THE GREEK ALPHABET

	Upper Case	Lower Case
alpha	A	α
beta	B	β
gamma	Γ	γ
delta	Δ	δ
epsilon	E	ϵ
zeta	Z	ζ
eta	H	η
theta	Θ	θ
iota	I	ι
kappa	K	κ
lambda	Λ	λ
mu	M	μ
nu	N	ν
xi	Ξ	ξ
omicron	O	o
pi	Π	π
rho	P	ρ
sigma	Σ	σ
tau	T	τ
upsilon	Υ	υ
phi	Φ	ϕ
chi	X	χ
psi	Ψ	ψ
omega	Ω	ω

Physics

Physics

RICHARD WOLFSON

Middlebury College

JAY M. PASACHOFF

Williams College

LITTLE, BROWN AND COMPANY

Boston   Toronto

Library of Congress Cataloging-in-Publication Data

Wolfson, Richard.
   Physics.

   Includes index.
   1. Physics. I. Pasachoff, Jay M. II. Title.
   QC21.2.W65 1987   530   86–21493
   ISBN 0–316–95183–8

Library of Congress Catalog Card No. 86–21493

ISBN 0-316-95183-8

9 8 7 6 5 4 3 2 1

KP

Published simultaneously in Canada
by Little, Brown & Company (Canada) Limited

Printed in the United States of America

The cover photo shows a test firing of the Particle Beam Fusion Accelerator II at Sandia
National Laboratories. The electrical discharges are a result of air breakdown at the surface of
the water that covers the pulse-forming section of the PBFA, and are due to extremely high
electromagnetic fields. Photo by Walter Dickenman, courtesy of Sandia National Laboratories.

Credits appear on pages 1065–1070

PREFACE

Physics is fundamental. To understand physics is to understand how the world works, both at the everyday level and on scales of time and space so small and so large as to defy our intuition. This book is meant to provide an understanding of physics for students of science and engineering.

To the student, physics can be at once fascinating, challenging, subtle, and yet simple. Physics fascinates in its myriad applications throughout science and engineering, and in its revelation of unexpected phenomena like superconductivity and black holes. It challenges with its needs for precise thinking and for the application of mathematics in the solution of physical problems. It can be subtle in its interpretation, especially where the phenomena it describes seem at odds with our everyday ways of thinking. Perhaps most important, though, physics is simple. Its few fundamental laws are stated in the simplest of terms, yet they encompass a universe of physical phenomena, both natural and technological.

Students who recognize the simplicity of physics develop a confidence that stems from understanding the physical universe at a fundamental level. Too often, though, students see physics as a complex and forbidding subject that presents a bewildering array of new and seemingly unrelated concepts. These students suffer both in their appreciation of physics and in their ability to apply physical principles in new situations. With this book, we hope to build student competence that is based on a flexible understanding of fundamental principles rather than on rigid memorization of details.

This is a textbook for science and engineering students. It covers the traditional material of a full-year introductory physics course, and does so in the traditional sequence. But within that traditional framework we have made a conscious effort to stress the underlying simplicity and unity of physics, and to make sure that students grasp the full meaning of physical laws as they learn to apply them. We use a variety of techniques to help students achieve a full and coherent understanding of physics. We follow the introduction of new concepts with worked examples that stress different aspects and applications of those concepts, then summarize by showing how these diverse examples are really manifestations of the same underlying concept or principle. Throughout the text, we avoid reinforcing artificial distinctions between phenomena that are really similar. For example, we treat circular motion on a par with all other forms of accelerated motion, stressing that this special case is really the same as any other application of Newton's second law and shunning special terminology that might make circular motion seem somehow different.

This is a calculus-based text designed for students who have had calculus or who are taking their first calculus course concurrently with physics. At the beginning we provide frequent reviews to refresh students' knowledge and to build mathematical confidence. As the text advances, we allow for steady growth in students' mathematical skills. We introduce calculus and vectors not in separate sections but as needed throughout the text. We make frequent references to the mathematical appendices that contain not only equations but also conceptual reviews. Our mathematical expositions stress the intuitive understanding of mathematical concepts, especially in the context of their physical applications. Even students well grounded in calculus should find their understanding of calculus enhanced by its use in this physics book.

The relation between physics and mathematics presents a special challenge. Some students suppose physics to be difficult because it is mathematical. For these students we are careful to distinguish those aspects of problem solving that relate to the concepts of physics from those aspects that are more computational, at the same time building skill in the use of mathematical tools. Other students might be highly skilled in mathematics but find physics difficult because they have trouble casting physics problems in mathematical terms. For these students we stress the relation between the physical and mathematical content of basic principles, making clear how equations are simply more precise renditions of ordinary-language statements. For all students, we refuse to let mathematical derivations go dangling; we ask frequently after the derivation of a new equation or solution of an example problem, "does this result make sense?" We show that it does by appeal to physical insight in easily grasped special cases. And we illustrate the meaning of equations and principles both verbally and through diagrams, making sure that interpretations are conceptually clear as students begin to use the material quantitatively.

We want this to be a book that students can learn from, without professors' needing continually to interpret what we are saying. Our consequent emphasis on careful and thorough explanations of physics and mathematics results in a larger-than-usual ratio of words to equations, but this does not entail any reduction in the level of mathematical sophistication. And our clearer but lengthier explanations often pave the way for examples and end-of-chapter problems involving more sophisticated applications of calculus and vector algebra.

Students in the introductory physics course for scientists and engineers learn most of their physics by working problems. This text contains 2530 end-of-chapter questions and problems. The problems range from simple exercises that familiarize students with basic equations and build confidence, to more complex and realistic problems involving the application of multiple concepts, and finally to physically or mathematically difficult problems that will challenge the best students. The problems for each chapter are grouped to correspond to the major text sections, and within each section they are in roughly increasing order of difficulty. A supplementary section contains problems that extend the ideas of the text, or require more substantial synthesis of concepts from other chapters, or are especially challenging. We have made a special effort to develop new and realistic problems based on contemporary applications in science and technology. With our emphasis on the conceptual understanding of physics, we also include thought-provoking questions intended to foster a deeper understanding of fundamental principles and their applications.

Students have a wide variety of reasons for taking introductory physics. We have tried to capture and maintain the interest of all students through a broad range of examples, applications, and problems drawn from contemporary science and engineering. We have been especially careful to illustrate the principles of classical physics with up-to-date examples from high technology and contemporary scientific research. Thus the student will find, for example, energy storage in capacitors made more lively through its application to laser fusion. The restoration of the Statue of Liberty is used as an example of torque and static equilibrium. In the thermodynamics section, earth's energy balance is discussed in the context of the latest calculations on the climatic effects of nuclear war. Space exploration, communications, biotechnology, geophysics, liquid crystals, computerized automobile braking systems, and jet aircraft technology are among many other applications used to illustrate basic concepts of physics. We integrate these applications directly into the text, where they are most likely to be read by students and to motivate interest in the underlying concepts. The applications incorporate a great many photographs, charts, and diagrams.

We have worked with our publisher to make this text easy to study from. Examples are singled out using color, as are important equations. The many photographs suggest realistic applications of the material. Chapter summaries highlight major ideas and their conceptual relationships. Appendices contain reviews of calculus and trigonometry as well as tables of physical data. Our appendices are referenced throughout the text and problems, to remind students that this background material is readily available. Two levels of subheadings within each chapter provide organizational clarity, and a comprehensive index allows ready access to the material. At the back of the book we also provide answers to most odd-numbered problems. Separate aids include a Student Study Guide by Jeffrey Braun, a manual of worked-out solutions to selected problems by Timothy Halpin-Healy, and an instructor's manual including answers to all problems.

ACKNOWLEDGMENTS

A project of this magnitude is not possible without considerable help and support beyond the authors themselves, and we would like to acknowledge many others whose assistance is greatly appreciated.

From Middlebury, Rich Wolfson would like to thank first all the students in the introductory physics courses Physics 109–110, on whom I have honed the teaching methods that I hope have been captured in this text. Special thanks go to those who went on to become physics majors during the five years this book was in preparation, for their frequent discussions and advice from the student perspective. My colleagues in the sciences at Middlebury have been more than willing to discuss applications in their particular fields and subfields, and many have cheerfully supplied references, data, or illustrations on rather short notice. I am grateful to them all, especially to my immediate colleagues in Physics into whose professional lives this project has made definite intrusions. Bob Prigo, with whom I have alternated teaching the different semesters of the introductory course, gave especially valuable insights into the needs of introductory students, and provided suggestions for many examples and demonstrations. Bob Gould was helpful with material on mechanics, waves, and fluid dynamics, and engaged in vigorous discussion of some more subtle points in electromagnetism. Frank Winkler and Steve Ratcliff supplied astrophysical data and examples, and Jeff Dunham put me in touch with contemporary work in atomic and nuclear physics. Many scientists and engineers at other institutions also supplied information, references, or photographs, often in response to last-minute telephone requests. Several others deserve special thanks. Ann Broughton and Sherry Mahady provided much clerical support. Vijaya Wunnava did much of the research that made possible realistic examples based on data from contemporary applications. The reference staff at Middlebury's Starr Library proved willingly resourceful in tracking down obscure information. Scot Gould, while a Middlebury undergraduate, wrote the software that allowed equations to be set easily in the manuscript. Later, as a graduate student in physics, Scot worked all the nearly 1700 problems and made many helpful suggestions for changes which I have incorporated into the problem sets. Laboratory Supervisor Cris Butler discussed proposed problems and examples, and prepared laboratory demonstrations featured in photographs supplied from Middlebury. Photographer Erik Borg captured those demonstrations on film, and often suggested the best ways for visual presentation. Finally, I am deeply grateful to my wife Artley and daughters Sarah and Carrie for their support and patience during the five long years that it took to bring this project to completion.

From Williams, Jay Pasachoff thanks David Park for his comments on modern physics and on historical aspects, and Stuart B. Crampton, William Wootters, Kevin Jones, Jefferson Strait, and C. Ballard Pierce for useful conversations. I thank also Hugh Kirkpatrick for setting up many lecture demonstrations to be photographed, and Timothy Halpin-Healy for his work on solutions and his contributions to the optics chapters, especially the section on rainbows. My wife, Naomi, and I exchange proofreading on each other's books, and I thank her for her excellent work on this one. Our daughters Eloise and Deborah continue to help me in many ways. The comments of Martine H. Westermann on early drafts of my material were very helpful. Elizabeth Stell has done outstanding work on photographs and other aspects of production.

Both authors would also like to thank a number of individuals without whose professional efforts this book would not have been possible. At Little, Brown and Company, Ian Irvine helped develop the idea of this text, and strongly influenced its direction through his interpretation of initial reviews and survey results. Ron Pullins capably guided the book during most of the years it was in preparation. Garret White supervised the project throughout, while Sally Stickney oversaw the final production. These editors were assisted by Mary Lou Wilshaw, Ann d'Entremont, and Bonnie Wood, who kept track of countless organizational details. We thank Nancy P. Kutner of Rensselaer Polytechnic Institute for preparing the index. Her experience as an indexer and her training in physics and computer science proved helpful in this always-difficult task. Finally, the content, level, and structure of this book were significantly influenced by the many physicists who reviewed the manuscript in its various stages. These reviewers also made a great many valuable suggestions for changes to enhance clarity, to correct conceptual and mathematical errors, and to ensure an authoritative text. We are grateful to them all: Angelo Armenti, Jr., Villanova; F. Todd Baker, University of Georgia; Robert Bearse, University of Kansas; James D. Finley, III, University of New Mexico; A. Lewis Ford, Texas A&M University; Anthony French, MIT; James B. Gerhart, University of Washington; Robert Hallock, University of Massachusetts, Amherst; William M. Hartmann, Michigan State; William H. Ingham, James Madison University; Carl A. Kocher, Oregon State University; Walter H. Kruschwitz, University of South Florida; M. A. K. Lodhi, Texas Tech; David Markowitz, University of Connecticut, Storrs; Terrill W. Mayes, UNC–Charlotte; Howard McAllister, University of Hawaii; Herbert R. Muether, SUNY–Stony Brook; Robert Reynolds, Reed College; John W. Stewart, University of Virginia; Fredrick Thomas, Sinclair Community College; Alan Van Heuvelen, New Mexico State University; and George A. Williams, Utah–SLC.

We have worked to make this book as interesting, as reliable, and as error-free as possible. We would be glad to hear from any readers—faculty, students, and others—with any corrections or suggestions. We are able to make some changes between subsequent printings and would appreciate prompt comments of any sort. We promise a speedy reply to each writer.

Richard Wolfson
Department of Physics
Middlebury College
Middlebury, Vermont 05753

Jay M. Pasachoff
Department of Physics and Astronomy
Williams College
Williamstown, Massachusetts 01267

July, 1986

CONTENTS

xii   Contents

Physics

1

DOING PHYSICS

physics

When you ride a bicycle, fly in a jet plane, or watch a thunderstorm, you experience physical laws. **Physics** is the science that tells us how and why things work. Knowledge of physics will help you understand both natural phenomena and the technologies that increasingly pervade our lives. In this book, we set forth the basic laws of physics and show how they apply to a myriad of phenomena from atoms and molecules to cars and airplanes, rockets and computers, stars and galaxies.

The dividing line between practical and fanciful applications of physics is eroding. Just a few decades ago, only science fiction writers could suggest that laws conceived by Isaac Newton in the 1600's would be used to guide spacecraft to the planets. Now that is history; exploration of the outer planets by Voyager spacecraft and the rendezvous with Halley's Comet by an international flotilla of spacecraft in 1986 represent triumphs of human ingenuity and of our understanding of the laws of physics (Fig. 1–1).

Fig. 1–1
The European Space Agency's Giotto spacecraft approaches Halley's Comet in March 1986. The spacecraft is an artist's rendition superimposed on an actual photo of the comet.

1.1
FIELDS OF PHYSICS

mechanics

classical mechanics

The branch of physics governing the motion of bodies is called **mechanics.** You use the laws of mechanics when you drive a car, ride a skateboard, build a skyscraper, and many other times each day. Through most of this book, we deal with **classical mechanics,** which applies to objects from molecules to galaxies that are moving at speeds small compared with the speed of light. In later chapters, we will see that classical mechanics is but an approximation to a more comprehensive set of physical laws that includes Einstein's theory of relativity and the theory of quantum mechanics.

In the first eight chapters we deal with the motion of single objects, often called particles. Then, in Chapters 9 to 13, we generalize to the motion of many-particle systems. The wave of standing people that sweeps

1

wave motion

through a football stadium, ocean waves that pound the coastlines, and the sound waves by which we communicate are examples of a special kind of many-particle motion called **wave motion,** which we examine in Chapter 14.

Our earth is five-eighths covered by liquid water, and is surrounded entirely by a gaseous atmosphere. Even the continents on which we live float around on an underlying fluid layer. The cells of our bodies are nurtured by the motion of fluid through our circulatory systems. And the bulk of the matter in the universe—stars and interstellar gas—is in the fluid state. In Chapter 15, we apply the laws of mechanics to fluids.

Nearly all life on earth thrives on a constant supply of energy from the sun, and a delicate balance of energy-transfer processes keeps our planet at a habitable temperature; we worry increasingly that human activity may upset that balance and bring severe climatic changes. For our technological society, we have developed nuclear power plants, gasoline engines, solar collectors, oil and gas furnaces, and a host of other energy sources. Worries about our planet's energy balance, and the operation of our energy technol-

thermodynamics

ogies, are both concerns of **thermodynamics**—the study of heat and its interaction with matter. We follow our study of mechanics with four chapters on thermodynamics.

Following, we deal extensively with another major branch of physics:

electromagnetism

electromagnetism. The Greeks knew something of the electric and magnetic properties of matter; the words electricity and magnetism themselves originate in Greek words for natural materials displaying electric and magnetic effects. The Chinese invention of the magnetic compass resulted by the twelfth century A.D. in widespread use of earth's natural magnetism to aid navigation. In 1800, Alessandro Volta invented the electric battery, paving the way for a series of experiments involving electric current and its magnetic effects. By the late 1800's, application of electromagnetism led to practical technologies including telegraph, telephone, and electric power distribution. In one of the most sweeping syntheses in the history of science, James Clerk Maxwell in the 1860's formulated a complete description of electromagnetism, showing that electricity and magnetism are intimately related. A startling prediction of Maxwell's theory was that light is an electromagnetic wave phenomenon, and that other electromagnetic waves should be possible as well. Radio and television follow directly from that prediction. Today, electromagnetic technology dominates our civilization, as evidenced by our digital watches, video recorders, microwave ovens, and especially computers (Fig. 1–2).

Fig. 1–2
This tiny silicon chip forms the heart of a personal computer. The chip contains the equivalent of over 100,000 electronic parts. Here we see an electron micrograph of an ant carrying the chip.

Electromagnetic theory led directly to many practical technologies, but it also raised bafflingly deep questions about the nature of light, motion, space, and time. With a radically simple statement whose implications seem to violate our common sense, Albert Einstein in 1905 confronted these questions (Fig. 1–3). His **special theory of relativity** gives a surpris-

special theory of relativity

ingly simple picture of reality, in the context of radically altered notions of space and time. Although relativity provides a more correct description of physical reality than does classical mechanics, the two theories differ significantly only at speeds approaching that of light—seven times around the earth in one second. In our everyday existence we need not worry about relativity. But in particle accelerators probing the basic structure of matter;

in studies of exploding supernovas, neutron stars, and the early universe; and even in designing a color television, we must take account of relativity. We explore the theory of relativity in Chapter 33.

optics

The behavior of light is intimately tied to electromagnetism and relativity. In Chapters 34 and 35 we explore the branch of physics called **optics,** building an understanding of light and of optical devices like lenses, prisms, diffraction gratings, and the instruments we build from them—microscopes, telescopes, spectrometers, and the like (Fig. 1–4).

quantum mechanics

But what is light? This question leads to one of the major developments of twentieth-century thought: **quantum mechanics.** Dealing with the smallest objects in the universe, quantum mechanics blurs the distinction between light and matter, giving a description of matter on the atomic scale for which our macroscopic intuition is simply inadequate. We give a brief introduction to quantum mechanics and its philosophical implications in Chapter 36. We also discuss our evolving knowledge of the elementary particles that are the fundamental constituents of matter. Finally, we show how studies of the largest and smallest components of the universe are uniting to give us a picture of the universe only a fraction of a second after its creation.

Fig. 1–3
Albert Einstein at age 26, when he developed the special theory of relativity.

1.2
THE SIMPLICITY OF PHYSICS

Physics is the fundamental science. Its laws and theories describe the workings of the universe at the most basic level. For that reason physics is immensely powerful; the same laws describe the behavior of molecules, of airplanes, and of galaxies. Although no one has done so, most scientists believe in principle that it would be possible to describe the operation of a living cell or organism using only the fundamental laws of physics.

Applying the laws of physics can give rise to challenging problems whose solutions call for clever insight and mathematical agility. The challenge of problem solving is what gives physics some of its intellectual interest and also its reputation as a difficult subject. But if you approach this course thinking that physics presents you with numerous difficult things to learn, you are missing the point. Because it is so fundamental, physics is inherently simple. There are only a few basic laws to learn; if you really understand those laws, you can readily apply them in a wide variety of situations. We wrote this book in a spirit that emphasizes the underlying simplicity of physics by reminding you how diverse examples are really manifestations of the *same* underlying physical laws. You should come to understand the basic laws thoroughly so you can apply them confidently in new situations. As you read the text and work physics problems, remember the simplicity of the underlying physical principles. Ask yourself how each problem you approach is really similar to other problems and to the text examples. And you will find that similarity, because the many problems and examples really do involve only a few underlying laws. So physics is simple—challenging, too—but with an underlying simplicity that reflects the scope and power of this fundamental science.

Fig. 1–4
The Hubble Space Telescope will use optical principles to bring us information about the most remote objects in the universe. Here we see it being assembled; the tube inserted into a central hole in the 2.4-m-diameter main mirror is to block internal stray light.

Fig. 1–5
A standard foot and yard were placed on a wall in the courtyard of the Royal Observatory in Greenwich, England, in the mid-1800's.

1.3
MEASUREMENT SYSTEMS

"A long way" means different things to a sedentary person, to a marathon runner, to a pilot, and to an astronaut. We need to quantify our measurements. In the United States, we still use a unit derived from the length of a person's foot—the foot, a unit that can be traced back through Greek and Roman to Egyptian times. The inch and the yard share that lineage. The Roman mile of 5000 feet was changed by Queen Elizabeth in the late sixteenth century to 5280 feet, or 8 furlongs. But though we think of inches, feet, yards, miles, and pounds as "English units" (Fig. 1–5), they are being phased out even in England. Every country in the world, except the United States, Burma, and Brunei (a sultanate on the island of Borneo), now uses the metric system, in which the unit of length is based on the **meter** (from the Greek word "metron," meaning "a measure"). The meter was first defined at the time of the French Revolution to be one ten-millionth of the distance from the equator to the north pole through Paris. This distance, of course, is not trivial to determine—surveyors sent out at the time to measure a portion of the path were nearly guillotined—and more readily accessible measuring standards proved necessary.

meter

In 1889, a standard meter was carefully made; it is the distance between two lines on a certain bar made of a platinum-iridium alloy when examined at the temperature of the melting point of ice. It is kept in a temperature-controlled room at the Bureau International des Poids et Mesures in Sèvres, France, where the international treaty on measurement signed in 1875 is monitored. Thirty secondary standards were also made from the same ingot of alloy; the United States Bureau of Standards in Washington, D.C., has numbers 21 and 27. (Comparisons over 15 years showed that the standard meters remained the same to within 2 tenmillionths of a meter.) But as new methods developed, even the standard meter was not sufficiently accurate. In 1960, the meter was redefined as 1,650,763.73 wavelengths of orange-red light emitted by the isotope krypton-86.

In recent years, even this standard became insufficient. The speed of light is now one of the most accurately determined quantities (Fig. 1–6). In 1983, the meter was redefined in terms of the speed of light, which, in vacuum, has been measured at 299,792.458 kilometers per second (km/s). The 1983 Conférence Général des Poids et Mesures inverted this result and defined the meter as "the length of the path travelled by light in a vacuum during a time interval of 1/299,792,458 of a second."

In science, we now use the version of the metric system known as SI, for Système International d'Unités (International System of Units). Its basic version was developed in 1960, but it has been modified several times since. Engineers in the United States still often use the English system. If you were in England, though, you would find that bottles labelled with the English unit "1 gallon" actually contain 5 liters, a metric unit. In the United States, the use of the metric system was legalized in 1866, and the yard and the pound have been defined in terms of the meter and the kilogram since 1893. The definitions were changed in 1959 to make 1 yard exactly equal

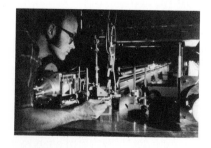

Fig. 1–6
A helium-neon laser at the National Bureau of Standards laboratory in Boulder, Colorado, is used to make accurate determinations of the speed of light. Here, Kenneth Evenson calibrates the instrument.

to 0.9144 m—slightly less than its previous value of 36/39.37 m—and to make a one-pound mass exactly equal to 0.453 592 37 kg.

SI includes 7 independent base units: the meter (m) for length, the kilogram (kg) for mass, the second (s) for time, the ampere (A) for electric current, the kelvin (K) for temperature, the mole (mol) for the amount of substance, and the candela (cd) for luminosity. Two supplementary units are used to measure angle—the radian (rad) for ordinary angles and the steradian (sr) for solid angles (Fig. 1–7). Units for all other quantities are derived from these base units. In mechanics, we use only the first three of the base units—meter, kilogram, and second—along with measures of angle.

The metric system is a decimal system. Its base units are used with the prefixes listed in Table 1–1 to indicate multiplication by powers of 10. For example, k is the SI symbol for the prefix "kilo-," which means "times 1000." 1 km is thus 1000 m.

When two units are used together, a hyphen appears between them: newton-meter. Each unit has a symbol, such as m for meter or km for kilometer. Since these are symbols, rather than abbreviations, they are not followed by periods. Neither are the plural forms followed by an "s." Symbols are ordinarily lowercase; only those named after people are uppercase (with one exception). Thus the symbol for gram is a lowercase g; the unit "newton" is written with a small "n" but its symbol is a capital N. The only exception is for the volume unit, the liter (spelled "litre" in most other countries), defined as the volume of a cube one-tenth of a meter on each side. Since the lowercase "l" is easily confused with the number one, the symbol for liter is written with a capital L. When two units are multiplied, their symbols are separated by a centered dot: N·m for newton-meter. Division of units is expressed using the slash (/), or by writing the symbol for the denominator unit raised to the −1 power. Thus the SI unit of speed is the meter per second, written m/s or m·s^{-1}. When four or more digits are used together, SI recommends the use of spaces to separate groups of three:

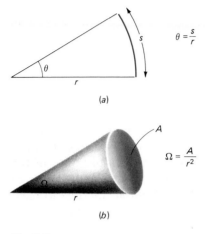

$$\theta = \frac{s}{r}$$

$$\Omega = \frac{A}{r^2}$$

Fig. 1–7
SI units of angle. *(a)* The radian measure of an angle is defined as the ratio of the subtended arc length to the radius. Since the circumference of a circle is $2\pi r$, there are 2π radians in a complete circle. One radian is then equal to $360°/2\pi$ or $57.3°$. *(b)* The measure of a solid angle is defined as the ratio of the subtended area to the radius squared.

TABLE 1–1
SI Prefixes

Prefix	Symbol	Power	
exa	E	10^{18}	= 1 000 000 000 000 000 000
peta	P	10^{15}	= 1 000 000 000 000 000
tera	T	10^{12}	= 1 000 000 000 000
giga	G	10^{9}	= 1 000 000 000
mega	M	10^{6}	= 1 000 000
kilo	k	10^{3}	= 1 000
hecto	h	10^{2}	= 100
deca	da	10^{1}	= 10
—	—	10^{0}	= 1
deci	d	10^{-1}	= 0.1
centi	c	10^{-2}	= 0.01
milli	m	10^{-3}	= 0.001
micro	μ	10^{-6}	= 0.000 001
nano	n	10^{-9}	= 0.000 000 001
pico	p	10^{-12}	= 0.000 000 000 001
femto	f	10^{-15}	= 0.000 000 000 000 001
atto	a	10^{-18}	= 0.000 000 000 000 000 001

34 365.456 32; we tend to use commas instead of spaces in the United States, though some countries use a comma as we use a period to indicate a decimal point.

The study of the fundamental units is itself an interesting field of physics. There are, for example, at least a half-dozen definitions of units of time, each different. The division of time into hours, minutes, and seconds goes back to the Babylonians. The **second** used to be defined by the earth's rotation, as $1/(24 \times 60 \times 60)$ of a mean solar day, but then it was realized that the earth's rotation is slowing slightly. The second was then redefined as a specific fraction of the year 1900. Later, the second was defined in terms of the vibrations of a certain type of cesium atom (Fig. 1–8); the device used was called an "atomic clock." More recently, devices called hydrogen masers have been used as even more stable time standards. And now some scientists are anticipating that even more accurate time standards will be provided by pulsars—celestial objects that pulse at an extremely steady rate. One pulsar in particular, when its minuscule rate of slowing down is taken into account, can give us time accurate to one-billionth of a second per year. It is a good thing that we have better standards of time than the rotation of the earth; new research in 1984 revealed that the earth's rotation slows down and speeds up slightly as the wind blows. To keep atomic time and earth-spin time in agreement, "leap seconds" are added to our terrestrial clocks when necessary, usually about once a year.

The standard of mass is now the least satisfactory. Unlike the standards of length and time, which are based on measurement procedures that can be repeated by scientists anywhere, the unit of mass is still defined in terms of a particular object—the international prototype **kilogram** kept at the International Bureau of Weights and Measures at Sèvres, France (Fig. 1–9). It is made out of a special platinum-iridium alloy that is very hard, not subject to corrosion, and very dense. Nevertheless, it could conceivably change, and in any event comparison with such a standard is less convenient than a procedural standard that can be checked in a laboratory. So scientists are now working on techniques to measure the spacing of atoms in a crystal of silicon, essentially counting the atoms in a given volume, and therefore to scale up from the mass of a single silicon atom to a new definition of the kilogram.

second

Fig. 1–8
This cesium atomic-beam clock is the primary standard for time in the United States. The SI second is defined as the duration of 9,192,631.770 periods of the waves emitted as the cesium atoms undergo a particular transition.

kilogram

Changing Units

Frequently, we need to change units from one system to another—for example, from English to SI units. Appendix D contains extensive tables for converting among unit systems; you should familiarize yourself with this and the other appendices, and refer to them often.

In changing units, it is helpful to treat the units as algebraic quantities. Then we can derive ratios that are equivalent to 1 and have no dimensions.

Fig. 1–9
In this vault at the International Bureau of Weights and Measures at Sèvres, France, we find on the top shelf the international prototype meter bar of 1889 in its protective case. On the bottom shelf is the international prototype kilogram, its six comparison standards, two thermometers, and a hygrometer. The prototype kilogram has been used only three times: before 1889, in 1939, and in 1946. The standard meter has been superseded by measurement techniques based on the speed of light, but the international prototype kilogram remains the official standard of mass.

Since we can always multiply anything by 1 without changing its value, we can manipulate units in an expression by multiplying by these ratios.

For example, we know that 60 min = 1 h. Dividing both sides of this equation by 1 h, we have

$$\frac{60 \text{ min}}{1 \text{ h}} = 1.$$

Then, if we want to transform 24 hours into minutes, we multiply by this equivalent-to-1 ratio; the h in the denominator of our ratio cancels the h in the top of our original number:

$$(24 \cancel{h}) \left(\frac{60 \text{ min}}{1 \cancel{h}}\right) = 1440 \text{ min}.$$

We could equally well have divided the equation "60 min = 1 h" by 60 min to get 1 = (1 h)/(60 min), which puts the minutes into the denominator but also gives a ratio equivalent to 1. If we were changing minutes to hours, we would want this alternative form.

This method of changing units can also be used to change between metric and English units. For example, with 1 m = 39.37 in, the 400-inch-diameter mirror of the Keck telescope in Hawaii—to be the world's largest optical telescope—will be

$$(400 \text{ in}) \left(\frac{1 \text{ m}}{39.37 \text{ in}}\right) = 10.2 \text{ m}$$

Fig. 1–10
A model of the Keck Ten-Meter Telescope.

across (see Figure 1–10).

When you change units in some complicated fashion, a chain of such multiplications helps you keep track of units in a mechanical way, and prevents you from carelessly inverting one of the conversion factors.

Example 1–1
Changing Units

Express the United States' 55 mi/h speed limit in meters per second.

Solution

In Appendix D, we find that 1 mi = 1609 m, so we can multiply miles by the ratio 1609 m/mi to get meters. Similarly, we use the conversion factor 3600 s/h to convert hours to seconds. So we have

$$55 \text{ mi/h} = \frac{(55 \text{ mi})(1609 \text{ m/mi})}{(1 \text{ h})(3600 \text{ s/h})} = 24.6 \text{ m/s}.$$

1.4
EXPONENTIAL NOTATION

The range of measured quantities in the universe is enormous; lengths alone range from about 1/1 000 000 000 000 000 m for the radius of the proton to 10 000 000 000 000 000 000 000 m for the size of a galaxy; our telescopes observe to distances 10,000 times farther still. It is easy to lose track with so many zeroes involved. We thus often write numbers as decimals with one digit before the decimal point multiplying 10 to some power.

Fig. 1–11
A galaxy can be 10^{21} m across.

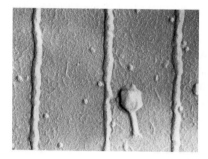

Fig. 1–12
Taken with an electron microscope, this photograph shows both artificial and natural structures on the small scale. Vertical lines are 20-nm-wide strips of polymethyl methacrylate, laid down on a silicon substrate by a process called x-ray lithography. The tailed structure is a T-4 bacteriophage virus, the head of which measures about 100 nm across.

TABLE 1–2
Some Typical Distances

Distance to farthest quasar detected	1×10^{26}	m
Distance to nearest large galaxy (Andromeda)	2×10^{22}	m
Diameter of our galaxy (Milky Way)	8×10^{20}	m
Distance to nearest star (Proxima Centauri)	4×10^{16}	m
Radius of Pluto's orbit	6×10^{12}	m
Radius of earth's orbit	1.5×10^{11}	m
Diameter of earth	1.3×10^{7}	m
Deepest ocean trench (Marianas Trench)	1.1×10^{4}	m
Skyscraper (Sears Tower, Chicago)	4.4×10^{2}	m
Height of person	2	m
Transistor on semiconductor chip	1×10^{-5}	m
Diameter of red blood cell	8×10^{-6}	m
Diameter of virus	1×10^{-7}	m
Smallest human-made structures	2×10^{-9}	m
Atomic diameter (hydrogen)	1×10^{-10}	m
Diameter of proton	2×10^{-15}	m

TABLE 1–3
Some Typical Times

Time since Big Bang	5×10^{17}	s
Age of earth	1.5×10^{17}	s
Time for sun to orbit galactic center	8×10^{15}	s
Existence of human species	6×10^{13}	s
Half-life of plutonium	8×10^{11}	s
Human lifespan	2×10^{9}	s
Orbital period of earth (1 year)	3×10^{7}	s
Rotation period of earth (1 day)	9×10^{4}	s
Flight time of intercontinental missile	1×10^{3}	s
Human heartbeat	1	s
Reaction time of human nervous system	1×10^{-1}	s
Period of highest audible sound	5×10^{-5}	s
Period of AM radio wave	1×10^{-6}	s
Time for fast computer to add two numbers	1×10^{-8}	s
Period of rotation of typical molecule	1×10^{-12}	s
Shortest light pulse produced in laboratory	1×10^{-15}	s
Half-life of neutral pion	2×10^{-16}	s
Time for light to cross a proton	7×10^{-24}	s
Earliest time after Big Bang to which known laws of physics apply	10^{-43}	s

For example, 14 can be written as 1.4×10^{1}, 1687 can be written as 1.687×10^{3}, and 0.023 can be written as 2.3×10^{-2}. (Note that a negative exponent means to take the inverse of the same quantity with a positive exponent, so that $10^{-2} = 1/10^{2} = 0.01$.) This way of writing numbers is called **exponential notation** or **scientific notation**. Exponential notation is often used in computers and calculators, with the letter E for exponential instead of the 10; the numbers above would then be written 1.4E1, 1.687E3, and 2.3E-2, respectively. Tables 1–2, 1–3, and 1–4 list some typical distance, time, and mass measurements in exponential notation (see also Figures 1–11, 1–12, and 1–13).

To add two quantities that have the same exponent, the distributive law of multiplication shows that we simply add the coefficients:

exponential notation
scientific notation

TABLE 1–4 ■■■■

Some Typical Masses

Galaxy (Milky Way)	4×10^{41}	kg
Sun	2×10^{30}	kg
Earth	6×10^{24}	kg
Mountain	2×10^{18}	kg
747 jetliner	4×10^{5}	kg
Compact car	1×10^{3}	kg
Human	65	kg
Raisin	1×10^{-3}	kg
Raindrop	1×10^{-6}	kg
Red blood cell	1×10^{-13}	kg
DNA molecule	2×10^{-18}	kg
Uranium atom	4×10^{-25}	kg
Proton	1.7×10^{-27}	kg
Electron	9.1×10^{-31}	kg

significant figures

Fig. 1–13
A 747 jetliner has a mass of 4×10^5 kg.

$$1.4 \times 10^1 \text{ kg} + 3.8 \times 10^1 \text{ kg} = (1.4 + 3.8) \times 10^1 \text{ kg} = 5.2 \times 10^1 \text{ kg}.$$

As in this example, most numbers we work with in physics represent physical quantities. It is important in working with such quantities to keep the numbers labelled with the appropriate unit symbol; the answer for a physical quantity is meaningful only if its units are specified, and furthermore, attention to units can alert you to mistakes in algebra. In particular, the only quantities that can be added meaningfully are quantities with the same units (adding 5 kilograms to 3 seconds, for example, makes no sense). We discuss the manipulation of units further in Chapter 2.

To add quantities with different exponents, we must first express them with the same exponent. After all, to add 3 pennies to 2 dimes, you would first convert the dimes to their equivalent number of pennies. Similarly,

$$6.8 \times 10^4 \text{ s} + 3.1 \times 10^5 \text{ s} = 6.8 \times 10^4 \text{ s} + 31 \times 10^4 \text{ s} = 37.8 \times 10^4 \text{ s}$$

$$= 3.78 \times 10^5 \text{ s},$$

where we converted 3.1×10^5 s to its equivalent, 31×10^4 s.

How accurate is the answer we just calculated? We only knew two digits of accuracy for each of the original numbers; we say that those numbers are accurate to two **significant figures.** The mere act of calculating cannot add accuracy, so the answer should not have more than two significant figures. We thus round it off to 3.8×10^5 s. Calculators and computers often give us numbers to many figures, but some or even most of those figures may be meaningless. If they are outside the range of accuracy set by our original numbers, they must be discarded.

Exponential notation is particularly useful when we multiply numbers, since then we merely multiply the digits and add the exponents. For example,

$$(3.0 \times 10^8 \text{ m/s})(1.5 \times 10^{-9} \text{ s}) = (3.0)(1.5) \times 10^{8+(-9)} \text{ m}$$

$$= 4.5 \times 10^{-1} \text{ m} = 45 \text{ cm}$$

is the distance that light, moving at 3.0×10^8 m/s, travels in 1.5 ns (nano is the prefix meaning one-billionth, or 10^{-9}; see Table 1.1). Notice in this example how we treated units as algebraic quantities, multiplying meters per second by seconds to get meters.

To raise a number written in exponential form to some power, we raise the digits preceding 10 to that power and multiply the exponent by the power:

$$(3 \times 10^4 \text{ m})^2 = 3^2 \times 10^{(4)(2)} \text{ m}^2 = 9 \times 10^8 \text{ m}^2$$

and

$$\sqrt{64 \times 10^6 \text{ m}^2/\text{s}^2} = (64 \times 10^6 \text{ m}^2/\text{s}^2)^{1/2} = 8.0 \times 10^3 \text{ m/s}.$$

When taking the square root of an odd power, we first convert it to an even power:

$$\sqrt{2.5 \times 10^7} = (25 \times 10^6)^{1/2} = 5.0 \times 10^3.$$

Note that raising a quantity to the one-half power is the same as taking the square root.

1.5
ESTIMATION

Some problems in physics and engineering call for very precise numerical answers. We need to know exactly how long to fire a rocket to put a space probe on course for a distant planet, or exactly what size to cut the tiny quartz crystal whose vibrations set the pulse of a digital watch. But for many other purposes, we need only a rough idea of the size of a given physical effect. Often a rough estimate will tell us that we can ignore a certain effect before we go to the trouble of calculating it exactly. In other cases, an estimate—even though it may be off by a factor of ten or more— may help decide among competing theories. And we can often make a rough estimate to check whether the results of more difficult calculations make sense.

When confronted with any physics problem, especially one involving estimation, think of what you know that might possibly pertain to the problem. Finding such links between what you already know and what the problem requires is more important than looking up formulas in a textbook. Sometimes a little thought will lead you to solutions that you didn't at first think you could achieve.

Example 1–2
Estimation: Is Earth Growing?

(a) Estimate the present mass of the earth.   (b) Earth is about 5 billion years old. Micrometeorites, roughly the size of dust grains, rain down on earth at the rate of about 10^6 kg per day (see Figure 1–14). Has this rain of dust contributed significantly to earth's present mass?

Solution

(a) How big is the earth? Maybe you know its diameter, but if you don't, consider how the United States looks on a globe. In its east-west dimension, the United States spans perhaps one-sixth of the globe. From the Atlantic to the Pacific coast is about 3000 miles, a little less than 5000 km. So earth is about 5000 km × 6 $= 3 \times 10^4$ km around at the latitude of the United States. At the equator, it is farther around—say about 4×10^4 km. The circumference of a circle is $2\pi r$, so earth's radius is 4×10^4 km$/2\pi$, or about 6×10^3 km. (The actual value, given in Appendix F, is 6.4×10^3 km, so our estimate is quite close.) Now the volume of a sphere is $\frac{4}{3}\pi r^3$, or, since π is close to 3, about $4r^3$. So the volume of the earth is

$$V_e \sim (4)(6 \times 10^6 \text{ m})^3 \sim 8 \times 10^{20} \text{ m}^3,$$

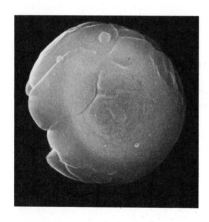

Fig. 1–14
A scanning-electron-microscope image of a micrometeorite.

where we use the sign $\sim$ to indicate a rough approximation. How dense is earth? It is made of rock, which is certainly denser than water. How much denser? Significantly denser, but on the other hand not ten times as dense. Say about twice as dense. Now how dense is water? Maybe you know that 1 cm^3 of water has a mass of 1 gram. But if not, maybe you've heard that "a pint's a pound," or maybe you can guess from experience that a gallon of water is around 10 lb, or 5 kg. Let's use the latter estimate. A gallon contains 4 quarts, and a quart is about a liter (look at the label on a 1-quart soda bottle to confirm this!). So a quart of water has a mass of about 5 kg/4, or about 1 kg. A liter is 1000 cm^3, or 10^{-3} m^3, so the density of water is about 1 kg$/10^{-3}$ m^3, or 1×10^3 kg/m^3. If rock is twice as dense, the density of rock is 2×10^3 kg/m^3. Then, using our value for earth's volume, we can estimate the mass of the earth as

$$M_e = (\text{volume})(\text{density}) = (8 \times 10^{20} \text{ m}^3)(2 \times 10^3 \text{ kg/m}^3) = 1.6 \times 10^{24} \text{ kg.}$$

Appendix F gives the correct value as 5.98×10^{24} kg; our answer is low by a factor of about 4, but is still quite adequate to our purpose of determining whether the rain of micrometeorites is significant. (Part of the reason for our low estimate is the fact that material in the core of the earth is compressed to much higher densities than the rocks with which we are familiar.)

(b) Coming down at the rate of 10^6 kg/day, the total mass of micrometeorites landing on earth in 5×10^9 years is

$$m = (10^6 \text{ kg/day})(365 \text{ days/year})(5 \times 10^9 \text{ years}) = 2 \times 10^{18} \text{ kg.}$$

Since this is far less than our estimate of earth's mass (only about $10^{18}/10^{24}$, or one-millionth), the rain of micrometeorites has made an insignificant contribution to earth's mass.

Our conclusion rests on an important assumption: that micrometeorites have fallen at roughly the same rate for 5 billion years. Actually, the rate was probably higher when the solar system was young. But except for the earliest times, when earth itself was forming from the matter of the primordial solar system, extraterrestrial matter probably has not contributed significantly to earth's mass.

Example 1–3

Estimation: Post Offices

How many post offices are there in the United States?

Solution

Are you tempted to ask for more information? You don't have to! What do you know about post offices? There is at least one in every city, and large cities have several. What is the size of a city that has only one? Let us estimate 20,000. The population of the United States is about 240 million. 240 million divided by 20 thousand is about 10 thousand. So that is one estimate of the number of post offices.

Can we get the answer another way? Consider the area served by a post office. Again, a large city will have several post offices, and small towns separated by 50 km will certainly have separate post offices, so let us estimate an area 25 km $\times$ 25 km, say 30 km $\times$ 30 km to make the multiplication easier. This is about 1000 square km for a single post office. What is the area of the United States? It is about 3000 miles from coast to coast, and 2000 miles from Maine to Florida. This translates into about (5000 km) (3000 km) = 15×10^6 km^2. At 1000 km^2 per post office, this estimate calls for 15×10^6 km^2/(1000 km/post office) = 15,000 post offices. This answer is within a factor of 2 of our earlier estimate, which is more consistency than we might expect when estimating.

How about a third method? Zip codes are 5 digits long, so there clearly must be fewer than 99,999 post offices.

What is the right answer? The U.S. Postal Service reports that there are about 30,000 post offices. Our estimates are all within a factor of about 3.

Example 1–4

Estimation: Nuclear-Electric Cars?

(a) Estimate the annual consumption of gasoline by cars in the United States. (b) Burned in a car, a gallon of gasoline is roughly equivalent to 30 kilowatt-hours (kWh) of electricity. If a large nuclear power plant produces electricity at the rate of 10^6 kWh per hour (1000 MW), how many such plants would have to be built if the United States converted to electric cars?

Solution

How far is your family car driven each year? Mine goes about 12,000 miles, or roughly 10^4 mi. A typical car might get 20 miles/gallon, so at 10^4 miles/year, it would use $(10^4 \text{ mi/y})/(20 \text{ mi/gal}) = 500$ gal/y. How many cars are there in the United States? There are over 200 million people, but certainly not everyone has a car. My family has four people and two cars, so perhaps there are about 100 million, or 10^8 cars. The annual gasoline consumption of all these cars is then $(500 \text{ gal/y/car})(10^8 \text{ cars}) = 5 \times 10^{10}$ gal/y. (Figures from the U.S. Department of Energy show that the total gasoline consumption in the mid-1980's was somewhat over 10^{11} gal/y, so our estimate is low by only a factor of about 2.)

At 30 kWh/gal, our estimated gasoline consumption is equivalent to $(30 \text{ kWh/gal})(5 \times 10^{10} \text{ gal/y}) \sim 10^{12}$ kWh/y. How many power plants would this require if we converted to electric cars? Each plant produces electricity at the rate of 10^6 kWh/h. But there are 24×365 or about 10^4 hours in a year, so each plant produces $(10^6 \text{ kWh/h})(10^4 \text{ h/y})$ or 10^{10} kWh/y. We need 10^{12} kWh/y; this requires $(10^{12} \text{ kWh})/(10^{10} \text{ kWh/plant})$ or 100 new nuclear plants. Since there are about 75 nuclear power plants in the United States, this would represent substantial new construction.

SUMMARY

1. **Physics** is the fundamental science, dealing with the most basic laws governing the behavior of the physical universe. This book deals mainly with **classical physics,** an approximation to the more accurate description provided by **quantum mechanics** and **relativity.** Classical physics is a valid approximation for dealing with all but the smallest systems and the most rapid motion.

2. Physics is a quantitative science. In this book and throughout the scientific community, we use the SI (Système International) of measurement units. Although seven SI base units are defined, the three most central to mechanics are those of length, time, and mass.

a. The SI unit of length is the **meter,** defined in terms of the distance light travels in a fixed time.
b. The SI unit of time is the **second,** defined in terms of the vibration period of a certain atom.
c. The SI unit of mass is the **kilogram,** defined by the international prototype kilogram at the International Bureau of Weights and Measures in France.

3. Some applications call for very precise numerical measurements or calculations; the number of **significant figures** in a number reflects how precisely we know that quantity.

4. In other applications, **estimation** may be all that is required to grasp the significance of a physical phenomenon.

SUGGESTED READINGS

Allen V. Astin, "Standards of Measurement," *Scientific American,* vol. 218, No. 6, p. 50, June 1968.

David T. Goldman and R. J. Bell, eds., "The International System of Units (SI)," NBS Special Publication 330, 1986 edition (U.S. Government Printing Office).

Lewis V. Judson, "Units and Systems of Weights and Measures: Their Origin, Development, and Present Status," U.S. Department of Commerce Letter Circular LC 1035, 1960, revised 1976 by L. E. Barbrow.

Robert A. Nelson, "Foundations of the International System of Units (SI)," *The Physics Teacher,* December 1981, pp. 596–613.

Francis M. Pipkin and Rogers C. Ritter, "Precision Measurements and the Fundamental Constants," *Science,* February 25, 1983, vol. 219, pp. 913–921.

Arthur L. Robinson, "Using Time to Measure Length," *Science,* June 24, 1983, p. 1367.

Margaret L. Silbar, "Measurement to the Limit," *Mosaic,* November/December 1981, pp. 27–32.

QUESTIONS

1. Explain why measurement standards based on laboratory procedures are preferable to those based on specific objects like the international prototype kilogram.
2. Which measurement standards are now defined procedurally? Which are not?
3. Why is it important that the international prototype kilogram be made from corrosion-resistant material?
4. Given present-day definitions of the fundamental units, is it meaningful to attempt a more accurate measurement of the speed of light?
5. Why are "leap seconds" necessary?
6. When a computer carrying 7 significant figures adds 1.000000 and 2.5E-15, what answer does it display? Why?
7. In what way are the sciences of biology and chemistry based in physics?
8. Why does earth's rotation not provide a suitable standard of time?
9. For which of the two coordinates—latitude or longitude—do navigators need a precise measure of time? Explain.
10. Astronomers distinguish two different measures of the length of a day—solar and sidereal. A solar day is the interval between times when the sun is at its highest position in the sky. A sidereal day is the interval between times when any star other than the sun is at its highest position in the sky. Why are they different? Which is longer?
11. To raise a power of ten to another power, you multiply the exponent by the power. Explain why this works.
12. A scientist and a creationist are arguing about the age of the earth. What facts might the scientist use in estimating this age?
13. How would you determine the length of a curved line?
14. How would you measure the thickness of a sheet of paper?

PROBLEMS

Section 1.3 *Measurement Systems*

1. What is your mass in (a) kg; (b) g; (c) Gg; (d) fg?
2. A year is very nearly $\pi \times 10^7$ s. By what percentage is this figure in error?
3. How many cubic centimeters (cm^3) are there in a cubic meter (m^3)?
4. I need a 14-mm wrench to remove the spark plugs in my car, but I have only an English set with sizes that are increments of one-sixteenth of an inch. What size should I use?
5. A gallon of paint covers 350 ft². What is its coverage in m²/L?
6. By what percentage do the 1500-m and 1-mile foot races differ?
7. Superhighways in Canada have speed limits of 100 km/h. Does this exceed the United States' 55 mi/h speed limit? If so, by how much?
8. One m/s is how many km/h?
9. A hydrogen atom is about 0.1 nm in diameter. How many hydrogen atoms lined up side-by-side would make a line 1 cm long?
10. How long a piece of wire would you need to form a circular arc subtending an angle of 1.4 rad, if the radius of the arc is 8.1 cm?

Section 1.4 *Exponential Notation*

11. Add 3.6×10^5 m and 2.1×10^3 km.
12. The volume of the moon is 2.21×10^{19} m³. What is its radius?
13. Divide 4.2×10^3 m/s by 0.57 ms, and express your answer in m/s².
14. Add 5.1×10^{-2} cm and 6.8×10^3 μm, and multiply the result by 1.8×10^4 N (1 N is the SI unit of force).
15. If there are 100,000 electronic components on a semiconductor chip that measures 5.0 mm by 5.0 mm, (a) how much area does each component occupy? (b) If the individual components are square, how long is each on a side?
16. What is the cube root of 2.7×10^{13}?

Section 1.5 *Estimation*

17. How many earths would fit inside the sun?
18. Estimate the number of people, standing with outstretched arms touching, needed to form a line from New York to Los Angeles.
19. If you set your watch to be exactly correct now, and if it runs slow by 1 s per month, in how many years would it again read exactly the right time?
20. (a) Estimate the volume of water going over Niagara Falls each second. (b) The falls provides the outlet for Lake Erie; if the falls were shut off, estimate how long it would take Lake Erie to rise 1 m.
21. (a) Estimate the volume of water in earth's oceans. (b) If scientists succeed in harnessing nuclear fusion as an energy source, each gallon of sea water will be equivalent to 400 gallons of gasoline. At our present rate of gasoline consumption (see Example 1–4), how long would the oceans supply our fuel needs?
22. Estimate the number of air molecules in your dormitory room.
23. The density of interstellar space is about 1 atom per cubic cm. Stars in our galaxy are typically a few light-years apart (one light-year is the distance light travels

in one year), and have typical masses of 10^{30} kg. Estimate whether there is more matter in the stars or in the interstellar gas.

24. Solar power roughly equivalent to the light from 10 100-watt light bulbs falls on each square meter of earth's surface. Using the fact that earth is 150 million km from the sun, and that the sun radiates energy equally in all directions, estimate how many 100-watt light bulbs are equivalent to the sun's total power output.

Supplementary Problems

25. A human hair is about 100 μm across. Estimate the number of hairs in a typical braid.
26. The density of bubble gum is about 1 g/cm^3. You blow a 50-g wad of gum into a bubble 10 cm in diameter. What is the thickness of the bubble? *Hint:* think about unrolling the bubble into a flat sheet. The surface area of a sphere is $4\pi r^2$.
27. The moon barely covers the sun at a solar eclipse. Given that the moon is 4×10^5 km from earth and that the sun is 1.5×10^8 km from earth, determine how much bigger the sun's diameter is than the moon's. If the moon's radius is 1800 km, how big is the sun?
28. The semiconductor chip at the heart of a personal computer is a square 4 mm on a side, and contains 10^5 electronic components. (a) If each component is a square, what is the distance across each component? (b) If a calculation requires that electrical impulses traverse 10^4 elements on the chip, each a million times, how many such calculations can the computer perform each second? The maximum speed of an electrical impulse is close to the speed of light, 3×10^8 m/s.

29. Estimate the number of (a) atoms and (b) cells in your body.
30. When we write the number 3.6 as typical of a number with two significant figures, we are saying that the actual value is closer to 3.6 than to 3.5 or 3.7. That is, the actual value lies between 3.55 and 3.65. Show that the percent uncertainty implied by two-significant-figure accuracy varies with the value of the number, being the least for numbers beginning with 9 and most for numbers beginning with 1. In particular, what is the percent uncertainty implied by the numbers (a) 1.1, (b) 5.0, and (c) 9.9?
31. Thermonuclear explosives have about ten million times the explosive power per unit mass of conventional chemical explosives. Estimate the mass of each of the ten 350-kiloton warheads carried on an MX missile. (One kiloton means the equivalent of one thousand tons of chemical explosive.)

2

MOTION IN A STRAIGHT LINE

Electrons swarming around atomic nuclei, cars speeding along a highway, a comet sweeping through the solar system, the galaxies themselves rushing apart in the expanding universe—all these are examples of matter in motion. In physics, the study of motion is called **kinematics** (from the Greek "kinema," motion, as in motion pictures). In this chapter, we deal with the simplest case of motion: a single particle moving along a straight line. Later, we generalize to motion in more dimensions and with more complicated objects. But the basic concepts we develop here remain fundamental to these more complex situations.

We define a **particle** as an isolated point of mass, having no size or structure, and incapable of rotation, vibration, or other internal motions. Whether we can treat a real object as a particle depends on the circumstances. We can consider the earth to be a particle when we describe its orbital motion around the sun, for its size and rotation are essentially negligible in that context. But we must consider earth's nonzero size and rotation when we show how the moon gives rise to tides. In later chapters, we will see how the behavior of complicated objects follows from the laws we develop for their constituent particles.

kinematics

particle

2.1
DISTANCE, TIME, SPEED, AND VELOCITY

average speed

We define the **average speed** of an object as the distance it travels divided by the travel time. The units of speed are therefore length per time, such as m/s (read "meters per second"), km/h, or mi/h (often written mph). For

15

example, a car that goes 7.2 km in 12 min has a speed of

$$\frac{7.2 \text{ km}}{12 \text{ min}} = 0.60 \text{ km/min} = \frac{0.60 \text{ km}}{(1 \text{ min})(1 \text{ h/60 min})} = 36 \text{ km/h}.$$

At this speed, the car would go $(36 \text{ km/h})(2.5 \text{ h}) = 90$ km in 2.5 h. Note how in these calculations we cancelled units just as we multiplied and divided numbers. Had our answers not come out with the correct units, we would have a sure indication of an error in the calculation.

Vectors, Scalars, and Displacement

Suppose you drive 15 minutes to a hamburger stand 10 miles away, grab your food at a drive-through window, and take another 15 minutes to drive home. You've gone a total of 20 miles in half an hour, giving an average speed of 40 mi/h. You certainly have been travelling, as the hamburger proves. But in another sense, you have gone nowhere at all, since you end up back at your starting point.

Physicists distinguish between the distance actually travelled and the overall change in position. This overall change—between the position at which you started and the position at which you finished—we call the **displacement**. In our hamburger example, the displacement is zero even though the overall distance travelled is 20 miles.

displacement

Starting from Boston, a 400-km displacement to the southwest gets you to Philadelphia. But head north-northwest the same distance and you find yourself in Montreal. Specifying displacement requires not only the overall distance between your final and initial positions, but also the direction you travel. A quantity for which we keep track of direction, like displacement, is a **vector**. A quantity not associated with a direction, like mass, is a **scalar.**

vector

scalar

In this book, we write scalars in regular type and vectors in bold type. When handwriting vectors, we often write an arrow over the symbol, as in $\vec{r}$ for the displacement vector. We write the magnitude of the displacement vector **r** as $|\mathbf{r}|$ or simply r. The magnitude of a vector is always greater than or equal to zero (≥ 0).

component

We define a **component** of a vector as the projection of the vector onto a given axis. Figure 2–1, for example, graphs our Boston-to-Montreal displacement on coordinate axes x and y that run east-west and north-south. We designate the components of an arbitrary vector **A** by the terms A_x, A_y, and A_z. The components of a vector are simply numbers. As shown in Fig. 2–1, they are the projections of the vector onto the coordinate axes, and can be positive or negative. (We normally define the axes as in Fig. 2–1, with the x-axis horizontal, the y-axis vertical, and the z-axis out of the page. In some circumstances, we will find other orientations more convenient.)

Velocity

In describing motion, we want to know not only the displacement an object undergoes, but also how rapidly that displacement occurs. Accordingly, we define **average velocity** as displacement divided by time interval.

average velocity

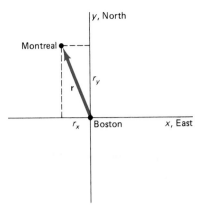

Fig. 2–1
The displacement vector **r** describing a trip from Boston to Montreal. The *y*-component r_y of the vector is the change in north-south position, while the *x*-component r_x is the change in the east-west position. With the positive *x*-direction eastward, the *x*-component is negative.

(To divide a vector by a scalar, we simply divide the magnitude of the vector by the scalar to get a new vector that lies along the same line as the original vector.) In our hamburger example, our average velocity was zero because our displacement was zero. If our 400-km Boston-to-Montreal trip took 4 hours, then our average velocity was 100 km/h north-northwest. Since displacement is a vector, so is velocity, and we must specify not only its magnitude but also its direction.

To see the relation among velocity, displacement, and time, consider a trip to a town 50 km east of home. If you travel eastward at a constant 50 km/h for 1 hour, your average velocity is 50 km/h eastward. If you paused for an hour in the midst of your trip, then you would have travelled 50 km eastward in 2 hours and your average velocity would have been eastward at 50 km/2 h = 25 km/h. Or suppose you got almost to the town, realized you forgot something, and roared rapidly home and back to the town. If the whole trip again took 2 hours, your average velocity would still be 25 km/h, even though you were actually moving much faster most of the time.

Using symbols, we can write the average velocity as

$$\bar{\mathbf{v}} = \frac{\Delta\mathbf{r}}{\Delta t},$$ (2–1)

where $\Delta\mathbf{r}$ is the vector displacement in the time interval Δt. The bar over the vector **v** indicates an average quantity (and is read "v bar"). We have here introduced the symbol Δ (the Greek capital delta) for "the change in . . . ." The individual components of velocity may be calculated from the components of the displacement; for example, the x-component of the average velocity, $\bar{v}_x$, is

$$\bar{v}_x = \frac{\Delta x}{\Delta t},$$ (2–2)

where Δx is the x-component of the displacement vector $\Delta\mathbf{r}$. In this chapter on one-dimensional motion, we will consider only motion with one non-zero component.

Example 2–1 —————————————————————

Flowing Lava

Geologists studying the flow of lava underground often drop a branch of wood in at one hole, and watch downstream a measurable distance away to see how long it takes for the wood to travel that known distance. At a recent volcanic eruption on the island of Hawaii, a branch travelled 100 meters in 12 seconds. What was the magnitude of its average velocity?

Solution

Applying Equation 2–1, we have

$$\bar{v} = \frac{\Delta r}{\Delta t} = \frac{100 \text{ m}}{12 \text{ s}} = 8.3 \text{ m/s}.$$

Since we were interested only in the **magnitude**, we have not written *v* or Δr as vectors, or specified a direction.

2.2
MOTION IN SHORT TIME INTERVALS

Although our geologists' experiment told them the average velocity of the lava flow, it did not give them all the details of the motion. For example, did the lava move faster at the beginning of the interval? Was there a region where the flow slowed down significantly? The geologists could answer these questions with a series of observations measuring the lava flow velocity over smaller intervals of time and distance (Fig. 2–2).

As the size of the intervals shrinks, and their number grows, an ever-more-detailed picture of the velocity emerges. In the limit of arbitrarily small time intervals, we achieve a description that gives a value for the velocity at essentially each instant. We call the velocity obtained through this limiting process the **instantaneous velocity, v.** The magnitude of the instantaneous velocity is called the **instantaneous speed.**

Physically, we can determine an approximate value for the instantaneous velocity by measuring the average velocity $\Delta \mathbf{r}/\Delta t$ over a very small time interval; we will soon see how we can calculate the instantaneous velocity mathematically given displacement as a function of time.

instantaneous velocity
instantaneous speed

Example 2–2 _____
Average and Instantaneous Velocity

You arrive at Boston's Logan Airport at 2:00 PM to check in for a trip to Detroit. Your plane departs at 3:00 PM and flies westward 1400 km to Chicago at 900 km/h. There, you catch a 5:30 flight to Detroit, which lies 390 km east of Chicago. Your plane lands in Detroit at 6:10. Assuming the planes maintain constant velocity while in the air, what are your instantaneous velocities at 2:30 PM, 4:00 PM, and 6:00 PM? What is your average velocity for the entire trip?

Fig. 2–2
Determining the velocity of a lava flow.
(a) Measuring the time Δt for a branch to go the distance Δx from A to B gives the average flow velocity $\bar{v} = \Delta x/\Delta t$.
(b) A series of measurements over shorter intervals gives a more detailed picture of the flow, by providing values for the velocities $\bar{v}_1 = \Delta x_1/\Delta t_1$, $\bar{v}_2 = \Delta x_2/\Delta t_2$, etc.

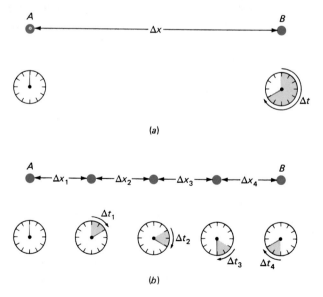

Solution

At 2:30, you are still waiting in Boston, so your instantaneous velocity is zero. At 900 km/h, the 1400-km trip to Chicago will take well over an hour, so that at 4:00 PM you are still en route to Chicago, with instantaneous velocity 900 km/h westward. At 6:00 PM, you are en route to Detroit. Since the 390-km trip from Chicago to Detroit takes 40 min or 0.67 h, the magnitude of your velocity is 390 km/0.67 h = 580 km/h. Because the plane's velocity is constant, your instantaneous velocity at 6:00 PM is also eastward at 580 km/h. The entire trip takes 4 h 10 min, or 4.17 h. Since you travel westward 1400 km, then eastward 390 km, your net displacement is 1400 km − 390 km = 1010 km westward. Then your average velocity is 1010 km/4.17 h = 242 km/h westward.

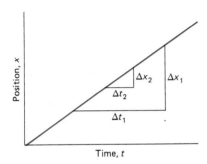

Fig. 2–3
A graph of position versus time for an object moving at constant velocity. Shown are two time intervals Δt and the corresponding displacements Δx. The average velocity over an interval is the ratio $\Delta x/\Delta t$, or the slope of a line connecting the ends of the interval. With constant velocity this slope is the same for all intervals.

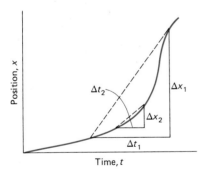

Fig. 2–4
A graph of position versus time for an object whose velocity is continually changing. Dotted lines connect the endpoints of two intervals; since the lines have different slopes, the average velocities $\Delta x/\Delta t$ have different values over the two intervals.

To make the notion of instantaneous velocity more precise, consider first motion at constant velocity, as shown in Fig. 2–3. The graph is drawn in the usual format for a graph describing motion: time along the horizontal axis and distance along the vertical axis. Note that if we shorten the length of the time interval we consider by any factor, the distance is shortened by the same factor. We get the same value for velocity no matter how short the time interval; thus for this constant-velocity situation, the average and instantaneous velocities are equal. Both velocities are, indeed, given by the slope of the line on the graph—that is, by distance divided by time.

In Fig. 2–4, we plot the motion of an object whose velocity is continually changing. Now as we consider different time intervals, we get different slopes. But if we examine shorter and shorter intervals around one specific time, we find that the part of the curve we are limited to gets closer and closer to a straight line (Fig. 2–5). The average velocity we compute then approaches a single value, which in the limit of arbitrarily small intervals is just the slope of the curve at a single point.

Thus, to compute the instantaneous velocity—the velocity at any given instant—we apply Equation 2–1 or 2–2 as the time interval Δt becomes arbitrarily small. In doing so, we are carrying out the mathematical equivalent of the physical procedure our geologists could have used to discover the detailed motion of their lava flow in Example 2–1. Of course, we cannot take Δt all the way to zero, for then $\Delta x = 0$, and the velocity would be undefined. But the velocity remains perfectly well defined for all nonzero values of Δt, no matter how small. So as we consider ever smaller time intervals, the average velocity we calculate comes ever closer to the instantaneous velocity. We can say this by writing

$$v_x = \lim_{\Delta t \to 0} \frac{\Delta x}{\Delta t}. \qquad (2\text{–}3a)$$

Read this equation as "v_x is the limit as Δt approaches zero of Δx divided by Δt." This statement is true even though the value of $\Delta x/\Delta t$ would be undefined if we evaluated it at $\Delta t = 0$. And it is true even for nonconstant velocity, since a sufficiently short segment of a curve comes arbitrarily close to being straight (Fig. 2–5).

Here physics diverges from mathematics in two significant ways. Mathematically, we can construct position-versus-time graphs including points

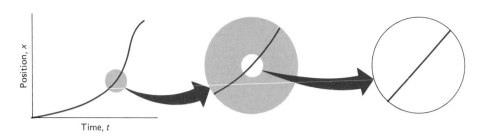

Fig. 2–5
As we examine an ever-smaller segment of the curve, it begins to resemble a straight line. The slope of this line is the magnitude of the instantaneous velocity.

where the limit in Equation 2–3a is undefined (Fig. 2–6). But such examples correspond physically to the impossible cases of an abrupt change in velocity or an infinite velocity. So in physics, at least, the limit in Equation 2–3a always exists. A second distinction arises because mathematically we take the limiting procedure to arbitrarily small time intervals Δt. But in attempting to measure an instantaneous velocity physically (or for that matter, in calculating such a velocity from position-versus-time data produced by a computer), there is a minimum time interval over which we can obtain accurate data. Our physical determinations of instantaneous velocities are therefore always approximations.

The limiting process we have just described occurs so frequently in physics and mathematics that we give the result a special name and symbol. We define the **derivative** of x with respect to t (symbol dx/dt) as

derivative

$$\frac{dx}{dt} = \lim_{\Delta t \to 0} \frac{\Delta x}{\Delta t}.$$

infinitesimals

The quantities dx and dt are called **infinitesimals;** they represent arbitrarily small quantities associated with the limiting process that defines the derivative. We can then write Equation 2–3a for the velocity component in the form

$$v_x = \frac{dx}{dt}. \tag{2–3b}$$

We have discussed this limiting procedure for the x-component of the velocity vector—the only component in one-dimensional motion. In the more general case (which we will discuss further in Chapter 3), we write

$$\mathbf{v} = \lim_{\Delta t \to 0} \frac{\Delta \mathbf{r}}{\Delta t} = \frac{d\mathbf{r}}{dt}, \tag{2–4}$$

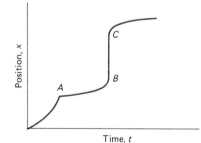

Fig. 2–6
A position-versus-time graph including an abrupt change in velocity (point *A*) and a time when the velocity is infinite (section *BC*). In both these cases, the limit in Equation 2–3a is undefined, but both are physically impossible.

where **r** is the displacement vector. Clearly, Equation 2–4 represents the limiting procedure applied to Equation 2–1. More generally, if some variable y is a function of some other variable x, then the derivative dy/dx measures the rate of change of y with respect to x—that is, the slope of the y versus x graph. Thus Equation 2–4 says that velocity is the rate of change of position.

Example 2–3

Derivative of a Square Law

The altitude of a space shuttle in the first half minute of its ascent is given approximately by a "square law," that is, $x = bt^2$, where the constant b is 2.9 m/s^2, x is in meters, and t is the time in seconds since lift-off. Using the process of taking limits, derive a general expression for the velocity, and determine its value when $t = 20$ s.

Solution

To find the instantaneous velocity v at some time t, we need to consider a small time interval Δt that includes t. In the limit $\Delta t \to 0$, the average velocity $\Delta x / \Delta t$ over that interval becomes the instantaneous velocity dx/dt. So consider an interval stretching from t to $t + \Delta t$; then evaluating $x = bt^2$ at $t + \Delta t$ and at t gives

$$v_x = \lim_{\Delta t \to 0} \frac{\Delta x}{\Delta t} = \lim_{\Delta t \to 0} \frac{b(t + \Delta t)^2 - bt^2}{\Delta t}$$

$$= \lim_{\Delta t \to 0} \frac{b(t^2 + 2t\Delta t + \Delta t^2 - t^2)}{\Delta t}$$

$$= \lim_{\Delta t \to 0} \frac{b(2t\Delta t + \Delta t^2)}{\Delta t} = \lim_{\Delta t \to 0} (2bt + 2b\Delta t) = 2bt.$$

For $t = 20$ s, $v_x = 2(2.9$ m/s$^2)(20$ s$) = 120$ m/s. Note that the units of b had to be m/s^2 for the expression $x = bt^2$ to be dimensionally correct.

We need not repeat the rather lengthy calculation of Example 2–3 every time we need the derivative of a simple function like t^2. The result of that example—that the derivative of bt^2 is $2bt$—is but a special case of a general rule for derivatives of variables that are raised to a power. We will not prove the general case here; that is usually done in introductory calculus courses. The result is that, for

$$x = bt^n,$$

where b and n are constants, then

$$\frac{dx}{dt} = bnt^{n-1}. \tag{2–5}$$

Equation 2–5 is true for all values of n, including both integer and non-integer values, positive and negative. The equation provides us directly with the result of the limiting process that defines the derivative; using it allows us to bypass the detailed limiting process we did in Example 2–3. You should verify that Equation 2–5, applied to $x = bt^2$, gives the result of Example 2–3. Derivatives of other functions, including trigonometric functions, exponentials, and logarithms, are given in Appendix B (see also Problem 63).

2.3
ACCELERATION

acceleration

In our space shuttle example, the velocity of the shuttle was changing. When velocity changes, an object is said to undergo **acceleration.** Since velocity is a vector, it has both a direction and a magnitude, and either can change. A car starting up from a stoplight is accelerating because the magnitude of its velocity is changing. A car going around a corner at constant speed is also accelerating because the direction of its velocity is changing; we will discuss such two-dimensional motion in the next chapter.

We define acceleration as the rate of change of velocity, just as we defined velocity as the rate of change of position. The average acceleration over a time interval Δt is

$$\bar{\mathbf{a}} = \frac{\Delta \mathbf{v}}{\Delta t},$$

where $\Delta \mathbf{v}$ is the change in velocity and where the bar over the **a** indicates that it is an average value. Since velocity is a vector, so is its rate of change. However, the velocity and acceleration vectors need not have the same direction. In one-dimensional motion, the acceleration is in the same direction as the velocity if the speed is increasing, but is in the opposite direction if the speed is decreasing. Just as we defined instantaneous velocity through a limiting procedure, so we define instantaneous acceleration as

$$\mathbf{a} = \lim_{\Delta t \to 0} \frac{\Delta \mathbf{v}}{\Delta t} = \frac{d\mathbf{v}}{dt}. \qquad (2\text{--}6)$$

In this chapter, we will always discuss a_x, the component of the acceleration along the x-axis, rather than **a,** the vector acceleration. Clearly, the x-component of Equation 2–6 is

$$a_x = \lim_{\Delta t \to 0} \frac{\Delta v_x}{\Delta t} = \frac{dv_x}{dt}. \qquad (2\text{--}7)$$

The word "deceleration" is often used to describe a slowing-down—that is, an acceleration whose direction is opposite to the velocity. When solving problems, though, it is usually easier to consider acceleration in one dimension as an algebraic quantity, and to find out whether a solution represents a decrease or increase in speed by comparing the sign of the acceleration with that of the velocity.

Since acceleration is the rate of change of velocity, the units of acceleration must be (unit of velocity) per (unit of time). Since the units of velocity are distance per time, the units of acceleration are (distance/time) per time, which can be written distance/time^2. For example, if you accelerate in such a way that your velocity changes by 5 m/s during each second, your acceleration is 5 (m/s)/s or 5 m/s^2. Sometimes acceleration is specified in mixed units; for example, a car going from 0 to 60 mi/h in 10 seconds has an average acceleration of 6 mi/h/s.

Example 2–4

Average Acceleration

A jetliner rolls down the runway with constant acceleration. From rest, it reaches its takeoff speed of 250 km/h in 1 min. What is its acceleration? Express in km/h².

Solution

The plane undergoes a velocity change Δv_x of 250 km/h in time $\Delta t = 1$ min or 1/60 h. The average acceleration is therefore

$$\bar{a}_x = \frac{\Delta v_x}{\Delta t} = \frac{250 \text{ km/h}}{(1/60 \text{ h})} = 15{,}000 \text{ km/h}^2.$$

Does this huge result make sense? Yes—the jet reaches 250 km/h in only 1 minute, so an increase of 15,000 km/h in 1 hour is not unreasonable. Of course, the plane never reaches such a high speed because it does not accelerate for a full hour.

Example 2–5

Acceleration and Deceleration

Two cars are moving eastward at 20 m/s. At time $t = 0$ both undergo accelerations of magnitude 2 m/s², with one car's acceleration eastward and the other westward. How fast is each car going 5 s later?

Solution

Each car's velocity changes at the rate of 2 m/s², or 2 (m/s)/s, so that in 5 s, each velocity has changed by (2 m/s²)(5 s) = 10 m/s. (Note how the units work out.) Both cars are *moving* eastward. For the car that is also *accelerating* eastward, this change represents a speeding up to 30 m/s. But for the car that is accelerating westward—opposite to the direction of its velocity—the change represents a slowing down, to 10 m/s.

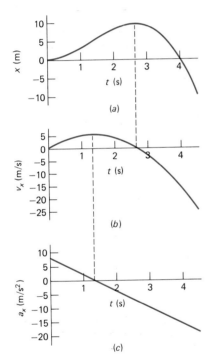

Fig. 2–7
Position, velocity, and acceleration versus time for one-dimensional motion with position given by $x = bt^2 - ct^3$, where $b = 4$ m/s² and $c = 1$ m/s³. Note that zero velocity occurs where the *slope* of the position graph is zero, and zero acceleration where the *slope* of the velocity graph is zero. In general, each graph gives the *slope* of the one above it; there is no particular relation between the actual *values* on one graph and those on a lower graph.

Just as velocity is the slope of the position-versus-time graph, so acceleration is the slope of the velocity-versus-time graph. Figure 2–7 shows graphs of position, velocity, and acceleration versus time for a particle undergoing one-dimensional motion with position x given by $x = bt^2 - ct^3$, where $b = 4$ m/s² and $c = 1$ m/s³. We obtained the velocity by differentiating the position: $v_x = d(bt^2 - ct^3)/dt = 2bt - 3ct^2$; a second differentiation gives the acceleration: $a_x = d(2bt - 3ct^2)/dt = 2b - 6ct$. Note that where the curve of Fig. 2–7a is steep—where the position x changes rapidly—the magnitude of the velocity in Fig. 2–7b is large. When x is decreasing with time, the velocity is negative. Similarly, where the slope of the velocity graph is greatest, the acceleration has its greatest magnitude. And when the velocity is decreasing—regardless of whether its value is positive or negative—the acceleration is negative. Finally, we have marked on Fig. 2–7 the points where velocity and acceleration are zero. The zero-velocity points appear where the *slope* of the position graph is zero—where that graph is flat. Similarly, the acceleration is zero where the *slope* of the velocity curve is zero.

Notice in Fig. 2–7 that the *value* of the velocity is unrelated to the *value* of the position. We could change all positions by some fixed amount (by simply moving the entire position-versus-time curve up or down on its graph), and this would have no effect on the velocity graph. The velocity is the *slope* of the position curve—and the slope depends on how the position is changing, not on its actual value. Similarly, the value of the acceleration depends only on the rate of change of velocity, not on the velocity itself. In particular, we can have zero velocity and still be accelerating! You can see this in Fig. 2–7, where the point of zero velocity does not correspond to zero acceleration.

How can something be accelerating if it's not moving? If this idea troubles you, remember that we are talking about *instantaneous* values. If the velocity stays zero for any length of time, then the acceleration must also be zero during that time. But if the velocity goes through zero instantaneously—as happens to a ball thrown straight up at the top of its flight—then the acceleration can remain nonzero. Just before the ball was stopped, it was moving upward. Just after it was stopped, it was moving downward. No matter how small a time interval Δt you choose around the instant the ball was stopped, you will find that on one side of that interval the ball was moving upward, and on the other side it was moving downward. There is always a change in velocity, and therefore always an acceleration—even in the limit $\Delta t \rightarrow 0$ that gives the acceleration at the instant the velocity is zero. Graphically, a plot of velocity versus time for the ball shows that the velocity is always changing—even as it passes through zero on its way to negative (downward) values (Fig. 2–8).

Since acceleration is the time derivative (the derivative with respect to time) of velocity and velocity is, in turn, the time derivative of position, we

second derivative

define a **second derivative** as a derivative taken twice. The second derivative of x with respect to t is written d^2x/dt^2; this is a symbol and does not mean that we actually square anything. We then write

$$\mathbf{a} = \frac{d\mathbf{v}}{dt} = \frac{d}{dt}\left(\frac{d\mathbf{r}}{dt}\right) = \frac{d^2\mathbf{r}}{dt^2} \tag{2–8}$$

for the acceleration vector and

$$a_x = \frac{dv_x}{dt} = \frac{d}{dt}\left(\frac{dx}{dt}\right) = \frac{d^2x}{dt^2} \tag{2–9}$$

for its x-component.

2.4
CONSTANT ACCELERATION

When the acceleration is constant, the mathematical description of motion takes an especially simple form. In this section, we derive and apply basic equations for the velocity and position of an object undergoing constant acceleration. We do so in two steps, going first from acceleration to velocity and then from velocity to position.

Since $a_x = dv_x/dt$, we must find an expression for v_x which, when differentiated, gives constant a_x. From Equation 2–5, we see that $v_x = a_x t$ is such an expression, since $d(a_x t)/dt = a_x$. If we add a constant to this

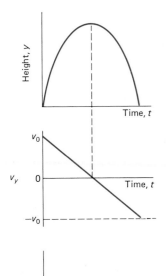

Fig. 2–8
Position, velocity, and acceleration versus time for a ball thrown straight up. At the peak of its flight, the ball is instantaneously at rest ($v = 0$). But even at that point, the velocity is in the process of changing—here from upward to downward—so that the acceleration is nonzero. In this case the acceleration is, in fact, constant and downward (i.e., negative).

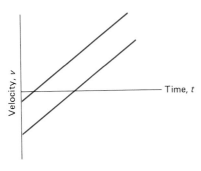

Fig. 2–9
Two velocity-versus-time curves that differ everywhere by the same constant. The two curves have the same slope, and therefore imply the same acceleration.

expression for velocity, the acceleration is still

$$\frac{d}{dt}(a_x t + \text{constant}) = \frac{d(a_x t)}{dt} + \frac{d}{dt}(\text{constant}) = a_x,$$

since the derivative—the rate of change—of a constant is zero. On a graph, adding the constant is equivalent to sliding the entire velocity-versus-time curve up or down by a fixed amount (Fig. 2–9). Such an action does not change the slope, so has no effect on the acceleration.

Evaluating the expression $v_x = a_x t + \text{constant}$ at time $t = 0$ shows that the velocity at time $t = 0$ is equal to the constant. We therefore call this constant v_{x0}, the initial velocity. Thus

$$v_x = v_{x0} + a_x t \tag{2–10}$$

is the general expression for velocity v_x as a function of time when an object undergoes constant acceleration a_x.

Just as we found an expression for velocity that gives constant acceleration, we now seek an expression for position that gives the velocity (Equation 2–10) associated with constant acceleration. Since $v_x = dx/dt$, this means that we want position x as a function of time that when differentiated gives Equation 2–10. We will get the constant term v_{x0} if we start with $v_{x0}t$, since $d(v_{x0}t)/dt = v_{x0}$. As in our reasoning to Equation 2–10, we can in fact start with $v_{x0}t + \text{constant}$ and still get the v_{x0} term when we differentiate. What about the term $a_x t$? Differentiating $a_x t^2$ with respect to t gives $2a_x t$, which is twice what we want, so we must start with $\frac{1}{2}a_x t^2$; again we can add any constant. Combining the two constants gives

$$x = v_{x0}t + \tfrac{1}{2}a_x t^2 + \text{constant}.$$

Evaluating this equation at time $t = 0$ shows that the constant is the initial position, which we call x_0. Then

$$x = x_0 + v_{x0}t + \tfrac{1}{2}a_x t^2 \tag{2–11}$$

is the general expression for the position of an object undergoing constant acceleration in one dimension, if its initial position is x_0 and its initial velocity is v_{x0}. The t^2 dependence in Equation 2–11 makes physical sense: with constant acceleration, the longer we travel the faster we go and therefore the more distance we cover in a given time.

We have been working backward, "anti-differentiating," to find the equation for x that would give the result we know we need for $a_x = d^2x/dt^2$. Later, we will use this process, formally known as "integration," more directly.

We can also understand Equation 2–11 by considering the average velocity of an object undergoing constant acceleration. Suppose that at time 0, the velocity is v_{x0}, and at some arbitrary later time t it is v_x. Since the acceleration is constant, the velocity rises steadily with time (Fig. 2–10a). Because the velocity rises steadily, the average velocity $\bar{v}_x$ in the interval from time 0 to time t lies midway between the initial velocity v_{x0} and the final velocity v_x:

$$\bar{v}_x = \tfrac{1}{2}(v_{x0} + v_x). \quad \text{(constant acceleration)} \tag{2–12}$$

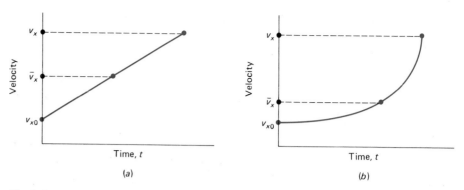

Fig. 2–10
(a) With constant acceleration, the velocity rises steadily, with slope equal to the acceleration. Because the rise is steady, the average velocity lies midway between the initial velocity v_{x0} and the final velocity v_x. *(b)* When the acceleration is not constant, the velocity-versus-time curve is not a straight line, and the average velocity is not midway between the initial and final velocities.

Note that this result is necessarily true *only* for the case of constant acceleration. When the acceleration varies with time, the velocity-versus-time curve is not straight, and the average velocity will, in general, lie closer to one of the two extremes (Fig. 2–10b).

From Equation 2–2, we know that the average velocity is the displacement divided by time interval, so that in time Δt, we travel a distance $\Delta x = \bar{v}_x \, \Delta t$. Calling x_0 the initial position at time 0, and x the position at time t, we have $\Delta x = x - x_0$. Our time interval Δt is the final time t minus the initial time 0, or simply t. Then we have

$$x - x_0 = \bar{v}_x t.$$

In the case of constant acceleration, we can use Equation 2–12 for the average velocity $\bar{v}_x$, giving

$$x - x_0 = \tfrac{1}{2}(v_{x0} + v_x)t. \qquad (2\text{–}13)$$

Equation 2–13 is a useful form if we know we have constant acceleration and are given the initial and final velocities but not the acceleration itself. From Equation 2–10, we can write $v_x = v_{x0} + a_x t$, so that Equation 2–13 becomes

$$x - x_0 = \tfrac{1}{2}(v_{x0} + v_{x0} + a_x t)t,$$

or

$$x = x_0 + v_{x0}t + \tfrac{1}{2}a_x t^2,$$

which is Equation 2–11.

In some cases, we want to relate position and velocity directly, without explicit mention of time. To do so, we solve Equation 2–10 for the time t, giving

$$t = \frac{v_x - v_{x0}}{a_x}.$$

Substituting this result for t in Equation 2–13 gives

TABLE 2–1 ▬▬▬▬▬▬▬▬▬▬▬▬▬▬▬▬▬▬▬▬▬▬▬▬▬▬▬▬▬
Equations of Motion for Constant Acceleration

Equation	*Contains*	*Number*
$v_x = v_{x0} + a_x t$	$v, a, t;$ no x	2–10
$x = x_0 + v_{x0}t + \frac{1}{2}a_x t^2$	$x, a, t;$ no v	2–11
$\bar{v}_x = \frac{1}{2}(v_{x0} + v_x)$	Average v	2–12
$x - x_0 = \frac{1}{2}(v_{x0} + v_x)t$	$x, v, t;$ no a	2–13
$v_x^2 = v_{x0}^2 + 2a_x(x - x_0)$	$x, v, a;$ no t	2–14

$$x - x_0 = \frac{1}{2}\frac{(v_{x0} + v_x)(v_x - v_{x0})}{a_x},$$

or, since $(a + b)(a - b) = a^2 - b^2$,

$$v_x^2 = v_{x0}^2 + 2a_x(x - x_0). \tag{2–14}$$

Equations 2–10 through 2–14 link all possible combinations of position, velocity, and acceleration for motion with constant acceleration. We summarize these equations in Table 2–1.

You should not memorize these equations. Instead, you will grow familiar with them as you work problems involving one-dimensional motion. As you use the equations, do not regard them as a set of separate laws, but instead see them as complementary descriptions of the same underlying phenomenon—one-dimensional motion with constant acceleration. We stress again that we derived Equations 2–10 to 2–14 under the assumption of *constant acceleration*. They do not hold if the acceleration vector is changing in either magnitude or direction.

Example 2–6 _____
Acceleration from Distance and Velocity

The electron beam in a color TV tube jumps from near zero velocity to 1×10^{10} cm/s in a distance of 4 cm near the back of the tube. What is its acceleration?

Solution

We are given the initial and final velocities, and the distance over which the acceleration occurs. Equation 2–14 links these quantities with the acceleration. Solving for a_x, with $v_{x0} = 0$, we have

$$a_x = \frac{v_x^2}{2(x - x_0)} = \frac{(1 \times 10^{10} \text{ cm/s})^2}{2(4 \text{ cm})} = 1 \times 10^{19} \text{ cm/s}^2,$$

where we keep only one significant figure in our answer because that is all we were given in the original data. Such a huge acceleration is possible because of the tiny mass of the electron. We will explore the relation between mass and acceleration in Chapter 4.

Example 2–7 _____
Distance from Velocity and Time

I am riding my bicycle at a leisurely 10 km/h, then speed up at a steady rate for 10 s to escape a dog. At the end of that time, I am going at 30 km/h. How far did I go during the 10-s interval?

Solution

We want to link v_x and t, which we know, with x, which we don't know. Equation 2–13 does this. Our answer is the distance $x - x_0$ travelled during the time t:

$$x - x_0 = \tfrac{1}{2}(v_{x0} + v_x)t = \tfrac{1}{2}(10 \text{ km/h} + 30 \text{ km/h})(10 \text{ s})(1 \text{ h}/3600 \text{ s})$$
$$= 0.028 \text{ km} = 28 \text{ m}.$$

Note that, with our time in seconds and our velocity in km/h, we had to convert seconds to hours for consistency of units.

Occasionally, we are interested in comparing the one-dimensional motions of two different objects, as the example below illustrates.

Example 2–8

Speed Trap

A speeding motorist zooms through a 50 km/h zone at 75 km/h, without noticing a stationary police car by the roadside. The police officer immediately heads after the speeder, accelerating at 9.0 km/h/s. When the officer pulls alongside the speeder, how far down the road are they, and how fast is the police car going?

Solution

In this example, we deal with two different one-dimensional motions, both described by Equation 2–11. We want to know when the two cars are in the same place—that is, when equations for the positions of the two cars give the same value. In working such a problem, it is important to define the initial positions and times consistently for both motions; otherwise a comparison between the two is meaningless.

Let $t = 0$ be the time the speeder first passes the police car, and let $x = 0$ be the position at which this occurs. Then $x_0 = 0$ for both cars. For the speeder, we have $v_{s0} = 75$ km/h $= 21$ m/s and $a_s = 0$. For the police car, $v_{p0} = 0$ but $a_p = 9.0$ km/h/s $= 2.5$ m/s^2. (Here we use subscripts s and p for quantities associated with the speeder and police, respectively, and have dropped the subscript x because it is clear we are dealing only with one-dimensional motion.) Then for the two cars Equation 2–11 becomes

$$x_s = v_{s0}t \quad \text{(speeder)}$$

and

$$x_p = \tfrac{1}{2}a_p t^2. \quad \text{(police)}$$

Equating these expressions tells when the speeder and police car are at the same place:

$$v_{s0}t = \tfrac{1}{2}a_p t^2,$$

so that $t = 0$ or $t = 2v_{s0}/a_p$. Why two answers? We asked the equations for *any* times when the two cars were in the same place. Our first answer is just the initial encounter as the speeder passes the stationary police car. The second answer is the one we want—the time when the officer catches up with the speeder. Where does this occur? Using the time $2v_{s0}/a_p$ in the speeder equation, we have

$$x = v_{s0}t = v_{s0}\frac{2v_{s0}}{a_p} = \frac{2v_{s0}^2}{a_p} = \frac{(2)(21 \text{ m/s})^2}{2.5 \text{ m/s}^2} = 350 \text{ m}.$$

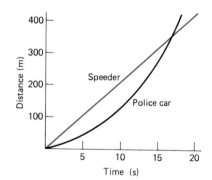

Fig. 2–11
Position versus time for the speeder and police car of Example 2–8.

We can calculate the police car's speed at this time from Equation 2–10, with $v_{p0} = 0$:

$$v_p = a_p t = a_p \frac{2v_{s0}}{a_p} = 2v_{s0} = 150 \text{ km/h}.$$

Does this result make sense? Yes—the police car and the speeder cover the same distance in the same time. Therefore, their average velocities must be the same. The speeder maintains constant velocity, but the police car accelerates from rest. Since the average velocity with constant acceleration is half the sum of the initial and final velocities, the police car must be going twice as fast as the speeder when they meet again. Note that this remains true no matter what the acceleration of the police car.

Figure 2–11 shows both motions plotted on the same graph. Sketching such a graph is often helpful in understanding problems involving two or more one-dimensional motions.

2.5
THE CONSTANT ACCELERATION OF GRAVITY

Gravity is a force with which we are very familiar, so it is suitable to discuss some aspects of it here even though we will not formally define force until Chapter 4, and will study gravity in detail in Chapter 8. Here we concern ourselves only with gravity near the surface of the earth, where its effects do not vary significantly with height.

When you drop an object, it falls at an ever-increasing rate (Fig. 2–12)—that is, it accelerates because the force of gravity is attracting it to the earth. Experimentally, we find that the acceleration remains constant for objects falling near the surface of the earth, and furthermore that this constant acceleration has the same value for all objects, independent of mass or other properties. The rate at which the speed of a falling object increases—the **acceleration of gravity**—is approximately 9.8 m/s^2 near earth's surface. (A more exact value is given in Appendix F, along with other properties of the earth and other planets.)

The acceleration of gravity applies strictly only to objects in **free fall**—that is, objects under the influence of gravity alone. The presence of other forces, in particular air resistance, may alter dramatically the statement that all objects fall with the same constant acceleration. Try dropping a flat piece of paper and a rock, for example. Their behavior is radically different. But if you crumple the paper, lessening its air resistance, it falls with an acceleration closer to that of the rock. An experiment illustrating that all objects fall with the same acceleration was carried out by astronauts on the moon in 1971, when they showed that a feather and a hammer fell together at the airless lunar surface (Fig. 2–13). The acceleration of gravity on the moon is different from that on earth, but the crucial concept is that all objects at the same location experience the same gravitational acceleration.

Galileo is reputed to have shown that all objects fall with the same acceleration by dropping objects off the Leaning Tower of Pisa in about 1600. There is no evidence that the Tower of Pisa played a role, but it is known that Galileo carried out careful experiments by rolling balls down

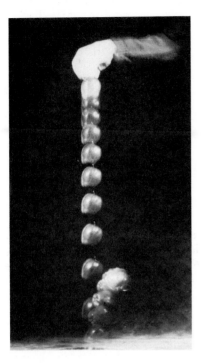

Fig. 2–12
Strobe photo of a falling apple, with the strobe flashing 30 times per second. Appearances of the apple are farther apart as it falls, showing that the apple is accelerating.

acceleration of gravity

free fall

Fig. 2–13
Apollo 15 astronaut David Scott dropping a feather and a hammer on the moon. Only video images exist, and the objects are hard to see in stills.

TABLE 2–2 ▬▬▬▬▬▬▬▬▬▬▬▬▬▬
A Freely Falling Object

Time (s)	Distance (m)	Speed (m/s)	Acceleration (m/s²)
0	0	0	9.8
1	4.9	9.8	9.8
2	19.6	19.2	9.8
3	44.1	29.4	9.8
4	78.4	39.2	9.8
5	123	49.0	9.8

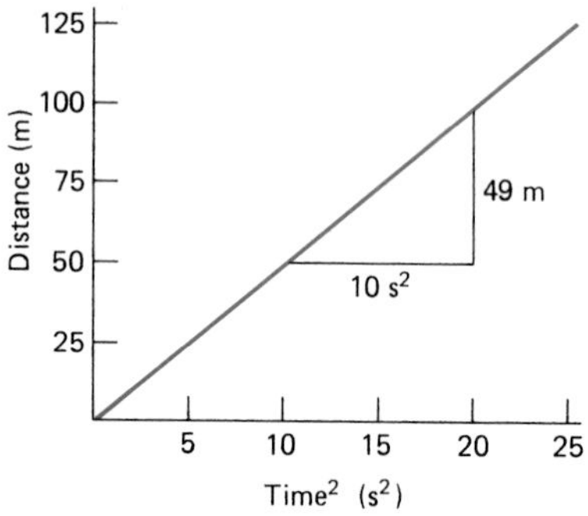

Fig. 2–14
A graph of distance versus time² for the data of Table 2–2. According to Equation 2–11, a straight-line graph shows that we have constant acceleration whose magnitude is twice the slope.

inclined planes. The gradual inclines "diluted" the effect of gravity, allowing Galileo to make more precise measurements. We now study vertically falling bodies with strobe lights or video cameras, equipment that was not available to Galileo.

That the acceleration of gravity is a constant 9.8 m/s² for objects in free fall near earth's surface means that the downward velocity component increases by 9.8 m/s every second. Table 2–2 gives the position, speed, and acceleration of a freely falling object like that of Fig. 2–12.

We can tell that the data of Table 2–2 represent constant acceleration because the distance increases as the *square* of the time. For example, the total distance after 2 s is 2^2 or 4 times that after 1 s; after 3 s the object has gone 3^2 or 9 times as far as in the first second. According to Equation 2–11, constant acceleration from rest should give distance $= \frac{1}{2}a(\text{time})^2$. Therefore, a graph of distance versus time² should be a straight line with slope $\frac{1}{2}a$. Figure 2–14 shows such a plot for the data of Table 2–2; that the plot is a straight line confirms that we have constant acceleration. The measured slope is 4.9 m/s²; thus the acceleration is 2(4.9 m/s²) or 9.8 m/s². Correspondingly, the velocity increases in direct proportion to the time, adding 9.8 m/s for every second. Although we work in SI units for the most part in this book, it is still worth knowing that the acceleration of gravity in English units is 32 (ft/s)/s, usually written 32 ft/s².

The magnitude of the gravitational acceleration is given the symbol g. Remember that g is relatively constant near earth's surface, but is not fundamentally a constant quantity. If we define the positive y-direction as vertically upward, then the downward acceleration enters Equations 2–10, 2–11, 2–13, and 2–14 with a minus sign, and these equations become

$$v_y = v_{y0} - gt \tag{2–15}$$

$$y = y_0 + v_{y0}t - \tfrac{1}{2}gt^2 \tag{2–16}$$

$$y - y_0 = \tfrac{1}{2}(v_y + v_{y0})t \tag{2–17}$$

$$v_y{}^2 = v_{y0}{}^2 - 2g(y - y_0), \tag{2–18}$$

where we have replaced x by y since we are describing motion along the vertical direction. Depending on our measurement system, we use g = 9.8 m/s² or g = 32 ft/s²; the minus signs in the equations take care of the fact that the acceleration is actually downward.

Fig. 2–15
A high dive (Example 2–9).

Fig. 2–16
Lava fountains at the 1983 eruption of
Kilauea volcano in Hawaii.

Example 2–9

A High Dive

A stunt diver drops from a 30-m-high platform (Fig. 2–15). At what speed does he enter the water? Give the answer in both m/s and mi/h. How long is he in the air? If a photographer's motor drive can take three frames per second, how many frames cover the dive?

Solution

To find the speed, given the distance, we use Equation 2–18. Since the diver starts from rest, $v_0 = 0$, and the equation becomes

$$v_y^2 = -2g(y - y_0).$$

Calling the water $y = 0$, the initial height is $y_0 = 30$ m. Then we have

$$v_y^2 = (-2)(9.8 \text{ m/s}^2)(0 \text{ m} - 30 \text{ m}) = 590 \text{ m}^2/\text{s}^2.$$

Taking the square root, we find that the speed—the magnitude of the velocity—at impact with the water is 24 m/s. (The actual velocity component v_y is -24 m/s, with the negative sign indicating the diver's downward motion.) Converting this result to mi/h gives

$$v = (24 \text{ m/s})(3600 \text{ s/h})(0.001 \text{ km/m})(0.62 \text{ mi/km}) = 54 \text{ mi/h}.$$

Note how the units cancel. Appendix D is useful for conversions such as this one.

Knowing the initial and final velocities, we can use Equation 2–15 to find out how long the diver is in the air. Solving that equation for time t gives

$$t = \frac{v_{y0} - v_y}{g} = \frac{0 \text{ m/s} - (-24 \text{ m/s})}{9.8 \text{ m/s}^2} = 2.5 \text{ s}.$$

Note here that we were careful to use the negative sign with v_y, indicating a downward velocity. Alternatively, we could have obtained the time from Equation 2–16, knowing the distance and acceleration. Solving for t^2 and setting $y_0 = 30$ m (top of platform), $y = 0$ (water), and $v_{y0} = 0$, we have

$$t^2 = \frac{2(y_0 - y)}{g} = \frac{2(30 \text{ m} - 0 \text{ m})}{9.8 \text{ m/s}^2} = 6.1 \text{ s}^2,$$

so that again, $t = 2.5$ s. At 3 frames per second, the diver will appear on (2.5)(3), or 7, complete frames after the dive starts.

Example 2–10

A Volcanic Eruption

The fountains of lava at the volcanic eruption shown in Fig. 2–16 are so hot that geologists can't get close. But they can watch a chunk of lava that has been thrown upward, and time its fall. Lava chunks at the peak of the fountain take 5.0 s to fall. How high are the fountains?

Solution

Here we know the time and the acceleration, so we can calculate the distance from Equation 2–16. Let $y_0 = h$, our unknown height at the peak of the fountain, and take $y = 0$ at the ground. With $v_{y0} = 0$ at the peak, Equation 2–15 becomes

$$0 = h - \tfrac{1}{2}gt^2,$$

so that

$$h = \tfrac{1}{2}gt^2 = \tfrac{1}{2}(9.8 \text{ m/s}^2)(5.0 \text{ s})^2 = 120 \text{ m}.$$

So far, we have treated only examples of objects that are actually moving downward. In that case, the magnitude of the velocity increases because the acceleration is in the same direction as the velocity. But the acceleration of an object subject only to gravity is 9.8 m/s² *downward* no matter what the direction of its motion. When you throw a ball straight up, for example, it is accelerating *downward* even while moving *upward*. Why? Because its upward velocity is *decreasing*—that is, the *change* in velocity is negative. Equations 2–15 to 2–18 continue to describe the situation no matter what the direction of this initial velocity.

Example 2–11
Upward Initial Velocity

You toss a ball straight up with an initial speed of 7.3 m/s. When it leaves your hand, it is 1.5 m above the floor. Use Equation 2–16 to determine when it hits the floor. Find also the maximum height it reaches and its speed when it passes your hand on the way down.

Solution

If we choose $y=0$ at the floor, then $y_0 = 1.5$ m and $v_{y0} = +7.3$ m/s. To find out when the ball hits the floor, we must ask of Equation 2–16: At what time t is the ball at $y=0$? That is, we want t such that

$$0 = y_0 + v_{y0}t - \tfrac{1}{2}gt^2.$$

Using the quadratic formula (see Appendix A), we have

$$t = \frac{v_{y0} \pm \sqrt{v_{y0}^2 + 2y_0g}}{g}.$$

The quantity under the square root is

$$v_{y0}^2 + 2y_0g = (7.3 \text{ m/s})^2 + 2(1.5 \text{ m})(9.8 \text{ m/s}^2) = 82.7 \text{ m}^2/\text{s}^2,$$

so that

$$t = \frac{7.3 \text{ m/s} \pm \sqrt{82.7 \text{ m}^2/\text{s}^2}}{9.8 \text{ m/s}^2} = 1.7 \text{ s or } -0.18 \text{ s}.$$

1.7 s is a reasonable answer, but what about the second answer, -0.18 s? Remember that Equations 2–15 to 2–18 were derived under the assumption that gravity alone provides the acceleration. The equations don't "know" anything about the upward acceleration of your hand or about the ball stopping when it hits the floor; they "assume" that the ball always undergoes a downward acceleration g. Our negative answer means simply this: had our hand not been involved, a ball travelling upward at 7.3 m/s when it was 1.5 m off the floor would have been on the floor 0.18 s *earlier*. Our positive answer, the one we expected, says that the ball will again hit the floor 1.7 s *later*.

At the peak of the ball's flight, its instantaneous velocity is zero since it is moving neither up nor down at that point. Setting $v_y^2 = 0$ in Equation 2–18, we can then solve for the peak height y:

$$0 = v_{y0}^2 - 2g(y - y_0),$$

so that

$$y = \frac{2gy_0 + v_{y0}^2}{2g} = y_0 + \frac{v_{y0}^2}{2g} = 1.5 \text{ m} + \frac{(7.3 \text{ m/s})^2}{(2)(9.8 \text{ m/s}^2)} = 4.2 \text{ m}$$

above the floor.

To determine the speed when the ball reaches 1.5 m on the way down, we can set $y = 1.5$ m in Equation 2–18 and solve for v_y. Since $y_0 = 1.5$ m, the term $y - y_0 = 0$, and Equation 2–18 becomes simply $v_y^2 = v_{y0}^2$. There are two possibilities: $v_y = v_{y0}$ and $v_y = -v_{y0}$. Again, our formulation of the problem does not really ask only about the downward flight; it simply asks for the velocity whenever the height is 1.5 m. The positive answer corresponds to the initial upward velocity and the negative answer to the velocity on the downward flight. In deriving this result, we have demonstrated the important point that an object thrown upward returns to its initial height with the same speed with which it left that height. (This result is exactly true only in the absence of air resistance.)

2.6
DIMENSIONAL ANALYSIS

We have seen that we can treat units as algebraic quantities, cancelling or multiplying them as needed. Each term of an equation, on either side of the equal sign, must have the same units in order for us to add and compare the terms.

It is often convenient to check the "dimensions" of the terms in an equation, ignoring the particular system of units. The basic dimensions in mechanics are length (L), time (T), and mass (M). Consider, for example, Equation 2–10: $v_x = v_{x0} + a_x t$. Here v_x has the dimensions of L/T. We express this by writing $[v_x] = L/T$, with the square brackets to indicate we are talking only about dimensions, not actual values. How about the term $a_x t$? Acceleration has units of velocity per time, so $[a_x] = (L/T)/T = L/T^2$. Thus $[a_x t] = (L/T^2)(T) = L/T$, matching the other terms in the equation.

Similarly, for Equation 2–11 we see that $[v_{x0} t] = (L/T)(T) = L$ and $[\frac{1}{2} a_x t^2] = (L/T^2)(T^2) = L$, matching the terms $[x]$ and $[x_0]$. Note that numerical constants like $\frac{1}{2}$ have no dimensions.

Dimensional analysis helps physicists and engineers see the form that new equations should take. It is also helpful in checking whether you have made a mistake in working a physics problem. All equations must be dimensionally consistent.

SUMMARY

1. **Kinematics** is the study of motion. In this chapter we limit ourselves to the one-dimensional motion of **particles**—objects that have no extent and therefore cannot rotate or vibrate. Real objects approximate particles to the extent we can ignore their rotation, vibration, or other internal motions.

2. **Displacement** is the net change in an object's position. Both distance and direction are needed to specify displacement, which is therefore a **vector** quantity—one that includes both direction and magnitude. We designate displacement from some fixed origin by the vector **r**.

3. **Velocity** is the rate of change of displacement. The **average velocity** $\bar{\mathbf{v}}$ is the displacement $\Delta \mathbf{r}$ divided by the time interval Δt over which it occurred:

$$\bar{\mathbf{v}} = \frac{\Delta \mathbf{r}}{\Delta t}.$$

Like displacement, velocity is a vector quantity. As with any vector, we can describe velocity in terms of **components**—the projections of the velocity vector onto the coordinate axes. In this chapter on one-dimensional motion, we deal with velocity that has only one nonzero component, for example, v_x. We then write

$$\bar{v}_x = \frac{\Delta x}{\Delta t}$$

for the x-component of the average velocity, where Δx is the x-component of the displacement vector $\Delta \mathbf{r}$.

4. To find the **instantaneous velocity**—the velocity at a specific instant of time—we take the limit of the average velocity as the time interval becomes arbitrarily small. Mathematically, this process defines the **derivative** $d\mathbf{r}/dt$:

$$\mathbf{v} = \lim_{\Delta t \to 0} \frac{\Delta \mathbf{r}}{\Delta t} = \frac{d\mathbf{r}}{dt}.$$

When a component of position can be written in the form $x = bt^n$, the derivative dx/dt, or v_x, is

$$v_x = \frac{dx}{dt} = bnt^{n-1}.$$

Graphically, the instantaneous velocity—the derivative of position with respect to time—is the slope of the position-versus-time curve.

5. **Acceleration** is a vector giving the rate of change of velocity. If $\Delta \mathbf{v}$ is the change in velocity in a time interval Δt, then the **average acceleration** over that interval is

$$\bar{\mathbf{a}} = \frac{\Delta \mathbf{v}}{\Delta t}.$$

Like the instantaneous velocity, the **instantaneous acceleration** is defined as the average acceleration taken in the limit of arbitrarily small time intervals Δt:

$$\mathbf{a} = \lim_{\Delta t \to 0} \frac{\Delta \mathbf{v}}{\Delta t} = \frac{d\mathbf{v}}{dt}.$$

In one dimension, the x-component of acceleration is given by $a_x = dv_x/dt$. The direction of the acceleration in one dimension is the same as that of the velocity when the speed is increasing; when the speed is decreasing, the acceleration is directed opposite to the velocity.

6. For **constant acceleration** (constant in both magnitude and direction), position is a quadratic function of time. The following equations relate displacement, time, velocity, and acceleration in one-dimensional motion with constant acceleration:

$$v_x = v_{x0} + a_x t$$

$$x = x_0 + v_{x0}t + \tfrac{1}{2}at^2$$

$$\bar{v}_x = \tfrac{1}{2}(v_{x0} + v_x)$$

$$x - x_0 = \tfrac{1}{2}(v_{x0} + v_x)t$$

$$v_x{}^2 = v_{x0}{}^2 + 2a_x(x - x_0).$$

7. **Gravity** is a force that gives rise to an acceleration that is the same for all objects at the same location. Near the surface of the earth, the acceleration of gravity is directed downward and is essentially constant at 9.8 m/s^2 or 32 ft/s^2.

8. **Dimensional analysis** provides a means of checking equations for consistency. An equation cannot be correct unless all its additive terms have the same dimensions.

QUESTIONS

1. Why can we consider the earth to be a particle in some applications but not in others?
2. Give three common examples of velocity units.
3. What is the difference between your displacement and the total distance you travel between the time you wake up and the time you go back to bed on a typical day?
4. You are driving straight at a steady 80 km/h, but stop a while for a picnic lunch. How does the stop affect your average velocity?
5. Does a speedometer measure speed or velocity?
6. You check your odometer at the beginning of a day's driving, and again at the end. Under what conditions would the difference between the two readings represent the magnitude of your displacement?
7. In the previous question, under what conditions would the difference in odometer readings, divided by the total time, equal the magnitude of your average velocity?
8. Two displacement vectors add to give zero net displacement. How must their magnitudes and their directions compare?
9. Consider two possible definitions of average speed: (a) average speed is the average of the values of the instantaneous speed over a time interval; (b) average speed is the magnitude of the average velocity. Are these definitions equivalent? Give examples to demonstrate your conclusion.
10. Is speed a scalar or a vector?
11. Is the x-component of velocity a scalar or a vector?
12. If your speedometer remains fixed at 55 mi/h, do you know that you are not accelerating?
13. Is it possible to be at the position $x = 0$ and still be moving?
14. Is it possible to have zero velocity and still be accelerating?
15. Is it possible to have zero average velocity over a 10-s interval and still be accelerating during the interval? If so, give an example. If not, why not?
16. Is it possible to have zero instantaneous velocity at all times in a 10-s interval and still be accelerating during the interval? If so, give an example. If not, why not?
17. Suppose your car's brakes provide nearly constant deceleration, independent of speed. If you double your initial speed, what does this do to the time it will take to stop your car? What does it do to the distance the car travels while stopping?

18. Starting from rest, an object undergoes an acceleration given by $a_x = bt$, where t is time and b is a constant. Can you use the expression bt for a_x in Equation 2–11 to predict the object's position as a function of time? Why or why not?

19. In which of the velocity-versus-time graphs shown in Fig. 2–17 would the average velocity over the interval shown equal the average of the velocities at the ends of the interval?

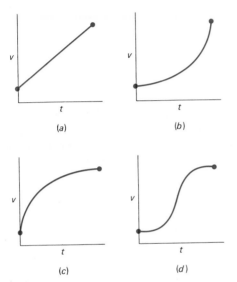

Fig. 2–17   Question 19.

20. What change should be made in Equations 2–15 to 2–18 if the positive y-direction were chosen to be downward?

21. At point A in Fig. 2–6, the slope of the position-versus-time curve changes abruptly. Comment on the velocity and acceleration at that point.

22. What would the third derivative of position signify? (The third derivative is sometimes called the "jerk." It is what can make you sick on an amusement park ride.)

23. What was the advantage in plotting distance versus time^2 rather than versus time in Fig. 2–14?

24. Standing on a roof, you simultaneously throw one ball straight up and drop another from rest. Which hits the ground first? Which hits the ground moving fastest?

25. Can the height of the lava fountains in Fig. 2–16 be used to estimate the masses of the lava chunks? Why or why not?

26. If you travel in a straight line at 50 km/h for one hour and at 100 km/h for another hour, is your average velocity 75 km/h? If not, is it more or less?

27. If you travel in a straight line at 50 km/h for 50 km, and then at 100 km/h for another 50 km, is your average velocity 75 km/h? If not, is it more or less?

28. (Zeno's paradox) If you try to reach the other side of the room, first you must cross half the distance, then half the remaining distance, then half the distance that still remains, and so forth. At any point, you still have half the distance to go. So it would seem that you can never reach the other side. What is wrong with this reasoning?

29. What relevance does the hammer/feather experiment on the moon have for our understanding of gravity on earth?

30. Why might crumpling a piece of paper make its air resistance less than that of a flat piece?

31. If you drop one rock from rest, and throw another down with velocity v_0, is the difference in their velocities when they hit the ground more than, less than, or equal to v_0? Don't apply any equations here; just think about how long each rock experiences the acceleration of gravity.

============ **PROBLEMS** ============

Section 2.1   *Distance, Time, and Velocity*

1. The world's record for the 200-m dash is 19.72 s, set by Italy's Pietro Mennea at the 1979 Mexico City Olympics. What was Mennea's average speed?

2. When races in a track meet are timed manually, timers start their watches when they see smoke from the starting gun, rather than when they hear the gun. How much error is introduced in timing a 200-m dash over a straight track if the watch is started on the sound rather than the smoke? The speed of sound is about 340 m/s.

3. The world's fastest marathon was the New York City Marathon of 1981, won by Alberto Salazar. Salazar covered the 26-mi 385-yd course in 2 h 8 min 13.0 s. What was his average speed?

4. Human nerve impulses travel at about 10^2 m/s. Estimate the minimum time that must elapse between the time you perceive a stalled car in front of you and the time you can activate the muscles in your leg to brake your car. (Your actual "reaction time" is much longer than this estimate.) Moving at 90 km/h, how far would your car travel in this time?

5. Starting from home, you bicycle 24 km north in 2.5 h, then turn around and pedal straight home in 1.5 h. What are your (a) displacement at the end of the first 2.5 h, (b) average velocity over the first 2.5 h, (c) average velocity for the homeward leg of the trip, (d) displacement for the entire trip, and (e) average velocity for the entire trip?

6. The European Space Agency's Giotto spacecraft en-

countered Halley's Comet in March 1986, when the comet was 93 million km from earth. How long did it take radio signals (travelling at the speed of light) to reach earth from Giotto?

7. A full-length triathlon is a gruelling event in which contestants first swim 2.4 miles, then bicycle 112 miles, then complete a 26-mile 385-yard marathon on foot. The world's triathlon record for women was set in 1983 by Canada's Sylviane Puntus. She completed the event in 10 h 43 min 36 s. What was her average speed? (Incidentally, her identical twin finished second.)

8. Find values, good to one significant figure, for the speed of light in (a) ft/ns (1 ns = 10^{-9} s) and (b) furlongs per fortnight, where there are 8 furlongs to a mile and 14 days per fortnight.

9. You allow yourself 40 min to drive 25 miles to the airport, but are caught in heavy traffic and average only 20 mi/h for the first 15 min. What must your average speed be on the rest of the trip if you are to get there on time?

10. Taking earth's orbit to be a circle of radius 1.5×10^8 km, determine the speed of earth's orbital motion in (a) m/s and (b) miles per second.

11. What is the conversion factor from m/s to mi/h?

12. If the average American driver goes 5000 miles each year on interstate highways, how much more time does the average driver spend on interstate highways each year as a result of the 1974 change in the speed limit from 70 mi/h to 55 mi/h?

13. A fast base runner can get from first to second base in 3.4 s. If he leaves first base as the pitcher throws a 90-mi/h fastball the 61-ft distance to the catcher, and if the catcher takes 0.45 s to catch and rethrow the ball, how fast does the catcher have to throw the ball to second base to make an out? Home plate to second base is the diagonal of a square 90 ft on a side.

14. Despite the fact that jet airplanes fly at about 600 mi/h, plane schedules and connections are such that the 3000-mi trip from Burlington, Vermont, to San Francisco ends up taking about 11 h. (a) What is the average speed of such a trip? (b) How much time is spent on the ground, assuming that the actual distance covered by the several aircraft involved in connecting flights is 4200 mi and that the planes maintain a steady 600 mi/h in flight?

15. If you drove the 4600 km from coast to coast of the United States at 55 mi/h (88 km/h), stopping an average of 30 min for rest and refueling every 2 h, (a) how long would it take? (b) What would be your average velocity for the entire trip?

16. I can run 9.0 m/s, 20 per cent faster than my kid brother. How much head start should I give him in order to have a tie race over 100 m?

17. A jetliner leaves San Francisco for New York, 4600 km away. With a strong tailwind, its speed is 1100 km/h. At the same time, a second jet leaves New York for San Francisco. Flying into the wind, it makes only 700 km/h. When and where do the two planes pass each other?

Section 2.2   *Motion in Short Time Intervals*

18. On a single graph, plot distance versus time for the three 50-km trips described on page 17. For each trip, identify graphically the average velocity and the instantaneous velocity for each segment of the trip.

19. By performing the limiting process as in Example 2–3, evaluate the instantaneous velocity as a function of time for an object whose position as a function of time is given by $x = bt^3$. Repeat, using Equation 2–5.

20. For the motion plotted in Fig. 2–18, estimate (a) the greatest velocity in the positive x-direction, (b) the greatest velocity in the negative x-direction, (c) any times when the object is instantaneously at rest, and (d) the average velocity over the interval shown.

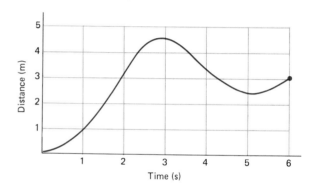

Fig. 2–18 Problem 20.

21. The position of an object is given by $x = bt^3 - ct^2 + dt$, with x in meters and t in seconds. The constants b, c, and d are $b = 3.0$ m/s^3, $c = 8.0$ m/s^2, and $d = 1.0$ m/s. (a) Find all times when the object is at position x = 0. (b) Determine a general expression for the instantaneous velocity as a function of time, and from it find (c) the initial velocity and (d) all times when the object is instantaneously at rest. (e) Graph the object's position as a function of time, and identify on the graph the quantities you found in (a) to (d).

22. In a drag race, the position of a car as a function of time is given by $x = bt^2$, with $b = 2.000$ m/s^2. In an attempt to determine the car's velocity midway down a 400-m track, two observers stand 20 m on either side of the 200-m mark and note the time when the car passes them. (a) What value do the two observers compute for the car's velocity? Give your answer to four significant figures. (b) By what percentage does this observed value differ from the actual instantaneous value at x = 200 m?

Section 2.3   *Acceleration*

23. A car roars away from a red light. Half a minute later, it screeches to a halt at a second light, one-fourth mile from the first. What is its average acceleration between the times it is stopped at the two lights?

24. The 1986 explosion of the space shuttle Challenger occurred 74 s after lift-off. At that time, mission con-

trol reported a shuttle speed of 2900 ft/s (880 m/s). What was the Challenger's average acceleration during its brief flight? Compare with the acceleration of gravity.

25. Under the influence of a radio wave, an electron in an antenna undergoes back-and-forth motion whose velocity as a function of time is described by Fig. 2–19. From the graph, estimate the electron's maximum acceleration.

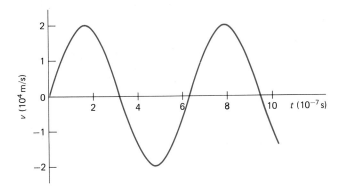

Fig. 2–19   Problem 25.

26. Determine the instantaneous acceleration as a function of time for the motion in Problem 21.

27. The position of an object is given by $x = bt^3$, where x is in meters, t in seconds, and where the constant b is 1.5 m/s^3. Determine (a) the instantaneous velocity and (b) the instantaneous acceleration at the end of 2.5 s. Find (c) the average velocity and (d) the average acceleration during the first 2.5 s.

Section 2.4   *Constant Acceleration*

28. In 1982, drag racer Gary Beck set a new record, travelling a quarter mile from rest in 5.484 s. Assuming constant acceleration, determine (a) Beck's acceleration, (b) the maximum speed achieved, and (c) the time to reach 55 mi/h.

29. Starting from rest, a car accelerates at a constant rate, reaching 88 km/h in 12 s. (a) What is its acceleration? (b) How far does it go in this time?

30. If you are running at 6 m/s and decelerate at 2 m/s^2, (a) how long will it take you to stop? (b) How far will you go while decelerating?

31. A car moving initially at 50 mi/h begins decelerating at a constant rate 100 ft short of a stoplight. If the car comes to a full stop just at the light, what is the magnitude of its deceleration?

32. In an x-ray tube, electrons are accelerated to a velocity of 10^8 m/s, then slammed into a tungsten target. When they collide with the tungsten atoms, the electrons undergo rapid deceleration, producing x-rays in the process. If the stopping time for an electron is on the order of 10^{-19} s, approximately how far does an electron move while decelerating? Assume constant deceleration.

33. A particle leaves its initial position x_0 at time $t = 0$, moving in the positive x-direction with speed v_0, but undergoing acceleration of magnitude a in the negative x-direction. Find expressions for (a) the time when it returns to the position x_0 and (b) its speed when it passes that point.

34. The Barringer meteor crater in northern Arizona is 180 m deep and 1.2 km in diameter. The fragments of the meteor lie just below the bottom of the crater. If these fragments decelerated at a constant rate of 4×10^5 m/s^2 as they ploughed through the earth in forming the crater, what was the speed of the meteor's impact at earth's surface?

35. A gazelle accelerates from rest at 4.1 m/s^2 over a distance of 60 m to outrun a predator. What is its final speed?

36. A frog's tongue flicks out 10 cm in 0.1 s to catch a fly. If it accelerates at a constant rate during this time, what are (a) its average velocity and (b) its acceleration?

37. Amtrak's 20th-Century Limited is en route from Chicago to New York at 110 km/h, when the engineer spots a cow on the track. The train brakes to a halt in 1.2 min with constant deceleration, stopping just in front of the cow. (a) What is the magnitude of the train's acceleration? (b) What is the direction of the acceleration? (c) How far was the train from the cow when the engineer first applied the brakes?

38. A jetliner touches down at 220 km/h, reverses its engines to provide braking, and comes to a halt 29 s later. What is the shortest runway on which this aircraft can land, assuming constant deceleration starting at touchdown?

39. A flea extends its 0.80-mm-long legs, propelling itself 10 cm into the air. Assuming that the legs provide constant acceleration as they extend from completely folded to completely extended, find the magnitude of that acceleration. Compare with the acceleration of gravity at earth's surface.

40. A motorist suddenly notices a stalled car and slams on the brakes, decelerating at the rate of 6.3 m/s^2. Unfortunately this isn't good enough, and a collision ensues. From the damage sustained, police estimate that the car was moving at 18 km/h at the time of the collision. They also measure skid marks 34 m long. (a) How fast was the motorist going when the brakes were first applied? (b) How much time elapsed from the initial braking to the collision?

41. The maximum acceleration that a human being can survive even for a short time is about 200g. In a highway accident, a car moving at 88 km/h slams into a stalled truck. The front end of the car is squashed by 80 cm on impact. If the deceleration during the collision is constant, will a passenger wearing a seatbelt survive?

42. A racing car undergoing constant acceleration covers 120 m in 2.7 s. (a) If it is moving at 53 m/s at the end of this interval, what was its speed at the beginning of the interval? (b) How far did it travel *from rest* to the end of the 120-m distance?

43. The maximum deceleration of a car on a dry road is about 8 m/s². If two cars are moving head-on toward each other at 88 km/h (55 mi/h), and their drivers apply their brakes when they are 85 m apart, will they collide? If so, at what relative speed? If not, how far apart will they be when they stop? On the same graph, plot distance versus time for both cars.

44. George, a physics student, leaves his dormitory at a speed of 1.2 m/s, heading for the physics building 95 m away. Just as he leaves his dorm, Amy, another physics student, leaves the physics building and heads toward George at a steady 1.6 m/s. George immediately spots her and begins accelerating at 0.075 m/s². Where and when do the two meet? Plot position-versus-time curves for both students on a single graph. *Hint:* Be sure to set the algebraic signs of positions, velocities, and accelerations consistently.

45. After 35 minutes of running, at the 9-km point in a 10-km race, you find yourself 100 m behind the leader and moving at the same speed. What should your acceleration be if you are to catch up by the finish line? Assume that the leader maintains constant speed throughout the entire race.

Section 2.5   *The Constant Acceleration of Gravity*

46. You drop a rock into a deep well, and 2.7 s later hear the splash. How far down is the water? Neglect the travel time of the sound.

47. Your friend is sitting 20 ft above you in a tree branch. How fast should you throw an apple so that it just reaches her?

48. A model rocket leaves the ground, heading straight up at 49 m/s. (*a*) What is its maximum altitude? What are its speed and altitude at (*b*) 1 s; (*c*) 4 s; (*d*) 7 s.

49. A foul ball leaves the bat going straight upward at 23 m/s. (*a*) How high does it rise? (*b*) How long is it in the air? Neglect the distance between the bat and the ground.

50. A Frisbee is lodged in a tree branch, 6.5 m above the ground. A rock thrown from below must be going at least 3 m/s to dislodge the Frisbee. How fast must such a rock be thrown upward, if it leaves the thrower's hand 1.3 m above the ground?

51. Space pirates kidnap an earthling and hold him imprisoned on one of the planets of the solar system. With nothing else to do, the prisoner amuses himself by dropping his watch from eye level (170 cm) to the floor. He observes that the watch takes 0.62 s to fall. On what planet is he being held? *Hint:* Consult Appendix F.

52. The earliest attempts to land instruments on the moon involved Ranger spacecraft that were to release instrument capsules at about 11 km above the lunar surface. After a retrorocket firing, the capsules were to fall freely to the moon. Surrounded by a balsa-wood sphere, the instrument capsule was designed to withstand an impact speed equivalent to a free fall from 150 m above earth's surface. From what height above the lunar surface could the capsule drop, assuming the retrorocket slowed it to zero speed at that

height? (In fact, none of the Ranger instrument packages ever made the intended cushioned landing.)

53. Anders, a physics student, is going to take an exam but has forgotten his calculator. He stops outside the physics building and yells for help. Charlie, another physics student 21 ft above him on the third floor, drops a calculator out the window. A third physics student, Torey, is on the fourth floor, 32 ft above Anders. She throws her calculator down at 15 ft/s at the same time Charlie drops his. Whose calculator reaches Anders first, and by how much?

54. A falling object travels one-fourth of its total distance in the last second of its fall. From what height was it dropped?

55. The defenders of a castle throw rocks down on their attackers from a 15-m-high wall. If the rocks are thrown with an initial speed of 10 m/s, how much faster do they hit the ground than if they were simply dropped?

56. A kingfisher is 30 m above a lake when it accidentally drops the fish it is carrying. A second kingfisher 5 m above the first dives toward the falling fish. What initial speed should it have if it is to reach the fish before the fish hits the water?

57. Two divers jump from a 3.00-m platform. One jumps upward at 1.80 m/s, and the second steps off the platform as the first passes it on the way down. (*a*) What are their speeds as they hit the water? (*b*) Which hits the water first, and by how much?

58. A balloon is rising at 10 m/s when its passenger throws a ball straight up at 12 m/s. How much time elapses before the passenger catches the ball?

59. A conveyer belt moves horizontally at 80 cm/s, carrying empty shoe boxes. Every 3 s, a pair of shoes is dropped from a chute 1.7 m above the belt. (*a*) How far apart should the boxes be spaced? (*b*) At the instant a pair of shoes drops, where should a box be in relation to a point directly below the chute?

Supplementary Problems

60. A penny dropped from a 90-m-high building embeds itself 3.6 cm into the ground. Compare its average acceleration on stopping with the acceleration of gravity. *Hint:* You need not compute any velocities or times.

61. You see the traffic light ahead of you is about to turn from red to green, so you slow to a steady speed of 10 km/h and cruise to the light, reaching it just as it turns green. You then accelerate to 60 km/h in the next 12 s, then maintain constant speed. At the light, you pass a Corvette that has stopped for the red light. Just as you pass (and the light turns green) the Corvette begins accelerating, reaching 65 km/h in 6.9 s, then maintaining constant speed. (*a*) Plot the motions of both cars on a graph showing the entire minute after the light turns green. (*b*) How long after the light turns green does the Corvette pass you? (*c*) How far are you from the light when the Corvette passes you? (*d*) How far ahead of you is the Corvette 1 min after the light turns green?

62. In the accident of Problem 41, calculate the relative speed with which a passenger not wearing a seatbelt collides with the dashboard. Assume the passenger undergoes no deceleration before striking the dashboard, and that the passenger is initially 1 m from the dashboard.

63. The position of a particle as a function of time is given by $x = x_0 \sin(\omega t)$, where x_0 and ω are constants. Using the limiting procedure as in Example 2–3, show that the instantaneous velocity is given by $v = x_0 \omega \cos(\omega t)$. *Hint:* Use the identity for the sine of a sum of angles (see Appendix A), along with appropriate approximations for the sine and cosine of small angles (see Appendix A).

64. A basketball player runs 12 m down court at 4.6 m/s, dribbling the ball so it has an initial downward velocity of 75 cm/s at 90 cm above the floor. How many times does the ball hit the floor during the 12-m run? Neglect any up-and-down motion of the player's hand. *Hint:* The vertical motion of the ball is independent of its horizontal motion.

65. You drop a rock into a well. 2.7 s later, you hear the splash. (*a*) How far down is the water? The speed of sound is 340 m/s. (*b*) What percentage error would be introduced by neglecting the travel time for the sound?

66. The depth of a well is such that an object dropped into the well hits the water going far slower than the speed of sound. Use the binomial theorem (see Appendix A) to show that, under these conditions, the depth of the well is given approximately by

$$d = \tfrac{1}{2}gt^2 \left(1 - \frac{gt}{v_s}\right),$$

where t is the time from when you drop the object until you hear the splash, and v_s is the speed of sound.

67. A student is staring idly out her dormitory window when she sees a water balloon fall past. If the balloon takes 0.22 s to cross the 130-cm-high window, from what height above the top of the window was it dropped?

68. A police radar has an effective range of 1.0 km, while a motorist's radar detector has a range of 1.8 km. The motorist is going 100 km/h in a 70 km/h zone when the radar detector beeps. At what rate must the motorist decelerate to avoid a speeding ticket?

3

MOTION IN MORE THAN ONE DIMENSION

Fig. 3–1
Landing at a major airport involves a complex pattern of three-dimensional motion. Here a jetliner passes over an experimental system designed to help pilots avoid dangerous wind shear.

Motion is not limited to moving back and forth along a single railroad track. As we walk, as we drive, as we sail, we move freely about the surface of the earth—that is, in two dimensions. As we fly from city to city, scale a mountain, or are launched into space, we move in a third dimension as well (Fig. 3–1). In this chapter, we formulate laws of motion for two and three dimensions. We show also how vectors provide a relatively simple way to analyze such multidimensional motion.

3.1
THE VECTOR DESCRIPTION OF MOTION

We recognized in the preceding chapter that the quantities characterizing motion—position, velocity, and acceleration—have both magnitude and direction, and are therefore vector quantities. In that chapter, we considered only one-dimensional motion, in which all vectors lie along a single line. Now we generalize to the case of arbitrary motion in two and three dimensions.

Position, Displacement, and Vector Addition

position vector

Motion involves changes of position; we specify position using the **position vector, r** (Fig. 3–2). We can characterize the position vector fully by giving both its length and its direction or, equivalently, by specifying its components—its projections on the coordinate axes (see Fig. 3–2).

What if we move the object in Fig. 3–2 1 m to the right, as shown in Fig. 3–3? The new position is described by a vector $\mathbf{r}_2$; we obtain $\mathbf{r}_2$ by adding to $\mathbf{r}_1$ a vector $\Delta\mathbf{r}$ that is 1 m long and points to the right. To accomplish the addition, we simply place $\Delta\mathbf{r}$ with its tail at the head of $\mathbf{r}_1$; the head of $\Delta\mathbf{r}$ then lies at the head of the sum $\mathbf{r}_1 + \Delta\mathbf{r}$. This procedure defines

40

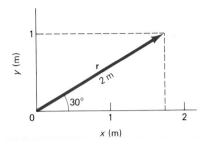

Fig. 3–2
The position vector **r** for an object located 2 m from the origin at 30° to the x-axis. The components of the vector are $x = (2\ m)(\cos 30°) = 1.7$ m and $y = (2\ m)(\sin 30°) = 1.0$ m. These expressions for the components follow from the definitions of sine and cosine as ratios of the opposite and adjacent sides of a right triangle, respectively, to the hypotenuse.

the general rule for adding any two vectors: place the vectors to be added head-to-tail; their vector sum is then a vector from the tail of the first vector to the head of the second. Remember that a vector is specified fully by giving its magnitude and direction. Where the vector starts doesn't matter, which is why we can slide vectors around—as long as we preserve their length and direction—to form vector sums.

In Fig. 3–3, we formed the vector sum $\mathbf{r}_1 + \Delta\mathbf{r}$ graphically; to specify the resultant vector $\mathbf{r}_2$, we could measure its angle and length on our graph. Or we could use the law of cosines to calculate the length of $\mathbf{r}_2$, and the law of sines to determine its angle. (The laws of cosines and of sines are described in Appendix A.) But we will often find it more convenient to handle vector arithmetic in terms of components. In Fig. 3–3, for example, it is obvious that the x-component of $\mathbf{r}_2$ is simply the sum $x_1 + \Delta x$, where x_1 and Δx are the x-components of the vectors $\mathbf{r}_1$ and $\Delta\mathbf{r}$, respectively. Similarly, the y-component of $\mathbf{r}_2$ is the sum of the y-components of $\mathbf{r}_1$ and $\Delta\mathbf{r}$, although in this case Δy, the y-component of $\Delta\mathbf{r}$, is zero.

Since the addition of vector components is commutative and associative, so is vector addition itself. That is, $\mathbf{A} + \mathbf{B} = \mathbf{B} + \mathbf{A}$ and $(\mathbf{A} + \mathbf{B}) + \mathbf{C} = \mathbf{A} + (\mathbf{B} + \mathbf{C})$. You can also see this graphically, as suggested in Fig. 3–4.

Example 3–1
Vector Addition

Determine the vector $\mathbf{r}_2$ in Fig. 3–3 (a) using the laws of cosines and sines and (b) using components.

Solution

The law of cosines (see Appendix A) is like a modified Pythagorean theorem, giving the third side, C, of a triangle in terms of the other two sides A and B and the angle θ between them:

$$C^2 = A^2 + B^2 - 2AB\cos\theta.$$

Applying the law of cosines to the lengths of the vectors in Fig. 3–3, we have

$$r_2{}^2 = r_1{}^2 + (\Delta r)^2 - 2r_1\Delta r \cos\theta.$$

We indicate the lengths—magnitudes—of the vectors by writing their symbols in italic type rather than boldface. Figure 3–5 shows the angle θ is 150°; therefore

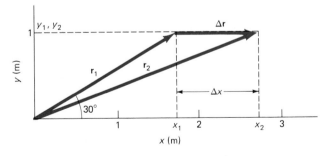

Fig. 3–3
Vector addition. The new position vector $\mathbf{r}_2$ defines the vector sum of the old position vector $\mathbf{r}_1$ and the change $\Delta\mathbf{r}$. Graphically, we sum the vectors by placing them head-to-tail; algebraically, we simply sum their components.

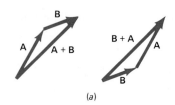

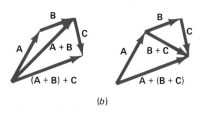

Fig. 3–4
(a) Vector addition is commutative; whether we form the sum **A**+**B** or the sum **B**+**A**, we get the same result. (b) Vector addition is also associative; that is, (**A**+**B**)+**C**=**A**+(**B**+**C**).

$$r_2{}^2 = (2.0 \text{ m})^2 + (1.0 \text{ m})^2 - (2)(2.0 \text{ m})(1.0 \text{ m})\cos 150° = 8.46 \text{ m}^2,$$

so that $r_2 = 2.9$ m. This value is the length of $\mathbf{r}_2$; what about its direction? Applying the law of sines (see Appendix A) to the triangle formed by the vectors $\mathbf{r}_1$, $\Delta\mathbf{r}$, and $\mathbf{r}_2$, we have

$$\frac{\sin\alpha}{\Delta r} = \frac{\sin 150°}{r_2},$$

where α is the angle between $\mathbf{r}_1$ and $\mathbf{r}_2$ (see Fig. 3–5). Then

$$\alpha = \sin^{-1}\left(\frac{\Delta r \sin 150°}{r_2}\right) = \sin^{-1}\left(\frac{(1.0 \text{ m})(0.50)}{2.9 \text{ m}}\right) = 9.9°.$$

Figure 3–5 then shows that $\mathbf{r}_2$ makes an angle $\beta = 30° - \alpha$, or 20°, with the x-axis. The full specification of the vector $\mathbf{r}_2$ includes both its 2.9-m magnitude and its 20° direction.

To determine $\mathbf{r}_2$ using components, we need first the components of $\mathbf{r}_1$ and $\Delta\mathbf{r}$. Using x_1 and y_1 as the components of the 2-m-long vector $\mathbf{r}_1$, we see from Fig. 3–2 that $x_1 = (2 \text{ m})(\cos 30°) = 1.73$ m and $y_1 = (2 \text{ m})(\sin 30°) = 1.0$ m. Since $\Delta\mathbf{r}$ points in the x-direction, only its x-component is nonzero, and is equal to the full 1-m length of the vector. [Formally, we could calculate its components Δx and Δy as we did for $\mathbf{r}_1$: $\Delta x = (1.0 \text{ m})(\cos 0°) = 1.0$ m and $\Delta y = (1.0 \text{ m})(\cos 90°) = 0.$] Adding the components of $\mathbf{r}_1$ and $\Delta\mathbf{r}$ gives the components x_2 and y_2 of $\mathbf{r}_2$:

$$x_2 = x_1 + \Delta x = 1.73 \text{ m} + 1.0 \text{ m} = 2.73 \text{ m}$$

and

$$y_2 = y_1 + \Delta y = 1.0 \text{ m} + 0 \text{ m} = 1.0 \text{ m}.$$

These two numbers give a full description of the vector $\mathbf{r}_2$; alternatively, we could determine the length and direction using the Pythagorean theorem and the definition of tangent as the ratio of the sides of a right triangle (see Fig. 3–5):

$$r_2 = \sqrt{x_2 + y_2} = [(2.73 \text{ m})^2 + (1.0 \text{ m})^2]^{1/2} = 2.9 \text{ m}$$

and

$$\beta = \tan^{-1}\left(\frac{y_2}{x_2}\right) = \tan^{-1}\left(\frac{1.0 \text{ m}}{2.73 \text{ m}}\right) = 20°,$$

in agreement with our graphical method.

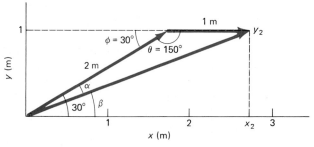

Fig. 3–5
We are given that the angle between $\mathbf{r}_1$ and the x-axis is 30°. Therefore, the angle ϕ is also 30° so that $\theta = 180° - 30° = 150°$. Knowing θ, we use the law of sines to get the angle α between $\mathbf{r}_1$ and $\mathbf{r}_2$; the angle β between $\mathbf{r}_2$ and the x-axis is $30° - \alpha$.

The two ways of specifying a vector—giving its length and direction or specifying its components—are completely equivalent. Using a vector diagram, the Pythagorean theorem, and definitions of the trig functions, we can always convert from one description to the other. In two dimensions, either description requires specifying two numbers. In three dimensions, we would need an additional angle or the z-component, for a total of three numbers. Whether we work only with length and direction, or only with components, or with some combination of the two, depends on what information we are given and on the form we need for our final answer.

In Example 3–1, we solved for the vector $\mathbf{r}_2$ satisfying the vector equation

$$\mathbf{r}_2 = \mathbf{r}_1 + \Delta\mathbf{r}. \tag{3-1}$$

Our second method—using components—involved solving the two scalar equations

$$x_2 = x_1 + \Delta x \tag{3-1a}$$
$$y_2 = y_2 + \Delta y \tag{3-1b}$$

for the components x_2 and y_2 of the vector $\mathbf{r}_2$. In general, we can think of a vector equation like Equation 3–1 as a shorthand for two (or in three dimensions, three) scalar equations for the vector components. Later in this chapter, we will see how the use of vector equations allows us to express the equations of two- and three-dimensional motion in compact form.

What if we had been given $\mathbf{r}_1$ and $\mathbf{r}_2$, and asked to find what vector $\Delta\mathbf{r}$ must be added to $\mathbf{r}_1$ to give $\mathbf{r}_2$? Then we could solve the vector equation 3–1 for the unknown $\Delta\mathbf{r}$:

$$\Delta\mathbf{r} = \mathbf{r}_2 - \mathbf{r}_1. \tag{3-2}$$

This is equivalent to the two scalar equations

$$\Delta x = x_2 - x_1 \tag{3-2a}$$

$$\Delta y = y_2 - y_1, \tag{3-2b}$$

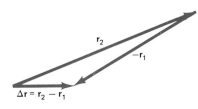

Fig. 3–6
To perform the subtraction $\mathbf{r}_2 - \mathbf{r}_1$, we add to $\mathbf{r}_2$ the vector $-\mathbf{r}_1$—that is, a vector the same length as $\mathbf{r}_1$ but pointing in the opposite direction. The result is the vector difference $\Delta\mathbf{r}$.

which give us the two components of the vector $\Delta\mathbf{r}$. Equivalently, we could solve Equation 3–2 graphically by subtracting the vector $\mathbf{r}_1$ from the vector $\mathbf{r}_2$—that is, by adding to $\mathbf{r}_2$ the vector $-\mathbf{r}_1$, whose length is that of $\mathbf{r}_1$ but which points in the opposite direction (Fig. 3–6).

It is cumbersome to say "a vector of length 2 m at 30° to the x-axis" or "a vector whose x-component is 1.73 m and whose y-component is 1.0 m." After all, a vector is a mathematical quantity, and we are used to writing mathematical quantities compactly, without a lot of words. We can express arbitrary vectors in compact form using **unit vectors**. We define three such vectors, each of length 1, pointing in the x, y, and z directions. These vectors are called $\hat{\mathbf{i}}$, $\hat{\mathbf{j}}$, and $\hat{\mathbf{k}}$, respectively. They have no dimensions, and serve only to indicate direction.

unit vectors

Before we can use the unit vectors, we must define what it means to multiply a vector by a scalar: the product of a scalar c with a vector $\mathbf{A}$ (written $c\mathbf{A}$) is a vector of length cA (where A is the length of $\mathbf{A}$), pointing in the same direction as $\mathbf{A}$. That is, multiplication by a scalar changes the length of a vector but not its direction (although multiplication by a negative scalar reverses the sense of the vector along the line on which it lies;

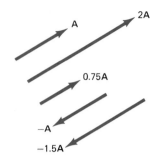

Fig. 3–7
Multiplication of a vector by a scalar. The result lies in the same direction as the original vector, although its sense is reversed if the scalar is negative. Note again that where a vector starts doesn't matter; its direction and magnitude fully characterize the vector.

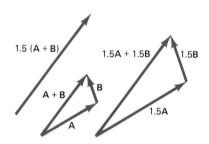

Fig. 3–8
Multiplication of a vector by a scalar is distributive.

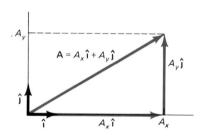

Fig. 3–9
An arbitrary vector expressed in terms of unit vectors $\hat{\imath}$ and $\hat{\jmath}$.

see Fig. 3–7). Multiplication by a scalar is distributive: $c(\mathbf{A}+\mathbf{B}) = c\mathbf{A}+c\mathbf{B}$ (Fig. 3–8).

Using unit vectors, we write an arbitrary vector $\mathbf{A}$, whose components are A_x, A_y, and A_z, as the sum of three vectors $A_x\hat{\imath}$, $A_y\hat{\jmath}$, and $A_z\hat{\mathbf{k}}$ that lie along the mutually perpendicular coordinate axes:

$$\mathbf{A} = A_x\hat{\imath} + A_y\hat{\jmath} + A_z\hat{\mathbf{k}}.$$

Figure 3–9 illustrates this process graphically for a vector that lies in the x-y plane. The unit vectors provide us a way of expressing compactly the three perpendicular directions of our coordinate system; weighted by the components along each axis and then summed, they give a full description of an arbitrary vector.

Adding and subtracting vectors expressed in unit vector notation is simple: we just add the individual terms. For example, if $\mathbf{A} = A_x\hat{\imath}+A_y\hat{\jmath}+A_z\hat{\mathbf{k}}$ and $\mathbf{B}=B_x\hat{\imath}+B_y\hat{\jmath}+B_z\hat{\mathbf{k}}$, the sum $\mathbf{A}+\mathbf{B}$ is

$$\begin{aligned}\mathbf{A} + \mathbf{B} &= (A_x\hat{\imath} + A_y\hat{\jmath} + A_z\hat{\mathbf{k}}) + (B_x\hat{\imath} + B_y\hat{\jmath} + B_z\hat{\mathbf{k}}) \\ &= (A_x+B_x)\hat{\imath} + (A_y+B_y)\hat{\jmath} + (A_z+B_z)\hat{\mathbf{k}}.\end{aligned} \tag{3-3}$$

Example 3–2

Unit Vector Notation

Express the vectors $\mathbf{r}_1$ and $\Delta\mathbf{r}$ of Example 3–1 in unit vector notation, and add them to get $\mathbf{r}_2$.

Solution

In Example 3–1, we found the components of $\mathbf{r}_1$ to be $x_1 = 1.73$ m and $y_1 = 1.0$ m, while those of $\Delta\mathbf{r}$ were $\Delta x = 1.0$ m and $\Delta y = 0$. In unit vector notation, then, these two vectors are

$$\mathbf{r}_1 = 1.73\hat{\imath} + 1.0\hat{\jmath} \text{ m}$$
$$\Delta\mathbf{r} = 1.0\hat{\imath} \text{ m}.$$

Adding these two, we have

$$\mathbf{r}_1 + \Delta\mathbf{r} = (1.73\hat{\imath}+1.0\hat{\jmath} \text{ m}) + (1.0\hat{\imath} \text{ m}) = 2.73\hat{\imath}+1.0\hat{\jmath} \text{ m}.$$

Unit vector notation allowed us to carry out this operation neatly and without a lot of words. Our final answer, $2.73\hat{\imath}+1.0\hat{\jmath}$ m, is a compact and complete description of the vector $\mathbf{r}_2$.

Velocity

In the preceding chapter, we defined average velocity $\bar{\mathbf{v}}$ as the ratio of change in position $\Delta\mathbf{r}$ to the time interval Δt over which that change occurs:

$$\bar{\mathbf{v}} = \frac{\Delta\mathbf{r}}{\Delta t}. \tag{3-4}$$

Taking the average velocity in the limit of arbitrarily short time intervals gave us the instantaneous velocity $\mathbf{v}$:

$$\mathbf{v} = \lim_{\Delta t\to 0}\frac{\Delta\mathbf{r}}{\Delta t} = \frac{d\mathbf{r}}{dt}, \tag{3-5}$$

The two ways of specifying a vector—giving its length and direction or specifying its components—are completely equivalent. Using a vector diagram, the Pythagorean theorem, and definitions of the trig functions, we can always convert from one description to the other. In two dimensions, either description requires specifying two numbers. In three dimensions, we would need an additional angle or the z-component, for a total of three numbers. Whether we work only with length and direction, or only with components, or with some combination of the two, depends on what information we are given and on the form we need for our final answer.

In Example 3–1, we solved for the vector $\mathbf{r}_2$ satisfying the vector equation

$$\mathbf{r}_2 = \mathbf{r}_1 + \Delta\mathbf{r}. \qquad (3\text{–}1)$$

Our second method—using components—involved solving the two scalar equations

$$x_2 = x_1 + \Delta x \qquad (3\text{–}1a)$$
$$y_2 = y_2 + \Delta y \qquad (3\text{–}1b)$$

for the components x_2 and y_2 of the vector $\mathbf{r}_2$. In general, we can think of a vector equation like Equation 3–1 as a shorthand for two (or in three dimensions, three) scalar equations for the vector components. Later in this chapter, we will see how the use of vector equations allows us to express the equations of two- and three-dimensional motion in compact form.

What if we had been given $\mathbf{r}_1$ and $\mathbf{r}_2$, and asked to find what vector $\Delta\mathbf{r}$ must be added to $\mathbf{r}_1$ to give $\mathbf{r}_2$? Then we could solve the vector equation 3–1 for the unknown $\Delta\mathbf{r}$:

$$\Delta\mathbf{r} = \mathbf{r}_2 - \mathbf{r}_1. \qquad (3\text{–}2)$$

This is equivalent to the two scalar equations

$$\Delta x = x_2 - x_1 \qquad (3\text{–}2a)$$

$$\Delta y = y_2 - y_1, \qquad (3\text{–}2b)$$

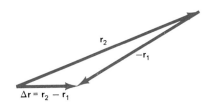

Fig. 3–6
To perform the subtraction $\mathbf{r}_2 - \mathbf{r}_1$, we add to $\mathbf{r}_2$ the vector $-\mathbf{r}_1$—that is, a vector the same length as $\mathbf{r}_1$ but pointing in the opposite direction. The result is the vector difference $\Delta\mathbf{r}$.

which give us the two components of the vector $\Delta\mathbf{r}$. Equivalently, we could solve Equation 3–2 graphically by subtracting the vector $\mathbf{r}_1$ from the vector $\mathbf{r}_2$—that is, by adding to $\mathbf{r}_2$ the vector $-\mathbf{r}_1$, whose length is that of $\mathbf{r}_1$ but which points in the opposite direction (Fig. 3–6).

It is cumbersome to say "a vector of length 2 m at 30° to the x-axis" or "a vector whose x-component is 1.73 m and whose y-component is 1.0 m." After all, a vector is a mathematical quantity, and we are used to writing mathematical quantities compactly, without a lot of words. We can express arbitrary vectors in compact form using **unit vectors.** We define three such vectors, each of length 1, pointing in the x, y, and z directions. These vectors are called $\hat{\mathbf{i}}$, $\hat{\mathbf{j}}$, and $\hat{\mathbf{k}}$, respectively. They have no dimensions, and serve only to indicate direction.

unit vectors

Before we can use the unit vectors, we must define what it means to multiply a vector by a scalar: the product of a scalar c with a vector $\mathbf{A}$ (written $c\mathbf{A}$) is a vector of length cA (where A is the length of $\mathbf{A}$), pointing in the same direction as $\mathbf{A}$. That is, multiplication by a scalar changes the length of a vector but not its direction (although multiplication by a negative scalar reverses the sense of the vector along the line on which it lies;

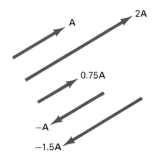

Fig. 3–7
Multiplication of a vector by a scalar. The result lies in the same direction as the original vector, although its sense is reversed if the scalar is negative. Note again that where a vector starts doesn't matter; its direction and magnitude fully characterize the vector.

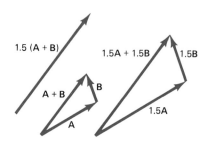

Fig. 3–8
Multiplication of a vector by a scalar is distributive.

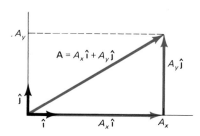

Fig. 3–9
An arbitrary vector expressed in terms of unit vectors $\hat{\imath}$ and $\hat{\jmath}$.

see Fig. 3–7). Multiplication by a scalar is distributive: $c(\mathbf{A} + \mathbf{B}) = c\mathbf{A} + c\mathbf{B}$ (Fig. 3–8).

Using unit vectors, we write an arbitrary vector $\mathbf{A}$, whose components are A_x, A_y, and A_z, as the sum of three vectors $A_x\hat{\imath}$, $A_y\hat{\jmath}$, and $A_z\hat{\mathbf{k}}$ that lie along the mutually perpendicular coordinate axes:

$$\mathbf{A} = A_x\hat{\imath} + A_y\hat{\jmath} + A_z\hat{\mathbf{k}}.$$

Figure 3–9 illustrates this process graphically for a vector that lies in the x-y plane. The unit vectors provide us a way of expressing compactly the three perpendicular directions of our coordinate system; weighted by the components along each axis and then summed, they give a full description of an arbitrary vector.

Adding and subtracting vectors expressed in unit vector notation is simple: we just add the individual terms. For example, if $\mathbf{A} = A_x\hat{\imath} + A_y\hat{\jmath} + A_z\hat{\mathbf{k}}$ and $\mathbf{B} = B_x\hat{\imath} + B_y\hat{\jmath} + B_z\hat{\mathbf{k}}$, the sum $\mathbf{A} + \mathbf{B}$ is

$$\begin{aligned}
\mathbf{A} + \mathbf{B} &= (A_x\hat{\imath} + A_y\hat{\jmath} + A_z\hat{\mathbf{k}}) + (B_x\hat{\imath} + B_y\hat{\jmath} + B_z\hat{\mathbf{k}}) \\
&= (A_x + B_x)\hat{\imath} + (A_y + B_y)\hat{\jmath} + (A_z + B_z)\hat{\mathbf{k}}.
\end{aligned} \tag{3–3}$$

Example 3–2
Unit Vector Notation

Express the vectors $\mathbf{r}_1$ and $\Delta\mathbf{r}$ of Example 3–1 in unit vector notation, and add them to get $\mathbf{r}_2$.

Solution

In Example 3–1, we found the components of $\mathbf{r}_1$ to be $x_1 = 1.73$ m and $y_1 = 1.0$ m, while those of $\Delta\mathbf{r}$ were $\Delta x = 1.0$ m and $\Delta y = 0$. In unit vector notation, then, these two vectors are

$$\mathbf{r}_1 = 1.73\hat{\imath} + 1.0\hat{\jmath} \text{ m}$$
$$\Delta\mathbf{r} = 1.0\hat{\imath} \text{ m}.$$

Adding these two, we have

$$\mathbf{r}_1 + \Delta\mathbf{r} = (1.73\hat{\imath} + 1.0\hat{\jmath} \text{ m}) + (1.0\hat{\imath} \text{ m}) = 2.73\hat{\imath} + 1.0\hat{\jmath} \text{ m}.$$

Unit vector notation allowed us to carry out this operation neatly and without a lot of words. Our final answer, $2.73\hat{\imath} + 1.0\hat{\jmath}$ m, is a compact and complete description of the vector $\mathbf{r}_2$.

Velocity

In the preceding chapter, we defined average velocity $\bar{\mathbf{v}}$ as the ratio of change in position $\Delta\mathbf{r}$ to the time interval Δt over which that change occurs:

$$\bar{\mathbf{v}} = \frac{\Delta\mathbf{r}}{\Delta t}. \tag{3–4}$$

Taking the average velocity in the limit of arbitrarily short time intervals gave us the instantaneous velocity $\mathbf{v}$:

$$\mathbf{v} = \lim_{\Delta t \to 0} \frac{\Delta\mathbf{r}}{\Delta t} = \frac{d\mathbf{r}}{dt}, \tag{3–5}$$

where the derivative notation dr/dt is a shorthand way of expressing the result of the limiting process. Since displacement is a vector, so is velocity.

Equation 3–4 shows us that the average velocity $\overline{\mathbf{v}}$ of an object moving between positions $\mathbf{r}_1$ and $\mathbf{r}_2$ in time Δt is given by the vector difference $\Delta\mathbf{r} = \mathbf{r}_2 - \mathbf{r}_1$ divided by the scalar Δt. (Dividing a vector by a scalar means multiplying by the reciprocal of the scalar; we have already defined such multiplication.) Suppose, for example, that the object of Examples 3–1 and 3–2 took 2.0 s to get from $\mathbf{r}_1$ to $\mathbf{r}_2$. We already found the difference $\Delta\mathbf{r}$ between these two vectors is $\Delta\mathbf{r} = 1.0\hat{\mathbf{i}}$ m, so that Equation 3–4 gives

$$\overline{\mathbf{v}} = \frac{\Delta\mathbf{r}}{\Delta t} = \frac{1.0\hat{\mathbf{i}}\ \text{m}}{2.0\ \text{s}} = 0.50\hat{\mathbf{i}}\ \text{m/s}.$$

Here an answer of 0.50 m/s alone would not be enough; velocity is a vector, and its specification requires both direction and magnitude. The answer $0.50\hat{\mathbf{i}}$ m/s includes both.

Example 3–3
LA to Denver via Salt Lake City

You leave Los Angeles on a flight to Denver, with a stop in Salt Lake City. Flying 55° north of east, the plane makes the 970-km trip to Salt Lake City in 72 min. After 40 min on the ground, the plane heads for Denver, which is 600 km away and in a direction 10° south of east. The plane reaches Denver 55 min after leaving Salt Lake City. Determine the average velocity of the plane between Los Angeles and Denver.

Solution

Figure 3–10 shows the situation graphically. The vector displacement $\Delta\mathbf{r}_{LD}$ from Los Angeles to Denver is the sum of the displacement $\Delta\mathbf{r}_{LS}$ from Los Angeles to Salt Lake City and the displacement $\Delta\mathbf{r}_{SD}$ from Salt Lake City to Denver. Choosing coordinate axes x and y running east-west and north-south, respectively, we can compute the components needed to write these vectors in unit vector notation:

$$\Delta\mathbf{r}_{LS} = (970\ \text{km})(\cos 55°)\hat{\mathbf{i}} + (970\ \text{km})(\sin 55°)\hat{\mathbf{j}} = 556\hat{\mathbf{i}} + 795\hat{\mathbf{j}}\ \text{km}$$

$$\Delta\mathbf{r}_{SD} = (600\ \text{km})[\cos(-10°)]\hat{\mathbf{i}} + (600\ \text{km})[\sin(-10°)]\hat{\mathbf{j}} = 591\hat{\mathbf{i}} - 104\hat{\mathbf{j}}\ \text{km}.$$

Note that the angle associated with the Salt Lake City to Denver displacement is negative, since the direction is south of east. Adding these vectors gives the displacement vector $\Delta\mathbf{r}_{LD}$ from Los Angeles to Denver:

$$\Delta\mathbf{r}_{LD} = \Delta\mathbf{r}_{LS} + \Delta\mathbf{r}_{SD} = (556\hat{\mathbf{i}} + 795\hat{\mathbf{j}}\ \text{km}) + (591\hat{\mathbf{i}} - 104\hat{\mathbf{j}}\ \text{km}) = 1147\hat{\mathbf{i}} + 691\hat{\mathbf{j}}\ \text{km}.$$

The total elapsed time for this trip includes the two flight times and the 40-min ground time, so that Δt = 72 min + 40 min + 55 min = 167 min or 2.78 h. The average velocity is then

$$\overline{\mathbf{v}} = \frac{\Delta\mathbf{r}_{LD}}{\Delta t} = \frac{1147\hat{\mathbf{i}} + 691\hat{\mathbf{j}}\ \text{km}}{2.78\ \text{h}} = 413\hat{\mathbf{i}} + 249\hat{\mathbf{j}}\ \text{km/h}.$$

The magnitude of this average velocity is given by the Pythagorean theorem:

$$\overline{v} = \sqrt{\overline{v}_x^2 + \overline{v}_y^2} = [(413\ \text{km/h})^2 + (249\ \text{km/h})^2]^{1/2} = 482\ \text{km/h}.$$

Since the actual path is not straight and is therefore longer than the magnitude of the vector $\Delta\mathbf{r}_{LD}$, the average *speed* would be greater than this.

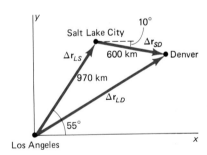

Fig. 3–10
Displacement vectors for a flight from Los Angeles to Denver via Salt Lake City.

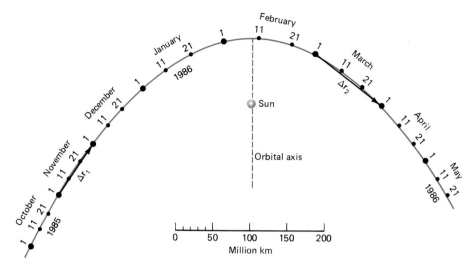

Fig. 3–11
The orbit of Halley's Comet in 1985–86, showing its position at several times each month.

Example 3–4

The Velocity of Halley's Comet

Figure 3–11 shows the orbit of Halley's Comet as it passed through the inner solar system in 1985–86. Determine the average velocity of the comet during November 1985 and again during March 1986.

Solution

On Fig. 3–11 we have drawn the displacement vectors $\Delta\mathbf{r}_1$ and $\Delta\mathbf{r}_2$ for the time intervals of interest. Measuring the length of $\Delta\mathbf{r}_1$ and using the scale factor shown, we find that $\Delta\mathbf{r}_1$ has a magnitude of 80×10^6 km. The time interval Δt_1 over which this displacement occurs is 30 days or 2.6×10^6 s, so that the average velocity $\overline{\mathbf{v}}_1$ over this interval has magnitude

$$\overline{v}_1 = \frac{\Delta r_1}{\Delta t_1} = \frac{80\times10^6 \text{ km}}{2.6\times10^6 \text{ s}} = 31 \text{ km/s}.$$

The velocity vector points in the same direction as $\Delta\mathbf{r}_1$. A similar analysis gives $\overline{v}_2 = 43$ km/s for the magnitude of the average velocity in March.

Acceleration

In the preceding chapter, we defined average acceleration as change in velocity divided by the time interval involved:

$$\overline{\mathbf{a}} = \frac{\Delta\mathbf{v}}{\Delta t}. \tag{3–6}$$

Taking the limit of arbitrarily small times gives the instantaneous acceleration:

$$\mathbf{a} = \lim_{\Delta t\to 0} \frac{\Delta\mathbf{v}}{\Delta t} = \frac{d\mathbf{v}}{dt}. \tag{3–7}$$

Velocity is a vector, so it can change in either *magnitude* or *direction*. You will understand two- and three-dimensional motion more clearly if you recognize that a change in *either* aspect of velocity necessarily involves acceleration. That's why a car going around a curve at constant speed is accelerating—its speedometer may read 45 mi/h all the while, but the *direction* of its velocity is nevertheless changing. In two and three dimensions, acceleration does not necessarily entail an increase or a decrease in speed. Whether the speed or direction, or both, change depends on the orientation of the acceleration vector relative to the velocity vector, as the examples below suggest.

Example 3–5

The Acceleration of Halley's Comet

Use the result of Example 3–4 to determine the average acceleration of Halley's Comet over the interval from November to March.

Solution

The initial and final velocities over the interval are approximately the average velocities $\bar{v}_1$ and $\bar{v}_2$ for November and March that we found in Example 3–4. Figure 3–12 shows a vector diagram of those two velocities and their difference Δv. Comparing the length of Δv with that of the 31-km/s vector v_1 shows that Δv represents a 54 km/s change in the magnitude of the velocity. The time interval over which this change occurs is 120 days or 1.04×10^7 s, so that the magnitude of the average acceleration is

$$\bar{a} = \frac{|\Delta v|}{\Delta t} = \frac{54 \text{ km/s}}{1.04 \times 10^7 \text{ s}} = 5.2 \times 10^{-6} \text{ km/s}^2 = 5.2 \times 10^{-3} \text{ m/s}^2.$$

The direction of $\bar{a}$ is the same as that of Δv, or 162° to the vertical of Fig. 3–11 and Fig. 3–12. In this example the acceleration involves changes in both the magnitude and the direction of the velocity.

As Equation 3–6 suggests, the average acceleration we have calculated is a rough approximation to the instantaneous acceleration near the middle of the November-March interval. Drawing the acceleration vector at the middle of the interval (Fig. 3–13) shows that the acceleration is directed approximately sunward; a more accurate calculation approaching the limit in Equation 3–7 would show that

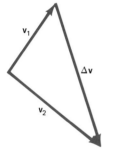

Fig. 3–12
Velocity vector diagram for determining the acceleration of Halley's Comet.

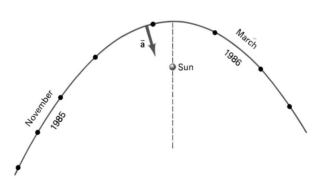

Fig. 3–13
Approximating the instantaneous acceleration at the midpoint of the November-March interval by the average acceleration suggests that the acceleration vector points roughly toward the sun. A more accurate calculation, using an arbitrarily small time interval instead of the 4-month interval of Example 3–5, would show that the acceleration is exactly sunward.

the direction is exactly sunward. This is not surprising, since the sun's gravity is what causes the comet's acceleration. We will explore gravitational acceleration further in Chapter 8.

Example 3–6

Acceleration Parallel and Perpendicular to Velocity

A plane is flying eastward at 250 m/s when it encounters a crosswind that gives it a southward acceleration of 1.2 m/s². What are the magnitude and direction of its velocity at the end of 1 minute? Repeat for the case of an eastward acceleration.

Solution

Choosing a coordinate system with the x-axis east-west and the y-axis north-south, the initial eastward velocity is $\mathbf{v}_0 = 250\hat{\mathbf{i}}$ m/s. Similarly, the southward acceleration is $\mathbf{a} = -1.2\hat{\mathbf{j}}$ m/s². Using this acceleration in Equation 3–6 gives the change in velocity over the 1-min interval:

$$\Delta\mathbf{v} = \mathbf{a}\Delta t = (-1.2\hat{\mathbf{j}} \text{ m/s})(60 \text{ s}) = -72\hat{\mathbf{j}} \text{ m/s}.$$

Then the velocity at the end of 1 min is given by

$$\mathbf{v} = \mathbf{v}_0 + \Delta\mathbf{v} = 250\hat{\mathbf{i}} \text{ m/s} - 72\hat{\mathbf{j}} \text{ m/s}.$$

Figure 3–14a shows the vectors $\mathbf{v}_0$, $\Delta\mathbf{v}$, and their sum $\mathbf{v}$. The magnitude of the vector $\mathbf{v}$ is given by the Pythagorean theorem:

$$v = \sqrt{v_x^2 + v_y^2} = [(250 \text{ m/s})^2 + (-72 \text{ m/s})^2]^{1/2} = 260 \text{ m/s},$$

while the tangent of the angle θ that $\mathbf{v}$ makes with the x-axis is the ratio of v_y to v_x:

$$\theta = \tan^{-1}\left(\frac{v_y}{v_x}\right) = \tan^{-1}\left(\frac{-72 \text{ m/s}}{250 \text{ m/s}}\right) = -16°,$$

or 16° south of east. Note that this acceleration—directed initially at right angles to the velocity—has relatively little effect on the magnitude of the velocity. The latter increases by only 10 m/s, even though the acceleration produces a 72 m/s change in the southward component of the velocity. We will find this to be generally true: an acceleration acting always at right angles to the velocity changes only the direction, not the magnitude of the velocity. (Here the acceleration is not always perpendicular to the velocity; as soon as the acceleration starts to act, subsequent velocity vectors are no longer exactly eastward.)

When the acceleration acts eastward, the situation is reduced to a one-dimensional problem. In vector notation, we still have $\mathbf{v}_0 = 250\hat{\mathbf{i}}$ m/s, but now $\mathbf{a} = 1.2\hat{\mathbf{i}}$ m/s, so that $\Delta\mathbf{v} = 72\hat{\mathbf{i}}$ m/s. Then the velocity after 1 min is

$$\mathbf{v} = \mathbf{v}_0 + \Delta\mathbf{v} = 250\hat{\mathbf{i}} \text{ m/s} + 72\hat{\mathbf{i}} \text{ m/s} = 322\hat{\mathbf{i}} \text{ m/s}.$$

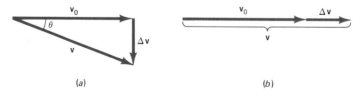

(a)   (b)

Fig. 3–14
Velocity vector diagrams for Example 3–6. (a) Eastward initial velocity with southward acceleration. (b) Eastward initial velocity with eastward acceleration. In (a), the acceleration results in a change in direction but only a small change in speed. In (b), the acceleration results in a large change in speed but no change in direction.

Here the acceleration—now in the same direction as the velocity—results in a large change in speed, but no change in direction. This, too, is generally true: an acceleration acting in the direction of the velocity changes only the magnitude, not the direction (an acceleration acting opposite to the velocity can reverse the sense of the velocity, but the new velocity still lies along the same line). Figure 3–14b shows velocity vector diagrams for this case.

3.2
CONSTANT ACCELERATION

When acceleration is constant—no change in either magnitude or direction—then the individual components of the acceleration vector must themselves be constants. The corresponding components of motion in the x- and y-directions are then described by the equations for constant acceleration that we developed in the preceding chapter, written once for each component:

$$v_x = v_{x0} + a_x t \qquad (3\text{–}8a) \qquad\qquad v_y = v_{y0} + a_y t \qquad (3\text{–}8b)$$

$$x = x_0 + v_{x0}t + \tfrac{1}{2}a_x t^2 \qquad (3\text{–}9a) \qquad\qquad y = y_0 + v_{y0}t + \tfrac{1}{2}a_y t^2 \qquad (3\text{–}9b)$$

$$\bar{v}_x = \tfrac{1}{2}(v_{x0} + v_x) \qquad (3\text{–}10a) \qquad\qquad \bar{v}_y = \tfrac{1}{2}(v_{y0} + v_y) \qquad (3\text{–}10b)$$

$$x - x_0 = \tfrac{1}{2}(v_{x0} + v_x)t \qquad (3\text{–}11a) \qquad\qquad y - y_0 = \tfrac{1}{2}(v_{y0} + v_y)t \qquad (3\text{–}11b)$$

$$v_x^{\,2} = v_{x0}^{\,2} + 2a_x(x - x_0) \qquad (3\text{–}12a) \qquad\qquad v_y^{\,2} = v_{y0}^{\,2} + 2a_y(y - y_0) \qquad (3\text{–}12b)$$

For fully three-dimensional motion, we could add a set of equations for constant acceleration in the z-direction. The first four pairs of equations above may be written more compactly in vector notation, using the position vector **r** whose components are the coordinates x and y:

$$\mathbf{v} = \mathbf{v}_0 + \mathbf{a}t \qquad\qquad (3\text{–}8)$$

$$\mathbf{r} = \mathbf{r}_0 + \mathbf{v}_0 t + \tfrac{1}{2}\mathbf{a}t^2 \qquad\qquad (3\text{–}9)$$

$$\bar{\mathbf{v}} = \tfrac{1}{2}(\mathbf{v}_0 + \mathbf{v}) \qquad\qquad (3\text{–}10)$$

$$\mathbf{r} - \mathbf{r}_0 = \tfrac{1}{2}(\mathbf{v}_0 + \mathbf{v})t \qquad\qquad (3\text{–}11)$$

Equations 3–8 to 3–11 describe fully the relations among the position, velocity, and acceleration vectors in two-dimensional motion with constant acceleration. Once we choose a coordinate system, we can consider separately the components of Equations 3–8 to 3–11 along our coordinate axes; these component equations are Equations 3–8a,b to 3–11a,b. (Although at this point we cannot combine Equations 3–12a and 3–12b into a single equation, we will later develop a form of vector multiplication that would make such a combination possible.)

Example 3–7 _____
A Spaceship Accelerates

A spaceship is travelling in a straight line at 6.2 km/s, when a rocket is fired that gives it an acceleration of 0.36 km/s² in a direction 60° to its velocity. If the rocket firing lasts 10 s, find the magnitude and direction of the ship's velocity at

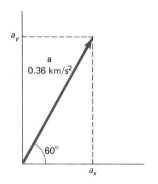

Fig. 3–15
The components of the acceleration vector are $a_x = a\,\cos60° = 0.18$ km/s^2 and $a_y = a\,\sin60° = 0.31$ km/s^2, where $a = 0.36$ km/s^2 is the magnitude of the acceleration vector.

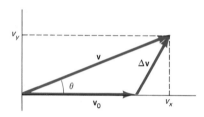

Fig. 3–16
Vector addition diagram for determining the spaceship's velocity after rocket firing. Note that the direction of the new velocity is given by $\tan\theta = v_y/v_x$.

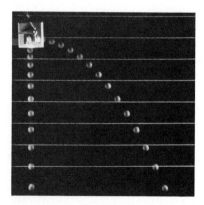

Fig. 3–17
Strobe photo showing one ball dropped and another projected horizontally at the same time. At each instant their vertical positions are identical, showing that the vertical and horizontal motions are independent. The time interval between images is 1/30 s, and the horizontal strings are 15.25 cm apart.

the end of the firing. At the end of the rocket firing, how far is the ship from its initial position?

Solution

Taking the x-axis along the initial direction of motion, the initial velocity is $\mathbf{v}_0 = 6.2\hat{\mathbf{i}}$ km/s and, as shown in Fig. 3–15, the acceleration is $0.18\hat{\mathbf{i}} + 0.31\hat{\mathbf{j}}$ km/s^2. To find the velocity after 10 s, we use Equations 3–8a and 3–8b:

$$v_x = v_{x0} + a_x t = 6.2 \text{ km/s} + (0.18 \text{ km/s}^2)(10 \text{ s}) = 8.0 \text{ km/s}$$

$$v_y = v_{y0} + a_y t = 0 \text{ km/s} + (0.31 \text{ km/s}^2)(10 \text{ s}) = 3.1 \text{ km/s},$$

so that the velocity is $\mathbf{v} = 8.0\hat{\mathbf{i}} + 3.1\hat{\mathbf{j}}$ km/s. The magnitude of this vector is given by the Pythagorean theorem:

$$v = \sqrt{v_x^2 + v_y^2} = [(8.0 \text{ km/s})^2 + (3.1 \text{ km/s})^2]^{1/2} = 8.6 \text{ km/s},$$

while its direction relative to the x-axis is, as shown in Fig. 3–16,

$$\theta = \tan^{-1}\left(\frac{v_y}{v_x}\right) = \tan^{-1}\left(\frac{3.1 \text{ km/s}}{8.0 \text{ km/s}}\right) = 21°.$$

We could also have handled this calculation graphically, as suggested in Fig. 3–16.

To find the change in position $\Delta\mathbf{r}$, we can use Equation 3–9. If we choose our origin at the point where the rocket is first fired, then $x_0 = y_0 = 0$, and the two components of Equation 3–9 become

$$x = v_{x0}t + \tfrac{1}{2}a_x t^2 = (6.2 \text{ km/s})(10 \text{ s}) + \tfrac{1}{2}(0.18 \text{ km/s}^2)(10 \text{ s})^2 = 71 \text{ km}$$

$$y = \tfrac{1}{2}a_y t^2 = \tfrac{1}{2}(0.31 \text{ km/s}^2)(10 \text{ s})^2 = 16 \text{ km},$$

where in writing the y-equation, we used the fact that $v_{y0} = 0$. The new position vector is then given by $\mathbf{r} = x\hat{\mathbf{i}} + y\hat{\mathbf{j}} = 71\hat{\mathbf{i}} + 16\hat{\mathbf{j}}$ km, so that the distance from the origin is

$$\Delta r = \sqrt{x^2 + y^2} = [(71 \text{ km})^2 + (16 \text{ km})^2]^{1/2} = 73 \text{ km}.$$

Because we had already calculated the velocity, we could equally well have used Equation 3–11 to determine the new position (see Problem 20).

As Example 3–7 shows, we handle problems of multidimensional motion by breaking them into two or three one-dimensional problems. These individual problems are solved using the methods of Chapter 2, and their solutions are combined to give a full vector description of the motion. Solving each one-dimensional problem separately implies that acceleration and velocity in one direction have no effect on the motion in a perpendicular direction. How do we know that this approach is correct? Ultimately, the answer comes from experiment. Figure 3–17 shows strobe photos of two balls, one dropped and one projected horizontally at the same time. The photos demonstrate that both experience the same downward acceleration, so that their different horizontal motions have no effect on their vertical motions. Similarly, if you toss a ball into the air on a horizontally moving airplane, the time of flight of the ball will be correctly given by the one-dimensional equations of Chapter 2, independent of the plane's horizontal motion.

3.3
PROJECTILE MOTION

projectile

A **projectile** is an object that is launched into the air and then moves predominantly under the influence of gravity. Examples are numerous: baseballs, streams of water, fireworks, missiles, ejecta from volcanoes, drops of ink in an ink-jet computer printer, and leaping dolphins are all projectiles (Fig. 3–18).

To treat projectile motion, we make two simplifying assumptions: (1) we neglect any variation in the direction or magnitude of the gravitational acceleration and (2) we neglect air resistance. Assumption 1 is equivalent to neglecting the curvature of the earth, and is valid for projectiles whose horizontal and vertical motions are small compared with earth's radius. Air resistance has a more variable effect; for dense objects moving relatively slowly, it is completely negligible, but for objects whose ratio of surface area to mass is large—like ping-pong balls and parachutes—or for high-speed objects like meteors entering earth's atmosphere, air resistance may dramatically alter the motion (Fig. 3–19). And the more subtle motions of a baseball—its curves and wobbles—can only be explained by invoking air resistance.

To describe projectile motion, it is usually convenient to choose a coordinate system with the y-axis vertically upward, and the x-axis horizontal and in the direction of the horizontal component of the projectile's initial velocity. Then there is neither velocity nor acceleration in the z-direction, and we have purely two-dimensional motion in the x-y plane. Furthermore, with the only acceleration provided by gravity, $a_x = 0$ and $a_y = -g$, so the components of Equations 3–8 and 3–9 become

$$v_x = v_{x0} \tag{3–13}$$

$$v_y = v_{y0} - gt \tag{3–14}$$

$$x = x_0 + v_{x0}t \tag{3–15}$$

$$y = y_0 + v_{y0}t - \tfrac{1}{2}gt^2 \,. \tag{3–16}$$

With $a_x = 0$, Equation 3–12a is identical to Equation 3–13, while Equation 3–12b is

$$v_y^2 = v_{y0}^2 - 2g(y - y_0). \tag{3–17}$$

In writing Equations 3–13 to 3–17, we take g to be a positive number, as we did in the preceding chapter, and account for the downward direction using minus signs. Equations 3–13 to 3–16 tell us mathematically what Fig. 3–17 tells us physically: that the horizontal motion of a projectile is not affected by gravity, and that the vertical motion is not affected by the horizontal motion.

Fig. 3–18
The 1986 explosion of a Titan rocket over Vandenburg Air Force Base in California. Each explosion fragment is a projectile, and each describes a parabolic trajectory.

Fig. 3–19
Parachutists descending from New York's Empire State Building. Air resistance has a dominant effect on the motion, and their trajectories are far from parabolic.

Example 3–8
Projectile Motion

Standing on the edge of a 120-m-high cliff, you throw a rock horizontally at 14 m/s. How far from the bottom of the cliff does it land? Neglect air resistance (although this assumption is not entirely justified for such a long fall).

Solution

Since there is no component of acceleration in the horizontal direction, the horizontal component of the velocity remains constant at $v_x = 14$ m/s. The answer we seek is the distance the rock travels horizontally; knowing the horizontal component of velocity, we could find this distance if we knew the flight time.

We can find the time by analyzing the vertical motion. Since the rock is thrown horizontally, $v_{y0} = 0$. If we choose our origin at the cliff bottom, then $y_0 = 120$ m. We want to find the time when the rock hits the ground—that is, when $y = 0$. Setting y and v_{y0} to zero in Equation 3–16, and solving for t, we have

$$t = \sqrt{\frac{2y_0}{g}} = \sqrt{\frac{(2)(120 \text{ m})}{9.8 \text{ m/s}^2}} = 4.95 \text{ s}.$$

During this time, the rock drops the 120-m distance to the bottom of the cliff. During the same time, it travels horizontally a distance given by Equation 3–15:

$$x - x_0 = v_{x0}t = (14 \text{ m/s})(4.95 \text{ s}) \doteq 69 \text{ m}.$$

Instead of solving for a numerical value of t, we could have used the expression $t = \sqrt{2y_0/g}$ in Equation 3–15 before substituting numerical values. It is usually easiest to carry calculations as far as possible in symbolic form.

In this example, we were asked only about the horizontal motion of the rock. But to find that, we needed the time—and that required us to analyze the vertical motion. This is a common situation with physics problems involving two dimensions. In such problems, it is often easiest to work backward from the answer, asking what quantities you would need to get that answer, and then formulating and solving other problems that will yield those quantities.

In Example 3–8 we were asked about the rock's horizontal position when it hit the ground. We didn't really care about the time of its flight, although we needed to calculate this time to solve the problem. More generally, we are often interested in describing the path, or **trajectory,** of a projectile without the details of where it is at each instant of time. A look at Fig. 3–18 suggests that objects undergoing projectile motion have similarly shaped trajectories. What is that shape?

Mathematically, we could specify a projectile's trajectory by giving its height y as a function of horizontal position x. To do so, consider a projectile that is launched from the origin ($x_0 = y_0 = 0$) at some angle θ to the horizontal, with initial speed v_0 (Fig. 3–20). As Fig. 3–20 suggests, the components of the initial velocity are $v_{x0} = v_0\cos\theta$ and $v_{y0} = v_0\sin\theta$. Then Equations 3–15 and 3–16 become

$$x = v_0\cos\theta\, t \qquad (3–18)$$

and

$$y = v_0\sin\theta\, t - \tfrac{1}{2}gt^2. \qquad (3–19)$$

Solving Equation 3–18 for the time t gives

$$t = \frac{x}{v_0\cos\theta};$$

using this result in Equation 3–19, we have

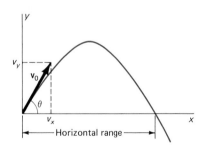

Fig. 3–20
Parabolic trajectory of a projectile.

trajectory

$$y = v_0 \sin\theta \left(\frac{x}{v_0 \cos\theta} \right) - \tfrac{1}{2}g \left(\frac{x}{v_0 \cos\theta} \right)^2,$$

or

$$y = x \tan\theta - \frac{g}{2v_0^2 \cos^2\theta} x^2. \qquad \text{(projectile trajectory)} \qquad (3\text{--}20)$$

In Equation 3–20 we have the mathematical description of the projectile's trajectory. Since y is a quadratic function of x, the trajectory is a parabola.

Example 3–9

The Trajectory of a Projectile

A construction worker is standing in a 2.6-m-deep cellar hole, 3.1 m from the side of the hole (Fig. 3–21). He tosses a hammer to a companion outside the hole. If the hammer leaves his hand 1.0 m above the bottom of the hole, at an angle of 35° to the horizontal, what is the minimum speed it must have to clear the edge of the hole, as shown in Fig. 3–21? How far from the edge of the hole does it land?

Solution

If it just clears the edge of the hole, the hammer must be on a trajectory that has $y = 1.6$ m when $x = 3.1$ m, where we take the origin at the point where the hammer leaves his hand, 1.0 m above the bottom of the 2.6-m-deep hole. Solving Equation 3–20 for v_0 and using these x and y values gives

$$v_0 = \left(\frac{gx^2}{2\cos^2\theta (x \tan\theta - y)} \right)^{1/2}$$

$$= \left(\frac{(9.8 \text{ m/s}^2)(3.1 \text{ m})^2}{(2)(\cos^2 35°)[(3.1 \text{ m})(\tan 35°) - 1.6 \text{ m}]} \right)^{1/2} = 11 \text{ m/s}.$$

To find where the hammer lands on the level ground outside the hole, we need to know the horizontal position when $y = 1.6$ m. Rearranging Equation 3–20 into the standard form $ax^2 + bx + c = 0$ for a quadratic equation for x gives

$$\frac{g}{2v_0^2 \cos^2\theta} x^2 - (\tan\theta)x + y = 0,$$

so that the coefficients a, b, and c are $a = g/2v_0^2\cos^2\theta = 0.0594$ m^{-1}, $b = -\tan\theta = -0.700$, and $c = y = 1.6$ m. Applying the quadratic formula (see Appendix A) then gives

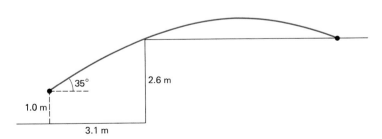

Fig. 3–21
Example 3–9

$$x = \frac{-b \pm \sqrt{b^2 - 4ac}}{2a} = \frac{0.700 \pm [0.700^2 - (4)(0.0594)(1.6)]^{1/2}}{(2)(0.0594)}$$

$$= 3.1 \text{ m or } 8.7 \text{ m},$$

where for clarity we have left off units in the calculation. These two answers give the two horizontal positions where the hammer is at ground level. The first provides a useful check on our work: it is just the horizontal position when the hammer clears the edge of the hole. The second answer is the one we want; it shows that the hammer lands 8.7 m − 3.1 m = 5.6 m to the right of the edge of the hole.

How far will a soccer ball go if I kick it at 12 m/s at 50° to the horizontal? If I can throw a rock at 15 m/s, can I get it across a 30-m-wide pond? At what angle should I throw it for maximum distance? How far off vertical can a rocket's trajectory be and still have it land within 50 km of its launch point? As in these examples, we are frequently interested in the **horizontal range** of a projectile—that is, in how far it moves horizontally over level ground.

horizontal range

Equation 3–20 describes the trajectory—height y versus horizontal position x—for a projectile. For a projectile launched on level ground, we can ask when the projectile will return to the ground by setting $y = 0$ in Equation 3–20:

$$0 = x \tan\theta - \frac{g}{2v_0{}^2 \cos^2\theta} x^2.$$

Factoring this equation gives

$$x \left(\tan\theta - \frac{gx}{2v_0{}^2 \cos^2\theta} \right) = 0,$$

so that either $x = 0$, corresponding to the launch point, or

$$x = \frac{2v_0{}^2}{g} \cos^2\theta \tan\theta = \frac{2v_0{}^2}{g} \sin\theta \cos\theta.$$

Here we have used $\tan\theta = \sin\theta/\cos\theta$; recalling further that $\sin2\theta = 2 \sin\theta \cos\theta$ (see Appendix A), we can write

$$x = \frac{v_0{}^2}{g} \sin2\theta. \tag{3-21}$$

We emphasize that Equation 3–21 gives the *horizontal* range—the distance a projectile travels horizontally before returning *to its starting height*. From the way it was derived—setting $y = 0$—you can see it does *not* give horizontal distance when the projectile returns to a different height (Fig. 3–22).

Does Equation 3–21 make sense? When $\theta = 0$, the range is zero: a projectile launched horizontally on level ground immediately hits the ground. When $\theta = 90°$, $\sin2\theta = \sin180° = 0$, and again the range is zero. Here the projectile is launched vertically upward, so of course it returns to the same point. When is the range a maximum? The largest value of $\sin2\theta$ is 1, which occurs when $\theta = 45°$. For a given initial speed v_0, then, the maximum range is attained by launching at 45°. Figure 3–23 shows the trajectories of projectiles launched at different angles with the same initial speed. At angles

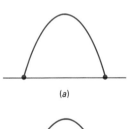

(a)

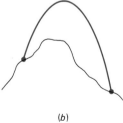

(b)

Fig. 3–22
(a) Equation 3–21 gives the horizontal range of a particle that returns to its starting height. (b) It does *not* apply otherwise.

Fig. 3–23
For a given initial speed, the range of a projectile is maximum at a launch angle of 45°. For launch angles equally spaced above and below 45°, the range is the same. Curves are for an initial speed of 50 m/s, with launch angles indicated.

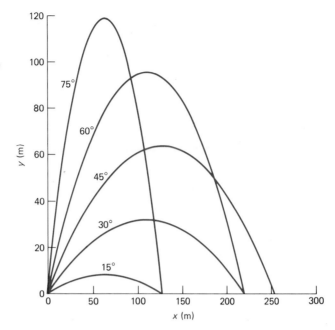

less than 45° the projectile has a greater horizontal component of velocity, but it doesn't get as high and therefore isn't in the air as long as the projectile launched at 45°; the net effect is a shorter range. At angles greater than 45°, the projectile rises higher and is therefore in the air longer, but its horizontal velocity is lower and the net effect is again a shorter range. In fact, as you can prove in Problem 39, the range is the same for angles equally spaced on either side of 45°. (These results are valid only in the absence of air resistance; when air resistance is taken into account, the maximum range angle turns out to be less than 45°.)

Example 3–10
A Sounding Rocket

A sounding rocket is launched from a remote desert site to probe the atmosphere. With a short firing of its engine, the rocket quickly reaches a speed of 4.6 km/s. If the rocket is to land within 50 km of its launch site, what is the maximum allowable deviation from a vertical trajectory? Neglect air resistance.

Solution

Here we neglect the short distance over which the rocket accelerates, considering that it left the ground with its initial speed of 4.6 km/s. Then the answer to the problem is the angle θ for which the range x is 50 km. Solving Equation 3–21 for $\sin 2\theta$ gives

$$\sin 2\theta = \frac{gx}{v_0{}^2} = \frac{(9.8 \text{ m/s}^2)(50 \times 10^3 \text{ m})}{(4.6 \times 10^3 \text{ m/s})^2} = 0.0232.$$

There are two solutions to this equation, corresponding to $2\theta = \sin^{-1}(0.0232) = 1.33°$ and $2\theta = 180° - 1.33°$. The second solution is the one we want; it represents a nearly vertical launch angle $\theta = 90° - 0.67°$, so that the launch angle must be within 0.67° of vertical.

3.4

CIRCULAR MOTION

uniform circular motion

An important case of accelerated motion in two dimensions is **uniform circular motion**—the motion of an object describing a circular path at constant speed. Although the speed is constant, the motion is accelerated because the *direction* of the velocity is changing. Velocity is a vector, and *any* change in that vector—whether in magnitude or direction or both—implies acceleration.

Examples of uniform circular motion are numerous. Many spacecraft are in circular orbits, and the orbits of the planets and their natural satellites are approximately circular. The earth's daily rotation carries you around in uniform circular motion. Pieces of rotating machinery describe uniform circular motion. In laboratory apparatus and astrophysical objects, electrons undergo uniform circular motion when they encounter magnetic fields. Even with the atom itself, we can gain some insight by picturing electrons in uniform circular motion about the nucleus—although this view is not fully consistent with modern quantum mechanics. Other situations involve uniform circular motion over a limited path. Driving around a curve at constant speed, you temporarily describe a circular arc and are therefore temporarily in uniform circular motion. On a back road in hilly country, you may undergo uniform circular motion as you negotiate valley bottoms and zoom over hillcrests. As you swing a baseball bat, golf club, or hockey stick, you produce motion that is circular and that may be approximately uniform.

In this section, we derive a relation among the acceleration, speed, and radius of uniform circular motion. Consider an object moving with constant speed v around a circle of radius r. Figure 3–24 shows the circular path and some velocity vectors for the object at several points on the path. Note that the velocity vectors are tangent to the circle, indicating the instantaneous direction of motion. Since the speed is constant, the velocity vectors all have the same length. But clearly the object is accelerating; the velocity vector $\mathbf{v}_C$ is not the same as the vector $\mathbf{v}_A$, since their directions differ. In fact, we can find the *average* acceleration between points A and C by dividing the change in velocity by the time interval involved. Figure 3–25 shows the vector $\Delta\mathbf{v}_{AC} = \mathbf{v}_C - \mathbf{v}_A$. This vector has magnitude $\sqrt{2}v$ and points diagonally downward and to the left at 45°. Between A and C, the object moves one-fourth of the way around its circle, or a distance $\frac{1}{2}\pi r$. At speed v, this motion takes a time $\Delta t = \frac{1}{2}\pi r/v$. The average acceleration $\Delta\mathbf{v}/\Delta t$ is then a vector whose magnitude is

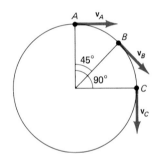

Fig. 3–24
Uniform circular motion, showing the velocity vectors at several points on the circle.

$$\overline{a} = \frac{|\Delta\mathbf{v}|}{\Delta t} = \frac{\sqrt{2}v}{\frac{1}{2}\pi r/v} = \frac{2\sqrt{2}v^2}{\pi r} = \frac{0.90v^2}{r}$$

and which points in the same diagonal direction as $\Delta\mathbf{v}$. Note that we used the magnitude of the *change in velocity*—not the change in speed (which is zero)—to compute the magnitude of the acceleration. We did this calculation in detail to show how to give a meaningful answer for the acceleration even in the case of uniform circular motion in which speed does not change. Clearly we could repeat the calculation for the average acceleration

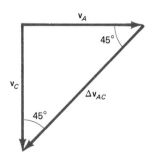

Fig. 3–25
The change in velocity from point A to point C is the vector $\Delta\mathbf{v}_{AC}$ that must be added to $\mathbf{v}_A$ to make $\mathbf{v}_C$. It has magnitude $\sqrt{2}v$ and points downward and to the left at 45°.

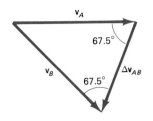

Fig. 3–26
The velocity change $\Delta \mathbf{v}$ from A to B is a vector of length $0.765v$, pointing downward and to the left at $67.5°$. In going from A to B, the object travels one-eighth of the circle or a distance $\frac{1}{4}\pi r$; at speed v, this takes $\Delta t = \frac{1}{4}\pi r/v$. The acceleration vector then has magnitude $|\Delta \mathbf{v}|/\Delta t = 0.97v^2/r$ and points in the same direction as $\Delta \mathbf{v}$.

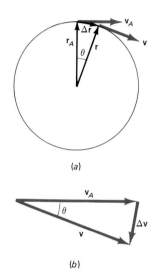

(a)

(b)

Fig. 3–27
(a) Velocity vectors $\mathbf{v}_A$ and $\mathbf{v}$ and (b) their difference $\Delta \mathbf{v}$. Also shown in (a) are the associated radius vectors and their difference $\Delta \mathbf{r}$. The angle θ in both figures is the same, so that the triangles formed by the radius vectors and their difference in (a) is similar to the triangle formed by $\mathbf{v}_A$, $\mathbf{v}$, and $\Delta \mathbf{v}$ in (b).

in the interval AB; analysis of Fig. 3–26 would show that the result is an average acceleration of magnitude $0.97v^2/r$, pointing downward and to the left at an angle of $67.5°$.

Now consider the average acceleration over ever shorter time intervals Δt; in the limit $\Delta t \to 0$, this quantity becomes the instantaneous acceleration. Figure 3–27a shows the velocity vector $\mathbf{v}_A$ and a nearby velocity vector $\mathbf{v}$. Now the vectors $\mathbf{v}_A$ and $\mathbf{r}_A$ are perpendicular; similarly, $\mathbf{v}$ and $\mathbf{r}$ are perpendicular. Therefore, the angle θ between the radii $\mathbf{r}_A$ and $\mathbf{r}$ in Fig. 3–27a is the same as the angle θ between the velocity vectors in Fig. 3–27b. Because the radii have the same length, and the velocity vectors have the same magnitude (*speed* is constant), the triangles shown in Fig. 3–27a and Fig. 3–27b are similar. Then we can write

$$\frac{|\Delta \mathbf{v}|}{v} = \frac{|\Delta \mathbf{r}|}{r}.$$

For short time intervals Δt, the angle θ is small, and the length of the vector $\Delta \mathbf{r}$ is approximately that of the circular arc joining the endpoints of the radius vectors. The length of this arc is just the distance travelled by the object in the time interval Δt, or $v\Delta t$, so that $|\Delta \mathbf{r}| \simeq v\Delta t$. Then we can write

$$\frac{|\Delta \mathbf{v}|}{v} \simeq \frac{v\Delta t}{r}.$$

The magnitude of the average acceleration is then approximately

$$\bar{a} = \frac{|\Delta \mathbf{v}|}{\Delta t} \simeq \frac{v^2}{r}.$$

Taking the limit $\Delta t \to 0$ gives the instantaneous acceleration; in this limit the circular arc and the vector $\Delta \mathbf{r}$ joining the ends of the radius vectors become completely indistinguishable, and the relation $|\Delta \mathbf{r}| \simeq v\Delta t$ becomes exact. So we have

$$a = \frac{v^2}{r} \qquad \text{(uniform circular motion)} \qquad (3\text{–}22)$$

for the magnitude of the instantaneous acceleration of an object moving in a circle of radius r at constant speed v. What about the direction of the acceleration vector? As Fig. 3–27b suggests, $\Delta \mathbf{v}$ is very nearly perpendicular to both velocity vectors; in the limit $\Delta t \to 0$, $\Delta \mathbf{v}$ and therefore the acceleration $\Delta \mathbf{v}/\Delta t$ becomes exactly perpendicular to the velocity. In the limit, then, the direction of the acceleration in Fig. 3–27b is downward—that is, toward the center of the circle.

Although we found the instantaneous acceleration for the point at the top of the circle, the geometrical arguments leading to our result are clearly independent of where the object is on its circular path. (You can rotate the book, for example, and Fig. 3–27 will still lead you to the same result.) We conclude therefore that the acceleration in uniform circular motion has constant magnitude v^2/r, and that it always points toward the center of the circle. Since the direction toward the center changes as the object moves around its circular path, the acceleration vector is not constant, even though its magnitude is. Therefore, uniform circular motion is *not* motion with constant acceleration, and our constant-acceleration equations do not apply. Indeed, we know that constant acceleration in two dimensions gives rise to a parabolic trajectory, not a circle.

Does Equation 3–22 make sense? Yes: an increase in speed v means that the time Δt for a given change in direction of the velocity becomes shorter. Not only that, but the associated change $\Delta\mathbf{v}$ in velocity is larger. These two effects combine to give an acceleration that depends on the *square* of the speed. On the other hand, an increase in radius with a fixed speed simply increases the time Δt associated with a given change in velocity, so that the acceleration is inversely proportional to the radius. Although the form of the acceleration in Equation 3–22 does not look familiar from our work with constant acceleration, we can check that its dimensions are correct:

$$[a] = [v^2]/[r] = (L/T)^2/L = L/T^2,$$

which are indeed the dimensions of acceleration. Incidentally, although we derived Equation 3–22 geometrically, we could have obtained the same results by writing the position in unit vector notation and then differentiating twice to get the acceleration. Problem 60 explores this approach.

Example 3–11
A Space Shuttle Orbit

A space shuttle is in a circular orbit at a height of 250 km, where the acceleration of earth's gravity is 93 per cent of its surface value. What is the period of its orbit? ("Period" means the time to complete one orbit.)

Solution

Here we are given the acceleration and radius and asked to find the period. Our equations for uniform circular motion relate acceleration, radius, and speed. But we can write the speed as the distance travelled in one orbit—the orbital circumference $2\pi r$—divided by the orbital period T:

$$v = \frac{2\pi r}{T}.$$

Using this result in Equation 3–22, we have

$$a = \frac{(2\pi r/T)^2}{r} = \frac{4\pi^2 r}{T^2}. \tag{3–23}$$

This equation is equivalent to Equation 3–22, but relates a, r, and T rather than a, r, and v. Here we are given that $a = 0.93g$ and that the shuttle is 250 km above earth's surface. But the r in Equation 3–23 is the distance to the center of the circular path—that is, to the center of the earth. So $r = R_{earth} + 250 \text{ km} = 6.6 \times 10^6$ m, where we used $R_{earth} = 6.37 \times 10^6$ m from Appendix F. Solving Equation 3–23 for the period T then gives

$$T = \left(\frac{4\pi^2 r}{a}\right)^{1/2} = \left(\frac{(4\pi^2)(6.6 \times 10^6 \text{ m})}{(0.93)(9.8 \text{ m/s}^2)}\right)^{1/2} = 5347 \text{ s} = 89 \text{ min},$$

or about an hour and a half. This value is approximately the orbital period of any object orbiting the earth at altitudes that are small compared with earth's radius. Scientists and engineers have no choice here; the orbital period is fixed by the size and mass of the earth. (At higher altitudes the decrease in the strength of gravity becomes more noticeable, and orbital periods lengthen. The moon, 390,000 km distant, orbits earth with a period of 27 days. We will study gravity and orbits further in Chapter 8.)

Example 3–12
Designing a Road

A flat, horizontal road is being designed for an 80-km/h speed limit. If the maximum acceleration of a car travelling this road at the speed limit is to be 1.5 m/s², what is the minimum radius for curves in the road?

Solution

Here we are given speed and acceleration, so we can solve Equation 3–22 for the radius r:

$$r = \frac{v^2}{a} = \frac{[(80 \times 10^3 \text{ m/h})(1 \text{ h/3600 s})]^2}{1.5 \text{ m/s}^2} = 330 \text{ m}.$$

A limit on acceleration is necessary to keep the occupants of a car in their seats, as we will see more clearly in Chapters 4 and 5.

Uniform circular motion represents the special case in which acceleration is always *perpendicular* to velocity. In that case, the acceleration changes only the *direction* of the velocity, and not its magnitude (speed). In this sense, uniform circular motion is the converse of one-dimensional accelerated motion that we examined in Chapter 2; there, acceleration was always *parallel* to velocity, and changed only the *magnitude* of the velocity. The case of constant acceleration in two dimensions lies between these two extremes; in projectile motion, for example, the acceleration is generally neither parallel nor perpendicular to the velocity. Then both the magnitude and direction of the velocity change. Figure 3–28 summarizes these three cases.

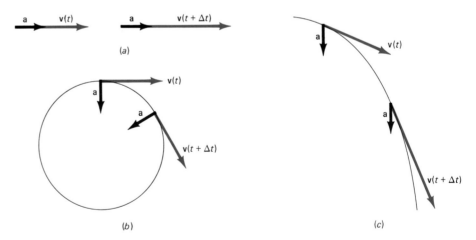

Fig. 3–28
(a) In one-dimensional motion, acceleration is always parallel to velocity and changes only the magnitude of the velocity vector. (b) In uniform circular motion, acceleration is always perpendicular to velocity and changes only the direction of the velocity vector. (c) In the more general case, of which projectile motion is an example, acceleration is neither parallel nor perpendicular to velocity. Then both the magnitude and direction of the velocity vector change.

What about circular motion whose speed changes, like a record turntable spinning up from rest to full speed, or a ball whirling around in a vertical circle (Fig. 3–29)? Since both the direction and magnitude of the velocity are changing, the acceleration cannot be either parallel or perpendicular to the velocity. But we can break the acceleration vector into two components, one perpendicular to the velocity and one parallel. We call the former the **radial acceleration** a_r, since it points radially inward toward the center of the circular path; the latter is the **tangential acceleration** a_t, since it is tangent to the path. The radial acceleration changes only the direction of the velocity; it acts just like the acceleration in uniform circular motion, and therefore has magnitude $a_r = v^2/r$. The tangential acceleration changes only the magnitude of the velocity; it acts just like one-dimensional acceleration, and therefore has magnitude $a_t = dv/dt$. The net acceleration vector can be found by combining these two components, as Example 3–13 illustrates.

radial acceleration

tangential acceleration

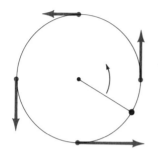

Fig. 3–29
A ball whirled about on a string in a vertical circle undergoes circular motion with nonconstant velocity; it moves faster near the bottom and more slowly near the top.

Example 3–13
Radial and Tangential Acceleration

A road makes a 90° bend with a radius of 190 m. A car enters the bend moving at 20 m/s. Finding this too fast, the driver decelerates at 0.92 m/s². Determine the acceleration of the car when its speed rounding the bend has dropped to 15 m/s.

Solution

Since it is rounding a curve, the car has a radial acceleration associated with its changing direction, in addition to the tangential deceleration that changes its speed. We are given that $a_t = 0.92$ m/s²; since the car is slowing down, the tangential acceleration is directed opposite the velocity. The radial acceleration is given by Equation 3–22:

$$a_r = \frac{v^2}{r} = \frac{(15 \text{ m/s})^2}{190 \text{ m}} = 1.2 \text{ m/s}^2.$$

Figure 3–30 shows both components of the acceleration and their vector sum, which is the net acceleration. From the figure, we see that this sum has magnitude

$$a = \sqrt{a_r^2 + a_t^2} = [(1.2 \text{ m/s})^2 + (0.92 \text{ m/s})^2]^{1/2} = 1.5 \text{ m/s}^2$$

and points at an angle

$$\theta = \tan^{-1}\left(\frac{a_r}{a_t}\right) = \tan^{-1}\left(\frac{1.2 \text{ m/s}^2}{0.92 \text{ m/s}^2}\right) = 53°$$

relative to the tangent line to the circle, as shown.

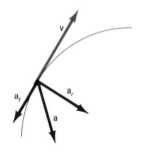

Fig. 3–30
The radial and tangential accelerations add vectorially to give the net acceleration.

3.5
MOTION IS RELATIVE

What does it mean to say that a car is moving at 80 km/h? Usually, it means 80 km/h relative to the earth. But earth is moving at 30 km/s relative to the sun, so that the car's speed relative to the sun is much greater than 80 km/h. Statements about velocity and speed are meaningful only when we have an answer to the question "velocity relative to what?" The object or system with respect to which velocity is measured is called a **frame of reference;** in most of the examples we have considered, the earth is our

frame of reference

frame of reference. But playing tennis on the deck of a moving ship, you would find the ship's frame of reference more appropriate for describing the motion of the tennis ball. Similarly, the motion of a comet or interplanetary space probe is described more simply in the reference frame of the sun rather than of the earth.

Suppose a car is moving at 50 mi/h. Then its speed relative to another car moving in the same direction at 30 mi/h is 50 mi/h − 30 mi/h = 20 mi/h. More generally, if an object is moving at velocity **v** relative to some frame of reference S, and if another frame of reference S′ moves with velocity **V** relative to S, then the velocity **v′** of the object with respect to the frame S′ is given by

$$\mathbf{v'} = \mathbf{v} - \mathbf{V}. \tag{3–25}$$

In our simple example, frame S is the earth, frame S′ is the car moving at 30 mi/h, $v = 50$ mi/h, $V = 30$ mi/h, and $v' = 20$ mi/h. Here all the velocities are in the same direction, so we deal only with their magnitudes. More generally, questions of relative velocity involve vectors, as in the following example.

Example 3–14
Relative Velocity

A jetliner has an airspeed (speed relative to the air) of 600 mi/h. It embarks on a flight from Houston to Omaha, a distance of 800 miles northward. In what direction should it fly if there is a steady 120-mi/h wind from the west at its cruising altitude? What will be its groundspeed?

Solution

Let S be the frame of reference of the ground, and choose x and y axes pointing east and north, respectively. Let S′ be the frame of reference of the air, so that the velocity **V** is 120î mi/h. The velocity **v′** of the plane relative to the air has magnitude 600 mi/h, but we don't know its direction. We can write this velocity as

$$\mathbf{v'} = v'\cos\theta\,\hat{\imath} + v'\sin\theta\,\hat{\jmath},$$

where v' is 600 mi/h and θ is the angle **v′** makes with the positive x-axis (see Fig. 3–31). The velocity **v** points northward, so it can be expressed in the form

$$\mathbf{v} = v\hat{\jmath},$$

where we don't know the magnitude v. We can now write the two components of Equation 3–25:

$$v_x' = v_x - V_x$$
$$v_y' = v_y - V_y.$$

Using our expressions for various velocities, these equations become

$$v' \cos\theta = 0 - V,$$

where $V = 120$ mi/h is the magnitude of **V**, and

$$v' \sin\theta = v - 0.$$

We can solve the first of these equations for θ:

$$\theta = \cos^{-1}\left(\frac{-V}{v'}\right) = \cos^{-1}\left(\frac{-120 \text{ mi/h}}{600 \text{ mi/h}}\right) = 102°.$$

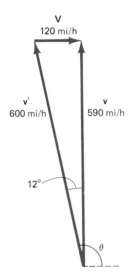

Fig. 3–31
Vector diagram for Example 3–14.

Using this result in the second equation gives

$$v = v' \sin\theta = (600 \text{ mi/h})(\sin 102°) = 590 \text{ mi/h}.$$

So the plane should fly at 102° to the x-axis, or 12° west of north, and its ground-speed will be 590 mi/h. Figure 3–31 is a vector diagram relating the velocities **v'**, **v**, and **V**. Before the advent of sophisticated navigation equipment, pilots routinely carried out such calculations, often using graphical techniques.

What about the acceleration of an object viewed in two different reference frames? Suppose **a** is the acceleration in some frame S; that is, **a** = $d\mathbf{v}/dt$, where **v** is the velocity relative to frame S. Equation 3–25 gives the velocity **v'** in a frame S' moving with velocity **V** relative to S. Differentiating this equation then gives the acceleration **a'**:

$$\mathbf{a}' = \frac{d\mathbf{v}'}{dt} = \frac{d\mathbf{v}}{dt} - \frac{d\mathbf{V}}{dt}.$$

We generally consider only reference frames moving with *constant* relative velocity **V**, so that $d\mathbf{V}/dt = 0$. Furthermore, $d\mathbf{v}/dt$ is just the acceleration **a** in the frame S, so that

$$\mathbf{a}' = \mathbf{a}. \tag{3–26}$$

That is, the acceleration of an object is the same in all frames of reference moving at constant velocity relative to one another. This simple result is true because acceleration depends on *changes* in velocity, and the addition of any constant velocity does not alter those changes.

That acceleration does not change from one uniformly moving reference frame to another has a deep significance in physics. Consider the simple experiment of tossing a ball into the air. The experiment will have exactly the same outcome—the ball will go straight up and down in accordance with our equations for one-dimensional motion—both on an airplane moving at a constant 600 mi/h and on the ground. Why? Because in both cases the ball experiences a downward acceleration g, which is unaffected by the uniform motion of the plane relative to the earth.

If you were on a plane flying through perfectly smooth air, there is in fact no experiment involving motion that you could do to tell that the plane was moving. All experiments would come out exactly the same as they would on the ground. This fact is expressed in a simple statement, called the **principle of Galilean relativity:**

principle of Galilean relativity

> The laws of motion are the same in all frames of reference in uniform motion.

What this means is that there is no way of using the laws of motion—the only laws of physics we will consider until Chapter 20—to answer the question "am I moving?" With respect to the laws of motion, the question is meaningless; only *relative motion* matters.

Although our discussion of relative motion and the principle of Galilean relativity may seem almost obvious and straightforward, in fact that discussion rests on deep-seated notions about the nature of time and space.

In Chapter 33, after we have completed our study of electromagnetism, we will see how Albert Einstein was able to extend the principle of Galilean relativity to all of physics; the result is Einstein's special theory of relativity. In the process, Einstein showed that our commonsense notions of space and time are not quite right. As a result, Equations 3–25 and 3–26—and indeed all the equations we develop in Chapters 1 through 19—are really only approximately correct; they work well for our everyday experience, and even for spacecraft probing the solar system, but they break down when relative speeds approach the speed of light.

SUMMARY

1. **Vectors** are used to describe motion in two or three dimensions. Vector quantities have magnitude and direction; we visualize vectors using arrows whose length represents magnitude and whose direction is that of the vector. A vector is described fully by giving its magnitude and direction or, equivalently, its components on a chosen set of coordinate axes.

   a. Vectors may be added: the vector sum $\mathbf{A}+\mathbf{B}$ is formed by placing the tail of vector $\mathbf{B}$ at the head of $\mathbf{A}$; the sum is then a vector from the tail of $\mathbf{A}$ to the head of $\mathbf{B}$. Alternatively, vectors may be added by adding their components to get the components of the sum: the x-component of the vector $\mathbf{A}+\mathbf{B}$, for example, is just A_x+B_x. Vector addition is commutative: $\mathbf{A}+\mathbf{B}=\mathbf{B}+\mathbf{A}$. It is also associative: $(\mathbf{A}+\mathbf{B})+\mathbf{C}=\mathbf{A}+(\mathbf{B}+\mathbf{C})$.

   b. Vector subtraction $\mathbf{A}-\mathbf{B}$ means adding to $\mathbf{A}$ a vector whose magnitude is that of $\mathbf{B}$, but which points in the opposite direction.

   c. A vector may be multiplied by a scalar. If c is a scalar and $\mathbf{A}$ a vector of length A, the vector $c\mathbf{A}$ is a vector of length cA having the same direction as $\mathbf{A}$. Multiplication by a scalar is distributive: $c(\mathbf{A}+\mathbf{B})=c\mathbf{A}+c\mathbf{B}$.

2. **Velocity** is a vector describing the rate of change of position. The average velocity over a time interval Δt is given by the change in position divided by the time interval:

$$\overline{\mathbf{v}} = \frac{\Delta \mathbf{r}}{\Delta t},$$

   where $\Delta\mathbf{r}=\mathbf{r}_2-\mathbf{r}_1$, with $\mathbf{r}_1$ and $\mathbf{r}_2$ the positions at the beginning and end of the time interval Δt. The **instantaneous velocity** is obtained by taking the limit of arbitrarily small time intervals:

$$\mathbf{v} = \lim_{\Delta t \to 0} \frac{\Delta \mathbf{r}}{\Delta t} = \frac{d\mathbf{r}}{dt}.$$

3. **Acceleration** is a vector describing the rate of change of velocity; the average acceleration over a time interval Δt is

$$\overline{\mathbf{a}} = \frac{\Delta \mathbf{v}}{\Delta t},$$

   while the instantaneous acceleration is

$$\mathbf{a} = \lim_{\Delta t \to 0} \frac{\Delta \mathbf{v}}{\Delta t} = \frac{d\mathbf{v}}{dt}.$$

   Since velocity is a vector, it can change in either magnitude or direction or both; any change in velocity represents an acceleration. When the acceleration vector is parallel to the velocity vector, only the magnitude of the velocity changes. When the acceleration vector is perpendicular to the velocity, only the direction of the velocity changes.

4. When an object undergoes **constant acceleration**, its motion is described by the vector equations

$$\mathbf{v} = \mathbf{v}_0 + \mathbf{a}t$$
$$\mathbf{r} = \mathbf{r}_0 + \mathbf{v}_0 t + \tfrac{1}{2}\mathbf{a}t^2.$$

   These equations may be broken into components along any chosen coordinate axes; the result is a set of equations describing independently the components of the motion along the different axes.

5. **Projectile motion** is an important case of motion with constant acceleration. Launched from the origin ($x=y=0$) with speed v_0 at an angle θ to the horizontal, a projectile follows a parabolic trajectory described by

$$y = x \tan\theta - \frac{g}{2v_0^2\cos^2\theta}x^2.$$

   On level ground, the horizontal range of the projectile is given by

$$x = \frac{v_0^2}{g} \sin 2\theta.$$

6. An object moving around a circle of radius r at constant speed v is in **uniform circular motion**. Since the direction of its velocity is changing, it is accelerating. The magnitude of its acceleration is

$$a = \frac{v^2}{r}, \qquad \text{(uniform circular motion)}$$

and the acceleration vector always points toward the center of the circle. When an object moves in a circular path with varying speed, this expression gives the **radial acceleration**—the component of the acceleration perpendicular to the velocity.

7. The concept of velocity is meaningful only in relation to a particular frame of reference; the statement "I am moving" is incomplete unless the answer to the question "moving relative to what?" is understood. Only **relative motion** matters. This idea is summarized in the **principle of Galilean relativity:**

The laws of motion are the same in all frames of reference in uniform motion.

In the context of Galilean relativity, we can determine the velocity $\mathbf{v}'$ of an object relative to some frame of reference S' moving with constant velocity $\mathbf{V}$ relative to a second frame S:

$$\mathbf{v}' = \mathbf{v} - \mathbf{V},$$

where $\mathbf{v}$ is the velocity of the object with respect to the frame S. The acceleration of the object is the same in either frame of reference: $\mathbf{a}' = \mathbf{a}$.

QUESTIONS

1. Under what conditions is the magnitude of the vector sum $\mathbf{A} + \mathbf{B}$ equal to the sum of the magnitudes of the two vectors?

2. Which of the following are valid expressions? Explain. (a) $\mathbf{C} = \mathbf{A} - \mathbf{B}$; (b) $a = \mathbf{B} + c$; (c) $\mathbf{B} = c\mathbf{A}$; (d) $\mathbf{B} = \mathbf{A}/c$; (e) $\mathbf{B} = c/\mathbf{A}$.

3. Can two vectors of equal magnitude sum to zero? Can two vectors of unequal magnitude?

4. Repeat the preceding question for three vectors.

5. One way of adding vectors is to place them tail-to-tail, and then form a parallelogram as shown in Fig. 3–32. The sum of the two vectors is then a vector along the diagonal of the parallelogram. Why is this method equivalent to the head-to-tail method described in the text?

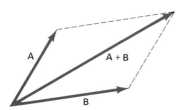

Fig. 3–32 Parallelogram method of adding vectors (Question 5).

6. Can an object have a southward acceleration while moving northward? A westward acceleration while moving northward?

7. Can an object move northward for 10 minutes, all the while having a southward acceleration?

8. Can an object move northward for 10 minutes, all the while having a westward acceleration?

9. Why can we usually use Cartesian coordinates on earth's surface even though earth is round?

10. Is the acceleration of an object in uniform circular motion constant? Explain.

11. A space shuttle completes one full orbit in 89 min. What are its average velocity and average acceleration over this interval? Is its average speed equal to its average velocity?

12. A fly heads west 5 ft, then north 3 ft, then straight up 6 ft. A mosquito starts from the same place, flies straight up 6 ft, west 5 ft, and north 3 ft. Compare their final positions.

13. Is it possible to form the unit vector $\hat{\mathbf{k}}$ from a sum of $\hat{\mathbf{i}}$ and $\hat{\mathbf{j}}$, each multiplied by appropriate scalars? Explain.

14. Is the speed of a projectile constant throughout its parabolic trajectory? Explain.

15. In this chapter, we found that the path of a projectile is a parabola. But what about a projectile launched vertically? It follows a straight up-and-down trajectory, as we described in the preceding chapter. Is there a contradiction here? Explain in terms of limiting cases of near-vertical motion.

16. Three objects are released from the same height, one from rest, one with horizontal velocity component v, and another with horizontal velocity component $2v$. Compare the times each is in the air. Compare the horizontal distances travelled by each.

17. Can there be a true "line drive" in baseball? What is required for the ball to travel a nearly straight horizontal trajectory?

18. Is there any point in the trajectory of a projectile where velocity and acceleration are perpendicular?

19. Explain why the acceleration in uniform circular motion is a vector pointing toward the center of the circle.

20. You are sitting in an airplane cruising at a steady 600 mi/h, and you throw a ball straight up. Describe its trajectory relative to you. Repeat for the case of an airplane accelerating down the runway.

21. A 45° angle maximizes the range of a ground-to-ground projectile. But suppose you are at the base of a slope inclined at some angle to the horizontal. If you want your projectile to land as far as possible up the slope, is 45° with respect to the horizontal still the optimum angle?

22. Why is the statement "I am moving" meaningless?

PROBLEMS

Section 3.1 *The Vector Description of Motion*

1. You walk west 220 m, then turn 45° toward the north and walk another 50 m. How far and in what direction from your starting point do you end up? Solve both with a diagram and by adding vector components.

2. For the vectors shown in Fig. 3–33, evaluate the vectors $\mathbf{A} + \mathbf{B}$, $\mathbf{A} - \mathbf{B}$, $\mathbf{A} + \mathbf{C}$, $\mathbf{A} + \mathbf{B} + \mathbf{C}$.

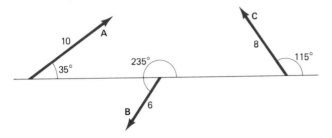

Fig. 3–33 Vectors for Problem 2; lengths are in arbitrary units.

3. A vector $\mathbf{A}$ is 10 units long and points 30° counterclockwise (CCW) from horizontal. What are its x and y components on a coordinate system (a) with the x-axis horizontal and the y-axis vertical; (b) with the x-axis at 45° CCW from horizontal and the y-axis 45° CCW from vertical; and (c) with the x-axis at 30° CCW from horizontal and the y-axis 90° CCW from the x-axis?

4. Express the sum of the unit vectors $\hat{\imath}$, $\hat{\jmath}$, and $\hat{k}$ in unit vector notation, and determine its magnitude.

5. The vector $\mathbf{A}$ is 12 units long and points 30° north of east. The vector $\mathbf{B}$ is 18 units long and points 45° west of north. Find a vector $\mathbf{C}$ such that $\mathbf{A} + \mathbf{B} + \mathbf{C} = 0$.

6. In a Hawaiian foot race, runners travel a twisting 58-km path, climbing from the ocean shore to an altitude of 3300 m at the summit of Haleakala volcano. They wind up a horizontal distance 15 km northwest of their starting point. (a) At what angle must someone at the starting line beam a laser to send a signal to someone at the finishing line? (b) Moving at the speed of light, how long does it take the laser beam to go from start to finish line?

7. Let $\mathbf{A} = 15\hat{\imath} - 40\hat{\jmath}$ and $\mathbf{B} = 31\hat{\jmath} + 18\hat{k}$. Find a vector $\mathbf{C}$ such that $\mathbf{A} + \mathbf{B} + \mathbf{C} = 0$.

8. A mountain expedition starts a base camp at an altitude of 5500 m. Four climbers then establish an advance camp at an altitude of 7400 m; the advance camp is southeast of the base camp, at a horizontal distance of 8.2 km. From the advance camp, two climbers head directly north to a 8900-m summit, a horizontal distance of 2.1 km. Using a coordinate system with the x-axis eastward, the y-axis northward, and the z-axis upward, and with origin at the base camp, express the positions of the advance camp and summit in unit vector notation, and determine the straight-line distance from base camp to summit.

9. An object is moving at 18 m/s at an angle of 220° counterclockwise from the x-axis. What are the x- and y-components of its velocity?

10. A car drives north at 40 mi/h for 10 min, then turns east and goes 5.0 mi at 60 mi/h. Finally, it goes southwest at 30 mi/h for 6.0 min. Draw a vector diagram and determine (a) the car's displacement and (b) its average velocity for this trip.

11. For the car of the preceding problem, determine the average acceleration between the end of the first leg and the beginning of the third leg of the trip.

12. A biologist studying the motion of bacteria notes a bacterium at position $\mathbf{r}_1 = 2.2\hat{\imath} + 3.7\hat{\jmath} - 1.2\hat{k}$ μm $(1\ \mu\text{m} = 10^{-6}\ \text{m})$. 6.2 s later the bacterium is at $\mathbf{r}_2 = 4.6\hat{\imath} + 1.9\hat{k}$ μm. What is its average velocity? Express in unit vector notation, and calculate the magnitude.

13. A supersonic aircraft is travelling east at 2100 km/h. It then begins to turn southward, emerging from the turn 2.5 min later heading due south at 1800 km/h. What are the magnitude and direction of its average acceleration during the turn?

14. For the preceding problem, write the initial and final velocities in unit vector notation, and from them calculate the average acceleration in unit vector notation. Verify that your result is consistent with the answer to the preceding problem.

15. An object undergoes acceleration of $2.3\hat{\imath} + 3.6\hat{\jmath}$ m/s^2 over a 10-s interval. At the end of this time, its velocity is $33\hat{\imath} + 15\hat{\jmath}$ m/s. (a) What was its velocity at the beginning of the 10-s interval? (b) By how much did its speed change? (c) By how much did its direction change? (d) Show that the speed change is *not* given by the magnitude of the acceleration times the time. Why not?

16. An object's position as a function of time is given by $\mathbf{r} = (bt^3 + ct)\hat{\imath} + dt^2\hat{\jmath} + (et + f)\hat{k}$, where b, c, d, e, and f are constants. Determine the velocity and acceleration as functions of time.

17. The position of an object is given by $\mathbf{r} = (ct - bt^3)\hat{\imath} + dt^2\hat{\jmath}$, where $c = 6.7$ m/s, $b = 0.81$ m/s^3, and $d = 4.5$ m/s^2. (a) Determine the object's velocity at time $t = 0$. (b) How long does it take for the direction of motion to change by 90°? (c) By how much does the speed change during this time?

Section 3.2 *Constant Acceleration*

18. An airplane heads northeastward down a runway, accelerating from rest at the rate of 2.1 m/s^2. Express the plane's velocity and position at $t = 30$ s in unit vector notation, using a coordinate system with x-axis eastward and y-axis northward, and with origin at the start of the plane's take-off roll.

19. A particle leaves the origin with initial velocity $\mathbf{v_0} = 11\hat{i} + 14\hat{j}$ m/s. It undergoes a constant acceleration given by $\mathbf{a} = -1.2\hat{i} + 0.26\hat{j}$ m/s².) (a) When does the particle cross the y-axis? (b) What is its y-coordinate at that time? (c) How fast is it moving, and in what direction, at that time?

20. Use Equation 3–11, along with the velocity calculated in the first part of Example 3–7, to determine the final position of the rocket in that example.

21. Figure 3–34 shows a cathode-ray tube, used to display electrical signals in oscilloscopes and other scientific instruments. Electrons are accelerated by the electron gun, then move down the center of the tube at 2.0×10^9 cm/s. In the 4.2-cm-long deflecting region they undergo an acceleration directed perpendicular to the long axis of the tube. The acceleration "steers" them to a particular spot on the screen, where they produce a visible glow. (a) What acceleration is needed to deflect the electrons through 15°, as shown in the figure? (b) What is the shape of an electron's path in the deflecting region?

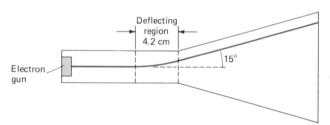

Fig. 3–34 A cathode-ray tube (Problem 21).

22. A particle starts from the origin with initial velocity $\mathbf{v_0} = v_0\hat{i}$ and constant acceleration $\mathbf{a} = a\hat{j}$. Show that the particle's distance from the origin and its direction relative to the x-axis are given by $d = t\sqrt{v_0^2 + \frac{1}{4}a^2t^2}$ and $\theta = \tan^{-1}(at/2v_0)$.

Section 3.3   *Projectile Motion*

23. A carpenter tosses a shingle off a 9.4-m-high roof, giving it an initially horizontal velocity of 7.2 m/s. (a) How long does it take to reach the ground? (b) How far does it move horizontally in this time?

24. You are trying to roll a ball off a 80.0-cm-high table to squash a bug on the floor 50.0 cm from the table's edge. How fast should you roll the ball?

25. Repeat the preceding problem, now with the bug moving away from the table at 30 mm/s and 50.0 cm from the table when the ball leaves the table edge.

26. Mike is standing outside the physics building, 15 ft from the wall. Debbie, at a window 12 ft above, tosses a physics book horizontally. What speed should she give it if it is to reach Mike?

27. You are on the ground 3.0 m from the wall of a building, and want to throw a package from your 1.5-m shoulder level to someone in a second floor window

4.2 m above the ground. At what speed and angle should you throw it so it just barely reaches the window?

28. Derive a general formula for the horizontal distance covered by a projectile launched horizontally at speed v_0 from a height h.

29. A submarine-launched missile has a range of 4500 km. (a) Assuming the missile reaches its full speed within a short distance of the ocean surface, what speed is needed for this range when the missile is launched at a 45° angle? (b) What is the total flight time for the missile? (The shortness of your answer is one reason that your survival in the nuclear age hangs in such precarious balance.) (c) To avoid destruction by space-based antimissile weapons, the missile's trajectory can be "depressed" by giving it a lower launch angle. What is the minimum launch speed needed to give the same 4500-km range if the launch angle is 20°? Neglect air resistance and the curvature of the earth.

30. A rescue airplane is flying horizontally at speed v_0 at an altitude h above the ocean, attempting to drop a package of medical supplies to shipwreck victims in a lifeboat. At what line-of-sight angle α (Fig. 3–35) should the pilot release the package?

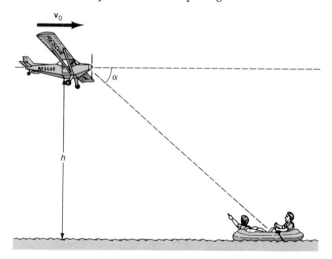

Fig. 3–35 Problem 30

31. At a circus, a human cannonball is shot from a cannon at 35 km/h at an angle of 40°. If he leaves the cannon 1.0 m off the ground, and lands in a net 2.0 m off the ground, how long is he in the air?

32. As you stand on the rim of the Grand Canyon, a friend 10 m back from the rim throws you a baseball at 60 km/h and an initial angle of 10°. You miss the ball, and it winds up dropping 500 m below the rim before it hits a trail descending into the canyon. Neglecting air resistance, (a) what horizontal distance from the rim would it travel and (b) at what speed would it hit the trail? (Neglect of air resistance here

is actually not justified; with air resistance, your answers would be quite different.)

33. If you can hit a golf ball 180 m on earth, how far can you hit it on the moon? (Your answer is an underestimate, because the distance on earth is restricted by air resistance as well as by a larger g.)

34. Prove that a projectile launched on level ground reaches its maximum height midway along its trajectory.

35. A projectile launched at an angle θ to the horizontal reaches a maximum height h. Show that its horizontal range is 4h/tanθ.

36. What is the minimum speed for throwing a ball over the house of Fig. 3–36 if the thrower is 15 ft from the wall? Assume the ball is thrown from essentially ground level.

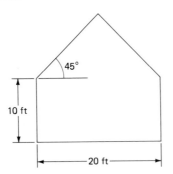

Fig. 3–36   Problem 36

37. A circular fountain has jets of water directed from the circumference inward at an angle of 45°. Each jet reaches a maximum height of 2.2 m. (a) If all the jets converge in the center of the circle and at their initial height, what is the radius of the fountain? (b) If one of the jets is aimed at 10° too low, how far short of the center does it fall?

38. When the Olympics were held in Mexico City in 1968, many sports fans feared that the high altitude would result in poor athletic performances due to reduced oxygen. To their surprise, new records were set in track and field events, probably as a result of lowered air resistance and a decrease in g to 9.786 m/s², both ultimately associated with the high altitude. In particular, Robert Beamon set a new world record of 8.90 m in the long jump. Photographs suggest that Beamon started his jump at a 25° angle to the horizontal. If he had jumped at sea level, where g = 9.81 m/s², at the same angle and initial speed as in Mexico City, how far would Beamon have gone? Neglect air resistance in both cases (although its effect is actually more significant than the change in g).

39. Show that, for a given initial speed, the horizontal range of a projectile is the same for launch angles 45° + α and 45° − α, where α is between 0 and 45°.

40. A typical intercontinental ballistic missile (ICBM) has a range of 10,000 km and reaches a maximum altitude of 1200 km. Assuming its rocket engine fires for only a short time at the beginning of the flight, determine (a) its launch angle and (b) its flight time. Neglect the curvature of the earth (this is not a particularly good approximation here).

41. A basketball player is 15 ft horizontally from the center of the basket, which is 10 ft off the ground. At what angle should the player aim the ball if it is thrown from a height of 8.2 ft with a speed of 26 ft/s?

Section 3.4   Circular Motion

42. How fast would a car have to round a turn 50 m in radius in order for its acceleration to be numerically equal to that of gravity?

43. Estimate the acceleration of the moon, which completes a nearly circular orbit of 390,000 km radius in 27 days.

44. An object is in uniform circular motion. Make a graph of its average acceleration, measured in units of v^2/r, versus angular separation for points on the circular path spaced 40°, 30°, 20°, and 10° apart. Your graph should show the average acceleration approaching the instantaneous value v^2/r as the angular separation decreases.

45. When Apollo astronauts landed on the moon, they left one astronaut behind in a circular orbit around the moon. For the half of each orbit that took him around the back side of the moon, this astronaut was "the loneliest man in the universe," who could neither see earth nor communicate by radio with earth or with his fellow astronauts on the moon. Estimate the duration of this total separation from the rest of humanity, assuming a sufficiently low orbit that you can neglect the distance above the lunar surface (see Appendix F for the acceleration of the moon's gravity).

46. A 10-in-diameter circular saw blade rotates at 3500 revolutions per minute. What is the acceleration of one of the saw teeth? Compare with the acceleration of gravity.

47. A beetle can cling to a phonograph record only if the acceleration is less than 0.25 g. How far from the center of a 33⅓-rpm record can the beetle stand?

48. A jet is diving vertically downward at 1200 km/h (see Fig. 3–37). If the pilot can withstand a maximum ac-

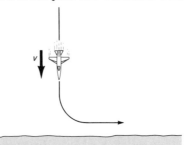

Fig. 3–37   Problem 48

celeration of 5g before losing consciousness, how close to the ground can the plane get before starting a quarter turn to pull out of the dive? Assume the plane's speed remains constant.

49. Electrons in a TV tube are deflected through a 55° angle as shown in Fig. 3–38. During the deflection they move at constant speed in a circular path of radius 4.30 cm. If they experience an acceleration of 3.35×10^{17} m/s², how long does the deflection take?

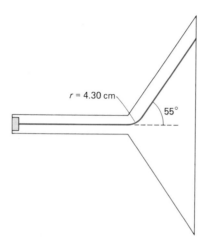

Fig. 3–38   Problem 49

50. How fast would the earth have to be rotating for the radial acceleration of an object on the equator to equal g? Compare with earth's actual rotation rate.

51. A record turntable takes 14 s to reach full speed of $33\frac{1}{3}$ revolutions per minute. With constant tangential acceleration, what are the tangential and radial components of the acceleration of a point on the rim of a 12-in record 10 s after the turntable is turned on?

52. A plane is heading northward when it begins to turn eastward on a circular path of radius 9.10 km. At the instant it begins to turn, its acceleration vector points 22.0° north of east and has magnitude 2.60 m/s². (a) What is the plane's speed? (b) At what rate is its speed increasing?

Section 3.5   *Motion Is Relative*

53. A dog paces around the perimeter of a rectangular barge that is headed up a river at 14 km/h relative to the riverbank. The current in the water is at 3.0 km/h. If the dog walks at 4.0 km/h, what are its speeds relative to (a) the shore and (b) the water as it walks around the barge?

54. A jetliner with an airspeed of 1000 km/h sets out on a 1500-km flight due south. To maintain a southward direction, however, the plane must be pointed 15° west of south. If the flight takes 100 min, what is the wind velocity?

55. A car is moving at a constant 19 m/s. Its 72-cm-diameter wheels rotate at 8.4 revolutions per second.

What are the instantaneous velocity and acceleration of (a) the bottom of the wheel, (b) the top of the wheel, and (c) the forward edge of the wheel? Determine these quantities both in the car's frame of reference and in the ground frame of reference.

Supplementary Problems

56. A juggler's hands are 80 cm apart, and the balls being juggled reach a maximum height of 100 cm above the juggler's hands. (a) At what velocity do the balls leave the juggler's hands? (b) If four balls are being juggled, how often must the juggler catch a ball?

57. A monkey is hanging from a branch a distance h above the ground. A hunter stands a horizontal distance d from a point directly below the monkey. The hunter wants to shoot the monkey, but knows that the monkey will let go of the branch at the instant the gun fires. (a) Show that the bullet will hit the falling monkey if the gun is aimed directly at the monkey's initial position. (b) Your result in (a) is independent of the speed of the bullet as it leaves the gun, provided that speed is above a certain minimum value that will ensure the bullet's reaching the monkey before the monkey hits the ground. Show that this minimum value is $\sqrt{(d^2+h^2)g/2h}$.

58. A child tosses a ball over a flat-roofed house 3.2 m high and 7.4 m wide, so that it just clears the corners on both sides, as shown in Fig. 3–39. If the child stands 2.1 m from the wall, what are the ball's initial speed and launch angle? Assume the ball is launched essentially from ground level.

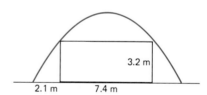

Fig. 3–39   Problem 58

59. A diver leaves a 3-m board on a trajectory that takes her 2.5 m above the board, and then into the water a horizontal distance of 2.8 m from the end of the board. At what speed and angle did she leave the board?

60. An object moves at constant speed v in the x-y plane, describing a circle of radius r centered at the origin. It is on the positive x-axis at time $t=0$. Show that the position of the object as a function of time can be written $\mathbf{r} = r[\cos(vt/r)\mathbf{\hat{i}} + \sin(vt/r)\mathbf{\hat{j}}]$, where the argument of the sine and cosine is in radians. Differentiate this expression once to obtain an expression for the velocity and again for the acceleration. Show that the acceleration has magnitude v^2/r and is directed radially inward (that is, opposite to $\mathbf{r}$).

61. In the Olympic hammer throw, contestants whirl a 7.3-kg ball on the end of a 1.2-m-long steel wire before releasing it. In a particular throw, the hammer is released horizontally from a height of 2.4 m and travels 84 m horizontally before hitting the ground. What is its radial acceleration just before release?

62. A projectile is launched at an angle θ to the horizontal, with sufficient speed to give it a horizontal range x. Show that the radius of curvature at the top of its trajectory is given by $r = x/2\tan\theta$.

63. While increasing its speed, a train enters a 90° circular turn of radius r with speed v_0 and with tangential acceleration equal to half its radial acceleration. If its tangential acceleration remains constant, show that when the train leaves the turn its tangential and radial accelerations are related by $a_t = a_r/(2 + \pi)$.

64. A convertible is speeding down the highway at 130 km/h, when the driver spots a police airplane 600 m back at an altitude of 250 m. The driver decelerates at 2.0 km/h/s. If the plane is flying horizontally at a steady 210 km/h, where should the plane be in relation to the car for the police officer to drop a speeding ticket into the car? Assume the ticket is dropped with no initial vertical motion, and it is in a heavy capsule that experiences negligible air resistance. Will the car still be speeding when the ticket reaches it?

65. A space shuttle orbits the earth at 27,000 km/h, while at the equator earth rotates at 1300 km/h. The two motions are in roughly the same direction (west to east), but the shuttle orbit is inclined at 25° to the equator. What is the shuttle's velocity relative to scientists tracking it from the equator?

4

DYNAMICS: WHY DO THINGS MOVE?

Fig. 4–1
This Voyager spacecraft, on its mission to the outermost planets, travels tens of kilometers each second without any need for rocket engines.

A spacecraft moves noiselessly, effortlessly, and smoothly through space. No rocket firing is necessary to maintain its motion (Fig. 4–1). Pioneer and Voyager spacecraft that have visited the outer planets are now on their way out of the solar system, and will continue onward forever—or until they run into something.

Until recent years we had trouble duplicating on earth situations that exist naturally in outer space. If we slide a book across a table top, it soon comes to rest because of friction. If we polish the table top to make it smoother, the book slides farther before friction inevitably stops it. In physics labs, we often use "air tables," in which air forced upward through small holes in the table top supports objects on a nearly frictionless cushion of air (Fig. 4–2). But even here, resistance of the air itself eventually stops the motion.

In this chapter, we seek the fundamental laws of motion, unobscured by the confusing effects of friction and air resistance. In earlier chapters, we studied only **kinematics**—the description of motion without reference to its causes. Here we also want to know *why* things move. The study of motion and its causes is called **dynamics,** from the Greek word "dynami-

kinematics

dynamics

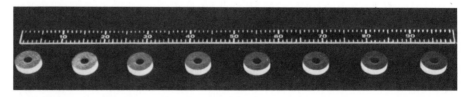

Fig. 4–2
Strobe photo of a puck sliding on a nearly frictionless air table. Time interval between images is the same; since the images are nearly equal distances apart, the puck's speed is essentially constant.

kos,'' meaning force or power. We will see how, building on work of Galileo, Isaac Newton in the seventeenth century laid down basic laws of motion that we still apply today. This chapter describes Newton's laws themselves and their immediate consequences. In the following chapter, we will apply Newton's laws in a variety of practical situations.

4.1
FINDING THE LAWS OF MOTION

What makes things move? For nearly two thousand years following the work of Aristotle (384–322 B.C.), most people believed that a force—a push or a pull—was needed to keep something moving. This idea made a lot of sense: when an ox stopped pulling an ox-cart, for example, the cart quickly came to a stop. Aristotelians extended their thinking to projectiles by saying that an arrow, for instance, could continue moving through the air because air rushed from the front to the back of the arrow to push it forward.

Actually, the very question ''what makes things move?'' is misleading. In the early 1600's, Galileo Galilei carried out a series of experiments that convinced him that a moving object has an intrinsic ''quantity of motion'' that needs no cause to maintain itself (Fig. 4–3). Instead of answering the question ''what makes things move?'' Galileo asserted that the question needs no answer. In doing so, he set the stage for centuries of progress in physics, beginning with the achievements of Isaac Newton.

Newton, son of a modest farming family, attended Trinity College of Cambridge University. Although the Aristotelian theories were taught, advanced students were allowed to explore newer ideas emerging on the European continent. Newton immersed himself in the work of Galileo, Descartes, and others who were breaking with Aristotelian tradition. In 1665–1666, young Isaac Newton (Fig. 4–4) had just graduated from Trinity College, when the plague swept across England and the universities were shut down. Newton returned home to Woolsthorpe in Lincolnshire (Fig. 4–5), where in two years of quiet contemplation he continued to lay the groundwork for much of his later work in physics. During this time Newton also began the development of calculus, a branch of mathematics that was needed for the full application of his laws of gravity and motion. Twenty years later, in completing his work on gravity, Newton formulated his three laws of motion.

Newton's laws of motion are the basis of mechanics. They involve three important concepts: **mass, force,** and **momentum.** Newton thought of mass as a measure of the quantity of matter in an object. Although this description is a valid one, we will see that it is more practical to give an operational definition of mass that arises from Newton's laws. We have an intuitive sense of **force** as a push or a pull, a sense that is consistent with Newton's concept of force. Since pushes and pulls have direction, force is a vector quantity. **Momentum** is a modern word for what Newton called ''quantity of motion.'' Seeking a quantity that reflected both the velocity of a moving object as well as the amount of matter in motion, Newton hit upon the product of mass and velocity as the best measure of the ''quantity

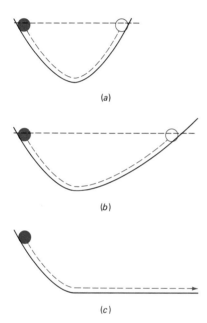

Fig. 4–3
(a) Galileo recognized that a ball rolling down an incline would rise to nearly its starting height on a second incline.
(b) Making the second incline more gradual would result in the ball's travelling farther in the horizontal direction. (c) If the second incline were reduced to a horizontal surface, Galileo reasoned, the ball's motion should continue forever.

mass, force, and momentum

force

momentum

of motion'' or momentum. Thus the momentum of an object of mass m moving with velocity $\mathbf{v}$ is $m\mathbf{v}$; we use the symbol $\mathbf{p}$ for momentum:

$$\mathbf{p} = m\mathbf{v}. \quad \text{(momentum)} \tag{4–1}$$

Since velocity is a vector, so is momentum.

We now state Newton's three laws, and then discuss each in detail:

Newton's first law of motion: A body in motion tends to remain in motion, or to remain stopped if stopped, unless acted on by an outside force.

Newton's second law of motion: The rate at which a body's momentum changes is equal to the net force acting on the body; mathematically

$$\mathbf{F} = \frac{d\mathbf{p}}{dt}. \quad \text{(Newton's 2nd law)} \tag{4–2}$$

When the mass of a body is constant, we can use the definition of momentum, $\mathbf{p} = m\mathbf{v}$, to write

$$\mathbf{F} = \frac{d(m\mathbf{v})}{dt} = m\frac{d\mathbf{v}}{dt}.$$

But $d\mathbf{v}/dt$ is the acceleration, $\mathbf{a}$, so that

$$\mathbf{F} = m\mathbf{a}. \quad \text{(Newton's 2nd law, constant mass)} \tag{4–3}$$

Although Newton originally wrote his second law in the form 4–2, which remains the most general form, the form 4–3 is more widely recognized. This form also provides an operational definition of mass: if the same force F is applied to two different objects, then the resulting accelerations are inversely proportional to their masses. Once a standard of mass has been established—like the standard kilogram described in Chapter 1—then this procedure can be used to compare an unknown mass with the standard.

Fig. 4–4
Isaac Newton at age 47, two years after he published the *Principia*.

Newton's third law of motion: For every action there is always an equal and opposite reaction.

Figure 4–6 shows the laws as stated by Newton in the first edition of his book *Philosophiae Naturalis Principia Mathematica* (*The Mathematical Principles of Natural Philosophy*), published in London in 1687. What we now call physical science was then known as natural philosophy.

Newton's First Law and Inertia

Fig. 4–5
Newton's family cottage at Woolsthorpe is still standing. In two years here, he laid the groundwork for his great achievements in science and mathematics.

Newton's first law is a more precise statement of what Galileo had earlier realized—that, in the absence of external complications like friction or other forces, a moving body will keep moving steadily forward. In other words, its velocity (a vector quantity) will not change in either direction or magnitude. We use the term **inertia** to describe a body's resistance to change in motion. So Newton's first law is equivalent to saying that a body has inertia; for this reason, the first law is also known as the law of inertia.

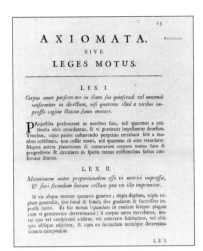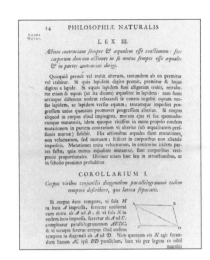

Fig. 4–6
Newton's three laws as published in his *Principia*. Literally translated, they read: "*Law I.* Every body remains in a state resting or moving uniformly in a straight line, except insofar as forces on it compel it to change its state. *Law II.* The force is proportional to the change of motion, which is in the direction of the force. *Law III.* Opposite actions always have opposite and equal reactions: two actions of a body are always equal and in opposite directions."

We can treat the case of a body at rest as a body with zero velocity. A stationary body's tendency to stay at rest is thus the same as a moving body's tendency to keep moving. Indeed, in Section 3.5, we showed that statements about velocity are meaningful only in the context of a particular frame of reference. A body that is stationary with respect to one frame of reference is moving with respect to another frame. But we also showed that the acceleration of an object is the same in *all* frames of reference that are themselves in uniform motion. So if a body exhibits no acceleration—no change in motion—in one uniformly moving frame of reference, its acceleration must be zero in *all* uniformly moving frames of reference. Newton's first law tells us that a body not subject to outside influences exhibits no change in motion; this statement is therefore true in all frames of reference in uniform motion. Such frames are often called **inertial frames,** since the law of inertia (Newton's first law) is valid in them.

But we do find accelerations; the motions of bodies obviously do change. Newton's first law says that, when motion changes, something must be acting to change it. Whatever is changing the motion is known as a **force.** Newton's first law is not quantitative; it does not assign numerical values to inertia or to forces. Still, it is extremely powerful, for it says that whenever we see the velocity of an object changing, we know that a force must be present. Often, change in velocity is the only visible manifestation of the force. If we see a car slowing, then some force must be acting. If we see a car turning even though its speed remains constant, its velocity is still changing, so again we know that a force must be acting. We observe the moon moving about the earth in a circular path; since the moon's velocity is changing direction, a force must be acting.

For Aristotle, the natural state of a body was to be at rest. This is what made meaningful the question "what makes things move?" But for Newton, the natural state of a body is motion at constant velocity. In the context of

inertial frames

force

Newton's laws, the meaningful question about motion is not "what makes things move?" but "what makes motion change?" The answer, embodied in the first law, is that forces make motion change. In the next few chapters, we will see the power of Newton's new idea, as we calculate the motions of objects from molecules to planets. Much later, in Chapter 33, we will see how the progression from Aristotelian to Newtonian thinking laid the foundation for another scientific revolution: Albert Einstein's theory of relativity.

4.2
FORCE

We are familiar with the forces in our everyday lives, and our usual commonsense notion of force is pretty close to the scientific meaning. We push, pull, or lift objects and thus exert forces on them. We experience the pull of gravity toward the center of the earth. We see objects slow down, and infer the existence of frictional forces. We know that ropes and cables provide forces that can support or accelerate heavy objects. We have used magnets enough to sense that they, too, exert forces. Finally, scientists' nuclear experiments infer the existence of strong forces that bind the atomic nucleus together. In some cases, like your hand pushing a book, the force seems associated with contact between two objects. In other cases, like the earth's gravitational force on the moon or the pull of a magnet, the force

action at a distance   seems to act over a great distance. This idea of **action at a distance** has bothered scientists and philosophers over the ages, and it may bother you. How can the moon, way out there, "know" about the earth so far away? Later, in Chapters 8 and 21, we will introduce the concept of a "force field" as an alternative way of describing forces like gravity, electricity, and magnestism that act over long distances. The field concept is a cornerstone of the modern description of matter and its interactions.

The Fundamental Forces

We have already alluded to gravitational forces, contact forces, frictional forces, electrical forces, tension forces, magnetic forces, and nuclear forces. How many kinds of forces are there? New theoretical work of the 1980's indicates that there is probably only one fundamental force, although we do not yet know how to describe it. At present, physicists identify three basic types of forces, each subsuming other forces once thought to be fundamental. The unification of seemingly unrelated forces has been a major theme in the history of physics, and is perhaps the central problem in contemporary physics (Fig. 4–7). Later we will explore one aspect of this unification in detail, as we come to understand how electricity and magnetism are fundamentally related. For now, we merely introduce the three fundamental forces with a brief description of each.

gravitational force   We experience the **gravitational force** in our everyday lives. Yet the gravitational force is the weakest of the fundamental forces, and perhaps the least understood theoretically. But though the gravitational force of an individual bit of matter is extremely weak, this force is cumulative. No

Fig. 4–7
Unification of seemingly unrelated forces has been a major theme in the history of physics.

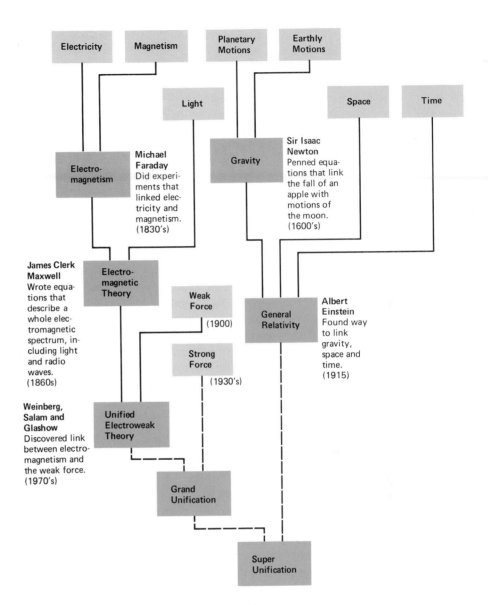

antigravity is known. So the gravity of a massive object like a star or planet is very large, and on a large scale gravitation is the dominant force in the universe. Our best theory of gravity is now the general theory of relativity, advanced by Albert Einstein in 1916. But physicists have not yet figured out how to combine general relativity with the quantum theory that describes matter on the atomic scale. The lack of a quantum theory of gravity has greatly hindered efforts to unify gravity with the other forces.

electroweak force

The **electroweak force** is also important in our everyday lives. The electroweak force includes the forces of electromagnetism and the so-called weak nuclear force. Just a few years ago, electromagnetism and the weak force were listed separately, and physicists counted four fundamental forces. For their theory linking the weak and electromagnetic forces, Sheldon Glashow, Steven Weinberg, and Abdus Salam received the 1979 Nobel

Fig. 4–8
Carlo Rubbia and Simon van der Meer, who won the 1984 Nobel Prize for Physics for their experimental work that confirmed the unification of electromagnetism and the weak force.

strong force

quark
color force

grand unified force
super unified force
supergravity

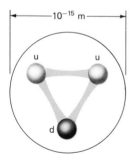

Fig. 4–9
A proton consists of three quarks, two of the so-called "up" variety and one a "down" quark. The quarks are bound by the color force shown symbolically as a link between the quarks. (In a proton, the strength of a color force is about 10,000 lb, or 50,000 N.) Although the quarks are much smaller than the 10^{-15}-m proton diameter, they are in constant motion within that region. The color force actually strengthens with increasing separation, preventing the quarks from escaping the proton volume.

Prize for Physics. That theory was confirmed experimentally in 1983 by a group working under the Italian physicist Carlo Rubbia, a Harvard professor working at CERN, the European Center for Nuclear Research (Fig. 4–8). The 1984 Nobel Prize for Physics went quickly to Rubbia and his Dutch colleague Simon van der Meer. Earlier still, electric and magnetic forces had been considered separate until they were linked by the theory of James Clerk Maxwell in the 1860's. We will study Maxwell's synthesis of electricity and magnetism in Chapters 27 through 32.

Virtually all the nongravitational forces we encounter in everyday life are manifestations of the electroweak force. Contact forces, friction, tension forces, the forces in your muscles, and the forces that hold everyday matter together are all essentially electrical forces. The forces associated with the interaction of light and matter are electromagnetic. Processes involving the weak force occur in the core of the sun and are partially responsible for the generation of solar energy that keeps us alive.

The third basic force of nature, the **strong force,** was advanced in the 1930's to explain how atomic nuclei are held together. Current theories show that the protons and neutrons that make up atomic nuclei are themselves composed of more fundamental particles called **quarks** (Fig. 4–9). The force acting between quarks is known as the **color force**. The color force links the quarks into particles like protons, neutrons, and others, and also links those particles together to form nuclei. What we call the strong force really results from the color force between quarks.

Theories for the gravitational, electroweak, and strong forces are fairly well established. Attempts are being made to unify the electroweak and strong forces, describing them as manifestations of a single **grand unified force.** Further behind are attempts to unify the grand unified force with gravity, to give a **super unified force,** for which a theory called **supergravity** is currently a leading candidate. Such further unifications are tasks for the rest of this century, if not the next. But scientists have realized that unified force theories can be studied by analyzing the earliest fraction of a second of the universe, immediately after the big bang in which the universe was formed. Theory suggests that forces now seemingly distinct would then have had the same strength, and been essentially identical. So studies of the early universe have joined with studies of nuclear physics to probe the fundamental forces of nature. We will look further at unification theories in Chapter 36.

4.3
NEWTON'S SECOND LAW

Why are we so interested in knowing about forces? Because forces cause changes in motion: if we know the force acting on a body, we can use Newton's second law to calculate its acceleration, and thus determine its subsequent motion.

We introduced Newton's second law in Section 4.1; for an object of constant mass, we saw that the law takes the form

$$\mathbf{F} = m\mathbf{a}. \tag{4–3}$$

Just what does this equation mean? We can think of it in two ways.

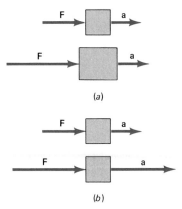

Fig. 4–10
(a) For two objects to have the same acceleration, a greater force must be exerted on the more massive object. (b) When two objects have the same mass, the one with the greater force experiences the greater acceleration.

First, Equation 4–3 provides a measure of force. If we observe a body of mass m undergoing an acceleration of magnitude a, we know that there is a force of magnitude ma acting on the body. The direction of the force is the same as that of the acceleration. Equation 4–3 shows that the units of force are kg·m/s². If we observe a 2-kg object accelerating at 3 m/s², then we know that a force of (2 kg)(3 m/s²) or 6 kg·m/s² is acting on the object. To honor Isaac Newton, we call one kg·m/s² a newton (symbol N). From Equation 4–3, we see that a force of 1 N will cause a 1-kg mass to accelerate at 1 m/s². Similarly, our 2-kg object accelerating at 3 m/s² experiences a force of 6 N. A 1-N force is about a quarter pound—a fairly small force on the human scale. You readily exert forces of tens to hundreds of newtons on the objects you encounter in everyday life.

Newton's law makes good sense in light of our everyday notions of force. We expect it should take a larger force to produce a given acceleration in a more massive object (Fig. 4–10a). And with a given mass, a larger force should be needed for a larger acceleration (Fig. 4–10b). Newton's second law, with force proportional to the product of mass and acceleration, reflects our intuitive understandings suggested in Fig. 4–10.

Example 4–1 _____

Force, Mass, and Acceleration

A 1200-kg car accelerates from rest to 20 m/s in 7.8 s, moving in a straight line. What force is acting on it? If the car then rounds a bend 85 m in radius at a steady 20 m/s, what force acts on it?

Solution

To calculate the force from Newton's second law, we need to know the acceleration and the mass being accelerated. When the car accelerates from rest to 20 m/s in 7.8 s, it undergoes an acceleration of magnitude

$$a = \frac{\Delta v}{\Delta t} = \frac{20 \text{ m/s}}{7.8 \text{ s}} = 2.6 \text{ m/s}^2.$$

The magnitude of the force causing this acceleration is then

$$F = ma = (1200 \text{ kg})(2.6 \text{ m/s}^2) = 3100 \text{ kg·m/s}^2 = 3100 \text{ N}.$$

Force is a vector, and the full vector form of Newton's second law—$\mathbf{F} = m\mathbf{a}$—shows that the force is in the same direction as the acceleration, which in this case also happens to be in the direction of motion. What is the physical origin of the force? Newton's law does not tell us that; it only tells us the magnitude and direction of the force. *Any* physical mechanism providing the same force on the car would produce the same acceleration. In the case of the accelerating car, the force is in fact provided by friction between tires and road—a force that is ultimately electrical in origin. If the driver tried to accelerate on frictionless ice, the wheels would simply spin and the car would not accelerate.

When the car rounds a circular bend at constant speed, we know from Section 3.4 that it undergoes an acceleration of magnitude v^2/r directed toward the center of the circle. Newton's second law therefore requires a force acting toward the center of the circle; the magnitude of this force is

$$F = ma = \frac{mv^2}{r} = \frac{(1200 \text{ kg})(20 \text{ m/s})^2}{85 \text{ m}} = 5600 \text{ N}.$$

Note that Newton's law does not distinguish between forces that result in a change

Fig. 4–11
Forces acting on the car of Example 4.1. (a) Accelerating in straight line. (b) Rounding a curve. (Only forces in the horizontal plane are shown; there are also vertical forces from gravity and from the road, but these add to zero and therefore do not accelerate the car.)

in speed (car accelerating in a straight line) and those that result in a change in direction (car accelerating in a circular path at constant speed). Newton's second law relates force, mass, and acceleration in *all* instances; there is nothing special about circular or any other form of motion in the eyes of Newton's second law.

What provides the accelerating force in the car's circular motion? Again, it is the force of friction between tires and road. On an icy road, the car may skid off the road in a straight line, since the frictional force is virtually absent.

Figure 4–11 shows the force vectors for the two cases in this example.

Example 4–2

Electrons in a TV Tube

Electrons in a TV tube are accelerated essentially from rest to a speed of 8.4×10^7 m/s in the "electron gun" at the back of the tube. If the electron gun is 8.2 cm long and if the acceleration in the gun is constant, what force is exerted on each electron?

Solution

To calculate the force, we need the acceleration and mass of the electron. Equation 2–14 relates acceleration, velocity, and distance in one-dimensional motion. With $v_0 = 0$, this equation becomes

$$v^2 = 2a(x - x_0),$$

so that

$$a = \frac{v^2}{2(x - x_0)}.$$

Here $x - x_0$ is the 8.2-cm length of the electron gun and v is the 8.4×10^7 m/s velocity of electrons emerging from the gun. Then Newton's second law gives

$$F = ma = \frac{mv^2}{2(x - x_0)} = \frac{(9.1 \times 10^{-31} \text{ kg})(8.4 \times 10^7 \text{ m/s})^2}{(2)(0.082 \text{ m})} = 3.9 \times 10^{-14} \text{ N},$$

where we found the electron mass in the table of fundamental constants inside the front cover. That such a small force can accelerate electrons to such high speed in only a few centimeters is possible because of their minuscule mass. (Our answer in this case is only approximate because effects of relativity become important.)

Example 4–3

A Variable Force

An object of mass m is observed to move in a straight line with velocity given by $v_x = bt^2 - ct$, where b and c are positive constants with appropriate units. Find an expression for the force on the object as a function of time, and find a time when the force is zero.

Solution

Acceleration is the rate of change of velocity, so that Newton's second law may be written

$$F = ma = m\frac{dv}{dt} = m\frac{d}{dt}(bt^2 - ct) = 2mbt - mc.$$

The force is zero when $2mbt - mc = 0$, or at time $t = c/2b$.

We have just used Newton's second law as a prescription for measuring forces. Conversely, given the force on an object, we can calculate the object's acceleration and thus predict its motion. It is this use of Newton's law that lets us send a spacecraft to Jupiter, design engines appropriate to a new aircraft, or determine the safe distance between cars on a highway. It also helps us analyze a skyscraper's response to gale-force winds, predict the positions of the planets and the timing of eclipses, and develop better tennis rackets. It is only a slight exaggeration to say that the equation $\mathbf{F} = m\mathbf{a}$ covers all of classical physics.

Example 4–4
An Airport Runway

Fully loaded, a 747 jetliner has a mass of 3.6×10^5 kg. Its four engines provide a total thrust (force) of 7.7×10^5 N. How long a runway is needed for a 747 to achieve takeoff speed of 310 km/h (86 m/s)?

Solution

We are given force and mass, from which we can calculate acceleration using Newton's second law:

$$a = \frac{F}{m}.$$

We need then to relate acceleration to speed and distance; Equation 2–14 provides this relation:

$$v^2 = 2a(x - x_0).$$

Solving for the runway length $x - x_0$ and using $a = F/m$, we have

$$x - x_0 = \frac{v^2}{2a} = \frac{v^2}{2(F/m)} = \frac{mv^2}{2F} = \frac{(3.6 \times 10^5 \text{ kg})(86 \text{ m/s})^2}{(2)(7.7 \times 10^5 \text{ N})} = 1700 \text{ m}.$$

Example 4–5
Predicting Motion Using Newton's Second Law

A 1.2-kg object initially at rest at the origin is acted on by a force $\mathbf{F} = 2.4\hat{\imath} + 1.7\hat{\jmath}$ N. What is the object's acceleration? Where is the object and how fast is it moving 3.5 s after the force is first applied?

Solution

Solving Newton's second law for the acceleration gives

$$\mathbf{a} = \frac{\mathbf{F}}{m} = \frac{2.4\hat{\imath} + 1.7\hat{\jmath} \text{ N}}{1.2 \text{ kg}} = 2.0\hat{\imath} + 1.4\hat{\jmath} \text{ m/s}^2.$$

Given the acceleration, we use methods of the previous chapters to find the position and speed. Equation 3–9 gives the position of an object undergoing constant acceleration:

$$\mathbf{r} = \mathbf{r}_0 + \mathbf{v}_0 t + \tfrac{1}{2}\mathbf{a}t^2 = \tfrac{1}{2}(2.0\hat{\mathbf{i}}+1.4\hat{\mathbf{j}}\ \text{m/s}^2)(3.5\ \text{s})^2 = 25\hat{\mathbf{i}}+17\hat{\mathbf{j}}\ \text{m}.$$

Here we have set $\mathbf{r}_0$ and $\mathbf{v}_0$ to zero since the object is initially at rest at the origin. The velocity after 3.5 s is given by Equation 3–8:

$$\mathbf{v} = \mathbf{v}_0 + \mathbf{a}t = (2.0\hat{\mathbf{i}}+1.7\hat{\mathbf{j}}\ \text{m/s})(3.5\ \text{s}) = 7.0\hat{\mathbf{i}}+6.0\hat{\mathbf{j}}\ \text{m/s},$$

so that the speed is

$$v = \sqrt{v_x{}^2 + v_y{}^2} = [(7.0\ \text{m/s})^2 + (6.0\ \text{m/s})^2]^{1/2} = 9.2\ \text{m/s}.$$

A third use of Newton's law involves determining unknown masses from measured forces and accelerations. This approach is actually used in orbiting spacecraft, where weightlessness (see Section 4.4) precludes more conventional measurements.

Fig. 4–12
Mass-measuring device used on Skylab.

Example 4–6

The Mass of an Astronaut

To aid doctors studying the physiological effects of spaceflight, astronauts on the long Skylab missions of the 1970's measured their masses using a chair subject to a known force exerted by a spring (Fig. 4–12). With astronaut Jack Lousma strapped in the chair, the 15-kg chair underwent an acceleration of 2.04×10^{-2} m/s^2 when the spring force was 2.07 N. What was Lousma's mass?

Solution

Solving Newton's second law for mass, we have

$$m = \frac{F}{a} = \frac{2.07\ \text{N}}{2.04 \times 10^{-2}\ \text{m/s}^2} = 101\ \text{kg}.$$

This is the combined mass of the chair and astronaut; subtracting the 15-kg chair mass then gives 86 kg for astronaut Lousma.

4.4

THE FORCE OF GRAVITY: MASS AND WEIGHT

Newton's second law shows that mass is a measure of a body's resistance to changes in motion—that is, of its inertia. The more massive a body, the greater the force needed to give it a particular acceleration. The mass of a body is an intrinsic property of the body; it doesn't depend on the body's location or other circumstances. If my mass is 65 kg, it is 65 kg whether I am on earth, in an orbiting spacecraft, on the moon, or in the most remote reaches of intergalactic space. That means that no matter where I am, a net force of 65 N is needed to give me an acceleration of 1 m/s^2.

We commonly speak of the weight of a body, using that term in the same way we use the term mass. In physics, though, weight has a very specific meaning that depends on the frame of reference in which the weight is being measured. Before defining weight, we rephrase the defini- tion of free fall given in Chapter 2: **free fall** is the state of motion of a body subject only to the gravitational force. The **weight** of a body in a particular reference frame is then defined as that force that, when applied to the body, would give it the same acceleration relative to the reference frame that it would have in free fall. For an object near the surface of the earth, the

free fall

weight

acceleration in free fall has magnitude g and points downward. In a frame of reference at rest with respect to the earth, the weight of a body has magnitude mg, where m is the body's mass. With my mass of 65 kg, I weigh (65 kg)(9.8 m/s^2) or 640 N. On the moon, the acceleration of gravity is only 1.6 m/s^2; although my mass is everywhere 65 kg, my weight on the moon would be only (65 kg)(1.6 m/s^2) or 100 N. And in the remote reaches of intergalactic space, far from any gravitating object, my weight would be essentially zero.

Example 4–7
Mass and Weight

The two Viking spacecraft that landed on Mars in 1976 had weights of 5880 N on earth. What were their masses on earth and Mars, and their weights on Mars?

Solution

Weight in a planet's frame of reference is the force that would give the local acceleration of gravity: $W = mg$. Solving for m and using the weight on earth and the acceleration of gravity on earth, we have

$$m = \frac{W}{g} = \frac{5880 \text{ N}}{9.8 \text{ m/s}^2} = 600 \text{ kg}.$$

The mass is the same everywhere; on Mars the weight is then given by

$$W = mg_{\text{Mars}} = (600 \text{ kg})(3.74 \text{ m/s}^2) = 2240 \text{ N},$$

where we obtained the acceleration of gravity on Mars from Appendix F.

What about my weight in a frame of reference that itself is falling freely—like a falling elevator or an orbiting spacecraft? Since all objects experience the same gravitational acceleration, with no force but gravity acting on me I experience no acceleration *relative to the freely falling frame* (Fig. 4–13). In that frame, I am therefore **weightless.** Weightlessness is not a mere mathematical condition; I really *feel* weightless. (The "weight" I experience standing on earth is actually the force of my muscles

weightlessness

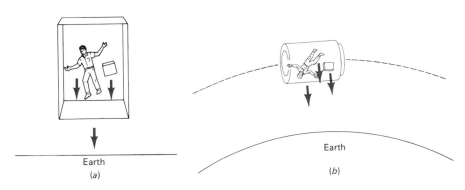

Fig. 4–13
Objects in a freely falling frame of reference are weightless because they experience the same acceleration as the reference frame. Weightlessness would occur in a freely falling elevator near earth's surface just as it does in an orbiting spacecraft. In both cases, the person, book, and elevator or spaceship all have the same acceleration toward earth so that none accelerates relative to its surroundings.

Fig. 4–14
(Left) Astronaut trainees are weightless as the NASA training aircraft executes a parabolic trajectory. *(Right)* Astronauts Richard Truly and Guy Bluford in the space shuttle.

acting to oppose gravity's tendency to accelerate me; in free fall, gravity acts unopposed and my muscles need not exert any force.) If I drop a book in a freely falling elevator, the book stays right next to me as we accelerate downward together; it too is weightless (see Fig. 4–13). If I jump off the floor of the elevator, I do not fall back down to it, but continue in uniform motion relative to the elevator until I hit the ceiling. This weightlessness is identical to that experienced by astronauts in an orbiting spacecraft; the only problem is that my experience of weightlessness would end all too soon when the elevator struck the ground and strong nongravitational forces acted. Weightlessness is not a condition peculiar to outer space; it occurs in any freely falling frame of reference. (Don't be misled by the word "fall." Free fall means that only gravity is acting; a spacecraft in circular orbit is as much in free fall as an elevator actually dropping toward earth's surface.) Practically, though, it is much easier to achieve weightlessness in space, where the absence of air eliminates the nongravitational force of air resistance and permits closed orbits that never intersect the earth. Astronaut training includes flights on aircraft that describe trajectories approximating the parabolas of projectiles moving under the influence of gravity alone; on these parabolic flights, astronauts are every bit as weightless as they are in orbit (Fig. 4–14).

Weightlessness does *not* imply the absence of gravitational force. The acceleration of gravity in low orbit is nearly the same as at earth's surface, so that the gravitational force on an orbiting astronaut is nearly the astronaut's surface weight mg. This had better be the case! It is the gravitational force that keeps the astronaut in orbit around the earth. If the gravitational force were zero, astronaut and spacecraft would obey Newton's first law and move in a straight line, quickly leaving earth's vicinity. Weight depends on frame of reference, and the astronaut is weightless only in the freely falling reference frame of the spacecraft; in a frame of reference at rest with respect to earth, the astronaut is not at all weightless.

The concepts of weight and mass are further confused by the common use of the SI unit kilogram to describe "weight." At the doctor's office, you may be told that you "weigh" 55 kg. Strictly speaking, this is not correct; the kilogram is the unit of mass, not of force or weight. A one-kilogram mass actually has a weight mg at earth's surface of $(1 \text{ kg})(9.8 \text{ m/s}^2) = 9.8 \text{ N}$, and your 55-kg "weight" is really the mass associated with a weight of $(55 \text{ kg})(9.8 \text{ m/s}^2) = 540 \text{ N}$. In the English system, in contrast, we customarily give weights in pounds. This is correct; the pound, like the newton, is a unit of force and therefore of weight. One pound is equivalent to 4.448 N. The English unit for mass is the **slug,** which is rarely used. On earth, the relation between mass in slugs and weight in pounds is again $W = mg$, where g in the English system is 32 ft/s^2 near earth's surface.

slug

In the reference frame of the earth, objects' weights are proportional to their masses. It is this proportionality that allows us to use the words "weight" and "mass" interchangeably in our everyday lives. But in physics, weight and mass mean very different things. Mass measures an object's resistance to acceleration, without any reference to gravity. Weight measures the force needed to give the acceleration of free fall. That we confuse them at all is the result of the remarkable fact that the free-fall acceleration of all objects at a given location is the same. This means that the *weight* of an object—a property defined in terms of free fall and gravity—is proportional to its *mass*—a property that seems to have nothing to do with gravitation. First inferred by Galileo in his experiments with falling bodies, this relation between gravitation and inertia seemed a coincidence until the early twentieth century. Finally, in his general theory of relativity, Albert Einstein showed how that simple relation reflects the underlying geometry of space and time in a way that links intimately the phenomena of accelerated motion and gravitation.

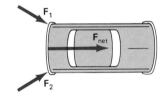

Fig. 4–15
When two people push a stalled car, the force they exert is the vector sum of their individual forces.

4.5
ADDING FORCES

In our previous examples, we considered only one force acting on an object. But often we can identify several forces acting (Fig. 4–15). The flight of a jet involves the forces associated with engine thrust, the force of air on the wings and body of the plane, and the force of gravity (Fig. 4–16). Indeed, for any object near earth, gravity is always among the forces acting.

Fig. 4–16
Forces on a jet include the force of the engines, the force of the air that provides both lift and drag, and the force of gravity. When the plane moves with constant velocity, these forces sum to zero.

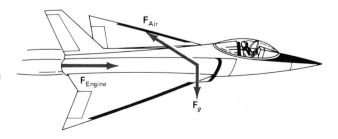

net force

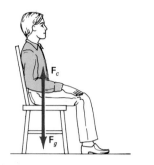

Fig. 4–17
A seated person is not accelerating, so that the net force on the person must be zero. Since the gravitational force F_g acts, there must be another force to counteract gravity and make zero net force. That force is the upward contact force F_c from the chair. (Ultimately, this contact force arises from electrical interactions between the molecules of the person and the chair.)

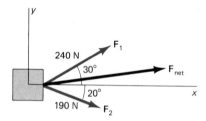

Fig. 4–18
Example 4–8. (View is from above; both x- and y-axes are in horizontal plane.)

How are we to apply Newton's law in these multiple-force cases? The answer is given ultimately by experiment: we add vectorially the individual forces to find the **net force** on an object. Newton's second law then relates the object's mass and acceleration to this net force.

The *net force* is all-important! It is only the *net force* that matters in Newton's second law. There may be all sorts of forces acting on a body, but if they sum to zero, then the net force is zero and the body is not accelerating (Fig. 4–17). More generally, when the net force on a body is not zero, then that body must be accelerating with acceleration given by Newton's second law: $\mathbf{a} = \mathbf{F}_{net}/m$.

Example 4–8 _____

The Net Force

Two fishermen are pulling a 260-kg ice-fishing shanty on essentially frictionless runners, with forces shown in Fig. 4–18. What is the net force on the shanty? What is its acceleration?

Solution

The net force is the vector sum of all the forces acting. In the coordinate system shown in Fig. 4–18, we can write

$$F_{x1} = (240 \text{ N})(\cos 30°) = 208 \text{ N},$$

$$F_{y1} = (240 \text{ N})(\sin 30°) = 120 \text{ N},$$

$$F_{x2} = (190 \text{ N})[\cos(-20°)] = 179 \text{ N},$$

$$F_{y2} = (190 \text{ N})[\sin(-20°)] = -65 \text{ N},$$

where the negative sign shows that $\mathbf{F}_2$ lies clockwise from the x-axis. Then the net force may be written

$$\mathbf{F} = (208 \text{ N} + 179 \text{ N})\hat{\imath} + (120 \text{ N} - 65 \text{ N})\hat{\jmath} = 387\hat{\imath} + 55\hat{\jmath} \text{ N}.$$

This is a vector of magnitude

$$F = \sqrt{F_x^2 + F_y^2} = [(387 \text{ N})^2 + (55 \text{ N})^2]^{1/2} = 391 \text{ N}$$

that points at an angle

$$\theta = \tan^{-1}\left(\frac{F_y}{F_x}\right) = \tan^{-1}\left(\frac{55 \text{ N}}{387 \text{ N}}\right) = 8.1°$$

to the x-axis. The acceleration then has magnitude

$$a = \frac{F}{m} = \frac{391 \text{ N}}{260 \text{ kg}} = 1.5 \text{ m/s}^2,$$

and points in the same direction as the force.

Have we accounted for all forces in this example? Not quite: there are also a downward gravitational force and an upward contact force from the ice. But the shanty is not accelerating in the vertical direction, so that these two forces sum to zero. The force we have calculated is then the true net force on the shanty.

Example 4–9 _____

An Elevator

A 740-kg elevator is accelerating upward at 1.1 m/s², pulled by a cable of negligible mass, as shown in Fig. 4–19. What is the tension force in the elevator cable?

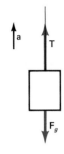

Fig. 4–19
Forces on the elevator include the cable tension **T** and the gravitational force **F**$_g$; the net force is their vector sum. Since the elevator is accelerating upward, this net force must be upward.

Solution

This is an important example, and understanding it thoroughly will help you apply Newton's second law correctly in more complicated situations. We are given the acceleration and mass and asked for the force in the cable. Can't we just write **F** = m**a** for that force? No! **F** = m**a** is true only for the *net force*, and here the cable tension is not the only force acting. There is also the force of gravity which has magnitude mg and points downward. An important step in working a Newton's law problem is to identify *all* the forces acting on the object of interest. It's a good idea to draw all the force vectors so you can picture all the forces contributing to the net force. Here only the two forces of tension and gravity are acting; we have shown them in Fig. 4–19. Calling the tension force **T** and the gravitational force **F**$_g$, Newton's second law becomes

$$\mathbf{F} = \mathbf{T} + \mathbf{F}_g = m\mathbf{a}. \tag{4–4}$$

You might be tempted to put minus signs in Equation 4–4 because the forces point in opposite directions. But not yet! Equation 4–4 is a *vector* equation stating that the *vector sum* of the forces acting is equal to the mass times the acceleration. Skipping the step of writing Newton's law in vector form will often get you into trouble with signs; that's why it's always best to write the full vector equation first. And in writing the vector equation, you don't need to worry about signs—they're built into the directional nature of the vectors.

Having written the vector form of Newton's second law for our problem, we next choose a coordinate system and write the components of Newton's law in that system. Here all the forces are vertical, so we choose our y-axis pointing vertically upward (vertically downward would do just as well, and would ultimately result in the same answer). Now we write the components of our vector equation 4–4. In this one-dimensional problem, there are no components of force or acceleration in the x- or z-directions (both horizontal), so that the x- and z-component equations read $0 = 0$. Only the y-equation is interesting; formally, it reads

$$T_y + F_{gy} = ma_y. \tag{4–5}$$

Now the tension force **T** points upward, or in the $+y$-direction; its vertical component T_y is positive and is equal to the magnitude T of **T**. The gravitational force **F**$_g$ has magnitude mg and points vertically downward, in the $-y$ direction. Its vertical component F_{gy} is therefore $-mg$. Then Equation 4–5 is

$$T - mg = ma_y,$$

so that

$$T = ma_y + mg = m(a_y + g). \tag{4–6}$$

Before putting in specific numbers, let us see if this result makes sense. Suppose the acceleration a_y were zero. Then the net force on the elevator would have to be zero. In this case, Equation 4–6 tells us that $T = mg$, showing that the tension force and gravity have the same magnitude. Since they point in opposite directions, they indeed sum to zero.

If, on the other hand, the elevator is accelerating upward, so that a_y is positive, then Equation 4–6 requires that the magnitude of the tension force exceed that of the gravitational force by an amount ma_y. Why is this? Because in this case the cable not only supports the elevator against gravity, but also provides the upward acceleration. For the numbers of this example, we have

$$T = m(a_y + g) = (740 \text{ kg})(1.1 \text{ m/s}^2 + 9.8 \text{ m/s}^2) = 8100 \text{ N}.$$

What if the elevator were accelerating downward? Then a_y is negative and Equation 4–6 shows that the tension force is less than mg. Does this make sense?

Yes: for a downward acceleration, the net force must be downward so that the magnitude of the downward gravitational force exceeds that of the upward tension force. Were the elevator in free fall, so that $a_y = -g$, the tension force would be exactly zero.

You probably could have reasoned out the answer to this problem in your head. But by applying the steps illustrated here—specifically, writing Newton's second law as a vector equation and then breaking it into components—you will be able to handle more complicated situations without confusion. We will outline the steps in solving a Newton's law problem more formally in the next chapter.

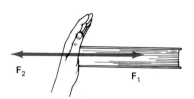

Fig. 4–20
When you push on a book, the book pushes back on you; otherwise you wouldn't feel its presence. The forces F_1 and F_2 have equal magnitudes and constitute an action-reaction pair.

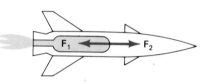

Fig. 4–21
The combustion chamber of a rocket engine exerts a force F_1 on the hot gases, expelling them through the nozzle. In return, the gases exert a force F_2 of equal magnitude on the rocket, accelerating it forward.

action-reaction pair

third law of motion

Fig. 4–22
The horse exerts a force on the cart, and the cart exerts an equal but opposite force on the horse. So how can the two start moving? The answer comes from looking at the *net* force on the horse. This includes not only the backward-pointing force of the cart, but also a forward-pointing force that the road exerts on the horse in reaction to the horse's pushing against the road. As long as that force exceeds the pull of the cart, the pair will accelerate forward.

4.6
NEWTON'S THIRD LAW

Try pushing your book across the table. As you exert a force on the book, you feel the book pushing back on your hand (Fig. 4–20). Kick a ball with bare feet and your toes will hurt. Why? You've exerted a force on the ball, and the ball has also exerted a force back on you. Or imagine a rocket engine, exerting forces that expel hot gases out of its nozzle. Something else happens too: the hot gases exert a force on the rocket, accelerating it forward (Fig. 4–21).

Experimentally, we find that whenever one object exerts a force on the other, the second object also exerts a force on the first. The two forces are in opposite directions, but they have equal magnitudes. The familiar expression that "for every action there is an equal and opposite reaction" is Newton's statement of this fact in seventeenth-century language. By "action" Newton means an applied force. "Reaction" is the second force, equal in magnitude but opposite in direction. Technically, it never makes any difference which force we call the action and which the reaction; they are both always present as an **action-reaction pair**. Newton's statement about action and reaction constitutes his **third law of motion.** In more modern language, we express the third law by stating that

> If object *A* exerts a force on object *B*, then object *B* exerts an oppositely directed force of equal magnitude on object *A*.

Notice that Newton's third law deals with *two forces* and also with *two objects*. The two forces of the action-reaction pair act on different objects. Therefore they do not cancel, even though they are of equal magnitude but opposite direction. Misunderstanding this point leads to a contradiction, embodied in the famous horse and cart dilemma: a horse pulls on a cart with a force **F**. But by Newton's third law, the cart pulls back on the horse with an equal but opposite force. So how can the horse-cart combination ever start moving (Fig. 4–22)? The answer lies in looking at *all* the forces on the horse. In addition to the backward-pointing force of the cart, there is also a forward-pointing force that occurs in reaction to the horse's feet pushing against the road. The forces of the road and the cart are *not* an action-reaction pair since one involves the horse-cart pair and the other the horse-road pair. So they need not have the same magnitude; as long as the road force exceeds the cart force, the pair will accelerate forward.

Example 4–10
Newton's Third Law

On a surface with negligible friction, you push with force **F** on a book of mass m_1 that in turn pushes on a book of mass m_2 (Fig. 4–23). What is the force exerted by the second book on the first?

Solution

Newton's *third* law tells us that the force of the second book on the first is equal in magnitude to that of the first book on the second. But what is that force? We can find it by applying Newton's *second* law to the combination of two books, then individually to the second book.

The total mass m of the two books is $m_1 + m_2$, and the net force applied to this combination is F (actually, this is only the horizontal component; the downward force of gravity and an upward force from the table also act, but they cancel). Then Newton's second law, $F = ma$, gives

$$a = \frac{F}{m} = \frac{F}{m_1 + m_2}.$$

This is the acceleration of the two books as they move together; since they are moving together, it is also the acceleration of each book individually. Now the only (horizontal) force on the second book is the force F_{12} exerted on it by the first book. So we can apply Newton's second law again to find this force:

$$F_{12} = m_2 a = m_2 \frac{F}{m_1 + m_2} = \frac{m_2}{m_1 + m_2} F.$$

Notice how, in applying Newton's second law, we always relate the *net* force on an object to the mass and acceleration *of that object*. Even though the applied force F acts on the first book, the acceleration of the first book is *not* F/m_1, because F is not the *net* force on that book. By Newton's third law, there is also a reaction force F_{21} exerted *by* the second book *on* the first; this force has the same magnitude as the force F_{12} that we just calculated, but it points in the opposite direction:

$$F_{21} = -F_{12} = -\frac{m_2}{m_1 + m_2} F.$$

Does all this make sense? We can see that it does by considering the motion of the first book. It too undergoes an acceleration $a = F/(m_1 + m_2)$. And what is the *net* force on it? For the first book, there are *two* forces acting in the horizontal plane: the applied force F and the reaction force F_{21} exerted by the second book. So the net force on the first book is

$$F_1 = F + F_{21}.$$

Using our expression for F_{21} then gives

$$F_1 = F - \frac{m_2}{m_1 + m_2} F = \frac{F(m_1 + m_2) - m_2 F}{m_1 + m_2} = \frac{m_1}{m_1 + m_2} F = m_1 a.$$

This is consistent with Newton's second law: we know that both books have the same acceleration $a = F/(m_1 + m_2)$, and indeed we find that the force on the first book is the product of that acceleration with the book's mass. Our result shows how Newton's second and third laws are both necessary for a fully consistent description of the motion.

Understanding this example thoroughly will give you confidence in applying Newton's laws of motion correctly.

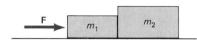

Fig. 4–23
Example 4–10

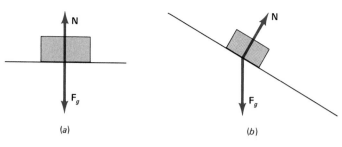

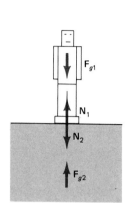

Fig. 4–24 ▲
Normal forces. *(a)* The upward force of a table on a block. *(b)* The normal force of a sloping surface on a block. Also shown in each case is the gravitational force on the block; since both the normal force and the gravitational force act on the same object, they do not constitute an action-reaction pair. In *(a)* the two forces balance, but in *(b)* they add to give a net force that accelerates the block down the slope.

◄ **Fig. 4–25**
Standing on the floor involves two action-reaction pairs: gravitational forces that the earth exerts on me and that I exert on the earth, and contact forces between my feet and the floor. The net force on each object is zero and therefore neither accelerates.

A contact force like the force between books in Example 4–10 acts at right angles to the surfaces of the objects that are pushing on each other. Such a force is called a **normal force.** ("Normal" comes from the Latin for "according to the carpenter's square.") Other examples of normal forces include the upward force of a table on an object, or the force normal to a sloping surface supporting an object (Fig. 4–24). (When friction is present, the force between two objects also has a component along the contacting surfaces. In that case, the term normal force refers only to the normal component—the component perpendicular to the surfaces.)

normal force

Newton's third law applies even for a force like gravity that doesn't involve direct contact. Since earth exerts a downward force on me, Newton's third law says that I must exert an upward force of equal magnitude on the earth. When I stand on the floor, the force of gravity $\mathbf{F}_{g1}$ on me is balanced by an upward normal force $\mathbf{N}_1$ from the floor. (These two forces don't constitute an action-reaction pair, since they both act on the same object—me.) Similarly, the upward force $\mathbf{F}_{g2}$ that I exert on earth—the reaction to earth's gravitational force on me—is balanced by a downward normal force $\mathbf{N}_2$ from my feet. All these forces are shown in Fig. 4–25. What if I jump off a table? Then the only force acting on me is earth's gravity, and I accelerate downward. But Newton's third law asserts that I exert an upward force on earth; since this force, too, is not cancelled by any other force, earth must accelerate upward as I fall. Does this really happen? Yes—but earth's acceleration is unnoticeably small. Why? Because, according to Newton's third law, the force I exert on earth has the same magnitude mg as the force earth exerts on me (Fig. 4–26). Applying Newton's second law shows that earth's acceleration has magnitude

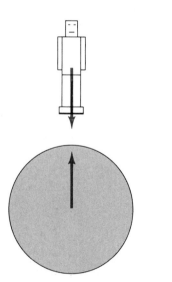

Fig. 4–26
Forces on me and on earth as we fall freely toward one another. The forces constitute an action-reaction pair, but since each acts on a different object and is the only force acting on that object, both objects accelerate. By Newton's third law, the force on each object has the same magnitude, but earth is so much more massive that its acceleration is negligibly small.

$$a = \frac{F}{M_e} = \frac{mg}{M_e} = \frac{m}{M_e}\, g,$$

where m is my mass and M_e is the mass of the earth. Because the ratio m/M_e is about 10^{-23}, earth's acceleration toward me is negligibly small.

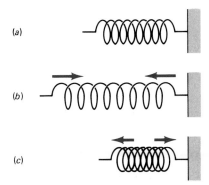

Fig. 4–27
(a) An unstretched spring attached to a wall. *(b)* When stretched, the spring exerts **tension forces** that oppose the stretching. *(c)* When compressed, it exerts **compression forces** that oppose the compression.

When we first listed Newton's laws in Section 4.1, we noted that Newton himself wrote his second law as a relation between force and the rate of change of "quantity of motion," or momentum, defined by $\mathbf{p} = m\mathbf{v}$. In Chapter 9, we will see how Newton's second law in this form combines with the third law to make a powerful statement about the momentum of systems consisting of more than one particle.

4.7
MEASURING FORCE

We have talked a lot about force in this chapter, but have not said much about how to measure this important physical quantity. Newton's second law provides us a direct way of determining force: apply an unknown force to a known mass and measure the resulting acceleration; the force is then the product of mass and acceleration.

Often, though, it is more convenient to use the properties of elastic objects like springs to measure forces. A spring, when stretched or compressed, exerts forces on whatever is deforming it (Fig. 4–27). In the case of a stretched spring, these forces are called **tension forces;** for a compressed spring, they are **compression forces.** In either case, Newton's third law requires that the forces be oppositely directed but of equal magnitude to the external forces acting on the ends of the spring to cause the stretching or compression.

tension forces

compression forces

Over a limited range of stretching and compression, the force exerted by a spring is directly proportional to the distance stretched or compressed. When this direct proportionality holds, the spring is said to obey **Hooke's law.** We then write

Hooke's law

$$F = -kx, \qquad \text{(Hooke's law; ideal spring)} \qquad (4\text{–}7)$$

where x is the amount by which the spring is stretched or compressed from its normal state, and where the minus sign indicates that the spring force is directed oppositely to the stretching or compression. The constant k is called the **spring constant;** its units are obviously N/m. A spring with $k = 200$ N/m (typical of a spring a few inches long that you might hold in your hand) would exert a force of 2 N when stretched or compressed 1 cm. If it were an **ideal spring**—one that obeyed Hooke's law (Equation 4–7) no matter how much it was deformed—it would exert a force of 200 N when stretched 1 m. A real spring a few inches long would break before reaching 1 m in length, and at any rate would cease to obey Hooke's law well before that point (Fig. 4–28).

spring constant

ideal spring

spring scale

We can use a spring to make a **spring scale** by attaching an indicator and a scale calibrated in force units (Fig. 4–29). Common bathroom scales, hanging scales in supermarkets, and laboratory spring scales are all examples of such scales. Even electronic scales widely used at supermarket checkouts are spring scales, with their "spring" a material that produces an electrical signal when it is deformed by an applied force.

We can use a spring scale to weigh objects by attaching one end of the scale to a fixed point and hanging the unknown weight from the other end (Fig. 4–30). When the object hangs at rest on the scale, the scale force is

Fig. 4–28
Force curve for ideal and real springs. The slopes are negative, indicating that the force is opposite to the deformation. Negative values of *x* mean compression; positive values mean stretching.

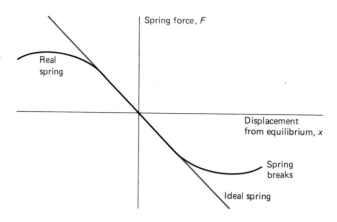

Fig. 4–29
A simple spring scale, made from a spring of known spring constant *k*, an indicating needle, and a scale calibrated in force units.

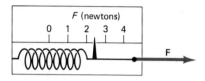

Fig. 4–30
Using a spring scale to determine weight. The scale is a special demonstration model, with the spring mounted outside the scale body.

just enough to keep it from accelerating. By definition, the force of the scale on the object is then equal to the object's weight in the scale's frame of reference. Since mass is proportional to weight in a given frame of reference, the spring scale also provides a measure of the object's mass.

Example 4–11
Weighing a Fish

When a fish is suspended from a spring of spring constant $k = 320$ N/m, the spring stretches 15 cm. What are the weight and mass of the fish?

Solution

From Equation 4–7, a 15-cm stretch corresponds to a spring force of magnitude

$$F = kx = (320 \text{ N/m})(0.15 \text{ m}) = 48 \text{ N}.$$

Since the net force on the fish is zero, the magnitude of the fish's weight is also 48 N. But weight in earth's frame of reference is just mg, so that $m = W/g = (48 \text{ N})/(9.8 \text{ m/s}^2) = 4.9$ kg.

When a spring scale is accelerating, then the weight it reads is that of the object in the accelerating reference frame of the scale. Newton's law shows that this weight must differ from the weight in an inertial (nonaccelerating) frame of reference. Why? Because only in the inertial frame is the net force on the object zero when it is at rest on the scale. In an accelerating frame, there must be a nonzero net force on the object for it to accelerate with the scale; therefore, the scale force and the gravitational force do not balance. When the object and scale are accelerating, the situation is like that of Example 4–9, with the scale playing the role of the elevator cable. The scale reading will be more or less than the actual weight depending on whether the acceleration is upward or downward.

Example 4–12
A Helicopter Ride

A helicopter is rising vertically, carrying a load of concrete to the top of a mountain to make the foundation for a ski lift. A 35-kg bag of concrete is resting on a scale in the helicopter. A construction worker in the copter notes that the scale reads 280 N. What is the acceleration of the copter?

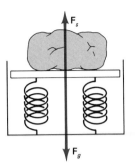

Fig. 4–31
Forces on the concrete of Example 4–12.

Solution

This problem is similar to Example 4–9. The two forces acting on the bag of concrete are the force of gravity $\mathbf{F}_g$ and the force of the scale $\mathbf{F}_s$ (Fig. 4–31). Newton's second law tells us that the net force is the product of mass and acceleration:

$$\mathbf{F} = \mathbf{F}_g + \mathbf{F}_s = m\mathbf{a}.$$

Since the motion is entirely vertical, we need work only with the vertical component of this equation. With the positive y-direction upward, the vertical component of the downward gravitational force is $-mg$; that of the scale force is positive and is equal to the magnitude F_s of the scale force. Then the y-component of Newton's second law for this case becomes

$$-mg + F_s = ma_y,$$

so that the vertical acceleration is

$$a_y = \frac{-mg + F_s}{m} = -g + \frac{F_s}{m} = -9.8 \text{ m/s}^2 + \frac{280 \text{ N}}{35 \text{ kg}} = -1.8 \text{ m/s}^2.$$

What does this negative result mean? It means that the direction of the helicopter's *acceleration* is downward—even though the copter is *moving* upward. That is, its speed is decreasing at the rate of 1.8 m/s² even as it rises. The result of this example is consistent with that of Example 4–9. There, a downward acceleration would result in a cable tension lower than the weight mg in the earth frame of reference. Here, a downward acceleration results in a scale reading lower than the weight in the earth frame as suggested by the length of the force vectors in Fig. 4–31. The magnitude and direction of the *velocity* are completely irrelevant.

We can also use spring scales to exert known forces on objects, as shown in Fig. 4–32. We attach the scale to the object and pull on the spring scale in such a way that the scale reading remains constant. In the absence of friction or other forces, the scale reading is then the net (horizontal) force on the object. If we measure the resulting acceleration, we can use Newton's law to calculate the object's mass. In a frame of reference at rest on earth, it is much easier to determine mass by hanging an object vertically from a scale. But in a freely falling reference frame like an orbiting spacecraft, this second approach must be used (see Examples 4–6 and 4–13).

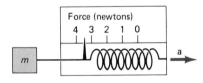

Fig. 4–32
Using a spring scale to apply a known force to an unknown mass. In the absence of friction, the force on the mass is equal to the scale reading. (The force applied at the other end of the scale is greater, though, because it also accelerates the scale.)

Example 4–13
Determining Mass in Orbit

A biologist is studying the growth of rats in an orbiting space laboratory. To determine a rat's mass, she puts it in a 320-g cage, attaches a spring scale, and pulls so that the scale reads 0.46 N. If the resulting acceleration of rat and cage is 0.40 m/s², what is the mass of the rat?

Solution

The spring scale reading F_s is the force on the cage and rat, so that Newton's second law gives

$$F_s = (m_r + m_c)a,$$

where m_r and m_c are the rat and cage mass, respectively. Then

$$m_r = \frac{F_s}{a} - m_c = \frac{0.46 \text{ N}}{0.40 \text{ m/s}^2} - 0.32 \text{ kg} = 0.83 \text{ kg}.$$

In pulling on the scale, the biologist must exert a greater force than the 0.46-N scale force. Why? Because in addition to the rat and cage, she is also accelerating the mass of the scale.

accelerometer   Finally, if we attach a known mass to a spring scale and fasten the other end of the scale to a vehicle or other moving object, we then have an **accelerometer** that can be used to measure the acceleration of the object. This application is explored in Problem 28.

SUMMARY

1. Newton's **first law of motion** states that a body continues in uniform motion unless acted on by a force. Philosophically, the first law represents a shift from the Aristotelian notion that the natural state of motion for objects on earth is rest. In Newton's view, uniform motion is a natural state; the study of motion then emphasizes not the cause of motion itself, but the cause of *changes* in motion.

2. Newton's **second law of motion** quantifies the change in motion brought about by a force. The law states that the rate of change of a body's momentum (product of mass and velocity; symbol **p**) is equal to the net force on the body:

$$\mathbf{F} = \frac{d\mathbf{p}}{dt}.$$

As long as the mass of the body does not change, Newton's second law can also be written in the form

$$\mathbf{F} = m\mathbf{a}.$$

Newton's second law is valid only when **F** is the **net force** on a body—that is, the vector sum of all forces acting on the body. The SI unit of force is the **newton**, defined as the force that gives a 1-kg mass an acceleration of 1 m/s².

3. The **weight** of a body is defined as the force that would be required to give the body the acceleration of free fall in a particular frame of reference. In a reference frame fixed with respect to earth or another gravitating object, the magnitude of a body's weight is just the body's mass multiplied by the local acceleration of gravity: $W = mg$. Objects in a freely falling frame of reference have no acceleration relative to their reference frame; therefore they are **weightless.** Weightlessness does not imply the absence of gravitational force; instead, weightlessness arises because all objects at the same place experience the same gravitational acceleration and therefore fall freely without accelerating relative to one another.

4. Newton's **third law of motion** states that forces always come in **action-reaction pairs.** When one object exerts a force on another, the second object exerts an oppositely directed force of equal magnitude back on the first. Newton's second and third laws together permit a consistent description of the motions of interacting objects.

5. Elastic objects like springs provide a practical way of measuring forces. When the force exerted by a spring is directly proportional to the amount of stretch or compression, the spring is said to obey **Hooke's law;** the force is then given by

$$F = -kx, \quad \text{(Hooke's law)}$$

where k is the **spring constant,** with SI units of N/m.

QUESTIONS

1. Distinguish between the Aristotelian and Galilean/Newtonian view of the natural state of motion.

2. Use Newton's laws of motion to explain the function of a padded dashboard in a car.

3. A ball bounces off a wall with the same speed it had before it hit the wall. Has its momentum changed? Has a force acted on the ball? Has a force acted on the wall? Relate your answers to Newton's laws of motion.

4. Give several examples from the history of physics in which seemingly unrelated forces were found to be related.

5. We often use the term "inertia" to describe human sluggishness. How is this usage related to the meaning of inertia in physics?

6. Why do subway riders lurch?

7. My high school physics teacher defined mass as "inverse pushability aroundness." Comment.

8. Which of the fundamental forces do you deal with in everyday life? Give several examples.

9. Does a body necessarily move in the direction of the net force on it?

10. Given that $\mathbf{F} = m\mathbf{a}$, is it possible to have a net force acting on a body that is at rest? Explain.

11. Can a motorcycle and a truck have the same momentum? Explain.
12. How would you weigh yourself (a) on the moon and (b) in an orbiting spacecraft?
13. Does the length of time that a force is applied to an object affect the change in the object's momentum? Explain.
14. Does the length of time that a force is applied to an object affect the object's acceleration? Explain.
15. An elevator in the physics building at the University of Colorado includes a built-in physics demonstration consisting of a 98-N weight suspended from a spring scale. Is the scale reading more than, less than, or equal to 98 N when the elevator (a) is stationary; (b) first starts moving upward; (c) moves steadily upward; (d) slows down on its way up; (e) starts moving downward; (f) moves steadily downward; (g) slows down on its way down?
16. The surface gravity of Jupiter's moon Io is one-fifth that of earth. What would happen to your weight and to your mass if you were to travel to Io?
17. What does it mean to say that someone "weighs" 50 kilograms?
18. Consider two teams in a tug-of-war (Fig. 4–33). Newton's third law says that the force team A exerts on team B is equal in magnitude to the force team B exerts on team A. How, then, can either team win?
19. Describe the action and reaction forces involved in kicking a football.
20. What is the force that ultimately stops the movement of a braking car?

Fig. 4–33 A tug-of-war (Question 18).

21. If you move from bow to stern of a canoe, the canoe moves in the opposite direction. Why?
22. As you sit in your chair, the chair exerts an upward force on you. Why don't you accelerate upward?
23. As you sit in a car, what direction is the force exerted on you by the car seat (a) when the car moves at constant speed; (b) when the car accelerates in the forward direction; (c) when the car brakes while moving in a straight line; (d) while the car rounds a corner at constant speed?
24. As your plane accelerates down the runway, you take your keys from your pocket and suspend them by a thread. Do they hang vertically? Explain.
25. Since every force has an oppositely directed reaction force of equal magnitude, how can there ever be a net force on an object?
26. A friend thinks that astronauts are weightless because there is no gravity in outer space. Criticize this statement, and supply your friend with a correct explanation for weightlessness.

PROBLEMS

Section 4.3 *Newton's Second Law*

1. A subway train has a mass of 1.5×10^6 kg. What force is required to accelerate the train at 2.5 m/s²?
2. A railroad locomotive with a mass of 6.1×10^4 kg can exert a force of 1.2×10^5 N. At what rate can it accelerate (a) by itself and (b) when pulling a 1.4×10^6-kg train?
3. The maximum braking force of a 1400-kg car is about 8000 N. Estimate the stopping distance when the car is travelling (a) 40 km/h; (b) 60 km/h; (c) 80 km/h; (d) 55 mi/h.
4. A car leaves the road travelling at 95 km/h and hits a tree, coming to a complete stop in 0.16 s. What force does a seatbelt exert on a 55-kg passenger during this collision?
5. Suppose the passenger in the previous problem were not wearing a seatbelt. What force would the passenger experience in striking a padded dashboard that compresses 8.0 cm on impact? Assume that the car had come to a full stop by the time the passenger struck the dashboard. (Is this a consistent assumption?)

6. As a function of time, the velocity of an object of mass m is given by $\mathbf{v} = bt^2\hat{\mathbf{i}} + (ct + d)\hat{\mathbf{j}}$, where b, c, and d are constants with appropriate units. What is the force acting on the object, as a function of time?
7. In an x-ray tube, electrons are accelerated to speeds on the order of 10^8 m/s, then slammed into a target where they come to a stop in about 10^{-18} s. Estimate the average stopping force on each electron.
8. A 3800-kg jet touches down at 240 km/h on the deck of an aircraft carrier, and immediately deploys a parachute to slow itself down. If the plane comes to a stop in 170 m, what is the average force of air on the parachute? Assume the parachute provides essentially all the stopping force.

Section 4.4 *The Force of Gravity*

9. My spaceship crashes on one of the sun's nine planets. Fortunately, the ship's scales are intact, and show that my weight is 532 N. If I know my mass to be 60 kg, where am I? *Hint:* Consult Appendix F.
10. If I can barely lift a 50-kg concrete block on earth, how massive a block can I lift on the moon?

11. The frictional force between a car's tires and the road has a magnitude that is 75 percent of the car's weight. What is the maximum possible acceleration for this car?

Section 4.5   *Adding Forces*

12. A 65-kg parachute jumper descends at a steady 40 km/h. What is the force of the air on the parachute?

13. A 10-kg mass is suspended at rest by two strings attached to walls, as shown in Fig. 4–34. What are the tension forces in the two strings?

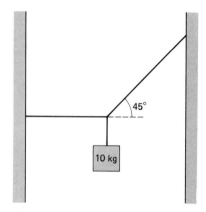

Fig. 4–34   Problem 13

14. A 3700-kg barge is being pulled along a canal by two mules, as shown in Fig. 4–35. The tension in each tow rope is 1100 N, and the ropes make 25° angles with the forward direction. What force does the water exert on the barge (*a*) if it moves with constant velocity and (*b*) if it accelerates forward at 0.16 m/s²?

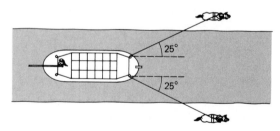

Fig. 4–35   Problem 14

15. At lift-off, a space shuttle with 2.0×10^6 kg total mass undergoes an upward acceleration of 0.60 g. (*a*) What is the total thrust (force) developed by its engines? (*b*) By what factor is the weight of an astronaut increased in the reference frame of the accelerating shuttle?

16. An airplane accelerates down the runway at 3.3 m/s². A passenger suspends a 25-g watch from a thread. (*a*) What angle does the thread make with the vertical? (*b*) What is the tension in the thread?

17. A 15-kg monkey hangs from the middle of a massless rope as shown in Fig. 4–36. What is the tension in the rope? Compare with the monkey's weight.

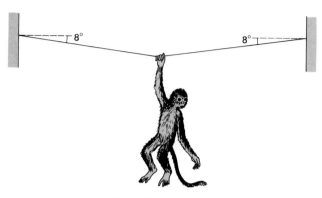

Fig. 4–36   Problem 17

18. Two forces act on a 3.1-kg mass, which undergoes acceleration $\mathbf{a} = 0.91\hat{\imath} - 0.27\hat{\jmath}$ m/s². If one of the forces is $\mathbf{F}_1 = -1.2\hat{\imath} - 2.5\hat{\jmath}$ N, what is the other force?

19. A block of mass m slides with acceleration a down a frictionless slope that makes an angle θ to the horizontal; the only forces acting on it are the force of gravity $\mathbf{F}_g$ and the normal force $\mathbf{N}$ of the slope, as shown in Fig. 4–37. Show that the magnitude of the normal force is given by $N = m\sqrt{g^2 - a^2}$.

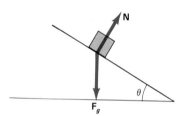

Fig. 4–37   Problem 19

Section 4.6   *Newton's Third Law*

20. In a tractor pulling contest, a 2300-kg tractor pulls a 4900-kg sledge with an acceleration of 0.61 m/s². If the tractor exerts a horizontal force of 7700 N on the ground, determine the magnitudes of (*a*) the force of the tractor on the sledge, (*b*) the force of the sledge on the tractor, and (*c*) the frictional force exerted on the sledge by the ground.

21. A 2200-kg airplane is pulling two gliders, the first of mass 310 kg and the second of mass 260 kg, down the runway with an acceleration of 1.9 m/s² (Fig. 4–38). Neglecting the mass of the two ropes and any frictional forces, determine (*a*) the horizontal thrust of the plane's propeller; (*b*) the tension force in the first rope; (*c*) the tension force in the second rope; and (*d*) the net force on the first glider.

Fig. 4–38 Problem 21

22. A 68-kg astronaut pushes off a 420-kg satellite, exerting a force of 120 N for the 2.9 s they are in contact. How far apart are the two after 1 minute?

23. I have a mass of 65 kg. If I jump off a 120-cm-high table, how far toward me does earth move during the time I fall?

Section 4.7   *Measuring Force*

24. A spring with spring constant $k = 340$ N/m is used to weigh a 6.7-kg fish. How far does the spring stretch?

25. Two springs have the same unstretched length but different spring constants k_1 and k_2. (*a*) If they are connected side-by-side and stretched a distance x, as shown in Fig. 4–39*a*, show that the force exerted by the combination is $(k_1 + k_2)x$. (*b*) If they are now connected end-to-end and the combination is stretched a distance x (Fig. 4–39*b*), show that they exert a force $2k_1k_2x/(k_1 + k_2)$.

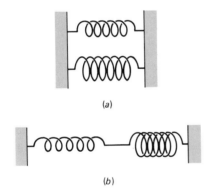

(a)

(b)

Fig. 4–39 Problem 25

26. An elastic tow rope has a spring constant of 1300 N/m. It is connected between a truck and a 1900-kg car. As the truck tows the car, the rope stretches 55 cm. Starting from rest, how far do the truck and car move in 1 min? Assume the car experiences negligible friction.

27. A 7.2-kg mass is hanging from the ceiling of an elevator by a spring of spring constant 150 N/m whose unstretched length is 80 cm. What is the overall length of the spring when the elevator (*a*) starts moving upward with acceleration 0.95 m/s²; (*b*) moves upward at a steady 14 m/s; (*c*) comes to a stop while moving upward at 14 m/s, taking 9.0 s to do so? (*d*) If the elevator measures 3.2 m from floor to ceiling,

what is the maximum acceleration it could undergo without the 7.2-kg mass hitting the floor?

28. An accelerometer consists of a spring of spring constant $k = 1.25$ N/m and unstretched length $\ell_0 = 10.0$ cm fastened to a frictionless surface by a pivot that allows it to swivel in any direction in a horizontal plane. A 50.0-g mass is attached to the other end of the spring, as shown in Fig. 4–40. The whole system is mounted securely in an automobile. When the vehicle accelerates, the spring provides a force to keep the 50-g mass accelerating with the vehicle; by measuring the stretch of the spring, the acceleration can then be determined. To calibrate the accelerometer, circles marked with values of acceleration can be drawn on its frictionless surface. (*a*) How far apart should the circles be if each represents an acceleration of 0.250 m/s² larger than the next smaller circle? (*b*) What should be the radius of the circle marked 2.0 m/s²? (*c*) How do you read the direction of the car's acceleration from this device?

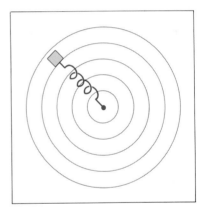

Fig. 4–40 Top view of accelerometer for Problem 28.

Supplementary Problems

29. Tarzan (mass 73 kg) slides down a vine that can withstand a maximum tension force of 580 N before breaking. What minimum acceleration must Tarzan have?

30. A 90-kg weightlifter is standing on a scale calibrated in newtons, holding a 140-kg barbell. (*a*) What does the scale read? (*b*) The lifter then gives the barbell a constant upward acceleration. If the scale reads 2400 N, what is the acceleration of the barbell? (*c*) Suppose the lifter were in an orbiting spaceship with his feet contacting an identical scale. If he gave the barbell the same acceleration as in (*b*), what would the scale read? Assume in (*b*) and (*c*) that the lifter keeps his body rigid, bending only his arms, and neglect the mass of the arms and any downward acceleration of the lifter.

31. To escape a dog, a 1.3-kg squirrel runs up a vertical tree trunk, accelerating at 0.82 m/s^2. What is the average vertical force the tree exerts on the squirrel?
32. The second floor of a house can safely carry a load of 3500 N for each square meter of floor. Can it support a waterbed measuring 1.4 m wide by 1.8 m long by 25 cm thick? (The density of water is 1 g/cm^3.)
33. A string is draped over a pulley, and masses m_1 and m_2 are tied to its two ends, as shown in Fig. 4–41. String and pulley have negligible mass, and the pulley is lubricated so that friction is negligible. Show that the magnitude of the two blocks' acceleration is given by

$$a = \frac{m_2 - m_1}{m_2 + m_1} g,$$

where a positive result corresponds to an upward acceleration of m_1.

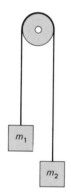

Fig. 4–41 Problem 33

34. A child sends a 2.3-kg stone sliding across a frozen pond at 11 m/s. If the stone slides 67 m before stopping, what is the average frictional force acting on the stone?
35. Three identical massless springs of unstretched length ℓ and spring constant k are connected to three equal masses m as shown in Fig. 4–42. A force is applied at the top of the upper spring to give the whole system the same acceleration a. Determine the length of each spring.
36. A block of mass m_1 on a frictionless table top is connected by a massless string over a massless, frictionless pulley to a mass m_2 (Fig. 4–43). Determine (a) the string tension and (b) the magnitude of the blocks' acceleration.
37. A block of mass M hangs from a rope of length ℓ and mass m. Find an expression for the tension in the rope as a function of the distance y measured vertically downward from the top of the rope.
38. A tow truck exerts a 750-N force on a 54.4-kg tow chain. The other end of the chain is attached to a car, and the whole system accelerates at 0.436 m/s^2. Find

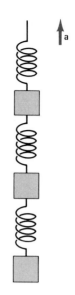

Fig. 4–42 Problem 35

(a) the force exerted on the truck by the cable, (b) the force exerted on the car by the cable, and (c) the mass of the car. Assume there are no horizontal forces acting on the car except for the cable. Why are your answers to (a) and (b) different?
39. In throwing a 200-g ball, your hand exerts a constant upward force of 9.4 N for 0.32 s. How high does the ball rise after leaving your hand?
40. What downward force is exerted on the air by the blades of a 4300-kg helicopter when it is (a) hovering at constant altitude; (b) dropping at 21 m/s with speed decreasing at 3.2 m/s^2; (c) rising at 17 m/s with speed increasing at 3.2 m/s^2; (d) rising at a steady 15 m/s; (e) rising at 15 m/s with speed decreasing at 3.2 m/s^2?
41. What engine thrust (force) is needed to accelerate a rocket of mass m (a) downward at 1.40g near earth's surface; (b) upward at 1.40g near earth's surface; (c) at 1.40g in interstellar space far from any star or planet?

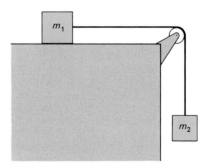

Fig. 4–43 Problem 36

42. An elevator cable can withstand a maximum tension of 19,500 N before breaking. The elevator has a mass of 490 kg, and a maximum acceleration of 2.24 m/s². Engineering safety standards require that the cable tension never exceed two-thirds of the breaking tension. How many 65-kg people can the elevator safely accommodate?

43. You have a mass of 60 kg, and you jump from a 78-cm-high table onto a hard floor. (a) If you keep your legs rigid, you come to a stop in a distance of 2.9 cm, as your body tissues compress slightly. What force does the floor exert on you? (b) If you bend your knees when you land, the bulk of your body comes to a stop over a distance of 0.54 m. Now estimate the force exerted on you by the floor. Neglect the fact that your legs stop in a shorter distance than the rest of you.

44. An electric winch is lifting a 230-kg engine out of a car using a steel cable and pulley, as shown in Fig. 4–44. Find the force of the cable on the pulley if the engine accelerates upward at 0.24 m/s². Neglect friction and the mass of the cable.

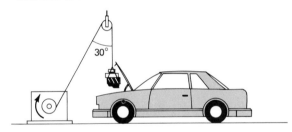

Fig. 4–44 Problem 44

45. The force of air resistance is roughly proportional to the square of an object's speed, and is directed oppositely to the velocity: $F_{air} = -cv^2$, where c is a constant that depends largely on the shape of the falling body. As a result of air resistance, a falling object eventually reaches a constant **terminal speed,** at which it ceases to accelerate. The constant c for a 70-kg skydiver is approximately 0.30 N·s²/m² (this depends on how the diver's body is oriented). Find (a) the skydiver's acceleration while falling at 25 m/s and (b) find the terminal speed.

46. An F-14 jet fighter has a mass of 1.6×10^4 kg and an engine thrust of 2.7×10^5 N. A 747 jumbo jet has a mass of 3.6×10^5 kg and a total engine thrust of 7.7×10^5 N. Is it possible for either plane to climb vertically, with no lift from its wings? If so, what vertical acceleration could it achieve?

47. A spider of mass m_s drapes a silk thread of negligible mass over a stick with its far end a distance h off the ground as shown in Fig. 4–45. The stick is lubricated by a drop of dew, so that there is essentially no friction between silk and stick. The spider waits on the ground until a fly of mass m_f lands on the other end of the silk and sticks to it. The spider immediately begins to climb her end of the silk. (a) With what acceleration must she climb to keep the fly from falling? (b) If she climbs with acceleration a_s, at what height y above the ground will she encounter the fly?

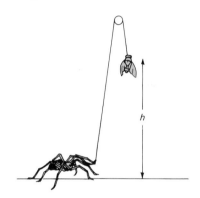

Fig. 4–45 Problem 47

48. Each of the Viking spacecraft that landed on Mars used retrorockets to begin its descent to the Martian surface. At the beginning of the rocket firing, the craft had a mass of 1070 kg, and its rocket engine developed a thrust of 2840 N. If it dropped vertically with its retrorocket firing downward to slow the fall, what was its acceleration? Assume the craft was close enough to the Martian surface to neglect any variation in the local value of g.

5

USING NEWTON'S LAWS

With his three laws of motion, Newton gave us the means to understand most aspects of motion throughout the entire universe. The important concepts he codified—such as force and inertia—provide our basic knowledge of how objects move. By quantifying the relation between force and motion, Newton showed how we can solve in detail for the motions of objects in specific cases (Fig. 5–1).

In this chapter, we apply Newton's laws in a variety of physical situations. The basic methods we use are at the heart of all applications of Newton's laws, from the simplest textbook problems to the most complicated computer programs that guide space probes to distant planets. As you read through the examples, don't look on each one as an entirely new thing to learn about. Rather, think about how the examples are related and how they all represent applications of the same basic physical laws.

5.1
USING NEWTON'S SECOND LAW

Newton's second law, $\mathbf{F} = m\mathbf{a}$, is the cornerstone of mechanics. That law relates the *net force* on an object to the acceleration *of that object*. To apply Newton's law, we must identify the object whose motion interests us and all the forces that together provide the net force acting on that object. Only then can we write Newton's law and solve the equations it provides. In working Newton's law problems, you will find the following steps helpful in correctly applying the law and developing from it a mathematical description of each problem:

1. Identify the object of interest.
2. Identify all the forces acting on the object, and draw them on a vector diagram. (This diagram is often called a "free-body diagram.")

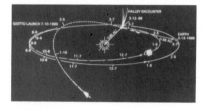

Fig. 5–1
The orbit of Halley's Comet and the trajectory of the European Space Agency's Giotto spacecraft from earth to its encounter with the comet in March, 1986. Careful application of Newton's laws to the motions of both objects made possible Giotto's approach to within 605 km of the comet's nucleus.

3. Write Newton's second law, **F** = m**a**, *in vector form, using for* **F** *the vector sum of the forces in your diagram.*

4. Choose a convenient coordinate system, and draw the coordinate axes on your force diagram.

5. Write the components of your vector equation in your chosen coordinate system. Be guided in this by the geometry of the force diagram, and by any constraints on the motion (for example, the constraint that an object move along a given surface).

6. Solve the equations symbolically for whatever the problem asks you to find.

7. Ask whether your results make sense. In particular, ask about extreme cases such as purely horizontal or purely vertical motion to see if they lead to results that you have seen before or that are physically obvious.

8. Finally, insert numerical values and work the arithmetic, if numerical answers are called for.

Don't memorize these steps. Rather, you should refer to them as you begin to work problems, and will gradually come to realize that they simply reflect the full meaning of Newton's second law. Later, as you develop confidence in problem solving, you can take short cuts. But whenever you deal with a complicated problem, following these steps will help avoid confusion.

Example 5–1

An Inclined Surface

A block of mass m slides without friction on a surface tilted at an angle θ to the horizontal. Find its acceleration and the magnitude of the force the block exerts on the surface.

Solution

Here we outline the solution in the context of our eight steps:

1. The object of interest is the block.
2. There are two forces on the block: the gravitational force **F**$_g$ and the normal force **N** from the surface. Both are shown in Fig. 5–2.
3. In vector form, Newton's second law becomes

$$\mathbf{F}_g + \mathbf{N} = m\mathbf{a}. \tag{5–1}$$

4. Now we choose a coordinate system. Since gravity points vertically downward, you might pick a system with horizontal and vertical axes. But a better choice has axes parallel and perpendicular to the slope, for then two of the three vectors in Equation 5–1—the acceleration and the normal force—have only one nonzero component. The individual component equations will be simpler in such a coordinate system, which we show superimposed on Fig. 5–2. (In Example 5–1b we repeat this example using a horizontal/vertical coordinate system to show that the choice of coordinate system affects only the ease of computation, not the results.)

5. We now write the components of Equation 5–1. In our tilted coordinated system, the normal force has no x-component, but the gravitational force does. From Fig. 5–3, we see that the x-component of the gravitational force is $F_g \sin\theta$; since the magnitude F_g is mg, we have $F_{gx} = mg \sin\theta$. Are the signs right here? Yes: the gravitational force points vertically downward, and our positive x-direction is downslope, so that the x-component of the gravitational force must be positive.

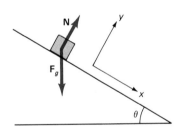

Fig. 5–2
Force diagram for block sliding on incline.

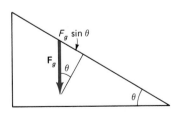

Fig. 5–3
The two triangles shown are similar, so that the angles marked θ are equal. Since the sine of an angle is the ratio of opposite side to hypotenuse, the x-component of the gravitational force **F**$_g$ is $F_g\sin\theta$.

What about the acceleration? The block is constrained to move along the slope, so that there is no y-component of acceleration. Since the x-component of force is positive, so is the x-component of acceleration, which is therefore just the magnitude a. Then the x-component of Newton's law is

$$mg \sin\theta = ma.$$

Similarly, the y-component of the gravitational force is $-mg \cos\theta$; with no acceleration perpendicular to the slope, the y-component of Newton's law reads

$$N - mg \cos\theta = 0.$$

6. Now we solve these equations for the quantities of interest. The x-equation gives immediately

$$a = g \sin\theta,$$

while the y-equation gives

$$N = mg \cos\theta.$$

This is the magnitude of the force that the surface exerts on the block; by Newton's third law, it is also the magnitude of the force that the block exerts on the surface.

7. To check our results, we can consider extreme values for the angle θ. When $\theta = 0$, corresponding to a horizontal surface, we have $\sin\theta = 0$ so that $a = 0$, as we expect. In this case $N = mg$, and the surface supports the entire weight of the block. When $\theta = 90°$, $\sin\theta = 1$ and $a = g$. In this case the block falls freely; the normal force is $mg \cos 90° = 0$, so that the presence of the surface is irrelevant.

8. There are no numerical values in this example.

Our result in this example shows that a frictionless, sloping surface "dilutes" the acceleration of gravity. It was through the use of such "diluted" gravity on inclined planes that Galileo first probed the laws of motion.

Example 5–1a

Leveling an Air Track

An air track used in a physics lab is supported on adjustment screws 88 cm apart (Fig. 5–4). When the track is not quite level, a glider released from rest at one end reaches the other end, 120 cm away, in 6.8 s. By how much should the support near the higher end be dropped in order to level the track?

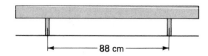

Fig. 5–4
An air track (Example 5–1a).

Solution

From Example 5–1, we know that the acceleration down a sloping track is given by $a = g \sin\theta$. Here we are given the distance the accelerating glider travels in a known time; from this we can use Equation 2–11 to get the acceleration:

$$x - x_0 = \tfrac{1}{2}at^2,$$

so that

$$a = \frac{2(x - x_0)}{t^2},$$

where $x - x_0$ is the 120-cm length of the track. Since $a = g \sin\theta$, the track is tilted at an angle θ given by

$$\sin\theta = \frac{a}{g} = \frac{2(x - x_0)}{gt^2}.$$

In Fig. 5–5, we see an exaggerated view of the sloping track; from the figure, we see that the high end must be dropped by an amount $h = d \tan\theta$, where d is the 88-cm

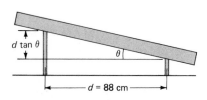

Fig. 5–5
Exaggerated view of the sloping track. Angle θ is actually so small that $\sin\theta$ is very nearly equal to $\tan\theta$.

spacing between supports. But for very small angles, sine and tangent are essentially equal (see Appendix A), so that the high end must be dropped by

$$h = d \tan\theta \simeq d \sin\theta = \frac{2d(x - x_0)}{gt^2}$$

$$= \frac{2(0.88 \text{ m})(1.2 \text{ m})}{(9.8 \text{ m/s}^2)(6.8 \text{ s})^2} = 4.7 \times 10^{-3} \text{ m} = 4.7 \text{ mm}.$$

Example 5–1b _____
Another Coordinate System

Rework Example 5–1, solving for the acceleration using a coordinate system with x- and y-axes horizontal and vertical, respectively.

Solution

Steps 1 to 3 of our solution procedure are independent of the choice of coordinate system, so the forces are still as shown in Fig. 5–2 and Newton's second law is still given by Equation 5–1. The essential physics is summed up in that equation; the rest of the problem is algebra.

In Fig. 5–6 we show the forces on a horizontal/vertical coordinate system. The gravitational force is simpler in this system; it has only a vertical component given by $F_{gy} = -mg$. But now the normal force has two components; from Fig. 5–6, we see that they are $N_x = N \sin\theta$ and $N_y = N \cos\theta$. What about the acceleration? The block is constrained to accelerate down the slope, so from Fig. 5–7 we then have $a_x = a \cos\theta$ and $a_y = -a \sin\theta$, where a is the magnitude of the acceleration. Having resolved the forces and acceleration into components, we rewrite Equation 5–1 and then give its components in our coordinate system:

$$\mathbf{F_g} + \mathbf{N} = m\mathbf{a} \tag{5–1}$$

x-component: $\qquad\qquad N \sin\theta = ma \cos\theta \tag{5–2}$

y-component: $\qquad -mg + N \cos\theta = -ma \sin\theta. \tag{5–3}$

With our sensible choice of coordinate system in Example 5–1, we needed only one of the component equations to get the acceleration; here, with a less appropriate coordinate system, we must deal with both equations. Solving Equation 5–2 for the unknown normal force N gives

$$N = \frac{ma \cos\theta}{\sin\theta};$$

using this result in Equation 5–3, we have

$$-mg + \frac{ma \cos^2\theta}{\sin\theta} = -ma \sin\theta.$$

Multiplying through by $\sin\theta$ and rearranging the equation gives

$$ma(\cos^2\theta + \sin^2\theta) = mg \sin\theta.$$

But $\cos^2\theta + \sin^2\theta = 1$, so we recover our previous result,

$$a = g \sin\theta.$$

Of course we *had* to get the same result, because we described the same physical situation. But our poor choice of coordinate system resulted in more complicated algebra. Comparison of Examples 5–1 and 5–1b shows that a little thought about coordinate systems can save a good deal of mathematical effort.

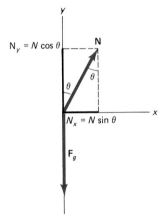

Fig. 5–6
Forces of Example 5–1 on a horizontal/vertical coordinate system. Also shown is the breakdown of the normal force **N** into components $N_x = N \sin\theta$ and $N_y = N \cos\theta$. (The triangle formed by N_x, N_y, and the vector **N** is a right triangle, and the trig functions sine and cosine are defined as ratios of opposite and adjacent sides to the hypotenuse of a right triangle.)

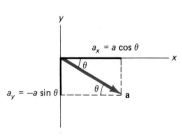

Fig. 5–7
The acceleration vector points downslope, so its components are $a_x = a \cos\theta$ and $a_y = -a \sin\theta$.

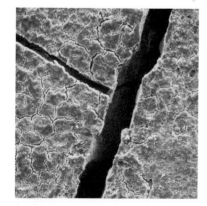

Fig. 5–8
Forces on the block of Example 5–2.

Example 5–2

An Equilibrium Situation

A horizontal force of what magnitude is needed to keep the block of Example 5–1 from sliding?

Solution

Now there are three forces acting: gravity, the normal force, and the horizontal restraining force $\mathbf{F}_h$, as shown in Fig. 5–8. Newton's second law then reads

$$\mathbf{F}_g + \mathbf{N} + \mathbf{F}_h = m\mathbf{a} = 0, \tag{5–4}$$

where we set the right-hand side to zero because we want $\mathbf{F}_h$ to keep the block from accelerating. Now two of the three nonzero vectors in our equation point in the horizontal or vertical direction, so that we are best off with the horizontal/vertical coordinate system of Example 5–1b. In this system, the gravitational force has only a vertical component $-mg$, and the normal force breaks down as in Example 5–1b into components $N_x = N\sin\theta$ and $N_y = N\cos\theta$. What about $\mathbf{F}_h$? It points in the negative x-direction, so it has only an x-component $-F_h$. Then the components of Newton's second law (Equation 5–4 in this case) become

x-component: $N\sin\theta - F_h = 0$ \hfill (5–5)

and

y-component: $-mg + N\cos\theta = 0.$ \hfill (5–6)

Solving Equation 5–6 for the normal force gives

$$N = \frac{mg}{\cos\theta}.$$

Using this result in Equation 5–5, we have

$$F_h = \frac{mg}{\cos\theta}\sin\theta = mg\tan\theta.$$

Does this answer make sense? When $\theta = 0$, corresponding to a horizontal surface, then $\tan\theta = 0$ and therefore $F_h = 0$. Of course: it takes no force to prevent acceleration on a horizontal surface. As the slope approaches vertical, however, $\tan\theta \to \infty$, and it becomes impossible for a purely horizontal force $\mathbf{F}_h$ to prevent the acceleration.

5.2
FRICTION

We have seen how the presence of friction prevented scientists before Galileo and Newton from recognizing the true nature of motion. Further, we have generally considered frictionless systems in our examples and problems. But friction does occur in real systems. What is friction, and how can we deal with it?

Friction is a force that acts between two surfaces to oppose their relative motion. Even the smoothest of surfaces is highly irregular on a microscopic scale (Fig. 5–9). When two surfaces are placed in contact, their irregularities adhere because of electrical forces between their molecules (Fig. 5–10). The result is a force that opposes relative motion of the sur-

Fig. 5–9
Electron micrograph of a cracked eggshell shows that the apparently smooth surface is actually rough on a microscopic scale.

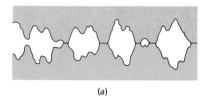

(a)

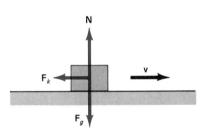

(b)

Fig. 5–10
(a) When two surfaces are in contact, microscopic irregularities in the two surfaces adhere, giving rise to a force that opposes their relative motion. Only the microscopic irregularities actually touch, so that the actual contact area is much less than the measured area. (b) When the normal force between the surfaces increases, the microscopic contact area increases as the irregularities are crushed together. This increases the frictional force.

faces; in order to move them, an external force must be applied to break the microscopic electrical bonds.

On earth, friction can rarely be ignored. As much as 20 per cent of the gasoline burned in a car is used to overcome friction in the engine. But frictional forces are also beneficial to the car; without friction between tires and road, the car could not stop, start, or turn corners. And it is friction between your feet and the floor that lets you walk.

We distinguish two types of friction: (1) kinetic, or sliding, friction and (2) static friction.

Kinetic Friction

kinetic friction

Kinetic friction is the friction between surfaces in relative motion. It is kinetic friction that slows down a book you shove across a table, or makes it hard to push a heavy trunk even at constant speed. Experimentally, we find that the force of kinetic friction is independent of the surface areas that seem to be in contact, but is proportional to the normal force acting between the surfaces. Intuitively, you might have expected the frictional force to increase with increasing contact area. On a microscopic scale, your intuition is correct: the greater the area *in actual contact,* the greater the friction. But microscopically, only a small fraction of the area you measure macroscopically is actually in contact with the other surface (Fig. 5–10a). As the normal force between the surfaces increases, the surface irregularities are crushed together and the actual contact area increases, so that the frictional force increases (Fig. 5–10b).

We can characterize the force of kinetic friction mathematically by writing

$$F_k = \mu_k N, \tag{5–7}$$

where F_k is the force of kinetic friction, N the normal force between the two surfaces, and μ_k the **coefficient of kinetic friction,** a quantity that depends on the properties of the two surfaces. Equation 5–7 relates only the magnitudes; the direction of the frictional force is opposite to the relative motion of the surfaces, and is therefore perpendicular to the normal force (Fig. 5–11).

Since both F_k and N have the units of force, the coefficient of kinetic friction is a dimensionless quantity; it has no units. μ_k ranges from about 0.01 for very smooth surfaces to 1.5 for the roughest surfaces. That is, the force of friction on a smooth surface is about 1 per cent of the normal force; if you push an object on a smooth horizontal surface, you must apply a force about 1 per cent of the object's weight. On a rough surface, on the other hand, you may have to push an object with a force greater than its weight just to keep it moving horizontally at constant speed. No wonder Aristotle was confused about the laws of motion!

Fig. 5–11
The force of kinetic friction F_k acts to oppose relative motion **v,** and is therefore parallel to the surfaces in contact and perpendicular to the normal force. In the absence of an applied force to the right, the block shown here would be decelerated by the frictional force.

coefficient of kinetic friction

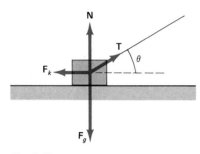

Fig. 5–12
Forces on a trunk being pulled across a level floor include the rope tension **T**, the force of kinetic friction **F**$_k$, the normal force **N**, and the gravitational force **F**$_g$. Only the tension force has nonzero components in both the horizontal and vertical directions; the figure shows that these are $T_x = T \cos\theta$ and $T_y = T \sin\theta$.

Example 5–3
Pulling a Trunk

A trunk of mass m is being pulled across a heavy floor by means of a massless rope that makes an angle θ with the horizontal. The coefficient of friction between trunk and floor is μ_k. What rope tension is required to move the trunk at constant velocity?

Solution

The trunk is our object of interest; Fig. 5–12 shows that the forces include the rope tension **T**, the force of kinetic friction **F**$_k$, the normal force **N**, and the gravitational force **F**$_g$. Since the trunk moves with constant velocity, its acceleration is zero and therefore the net force is zero:

$$\mathbf{T} + \mathbf{F}_k + \mathbf{N} + \mathbf{F}_g = 0.$$

The frictional force acts only in the horizontal direction, while the normal and gravitational forces act only in the vertical direction. So we choose a coordinate system with x-axis horizontal and in the direction of the trunk's motion, and with y-axis pointing vertically upward. From Fig. 5–12, we see that the tension force has horizontal and vertical components $T_x = T \cos\theta$ and $T_y = T \sin\theta$, respectively. Equation 5–7 shows that the magnitude of the frictional force is $\mu_k N$; this force points in the negative x-direction to oppose the trunk's motion. Finally, the normal and gravitational forces are entirely vertical, with the vertical component of the gravitational force given by $-mg$ since it points downward. Then the two components of Newton's law are

x-component: $$T \cos\theta - \mu_k N = 0$$

y-component: $$T \sin\theta + N - mg = 0.$$

Solving the y-equation for the normal force N gives $N = mg - T \sin\theta$; using this result in the x-equation, we have

$$T \cos\theta - \mu_k(mg - T \sin\theta) = 0,$$

or

$$T(\cos\theta + \mu_k\sin\theta) = \mu_k mg.$$

Then the rope tension T is

$$T = \frac{\mu_k mg}{\cos\theta + \mu_k\sin\theta}.$$

Does this result make sense? As $\mu_k \to 0$, the frictional force vanishes and the force needed to pull the trunk at constant velocity becomes zero, in accordance with Newton's first law. And as the frictional force increases, it becomes more difficult to pull the trunk. We can also look at extreme possibilities for the angle θ: for $\theta = 0$, we have $\cos\theta = 1$ and $\sin\theta = 0$, so that $T = \mu_k mg$. When we pull with a horizontal force on a horizontal surface, the force we must provide is just the coefficient of friction times the object's weight. But in the more general case, the upward component of the tension force helps balance the gravitational force, so that the normal force and therefore the frictional force are lowered. Finally, when $\theta = 90°$, our result gives $T = mg$, independent of μ_k. And this makes sense, too: here we are lifting the trunk vertically; with no acceleration, we provide a force equal to the trunk's weight. Friction plays no role because the trunk is no longer sliding along the surface.

Example 5–4

An Incline with Friction

A child sleds down a 20° snow-covered slope. If the mass of the child and sled together is 38 kg, and the coefficient of kinetic friction is $\mu_k = 0.085$, what is the child's acceleration? At what angle would the child slide with constant speed?

Solution

Here the forces on the child and sled together include only gravity, the normal force, and the frictional force, as shown in Fig. 5–13. Newton's second law then reads

$$\mathbf{F}_g + \mathbf{N} + \mathbf{F}_k = m\mathbf{a}.$$

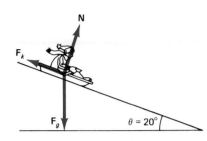

Fig. 5–13
Forces on the child and sled.

As in Example 5–1, the simplest coordinate system is one with axes parallel and perpendicular to the slope. In this system the gravitational force has an x-component $mg \sin\theta$ (see Fig. 5–14), while the frictional force points upslope, or in the negative x-direction; $F_{kx} = -\mu_k N$. So the x-component of Newton's law becomes

x-component: $$mg \sin\theta - \mu_k N = ma.$$

There is no acceleration in the y-direction, perpendicular to the slope; forces in the y-direction include the normal force N and the y-component $-mg \cos\theta$ of the gravitational force (see Fig. 5–14), so that the y-component of Newton's law is

y-component: $$N - mg \cos\theta = 0.$$

Solving the y-equation gives $N = mg \cos\theta$; using this result in the x-equation, we have

$$a = g \sin\theta - \mu_k g \cos\theta,$$

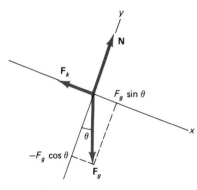

Fig. 5–14
Resolving the forces into components. The magnitude F_g of the gravitational force is mg, so the components become $F_{gx} = mg \sin\theta$ and $F_{gy} = -mg \cos\theta$.

where the mass m cancels from all terms. Note that if $\mu_k = 0$, we recover the result $a = g \sin\theta$ for acceleration down a frictionless slope.

For the sled to slide with constant velocity, its acceleration must be zero; then

$$g \sin\theta = \mu_k g \cos\theta,$$

so that

$$\mu_k = \frac{\sin\theta}{\cos\theta} = \tan\theta.$$

Does this make sense? As the slope becomes very steep, $\tan\theta$ increases without bound and it takes a very large frictional force to counteract gravity. For a horizontal slope, on the other hand, $\tan\theta = 0$ and the sled slides with constant velocity only in the absence of friction: $\mu_k = 0$.

For the numbers of this example, we have

$$a = g(\sin\theta - \mu_k \cos\theta) = (9.8)[\sin 20° - (0.085)(\cos 20°)] = 2.6 \text{ m/s}^2$$

for the acceleration on a 20° slope, and

$$\theta = \tan^{-1}\mu_k = \tan^{-1}(0.085) = 4.9°$$

when the sled undergoes no acceleration.

Note that in working this problem we could not write the frictional force as μ_k times the weight mg. Why not? Because the frictional force depends on the *normal force*, whose magnitude is not equal to the weight when the object is on a sloping surface.

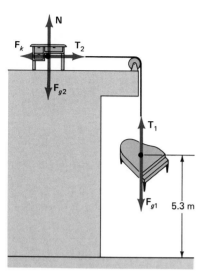

Fig. 5–15
Example 5–5

Example 5–5

Losing a Piano

Workers are pulling a 190-kg piano up a building using a massless rope over a frictionless corner, as shown in Fig. 5–15. When lunchtime arrives, the workers tie the rope to a 210-kg desk, leaving the piano dangling 5.3 m above the ground. Unfortunately, the desk begins to slide; the coefficient of kinetic friction is $\mu_k = 0.71$. What is the acceleration of the piano? At what speed does it hit the ground?

Solution

In this example, we must solve two related Newton's law problems: one for the motion of the piano and one for the desk. Let m_1 be the mass of the piano and m_2 that of the desk. Figure 5–15 shows the forces acting on each. For the piano they are the gravitational force $\mathbf{F}_{g1}$ and the tension force $\mathbf{T}_1$. For the desk they are the gravitational force $\mathbf{F}_{g2}$, the normal force $\mathbf{N}$, the tension force $\mathbf{T}_2$, and the frictional force $\mathbf{F}_k$. So we write Newton's law for each object:

piano:
$$\mathbf{F}_{g1} + \mathbf{T}_1 = m\mathbf{a}_1$$

desk:
$$\mathbf{F}_{g2} + \mathbf{N} + \mathbf{T}_2 + \mathbf{F}_k = m\mathbf{a}_2,$$

where $\mathbf{a}_1$ and $\mathbf{a}_2$ are the respective accelerations.

Given the geometry of the situation, we can use the same horizontal/vertical coordinate system for each object. (More generally, we might want two different coordinate systems; see Problem 4.) For the piano, the gravitational force points downward and has magnitude $m_1 g$; the tension force points upward and has magnitude T_1. Clearly the acceleration points downward, so that the y-component of Newton's law for the piano becomes

$$-m_1 g + T_1 = -m_1 a_1, \tag{5–8}$$

where a_1 is the magnitude of the acceleration. (Had the direction of the acceleration not been obvious, we could have written ma_{y1} on the right-hand side and let the algebra determine the correct sign.)

The desk experiences forces in both horizontal and vertical directions; with its acceleration strictly horizontal, components of Newton's law for the desk are

y-component:
$$-m_2 g + N = 0 \tag{5–9}$$

x-component:
$$T_2 - \mu_k N = m_2 a_2. \tag{5–10}$$

Solving Equation 5–9 for the normal force and using the result in Equation 5–10, we have

$$T_2 - \mu_k m_2 g = m_2 a_2. \tag{5–11}$$

So far we have not used the fact that the two masses are, in fact, connected by a single rope. Because of this connection, the magnitudes of their accelerations must be the same. Furthermore, the rope is massless and passes over a frictionless corner, so that no force is required to accelerate it; therefore, the tensions at the two ends of the rope are of equal magnitude. So $a_1 = a_2 = a$ and $T_1 = T_2 = T$, so we can drop the subscripts on a and T. Equations 5–8 and 5–11 then become

$$-m_1 g + T = -m_1 a$$

$$T - \mu_k m_2 g = m_2 a.$$

These are two equations in the two unknowns T and a. Solving the first equation for T gives $T = m_1(g-a)$; using this result in the second equation, we have

$$m_1(g-a) - \mu_k m_2 g = m_2 a.$$

Solving for the acceleration a then gives

$$a = \frac{(m_1 - \mu_k m_2)g}{m_1 + m_2}.$$

Does this result make sense? Consider first the frictionless case $\mu_k = 0$. Then $a = m_1 g/(m_1 + m_2)$. This is the acceleration we would expect for an object of mass $m_1 + m_2$ subject to a force $m_1 g$. Next consider the case $m_2 = 0$. Then there is no friction and no force needed on m_2; the rope tension is zero and mass m_1 falls freely with acceleration g. On the other hand, if m_2 or the coefficient of friction become large enough that $m_1 - \mu_k m_2 = 0$, then there is no acceleration and the masses move with constant speed or, if at rest, remain at rest.

For the numbers of this example, we have

$$a = \frac{(m_1 - \mu_k m_2)g}{m_1 + m_2} = \frac{[190 \text{ kg} - (0.71)(210 \text{ kg})](9.8 \text{ m/s}^2)}{190 \text{ kg} + 210 \text{ kg}} = 1.0 \text{ m/s}^2.$$

The piano falls a distance $h = 5.3$ m; Equation 2–14 then shows that it hits the ground with speed

$$v = \sqrt{2ah} = \sqrt{(2)(1.0 \text{ m/s}^2)(5.3 \text{ m})} = 3.3 \text{ m/s}.$$

Smash!

Static Friction

static friction

Kinetic or sliding friction is associated with microscopic bonds that continually form and break as the surfaces slide past each other. In contrast, **static friction** describes the frictional force between two surfaces at rest with respect to each other. When two surfaces are stationary, more bonds have time to form, so that a greater force is needed to initiate relative motion.

coefficient of static friction

The static friction between two surfaces is characterized by the **coefficient of static friction**, μ_s, which relates the frictional force to the normal force. Consider a typical situation involving static friction: a book resting on a table (Fig. 5–16). With no externally applied forces in the horizontal

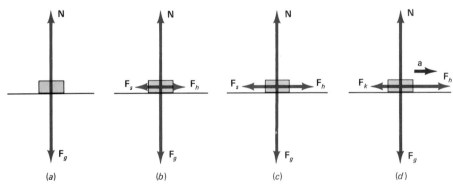

Fig. 5–16
A book on a table. In the vertical direction, the normal force balances the gravitational force. *(a)* With no externally applied horizontal force, the force of static friction is also zero. *(b)* When a horizontal force $\mathbf{F}_h$ of magnitude less than $\mu_s N$ is applied, a static frictional force of the same magnitude acts to oppose the applied force. *(c)* An applied force of magnitude $\mu_s N$ is the largest that the static frictional force can balance. *(d)* For applied forces of magnitude greater than $\mu_s N$, the book begins to accelerate; its subsequent motion is determined by the applied force and the force of kinetic friction.

direction, the frictional force must be zero; otherwise there would be an unbalanced frictional force on the book and it would accelerate. If we apply a small force in the horizontal direction, we find that the book still does not move; evidently the force of static friction has the same magnitude as the applied force. As the applied force is increased, there comes a point when the force of static friction can no longer balance the applied force; then the book begins to slide and its subsequent motion is governed by the applied force and the force of kinetic friction. Experimentally, we find that the maximum value for the static frictional force is proportional to the normal force between the surfaces. We are able to describe static friction by writing

$$F_s \leq \mu_s N. \tag{5–12}$$

Because bonds between stationary surfaces are harder to break, the coefficient of static friction μ_s is generally higher than the coefficient of kinetic friction μ_k.

Example 5–6
Static Friction

A 2.5-kg block is placed on a horizontal board, and the board is gradually tilted. The block remains at rest until the board makes an angle of 38° to the horizontal, at which point it begins to slide. What is the coefficient of static friction?

Solution

Figure 5–17 shows the situation, which is identical to that of Example 5–4 except that static friction replaces kinetic friction until the block begins to slide. Following Example 5–4, we can immediately write

$$\mathbf{F}_g + \mathbf{N} + \mathbf{F}_s = 0$$

for Newton's second law before the block starts to slide. Then, using the tilted coordinate system of Example 5–4, and setting the frictional force to its maximum value $\mu_s N$, the components of Newton's law become

x-component: $mg \sin\theta - \mu_s N = 0$

y-component: $-mg \cos\theta + N = 0,$

where θ is the maximum angle at which the block remains at rest. Solving the y-equation for N and using the result in the x-equation gives

$$mg \sin\theta - \mu_s mg \cos\theta = 0,$$

so that

$$\mu_s = \frac{\sin\theta}{\cos\theta} = \tan\theta.$$

Does this make sense? When $\theta = 0$, μ_s must be zero, showing that the block can slide freely without friction. As $\theta \to 90°$, the coefficient of friction must become very large. This is partly because the downslope component of the gravitational force becomes larger, but more importantly it is because the normal force gets smaller, so that a much larger coefficient of friction is needed to counteract the increasing downslope component of gravity. For the numbers of this example, we have

$$\mu_s = \tan 38° = 0.78.$$

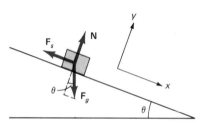

Fig. 5–17
Example 5–6

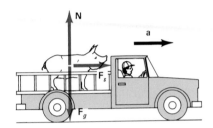

Fig. 5–18
Forces on the hog. The static friction force provides the acceleration of the hog; it points forward to oppose the tendency of the hog to slip backward with respect to the truck.

Fig. 5–19
Forces involved in walking. The foot in contact with the ground is temporarily at rest. It pushes back against the ground, exerting the static frictional force F_1 on the ground. In reaction, the ground pushes forward on the foot with the equal but opposite frictional force F_2. Transmitted through muscle and bone, this is the force that accelerates your upper body forward.

Fig. 5–20
Walking is difficult on a frictionless surface!

Example 5–7 _____

Transporting a Hog

A farmer loads a 170-kg hog onto a flatbed truck. If the coefficient of static friction between hog and truck is 0.21, what is the maximum acceleration the truck should undergo if the hog is not to slide out the back?

Solution

Here the object of interest is the hog; as long as the hog is not slipping relative to the truck, the forces on it include static friction, gravity, and a normal force (Fig. 5–18). Are you tempted to draw another force so that the forces on the hog sum to zero? DON'T! If it stays with the truck, the hog is *accelerating*, and there had better be a nonzero net force to provide that acceleration. Static friction ($\mathbf{F}_s$) is that force.

Newton's second law for the hog reads

$$\mathbf{F}_s + \mathbf{F}_g + \mathbf{N} = m\mathbf{a},$$

where $\mathbf{a}$ is the acceleration of the hog, which is equal to that of the truck as long as the hog does not slip. In a coordinate system with y-axis pointing vertically upward, and x-axis in the direction of the acceleration, the components of Newton's law become

y-component: $\qquad\qquad -mg + N = 0$

x-component: $\qquad\qquad \mu_s N = ma.$

Setting the frictional force to its maximum value $\mu_s N$ gives the maximum acceleration; the frictional force enters with a positive sign because it points in the $+x$-direction, opposing not the motion of the truck but any tendency of the hog to slip backward relative to the truck. Solving the y-equation for the normal force N and using the result in the x-equation, we find that the mass cancels, leaving

$$a = \mu_s g.$$

Does this make sense? Yes: with purely horizontal motion, the normal force is equal to the hog's weight mg, so the maximum frictional force is just the fraction μ_s of the hog's weight. According to Newton's second law, the corresponding maximum acceleration is $a = F/m$, or $\mu_s g$. For this problem, $a = (0.21)(9.8 \text{ m/s}^2) = 2.1 \text{ m/s}^2$ is the maximum acceleration the farmer should give the truck.

Example 5–7 shows that the force of static friction can be useful in accelerating objects. Other cases where static friction plays this accelerating role including walking and driving. When you walk, the foot in contact with the ground is temporarily at rest, pushing back against the ground. The reaction force to this backward push is the forward-pointing force of static friction that accelerates you forward (Fig. 5–19). On a near-frictionless surface, walking becomes very difficult (Fig. 5–20). An accelerating car

By permission of Johnny Hart and News America Syndicate.

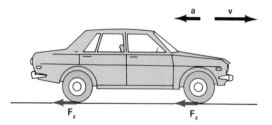

is in a similar situation. If the tires are not slipping, the bottom of each tire—the part in contact with the road—is instantaneously at rest. The tire pushes back against the road, and in reaction the force of static friction pushes forward to accelerate the car. When the car brakes, the wheel pushes forward against the road, and the reaction force of friction points backward to stop the car (Fig. 5–21). Note that the brakes only stop the wheel's spinning; it is friction that stops the car. You know this if you've ever applied your brakes on ice!

Application

Automotive Safety

Why do we talk about *static* friction in connection with an accelerating car? Because the part of the tire in contact with the road is instantaneously at rest as long as there is no slipping or skidding. That the coefficient of static friction is higher than that of kinetic friction has important implications for automobile safety (Fig. 5–22). When you slam on your brakes so hard that the wheels lock and slide along the road, the force between tires and road is *kinetic* friction. When you pump the brakes so that the wheels keep rolling without slipping, the bottoms of the tires are instantaneously at rest with respect to the road, and the force is that of *static* friction. Since $\mu_s > \mu_k$, the deceleration is greater and therefore the stopping distance is shorter.

To provide the safest possible braking, several manufacturers have developed computer-controlled braking systems. Sensors check the wheel motion many times each second and feed their information to a computer. Just as the car begins to skid, the computer responds by momentarily releasing pressure on the brakes, thereby maintaining a static friction situation and the greatest possible stopping force. Figure 5–23 shows the decrease in stopping distance achieved with such a system.

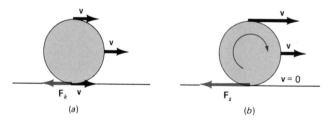

(a)   (b)

Fig. 5–22
(a) In a wheel that is locked and skidding, all points move with the same velocity. Kinetic friction then acts between the wheel and road. *(b)* When the brakes are pumped so that the wheel keeps rolling, the bottom of the tire is instantaneously at rest so the force acting is that of static friction. Because the coefficient of static friction is greater, the stopping distance is reduced when the wheels are kept rolling.

$$\Delta x_{\text{no skid}} = \frac{v_0^2}{\mu_s g} = \frac{(24 \text{ m/s})^2}{(0.89)(9.8 \text{ m/s}^2)} = 66 \text{ m}$$

and

$$\Delta x_{\text{skid}} = \frac{v_0^2}{\mu_k g} = \frac{(24 \text{ m/s})^2}{(0.81)(9.8 \text{ m/s}^2)} = 73 \text{ m}.$$

The 7-m difference could well be enough to prevent an accident.

5.3
CIRCULAR MOTION

The giant planet Jupiter, over 300 times more massive than earth, circles the sun at a speed of 13 km/s. A nimble sports car rounds a tight curve at 120 km/h. Protons circle the 2-km Fermilab accelerator at nearly the speed of light.

What keeps these objects in their circular paths? The answer, according to Newton's first law, is a force. For Jupiter, it is a gravitational force; for the sports car, a frictional force between tires and road; and for the proton, a magnetic force. And how big is that force? Newton's second law tells us: the force required to give an object of mass m an acceleration $\mathbf{a}$ is $\mathbf{F} = m\mathbf{a}$. For an object in circular motion at constant speed v, we know from Section 3.4 that the object accelerates—because of its changing direction—at the rate $a = v^2/r$. The direction of the acceleration is toward the center of the circle. So for an object of mass m to be in uniform circular motion, a net force of magnitude

$$F = ma = \frac{mv^2}{r} \qquad \text{(force in uniform circular motion)} \qquad (5\text{–}13)$$

must act on the object; the force vector must be directed toward the center of the circular path. Whenever we see an object in uniform circular motion, we know that a net force of this magnitude must be acting. Something—gravity, tension in a string, an electric or magnetic force, friction, or some other physical mechanism must provide that force.

Newton's second law describes circular motion in exactly the same way that it does any other motion: by relating the net force, the mass, and the acceleration. We can analyze circular motion using exactly the same steps we outlined in Section 5.1, for we are dealing with exactly the same law of motion.

Example 5–9
A Mass on a String

A ball of mass m is whirled around in a horizontal circle at the end of a string of length ℓ. The string makes an angle θ with the horizontal. Determine (a) the speed of the ball and (b) the tension in the string.

Solution

To see that this circular motion problem is like any other Newton's law problem, we apply rigorously the steps outlined in Section 5.1:

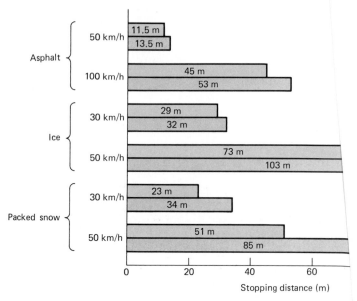

Fig. 5–23

Stopping distances for computerized braking system (colored bars) and for ⟨conventional⟩ (gray bars). Maximum brake pedal pressure is applied in both cases. Comp⟨uterized⟩ system keeps wheels rolling so that the coefficient of static friction remains ⟨. In the⟩ conventional system wheels skid, and the lower coefficient of sliding friction ⟨increases⟩ stopping distance. (Data from Robert Bosch Automotive Equipment Division⟨.⟩

Example 5–8

Stopping a Car

The kinetic and static coefficients of friction between a car⟨'s and a⟩ road are 0.81 and 0.89, respectively. If the car is travelling at 8⟨8 km/h when the⟩ brakes are applied, determine (a) the minimum stopping distance ⟨and (b) the stop⟩ping distance with the wheels fully locked.

Solution

Figure 5–24 shows the forces on the car; they include gravity, ⟨the normal force,⟩ and a frictional force $\mathbf{F}_f$. We have worked enough examples to see ⟨that the normal⟩ force in this case of horizontal motion is equal in magnitude to the ⟨weight, so that⟩ the horizontal component of Newton's law reads

$$-\mu mg = ma_x,$$

or

$$a_x = -\mu g,$$

where the negative sign appears because we have taken the positive ⟨x-direction as⟩ the direction of the car's motion. Equation 2–14 relates distance an⟨d velocity:⟩ $v_x^2 = v_{x0}^2 + 2a_x\Delta x$. Putting $v_x^2 = 0$ to get the stopping distance, and using ⟨$a_x = -\mu g$, the⟩ stopping distance Δx becomes

$$\Delta x = \frac{v_{x0}^2}{-2a_x} = \frac{v_0^2}{\mu g}.$$

The car's initial speed is 88 km/h or 24 m/s; using the appropriate c⟨oefficients of⟩ μ, the stopping distances are

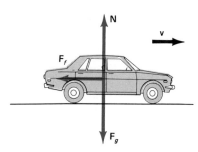

Fig. 5–24
Forces on a braking car.

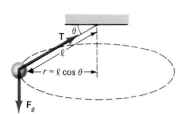

Fig. 5–25
Forces on a ball being whirled about at
the end of a string.

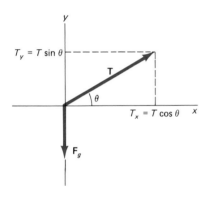

Fig. 5–26
Force components on a
horizontal/vertical coordinate system.

1. The object of interest is the ball.
2. There are two forces acting on the ball: the string tension and the gravitational force, as shown in Fig. 5–25. Are you tempted to draw a third force in Fig. 5–25, perhaps pointing outward to balance the other two? DON'T! The tension and gravitational forces *DO NOT BALANCE*. Why not? Because the ball is accelerating, so that there must be a nonzero net force acting on it! Or maybe you're tempted to draw a third force pointing inward, the force mv^2/r. DON'T! mv^2/r is *not* another physical force; it is the magnitude of the mass times the acceleration, and gives the magnitude of the net physical force that must be applied to provide that acceleration. Here, that net force is provided by gravity and the tension force.
3. Newton's law tells us that $\mathbf{F} = m\mathbf{a}$; here we have

$$\mathbf{T} + \mathbf{F}_g = m\mathbf{a}.$$

4. Now we choose a coordinate system. The tension force is at some angle θ to the horizontal. The gravitational force is purely vertical. What about the acceleration? It points toward the center of the circular path—that is, horizontally. So if we choose a horizontal/vertical coordinate system, then two of the three terms in Newton's law will have only one nonzero component. Figure 5–26 shows the forces with appropriate coordinate axes.
5. The downward-pointing gravitational force $\mathbf{F}_g$ has a vertical component $F_{gy} = -mg$; from Fig. 5–26 we see that the tension force $\mathbf{T}$ has a vertical component $T_y = T\sin\theta$, where T is the magnitude of $\mathbf{T}$. The acceleration has no component in the vertical direction, so that the y-component of Newton's law becomes

y-component: $T\sin\theta - mg = 0.$

The gravitational force has no horizontal component, while Fig. 5–26 shows that the horizontal component of the tension force is $T_x = T\cos\theta$. At the point where we show the ball in Fig. 5–25, the acceleration points entirely in the x-direction. The constraint that the ball move in a circle tells us that the magnitude of the acceleration is mv^2/r, where r is the radius of the circular path. From Fig. 5–25, we see that this radius is $\ell\cos\theta$. Then the x-component of the acceleration is $mv^2/\ell\cos\theta$, so that the x-component of Newton's law becomes

x-component: $T\cos\theta = \dfrac{mv^2}{\ell\cos\theta}.$

6. Now we're ready to solve for the tension and speed. The vertical equation gives the tension:

$$T = \frac{mg}{\sin\theta}.$$

Using this result in the horizontal equation, we have

$$\frac{mg}{\sin\theta}\cos\theta = \frac{mv^2}{\ell\cos\theta},$$

or

$$v = \sqrt{\frac{g\ell\cos^2\theta}{\sin\theta}}.$$

7. Do these results make sense? When $\theta = 90°$, so that the string hangs vertically downward, $\cos\theta = 0$ and therefore $v = 0$. In this case there is no acceleration, and the tension is vertically upward and equal in magnitude to the weight. But as $\theta \to 0°$, $\sin\theta \to 0$ and the speed v becomes very large. This tells us that it is not possible to whirl the ball around with the string exactly horizontal. Why not?

Because there must always be a vertical component of string tension to balance the ball's weight. As the speed increases, so does the string tension, and therefore the vertical component can be a smaller fraction of the total tension. But it can never be zero.

8. There are no numerical values in this example.

Example 5–10
Steering a Car

A level road makes a 90° turn with a 95-m radius of curvature. What is the maximum speed with which a car can negotiate this turn (a) when the road is dry and the coefficient of static friction is 0.88 and (b) when the road is snow-covered and the coefficient of static friction is 0.21?

Solution

Figure 5–27 shows the forces on the car as it rounds the corner. The only force acting in the horizontal direction is the force of static friction between tires and road. Why *static* friction? Because there is no motion of the tire in the radial direction, perpendicular to the car (Fig. 5–28). This frictional force is what provides the car's acceleration; with the car in a circular path of radius r, we can write

$$\mu_s N = \frac{mv^2}{r}$$

for the horizontal component of Newton's second law. Here we have used the maximum possible value, $\mu_s N$, for the frictional force because we want the maximum possible speed v. In the vertical direction there is no acceleration, so that the normal force balances gravity: $N = mg$. Substituting this result in the horizontal equation, we have

$$\mu_s mg = \frac{mv^2}{r},$$

so that

$$v = \sqrt{\mu_s gr}.$$

Does this make sense? On a frictionless road, $v = 0$: you can't corner without a force to deflect you from a straight-line path. With friction, the maximum safe speed increases as the curve becomes more gradual—that is, with increasing r.

For the numbers of this example, we have

$$v_{\text{dry road}} = [(0.88)(9.8 \text{ m/s}^2)(73 \text{ m})]^{1/2} = 25 \text{ m/s} = 90 \text{ km/h}$$

and

$$v_{\text{snow}} = [(0.21)(9.8 \text{ m/s}^2)(73 \text{ m})]^{1/2} = 12 \text{ m/s} = 44 \text{ km/h}.$$

If you exceed these speeds, your car must inevitably move in a path of greater radius—and that means going off the road!

Example 5–11
Designing a Road

Curves in roads designed for safe, high-speed travel are usually banked so that the normal force of the road has a component toward the center of the curve; this allows vehicles to turn without relying on the force of friction. At what angle should the road of the previous example be banked, if it is designed for travel at 55 km/h (15 m/s) without relying on friction?

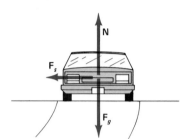

Fig. 5–27
Forces on a car rounding a corner include gravity, the normal force, and the force of static friction between tires and road.

Fig. 5–28
Cornering at high speed. The force of static friction acts on the bottom of the tire and is directed toward the center of the turn. Tension in the tire, evident in this picture through the distortion of the tire, exerts a force on the rest of the car.

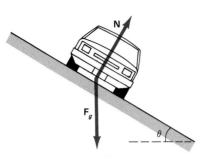

Fig. 5-29
Forces on a car rounding a banked turn. The horizontal component of the normal force provides the acceleration.

Solution

Figure 5–29 shows the forces acting on a car as it rounds the banked turn. They include only gravity and the normal force; we do not show a frictional force because we want to calculate the angle at which the frictional force can be zero. Newton's second law then reads

$$\mathbf{F_g} + \mathbf{N} = m\mathbf{a}.$$

The gravitational force points vertically downward, and the vertical component of the normal force is $N\cos\theta$. There is no acceleration in the vertical direction, so the vertical component of Newton's second law is

$$N\cos\theta - mg = 0.$$

The only force component in the horizontal direction is $N\sin\theta$; since the car describes a circular path, the horizontal component of Newton's law becomes

$$N\sin\theta = \frac{mv^2}{r}.$$

Substituting $N = mg/\cos\theta$ from the vertical equation gives

$$\frac{mg}{\cos\theta}\sin\theta = \frac{mv^2}{r},$$

so that

$$\tan\theta = \frac{v^2}{gr}.$$

Does this make sense? At very low speeds, banking is hardly needed. But at high speeds, the force needed to maintain circular motion increases, and so does the banking angle.

For the numbers of this example,

$$\theta = \tan^{-1}\left(\frac{v^2}{gr}\right) = \tan^{-1}\left[\frac{(15\text{ m/s})^2}{(9.8\text{ m/s}^2)(73\text{ m})}\right] = 17°.$$

Example 5-12
Weight at the Equator and Poles

A 60.00-kg person stands on a spring scale. What does the scale read (a) at the north pole and (b) at the equator? Assume (unrealistically!) that g has the value 9.810 m/s^2 everywhere on earth's surface.

Solution

Why should the scale give a different reading at the equator than at the pole? Because earth rotates, and therefore a person standing at the equator is accelerating, while a person standing at the pole is not. Figure 5–30 shows the forces on the person at both locations. They include the upward force $\mathbf{F_s}$ of the spring scale and the downward force $\mathbf{F_g}$ of gravity. According to Newton's second law, the vector sum of the two forces is equal to the product of the mass and acceleration:

$$\mathbf{F_s} + \mathbf{F_g} = m\mathbf{a}.$$

In a coordinate system with the positive direction upward, we have at the pole

$$F_s - mg = 0,$$

so that the spring force is equal to the gravitational force mg, or

pole: $\qquad F_s = mg = (60.00\text{ kg})(9.810\text{ m/s}^2) = 588.6\text{ N}.$

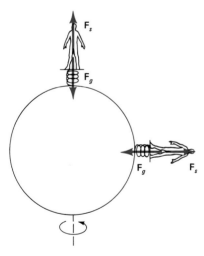

Fig. 5-30
Forces on a person standing on a spring scale at the pole and the equator. In both cases, the only forces acting are the spring force and gravity. At the pole they sum to zero because the person is not accelerating. At the equator they sum to $m\mathbf{a}$, where $\mathbf{a}$ is the acceleration associated with the circular motion of the rotating earth.

Fig. 5–31
A loop-the-loop roller coaster. At the top of the loop, net force on the train and its passengers is downward; that force provides the acceleration that keeps them in their circular path.

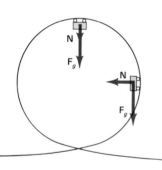

Fig. 5–32
Forces on the train include gravity and the normal force. At the top of the loop, both point downward.

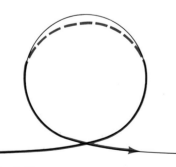

Fig. 5–33
If the train's speed is too low, it leaves the track on a parabolic trajectory, subject only to the gravitational force.

At the equator, however, the person moves in a circular path as earth rotates, and therefore experiences an acceleration directed toward the center of the earth—that is, downward. So at the equator the two forces don't balance; in a coordinate system with the upward direction positive, we have

equator:
$$F_s - mg = -\frac{mv^2}{r},$$

where the minus signs show that the gravitational force and the acceleration are both toward the center of the earth. Now earth takes a time $T = 1$ day to rotate about its axis; in this time a point on the equator moves one earth circumference, or $2\pi R_e$, where R_e is earth's radius. So $v = 2\pi R_e / T$, and Newton's law becomes

$$F_s - mg = -\frac{m(2\pi R_e / T)^2}{R_e} = -\frac{4\pi^2 m R_e}{T^2}.$$

Then the spring force—that is, the scale reading—is

$$F_s = mg - \frac{4\pi^2 m R_e}{T^2}$$

$$= (60.00 \text{ kg})(9.810 \text{ m/s}^2) - \frac{(4\pi^2)(60.00 \text{ kg})(6.37 \times 10^6 \text{ m})}{[(24 \text{ h})(3600 \text{ s/h})]^2}$$

$$= 586.6 \text{ N},$$

or about 0.34 per cent below the weight at the pole. For points between the equator and pole, the difference is correspondingly smaller (see Problem 65). Actually, the difference is about 50 per cent greater because g varies due to the nonspherical shape of the earth.

Example 5–13
Loop-the-Loop

The "Great American Revolution" roller coaster at Valencia, California, includes a loop-the-loop section whose radius is 6.3 m (Fig. 5–31). What is the minimum speed of a train at the top of the loop if it is not to leave the track?

Solution

What does it mean for the train to stay on the track? It means that there must be a nonzero normal force between train and track; if that force goes to zero, then the train and track are no longer in contact. The only forces acting on the train are gravity and the normal force of the track; Fig. 5–32 shows these forces at two points on the loop. Newton's second law relates the net force to the acceleration:

$$\mathbf{F}_g + \mathbf{N} = m\mathbf{a}.$$

At the top of the loop, both forces point downward. In a coordinate system with the positive direction downward, the vertical component of Newton's law then becomes

$$mg + N = ma = \frac{mv^2}{r},$$

so that

$$v^2 = gr + \frac{N}{m}.$$

At the minimum speed, the normal force barely reaches zero as the train passes the top of the loop; at that instant the gravitational force alone provides the acceleration keeping the train in its circular path. The minimum speed is then given by setting $N=0$ at the top of the loop:

$$v_{min} = \sqrt{gr}.$$

For speeds lower than v_{min}, the normal force goes to zero before the train reaches the top of the loop, and the train leaves the track travelling in the parabolic trajectory of a projectile (Fig. 5–33). For speeds greater than v_{min}, the normal force is everywhere nonzero, and the train remains in contact with the track. For the "Great American Revolution," $r=6.3$ m so that $v_{min}=\sqrt{gr}=[(9.8$ m/s$^2)(6.3$ m$)]^{1/2}=7.9$ m/s. The actual speed of the train at the top of the loop is 9.7 m/s to maintain a margin of safety.

SUMMARY

1. Newton's second law provides a universal description of motion in the realm of classical physics. The following steps will help you apply Newton's law correctly in a variety of situations:
   1. Identify the object of interest.
   2. Identify all forces acting on the object, and draw the force vectors on a single diagram.
   3. Write Newton's law in vector form: $\mathbf{F}=m\mathbf{a}$, where $\mathbf{F}$, the net force, is the vector sum of the forces you have identified.
   4. Choose a suitable coordinate system.
   5. Write the components of Newton's law in your coordinate system.
   6. Solve algebraically for the quantities of interest.
   7. Check some special cases to see if your answers make sense.
   8. Insert numerical values into the equations to obtain numerical answers.
2. **Friction** is a force associated with microscopic bonds between two contacting surfaces. The frictional force always acts to oppose motion. Its magnitude depends on the normal force between the two surfaces. When the surfaces are moving relative to one another, the frictional force is the force of **kinetic friction**, $\mathbf{F}_k$ (also called **sliding friction**); in that case the relation between friction and normal forces is quantified by the **coefficient of kinetic friction**, μ_k:

$$F_k = \mu_k N.$$

When the surfaces are at rest, the force of **static friction** opposes an applied force:

$$F_s \leq \mu_s N,$$

where μ_s is the **coefficient of static friction**. In general, $\mu_s > \mu_k$. Once an applied force exceeds the maximum static frictional force $\mu_s N$, the surfaces start moving relative to one another and the subsequent motion is governed by the applied force and the force of kinetic friction.
3. When an object undergoes circular motion at speed v in a circle of radius r, it accelerates at the rate v^2/r in the direction toward the center of the circle. Newton's second law shows that a force of magnitude mv^2/r must then be acting in the same direction.

QUESTIONS

1. Compare the net force on a heavy trunk when it is (a) at rest on the floor; (b) being slid across the floor at constant speed; (c) being pulled upward in an elevator whose cable tension equals the combined weight of the elevator and trunk; (d) sliding down a frictionless ramp.
2. Can an object move if there is no net force on it? Can it start moving if there is no net force?
3. If you take your foot off the gas pedal while driving on a horizontal road, your car soon coasts to a stop. Is there a net force on the car during this time? If not, why not? If so, what forces might contribute to it?
4. The force of static friction acts only between surfaces at rest. Yet that force is essential in walking and in accelerating or braking a car. Explain.

5. A jet plane flies at constant speed in a vertical circular loop (Fig. 5–34). At what point in the loop does

Fig. 5–34 Question 5

the seat exert the greatest force on the pilot? The least force?

6. In cross-country skiing, skis should easily glide forward but should remain at rest when the skier pushes back against the snow. What frictional properties should the ski wax have to achieve this goal?

7. Why is it easier for a child to stand nearer the inside of a rotating merry-go-round?

8. Can a coefficient of friction exceed 1?

9. By pushing horizontally with your hand, you can hold a book at rest on a perfectly vertical wall (Fig. 5–35). How is this possible?

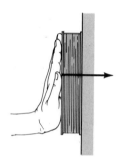

Fig. 5–35   Question 9

10. In general, a mass suspended from a string will not point exactly toward the center of the earth. Explain in terms of earth's rotation. Are there places on earth where the string will point toward the center of the earth?

11. Earth's gravity pulls a satellite toward the center of the earth. So why doesn't the satellite actually fall to earth?

12. Why can front-wheel-drive cars corner better on slippery roads than do rear-wheel-drive cars? Explain in terms of the forces acting on the wheels.

13. Explain why stepping as hard as you can on the brake pedal does not necessarily result in the shortest stopping distance for your car.

14. A fishing line has a breaking strength of 20 lb. Is it possible to break the line while reeling in a 15-lb fish? Explain.

15. Two blocks rest on slopes of unequal angle, connected by a rope passing over a pulley (Fig. 5–36). If the blocks have equal masses, will they remain at rest? Why? Neglect friction.

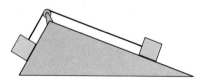

Fig. 5–36   Question 15

16. Why do a car's front brakes usually wear out before the rear brakes?

17. The dominant aerodynamic force on an airplane is the force of air approximately perpendicular to the wings. Why do airplanes bank when turning?

18. Someone once tried to describe the orbital motion of a satellite by saying that "the gravitational force balances the force mv^2/r." Criticize this statement.

19. In what sense is the force of friction between two surfaces independent of their contacting surface areas? In what sense does the force depend on surface area?

20. On landing, jet aircraft brake by reversing their engine thrust. Is the stopping distance affected by the coefficient of friction between the plane's tires and the runway?

21. If surfaces are very highly polished, friction actually increases. Why might this be?

22. A car rounds a banked turn at a speed lower than the design speed. Which way does the frictional force point?

23. For safety, automobile racetracks are banked as shown in Fig. 5–37. Explain why this improves the safety of the track.

Fig. 5–37   Banking of an automobile racetrack. Inside edge of turn is to left (Question 23).

24. In contrast to the automobile racetrack of the previous question, well-designed running tracks are sometimes banked as shown in Fig. 5–38. Explain why, neglecting safety considerations, this is a more appropriate banking scheme.

Fig. 5–38   Banking of a running track. Inside edge of turn is to left (Question 24).

25. Would the difference between apparent weights at the equator and pole be measurable with a pan balance? Why or why not?

PROBLEMS

Section 5.1   *Using Newton's Second Law*

1. A 20-kg fish at the end of a fishing line is being lifted vertically into a boat. When its acceleration reaches 1.2 m/s^2, the line breaks and the fish goes free. What is the maximum allowable tension in the line?

2. At what angle should you tilt an air table to simulate motion on the moon's surface, where $g = 1.6$ m/s^2?

3. A tow truck is connected to a 1400-kg car by a cable that makes a 25° angle to the horizontal, as shown in Fig. 5–39. If the truck accelerates at 0.57 m/s^2, what is the magnitude of the cable tension? Neglect friction and the mass of the cable.

Fig. 5–39   Problem 3

4. A 2.0-kg mass and an unknown mass are on friction-less surfaces and are connected by a massless string over a frictionless pivot, as shown in Fig. 5–40. If the 2.0-kg mass accelerates downslope at 1.8 m/s^2, what is the unknown mass?

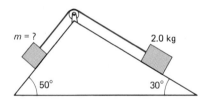

Fig. 5–40   Problem 4

5. Your 12-kg baby sister is pulling on the bottom of the tablecloth with all her weight. In the middle of the table, 60 cm from each edge, is a 6.8-kg roast turkey. (a) What is the acceleration of the turkey? (b) From the time she starts pulling, how long do you have to intervene before the turkey goes over the edge of the table?

6. Find expressions for the acceleration of the blocks in Fig. 5–41, where the string is fastened securely to the ceiling. Neglect friction and assume that the masses of pulley and string are negligible.

7. A block is launched up a frictionless ramp that makes an angle of 35° to the horizontal. If the block's initial speed is 2.2 m/s, how far up the ramp does it slide?

8. A 1300-kg car is being pulled up a 20° incline using a tow rope that makes a 30° angle with the incline, as shown in Fig. 5–42. If the car moves with constant speed, what is the tension in the rope? Neglect friction.

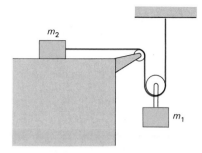

Fig. 5–41   Problem 6

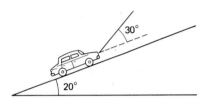

Fig. 5–42   Problem 8

9. A skier starts from rest at the top of a 24° slope 1.3 km long. Neglecting friction, how long does it take to reach the bottom?

10. A 73,200-kg space shuttle is in circular orbit at an altitude of 250 km above earth, where the acceleration of gravity is 0.93 times its surface value. With what magnitude and direction should rocket thrust be applied to the shuttle so that it begins moving in a straight line tangent to its orbit?

11. At the start of a race, a 70-kg swimmer pushes off the starting block with a force of 950 N directed at 15° below the horizontal. (a) What is the swimmer's horizontal acceleration? (b) If the swimmer is in contact with the starting block for 0.29 s, what is the horizontal component of his velocity when he hits the water?

Section 5.2   *Friction*

12. Eight rugby players push on a scrum machine—a device with vertically mounted pads for pushing and a horizontal surface where additional players stand to increase the total weight. The scrum machine has a mass of 70 kg, and holds eight more 80-kg players. The coefficient of friction between the scrum machine and ground is 0.78. If the pushing players push the scrum machine at constant speed, what total force do they exert?

13. The handle of a 22-kg lawnmower makes a 35° angle with the horizontal. If the coefficient of friction between lawnmower and ground is 0.68, what magnitude of force is required to push the mower at constant speed? Assume the force is applied in the

direction of the handle. Compare with the mower's weight.

14. The coefficient of friction between a suction cup and window glass is 1.3. If the suction cup is mounted on a vertical window, and if the net force of air on the cup is 85 N, how much mass can the cup support?

15. During an ice storm, the coefficients of friction between car tires and road are reduced to $\mu_k = 0.088$ and $\mu_s = 0.14$. (a) What is the maximum slope on which a car can be parked without sliding? (b) On a slope just steeper than this maximum, with what acceleration will a car slide down the slope?

16. A bat crashes into the vertical front of an accelerating subway train. If the coefficient of friction between bat and train is 0.86, what is the minimum acceleration of the train that will allow the bat to remain in place?

17. In a factory, boxes drop vertically onto a conveyor belt moving horizontally at 1.7 m/s. If the coefficient of kinetic friction is 0.46, how long does it take each box to come to rest with respect to the belt?

18. The coefficient of static friction between steel train wheels and steel rails is 0.58. The engineer of a train moving at 140 km/h spots a stalled car on the tracks 150 m ahead. If he applies the brakes so that the wheels do not slip, will the train stop in time?

19. If you neglect to fasten your seatbelt, and if the coefficient of friction between you and your car seat is 0.42, what is the maximum deceleration for which you can remain in your seat? Compare with the deceleration in an accident that brings a 60-km/h car to rest in a distance of 1.6 m.

20. A 310-g paperback book rests on a 1.2-kg textbook. A force is applied to the textbook, and the two books accelerate together from rest to 96 cm/s in 0.42 s. The textbook is then brought to a stop in 0.33 s, during which time the paperback slides off. Within what range does the coefficient of static friction between the two books lie?

21. A 2.5-kg block and a 3.1-kg block slide down a 30° incline as shown in Fig. 5–43. The coefficient of kinetic friction between the 2.5-kg block and the slope is 0.23; between the 3.1-kg block and the slope it is 0.51. Determine the (a) acceleration of the pair and (b) force the lighter block exerts on the heavier one.

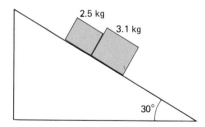

Fig. 5–43 Problem 21

22. Children sled down a 41-m-long hill inclined at 25°. At the bottom the slope levels out. If the coefficient of friction is 0.12, how far do the children slide on the level?

23. In a typical front-wheel-drive car, 70 per cent of the car's weight rides on the front wheels. If the coefficient of friction between tires and road is 0.61, what is the maximum acceleration of the car?

24. Repeat the previous problem for a rear-wheel-drive car with the same portion of its weight over the front wheels.

25. A police officer investigating an accident estimates from the damage done that a moving car hit a stationary car at 25 km/h. If the moving car left skid marks 47 m long, and if the coefficient of kinetic friction is 0.71, what was the initial speed of the moving car?

26. A skier finds she must give herself a push to get started on slopes of less than 8°. What is the coefficient of static friction?

27. Starting from rest, the skier of the previous problem traverses a 1.8-km-long trail in 65 s. If the trail is inclined at 12°, what is the coefficient of kinetic friction?

28. A slide inclined at 35° takes bathers into a swimming pool. With water sprayed onto the slide to make it essentially frictionless, a bather spends only one-third as much time on the slide as when it is dry. What is the coefficient of friction on the dry slide?

29. Planes operated by USAir have tables at their forward-most seats. (a) Sitting in the rear-facing seat at the forward side of the table, I place a book on the table during take-off. If the coefficient of static friction is 0.33, and if the plane accelerates at 2.4 m/s² during its take-off roll, will my book stay in place? (b) Once the plane is airborne, it heads upward at an 18° angle but with constant speed. Now will the book stay in place?

30. You try to push a heavy trunk, exerting a force at an angle of 50° below the horizontal (Fig. 5–44). Show that, no matter how hard you try to push, it is impossible to budge the trunk if the coefficient of static friction exceeds 0.84.

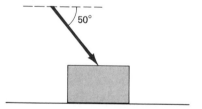

Fig. 5–44 Problem 30

31. A block of mass m is being pulled at constant speed v down a slope that makes an angle θ with the horizontal. The pulling force is applied through a horizontal rope, as shown in Fig. 5–45. If the coefficient of kinetic friction is μ_k, what is the rope tension?

32. A block is shoved down a 22° slope with an initial speed of 1.4 m/s. If it slides 34 cm before stopping, what is the coefficient of friction?

33. If the block in the previous problem were shoved up the slope with the same initial speed, (a) how far

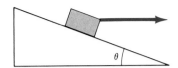

Fig. 5–45 Problem 31

would it go? (b) Once it stopped, would it slide back down?

Section 5.3 *Circular Motion*

34. Suppose the moon were held in its orbit not by gravity but by the tension in a massless cable. Estimate the magnitude of the cable tension. (See Appendix F for relevant data.)

35. Show that the force needed to keep a mass m in a circular path of radius r with period T is $4\pi^2 mr/T^2$.

36. A mass m_1 undergoes circular motion of radius R on a horizontal frictionless table, connected by a massless string through a hole in the table to a second mass m_2 (Fig. 5–46). If m_2 is stationary, find (a) the tension in the string and (b) the period of the circular motion.

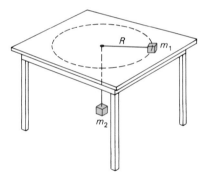

Fig. 5–46 Problem 36

37. A bug starts walking outward from the center of a $33\frac{1}{3}$-rpm record. If the coefficient of friction between bug and record is 0.15, how far does the bug get before it starts to slip?

38. A 340-g rock is whirled in a horizontal circle at the end of a 1.3-m-long string. If the breaking strength of the string is 250 N, what is the maximum speed of the rock? Assume that the string is very nearly horizontal at this speed. (How do you know that this assumption is reasonable?)

39. If the rock of the previous problem is whirled in a vertical circle, what is the minimum speed necessary at the top of the circle in order that the string remain taut?

40. A subway train rounds an unbanked curve at 67 km/h. A passenger hanging onto a strap notices that an adjacent unused strap makes an angle of 15° to the vertical. What is the radius of the turn?

41. An Olympic hammer thrower whirls a 7.3-kg hammer on the end of a 120-cm chain. If the chain makes a 10° angle with the horizontal, what is the speed of the hammer?

42. A Volvo model 240 sedan has a minimum turning radius of 9.8 m. If the coefficient of static friction between tires and road is 0.81, what is the maximum speed the car can have without skidding if the steering wheel is turned fully to the right?

43. A drum major whirls a ball of mass m in a vertical circle at the end of a massless rope of length ℓ. The ball's speed is constant, and is just fast enough to keep the rope always taut. If the drum major releases the rope when the ball is headed vertically upward, to what height h does it rise above its release point?

44. The world's largest Ferris wheel, at the Tsukuba Exposition in Japan, is 100 m in diameter and rotates once every 15 min. By what percentage would your weight in the frame of reference of the rotating wheel differ from your weight in earth's frame of reference when you are (a) at the top and (b) at the bottom of the wheel?

45. A space station is in the shape of a hollow ring with an outer diameter of 150 m. How fast should it rotate to simulate earth's surface gravity—that is, so that the force exerted on an object by the outside wall is equal to the object's weight on earth's surface?

46. A 45-kg skater rounds a 5.0-m-radius turn at 6.3 m/s. (a) What are the horizontal and vertical components of the force the ice exerts on her skate blades? (b) At what angle can she lean without falling over?

47. An indoor running track is square-shaped with rounded corners; each corner has a radius of 6.5 m on its inside edge. The track includes six 1.0-m-wide lanes. What should be the banking angles on (a) the innermost and (b) the outermost lanes if the design speed of the track is 24 km/h?

48. A jetliner flying horizontally at 850 km/h banks at 32° to make a turn. What is the radius of the turn? Note: the main aerodynamic force on an airplane is normal to the wing surfaces.

49. A 550-g rock is whirled in a horizontal circle at the end of a string 80 cm long. If the period of the circular motion is 1.6 s, what angle does the string make with the vertical?

50. A bucket of water is whirled in a vertical circle of radius 85 cm. What is the minimum speed that will keep the water from falling out?

51. A child is sitting on the edge of a merry-go-round 2.8 m in diameter. When the speed of the merry-go-round reaches one revolution in 3.2 s, the child slides off. What is the coefficient of static friction?

52. Riders on the "Great American Revolution" loop-the-loop roller coaster of Example 5–13 wear seatbelts as the roller coaster negotiates its 6.3-m-radius loop with a speed of 9.7 m/s. At the top of the loop, what are the magnitude and direction of the force exerted on a 60-kg rider (a) by the roller-coaster seat and (b) by the seatbelt? (c) What would happen if the rider unbuckled at this point?

53. A 1200-kg car drives on the country road shown in Fig. 5–47. The radius of curvature of the crests and dips is 31 m. (a) What is the maximum speed at which the car will always maintain contact with the road? (b) If the maximum rated load for the car is 450 kg, and if it carries four 65-kg people, what is the maximum speed for which it will not exceed its rated load?

Fig. 5–47   Problem 53

54. The Tethered Satellite System (TSS) is a proposed NASA experiment consisting of a 500-kg satellite connected to the space shuttle by a 20-km long cable whose mass is negligible compared with the satellite's mass. Suppose the shuttle is placed in a 250-km high circular orbit, where the acceleration of gravity is 0.926 times its value at earth's surface. The TSS hangs vertically on its tether (Fig. 5–48), and at its 230-km altitude the acceleration of gravity is 0.932 times its surface value. What is the tension in the cable? Assume that the shuttle's orbit is not significantly affected by the TSS, but that the TSS is forced by the cable to have the same orbital period as the shuttle. (Why is this assumption reasonable?)

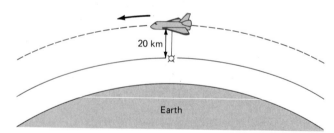

Fig. 5–48   NASA's Tethered Satellite System (Problem 54).

55. A block of mass m slides in a circle around the inside of a frictionless cone, as shown in Fig. 5–49. Find an expression for the speed of the block, if it is a vertical distance h above the apex of the cone.

Supplementary Problems

56. An astronaut is training in a centrifuge that consists of a small chamber whirled around horizontally at the end of a 5.1-m-long shaft. The astronaut places a notebook on the vertical wall of the chamber and it stays in place. If the coefficient of static friction is 0.62, what is the minimum rate at which the centrifuge must be revolving?

57. A block of mass m is given an initial speed v_0 on a horizontal table of height h. It slides a distance x_1

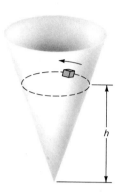

Fig. 5–49   Problem 55

across the table and lands on the floor a horizontal distance x_2 from the edge of the table. Find an expression for the coefficient of friction between block and table.

58. Driving in thick fog on a horizontal road, a driver spots a tractor-trailer truck jackknifed across the road, as in Fig. 5–50. To avert a collision, the driver could brake to a stop or swerve in a circular arc, as suggested in Fig. 5–50. Which offers the greater margin of safety? Assume that the same coefficient of static friction is operative in both cases, and that the car maintains constant speed if it swerves.

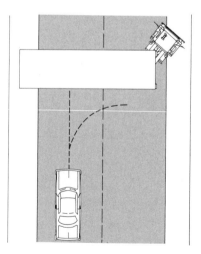

Fig. 5–50   Problem 58

59. A highway turn of radius R is banked for a design speed v_d. If a car enters the turn at speed $v = v_d + \Delta v$, where Δv can be positive or negative, show that the minimum coefficient of static friction needed to prevent slipping is

$$\mu_s = \frac{|\Delta v|}{gR} (2v_d + \Delta v).$$

Show that your result is consistent with Example 5–6 for $\Delta v = -v_d$—that is, for a car at rest on the banked highway.

60. Suppose the coefficient of friction between a block and a horizontal surface is proportional to the block's speed: $\mu = \mu_1 v/v_1$, where μ_1 and v_1 are constants. If the block is given an initial speed v_0, show that it comes to rest in a distance $x = v_0 v_1/\mu g$.

61. A block is projected up an incline making an angle θ with the horizontal. It returns to its initial position with half its initial speed. Show that the coefficient of friction is $\mu = \frac{3}{5}\tan\theta$.

62. A conical pendulum consists of a mass m whirled in a horizontal circle at the end of a massless string of length ℓ. Show that in the limit when the string is nearly vertical, the period of the circular motion is $2\pi\sqrt{\ell/g}$.

63. A 2.1-kg mass is connected to a spring of spring constant $k = 150$ N/m and unstretched length 18 cm. The pair are mounted on a frictionless air table, with the free end of the spring attached to a frictionless pivot. The mass is set into circular motion at 1.4 m/s. Find the radius of its path.

64. The victim of a political kidnapping is forced into a north-facing car and then blindfolded. The car pulls into traffic and, from the sound of the surrounding traffic, the victim knows that the car is moving at about the legal speed limit of 85 km/h. The car then turns to the right; the victim estimates that the force the seat exerts on him is one-fifth of his weight. The victim experiences this force for 28 s. At the end of that time, in what direction can the victim conclude that he is heading?

65. A mass m is suspended from a spring scale at latitude θ on earth's surface. (a) Show that the spring and mass make an angle ϕ with the radial direction to earth's center, where ϕ in radians is given by

$$\phi = \frac{\pi}{2} - \theta - \tan^{-1}\left[\cot\theta\left(1 - \frac{4\pi^2 R_e}{gT^2}\right)\right],$$

with R_e the earth's radius and T its rotation period. (b) Show that the weight read by the spring scale is

$$F_s = \frac{mg\cos\theta}{\sin\left(\frac{\pi}{2} - \theta - \phi\right)}.$$

(c) Show that the equations you write in deriving these results are consistent with $\phi = 0$ at poles and equator, and that in these cases the spring scale force is as given in Example 5–12. (d) Evaluate ϕ and F_s for the case when you weigh yourself at your present latitude. Assume throughout that g is constant.

6

WORK, ENERGY, AND POWER

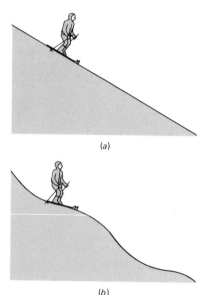

Fig. 6–1
Skier on (a) a uniform slope and (b) a nonuniform slope.

In Fig. 6–1a, a skier starts from rest at the top of a uniform slope. What is the skier's speed at the bottom? You could easily solve this problem by applying Newton's second law to find the acceleration, then using the appropriate constant acceleration equation to find the speed. But what about the skier in Fig. 6–1b? Here the slope is continuously changing, and so is the acceleration. Constant acceleration equations do not apply, so that solving for the details of the skier's motion would prove very difficult. Nevertheless, by introducing two important new concepts—work and energy—we will see how to "shortcut" the detailed application of Newton's law to arrive very easily at the answers to this and many other practical problems.

6.1
WORK

work

We all have an intuitive sense of the term "work." Carrying a piece of furniture upstairs involves work. The heavier the furniture, or the higher the stairs, the greater the work we must do. Pushing a stalled car involves work. Again, the heavier the car or the farther we push it, the more work we must do. Our precise definition of work reflects this intuition. For an object moving in one dimension—say, along the x-axis—the **work** W done *on* the object *by* an applied force $\mathbf{F}$ is

$$W = F_x \Delta x, \tag{6–1}$$

where F_x is the component of the force in the direction of the object's motion and where Δx is the distance moved. The force $\mathbf{F}$ need not be the net force; if we are interested, for example, in how much work *we* must do to drag a heavy box across the floor, then $\mathbf{F}$ is the force *we apply* and W is the work *we do*.

124

joule

foot-pound

Fig. 6–2
James Joule.

Fig. 6–3
When pushing in the direction of the car's motion, the person does work equal to the force he applies times the distance the car moves.

Fig. 6–4
Here only the component of force in the direction of motion contributes to the work.

Equation 6–1 shows that the SI unit of work is the newton-meter (N·m). One newton-meter is given the name **joule**, in honor of the nineteenth-century British physicist and brewer James Joule (Fig. 6–2). Since a newton is a kilogram-meter/second^2, the units of work can also be expressed as kg·m^2/s^2. In the English system, where the unit of force is the pound and the unit of distance is the foot, work is measured in **foot-pounds** (ft·lb).

Equation 6–1 shows that the person pushing the car in Fig. 6–3 does work equal to the force he applies times the distance the car moves. But the person pulling the suitcase in Fig. 6–4 does work equal only to the horizontal component of the force she applies times the distance the suitcase moves.

Furthermore, by our definition, the waiter of Fig. 6–5 does no work on the tray. Why not? Because the direction of the force he applies is perpendicular to the horizontal motion of the tray; there is no component of force along the direction of motion. Similarly, the weightlifter of Fig. 6–6 does no work while he holds the barbell steady over his head. Now you know both the waiter and weightlifter get tired. What is going on is that their muscles are undergoing minuscule contractions and lengthenings, so that on a small scale the muscles are doing work. But according to our definition, no work is being done on the tray or barbell. Similarly, a normal force from the road acts to support a moving car against gravity. But because that force is always perpendicular to the motion, it does no work on the car. Only the component of force in the direction of motion contributes to the work.

Example 6–1

Pushing a Trunk

You push a 75-kg trunk 6.4 m across a floor, exerting a horizontal force of 540 N. How much work do you do?

Solution

Since the force is in the direction of motion, Equation 6–1 gives simply

$$W = F\Delta x = (540 \text{ N})(6.4 \text{ m}) = 3500 \text{ N·m} = 3500 \text{ J}.$$

If the trunk is pushed at constant speed, then this work is expended to overcome the frictional force.

Example 6–2

A Garage Lift

A garage lift raises a 1300-kg car a vertical distance of 2.3 m at constant speed. How much work does it do?

Solution

To find the work, we need the force the lift exerts on the car. Since the car is raised at constant speed, it is not accelerating and therefore the *net* force on it is zero. The two forces acting are the upward lift force and the downward force of gravity; with no acceleration, they have the same magnitude, mg. The lift force is in the same direction as the motion, so that the work done by the lift is

$$W = F\Delta x = mg\Delta x = (1300 \text{ kg})(9.8 \text{ m/s}^2)(2.3 \text{ m})$$
$$= 2.9 \times 10^4 \text{ kg·m}^2/\text{s}^2 = 2.9 \times 10^4 \text{ J}.$$

Fig. 6–5
The waiter applies a vertical force to the tray; since this is perpendicular to the tray's horizontal motion, the waiter does no work on the tray.

Fig. 6–6
This weightlifter, strain though he may, is not doing work in the sense the word is used in physics.

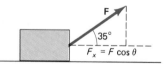

Fig. 6–7
Example 6–3

In the previous two examples, the force we were interested in was in the same direction as the motion. When that is not the case, we must first determine the component of force along the direction of motion, as the next example shows.

Example 6–3
Pulling a Suitcase

An airline passenger pulls a wheeled suitcase, exerting a 60-N force at a 35° angle to the horizontal. How much work does she do in pulling the suitcase 45 m from a taxi to the check-in counter?

Solution

Figure 6–7 shows that the horizontal component of the force is $F_x = F \cos\theta$, so that the work is

$$W = F_x \Delta x = F \cos\theta \, \Delta x = (60 \text{ N})(\cos 35°)(45 \text{ m}) = 2200 \text{ J}.$$

Work can be positive or negative. When a force acts in the same general direction as the motion, the work it does is positive. Acting at 90° to the direction of motion, a force does no work. And when a force acts to oppose motion, it does negative work (Fig. 6–8).

Example 6–4
Stopping a Car

The coefficient of kinetic friction between a wet, horizontal road and the wheels of a skidding car is 0.21. The mass of the car is 1500 kg. How much work is done on the car by the frictional force if the car comes to a stop in 54 m?

Solution

To find the work, we need to know the frictional force. On a horizontal road, the normal force balances gravity, so that $N = mg$. Then the magnitude of the frictional force is $\mu_k N = \mu_k mg$. The direction of this force is opposite the car's motion. If we take the positive x-direction in the direction of the car's motion, then the component of force along this direction is negative: $F_x = -\mu mg$. So the work done by friction is

$$W = F_x \Delta x = (-\mu_k mg)(\Delta x) = -(0.21)(1500 \text{ kg})(9.8 \text{ m/s}^2)(54 \text{ m})$$
$$= -1.7 \times 10^5 \text{ kg·m}^2/\text{s}^2 = -1.7 \times 10^5 \text{ J}.$$

6.2
WORK AND THE SCALAR PRODUCT

Work is a *scalar* quantity: it is specified completely by a single number, and has no direction associated with it. But Fig. 6–8 shows clearly that work involves a relation between two *vectors*—the force vector **F** and the displacement vector **Δr**. If θ is the angle between those two vectors, then the component of the force along the direction of motion is $F \cos\theta$. If the length of the displacement vector is Δr, then the work is simply

$$W = (F \cos\theta)(\Delta r) = F \Delta r \cos\theta. \qquad (6\text{–}2)$$

(Equation 6–2 is really the same as our definition 6–1, written in slightly more general terms. If we choose the x-axis along the displacement vector $\Delta \mathbf{r}$, then $F_x = F \cos\theta$ and $\Delta r = \Delta x$, and we recover Equation 6–1.)

Equation 6–2 shows that the work is the product of the magnitudes of the vectors $\mathbf{F}$ and $\Delta \mathbf{r}$ and the cosine of the angle between them. This combination—the product of vector magnitudes with the cosine of the angle between the vectors—occurs so often in physics and mathematics that it is given a special name: the **scalar product** of two vectors. If $\mathbf{A}$ and $\mathbf{B}$ are any two vectors, their scalar product is defined as

scalar product

$$\mathbf{A}\cdot\mathbf{B} = AB \cos\theta, \qquad (6\text{–}3)$$

where A and B are the magnitudes of the vectors and θ the angle between them. Note that we use a centered dot to designate the scalar product; for this reason the scalar product is often called the **dot product.** Although the scalar product $\mathbf{A}\cdot\mathbf{B}$ is formed from two vectors, the product itself is a scalar.

dot product

Since scalar multiplication is commutative, we can rearrange terms on the right-hand side of Equation 6–3 to get $\mathbf{A}\cdot\mathbf{B} = BA\cos\theta$. But $BA\cos\theta$ is, by definition, the scalar product $\mathbf{B}\cdot\mathbf{A}$, so that the scalar product is itself commutative: $\mathbf{A}\cdot\mathbf{B} = \mathbf{B}\cdot\mathbf{A}$. Problems 11 and 12 ask you to confirm that the scalar product is also distributive:

$$\mathbf{A}\cdot(\mathbf{B} + \mathbf{C}) = \mathbf{A}\cdot\mathbf{B} + \mathbf{A}\cdot\mathbf{C}$$

and that it has the property

$$(c\mathbf{A})\cdot\mathbf{B} = \mathbf{A}\cdot(c\mathbf{B}),$$

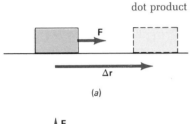

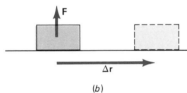

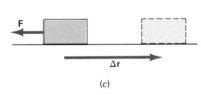

where c is a scalar. (In Chapter 11 we will define a way of multiplying vectors to form another vector; we will find that that operation is not commutative.)

Comparing Equation 6–2 with the definition 6–3 of the scalar product, we see immediately that work is given by

$$W = \mathbf{F}\cdot\Delta\mathbf{r}. \qquad (6\text{–}4)$$

This vector notation expresses the work in a compact way that accounts for both direction and magnitude of the vectors involved.

The forms 6–3 and 6–4 are convenient when we are given the magnitudes and directions of both vectors. But often vectors are expressed in unit vector notation. In that case, we can apply the distributive law to derive a simple form for the scalar product. Let $\mathbf{A} = A_x\hat{\mathbf{i}} + A_y\hat{\mathbf{j}} + A_z\hat{\mathbf{k}}$ and $\mathbf{B} = B_x\hat{\mathbf{i}} + B_y\hat{\mathbf{j}} + B_z\hat{\mathbf{k}}$. Then the scalar product $\mathbf{A}\cdot\mathbf{B}$ is

$$\mathbf{A}\cdot\mathbf{B} = (A_x\hat{\mathbf{i}} + A_y\hat{\mathbf{j}} + A_z\hat{\mathbf{k}})\cdot(B_x\hat{\mathbf{i}} + B_y\hat{\mathbf{j}} + B_z\hat{\mathbf{k}}).$$

Multiplying this out would give a sum of nine terms. The first term, for example, is $A_x\hat{\mathbf{i}}\cdot B_x\hat{\mathbf{i}}$. This can be written $A_x B_x\hat{\mathbf{i}}\cdot\hat{\mathbf{i}}$. But what is the scalar product $\hat{\mathbf{i}}\cdot\hat{\mathbf{i}}$? The vector $\hat{\mathbf{i}}$ is a *unit* vector—it has length 1—and the angle between any vector and itself is zero, so that

Fig. 6–8
(a) A force applied in the same general direction as the motion does positive work. (b) A force acting at right angles to the motion—like the normal force of the floor—does no work. (c) A force acting to oppose the motion—like the frictional force—does negative work.

$$\hat{\mathbf{i}} \cdot \hat{\mathbf{i}} = (1)(1)(\cos 0°) = 1.$$

Thus the term $A_x \hat{\mathbf{i}} \cdot B_x \hat{\mathbf{i}}$ becomes $A_x B_x$. Clearly there are two similar terms, $A_y B_y$ and $A_z B_z$. What about the other six terms? They involve expressions like $A_x \hat{\mathbf{i}} \cdot B_y \hat{\mathbf{j}} = A_x B_y \hat{\mathbf{i}} \cdot \hat{\mathbf{j}}$. But the different unit vectors are perpendicular, so $\hat{\mathbf{i}} \cdot \hat{\mathbf{j}} = 0$ and the six "cross terms" vanish. Then the scalar product becomes simply

$$\mathbf{A} \cdot \mathbf{B} = A_x B_x + A_y B_y + A_z B_z. \tag{6-5}$$

Since Equation 6–5 follows directly from the definition 6–3, either form may be used in computing the scalar product of two vectors.

Example 6–5

Work and Unit Vectors

A force $\mathbf{F} = 33\hat{\mathbf{i}} + 14\hat{\mathbf{j}}$ N acts to move an object on a straight path from the origin to the point $x = 4.0$ m, $y = 2.0$ m. How much work is done by the force?

Solution

The work is given by Equation 6–4: $W = \mathbf{F} \cdot \Delta\mathbf{r}$. Here the displacement vector from the origin to the point (4.0 m, 2.0 m) is $\Delta\mathbf{r} = 4.0\hat{\mathbf{i}} + 2.0\hat{\mathbf{j}}$ m, so that Equation 6–5 gives immediately

$$W = \mathbf{F} \cdot \Delta\mathbf{r} = F_x \Delta x + F_y \Delta y = (33 \text{ N})(4.0 \text{ m}) + (14 \text{ N})(2.0 \text{ m}) = 160 \text{ J},$$

where Δx and Δy are the components of the displacement vector.

Equation 6–5 provides the best approach here since the vectors are given in unit vector notation. If you insist on using Equation 6–3, you must first compute the magnitudes of both vectors and the angle between them. Problem 15 explores this approach.

6.3

A VARYING FORCE

Often the force applied to an object varies with position. Especially important examples include the fundamental forces of nature; the electrical and gravitational forces, for example, vary as the inverse square of the distance between interacting objects. The force of a spring that we encountered in Chapter 4 provides another example; as the spring stretches, the force increases.

How are we to calculate the work done by a varying force? We consider first the case of one-dimensional motion, with the force in the same direction as the motion. Figure 6–9 is a plot of a force F that varies with position x. We would like to find the work done by the force as an object moves from x_1 to x_2. We can't simply write $F(x_2 - x_1)$, since the force varies so that there is no single value for F. What we can do, though, is to divide the region into a series of small rectangles of width Δx, as shown in Fig. 6–10. If we make the width Δx small enough, the force will be nearly constant over the width of each rectangle. Then the work ΔW done in moving the width Δx of one such rectangle is approximately $F(x)\Delta x$, where $F(x)$ is the

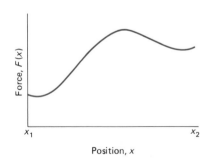

Fig. 6–9
A force $F(x)$ that varies with position x.

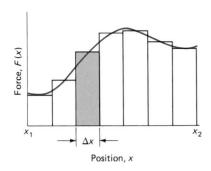

Fig. 6–10
The work done by the varying force is the area under the force versus position curve, and is approximated as the sum of the areas of many narrow rectangles.

definite integral

magnitude of the force at the midpoint x of that rectangle; we write $F(x)$ to show explicitly that the force is a function of position. Note that the quantity $F(x)\Delta x$ is simply the area of the rectangle, expressed in the appropriate units (N·m, or J). Suppose there are N rectangles. Let x_i be the midpoint of the ith rectangle. Then the total work done in moving from x_1 to x_2 is given approximately by the sum of the individual amounts of work ΔW_i associated with each rectangle, or

$$W \simeq \sum_{i=1}^{N} W_i = \sum_{i=1}^{N} F(x_i)\Delta x. \qquad (6\text{–}6)$$

How good is this approximation? That depends on how small we make the rectangles. Suppose we let them get arbitrarily small. Then the number of rectangles must grow arbitrarily large. In the limit of infinitely many infinitesimally small rectangles, the approximation in Equation 6–6 becomes exact. Then we have

$$W = \lim_{\Delta x \to 0} \sum_i F(x_i)\Delta x, \qquad (6\text{–}7)$$

where the sum is over all the infinitesimal rectangles between x_1 and x_2. The sum of infinitely many infinitesimal quantities occurs frequently in physics and mathematics, and is called a **definite integral.** The quantity on the right-hand side of Equation 6–7 is the definite integral of the function $F(x)$ over the interval from x_1 to x_2. We introduce a special symbolism for the limiting process of Equation 6–7, writing

$$W = \int_{x_1}^{x_2} F(x)\, dx. \quad \left(\begin{array}{l}\text{work done by a varying}\\\text{force in one dimension}\end{array}\right) \qquad (6\text{–}8)$$

What does expression 6–8 mean? It means, by definition, exactly the same thing as Equation 6–7: it tells us to break the interval from x_1 to x_2 into many small rectangles of width Δx, to multiply the value of the function $F(x)$ at each rectangle by the width Δx, and to sum those products. As we take arbitrarily many arbitrarily small rectangles, the result of this process gives us the value of the definite integral. You can think of the symbol $\int$ in Equation 6–8 as standing for "sum," and of the symbol dx as a limiting case of arbitrarily small Δx. The definite integral has a simple geometrical interpretation: as the sum of the areas of the many rectangles, it is the area under the curve $F(x)$ between the limits x_1 and x_2.

How are we ever to evaluate the infinite sum that is implied in Equation 6–8? One way is to back off and take a sum over a large but finite number of small but not infinitesimally small rectangles. This method gives an approximation that can be very accurate if the rectangle width Δx is small enough. It is widely used in computer calculations and for integrals that cannot be handled in any other way.

Calculus provides us with a powerfully simple way to evaluate many integrals. In the box below, we show how and why this is so. If you are familiar with integrals from your calculus course, you can skip the box and move on to more physics. But even if you are comfortable with integrals, you may find the box useful in refreshing your sense of just what an integral means.

Box 6–1

Integrals and Derivatives

The limiting process defining the integral, and even the symbol dx in the expression 6–8, should remind you of our definition of the derivative in Chapter 2. Integration and differentiation are related processes—in fact, they are inverses. You either have proved or will soon prove this rigorously in your calculus course; here we give a graphical illustration of the relation between differentiation and integration.

Figure 6–11a again shows our force function $F(x)$, with the area under it divided into small rectangles. We define a new function, called $A(x)$, which gives the area under the $F(x)$ curve between x_1 and some arbitrary point x. That is,

$$A(x) = \int_{x_1}^{x} F(x)\, dx. \qquad (6\text{–}9)$$

(Setting $x = x_2$ in the upper limit of the integral gives the area from x_1 to x_2.) In Fig. 6–11b, we plot the function $A(x)$. At the end of the first interval in Fig. 6–11a, the total area under the curve is the area marked A_1; we plot the value of this area at the appropriate point in Fig. 6–11b. At the end of the second interval in Fig. 6–11b, the total area under the curve is the sum of the areas A_1 and A_2; we plot this area on Fig. 6–11b as well. We continue this process for the intervals of Fig. 6–11a, obtaining the points shown on Fig. 6–11b. We could fill in intermediate points by considering smaller intervals in Fig. 6–11a; the result is the smooth curve shown in Fig. 6–11b.

What is the *rate* at which the function $A(x)$ increases with increasing x? That depends on the *value* of the function $F(x)$ at the point x. If $F(x)$ were everywhere constant, $A(x)$ would increase at a steady rate (Fig. 6–12). What does all this mean? Simply this: the rate of change of the area $A(x)$ under a curve $F(x)$ is given by the value of $F(x)$. But, as we found in Chapter 2, the rate of change of a function is the *derivative* of that function. So we have

$$\frac{dA(x)}{dx} = F(x). \qquad (6\text{–}10)$$

Equations 6–9 and 6–10 are both relations between $A(x)$ and $F(x)$. Equation 6–9 tells us that $A(x)$ is the integral of $F(x)$; conversely, Equation 6–10 tells us that $F(x)$ is the derivative of $A(x)$. Differentiation and integration are therefore inverse operations.

In particular, the integral of $F(x)$ must be a function that, when differentiated, gives the function $F(x)$. Now there are many such functions; if $A(x)$ is one of them, so that $dA(x)/dx = F(x)$, then so is $A(x) +$ constant, since the derivative of any constant is zero. So how are we to choose? In calculus, we distinguish between the **indefinite integral** of $F(x)$, defined as any function $A(x)$ whose derivative is $F(x)$, and the **definite integral,** which we defined earlier as the area under the curve $F(x)$ between two specific limits. If we want the area between some point x_1 and another point x_2, then our result should reduce to zero when $x_2 = x_1$. We can ensure this by adding to $A(x)$ a constant equal to $-A(x_1)$. Then our definite integral is

$$\int_{x_1}^{x_2} F(x)\, dx = A(x_2) - A(x_1) \equiv A(x)\Big|_{x_1}^{x_2}, \qquad (6\text{–}11)$$

where the term $A(x)\Big|_{x_1}^{x_2}$ is a shorthand for $A(x_2) - A(x_1)$. The expression 6–11 has all the features we want: its derivative (with respect to x_2) is the function F, and it correctly gives zero when $x_2 = x_1$.

We learned in Chapter 2 that the derivative of the function x^n is nx^{n-1}. Since we have just shown that integration and differentiation are inverse relations, this means that x^n (plus any constant) is the indefinite integral of nx^{n-1}. That is, to integrate a power of x, we increase the exponent by one and divide by the new exponent. If we start with x^n, we end up, after integrating, with

$$\int x^n\, dx = \frac{x^{n+1}}{n+1} + \text{constant.} \qquad (6\text{–}12)$$

You can verify this equation by differentiating the right-hand side; the result is just x^n. Equation 6–12 gives the indefinite integral of x^n; the definite integral is given by applying Equation 6–11:

$$\int_{x_1}^{x_2} x^n\, dx = \frac{x^{n+1}}{n+1}\Big|_{x_1}^{x_2} = \frac{x_2^{n+1}}{n+1} - \frac{x_1^{n+1}}{n+1}, \qquad (6\text{–}13)$$

and is equal to the area under the curve x^n between x_1 and x_2. In your calculus course, you either have learned or will learn how to integrate other functions. Appendix B lists integrals of some common functions.

Example 6–6

Stretching a Spring

A spring of constant k is stretched a distance x_1 from its equilibrium position. How much work does this take?

Solution

We learned in Section 4.7 that an ideal spring exerts a force $-kx$, where x is the distance it has been stretched or compressed and where the minus sign indicates that the spring opposes the stretching or compression. To stretch a spring, we must exert a force $+kx$ to overcome the spring force. The work we do is then given by the definite integral in Equation 6–8:

$$W = \int_0^{x_1} kx \, dx = \left.\frac{kx^2}{2}\right|_0^{x_1} = \tfrac{1}{2}kx_1^2 - \tfrac{1}{2}k(0)^2 = \tfrac{1}{2}kx_1^2.$$

Since x_1 is quite arbitrary, we can drop the subscript and say that the work needed to stretch a spring of constant k a distance x from its equilibrium position is

$$W = \tfrac{1}{2}kx^2. \qquad \text{(work to stretch spring)} \qquad (6\text{–}14)$$

In this simple case, we can also solve the problem graphically. Figure 6–13 graphs the force kx we apply to stretch the spring. The work required is the area under the line $F = kx$ between $x = 0$ and $x = x_1$. Here the area is that of a triangle: one-half of the base times the height, or $(\tfrac{1}{2})(x_1)(kx_1) = \tfrac{1}{2}kx_1^2$, as before.

Notice that in this example we could not simply multiply the force kx_1 by the stretch x_1 to get kx_1^2 for the work. Why not? Because the force does not always have the value kx_1; rather, it increases steadily from zero. To evaluate the work, we must go through the procedure of adding up the small amounts of work done as we stretch the spring over small intervals, each of which has a *different* value of the force. The definite integral we used provides us the exact results of carrying out this procedure in the limit of arbitrarily small intervals. You will find integrals appearing in physics any time you are involved with quantities that vary continuously over an interval.

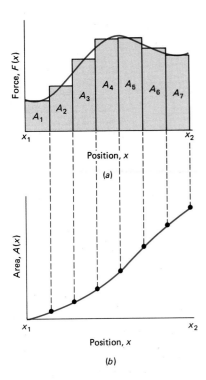

Fig. 6–11
(a) The force function $F(x)$, showing the area under the curve divided into small rectangles. (b) The function $A(x)$ is the area under the curve $F(x)$ from the point x_1 to an arbitrary point x.

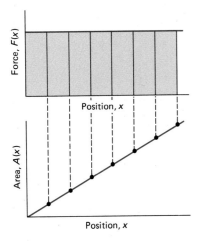

Fig. 6–12
With constant $F(x)$, the area $A(x)$ under the $F(x)$ curve increases at a constant rate; that rate depends on the *value* of the function F. The $F(x)$ curve is therefore the slope, or derivative, of the $A(x)$ curve.

Example 6–7
Rough Sliding

Workers pushing a 180-kg crate across a level floor encounter a 9.5-m-long region where the floor becomes increasingly rough; the coefficient of friction increases from 0.17 to 0.87, according to the expression

$$\mu_k = ax^2 - bx^3 + \mu_0,$$

where $a = 0.022 \text{ m}^{-2}$, $b = 1.5 \times 10^{-3} \text{ m}^{-3}$, $\mu_0 = 0.17$, and where x is the distance in meters from the beginning of the rough region. How much work does it take to push the crate at constant speed across this region?

Solution

Pushing at constant speed on a level floor, the workers must supply just enough force to balance the frictional force $\mu_k N$; here the normal force balances gravity, so that $N = mg$. Then the frictional force is $\mu_k mg = (ax^2 - bx^3 + \mu_0)mg$. As in the previous example, the force varies with position, so that we must effectively add the contributions from many small intervals over which we can consider the force roughly constant. Equation 6–8 expresses this procedure and Equation 6–13 lets us evaluate the result:

$$W = \int_{x_1}^{x_2} F(x) \, dx = \int_{x_1}^{x_2} mg\,(ax^2 - bx^3 + \mu_0) \, dx = mg\left.\left(\frac{ax^3}{3} - \frac{bx^4}{4} + \mu_0 x\right)\right|_{x_1}^{x_2}.$$

Here $x_1 = 0$, the start of the rough interval, and $x_2 = 9.5$ m. Putting in the values for a and b, we then have

$$W = mg\left(\frac{ax_2^3}{3} - \frac{bx_2^4}{4} + \mu_0 x_2\right) - mg\left(\frac{ax_1^3}{3} - \frac{bx_1^4}{4} + \mu_0 x_1\right)$$

$$= (180 \text{ kg})(9.8 \text{ m/s}^2)\left[\frac{(0.022 \text{ m}^{-2})(9.5 \text{ m})^3}{3} - \frac{(1.5 \times 10^{-3} \text{ m}^{-3})(9.5 \text{ m})^4}{4}\right.$$

$$\left. + (0.17)(9.5 \text{ m})\right] - 0 = 8.6 \times 10^3 \text{ J}.$$

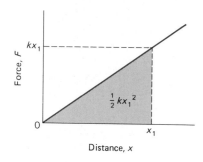

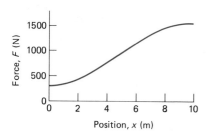

Fig. 6–13
The force required to stretch a spring increases linearly with distance x. The work done in stretching the spring an amount x_1 is the area under the force-distance curve, or $\frac{1}{2}kx_1^2$.

Fig. 6–14
Force versus position for Example 6–7.

Does this result make sense? It is well above the 2.8×10^3 J we get by calculating $\mu_k mg\Delta x$ with μ_k taking its value 0.17 at the beginning of the rough region. On the other hand, it's well below the 1.5×10^4 J we get using the value $\mu_k = 0.87$ appropriate to the end of the interval. Since the actual coefficient varies between these limits, we expect the work to lie between those extremes.

Figure 6–14 shows the force $\mu_k mg$ as a function of position. Unlike the previous example, we cannot calculate the work from the graph using a simple geometrical formula; to get the exact area under the curve we must do the integral. However, we can approximate the work by estimating the area graphically; Problem 24 explores this approach.

6.4
FORCE AND WORK IN THREE DIMENSIONS

What if a force varies not only in magnitude but also in direction? Or what if an object's path is not straight but curved? Then how do we calculate work? Our basic definition $W = \mathbf{F} \cdot \Delta \mathbf{r}$ provides the answer. Figure 6–15a shows the curved path of an object and several vectors of an applied force for which we would like to know the work. In this most general case, both the direction of the path and the direction and magnitude of the force vary. But if we magnify a small section $\Delta \mathbf{r}$ of the path (Fig. 6–15b), it will appear essentially straight and the force will not vary significantly over the segment. What if the path is sharply bent, or the force is changing rapidly? Then just make the segment smaller. Eventually it will be small enough to look approximately straight. Over this short segment, with the force a nearly constant quantity $\mathbf{F}$, the work ΔW is approximately

$$\Delta W = \mathbf{F} \cdot \Delta \mathbf{r}.$$

If we determine the work ΔW associated with each small segment of the path from some point $\mathbf{r}_1$ to another point $\mathbf{r}_2$, then the total work is given by

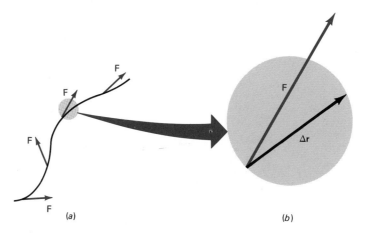

Fig. 6–15
(a) An object moves on a curved path. We want to know the work done by a force that may vary in direction and/or magnitude along the path. (b) A very small segment $\Delta \mathbf{r}$ of the path appears straight and the force does not vary significantly over that segment. The work involved in traversing the short segment is $\Delta W = \mathbf{F} \cdot \Delta \mathbf{r}$.

summing all the ΔW's:

$$W \simeq \sum \Delta W = \sum \mathbf{F} \cdot \Delta \mathbf{r}.$$

If we let the size of the segments grow arbitrarily small and their number increase without bound, this approximation becomes increasingly accurate. In the limit $\Delta \mathbf{r} \to 0$, the sum becomes an integral and we write

$$W = \int_{\mathbf{r}_1}^{\mathbf{r}_2} \mathbf{F} \cdot d\mathbf{r}, \tag{6-15}$$

line integral

where the limits $\mathbf{r}_1$ and $\mathbf{r}_2$ imply that the integral is taken over the specified path from point $\mathbf{r}_1$ to $\mathbf{r}_2$. An integral like that in Equation 6–15, involving the dot product of a vector with a particular path, is called a **line integral.** You will encounter line integrals again when you study electromagnetism.

What does the line integral 6–15 mean? Its meaning comes from the procedure through which we arrived at it: form the many scalar quantities $\mathbf{F} \cdot \Delta \mathbf{r}$ over the path of interest and sum them, taking the limit as $\Delta \mathbf{r}$ becomes arbitrarily small. Example 6–8 illustrates the use of a line integral.

Example 6–8

A Line Integral and the Gravitational Force

A car drives down the hill shown in Fig. 6–16a. How much work does the force of gravity do on the car as it drives from the top of the hill ($y = h$) to the bottom ($y = 0$)?

Solution

Figure 6–16b shows a small segment $\Delta \mathbf{r}$ of the path, which can be written $\Delta \mathbf{r} = \Delta x \hat{\imath} + \Delta y \hat{\jmath}$ in a coordinate system with x-axis horizontal and y-axis vertically upward. The force of gravity is always downward: $\mathbf{F}_g = -mg\hat{\jmath}$. Then the amount of work ΔW done by gravity as the car undergoes the displacement $\Delta \mathbf{r}$ is

$$\Delta W = \mathbf{F} \cdot \Delta \mathbf{r} = -mg\hat{\jmath} \cdot (\Delta x \hat{\imath} + \Delta y \hat{\jmath}) = -mg\Delta y.$$

Summing over the entire path and taking the limit as $\Delta y \to 0$ gives

$$W = \int_{y=h}^{y=0} (-mg \, dy) = -mgy \Big|_h^0 = -mg(0) - (-mgh) = mgh.$$

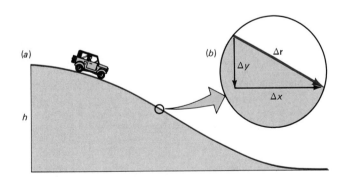

Fig. 6–16
(a) Hill of Example 6–8; (b) shows a small portion of the hill where the car undergoes displacement $\Delta \mathbf{r} = \Delta x \hat{\imath} + \Delta y \hat{\jmath}$.

In this case the details of the path don't matter at all; the work depends only on the vertical distance h. This illustrates a property of the gravitational force that we will explore further in the next chapter.

An interesting case involving work as a line integral is provided by uniform circular motion. Here the force **F** is always perpendicular to the displacement vector Δ**r**; the product **F**·Δ**r** is therefore always zero and no work is done, no matter how far the object moves in its circular path. Thus it takes no work to whirl a ball around on the end of a horizontal string; similarly, earth's gravity does no work on a satellite in a circular orbit.

6.5
KINETIC ENERGY

So far we have discussed only the work done by a particular force acting on an object. We didn't distinguish between the case when that was the only force acting—in which case the object must be accelerating—and the case where other forces act as well, perhaps giving zero net force. We did not need to make a distinction because the work done by the one force of interest is $\int$**F**·d**r** regardless of what other forces may be acting.

We now turn our attention to the *net* force on an object. The *total* work done *on* an object is given by summing the work associated with each individual force. Since the sum of the forces is the net force, this amounts to using the net force in our expression for work:

$$W_{total} = \int \mathbf{F}_{net} \cdot d\mathbf{r}. \tag{6–16}$$

For example, if you push a heavy crate across the floor at constant speed, you do work on the crate. But the frictional force does negative work of equal magnitude; the net force is zero and so is the total mechanical work done on the crate.

energy   Considering the total work in relation to the net force leads to the important concept of **energy**—a concept that will become increasingly vital as your understanding of physics develops. Here we consider the case of one-dimensional motion; the general case is readily treated by retaining all three terms that form the dot product in Equation 6–16. In one dimension, Equation 6–16 becomes

$$W = \int F \, dx,$$

where we have dropped the subscripts "total" and "net" because it is understood we are talking about total work and net force. But the net force can be written in terms of Newton's second law: $F = ma$, or $F = mdv/dt$, so that

$$W = \int m \frac{dv}{dt} dx. \tag{6–17}$$

The quantities dv, dt, and dx arose as the limits of small numbers Δv, Δt, and Δx. In your calculus class, you have by now shown that the limit of a product or quotient is just the product or quotient of the individual terms involved. For these reasons, we can rearrange the symbols dv, dt, and dx

to rewrite Equation 6–17 in the form

$$W = \int m \, dv \, \frac{dx}{dt}.$$

But $dx/dt = v$, so we have

$$W = \int mv \, dv.$$

The integral is simple; it is just like the form $\int x \, dx$, which we integrate by raising the exponent and dividing by the new exponent. What about the limits? Suppose our object starts out at some speed v_1 and ends up at v_2. Then we have

$$W = \int_{v_1}^{v_2} mv \, dv = \tfrac{1}{2}mv^2 \Big|_{v_1}^{v_2} = \tfrac{1}{2}mv_2^2 - \tfrac{1}{2}mv_1^2. \qquad (6\text{–}18)$$

kinetic energy

What is the significance of Equation 6–18? It shows that an object has associated with it a quantity $\tfrac{1}{2}mv^2$ that changes when, and only when, work is done on the object. The quantity $\tfrac{1}{2}mv^2$ is called the **kinetic energy,** K. Equation 6–18 shows that the change in an object's kinetic energy is equal to the total work done on the object; we often rewrite Equation 6–18 in the form

$$\Delta K = W, \qquad (6\text{–}19)$$

work-energy theorem

where ΔK is the change in kinetic energy. Equations 6–18 and 6–19 are equivalent forms of the **work-energy theorem.**

In particular, an object accelerated from rest to speed v experiences a change $\tfrac{1}{2}mv^2$ in kinetic energy; we then say that an object moving at speed v has kinetic energy $\tfrac{1}{2}mv^2$ by virtue of its motion. Like velocity, kinetic energy is a relative term; its value depends on the reference frame in which it is measured. Unlike velocity, though, kinetic energy is a scalar. Since it depends on the square of the speed, it is always positive.

What if we stop a moving object? Then the kinetic energy changes from $\tfrac{1}{2}mv^2$ to zero—a change of $-\tfrac{1}{2}mv^2$. So we must do negative work on the object to stop it—that is, we must apply a force directed oppositely to the motion. By Newton's third law, the object exerts an equal but oppositely directed force on us, therefore doing positive work $\tfrac{1}{2}mv^2$ on us. So an object of mass m moving at speed v can do work equal to its initial kinetic energy if it is brought to rest.

Example 6–9
Work-Energy Theorem

A biologist uses a spring-loaded gun to shoot tranquilizer darts into an elephant. The gun's spring has spring constant $k = 940$ N/m, and is compressed 25 cm from its equilibrium position before firing a 38-g tranquilizer dart. At what speed does the dart leave the gun?

Solution

From Example 6–6, we know that the work required to compress the spring is $\tfrac{1}{2}kx^2$. When the spring is released, it exhibits the same force-distance relation as when it was compressed, so it now does positive work $\tfrac{1}{2}kx^2$ on the dart. The work-energy theorem tells us that this is also the change in the dart's kinetic energy; since the dart starts from rest, we then have

$$\tfrac{1}{2}kx^2 = \tfrac{1}{2}mv^2,$$

so that

$$v = \sqrt{\frac{k}{m}}\, x = \left(\sqrt{\frac{940 \text{ N/m}}{0.038 \text{ kg}}}\right)(0.25 \text{ m}) = 39 \text{ m/s}.$$

Does this result make sense? Yes: in algebraic form, our answer $v = \sqrt{k/m}\, x$ shows that the stiffer the spring, the higher the speed; the greater the mass, the lower the speed, as we expect from Newton's second law. And the further we compress the spring, the greater the speed as well.

Notice how easy the work-energy theorem made this problem. Had we tried to calculate the speed using Newton's second law, we would have had a hard time because the force, and therefore the acceleration, is not constant.

As Example 6–9 shows, the work-energy theorem gives us a quick way to solve problems that might be difficult using Newton's second law. Of course, the work-energy theorem is itself an application of Newton's second law; we used Newton's law to replace the net force with ma in deriving that theorem. But in our calculus derivation of the work-energy theorem, we developed a shortcut way of applying Newton's law that bypasses the details we would have to work out if we applied the law directly.

6.6
ENERGY UNITS

Since work is equal to the change in kinetic energy, the units of kinetic energy are the same as those of work. In SI, the unit of kinetic energy is the joule, equal to one newton-meter.

erg

In the centimeter-gram-second (cgs) system of units still commonly used in several branches of physics, the unit of energy (and work) is the **erg,** equal to one dyne-centimeter (the dyne is the cgs force unit, and is equal to 10^{-5} N). One erg is equal to 10^{-7} J.

foot-pound

We have already seen that the unit of work in the English system is the **foot-pound.** This is therefore also the English unit of energy, and is equal to 1.356 J.

kilowatt-hour

When you pay your electric bill, you are charged for your energy usage in **kilowatt-hours** (kW·h or kWh). One kWh is the amount of energy consumed by a 100-watt light bulb burning for 10 hours and is equal to 3.6×10^6 J.

electron-volt

All these energy units are most useful on the macroscopic scale. In atomic physics and molecular chemistry, scientists often use the **electron-volt** (eV), an energy unit equal to 1.60×10^{-19} J. The electron-volt is based on the energy gained by an electron under specified circumstances that we will discuss in Chapter 23. Chemical reactions between individual atoms and molecules typically involve energies on the order of 1 eV. Nuclear reactions involve energies on the order of 10 MeV (1 MeV $= 10^6$ eV; see Table 1–1). The factor of 10 million between chemical and nuclear reactions shows why the invention of nuclear weapons so radically changed our world. Our largest particle accelerators, probing the fundamental structure of matter, accelerate elementary particles to energies measured in TeV (1 TeV $= 10^{12}$ eV). And physicists seeking a single unified force law governing all of physics consider elementary particle energies of 10^{24} eV—about the kinetic energy of a car moving at 50 km/h.

Appendix D contains an extensive table of energy conversion factors.

Example 6–10
Kinetic Energy

Evaluate and compare the kinetic energies of a 1200-kg car moving at 50 km/h, 100 km/h, and 150 km/h.

Solution

To work in SI units, we convert km/h to m/s; the conversion factor is 1 km/h $=(1000 \text{ m})/(3600 \text{ s})=0.278$ m/s, so that the speeds become 14 m/s, 28 m/s, and 42 m/s. Then our energies are

$$K_{50 \text{ km/h}} = \tfrac{1}{2}(1200 \text{ kg})(14 \text{ m/s})^2 = 1.2 \times 10^5 \text{ J},$$

$$K_{100 \text{ km/h}} = \tfrac{1}{2}(1200 \text{ kg})(28 \text{ m/s})^2 = 4.7 \times 10^5 \text{ J},$$

$$K_{150 \text{ km/h}} = \tfrac{1}{2}(1200 \text{ kg})(42 \text{ m/s})^2 = 1.1 \times 10^6 \text{ J}.$$

Since the speed is squared, the kinetic energy is four times as high when the speed is doubled, and more than double when the speed is increased by 50 per cent. Since all the kinetic energy must be removed to stop the vehicle, this example suggests the hazard of high-speed driving.

Example 6–11
A Falling Coconut

What is the kinetic energy of a 1.5-lb coconut landing on your head after falling 30 ft starting from rest? Express in foot-pounds and joules. Determine also the final speed of the coconut.

Solution

The only force acting on the falling coconut is the constant downward force of gravity, which is the weight of the coconut. Acting over a distance $\Delta y = 30$ ft, this force does work

$$W = F_g \Delta y = (1.5 \text{ lb})(30 \text{ ft}) = 45 \text{ ft·lb}.$$

Since the gravitational force is the net force, the work-energy theorem tells that this is the kinetic energy gained by the coconut. With 1 ft·lb$=1.356$ J, the kinetic energy of the coconut is (45 ft·lb)(1.356 J/ft·lb)$=61$ J.

The speed is given by solving $K=\tfrac{1}{2}mv^2$ for v. Here the *weight* of the coconut is 1.5 lb, or 6.7 N (see Appendix D for conversion tables); the mass is then (6.7 N)/(9.8 m/s^2) $=0.68$ kg. So

$$v = \sqrt{\frac{2K}{m}} = \sqrt{\frac{(2)(61 \text{ J})}{0.68 \text{ kg}}} = 13 \text{ m/s}.$$

Example 6–12
A Falling Snowflake

What is the kinetic energy of a 10-mg snowflake settling on your hand at 90 cm/s? Express in joules, ergs, and eV.

Solution

In SI, we have

$$K = \tfrac{1}{2}mv^2 = \tfrac{1}{2}(10 \times 10^{-6}\text{ kg})(0.90\text{ m/s})^2 = 4.1 \times 10^{-6}\text{ J}.$$

With 10^7 erg/J, this is $(4.1 \times 10^{-6}\text{ J})(10^7\text{ erg/J}) = 41$ erg. Or, dividing by the conversion factor 1.6×10^{-19} J/eV, it is also equivalent to $(4.1 \times 10^{-6}\text{ J})/(1.6 \times 10^{-19}\text{ J/eV}) = 2.6 \times 10^{13}$ eV. This is only slightly higher than the energy given to individual protons in the largest particle accelerators.

6.7
POWER

Climbing stairs at constant speed requires the same amount of work no matter how fast you go. Why? Because at constant speed, you exert an average force exactly equal to your weight, and do so over a fixed distance; work is the product of force and distance. But of course it's much harder to run up the stairs than to walk. In a case like this, we are concerned not so much with the total work done as with the *rate* at which we do work.

power
average power

Power is defined as the rate of doing work. If an amount of work ΔW is done in a time Δt, then the **average power $\overline{P}$** is

$$\overline{P} = \frac{\Delta W}{\Delta t}. \tag{6–20}$$

instantaneous power

Often the rate of doing work varies with time. Then we define the **instantaneous power** as the average power taken in the limit of arbitrarily small time interval Δt:

$$P = \lim_{\Delta t \to 0} \frac{\Delta W}{\Delta t} = \frac{dW}{dt}. \tag{6–21}$$

Equations 6–20 and 6–21 both show that the units of power are joules/second. One J/s is given the name watt (W) in honor of James Watt, a Scottish engineer and inventor who was instrumental in developing the steam engine as a practical power source. It is ironic that the SI unit, the watt, supersedes the English unit of power suggested by Watt himself, the horsepower. Watt defined one horsepower as 550 ft·lb/s; he chose this figure—somewhat greater than the experimentally measured output of a horse—so that buyers of his steam engines would not feel short-changed. Since a ft·lb is 1.356 J, one horsepower is about 746 J/s or 746 W.

Example 6–13
Climbing Mt. Washington

A 55-kg hiker ascends New Hampshire's Mount Washington, a vertical rise of 1300 m from the base elevation. The hike takes 2 hours. A 1500-kg car drives up the Mount Washington Auto Road, the same vertical rise, in half an hour. What is the average power output in each case? Assume the hiker and car maintain constant speed and neglect friction, so that each does work only against the gravitational force.

Solution

In Example 6–8, we found that the work done by gravity depends only on the overall change in vertical position. Here the hiker and car do positive work $\Delta W = mgh$ to overcome the negative work $-mgh$ done by gravity. So the average power outputs are

$$\overline{P}_{\text{hiker}} = \frac{\Delta W}{\Delta T} = \frac{(55 \text{ kg})(9.8 \text{ m/s}^2)(1300 \text{ m})}{(2.0 \text{ h})(3600 \text{ s/h})} = 97 \text{ W}$$

and

$$\overline{P}_{\text{car}} = \frac{\Delta W}{\Delta t} = \frac{(1500 \text{ kg})(9.8 \text{ m/s}^2)(1300 \text{ m})}{(0.50 \text{ h})(3600 \text{ s/h})} = 1.1 \times 10^4 \text{ W}.$$

These values correspond to 0.13 hp and 15 hp, respectively. The figure of 97 W is typical of the sustained long-term power output of the human body. Remember that next time you leave a 100-W light bulb burning! The power plant supplying the electricity is doing work at about the rate your body can (actually, it's doing work about three times this rate, for reasons we will examine in Chapter 19). You may be surprised at the low output of the car, given that it probably has an engine rated at several hundred horsepower. Actually, only a small fraction of a car engine's rated horsepower is available in mechanical power to the wheels. The rest is lost in friction and heating.

When the power is constant, so that the average and instantaneous power are the same, then Equation 6–20 shows that the amount of work W done in a time Δt is just

$$W = P \Delta t. \tag{6–22}$$

When the power is not constant, we can consider small amounts of work ΔW, each taken over so small a time interval Δt that the power is nearly constant. Adding all these small amounts of work, and taking the limit as Δt becomes arbitrarily small, we have

$$W = \lim_{\Delta t \to 0} \sum P \Delta t = \int_{t_1}^{t_2} P \, dt, \tag{6–23}$$

where t_1 and t_2 are the beginning and end of the time interval over which we calculate the power.

Example 6–14

An Electric Bill: Yankee Stadium

Each of the 500 floodlights at Yankee Stadium uses electrical energy at the rate of 1.0 kW. How much does it cost to run these lights during a 4-hour night game, if electricity costs 9.5 cents per kilowatt-hour?

Solution

The total power consumption of the 500 lights is 500 kW. Since the power is constant, the total work done to run the lamps is given by Equation 6–22:

$$W = P \Delta t = (500 \text{ kW})(4.0 \text{ h}) = 2000 \text{ kWh}.$$

The cost is then (2000 kWh)(9.5 cents/kWh) = $190.

Example 6–15

Variable Power

Figure 6–17 shows the rate of electrical energy use in a typical home between the hours of 3 PM and 9 PM. The curve shown is given by

$$P = at^2 + bt + c,$$

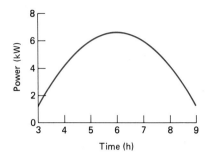

Fig. 6–17
Electrical power consumption in a home (Example 6–15).

where t is the time in hours and where $a = -0.60$ kW/h², $b = 7.2$ kW/h, and $c = -15$ kW. Find the total energy consumption during this interval.

Solution

Now the power varies with time, so we must evaluate the integral in Equation 6–23:

$$W = \int_{t_1}^{t_2} P\, dt = \int_{t_1}^{t_2} (at^2 + bt + c)\, dt$$

$$= \left(\frac{at^3}{3} + \frac{bt^2}{2} + ct \right)\Bigg|_{t_1}^{t_2} = \frac{a(t_2^3 - t_1^3)}{3} + \frac{b(t_2^2 - t_1^2)}{2} + c(t_2 - t_1),$$

where we have used the usual rule for integrating a power of a variable: raise the exponent by 1 and divide by the new exponent. Putting in numbers including the limits $t_1 = 3$ h and $t_2 = 9$ h, we have

$$W = \frac{(-0.60)(9^3 - 3^3)}{3} + \frac{(7.2)(9^2 - 3^2)}{2} + (-15)(9-3) = 29 \text{ kWh},$$

where we have left off the units in doing the calculation. Note that with power in kW and time in hours, our answer must come out in kWh.

Does our result make sense? Figure 6–17 shows a maximum power of about 6.5 kW; at this rate we would use about 40 kWh in the 6-hour period. The minimum power in Fig. 6–17 is a little over 1 kW; at this rate we would use 6 kWh in the period. Our answer, 29 kWh, lies in between and is toward the high end because the curve rises rapidly. The average power consumption during the interval is just 29 kWh/6 h or 4.8 kW.

We can derive an expression relating power, applied force, and velocity by noting that the work ΔW done by a force **F** acting on an object that undergoes a small displacement $\Delta \mathbf{r}$ is given by Equation 6–4:

$$\Delta W = \mathbf{F} \cdot \Delta \mathbf{r}.$$

Dividing both sides by the time interval Δt in which the displacement $\Delta \mathbf{r}$ occurs, and taking the limit as $\Delta t \to 0$, we have

$$\lim_{\Delta t \to 0} \frac{\Delta W}{\Delta t} = \lim_{\Delta t \to 0} \mathbf{F} \cdot \frac{\Delta \mathbf{r}}{\Delta t} = \mathbf{F} \cdot \frac{d\mathbf{r}}{dt}.$$

But the limit of $\Delta W / \Delta t$ is just the instantaneous power, P, and $d\mathbf{r}/dt$ is the velocity, **v**, so that

$$P = \mathbf{F} \cdot \mathbf{v}. \tag{6–24}$$

Example 6–16
Bicycling

As you ride your 14-kg bicycle on level ground, you must overcome a frictional force of 30 N. If your mass is 65 kg, what is the power you must supply to maintain a steady 25 km/h (a) on level ground and (b) going up a 5° incline?

Solution

The force of friction opposes your motion; to overcome friction, you must apply a force of equal magnitude in the direction of motion. The 25-km/h speed is equivalent to $(25 \times 10^3$ m$)/(3600$ s$) = 6.9$ m/s, so the power you supply is

$$P = \mathbf{F} \cdot \mathbf{v} = Fv = (30 \text{ N})(6.9 \text{ m/s}) = 210 \text{ W},$$

or a little over one-fourth horsepower. (In writing $\mathbf{F} \cdot \mathbf{v} = Fv$, we used the fact that the force is in the same direction as the velocity, so the cosine in the dot product is 1.)

In climbing the hill, you must overcome not only friction but also the downslope component of the gravitational force. From Example 5–1, this downslope component is $mg\sin\theta$, here equal to $(14$ kg $+ 65$ kg$)(9.8$ m/s$^2)(\sin5°) = 67$ N. Also from Example 5–1, we know that the normal force is $mg\cos\theta$, so that the frictional force is reduced from its level-ground value by a factor $\cos\theta$. But here $\cos\theta = \cos5° = 0.996$, so that the frictional force is still essentially 30 N. Then the power you must supply is

$$P = Fv = (30 \text{ N} + 67 \text{ N})(6.9 \text{ m/s}) = 670 \text{ W},$$

or nearly one horsepower. Here again we wrote $P = Fv$ since the applied force is in the direction of the velocity.

SUMMARY

This chapter introduces the important concepts of work and energy, whose meanings in physics are similar but not identical to their meanings in everyday language.

1. **Work** is defined as the product of the component of force in the direction an object moves and the distance through which the object moves. Taking the x-direction as the direction of motion, work is

$$W = F_x \Delta x.$$

The SI unit of work is the newton-meter, given the special name **joule** (J).

2. The **scalar product** or **dot product** of two vectors provides a convenient shorthand for writing work in terms of the force and displacement vectors. The scalar product of any two vectors $\mathbf{A}$ and $\mathbf{B}$ is defined as the product of their magnitudes with the cosine of the angle between them:

$$\mathbf{A} \cdot \mathbf{B} = AB\cos\theta.$$

In terms of the dot product, the work done by a force $\mathbf{F}$ acting on an object undergoing displacement $\Delta\mathbf{r}$ is

$$W = \mathbf{F} \cdot \Delta\mathbf{r}.$$

3. When force varies with distance, work must be calculated by summing the terms $\mathbf{F} \cdot \Delta\mathbf{r}$ in the limit of arbitrarily small $\Delta\mathbf{r}$:

$$W = \lim_{\Delta\mathbf{r} \to 0} \sum \mathbf{F} \cdot \Delta\mathbf{r}.$$

This limiting process defines the **definite integral**:

$$W = \int_{\mathbf{r}_1}^{\mathbf{r}_2} \mathbf{F} \cdot d\mathbf{r}.$$

Written in terms of a dot product of force with the infinitesimal displacement $d\mathbf{r}$, this form of the definite integral is called a **line integral**. In the case of one-dimensional motion along the x-axis, the line integral reduces to

$$W = \int_{x_1}^{x_2} F \, dx.$$

Integration is the inverse of differentiation; the methods of calculus show how to evaluate integrals. In particular, the integral of a power of a variable is obtained by raising the exponent and dividing by the new exponent:

$$\int_{x_1}^{x_2} x^n \, dx = \frac{x^{n+1}}{n+1}\Big|_{x_1}^{x_2},$$

where the $\big|_{x_1}^{x_2}$ means to evaluate the expression at x_2 and subtract from it the same expression evaluated at x_1.

4. Work done by a nonzero net force results in a change in an object's **kinetic energy**, a scalar quantity given by $K = \frac{1}{2}mv^2$. The **work-energy theorem** shows that the change in kinetic energy is exactly equal to the work done by the net force:

$$W = \Delta K = \tfrac{1}{2}mv_2^2 - \tfrac{1}{2}mv_1^2,$$

where v_1 and v_2 are the speeds before and after the work is done, respectively. The units of energy are the same as those of work. In addition to the SI unit, the joule, other common units include the English foot-pound; the kilowatt-hour; the cgs unit erg; and the electron-volt, widely used in atomic, nuclear, and high-energy physics.

5. **Power** is the rate at which work is done. In SI units, power is measured in joules/second, or **watts;**

1 W $=$ 1 J/s. The average power $\overline{P}$ is given by

$$\overline{P} = \frac{\Delta W}{\Delta t},$$

while the instantaneous power is

$$P = \frac{dW}{dt}.$$

Inverting this last expression allows us to calculate

work as the integral of the power:

$$W = \int_{t_1}^{t_2} P \, dt.$$

When a force **F** acts on an object moving with velocity **v,** the instantaneous power supplied by that force is

$$P = \mathbf{F} \cdot \mathbf{v}.$$

QUESTIONS

1. Give two examples of situations in which you might think you are doing work but in which, in the technical sense, you do no work.
2. If the scalar product of two nonzero vectors is zero, what can you conclude about their relative directions?
3. Must you do work to whirl a ball around on the end of a string? Explain.
4. If you pick up a suitcase and put it down, how much total work have you done on the suitcase? Does your answer change if you pick up the suitcase and drop it?
5. You lift a book from your desk to a bookshelf. It is initially at rest and ends up at rest, so that its kinetic energy has not changed. Yet you have certainly done work on the book. Explain why this is not a violation of the work-energy theorem.
6. Would Equation 6–5 hold in a coordinate system whose axes were not perpendicular?
7. A given force F is applied to move a crate a fixed distance across the floor. Does the amount of work done by the force depend on the coefficient of friction? Does the total work done on the crate depend on the coefficient of friction?
8. Discuss the relation between a sum and an integral.
9. Two cross-country skiers race up opposite sides of the mountain shown in Fig. 6–18. If the skier coming up the steeper side wins, compare the work and the average power associated with each skier. Assume that no work is done by friction (although static friction, which does no work, is what makes the climb possible).

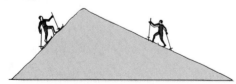

Fig. 6–18   Question 9

10. With a given nonzero coefficient of friction, compare the force required to slide a piano up a steep ramp

versus a gradual ramp. Compare also the work done in raising the piano the same height on each ramp.
11. Does the gravitational force of the sun do work on a planet in a circular orbit? On a comet in an elliptical orbit? Explain.
12. How does the dependence of a car's kinetic energy on speed affect its stopping distance, assuming a constant braking force?
13. Give an example showing that kinetic energy depends on frame of reference.
14. Can kinetic energy ever be negative? Explain.
15. A pendulum bob swings back and forth on the end of a string, describing a circular arc. Does the tension force in the string do any work?
16. In the preceding question, does the kinetic energy of the bob change as it swings back and forth? Why?
17. Does your car's kinetic energy change if you drive at constant speed for 1 hour?
18. List in order of increasing size: erg, kWh, eV, ft·lb, J. Which units are most appropriate in discussing the energetics of macroscopic systems? Which is most appropriate in atomic physics?
19. A watt-second is a unit of what quantity? Relate it to a more standard SI unit.
20. A body collides with a spring and compresses it an amount Δx. How will Δx change if you double the impact velocity? Half the mass of the body? Triple the spring constant?
21. In the absence of air resistance, it would take the same time for a fly ball to rise as to fall. In the presence of air resistance, however, the ball will take longer to fall. Use the work-energy theorem to explain this difference.
22. Two particles of different mass have the same momentum. Compare their kinetic energies.
23. A truck is moving northward at 55 mph. Later, it is moving eastward at the same speed. Has its kinetic energy changed? Has its momentum changed? Has work been done on the truck? Has a force acted on the truck? Explain.

Fig. 6–20 Problem 9

PROBLEMS

Section 6.1 *Work*

1. How much work do you do as you exert a 95-N force to push a 30-kg shopping cart through a 14-m-long supermarket aisle?

2. If the coefficient of kinetic friction is 0.21, how much work do you do when you slide a 50-kg box at constant speed across a 4.8-m-wide room?

3. A crane lifts a 500-kg beam vertically upward 12 m, then swings it eastward 6.0 m. How much work does the crane do? Neglect friction, and assume the beam moves with constant speed.

4. You lift a 45-kg barbell from the ground to a height of 2.5 m. (*a*) How much work do you do on the barbell? (*b*) You hold the barbell aloft for 2.0 min. How much work do you do on the barbell during this time? (*c*) You lower the barbell to the ground. Now how much work do you do on it?

5. You slide a box of books at constant speed up a 30° ramp, applying a force of 200 N directed up the slope. The coefficient of sliding friction is 0.18. (*a*) How much work have you done when the box has risen 1 m vertically? (*b*) What is the mass of the box?

6. Two people push a stalled car at its front doors, each applying a 330-N force at 25° to the forward direction, as shown in Fig. 6–19. How much work does each do in pushing the car 6.2 m?

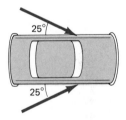

Fig. 6–19 Problem 6

7. A locomotive does 8.8×10^{11} J of work in pulling a 2×10^6-kg train 150 km. What is the average force in the coupling between the locomotive and the rest of the train?

8. An elevator of mass *m* rises a distance *h* up a vertical shaft with upward acceleration equal to one-tenth *g*. How much work does the elevator cable do on the elevator?

9. A 20-kg child lies on her back and scoots 7.4 m across the floor at constant speed by applying a 50-N force with the soles of her feet (Fig. 6–20). (*a*) How much work does she do? (*b*) What is the coefficient of friction between the child's back and the floor?

10. The mass of a hydrogen atom is 1.67×10^{-27} kg. How many eV of work does gravity do on the atom when it falls a distance of 10 cm? (See Section 6.6 for the definition of the eV.)

Section 6.2 *Work and the Scalar Product*

11. Show that the scalar product is distributive: $\mathbf{A} \cdot (\mathbf{B} + \mathbf{C}) = \mathbf{A} \cdot \mathbf{B} + \mathbf{A} \cdot \mathbf{C}$.

12. Show that the scalar product has the property that $(c\mathbf{A}) \cdot \mathbf{B} = \mathbf{A} \cdot (c\mathbf{B})$, where *c* is a scalar.

13. Given the following vectors:

> **A** has length 10 and points 30° above the x-axis
>
> **B** has length 4.0 and points 10° to the left of the *y*-axis
>
> $\mathbf{C} = 5.6\hat{\mathbf{i}} - 3.1\hat{\mathbf{j}}$
>
> $\mathbf{D} = 1.9\hat{\mathbf{i}} + 7.2\hat{\mathbf{j}},$

compute the scalar products (*a*) $\mathbf{A} \cdot \mathbf{B}$; (*b*) $\mathbf{C} \cdot \mathbf{D}$; (*c*) $\mathbf{B} \cdot \mathbf{C}$.

14. (*a*) Find the scalar product of the vectors $a\hat{\mathbf{i}} + b\hat{\mathbf{j}}$ and $b\hat{\mathbf{i}} - a\hat{\mathbf{j}}$, and (*b*) determine the angle between them. (*a* and *b* are arbitrary constants.)

15. Rework Example 6–5 using Equation 6–3 instead of Equation 6–5.

16. Use Equations 6–3 and 6–5 to show that the angle between the vectors $\mathbf{A} = a_x\hat{\mathbf{i}} + a_y\hat{\mathbf{j}}$ and $\mathbf{B} = b_x\hat{\mathbf{i}} + b_y\hat{\mathbf{j}}$ is

$$\theta = \cos^{-1}\left\{ \frac{a_x b_x + a_y b_y}{[(a_x^2 + a_y^2)(b_x^2 + b_y^2)]^{1/2}} \right\}$$

17. Given that $\mathbf{A} = 2\hat{\mathbf{i}} + 2\hat{\mathbf{j}}$, $\mathbf{B} = 5\hat{\mathbf{i}}$, and $\mathbf{C} = \sqrt{2}\hat{\mathbf{i}} - \pi\hat{\mathbf{j}}$, find the angles between (*a*) **A** and **B**; (*b*) **A** and **C**; (*c*) **B** and **C**. Hint: See previous problem.

18. A force $\mathbf{F} = 67\hat{\mathbf{i}} + 23\hat{\mathbf{j}} + 55\hat{\mathbf{k}}$ N is applied to a body as it moves in a straight line from $\mathbf{r}_1 = 16\hat{\mathbf{i}} + 31\hat{\mathbf{j}}$ to $\mathbf{r}_2 = 21\hat{\mathbf{i}} + 10\hat{\mathbf{j}} - 14\hat{\mathbf{k}}$ m. How much work is done by the force?

19. A rope pulls a box a horizontal distance of 23 m, as shown in Fig. 6–21. If the rope tension is 120 N, and if the rope does 2500 J of work on the box, what angle does it make with the horizontal?

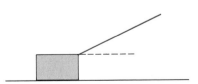

Fig. 6–21 Problem 19

Section 6.3   A Varying Force

20. Find the total work done by the force shown in Fig. 6–22, as the object on which it acts moves from $x=0$ to $x=5.0$ m.

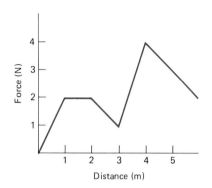

Fig. 6–22   Problem 20

21. A force F acts in the x-direction, its magnitude given by $F=ax^2$, where x is in meters and a is exactly 5 N/m². (a) Find an exact value for the work done by this force as it acts on a particle moving from $x=0$ to $x=6$ m. Now find approximate values for the work by dividing the area under the force curve into rectangles of width (b) $\Delta x=2$ m; (c) $\Delta x=1$ m; (d) $\Delta x=\frac{1}{2}$ m with height equal to the magnitude of the force in the center of the interval. Calculate the per cent error in each case.

22. A spring has spring constant $k=200$ N/m. How much work does it take to stretch the spring (a) 10 cm from equilibrium and (b) from 10 cm to 20 cm from equilibrium?

23. A certain amount of work is required to stretch spring A a certain distance. Twice as much work is required to stretch spring B half that distance. Compare the spring constants of the two springs.

24. On graph paper, draw an accurate force-distance curve for the force of Example 6–7, and obtain a solution to that example by determining graphically the area under the curve.

25. A force $\mathbf{F}$ acts in the x-direction, its magnitude given by $F=F_0\cos(x/x_0)$, where $F_0=51$ N and $x_0=13$ m. Calculate the work done by this force acting on an object as it moves from $x=0$ to $x=37$ m. *Hint:* Consult Appendix B for the integral of the cosine function, and treat the argument of the cosine as a quantity in radians.

26. Work the preceding problem graphically, by making an accurate plot of the force versus distance curve on graph paper, and determining the area (in units of work) under the curve.

27. A force given by $F=a\sqrt{x}$ acts in the x-direction, where $a=14$ N/m¹ᐟ². Calculate the work done by this force acting on an object as it moves (a) from $x=0$ to

$x=2.2$ m; (b) from $x=2.2$ m to $x=4.4$ m; (c) from 4.4 m to 6.6 m.

28. A force given by $F=b/\sqrt{x}$ acts in the x-direction, where b is a constant with the units N·m¹ᐟ². Show that even though the force becomes arbitrarily large as x approaches 0, the work done in moving from x_1 to x_2 remains finite even as x_1 approaches zero. Find an expression for that work in the limit $x_1\to0$.

29. The force exerted by a rubber band is given approximately by

$$F = F_0\left[\frac{\ell_0+x}{\ell_0} - \frac{\ell_0{}^2}{(\ell_0+x)^2}\right],$$

where ℓ_0 is the unstretched length, x the stretch, and F_0 is a constant (although F_0 varies with temperature). Find the work needed to stretch the rubber band a distance x.

Section 6.4   Force and Work in Three Dimensions

30. You put your little sister (mass m) on a swing whose chains have length ℓ, and pull slowly back until the swing makes an angle ϕ with the vertical. Show that the work you do is $mg\ell(1-\cos\phi)$.

31. A cylindrical log of radius R lies half buried in the ground, as shown in Fig. 6–23. An ant climbs the log at constant speed, supplying just enough force to balance gravity. Show by evaluating $\int\mathbf{F}\cdot d\mathbf{r}$ that the work required to reach the top is mgR.

Fig. 6–23   Problem 31

32. A particle of mass m moves from the origin to the point $x=3$, $y=6$ along the curve $y=ax^2-bx$, where $a=2$ m⁻¹ and $b=4$. It is subject to a force $\mathbf{F}=cxy\hat{\mathbf{i}}+d\hat{\mathbf{j}}$, where $c=10$ N/m² and $d=15$ N. Calculate the work done by the force.

33. Repeat the preceding problem for the case when the particle moves first along the x-axis from the origin to the point (3,0), then parallel to the y-axis until it reaches (3,6).

34. A playground structure is in the form of a spherical cap of radius R; an arc from the top of the cap to the ground subtends an angle θ, as shown in Fig. 6–24. A child of mass m slides from the top of the structure to the ground. If the coefficient of friction is μ, show that the work done on the child by the frictional force is $-\mu mgR\sin\theta$.

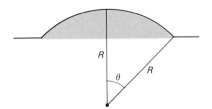

Fig. 6–24 Problem 34

Sections 6.5–6.6 *Kinetic Energy and Energy Units*

35. A 3.5×10^5-kg jumbo jet is cruising at 1000 km/h. (a) What is its kinetic energy relative to the ground? A 85-kg passenger strolls down the aisle at 2.9 km/h. What is the passenger's kinetic energy (b) relative to the ground and (c) relative to the plane?

36. Electrons in a color TV tube are accelerated to 25% of the speed of light. How much work does the TV tube do on each electron? (At this speed, relativity theory introduces small but measurable corrections; here you neglect these effects.) See inside front cover for the electron mass.

37. In a cyclotron used to produce radioactive isotopes for medical research at the Massachusetts General Hospital, deuterium nuclei (mass 3.3×10^{-27} kg) are given kinetic energies of 5.5 MeV. What is their speed? Compare with the speed of light.

38. At what speed must a 950-kg subcompact car be moving to have the same kinetic energy as a 3.2×10^4-kg truck going 20 km/h?

39. A 60-kg skateboarder comes over the top of a hill at 5.0 m/s, and reaches 10 m/s at the bottom of the hill. Find the total work done on the skateboarder between the top and bottom of the hill.

40. Two unknown elementary particles pass through a detection chamber. If they have the same kinetic energy and their mass ratio is 4:1, what is the ratio of their speeds?

41. A 300-kg boulder rolls after an intrepid explorer at a speed of 10 m/s (Fig. 6–25). The explorer is running away at 3.0 m/s. What is the kinetic energy of the boulder as it hits the explorer, measured in the explorer's frame of reference? (Photo from *Raiders of the Lost Ark* © MCMLXXXI by Lucasfilm, Ltd. All rights reserved.)

42. After a tornado, a 0.50-g drinking straw was found embedded 4.5 cm in a tree. Subsequent measurements showed that the tree would exert a stopping force of 70 N on the straw. What was the straw's speed when it hit the tree?

43. You drop a 150-g baseball from a sixth-story window 16 m above the ground. What are (a) its kinetic energy and (b) its speed when it hits the ground? Neglect air resistance.

44. A hospital patient was being wheeled in for x-ray when his leg slipped off the stretcher and his heel hit the concrete floor. As a physicist, you are called to testify about the forces involved in this accident. You estimate that the foot and leg had an effective mass of 8 kg, that they dropped freely a distance of 70 cm, and that the stopping distance was 2 cm. What force can you claim was exerted on the foot by the floor? Give your answer in pounds so the jury will have a feel for the size of the force.

45. From what height would you have to drop a car for its impact to be equivalent to a collision at 20 mph?

46. A 2.3-kg particle's position as a function of time is given by $x = bt^3 - ct$, where $b = 0.41$ m/s^3 and $c = 1.9$ m/s. Find the work done on the particle between the time $t = 0$ and $t = 2.0$ s.

47. In a switchyard, freight cars start from rest and roll down a 2.8-m incline and come to rest against a spring bumper at the end of the track (Fig. 6–26). If the spring constant is 4.3×10^6 N/m, how much is the spring compressed when hit by a 57,000-kg freight car?

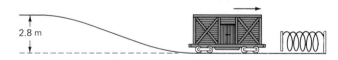

2.8 m

Fig. 6–26 Problem 47

48. Catapults run by high-pressure steam from the ship's nuclear reactor are used on the aircraft carrier Enterprise to launch jet aircraft to takeoff speed in only 76 m of deck space. A catapult exerts a 1.1×10^6 N force on a 3.3×10^4 kg aircraft. What are (a) the kinetic energy and (b) the speed of the aircraft as it leaves the catapult? (c) How long does the catapulting operation take? (d) What is the acceleration of the aircraft?

49. A block of mass m slides from rest without friction down the slope shown in Fig. 6–27. (a) How much work is done on the block by the normal force of the slope? (b) Show that the final speed is $\sqrt{2gh}$ regardless of the details of the slope.

Fig. 6–25 Problem 41

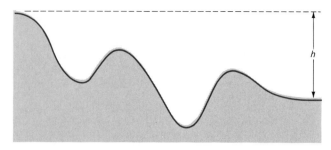

Fig. 6–27 Problem 49

Section 6.7 *Power*

50. A horse plows a 200-m-long furrow in 5.0 min, exerting a force of 750 N. What is its power output, measured in watts and in horsepower?
51. A typical car battery stores about 1 kWh of energy. What is its power output if it is drained completely in (a) 1 minute; (b) 1 hour; (c) 1 day?
52. Herschel Walker's mass is 90 kg. (a) When he is running at 7.0 m/s, what is his kinetic energy? (b) If a tackler grabs him, and slows him to 3.0 m/s over a 1.0-s interval, what power does Herschel supply?
53. A "mass driver" is designed to launch raw materials mined on the moon to a factory in lunar orbit. The driver can accelerate a 1000-kg package to 2.0 km/s (just under lunar escape speed) in 55 s. (a) What is its power output during a launch? (b) If the driver makes one launch every 30 minutes, what is its average power consumption?
54. An 85-kg long-jumper takes 3.0 s to reach a prejump speed of 10 m/s. What is his power output?
55. Estimate your power output as you do deep knee bends at the rate of one per second.
56. At what rate can a one-half horsepower well pump deliver water to a tank 60 m above the water level in the well? Give your answer in kg/s and gal/min.
57. A young boy, fortified by a hearty lunch, straps himself into his human-powered go-cart and begins to expend his bodily energy in such a way that his vehicle's speed increases quadratically with time. Show that his power output must increase in proportion to t^3.
58. With your car initially at rest, you depress the accelerator pedal in such a way that the power delivered by the car's engine is proportional to the time since you started. Neglecting losses to air and road friction, show that the car's speed also increases linearly with time.

Supplementary Problems

59. If the ant of Problem 31 climbs with constant speed v along the log surface, (a) what is its power output as a function of time? (b) Integrate your expression for power to show that the total work to climb the log is the same as that obtained in Problem 31.
60. You need to purchase a refrigerator, and are looking at an energy-efficient, 250-W model for $950 and a

standard, 450-W model for $800. If each refrigerator is actually running 20% of the time, and if electricity costs 9 cents/kWh, how long would you have to own the refrigerator for the energy-efficient model to be more economical? Work your answer (a) neglecting interest on your money and (b) assuming you can earn 7.4% interest on your savings, compounded continuously. *Note:* Part (b) calls for an iterative calculation.

61. A machine delivers power at a decreasing rate $P = W_0/(t+1)^2$, where W_0 is a constant with the units of work, and t is the time in seconds. The machine starts at time $t = 0$ and continues forever. Show that it nevertheless does only a finite amount of work, equal to W_0.

62. An unusual spring has the force-distance curve in Fig. 6–28. This curve is described analytically by $F = 100x^2$ for $0 \leq x \leq 1$ and $F = 100(4x - x^2 - 2)$ for $1 \leq x \leq 2$, where x is the displacement in meters from the equilibrium position and F is in newtons. Find the work done in stretching this spring (a) from equilibrium to 1 m and (b) from 1 m to 2 m. (c) Show that the work done in stretching the spring from equilibrium to 2 m is the same as for an ideal spring with $k = 100$ N/m. (d) Explain, in terms of the shape of the graph, why this is true. Is it true for any other distance stretched?

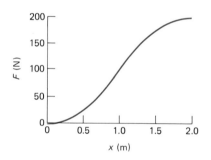

Fig. 6–28 Problem 62

63. The force of earth's gravity on an object of mass m is given by mgR_e^2/r^2, where g is the acceleration of gravity at earth's surface, R_e is earth's radius, and r is the distance in meters from the center of the earth. (a) Show that the work required to move an object from earth's surface to a distance r from the center of the earth is given by

$$W = mgR_e \left(1 - \frac{R_e}{r} \right).$$

Calculate and compare the work needed to boost an 850-kg satellite (b) to low earth orbit, 250 km above earth's surface; (c) to geosynchronous orbit, 36,000 km above earth's surface; (d) to the moon's orbit, 390,000 km from earth. (Your answers do not include additional work needed to achieve orbital speed.)

64. A spring scale hangs vertically. A mass is attached to the unstretched scale and released from rest. It bounces up and down awhile until friction eventually brings it again to rest, with the final scale reading equal to the weight. Show that the maximum scale reading is twice the object's weight.

65. A locomotive accelerates a freight train of total mass M, applying constant power P from rest. Determine the velocity and position of the train as functions of time.

66. A possible reaction for a nuclear fusion reactor involves the fusing of two deuterium nuclei (each consisting of one proton and one neutron) to form helium, a process that releases 24 MeV of energy. In the fusion process, the two nuclei are bound tightly by the strong nuclear force. But the nuclear force is effective only when the nuclei are closer than about 10^{-15} m; beyond this their interaction is dominated by a repulsive electrical force given in newtons by $F = 2.3 \times 10^{-28}/r^2$, where r is the distance in meters between the two nuclei. Consider one deuterium nucleus held at rest and another a long way from it. (a) Calculate the work needed to overcome the electrical force and bring the second nucleus to within 10^{-15} m of the first. (b) If the electric force alone acts between the nuclei, what initial speed should the second nucleus have if it is to fuse with the first? (c) A typical large nuclear fission power plant has an electrical power output of 1000 MW. If a fusion plant with the same power output converts fusion energy to electricity with 40% efficiency, at what rate must deuterium-deuterium fusion reactions occur in its reactor?

7

CONSERVATION OF ENERGY

(a)

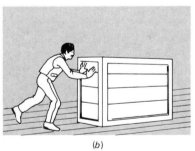

(b)

Fig. 7–1
Both the rock climber *(a)* and the mover
(b) do work, but only the climber can
recover that work as kinetic energy.

conservative force

nonconservative force

The rock climber of Fig. 7–1a does work as she ascends the vertical cliff. So does the mover of Fig. 7–1b, as he pushes a heavy chest across the floor. But there is a difference. If the rock climber lets go, down she goes; the work she put into the climb comes back as kinetic energy of her fall. If the mover lets go of the trunk, though, it just sits there; it has no tendency to return to its starting place.

The contrast between the two situations of Fig. 7–1 illustrates an important distinction between two types of forces. In this chapter, we explore that distinction and from it develop one of the most important principles in all of physics: conservation of energy. In later chapters, we will expand the conservation of energy principle to include new forms of energy we encounter. Ultimately, in Chapter 33, we will see how Einstein's theory of relativity subsumes two great principles—conservation of energy and conservation of matter—into a single statement.

7.1
CONSERVATIVE AND NONCONSERVATIVE FORCES

Both the climber and the mover of Fig. 7–1 are working against external forces—gravity for the climber and friction for the mover. The difference between the two situations is this: if the climber lets go, the gravitational force "gives back" the work that she did; that work manifests itself as a gain in her kinetic energy. But the frictional force does not "give back" the work that the mover did; that work has disappeared, and cannot be recovered as kinetic energy.

We use the term **conservative force** to describe a force that, like gravity, "gives back" work that has been done against it. A force like friction is **nonconservative.** We can give a more precise mathematical sense of this distinction by considering the work involved in moving an object over a closed path—that is, a path that ends up at the same point where it started. Suppose, for example, that our rock climber ascends a cliff of height *h* and then returns to her starting point. How much work has the gravitational force done on her? That force has magnitude *mg*, and points downward. Going up, the force is directed opposite to her motion, so that the work done by gravity is −*mgh*. Coming down, the force is in the same direction as the motion, so that the work done by gravity is +*mgh*. The total work that gravity does on the climber as she traverses the closed path from the bottom to the top of the cliff and back again is therefore zero.

(We're concerned here only with the work done by gravity, so other forces don't affect this result. For example, if the climber falls from the top of the cliff, then the only work done during her descent is that of gravity, so that she gains kinetic energy on the way down. But if she climbs carefully down, she exerts upward forces so that there is no net work done on the way down, and she arrives safely at the bottom. But in either case, the work done on her *by gravity* is *mgh*.)

Now suppose the mover of Fig. 7–1b pushes the chest a distance ℓ across a room, discovers it's the wrong room, and pushes it back to the door. Like the climber, the chest describes a closed path. How much work is done by the frictional force acting over this path? That force has magnitude *μN*, where the normal force in this case of horizontal motion is just *mg*. But the frictional force *always* acts to oppose the motion, so that it *always* does negative work. With frictional force *μmg* opposing the motion, the work done in crossing the width ℓ of the room is −*μmgℓ*. Coming back, it is also −*μmgℓ*, so that the total work done by the frictional force is −2*μmgℓ*.

The difference between our answers for the total work done by the gravitational and frictional forces acting over closed paths provides one precise definition of the distinction between conservative and nonconservative forces:

> When the total work done by a force acting as an object moves over any closed path is zero, then the force is conservative. Mathematically,

$$\oint \mathbf{F} \cdot d\mathbf{r} = 0. \qquad \text{(conservative force)} \qquad (7\text{–}1)$$

The circle on the integral sign indicates that the integral is taken over a *closed* path. A force for which the integral is nonzero, like the frictional force, is nonconservative.

You can see why the mathematical definition Equation 7–1 is equivalent to our more physical statement that a conservative force "gives back" work that was done against it. When we do work against a conservative force, it simultaneously does negative work on us. But for the total work done by the force to be zero, it must subsequently do positive work on us; during this time it "gives back" the work we did earlier.

Equation 7–1 suggests a related property of conservative forces. Suppose we move an object along the straight path between points *A* and *B* shown in Fig. 7–2, along which a conservative force acts; let the work done by the conservative force be W_{AB}. Since the work done by a conservative force over any closed path is zero, the work W_{BA} done in moving back from

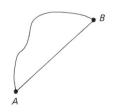

Fig. 7–2
Since the work done by a conservative force over a closed path is zero, the work done in moving from point *A* to point *B* must be independent of the path.

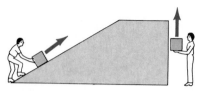

Fig. 7–3
Movers raise identical boxes a vertical distance *h*. The work done by gravity is the same for the box lifted vertically and for the box pushed up the ramp. (The horizontal location of the starting point doesn't matter here, since gravity does no work on an object moved horizontally.)

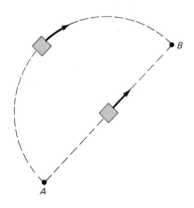

Fig. 7–4
Top view of boxes being pushed along a horizontal floor. The work done by friction is least for the straight-line path.

B to *A* must be $-W_{AB}$ whether we return along the straight path or take the curved path shown or any other path. That is, the work done by a conservative force as we move between two points is *independent of the path taken*: $\int_A^B \mathbf{F}\cdot d\mathbf{r}$ is path-independent. We saw this in Example 6–8, where we found that the work done by gravity does not depend on the details of the path taken (Fig. 7–3). In contrast, the work done by a nonconservative force is *not* path-independent. On a frictional surface, for example, the least work is done over a straight-line path; any other path involves more work (Fig. 7–4).

Important examples of conservative forces include gravity and the static electric force. The force of an ideal spring—fundamentally an electric force—is also conservative. Nonconservative forces include friction and the electric force in the presence of time-varying magnetic effects.

Example 7–1
Conservative and Nonconservative Forces

Two forces point in the y-direction; the first is given by $\mathbf{F}_1 = ay\hat{\mathbf{j}}$ and the second by $\mathbf{F}_2 = bx\hat{\mathbf{j}}$, where *a* and *b* are constants. For each force, determine the work done in moving from the origin to the point (ℓ,ℓ) by first moving up the y-axis to $(0,\ell)$, then moving straight over to (ℓ,ℓ). Compare with the work done in moving first along the x-axis to $(\ell,0)$, then straight up to (ℓ,ℓ). Comment on the conservative or nonconservative nature of the forces.

Solution

Figure 7–5*a* shows some force vectors for $\mathbf{F}_1$ on the two paths *ABC* and *ADC*. If we start at the origin and go upward, we have

$$W_{AB} = \int_A^B \mathbf{F}\cdot d\mathbf{r} = \int_{(0,0)}^{(0,\ell)} ay\hat{\mathbf{j}}\cdot d\mathbf{r}$$

for the work done by $\mathbf{F}_1$ along the vertical segment *AB*. But along this segment, $d\mathbf{r}$ is vertical; $d\mathbf{r} = \hat{\mathbf{j}}dy$. Then we have

$$W_{AB} = \int_{y=0}^{y=\ell} a\,y\,dy = \frac{ay^2}{2}\bigg|_0^\ell = \tfrac{1}{2}a\ell^2.$$

Going across the top of the square, from *B* to *C*, the force and displacement are perpendicular, so that no work is done; therefore, $W_{ABC} = \tfrac{1}{2}a\ell^2$ for force $\mathbf{F}_1$. Going

Fig. 7–5
Force vectors and paths for the forces $\mathbf{F}_1$ and $\mathbf{F}_2$ of Example 7–1.

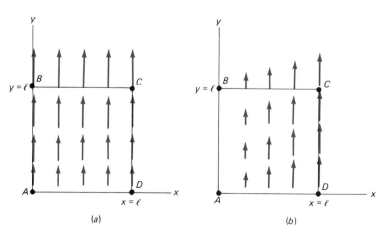

along the x-axis from A to D, the force and displacement are again perpendicular and no work is done. Going up from D to C, we again have $d\mathbf{r} = \hat{\jmath}\,dy$; then

$$W_{DC} = \int_{y=0}^{y=\ell} a\,y\,dy = \tfrac{1}{2}a\ell^2,$$

so that the work W_{ADC} is equal to the work W_{ABC}, and the force could be conservative. (Problem 2 asks you to verify that the work is the same along the diagonal path AC.)

For the force $\mathbf{F}_2$, on the other hand, $x = 0$ and therefore $\mathbf{F}_2 = 0$ along the segment AB. With $\mathbf{F}_2$ perpendicular to displacement along BC, the work $W_{ABC} = 0$ for this force. Along AD, the work is again zero, but going up from D to C we have $d\mathbf{r} = \hat{\jmath}\,dy$, so that with $x = \ell$ along this segment, the work becomes

$$W_{DC} = \int_{y=0}^{y=\ell} b\,\ell\,dy = b\ell y\Big|_0^\ell = b\ell^2.$$

Then the total work W_{ADC} is $b\ell^2$ for the force $\mathbf{F}_2$; since this differs from the work over the path W_{ABC}, the work is not path-independent and therefore the force is not conservative.

Although we calculated the work integrals in detail here, you should be able to see directly from Fig. 7–5 that $\mathbf{F}_1$ is conservative and $\mathbf{F}_2$ is not. Question 1 explores other examples like this one.

7.2
POTENTIAL ENERGY

Work done against a conservative force is somehow "stored," in the sense that we can get it back again in the form of kinetic energy. The climber of Fig. 7–1a is acutely aware of that "stored work"; it gives her the potential for a dangerous fall. So is the archer of Fig. 7–6; the work he did against the springiness of his bow has the potential to propel the arrow at high speed. "Potential" is an appropriate word here: we can consider the "stored work" as **potential energy,** potential in the sense that it can become actualized as kinetic energy.

potential energy

We define potential energy formally in terms of the the work done by a conservative force. Specifically, the change ΔU_{AB} in potential energy associated with moving an object from point A to point B is the negative of the work done by the conservative force acting on that object:

$$\Delta U_{AB} = -\int_A^B \mathbf{F}\cdot d\mathbf{r}. \qquad \text{(potential energy)} \qquad (7\text{--}2)$$

Why the minus sign? Because we want the potential energy to represent stored work. When you lift a book from the floor to your desk, for example, you do positive work on the book. But gravity—the conservative force of interest—does negative work; putting the minus sign in our definition of potential energy then gives a positive value for the "stored work."

Our definition speaks only of *changes* in potential energy. In fact, only changes in potential energy ever matter physically. The actual value of potential energy is physically meaningless. Often, though, it is convenient to

Fig. 7–6
The archer's taut bow stores potential energy that can become kinetic energy of the arrow.

establish a reference point at which the potential energy is defined to be zero. When we then speak of "the potential energy U," we really mean the potential energy difference ΔU between that reference point and some other point. The rock climber of Fig. 7–1a, for example, would find it convenient to take the zero of potential energy at the base of the cliff. When dealing with the gravity of the earth in relation to the sun and other planets, we will find the mathematics simplified if we take the zero of potential energy at an infinite distance from the earth. But the choice is purely for convenience; only potential energy *differences* ever matter, and no physically significant result ever depends on the choice for the zero of potential energy.

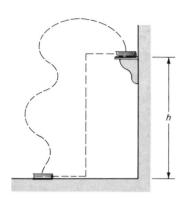

Fig. 7–7
The work done by gravity as the book moves from floor to shelf is the same for both paths shown, but is more easily calculated for the path composed of horizontal and vertical sections.

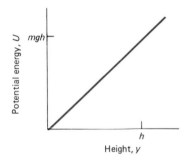

Fig. 7–8
Since the gravitational force is constant, the potential energy increases linearly with height.

Example 7–2
Gravitational Potential Energy

A book of mass m is moved from the floor to a shelf of height h. (a) What is the change in its potential energy? (b) Taking the zero of potential energy at the floor, what is the potential energy when the book is on the floor and when it is on the shelf? (c) Taking the zero of potential energy at the shelf, what is the potential energy when the book is on the floor and when it is on the shelf?

Solution

The potential energy difference is given by Equation 7–2. Here $\mathbf{F}$ is the gravitational force, $-mg\hat{\jmath}$. Over what path are we to evaluate the integral in Equation 7–2? It doesn't matter: gravity is a conservative force, so the work is path-independent. The simplest path involves only horizontal and vertical motion (Fig. 7–7); the work done by gravity on the horizontal sections is zero, and on the vertical section is just $-mgh$. So the potential energy change—the negative of the work done by gravity—is mgh. If we take the zero of potential energy at the floor, then the potential energy at the shelf is mgh. If we take the zero of potential energy at the shelf, then the potential energy at the floor must be such that its value plus the change $\Delta U = mgh$ gives zero. So the potential energy at the floor must be $-mgh$. We can get these results formally by applying Equation 7–2. With $U = 0$ at the shelf height $y = h$, for example, we have

$$U_{\text{floor}} = -\int_{y=h}^{y=0} (-mg\hat{\jmath}) \cdot \hat{\jmath}\, dy = -(-mgy)\Big|_{h}^{0} = -[0 - (-mgh)] = -mgh.$$

Figure 7–8 shows potential energy as a function of vertical position for this example, with the zero of potential energy taken at the floor. The *linear* increase in potential energy with height is a reflection of the *constant* gravitational force.

Just where is the "stored work" represented by the potential energy actually stored? We often speak loosely of the potential energy of a particular object, like the book of Example 7–2. But the book itself hasn't changed as we raise it to the shelf. What has changed is the configuration of a system consisting of the book and the earth. The potential energy change is associated with that change in configuration; potential energy resides not in the book but in the altered gravitational situation between earth and book. We will explore this notion further when we study gravity in the next chapter, and again when we examine the electric force. A spring provides a clearer example that potential energy is associated with a

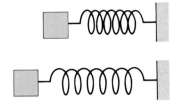

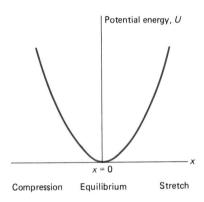

Fig. 7–9
A mass attached to a spring. When the spring is stretched, the increased potential energy is associated not with the mass but with the altered configuration of the spring.

Potential energy, U

$x = 0$

Compression   Equilibrium   Stretch

Fig. 7–10
The potential energy curve for a spring is a parabola.

change in configuration. In Fig. 7–9, the increase in potential energy is clearly associated not with the mass but with the altered configuration of the spring.

<hr>

Example 7–3
The Potential Energy of a Spring

Find the potential energy of a spring of spring constant k when it is stretched or compressed a distance x from equilibrium. Take $U = 0$ in the spring's equilibrium configuration.

Solution

The spring exerts a force $-kx$, so that Equation 7–2 becomes

$$U = -\int_0^x (-kx)\,dx = \int_0^x kx\,dx = \tfrac{1}{2}kx^2 \bigg|_0^x$$

$$= \tfrac{1}{2}kx^2. \quad \text{(potential energy of a spring)} \tag{7–3}$$

This result is the same as the expression we found in Example 6–6 for the work done by an applied force in stretching the spring. Of course: we defined potential energy so it would represent the "stored work" done by an applied force acting to oppose a given conservative force.

Figure 7–10 shows the potential energy as a function of position for the spring. The *parabolic* shape of the potential energy curve—U varying in proportion to x^2—reflects the *linear* relation between force and position.

<hr>

Example 7–4
Electric Potential Energy

In the vicinity of an electrically charged wire of radius R, a proton experiences a force given by

$$\mathbf{F} = F_0 \frac{R}{r}\hat{\mathbf{r}},$$

where $\hat{\mathbf{r}}$ is a unit vector directed radially outward from the center of the wire, r is the distance from the center of the wire, and F_0 is the force that would act on the proton at $r = R$ (F_0 is related to the electric charge on the wire and on the proton). Taking the zero of potential energy at the surface of the wire, calculate the potential energy at an arbitrary point outside the wire.

Solution

Moving radially outward from the wire, we can write our infinitesimal displacement $d\mathbf{r}$ as $\hat{\mathbf{r}}dr$, where dr is a scalar (in the same way, we wrote $d\mathbf{r} = \hat{\mathbf{j}}dy$ when we moved along the y-axis in Example 7–1). Then Equation 7–2 gives

$$U = -\int_R^r \mathbf{F}\cdot d\mathbf{r} = -\int_R^r F_0 \frac{R}{r}\hat{\mathbf{r}}\cdot\hat{\mathbf{r}}\,dr = -F_0 R \int_R^r \frac{dr}{r},$$

where the dot product of $\hat{\mathbf{r}}$ with itself is simply 1 and where we have taken constants outside the integral sign (we can always do this using the distributive law, remembering that an integral is just the limit of a sum). The integral of $1/r$, or r^{-1}, is the one exception to our rule for integrating powers. This integral, in fact, defines the natural logarithm: $\int r^{-1}dr \equiv \ln(r)$. So we have

$$U = -F_0 R \ln(r) \bigg|_R^r = -F_0 R\,[\ln(r) - \ln(R)] = -F_0 R \ln\!\left(\frac{r}{R}\right).$$

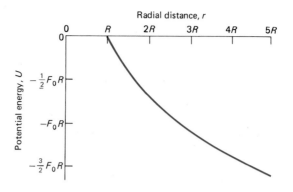

Fig. 7–11
Potential energy curve for Example
7–4. (Curve does not extend to $r < R$
since force was only given for $r \geq R$.)

Does this result make sense? For positive F_0, the force is directed away from the wire. We must then do work to move the proton toward the wire, thereby increasing the stored potential energy. So the potential energy must get lower as we move away from the wire. Since we chose $U = 0$ at $r = R$, the potential energy must be negative for $r > R$, as our expression indeed shows. Note again that only potential energy differences matter. There is nothing impossible or unusual about our negative answer for the potential energy itself; it is significant only in telling us that potential energy decreases as we move away from the wire.

Figure 7–11 shows the potential energy curve for this example, and is analogous to Fig. 7–8 and Fig. 7–10 for the gravitational and spring forces. The potential energy decreases rapidly near the wire, reflecting the strong force there, but soon decreases more gradually as the force and therefore the work done in moving a given distance also decrease.

We worked this example as though we were considering a path radially outward from the wire. But what if we had moved between the points A and D shown in Fig. 7–12? Since potential energy is path-independent (we are dealing with a conservative force whenever we talk about potential energy), we can use *any* path to evaluate the integral for potential energy. In Fig. 7–12, we show a path that extends first in a straight line AB along the wire surface, then in a circular arc BC around the wire, and finally radially outward along a line CD. Along the first two segments, the radially outward force is at right angles to the displacement, and the potential energy change $-\int \mathbf{F} \cdot d\mathbf{r}$ over these segments is zero. Over the segment CD, the potential energy change has the value we calculated in this example; since the changes over the other segments are zero, segment CD gives the overall change in potential energy from A to D. So the potential energy in this case depends only on the radial distance from the center of the wire, and not on how far we move along or around the wire.

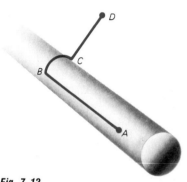

Fig. 7–12
Since potential energy is independent of the path taken, it can be calculated using any path. The potential energy change is zero over the two segments AB and BC of the path shown, and is given by the result of Example 7–4 for the segment CD.

7.3
CONSERVATION OF MECHANICAL ENERGY

In the preceding chapter, we introduced the work-energy theorem, which shows that the change ΔK in a body's kinetic energy is equal to the total work done on it:

$$\Delta K = W.$$

Consider separately the work W_c done by conservative forces and the work W_{nc} done by nonconservative forces, so that

$$\Delta K = W_c + W_{nc}.$$

But we have defined the change in potential energy, ΔU, as the negative of the work done by conservative forces. So we can write

$$\Delta K = -\Delta U + W_{nc},$$

or

$$\Delta K + \Delta U = W_{nc}. \tag{7–4}$$

mechanical energy

We define the sum of the kinetic and potential energy as the **mechanical energy**. Then Equation 7–4 can be written

$$\Delta(K + U) = W_{nc}, \tag{7–5}$$

showing that the change in mechanical energy is equal to the work done by nonconservative forces.

In the absence of nonconservative forces (or when those forces are perpendicular to displacement, so they do no work), then Equations 7–4 to 7–6 show that the mechanical energy is unchanged:

$$\Delta K + \Delta U = 0 \tag{7–6}$$

and, equivalently,

$$K + U = \text{constant}. \tag{7–7}$$

law of conservation of
mechanical energy

Equations 7–6 and 7–7 express the **law of conservation of mechanical energy.** They show that, in the absence of nonconservative forces, there is a quantity $K + U$ that remains always the same. The kinetic energy K may change, but that change is always compensated by an equal but opposite change in potential energy.

Conservation of mechanical energy is a powerful principle, one that is grounded in Newton's laws and in the nature of conservative forces. Throughout physics, from the subatomic realm through practical problems in engineering and on to astrophysics, the principle of energy conservation is widely used in solving problems that would be intractable without it.

Applying the conservation of energy principle is as simple as writing $K + U = \text{constant}$, with the form of U appropriate to the situation at hand and the constant arrived at from the statement of the problem. In classical physics, the kinetic energy K is given by $\frac{1}{2}mv^2$. But the form of the potential energy depends on the details of the conservative force; in Examples 7–2, 7–3, and 7–4 we worked out some specific cases. Usually we are given the speed at some position; then we can find the kinetic energy K_0 and potential energy U_0 at that position. Since mechanical energy is constant, the quantity $K_0 + U_0$ is the constant in Equation 7–7. In the examples below, we show how the conservation of energy principle is applied.

Example 7–5

Accelerating a Proton

The proton of Example 7–4 is released from rest at the surface of the wire. How fast is it moving when it is a distance r from the wire?

Solution

Since only the conservative electric force acts, the mechanical energy $K + U$ is conserved. So we can write

$$K + U = K_0 + U_0,$$

where K_0 and U_0 are the kinetic and potential energy at the surface of the wire. The proton is released from rest, so $K_0 = 0$. In Example 7–4, we choose $U = 0$ at the surface of the wire, so that $U_0 = 0$ as well [we can also get this by setting $r = R$ in our expression for U in Example 7–4, noting that $\ln(R/R) = \ln(1) = 0$]. So the mechanical energy with the proton at the wire surface is 0; since mechanical energy is conserved, this must be its value anywhere else. (Again, remember that only differences in potential energy matter; the actual value of the potential energy and therefore of the total mechanical energy has no significance, so you should not be bothered by zero total energy.) The kinetic energy anywhere is given by $K = \frac{1}{2}mv^2$. From Example 7–4, we know that the potential energy in this case is $U = -F_0R \ln(r/R)$. So we can write

$$K + U = \tfrac{1}{2}mv^2 - F_0R \ln(r/R) = 0.$$

This is the statement of mechanical energy conservation—Equation 7–7—specialized to this particular example. Solving for the speed v gives

$$v = \sqrt{\frac{2F_0R \ln(r/R)}{m}}.$$

Does this result make sense? Yes: Fig. 7–13 shows a plot of speed versus distance r from the center of the wire. At first the speed increases rapidly, as the proton experiences a strong repulsive force. But the force decreases with distance, and while the proton continues to accelerate, it does so at an ever-decreasing rate.

This example illustrates the power of the conservation of energy principle. Applying Newton's second law in this case of a force and therefore acceleration that varies as $1/r$ would prove quite difficult. But conservation of energy bypasses the details of the acceleration and gets us directly to an expression for speed as a function of position. What conservation of energy cannot do, though, is give us all the details; in particular, it can never give us position and speed as functions of *time*. Only the full application of Newton's second law—often requiring computer calculations—can do that.

Fig. 7–13
Speed as a function of position for the proton in Example 7–5.

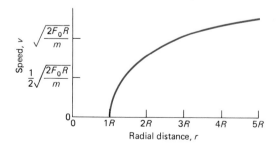

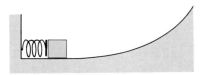

Fig. 7–14
Example 7–6

Example 7–6
Conservation of Energy: A Spring and Gravity

The spring in Fig. 7–14 has spring constant $k = 140$ N/m. A 50-g block is placed against the spring, which is then compressed 11 cm. When the block is released, how high up the slope does it rise? Both the horizontal surface and the slope are frictionless.

Solution

To solve this problem, we equate the total energy $K_0 + U_0$ in the initial position with the total energy $K_1 + U_1$ at the maximum height:

$$K_1 + U_1 = K_0 + U_0.$$

Initially, the block is at rest, so $K_0 = 0$. At maximum height on the slope, it is also at rest, so $K_1 = 0$. The initial potential energy is that of a compressed spring, or $\frac{1}{2}kx^2$ from Example 7–3. The final potential energy is the gravitational potential energy mgh, as we found in Example 7–2. So our statement of energy conservation reads simply

$$\tfrac{1}{2}kx^2 = mgh.$$

Solving for the height h and using the values given, we have

$$h = \frac{kx^2}{2mg} = \frac{(140 \text{ N/m})(0.11 \text{ m})^2}{(2)(0.050 \text{ kg})(9.8 \text{ m/s}^2)} = 1.7 \text{ m}.$$

You might have been tempted in working this example to solve for the speed after the block left the spring, and then equate $\frac{1}{2}mv^2$ to mgh to find the height. And you certainly could have done that. But you don't have to! Conservation of energy short-cuts all details of the motion, and allows you to equate directly the total energy at two points of interest. Here the speed is zero at both these points, so you need not be concerned with kinetic energy.

Example 7–7
A Pendulum

The pendulum of Fig. 7–15 is pulled back until its string makes an angle θ_0 with the vertical, then released from rest. (a) How fast is it going when it reaches its lowest point? (b) At its lowest point, the string catches on a nail located a distance a above the bottom of the string. Assuming the string remains taut, what is the maximum angle θ_1 that the string makes with the vertical?

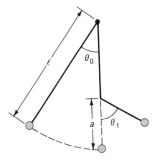

Fig. 7–15
Example 7–7

Solution

Here two forces are acting: the conservative force of gravity and the string tension. But the string tension is perpendicular to the motion, and so does no work. So the total energy remains constant, and the potential energy is associated with the gravitational force alone:

$$K + U = K_0 + U_0.$$

To write out the potential energy terms, we need an expression for the potential energy as a function of angle. Figure 7–16 shows that an angle θ corresponds to a height $h = \ell(1 - \cos\theta)$. Setting the zero of potential energy at the lowest point, we can then write $U_0 = mgh = mg\ell(1 - \cos\theta_0)$, where m is the mass of the pendulum bob (we assume the string has negligible mass). Since the pendulum is released from rest, $K_0 = 0$, so that $K_0 + U_0 = mg\ell(1 - \cos\theta_0)$.

(a) To find the speed at the bottom, we equate the energy at the bottom to our total energy $K_0 + U_0$. At the bottom, we have defined $U=0$, so that $K+U=\frac{1}{2}mv^2$ at the bottom. Then

$$\tfrac{1}{2}mv^2 = mg\ell(1-\cos\theta_0),$$

so that

$$v = \sqrt{2g\ell(1-\cos\theta_0)}.$$

(b) To find the maximum angle as the pendulum swings to the right, we equate the energy in the rightmost position to the total energy $K_0 + U_0$. On the right, the geometry is the same as in Fig. 7–16 but with the length a replacing ℓ. So the potential energy in the rightmost position is $mga(1-\cos\theta_1)$. Since the pendulum is instantaneously at rest in this position, $K=0$, and conservation of energy becomes

$$mga(1-\cos\theta_1) = mg\ell(1-\cos\theta_0).$$

Solving for θ_1 gives

$$\theta_1 = \cos^{-1}\left(\frac{\ell}{a}\cos\theta_0 + \frac{a-\ell}{a}\right).$$

Does this result make sense? When $a=\ell$, so that the nail is the pivot for the pendulum, then $\theta_1 = \theta_0$, and the pendulum swings symmetrically about its lowest position. And no matter where the nail, our conservation of energy statement shows that $a(1-\cos\theta_1) = \ell(1-\cos\theta_0)$; from Fig. 7–16, we see that this means the pendulum rises to the same height on either side, regardless of the presence of the nail. Of course it must! At its extreme positions, it has only gravitational potential energy. Since gravitational potential energy is mgh regardless of the details of the path taken to rise a distance h, the heights must be the same on both sides of the pendulum's arc.

Will the string really remain taut, as we have assumed? Not always! This question is explored in Problems 27 and 60.

How long does it take the pendulum to complete its swing? That is a much harder question, one that energy conservation cannot answer. We will solve Newton's second law for the pendulum in Chapter 14; only then can we determine the time-dependence of the motion.

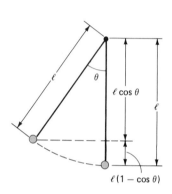

Fig. 7–16
When the pendulum makes an angle θ with the vertical, it is a distance $h = \ell(1-\cos\theta)$ above its lowest position.

Example 7–8

Stopping an Electron

An electron with kinetic energy of 25 keV is moving in the $+x$-direction. At $x=0$, it enters a region where it experiences a conservative force pointing in the $-x$-direction; the force is given by $F = -ax^2$, where $a=18$ eV/cm³. (You should verify that these are reasonable units for a.) How far into the region does the electron penetrate?

Solution

Again, mechanical energy is conserved, so that

$$K + U = K_0 + U_0.$$

If we take $U=0$ at $x=0$, then the right-hand side of the equation is known: $U_0=0$ by definition, and K_0 is given as 25 keV. We don't have an expression for U, but can get this by integrating the force as described by Equation 7–2:

$$U(x) = -\int_0^x F\,dx = -\int_0^x (-ax^2)dx = \int_0^x ax^2\,dx = \tfrac{1}{3}ax^3.$$

(Since a has units eV/cm^3, the units of U are eV when x is measured in cm.) With $K=0$ at maximum x, our conservation of energy statement becomes

$$\tfrac{1}{3}ax^3 = K_0,$$

or

$$x = \left(\frac{3K_0}{a}\right)^{1/3} = \left[\frac{(3)(25 \times 10^3 \text{ eV})}{18 \text{ eV/cm}^3}\right]^{1/3} = 16 \text{ cm.}$$

Note how the units work out without our ever needing to convert eV to joules. Atomic and high-energy physicists often work in the "mixed units" of this example.

7.4
POTENTIAL ENERGY CURVES

Figure 7–17 shows a frictionless roller-coaster track. How fast must a car be coasting at point A if it is to reach point D? What happens if it is going more slowly than this?

The answers to our questions are provided by the conservation of energy principle. If we take the zero of potential energy at the lowest point on the track, then the energy at point A is $\tfrac{1}{2}mv_A^2 + mgh_A$; since mechanical energy is conserved, this is the energy anywhere. To reach point D, the car must clear the highest peak C, where its potential energy is mgh_C. If it just barely clears the peak, its kinetic energy is arbitrarily close to zero on the peak; then the statement of energy conservation becomes $U_C = K_A + U_A$, or

$$mgh_C = \tfrac{1}{2}mv_A^2 + mgh_A,$$

so that

$$mg(h_C - h_A) = \tfrac{1}{2}mv_A^2.$$

In other words, the initial kinetic energy must be at least equal to the potential energy difference between the highest point and the initial point. Solving for v_A would give our answer for the minimum speed required.

What if the car is moving only a little more slowly? Then it won't ever get to the top of the second peak, but will reverse direction before then. Where? At the height h where the kinetic energy becomes zero: $mgh = \tfrac{1}{2}mv_A^2 + mgh_A$. It will head back, clearing peak B, then down and up past point A until its kinetic energy is again zero. Where is this? Again, at a point where $mgh = \tfrac{1}{2}mv_A^2 + mgh_A$—that is, at the same height where it turned earlier. So the car will run back and forth between two **turning points** set by the value of its total energy. With still lower speed, the car won't clear peak B; then its motion will be confined to the first valley alone.

turning points

Fig. 7–17
A roller-coaster track.

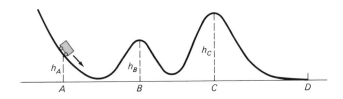

Figure 7–17 is a drawing of the actual roller-coaster track. But, because gravitational potential energy near earth's surface is directly proportional to height, we can also regard it as a plot of potential energy versus position: a **potential energy curve.** We can understand the car's motion graphically by plotting the car's total energy on the same graph as the potential energy curve. Since total energy is constant, the total energy curve is a straight, horizontal line. Figure 7–18 shows the potential energy curve and the total energy curve for several values of the total energy. These graphs tell us immediately about the motion of the car. In Fig. 7–18a, the total energy exceeds the potential energy at peak C; therefore, the car will reach C with kinetic energy to spare, and will make it all the way to D. In Fig. 7–18b, the total energy is less than the potential energy at C; the car must stop when its total energy is entirely potential. This happens when the total energy curve intersects the potential energy curve; the points of intersection are the turning points that bound the car's motion. In Fig. 7–18c, the energy is still lower, and the turning points closer together. We say that the car is **trapped** in the **potential well** between its turning points.

In both Fig. 7–18b and Fig. 7–18c, the car's total energy exceeds the potential energy in the rightmost region, so that motion in this region is energetically possible. But starting at point A, the car is blocked from this region by the **potential barrier** of peak C. In the case of Fig. 7–18c, peak B poses an additional potential barrier that keeps the car out of the valley between B and C where motion is also energetically possible. The terminology of this example—potential wells and barriers—is widely used in many fields of physics.

Although we phrased our discussion in terms of a potential energy curve corresponding directly to physical hills and valleys of a roller-coaster track, potential energy curves remain useful even when such a correspondence is absent. For example, Fig. 7–19 shows the potential energy versus separation distance for a pair of deuterium (heavy hydrogen) nuclei. The potential barrier is associated with the repulsive electric force, while the deep potential well arises from the strongly attractive but short-range nuclear force. Even though the forces here are not gravitational, you can picture the potential barrier as a "hill" with a deep hole on the other side. If you can get a nucleus over the hill, it will "fall" into the hole. The result is a fusion reaction in which two deuterium nuclei fuse to form a helium nucleus, releasing a great deal of energy in the process. (This classical description of fusion is only partially correct; quantum-mechanical effects actually permit "barrier penetration" by nuclei whose energies are below the barrier height.) If nuclear fusion could be harnessed to produce electricity, each gallon of seawater would be energetically equivalent to 400 gallons of gasoline. Since the 1950's, scientists have made slow progress toward the

potential energy curve

trapped
potential well

potential barrier

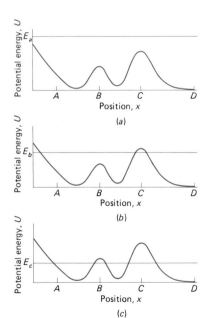

Fig. 7–18
Potential and total energy for a car on a roller coaster. *(a)* The total energy E_a exceeds the potential energy at peak C, and the car can move over the entire track. *(b)* The total energy E_b is lower than the potential energy at C, and the car's motion is confined between the turning points where the total and potential energy curves intersect. *(c)* The total energy E_c is lower than the potential energy at B, limiting the car's motion still further. Although motion between B and C is energetically possible, a car starting at A with this total energy cannot overcome the potential energy barrier of peak B to reach the region between B and C.

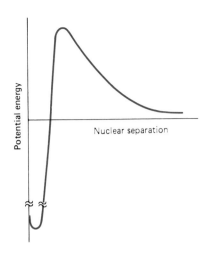

Fig. 7–19
Potential energy curve for a pair of deuterium nuclei. Break in curve indicates that nuclear potential energy well is really much deeper than shown.

goal of developing a controlled fusion reactor—a device in which nuclei are given the high energies needed for fusion, in an environment where there are enough nuclei to ensure frequent collisions that result in fusion reactions. Unfortunately, we've long known how to make fusion work in an uncontrolled fashion, by using fissioning plutonium to "ignite" the fusion reaction in thermonuclear weapons (hydrogen bombs).

Example 7–9

A Diatomic Molecule

The force between the atoms of a diatomic molecule arises from the electrical interactions of the electrons and nuclei in each atom. For simple diatomic molecules, the associated potential energy is given, to a good approximation, by the so-called Lennard-Jones potential:

$$U = U_0\left[\left(\frac{a}{x}\right)^{12} - \left(\frac{a}{x}\right)^6\right], \qquad (7\text{–}8)$$

where x is the distance between the two atoms and where U_0 and a are constants. For oxygen (O_2), $U_0 = 5.6 \times 10^{-21}$ J and $a = 3.5 \times 10^{-10}$ m. (a) Plot the potential energy curve for oxygen, and from it determine the spacing between atoms at the minimum possible energy for the molecule. (b) If the molecule has more than this minimum energy, it may vibrate between two extreme sizes. Find these extrema for the case when the molecule's total energy is 1.0×10^{-21} J over the minimum. (c) What will happen if the molecule is given an energy 2.0×10^{-21} J over the minimum?

Solution

Figure 7–20 shows a plot of Equation 7–8. The minimum possible energy is at the lowest point on the curve, or -1.4×10^{-21} J; if the molecule has this minimum value, its energy is all potential and the two atoms cannot move. At this point, Fig. 7–20 shows that the atomic separation is 3.9×10^{-10} m. Figure 7–20 also shows horizontal lines representing total energies of 1.0×10^{-21} J and 2.0×10^{-21} J above the minimum. From the graph, we see in the case of 1.0×10^{-21} J that the turning points—here the points of maximum and minimum atomic separation—occur at about 3.6×10^{-10} m and 5.4×10^{-10} m. But what happens when the total energy is 2.0×10^{-21} J over the minimum? Then there is a minimum atomic separation of about 3.5×10^{-10} m, but no maximum: the total energy exceeds the potential energy no matter how far apart the atoms get. What does this mean? As would be the case

Fig. 7–20
Potential energy curve for diatomic oxygen, showing three possible values for total energy.

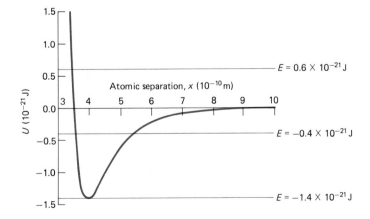

with a roller-coaster track shaped like Fig. 7–20, it means that the atoms get arbitrarily far apart. In other words, they do not stay together to form a molecule. High-speed collisions in a high-temperature gas, for example, might give a molecule enough energy to dissociate in this way.

Note that the zero of potential energy implicit in Equation 7–9 occurs when the atoms are a great distance apart. This is often a convenient choice when dealing with isolated but interacting particles. With this choice, total energies below zero result in two turning points and therefore correspond to the two atoms being bound together; total energies greater than zero correspond to unbound atoms.

7.5
FORCE AND POTENTIAL ENERGY

Figure 7–20 describes the potential energy of an oxygen molecule as a function of atomic separation. But it also tells us about the force acting between the atoms. As they get very close—specifically, to the left of the low point in Fig. 7–20—they experience a repulsive force. This is evident in Fig. 7–20 because the potential energy increases as the molecules move closer, indicating that the interatomic force does negative work on the atoms. Equivalently, if you tried to push the atoms together, you would have to do positive work to overcome the interatomic force. How strong is the repulsive force? That depends on *how rapidly* the potential energy increases with decreasing atomic separation—that is, on the *slope* of the potential energy curve. This makes sense mathematically, too. We wrote the potential energy as the integral of the force; now we find that the force is related to the slope, or derivative, of the potential energy curve. Again, we see the inverse relation between differentiation and integration.

What, exactly, is the relation between potential energy and force? In Fig. 7–20, the repulsive force to the left of the energy minimum points in the +x-direction; the slope of the curve in this region is negative. So our relationship must be

$$F_x = -\frac{dU}{dx}.$$

(7–9)

You can confirm that dU/dx has the units of force, so that no dimensional proportionality constant is needed in Equation 7–9. To the right of the energy minimum, the force is attractive and therefore points in the $-x$-direction; the slope dU/dx is positive, and therefore Equation 7–9 again correctly gives the force. What about the minimum point? Here the slope dU/dx is zero, so there must be no force. That is the one place where the atoms can remain at rest. We will explore such **equilibrium points** in more detail in Chapter 11.

equilibrium points

Example 7–10
Force and Potential Energy

Find an expression for the force between atoms in an oxygen molecule. Use your result to find the force when the atoms are separated by 5.0×10^{-10} m and to determine the equilibrium separation.

Fig. 7–21
Interatomic force in diatomic oxygen, as a function of atomic separation.

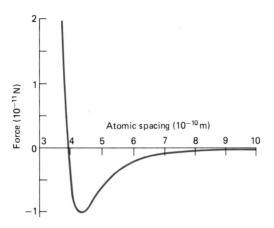

Solution

Equation 7–9 gives the potential energy as a function of separation:

$$U = U_0 \left[\left(\frac{a}{x}\right)^{12} - \left(\frac{a}{x}\right)^{6} \right].$$

Differentiating this expression with respect to x gives the force:

$$F = -\frac{dU}{dx} = -U_0 \left[-12\left(\frac{a^{12}}{x^{13}}\right) + 6\left(\frac{a^6}{x^7}\right) \right] = \frac{6U_0}{a} \left[2\left(\frac{a}{x}\right)^{13} - \left(\frac{a}{x}\right)^{7} \right]. \quad (7\text{–}10)$$

Evaluating this expression at $x = 5.0 \times 10^{-10}$ m, using values of U_0 and a from Example 7–9, gives $F = -6.0 \times 10^{-12}$ N. The minus sign indicates that at this separation the force is attractive.

Setting $F = 0$ in Equation 7–10 allows us to find the equilibrium separation:

$$2\left(\frac{a}{x}\right)^{13} - \left(\frac{a}{x}\right)^{7} = 0,$$

so that, on dividing by $(a/x)^7$, we have

$$\left(\frac{a}{x}\right)^{6} = \frac{1}{2},$$

or

$$x = 2^{1/6}a = (2^{1/6})(3.5 \times 10^{-10} \text{ m}) = 3.9 \times 10^{-10} \text{ m},$$

in agreement with our graphical result from Example 7–9. Figure 7–21 shows a plot of the interatomic force as a function of separation. The graph shows clearly that the force on either side of the equilibrium point tends to restore the molecule to equilibrium. We will explore the significance of such a restoring force further in Chapters 11 and 13.

We developed Equation 7–9, $F = -dU/dx$, for the case when the potential energy depends on only one position variable. In such a case, we may regard Equation 7–9 as the inverse of the potential energy equation $U = -\int F_x dx$. But what about the more general case when U may vary in more than one direction? Imagine you're standing on the side of a mountain. Your gravitational potential energy mgh is directly proportional to

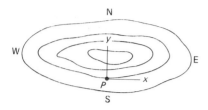

Fig. 7–22
Contour map of a mountain. Contours are lines of equal height and therefore of equal potential energy. At the point P, eastward (x) component of the gravitational force is zero. Gravitational force is a maximum in the southward (−y) direction.

your altitude h. Figure 7–22 shows a contour map of the mountain. The contour lines—lines of constant altitude—are also lines of constant potential energy. If you walk eastward from the point P—that is, along a contour line—your potential energy does not change. And the component of the gravitational force in the direction of your motion is also zero. Mathematically, we can write $F_x = -dU/dx = 0$. But if you walk southward—at right angles to the contour lines—your potential energy changes rapidly and the component of gravitational force in your direction of motion is large. Mathematically, we now write $F_y = -dU/dy$; the minus sign arises because the potential energy increases up the slope, while the force is downslope. More generally, if you walk in some direction along the mountainside, the component of gravitational force in that direction is the negative of the rate of change of your potential energy as you walk; we write

$$F_\ell = -\frac{dU}{d\ell},\qquad(7\text{–}11)$$

where ℓ designates an arbitrary direction and F_ℓ the component of force in that direction.

Can we write an expression for the actual force vector itself? Yes: we already wrote $F_x = -dU/dx$ and $F_y = -dU/dy$. Just what do the derivatives mean here? dU/dx means the rate of change of U as we move *in the x-direction only*; that is, the derivative of U with respect to x *while y is held constant*. Similarly, dU/dy means the derivative of U with respect to y *while x is held constant*. To avoid confusion in cases like this where a quantity (here U) depends on more than one variable (here x and y), we define a **partial derivative** as a derivative taken with respect to only one variable while all other variables are held constant. Partial differentiation is indicated by the special symbol ∂ instead of the usual d; we then write $F_x = -\partial U/\partial x$ and $F_y = -\partial U/\partial y$. So, using unit vector notation, we can write the vector force itself:

partial derivative

$$\mathbf{F} = -\left(\frac{\partial U}{\partial x}\hat{\mathbf{i}} + \frac{\partial U}{\partial y}\hat{\mathbf{j}}\right).\qquad(7\text{–}12)$$

The combination $\hat{\mathbf{i}}\partial/\partial x + \hat{\mathbf{j}}\partial/\partial y$ acting on a scalar function occurs often in physics and mathematics, and is given the special symbol $\boldsymbol{\nabla}$:

$$\boldsymbol{\nabla} \equiv \hat{\mathbf{i}}\frac{\partial}{\partial x} + \hat{\mathbf{j}}\frac{\partial}{\partial y}.\qquad(7\text{–}13)$$

[The extension to a three-dimensional dependence is obvious; we just add a term $(\partial U/\partial z)\hat{\mathbf{k}}$ to Equation 7–12 and a term $\hat{\mathbf{k}}\partial/\partial z$ to the definition 7–13.] Using the definition 7–13, we can write the force in terms of the potential energy in compact vector notation:

$$\mathbf{F} = -\boldsymbol{\nabla}U.\qquad(7\text{–}14)$$

gradient

The vector quantity $\boldsymbol{\nabla}U$ is called the **gradient** of U; it is a vector whose magnitude is the maximum rate of change of U with position, and whose direction is that in which U increases most rapidly. Equation 7–12 shows how to interpret the compact notation of Equation 7–14.

Example 7–11

The Gradient

The potential energy of a particle moving through a region is given by $U = ax^2y + bx$, where a and b are constants. Obtain an expression for the force as a function of position, and show that there is no point where the force is zero.

Solution

The force is given by Equation 7–14, whose interpretation is Equation 7–12:

$$\mathbf{F} = -\boldsymbol{\nabla}U = -\left(\hat{\mathbf{i}}\frac{\partial U}{\partial x} + \hat{\mathbf{j}}\frac{\partial U}{\partial y}\right) = -[(2axy + b)\hat{\mathbf{i}} + ax^2\hat{\mathbf{j}}].$$

Note that, in taking derivatives, we treated y as a constant in the x-differentiation and vice versa. For a vector to be zero, all its components must be zero. So we must have $x = 0$ for the y-component, $F_y = -ax^2$, to be zero. But then the x-component is the nonzero constant $-b$, so that the force cannot be zero anywhere.

7.6

NONCONSERVATIVE FORCES

We developed the conservation of energy principle and the notion of potential energy on the assumption that only conservative forces were acting. But with nonconservative forces like friction, mechanical energy is not conserved and we must go back to Equation 7–5:

$$\Delta(K + U) = W_{nc}, \tag{7–5}$$

where K is the kinetic energy, U the potential energy associated with any conservative forces acting, and W_{nc} the work done by nonconservative forces. It makes no sense to define a potential energy associated with nonconservative forces, since these forces do not "give back" work done against them. When you do work to push a trunk along a floor, there is no tendency for the trunk to slide back when you let go; the trunk gains no potential for doing work.

If we have an expression for W_{nc} in Equation 7–5, we can use that equation as we did the conservation of energy principle to provide a shortcut to the solution of many practical problems.

Example 7–12

A Nonconservative Force

A cross-country skier moving at 4.8 m/s on level ground encounters a nearly frictionless downward slope 6.1 m high. On the level ground below, the snow has been worn thin and consequently the coefficient of friction is 0.27. After coasting down the hill, how far will the skier glide across the level stretch?

Solution

The skier starts with mechanical energy $K + U = \frac{1}{2}mv_0^2 + mgh$, where v_0 is the speed at the top of the slope and h the height of the hill, and where we have taken the zero of potential energy at the bottom of the slope. When the skier has stopped on the level stretch, both the kinetic and potential energy are zero; the total change in mechanical energy is therefore $\Delta(K+U) = -(\frac{1}{2}mv_0^2 + mgh)$. The work done by friction is $\mathbf{F}_f \cdot \Delta\mathbf{r}$ where $\mathbf{F}_f$ is the frictional force and $\Delta\mathbf{r}$ the displacement. On level ground, the frictional force has magnitude μmg; if we take the x-direction as the direction of motion, the work is then $-\mu mg\Delta x$, where the minus sign arises because the force is opposite the direction of motion. Then Equation 7–5, $\Delta(K+U) = W_{mc}$, becomes

$$-(\tfrac{1}{2}mv_0^2 + mgh) = -\mu mg\Delta x.$$

so that

$$\Delta x = \frac{\frac{1}{2}v_0^2}{\mu g} + \frac{h}{\mu} = \frac{(\frac{1}{2})(4.8 \text{ m/s})^2}{(0.27)(9.8 \text{ m/s}^2)} + \frac{6.1 \text{ m}}{0.27} = 27 \text{ m}.$$

7.7
CONSERVATION OF ENERGY AND MASS-ENERGY

When a conservative force acts, work done against that force is "stored" as potential energy. Potential energy comes in many forms. Gravity provides one example of a conservative force and its associated potential energy. So does the force of a spring—ultimately electrical—and the force—also electrical—associated with the interactions of atoms and molecules. The energy contained in a fuel like gasoline is, in fact, electrical potential energy associated with the particular configuration of atoms in the gasoline molecules. When you burn gasoline, those molecules are reconfigured into substances—mostly carbon dioxide and water—whose potential energy is lower. The excess energy becomes kinetic energy of the individual molecules, which in turn may do work as they collide with pistons in an automobile engine. The forces binding atomic nuclei together provide yet another example of stored potential energy. Liberation of this nuclear potential energy powers the sun as well as our nuclear power plants and nuclear weapons.

But what about nonconservative forces? Is energy really lost when they act? No: but it is converted to forms that we don't normally associate with mechanical energy. Friction, for example, converts the kinetic energy of a moving object like a car into randomly directed kinetic energies of individual molecules in the frictional surfaces. The energy is still there; it's just in a form where it's not as obvious or, for reasons we detail in Chapter 19, as useful. (The term "heat" is sometimes used to describe this random molecular kinetic energy, although in Chapter 16 we will find that the term "internal energy" is more accurate.) And what about the electrical energy we supply to a light bulb? It, too, turns into random kinetic energy of atoms in the bulb's filament. Those atoms, in turn, convert their energy into electromagnetic radiation, a form of energy that travels through space at 3×10^8 m/s. We'll see just how that comes about in Chapter 32; leading up to our study of electromagnetic radiation, we will see also how energy can be stored in both electrical and magnetic forms.

Energy can take a variety of forms. Much of physics and engineering involves identifying, following, or controlling the interchanges of energy among its different forms. In classical physics, one overriding principle conservation of energy

governs those interchanges: the principle of **conservation of energy.** If we keep track of all forms of energy involved, we find that energy is never lost in any interchange; it is just converted to another form. The interchange of potential and kinetic energy is just one example that we examined in detail in this chapter. If we account for the random kinetic energy of molecular motion, we find that even a "nonconservative" force like friction conserves energy; we examine this situation further in Chapters 16 to 19. If we account for electromagnetic radiation, we find that processes like the operation of a radio transmitter, the cooling of a hot stove burner, or the death of a star also conserve energy; Chapters 20 to 32 amount to a detailed study of electromagnetic energy.

The conservation of energy principle stands at the heart of classical physics. Alongside it is the principle of conservation of mass. But if we look closely, we find that neither principle really stands by itself. If, for example, we determine the mass of two deuterium nuclei before they fuse to form helium, we find their total mass to be slightly greater than the mass of the helium. Some mass seems to have disappeared; at the same time, energy has been released. Albert Einstein explained this phenomenon as a consequence of his special theory of relativity. Mass and energy, said Einstein, are interchangeable. His equation $E = mc^2$ describes that interchangeability; it says that a quantity of matter m is equivalent to a quantity of energy mc^2 (check the units: they work out!). In deuterium fusion, the energy released is exactly Δmc^2, where Δm is the mass difference between the deuterium nuclei before they react and the helium nucleus after the reaction. We can describe the reaction by saying that a small fraction of the original mass ends up after the reaction as kinetic energy. In more extreme cases, all the mass of a pair of reacting particles may change form. When an electron encounters its antimatter opposite, a positron, the pair annihilate completely; all the mass disappears in a burst of electromagnetic radiation. So much energy is produced in this total annihilation that the mc^2 energy of a single raisin could supply the energy needs of a major city for a day! And the opposite process occurs, too: under suitable conditions, a pair of particles can form out of pure energy (Fig. 7–23). Such total interchange of matter and energy is rare on earth; it occurs only in our large particle accelerators and in some reactions involving cosmic rays. But in some astrophysical situations, and especially in the very early universe, these events play a major role.

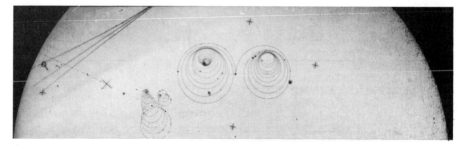

Fig. 7–23
Pair creation. This photograph, taken in the bubble chamber of the Stanford Linear Accelerator, shows the spiral tracks of an electron and positron that materialized out of pure electromagnetic energy. They are curving oppositely in the strong magnetic field that is present.

Conversion of matter to energy is popularly associated with nuclear energy only. That misconception does an injustice to Einstein's universal statement of mass-energy equivalence. Any time there is a net energy loss from a system, its mass goes down by an amount E/c^2. This occurs in burning gasoline just as it does in fissioning uranium; it's just that nuclear reactions convert a much greater portion of their mass to energy than do chemical reactions. Even a stretched spring is more massive than the same spring at equilibrium. The extra mass is given by $\frac{1}{2}kx^2/c^2$, since $\frac{1}{2}kx^2$ is the potential energy stored in the spring. It is only because the speed of light is so large that we do not notice the general equivalence of mass and energy.

conservation of mass-energy

Einstein's mass-energy equivalence replaces the two conservation statements for mass and energy separately with a single statement: the **conservation of mass-energy.** Whatever interchanges occur among different forms of energy, or among energy and matter, we find that the total mass-energy of a closed system remains the same. In this sense, matter and energy are essentially the same basic "stuff," just manifesting itself in different forms. We will see how mass-energy equivalence arises when we examine the theory of relativity in Chapter 33; in Chapter 36 we will see how, at the atomic level, the theory of quantum mechanics further erodes the distinction between matter and energy.

SUMMARY

1. A **conservative force** is one that "stores" work done against it. Mathematically, a conservative force is one that does zero work on an object moved around any closed path:

$$\oint \mathbf{F} \cdot d\mathbf{r} = 0. \quad \text{(conservative force)}$$

A corollary is that the work done by a conservative force as an object is moved between two points is independent of the path taken. Gravity is a familiar example of a conservative force.

2. With a **nonconservative force,** like friction, the work done by the force is not path-independent, and the total work done on an object describing a closed path need not be zero.

3. **Potential energy** describes the "stored work" associated with a conservative force. Mathematically, the potential energy difference as an object moves from point A to point B is defined as the negative of the work done on the object by the conservative force:

$$\Delta U_{AB} = \int_A^B \mathbf{F} \cdot d\mathbf{r}. \quad \text{(potential energy)}$$

Only potential energy differences have physical significance. We are free to assign zero potential energy to any point we choose; when we then speak of the potential energy U at some other point, we really mean the potential energy difference between that point and the point where $U=0$.

The form of the potential energy function depends on the specific conservative force involved. Two commonly encountered cases include gravitational potential energy near earth's surface:

$$U = mgh, \quad \text{(gravitational potential energy)}$$

and the potential energy of an ideal spring:

$$U = \tfrac{1}{2}kx^2. \quad \text{(ideal spring)}$$

4. When only conservative forces act, the **mechanical energy**—the sum $K+U$ of the kinetic and potential energies—remains constant. This statement is the principle of **conservation of mechanical energy.** The conservation of energy principle allows us to solve easily problems that would be difficult to solve using Newton's second law.

5. Graphs of potential energy as a function of position—**potential energy curves**—reveal features of an object's motion. In particular, from such curves we can identify **turning points** where an object's kinetic energy goes to zero; when there are two such points, the object is confined to a **potential well** by **potential barriers** on either side.

6. Potential energy is related to the integral of force over distance; inversely, force can be expressed as the derivative of potential energy. Specifically, the component of force in some direction designated by ℓ is

$$F_\ell = -\frac{dU}{d\ell},$$

where the derivative indicates the rate of change of the potential energy U as one moves in the ℓ-direction.

The force vector may be written in terms of the **gradient** of the potential energy:

$$\mathbf{F} = -\nabla U,$$

where

$$\nabla U = \hat{\mathbf{i}}\frac{\partial U}{\partial x} + \hat{\mathbf{j}}\frac{\partial U}{\partial y} + \hat{\mathbf{k}}\frac{\partial U}{\partial z}.$$

The **partial derivatives** in this expression mean differentiation with respect to one variable while the others are held constant.

7. Energy exists in many forms besides kinetic and potential. These include the internal energy associated with random molecular motions, electrical energy, magnetic energy, and the energy of electromagnetic radiation. The **conservation of energy** principle states that, when all forms are taken into account, energy is neither created nor destroyed; it only changes form.

8. Einstein showed that energy and matter are interchangeable according to the equation $E = mc^2$. His theory of relativity therefore replaces the classical principles of conservation of energy and conservation of matter with the single **conservation of mass-energy** principle.

QUESTIONS

1. Figure 7–24 shows force vectors associated with four different forces. Which are conservative?

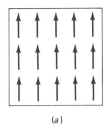 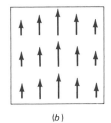

(a)   (b)

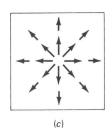 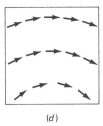

(c)   (d)

Fig. 7–24 Question 1

2. Is the conservation of energy principle related to Newton's laws, or is it an entirely separate physical principle? Discuss.
3. Why can't we define a potential energy associated with friction?
4. Why are you free to choose the zero of potential energy?
5. Can potential energy be negative? Can kinetic energy? Can total mechanical energy? Explain.
6. If the potential energy is zero at a point, must the force also be zero at that point? Give an example.
7. If the force is zero at a point, must the potential energy also be zero at that point? Give an example.
8. Explain how path-independence of the work done by a conservative force follows from the fact that the work done over a closed path is zero.
9. If the difference in potential energy between two points is zero, does that necessarily mean that an object moving between those points experiences no force?
10. A tightrope walker follows an essentially horizontal rope between two mountain peaks of equal altitude. A climber descends from one peak and climbs the other. Compare the work done by the gravitational force on the tightrope walker and on the climber.
11. A bowling ball is tied to the end of a long rope and suspended from the ceiling. A student climbs a stool at one side of the room and holds the ball to her nose, then releases it from rest (Fig. 7–25 left). Should she duck as it swings back (Fig. 7–25 right)? Argue from conservation of energy.
12. An avalanche thunders down a mountainside. Discuss the various energy transfers that resulted in water from the ocean eventually ending up as snow in the avalanche.

Fig. 7–25 Question 11

13. Could you define a potential energy function for a velocity-dependent force?

14. A block of mass m is held against a spring of constant k while the spring is compressed a distance x. Discuss how the speed of the block when released scales with k, x, and m.

15. Figure 7–26 shows a potential energy curve along with total energies for three particles. If the particles are all initially at x_0 and moving to the right, discuss qualitatively their subsequent motions.

16. A block *slides* down a frictionless incline. (a) Is its mechanical energy conserved? (b) If the incline is not frictionless, is mechanical energy conserved? (c) Now consider a cylinder that *rolls* without slipping down the incline. Note that friction is necessary to make the cylinder roll instead of sliding. Nevertheless, energy is conserved in this case. Why?

17. If potential energy is an even function, meaning that $U(x) = U(-x)$, what can you conclude about the force acting at the origin?

18. Why is it a misconception to apply $E = mc^2$ only to nuclear reactions?

19. High-energy physicists often give the mass of elementary particles in eV; for example, the "mass" of an electron is 511 keV. Explain.

20. You use a microwave oven to boil a cup of coffee.

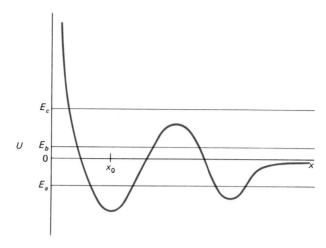

Fig. 7–26 Question 15

Assuming that your electricity comes from a hydroelectric power plant, trace the energy of your hot coffee as far back as you can.

21. In terms of potential energy, how do a new and a rundown battery differ?

PROBLEMS

Section 7.1   *Conservative and Nonconservative Forces*

1. Determine the work done by the frictional force in moving a block of mass m from point 1 to point 2 over the two paths shown in Fig. 7–27. The coefficient of friction has the constant value μ over the surface.

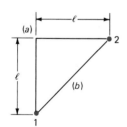

Fig. 7–27 Problem 1

2. Calculate the work done by each of the forces in Example 7–1 as the object is moved along a diagonal straight line from A to C in Fig. 7–5, and compare with the results of Example 7–1.

3. Force $\mathbf{F}_1$ points only in the x-direction, and its magnitude varies with x only. Force $\mathbf{F}_2$ points only in the x-direction, and its magnitude varies with y only. Draw diagrams showing some force vectors for forces with these general properties, and by drawing appropriate closed paths show that one force is conservative and the other nonconservative.

4. A force is given by $\mathbf{F} = F_0 (-\hat{\mathbf{i}} \sin\theta + \hat{\mathbf{j}} \cos\theta)$, where θ is the angle from the x-axis to the point where the force is being evaluated, and F_0 is a constant. Find the work done by the force as an object moves around a circle of radius R centered on the origin. Is the force conservative?

Section 7.2   *Potential Energy*

5. Show using Equation 7–2 that the potential energy difference between the ground and a distance h above the ground is mgh regardless of whether you choose the y-axis upward or downward.

6. An incline makes an angle θ with the horizontal. Find the gravitational potential energy associated with a mass m located a distance x measured along the incline. Take the zero of potential energy at the bottom of the incline.

7. The force exerted by a rubber band is given approximately by

$$F = F_0 \left[\frac{\ell + x}{\ell} - \frac{\ell^2}{(\ell + x)^2} \right],$$

where ℓ is the unstretched length and F_0 a constant. Find the potential energy of the rubber band as a function of the distance x it is stretched. Take the zero of potential energy in the unstretched position.

8. The force on a particle is given by $\mathbf{F} = A\hat{\mathbf{i}}/x^2$, where A is a positive constant. (a) Find the potential energy difference between two points x_1 and x_2, where $x_1 > x_2$. (b) Show that the potential energy difference remains finite even when $x_1 \rightarrow \infty$.

9. A 1.50-kg brick measures 20.0 cm × 8.00 cm × 5.50 cm. Taking the zero of potential energy when the brick lies on its broadest face, what is the potential energy (a) when the brick is standing on end and (b) when it is balanced on its 8-cm edge, with its center directly above that edge? Note: You can treat the brick as though all its mass is concentrated at its center.

10. Can you define a potential energy associated with a force $\mathbf{F} = -bx^4\hat{\mathbf{i}}$, where b is a positive constant? If so, give the potential energy as a function of position. If not, why not?

11. Can you define a potential energy associated with a force $\mathbf{F} = -bx^4\hat{\mathbf{j}}$, where b is a positive constant? If so, give the potential energy as a function of position. If not, why not?

12. The top of the volcano Haleakala on Maui, Hawaii, is 3050 m above sea level and 18 km inland from the sea. By how much does your gravitational potential energy change as you come down from the mountaintop observatory to swim in the ocean? Assume your mass is 75 kg.

13. A particle moves along the x-axis under the influence of a force $F = ax^2 + b$, where a and b are constants. Find its potential energy as a function of position, taking $U = 0$ at $x = 0$.

14. A 60-kg hiker ascending 1250-m-high Camel's Hump mountain in Vermont has potential energy -2.4×10^5 J; the zero of potential energy is taken at the mountain top. What is her altitude?

15. A 3.0-kg fish is hanging from a spring scale whose spring constant is 240 N/m. (a) What is the potential energy of the spring? (b) If the fish were moved slowly upward to the equilibrium position of the spring, by how much would its gravitational potential energy change? (c) In case (b), by how much would the spring's potential energy change? Explain any apparent discrepancies.

Section 7.3 *Conservation of Mechanical Energy*

16. A skier starts down a frictionless 30° slope. After a vertical drop of 22 m, the slope temporarily levels out, then drops at 20° an additional 31 m vertically before levelling out again. What is the skier's speed on the two level stretches?

17. A Navy jet of mass 10,000 kg lands on an aircraft carrier and snags a cable to slow it down. The cable is attached to a spring with spring constant 40,000 N/m. If the spring stretches 25 m to stop the plane, what was the landing speed of the plane?

18. A child is on a swing whose 3.2-m-long chains make a maximum angle of 50° with the vertical. What is the child's maximum speed?

19. Derive Equation 2–18 using the conservation of energy principle.

20. A 200-g block slides back and forth on a frictionless surface between two springs, as shown in Fig. 7–28. The left-hand spring has $k = 130$ N/m and its maximum compression is 16 cm. The right-hand spring has $k = 280$ N/m. Find (a) the maximum compression of the right-hand spring and (b) the speed of the block as it moves between the springs.

Fig. 7–28 Problem 20

21. A ball of mass m is being whirled around on a string of length R in a vertical circle; the string does no work on the ball. (a) Show from force considerations that the speed at the top of the circle must be at least $\sqrt{Rg}$ if the string is to remain taut. (b) Show that, as long as the string remains taut, the speed at the bottom of the circle can be no more than $\sqrt{5}$ times the speed at the top.

22. An 840-kg roller-coaster car is launched from a giant spring of constant $k = 31,000$ N/m into a frictionless loop-the-loop track of radius 6.2 m, as shown in Fig. 7–29. What is the minimum amount that the spring must be compressed if the car is to stay on the track? *Hint:* See preceding problem.

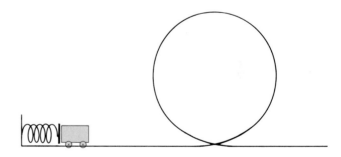

Fig. 7–29 Problem 22

23. An initial speed of 2.4 km/s (the "escape speed") is required for an object launched from the moon to get arbitrarily far from the moon. At a mining operation on the moon, 1000-kg packets of ore are to be launched to a smelting plant in orbit around the earth. If they are launched with a large spring whose maximum compression is 15 m, what should be the spring constant of the spring?

24. A runaway truck lane heads uphill at 30° to the horizontal. If a 16,000-kg truck goes out of control and enters the lane going 110 km/h, how far along the ramp does it go? Neglect friction.

25. A low-damage bumper on a 1500-kg car is mounted on springs whose total effective spring constant is 8.0×10^5 N/m. The springs can undergo a maximum compression of 18 cm without damage to the bumper,

springs, or car. What is the maximum speed at which the car can collide with a stationary object without sustaining damage?

26. A block slides on the frictionless loop-the-loop track shown in Fig. 7–30. What is the minimum height h at which it can start from rest and still make it around the loop?

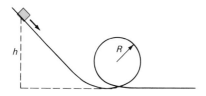

Fig. 7–30   Problem 26

27. Show that the pendulum string in Example 7–7 can remain taut all the way to the top of its smaller loop only if $a \le \frac{2}{5}\ell$. (Note that the maximum release angle is 90° for the string to be taut on the way down.)

28. The maximum speed of the pendulum bob in a grandfather clock is 0.55 m/s. If the pendulum makes a maximum angle of 8.0° with the vertical, what is the length of the pendulum?

29. A 2.0-kg mass rests on a frictionless table and is connected over a frictionless pulley to a 4.0-kg mass, as shown in Fig. 7–31. Use conservation of energy to calculate the speed of the masses after they have moved 50 cm.

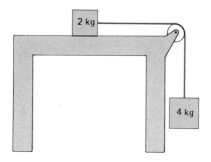

Fig. 7–31   Problem 29

30. Your shot in a pinball game falls short; you would have wanted the 100-g ball to move 15.0 cm farther along the game. You had drawn a spring of constant $k = 140$ N/m back 6.6 cm. If the game is inclined at 20°, how much farther should the next player pull back the spring?

31. Boyoma Falls in Zaire has the largest flow of any waterfall on earth, with 2×10^7 kg of water plunging over the 60-m-high falls each second. If all the energy gained in this fall could be converted to electricity in a hydroelectric generating plant, what would be the power output of the plant? Compare with the 1000-MW output of a typical large nuclear plant.

32. A mass m is dropped from a height h above the top of a spring of constant k that is mounted vertically on the floor (Fig. 7–32). Show that the maximum compression of the spring is given by $(mg/k)(1 + \sqrt{1 + 2kh/mg})$. What is the significance of the other root of the quadratic equation?

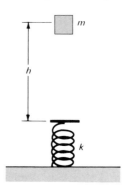

Fig. 7–32   Problem 32

33. A uranium nucleus has a radius of 1.43×10^{-10} m. An alpha particle (mass 6.7×10^{-27} kg) leaves the surface of the nucleus with negligible speed, subject to a repulsive force whose magnitude is $F = A/x^2$, where $A = 4.1 \times 10^{-26}$ N·m², and where x is the distance from the alpha particle to the center of the nucleus. What is the speed of the alpha particle when it is (a) 4 nuclear radii from the nucleus; (b) 100 nuclear radii from the nucleus; (c) very far from the nucleus $(x \rightarrow \infty)$?

Section 7.4   *Potential Energy Curves*

34. Derive an expression for the potential energy of an object subject to a force $\mathbf{F} = (ax - bx^3)\hat{\mathbf{i}}$, where $a = 5$ N/m and $b = 2$ N/m³. Graph the potential energy curve, and use your graph to determine the turning points and the points of maximum speed.

35. The potential energy associated with a conservative force is shown in Fig. 7–33. Consider particles with

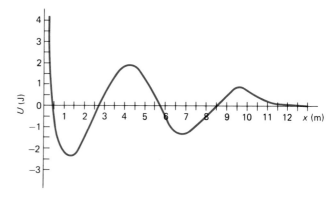

Fig. 7–33   Problem 35

total energies $E_1 = -1.5$ J, $E_2 = -0.5$ J, $E_3 = 0.5$ J, $E_4 = 1.5$ J, and $E_5 = 3.0$ J. Discuss the subsequent motion, including the approximate location of any turning points, if the particles are initially at point $x = 1$ m and moving in the $-x$-direction.

36. Make an accurate potential energy curve, covering the region -8 m$<x<8$ m, for potential energy $U = (ax^2 - b)e^{-x^2/c^2}$, where $a = 1.5$ J/m^2, $b = 5.0$ J, and $c = 3.0$ m. Discuss the subsequent motion of 1-kg particles starting from the origin and moving initially in the $+x$-direction with total energies of -3 J, 1 J, and 4 J. Include the location of any turning points. Determine also the speed of the highest-energy particle when it is a great distance from the origin.

Section 7.5 *Force and Potential Energy*

37. (a) Find an expression for the force in the preceding problem as a function of position. (b) Plot your result and compare with the graph of the preceding problem.

38. (a) Find the force as a function of position for potential energy given by $U = Axy + Bx^2$, where $A = 2$ J/m^2 and $B = 3$ J/m^2. (b) Find all points where the force is directed at 45° counterclockwise from the positive x-axis.

39. The potential energy associated with a conservative force is given by $U = bx^2$, where b is a constant. Show that the force always tends to accelerate a particle toward the origin if b is positive, and away from the origin if b is negative.

40. The potential energy of a spring is given by $U = ax^2 - bx + c$, where $a = 5.20$ N/m, $b = 3.12$ N, and $c = 0.468$ J, and where x is the *overall* length of the spring (not the stretch!). (a) What is the equilibrium length of the spring? (b) What is the spring constant?

Section 7.6 *Nonconservative Forces*

41. Repeat Problem 16 for the case when the coefficient of friction is 0.11.

42. A spring of constant $k = 280$ N/m is used to launch a 2.5-kg block along a horizontal surface whose coefficient of sliding friction is 0.24. If the spring is compressed 15 cm, how far does the block slide?

43. A 2.5-kg block strikes a horizontal spring at a speed of 1.8 m/s, as shown in Fig. 7–34. The spring constant is 100 N/m. If the maximum compression of the spring is 21 cm, what is the coefficient of friction between the block and the surface on which it is sliding?

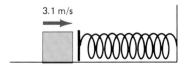

3.1 m/s

Fig. 7–34 Problem 43

44. A meteorite strikes earth and embeds itself 1.7 m into the ground. Scientists dig up the meteorite and find that its mass is 400 g; they estimate that the ground exerted a retarding force of 10^6 N on the meteorite. Estimate the impact speed of the meteorite.

45. In the presence of time-varying magnetic effects, an electron experiences a nonconservative force. In a particular situation, the force vectors lie tangent to concentric circles (Fig. 7–35), and have magnitude given by $F = \frac{1}{2}e\dot{B}r$, where e and $\dot{B}$ are constants (they are the electron charge and the rate of change of the magnetic field), and r the distance from the origin. If an electron starts from rest a distance R from the origin and is constrained by some force to move on a circle of this radius, find its speed after it has made one complete circle.

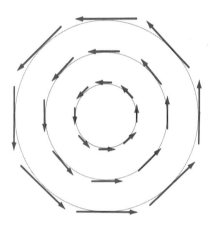

Fig. 7–35 Problem 45

46. A biologist uses a spring-loaded dart gun to shoot a 50-g tranquilizing dart into an elephant 21 m away. The gun's spring has spring constant $k = 690$ N/m, and is pulled back 14 cm to launch the dart. The dart embeds itself 2.2 cm in the elephant. (a) What is the average stopping force exerted on the dart by the elephant's flesh? (b) How long does it take the dart to reach the elephant? Assume the dart's trajectory is nearly horizontal.

47. A surface is frictionless except for a region between $x = 1$ m and $x = 2$ m, where the coefficient of friction is given by $\mu = ax^2 + bx + c$, with $a = -2$ m^{-2}, $b = 6$ m^{-1}, and $c = -4$. A block is sliding in the $+x$-direction when it encounters this region. What is the minimum speed it must have to get all the way across the region?

48. An object starts at rest from point A as shown in Fig. 7–36 and slides down the straight incline (path 1), reaching point B with speed v. The coefficient of friction on the incline is μ. If the coefficient of friction has the same value on the other paths shown, (a) show that the object will reach point B with the same speed v on either path 2 or path 3. (Assume there is a gradual bend at the bottom, but neglect the

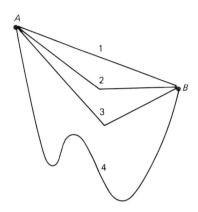

Fig. 7–36   Problem 48

details of this turn.) (*b*) Make a general argument, based on your result in part (*a*), to show that the speed will be the same for an arbitrary path that lies in the plane of the page, like path 4. (*c*) Explain why, in this problem, the work done by friction seems to be path-independent, even though friction is a non-conservative force. Give an example of a path for which your result (*b*) does not hold.

49. Two mountain peaks are a horizontal distance of 3400 m apart. One peak rises 1100 m above the valley between them, the other 900 m. Find the maximum coefficient of friction that will allow a skier starting from rest at the higher peak to reach the lower peak.

50. A bug slides back and forth in a hemispherical bowl of 11 cm radius, starting from rest at the top, as shown in Fig. 7–37. The bowl is frictionless except for a 1.5-cm-wide sticky patch at the bottom, where the coefficient of friction is 0.87. How many times does the bug cross the sticky region?

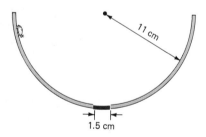

Fig. 7–37   Problem 50

51. A 190-g block is launched by compressing a spring of constant $k = 200$ N/m a distance of 15 cm. The spring is mounted horizontally and the surface directly under it is frictionless. But beyond the equilibrium position of the spring end, the surface has coefficient of friction $\mu = 0.27$. This frictional surface extends 85 cm, followed by a frictionless curved rise, as shown

in Fig. 7–38. After launch, where does the block finally come to rest? Measure from the left end of the frictional zone.

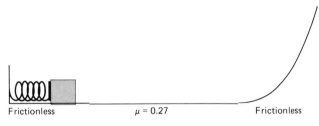

Fig. 7–38   Problem 51

Section 7.7   *Conservation of Energy and Mass-Energy*

52. Particle physicists often describe the "mass" of the proton as 938 MeV. To what mass in kg does this correspond?

53. A hypothetical power plant converts matter entirely into electrical energy. Each year, a worker at the plant buys a box of 1-g raisins at the grocery store; each day, he drops one raisin into the plant's energy conversion unit. Estimate the average power output of the plant. Compare with that of a 500-mW coal-burning power plant, which consumes a 100-car trainload of coal every 3 days.

54. The sun's total power output is 4×10^{26} W. What is the associated rate at which the sun loses mass?

Supplementary Problems

55. Show that, for small displacements from the origin, the potential energy associated with the mass-spring system of Fig. 7–39 is given approximately by $U = \frac{1}{2}kr^2$, where $r^2 = x^2 + y^2$. Take the origin and the zero of potential energy when the mass is centered between the springs. All four springs have the same spring constant k. Why is your result only approximately correct?

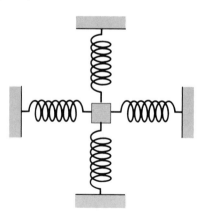

Fig. 7–39   Problem 55

56. A mass m is attached to a spring of constant k that is hanging from the ceiling. (a) Taking the zero of potential energy when the spring is in its normal unstretched position, derive an expression for the total potential energy (gravitational plus spring) as a function of distance y taken positive downward. (b) Find the point where the potential energy is a minimum, and explain its significance. (c) Find a second point where the total potential energy is zero. Discuss its significance in terms of an experiment where you attach the mass to the unstretched spring and let go.

57. With the brick of Problem 9 standing on end, what is the minimum kinetic energy that can be given the brick to make it fall over?

58. A bug lands on top of the frictionless, spherical head of a bald man. It begins to slide down the head (Fig. 7–40). Show that the bug leaves the head when it has dropped a vertical distance one-third the radius of the head.

Fig. 7–40 Problem 58

59. Together, the springs in a 1200-kg car have an effective spring constant of 110,000 N/m, and can compress a maximum distance of 40 cm. What is the maximum abrupt drop in road level (Fig. 7–41) that the car can tolerate without "bottoming out"—that is, without its springs reaching maximum compression? Assume the car is driving fast enough that it becomes temporarily airborne.

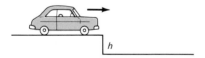

Fig. 7–41 Problem 59

60. Show that the pendulum string in Example 7–7 will cease to be taut when the string has caught on the nail and makes an angle

$$\theta = \cos^{-1}\left[\frac{2\ell}{3a}\left(\cos\theta_0 + \frac{a}{\ell} - 1\right)\right]$$

with the vertical. Show that your answer is consistent with that of Problem 27, in that when $\theta_0 = 90°$ and $a = \frac{2}{5}\ell$, the string remains taut all the way to the top of the small circle.

61. An electron with kinetic energy of 10 keV enters a region where its potential energy is $U = ax^2 - bx$, where $a = 1.7$ keV/cm², $b = 2.6$ keV/cm, and x is the distance in cm from the start of the region. (a) How far into the region does the electron penetrate? (b) What is the maximum speed of the electron? (c) At what point does this maximum speed occur?

62. A particle of mass m is subject to a force $\mathbf{F} = (a\sqrt{x})\hat{\mathbf{i}}$, where a is a constant. The particle is initially at rest at the origin, and is given a slight nudge in the positive x-direction. Find an expression for the particle's speed as a function of position x.

63. (a) Repeat the previous problem for a force $\mathbf{F} = (ax - bx^3)\hat{\mathbf{i}}$, where a and b are positive constants. (b) What is the significance of the negative square root that can occur for some values of x? (c) Find an expression for the particle's maximum speed.

64. A 17-m-long vine hangs vertically from a tree on one side of a 10-m-wide gorge, as shown in Fig. 7–42. Tarzan runs up, hoping to grab the vine, swing over the gorge, and drop vertically off the vine to land on the other side. At what minimum speed must he be running?

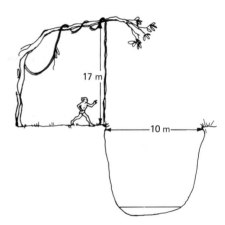

Fig. 7–42 Problem 64

65. A force points in the $-x$-direction with magnitude given by $F = ax^b$, where a and b are constants. Evaluate the potential energy as a function of position, taking $U = 0$ at some point $x_0 > 0$. Use your result to show that an object of mass m released infinitely far from x_0 will reach x_0 with finite velocity provided $b < -1$. Find the velocity for this case.

66. A block slides on a horizontal surface with coefficient of sliding friction $\mu_k = 0.37$. It collides with a spring and stops at the point of maximum compression. If the block hit the spring at 1.79 m/s, and if the spring compressed 22 cm, and if these are the maximum speed and compression for which the block stops, show that the coefficient of static friction is twice the coefficient of sliding friction.

8

GRAVITATION

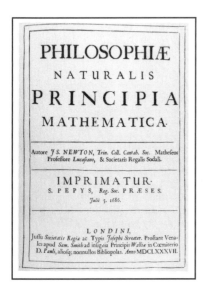

Fig. 8–1
Title page of Newton's *Principia*.

In 1684, Isaac Newton was a professor at Cambridge University when he received a visit from the young astronomer Edmond Halley. Halley had been discussing orbits with colleagues at London's Royal Society, a scientific organization. The London scientists had reasoned from Johannes Kepler's description of planetary orbits that the force acting on the planets must decrease with the square of their distance from the sun.

Halley asked Newton whether he knew what shape orbit such a force would imply. Though Kepler had shown empirically that planetary orbits were ellipses, Halley and his colleagues had been unable to prove this from their conjecture about the inverse-square dependence of the force. "Sr Isaac replied immediately that it would be an Ellipsis [an ellipse]," as was later reported by a person to whom Newton told the story. "The Doctor struck with joy & amazement asked him how he knew it, why saith he I have calculated it."*

After some months, Newton sent Halley a 9-page piece called *De Moto Corporum in Gyrum* (On the Motion of Bodies in an Orbit). In it, he proved all three of Kepler's three empirical rules of planetary motion. Newton became consumed with his work, and by 1686 he submitted to the Royal Society his masterwork *Philosophiae Naturalis Principia Mathematica* (The Mathematical Principles of Natural Philosophy). Halley saw the book through the press, and it was published in 1687 (Fig. 8–1). It was soon recognized as a monumental work that changed the course of science.

In addition to the three laws of motion that we introduced in Chapter 4, Newton advanced a law describing a force that tended to pull masses together. In his successive works in the 1680's he called it "gravity," from the Latin word *gravitas*, meaning "heaviness." We now know gravity as one of the fundamental forces of the universe (Section 4.2). In this chapter, we explore Newton's law of gravity and its application in the context of Newton's laws of motion and the concepts of work and energy.

---

*See R. Westfall, *Never at Rest* (Cambridge University Press, 1980) for an authoritative account of Newton's life and work.

8.1

TOWARD THE CONCEPT OF UNIVERSAL GRAVITATION

Newton's theory of gravity was the culmination of two centuries of scientific revolution that began in 1543 when the Polish astronomer Nicolaus Copernicus made his radical suggestion that the planets orbit not the earth but the sun. Before that time, theories of motion and astronomy had been dominated by the views of the Greek philosopher-scientists Aristotle (ca. 350 B.C.) and Ptolemy (ca. A.D. 140). The Greeks held that the earth was the center of the universe, and the realm of imperfection. The natural tendency of terrestrial objects was to move toward the earth. The heavens, in contrast, were the realm of perfection, the home of the gods and of perfect celestial spheres like stars and planets. In the celestial realm, objects moved in the most perfect way possible—in perfect circles (Fig. 8–2). Even Copernicus maintained the sharp distinction between terrestrial and celestial motion; although he correctly placed the sun at the center of the solar system, he still insisted that planetary motion be described entirely in terms of circles (Fig. 8–3).

Fifty years after Copernicus' work was published, the Danish noble Tycho Brahe began a program of accurate observations of planetary motion (Fig. 8–4). After Tycho's death in 1601, his assistant Johannes Kepler worked for years to make sense of these observations. Success came when Kepler took a radical step: he gave up the requirement that planetary motion involve only perfect circles. In 1609 Kepler published his first two laws:

Fig. 8–2
In the Aristotelian scheme, planets moved about the earth in perfect circles. To match the observed planetary behavior, though, additional circular motions about the main circular path had to be postulated. Ptolemy made this addition.

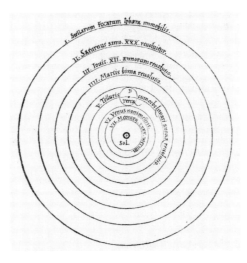

Fig. 8–3
The Copernican theory shifted the center from earth to sun, but maintained that celestial motion involved only circles. We see here the original diagram from Copernicus' book *De Revolutionibus,* published in 1543. Note *sol* (sun) at the center with *telluris* (earth) moving around it.

Fig. 8–4
Before the invention of the telescope, Tycho Brahe developed instruments for accurate measurements of planetary motion. Here Tycho demonstrates a mural quadrant used to measure the altitudes of stars and planets as they crossed the meridian.

Kepler's laws

Kepler's first law: The planets orbit the sun in ellipses, with the sun at one focus.

Kepler's second law: The line joining the sun and a planet sweeps out equal areas in equal times.

The first law sets the shape of the orbit (Fig. 8–5), while the second law describes the speed at which the planet moves along its orbit (Fig 8–6). In 1618 Kepler added his third law, relating the period (the time to complete one orbit) and the semimajor axis of the orbit (half the length of the major axis shown in Fig. 8–5):

Kepler's third law: The square of a planet's orbital period is proportional to the cube of the semimajor axis of its orbit.

Fig. 8–5
Kepler's first law asserts that planetary orbits are ellipses with the sun at one focus. The ellipse shown here is highly exaggerated; the orbits of all the planets but Pluto are nearly but not exactly circular. (A circle is an ellipse with both foci at the center.) Straight line is the major axis of the ellipse; the points are the foci.

These laws were all empirical—worked out to describe the data—and had no theoretical basis. So Kepler knew *how* the planets moved in their orbits, but not *why* they did so.

Shortly after Kepler published his first two laws, Galileo became the first to use a telescope to observe the heavens. With his first telescopes, no more powerful than today's binoculars, he discovered that Jupiter had moons orbiting it, that earth's moon was not perfect but pocked with craters, that the sun was blemished by sunspots, that the Milky Way was made of uncountable stars, and a host of other facts that are now common astronomical knowledge. His observations of Venus (Fig. 8–7) provided strong support for the Copernican theory. For favoring Copernican theory, the church condemned Galileo to life imprisonment, later commuted to house arrest.

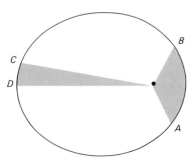

Fig. 8–6
Kepler's equal-area law. If the planet takes the same time to go from *A* to *B* and from *C* to *D*, then the shaded areas are the same. Note that this means the planet must be moving faster when it is closer to the sun.

By Newton's time the intellectual climate was ripe for the culmination of the revolution that had begun with Copernicus. The popular story is that Newton was sitting under an apple tree when a falling apple struck him on the head and caused him to discover gravity. That story is a myth, but if it were true the other half would be that Newton was staring at the moon when the apple hit him (Fig. 8–8). And Newton's stroke of genius was this: he realized that **the motion of the apple and the motion of the moon were the *same* motion; that both were "falling" toward earth under the influence of the same force.** In one of the most sweeping syntheses in all of human thought, Newton eliminated forever the distinction between terrestrial and celestial realms, suggesting that the behavior of everything in the universe follows a single set of physical laws.

8.2

THE LAW OF UNIVERSAL GRAVITATION

law of universal gravitation

In his *Principia,* Newton quantified his three laws of motion, and proposed further his **law of universal gravitation.** Newton suggested that any two particles in the universe exert an attractive force on each other, with the magnitude of the force given by

$$F = \frac{Gm_1m_2}{r^2}, \quad \text{(universal gravitation)} \tag{8–1}$$

constant of universal
gravitation

Fig. 8–7
The phases of Venus. In an earth-centered universe, Venus would not show the full range of phases from crescent to full, and its apparent size would remain constant because of its constant distance from earth. Galileo's observations of the phases of Venus and the associated variation in its apparent size were strong evidence for Copernicus' heliocentric theory.

where m_1 and m_2 are the masses of the particles, r the distance between them, and G a constant. The value of G—called the **constant of universal gravitation**—is 6.67×10^{-11} N·m²/kg², although this value was not determined until after Newton's time. G is truly universal; both observation and theory suggest that it has the same value throughout the universe.

The force of gravity acts *between* the two particles m_1 and m_2. That is, m_1 exerts an attractive force on m_2, and m_2 exerts an equal but oppositely directed force on m_1. The two gravitational forces therefore obey Newton's third law, and constitute an action-reaction pair.

Newton's law of universal gravitation applies strictly only to pointlike particles that have no extent. But, as Newton showed and as we will prove in Section 8.5, it also holds for spherically symmetric objects of any size, provided that the distance r is measured from the centers of the objects. Thus the law 8–1 applies to planets and stars. It also applies approximately to arbitrarily shaped objects provided that the distance between them is large compared with their sizes (see Problem 29). For example, the gravitational force of earth on a space shuttle is given accurately by Equation 8–1 because (1) the earth is spherical and (2) the space shuttle, although irregular in shape, is much smaller than its distance from the center of the earth. To apply Equation 8–1 accurately to closely spaced irregular objects, we add vectorially the gravitational forces between all the individual particles making up the objects. We discuss this process further in Section 8.5.

Example 8–1
The Acceleration of Gravity

(a) Use Newton's law of universal gravitation to find an expression for the force of earth's gravity on a mass m, and apply Newton's second law of motion to calculate the corresponding acceleration. (b) Evaluate this acceleration at earth's surface and the 250-km altitude of a space shuttle. (c) Adapt your calculations to find the acceleration due to the moon's gravity at the lunar surface.

Solution

Since earth is spherical, Equation 8–1 gives the gravitational force of earth on a mass m:

Fig. 8–8
Newton's stroke of genius was the realization that terrestrial and celestial objects move under the influence of the same force: gravity.

$$F = \frac{GM_e m}{R_e^{\ 2}},$$

where M_e and R_e are the mass and radius of the earth. We use R_e because the mass m is at earth's surface, and the distance r in Equation 8–1 is measured from the *center* of the gravitating object.

Newton's second law, $F = ma$, shows that the acceleration is F/m. So

$$a = \frac{F}{m} = \frac{GM_e}{R_e^{\ 2}}. \tag{8–2}$$

Taking values of M_e and R_e from Appendix F gives

$$a = \frac{(6.67 \times 10^{-11}\ \text{N·m}^2/\text{kg}^2)\,(5.97 \times 10^{24}\ \text{kg})}{(6.37 \times 10^6\ \text{m})^2} = 9.8\ \text{m/s}^2.$$

This, of course, is just the value of g—the acceleration of gravity at earth's surface. Our calculation shows clearly the difference between the two numbers G and g commonly associated with gravity. G is a truly universal constant, and is not a measure of acceleration; g is the acceleration of gravity at a particular place in the universe—the surface of the earth—and happens to have the value it does because of the size and mass of the earth.

At the altitude of the space shuttle, $r = R_e + 250$ km, so

$$a = \frac{F}{m} = \frac{GM_e}{r^2} = \frac{(6.67 \times 10^{-11}\ \text{N·m}^2/\text{kg}^2)\,(5.97 \times 10^{24}\ \text{kg})}{(6.37 \times 10^6\ \text{m} + 250 \times 10^3\ \text{m})^2} = 9.1\ \text{m/s}^2,$$

or about 93 per cent of its surface value. This calculation supports our contention in Chapter 4 that weightlessness does not mean the absence of gravitational force. The gravitational force on the orbiting astronauts is about 93 per cent of its value at earth's surface. But, as Equation 8–2 shows, the acceleration of an object arising from the gravitational force is independent of the object's mass. So the astronauts experience the same acceleration as their spacecraft, and are therefore weightless in the spacecraft's frame of reference.

Finally, we can find lunar surface gravity by replacing M_e and R_e with the appropriate values for the moon, also listed in Appendix F:

$$g_{\text{moon}} = \frac{GM_m}{R_m^{\ 2}} = \frac{(6.67 \times 10^{-11}\ \text{N·m}^2/\text{kg}^2)\,(7.35 \times 10^{22}\ \text{kg})}{(1.74 \times 10^6\ \text{m})^2} = 1.6\ \text{m/s}^2,$$

or about one-sixth the value at earth's surface. How can this be, when the moon has only about one one-hundredth of earth's mass? Surface gravity depends on both the mass and the radius of the gravitating object; although the moon is less massive than earth, it is also smaller. An object on the lunar surface is therefore closer to the center, and this effect increases the gravitational acceleration even as the effect of lower mass decreases it.

Although we treated earth and moon as perfectly spherical in Example 8–1, they are actually not. Deviations of earth from spherical symmetry are associated with the overall shape of the planet as well as with geological features like mountains and subsurface rock formations. Using sensitive instruments called gravimeters, geologists map minute variations in g, and interpret their results to help locate mineral and oil deposits (Fig. 8–9). Tracking spacecraft in lunar orbit has led similarly to the discovery of "mascons"—concentrations of mass beneath the lunar surface.

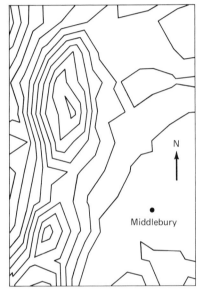

Fig. 8-9
(Above) Student using a gravimeter to determine the local value of g. (Below) Contour map produced from gravimeter data shows variation of g in the vicinity of Middlebury, Vermont. The area shown measures 10 by 15 km; the value of g on adjacent contour lines differs by one milligal or 0.001 cm/s². Structure at the left is a trough of decreased gravitational acceleration associated with the Champlain Thrust, where low-density layers overlie the deeper bedrock.

Example 8-2 _____

The Gravitational Force between Small Objects

(a) Estimate the gravitational force between a 55-kg woman and a 75-kg man when they are 1.6 m apart. (b) Find the gravitational force between two electrons 1.0 cm apart, and compare with the 2.3×10^{-24} N repulsive electrical force that the two electrons also experience.

Solution

The people are hardly spherical, and their separation is not large compared with their sizes, so that Equation 8-1 can provide at best an order-of-magnitude estimate of the force:

$$F \sim \frac{Gm_1m_2}{r^2} = \frac{(6.67 \times 10^{-11} \text{ N·m}^2/\text{kg}^2)\ (55 \text{ kg})\ (75 \text{ kg})}{(1.6 \text{ m})^2} = 10^{-7} \text{ N}.$$

This force is far smaller than typical forces we usually discuss in connection with people-sized objects.

Electrons are essentially point particles, so Equation 8-1 applies exactly. Taking the electron mass from the list of fundamental constants inside the front cover, we have for two electrons

$$F = \frac{Gm_e^2}{r^2} = \frac{(6.67 \times 10^{-11} \text{ N·m}^2/\text{kg}^2)(9.1 \times 10^{-31} \text{ kg})^2}{(0.010 \text{ m})^2} = 5.5 \times 10^{-67} \text{ N},$$

or about 10^{42} times smaller than the electrical force.

Example 8-2 suggests that the gravitational force between subatomic particles and even ordinary-sized objects like people is very small compared with other forces that usually act. Why, then, is gravity the dominant force on the large scale? Because, unlike the electrical force, gravity is always attractive. No antigravity, or negative mass, has ever been found.* So the gravitational force is always cumulative, and large concentrations of matter like stars and planets give rise to substantial gravitational forces. Electrical forces, in contrast, can be attractive or repulsive because there are two kinds of electric charge. A large object like the earth contains nearly equal amounts of positive and negative charge, and therefore exerts no significant electrical force. We will discuss this fundamental difference between the gravitational and electrical forces further in Chapter 20.

The Cavendish Experiment: Weighing the Earth

We showed in Example 8-1 that the acceleration of gravity at earth's surface is given by $g = GM_e/R_e^2$. Given the radius and mass of the earth, and the measured value of g, we could determine G, the constant of universal gravitation. Or we might determine G using the parameters of earth's orbit in conjunction with the mass of the sun. Unfortunately, though, there is no way to determine accurately the masses of planets and stars except by ob-

_____

*Even antimatter should be attracted by the gravitational force of ordinary matter.

serving their gravitational effects and then using Newton's law of gravity to calculate the mass. For this we need to know G.

The only way to determine G is to measure the gravitational force arising from a *known* mass. Given the very weak gravitational force of ordinary-sized objects, though, this is a difficult task. That task was accomplished in 1798 through an ingenious experiment by the British physicist Henry Cavendish. Cavendish mounted two 5-cm-diameter lead spheres on the ends of a rod that was suspended from a thin fiber. He then brought two 30-cm lead spheres near the smaller spheres (Fig. 8–10). The gravitational attraction caused a slight twisting of the fiber. Knowing the properties of the fiber, Cavendish could determine the force involved. From the known masses and separation of the lead spheres, he then used Equation 8–1 to determine G. His result was within 1 per cent of the currently accepted value. Cavendish used his value of G to compute the mass of the earth; indeed, his published paper was titled "On Weighing the Earth."

Improved versions of the Cavendish apparatus are used today both in introductory physics laboratories and in experiments aimed at providing more accurate values of G. Modern experiments often use a laser beam reflected from a mirror mounted on the twisting fiber.

After his death, Cavendish left his considerable fortune to relatives, who later endowed the Cavendish Laboratory at Cambridge University. In "the Cavendish," many of the key discoveries about the nature of matter were made, including the discovery of the electron in 1897, of alpha and beta particles from radioactive decay in 1898, and of the neutron in 1932. Research on the nature of DNA by James Watson and Francis Crick in 1953 was done in the Cavendish as well.

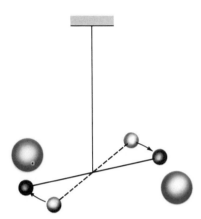

Fig. 8–10
A Cavendish apparatus, used to determine G. The original position of the rod and small spheres is indicated by the dashed line.

8.3
ORBITAL MOTION

orbital motion

Orbital motion refers not just to spacecraft and planets, but quite generally to objects moving under the influence of gravity alone. Minor constituents of the solar system—comets, asteroids, meteors—are, like the planets, in orbital motion about the sun. An individual astronaut, floating outside a spacecraft, is in earth orbit. And the sun itself orbits the center of the galaxy, taking about 250 million years to complete one revolution. Neglecting air resistance, even a baseball is temporarily in orbit. Here we discuss quantitatively the special case of circular orbits, then describe qualitatively the general case.

Circular Orbits

Newton's genius was to recognize that the moon is held in its circular orbit by the same force that pulls an apple to the ground. From there, it was a short step for Newton to grasp the possibility that human-made objects could be put into orbit. Nearly 300 years before the first artificial satellites were launched, he imagined a projectile launched horizontally from a high mountain (Fig. 8–11). If the projectile is simply dropped, it falls vertically like an apple from a tree. But given an initial horizontal velocity, the projectile describes a curve as the gravitational force pulls it from a

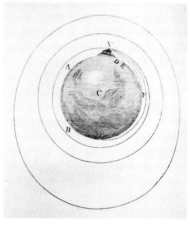

Fig. 8–11
Newton's thought experiment, showing the relation between projectile motion and orbital motion. This is from his *System of the World,* published in 1728.

straight-line path. As the initial horizontal speed is increased, the projectile travels farther before striking earth. But always gravity pulls it out of a straight-line path. Finally, there comes a speed for which the rate at which the projectile is pulled out of its straight-line path is exactly equal to the rate at which earth's surface curves away beneath it. It is then in **circular orbit**, returning eventually to its starting point and continuing forever unless a nongravitational force acts.

circular orbit

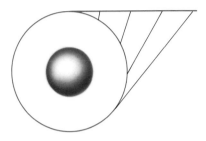

Fig. 8–12
An object in circular orbit is falling away from the straight-line path it would follow in the absence of gravity.

Why doesn't the orbiting object fall toward earth? It does! It falls toward earth in the sense that its motion deviates from a straight line, and the direction of that deviation is toward earth. Under the influence of earth's gravity, it is always getting closer to earth than it would be on a straight-line path (Fig. 8–12). It is behaving exactly as Newton's second law suggests it should. Remember that Newton's laws are not so much about *motion* as they are about *changes in motion*. That emphasis on change is what distinguishes Newton's view from the earlier and misleading Aristotelian view. If you ask why a satellite doesn't fall down, you are adopting the archaic Aristotelian view. The Newtonian question is: why doesn't the satellite move in a straight line? And the answer is simple: because a force is acting. That gravitational force is exactly analogous to the tension force that keeps a ball on a string whirling in a circular path (Fig. 8–13).

We can analyze circular orbits quantitatively because we know that any object in uniform circular motion is undergoing an acceleration v^2/r, where v is its speed and r the radius of the circular path. Newton's first law of motion tells us that a force is necessary to provide this acceleration. Newton's second law tells us the magnitude of that force:

$$F = ma = \frac{mv^2}{r}.$$

Fig. 8–13
A ball whirling on a string is like a satellite in circular orbit. The tension force pulls the ball toward the center of the circle, but the ball never gets closer to the center. Similarly, the gravitational force pulls the satellite toward earth's center, but the satellite never gets closer to the earth.

Some physical force of this magnitude must act to maintain the circular motion. In the case of an orbit, that force is gravity. Newton's law of gravity tells the strength of the gravitational force:

$$F = \frac{GMm}{r^2},$$

where M is the mass of the gravitating object and m the mass of the orbiting object. (We assume here that $M \gg m$, so we can consider the gravitating object to be at rest, essentially unaffected by the gravitational force.) Why do we have two equations for the same force F? The first, $F = mv^2/r$, is Newton's second law. It tells us nothing about the *physical nature* of the force, but only what its magnitude must be to keep the object in its circular path. The second, $F = GMm/r^2$, is a description of the particular physical force acting, in this case the gravitational force. The two describe the same force; equating them gives a relation between v and r that is necessary for a circular orbit:

$$\frac{GMm}{r^2} = \frac{mv^2}{r},$$

or

$$v^2 = \frac{GM}{r}. \tag{8–3}$$

orbital period

Often we're interested in the **orbital period,** or time to complete one orbit. In one period T, the orbiting object moves the orbital circumference $2\pi r$, so that its speed is $2\pi r/T$. Using this expression for v in Equation 8–3 gives

$$\frac{4\pi^2 r^2}{T^2} = \frac{GM}{r},$$

or

$$T^2 = \frac{4\pi^2 r^3}{GM}. \qquad \text{(orbital period, circular orbit)} \qquad (8\text{–}4)$$

In deriving Equation 8–4, we have proved Kepler's third law—that the square of the orbital period is proportional to the cube of the semimajor axis—for the special case of a circular orbit, whose semimajor axis is just its radius.

Note that neither the orbital speed nor the orbital period depends on the mass m of the orbiting object. This is another indication that all objects experience the same gravitational acceleration, and is what allows an astronaut to remain motionless with respect to an orbiting spacecraft (Fig. 8–14).

Fig. 8–14
Astronaut Robert L. Stewart floats motionless with respect to the Space Shuttle from which this picture was taken. Because orbital parameters are independent of the orbiting mass, Stewart remains in the same orbit as the Shuttle, both circling earth at nearly 8 km/s.

Example 8–3
A Space Shuttle Orbit

A space shuttle is in circular orbit 250 km above earth's surface. What are its orbital speed and period?

Solution

Remember that the r appearing in Equations 8–3 and 8–4 is the distance from the *center* of the gravitating object. So here r is the earth's radius plus the 250-km altitude, or $r = 6.37 \times 10^6$ m $+ 0.25 \times 10^6$ m $= 6.62 \times 10^6$ m. Then, taking the mass of the earth from Appendix F, Equation 8–3 gives

$$v^2 = \frac{GM_e}{r} = \frac{(6.67 \times 10^{-11} \text{ N·m}^2/\text{kg}^2)(5.97 \times 10^{24} \text{ kg})}{6.62 \times 10^6 \text{ m}} = 6.0 \times 10^7 \text{ m}^2/\text{s}^2,$$

so that $v = 7.8$ km/s, or about 17,500 mph. Astronauts have no choice; if they want a circular orbit 250 km up, they must achieve this speed.

We can get the orbital period either from the speed and radius, or directly from Equation 8–4:

$$T^2 = \frac{4\pi^2 r^3}{GM_e} = \frac{(4\pi^2)(6.62 \times 10^6 \text{ m})^3}{(6.67 \times 10^{-11} \text{ N·m}^2/\text{kg}^2)(5.97 \times 10^{24} \text{ kg})} = 2.88 \times 10^7 \text{ s}^2,$$

so that $T = 5363$ s, or about 90 minutes. Again, there is no choice: as long as gravity is the only force acting, an orbit at 250 km altitude must take this much time.

Since 250 km is small compared with earth's radius, the orbital speed and 90-minute period found in this example are approximately correct for any near-earth orbit.

Example 8–3 showed that the orbital period in near-earth orbit is about 90 minutes. The moon, on the other hand, takes 27.3 days to complete its nearly circular orbit. (Comparison of the moon's acceleration with that of an object falling at earth's surface helped Newton verify the inverse-square

Fig. 8–15
Weather over Africa and the South
Atlantic as photographed by the
European Space Agency's METEOSAT
satellite from its geosynchronous orbit.

dependence of the gravitational force; see Problem 8.) So there must be a distance where the orbital period is 24 h—the same as the rotation period of the earth. A satellite at this altitude whose orbital motion is parallel to earth's equator will remain at rest with respect to the earth. For this reason, weather and communications satellites are placed in geosynchronous orbit (Fig. 8–15). The increasingly popular home TV dish antennas are trained on geosynchronous satellites (Fig. 8–16).

Example 8–4
Geosynchronous Orbit

At what altitude may a satellite be placed in geosynchronous orbit?

Solution

We want the period T to be 24 h or 8.64×10^4 s. Solving Equation 8–4 for r gives

$$
r = \left(\frac{GM_eT^2}{4\pi^2} \right)^{1/3}
$$

$$
= \left(\frac{(6.67 \times 10^{-11}\ \mathrm{N \cdot m^2/kg^2})\ (5.97 \times 10^{24}\ \mathrm{kg})\ (8.64 \times 10^4\ \mathrm{s})^2}{4\pi^2} \right)^{1/3}
$$

$$
= 4.22 \times 10^7\ \mathrm{m},
$$

or 42,000 km. This is the distance from earth's center; the altitude above earth's surface is then 36,000 km or about 22,000 miles.

Fig. 8–16
This TV satellite dish antenna is trained on a communications satellite 36,000 km above the equator.

Elliptical Orbits

Using his laws of gravity and motion, Newton was able to prove Kepler's first law: that the planets move in elliptical paths with the sun at one focus. Since Newton's law of gravity is truly universal, Kepler's law applies to objects orbiting any spherical, gravitating body (here we assume the gravitating body is much more massive than its satellite; if this is not true, both bodies describe ellipses with one focus at the so-called center of mass of the system; see Fig. 8–17).

The circular orbits we discussed in the preceding section represent the special case when the two foci of the ellipse coincide, and the distance from the gravitating body remains constant. In the more general case, the orbit is distinctly noncircular. For bodies orbiting the sun, the point of closest approach is the **perihelion;** the most distant point is the **aphelion.** For bodies orbiting earth, the corresponding points are **perigee** and **apogee.** The orbits of the planets are nearly circular; earth's distance from the sun, for example, varies by only 3 per cent throughout the year. But the orbits of

perihelion
aphelion
perigee
apogee

Fig. 8–17
(a) Two masses interacting by gravity alone describe elliptical orbits with one focus at the center of mass of the system (small dot). (b) When one object is much more massive, it remains essentially fixed and the other orbits in an ellipse with one focus very near the center of the massive object.

(a) (b)

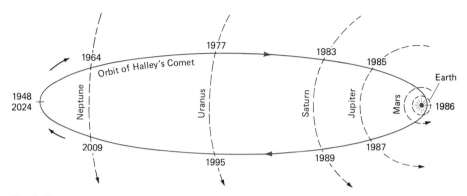

Fig. 8–18
Orbits of Halley's Comet and the planets, showing the years when the comet crosses each planetary orbit.

Fig. 8–19
Halley's Comet as imaged by the European Space Agency's Giotto spacecraft from a distance of 18,000 km. The comet's nucleus is the dark object at left.

comets are typically quite elongated. Figure 8–18 is an accurately scaled diagram showing the orbit of Halley's Comet in relation to the planetary orbits. From the dates of each orbit crossing, you can see the comet falling toward the sun, picking up speed on its inward journey, then whipping around the sun under the influence of the strong gravitational force well inside earth's orbit, and finally slowing as it climbs away from the sun. Halley's most recent dash through the inner solar system occurred in the winter of 1986, with perihelion on February 9. The comet is now outbound, and will spend most of its 76-year period moving slowly through the outer solar system. Although Halley's 1986 journey did not take it as close to earth as in 1910, the comet was visited by an international armada of spacecraft that returned spectacular pictures along with much scientific data (Fig. 8–19).

When we discussed projectile motion in Chapter 3, we concluded that the trajectory of a projectile is a parabola. But in our derivation, we neglected the curvature of the earth and the associated decrease in the acceleration of gravity with altitude. In fact, a projectile launched from earth is just like any other orbiting body. Neglecting air resistance, it too obeys Kepler's first law, describing an elliptical orbit with earth's center at one focus. For trajectories small compared with earth's radius, the true elliptical path and the parabolic path of Chapter 3 are indistinguishable (Fig. 8–20). Un-

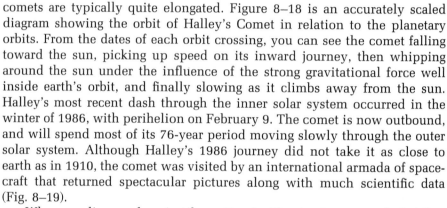

(a)   (b)

Fig. 8–20
(a) For projectiles of limited range, the true elliptical trajectory (solid line) differs only slightly from the parabolic trajectory calculated on the assumption of a flat earth. (b) Long-range trajectories, like those of intercontinental ballistic missiles, are clearly elliptical. Were earth's mass concentrated at the center, such a projectile would continue forever in its elliptical orbit.

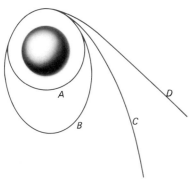

Fig. 8–21
As the projectile speed is increased above that required for circular orbit (*A*), the trajectory becomes first an elongated ellipse (*B*), then a parabola (*C*), and finally a hyperbola (*D*).

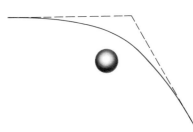

Fig. 8–22
Hyperbolic orbit of an object starting far from the sun with nonzero velocity. Only near the sun does the orbit deviate significantly from a straight line; at great distances the orbit approaches the asymptotes of the hyperbola (dashed lines).

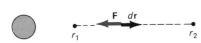

Fig. 8–23
The change in gravitational potential energy between points r_1 and r_2 is found by integrating the gravitational force; along the path of integration the force **F** and path element *dr* are oppositely directed.

like a spacecraft orbit, the orbit of a projectile intersects the earth; at that point, nongravitational forces quickly put an end to the orbital motion. Even a baseball is in an elliptical orbit, and would continue forever in this orbit if earth were suddenly shrunk to the size of a grapefruit. Newton's ingenious insight was correct: barring air resistance, there is truly no difference between the motion of everyday objects near earth and the motion of celestial objects. A continuous range of orbits—from the straight-line path of a falling apple to the circular orbit of the moon—includes all such motions.

Open Orbits

The elliptical and circular orbits we have just discussed are all closed—that is, the same motion is repeated over and over because the orbital path closes back on itself. But closed orbits are not the only ones possible. Imagine again Newton's thought experiment (Fig. 8–21). Firing the projectile at the appropriate speed gives a circular orbit. Firing it faster causes it to rise higher above the earth, giving an elongated ellipse. But for great enough speeds—which we will derive quantitatively in the next section—the projectile travels in a path that takes it ever farther from earth. Eventually it gets so far that the influence of the gravitating mass becomes negligible, and the trajectory approaches a straight line. Mathematically, the shape of the trajectory is a hyperbola (Fig. 8–22). At the dividing line between closed elliptical and open hyperbolic orbits is the parabolic orbit *C* of Fig. 8–21, in which the initial velocity is just barely enough to prevent the orbit from closing. In the next section, we show how orbital energy determines these possibilities.

8.4
GRAVITATIONAL POTENTIAL ENERGY

How much work does it take to boost a satellite from earth's surface to geosynchronous orbit? Equivalently, what is the satellite's gravitational potential energy as a function of height? Our simple answer *mgh* won't do here, since the acceleration of gravity varies with height over the substantial distances involved. The expression *mgh* for potential energy, like the parabolic projectile trajectories of Chapter 3, is an approximation based on the assumption that g is constant—an approximation that is valid only for heights much smaller than the radius of the earth or other gravitating body.

Figure 8–23 shows two points at distances r_1 and r_2 from the center of a gravitating mass *M*. To find the change in potential energy associated with moving a mass *m* from r_1 to r_2, we use Equation 7–2:

$$U_{12} = -\int_{r_1}^{r_2} \mathbf{F} \cdot d\mathbf{r}.$$

Here the force points radially inward, and has magnitude GMm/r^2, while the path element d**r** points radially outward. Then $\mathbf{F} \cdot d\mathbf{r} = -(GMm/r^2)\,dr$, and the potential energy becomes

$$\Delta U_{12} = \int_{r_1}^{r_2} \frac{GMm}{r^2}\, dr = GMm \int_{r_1}^{r_2} r^{-2}\, dr$$

$$= GMm \left. \frac{r^{-1}}{-1} \right|_{r_1}^{r_2} = GMm \left(\frac{1}{r_1} - \frac{1}{r_2} \right). \tag{8-5}$$

Does this result make sense? Yes: for $r_1 < r_2$, the potential energy change is positive, showing that potential energy increases with increasing height. The result 8–5 is thus consistent with our result $\Delta U = mg\,\Delta y$ for the potential energy change associated with a change Δy in vertical position near earth's surface. Problem 34 shows that $\Delta U = mg\,\Delta y$ is a special case of the form 8–5 when $r_1 \approx r_2$. Although we derived Equation 8–5 for two points along a radial line, it holds for any two points at distances r_1 and r_2 from the gravitating center, as suggested by Fig. 8–24.

The potential energy difference 8–5 has an interesting feature: it remains finite even when the points are infinitely far apart. As a result, it takes only a finite amount of work to propel a spacecraft arbitrarily far from earth. Even though the gravitational force is always doing work on the spacecraft as it moves away, that force drops off so rapidly that its cumulative effect remains finite.

This property of the gravitational potential energy makes it convenient to set the zero of potential energy an infinite distance from the gravitating mass. Setting $r_1 = \infty$, and dropping the subscript on r_2, we then have an expression for the potential energy at an arbitrary distance r from the gravitating center:

$$U(r) = -\frac{GMm}{r}. \qquad \text{(gravitational potential energy)} \tag{8-6}$$

The potential energy is negative because we chose $U = 0$ at $r = \infty$; any point $r < \infty$ is closer to the gravitating center and therefore has lower potential energy.

Knowing the form 8–6 for the gravitational potential energy allows us to apply the powerful conservation of energy principle. Figure 8–25 shows the potential energy curve for a gravitating mass. Superimposing several total energy curves shows immediately that an object with $E_{\text{total}} < 0$ is in a bound orbit, for there is a maximum distance beyond which it cannot move. In general, the orbit for $E_{\text{total}} < 0$ is an ellipse. But for $E_{\text{total}} > 0$, the object is not bound and can move infinitely far from the gravitating center. This corresponds to the case of a hyperbolic orbit. The case $E_{\text{total}} = 0$ is the intermediate case of a parabolic orbit; here the object has just barely enough energy to move infinitely far from the gravitating center.

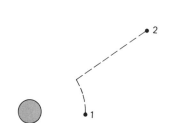

Fig. 8–24
The path between two arbitrary points can be broken into an arc at fixed radius and a radial line. Since the gravitational force is perpendicular to the arc, there is no potential energy change along the arc. Potential energy change along the line is given by Equation 8–5.

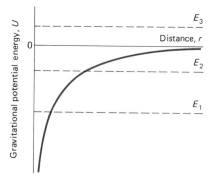

Fig. 8–25
Potential energy curve—a graph of Equation 8–6—for a mass m moving under the gravitational influence of a fixed mass M. Total energies E_1 and E_2, both less than zero, represent bound orbits; positive total energy E_3 represents an unbound orbit.

Example 8–5
Steps to the Moon

Materials to construct a 11,000-kg lunar base module are boosted from earth to geosynchronous orbit. There they are assembled, then launched to the moon, 390,000 km from earth. Compare the work that must be done against earth's gravity on the two steps of the trip.

Solution

As we saw in the preceding chapter, the work that must be done against a conservative force is equal to the associated potential energy difference. That difference is given by Equation 8–5. For the first step we have $r_1 = R_e$ and, from Example 8–4, $r_2 = 42,000$ km. So

$$W = \Delta U_{12} = GM_e m \left(\frac{1}{r_1} - \frac{1}{r_2} \right)$$

$$= (4.38 \times 10^{18} \text{ N·m}^2) \left(\frac{1}{6.37 \times 10^6 \text{ m}} - \frac{1}{4.2 \times 10^7 \text{ m}} \right)$$

$$= 5.8 \times 10^{11} \text{ J},$$

where the quantity 4.38×10^{18} N·m^2 is the value of $GM_e m$ calculated from G, M_e, and $m = 11,000$ kg. From geosynchronous orbit to the moon,

$$W = (4.38 \times 10^{18} \text{ N·m}^2) \left(\frac{1}{4.2 \times 10^7 \text{ m}} - \frac{1}{3.9 \times 10^8 \text{ m}} \right)$$

$$= 9.3 \times 10^{10} \text{ J}.$$

Even though the trip from geosynchronous orbit to the moon is much longer, the rapid drop-off of the gravitational force means that more work is required for the boost from earth to geosynchronous orbit.

The work calculated in this problem is solely the work done against earth's gravity. It does not include the work done to accelerate the lunar module material to orbital speed, and does not take into account work done by the moon. In Section 8.5 we show how to handle the moon's effect as well, while Problem 24 addresses the work associated with orbital speed.

Example 8–6
Halley's Comet

At perihelion on February 9, 1986, Halley's Comet was 8.79×10^7 km from the sun and was moving at 54.6 km/s relative to the sun. Find its speed at its March 13 encounter with the Giotto spacecraft, when the comet was 1.16×10^8 km from the sun, and its speed at its next aphelion in the year 2024, when the comet will be 5.28×10^9 km from the sun.

Solution

The comet's total energy is conserved, so we can write

$$K + U = K_p + U_p,$$

where the subscript p denotes perihelion. The kinetic energy is $\frac{1}{2}mv^2$, while the potential energy is given by Equation 8–6, so that

$$\frac{1}{2}mv^2 - \frac{GM_\odot m}{r} = \frac{1}{2}mv_p{}^2 - \frac{GM_\odot m}{r_p},$$

where $M_\odot$ is the sun's mass ($\odot$ is used by astronomers to designate the sun) and m the comet's mass. The comet mass cancels, and we can solve for the speed v:

$$v = \left[v_p{}^2 + 2GM_\odot \left(\frac{1}{r} - \frac{1}{r_p} \right) \right]^{1/2}.$$

Evaluating this expression at the Giotto encounter gives

$$v_{encounter} = \left[(5.46 \times 10^4 \text{ m/s})^2 + (2)(6.67 \times 10^{-11} \text{ N·m}^2/\text{kg}^2)(1.99 \times 10^{30} \text{ kg}) \right.$$
$$\left. \left(\frac{1}{1.16 \times 10^{11} \text{ m}} - \frac{1}{8.79 \times 10^{10} \text{ m}} \right) \right]^{1/2}$$
$$= 4.74 \times 10^4 \text{ m/s} = 47.4 \text{ km/s}.$$

A similar calculation gives $v_{aphelion} = 2.92$ km/s. The substantial difference between speeds at perihelion and aphelion results from Halley's highly elliptical orbit (see Fig. 8–18). As the comet climbs to aphelion well beyond Neptune's orbit, its gravitational potential energy increases greatly, leaving very little kinetic energy.

Escape Speed

Throw a ball straight up and eventually it comes down. Is this always true? No! Since there is a finite difference in gravitational potential energy between earth's surface and infinity, an object with enough total energy can escape all the way to infinity, and will never come back.

How much energy is needed for escape? At infinity, the potential energy is zero. If an object has barely enough energy, it will reach infinity with no kinetic energy remaining. So zero total energy is required (we saw this graphically in our analysis of Fig. 8–25):

$$K + U = 0.$$

If v is the speed of a mass m at some distance r from a gravitating mass M, then $K = \frac{1}{2}mv^2$ and $U = -GMm/r$, and we have

$$\frac{1}{2}mv^2 - \frac{GMm}{r} = 0,$$

so that

$$v = \sqrt{\frac{2GM}{r}}. \qquad \text{(escape speed)} \qquad (8-7)$$

escape speed   The speed v in Equation 8–7 is known as the **escape speed.** It is the speed that must be given an object at the distance r in order for it to escape all the way to infinity. At earth's surface, $v_{esc} = 11.2$ km/s. Earth-orbiting spacecraft are given somewhat lower speeds, depending on the desired orbital parameters. Moon-bound spacecraft are given speeds just under escape speed, so that if anything goes wrong (as happened with Apollo 13) they will return to earth. Spacecraft headed for other planets have speeds greater than v_{esc}; Pioneer and Voyager missions to the outer solar system left earth's vicinity at 14 km/s. In their encounters with Jupiter these spacecraft gained additional energy, giving them escape speed relative to the sun as well. Leaving the solar system altogether, these craft will coast through the galaxy for the foreseeable future (Fig. 8–26).

On a much grander scale, the concept of escape speed applies to the universe itself. Our universe is known to be expanding; cosmologists ask whether that expansion will continue forever or whether the universe will

Fig. 8–26
In 1983, Pioneer 10 became the first artifact to leave the solar system. With its speed in excess of solar escape speed, Pioneer will travel forever through the galaxy.

eventually collapse on itself. Although detailed analysis requires the general theory of relativity (Section 8.8), the question is basically whether or not the expansion speed exceeds the escape speed associated with the gravitational attraction of all the matter in the universe.

Energy and Circular Orbits

In the special case of a circular orbit, the potential and kinetic energies are related in a very simple way. In Section 8.3, we found that the speed in a circular orbit is given by

$$v^2 = \frac{GM}{r},$$

where r is the distance from a gravitating center of mass M. So the kinetic energy of the object is

$$K = \tfrac{1}{2}mv^2 = \frac{GMm}{2r}.$$

The potential energy is given by Equation 8–6:

$$U = -\frac{GMm}{r}.$$

Comparison of these two expressions shows that

$$U = -2K. \qquad \text{(circular orbit)}$$

The total energy $U + K$ is therefore given by

$$E_{\text{total}} = U + K = -2K + K = -K, \qquad \text{(circular orbit)} \qquad (8\text{–}8)$$

or, equivalently,

$$E_{\text{total}} = \tfrac{1}{2}U. \qquad \text{(circular orbit)} \qquad (8\text{–}9)$$

Equation 8–8 shows that for circular orbits, higher kinetic energy corresponds to lower total energy. This surprising result arises because *higher orbital speeds correspond to lower* orbits. The implications for space flight

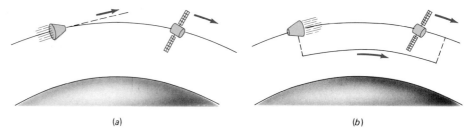

Fig. 8–27
(a) Thrusting toward the target only takes the spacecraft into a higher-energy—and therefore higher but slower—orbit. (b) The correct maneuver is counterintuitive. The spacecraft thrusts *away from* its target, decreasing its total energy and dropping into an elliptical orbit. A second thrust circularizes the orbit; the lower circular orbit is faster than the target's orbit. After coming ahead of the target, the spacecraft reverses this series of maneuvers.

are counterintuitive. To get to a faster orbit, a spacecraft must *lose* energy—as if a car, to speed up, had to apply its brakes. Astronauts Ed White and Jim McDivitt, attempting the first orbital rendezvous in 1965, experienced this firsthand. As they fired their thruster rockets in a direction they thought would take them toward their target, they found that they only moved farther away (Fig. 8–27a). Figure 8–27b shows the correct maneuvers, which involve thrusting opposite the desired direction to achieve a lower but faster orbit.

Example 8–7

Maneuvering a Spacecraft

A shuttle crew is sent to service the Hubble Space Telescope in its 600-km-altitude circular orbit. They maneuver into the same orbit, but find themselves 1.8 km behind the telescope. With two quick thrusts, they drop into an elliptical and then a circular orbit 250 m below the telescope. How long should they stay in this orbit before reversing the procedure to rendezvous with the telescope? Neglect the thrusting time and the time spent in the elliptical transfer orbits. How many orbits does the spacecraft complete during this maneuver?

Solution

Equation 8–3 shows that the speed in circular earth orbit is $v = \sqrt{GM_e/r}$. The telescope's speed v_t is therefore

$$v_t = \left(\frac{GM_e}{r_t}\right)^{1/2} = \left[\frac{(6.67 \times 10^{-11}\ \text{N·m}^2/\text{kg}^2)(5.97 \times 10^{24}\ \text{kg})}{6.97 \times 10^6\ \text{m}}\right]^{1/2} = 7560\ \text{m/s},$$

where $r_t = R_e + 600\ \text{km} = 6.97 \times 10^6\ \text{m}$ is the telescope's orbital radius. Speed in the lower orbit is given by the same expression, with r lower by an amount $\Delta r = 250\ \text{m}$. Although we could solve directly for the speed in the lower orbit, it is more convenient with such a small Δr to use the binomial approximation, $(1+x)^n \simeq 1 + nx$, which holds for $|nx| \ll 1$, and is described in Appendix A. We can write the shuttle speed v_s as

$$v_s = \left(\frac{GM_e}{r_t - \Delta r}\right)^{1/2} = \left(\frac{GM_e}{r_t}\right)^{1/2} \left(\frac{1}{1 - \Delta r/r_t}\right)^{1/2}.$$

The term $(GM_e/r_t)^{1/2}$ is just the telescope speed v_t; rewriting the second term as a quantity to the $-\frac{1}{2}$ power, we have

$$v_s = v_t \left(1 - \frac{\Delta r}{r_t}\right)^{-1/2}.$$

The term in parentheses has the form $(1+x)^n$, with $x = -\Delta r/r_t$ and $n = -\frac{1}{2}$. Since $\Delta r \ll r_t$, we can apply the binomial approximation and write

$$v_s \simeq v_t \left(1 + \frac{\Delta r}{2r_t}\right).$$

The difference between the two speeds—that is, the shuttle speed *relative* to the telescope—is therefore

$$v_{rel} = v_s - v_t = \frac{v_t \Delta r}{2r_t} = \frac{(7.56 \times 10^3 \text{ m/s})(250 \text{ m})}{(2)(6.97 \times 10^6 \text{ m})} = 0.136 \text{ m/s}.$$

Since the upper and lower orbits have nearly the same radius, the shuttle must travel very nearly the entire 1.8 km relative to the telescope in order to catch up. With relative speed v_{rel}, this takes a time Δt given by

$$\Delta t = \frac{\Delta x}{v_{rel}} = \frac{1800 \text{ m}}{0.136 \text{ m/s}} = 1.32 \times 10^4 \text{ s} = 221 \text{ min}.$$

Since the period in near-earth orbit is about 90 min, the spacecraft completes more than two orbits during the maneuver.

8.5
THE SUPERPOSITION PRINCIPLE

So far we have considered only the effect of a single gravitating mass. But things are more complicated than that! The moon, for example, experiences substantial gravitational forces from both earth and sun, as well as smaller forces from the other planets. How are we to handle such cases? Experimentally, we find that gravitational forces given by Equation 8–1 simply add vectorially (Fig. 8–28). This experimental fact is an instance of the **superposition principle.**

superposition principle

To determine the motion of several gravitating masses, we must first add vectorially the gravitational forces acting on each, and then solve Newton's second law for the motion. So complicated is this process that, in general, it is impossible to solve exactly for the motions of even three objects. In practice, computer calculations are used to provide highly accurate approximation in these cases. Before the age of computers, orbital calculations often involved years of tedious work (Fig. 8–29).

Fig. 8–28
The net force on the moon is the sum of the gravitational forces of earth and sun.

Fig. 8–29
Halley's original observations of his comet in 1682. The left-hand page shows observations of September 4, 1682, followed by calculations of the orbit. The right-hand page shows a sketch of the inner orbit.

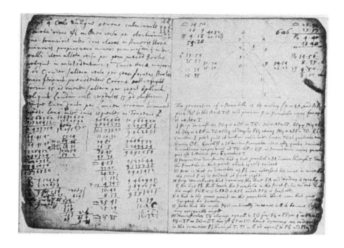

Much of the difficulty in treating complicated gravitational situations involves the vector addition of forces that vary in both direction and magnitude. But because forces add, so do the potential energies associated with these forces. Then the gravitational potential energy of a mass m a distance r_1 from a gravitating mass M_1 and a distance r_2 from a second gravitating mass M_2 is just the *scalar* sum

$$U = -\frac{GM_1 m}{r_1} - \frac{GM_2 m}{r_2} = -Gm\left(\frac{M_1}{r_1} + \frac{M_2}{r_2}\right), \qquad (8\text{–}10)$$

regardless of the directions of the *vectors* $\mathbf{r_1}$ and $\mathbf{r_2}$. In the preceding chapter, we found that the force is given by the gradient of the potential energy, so that we can in principle calculate the gravitational force from the associated potential energy function.

Example 8–8

Potential Energy and Superposition

(a) Find an expression for the gravitational potential energy of a mass m located a distance r from earth's center along a line joining earth and moon. (b) Revise accordingly the calculation in Example 8–5 for the work done in boosting the 11,000-kg lunar base module from geosynchronous orbit to the moon's surface. (c) Use your potential energy function to find an expression for the net gravitational force, and find a point where the force vanishes.

Solution

(a) Letting R be the 390,000-km distance from earth to moon, Equation 8–10 becomes

$$U = -Gm\left(\frac{M_e}{r} + \frac{M_m}{R-r}\right).$$

(b) At geosynchronous orbit, $r = 42{,}000$ km and $R-r = 350{,}000$ km, so that for the 11,000-kg lunar base module

$$U_{geo} = (-6.67 \times 10^{-11} \text{ N·m}^2/\text{kg}^2)(11 \times 10^3 \text{ kg})\left(\frac{5.97 \times 10^{24} \text{ kg}}{42 \times 10^6 \text{ m}} + \frac{7.35 \times 10^{22} \text{ kg}}{350 \times 10^6 \text{ m}}\right)$$

$$= -1.0 \times 10^{11} \text{ J}.$$

At the moon's surface, r is nearly the full 390,000 km and $R - r$ is one lunar radius, or 1740 km, so that

$$U_{moon} = -(6.67 \times 10^{-11} \text{ N·m}^2/\text{kg}^2)(11 \times 10^3 \text{ kg})\left(\frac{5.97 \times 10^{24} \text{ kg}}{390 \times 10^6 \text{ m}} + \frac{7.35 \times 10^{22} \text{ kg}}{1.74 \times 10^6 \text{ m}}\right)$$

$$= -4.2 \times 10^{10} \text{ J}.$$

Then the work needed to lift the module from geosynchronous orbit to the moon is

$$W = U_{moon} - U_{geo} = (-4.2 \times 10^{10} \text{ J}) - (-1.0 \times 10^{11} \text{ J}) = 6 \times 10^{10} \text{ J},$$

about 35 per cent lower than the 9.3×10^{10} J we found in Example 8–5 when we neglected the moon's gravity. (c) In this one-dimensional situation, the gravitational force lies along the earth-moon line; Equation 7–11 then shows that $F = -dU/dr$, or

$$F = Gm\frac{d}{dr}\left(\frac{M_e}{r} + \frac{M_m}{R-r}\right) = -\frac{GmM_e}{r^2} + \frac{GmM_m}{(R-r)^2}.$$

Here $R - r$ is the distance from the moon, so that the second term is the force of the moon's gravity, while the first is that of earth's gravity. The positive term indicates a force in the direction of increasing r—that is, toward the moon—while the negative term indicates an earthward force. The force is zero when

$$\frac{M_m}{(R-r)^2} = \frac{M_e}{r^2},$$

or

$$r\sqrt{M_m} = (R-r)\sqrt{M_e},$$

so that

$$r = R\frac{\sqrt{M_e}}{\sqrt{M_e} + \sqrt{M_m}}.$$

Using the appropriate values shows that the gravitational force vanishes at a point 350,000 km from earth's center. (Because earth and moon are in orbit about each other, this zero-gravity point does not stay fixed, so a particle would not remain at rest there.)

The Gravitational Effect of a Spherical Mass

Newton was delayed 20 years in the publication of his ideas about gravity, until he could prove that a spherical body's gravity acted as though all the mass of the body were concentrated at its center. Here we determine the gravitational effect of a spherical mass by calculating the gravitational potential energy.

We begin by considering a thin spherical shell of mass M and radius R, with a small mass m located some distance r from the center of the shell (Fig. 8–30). We seek an expression for the gravitational potential energy of this configuration, as a function of the distance r. Spherical symmetry ensures that the potential energy depends only on r, and not on angular position, so that the position of our "test mass" m is quite general.

Now imagine dividing the shell into many mass elements ΔM_i. If the mass elements are small enough, they behave essentially like point particles, and each gives rise to a gravitational potential energy given by Equation 8–6:

Fig. 8–30
A small mass m located a distance r from the center of a spherical shell of mass M. Also shown is a hoop-shaped mass element, all of which is a distance s from the test mass m.

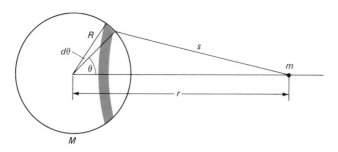

$$\Delta U = -\frac{Gm\,\Delta M_i}{s_i},$$

where s_i is the distance from the mass element ΔM_i to the test mass m. Generalizing Equation 8–10, we can write the total potential energy of the mass m as the sum of the potential energies associated with each of the mass elements:

$$U = -\sum \frac{Gm\,\Delta M_i}{s_i}.$$

In the limit $\Delta M_i \rightarrow 0$, our approximation of the mass elements as point particles becomes exact, and the sum becomes an integral:

$$U = -\int \frac{Gm\,dM}{s}, \qquad (8\text{–}11)$$

where the integration is taken over all mass elements making up the shell.

What constitutes a suitable mass element? Applying Equation 8–6 to write the potential energy of a mass element as $\Delta U = -Gm\Delta M/s$ did not really require that each mass element resemble a point particle, but only that it be small enough so the distance s is the same for all parts of the mass element. Suitable mass elements with this property are narrow hoops like the one shown in Fig. 8–30. In the limit that led to the integral 8–11, such a hoop has infinitesimal mass dM; since the shell is uniform, the ratio dM/M is the same as the ratio of the hoop area to the total area of the shell. Figure 8–30 shows that the hoop subtends an angle $d\theta$, so that its width is $R\,d\theta$. The radius of the hoop is $R\sin\theta$, where θ is shown in Fig. 8–30. Now imagine "unrolling" the hoop; the result is a thin, nearly rectangular strip of area $2\pi R^2 \sin\theta\,d\theta$ (Fig. 8–31). Since the area of the entire shell is $4\pi R^2$, we have

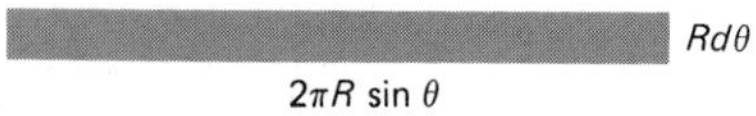

$$\frac{dM}{M} = \frac{\text{hoop area}}{\text{shell area}} = \frac{2\pi R^2 \sin\theta\,d\theta}{4\pi R^2} = \tfrac{1}{2}\sin\theta\,d\theta.$$

Solving for dM and using the result in Equation 8–11 gives

$$U = -\int \frac{GMm}{2s} \sin\theta\,d\theta.$$

To evaluate this integral, we must relate the variables s and θ. Such a relationship is provided by the law of cosines (see Appendix A); applying that law to the triangle shown in Fig. 8–30 gives

$$s^2 = R^2 + r^2 - 2rR\cos\theta.$$

$Rd\theta$

$2\pi R \sin\theta$

Fig. 8–31
Unrolling a hoop-shaped mass element gives a rectangular strip of length $2\pi R \sin\theta$ and width $R\,d\theta$; the area of the strip is $2\pi R^2 \sin\theta\,d\theta$.

Both the shell radius R and the distance r to the test mass are independent of θ, so that differentiating this expression with respect to θ gives

$$2s \frac{ds}{d\theta} = 2rR\sin\theta,$$

since $d(\cos\theta)/d\theta = -\sin\theta$. We can rearrange this expression to get

$$\frac{\sin\theta \, d\theta}{s} = \frac{ds}{rR};$$

using this result in our integral then gives

$$U = -\frac{GMm}{2rR} \int ds.$$

The limits of this integral must be chosen to cover the entire shell; that is, they must include all values of the distance s corresponding to all the hoops making up the shell. Calling s_1 the smallest such value and s_2 the largest, our integral becomes

$$U = -\frac{GMm}{2rR} \int_{s_1}^{s_2} ds = -\frac{GMm}{2rR} (s_2 - s_1). \tag{8–12}$$

We now consider the two distinct cases of (1) the test mass m outside the shell and (2) the test mass m inside the shell.

   Test mass outside shell: In this case, Fig. 8–30 shows that the smallest value of s is $s_1 = r - R$, while the largest is $s_2 = r + R$. Then Equation 8–12 becomes

$$U = -\frac{GMm}{2rR} [(r+R) - (r-R)] = -\frac{GMm}{r}.$$

This result is identical to Equation 8–6 for the potential energy of a mass m a distance r from a point mass M. Thus for a test mass outside the shell, the shell's gravitational effect is the same as if all the shell's mass were concentrated at its center.

   Since a solid sphere can be considered to be composed of a nested set of shells (Fig. 8–32), with each shell acting as if its mass were concentrated at the center, the total gravitational effect of the sphere is still that of a mass point at the sphere's center. Note that the density does not even have to be constant from shell to shell. As long as there is only a radial variation in density, so that the sphere remains spherically symmetric, then the gravitational effect remains that of a mass point concentrated at the center.

   Test mass inside shell: Figure 8–33 shows that in this case $s_1 = R - r$ while $s_2 = R + r$. Then Equation 8–12 gives

$$U = -\frac{GMm}{2rR} [(R+r) - (R-r)] = -\frac{GMm}{R}.$$

Since all the quantities in this expression are constants, the potential energy has the same value everywhere inside the shell. That means no work is involved in moving the test mass around inside the shell, and therefore the force must be zero everywhere inside the shell! How can this be? The answer lies in the inverse-square dependence of the force associated with

Fig. 8–32
A solid sphere can be considered to be composed of a nested set of shells.

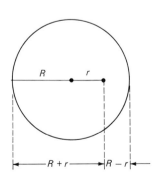

Fig. 8–33
With the test mass inside the shell, values of s range from $R - r$ to $R + r$.

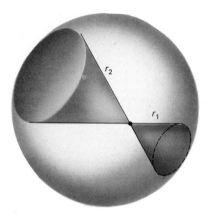

Fig. 8–34
Within any cone of arbitrary angle, the amount of mass on the more distant part of the shell is larger than that on the nearer part by a factor r_2^2/r_1^2. But the force from each mass point on the more distant part is weaker by r_1^2/r_2^2. The result is equal forces in two opposite directions, so that there is no net force anywhere inside the shell.

a point mass. At any point inside the shell, the force from nearby portions of the shell is exactly compensated by the greater mass associated with more distant parts of the shell (Fig. 8–34).

Our results for the gravitational force within and without a shell can be combined to give the force inside a solid sphere like the earth. Since a shell exerts no force on a test mass inside it, the overlying layers have no effect on a test mass inside the sphere (Fig. 8–35). The underlying layers, meanwhile, act as though their mass were concentrated at the center, and provide the total force on the test mass. So we can write

$$F = \frac{GM(r)m}{r^2}, \qquad (8–13)$$

where $M(r)$ signifies the total mass interior to the radius r. The direction of this force is, of course, toward the center. In the case of a *uniform* solid sphere (a poor approximation to the earth, whose density is much greater at the center), the mass within a given radius is proportional to the volume within that radius, so that

$$M(r) = \frac{r^3}{R^3} M,$$

with M the total mass and R the radius of the sphere. Using this expression in Equation 8–13 gives

$$F = G\left(\frac{Mr^3}{R^3}\right)\left(\frac{m}{r^2}\right) = \frac{GMmr}{R^3},$$

so that the gravitational force increases linearly from the center to the surface, after which it decreases with the $1/r^2$ dependence characteristic of a point mass (Fig. 8–36).

8.6
THE GRAVITATIONAL FIELD

Our description of gravity so far has been one in which a massive object like the earth somehow "reaches out" across empty space and tugs on external objects like the moon or a baseball. Such a description is termed **action at a distance** because of its obvious assumption that earth can somehow exert its influence on distant objects.

action at a distance

There is an alternative way of looking at gravity that avoids explicit reference to a force acting between distant objects. A 1-kilogram object placed at some point in earth's vicinity experiences a force of 9.8 newtons, directed downward. A 2-kilogram object placed at the same point experiences twice the force. We ascribe this force not to a direct interaction between earth and each object but rather to a **gravitational field** that exists in the vicinity of earth. Near earth, the field produces a force of 9.8 newtons on every kilogram, directed downward. Since we need to specify both magnitude and direction, the gravitational field is a vector. At every point in space there is a gravitational field vector, giving the magnitude and direction of the gravitational force on a 1-kilogram mass placed at that point. Mathematically, we can write the gravitational field **g** as

gravitational field

Fig. 8–35
A test mass inside a solid sphere. Matter outside the radius r where the test mass is located exerts no force on the test mass, since that matter is composed of spherical shells of which the test mass is on the inside. The gravitational force arises entirely from the mass interior to the radius r, and that mass behaves as though it were concentrated at the center.

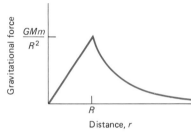

Fig. 8–36
Gravitational force on a test mass m inside and outside a uniform sphere of mass M and radius R. Inside the sphere, force increases with radius because the r^3 increase in enclosed mass overcomes the $1/r^2$ drop-off in strength. Once outside, the enclosed mass remains constant and the force drops as $1/r^2$.

$$\mathbf{g} = \frac{\mathbf{F}_g}{m}. \quad \text{(gravitational field)} \tag{8–14}$$

where $\mathbf{F}_g$ is the gravitational force on a mass m (we assume the mass m is small enough that its own gravitational effects do not alter the arrangement of matter giving rise to the field). Since the force on a mass m in the field of a point particle or spherical mass M has magnitude GMm/r^2 and is directed toward the mass M, we have

$$\mathbf{g} = -\frac{GM}{r^2}\hat{\mathbf{r}}. \quad \left(\begin{array}{c}\text{gravitational field} \\ \text{of a spherical mass}\end{array}\right) \tag{8–15}$$

for the associated gravitational field, where $\hat{\mathbf{r}}$ is a unit vector pointing radially outward and where the minus sign shows that the field points toward the gravitating center. Both the definition 8–14 and Equation 8–15 show that the magnitude of the gravitational field $\mathbf{g}$ is just the local acceleration of gravity, g. The units of $\mathbf{g}$—N/kg—are equivalent to those of acceleration—m/s². The difference is one of interpretation; g reminds us of the acceleration of a mass falling freely under the influence of gravity, while $\mathbf{g}$ tells us the gravitational force per unit mass that arises whether or not the object is allowed to accelerate.

Near earth's surface, g is approximately constant and Equation 8–14 gives $\mathbf{g} = -g\hat{\mathbf{j}}$, where the unit vector $\hat{\mathbf{j}}$ points upward. The gravitational field is uniform, with field vectors of the same magnitude and direction everywhere (Fig. 8–37a). On a larger scale, the field is given by Equation 8–15, and the field vectors differ in magnitude and direction from point to point (Fig. 8–37b).

What is this field? Is it really "something," or just a mathematical construct we use to describe the gravitational interaction? At present you are perfectly justified in arguing the latter point of view. As you progress in

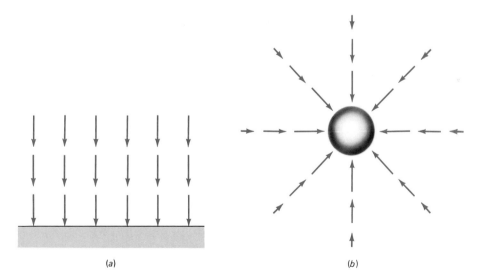

(a)   (b)

Fig. 8–37
(a) Gravitational field vectors in a small region near earth's surface have the same magnitude and direction. (b) On a larger scale, the field vectors vary in both direction and magnitude.

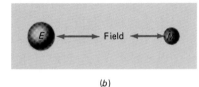

Fig. 8–38
(a) In the action-at-a-distance view, the earth interacts directly with a distant object, for example, the moon. (b) In the field view, earth gives rise to a field everywhere in space. The moon, in turn, responds to the field in its immediate vicinity.

your study of physics, though, you will encounter other forces—particularly those of electricity and magnetism—that are most easily described using the field concept. Gradually that concept will become almost indispensable, and the fields will become increasingly real to you. To a physicist, fields are every bit as real as matter.

Using the field concept, our description of the force on the moon becomes slightly more involved. We now say that the moon interacts only with the field in its immediate vicinity. The field, in turn, is produced by the earth. The field becomes an intermediary between earth and moon (Fig. 8–38).

What do we gain with this more complicated description? If we know the field at a point, we can readily calculate the force on *any* object placed in that field. We gain a more general description that applies to any object. But there is a far more important reason for speaking of fields rather than of action at a distance. It might bother you to think that the moon can somehow "know" that the earth is there to pull on it. It should bother you even more to imagine what would happen if the earth suddenly disappeared. Would the moon know immediately to stop feeling a force? Would it immediately fly off in the straight-line path required by Newton's first law? How could something that happens elsewhere—namely, at earth—instantaneously affect what goes on where the moon is? With the field concept these problems are avoided. The moon responds only to the field in its immediate vicinity. There is no question of its needing to know instantaneously what is going on in some distant place. The field concept makes the interaction a local one, in which what happens at some point depends only on conditions (that is, the field vector) right at that point.

As long as we deal with situations in which nothing changes, the field concept and the action-at-a-distance explanation give the same results. The choice of which to use is a matter of convenience or philosophical inclination. But as soon as we allow objects to move, only the field description can be correct. The reason lies in Einstein's special theory of relativity, which asserts that information cannot be transmitted instantaneously. If the earth disappears or simply jumps sideways so that "down" becomes a new direction, the field in the moon's vicinity cannot change instantaneously. Instead, a change in the field propagates outward from earth at a high, but finite, speed (in fact, it is the speed of light). The moon experiences the changed field only after a small but nonzero time has elapsed. Only when the field has changed will the direction or magnitude of the force on the moon also change. This field description is consistent with the requirement that information not be transmitted instantaneously. The action-at-a-distance view is not.

8.7
TIDAL FORCES AND THE ROCHE LIMIT

When the gravitational field is uniform, all parts of a freely falling object experience exactly the same gravitational acceleration (Fig. 8–39a). But when the field is nonuniform, the acceleration of gravity varies in magnitude and/or direction from place to place (Fig. 8–39b). The result is a

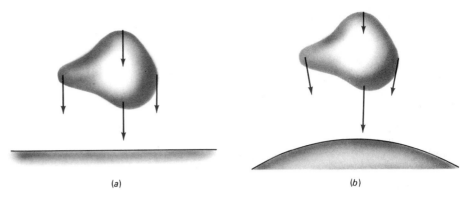

Fig. 8–39
(a) In a uniform gravitational field, all parts of an object experience the same gravitational acceleration. (b) When the field is not uniform, gravity acts differently on the different parts of the object. The result is a differential force that tends to stretch an object along the field and compress it at right angles to the field.

differential force

tidal forces

differential force—depending not on the strength of gravity but on how gravity varies from place to place—that tends to stretch or compress the object.

Ocean tides are an important manifestation of this differential force (Fig. 8–40). Because of the association with ocean tides, the differential forces of gravity are usually called **tidal forces.** In reality, the effects of continents and sea floor topography greatly alter the simplified tidal picture of Fig. 8–40. And the differential force of the sun, although weaker than that of the moon, still has a significant effect. As a result, tides are highest when the sun, moon, and earth are in a line and weakest when they make a right angle.

The force of the sun's gravity on earth is far greater than that of the moon (see Problem 37). Why, then, should the moon's gravity be the dominant influence on the tides? The answer lies in the very rapid drop-off of tidal forces with distance. Consider an object consisting of two small masses m separated by a distance $2a$, and located some distance r from a gravitating mass M (Fig. 8–41). The tidal force arises from the *difference* in gravitational field across the object. With the object oriented as in Fig. 8–41, the gravitational force points in the same direction at both ends of the object; its magnitude at either end is given by Equation 8–1, using $r \pm a$ for the distance. So the tidal force on the object—the difference between

Earth   Moon

Fig. 8–40
Ocean tides result primarily from the differential force of the moon's gravity. The gravitational force is strongest on the moonward ocean, weaker on the solid earth, and weakest on the distant ocean. Thus the water on the moonward side is pulled away from the solid earth, while earth itself is pulled away from the distant water. Two tidal bulges result, so that a given location experiences two high tides each day as earth rotates. (Not shown is the differential force associated with the sun's gravity, which is weaker but also important.)

Fig. 8–41
The tidal force on an object arises from the difference in gravitational field strength across the object.

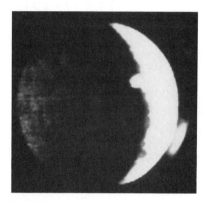

Fig. 8–42
Two volcanic eruptions on Jupiter's moon Io are visible in this image from Voyager I. The volcanic plume at right extends for 250 km. The bright spot on the terminator—the line between night and day—is another erupting volcano. Io's volcanism is thought to arise from heat generated internally by tidal stresses.

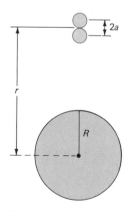

Fig. 8–43
Two small spheres attracted by their own gravity, but pulled apart by the tidal force of the planet they orbit. Which dominates depends on how close they are to the planet.

the gravitational forces on either end—is

$$F_T = \frac{GMm}{(r-a)^2} - \frac{GMm}{(r+a)^2} = GMm\frac{(r+a)^2 - (r-a)^2}{(r-a)^2(r+a)^2}.$$

Expanding both terms in the numerator, and noting that $(r-a)(r+a) = r^2 - a^2$, we have

$$F_T = GMm\frac{(r^2 + 2ra + a^2) - (r^2 - 2ra + a^2)}{(r^2 - a^2)^2} = \frac{4GMmra}{(r^2 - a^2)^2}.$$

If the object's size $2a$ is small compared with the distance r from the gravitating object—as is the case with earth's diameter in relation to the earth-moon distance—then we can neglect a^2 compared with r^2 in the denominator, so that the tidal force becomes very nearly

$$F_T = \frac{4GMma}{r^3}. \qquad \text{(tidal force)} \qquad (8\text{–}16)$$

This equation shows that the tidal force drops off as the inverse cube of the distance—much more rapidly than the gravitational force itself. For this reason the tidal force of the moon is greater than that of the more massive but more distant sun.

Tidal forces exert stresses on any extended object in a nonuniform gravitational field, and the astronomical consequences of tidal forces are therefore numerous. The extensive volcanic activity of Jupiter's moon Io (Fig. 8–42) is thought to arise from heat generated by tidal stresses. Tidal forces can even become strong enough to break an object apart. This happened to the Comet Biela in 1846, when it was split into two pieces by the strong tidal forces at perihelion.

Since astronomical objects are generally held together by their own gravity, we can ask when tidal forces might become strong enough to overcome the internal gravitational force. Figure 8–43 shows an object consisting of two small spheres of mass m and radius a in free fall a distance r from a planet of mass M. Since the small masses are spherical, they attract each other gravitationally as though they were point masses located at their centers, a distance $2a$ apart. So the magnitude of the attractive force between them is

$$F_g = \frac{Gm^2}{(2a)^2} = \frac{Gm^2}{4a^2}.$$

In the configuration of Fig. 8–43, the tidal force associated with differences in the planet's gravity tends to pull the spheres apart; since the spheres are small compared with their distance to the planet's center, and their own centers are a distance $2a$ apart, the magnitude F_T of the tidal force is given by Equation 8–16:

$$F_T = \frac{4GMma}{r^3}.$$

The object will be pulled apart if $F_T > F_g$, so that

$$\frac{4GMma}{r^3} > \frac{Gm^2}{4a^2}$$

Fig. 8–44
The rings of a planet lie within the planet's Roche limit, where tidal forces overcome the gravitational forces that would hold a moon together. Here, left to right, are the rings of Jupiter, Saturn, and Uranus.

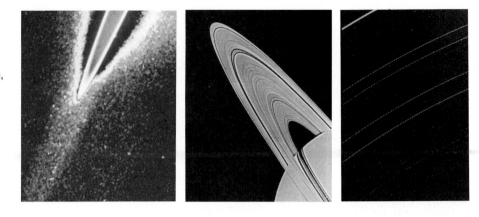

or when

$$r^3 < \frac{16Ma^3}{m}.$$

If we let ρ_p be the density of the planet, R its radius, and ρ_s the density of its satellite—the composite object we are considering—then $M = \frac{4}{3}\pi R^3 \rho_p$ and $m = \frac{4}{3}\pi a^3 \rho_s$. Then our condition for tidal break-up becomes

$$r < \left(\frac{16\rho_p}{\rho_s}\right)^{1/3} R. \qquad (8\text{--}17)$$

Roche limit

The critical radius r for tidal break-up is called the **Roche limit,** after the French scientist Edouard Roche, who suggested in 1848 that a moon would break up within this distance of a planet's center. Equation 8–17 shows that the Roche limit occurs at about 2.5 times the planet's radius when planet and satellite densities are comparable. A less simplified analysis would show the Roche limit depends also on the composition of the planet and satellite and whether the satellite material is liquid or solid.

Within the Roche limits of the planets Jupiter, Saturn, Uranus, and Neptune we find rings rather than moons (Fig. 8–44), showing that the tidal force prevents material within that limit from collapsing gravitationally to form moons. Beyond their Roche limits, all these planets have moons. We regularly place artificial satellites in orbit within earth's Roche limit, but these are held together by bolts, welds, and the strengths of materials rather than by gravitational forces, so that our force analysis does not apply.

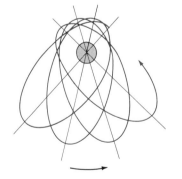

Fig. 8–45
The shift in Mercury's perihelion results in an orbit that does not quite close on itself. Both the shift and ellipticity of the orbit are greatly exaggerated; the actual orbit is nearly circular, and the perihelion shift amounts to only 43 seconds of arc per century.

8.8
GRAVITY AND THE GENERAL THEORY OF RELATIVITY

Although Newton's theory of gravity provided a brilliantly successful explanation of planetary motion, and today guides space probes like Voyager on billion-mile journeys to the outer planets, that theory is not perfectly correct. In very strong gravitational fields deviations from Newtonian predictions become evident. For decades the only evidence for this was a very slight shift in the position of the perihelion of Mercury's orbit (Fig. 8–45),

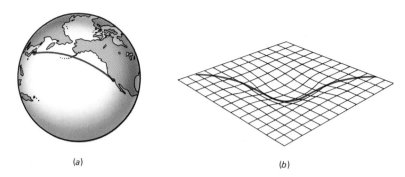

(a)   (b)

Fig. 8–46
(a) The "straightest" path between San Francisco and Tokyo is a great circle skirting the Aleutian Islands. (b) Similarly, the "straightest" path in the curved space and time near a massive object is not a straight line in the Newtonian sense. Depression in the surface shown represents the effect of a massive object on space and time.

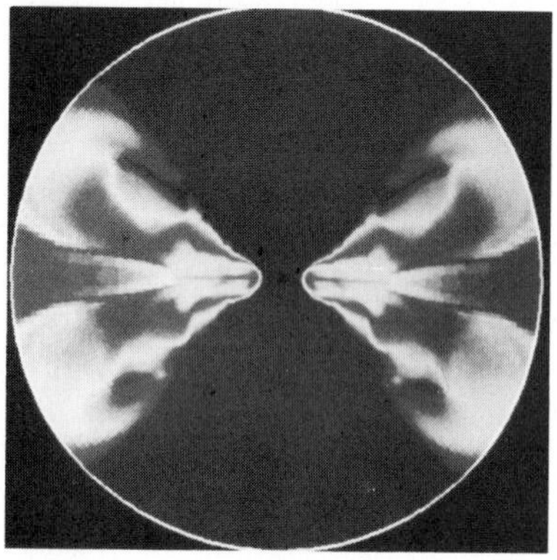

Fig. 8–47
This supercomputer simulation shows the results of a calculation of the motion of gas as it spirals into a black hole. The black hole itself is invisible at the center, but we see the disk of gas that forms around it. X-rays emitted by the hot gas allow astronomers to study such systems.

amounting to only 43 seconds of arc per century (one second is 1/3600 of a degree).

That shift is consistent with the general theory of relativity, a radically different description of gravity completed by Albert Einstein in 1916. Unlike his special theory of relativity, which we discuss in detail in Chapter 33, Einstein's general theory is largely beyond the scope of this book. Here we note only that Einstein describes gravity as a geometrical effect resulting from the curvature of space and time caused by the presence of matter. In Einstein's theory, the natural state of motion is not uniform motion but free fall. An object moving under the influence of gravity alone follows the "straightest" path in curved space and time. But that straightest path need not be a true straight line, just as the straightest path on the surface of the round earth is not a straight line (Fig. 8–46).

The general theory predicts deviations from the Newtonian description of space, time, and gravity that become more significant as the gravitational field increases. Within our solar system, careful measurements of Mercury's orbit along with more recent precision spacecraft experiments have confirmed Einstein's remarkable theory. So have earth-based experiments that measure the minute effect of earth's gravity on time itself. And sensitive measurements of star positions during solar eclipses have confirmed Einstein's prediction of the influence of gravity on light.

Only with the past two decades' advances in astrophysics, though, have the effects of general relativity provided more than minor corrections to Newtonian theory. Today we observe extraordinarily dense collapsed stars, some in such close orbits about each other that relativistic effects on their

Fig. 8–48
This massive aluminum bar was built at Bell Laboratories to detect gravitational waves. Sensitive instruments on the bar could detect vibrations whose amplitude is no larger than an atomic nucleus. Such vibrations might be caused by gravitational waves from an exploding star elsewhere in our galaxy.

black holes

gravitational waves

orbits are quite obvious. Other collapsed stars, we speculate, may involve **black holes**—objects so dense that they warp space and time until not even light can escape (Fig. 8–47). Finally, scientists are testing general relativity's prediction that **gravitational waves** can exist. Literally ripples in the fabric of space and time, these waves could be generated in violent stellar explosions, and might be detected by the minute vibrations they set up in massive objects (Fig. 8–48).

SUMMARY

1. The concept of **universal gravitation** represents a major philosophical shift in human understanding of the universe. In contrast to the ancients' view that the celestial and terrestrial realms were entirely distinct and governed by different laws, Newton's universal gravitation provides, in conjunction with his laws of motion, a coherent description of motion throughout the entire universe. In particular, Kepler's laws of planetary motion follow as a natural consequence of universal gravitation.

2. Newton's **law of universal gravitation** states that any two particles in the universe exert a mutually attractive force, whose magnitude is given by

$$F = \frac{Gm_1m_2}{r^2},$$

where m_1 and m_2 are the particle masses and r the distance between them. Although the law applies strictly only to point particles, Newton showed that it is also exactly true for spherically symmetric masses of any size. Furthermore, it remains approximately true for arbitrarily shaped objects whose separation is much greater than their size.

3. The **constant of universal gravitation,** G, was first determined by Cavendish, who measured the gravitational attraction of lead spheres in the laboratory. The value of G is approximately 6.67×10^{-11} N·m²/kg².

4. An **orbit** is the path of an object moving under the influence of gravity. Orbital motion is generally very complicated, but the case of two isolated objects interacting gravitationally is more simple.
   a. Objects whose total energy is negative travel in closed, **elliptical orbits.**
   b. A special case of an ellipse is the **circular orbit,** for which gravity provides the mv^2/r force needed for circular motion. For an object in orbit about a much more massive object M, the orbital period and radius are related by

   $$T^2 = \frac{4\pi^2 r^3}{GM}. \quad \text{(circular orbit)}$$

   This expression is a special case of Kepler's third law, which states that the square of the period of an elliptical orbit is proportional to the cube of the semimajor axis. An especially important earth orbit

is the equatorial **geosynchronous orbit,** where a satellite's orbital period equals one day so that the satellite remains fixed over a point on earth's equator. Geosynchronous orbit occurs at about 36,000 km above earth's surface.
   c. Objects whose total energy is greater than zero travel in open, **hyperbolic orbits** that take them infinitely far from the gravitating center. The intermediate case of zero total energy corresponds to a parabolic orbit.

5. The gravitational potential energy of two masses M and m is given by

$$U = -\frac{GMm}{r}, \quad \text{(gravitational potential energy)}$$

where r is the distance between their centers and where the zero of potential energy is taken at infinity.
   a. Because the potential energy difference remains finite even over infinite distances, it is possible to launch an object with sufficient speed that it will never return to its starting point. The required speed is the **escape speed,** given by

   $$v_{\text{esc}} = \sqrt{\frac{2GM}{r}}. \quad \text{(escape speed)}$$

   b. For an object in circular orbit, the potential and kinetic energies are related in a simple way:

   $$U = -2K, \quad \text{(circular orbit)}$$

   so that the total energy is

   $$E_{\text{total}} = \tfrac{1}{2}U = -K. \quad \text{(circular orbit)}$$

6. The **superposition principle** states that gravitational forces add vectorially. Consequently, the scalar potential energies also add. Integrating the potential energy associated with a spherical mass proves Newton's assertion that the gravitational force outside a spherically symmetric body is the same as if the body's mass were concentrated at its center.

7. The **gravitational field** describes the force per unit mass at each point in space. The field concept avoids the action-at-a-distance view that the earth or other massive object somehow "reaches out" across empty space to pull on distant objects. Instead, earth gives

rise to a gravitational field that pervades space, and a distant object responds to the field in its vicinity.

8. **Tidal forces** result from the difference in gravitational force over an extended body. Tidal forces give rise to stresses within astronomical objects; within the so-called **Roche limit,** tidal forces are strong enough to break apart objects held together by their own gravitation.

9. The **general theory of relativity** is Einstein's description of gravity in terms of the geometry of space and time. The predictions of general relativity differ only very slightly from those of Newtonian theory in the weak gravitational fields of the solar system, but in more exotic astrophysical situations the general relativistic description of gravity must be used.

QUESTIONS

1. What is the difference between Kepler's and Newton's assertions that planetary orbits are elliptical?
2. Explain the difference between G and g.
3. Earth's orbital motion is fastest in January and slowest in June. When is earth closest to the sun? Explain.
4. Newton was unable to determine the value of G, yet he managed to show that Kepler's laws followed from his law of universal gravitation. Why was this possible?
5. When you stand on the earth, the distance between you and earth is zero. So why isn't the gravitational force infinite?
6. The force of gravity on an object is proportional to the object's mass, and yet all objects fall with the same acceleration. Why?
7. A friend who knows nothing about physics asks what keeps an orbiting space shuttle from falling to earth. Give an answer that will satisfy your friend.
8. Could you put a satellite in an orbit that kept it stationary over the south pole? Explain.
9. It takes more energy to launch a satellite into the polar orbit shown in Fig. 8–49 than into the equatorial orbit shown. Why?

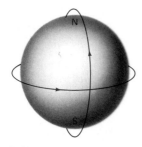

Fig. 8–49 Polar and equatorial orbits (Question 9).

10. Why are satellites generally launched to the east?
11. Can you launch a projectile that will travel three-fourths of the way around the earth, as suggested in Fig. 8–50? Why or why not?
12. Why was the United States' first satellite launching facility (at Cape Canaveral, Florida) in the southern part of the country?
13. Given the mass of the earth, the distance to the moon, and the period of the moon's orbit, could you calculate the mass of the moon? How or why not?

Fig. 8–50 Question 11

14. How should a satellite be launched so that its orbit takes it over every point on earth?
15. Does escape speed depend on the angle at which an object is launched? See Fig. 8–51.

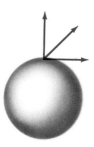

Fig. 8–51 Question 15

16. Name some advantages and disadvantages of choosing the zero of gravitational potential energy at earth's surface rather than at infinity.
17. Is it possible for a spacecraft to reach Jupiter if launched from earth with less than escape speed relative to earth? Explain.
18. Is it possible to launch a projectile to the moon using a high-speed slingshot mounted on the earth?
19. Is it possible to launch a projectile into circular earth orbit using a high-speed slingshot mounted on the earth?
20. Describe the maneuvers necessary for a spacecraft to catch up with another that is ahead of it in the same circular orbit.

21. Are there any points not on a line between two masses where a third mass would experience zero gravitational force?

22. If the sun suddenly shrank to a white dwarf star the size of the earth, but with no loss of mass, what would happen to earth's orbit?

23. Newton asserted that a spherical object of radius R behaves gravitationally as though all its mass were concentrated at the center. Is this an approximation valid only at distances r large compared with the object's radius ($r \gg R$), or is it exactly true for any distance $r > R$?

24. Why is it significant whether the total energy of a comet is greater or less than zero?

25. Satellites in orbits lower than a few hundred km experience significant frictional drag due to residual air at these altitudes. Is the ultimate effect of this frictional drag to slow down or speed up a satellite? Explain.

26. Inside a spherical shell of matter, the gravitational force of the shell is zero. Does this mean that you can shield yourself from all gravitational forces by going inside such a shell?

PROBLEMS

Section 8.2   *The Law of Universal Gravitation*

1. Space explorers land on a planet with the same mass as the earth, but find that they weigh twice what they would on earth. What is the radius of the planet?

2. Jupiter's satellites Himalia and Io have masses of 9×10^{18} kg and 8.9×10^{22} kg, respectively. Their radii are 93 km and 1820 km, respectively. Determine the acceleration of gravity on the surface of each.

3. If you are standing on the ground 15 m directly below the center of a spherical water tank containing 4×10^6 kg of water, by what fraction is your weight reduced?

4. Compare the gravitational attraction of the earth for an astronaut on the surface of the moon with the gravitational attraction of the moon for the astronaut.

5. A sensitive gravimeter is carried to the top of Chicago's Sears Tower, where its reading of the acceleration of gravity is 0.00136 m/s² lower than at street level. What is the height of the building?

6. The density of lead is 11 g/cm³. What was the gravitational attraction between the 5.0-cm- and 30-cm-diameter lead spheres in Cavendish's experiment when they were nearly touching?

Section 8.3   *Orbital Motion*

7. Mars' orbit has a diameter 1.52 times that of earth's orbit. How long does it take Mars to orbit the sun?

8. Calculate the acceleration of the moon in its circular orbit (radius 390,000 km, period 27.3 days), and compare with the acceleration of gravity at earth's surface. Show that the acceleration of the moon is smaller by a factor (radius of moon's orbit)²/(radius of earth)², thereby confirming the inverse-square nature of the gravitational force.

9. Comets are thought to originate in a cloud of ice chunks orbiting in roughly circular orbits at a distance of almost one light-year (the distance light travels in one year) from the sun. Effects of collisions or gravitational interactions with other stars occasionally send a chunk out of its circular orbit and into the elliptical orbit of the comets we observe. What is the orbital period of the cometary ice chunks in their distant circular orbits?

10. The asteroid Icarus has an orbital period of 410 days. What is the semimajor axis of its orbit? Compare with earth's orbital radius.

11. During the Apollo moon landings, one astronaut remained with the command module in lunar orbit, about 130 km above the surface. For half of each orbit, this astronaut was the "loneliest man in the universe," as his spacecraft rounded the back side of the moon and he was cut off from visual or radio contact with earth and with his fellow astronauts. How long did this period last?

12. Derive an expression for the period of a circular orbit in terms of the orbital speed.

13. We derived Equation 8–4 for the period of a circular orbit on the assumption that the massive gravitating center remained fixed. Now consider the case of two objects of equal mass M orbiting about each other as shown in Fig. 8–52. Show that the orbital period is given by

$$T^2 = \frac{16\pi^2 r^3}{GM},$$

where r is the orbital radius, or half the distance between the objects.

Fig. 8–52 Problem 13

14. Use the result of the preceding problem to determine the mass of two equal-mass stars in a binary star system, if the stellar separation is 2.4 AU (one AU, or

astronomical unit, is the distance from earth to sun) and the orbital period is 180 days.

15. A white dwarf is a collapsed star with roughly the mass of the sun but the radius of the earth. What would be the orbital period of a spaceship in low orbit around such a white dwarf?

16. Our Milky Way galaxy belongs to a large group of galaxies known as the Virgo Cluster, which contains about 1000 times the mass of our galaxy, which in turn contains about 10^{11} times the mass of the sun. We are roughly 50 million light-years (1 light-year is the distance light travels in 1 year) from the center of this approximately spherical distribution of galaxies. Could our galaxy have completed a full orbit of the cluster in the time since the universe began, some 15 billion years ago?

Section 8.4   *Gravitational Potential Energy*

17. Show that an object released from rest very far from earth $(r \gg R_e)$ reaches earth's surface at nearly escape speed.

18. What is the total mechanical energy associated with earth's orbital motion?

19. By what factor must the speed of an object in circular orbit be increased to reach escape speed from its orbital altitude?

20. Two meteoroids are 250,000 km from earth and are moving at 2.1 km/s. One is headed straight for earth, while the other is on a path that will miss earth by a distance of 8500 km from earth's center (Fig. 8–53). (a) What is the speed of the first meteoroid when it strikes earth? (Neglect air resistance.) (b) What is the speed of the second meteoroid at its closest approach to earth? (c) Will the second meteoroid return to earth's vicinity again?

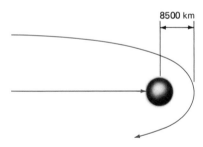

Fig. 8–53   Partial trajectories of two meteoroids (Problem 20).

21. Neglecting air resistance, to what height would you have to fire a rocket for the constant acceleration equations of Chapter 3 to give a height that is in error by 1%? Would those methods provide an overestimate or an underestimate?

22. Recent estimates suggest that efficient rockets of the twenty-first century may be able to place payloads in low earth orbit (altitude 300 km) for as little as

$700/kg. (This compares with 1985 space shuttle costs of about $18,000/kg.) If most of this cost is for fuel, and if fuel efficiency is independent of altitude, what would be the twenty-first century cost of getting 1 kg into geosynchronous orbit?

23. One component of the proposed Strategic Defense Initiative ballistic missile defense calls for powerful ground-based lasers aimed at their targets by space-based mirrors. Tentative plans call for 30-m-diameter mirrors in geosynchronous orbit to reflect the laser beam to low-orbit "battle mirrors." One simple means of foiling this delicate technology is to put a load of rocks in the same orbit as the mirrors, but going in the opposite direction. At what relative speed would the rocks hit a mirror in geosynchronous orbit?

24. Neglecting earth's rotation, show that the energy needed to launch a satellite of mass m into a circular orbit of altitude h is

$$\frac{GM_e m}{R_e} \frac{R_e + 2h}{2(R_e + h)}.$$

25. Satellites are usually launched eastward to take advantage of earth's rotation. Show that 0.30% more energy is needed to launch a satellite into a 1000-km-high equatorial orbit than into a polar orbit at the same altitude.

26. Late twentieth-century humanity consumes energy at the rate of about 10^{13} W. Suppose a method were invented to extract usable energy from the moon's orbital motion. If all our energy were supplied from this source, and our energy use remained constant, by how much would (a) the moon's orbital radius and (b) its orbital period change in a century?

27. A projectile is launched vertically upward from a planet of mass M and radius R; its initial speed is twice the escape speed. Derive an expression for its speed as a function of distance r from the center of the planet.

28. A spacecraft is in circular earth orbit at an altitude of 5500 km. By how much will its altitude decrease if it moves to a new circular orbit where (a) its orbital speed is 10% higher or (b) its orbital period is 10% shorter?

Section 8.5   *The Superposition Principle*

29. Two particles of equal mass M are located a distance ℓ apart. (a) Determine the gravitational potential energy of a third mass m at an arbitrary point on the perpendicular bisector shown in Fig. 8–54. (Take the bisector to be the x-axis, and give your answer as a function of x.) (b) Show that your result approaches the potential energy associated with the gravity of a point particle of mass $2M$ when $x \gg \ell$. (c) Show with a simple sketch that the gravitational force on the mass m must point in the x-direction, and differentiate your expression for potential energy to get an expression for the force. (d) Show that the gravita-

tional force is zero midway between the two masses, and that it drops off as $1/x^2$ for $x \gg \ell$.

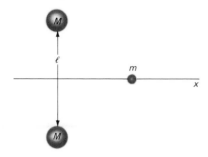

Fig. 8-54 Problem 29

30. An advanced galactic civilization constructs a solid shell of mass M_s and radius R_s around a planet of mass M_p and radius R_p. The shell has negligible thickness. Find expressions for the acceleration of gravity (a) for $R_p < r < R_s$ and (b) for $r > R_s$, where r is the radial distance from the center of the planet. Make a sketch of the gravitational acceleration as a function of r.

31. Show, by integrating the gravitational force, that the work needed to move a mass m from the center to the surface of a uniform sphere of mass M and radius R is $GMm/2R$.

32. Suppose earth's density were uniform, and that a small hole was drilled to the center of the earth. If an object were dropped from rest at the top of the hole, with what speed would it reach the center of the earth? Compare with the speed in a circular orbit just above earth's surface. *Hint:* See result of preceding problem.

33. (a) Use the result of Problem 31 to show that the potential energy of a mass m inside a uniform sphere of mass M and radius R is

$$U = -\frac{GMm}{2R}\left[3 - \left(\frac{r}{R}\right)^2\right],$$

where r is the distance from the center and where $U = 0$ at $r = \infty$. (b) Sketch the potential energy as a function of position both inside and outside the sphere.

Supplementary Problems

34. Show that the form $\Delta U = mg\Delta r$ follows from Equation 8-5 for the case $r_2 - r_1 \ll r_1$.

35. Although the study of black holes necessarily involves general relativity, one way to think about a black hole is to consider it an object so dense that the escape speed is the speed of light. (a) Using this criterion, show that the radius R_s of a black hole of mass M is given by $R_s = 2GM/c^2$, where c is the speed of light. What are the radii of black holes with (b) the mass of the earth and (c) the mass of the sun?

36. When a star with a mass twice that of the sun exhausts its nuclear fuel, it collapses into a neutron star with a radius of about 10 km. (a) What would be the acceleration of gravity on the surface of such a neutron star? Neutron stars are usually in rapid rotation, for reasons we will examine in Chapter 12. (b) If your mass is 70 kg, what would be your weight at the pole of the neutron star? (c) If the star is rotating once per second, by what fraction would your weight at the equator differ from your weight at the pole?

37. Show that the force of the sun's gravity on earth is nearly 200 times that of the moon's gravity, but that the tidal force associated with the sun's gravity is about half that associated with the moon's gravity.

38. The innermost satellites of Mars, Jupiter, and Saturn lie at distances of 9400 km, 128,000 km, and 138,000 km, respectively, from their respective planets' centers. Use data from Appendix F to compare these orbital radii with the appropriate Roche limits.

39. A uniform solid sphere of radius R and initial mass M is hollowed out to make a cavity of radius $R/2$ that just touches the center of the sphere, as shown in Fig. 8-55. Use the superposition principle to show that the gravitational field a distance $r > R$ to the right of the center of the sphere is

$$\frac{GM}{r^2}\left(1 - \frac{1}{8(1 - R/2r)^2}\right),$$

where r is the distance from the large sphere.

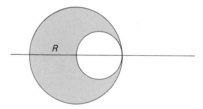

Fig. 8-55 Problem 39

40. A straight tunnel is drilled from New York to San Francisco, a surface distance of 4900 km, as shown in Fig. 8-56. If a vehicle is released from rest at New York and if friction is negligible, what is the maximum speed it achieves? *Hint:* See Problem 33.

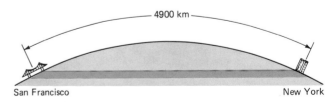

Fig. 8-56 Problem 40

41. Two satellites are in geosynchronous orbit, but in diametrically opposite positions (Fig. 8–57). Into how much lower a circular orbit should one spacecraft descend if it is to catch up with the other after 10 complete orbits? Neglect rocket firing time and time spent in transfer orbits.

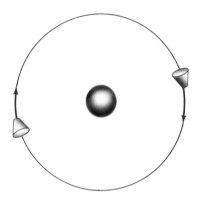

Fig. 8–57   Problem 41

42. A black hole (see Problem 35) has the same mass as the sun. Estimate how close you could get to the black hole before the tidal force stretching you apart had a magnitude of 1 ton.

9

SYSTEMS OF PARTICLES

In the previous chapters, we treated most objects as point particles. That is, we neglected any effects associated with the nonzero size of objects. But in reality, the sizes of objects may greatly affect their physical interactions. We have seen, for example, how the fact that earth is essentially a large sphere rather than a point delayed Newton for many years when he was considering his law of universal gravitation.

rigid bodies

In this chapter, we deal with systems composed of many particles. These include **rigid bodies**—objects like baseballs, cars, and planets composed of many particles linked together at fixed distances and with fixed orientations, as well as systems whose constituent parts move with respect to one another, like a gymnast, an exploding star, or a cup of coffee being stirred. In Chapter 12, we will concentrate specifically on the rotational motion of rigid bodies, and in Chapter 15 will deal with fluid systems. Our work in the present chapter is an important stage on the route to understanding these more complex and realistic systems. Some systems, like the air masses that determine earth's weather, are still too complicated for us to predict their motion. The advance of supercomputers, though, is enhancing our ability to apply the laws of physics in such situations.

Fig. 9–1
Most parts of the hammer move in complicated paths. But one point—the center of mass—follows the simple parabolic trajectory of a projectile.

center of mass

9.1
CENTER OF MASS

Figure 9–1 shows a hammer tossed into the air. The motion of most parts of the hammer is quite complex. But the curve superimposed on the photograph shows that one point seems to follow the parabola that we expect of a point projectile (Section 3.3). This point is known as the **center of mass,** since the force of gravity acts as though all the mass were concentrated there.*

---

*Actually, the point where gravity seems to act is called the **center of gravity.** For objects small enough that the acceleration of gravity doesn't vary over their size, the center of mass and center of gravity coincide. We discuss this point further in Section 11.3.

211

How do we find out where the center of mass is? Figure 9–1 shows that the center of mass obeys Newton's second law just as would a point particle. If we can find a point whose acceleration **A** obeys Newton's second law **F**=M**A**, where **F** is now the net force on an entire system and M the mass of that system, then we will have found the center of mass. (Note that we use capital letters for quantities associated with the center of mass.)

Consider a system consisting of many particles. Our system could be a rigid body, or its constituent particles could be moving relative to one another; we only require that the total mass remain constant. To find the center of mass, we want an equation that looks like Newton's second law but involves the total mass of the system and the net force on the entire system. If we apply Newton's second law to the ith particle in the system, we have

$$\mathbf{F}_i = m_i\mathbf{a}_i = m_i\frac{d^2\mathbf{r}_i}{dt^2} = \frac{d^2 m_i\mathbf{r}_i}{dt^2},$$

where $\mathbf{F}_i$ is the net force on the particle, m_i its mass, and where we have written the acceleration $\mathbf{a}_i$ as the second derivative of the position $\mathbf{r}_i$. Now consider the entire object. The net force is $\mathbf{F}=\Sigma\mathbf{F}_i$, where we sum the forces acting on all the particles making up the object. Thus

$$\mathbf{F} = \sum \frac{d^2 m_i\mathbf{r}_i}{dt^2},$$

where the sum is taken over all values of i that label the individual particles. But the sum of derivatives is the same as the derivative of the sum, so that

$$\mathbf{F} = \frac{d^2(\sum m_i\mathbf{r}_i)}{dt^2}.$$

We can now put this equation in the form of Newton's second law. Multiplying and dividing the right-hand side by the total mass $M=\Sigma m_i$, and distributing this constant M through the differentiation, we have

$$\mathbf{F} = M\frac{d^2}{dt^2}\left(\frac{\sum m_i\mathbf{r}_i}{M}\right). \tag{9–1}$$

Equation 9–1 has the desired form $\mathbf{F}=M\mathbf{A}=M(d^2\mathbf{R}/dt^2)$ if we define

$$\mathbf{R} = \frac{\sum m_i\mathbf{r}_i}{M}. \quad \text{(center of mass)} \tag{9–2}$$

We have almost reached our goal of showing that the point **R** moves according to Newton's law $\mathbf{F}=Md^2\mathbf{R}/dt^2$. Why aren't we there now? Because the force **F** in Equation 9–1 is the sum of *all* the forces acting on all the particles of the system, including the **internal forces** that act between the particles. We would like the force **F** in Newton's law to be the net **external force**—the net force applied from outside the system. We can write the force **F** in Equation 9–1—the total force on all the particles of the system—as

$$\mathbf{F} = \sum\mathbf{F}_{\text{ext}} + \sum\mathbf{F}_{\text{int}},$$

internal forces

external force

where $\Sigma\mathbf{F}_{ext}$ is the sum of all the external forces and $\Sigma\mathbf{F}_{int}$ the sum of the internal forces. According to Newton's third law of motion, each of the internal forces has an equal but oppositely directed reaction force that itself acts on a particle of the system and is therefore included in the sum $\Sigma\mathbf{F}_{int}$. (Each external force also has a reaction force, but these reaction forces act *outside* the system and are therefore not included in the sum.) Added vectorially, the reaction forces therefore cancel in pairs, so that $\Sigma\mathbf{F}_{int} = 0$, and the force $\mathbf{F}$ in Equation 9–1 is just the net *external* force applied to the system. So the point $\mathbf{R}$ defined in Equation 9–2 does obey Newton's law, written in the form

$$\mathbf{F} = M\frac{d^2\mathbf{R}}{dt^2}, \tag{9–3}$$

where $\mathbf{F}$ is the net external force applied to the system and M the total mass.

We have defined the center of mass $\mathbf{R}$ so that we can apply Newton's second law to the entire system rather than to each individual particle. As far as its overall motion is concerned, a complex system acts as though all its mass were concentrated at the center of mass. Physically, the center of mass represents a kind of "average position," determined from the positions of all the individual particles weighted according to their masses.

Finding the Center of Mass

Equation 9–2 expresses the center of mass in vector form; taking components gives three separate scalar equations for the xyz-coordinates of the center of mass. If we consider only a one-dimensional situation, in which the constituent particles lie along a single line, then the center of mass must lie along that line. Similarly, for a two-dimensional object, the center of mass lies in the plane of the object. For the simple case of two point masses, we can take the x-axis along the line joining the masses, and write Equation 9–2 in the form

$$X = \frac{m_1x_1 + m_2x_2}{m_1 + m_2}. \tag{9–4}$$

Example 9–1
Weightlifter

You want to lift a barbell at its center of mass. (a) Suppose there are equal 60-kg masses at either end of the 1.5-m-long bar. (b) Suppose someone put unequal masses of 50 kg and 80 kg on the ends. Where is the center of mass in each case? Assume that the size of each of the masses is negligible compared with the distance between them.

Solution

In (a) it's obvious that the center of mass must lie at the center of the barbell. If we choose the origin to lie at one of the masses, then $x_1 = 0$ m and $x_2 = 1.5$ m; applying Equation 9–4 then gives

$$X = \frac{m_1x_1 + m_2x_2}{m_1 + m_2} = \frac{(60 \text{ kg})(0 \text{ m}) + (60 \text{ kg})(1.5 \text{ m})}{60 \text{ kg} + 60 \text{ kg}} = 0.75 \text{ m},$$

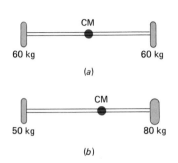

Fig. 9–2
Center of mass of a barbell. (*a*) Two equal masses. (*b*) 50-kg and 80-kg masses.

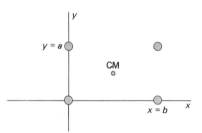

Fig. 9–3
Example 9–2

as expected. In case (*b*), with the origin at the 50-kg mass, we have

$$X = \frac{m_1x_1 + m_2x_2}{m_1 + m_2} = \frac{(50 \text{ kg})(0 \text{ m}) + (80 \text{ kg})(1.5 \text{ m})}{50 \text{ kg} + 80 \text{ kg}} = 0.92 \text{ m}.$$

Again, the result makes sense: the center of mass is closer to the more massive constituent of the system (Fig. 9–2). Our choice of origin at one of the masses was convenient because it made one of the terms in the sum $\Sigma m_i x_i$ zero. But that choice is purely for convenience; it cannot influence the actual physical location of the center of mass (see Problem 2).

Example 9–2
Center of Mass in a Plane

Four equal masses m lie at the corners of a rectangle of width a and length b. Where is the center of mass?

Solution

A convenient coordinate system has two sides of the rectangle along the x- and y-axes, with one corner at the origin (Fig. 9–3). Then the x- and y-components of Equation 9–2 are

$$X = \frac{\sum m_i x_i}{M} = \frac{m(0 + b + b + 0)}{4m} = \frac{b}{2}$$

and

$$Y = \frac{\sum m_i y_i}{M} = \frac{m(0 + 0 + a + a)}{4m} = \frac{a}{2},$$

where in both cases we summed going counterclockwise around the rectangle, starting at the origin. Our result is hardly surprising: the center of mass lies at the center of the rectangle.

It is no accident that the center of mass in Example 9–2 lies in the center of the rectangle. For any object with an axis of symmetry, the center of mass must lie along that axis. This is because particles on opposite sides of the symmetry axis cancel each other's effect on the component of the center-of-mass position in a direction perpendicular to the axis. The rectangle of Example 9–2 has *two* symmetry axes; they are vertical and horizontal lines through its center. The center of mass must lie on *both* symmetry axes, and is therefore at their one point of intersection, the center. Attention to symmetry here and in many other areas of physics can save a great deal of calculation.

We have expressed the center of mass in terms of sums over individual particles. Ultimately, matter is composed of individual particles. But it is often a convenient approximation to consider that matter is continuously distributed; after all, we don't want to deal with 10^{23} or so atoms to find the center of mass of a book or other macroscopic object. We can consider a chunk of continuous matter to be composed of individual pieces of mass Δm_i, with position vectors $\mathbf{r}_i$; we call these pieces **mass elements** (Fig. 9–4). The center of mass of the entire chunk is then given by Equation 9–2:

$$\mathbf{R} = \frac{\sum \Delta m_i \mathbf{r}_i}{M},$$

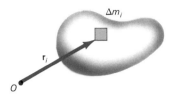

Fig. 9–4
A chunk of continuous matter, showing one mass element Δm_i and its position vector $\mathbf{r}_i$ from an arbitrary origin.

where $M = \Sigma \Delta m_i$ is the total mass. This expression is slightly vague, for we don't know exactly where within a mass element its position vector should terminate. But in the limit as the mass elements become arbitrarily small, like point particles, the expression becomes exact. As in Chapter 6, that limiting process defines an integral, giving

$$\mathbf{R} = \lim_{\Delta m_i \to 0} \frac{\sum \Delta m_i \mathbf{r}_i}{M} = \frac{1}{M} \int \mathbf{r}\, dm, \quad \left(\begin{array}{c} \text{center of mass,} \\ \text{continuous matter} \end{array} \right) \qquad (9\text{–}5)$$

where the integration is over the entire volume of the object. Like the sum 9–2, the vector form 9–5 stands for three separate integrals for the three components of the center-of-mass position. Example 9–3 shows how to evaluate these integrals.

Example 9–3

An Aircraft Wing

The wing of a supersonic aircraft is in the form of an isosceles triangle of length ℓ, width w, and negligible thickness (Fig. 9–5). It has total mass M, distributed uniformly over the wing. Where is its center of mass?

Solution

Figure 9–5 shows the wing with a suitable coordinate system. Since the wing is flat and of negligible thickness, its center of mass must lie in the x-y plane. The wing is symmetric about the x-axis, so that the center of mass must lie on the x-axis; that is, $Y = 0$. So we need to integrate only to find the x-coordinate. Figure 9–5 shows an appropriate mass element, here in the form of a strip of width dx located a distance x along the axis. To find the mass dm of the strip, we first find its area. Figure 9–5 shows that the equations describing the sloping sides of the wing are $y = \pm wx/2\ell$, so that our strip extends from $x = -wx/2\ell$ to $+wx/2\ell$; its length is therefore wx/ℓ. Since its width is dx, its area is then $dA = (wx\, dx)/\ell$. [Here we've treated the strip as a rectangle, which is not quite right because of its sloping sides. But in the limit of arbitrarily small dx, the area of the strip and the rectangle area $(wx\, dx)/\ell$ become the same (Fig. 9–6).] Since mass is distributed uniformly over the wing, the mass dm of the strip is the same fraction of the total mass as its area is of the total area:

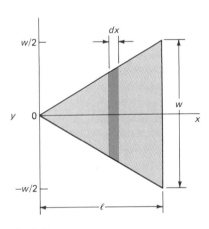

Fig. 9–5
A supersonic aircraft wing (Example 9–3). Shaded area is a mass element of width dx.

$$\frac{dm}{M} = \frac{dA}{A} = \frac{(wx\, dx)/\ell}{\frac{1}{2}w\ell} = \frac{2x\, dx}{\ell^2},$$

so that $dm = (2Mx\, dx)/\ell^2$. The x-component of the center-of-mass position is given by the x-component of Equation 9–5:

$$X = \frac{1}{M} \int x\, dm.$$

Using our expression for dm, and integrating over the entire triangle—that is, from $x = 0$ to $x = \ell$—we have

$$X = \frac{1}{M} \int_0^\ell \frac{2M}{\ell^2} x^2\, dx = \frac{2}{\ell^2} \left. \frac{x^3}{3} \right|_0^\ell = \tfrac{2}{3}\ell.$$

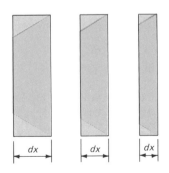

Fig. 9–6
As the width dx of the strip becomes arbitrarily small, its area approaches that of the rectangle.

Does this make sense? Yes: more of the wing's mass is concentrated near the wide end at $x = \ell$, so the center of mass should be located closer to that end.

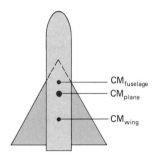

Fig. 9–7
The center of mass of the airplane is found by treating the wing and fuselage as point particles located at their respective centers of mass. Given the center-of-mass location, is the wing or the fuselage more massive?

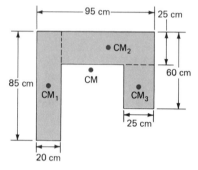

Fig. 9–8
An irregular object, showing centers of mass of the constituent rectangles and of the object itself. Note that the center of mass lies outside the object.

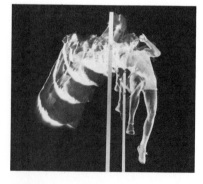

Fig. 9–9
The high jumper's center of mass does not clear the bar, even though his entire body does. This is possible because the entire body is not above the bar at the same time.

With still more complex objects, it is convenient to find the centers of mass of major subpieces, then treat those subpieces as point particles to find the center of mass of the entire object. If the wing of Example 9–3 is placed on a cylindrical airplane fuselage, for example, then we can find the center of mass of the entire plane by treating the wing and fuselage as point particles located at their respective centers of mass (Fig. 9–7). And an irregular object like that of Fig. 9–8 is treated by locating the centers of mass of its three constituent rectangles (see Problem 5).

Figure 9–8 shows that the center of mass need not lie within the object. A high jumper (Fig. 9–9) takes advantage of this fact by arching his body into a shape qualitatively like that of Fig. 9–8. The jumper's center of mass never has to get as high as the bar, thereby reducing the amount of work he must do against gravity. Similarly, a ballet dancer can appear to float through the air, her head and torso following a horizontal trajectory, by raising her legs and therefore the position of the center of mass relative to her torso. The center of mass, of course, follows a parabolic trajectory, but that's not what the audience notices (Fig. 9–10).

Motion of the Center of Mass

We defined the center of mass (Equation 9–2) so that its motion would obey Newton's law $\mathbf{F} = M\mathbf{A}$, where $\mathbf{F}$ is the net external force applied to a system and M the total mass. When, for example, gravity is the only external force acting on a system, then its center of mass follows the trajectory appropriate to a point particle. We saw this in Fig. 9–1 and Fig. 9–10, and see it again with the exploding fireworks rocket of Fig. 9–11. An example similar to Fig. 9–11 is provided on a grander scale by the breakup of comet Biela by tidal forces as it passed the sun in the mid-1800's. Today, the Andromedid swarm of meteors orbits the sun in similar but not identical orbits. But the center of mass of the swarm remains in the orbit of the original comet, despite the fact that there is no longer any matter there!

When there is no net external force on a system, then the center of mass acceleration $\mathbf{A}$ is zero, and the center of mass moves with constant velocity. In the special case of a system at rest and subject to no net external force, the center of mass remains at rest despite any internal motions of the system's constituents.

Example 9–4
Leaping to Shore

You are standing 1.8 m from the shoreward end of a 4.2-m-long boat; from the end to shore is another 1.1 m (Fig. 9–12). Your mass is 70 kg and the boat's mass is 150 kg. The center of mass of the boat lies at its center. You walk to the shoreward end of the boat, preparing to leap to shore. How far do you have to leap? Where is the boat at the instant you reach shore? Neglect any frictional effects.

Solution

The net external force on the system consisting of you and the boat is zero, so that the center of mass of the system remains fixed. Letting m_1 and x_1 be your mass and position, and m_2 and x_2 the mass and position of the boat (that is, of the boat's

Fig. 9–10
Performing a grand jeté, Jennifer Davis of the Pittsburgh Ballet Theatre keeps her head and torso moving nearly horizontally while her center of mass describes a parabolic trajectory.

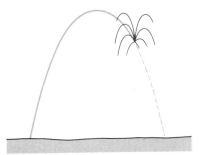

Fig. 9–11
An exploding fireworks rocket. The center of mass of the system follows a parabolic trajectory despite the strong internal forces of the explosion.

center of mass), and designating the initial and final values by the subscripts i and f, respectively, we equate the initial and final positions of the center of mass:

$$\frac{m_1x_{1i} + m_2x_{2i}}{m_1+m_2} = \frac{m_1x_{1f} + m_2x_{2f}}{m_1+m_2}.$$

The total mass m_1+m_2 cancels, so that

$$m_1x_{1i} + m_2x_{2i} = m_1x_{1f} + m_2x_{2f}. \tag{9-6}$$

Since we're interested in the final position with respect to the shore, an appropriate choice of coordinate system has the origin at the shore. Then Fig. 9–12 shows that $x_{1i}=1.1$ m$+1.8$ m$=2.9$ m and $x_{2i}=1.1$ m$+2.1$ m$=3.2$ m. x_{1f} and x_{2f} are still unknown. But they are related: when you reach the end of the boat, you are half a boat length closer to shore than the center of the boat. Therefore, $x_{2f}=x_{1f}+\frac{1}{2}L$, where L is the 4.2-m length of the boat. Using this result in Equation 9–6 gives

$$m_1x_{1i} + m_2x_{2i} = m_1x_{1f} + m_2(x_{1f}+\tfrac{1}{2}L).$$

Solving for x_{1f}, we then have

$$x_{1f} = \frac{m_1x_{1i} + m_2x_{2i} - \tfrac{1}{2}m_2L}{m_1+m_2}$$

$$= \frac{(70\text{ kg})(2.9\text{ m}) + (150\text{ kg})(3.2\text{ m}) - \tfrac{1}{2}(150\text{ kg})(4.2\text{ m})}{70\text{ kg} + 150\text{ kg}} = 1.7\text{ m}.$$

This is the distance you have to leap from the end of the boat to shore. When you reach shore, you are at the origin: $x_{1f}=0$. Solving Equation 9–6 for the boat's position then gives

Fig. 9–12
Example 9–4

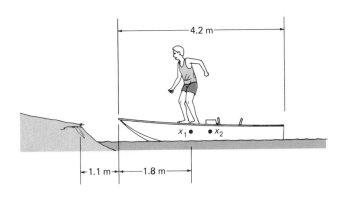

$$x_{2f} = \frac{m_1 x_{1i} + m_2 x_{2i}}{m_2} = \frac{(70 \text{ kg})(2.9 \text{ m}) + (150 \text{ kg})(3.2 \text{ m})}{150 \text{ kg}} = 4.6 \text{ m}.$$

After the instant you reach shore, we can no longer argue that the center of mass remains fixed. Why not? Because an external force—the stopping force of the ground—has acted on part of the system, so that the center of mass is no longer fixed. Indeed, the boat continues to drift away from shore. Question 10 explores this situation further.

9.2
MOMENTUM

When we introduced Newton's second law of motion in Chapter 4, we defined the linear momentum $\mathbf{p} = m\mathbf{v}$ of a particle, and first wrote Newton's law in the form $\mathbf{F} = d\mathbf{p}/dt$. We subsequently applied the law to single particles, using the form $\mathbf{F} = m\mathbf{a}$, but suggested that the momentum concept would play an important role in many-particle systems.

Consider the total momentum of a system of particles:

$$\mathbf{P} = \sum \mathbf{p}_i = \sum m_i \mathbf{v}_i,$$

where m_i and $\mathbf{v}_i$ are the masses and velocities of the individual particles, and $\mathbf{p}_i$ their corresponding momenta. But $\mathbf{v} = d\mathbf{r}/dt$, so that

$$\mathbf{P} = \sum m_i \frac{d\mathbf{r}_i}{dt} = \frac{d}{dt} \sum m_i \mathbf{r}_i,$$

where the last step follows because the individual particle masses are constant and because the sum of derivatives is the derivative of the sum. In Section 9.1 we defined the center-of-mass position $\mathbf{R}$ as $\sum m_i \mathbf{r}_i / M$, where M is the total mass. So the total momentum can be written

$$\mathbf{P} = \frac{d}{dt} M\mathbf{R},$$

or, assuming the system mass M remains constant,

$$\mathbf{P} = M\frac{d\mathbf{R}}{dt} = M\mathbf{V}, \tag{9-7}$$

where $\mathbf{V} = d\mathbf{R}/dt$ is the center-of-mass velocity. So the momentum of a system is given by an expression similar to that of a single particle; it is the product of the system mass with the system velocity—that is, with the velocity of the center of mass. If this seems so obvious as not to need deriving, watch out! We'll see in Section 9.3 that the same is *not* true for the system's total energy.

If we differentiate Equation 9–7 with respect to time, we have

$$\frac{d\mathbf{P}}{dt} = M\frac{d\mathbf{V}}{dt} = M\mathbf{A},$$

where $\mathbf{A}$ is the center-of-mass acceleration. But we defined the center of mass so that its motion obeyed Newton's second law, $\mathbf{F} = M\mathbf{A}$, with $\mathbf{F}$ the net external force on the system. So we can write simply

$$\mathbf{F} = \frac{d\mathbf{P}}{dt}, \qquad (9\text{--}8)$$

showing that the momentum of a system of particles changes only if there is a net external force on the system. Remember the hidden role of Newton's third law in all this: only because the forces *internal* to the system cancel in pairs can we ignore them and consider just the *external* force.

Conservation of Momentum

In the special case when the net external force is zero, Equation 9–8 gives

$$\frac{d\mathbf{P}}{dt} = 0, \qquad \text{(no net external force)}$$

so that

$$\mathbf{P} = \text{constant.} \qquad \text{(conservation of momentum)} \qquad (9\text{--}9)$$

law of conservation of momentum

Equation 9–9 expresses the **law of conservation of momentum.** It says that, in the absence of a net external force, the total momentum **P** of a system—the vector sum $\Sigma m_i \mathbf{v}_i$ of the momenta of the individual particles—remains constant. (More explicitly, we have written the "law of conservation of linear momentum." We will deal with "angular momentum" in Chapter 12.)

Equation 9–9 holds no matter how many particles are involved, and no matter how they may be moving. It is valid for a set of pool balls both before and after they are struck by the cue ball (Fig. 9–13). It is also valid for a globular cluster containing 1,000,000 stars interacting by their mutual gravity (Fig. 9–14), to the extent that we can ignore the external force from the rest of the galaxy (since the cluster takes hundreds of millions of years to orbit the galaxy, that is a reasonable approximation for an astrophysicist studying the much more rapid internal dynamics of the cluster). And it is valid for a system consisting of 10^{23} air molecules and a closed vessel containing them. We will, in fact, use momentum conservation in Chapter 17 to show how the behavior of gases follows from Newton's laws.

We derived the conservation of momentum principle from Newton's laws. But in a fundamental sense, momentum conservation is more basic than Newton's laws. Momentum is conserved even in subatomic and nuclear systems where the laws and even language of Newtonian physics are hopelessly inadequate. The examples below show the wide range and power of the momentum conservation principle.

Fig. 9–13
(Above) The momentum **P** of the system of pool balls is that of the cue ball alone. (Below) After the cue ball has struck the others, the total momentum $\Sigma m_i \mathbf{v}_i$ is still equal to the original **P**.

Example 9–5
Momentum Conservation in a Canoe

Amy (mass 55 kg) and George (mass 70 kg) are sitting at opposite ends of a canoe (mass 22 kg) at rest on frictionless water. She tosses him a 14-kg pack, giving it a speed of 3.1 m/s relative to the water. What is the speed of the canoe (*a*) while the pack is in the air and (*b*) after George catches it?

Solution

We're interested here in the horizontal motion of the canoe. With no external forces acting in the horizontal direction, the horizontal component of the momentum is conserved. Originally, everything is at rest and the total momentum is zero.

Fig. 9–14
A globular cluster, containing about 10^6 stars interacting via gravitational forces. In the absence of external forces, the total momentum of the system remains constant.

While the pack is in the air, it has horizontal momentum $m_p v_p$. The canoe and its passengers are all moving together; they have momentum $(m_a + m_g + m_c)v_c$, where m_a, m_g, and m_c are the masses of Amy, George, and the canoe, respectively, and where v_c is the canoe's horizontal velocity. Since horizontal momentum is conserved, it must still be zero when the pack is in the air, so that

$$m_p v_p + (m_a + m_g + m_c)v_c = 0,$$

or

$$v_c = -\frac{m_p v_p}{m_a + m_g + m_c} = -\frac{(14 \text{ kg})(3.1 \text{ m/s})}{55 \text{ kg} + 70 \text{ kg} + 22 \text{ kg}} = -0.30 \text{ m/s}.$$

The minus sign tells us that the canoe moves in the opposite direction from the pack.

There is no need to calculate the speed after George catches the pack: canoe, passengers, and pack all have the same velocity then. For the total momentum to remain zero, that velocity must be zero.

Example 9–6
The Speed of a Hockey Puck

A hockey coach enlists a physics student on his team to measure the speed of a hockey puck. The student loads a small styrofoam chest with sand, giving it a total mass of 6.4 kg. He places the chest, at rest on frictionless ice, in the path of the hockey puck, whose mass is 160 g. The puck embeds itself in the styrofoam, and the chest moves off at 1.2 m/s. What was the speed of the puck?

Solution

Since there is no net external force, momentum is conserved. Originally, all the momentum of the puck+chest system is in the puck: $P = m_p v_p$, where we do not need vector notation because the motion is in one dimension. After the puck hits the chest, the two move off together at the same speed: $P = (m_p + m_c)v_c$. So the statement that momentum is conserved becomes

$$m_p v_p = (m_p + m_c)v_c,$$

so that

$$v_p = \frac{(m_p + m_c)v_c}{m_p} = \frac{(0.16 \text{ kg} + 6.4 \text{ kg})(1.2 \text{ m/s})}{0.16 \text{ kg}} = 49 \text{ m/s}.$$

Variations of this technique are routinely used to measure the speeds of rapidly moving objects like bullets.

Example 9–7
Radioactive Decay

A lithium-5 nucleus (^{5}Li) is moving at 1.6×10^6 m/s when it decays into a proton (^{1}H) and an alpha particle (^{4}He). [The superscripted numbers are the total number of nucleons (neutrons and protons) in a nucleus, and give the approximate nuclear mass in atomic mass units.] The alpha particle is detected moving at 1.4×10^6 m/s, at 33° to the original velocity of the ^{5}Li nucleus. What are the magnitude and direction of the proton's velocity?

Solution

Again, momentum is conserved. Before the decay, we have only the lithium nucleus. After the decay, there are two particles whose individual momenta contribute to the total. So the statement that momentum is conserved becomes

$$m_{Li}\mathbf{v}_{Li} = m_H\mathbf{v}_H + m_{He}\mathbf{v}_{He},$$

where we keep the full vector notation because the motion involves two dimensions. If we choose the x-axis along the direction of $\mathbf{v}_{Li}$, then we can write separately the two components of the momentum conservation equation:

x-component: $m_{Li}v_{Li} = m_Hv_H\cos\theta + m_{He}v_{He}\cos\phi$

y-component: $0 = m_Hv_H\sin\theta + m_{He}v_{He}\sin\phi,$

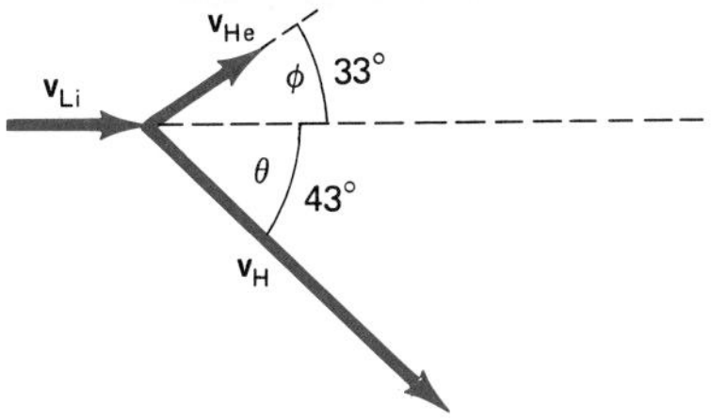

where θ and ϕ are the angles that the velocities $\mathbf{v}_H$ and $\mathbf{v}_{He}$ make with the positive x-axis (Fig. 9–15). Writing these two equations completes the physics of the problem; the rest is the mathematics of solving them. Moving the terms involving the unknown proton speed v_H to one side of each equation, we can write

$$m_Hv_H\cos\theta = m_{Li}v_{Li} - m_{He}v_{He}\cos\phi$$
$$m_Hv_H\sin\theta = -m_{He}v_{He}\sin\phi.$$

Squaring both sides of each equation and adding the resulting equations gives

$$(m_Hv_H)^2 = (m_{Li}v_{Li} - m_{He}v_{He}\cos\phi)^2 + (m_{He}v_{He}\sin\phi)^2$$

where we used the fact that $\cos^2\theta + \sin^2\theta = 1$. Then

$$v_H = \left[\left(\frac{m_{Li}}{m_H}v_{Li} - \frac{m_{He}}{m_H}v_{He}\cos\phi\right)^2 + \left(\frac{m_{He}}{m_H}v_{He}\sin\phi\right)^2\right]^{1/2}.$$

Since the proton mass is 1 u, the mass ratios m_{Li}/m_H and m_{He}/m_H are equal to the nucleon numbers of ^{5}Li and ^{4}He, respectively. So

$$v_H = [(5v_{Li} - 4v_{He}\cos\phi)^2 + (4v_{He}\sin\phi)^2]^{1/2}$$
$$= \{[(5)(1.6\times10^6 \text{ m/s}) - (4)(1.4\times10^6 \text{ m/s})(\cos33°)]^2$$
$$+ [(4)(1.4\times10^6 \text{ m/s})(\sin33°)]^2\}^{1/2}$$
$$= 4.5\times10^6 \text{ m/s}.$$

Knowing the speed of the proton, we can solve the y-component of the momentum conservation equation for the angle θ:

$$\theta = \sin^{-1}\left(\frac{-m_{He}v_{He}\sin\phi}{m_Hv_H}\right) = \sin^{-1}\left[\frac{-(4\text{ u})(1.4\times10^6 \text{ m/s})(\sin33°)}{(1\text{ u})(4.5\times10^6 \text{ m/s})}\right] = -43°,$$

where the minus sign indicates that the proton's velocity vector points below the x-axis. We can see graphically that our results make sense by redrawing Fig. 9–15 as a momentum vector diagram—that is, by multiplying each velocity vector by the appropriate mass. The result, Fig. 9–16, shows that the y-components of the He and H momenta cancel, and that their x-components sum to the momentum of the original nucleus.

With no external forces acting, momentum had to be conserved in this decay. But kinetic energy is not conserved, as you will find if you evaluate the kinetic energies before and after decay (see Problem 39). Internal potential energy of the ^{5}Li nucleus is converted to kinetic energy in the decay process.

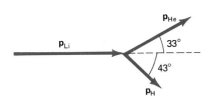

Fig. 9–15
Decay of a ^{5}Li nucleus (Example 9–7).

Fig. 9–16
Momentum diagram for Example 9–7.

How do we decide whether or not the forces acting on a system are external? That depends on our choice of what to include in the system. In Example 9–6, for example, we chose to include both the hockey puck and the styrofoam chest in our system; the force exerted on the chest by the puck was then an internal force and we did not consider it explicitly. On the other hand, we could have chosen our system to include just the chest;

then the force of the puck would be an external force, and the momentum of the system—the chest alone—would change. Of course, our definition of the system cannot affect what actually happens. Either description of the puck-chest interaction is equally valid; our choice was made so we could apply the conservation of momentum principle.

Sometimes we're interested in the force exerted on a particular object. Then it is convenient to redefine our system to include only that object, so that the force of interest is an external force. The example below illustrates this situation.

Example 9–8

Fighting a Fire

A firefighter directs a stream of water against the window of a burning building, hoping to break it so the water can get to the fire. The hose delivers water at the rate of 45 kg/s; the water hits the window moving horizontally at 32 m/s. After hitting the window, the water drops vertically downward. What is the horizontal force exerted on the window?

Solution

The horizontal motion of the water stops abruptly at the window; since the water is moving initially at 32 m/s, each kg of water loses 32 kg·m/s of momentum. But water strikes the window at the rate of 45 kg/s, so that the rate of momentum loss is

$$\frac{dP}{dt} = (45 \text{ kg/s})(32 \text{ kg·m/s/kg}) = 1400 \text{ kg·m/s}^2.$$

If we regard the water alone as our system, then this change in the system's momentum means that an external force of magnitude 1400 N is acting. That force is provided by the window. By Newton's third law, the water must therefore exert an oppositely directed force of equal magnitude on the window. Because the window is attached to the building, and the building to the earth, the force does not give rise to any significant acceleration. Once the window breaks, though, the glass fragments will be accelerated violently.

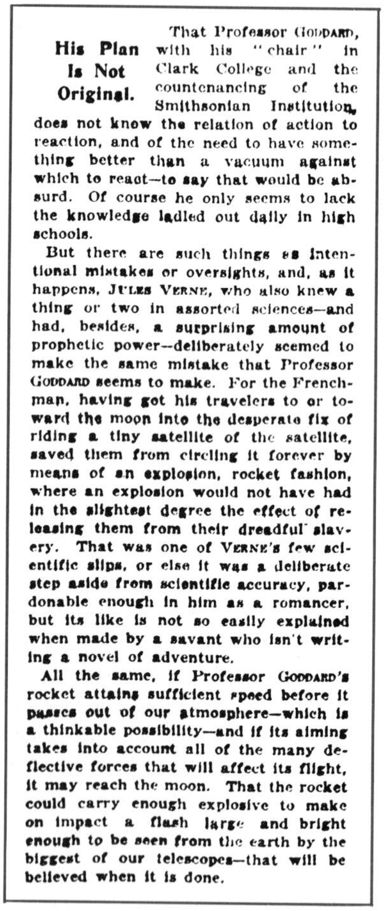

Rockets

Almost everyone knows that a rocket moves forward by ejecting gas backward. Its motion is a consequence of Newton's third law. Yet in the 1920's, *The New York Times* criticized the American rocket pioneer Robert Goddard, claiming that his rockets would never work in space because they had nothing to push against (Fig. 9–17). The *Times* was wrong; today we routinely use rocket engines in space to deploy satellites from space shuttles, to point space-based telescopes at their targets, to make midcourse corrections to planetary probes, and to return manned spacecraft from orbit.

Consider a rocket in deep space, where we can neglect gravity. The rocket has a total mass M including fuel, and is moving with velocity $\mathbf{v}$ relative to some nonaccelerating frame of reference. Its engine is then fired for a time Δt, during which time its mass changes by an amount ΔM, where ΔM is a *negative* quantity. The mass lost by the rocket appears as exhaust gas, moving with velocity $\mathbf{v}_{\text{ex}}$ relative to the rocket. At the end of the interval, the rocket is moving with velocity $\mathbf{v} + \Delta \mathbf{v}$ (Fig. 9–18). Before

His Plan Is Not Original. That Professor GODDARD, with his "chair" in Clark College and the countenancing of the Smithsonian Institution, does not know the relation of action to reaction, and of the need to have something better than a vacuum against which to react—to say that would be absurd. Of course he only seems to lack the knowledge ladled out daily in high schools.

But there are such things as intentional mistakes or oversights, and, as it happens, JULES VERNE, who also knew a thing or two in assorted sciences—and had, besides, a surprising amount of prophetic power—deliberately seemed to make the same mistake that Professor GODDARD seems to make. For the Frenchman, having got his travelers to or toward the moon into the desperate fix of riding a tiny satellite of the satellite, saved them from circling it forever by means of an explosion, rocket fashion, where an explosion would not have had in the slightest degree the effect of releasing them from their dreadful slavery. That was one of VERNE's few scientific slips, or else it was a deliberate step aside from scientific accuracy, pardonable enough in him as a romancer, but its like is not so easily explained when made by a savant who isn't writing a novel of adventure.

All the same, if Professor GODDARD's rocket attains sufficient speed before it passes out of our atmosphere—which is a thinkable possibility—and if its aiming takes into account all of the many deflective forces that will affect its flight, it may reach the moon. That the rocket could carry enough explosive to make on impact a flash large and bright enough to be seen from the earth by the biggest of our telescopes—that will be believed when it is done.

Fig. 9–17
New York Times editorial of January 13, 1920, criticizing Robert Goddard's suggestion that rockets might be used for space travel.

engine firing, the total momentum is

$$P_{\text{initial}} = Mv,$$

where we drop the vector notation for the case of one-dimensional motion. What is the momentum after firing? The rocket's mass has changed by ΔM, and its speed by Δv, so its new momentum is $(M+\Delta M)(v+\Delta v)$. (Remember that ΔM is a negative quantity, so that the mass has actually decreased.) The gas is exhausted backward at speed v_{ex} relative to the rocket, so that the exhaust velocity in our reference frame is $v - v_{\text{ex}}$. (Actually, this is only approximately true, since the rocket speed changes during the firing. But when we take the limit $\Delta t \rightarrow 0$, this approximation will become exact.) The mass of the ejected gas is the positive quantity $-\Delta M$, where ΔM—a negative quantity—is the change in the rocket's mass as a result of its ejecting the gas. So the momentum of the gas is given by $p_{\text{gas}} = -\Delta M(v - v_{\text{ex}})$, and therefore the total momentum after firing is

$$P_{\text{final}} = p_{\text{rocket}} + p_{\text{gas}} = (M+\Delta M)(v+\Delta v) - \Delta M(v - v_{\text{ex}}).$$

There are no external forces acting on the rocket-gas system, so the initial and final momenta are the same. Expanding the products in the expression for the final momentum, and equating to the initial momentum, we have

$$Mv = Mv + v\Delta M + M\Delta v + \Delta M\Delta v - v\Delta M + v_{\text{ex}}\Delta M,$$

or

$$0 = M\Delta v + \Delta M\Delta v + v_{\text{ex}}\Delta M.$$

Rearranging and dividing by the time interval Δt gives

$$M\frac{\Delta v}{\Delta t} = -(v_{\text{ex}} + \Delta v)\frac{\Delta M}{\Delta t}.$$

In the limit $\Delta t \rightarrow 0$, the ratios $\Delta v/\Delta t$ and $\Delta M/\Delta t$ become derivatives, while $\Delta v \rightarrow 0$, so that

$$M\frac{dv}{dt} = -v_{\text{ex}}\frac{dM}{dt}. \qquad \text{(rocket equation)} \qquad (9\text{--}10)$$

thrust

The quantity $-v_{\text{ex}}dM/dt$ is defined by rocket engineers to be the **thrust** of the rocket. Thrust has the dimensions of force; physically, it is the force exerted on the interior of the rocket engine by the exhaust gases. We see from Equation 9–10 that the thrust can be increased by ejecting the gases at a higher exhaust speed, by ejecting gas at a greater rate, or both. The minus sign in Equation 9–10 shows that an *increase* in speed—positive dv/dt—corresponds to a *decrease* in rocket mass—negative dM/dt. In practice, the exhaust gas plume does not point straight backward, but fans out somewhat, thereby reducing the forward thrust (Fig. 9–19).

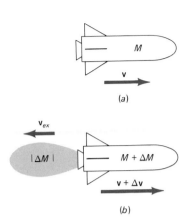

Fig. 9–18
A rocket (a) before and (b) after a short firing of its engine. The change ΔM in the rocket's mass is negative, so that $M+\Delta M < M$. Rocket speed is relative to a nonaccelerating frame of reference; exhaust speed v_{ex} is relative to the rocket.

Example 9–9
On to Neptune!

Several weeks after its rendezvous with Uranus in January 1986, the 815-kg Voyager 2 spacecraft was moving at 19.698 km/s relative to the sun. To set its course for an August 1989 rendezvous with Neptune, Voyager's rocket thrusters

Fig. 9–19
The exhaust plume of a rocket fans out slightly, reducing thrust from the theoretical value given by Equation 9–10. The photo shows the launch of the Solar Maximum Mission, a satellite that has carried out extensive studies of the sun.

were fired for a period of 2 h 33 min, to give the craft an acceleration in the same direction as its velocity. During the rocket firing 9.59 kg of fuel was ejected at 1448 m/s relative to the spacecraft. By how much did Voyager's speed increase?

Solution

For this firing $|\Delta M| \ll M$, so that we can "back off" just slightly from the limit $\Delta t \rightarrow 0$ that led to Equation 9–10, and write approximately

$$M\frac{\Delta v}{\Delta t} = -v_{ex}\frac{\Delta M}{\Delta t},$$

so that

$$\Delta v = -v_{ex}\frac{\Delta M}{M} = -(1448 \text{ m/s})\left(\frac{-9.59 \text{ kg}}{815 \text{ kg}}\right) = 17 \text{ m/s}$$

is the increase in Voyager's speed. That $\Delta v \ll v_{ex}$ is another indication that our approximation is reasonable.

What if the rocket engine is fired for so long a time that the mass changes substantially? Then we cannot use the approximation in Example 9–9. But if we multiply both sides of Equation 9–10 by dt/M, we obtain

$$dv = -v_{ex}\frac{dM}{M}.$$

Integrating this expression from some initial speed v_0 to some final speed v, during which the mass goes from M_0 to M, gives

$$\int_{v_0}^{v} dv = -v_{ex}\int_{M_0}^{M}\frac{dM}{M}.$$

The integral on the left is just $v - v_0$, while the integral on the right is the one we encountered in Example 7–4; it defines the natural logarithm. So we have

$$v - v_0 = -v_{ex}[\ln(M) - \ln(M_0)] = v_{ex}[\ln(M_0) - \ln(M)].$$

But $\ln(a) - \ln(b) = \ln(a/b)$, so that

$$v = v_0 + v_{ex}\ln\left(\frac{M_0}{M}\right). \tag{9–11}$$

Since $M_0 > M$, this equation gives the expected result that speed increases as the rocket ejects fuel.

A rocket carries only a finite amount of fuel, so that the mass ratio from fully loaded to fully empty fuel tanks sets the maximum achievable speed—the **terminal speed** of the rocket. For a given mass ratio, the terminal speed can be increased by increasing the velocity of the exhaust gases.

terminal speed

Example 9–10
Terminal Speed

A spacecraft is being designed to probe the interstellar medium beyond the solar system. The craft is to be boosted from earth and sent on an outbound trajectory by conventional rockets. At the orbit of Pluto, it is to be moving away from the sun at 34 km/s; at that point an advanced rocket engine will fire to accelerate the craft to its terminal speed of 150 km/s. If the mass of the spacecraft alone is 750 kg, and if the advanced rocket exhausts fuel at 47 km/s, how much fuel must be carried to Pluto's orbit? Plot the speed of the craft as a function of time during the rocket firing, assuming that fuel is exhausted at the rate of 19 kg/s. At Pluto, the sun's gravity has a negligible effect on the spacecraft.

Solution

Let M be the spacecraft mass and M_0 the mass of the spacecraft plus fuel. To solve Equation 9–11 for the fuel mass, we first write

$$\ln\left(\frac{M_0}{M}\right) = \frac{v - v_0}{v_{ex}} = \frac{150 \text{ km/s} - 35 \text{ km/s}}{47 \text{ km/s}} = 2.45,$$

so that

$$\frac{M_0}{M} = e^{2.45} = 11.6,$$

where we used the fact that natural logarithm and exponential are inverse functions: $e^{\ln x} = x$. So the total mass before the rocket firing is 11.6M, or 8700 kg. Subtracting the 750-kg spacecraft mass leaves 7950 kg of fuel. This example shows why the achievement of very high exhaust speeds is crucial to long-distance space travel; even in this case the useful payload is dwarfed by the fuel mass, and the situation would be far worse with present-day rockets whose exhaust speeds are typically several km/s.

If fuel is burned at a steady rate dM/dt, the mass at time t is given by $M_0 - t(dM/dt)$, so that the speed is

$$v = v_0 + v_{ex} \ln\left[\frac{M_0}{M_0 - t(dM/dt)}\right].$$

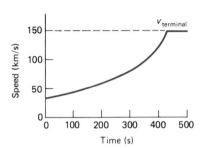

Fig. 9–20
Speed versus time for the spacecraft of Example 9–10. The craft coasts at its terminal speed once all its fuel is exhausted. Note that the acceleration increases as the craft loses mass.

The general principle of rocket propulsion applies as well to jet aircraft engines. In analyzing a jet engine, however, we must account also for the momentum of the air that is drawn into the intake of the jet (see Problem 47).

9.3
KINETIC ENERGY IN MANY-PARTICLE SYSTEMS

We have seen in this chapter how the center-of-mass concept allows us to treat the overall motion of many-particle systems using the laws we introduced earlier for individual point particles. In particular, we've seen that the total momentum of a system is determined completely by the motion of its center of mass; motions and interactions of the constituent particles don't matter. For example, a firecracker sliding on ice has the same total momentum before and after it explodes.

The same, however, is not true of a system's kinetic energy. Energetically, our firecracker is very different after it explodes; internal potential energy has been converted to kinetic energy of the fragments. Nevertheless, the center-of-mass concept remains useful in categorizing the kinetic energy associated with a system of particles.

The total kinetic energy of a system is simply the sum of the kinetic energies of the constituent particles:

$$K = \sum \tfrac{1}{2} m_i v_i^2.$$

But the velocity $\mathbf{v}_i$ of a particle can be written as the vector sum of the center-of-mass velocity $\mathbf{V}$ and a velocity $\tilde{\mathbf{v}}_i$ of that particle relative to the center of mass:

$$\mathbf{V}_i = \mathbf{V} + \tilde{\mathbf{v}}_i.$$

Then the total kinetic energy of the system is

$$\begin{aligned} K &= \sum \tfrac{1}{2} m_i (\mathbf{V} + \tilde{\mathbf{v}}_i) \cdot (\mathbf{V} + \tilde{\mathbf{v}}_i) \\ &= \sum \tfrac{1}{2} m_i V^2 + \sum m_i \mathbf{V} \cdot \tilde{\mathbf{v}}_i + \sum \tfrac{1}{2} m_i \tilde{v}_i^2, \end{aligned} \tag{9-12}$$

where in the middle term the $\tfrac{1}{2}$ cancelled the 2 from the cross terms in multiplying out the dot product.

Let us examine the three sums making up the total kinetic energy. Since the center-of-mass speed V is common to all particles, it can be factored out of the first sum, so that

$$\sum \tfrac{1}{2} m_i V^2 = \tfrac{1}{2} V^2 \sum m_i = \tfrac{1}{2} M V^2,$$

where M is the total mass. So what is this term? It is just the kinetic energy associated with a particle of mass M moving with speed V; we call this term K_{cm}, the **kinetic energy of the center of mass.**

kinetic energy of the center of mass

The center-of-mass velocity can also be factored out of the second term in Equation 9-12, giving

$$\sum m_i \mathbf{V} \cdot \tilde{\mathbf{v}}_i = \mathbf{V} \cdot \sum m_i \tilde{\mathbf{v}}_i.$$

But what is $\sum m_i \tilde{\mathbf{v}}_i$? Because the $\tilde{\mathbf{v}}_i$'s are the particle velocities relative to the center of mass, that sum is just the total momentum measured in a frame of reference moving with the center of mass. But we found earlier that the total momentum is $\mathbf{P} = m\mathbf{V}$, with $\mathbf{V}$ the center of mass velocity. In a frame of reference where $\mathbf{V} = 0$—that is, a frame of reference moving with the center of mass—the total momentum must be zero. So the entire second term in Equation 9-12 is zero.

internal kinetic energy

The third term in Equation 9-12, $\sum \tfrac{1}{2} m_i \tilde{v}_i^2$, is just the sum of the individual kinetic energies measured in a frame of reference moving with the center of mass. We call this term K_{int}, the **internal kinetic energy.**

With the middle term gone, Equation 9-12 shows that the kinetic energy of a system breaks down into two terms:

$$K = K_{cm} + K_{int}. \tag{9-13}$$

The first term, the kinetic energy of the center of mass, depends only on the motion of the center of mass. In our firecracker example, K_{cm} cannot change when the firecracker explodes. The second term, the internal kinetic energy, depends only on the motions of the constituent particles relative to the center of mass. It is therefore unchanged by external forces acting on the system, but it can be changed by internal forces. The internal kinetic energy of our firecracker increases dramatically when it explodes.

SUMMARY

1. A system of particles of total mass M has associated with it a point called the **center of mass** to which Newton's laws of motion apply as they do to a point particle:

$$\mathbf{F} = M\frac{d^2\mathbf{R}}{dt^2} = M\mathbf{A},$$

where $\mathbf{F}$ is the net external force on the system and $\mathbf{A}$ the acceleration of the center of mass. The position $\mathbf{R}$ of the center of mass is given by

$$\mathbf{R} = \frac{\sum m_i \mathbf{r}_i}{M}, \qquad \text{(center of mass)}$$

where m_i and $\mathbf{r}_i$ represent the masses and positions of the individual particles in the system. For continuously distributed matter, the center-of-mass position is given by an integral:

$$\mathbf{R} = \frac{1}{M}\int \mathbf{r}\,dm,$$

where the integration is taken over the entire system. That the center-of-mass concept is useful is a consequence of Newton's third law, which requires that internal forces cancel in pairs, leaving the overall system motion determined only by external forces.

2. In the absence of a net external force, the center-of-mass velocity $\mathbf{V} = \sum m_i \mathbf{v}_i$ remains constant. In particular, if the center of mass is at rest, then it remains at rest regardless of the motions of the constituent particles.

3. The total momentum $\mathbf{P}$ of a system is the vector sum of the momenta $\mathbf{p}_i = m_i \mathbf{v}_i$ of the constituent particles:

$$\mathbf{P} = \sum m_i \mathbf{v}_i.$$

Application of Newton's third law, as embodied in the center-of-mass concept, shows that the system obeys Newton's second law in the form

$$\mathbf{F} = \frac{d\mathbf{P}}{dt},$$

where $\mathbf{F}$ is the net external force applied to the system.

4. In the absence of a net external force, the momentum of a system remains constant. This statement is known as the **law of conservation of momentum.** It is a fundamental principle of physics, extending even into realms where Newtonian mechanics is not applicable.

5. The motion of a rocket is a manifestation of the momentum principle. The rate of change of a rocket's speed v is given by

$$M\frac{dv}{dt} = -v_{\text{ex}}\frac{dM}{dt},$$

where M is the rocket's mass, v_{ex} the speed of the exhaust gas relative to the rocket, and dM/dt the rate of change of the rocket's mass due to the ejection of gas.

6. The total kinetic energy of a system of particles includes the kinetic energy of the center of mass, given by

$$K_{\text{cm}} = \tfrac{1}{2}MV^2,$$

where M is the total mass and V the speed of the center of mass, and the **internal kinetic energy,** given by

$$K_{\text{int}} = \sum \tfrac{1}{2}m_i \tilde{v}_i^{\,2},$$

where m_i is the mass of an individual particle and $\tilde{v}_i$ its speed relative to the center of mass. The total kinetic energy is the sum of these two:

$$K_{\text{total}} = K_{\text{cm}} + K_{\text{int}}.$$

QUESTIONS

1. Roughly where is your center of mass when you are standing?
2. Can you form your body into a shape so that your center of mass is not within your body? If not, why not? If so, describe.
3. Where is the center of mass of a solid wooden wheel? Of a wagon wheel, with spokes?
4. Earth and moon are both in nearly circular orbits about a point 4600 km from the center of the earth. Explain in terms of the center-of-mass concept as applied to the earth-moon system.
5. Hunters drive a herd of bison over the edge of a cliff. As they fall, what happens to the center of mass of the earth + bison system? What happens to the centers of mass of the herd and the earth separately? Explain.
6. Explain why a high jumper's center of mass need not clear the bar.
7. Where would our derivation of the center-of-mass position $\mathbf{R}$ fail if the number of particles in the system were not constant?
8. The center of mass of a solid sphere is clearly at its center. If the sphere is cut in half, and the two halves are stacked as in Fig. 9–21, is the center of mass at the point where they touch? If not, roughly where is it? Explain.

Fig. 9–21 Question 8

9. How can a jet engine work when it takes air in the front as well as exhausting it out the back?

10. When you reach shore in Example 9–4, the center of mass of the boat+you system suddenly begins moving. What force accelerates it?

11. In Example 9–4, does the motion of the boat change when you reach shore?

12. Does the momentum of a basketball change as it rebounds off a backboard? What happens to the momentum of the backboard?

13. The momentum of a system of pool balls is the same before and after they are hit by the cue ball. Is it still the same after one of the balls strikes the edge of the table? Explain.

14. If all the cars in the world started driving eastward, what would happen to the length of the day?

15. When the chest of Example 9–6 hits the wall of the hockey rink, its momentum is no longer conserved. How would you redefine the system so that momentum is still conserved?

16. Two unequal masses are attached to the ends of an ideal spring. The spring is then compressed and the system placed at rest on a frictionless horizontal surface. When the masses are released, they oscillate back and forth, alternately stretching and compressing the spring. (a) Does the center of mass move during this process? (b) Now suppose the spring is not ideal, so that energy is dissipated in the spring and the motion eventually stops. Has the center of mass moved from its original position? (c) Now suppose the surface is not frictionless, so that again the motion eventually stops. Has the center of mass moved from its original position? Explain your answers.

17. Consider the system consisting of a shot putter and his shot; the mass of the shot is about 10% that of the shot putter. Describe the motion of the system's center of mass (a) if the shot putter is standing on a concrete pad with high coefficient of friction and (b) if he is standing on a frictionless surface.

18. An asteroid 100 km in diameter is discovered to be on collision course with earth. In a show of international cooperation and mutual disarmament, the superpowers launch all their nuclear missiles at the asteroid and succeed in blowing it to smithereens. Will the center of mass of the asteroid still hit earth?

19. Soon after the event of the previous question, a smaller asteroid is discovered heading for earth. Having used all their nuclear missiles on the first asteroid, the superpowers now cooperate in the manufacture of a huge rocket engine, which they transport to the asteroid. The rocket is strapped to the asteroid and fired long enough to move the asteroid off its collision course. Now does the center of mass of the asteroid hit earth?

20. Can you make a sailboat go by blowing air onto the sails with a fan mounted on the boat? With air from a compressed-air cylinder mounted on the boat? [See *The Physics Teacher*, vol. 24, pp. 38–39, 392–393 (1986).]

21. Criticize the editorial shown in Fig. 9–17.

22. An hourglass is inverted and placed on a scale. Compare scale readings (a) before sand begins to hit the bottom; (b) while sand is hitting the bottom; (c) when all the sand is on the bottom.

23. The momentum of the center of mass of a system is the sum of the momenta of the individual particles, but the same is not true for kinetic energy. Why not?

PROBLEMS

Section 9.1   *Center of Mass*

1. Four masses lie at the corners of a square 1.0 m on a side, as shown in Fig. 9–22. Where is the center of mass?

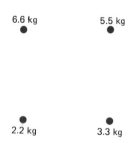

6.6 kg   5.5 kg

2.2 kg   3.3 kg

Fig. 9–22   Problem 1

2. Rework Example 9–1 with the origin at the center of the barbell, showing that the physical location of the center of mass is independent of the choice of coordinate system.

3. Three equal masses lie at the corners of an equilateral triangle of side ℓ. Where is the center of mass?

4. How far from the center of the earth is the center of mass of the earth-moon system? *Hint:* Consult Appendix F.

5. Find the center of mass of the plane object shown in Fig. 9–8, assuming uniform density.

6. Find the center of mass of the solar system at a time when all the planets are lined up on the same side of the sun. *Hint:* Consult Appendix F.

7. Find the center of mass of a pentagon of side a with one triangle missing, as shown in Fig. 9–23. *Hint:* See Example 9–3.

8. Find the center of mass of a uniformly solid cone of height h and base radius a.

9. A water molecule (H_2O) consists of two hydrogen atoms, each of mass 1.0 u, and one oxygen atom of mass 16 u. The hydrogen atoms are 0.96 Å ($1 \text{ Å} = 10^{-10}$ m) from the oxygen and are separated by an angle of 105° (Fig. 9–24). Where is the center of mass of the molecule?

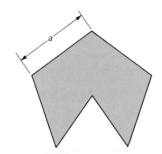

Fig. 9–23 Problem 7

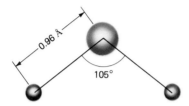

Fig. 9–24 A water molecule (Problem 9).

10. A flat, T-shaped structure is made from two identical rectangles of length ℓ and width w (Fig. 9–25). Find its center of mass.

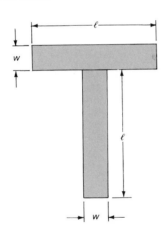

Fig. 9–25 Problem 10

11. Consider a system of three equal mass particles moving in a plane; their positions are given by $a_i\hat{i}+b_i\hat{j}$, where a_i and b_i are functions of time with the units of position. Particle 1 has $a_1=3t^2+5$ and $b_1=0$; particle 2 has $a_2=7t+2$ and $b_2=2$; particle 3 has $a_3=3t$ and $b_3=2t+6$. Find the position, velocity, and acceleration of the center of mass as functions of time.

12. A 50-kg woman on a 650-kg boat hooks a 150-kg blue marlin fish on the end of a nearly horizontal line 35 m long. If fish and boat are originally at rest on a frictionless sea, how far does the boat move as she reels in the fish?

13. Estimate how much the earth would move if its entire human population of 4.5 billion jumped simultaneously into the 1-mile-deep Grand Canyon.

14. You are with 19 other people on a boat at rest in frictionless water. Your total mass is 1500 kg, and the mass of the boat itself is 12,000 kg. The entire party walks the 6.5-m distance from bow to stern. How far does the boat move?

15. A hemispherical bowl is at rest on a frictionless kitchen counter. A mouse drops onto the rim of the bowl from a cabinet directly overhead. The mouse climbs down the inside of the bowl to eat the crumbs at the bottom. If the bowl moves along the counter a distance equal to one-tenth of its diameter, how does the mouse's mass compare with the bowl's mass?

16. An 850-kg elevator is balanced perfectly by a counterweight connected to the elevator by a cable running over a pulley (Fig. 9–26). When the elevator is at the first floor, the counterweight is at the seventeenth floor. (a) Where is the center of mass of the elevator and counterweight system? (b) If a 60-kg passenger boards the elevator when it is at rest on the fifth floor, what is the subsequent motion of the center of mass of the system consisting of elevator, counterweight, and person?

Fig. 9–26 Problem 16

Section 9.2 *Momentum*

17. The Olympic-champion ice dancers Torville and Dean start from rest and push off each other. Torville's mass is 70 kg and Dean's mass is 50 kg, and the ice is essentially frictionless. If Torville's speed is 6.2 m/s, how far apart are they after 2.0 s?

18. A plutonium-239 nucleus at rest decays into a uranium-235 nucleus by emitting an alpha particle (^{4}He) with kinetic energy of 5.15 MeV. What is the speed of the uranium nucleus?

19. A runaway toboggan of mass 8.6 kg is moving horizontally at 23 km/h. As it passes under a tree, 15 kg of snow drop onto it. What is its subsequent speed?

20. A 950-kg airplane touches down on an icy runway, heading northward at 150 km/h. It collides with a 1200-kg plane taxiing eastward at 85 km/h. Find the speed and direction of the combined wreckage.

21. During a heavy storm, rain falls at the rate of 2.0 cm/hour; the speed of the individual raindrops is 25 m/s. (a) If the rain strikes a flat roof and then flows off the roof with negligible speed, what is the force exerted per square meter of roof area? (b) How much water would have to stand on the roof to exert the same force? The density of water is 1.0 g/cm^3.

22. A circle of Ice Capades clowns contains six individuals with masses of 50, 55, 60, 65, 70, and 75 kg. If they are at rest and evenly spaced around a circle, then start skating radially outward each with the same momentum $p = 200$ kg·m/s, describe the subsequent motion of their center of mass.

23. An 11,000-kg freight car is resting against a spring bumper at the end of a railroad track. The spring has constant $k = 3.2 \times 10^5$ N/m. The car is hit by a second car of 9400 kg mass moving at 8.5 m/s, and the two cars couple together. (a) What is the maximum compression of the spring? (b) What is the speed of the two cars together when they rebound from the spring?

24. On an icy road, a 1200-kg car moving at 50 km/h strikes a 4400-kg truck moving in the same direction at 35 km/h. The combination is soon hit from behind by a 1500-kg car speeding at 65 km/h. If all three vehicles stick together, what is the speed of the wreckage?

25. A 1600-kg automobile is resting at one end of a 4500-kg railroad flatcar that is also at rest. The automobile then drives along the flatcar at 15 km/h relative to the flatcar. Unfortunately, the flatcar brakes are not set. How fast does it move?

26. One plan for a space-based anti-ballistic-missile weapon involves shooting a beam of 100-MeV hydrogen atoms at an approaching missile. The power delivered by the beam is 100 MW. To destroy a missile, the beam should be turned on for 2 s, and should cover an area of 1 m^2. Determine (a) the momentum in the beam and (b) the average force exerted on the missile. Assume the beam is absorbed by the missile. (c) If the beam hits a 100-kg warhead moving at 8 km/s, will the warhead's trajectory be affected significantly?

27. An alternative to the anti-missile weapon of the previous problem is to hit the missile with objects of about 1-kg mass travelling at 20 km/s. Assuming that such a "kinetic kill weapon" embeds itself in the missile, (a) how much momentum does it impart to the missile; (b) how much kinetic energy does it carry; (c) does it significantly affect the trajectory of a 100-kg warhead moving at 8 km/s? (d) Compare the energy and momentum in this problem with that of the preceding problem. Explain any similarities and differences.

28. A ^{238}U nucleus is moving in the x-direction at 5.0×10^5 m/s, when it decays into an alpha particle (^{4}He) and a ^{234}Th nucleus. If the alpha particle moves off at 22° above the x-axis with a speed of 1.4×10^7 m/s, what is the recoil velocity of the thorium nucleus?

29. An ideal spring of spring constant k rests on a frictionless surface. Blocks of mass m_1 and m_2 are pushed against the two ends of the spring until it is compressed a distance x from its equilibrium length. The blocks are then released. What are their speeds when they leave the spring?

30. A car of mass M is initially at rest on a frictionless surface. A jet of water carrying mass at the rate dm/dt and moving horizontally at speed v_0 strikes the rear window of the car, which makes a 45° angle with the horizontal; the water bounces off at the same relative speed with which it hit the window, as shown in Fig. 9–27. (a) Find an expression for the initial acceleration of the car. (b) What is the maximum speed reached by the car?

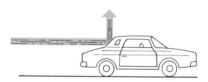

Fig. 9–27   Problem 30

31. A 950-kg compact car is moving with velocity $\mathbf{v}_1 = 32\hat{\imath} + 17\hat{\jmath}$ m/s. It skids on a frictionless icy patch, and collides with a 450-kg hay wagon moving with velocity $\mathbf{v}_2 = 12\hat{\imath} + 14\hat{\jmath}$ m/s. If the two stay together, what is their velocity?

32. The 975-kg Giotto spacecraft, when it was close to Halley's Comet in 1986, passed through cometary dust of density 0.10 g/m^3 at a speed of 200 m/s. Its frontal surface area was 10 m^2. Assuming that the dust stuck to the front of the spacecraft, at what rate was the spacecraft decelerated?

33. A biologist fires 20-g rubber bullets at a rhinoceros that is charging at 1.9 m/s. If the gun fires 5 bullets per second, with a speed of 1600 m/s, and if the biologist fires for 13 s to stop the rhino in its tracks, what is the mass of the rhino? Assume the bullets drop vertically after striking the rhino.

34. A wedge-shaped block of mass M, length ℓ, and angle θ is at rest on a frictionless surface. A small block of mass m is released from rest at the top of the wedge, and slides without friction to the bottom and onto the horizontal surface (Fig. 9–28). Show that the speed of the wedge after the block leaves it is

$$v = m \sqrt{\frac{2g\ell \, \sin\theta \, \cos\theta}{(M+m)(M+m \, \sin^2\theta)}}.$$

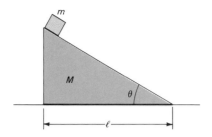

Fig. 9–28 Problem 34

42. The triangle of Fig. 9–29 has a density that varies in proportion to y^α, where α is a constant. Show that its center of mass is located at $y = (\alpha + 2)h/(\alpha + 3)$.

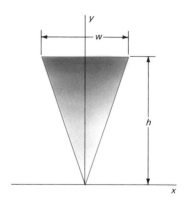

Fig. 9–29 Problem 42

35. An Ariane rocket ejects 1.0×10^5 kg of fuel in the 90 s after launch. (a) How much thrust is developed if the fuel is ejected at 3.0 km/s with respect to the rocket? (b) What is the maximum total mass of the rocket if it is to get off the ground?

36. 90% of a rocket's initial mass is fuel. If the fuel is exhausted at 3500 m/s, what is the terminal speed of the rocket?

37. If a rocket's exhaust speed is 200 m/s, what fraction of its initial mass must be ejected to increase the rocket's speed by 50 m/s?

38. A space shuttle's main engines develop a thrust of 35×10^6 N as they eject gas at 2500 m/s. (a) How much fuel must the shuttle carry to permit a 5-minute engine firing? (b) At 200 kg/m^3, how large a cubical fuel tank would be required?

Section 9.3   *Kinetic Energy in Many-Particle Systems*

39. Determine the center of mass and internal kinetic energies before and after decay of the lithium nucleus of Example 9–7. Treat the individual nuclei as point particles.

40. A 150-g trick baseball is thrown at 60 km/h. It explodes in flight into two pieces, with a 38-g piece continuing straight ahead at 85 km/h. How much energy is released in the explosion?

41. The solar corona is a gas containing predominantly electrons and protons. The electrons are moving in random directions at speeds on the order of 10^7 m/s; this speed corresponds to a temperature of about 2 million K. The average coronal density near the sun is about 10^9 electrons/cm^3, with an equal density of protons. Large volumes of coronal material are sometimes ejected into space at speeds on the order of 300 km/s. Assuming that the protons partake only in this overall motion, while the electrons have superimposed their random 10^7 m/s motions, determine the center of mass and internal kinetic energies in a 1000 km^3 volume of such ejected coronal matter. Which particles make the dominant contribution to the center of mass and internal energies?

43. While standing on frictionless ice, you (mass 65.0 kg) toss a 4.50-kg rock with initial speed of 12.0 m/s. If the rock is 15.2 m from you when it lands, (a) at what angle did you toss it? (b) How fast are you moving?

44. A drunk driver in a 1600-kg car plows into a 1300-kg parked car with its brake set. Police measurements show that the two cars skid together a distance of 25 m before stopping. If the effective coefficient of friction is 0.77, how fast was the drunk going just before the collision?

45. A fireworks rocket is launched vertically upward at 40 m/s. At the peak of its trajectory, it explodes into two equal-mass fragments. One reaches the ground 2.87 s after the explosion. When does the second reach the ground?

46. A fire hose delivers 50 kg/s of water at 30 m/s. How many 75-kg firefighters are needed to hold the hose on muddy ground for which the coefficient of static friction is 0.35?

47. (a) Derive an expression for the thrust of a jet aircraft engine. Moving through the air with speed v, the engine takes in air at the rate dM_{in}/dt. It uses the air to burn fuel with a fuel/air ratio f (that is, f kg of fuel burned for each kg of air), and ejects the exhaust gases at speed v_{ex} with respect to the engine. (b) Use your result to find the thrust of a JT-8D engine on a Boeing 727 jetliner, for which $v_{ex} = 1034$ ft/s and $dM_{in}/dt = 323$ lb/s, and which consumes 3760 lb of fuel per hour while cruising at 605 mi/h.

48. Two blocks are sliding along a one-dimensional frictionless surface with speeds v_1 and v_2, when they collide. Use the conservation of momentum principle to show that the most kinetic energy is lost if they stick together.

10

COLLISIONS

On the highway, we try to avoid colliding with other cars. On the pool table, we tailor collisions to suit our purposes. In particle accelerators, we interpret collisions to reveal the innermost structure of matter. On a cosmic scale, we wonder if a collision with an asteroid led to the extinction of the dinosaurs, and whether such occasional collisions will have drastic effects in the future. Collisions are important features of our physical universe (Fig. 10–1).

collision

A **collision** is a relatively violent interaction of objects that lasts for a short time. By "violent interaction" we mean that forces associated with the collision are much greater than any other forces that may be acting on the objects. By "short time" we mean short compared with time scales appropriate in describing the overall motion of the objects. A collision between two airplanes, for example, involves forces much greater than the gravitational and aerodynamic forces; the time interval over which the collision occurs is short compared with the overall flight times. Similarly, a collision between two galaxies (Fig. 10–2) may take millions of years, but that period is still short compared with the ages of the galaxies.

Fig. 10–1
Artist's conception of an asteroid colliding with the earth. Statistics show that an asteroid larger than 1 km probably hits our planet every few hundred million years. Some scientists have suggested that such a collision 65 million years ago created a cloud of dust and soot that led to the extinction of the dinosaurs.

impulsive force

10.1
IMPULSE AND COLLISIONS

In a collision of two objects—like the foot and football in Fig. 10–3—the velocity of one or both objects is abruptly changed. We know, from Newton's laws of motion, that such a change means a force has been applied. The strong but short-duration force associated with a collision is called an **impulsive force**. Figure 10–4 shows force as a function of time for the football of Fig. 10–3. Although the gravitational force also acts on the ball, during the collision the impulsive force is overwhelmingly dominant.

We can relate the impulsive force to the change in an object's momen-

232

Fig. 10–2
Colliding galaxies.

Fig. 10–3
High-speed photo of a collision between a foot and a football. The strong impulsive force is evident in the deformation of the football. This force causes a rapid change in the ball's velocity.

impulse

tum through Newton's second law:

$$\mathbf{F} = \frac{d\mathbf{p}}{dt}.$$

Multiplying this equation by dt and integrating from some time t_1 before the collision to a time t_2 after the collision, we have

$$\int_{t_1}^{t_2} \mathbf{F}\,dt = \int_{t_1}^{t_2} d\mathbf{p}.$$

The right-hand integral is just the change $\Delta\mathbf{p}$ that occurs during the collision. The integral on the left is known as the **impulse, I,** associated with the collision. Our result then shows that the change in an object's momentum is equal to the impulse:

$$\mathbf{I} \equiv \int_{t_1}^{t_2} \mathbf{F}\,dt = \Delta\mathbf{p}. \qquad (10\text{–}1)$$

The units of impulse are the same as those of momentum: kg·m/s, or, equivalently, N·s. Since the definite integral corresponds to the area under a curve, Equation 10–1 shows that the impulse is given geometrically by the area under the force versus time curve. It doesn't much matter whether the force curve includes any non-impulsive forces that may be present, for our definition of a collision ensures that the change in momentum associated with non-impulsive forces is small during the time of the collision.

Although the impulsive force in a collision may be a very complicated function of time (Fig. 10–5), the overall effect of a collision on the motion of an object is summed up by the change in the object's momentum. We can work backward from that change to determine the **average impulsive force** $\overline{\mathbf{F}}$—the force that, if applied constantly during the collision time Δt, would give the same change in momentum. That is,

$$\overline{\mathbf{F}} = \frac{\mathbf{I}}{\Delta t} = \frac{\Delta\mathbf{p}}{\Delta t}. \qquad (10\text{–}2)$$

average impulsive force

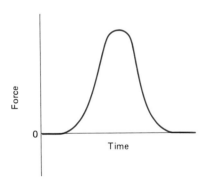

Fig. 10–4
Force as a function of time for the football of Fig. 10–3. The impulse is given by the area under the force versus time curve, and is equal to the change in the football's momentum.

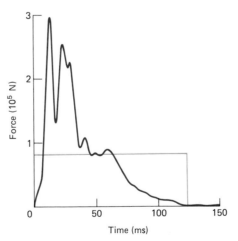

Fig. 10–5
Impulsive force in the crash of a 1986 Suzuki Samuri into a rigid barrier. Horizontal line is the average force; area under the rectangle is the same as under the jagged curve.

Graphically, the average force is represented by the height of a rectangle whose area is the same as the actual impulse area $\int F dt$ (see Fig. 10–5).

Example 10–1
Impulse

A 150-g baseball is moving horizontally at 30 m/s just before reaching the bat. Bat and ball are then in contact for 2.0 ms (Fig. 10–6). Find the impulse and the average impulsive force (a) if the hit is a line drive to center field at 45 m/s and (b) if the hit is a pop foul that starts straight up at 22 m/s.

Solution

In a coordinate system with x-axis along the line from home plate to center field, the incoming momentum is

$$\mathbf{p}_1 = m\mathbf{v}_1 = (0.15 \text{ kg})(-30\hat{\imath} \text{ m/s}) = -4.5\hat{\imath} \text{ kg·m/s}.$$

For the line drive, the outgoing momentum is

$$\mathbf{p}_2 = m\mathbf{v}_2 = (0.15 \text{ kg})(45\hat{\imath} \text{ m/s}) = 6.8\hat{\imath} \text{ kg·m/s},$$

giving an impulse

$$\mathbf{I} = \Delta\mathbf{p} = (6.8\hat{\imath} \text{ kg·m/s}) - (-4.5\hat{\imath} \text{ kg·m/s}) = 11\hat{\imath} \text{ kg·m/s}$$

and an average force

$$\overline{\mathbf{F}} = \frac{\Delta\mathbf{p}}{\Delta t} = \frac{11\hat{\imath} \text{ kg·m/s}}{2.0 \times 10^{-3} \text{ s}} = 5.5 \times 10^3 \hat{\imath} \text{ N}.$$

This is a large force, nearly 4000 times the gravitational force on the ball.

For the pop foul, the outgoing momentum is in the y-direction:

$$\mathbf{p}_2 = m\mathbf{v}_2 = (0.15 \text{ kg})(22\hat{\jmath} \text{ m/s}) = 3.3\hat{\jmath} \text{ kg·m/s},$$

so that

$$\mathbf{I} = \Delta\mathbf{p} = (3.3\hat{\jmath} \text{ kg·m/s}) - (-4.5\hat{\imath} \text{ kg·m/s}) = 4.5\hat{\imath} + 3.3\hat{\jmath} \text{ kg·m/s}$$

and

$$\overline{\mathbf{F}} = \frac{\Delta\mathbf{p}}{\Delta t} = \frac{4.5\hat{\imath} + 3.3\hat{\jmath} \text{ kg·m/s}}{2.0 \times 10^{-3} \text{ s}} = 2.3 \times 10^3 \hat{\imath} + 1.7 \times 10^3 \hat{\jmath} \text{ N},$$

or an average impulse force of magnitude 2800 N. Large forces like this are typical of collisions, and explain why collisions often involve so much damage.

Fig. 10–6
High-speed photo of ball and bat. The two are in contact for only about 2 ms.

10.2
COLLISIONS AND THE CONSERVATION LAWS

Often we know no more about the impulsive force of a collision than that it acts for a short time and produces abrupt changes in motion. Nevertheless, we can use the conservation laws developed in earlier chapters to relate the interacting objects' motions before and after collision. As always, conservation laws provide a "shortcut," giving us an overall picture lacking in some details. In the case of a collision involving an unknown impulsive force, those missing details are confined to the very short time of the collision.

Momentum

One thing we do know about impulsive forces between colliding objects is that they are *internal* to a system comprising those objects. We found in the preceding chapter that the momentum of a system can be changed only by external forces. Therefore, the impulsive forces of a collision leave the total momentum of the colliding objects unchanged. What about any external forces that may also act during a collision, like the gravitational force on a tennis ball as it collides with the racket? Our definition of a collision requires that the collision take place in a very short time, so that external forces do not have time to alter the system momentum significantly during the collision. To a very good approximation, therefore, the total momentum of a colliding system is conserved during a collision. This conclusion is valid regardless of the details of the impulsive force.

Energy

What about mechanical energy? It may or may not be conserved during a collision. If mechanical energy is conserved, the collision is termed elastic. Like a completely frictionless surface, an **elastic collision** is an impossible idealization in the macroscopic world. In a macroscopic collision, some mechanical energy is inevitably transformed to other forms like sound, random molecular motions, or the energy associated with permanently deforming materials (Fig. 10–7). But some interactions in the atomic, nuclear, and elementary particle realms are completely elastic. And macroscopic objects like billiard balls, trampolines, and hockey pucks undergo nearly elastic collisions. The important criterion is that in an elastic collision, the internal forces must be conservative. Then kinetic energy is stored temporarily as potential energy, all of which is again released as kinetic energy before the collision is over.

When mechanical energy is lost during a collision, the collision is **inelastic.** In a **totally inelastic collision,** the colliding objects stick together to form one composite object. Even then, mechanical energy is usually not

elastic collision

inelastic collision

totally inelastic collision

Fig. 10–7
(Left) The superball on the left undergoes a nearly elastic collision with the ground, rebounding with much greater speed than the squash ball on the right. (Above) An automobile collision is invariably inelastic, with much of the kinetic energy lost to permanent deformation of the colliding vehicles.

all lost; complete loss of mechanical energy in most cases would violate conservation of momentum. But a totally inelastic collision involves the maximum loss of mechanical energy that is consistent with momentum conservation. In general, we can determine the motion after collision only for the case of completely elastic or totally inelastic collisions; in intermediate cases we must know just how much energy is lost before we can solve for the post-collision motion (see Problem 47).

10.3
INELASTIC COLLISIONS

The motion after a totally inelastic collision is determined entirely by the conservation of momentum principle. We analyzed some totally inelastic collisions in the previous chapter; for instance, measuring the speed of the hockey puck in Example 9–6 involved a totally inelastic collision.

Consider two objects with masses m_1 and m_2 and initial velocities $\mathbf{v}_{1i}$ and $\mathbf{v}_{2i}$ that undergo a totally inelastic collision. That is, after collision they stick together to form a single composite body of mass $m_1 + m_2$ and final velocity $\mathbf{v}_f$. Conservation of momentum states that the initial and final momenta of this two-particle system must be the same:

$$m_1\mathbf{v}_{1i} + m_2\mathbf{v}_{2i} = (m_1 + m_2)\mathbf{v}_f. \quad \text{(inelastic collision)} \quad (10\text{–}3)$$

Given four of the five quantities m_1, $\mathbf{v}_{1i}$, m_2, $\mathbf{v}_{2i}$, and $\mathbf{v}_f$, we can solve for the fifth. Often but not always, the unknown is the final velocity:

$$\mathbf{v}_f = \frac{m_1\mathbf{v}_{1i} + m_2\mathbf{v}_{2i}}{m_1 + m_2}. \quad (10\text{–}4)$$

Calculating the total kinetic energy before and after the collision will convince you that this collision is indeed inelastic (see Problem 13).

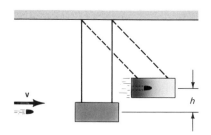

Fig. 10–8
A ballistic pendulum (Example 10–2).

Example 10–2
The Ballistic Pendulum

The ballistic pendulum is a device used to measure the speeds of fast-moving objects like bullets. It consists of a wooden block of mass M suspended from vertical strings (Fig. 10–8). When a bullet strikes the block, it undergoes an inelastic collision and embeds itself in the block. If the bullet has mass m, and if the block rises a maximum height h after the collision, show that the speed of the bullet is $(m + M)\sqrt{2gh}/m$.

Solution

To solve this problem, we deal separately first with the collision and then with the subsequent rise of the block. Understanding the difference in the way we handle these two steps will strengthen your confidence in applying the conservation laws.

1. The incoming bullet undergoes an inelastic collision with the block. Mechanical energy is not conserved. But because the collision takes such a short time, external forces—here gravity and the string tension—have little effect and the momentum of the bullet + block system is essentially conserved. So we can write:

$$mv = (m + M)V,$$

Momentum

One thing we do know about impulsive forces between colliding objects is that they are *internal* to a system comprising those objects. We found in the preceding chapter that the momentum of a system can be changed only by external forces. Therefore, the impulsive forces of a collision leave the total momentum of the colliding objects unchanged. What about any external forces that may also act during a collision, like the gravitational force on a tennis ball as it collides with the racket? Our definition of a collision requires that the collision take place in a very short time, so that external forces do not have time to alter the system momentum significantly during the collision. To a very good approximation, therefore, the total momentum of a colliding system is conserved during a collision. This conclusion is valid regardless of the details of the impulsive force.

Energy

What about mechanical energy? It may or may not be conserved during a collision. If mechanical energy is conserved, the collision is termed elastic. Like a completely frictionless surface, an **elastic collision** is an impossible idealization in the macroscopic world. In a macroscopic collision, some mechanical energy is inevitably transformed to other forms like sound, random molecular motions, or the energy associated with permanently deforming materials (Fig. 10–7). But some interactions in the atomic, nuclear, and elementary particle realms are completely elastic. And macroscopic objects like billiard balls, trampolines, and hockey pucks undergo nearly elastic collisions. The important criterion is that in an elastic collision, the internal forces must be conservative. Then kinetic energy is stored temporarily as potential energy, all of which is again released as kinetic energy before the collision is over.

When mechanical energy is lost during a collision, the collision is **inelastic.** In a **totally inelastic collision,** the colliding objects stick together to form one composite object. Even then, mechanical energy is usually not

elastic collision (margin note)

inelastic collision (margin note)
totally inelastic collision (margin note)

Fig. 10–7
(Left) The superball on the left undergoes a nearly elastic collision with the ground, rebounding with much greater speed than the squash ball on the right. (Above) An automobile collision is invariably inelastic, with much of the kinetic energy lost to permanent deformation of the colliding vehicles.

all lost; complete loss of mechanical energy in most cases would violate conservation of momentum. But a totally inelastic collision involves the maximum loss of mechanical energy that is consistent with momentum conservation. In general, we can determine the motion after collision only for the case of completely elastic or totally inelastic collisions; in intermediate cases we must know just how much energy is lost before we can solve for the post-collision motion (see Problem 47).

10.3
INELASTIC COLLISIONS

The motion after a totally inelastic collision is determined entirely by the conservation of momentum principle. We analyzed some totally inelastic collisions in the previous chapter; for instance, measuring the speed of the hockey puck in Example 9–6 involved a totally inelastic collision.

Consider two objects with masses m_1 and m_2 and initial velocities $\mathbf{v}_{1i}$ and $\mathbf{v}_{2i}$ that undergo a totally inelastic collision. That is, after collision they stick together to form a single composite body of mass $m_1 + m_2$ and final velocity $\mathbf{v}_f$. Conservation of momentum states that the initial and final momenta of this two-particle system must be the same:

$$m_1\mathbf{v}_{1i} + m_2\mathbf{v}_{2i} = (m_1 + m_2)\mathbf{v}_f. \qquad \text{(inelastic collision)} \qquad (10\text{--}3)$$

Given four of the five quantities m_1, $\mathbf{v}_{1i}$, m_2, $\mathbf{v}_{2i}$, and $\mathbf{v}_f$, we can solve for the fifth. Often but not always, the unknown is the final velocity:

$$\mathbf{v}_f = \frac{m_1\mathbf{v}_{1i} + m_2\mathbf{v}_{2i}}{m_1 + m_2}. \qquad (10\text{--}4)$$

Calculating the total kinetic energy before and after the collision will convince you that this collision is indeed inelastic (see Problem 13).

Example 10–2
The Ballistic Pendulum

The ballistic pendulum is a device used to measure the speeds of fast-moving objects like bullets. It consists of a wooden block of mass M suspended from vertical strings (Fig. 10–8). When a bullet strikes the block, it undergoes an inelastic collision and embeds itself in the block. If the bullet has mass m, and if the block rises a maximum height h after the collision, show that the speed of the bullet is $(m + M)\sqrt{2gh}/m$.

Solution

To solve this problem, we deal separately first with the collision and then with the subsequent rise of the block. Understanding the difference in the way we handle these two steps will strengthen your confidence in applying the conservation laws.

1. The incoming bullet undergoes an inelastic collision with the block. Mechanical energy is not conserved. But because the collision takes such a short time, external forces—here gravity and the string tension—have little effect and the momentum of the bullet + block system is essentially conserved. So we can write:

$$mv = (m + M)V,$$

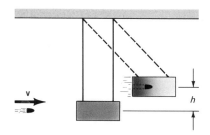

Fig. 10–8
A ballistic pendulum (Example 10–2).

where v is the initial speed of the bullet and V the speed of the bullet/block combination after the collision. Solving for V gives $V = mv/(m + M)$.

2. As the block swings upward, the external forces of gravity and string tension act to alter the momentum of the bullet + block system. Momentum is not conserved. But gravity is a conservative force, and the string tension acts at right angles to the motion and therefore does no work. So mechanical energy is conserved in the upward swing of the block. Taking the zero of potential energy in the block's initial position, we can then equate the kinetic energy just after the collision to the potential energy when the block is at its maximum height:

$$\tfrac{1}{2}(m + M)V^2 = (m + M)gh.$$

Now we relate the two parts of the problem. Using our expression for V in the statement of energy conservation, we have

$$\frac{1}{2}\left(\frac{mv}{m + M}\right)^2 = gh,$$

so that

$$v = \frac{(m + M)\sqrt{2gh}}{m}.$$

Example 10-3
A Fusion Reaction

In a fusion reaction, two deuterium nuclei (^{2}H) combine to form a helium nucleus (^{4}He). One of the incident nuclei is originally moving in the +x-direction at 3.50×10^6 m/s, and the helium nucleus is subsequently detected moving at 1.62×10^6 m/s at an angle of $\theta = 36°$ to the x-axis (Fig. 10–9). Find the initial velocity of the second deuterium nucleus.

Solution

Momentum conservation entirely determines the outcome of this totally inelastic collision. The statement of momentum conservation is given by Equation 10–3; solving for the initial velocity $\mathbf{v}_2$ of the second deuterium nucleus, we have

$$\mathbf{v}_2 = \frac{(m_1 + m_2)\mathbf{v} - m_1\mathbf{v}_1}{m_2}.$$

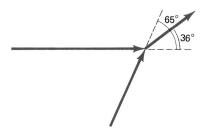

Fig. 10–9
Velocity vectors for Example 10–3.

Breaking this vector equation into its two components gives

x-component: $$v_2\cos\phi = \frac{(m_1 + m_2)v\cos\theta - m_1 v_1}{m_2}$$

y-component: $$v_2\sin\phi = \frac{(m_1 + m_2)v\sin\theta}{m_2},$$

where we have made use of the fact that $\mathbf{v}_1$ is entirely in the x-direction. Squaring and adding these two equations, and using $\cos^2\phi + \sin^2\phi = 1$, we get

$$v_2^2 = \frac{(Mv\cos\theta - m_1 v_1)^2 + M^2 v^2 \sin^2\theta}{m_2^2},$$

where $M = m_1 + m_2$ is the total mass. Using the given values of m_1, m_2, v_1, v_2, and θ, we find that $v_2 = 2.1 \times 10^6$ m/s. Solving the y-component equation for ϕ then gives

$$\phi = \sin^{-1}\left(\frac{Mv\sin\theta}{m_2 v_2}\right) = \sin^{-1}\left[\frac{(4\text{ u})(1.62 \times 10^6\text{ m/s})(\sin 36°)}{(2\text{ u})(2.1 \times 10^6\text{ m/s})}\right] = 65°.$$

Figure 10–9 shows the velocity vectors for the deuterium and helium nuclei.

Does this problem remind you of the radioactive decay of Example 9–7? It should: fusion and radioactive decay are essentially inverse processes. The same momentum conservation principle governs both, and even the same mathematical techniques apply.

10.4
ELASTIC COLLISIONS

We have seen that momentum is essentially conserved in any collision. In an elastic collision, kinetic energy is conserved as well. In the most general case of a two-body collision, we consider two objects of masses m_1 and m_2, moving initially with velocities $\mathbf{v}_{1i}$ and $\mathbf{v}_{2i}$, respectively. Their final velocities after collision are $\mathbf{v}_{1f}$ and $\mathbf{v}_{2f}$. Then the conservation statements for momentum and kinetic energy become

$$m_1\mathbf{v}_{1i} + m_2\mathbf{v}_{2i} = m_1\mathbf{v}_{1f} + m_2\mathbf{v}_{2f} \tag{10–5}$$

and

$$\tfrac{1}{2}m_1 v_{1i}^2 + \tfrac{1}{2}m_2 v_{2i}^2 = \tfrac{1}{2}m_1 v_{1f}^2 + \tfrac{1}{2}m_2 v_{2f}^2. \tag{10–6}$$

Given the initial velocities, we would like to be able to predict the outcome of a collision. In the preceding section, we did just that for an arbitrary inelastic collision. In a two-dimensional inelastic collision, we had two equations (the two components of the vector momentum conservation equation) and two unknowns (the magnitude and direction of the one final velocity). Mathematically, we had enough information to solve the system. Here, in the elastic case, we have the two components of the momentum conservation equation 10–5 and the single scalar equation for energy conservation 10–6. But we have four unknowns—the magnitudes and directions of both final velocities. With three equations and four unknowns, we do not have enough information to solve the general two-dimensional elastic collision. Later in this section we will see how other information can help us solve such problems. First, though, we look at the special case of a one-dimensional elastic collision.

Elastic Collisions in One Dimension

When two objects collide head-on, the internal forces act along the same line as the incident motion, and the objects' subsequent motion must therefore be along that same line (Fig. 10–10). Although such one-dimensional collisions are clearly a very special case, they do occur and they provide much insight into the more general case.

In the one-dimensional case, the momentum conservation equation 10–5 has only one nontrivial component:

$$m_1 v_{1i} + m_2 v_{2i} = m_1 v_{1f} + m_2 v_{2f}, \tag{10–5a}$$

where the v's stand for velocity components, rather than magnitudes, and can therefore be positive or negative. If we collect together the terms in Equations 10–5a and 10–6 that are associated with each mass, we have

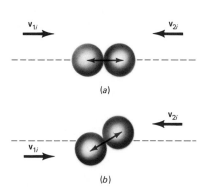

Fig. 10–10
(a) When two objects collide head-on, internal forces act along the same line as the incident motion, and the subsequent motion is also along this line. The collision is one-dimensional. (b) If the collision is not head-on, internal forces act in new directions, and the motion is necessarily two-dimensional.

$$m_1(v_{1i} - v_{1f}) = m_2(v_{2f} - v_{2i}) \qquad (10\text{--}5b)$$

and

$$m_1(v_{1i}^2 - v_{1f}^2) = m_2(v_{2f}^2 - v_{2i}^2). \qquad (10\text{--}6a)$$

But $a^2 - b^2 = (a+b)(a-b)$, so that Equation 10–6a can be written

$$m_1(v_{1i} - v_{1f})(v_{1i} + v_{1f}) = m_2(v_{2f} - v_{2i})(v_{2f} + v_{2i}). \qquad (10\text{--}6b)$$

Dividing the left and right sides of Equation 10–6b by the corresponding sides of Equation 10–5a then gives

$$v_{1i} + v_{1f} = v_{2f} + v_{2i}. \qquad (10\text{--}7)$$

(In performing this division, we assumed that we were not dividing by zero; that is, we assumed that the velocities were indeed changed by the collision.) Rearranging Equation 10–7 shows that

$$v_{1i} - v_{2i} = v_{2f} - v_{1f}. \qquad (10\text{--}8)$$

What does this equation tell us? Both sides describe the relative velocity between the two particles; the equation therefore shows that the relative speed remains unchanged after the collision, although the direction reverses. If one object is approaching another at a relative speed of 5 m/s, then after collision it will be receding at 5 m/s.

Continuing our search for the final velocities, we solve Equation 10–8 for v_{2f}:

$$v_{2f} = v_{1i} - v_{2i} + v_{1f},$$

and use this result in Equation 10–5a:

$$m_1 v_{1i} + m_2 v_{2i} = m_1 v_{1f} + m_2(v_{1i} - v_{2i} + v_{1f}).$$

Solving for v_{1f} then gives

$$v_{1f} = \frac{m_1 - m_2}{m_1 + m_2} v_{1i} + \frac{2m_2}{m_1 + m_2} v_{2i}. \qquad (10\text{--}9a)$$

Problem 32 asks you to show similarly that

$$v_{2f} = \frac{2m_1}{m_1 + m_2} v_{1i} + \frac{m_2 - m_1}{m_1 + m_2} v_{2i}. \qquad (10\text{--}9b)$$

Equations 10–9a and 10–9b are our desired result, expressing the final velocities in terms of the initial velocities alone. To see that these results make sense, we suppose that $v_{2i} = 0$. (This is really not a special case, since we can always work in a frame of reference in which m_2 is initially at rest.) Now consider several special cases.

Case 1: $m_1 \ll m_2$

For this case, picture a Ping-Pong ball colliding with a bowling ball, or any object colliding elastically with a perfectly rigid wall. Setting $v_{2i} = 0$ in Equations 10–9, and dropping m_1 as being negligible compared with m_2, Equations 10–9 become simply

$$v_{1f} = -v_{1i}$$

and

$$v_{2f} = 0.$$

That is, the lighter object rebounds with no change in speed, while the heavier object remains at rest. Does this make sense in light of the conservation laws that Equations 10–9 are supposed to reflect? Clearly energy is conserved: the kinetic energy of m_2 remains zero and the kinetic energy $\frac{1}{2}m_1v_1^2$ is unchanged. But what about momentum? The momentum of the lighter object has changed, from m_1v_{1i} to $-m_1v_{1i}$. But momentum *is* conserved; the momentum given up by the lighter object is absorbed by the heavier object. In the limit of an arbitrarily large m_2, though, the heavier object can absorb huge amounts of momentum without acquiring significant speed. If we "back off" from the extreme case that m_1 can be neglected altogether compared with m_2, we would find that the lighter object rebounds with slightly reduced speed and that the heavier object begins moving very slowly in the opposite direction (see Problem 41).

An important example of a case (1) collision is that of an electron and a proton, nearly 2000 times more massive than the electron. We have just seen that such collisions result in very little energy transfer from the lighter to the heavier object. For this reason, energy imparted to the electrons in a gas of electrons and protons—a plasma—is not readily transferred to the protons. Particle energy is associated with temperature, as we will see in Chapter 17, so that an electron-proton gas can actually have two quite different temperatures.

Case 2: $m_1 = m_2$

Again setting $v_{2i} = 0$, Equations 10–9 now give

$$v_{1f} = 0$$

and

$$v_{2f} = v_{1i}.$$

So the first object stops abruptly, transferring all its energy and momentum to the second object (Fig. 10–11). Case (2) collisions have important implications for the design of nuclear reactors. The fission reactions that power a reactor are most easily initiated by slowly moving neutrons, while each uranium atom that fissions releases fast-moving neutrons. The reactor's fuel rods are therefore surrounded by a moderator—a substance that slows down the neutrons. Our case (2) analysis shows that the most effective slowing-down is achieved in collisions of nearly equal masses. Most nuclear reactors use water as a moderator; the protons comprising the nuclei of the hydrogen atoms have almost the same mass as a neutron and are therefore efficient absorbers of neutron energy. Case (2) also shows why ordinary air does not have the dual-temperature complexity of the plasma described under case (1). In air, all the molecules have similar masses, and therefore energy is readily transferred in intermolecular collisions.

For purposes of energy transfer, two equal-mass particles are perfectly "matched." We will encounter analogous instances of energy transfer "matching" when we discuss wave motion and again in connection with electric circuits.

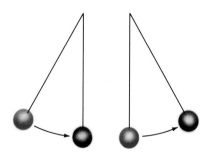

Fig. 10–11
During a collision between two pendulums of equal mass, the initially moving pendulum transfers all of its energy to the initially stationary one.

Case 3: $m_1 \gg m_2$

Now, with v_{2i} again set to zero, Equations 10–9 give

$$v_{1f} = v_{1i}$$

and

$$v_{2f} = 2v_{1i},$$

where we have neglected m_2 compared with m_1. So here the more massive object barrels right on with no change in motion, while the lighter one heads off in the same direction with twice the speed. How are momentum and energy conserved in this case? In the extreme limit where we neglect the mass m_2, the energy and momentum associated with m_2 are negligible. Essentially all the energy and momentum remain with the more massive object, and both these quantities are unchanged in the collision. If we "back off" from the extreme case where we neglect m_2 altogether, we find that the more massive object's speed is reduced slightly, while the lighter object heads off with a speed just under twice the initial speed of the heavier object (see Problem 42). The collision of a baseball bat and ball is a case (3) collision; in a reference frame moving with the ball's initial velocity, the massive bat hits the ball, which then heads out at a higher speed than that of the bat. The bat's motion is hardly altered.

Example 10–4
Car-Truck Collisions

A truck of mass $9M$ is moving at speed $v = 15$ km/h when it collides head-on with a parked car of mass M. Spring-mounted bumpers ensure that the collision is essentially elastic. Describe the subsequent motion of each vehicle. Repeat for the case of the car moving at speed $v = 15$ km/h and colliding with the stationary truck. What fraction of the incident vehicle's kinetic energy is transferred in each case?

Solution

For the first case, we can set $m_1 = 9M$, $v_{1i} = v = 15$ km/h, $m_2 = M$, and $v_{2i} = 0$. Then Equations 10–9 give

$$v_{1f} = \frac{9M - M}{9M + M}v = \tfrac{4}{5}v = 12 \text{ km/h}$$

and

$$v_{2f} = \frac{(2)(9M)}{9M + M}v = \tfrac{9}{5}v = 27 \text{ km/h}.$$

So both vehicles move off in the same direction, with the truck's speed reduced by 20 per cent and the car "kicked up" to 27 km/h. The initial energy of the truck is $\tfrac{9}{2}Mv^2$ or $4.5Mv^2$; the final energy of the car is $\tfrac{1}{2}M(\tfrac{9}{5}v)^2$ or $1.6Mv^2$. So 1.6/4.5, or 36 per cent of the truck's energy is transferred in this case.

For the second collision, we set $m_1 = M$ and $m_2 = 9M$, giving

$$v_{1f} = \frac{M - 9M}{M + 9M}v = -\tfrac{4}{5}v = -12 \text{ km/h}$$

and

$$v_{2f} = \frac{2M}{M + 9M}v = \tfrac{1}{5}v = 3.0 \text{ km/h},$$

so that the car rebounds with 80 per cent of its initial speed and the truck picks up a relatively small speed. Now the initial energy of the car is $\frac{1}{2}Mv^2$, while the final energy of the truck is $\frac{9}{2}M(\frac{1}{5}v)^2 = 0.18Mv^2$, so that again 36 per cent of the energy is transferred. Even though the mass ratio is quite large, neither collision approaches closely the case (1) or case (3) limit. Problem 26 explores the approach to those limits.

Example 10–5
"Weighing" an Elementary Particle

In a nuclear experiment, a particle of unknown mass moving at 460 km/s undergoes a head-on elastic collision with a carbon nucleus (^{12}C; mass approximately 12 u) moving at 220 km/s. After collision, the carbon continues in the same direction at 340 km/s. Find the mass and final velocity of the unknown particle.

Solution

If we designate the carbon as particle 2, we then know all quantities in Equation 10–9b except the unknown mass m_2. Multiplying that equation through by $(m_1 + m_2)$, we have

$$(m_1 + m_2)v_{2f} = 2m_1v_{1i} + (m_2 - m_1)v_{2i}.$$

Solving for m_1 then gives

$$m_1 = \frac{m_2(v_{2i} - v_{2f})}{v_{2f} - 2v_{1i} + v_{2i}} = \frac{(12\ \text{u})(220\ \text{km/s} - 340\ \text{km/s})}{340\ \text{km/s} - (2)(460\ \text{km/s}) + 220\ \text{km/s}} = 4\ \text{u}.$$

The unknown particle is therefore probably an alpha particle (^{4}He, or helium nucleus). Knowing m_1, we use Equation 10–9a to find v_{1f}:

$$v_{1f} = \frac{4\ \text{u} - 12\ \text{u}}{16\ \text{u}}(460\ \text{km/s}) + \frac{(2)(12\ \text{u})}{16\ \text{u}}(220\ \text{km/s}) = 100\ \text{km/s}.$$

Are these results consistent with our earlier finding that a less massive particle should rebound off a more massive one? After all, both particles now seem to continue in the same direction. But there is no inconsistency; our earlier conclusion was based on one particle's being initially at rest, while here both have nonzero initial velocities. If we look at the situation in a reference frame where the carbon is initially at rest, we have $v_{1i} = 460$ km/s $- 220$ km/s $= 240$ km/s, while $v_{1f} = 100$ km/s $- 220$ km/s $= -120$ km/s. The less massive nucleus does rebound, as expected. Figure 10–12 summarizes the collision in both reference frames; the change of reference frames is explored further in Problem 27.

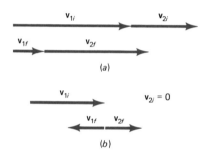

Fig. 10–12
The collision of Example 10–5 (a) in the original reference frame and (b) in a reference frame moving with the initial 220 km/s velocity of the carbon nucleus.

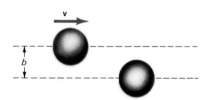

Fig. 10–13
The impact parameter is the perpendicular distance b between centers of the colliding spheres. In this simple case of equal-sized spheres, knowing the impact parameter allows us to find the line along which the internal forces act, and therefore to solve for the subsequent motion.

impact parameter

Elastic Collisions in Two Dimensions

To analyze an elastic collision in two dimensions, we must use the full vector statement of momentum conservation (Equation 10–5), along with the energy conservation equation 10–6. But we found earlier that these equations alone do not provide enough information to solve the problem. In a collision between reasonably simple macroscopic objects, that information may be provided by the so-called **impact parameter,** a measure of how much the collision differs from being head-on (Fig. 10–13). More typically, though, and especially with atomic and nuclear interactions, the needed information must be supplied by measurements done after the collision. Knowing the direction of motion of one particle after collision, for example, provides enough information to analyze the collision if the masses and initial velocities are also known.

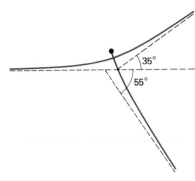

Fig. 10–14
A close encounter between two stars, in a frame of reference in which one star is initially at rest.

Example 10–6

Stars in a Close Encounter

A star of mass 1.4 $M_{\odot}$ ($M_{\odot}$ is the sun's mass, a convenient mass unit in astronomy) is a great distance from another star of equal mass, and is approaching the second star at 680 km/s. The two stars undergo close encounter, and much later the first star is moving at a 35° angle to its initial direction, as shown in Fig. 10–14. In the frame of reference in which the second star is initially at rest, find the final speeds of both stars and the direction of motion of the second star.

Solution

Since the force acting between the stars is the conservative gravitational force, the encounter is elastic.* Designating the initially moving star by the subscript 1, and taking the x-axis along the direction of its initial motion, the two components of the momentum conservation equation 10–5 become

x-component: $\qquad\qquad m_1 v_{1i} = m_1 v_{1f}\cos\theta_1 + m_2 v_{2f}\cos\theta_2$ $\qquad$ (10–10)

y-component: $\qquad\qquad 0 = m_1 v_{1f}\sin\theta_1 + m_2 v_{2f}\sin\theta_2,$ $\qquad$ (10–11)

where $\theta_{1,2}$ are the angles the final velocities make with the x-axis. A third equation is provided by energy conservation:

$$\tfrac{1}{2}m_1 v_{1i}^2 = \tfrac{1}{2}m_1 v_{1f}^2 + \tfrac{1}{2}m_2 v_{2f}^2, \qquad (10\text{–}12)$$

so we have three equations in the three unknowns v_{1f}, v_{2f}, and θ_2. Since the masses are equal, they cancel from all three equations. Isolating the unknown θ_2 on one side of Equations 10–10 and 10–11 gives

$$v_{2f}\cos\theta_2 = v_{1i} - v_{1f}\cos\theta_1$$

$$v_{2f}\sin\theta_2 = -v_{1f}\sin\theta_1.$$

Squaring and adding these two equations, and using $\cos^2\theta + \sin^2\theta = 1$ on both sides of the equation, we have

$$v_{2f}^2 = (v_{1i} - v_{1f}\cos\theta_1)^2 + v_{1f}^2\sin^2\theta_1 = v_{1i}^2 - 2v_{1i}v_{1f}\cos\theta_1 + v_{1f}^2.$$

Using this result in the energy conservation equation 10–12 gives

$$v_{1i}^2 = v_{1f}^2 + v_{2f}^2 = v_{1f}^2 + v_{1i}^2 - 2v_{1i}v_{1f}\cos\theta_1 + v_{1f}^2,$$

or

$$v_{1f}^2 = v_{1i}v_{1f}\cos\theta_1,$$

so that

$$v_{1f} = v_{1i}\cos\theta_1 = (680 \text{ km/s})(\cos 35°) = 557 \text{ km/s}.$$

From Equation 10–12, we can then write

$$v_{2f} = [v_{1i}^2 - v_{1f}^2]^{1/2} = [(680 \text{ km/s})^2 - (557 \text{ km/s})^2]^{1/2} = 390 \text{ km/s}.$$

Finally, the y-momentum equation 10–11 gives

$$\theta_2 = \sin^{-1}\left(\frac{-v_{1f}\sin\theta_1}{v_{2f}}\right) = \sin^{-1}\left[\frac{-(557 \text{ km/s})(\sin 35°)}{390 \text{ km/s}}\right] = -55°.$$

This angle is shown on Fig. 10–14. Note that the angle between the outgoing velocity vectors is 90°. Problem 29 asks you to prove that this is always true in an elastic collision of equal masses, one of which is initially at rest.

---

*This neglects tidal effects that might transform kinetic energy of stellar motion into internal energy.

Is this event really a collision? After all, the two stars interact only through the gravitational force which acts all the time. We certainly wouldn't consider a satellite in earth orbit—again interacting only via gravity—to be "colliding" with the earth. But our stellar encounter really is a collision. The two stars are in hyperbolic orbits; their energy exceeds escape energy, and they interact strongly only when they are very close. Far from each other, the stars travel in nearly straight lines, approaching the asymptotes of their hyperbolic orbits. So their interaction really does involve a strong force acting for only a short time, and therefore meets our definition of a collision. On the astronomical and subatomic scale, most collisions actually involve such "close encounters" in which particles interact without touching.

SUMMARY

1. A **collision** is a relatively violent interaction between two objects that takes place in a short time. An **impulsive force** acts between the colliding objects; this force is much stronger than the net external force acting on either object.
2. The **impulse** imparted to an object is given by integrating the impulsive force over time; Newton's second law shows that the impulse is equal to the change in the object's momentum:

$$I = \int_{t_1}^{t_2} F \, dt = \Delta p.$$

The average value $\overline{F}$ of the impulsive force is given by the impulse divided by the time interval Δt over which the force acts:

$$\overline{F} = \frac{I}{\Delta t} = \frac{\Delta p}{\Delta t}.$$

3. Because impulsive forces acting during a collision are internal to the system of colliding objects, they cancel in pairs and therefore cannot change the momentum of the system. Since a collision takes place in a short time, any external forces have little effect on the system momentum, so that to a good approximation **momentum is conserved during a collision.**
4. If kinetic energy is not conserved during a collision, the collision is **inelastic.** The maximum loss of kinetic energy occurs when the colliding objects stick to-

gether; then the collision is **totally inelastic.** In a totally inelastic collision, the final velocity v_f is determined entirely by the initial velocities through the momentum conservation equation:

$$m_1 v_{1i} + m_2 v_{2i} = (m_1 + m_2) v_f,$$

where m_1 and m_2 are the masses of the colliding objects.

5. In an **elastic collision,** kinetic energy is conserved. The relation between the initial and final velocities is governed by conservation of momentum and conservation of energy:

$$m_1 v_{1i} + m_2 v_{2i} = m_1 v_{1f} + m_2 v_{2f}$$

$$\tfrac{1}{2} m_1 v_{1i}^2 + \tfrac{1}{2} m_2 v_{2i}^2 = \tfrac{1}{2} m_1 v_{1f}^2 + \tfrac{1}{2} m_2 v_{2f}^2.$$

In the general case, these equations are not sufficient to solve for the final velocities. Additional information, such as the direction of one of the final velocities or the details of the collision, must also be used. In the special case of a one-dimensional collision, the momentum and energy conservation equations combine to determine completely the final velocities:

$$v_{1f} = \frac{m_1 - m_2}{m_1 + m_2} v_{1i} + \frac{2m_2}{m_1 + m_2} v_{2i}$$

$$v_{2f} = \frac{2m_1}{m_1 + m_2} v_{1i} + \frac{m_2 - m_1}{m_1 + m_2} v_{2i}.$$

QUESTIONS

1. Halley's Comet spends most of its 76-year orbital period moving slowly a long way from the sun. Does it make sense to treat its roughly 6-month close encounter with the sun as a collision? Discuss.
2. Why do we require a short time interval and strong impulsive forces for an interaction to qualify as a collision? What is meant by "short" and "strong" in this context?
3. Can the impulse associated with an impulsive force be zero even if the force is not always zero?
4. When a ball bounces off a wall, its momentum is reversed. How can this be, in view of the law of conservation of momentum?
5. A high jumper runs to the takeoff point with horizontal but no vertical momentum. When he jumps, where does his vertical momentum come from?

6. As you jump off a table, you bend your knees on landing. How does this lessen the shock to your body?

7. To what height does a bouncing ball return if the collision is elastic? Explain.

8. Must the analysis of a collision between two objects ever require consideration of all three dimensions? Explain.

9. Why are cars designed so that their front ends crush during an accident?

10. Discuss the relative advantages and disadvantages of air bags and seat belts for automotive safety during collisions.

11. Give three everyday examples of inelastic collisions.

12. Is it possible to have an inelastic collision in which *all* the kinetic energy of the colliding objects is lost? If not, why not? If so, give an example.

13. In our discussion of equal-mass elastic collisions, we argued that hydrogen-containing substances should make good moderators for nuclear reactors. Yet some reactors use graphite (a form of carbon) moderators. What can you say about the amount of graphite needed versus the amount of water in a water-moderated reactor? (The reactor that exploded at Chernobyl in the Soviet Union in 1986 used graphite as its moderator; the large amounts of burning graphite made the accident especially serious.)

14. If you want to stop the neutrons in a reactor, why not use massive nuclei like lead?

15. Photons—particles of light—carry momentum and energy but have no intrinsic mass. When an electron and a positron (the electron's antiparticle) collide, they annihilate to form a pair of photons. If the electron and positron were initially moving straight toward each other at the same speed, what can you conclude about the directions of the two photons? Would it be possible for just one photon to be created in this annihilation? Why or why not? Is mass conserved in the annihilation? Could energy be conserved?

16. Discuss the conservation principles used in the analysis of the ballistic pendulum. When is each valid?

17. A truck collides with an identical-looking truck initially at rest. The two move off together with more than half the original speed of the moving truck. What can you conclude about their relative loads?

18. Is it possible to have a collision during which the kinetic energy of the colliding particles increases? Explain.

19. How could you generalize the concept of energy to reinstitute the conservation of energy principle for inelastic collisions?

20. Ty Cobb slides into second base. His kinetic energy has been dissipated as heat, but where did his momentum go?

21. Does a batted ball go farther if the pitch is thrown faster, all other factors being equal? Explain.

22. Why is it relatively safe (though not recommended, of course!) to drop a lead weight on your foot if your foot is not in direct contact with the ground?

23. A downward-moving gas molecule collides with a downward-moving piston in an automobile engine. Comment on the speeds of each after the collision, and on any transfer of energy that might take place. How are such collisions important to the operation of the engine?

24. Two identical satellites are going in opposite directions in the same circular orbit, when they collide head-on. Describe their subsequent motion if the collision is (*a*) elastic or (*b*) inelastic.

PROBLEMS

Section 10.1   *Impulse and Collisions*

1. While doing aerobic dancing, you jump up 32 cm (that is, your center of mass rises that far). On the way down, you bend your knees as your feet hit the ground, lowering your center of mass an additional 12.0 cm in the 0.050 s it takes for your body to stop. If your mass is 60 kg, what are (*a*) the impulse and (*b*) the average impulsive force? Compare with the force of gravity on you.

2. A 59-g tennis ball is thrown straight up, and at the peak of its trajectory is hit by a racket that exerts a horizontal force given by $F = at - bt^2$, where t is the time in milliseconds from the instant the racket first contacts the ball, and where $a = 1200$ N/ms and $b = 400$ N/ms^2. The ball separates from the racket after 3.0 ms. Find (*a*) the impulse, (*b*) the average impulsive force, and (*c*) the ball's speed just after it leaves the racket.

3. What is the impulse imparted by the force shown in Fig. 10–15?

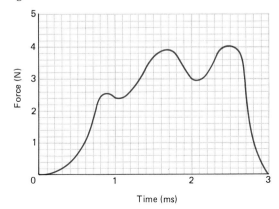

Fig. 10–15   Problem 3

4. Explosive bolts separate a 950-kg communications satellite from its 640-kg booster rocket. If the impulse of the explosion is 350 N·s, at what relative speed do the satellite and booster separate?

5. A 727 jetliner in level flight with a total mass of 8.6×10^4 kg encounters a downdraft lasting 1.3 s. During this time, the plane acquires a downward velocity component of 85 m/s. Find (a) the impulse and (b) the average impulsive force on the plane.

6. Find the magnitude of the impulse imparted to the stars in Example 10–6.

Section 10.2   *Collisions and the Conservation Laws*

7. At the peak of its trajectory, a 1.0-kg projectile moving horizontally at 15 m/s collides with a 2.0-kg projectile at the peak of a vertical trajectory. If the collision takes 0.10 s, how good is the assumption that momentum is conserved during the collision? To find out, compare the change in momentum of the colliding system with the system's total momentum.

8. An object like a ball used in a sporting event can be characterized by its **coefficient of restitution,** defined as the ratio of outgoing to incident speed when the ball collides with a rigid surface. The coefficient of restitution of a typical tennis ball is about 0.7. What fraction of the ball's kinetic energy is lost at each bounce?

Section 10.3   *Inelastic Collisions*

9. In a railroad switchyard, a 45-ton freight car is sent at 8.0 mi/h toward a 28-ton car that is moving in the same direction at 3.4 mi/h. (a) What is the speed of the pair after they couple together? (b) What fraction of the initial kinetic energy was lost in the collision?

10. In a totally inelastic collision between two equal masses, one of which is initially at rest, show that half the initial kinetic energy is lost.

11. A sled and child with a total mass of 33 kg are moving horizontally at 10 m/s when a second child leaps on with negligible speed. If the sled's speed drops to 6.4 m/s, what is the mass of the second child?

12. In an ice-show stunt, a 70-kg skater dressed as a baseball player catches a 150-g baseball moving at 23 m/s. (a) If the skater was initially at rest, what is his final speed? (b) If the catch takes 36 ms, what is the average impulsive force exerted by the ball?

13. A mass m collides totally inelastically with a mass M initially at rest. Show that a fraction $M/(m+M)$ of the initial kinetic energy is lost in the collision.

14. A 1200-kg Toyota and a 2200-kg Buick collide at right angles in an intersection. They lock together and skid 22 m; the coefficient of friction is 0.91. Show that at least one car must have exceeded the 25 km/h speed limit in effect at the intersection.

15. Astronomers warn that there is a nonzero chance of earth colliding inelastically with a substantial asteroid (Fig. 10–1). Impact speed of the asteroid might be 10 km/s. Estimate the mass of an asteroid needed (a) to alter earth's orbital speed by 0.01%; (b) to re-

lease energy equivalent to all the world's nuclear arsenal, about 16,000 megatons. (See Appendix D.)

16. Two identical trucks have mass 5500 kg when empty. One truck carries a 9500-kg load and is moving at 65 km/h. It collides inelastically with the second truck, which is initially at rest, and the pair moves off at 40 km/h. What is the load of the second truck?

17. Two pendulum bobs of equal length are suspended from the same point, and one is released from a distance h as shown in Fig. 10–16. When the first bob hits the second, the two stick together. Show that the maximum height to which the combination rises is $\frac{1}{4}h$.

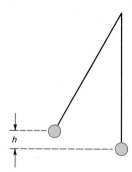

Fig. 10–16   Problem 17

18. A 5.4-kg hawk is diving at 18 m/s on a northward path at 65° to the horizontal when it grabs a 1.2-kg pigeon in its talons. If the pigeon was flying horizontally at 14 m/s in the same direction as the hawk's horizontal motion, what is the velocity of the pair just after the catch?

19. A 400-mg popcorn kernel is skittering across a nonstick frying pan at 8.2 cm/s when it pops and breaks into two equal-mass pieces. If one piece ends up at rest, how much energy was released in the popping?

20. A 1300-kg car moving at 10 km/h collides with a 1600-kg car moving in the same direction at 6.6 km/h. The first car is equipped with spring-loaded bumpers to prevent damage. If the spring constant is 28,000 N/m, find the maximum compression of the spring.

21. Two identical objects with the same initial speed collide and stick together. If the composite object moves with half the initial speed of either object, what was the angle between the initial velocities?

Section 10.4   *Elastic Collisions*

22. An alpha particle (^{4}He) strikes a stationary gold nucleus (^{197}Au) head-on. What fraction of the alpha particle's kinetic energy is transferred to the gold nucleus? Assume the collision is totally elastic.

23. While playing ball in the street, a child accidentally tosses a ball at 18 m/s toward the front of a car moving toward him at 14 m/s. What is the speed of the ball after it rebounds elastically from the car?

24. A block of mass m undergoes a one-dimensional elas-

tic collision with a block of mass M initially at rest. If both blocks have the same speed after the collision, how are their masses related?

25. Blocks B and C have masses $2m$ and m, respectively, and are at rest on a frictionless surface. Block A, also of mass m, is heading at speed v toward block B, as shown in Fig. 10–17. If all subsequent collisions are elastic, determine the final velocity of each block.

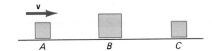

Fig. 10–17   Problem 25

26. A block of mass m_1 undergoes a one-dimensional elastic collision with an initially stationary block of mass m_2. Find an expression for the fraction of the initial kinetic energy transferred to the second block, and plot your result for mass ratios m_1/m_2 from 0 to 20. Show also that, for a given mass ratio, the energy transfer is the same no matter which mass is initially at rest.

27. Rework Example 10–5 using a frame of reference in which the carbon nucleus is initially at rest.

28. Two buses are approaching each other at 60 km/h relative to the road. The driver of one bus tosses a tennis ball straight ahead at 20 km/h relative to his bus. The ball bounces elastically off the vertical windshield of the second bus, and back to the driver who threw it. At what speed relative to the driver is it going when caught? Neglect gravity.

29. An object collides elastically with an equal-mass object initially at rest. If the collision is not head-on, show that the final velocity vectors are perpendicular.

30. On ice, a 3.2-kg rock moving at 1.0 m/s collides elastically with a 0.35-kg rock initially at rest. The smaller rock goes off at an angle of 45° to the larger rock's initial motion. What are the final speeds of the two rocks? What is the direction of the larger rock?

31. Two pendulums of equal length $\ell = 50$ cm are suspended from the same point. The pendulum bobs are steel spheres with masses of 140 and 390 g. The more massive bob is drawn back to make a 15° angle with the vertical (Fig. 10–18). When it is released the bobs

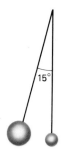

Fig. 10–18   Problem 31

collide elastically. What is the maximum angle made by the less massive pendulum?

32. Derive Equation 10–9b.

33. A particle of mass m collides elastically with a particle of mass $3m$. The more massive particle moves off at 1.2 m/s at 17° to the original direction of the less massive particle. The less massive particle moves off at 0.92 m/s at 48° to its original direction. Find the initial speeds of both particles.

34. A proton of mass 1.007 u is travelling through the bubble chamber of a particle accelerator at 2.4×10^5 m/s when it strikes a stationary deuteron (mass 2.014 u). The proton is deflected through an angle of 37°. Assuming the collision is elastic, what are the final velocities of the two particles?

35. A tennis ball moving at 18 m/s strikes the 45° hatchback of a car moving away at 12 m/s, as shown in Fig. 10–19. Both speeds are given with respect to the ground. What is the velocity of the ball with respect to the ground after it rebounds elastically from the car? *Hint:* Work in the frame of reference of the car; then transform to the ground frame.

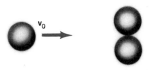

Fig. 10–19   Problem 35

36. Rework Example 10–6 for the case when the star initially at rest is three times as massive as the other star.

37. Two identical billiard balls are initially at rest when they are struck symmetrically by a third identical ball moving with velocity $\mathbf{v}_0 = v_0\hat{\mathbf{i}}$, as shown in Fig. 10–20. Find the velocities of all three balls after they undergo an elastic collision.

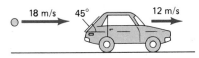

Fig. 10–20   Problem 37

Supplementary Problems

38. You set a small ball of mass m atop a large ball of mass $M \gg m$ and drop the pair from a height h. Assuming the balls are perfectly elastic, show that the smaller ball rebounds to a height $9h$.

39. A 1400-kg car moving at 75 km/h runs into a 1200-kg car moving in the same direction at 50 km/h. The two cars lock together and both drivers immediately slam on their brakes. If the cars come to rest in a distance of 18 m, what is the coefficient of friction?

40. To show that elastic collisions among electrons and

protons are not very effective at transferring energy, estimate the number of head-on collisions an electron must make with stationary protons ($m_p = 1800m_e$) before it has lost half its *initial* energy. *Hint:* Determine the fractional energy *remaining* with the electron. With each collision, the energy is further reduced by that factor. See Problem 26.

41. Consider a one-dimensional elastic collision with $m_1 \ll m_2$ and m_2 initially at rest. In discussing this extreme case, we neglected m_1 altogether and showed that m_1 then rebounds with its initial speed. Now use the binomial theorem (see Appendix A) to show that a better approximation gives a rebound speed that is less than the incident speed by an amount $2m_1 v_1/m_2$ for the case $m_1 \ll m_2$. In applying the binomial theorem, keep terms of order m_1/m_2, but neglect terms of order m_1^2/m_2^2.

42. Repeat the preceding problem for the case $m_1 \gg m_2$, and show that the less massive object moves off with speed given approximately by $2v_1(1 - m_2/m_1)$, where v_1 is the initial speed of the more massive object.

43. How many head-on collisions must a neutron (mass 1.0087 u) make with stationary ^{12}C nuclei (mass 11.9934 u) in a graphite-moderated nuclear reactor in order to lose as much energy as it would in a water-moderated reactor where it collides with a single proton (mass 1.0073 u)? *Hint:* See hint in Problem 40.

44. A 200-g block is released from rest 25 cm high on a frictionless 30° incline. It slides down the incline, then along a frictionless surface until it collides elastically with an 800-g block at rest 1.4 m from the bottom of the incline (Fig. 10–21). How much later do the two blocks collide again?

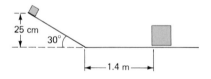

Fig. 10–21   Problem 44

45. A block of mass M is moving at speed v_0 on a frictionless surface that ends in a rigid wall. Farther from the wall is a more massive block of mass αM, initially at rest (Fig. 10–22). The less massive block undergoes elastic collisions with the other block and with the wall, and the motion of both blocks is confined to one dimension. (*a*) Show that the two blocks will undergo only one collision if $\alpha \leq 3$. (*b*) Show that the two blocks will undergo two collisions if $\alpha = 4$, and determine their final speeds. (*c*) Find out how many collisions the two blocks will undergo if $\alpha = 10$, and determine their final speeds.

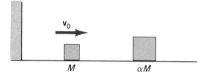

Fig. 10–22   Problem 45

46. Use a computer to investigate the number of collisions in the previous problem as a function of the mass ratio α, and plot the collision number versus α for values of α from 1 to 500.

47. A 1200-kg car moving at 25 km/h undergoes a one-dimensional collision with an 1800-kg car initially at rest. The collision is neither elastic nor totally inelastic; the kinetic energy lost is 5800 J. Find the speeds of both cars after the collision.

11

TORQUE AND STATIC EQUILIBRIUM

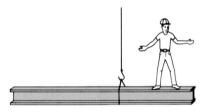

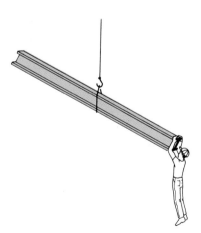

Fig. 11–1
A steelworker in static equilibrium.

What keeps the steelworker of Fig. 11–1 alive? Two things: First, the cable is strong enough to provide the tension force necessary to counteract the weight of the beam and the steelworker. Second, the forces on the beam are applied in such a way that there is no tendency for the beam to rotate. The steelworker remains in static equilibrium, neither accelerating nor rotating. Both conditions—force balance and no tendency to rotate—are necessary to maintain static equilibrium. In this chapter, we explore these conditions and their application in practical situations.

11.1
TORQUE

The first condition for static equilibrium—zero net force on a body—is familiar from our study of Newton's laws. If there is no net force on a body, then its center of mass cannot accelerate. If at rest, the center of mass will remain at rest. But the body could still rotate, as in Fig. 11–2. To prevent such rotation, it is not enough that the forces on a body sum to zero; those forces must also be applied in such a way that they cancel each other's tendency to rotate the body.

Figure 11–3 shows a beam supported by a cable. The beam, in turn, supports two equal weights hung from its ends. The tension in the cable is equal to the total weight, so that there is no net force on the beam. The right-hand weight tends to rotate the beam clockwise, but this tendency is counteracted by an equal counterclockwise effect from the left-hand weight, so that the system is in static equilibrium. If we move one of the weights, the system will begin to rotate (Fig. 11–4).

This simple example suggests that the tendency of a system to rotate is determined not only by the magnitude and direction of applied forces, but also by *where* those forces are applied. The concept of **torque** quantifies the

Fig. 11–2
Even though there is no net force on the beam, it may still rotate!

torque

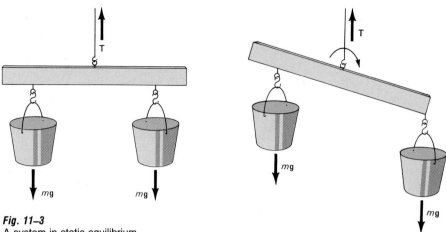

Fig. 11–3
A system in static equilibrium.

Fig. 11–4
If one of the weights in Fig. 11–3 is moved,
the system goes out of equilibrium.

relation between applied force and the tendency of a body to rotate. As Fig.
11–5 suggests, the torque—the tendency to rotate—is greater for greater distances from the pivot point to the point where the force is applied. Figure
11–6 suggests further that for a given force, the important quantity is the
perpendicular distance from the pivot point to the line along which the
force acts. We call this distance the **lever arm,** $r_\perp$, and write

lever arm

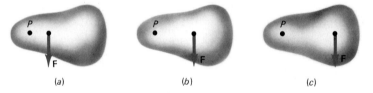

(a)   (b)   (c)

Fig. 11–5
Torque increases with distance from pivot P to the force application point. The same force **F**
produces the greatest torque in (c).

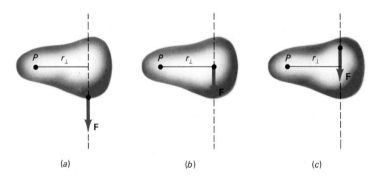

(a)   (b)   (c)

Fig. 11–6
Torque depends on the perpendicular distance $r_\perp$—called the lever arm—from the pivot P to the
line along which the force acts. Because $r_\perp$ and **F** are the same in cases (a) through (c), the
torque in each case is the same.

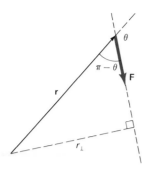

Fig. 11-7
The lever arm is given by
$r \sin(\pi - \theta) = r \sin\theta$.

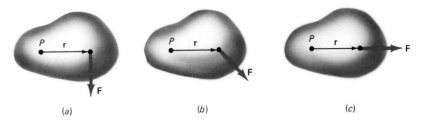

Fig. 11-8
The torque is greatest with **F** and **r** at right angles, and is zero when they are collinear.

$$\tau = r_\perp F \tag{11-1}$$

for the magnitude τ of the torque about a pivot point arising from a force F with lever arm $r_\perp$. Equation 11–1 shows that the units of torque are newton-meters (N·m), the same as those of energy. However, torque and energy are two different things, so that we call one newton-meter a joule only when it refers to energy.

Figure 11–7 shows that the lever arm $r_\perp$ is given by $r_\perp = r \sin\theta$, where r is the length of the vector **r** from the pivot point to the force application point, and θ the angle between **r** and the force vector **F**. Then Equation 11–1 for the magnitude of the torque may be written.

$$\tau = rF \sin\theta. \tag{11-2}$$

For a given force F and distance r, the torque is greatest when the vectors **F** and **r** are at right angles, and goes to zero as the two vectors become collinear (Fig. 11–8). Physically, you can see that no matter how hard you pull to the right in Fig. 11–8c, you can't cause the body to rotate about P because you are pulling in line with the pivot point. We emphasize that the vector **r** always points straight from the pivot point to the force application point. Although this vector often lies entirely within the body, this need not be the case (Fig. 11–9).

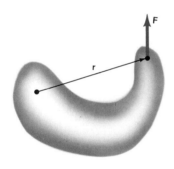

Fig. 11-9
The vector **r** always extends from the pivot point to the force application point, even for a concave body.

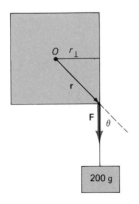

Fig. 11-10
Example 11-1

Example 11-1
Torque

A square slab of wood 20 cm on a side is free to pivot about a horizontal axis through its center. A 200-g mass is suspended from the lower right corner of the body, as shown in Fig. 11–10. Find the torque exerted by the weight of the 200-g mass about the axis of the slab.

Solution

To find the torque we must evaluate the right-hand side of Equation 11–1 or 11–2. For Equation 11–2 the distance r from the pivot point to the force application point is half a diagonal, or $10\sqrt{2}$ cm, while the angle θ is $\pi/4$ radians, or 45°, whose sine is $\sqrt{2}/2$. The mass exerts a force mg, so that the torque is

$$\tau = r \, F \, \sin\theta = (0.10\sqrt{2} \text{ m})(0.20 \text{ kg})(9.8 \text{ N/kg})(\sqrt{2}/2) = 0.20 \text{ N·m}.$$

We could solve this problem more easily by noting from Fig. 11–10 that the lever arm $r_\perp$ is just 10 cm, so that the torque is

$$\tau = r_\perp F = (0.10 \text{ m})(0.20 \text{ kg})(9.8 \text{ N/kg}) = 0.20 \text{ N·m}.$$

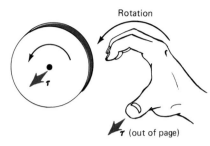

Fig. 11–11
The right-hand rule. Curl the fingers of your right hand in the direction that a torque would tend to rotate a system. Then your right thumb points in the direction of the torque.

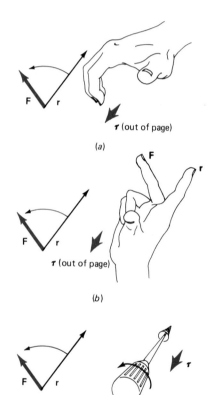

Fig. 11–12
Right-hand rules for the direction of torque and, more generally, for the cross product of two vectors. Rules (a), (b), and (c) are equivalent.

Does torque have direction? The torque in Example 11–1 tends to rotate the slab clockwise. The torques due to the weights in Fig. 11–3 are in opposite directions, and therefore cancel. We could specify the direction of a torque by its tendency to rotate a system either clockwise or counterclockwise (we call this the "sense" of the rotation), and it is often simplest to visualize torques in this way. But it is mathematically more convenient to treat torque like any other quantity with magnitude and direction—like a vector. We define the direction of the torque vector to be at right angles to the plane in which the torque tends to rotate a body, and in a sense given by the right-hand rule (Fig. 11–11). This definition may seem a little strange, since all the action is in the plane perpendicular to the torque. But there is no unique vector direction within the plane of rotation; in fact, only the direction perpendicular to that plane specifies the orientation of the plane. Torque is only the first of several vector quantities that we will define in this way; we do so because we obtain a unique vector representation that allows us to state physical laws in compact, powerful mathematical form.

Our right-hand rule definition for the direction of the torque is based on a physical consideration—the sense in which the torque would tend to rotate a body. We can form an equivalent but more mathematical right-hand rule for the torque vector by placing the vectors **r** and **F** tail-to-tail and curling the fingers of the right hand in a direction that would rotate **r** until it is aligned with **F**. The right thumb then points in the direction of the torque (Fig. 11–12a). Figure 11–12b and c show equivalent right-hand rules for the direction of the torque vector. Using our right-hand rule and Equation 11–2, we can say that the torque vector has magnitude $\tau = rF\sin\theta$, where θ is the angle between the vectors **r** and **F**, and that its direction is given by applying the right-hand rule to the vectors **r** and **F**. This definition makes it explicit that torque is always specified *with respect to some point*, for the vector **r** is drawn from that point.

Example 11–2

Torque as a Vector

Figure 11–13 shows a 1.8-N force applied 2.5 m from a point O and 1.3 m from a point P. What are the torques about O and P due to **F**?

Solution

Equation 11–2 gives the magnitudes of the torques:

$$\tau_O = r\,F\,\sin\theta = (2.5\text{ m})(1.8\text{ N})(\sin 30°) = 2.3\text{ N·m}$$

and

$$\tau_P = r\,F\,\sin\theta = (1.3\text{ m})(1.8\text{ N})(\sin 45°) = 1.7\text{ N·m}.$$

To determine the direction of the torque about O, we place the two vectors **r** and **F** tail-to-tail (Fig. 11–14) and apply the right-hand rule of Fig. 11–12, showing that the torque is into the page. Equivalently, we could imagine the force **F** rotating a physical object about the point O. Curling our fingers in the clockwise direction of such a rotation, we find again that the torque vector is into the page. A similar analysis shows that the torque about P is out of the page.

The Vector Cross Product

Torque is a vector that is determined entirely by the two other vectors **r** and **F** as described by Equation 11–2 and the right-hand rule. This operation—forming from two vectors **A** and **B** a third vector **C** of magnitude $C = AB\sin\theta$ and direction given by the right-hand rule—occurs frequently in physics. We will encounter it in discussing rotational motion in the next chapter, and later in connection with magnetism. We therefore give this operation a name: the **vector cross product**, often called simply the **cross product**. The vector cross product **C** of two vectors **A** and **B** is written

$$\mathbf{C} = \mathbf{A} \times \mathbf{B}.$$

This expression means to form the vector **C** with magnitude $AB\sin\theta$, where θ is the angle between **A** and **B**, and with **C** perpendicular to both **A** and **B**, as described by the right-hand rule of Fig. 11–12.

Torque is a particular example of a vector cross product, and we can write the torque vector simply as

$$\boldsymbol{\tau} = \mathbf{r} \times \mathbf{F}. \tag{11–3}$$

Having once defined the cross product, we need not give separate and wordy descriptions of the magnitude and direction of torque; both aspects of the torque vector are contained succinctly in Equation 11–3.

The vector cross product $\mathbf{A} \times \mathbf{B}$ bears some similarity to the dot product $\mathbf{A} \cdot \mathbf{B} = AB\cos\theta$ that we encountered in Chapter 7. Both depend on the product of the magnitudes of the vectors **A** and **B**, and on the angle between them. But where the dot product depends on the cosine of the angle, and is therefore maximum when $\theta = 0$, the cross product depends on the sine, and is maximum for two vectors at right angles. And most significantly, the dot product is a *scalar*—a single number, without direction—while the cross product is a *vector*. Because of this distinction, the two products are often called the scalar product of two vectors and the vector product of two vectors, respectively.

The vector cross product obeys some of the common rules for ordinary multiplication. For example, it is distributive over addition:

$$\mathbf{A} \times (\mathbf{B} + \mathbf{C}) = \mathbf{A} \times \mathbf{B} + \mathbf{A} \times \mathbf{C}.$$

But it is *not* commutative: the order of multiplication *does* matter. In fact

$$\mathbf{B} \times \mathbf{A} = -\mathbf{A} \times \mathbf{B}.$$

You can see this result from the right-hand rule, where rotating **B** into **A** points your thumb in the opposite direction from rotating **A** into **B**.

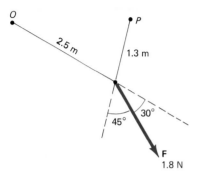

Fig. 11–13
Example 11–2

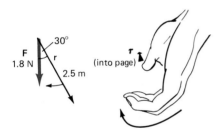

Fig. 11–14
Right-hand rule for the vectors of Example 11–2.

Example 11–3
The Vector Cross Product

Find the cross product of the vectors $\mathbf{A} = 3\hat{\imath}$ and $\mathbf{B} = \sqrt{3}\hat{\imath} + \hat{\jmath}$. Do so in two ways: (a) by first finding the directions and magnitudes of the two vectors, and (b) by working in unit vector notation and using the distributive law for cross products.

Solution

To calculate the vector cross product, we need the magnitudes of two vectors and the angle between them. Applying the Pythagorean theorem, we find that

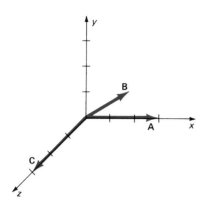

Fig. 11–15
The cross product **C** = **A** × **B** points out of the page, or in the + z direction.

$$B = \sqrt{B_x^2 + B_y^2} = \sqrt{3+1} = 2.$$

The angle **B** makes with the x-axis is

$$\theta = \tan^{-1}\left(\frac{1}{\sqrt{3}}\right) = 30°.$$

The vector **A** lies along the x-axis, so that 30° is the angle between the two vectors. **A** has magnitude 3, so that the magnitude of the cross product is

$$|\mathbf{A} \times \mathbf{B}| = AB\sin\theta = (3)(2)\sin30° = 3.$$

Rotating **A** into **B** shows that the direction of **A** × **B** is out of the page, or in the + z direction (Fig. 11–15). Then in compact vector notation, we can write

$$\mathbf{A} \times \mathbf{B} = 3\hat{\mathbf{k}}.$$

We could also obtain this result using unit vectors:

$$\mathbf{A} \times \mathbf{B} = (3\hat{\imath}) \times (\sqrt{3}\hat{\imath} + \hat{\jmath}) = 3\sqrt{3}(\hat{\imath} \times \hat{\imath}) + 3(\hat{\imath} \times \hat{\jmath}).$$

What about the cross products of the unit vectors? The angle between $\hat{\imath}$ and itself is zero, so that the first cross product is simply zero. The angle between $\hat{\imath}$ and $\hat{\jmath}$ is 90°, and each unit vector has length 1. Therefore, the magnitude of $\hat{\imath} \times \hat{\jmath}$ is 1 and from the right-hand rule we see that $\hat{\imath} \times \hat{\jmath}$ points in the + z direction (Fig. 11–16). Therefore, $\hat{\imath} \times \hat{\jmath}$ is simply the unit vector $\hat{\mathbf{k}}$. In general, the cross product of any two different unit vectors is ± the third unit vector, with the + sign taken when the vectors are in cyclic order $\hat{\imath}\hat{\jmath}\hat{\mathbf{k}}\hat{\imath} \ldots$.* Our product **A** × **B** is therefore simply $3\hat{\mathbf{k}}$, as we found before.

11.2
STATIC EQUILIBRIUM

Having defined torque, we are now ready to make a rigorous statement of the conditions for a body to be in static equilibrium:

static equilibrium

> A body is in static equilibrium when the net force and the net torque on the body are both zero.

Zero net force ensures that the center of mass does not accelerate; zero net torque ensures that the body has no tendency to rotate.

To determine whether a body is in static equilibrium, we draw a diagram showing all forces acting on the body. The first condition for equilibrium is met by requiring that there be no net force:

$$\sum \mathbf{F}_i = 0. \tag{11–4}$$

The second condition requires zero net torque:

$$\sum \boldsymbol{\tau}_i = \sum (\mathbf{r}_i \times \mathbf{F}_i) = 0. \tag{11–5}$$

The subscripts i in Equations 11–4 and 11–5 label the different forces that act on a body, the radius vectors from a pivot point to the different force application points, and the associated torques.

right-handed coordinate system

---

*If you ever wonder which way to put the axes in an xyz coordinate system, just choose their directions so that these unit vector cross products work out. Then you have a **right-handed coordinate system,** consistent with all the right-hand rules used in physics.

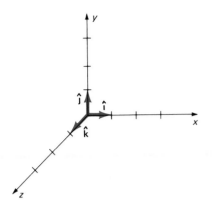

Fig. 11–16
Cross product of any two unit vectors is the third unit vector when the two are taken in cyclic order $\hat{\imath}\hat{\jmath}\hat{k}\hat{\imath}$ . . . and is the opposite of the third unit vector when taken in the reverse order.

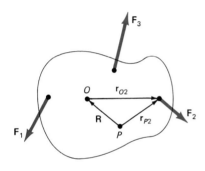

Fig. 11–17
The forces sum to zero, and the torques about O sum to zero. Is the net torque about P also zero?

There seems to be an ambiguity in Equation 11–5, for we have not specified the point to be used in evaluating the radius vectors $\mathbf{r}_i$. But a body in equilibrium must not rotate about *any* point, so that Equation 11–5 must hold no matter what point we choose to evaluate the torques. Must we then apply Equation 11–5 to every possible point? No! If the first equilibrium condition ($\Sigma \mathbf{F}_i = 0$) is satisfied, then it suffices to show that the sum of the torques about any arbitrary point is zero. We can prove this fact using Fig. 11–17, which shows an object subject to several forces whose vector sum is zero. Suppose also that the net torque about point O is zero. What about the torque about point P? This torque is given by

$$\boldsymbol{\tau}_P = \sum (\mathbf{r}_{Pi} \times \mathbf{F}_i), \qquad (11\text{–}6)$$

where $\mathbf{r}_{Pi}$ is the vector from P to the application point of $\mathbf{F}_i$. Let $\mathbf{R}$ be the vector from P to O. Then, as suggested in Fig. 11–17 for the case $i = 2$, we can write

$$\mathbf{r}_{Pi} = \mathbf{R} + \mathbf{r}_{Oi},$$

where $\mathbf{r}_{Oi}$ is the vector from O to the application point of $\mathbf{F}_i$. Using this expression in Equation 11–6, we have

$$\boldsymbol{\tau}_P = \sum (\mathbf{r}_{Oi} + \mathbf{R}) \times \mathbf{F}_i = \sum (\mathbf{r}_{Oi} \times \mathbf{F}_i + \mathbf{R} \times \mathbf{F}_i)$$
$$= \sum (\mathbf{r}_{Oi} \times \mathbf{F}_i) + \sum (\mathbf{R} \times \mathbf{F}_i) = \sum (\mathbf{r}_{Oi} \times \mathbf{F}_i) + \mathbf{R} \times \sum \mathbf{F}_i, \qquad (11\text{–}7)$$

where, in the last step, we have taken $\mathbf{R}$ outside the summation since it is the same vector for each term in the sum. The first term on the rightmost side of Equation 11–7 is just the net torque about O, which we have assumed to be zero. The second term is the cross product of the vector $\mathbf{R}$ with the net force on the body. But we have assumed the net force to be zero. Thus the net torque about P is also zero. Since the location of P is quite arbitrary, our result shows that if the net force on an object is zero, and the net torque about *some* point is zero, then the net torque about *any* point is also zero.

This result is useful in solving equilibrium problems, for it allows us to choose any convenient point about which to evaluate the torques. An appropriate point is usually the point of application of one of the forces, for then the torque due to that force is zero. The problem is thus simplified, since the torque sum in Equation 11–5 has one fewer term than it would otherwise.

Example 11–4 _____

Equilibrium and the Choice of Pivot Point

A 5.0-m-long canoe is tied at bow and stern to a pair of trees so that it rides parallel to the shore, as shown in Fig. 11–18. The tie ropes are perpendicular to the shore. A paddler boards the canoe 1.0 m from the stern, in the process exerting an outward force of 200 N at right angles to the shore. Find the tensions in the two ropes, if the canoe remains in static equilibrium. Assume that buoyancy of the water balances the gravitational force and that the torques resulting from these two forces exactly cancel, so that you need only consider equilibrium in the horizontal plane.

Fig. 11–18
Top view of canoe in Example 11–4.

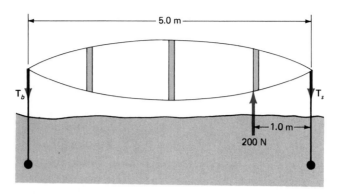

Solution

All three horizontal forces are perpendicular to the shore, so that the first equilibrium condition, Equation 11–4, becomes the single scalar equation

$$T_b + T_s - 200 \text{ N} = 0. \tag{11-8}$$

We can choose any point for our pivot; if we choose one of the force application points, then that force will not appear in the torque equation. Since the tensions are the unknowns, choosing either end of the canoe as our pivot will eliminate one unknown from the torque equation. Choosing the stern, the torque equation 11–5 becomes

$$(5.0 \text{ m})(T_b) - (200 \text{ N})(1.0 \text{ m}) = 0,$$

where the first term is the torque due to the force of the paddler and the second is the torque due to the bow-rope tension T_b. Here we take as positive a torque tending to produce counterclockwise rotation, so that the bow-rope torque has a negative sign. Solving for T_b gives

$$T_b = \frac{(200 \text{ N})(1.0 \text{ m})}{5.0 \text{ m}} = 40 \text{ N}.$$

Using this result in Equation 11–8 gives

$$T_s = 200 \text{ N} - T_b = 200 \text{ N} - 40 \text{ N} = 160 \text{ N}.$$

Suppose we had chosen the center of the canoe as our pivot point. The force-balance equation 11–8 would be unchanged, but the torque equation would now contain three terms:

$$(2.5 \text{ m})(T_b) + (1.5 \text{ m})(200 \text{ N}) - (2.5 \text{ m})(T_s) = 0,$$

where again we adopt the convention that a torque tending to produce counterclockwise rotation is positive. The three terms are due to the bow rope, the paddler, and the stern rope, respectively, and the distances are measured from the canoe center. Note that now the bow-rope and paddler torques have the same sign, since both tend to produce counterclockwise rotation about the canoe center. We can solve Equation 11–8 for T_s to get

$$T_s = 200 \text{ N} - T_b.$$

Using this result in the torque equation gives

$$(2.5 \text{ m})(T_b) + (1.5 \text{ m})(200 \text{ N}) - (2.5 \text{ m})(200 \text{ N} - T_b) = 0,$$

so that

$$T_b = \frac{(2.5\ \text{m})(200\ \text{N}) - (1.5\ \text{m})(200\ \text{N})}{5.0\ \text{m}} = 40\ \text{N},$$

as before. Although we get the same answer, the algebra is considerably more complicated with this choice of pivot point.

In Example 11–4, all the forces are perpendicular to the shore, and, as the right-hand rule shows, all the torque vectors point either vertically upward or downward. As a result, the two vector equations 11–4 and 11–5 reduce to two scalar equations in two unknowns. More generally, though, the forces can have components along any of three mutually perpendicular directions. With all three force components present, there will also be three components of torque that must sum to zero. In this most general case, then, Equations 11–4 and 11–5 become six scalar equations in six unknowns.

Often we have a situation in which all the forces lie in a single plane, so that Equation 11–4 has only two components. If we choose our pivot point to lie in the plane defined by the force vectors, then all torques will be perpendicular to that plane, so that the torque equation has only one component, making a total of three equations in three unknowns.

11.3
CENTER OF GRAVITY

Fig. 11–19
The gravitational force acts everywhere on an extended body.

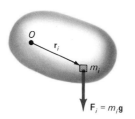

Fig. 11–20
Calculating the torque produced by an individual mass element about the pivot point O.

One force often present is gravity, which acts on all parts of a body (Fig. 11–19). We already know how to calculate the vector sum of these gravitational forces—it is just $M\mathbf{g}$. But what about the torques due to all the gravitational forces acting on different parts of the body? Summing these torques, we have

$$\boldsymbol{\tau} = \sum \mathbf{r}_i \times \mathbf{F}_i = \sum \mathbf{r}_i \times m_i\mathbf{g} = \left(\sum m_i \mathbf{r}_i\right) \times \mathbf{g}, \qquad (11\text{–}9)$$

where $\mathbf{F}_i$ is the gravitational force on the ith mass element, and $\mathbf{r}_i$ is the vector from some pivot point to that mass element (Fig. 11–20). We can rewrite Equation 11–9 by multiplying the right-hand side by M/M, where M is the total mass of the body:

$$\boldsymbol{\tau} = \left(\frac{\sum m_i \mathbf{r}_i}{M}\right) \times M\mathbf{g}.$$

The term in parentheses is just the location of the center of mass (see Section 9.1), while the right-hand term is the total weight of the body. Therefore, the net torque on the body due to gravity is just that of the gravitational force $M\mathbf{g}$ acting at the center of mass. In general, the point at which the gravitational force seems to act is called the **center of gravity.** We have just proven that the center of gravity coincides with the center of mass when the gravitational field is uniform—that is, when $\mathbf{g}$ is the same for all mass elements in the body. If a body is large enough that the variation in magnitude or direction of $\mathbf{g}$ is significant over the object, then the two

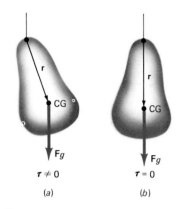

Fig. 11–21
The object is in equilibrium only when its center of gravity lies directly below the suspension point.

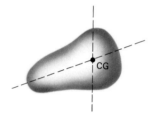

Fig. 11–22
The intersection of vertical lines from each suspension point is at the center of gravity.

points may not coincide (see Problem 35), although this is not a situation we usually encounter.

We can easily locate the center of gravity by suspending an object from a string or rope attached to its edge. The torque about the suspension point is due entirely to gravity, and is zero only if the center of gravity lies directly beneath that point (Fig. 11–21). In equilibrium, a vertical line from the suspension point must therefore pass through the center of gravity. If we now change the suspension point, we can determine another line that passes through the center of gravity. These two lines intersect in a single point, which must be the center of gravity (Fig. 11–22).

Example 11–5
Static Equilibrium

A 25-kg traffic signal is suspended over the center of a road by the structure shown in Fig. 11–23. The structure is secured to a concrete foundation with two bolts, as shown. What is the tension in the left-hand bolt?

Solution

Figure 11–24 shows a force diagram for the structure. There are four downward forces, which are the weights of each of the components of the structure. These forces are applied at the centers of gravity of the various components. There is also a downward force **T**, exerted by the tension in the left-hand bolt, and an upward contact force $\mathbf{F}_c$, exerted by the foundation at the point where the right-hand bolt attaches. The condition of force balance (Equation 11–4) becomes

$$m_1g + m_2g + m_3g + m_4g + T - F_c = 0, \qquad (11\text{--}10)$$

where we have adopted the convention that a downward force is positive. We are free to evaluate torques about any point. A convenient choice is the point O, where the axis of the vertical member intersects the concrete base. Then the lever arm, $r \sin\theta$, for each force is the horizontal distance x_i from the vertical axis to the force application point, so the torque-balance condition (Equation 11–5) becomes

$$m_1gx_1 + m_2gx_2 + m_3gx_3 + m_4gx_4 - Tx_5 - F_cx_6 = 0. \qquad (11\text{--}11)$$

Here we adopt the convention that a torque tending to produce clockwise rotation is positive, and that all distances x_i are positive numbers, so that the last two terms are negative. The fourth term, m_4gx_4, is actually zero since $x_4 = 0$, but we retain this term in symbolic form for mathematical convenience.

We can solve Equation 11–10 for the contact force F_c to get

$$F_c = g\sum m_i + T,$$

where we have factored out g and used the summation symbol, $\sum$, to indicate the sum of the four masses. Using this expression in Equation 11–11 gives

$$g\sum m_ix_i - Tx_5 - (g\sum m_i + T)x_6 = 0,$$

where again we have factored out g and used the summation symbol. We can now solve for T:

$$g\sum m_ix_i - gx_6\sum m_i - T(x_5 + x_6) = 0,$$

so that

$$T = \frac{g\left(\sum m_ix_i - x_6\sum m_i\right)}{x_5 + x_6}.$$

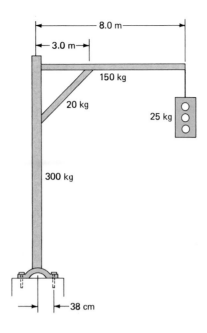

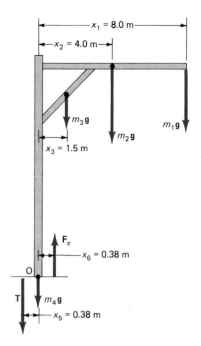

Fig. 11–23
Example 11–5

Fig. 11–24
Force diagram for structure of Fig. 11–23. Force vectors are not drawn to scale.

The second sum is just the total mass, 495 kg, while the first is

$$\sum m_i x_i = (25 \text{ kg})(8.0 \text{ m}) + (150 \text{ kg})(4.0 \text{ m}) + (20 \text{ kg})(1.5 \text{ m}) + 0 = 830 \text{ kg·m},$$

where we have neglected the relatively small radius of the vertical member in figuring the distances to the centers of mass of the horizontal and diagonal members. The fourth term in this sum is zero because the vertical member stands directly above the point about which the torques are evaluated (that is, $x_4 = 0$), and thus contributes nothing to the net torque. Then the bolt tension is

$$T = \frac{(9.8 \text{ m/s}^2)[830 \text{ kg·m} - (0.38 \text{ m})(495 \text{ kg})]}{0.38 \text{ m} + 0.38 \text{ m}} = 8.3 \times 10^3 \text{ N},$$

or nearly one ton! Engineers are responsible for ensuring that bolts and fittings on structures like this are sufficiently strong.

Example 11–6

Static Equilibrium

A board of mass m and length L is resting against a wall, as shown in Fig. 11–25. The wall is frictionless, while the coefficient of static friction between board and floor is μ. What is the minimum angle θ at which the board can be placed without slipping? Evaluate for the case $\mu = 0.27$, $L = 2.4$ m, $m = 5.0$ kg.

Solution

Figure 11–26 shows a force diagram for the board. In addition to the board's weight, there are normal forces F_1 and F_2 at the floor and wall, and a frictional force at the floor. We have expressed the frictional force in terms of the normal force of the floor, F_1, and the coefficient of static friction, μ (see Section 5.2), because the

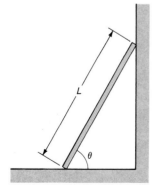

Fig. 11–25
Example 11–6

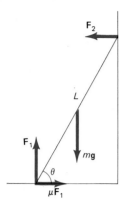

Fig. 11–26
In addition to its weight, two normal forces and a frictional force act on the board.

minimum angle occurs when the force of static friction takes its maximum value μF_1. This frictional force acts to oppose slipping of the board, and so points toward the wall. The equation of force balance now has both horizontal and vertical components:

$$F_1 - mg = 0; \quad \text{(vertical)} \tag{11–12}$$

$$\mu F_1 - F_2 = 0. \quad \text{(horizontal)} \tag{11–13}$$

Two forces are applied at the bottom of the board, so that if we choose this point to calculate the torques, we eliminate two of the forces from the torque equation. The remaining torques arise from the normal force $\mathbf{F}_2$ acting through the lever arm $L\sin\theta$ and from the gravitational force $m\mathbf{g}$ acting through the lever arm $\frac{1}{2}L\cos\theta$, as suggested in Fig. 11–26. Taking counterclockwise rotation as positive, the torque balance equation may then be written

$$LF_2\sin\theta - \tfrac{1}{2}Lmg\cos\theta = 0.$$

Having written the equations of force and torque balance, we have completed the essential physics of the problem. The rest is just algebra. We rewrite the torque equation in the form

$$\tan\theta = \frac{\sin\theta}{\cos\theta} = \frac{mg}{2F_2}. \tag{11–14}$$

We can find F_2 by solving Equations 11–12 and 11–13 for both normal forces F_1 and F_2. Equation 11–12 gives

$$F_1 = mg,$$

so that Equation 11–13 can be written

$$\mu mg = F_2.$$

Using this result in Equation 11–14 gives

$$\tan\theta = \frac{mg}{2\mu mg} = \frac{1}{2\mu},$$

so that

$$\theta = \tan^{-1}\left(\frac{1}{2\mu}\right).$$

This is a remarkably simple result—it doesn't depend on the length or mass of the board, or on g. Does it make sense? As the coefficient of friction goes to zero, the tangent of θ grows without bound, and θ approaches 90°. With no friction, we can lean the board against the wall only if it is vertical. Increasing the coefficient of friction allows us to lean the board at ever lower angles. When $\mu = 0.27$, we have

$$\theta = \tan^{-1}\left[\frac{1}{(2)(0.27)}\right] = 62°.$$

What if the wall had not been frictionless? Then we would have had an additional frictional force at the top. Although the physics of the situation would have been nearly the same, the algebra would be more complicated. Problem 20 explores this more realistic case.

Example 11–7

Equilibrium in the Human Body

Figure 11–27 depicts a section of a human arm. The forearm measures 32 cm from elbow joint to center of palm and has a mass $m = 2.7$ kg, with center of mass located 14 cm from the elbow joint. The biceps muscle attaches to the radius (fore-

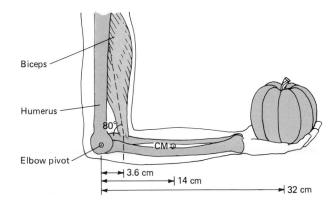

Fig. 11–27
A human arm.

arm bone) 3.6 cm from the elbow joint. With the humerus (upper arm bone) vertical and the forearm horizontal, the biceps muscle makes an 80° angle with the horizontal. If a pumpkin of mass $M = 4.5$ kg rests on the palm, what are the tensions in the biceps and the forces exerted on the elbow joint?

Solution

To solve this problem, we redraw the arm in schematic form, showing all the forces acting on it (Fig. 11–28). In addition to the weights m**g** and M**g** of the arm and pumpkin, the tension force **T** in the biceps and the contact force **F**$_c$ at the elbow joint also act on the arm. Denoting the horizontal and vertical components of the contact force by F_1 and F_2, the equations of force balance become

$$T \cos\theta - F_1 = 0, \quad \text{(horizontal)} \tag{11–15}$$

and

$$T \sin\theta - F_2 - mg - Mg = 0. \quad \text{(vertical)} \tag{11–16}$$

We now have two equations in the three unknowns T, F_1, and F_2. A third equation is provided by the condition that there be no net torque on the arm. Choosing the elbow joint as our pivot to eliminate the contact force from the torque equation gives

$$\ell_1 T \sin\theta - \ell_2 mg - \ell_3 Mg = 0$$

for the condition of torque balance. Here the ℓ's are the distances from the elbow to the biceps, the center of mass of the arm, and the pumpkin, respectively. We have taken a counterclockwise rotation as positive. Solving the torque equation for the biceps tension T gives

$$T = \frac{(\ell_2 m + \ell_3 M)g}{\ell_1 \sin\theta}$$

$$= \frac{[(0.14 \text{ m})(2.7 \text{ kg}) + (0.32 \text{ m})(4.5 \text{ kg})](9.8 \text{ N/kg})}{(0.036 \text{ m})(\sin 80°)}$$

$$= 500 \text{ N}.$$

Equations 11–15 and 11–16 then give the components of the elbow contact force:

$$F_1 = T \cos\theta = (500 \text{ N})(\cos 80°) = 87 \text{ N},$$

and

$$F_2 = T \sin\theta - (m+M)g$$

$$= (500 \text{ N})(\sin 80°) - (2.7 \text{ kg} + 4.5 \text{ kg})(9.8 \text{ N/kg}) = 420 \text{ N}.$$

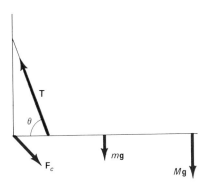

Fig. 11–28
A simplified picture of the arm, showing the forces acting.

This example shows that the human body often sustains forces far larger than the weights of objects it may be lifting.

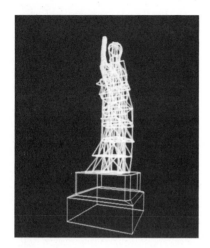

Fig. 11–29
The interior skeleton of the Statue of Liberty counteracts forces and torques associated with the statue itself as well as with the wind. This computer drawn image of Liberty's skeleton aided architects planning the restoration. (See color plate 1.)

Fig. 11–30
The Statue of Liberty's head and arm are offset from their originally planned positions, resulting in greater than expected forces and torques on structural members.

Application

Restoring the Statue of Liberty

In 1986, workers completed major renovation of the Statue of Liberty, France's famous gift to the United States that was first dedicated in 1886. Although its designer, French sculptor Frédéric-Auguste Bartholdi, suggested that his creation should last as long as Egypt's pyramids, many factors conspired to make major renovation necessary after only one century. These include corrosion from air pollution and from a chemical reaction between the iron framework and the copper skin, as well as an assembly change that resulted in excess torques on the statue's structural members.

Sculptor Bartholdi was no engineer, and without the work of the French engineer Eiffel (of tower fame) the statue could not have maintained itself in static equilibrium. Eiffel designed an inner skeleton of iron to provide the forces necessary to counteract the weights and torques associated with components of the statue, as well as with the force exerted by wind (Fig. 11–29). The statue was constructed in France in 300 separate pieces, then shipped to New York. During assembly, probably as a conscious aesthetic decision, Liberty's head and upraised arm were mounted 2 feet from their locations on Eiffel's plan, with the arm making a

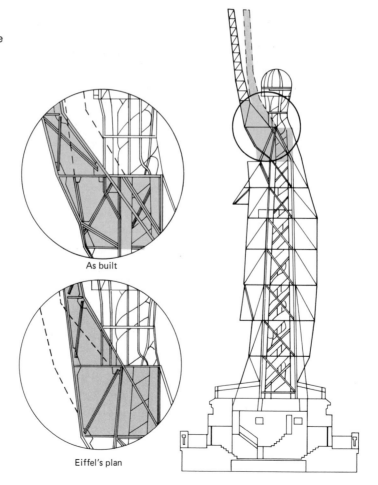

As built

Eiffel's plan

greater angle to the vertical than planned (Fig. 11–30). As a result, the statue's head had a nonzero lever arm about the central support, resulting in a significant torque. More importantly, the arm exerted a much greater torque about the shoulder than planned, resulting in forces greater than anticipated on structural components. For historical integrity, renovators did not return to Eiffel's original design; instead, they reinforced the support structure to withstand better the excess forces and torques.

11.4
STABILITY OF STATIC EQUILIBRIA

Fig. 11–31
Stable (left) and unstable (right) equilibria.

What if a body is disturbed from static equilibrium? Then, in general, it will experience nonzero net forces or torques, and it will accelerate or begin to rotate. Figure 11–31 shows two very different possibilities for the subsequent behavior of a body displaced slightly from equilibrium. On the left, a slight displacement results in a torque that moves the body back toward its equilibrium position. Although the body may rock back and forth for a while, any friction or other dissipation will soon sap the body's energy and it will settle back into its original equilibrium. On the right in Fig. 11–31, the torque arising from a slight displacement swings the body away from its original equilibrium. The former situation is an example of **stable equilibrium**, the latter of **unstable equilibrium**. Nearly all the equilibria we encounter in nature are stable, for a body in unstable equilibrium will not remain that way for any significant time. The slightest disturbance will set it in motion, and if it eventually stops, it will be in a very different equilibrium state from its original one.

stable equilibrium
unstable equilibrium

neutral stability

metastability

The distinction between stable and unstable equilibria is often subtler than Fig. 11–31 would imply. Figure 11–32 shows a ball in four different equilibrium situations. Clearly situation (a) is stable, while (b) is unstable. Situation (c) is neither stable nor unstable; it is termed **neutrally stable.** But what about situation (d)? For very small disturbances, the ball will return to its original state, so that the equilibrium is stable. But for larger disturbances—large enough to push the ball just over the highest points on the hill—it is unstable. Such an equilibrium is called **conditionally stable** or **metastable.** Most equilibria are conditionally stable. Figure 11–33 shows a rectangular block in two equilibrium situations. Although both are stable against small disturbances, the vertical orientation is less stable in the sense that it is easier to dislodge it from its stable situation and send it crashing down on its side.

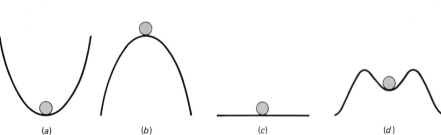

(a)   (b)   (c)   (d)

Fig. 11–32
Stable, unstable, neutrally stable, and conditionally stable equilibria.

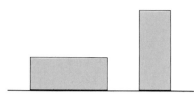

Fig. 11–33
Both equilibria are stable against very small disturbances, but even a modest disturbance will bring the vertical block into an unstable situation.

The stability of an equilibrium is closely associated with the potential energy of the system in equilibrium. In Fig. 11–32, for example, the shapes of the hills and valleys reflect the gravitational potential energy of the ball as a function of position. In all cases of equilibrium, the ball is at a minimum or maximum of the potential energy curve—at a place where the derivative of potential energy with respect to position is zero. For the stable and metastable equilibria, the potential energy at equilibrium is a minimum, at least with respect to positions immediately adjacent to equilibrium. A deviation from equilibrium requires that work be done against a net force that tends to restore the ball to its equilibrium position. The unstable equilibrium, in contrast, occurs at a maximum in the potential energy. Here, a deviation from equilibrium results in lower potential energy and in a net force that tends to accelerate the ball farther from equilibrium. For the neutrally stable equilibrium, there is no change whatever in potential energy as we move away from equilibrium; consequently the ball experiences no force.

The potential energy curve for the rectangular block of Fig. 11–33 is less obvious. But we can develop that curve by considering the height of the center of mass as the block is tipped. In both cases, tipping the block requires that the center of mass be raised, thereby raising the potential energy. Figure 11–34 shows the location of the center of mass as we tilt the block through various orientations. The path traced out by the center of mass is the potential energy curve for the block. From this curve we see immediately that there are two stable equilibria of lowest potential energy corresponding to the block lying on either side. Similarly, there are two metastable states corresponding to the block standing vertically on either end. In between are unstable equilibria that occur when the block is balanced on a corner so that its center of mass is just over the balance point. Disturbed from these unstable states, the block experiences a net torque that rotates it toward one of the stable or metastable states.

We can sum up our understanding of equilibrium and potential energy in two simple mathematical statements. First, equilibrium requires a local maximum or minimum in the potential energy, so that

$$\frac{dU}{dx} = 0, \qquad \text{(equilibrium condition)} \qquad (11\text{–}17)$$

where U is the potential energy of a system and x is a variable describing the configuration of that system. For the simple systems we have been considering, x measures the position or orientation of an object, but with more complicated systems it may represent other quantities such as the system's volume or even its composition. For a stable equilibrium, we require a local

Fig. 11–34
As the block is tilted, its center of mass traces out a potential energy curve. Shown are two stable equilibria (ends), a metastable equilibrium (center), and two unstable equilibria (tilted blocks).

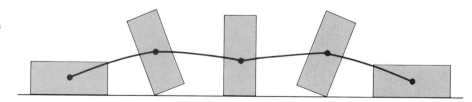

minimum, so that the potential energy curve is concave upward. Mathematically,

$$\frac{d^2U}{dx^2} > 0. \quad \text{(stable equilibrium)} \quad (11\text{–}18)$$

This condition applies to metastable equilibria as well, for they are *locally* stable. In contrast, unstable equilibrium occurs where the potential energy has a local maximum, so that

$$\frac{d^2U}{dx^2} < 0. \quad \text{(unstable equilibrium)} \quad (11\text{–}19)$$

The intermediate case $d^2U/dx^2 = 0$ corresponds to neutral stability.

Example 11–8

Stability of Equilibria

Solid-state physicists developing a new semiconductor device theorize that the potential energy of an electron in a region of the device should be given by

$$U(x) = ax^2 - bx^4,$$

where x is the electron's position measured in angstrom units (1 Å $= 10^{-10}$ m), U is its energy in electron volts (1 eV $= 1.6 \times 10^{-19}$ J), and constants a and b are given by $a = 8$ eV/Å^2 and $b = 1$ eV/Å^4. Locate the equilibrium positions for the electron, and describe their stability.

Solution

Equilibrium occurs where the potential energy has a minimum or maximum, so that $dU/dx = 0$. Taking the derivative of the potential energy and setting it to zero gives

$$0 = \frac{dU}{dx} = 2ax - 4bx^3 = 2x(a - 2bx^2) = 2x(8 - 2x^2) = 4x(4 - x^2) \text{ eV/Å},$$

where for mathematical clarity we have not bothered to write the units except in the final answer. This equation is satisfied if $x = 0$ or if $4 - x^2 = 0$, so that $x = \pm 2$. Therefore, the electron is in equilibrium at $x = -2$ Å, at $x = 0$ Å, or at $x = 2$ Å. To determine the stability of these equilibria, we evaluate the second derivative at each equilibrium position:

$$\frac{d^2U}{dx^2} = 2a - 12bx^2 = 16 - 12x^2 \text{ eV/Å}^2,$$

so that

$$\left(\frac{d^2U}{dx^2}\right)_{x=\pm 2} = 16 - (12)(4) = -32 \text{ eV/Å}^2$$

and

$$\left(\frac{d^2U}{dx^2}\right)_{x=0} = 16 - 0 = 16 \text{ eV/Å}^2.$$

From conditions 11–18 and 11–19, we see that the equilibria at $x = \pm 2$ Å are unstable, while the equilibrium at $x = 0$ Å is locally stable. Figure 11–35 shows a plot of the potential energy curve, which makes clear the location and stability of the equilibria.

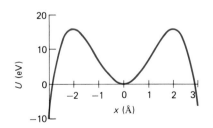

Fig. 11–35
Potential energy curve for Example 11–8.

Questions of stability apply not only to the physical locations of individual particles but also to the overall arrangements of the basic constituent particles that comprise matter. A mixture of hydrogen and oxygen, for example, is in metastable equilibrium at room temperature—the individual atoms are at local minima in their potential energy curves, as in Fig. 11–32d or Fig. 11–35. Applying a lighted match to the mixture adds enough energy to bring some of the atoms over the maxima in their potential energy curves, at which point the atoms can rearrange into a state of much lower potential energy—the state we call water (H_2O). Energy released in the process maintains the reaction, until virtually all the atoms have rearranged themselves into the lower-energy state. Since no state of lower potential energy is available to its constituent particles, the water molecule is quite stable, as opposed to the explosively metastable mixture of hydrogen and oxygen. Similarly, a uranium nucleus is at a local minimum in its potential energy curve—a minimum determined by the interplay of nuclear and electrical forces. But if it gains a small amount of excess energy, the nucleus may fission—split—into two smaller nuclei whose total potential energy is much lower. This transition from a less stable to a more stable equilibrium describes the basic physics of nuclear fission.

Potential energy curves for complex structures like molecules or skyscrapers cannot be described fully by simple one-dimensional graphs, for the configurations of these structures can be altered in many different ways. If potential energy varies with position in several different directions, for example, then we must consider all possible changes in potential energy in order to determine stability. A snowbank sitting on a saddle-shaped pass between two mountain peaks, or any other system with a saddle-shaped potential energy curve (Fig. 11–36), is stable with respect to displacements in one direction but not in another. Stability analysis of complex physical systems, ranging from nuclei and molecules to bridges and fusion reactors and on to stars and galaxies, is an important part of contemporary work in engineering and science.

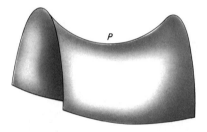

Fig. 11–36
A saddle-shaped potential energy curve. Saddle point P is an equilibrium position that is stable in one direction but not in another.

SUMMARY

1. **Torque** describes the tendency of a force to produce a rotation of an object about a point. The torque about a given pivot point depends on the magnitude of the force and on the perpendicular distance, or **lever arm,** from the pivot point to the line along which the force vector lies. Mathematically, the torque about a point O is a vector given by

$$\boldsymbol{\tau} = \mathbf{r} \times \mathbf{F},$$

where $\mathbf{r}$ is a vector from point O to the point where the force $\mathbf{F}$ is applied. The product in this expression is the **vector cross product,** defined as a vector whose magnitude is given by

$$\tau = rF \sin\theta$$

and whose direction is determined by the right-hand rule: curl the fingers of your right hand in a way that rotates the vector $\mathbf{r}$ onto the vector $\mathbf{F}$, and your right thumb points in the direction of the cross product $\boldsymbol{\tau} = \mathbf{r} \times \mathbf{F}$.

2. A body at rest is in **static equilibrium** only when there is no net force on the body and no net torque about any point:

$$\sum \mathbf{F} = 0$$

and

$$\sum \boldsymbol{\tau} = 0.$$

If the first condition is met, and the second holds for some point, then it holds for all points.

3. When the gravitational force acts on an extended body, its effect in producing a torque is that of a single force acting at a point that is called the **center of gravity.** If the gravitational field is the same throughout the body, the center of gravity coincides with the center of mass.

4. An equilibrium state is **stable** if a small displacement from equilibrium results in forces and/or torques that tend to restore the equilibrium. It is **unstable** if small displacements result in forces and/or torques that move the system farther from its original equilibrium. A **metastable** equilibrium is stable against very small displacements but becomes unstable if larger displacements occur. Stability properties of a system are closely related to its potential energy curve. Stable and metastable equilibria occur at minima of the potential energy curve, unstable equilibria at maxima. Mathematically,

$$\frac{dU}{dx} = 0, \quad \text{(equilibrium condition)}$$

$$\frac{d^2U}{dx^2} > 0, \quad \text{(stable or metastable equilibrium)}$$

and

$$\frac{d^2U}{dx^2} < 0. \quad \text{(unstable equilibrium)}$$

The intermediate case $d^2U/dx^2 = 0$ is termed **neutral stability.**

QUESTIONS

1. Why is the direction of the torque vector taken to be perpendicular to the plane in which the torque tends to rotate a body?
2. Figure 11–37 shows four forces acting on a body. What are the directions of each of the associated torques about the point O? About the point P?

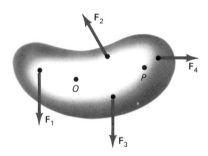

Fig. 11–37   Question 2

3. The best way to lift a heavy weight is to squat with your back vertical, rather than to lean over (Fig. 11–38). Why?

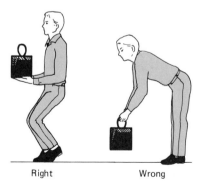

Right   Wrong

Fig. 11–38   Right and wrong ways to lift a heavy weight (Question 3).

4. Give an example of an object on which the net force is zero but which is not in static equilibrium.

5. As the beam of Fig. 11–4 rotates, does the torque on it remain constant? Explain.
6. Give an example of an object on which the net torque about the center of gravity is zero but which is not in static equilibrium. Is the net torque about all points zero in this case? Explain.
7. Estimate the torque you apply when tightening the wheel nuts of a car after changing a flat tire.
8. Some screwdrivers have flat shanks to which a wrench can be attached. How does this help increase the torque applied to a screw?
9. Pregnant women often assume a posture in which the shoulders are held well back of their normal position. Explain in terms of torque and center of gravity.
10. A large, irregularly shaped asteroid of uniform density is falling toward earth in the orientation shown in Fig. 11–39. Which is closer to earth, the asteroid's center of gravity or its center of mass?

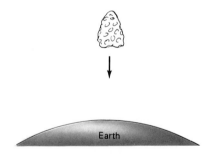

Earth

Fig. 11–39   Question 10

11. When you carry a bucket full of water with one hand, you often find it easier if you extend your opposite arm. Why?
12. Is a ladder more likely to slip when you stand near the top or the bottom? Explain.
13. How does a heavy keel on a boat help prevent the boat from tipping over?
14. In addition to wings, most airplanes have a smaller set of horizontal surfaces near the tail. Why? What

does this configuration suggest about the location of the wings relative to the center of gravity of the plane?

15. Why is a beer bottle less stable if stood on its top rather than on its base?

16. Stars end their lives containing a greater proportion of helium and other heavier elements than they had originally. What does this change say about the relative stability of matter arranged as helium and heavier elements versus the same matter arranged as hydrogen?

17. Why is a "balancing rock" (Fig. 11–40) such an unusual phenomenon?

Fig. 11–40 A balancing rock (Question 17).

PROBLEMS

Section 11.1 *Torque*

1. A rod is free to pivot about one end, and a force is applied at the other end, as shown in Fig. 11–41. What is the torque about the pivot point?

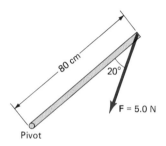

Fig. 11–41 Problem 1

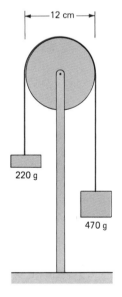

Fig. 11–42 Problem 2

2. A pulley 12 cm in diameter is free to rotate about a horizontal axle. A 220-g mass and a 470-g mass are tied to either end of a massless string, and the string is hung over the pulley, as shown in Fig. 11–42. If the string does not slip, what torque must be applied to keep the pulley from rotating?

3. A car tune-up manual calls for tightening the spark plugs to a torque of 25 ft·lb. If the plugs are tightened by pulling at right angles to the end of a wrench 10 in long, how much force must be applied to the wrench?

4. Two arm wrestlers are deadlocked in the position shown in Fig. 11–43. Each arm is 34 cm long and makes a 60° angle with the horizontal; the contact force at the hands is 180 N. What is the torque each exerts about the other's elbow? Neglect the weights of the wrestlers' arms.

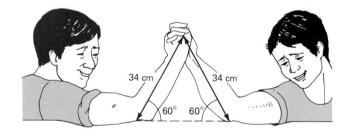

Fig. 11–43 Problem 4

5. Figure 11–44 shows the forces acting on a sailboat. These include the weight of the boat, 2200 N, acting at the center of gravity C; the upward buoyancy force of the water, acting at point B; and the force of the wind, acting horizontally at point A. Along an axis through the mast of the boat, point C lies 1.4 m below point B, and point A lies 4.2 m above B. The boat is heeled with its mast 30° to the vertical. What is the force of the wind?

6. What are the net force and the net torque about the origin due to the following forces?

$\mathbf{F}_1 = 2.0\hat{\mathbf{i}} - 0.80\hat{\mathbf{j}}$ N,
applied at $\mathbf{r}_1 = -1.3\hat{\mathbf{i}}$ m

$\mathbf{F}_2 = 1.0\hat{\mathbf{i}} + 2.6\hat{\mathbf{j}} - 1.4\hat{\mathbf{k}}$ N,
applied at $\mathbf{r}_2 = -0.70\hat{\mathbf{i}} + 1.1\mathbf{j}$ m

$\mathbf{F}_3 = 1.7\hat{\mathbf{j}}$ N,
applied at $\mathbf{r}_3 = -2.6\hat{\mathbf{k}}$ m

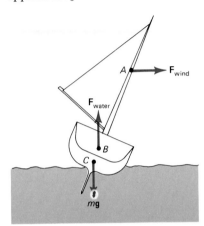

Fig. 11–44   Problem 5

7. What is the cross product of the vectors $\mathbf{A} = \hat{\mathbf{i}} - 2\hat{\mathbf{j}} + \pi\hat{\mathbf{k}}$ and $\mathbf{B} = 2\hat{\mathbf{i}} + 4\hat{\mathbf{j}} - 2\pi\hat{\mathbf{k}}$?
8. Vector $\mathbf{A}$ has length 1 and points along the positive x-axis. Vector $\mathbf{B}$ extends from the origin to the point (2,2). Evaluate the vector cross product $\mathbf{A} \times \mathbf{B}$.
9. Given the vectors $\mathbf{A} = 2\hat{\mathbf{i}} + 3\hat{\mathbf{j}}$, $\mathbf{B} = \hat{\mathbf{i}} - 2\hat{\mathbf{j}}$, and $\mathbf{C} = 4\hat{\mathbf{i}} + 2\hat{\mathbf{j}}$, find the following: (a) $\mathbf{A} \times \mathbf{B}$; (b) $\mathbf{A} \times \mathbf{C}$; (c) $\mathbf{B} \times \mathbf{C}$; (d) $\mathbf{A} \times (\mathbf{B} \times \mathbf{C})$; (e) $(\mathbf{A} \times \mathbf{B}) \times \mathbf{C}$; (f) $\mathbf{C} \cdot (\mathbf{A} \times \mathbf{B})$.
10. A rod of length ℓ has forces acting everywhere perpendicular to its length, as shown in Fig. 11–45. The rod lies along the x-axis, and the force per unit length is given by

$$\frac{dF}{dx} = \frac{F_0}{\ell}\left[1 - \left(\frac{x}{\ell}\right)^2\right].$$

Determine the magnitude of (a) the total force on the rod; (b) the torque about the end $x = 0$; (c) the torque about the end $x = \ell$.

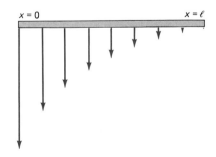

Fig. 11–45   Problem 10

Sections 11.2 and 11.3   *Static Equilibrium and Center of Gravity*

11. A 4.2-m-long beam is supported by a cable at its center. A 65-kg steelworker stands at one end of the beam. Where should a 190-kg bucket of concrete be suspended if the beam is to be in static equilibrium?
12. Two pulleys are mounted on a horizontal axle, as shown in Fig. 11–46. The inner pulley has a diameter of 6.0 cm, the outer a diameter of 20 cm. Cords are wrapped around both pulleys so that they do not slip. In the configuration shown, with what force must a person pull on the outer rope in order to support the 40-kg mass?

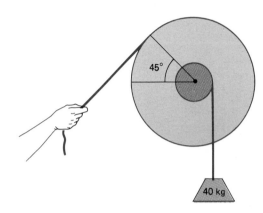

Fig. 11–46   Problem 12

13. A 23-m-long log of irregular cross section is lying horizontally, supported by a wall at one end and a cable attached 4.0 m from the other end, as shown in Fig. 11–47. The log weighs 7.5×10^3 N, and the tension in the cable is 6.2×10^3 N. Where is the log's center of gravity?

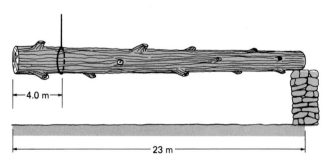

Fig. 11–47   Problem 13

14. Figure 11–48 shows an outstretched arm with a mass of 4.2 kg. The arm is 56 cm long, and its center of gravity is 21 cm from the shoulder. The hand at the end of the arm holds a 6.0-kg mass. (a) What is the torque about the shoulder due to the weights of the arm and the 6.0-kg mass? (b) If the arm is held in

equilibrium by the deltoid muscle, whose force on the arm acts 5.0° below the horizontal at a point 18 cm from the shoulder joint (Fig. 11–49), what is the force exerted by the muscle?

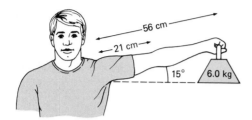

Fig. 11–48   Problem 14

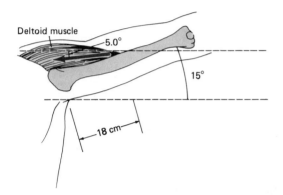

Fig. 11–49   Deltoid muscle (Problem 14).

15. Figure 11–50 shows how a scale with a capacity of only 250 N can be used to weigh a much heavier person. The board is 3.0 m long, has a mass of 3.4 kg, and is of uniform density. It is free to pivot about the end farthest from the scale. What is the weight of a person standing 1.2 m from the pivot end, if the scale reads 210 N? Assume that the beam remains nearly horizontal.

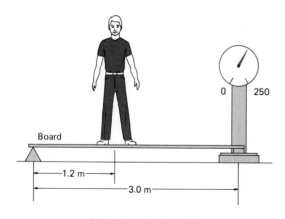

Fig. 11–50   Problem 15

16. Figure 11–51 shows a portable infant seat that is supported by the edge of a table. The mass of the seat is 1.5 kg, and its center of mass is located 16 cm from the table edge. A 12-kg baby is sitting in the seat with her center of mass over the seat's center of mass. Find the forces F_A and F_B that the seat exerts on the table.

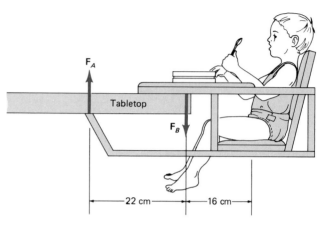

Fig. 11–51   Problem 16

17. A 15.0-kg door measures 2.00 m high by 75.0 cm wide. It hangs from hinges mounted 18.0 cm from top and bottom. Assuming that each hinge carries half the door's weight, determine the horizontal and vertical forces that the door exerts on each hinge.

18. Figure 11–52 shows a popular system for mounting bookshelves. An aluminum bracket is mounted on a vertical aluminum support by small tabs inserted into vertical slots. If each bracket in a shelf system supports 32 kg of books, with center of gravity 12 cm out from the vertical support, what is the horizontal component of the force exerted on the upper of the two bracket tabs? Assume contact between the bracket and support occurs only at the upper tab and at the bottom of the bracket, 4.5 cm below the upper tab.

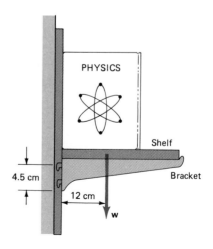

Fig. 11–52   Problem 18

19. Figure 11–53 shows a house designed to have high "cathedral" ceilings. Following a heavy snow, the total mass supported by each diagonal roof rafter is 170 kg, including building materials as well as snow. Under these conditions, what is the force in the horizontal tie beam near the roof peak? Is this force a compression or a tension? Neglect any horizontal component of force due to the vertical walls below the roof. Ignore the widths of the various structural components, treating contact forces as though they were concentrated at the roof peak and at the outside edge of the rafter/wall junction.

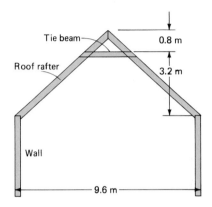

Fig. 11–53   Problem 19

20. Repeat Example 11–6, now assuming that the coefficient of friction at the floor is μ_1 and that at the wall μ_2. Show that the minimum angle at which the ladder will not slip is now given by

$$\theta = \tan^{-1}\left(\frac{1-\mu_1\mu_2}{2\mu_1}\right).$$

21. A uniform sphere of radius R is supported by a rope attached to a vertical wall, as shown in Fig. 11–54. The point where the rope is attached to the sphere is

located so that a continuation of the rope would intersect a horizontal line through the sphere's center a distance $R/2$ beyond the center, as shown. What is the smallest possible value for the coefficient of friction between wall and sphere?

22. Show that if the wall in the previous problem is frictionless, then a continuation of the rope line must pass through the center of the sphere.

23. A garden cart loaded with firewood is being pushed horizontally when it encounters a step 8.0 cm high, as shown in Fig. 11–55. The mass of the cart and its load is 55 kg, and the cart is balanced so that its center of mass is directly over the axle. The wheel diameter is 60 cm. What is the minimum horizontal force that will get the cart up the step?

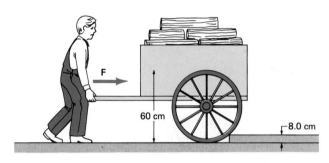

Fig. 11–55   Problem 23

24. A 3.0-m-long ladder is leaning against a frictionless vertical wall, with which it makes a 15° angle. The mass of the ladder is 5.0 kg, uniformly distributed over its length. The coefficient of friction between ladder and ground is 0.42. Can a 65-kg person climb to the top of the ladder without its slipping? If not, how high can the person climb? If so, how massive a person would make the ladder slip?

25. The boom in the crane of Fig. 11–56 is free to pivot about point P and is supported by the cable that joins halfway along its 18-m total length. The cable passes over a pulley and is anchored at the back of the crane. The boom has a mass of 1700 kg, distributed uniformly along its length, and the mass hanging from

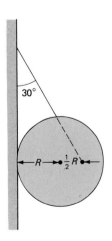

Fig. 11–54   Problem 21

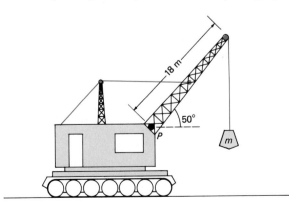

Fig. 11–56   Problem 25

the end of the boom is 2200 kg. The boom makes a 50° angle with the horizontal. What is the tension in the cable that supports the boom?

26. A uniform board of length ℓ and weight W is suspended between two vertical walls by ropes of length $\ell/2$ each. When a weight w is placed on the left end of the board, it assumes the configuration shown in Fig. 11–57. Find weight w in terms of the board weight W.

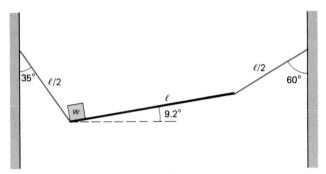

Fig. 11–57   Problem 26

Section 11.4   *Stability of Static Equilibria*

27. A roly-poly toy clown is made from part of a sphere topped by a cone. The sphere is truncated at just the right point so that there is no discontinuity in angle as the surface changes from sphere to cone (Fig. 11–58a). If the clown always returns to an upright position, what is the maximum possible height for its center of mass? Would your answer change if the continuity-of-angle condition were not met, as in Fig. 11–58b?

Fig. 11–58   Problem 27

28. The potential energy as a function of position for a certain electron is given by

$$U = U_0[1 - e^{(x-x_1)^2/x_2^2}],$$

where x is the electron's position and x_1 and x_2 are constants. Find any equilibria of this electron, and determine their stability.

29. A uniform rectangular block is twice as long as it is wide. Letting θ be the angle that the long dimension makes with the horizontal (Fig. 11–59), determine the angular positions of any static equilibria, and comment on their stability.

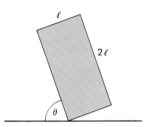

Fig. 11–59   Problem 29

30. The potential energy as a function of position for a certain particle is given by

$$U(x) = U_0\left(\frac{x^3}{x_0^3} + a\frac{x^2}{x_0^2} + 4\frac{x}{x_0}\right),$$

where U_0, x_0, and a are constants. For what values of a will there be two static equilibria? Comment on the stability of these equilibria.

Supplementary Problems

31. Suppose that the forces of Problem 6 are applied to a body. Show that it is not possible using a single additional force to put the body into static equilibrium. Find two additional forces, and their application points, that will result in static equilibrium. *Hint:* Think about the geometrical relationship of vectors in the cross product. You may also find the vector dot product useful.

32. Show that the cross product $\mathbf{A} \times \mathbf{B}$ of vectors $\mathbf{A} = a_x\hat{\mathbf{i}} + a_y\hat{\mathbf{j}} + a_z\hat{\mathbf{k}}$ and $\mathbf{B} = b_x\hat{\mathbf{i}} + b_y\hat{\mathbf{j}} + b_z\hat{\mathbf{k}}$ is given by

$$\mathbf{A} \times \mathbf{B} = (a_yb_z - a_zb_y)\hat{\mathbf{i}} + (a_zb_x - a_xb_z)\hat{\mathbf{j}} + (a_xb_y - a_yb_x)\hat{\mathbf{k}}.$$

33. Show that $\mathbf{A} \cdot (\mathbf{A} \times \mathbf{B})$ is zero for any two vectors $\mathbf{A}$ and $\mathbf{B}$.

34. A uniform solid cone of height h and base diameter $\frac{1}{3}h$ is placed on an incline, as shown in Fig. 11–60. The coefficient of static friction between cone and in-

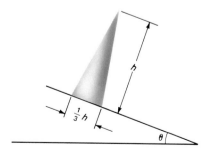

Fig. 11–60   Problem 34

cline is 0.63. As the slope of the incline is increased, will the cone first begin to slip or to tip over? At what angle would each occur? *Hint:* Begin with an integration to locate the center of mass of the cone.

35. A huge interstellar spacecraft from an advanced civilization is hovering above earth, as shown in Fig. 11–61. The ship consists of two small pods, each of mass m, separated by a rigid shaft of negligible mass that is one earth radius (R_e) long. The ship is hovering with one pod a distance R_e directly above earth's north pole and with the main axis of the ship parallel to the equatorial plane. (*a*) Find the net gravitational force on the ship (magnitude and direction), expressed in terms of the pod mass m and earth's surface gravity g. (*b*) Find the net torque on the ship about its center of mass. (*c*) Locate the center of gravity of the ship, and compare with the location of the center of mass.

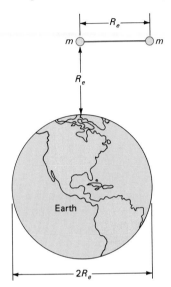

Fig. 11–61 Problem 35

12

ROTATIONAL MOTION

When a nonzero net force acts on an object, Newton's law tells us that the object accelerates. What happens when a nonzero net torque acts? Then the rotational motion of the object changes. We could describe this rotation without reference to torque, if we chose to consider the forces, accelerations, and velocities associated with each particle making up an object. That would be a difficult task, and would not give us much insight into the behavior of the object as a whole. Instead, we will develop a description of rotational motion that closely parallels what we know about motion as described by Newton's laws. We already established part of that description when we defined torque, which we can think of as the rotational analog of force.

12.1
ANGULAR VELOCITY AND ACCELERATION

Angular Speed

angular speed

Consider a record rotating on a turntable. If we described this motion in terms of linear velocity and acceleration, we would have to say that each point on the record has a different speed—depending on its distance from the center of the record—and a different acceleration to hold it in its circular path. But the record is a rigid body—all its parts maintain fixed positions relative to one another—so it is simpler to say that the entire record rotates at $33\frac{1}{3}$ revolutions per minute (rpm). When we say this, we are talking about **angular speed**—the rate at which the angular position of any point on the record changes. In this case, our unit of angle is one full revolution (360° or 2π radians), and our unit of time is the minute. We could equally well have expressed angular speed in revolutions per second (rev/s), degrees per second (°/s), or radians per second (rad/s). Because of the mathematically simple status of radian measure, we often use radians in calculations involving rotational motion.

average angular speed

We use the Greek symbol ω (omega) for angular speed, and define **average angular speed** $\overline{\omega}$ as:

$$\overline{\omega} = \frac{\Delta\theta}{\Delta t},$$ (12–1)

where $\Delta\theta$ is the change in angle occurring in the time Δt (Fig. 12–1). When the angular speed is constant, its value at any instant is the same as its average value. When angular speed is changing, we define **instantaneous angular speed** as the limit of the average angular speed taken over arbitrarily short time intervals:

instantaneous angular speed

$$\omega = \lim_{\Delta t \to 0} \frac{\Delta\theta}{\Delta t} = \frac{d\theta}{dt}.$$ (12–2)

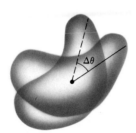

Fig. 12–1
A rigid body rotates through the angle $\Delta\theta$ in time Δt. Its average angular speed is $\Delta\theta/\Delta t$.

These definitions are analogous to our definitions of average and instantaneous linear speed introduced in Chapter 2. The only difference is the use of angular displacement $\Delta\theta$, rather than linear displacement Δx.

Knowing the angular speed of a rotating object, we can easily find the linear speed of any point on the object. Recall that an angle in radian measure is defined as the ratio of subtended arc length to radius (Fig. 12–2):

$$\theta = \frac{s}{r}.$$ (12–3)

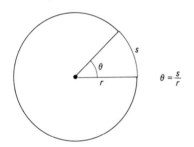

Fig. 12–2
The angle θ in radians is the ratio of the arc length s to the radius r.

Differentiating this expression with respect to time gives

$$\frac{d\theta}{dt} = \frac{1}{r}\frac{ds}{dt},$$

since the radius r is constant. But $d\theta/dt$ is the angular speed, and ds/dt is the linear speed, v, so that $\omega = v/r$, or

$$v = \omega r.$$ (12–4)

Thus the linear speed of any point on a rotating object is proportional both to the angular speed of the object and to the distance from that point to the axis of rotation (Fig. 12–3). A word of caution: Equation 12–4 was derived from the definition of angle in radians, and holds only for angular speed measured in radians per unit time. For other measures of angle, a proportionality constant must be included in the relation among v, ω, and r (see Problem 3). Furthermore, Equation 12–4 is valid only for rotation about a fixed rotation axis. In Section 12.3 we will consider the more general case of an object—like a rolling wheel—whose rotation axis is itself in motion.

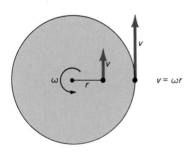

Fig. 12–3
The linear speed of a point in a rotating object is proportional to the angular speed of rotation and the distance from the rotation axis.

Example 12–1
Angular Speed

A record 30.5 cm in diameter is rotating at $33\frac{1}{3}$ rpm. Find its angular speed in rad/s, and determine the linear speed of a point on its outer edge.

Solution

One revolution is 2π radians, and one minute is 60 seconds, so that

$$\omega = 33\tfrac{1}{3} \text{ rpm} = \frac{(33\frac{1}{3}\text{ rev/min})(2\pi\text{ rad/rev})}{60\text{ s/min}} = 3.49 \text{ rad/s}.$$

This result makes sense, for at $33\frac{1}{3}$ rpm, the record turns a little more than half a revolution—π radians—in one second. From Equation 12–4, we then have

$$v = \omega r = (3.49 \text{ rad/s})\left(\frac{30.5 \text{ cm}}{2}\right) = 53.2 \text{ cm/s}.$$

Angular Acceleration

angular acceleration

If the angular speed of a rotating object changes, then the object undergoes an **angular acceleration,** α, defined analogously to linear acceleration:

$$\alpha = \lim_{\Delta t \to 0} \frac{\Delta \omega}{\Delta t} = \frac{d\omega}{dt}. \tag{12–5}$$

Taking the limit in Equation 12–5 gives the instantaneous angular acceleration; if we do not take the limit, then we have an average value over the time interval Δt. The SI units for angular acceleration are rad/s^2, although we will sometimes use other units such as rpm/s or rev/s^2. Having defined angular acceleration, we can differentiate Equation 12–4 to get a relation between angular acceleration and tangential acceleration a_t of a point a distance r from the axis of rotation:

$$a_t = \frac{dv}{dt} = r\frac{d\omega}{dt} = r\alpha. \tag{12–6}$$

This equation gives only the tangential acceleration—the component tangent to the circular path of a rotating point. There is also a radial acceleration a_r given by

$$a_r = \frac{v^2}{r} = \omega^2 r, \tag{12–7}$$

where we have used Equation 12–4 to express v in terms of ω.

Because angular speed and acceleration are defined in a way that is mathematically analogous to linear speed and acceleration, all the relations among linear position, speed, and acceleration automatically apply among angular position, angular speed, and angular acceleration. For example, angular acceleration is the second time derivative of angular position; angular speed is the integral of angular acceleration. And if angular acceleration is constant, then all our constant-acceleration formulas of Chapter 2 apply when we make the substitutions θ for x, ω for v, and α for a. Table 12–1 summarizes this direct analogy between linear and rotational quantities. With the analogies of Table 12–1, problems involving rotational motion are just like the one-dimensional linear problems you solved in Chapter 2.

Example 12–2

Constant Angular Acceleration

When the record of Example 12–1 finishes, the turntable shuts off and coasts to a halt in 17 s. If the angular acceleration is constant, how many revolutions does the turntable make during this time?

TABLE 12–1 ▬▬▬

Angular and Linear Position, Speed, and Acceleration

Linear Quantity or Equation	**Angular Quantity or Equation**
Position x	Angular position θ
Speed $v = \dfrac{dx}{dt}$	Angular speed $\omega = \dfrac{d\theta}{dt}$
Acceleration $a = \dfrac{dv}{dt} = \dfrac{d^2x}{dt^2}$	Angular acceleration $\alpha = \dfrac{d\omega}{dt} = \dfrac{d^2\theta}{dt^2}$

Equations for Constant Acceleration

$\bar{v} = \frac{1}{2}(v_0 + v)$	(2–12)	$\bar{\omega} = \frac{1}{2}(\omega_0 + \omega)$	(12–8)
$v = v_0 + at$	(2–10)	$\omega = \omega_0 + \alpha t$	(12–9)
$x = x_0 + v_0 t + \frac{1}{2}at^2$	(2–11)	$\theta = \theta_0 + \omega_0 t + \frac{1}{2}\alpha t^2$	(12–10)
$v^2 = v_0^2 + 2a(x - x_0)$	(2–14)	$\omega^2 = \omega_0^2 + 2\alpha(\theta - \theta_0)$	(12–11)

Solution

With constant angular acceleration, the average angular speed during the 17-s stopping time is just the average of the initial $33\frac{1}{3}$ rpm and the final speed of 0 rpm. Then the number of revolutions travelled in this time is

$$\theta = \bar{\omega}t = \frac{1}{2}(\omega_0 + \omega)t = \frac{1}{2}(33\frac{1}{3} \text{ rev/min} + 0)\left(\frac{17 \text{ s}}{60 \text{ s/min}}\right) = 4.7 \text{ revolutions.}$$

Alternatively, we could have solved for the angular acceleration using $\alpha = \Delta\omega/\Delta t$. Then Equation 12–10 would give the angular displacement $\theta - \theta_0$ during the stopping time. Or we could have used Equation 12–11, setting the final angular speed ω to zero and solving for $\theta - \theta_0$.

Example 12–3 ───────────────────────────────

Constant Angular Acceleration

A rotating wheel 80 cm in diameter is decelerating at 0.21 rad/s². What should be the initial angular speed if the wheel is to stop after exactly one revolution? What is the tangential linear deceleration of a point on the wheel's rim?

Solution

Can you see that this problem is exactly like one in which you are asked to find the initial vertical speed needed to throw a ball to a given height? We apply Equation 12–11, setting the final angular speed ω to zero and taking $\theta_0 = 0$ for convenience:

$$0 = \omega_0^2 + 2\alpha\theta.$$

Setting θ to 2π because we want exactly one revolution, and noting that α is negative because it is a deceleration, we solve for ω_0 to get

$$\omega_0 = \sqrt{-2\alpha\theta} = [(-2)(-0.21 \text{ rad/s}^2)(2\pi)]^{1/2} = 1.6 \text{ rad/s.}$$

The tangential acceleration of a point on the rim is then given by Equation 12–6:

$$a = r\alpha = (80 \text{ cm})(-0.21 \text{ rad/s}^2) = -17 \text{ cm/s}^2,$$

where the minus sign tells us that the tangential acceleration is directed oppositely to the linear velocity of the rim.

Fig. 12–4
The direction of the angular velocity
vector is given by the right-hand rule.

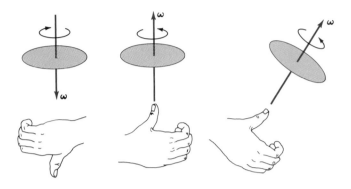

Angular Velocity and Acceleration Vectors

angular velocity

So far, we have treated angular quantities as scalars, with magnitude but not direction. But rotation of a wheel about a horizontal axis, for example, is not quite the same motion as rotation about a vertical axis. To distinguish these cases from each other and from all other possible orientations of the rotation axis, we define **angular velocity, ω,** as a vector whose magnitude is the angular speed ω and whose direction is parallel to the rotation axis. Specifically, the direction of the angular velocity is given by the right-hand rule: curl the fingers of your right hand so as to follow the rotation, and your right thumb points in the direction of the angular velocity (Fig. 12–4). Note that this definition of angular velocity is like our definition of torque, in that the vector direction is perpendicular to the plane of rotation.

By analogy with the linear acceleration vector, we define angular acceleration as the rate of change of the angular velocity vector:

$$\boldsymbol{\alpha} = \frac{d\boldsymbol{\omega}}{dt}.$$ (12–12)

This definition says that the angular acceleration vector points in the direction of the *change* in the angular velocity. If the angular velocity changes only in magnitude but not in direction, then **ω** simply grows or shrinks, so that **α** lies parallel or antiparallel to the rotation axis (Fig. 12–5). In this case, the magnitude of the angular acceleration is just the rate of change of

Fig. 12–5
(a) An increase in angular speed alone corresponds to a change Δω in the angular velocity that is parallel to the angular velocity vector. Therefore, the angular acceleration vector α is also parallel to angular velocity. *(b)* With a decrease in ω, Δω and α are antiparallel to **ω.**

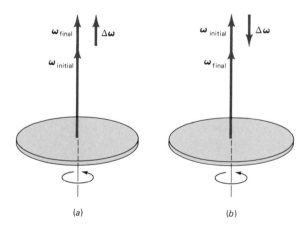

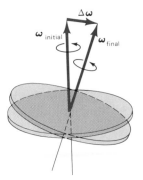

Fig. 12–6
When the angular velocity ω changes only in direction, the change Δω and therefore the angular acceleration vector α are perpendicular to the angular velocity.

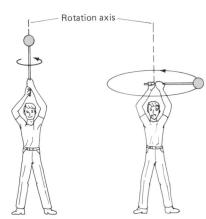

Fig. 12–7
A body's resistance to changes in rotational motion depends on the body itself and on the axis about which rotational motion is changed.

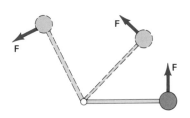

Fig. 12–8
A force applied perpendicular to the rod results in angular acceleration. (Here we assume no other external forces, such as gravity, are acting.)

angular speed, as in our scalar definition 12–5. But as with linear velocity, a change in the *direction* of angular velocity is also associated with acceleration. When ω changes only in direction, the angular acceleration vector is perpendicular to ω (Fig. 12–6). In the most general case, both the direction and magnitude of the angular velocity may change, and then α is neither parallel nor perpendicular to ω. These situations are exactly analogous to the linear case, where an acceleration that is parallel or antiparallel to velocity produces only a change in speed, while an acceleration that is perpendicular to velocity produces only a change in direction.

12.2
ROTATIONAL INERTIA AND THE ANALOG OF NEWTON'S LAW

Newton's second law, $\mathbf{F} = m\mathbf{a}$, proved very powerful in our study of linear motion. Ultimately, of course, Newton's law itself governs all motion, including rotation, but its application to every particle in a rotating object would be terribly cumbersome. Can we instead formulate an analogous law that deals directly with rotational quantities?

To develop such a law, we need rotational analogs of the three quantities $\mathbf{F}$, $\mathbf{a}$, and m. In torque and angular acceleration, we already have the analogs of force and linear acceleration. But what rotational quantity might be analogous to mass?

The mass m in Newton's law is a measure of a body's inertia—its resistance to changes in motion. We want a quantity that, analogously, describes a body's resistance to changes in rotational motion. Consider an object consisting of a small ball of mass m at the end of a rigid, massless rod of length R. This composite object is easy to start rotating about an axis that coincides with the rod, but much harder to start rotating about a perpendicular axis through the end of the rod (Fig. 12–7). Why is this? In the first case, no point is very far from the rotation axis, so that the linear speeds of all parts of the object remain small. But in the second case, the mass is far from the rotation axis, and to achieve a given change in angular speed requires a large change in linear speed. The rotational analog of mass must therefore depend both on the mass of the rotating object and on how far that mass is from the axis of rotation.

Suppose that we start our mass-rod object rotating by attaching one end to a frictionless pivot and applying a force $\mathbf{F}$ that is always at right angles to the rod (Fig. 12–8). That force results in an angular acceleration of the object about an axis perpendicular to the page. Starting from rest, it begins to rotate counterclockwise at an ever-increasing rate. Applying the right-hand rule, we see that the angular velocity vector ω is out of the page. Since ω is increasing in length, the angular acceleration vector α is also out of the page. We now develop a simple description of this situation in terms of angular quantities alone. Writing Newton's second law*, and using

---

*Actually, we are writing only the tangential component of Newton's law. There must also be a radial force acting at right angles to the applied force **F**. This force is provided automatically by the tension in the rod, and will not concern us further. Because it lies along a line from the mass to the rotation axis, the radial force contributes nothing to the torque.

Equation 12–6 to express the tangential acceleration a_t in terms of α and the distance R from the axis of rotation, we have

$$F = ma_t = mR\alpha. \tag{12–13}$$

Here we have assumed the ball small enough that all parts of it are essentially the same distance R from the pivot point. The force $\mathbf{F}$ produces a torque about the pivot point. From Equation 11–2, this torque has magnitude

$$\tau = RF,$$

and, by the right-hand rule, it points out of the page. Solving for F and using the result in Equation 12–13, we have

$$\frac{\tau}{R} = mR\alpha,$$

or

$$\tau = (mR^2)\alpha.$$

Here we have Newton's second law, $F = ma$, written in terms of rotational quantities. The torque—analogous to force—is the product of the angular acceleration with the quantity mR^2, which must therefore be the rotational analog of mass. We call this quantity mR^2 the **rotational inertia** or **moment of inertia,** and represent it by the symbol I. Rotational inertia has the units $\text{kg}\cdot\text{m}^2$, and accounts both for mass and for the location of mass relative to the axis of rotation. We can therefore write, in analogy with Newton's second law,

rotational inertia

moment of inertia

$$\boxed{\tau = I\alpha,} \tag{12–14}$$

where the fact that torque and angular acceleration are in the same direction allows us to write this as a vector equation.

Although we derived Equation 12–14 for a single mass point, we can apply it to extended objects if we interpret τ as the net torque on the object and I as the sum of the rotational inertias of all masses making up the object. The first subsection below shows how to calculate the rotational inertia for extended objects. The second subsection then explores the meaning of the net torque on such an object, and illustrates the practical application of Equation 12–14 to the dynamics of rotating objects.

Calculating the Rotational Inertia

When an object consists of a number of discrete mass points, its rotational inertia about an axis is just the sum of the rotational inertias of the individual mass points about that axis:

$$\boxed{I = \sum m_i r_i^2.} \tag{12–15}$$

Here m_i is the mass of the ith mass point, and r_i its distance from the rotation axis.

Example 12–4

Rotational Inertia

A dumbbell-shaped object consists of two equal masses $m = 0.64$ kg on the end of a massless rod of length $\ell = 85$ cm. Calculate the rotational inertia of this object about an axis one-fourth of the way from one end of the rod and perpendicular to it. If the object is accelerated from rest by a torque of 1.2 N·m pointing along the same axis, how many revolutions will it turn in 2.5 s?

Solution

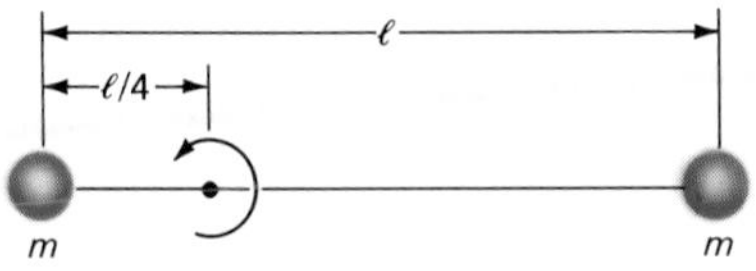

The situation is shown in Fig. 12–9. Summing the individual rotational inertias, we have

$$I = \sum m_i r_i^2 = m\left(\frac{1}{4}\ell\right)^2 + m\left(\frac{3}{4}\ell\right)^2 = \tfrac{5}{8}m\ell^2$$

$$= \tfrac{5}{8}(0.64 \text{ kg})(0.85 \text{ m})^2 = 0.29 \text{ kg·m}^2.$$

From Equation 12–14, the angular acceleration is

$$\alpha = \frac{\tau}{I}.$$

Using this result in Equation 12–10 then gives the angular displacement:

$$\theta = \frac{1}{2}\alpha t^2 = \frac{1}{2}\left(\frac{\tau}{I}\right)t^2 = \frac{1}{2}\left(\frac{1.2 \text{ N·m}}{0.29 \text{ kg·m}^2}\right)(2.5 \text{ s})^2 = 13 \text{ rad} = 2.1 \text{ rev}.$$

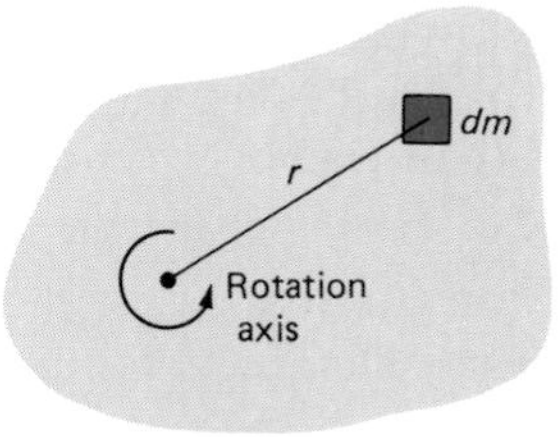

Fig. 12–9
Example 12–4

When an object consists of a continuous distribution of matter rather than a number of point masses, then we consider a large number of very small mass elements dm throughout the object, and sum the individual rotational inertias $r^2 dm$ over the entire object (Fig. 12–10). In the limit of an arbitrarily large number of very small mass elements, the sum becomes an integral, and we have

$$I = \int r^2 dm, \qquad (12\text{–}16)$$

where the limits of integration are chosen to include the entire object.

Fig. 12–10
To find the rotational inertia of a solid object, we imagine it divided into small elements dm, then integrate over the rotational inertias $r^2 dm$ of the individual mass elements.

Example 12–5

Rotational Inertia of a Rod

Find the rotational inertia of a uniform, narrow rod of mass M and length ℓ about an axis through its center and perpendicular to the rod.

Solution

Let the rod coincide with the x-axis, with origin at the center of the rod, and consider mass elements of mass dm and length dx, as shown in Fig. 12–11. The distance x from the center of the rod to the mass element dm plays the role of r in Equation 12–16, so that

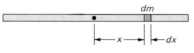

Fig. 12–11
A uniform rod, showing mass element of mass dm and length dx.

$$I = \int_{x=-\ell/2}^{x=\ell/2} x^2 \, dm, \qquad (12\text{–}17)$$

where we have chosen the limits to include the entire rod. To do the integration, we must relate x and dm so that we have a single variable under the integral sign. The length dx of the mass element is some tiny fraction of the total length ℓ. Since the rod is uniform, the mass dm is the *same* fraction of the total mass M. Therefore

$$\frac{dm}{M} = \frac{dx}{\ell}.$$

Solving for dm and using the result in Equation 12–17, we have

$$I = \int_{-\ell/2}^{\ell/2} \frac{M}{\ell} x^2 \, dx = \frac{M}{\ell} \int_{-\ell/2}^{\ell/2} x^2 \, dx$$

$$= \frac{M}{\ell} \frac{x^3}{3} \Big|_{-\ell/2}^{\ell/2} = \frac{1}{12} M\ell^2. \quad \text{(rod about bisector)}$$

(12–18)

Does this make sense? Yes: much of the rod's mass is near the axis of rotation, and therefore contributes little to the rotational inertia. The result is a much lower rotational inertia than if all the mass were at the ends of the rod.

Example 12–6
Rotational Inertia of a Ring

Find the rotational inertia of a thin ring of radius R and mass M about the ring's axis (Fig. 12–12).

Solution

We divide the ring into mass elements dm, one of which is shown in Fig. 12–12. All the mass elements in the ring are the same distance R from the rotational axis, so that r in Equation 12–16 is the constant R, and the equation becomes

$$I = \int R^2 \, dm = R^2 \int dm,$$

where the integration is over the ring. But the sum of mass elements over the ring is just the total mass, M, so that

$$I = MR^2. \quad \text{(thin ring)}$$

(12–19)

The rotational inertia of the ring is the same as if all the mass were concentrated in one place a distance R from the rotation axis; the angular distribution of the mass about the axis does not matter. Notice, too, that it does not matter whether the ring is narrow like a loop of wire or long like a section of hollow pipe, as long as it is thin enough that all of it is essentially equidistant from the rotation axis (Fig. 12–13).

Example 12–7
Rotational Inertia of a Disk

A disk of radius R and mass M has uniform density. Find the rotational inertia of the disk about an axis through its center and perpendicular to the disk.

Solution

Not all parts of the disk are the same distance from the axis, so that r is not constant in Equation 12–16. Since we already know the rotational inertia of a ring, we can divide the disk into mass elements that are themselves rings, as shown in Fig. 12–14. A given mass element is a ring of radius r, width dr, and mass dm. Its rotational inertia is $r^2 dm$, so that the total rotational inertia of the disk is

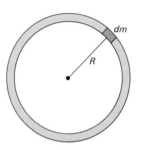

Fig. 12–12
A thin ring, showing one mass element dm.

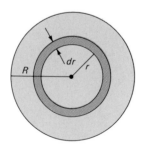

Fig. 12–13
The rotational inertia is MR^2 for any thin ring, whether it is narrow like a wire loop (left) or long like a section of pipe (right).

Fig. 12–14
A disk may be divided into ring-shaped mass elements of mass dm, radius r, and width dr.

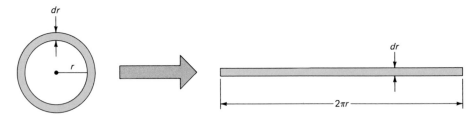

Fig. 12–15
The mass element *dm* may be "unwound" into a thin rectangle of length $2\pi r$ and width *dr*.

$$I = \int_{r=0}^{r=R} r^2 \, dm, \qquad (12\text{--}20)$$

where the limits are chosen to include all the ring-shaped mass elements in the entire disk. To evaluate this integral, we must relate the variables m and r. The mass element dm is a tiny fraction of the total mass M—but it is the *same* fraction of the total mass as its area is of the total disk area, πR^2. Since the ring dm is very thin, we can imagine "unwinding" it to get a thin rectangle of length $2\pi r$ and width dr (Fig. 12–15). Then the area of the ring is $2\pi r \, dr$, and we have

$$\frac{dm}{M} = \frac{2\pi r \, dr}{\pi R^2}.$$

Solving for dm and using the result in Equation 12–20 leaves only the variable r under the integral, allowing us to do the integration:

$$I = \int_{r=0}^{r=R} r^2 \, dm = \int_0^R r^2 \frac{2M}{R^2} r \, dr = \frac{2M}{R^2} \int_0^R r^3 \, dr$$

$$= \frac{2M}{R^2} \frac{r^4}{4} \bigg|_0^R = \tfrac{1}{2} M R^2. \qquad \text{(disk)} \qquad (12\text{--}21)$$

Does this make sense? In the disk, mass is distributed from the axis out to the edge. Mass nearer the axis contributes less to the rotational inertia, so that we expect a lower rotational inertia for the disk than for a ring of the same mass and radius.

The rotational inertias of other shapes about various axes may be found by integration as in these three examples. Table 12–2 lists results for some common shapes. Note that more than one rotational inertia is listed for some shapes, since the rotational inertia depends on the rotation axis.

If we know the rotational inertia, I_{cm}, about an axis through the center of mass of a body, a useful relation called the **parallel axis theorem** allows us to calculate the rotational inertia, I, through any parallel axis. The parallel axis theorem states that

parallel axis theorem

$$I = I_{cm} + Mh^2, \qquad (12\text{--}22)$$

where h is the distance from the center of mass axis to the parallel axis and M the total mass of the object. Problem 14 explores the proof of this theorem.

TABLE 12–2
Rotational Inertias

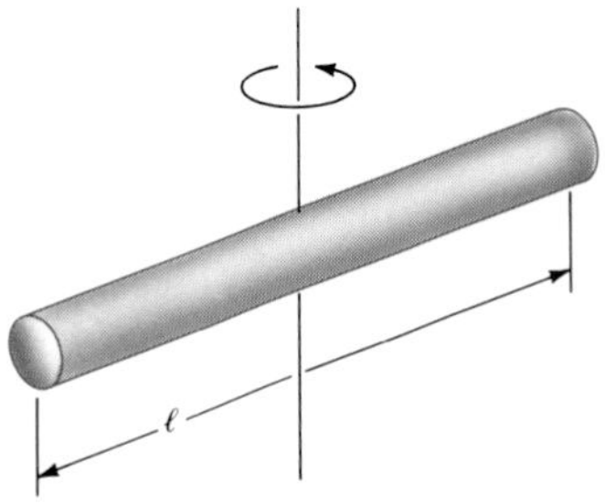

Thin rod about center
$I = \frac{1}{12}M\ell^2$

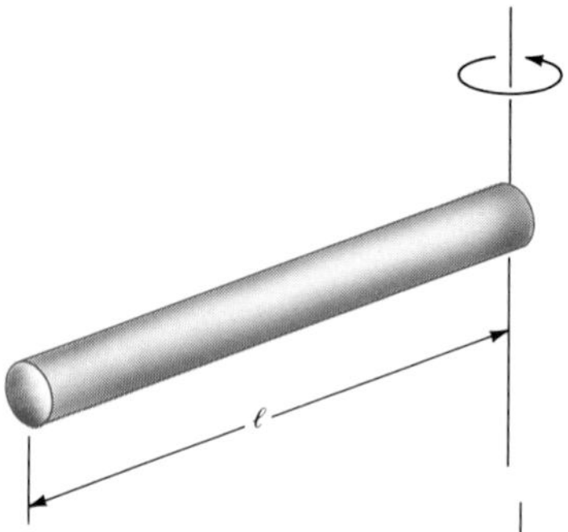

Thin rod about end
$I = \frac{1}{3}M\ell^2$

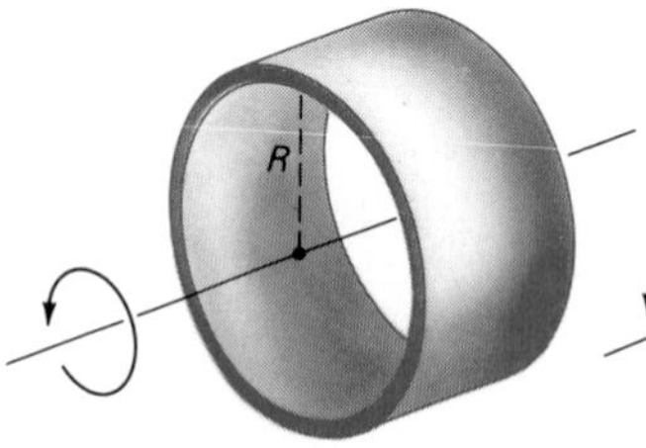

Thin ring or hollow cylinder
about its axis
$I = MR^2$

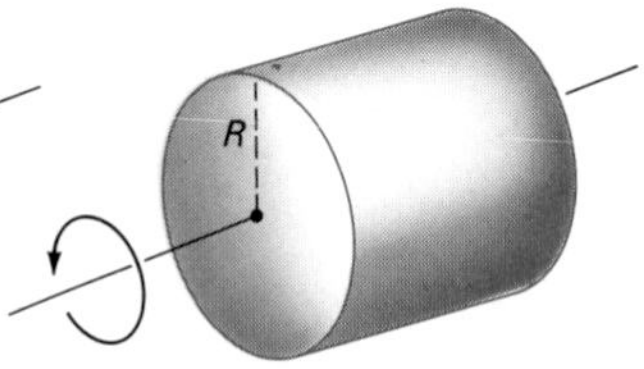

Disk or solid cylinder
about its axis
$I = \frac{1}{2}MR^2$

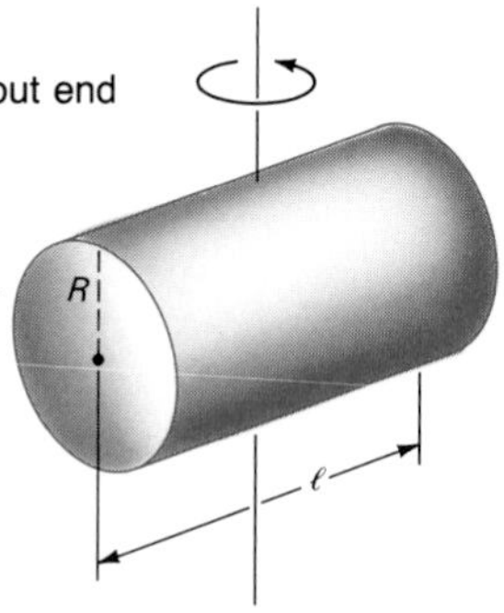

Solid cylinder about
perpendicular bisector
$I = \frac{1}{4}MR^2 + \frac{1}{12}M\ell^2$

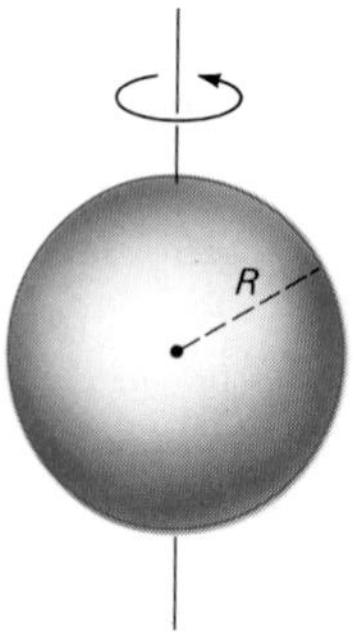

Solid sphere about diameter
$I = \frac{2}{5}MR^2$

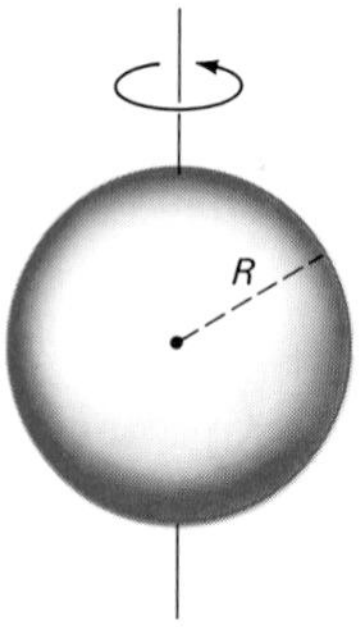

Hollow spherical shell about
diameter
$I = \frac{2}{3}MR^2$

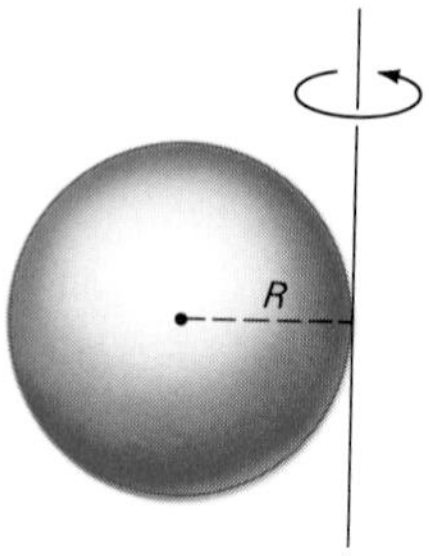

Sphere about tangent line
$I = \frac{7}{5}MR^2$

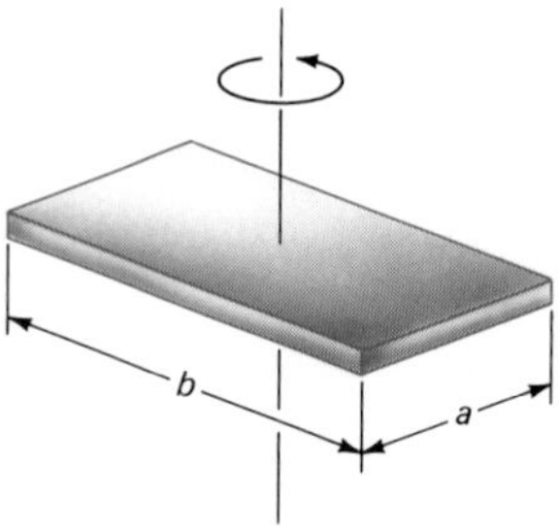

Flat plate about perpendicular
axis
$I = \frac{1}{12}M(a^2 + b^2)$

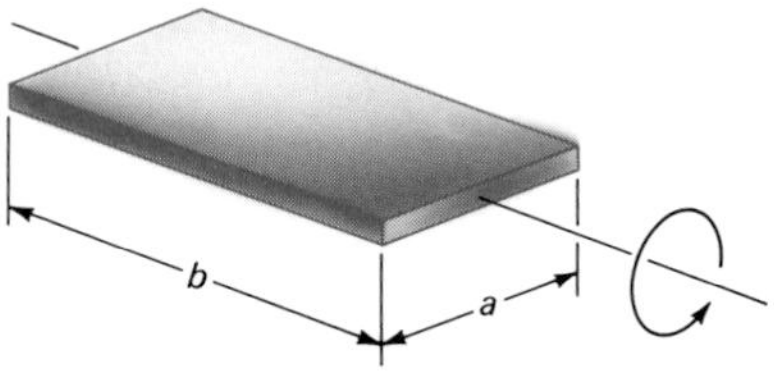

Flat plate about central axis
$I = \frac{1}{12}Ma^2$

Example 12-8

The Parallel Axis Theorem

Find the rotational inertia of a uniform, narrow rod of length ℓ and mass M about an axis through one end and perpendicular to the rod.

Solution

In Example 12–5, we found the rotational inertia through the center of mass of the rod to be $I_{cm} = M\ell^2/12$. Our new rotation axis is a distance $\ell/2$ from the axis through the center of mass, so that the parallel axis theorem becomes

$$I = I_{cm} + Mh^2 = \tfrac{1}{12}M\ell^2 + M\left(\frac{\ell}{2}\right)^2 = \tfrac{1}{3}M\ell^2.$$

Use of the parallel axis theorem can eliminate a great deal of integration!

Rotational Dynamics

Knowing the rotational inertia of a body, we can use the rotational analog of Newton's second law (Equation 12–14) to determine the behavior of the body just as we used $\mathbf{F} = m\mathbf{a}$ itself to determine the dynamics of linear motion. The torque in Equation 12–14 is the *net* torque on the object in question. As with forces in the linear case, we must be careful to isolate the object and determine *all* torques that act on it, then take their vector sum to get the net torque.

When we studied the linear motion of a composite system, we found that only external forces matter; internal forces cancel in pairs according to Newton's third law. Can we similarly disregard torques due to internal forces among the various parts of an object? The answer is yes, if we make one further assumption about those internal forces—that they act along a line joining the two mass elements in question (Fig. 12–16). Then the lever arm r sinθ for each torque is the same, as shown in Fig. 12–16. By Newton's third law, the forces are equal but opposite; since the lever arms r sinθ are the same, the torques are also equal and opposite. Therefore, internal torques cancel in pairs, and the torque in Equation 12–14 is the net *external* torque.

We now apply Equation 12–14 in two examples. These examples are analogous to one-dimensional linear motion, in that torque and angular

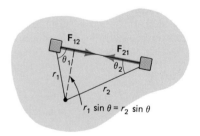

Fig. 12–16
When internal forces act along the line joining two mass elements, then the associated torques cancel in pairs.

velocity are in the same direction. Later, in Section 12.4, we consider the case in which a torque acts to change the direction of angular velocity.

Example 12–9

De-spinning a Satellite

A cylindrical satellite is 1.4 m in diameter, with its mass of 940 kg distributed approximately uniformly throughout its volume. The satellite is spinning at 10 rpm about its axis, but must be stopped so that a space shuttle crew can make repairs. Two small gas jets, each with a thrust of 20 N, are mounted on opposite sides of the satellite and are aimed tangent to the satellite's surface (Fig. 12–17). How long must the jets be fired in order to stop the satellite's rotation?

Solution

The satellite's angular speed must change by $\Delta\omega = 10$ rpm. At constant angular acceleration α, this takes time

$$\Delta t = \frac{\Delta\omega}{\alpha} = \frac{\Delta\omega I}{\tau}, \tag{12–23}$$

where we have used Equation 12–14 to write $\alpha = \tau/I$. From Table 12–2 the rotational inertia of the cylindrical satellite is $I = \frac{1}{2}MR^2$. The torque exerted by the two jets, each a distance R from the rotational axis and directed perpendicular to the radius, is $\tau = 2RF$, with F the thrust of one jet. Then Equation 12–23 becomes

$$\Delta t = \frac{(\Delta\omega)(\frac{1}{2}MR^2)}{2RF} = \frac{\Delta\omega MR}{4F} = \frac{(10 \text{ rpm})(2\pi \text{ rad/rev})}{60 \text{ s/min}} \frac{(940 \text{ kg})(0.70 \text{ m})}{(4)(20 \text{ N})} = 8.6 \text{ s},$$

where we have converted the change in angular speed from rpm to rad/s.

Often, rotational and linear motion are coupled in a single problem, as in Example 12–10.

Example 12–10

Rotational and Linear Dynamics

A solid cylinder of mass M and radius R is mounted on a frictionless horizontal axle over a well, as shown in Fig. 12–18. A rope of negligible mass is wrapped around the cylinder, and a bucket of mass m is suspended from the rope. Find an expression for the acceleration of the bucket as it falls down the empty well shaft.

Solution

Were it not connected to the cylinder, the bucket would, of course, accelerate downward with linear acceleration g. But now the rope exerts an upward tension force T on the bucket, reducing the net downward force and at the same time exerting a torque on the cylinder. The downward force on the bucket is $F = mg - T$ (Fig. 12–19), so that Newton's second law gives

$$ma = mg - T, \tag{12–24}$$

where a is the linear acceleration of the bucket. Looking at the cylinder end-on (Fig. 12–20), we see that the rope exerts a torque $\tau = RT$ on the cylinder, giving it an angular acceleration

$$\alpha = \frac{\tau}{I} = \frac{RT}{I}, \tag{12–25}$$

where I is the rotational inertia.

Fig. 12–17
Satellite with gas jets.

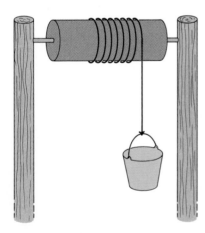

Fig. 12–18
Example 12–10

Since **L** and **ω** are in the same direction* we can write this equation vectorially:

$$\mathbf{L} = I\boldsymbol{\omega},$$

showing that Equations 12–33 and 12–34 are indeed equivalent for a rigid object.

For a rotating object, it is usually easier to use Equation 12–34 to calculate angular momentum. But the concept of angular momentum applies, more generally, even to systems of particles that do not share a common circular motion. Equation 12–33 remains the fundamental definition, and may be used to calculate angular momentum in all cases.

Example 12–16

Angular Momentum of a Particle in Uniform Motion

An electron of mass m moves in a straight line at constant speed v past a neutron, with b its distance of closest approach (Fig. 12–28). Determine the angular momentum of the electron about the neutron, and show that it remains constant.

Solution

From Equation 12–33, the magnitude of **L** is

$$L = rp \sin\theta = mvr \sin\theta.$$

But from Fig. 12–28, we see that $r \sin\theta = b$, so that

$$L = mvb,$$

which is independent of the electron's position on its straight-line trajectory. What about the direction of the angular momentum vector? From the right-hand rule, we see that $\mathbf{r} \times \mathbf{v}$ is everywhere into the page, so that the angular momentum is constant in magnitude and direction.

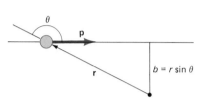

Fig. 12–28
The angular momentum of an object moving uniformly in a straight line is constant.

Torque and Angular Momentum

In Section 12.2, we considered the rotational analog of Newton's law $\mathbf{F} = m\mathbf{a}$. Newton's law in this form holds when the mass m of a system is constant; analogously, our equation $\boldsymbol{\tau} = I\boldsymbol{\alpha}$ holds when the rotational inertia I is constant. More generally, we saw in Chapter 9 that Newton's law applied to a system of particles has the form

$$\mathbf{F} = \frac{d\mathbf{P}}{dt},$$

where **F** is the sum of the external forces acting on the system, and **P** is the total momentum of the system. Can we write, by analogy,

$$\boldsymbol{\tau} = \frac{d\mathbf{L}}{dt}?$$

To see that we can, we write the angular momentum of a system as the sum of the angular momenta of its constituent particles:

---

*Actually, this is true only when the rotation axis is stationary. In more complicated situations involving asymmetric objects, the angular momentum and angular velocity may be in different directions. In that case the rotational inertia cannot be described by a single number but instead is described by a more complicated mathematical entity called a tensor.

$$\mathbf{L} = \sum \mathbf{L}_i = \sum \mathbf{r}_i \times \mathbf{p}_i,$$

where the subscript i refers to the ith particle of the system. Differentiating this equation with respect to time gives

$$\frac{d\mathbf{L}}{dt} = \sum \left(\mathbf{r}_i \times \frac{d\mathbf{p}_i}{dt} + \frac{d\mathbf{r}_i}{dt} \times \mathbf{p}_i \right),$$

where we have applied the product rule for differentiation, being careful to preserve the order of the cross product since it is not commutative. But $d\mathbf{r}_i/dt$ is the velocity of the ith particle, so that the second term in the sum is the cross product of velocity $\mathbf{v}$ and momentum $\mathbf{p} = m\mathbf{v}$. Since these two vectors are parallel, their cross product is zero, and we are left with only the first term in the sum:

$$\frac{d\mathbf{L}}{dt} = \sum \mathbf{r}_i \times \frac{d\mathbf{p}_i}{dt} = \sum \mathbf{r}_i \times \mathbf{F}_i,$$

where we have used Newton's law to write $d\mathbf{p}_i/dt = \mathbf{F}_i$. But $\mathbf{r}_i \times \mathbf{F}_i$ is the torque, $\boldsymbol{\tau}_i$, on the ith particle, so that

$$\frac{d\mathbf{L}}{dt} = \sum \boldsymbol{\tau}_i.$$

As we discussed in Section 12.2, the internal torques cancel in pairs as long as the internal forces in a system of particles act along the lines between the particles. Under these conditions, the sum of all the torques on all the particles reduces to the net *external* torque on the system, and we have

$$\frac{d\mathbf{L}}{dt} = \boldsymbol{\tau}, \qquad (12\text{–}35)$$

where $\boldsymbol{\tau}$ is the net external torque. Thus our analogy between linear and rotational motion holds for momentum as well as for the other quantities we have discussed.

Conservation of Angular Momentum

When there is no external torque on a system, Equation 12–35 tells us that angular momentum is constant. This statement—that the angular momentum of an isolated system cannot change—is of fundamental importance in physics, and applies to systems ranging from subatomic particles to galaxies (Fig. 12–29). In deriving Equation 12–35, we did not require that the system in question be a rigid object, so that conservation of angular momentum applies even to systems that undergo changes in configuration and therefore in rotational inertia. The classic example of such a change is a figure skater, who starts spinning relatively slowly with arms extended, then pulls her arms in to spin much more rapidly (Fig. 12–30). Why does this happen? As the skater's arms move in, her mass is concentrated more toward the rotation axis, lowering her rotational inertia, I. But her angular momentum $I\omega$ is conserved, so that her angular speed ω must increase. A much more dramatic example is illustrated in Example 12–17: the collapse

Fig. 12–29
The Andromeda galaxy, nearest large galaxy to our own Milky Way. The galaxy is a complex system of 100 billion stars and clouds of gas and dust, undergoing slow rotation. But the galaxy is essentially isolated, so that no matter how complex the interactions among its constituent parts, its total angular momentum remains constant.

Fig. 12–30
As the skater pulls in her arms, her rotational inertia I decreases. To conserve her angular momentum $I\omega$, her angular speed ω must increase.

of a star at the end of its lifetime. Again, the rotational inertia decreases, requiring an increase in the star's rotational speed.

Example 12–17
Pulsars and the Conservation of Angular Momentum

A certain star rotates about its axis once every 45 days. At the end of its life, when the star has exhausted its nuclear fuel, it undergoes a colossal outburst called a supernova explosion. During this event, much of the star's mass is flung outward into the interstellar medium, where it will eventually condense into new stars and planets and perhaps find its way into living creatures (Fig. 12–31). But the inner core of the star, whose initial radius was 2×10^7 m, collapses into an object only 6 km in radius. This **neutron star** is so dense that one teaspoonful of its material has a mass of a billion tons! Calculate the rotation rate of the neutron star, assuming that no torque acts on the stellar core during the collapse. Consider the core to be a uniform sphere both before and after collapse.

neutron star

Solution

This situation is exactly like that of the figure skater, in that a decrease in rotational inertia requires an increase in angular speed. Conservation of angular momentum states that

$$I_1\omega_1 = I_2\omega_2,$$

where the subscripts 1 and 2 refer to conditions before and after the collapse, respectively. For a uniform sphere, $I = \tfrac{2}{5}MR^2$, so that

$$\tfrac{2}{5}MR_1^2\omega_1 = \tfrac{2}{5}MR_2^2\omega_2.$$

Solving for ω_2 and using the known values of ω_1, R_1, and R_2 gives

$$\omega_2 = \omega_1\left(\frac{R_1}{R_2}\right)^2 = \left(\frac{1\ \text{rev}}{45\ \text{day}}\right)\left(\frac{2\times10^7\ \text{m}}{6\times10^3\ \text{m}}\right)^2 = 2.5\times10^5\ \text{rev/day} = 3\ \text{rev/s}.$$

This neutron star is a fantastic thing—an object with more mass than the entire sun, rotating three times a second. Often intense radio waves, light, or x-rays radiate in a narrow beam from the magnetic axis of the neutron star. As the neutron star

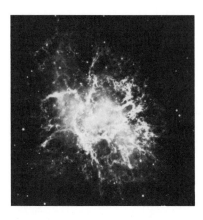

Fig. 12–31
The Crab Nebula is the remains of a supernova explosion as it looks 1000 years after the explosion.

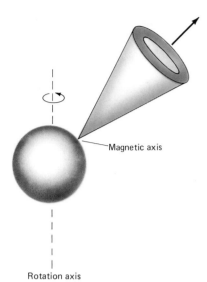

Magnetic axis

Rotation axis

Fig. 12–32
A rotating neutron star emits a beam of radiation, much like a searchlight.

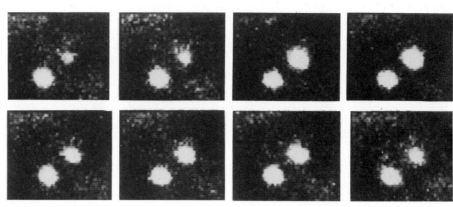

Fig. 12–33
This series of optical images shows the pulsar in the Crab Nebula at different times in its 33-ms cycle. The object on the left is a normal star and remains at constant brightness. The object on the right is the pulsar; its intensity is lowest in the first frame.

rotates, this beam of radiation sweeps through space like a searchlight beacon (Fig. 12–32). As this beam repeatedly sweeps past earth, we detect a series of regular pulses of radiation that give the rotating neutron star the name "pulsar" (Fig. 12–33). Over 300 pulsars have been identified, with rotation periods ranging from 1.6 ms to 4 s. Pulsar rotation rates actually decrease slowly, as magnetic forces exert slight torques on the neutron stars.

Example 12–18

Conservation of Angular Momentum

A merry-go-round of radius $R = 1.3$ m has a rotational inertia $I = 240$ kg·m², and is rotating freely at $\omega_1 = 11$ rpm. A child of mass $m_1 = 28$ kg leaps onto the edge of the merry-go-round, heading directly toward the center (Fig. 12–34). What is the new angular speed of the merry-go-round? Later a second child, of mass $m_2 = 32$ kg, leaps onto the merry-go-round by running tangentially at speed $v = 3.7$ m/s. If the direction of the child's motion is the same as that of the merry-go-round (see Fig. 12–34), what is the new angular speed?

Solution

Moving directly toward the rotation axis, the first child carries no angular momentum relative to the axis, so that the total angular momentum is $I\omega_1$. But sitting at the edge of the merry-go-round, the child adds rotational inertia m_1R^2 to the system. Then conservation of angular momentum may be written

$$I\omega_1 = (I + m_1R^2)\omega_2,$$

where ω_2 is the angular speed after the child is on the merry-go-round. Solving for ω_2 gives

$$\omega_2 = \frac{I\omega_1}{I + m_1R^2} = \frac{(240 \text{ kg·m}^2)(11 \text{ rpm})}{240 \text{ kg·m}^2 + (28 \text{ kg})(1.3 \text{ m})^2} = 9.2 \text{ rpm}.$$

Notice that in this calculation, where we dealt with a ratio of angular speeds, we did not need to convert from rpm to rad/s.

The second child, running tangentially to the merry-go-round, carries angular momentum m_2vR, in the same direction as the merry-go-round's angular momen-

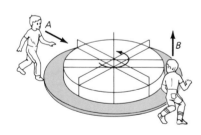

Fig. 12–34
Running toward the axis, child *A* brings no angular momentum relative to the axis of the merry-go-round. Running tangentially, child *B* brings angular momentum *mvR*.

tum (see Example 12–16), so that the total angular momentum of the system is now $I\omega_1 + m_2vR$. After the second child leaps on, the total rotational inertia includes that of the merry-go-round along with a contribution mR^2 from each child. Calling the final angular speed ω_3, conservation of angular momentum may then be written

$$I\omega_1 + m_2vR = (I + m_1R^2 + m_2R^2)\omega_3,$$

so that

$$\omega_3 = \frac{I\omega_1 + m_2vR}{I + (m_1 + m_2)R^2}.$$

With m_2 in kg, v in m/s, and R in m, the angular momentum m_2vR of the second child has the units of kg·m²·rad/s, while all our other angular momenta have been expressed implicitly in kg·m²·rpm. To solve for ω_3 in rpm, we convert the second child's angular momentum into the latter unit:

$$\omega_3 = \frac{(240 \text{ kg·m}^2)(11 \text{ rpm}) + (32 \text{ kg})(3.7 \text{ m/s})(1.3 \text{ m})(1 \text{ rev}/2\pi \text{ rad})(60 \text{ s/min})}{240 \text{ kg·m}^2 + (28 \text{ kg} + 32 \text{ kg})(1.3 \text{ m})^2} = 12 \text{ rpm}.$$

Is mechanical energy conserved as the children leap onto the merry-go-round? No! In each case frictional forces are involved in bringing children and merry-go-round to rest with respect to each other, so that the situation is like an inelastic collision.

Rotational Dynamics in Three Dimensions

Although we developed rotational analogs of Newton's law (Equations 12–14 and 12–35) in vector form, so far we have explored rotational dynamics only in cases where the angular momentum vector and torque are in the same direction. When these two vectors are not collinear, new and often startling behavior results. However strange this behavior may seem, you can always understand it if you really accept the rotational analog of Newton's law—that the rate of change of the angular momentum *vector* is equal to the applied torque *vector*.

A toy gyroscope provides a common example of rotational dynamics in three dimensions. This device seems to defy gravity, balanced on a single point with most of its mass clearly unsupported (Fig. 12–35). What is going on here? The downward force of gravity produces a torque whose direction is into the page for the orientation shown in Fig. 12–35. Were the gyroscope not spinning, this torque would cause it to fall over, gaining angular momentum about its pivot point (Fig. 12–36). The change in angular momentum—here an increase—is in the same direction as the torque, in agreement with the rotational analog of Newton's law. When the gyroscope is spinning, the gravitational torque is still into the page. But now the gyroscope already has angular momentum that points along its rotation axis. The rotational analog of Newton's law states that this angular momentum *vector* must change at a rate given by the torque, and that this change must be in the same direction as the torque—at right angles to the angular momentum vector. As a result the gyroscope moves around so that its rotation axis sweeps out a circle—a phenomenon called **precession.** The torque and angular momentum vectors remain always at right angles (Fig. 12–37), so that the torque changes only the direction but not the magnitude of the angular momentum vector.

Fig. 12–35
A gyroscope.

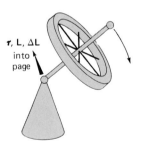

τ, L, ΔL
into
page

Fig. 12–36
Were it not spinning, the gyroscope would fall over, gaining angular momentum about its pivot point. The change in angular momentum points in the same direction as the torque, satisfying the rotational analog of Newton's law.

precession

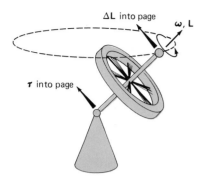

Fig. 12–37
With **L** and τ at right angles, the gyroscope precesses, its rotation axis describing a circle. The *change* in angular momentum is still in the same direction as the torque.

So why doesn't the gyroscope fall over? The reason is analogous to why a satellite doesn't fall to earth. There is a gravitational force on the satellite, but it is providing just the right force to hold the satellite in orbit. Analogously, there is a torque on the gyroscope, but it has all it can do to change the angular momentum at a rate required to maintain precession.

We can analyze the gyroscope's behavior quantitatively through Fig. 12–38, which shows a gyroscope spinning with its axis in a horizontal plane. The downward force of gravity on the unbalanced gyroscope produces a torque that is into the page in Fig. 12–38a—that is, at right angles to the angular momentum vector. In a short time dt, the angular momentum vector changes by an amount $d\mathbf{L}$ given by Equation 12–35:

$$d\mathbf{L} = \boldsymbol{\tau} dt. \tag{12–36}$$

This change is shown in Fig. 12–38b, and represents a rotation of the angular momentum vector through a small angle $d\phi$. For an arbitrarily small time interval dt, and correspondingly small angle $d\phi$, the magnitude dL of the change in angular momentum is essentially equal to the arc length through which the tip of the angular momentum vector swings in Fig. 12–38b. From the definition of angle (Equation 12–3), this arc length is just $Ld\phi$, so that

$$dL = Ld\phi.$$

Combining this result with Equation 12–36 gives

$$\tau dt = Ld\phi. \tag{12–37}$$

The angular speed of precession, called Ω, is the rate at which the angular momentum vector rotates, or $d\phi/dt$. Solving Equation 12–37 for this quantity gives

$$\Omega = \frac{d\phi}{dt} = \frac{\tau}{L}. \tag{12–38}$$

Although we derived Equation 12–38 for the case of a gyroscope whose angular momentum is horizontal,* it holds true for any angle the gyroscope

---

*Actually, **L** in this example is not quite horizontal. The gyroscope also has a small vertical component of angular momentum associated with its precession. But if the angular momentum associated with spin about the axis is large, the precession rate and vertical component of **L** are small.

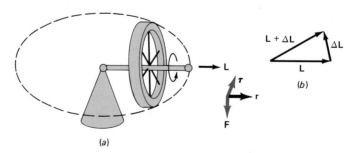

Fig. 12–38
(a) A gyroscope spinning with its axis in the horizontal plane. (b) Top view showing change in the angular momentum vector.

might make with the horizontal (see Problem 39). To use Equation 12–38 it is, of course, necessary to know both the angular momentum L and the torque τ. Problems 40 and 41 explore practical applications of this equation.

The motion of a gyroscope can be more complicated than we have described. The gyroscope's precession represents a component of angular momentum about a vertical axis. But unless it is given an initial impulse in the direction of precession, the gyroscope has initially no vertical component of angular momentum. Since the gravitational torque acts only in the horizontal direction, the vertical component of the gyroscope's angular momentum must remain zero. As a result, a gyroscope released with its angular momentum vector strictly horizontal will drop slightly as it begins to precess, gaining a downward component of angular momentum that compensates for the upward component associated with precession. In the absence of friction, it overshoots the position at which the two would balance, resulting in a slight oscillation that is called **nutation** (Fig. 12–39). If the angular momentum associated with the spin of the gyroscope is large, this nutation is very small. In any event, torques that are associated with friction at the gyroscope support usually damp out the nutation in a short time.

An important example of precession is provided by the earth itself. The earth is slightly bulged at its equator, and the gravitational force of the sun and moon on this bulge results in a slight torque on the tilted, spinning earth. In response, the rotation axis of the earth precesses, taking about 26,000 years to complete a full circle (Fig. 12–40). Although earth's axis points now toward the star Polaris, which we call the North Star, it will not always do so! When astronomers measure star positions in earth-based coordinates, they must take account of precession and, for extreme accuracy, of nutation as well.

Conservation of Angular Momentum in Three Dimensions

When no external torques act on a system, both the direction and magnitude of its angular momentum remain constant. For example, a spinning gyroscope that is supported at both ends maintains its rotation axis in the same direction even as it is moved about the earth (Fig. 12–41). Gyroscopes of this sort are used extensively in navigation systems. Changes in the orientation of the rotation axis relative to a submarine, ship, or aircraft are "remembered" by a computer that can consequently calculate where the craft is relative to its starting point.

Application _____

Pointing a Satellite

One way to steer a satellite (for example, to point at an astronomical object under study) is to apply torques using small gas jets. But eventually these jets run out of gas and the satellite can no longer be controlled. Another approach involves angular momentum conservation. The satellite is equipped with three internal

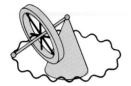

Fig. 12–39
Nutation.

nutation

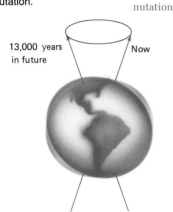

13,000 years in future · Now

Fig. 12–40
Torque due to the sun's gravity on earth's equatorial bulge (shown highly exaggerated) results in precession of the planet's rotation axis with a period of about 26,000 years.

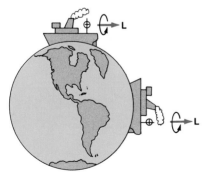

Fig. 12–41
A gyrocompass. When no external torques act, the rotation axis of the shipboard gyroscope always points in the same direction.

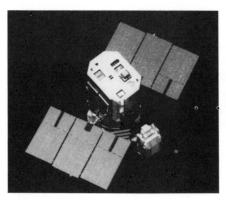

Fig. 12–42
The Solar Maximum Mission satellite is aimed at the sun by a reaction wheel system. An electrical failure crippled the reaction wheels about six months after launch in 1980. In 1984 Solar Max became the first satellite to be repaired in space. Here an astronaut attempts to link with the slowly spinning satellite prior to the actual repair.

Fig. 12–43
Close-up of the Solar Maximum Mission satellite reaction wheel system. Arrows at rear point to three reaction wheels with mutually perpendicular axes. At front is a fourth wheel mounted at an angle to provide redundancy should one of the others fail.

wheels whose axes are mutually perpendicular (Figs. 12–42 and 12–43). Each wheel has a motor and brakes to start and stop its rotation. When a wheel starts rotating, the satellite rotates in the opposite direction to conserve angular momentum. After the satellite has rotated through the desired angle, the wheel is stopped and the satellite also stops. The use of three wheels allows the satellite to be pointed in any direction. Electricity generated by solar panels powers the motors and brakes, so there is no fuel to run out.

SUMMARY

Rotational motion is described using quantities analogous to those of linear motion. These rotational quantities satisfy the same relations as their linear counterparts, making problems involving rotational motion similar to the linear motion problems of previous chapters.

1. The rotational analog of displacement is **angular displacement,** defined as the angle through which an object has rotated. The SI unit of angular displacement is the radian. The first and second time derivatives of angular displacement are **angular speed,** ω, and **angular acceleration,** α. **Angular velocity** ω is a vector whose magnitude is the angular speed and whose direction is along the rotation axis, as given by the right-hand rule. The **angular acceleration vector** α is the time derivative of the angular velocity vector, and points in the direction of the change in the angular velocity.

2. The rotational analog of mass is **rotational inertia,** a quantity that measures the resistance of a body to changes in rotational motion. Rotational inertia depends on the mass of a body and on the distribution of mass about the rotation axis, and is given by

$$I = \sum m_i r_i^2$$

for a body consisting of discrete masses, and by

$$I = \int r^2 dm$$

for a continuous distribution of matter.

3. **Torque** is the rotational analog of force; torque, rotational inertia, and angular acceleration are related by the rotational analog of Newton's second law:

$$\tau = I\alpha.$$

4. **Rotational kinetic energy** is calculated from rotational inertia and angular speed in the same way that translational kinetic energy is calculated from mass and linear speed:

$$K_{\text{rot}} = \tfrac{1}{2}I\omega^2.$$

The total kinetic energy of a rigid, rotating object may be written as the sum of the translational kinetic energy of its center of mass and the rotational kinetic energy about its center of mass.

5. The **angular momentum L** of a particle about some point is defined in terms of the particle's linear momentum **p** and its displacement **r** from that point:

$$\mathbf{L} = \mathbf{r} \times \mathbf{p}.$$

The angular momentum of a system of particles is the sum or integral of the angular momenta of the constituent particles. For a rigid, rotating object, the angular momentum may also be written in terms of strictly rotational quantities:

$$\mathbf{L} = I\boldsymbol{\omega},$$

in analogy with linear momentum $\mathbf{p} = m\mathbf{v}$.

The rotational analog of Newton's law may be written in terms of angular momentum as

$$\boldsymbol{\tau} = \frac{d\mathbf{L}}{dt}.$$

In the absence of external torques, the angular momentum of a system is conserved. The application of external torques to a rotating system results in a change in angular momentum. When the external torque is at right angles to the angular momentum, the angular momentum vector **precesses,** describing a circular path as its direction changes.

Table 12–3 summarizes the analogy between linear and rotational quantities.

TABLE 12–3
Linear and Angular Quantities

Linear Quantity or Equation	Angular Quantity or Equation	Relation Between Linear and Angular Quantities
Position x	Angular position θ	
Speed $v = dx/dt$	Angular speed $\omega = d\theta/dt$	$v_t = \omega r$
Velocity **v**	Angular velocity $\boldsymbol{\omega}$	
Acceleration **a**	Angular acceleration α	$a_t = \alpha r$
Mass m	Rotational inertia I	$I = \int r^2 \, dm$
Force **F**	Torque τ	$\boldsymbol{\tau} = \mathbf{r} \times \mathbf{F}$

Newton's second law (constant mass or rotational inertia):

$\mathbf{F} = m\mathbf{a}$ $\tau = I\alpha$

Kinetic energy:

$K_{\text{trans}} = \frac{1}{2}mv^2$ $K_{\text{rot}} = \frac{1}{2}I\omega^2$

Momentum:

$\mathbf{p} = m\mathbf{v}$ (linear) $\mathbf{L} = I\boldsymbol{\omega}$ (angular) $\mathbf{L} = \mathbf{r} \times \mathbf{p}$

Newton's second law in terms of momentum:

$\mathbf{F} = d\mathbf{p}/dt$ $\tau = d\mathbf{L}/dt$

QUESTIONS

1. Do all points on a rigid, rotating object have the same angular speed? linear speed? radial acceleration?
2. Suppose you mount tires on your car that are slightly larger in diameter than factory-supplied tires. If you are in a 55-mi/h zone and your speedometer says 55 mi/h, are you speeding? Explain.
3. A wheel undergoes a constant angular acceleration. How do each of the following depend on time: (a) tangential acceleration of a point on the rim? (b) radial acceleration of a point on the rim? (c) angular speed of the wheel?
4. Part of a train wheel extends below the point of contact with the rails, as shown in Fig. 12–44. It is often said that this part of the train moves backward. Explain.

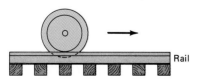

Fig. 12–44   Question 4

5. An electric circular saw takes a long time to stop rotating after the power is turned off. Without the saw blade mounted, the motor stops much more quickly when turned off. Why?
6. A solid sphere and a solid cube have the same mass, and the side of the cube is equal to the diameter of the sphere. Which has the greatest rotational inertia about an axis through the center of mass? The rotational axis for the cube is perpendicular to two of the cube faces.
7. A badly unbalanced wheel has its center of mass located closer to its rim than to its geometrical center. The wheel is released from rest on a slope where it is free to roll without slipping. Describe its initial motion when released from each of the orientations shown in Fig. 12–45.
8. The lower leg of a horse contains essentially no muscle. How does this help the horse to run fast? Explain in terms of rotational inertia.
9. You wish to store energy in a rotating flywheel. Given a fixed amount of material, what shape should you make the flywheel so that it will store the most energy at a given angular speed?

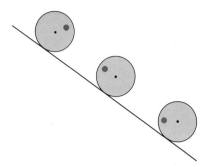

Fig. 12–45 The darker circle indicates the center of mass of the wheel (Question 7).

10. A ball starts from rest and rolls without slipping down an incline, then across a horizontal surface. It then starts up a *frictionless* incline (Fig. 12–46). Compare its maximum height on the frictionless incline with its starting height on the first incline.

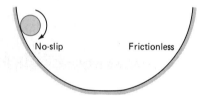

Fig. 12–46 Question 10

11. Two solid cylinders of the same mass but different radii are rolling with the same linear speed. How do their translational and rotational kinetic energies compare?

12. A raw egg and a hard-boiled egg, otherwise identical, are rolled down an incline. Which hits the bottom first? Explain.

13. A cloud of interstellar matter collapses to form a star. Under what conditions will the star *not* be rotating?

14. What happens to the rotational inertia of a washing machine tub and its contents when the spin portion of the cycle starts?

15. A bug, initially at rest on a stationary record turntable, begins walking around the turntable. After a while the bug stops. Must the turntable stop immediately? Or must it first return to its initial angular position? Or must it keep rotating?

16. Tornadoes in the northern hemisphere rotate counterclockwise, as viewed from above. A farfetched suggestion was once made that tornadoes occur more frequently in the northern hemisphere because most vehicles drive on the right-hand side of the road. Could there by any physical basis for this suggestion? Explain in terms of the angular momentum imparted to the air by vehicles passing in opposite directions.

17. As they are launched, rockets are often set rotating about a vertical axis. Why does this help stabilize the rockets?

18. A physics student is standing on a nonrotating platform that is free to rotate, and is holding a wheel that is rotating about a vertical axis (Fig. 12–47). Describe what happens if she turns the wheel upside down.

Fig. 12–47 Question 18

19. How does angular momentum conservation help explain why there are seasons on earth?

20. A gyroscope is spinning with its rotation axis horizontal, and is precessing about a vertical axis. If a weight is hung from the end of the gyroscope (Fig. 12–48), its precession rate increases. Why does this happen?

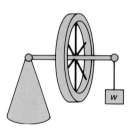

Fig. 12–48 The gyroscope precesses faster when a weight is hung from its end (Question 20).

21. A balanced gyroscope is spinning without precessing, its rotation axis horizontal. If you push with your finger in a horizontal direction at right angles to the rotation axis (Fig. 12–49), what happens?

Fig. 12–49 Question 21

PROBLEMS

Section 12.1 *Angular Velocity and Acceleration*

1. Determine the angular speed, in rad/s, of (*a*) earth about its axis; (*b*) the minute hand of a clock; (*c*) the hour hand of a clock; (*d*) an egg beater turning at 300 rpm.

2. What is the linear speed of a point on earth's equator? At your latitude?

3. Express Equation 12–4 with constants appropriate to angular speed in (*a*) rpm, (*b*) rev/s, (*c*) °/s.

4. A 25-cm-diameter circular saw blade spins at 3500 rpm. How fast would you have to push a straight hand saw to have the teeth move through the wood at the same rate as the circular saw teeth?

5. A wheel is turned through 2.0 revolutions while being accelerated at 18 rpm/s. (*a*) What is the final angular speed? (*b*) How long does it take to turn the 2.0 revolutions?

6. You switch a record turntable from 45 rpm to $33\frac{1}{3}$ rpm. The turntable takes 2.2 s to reach the new speed. Assuming that the deceleration is constant, how many revolutions does the turntable make during this time?

7. The angular acceleration of a wheel in rad/s^2 is given by $24t^2 - 16t^3$, where t is the time in seconds. The wheel starts from rest at $t=0$. (*a*) When is it again at rest? (*b*) How many revolutions has it turned between $t=0$ and when it is again at rest?

8. A circular saw blade completes 1200 revolutions in 40 s while coasting to a stop after being turned off. Assuming constant deceleration, what are (*a*) the angular deceleration and (*b*) the initial angular speed?

Section 12–2 *Rotational Inertia and the Analog of Newton's Law*

9. Each propeller on a King Air twin engine airplane consists of three blades, each of mass 10 kg and length 125 cm. The blades may be treated approximately as uniform, thin rods. (*a*) What is the rotational inertia of the propeller? (*b*) If the propeller is driven by an engine that develops a torque of 2700 N·m, how long will it take to change the propeller's angular speed from 1400 rpm to 1900 rpm? Neglect the rotational inertia of the engine and any aerodynamic forces on the propeller.

10. A neutron star is an extremely dense, rapidly spinning object that results from the collapse of a star at the end of its life (see Example 12–17). A neutron star of 1.8 times the sun's mass has an approximately uniform density of 1×10^{18} kg/m^3. (*a*) What is its rotational inertia? (*b*) The neutron star's spin rate slowly decreases as a result of torque associated with magnetic forces. If the spin-down rate is 5×10^{-5} rad/s^2, what is the magnetic torque?

11. Verify by direct integration the formula given in Table 12–2 for the rotational inertia of a flat plate about a central axis.

12. Use the parallel axis theorem to derive the rotational inertia of a sphere about a tangent line from that of a sphere about a diameter (see Table 12–2).

13. A square frame is made from four thin rods, each of length ℓ and mass m. Calculate its rotational inertia about the three axes shown in Fig. 12–50.

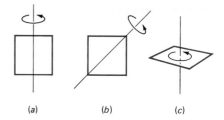

(*a*)   (*b*)   (*c*)

Fig. 12–50 Problem 13

14. Proof of the parallel axis theorem: Fig. 12–51 shows an object of mass M with axes through the center of mass and through an arbitrary point A. Both axes are perpendicular to the page. Let **h** be a vector from the axis through the center of mass to the axis through point A, $\mathbf{r}_{cm}$ a vector from the axis through the CM to an arbitrary mass element dm, and **r** a vector from the axis through point A to the mass element dm, as shown. (*a*) Use the law of cosines to show that

$$r^2 = r_{cm}^2 + h^2 - 2\mathbf{h}\cdot\mathbf{r}_{cm}.$$

(*b*) Use this result in the expression $I=\int r^2 dm$ to calculate the rotational inertia of the object about the axis through A. Each of the three terms in your expression for r^2 leads to a separate integral. Identify one as the rotational inertia about the CM, another as the quantity Mh^2, and show that the third is zero because it involves the position of the center of mass relative to the center of mass. Your result is then a statement of the parallel axis theorem.

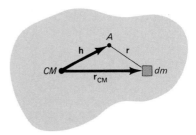

Fig. 12–51 Problem 14

15. Two blocks of mass m_1 and m_2 are connected by a massless string that passes over a solid cylindrical pulley, as shown in Fig. 12–52. The surface under the block m_2 is frictionless, and the pulley rides on frictionless bearings. The string passes over the pulley

without slipping. When released, the masses accelerate at $\frac{1}{8}$g. The tension in the lower half of the string is 2.7 N and that in the upper half, 1.9 N. What are the masses of the pulley and of the two blocks?

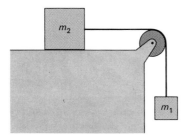

Fig. 12–52   Problem 15

16. A disk of radius R and thickness w has a mass density that increases from the center outward, being given by $\rho = \rho_0 r/R$, where r is the distance from the axis of the disk. (a) Calculate the total mass M of the disk. (b) Calculate the rotational inertia about the disk axis, in terms of M and R. Compare with the results for a solid disk of uniform density and for a ring.

17. A space station is constructed in the shape of a wheel 22 m in diameter, with essentially all of its 5.0×10^5-kg mass at the rim (Fig. 12–53). Once the station is completed, it is set rotating at a rate that requires an object at the rim to have radial acceleration g, thereby simulating earth's surface gravity. This is accomplished using two small rockets, each with 100 N thrust, that are mounted on the rim of the station as shown. (a) How long will it take to reach the desired spin rate? (b) How many revolutions will the station make in this time?

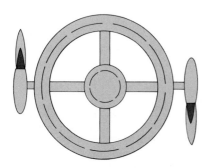

Fig. 12–53   Problem 17

18. A standard 108-g Frisbee is 24 cm in diameter and has about half its mass spread uniformly in a disk, and the other half concentrated in the rim. With a quarter-turn flick of the wrist, a student sets the Frisbee rotating at 550 rpm. (a) What is the rotational inertia of the Frisbee? (b) What is the magnitude of the torque, assumed constant, that the student applies?

19. The crane shown in Fig. 12–54 contains a hollow drum of mass 150 kg and radius 0.80 m that is driven by an engine to wind up a cable. The cable passes over a solid cylindrical 30-kg pulley 0.30 m in radius to lift a 2000-N weight. How much torque must the engine apply to the drum to lift the weight with an acceleration of 1.0 m/s^2? Neglect the rotational inertia of the engine and the mass of the cable.

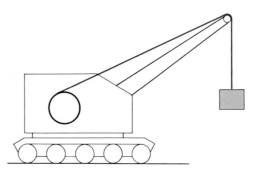

Fig. 12–54   Problem 19

20. A 2.4-kg block rests on a 30° slope, and is attached by a string of negligible mass to a solid drum of mass 0.85 kg and radius 5.0 cm, as shown in Fig. 12–55. When released, the block accelerates down the slope at 1.6 m/s^2. What is the coefficient of friction between block and slope?

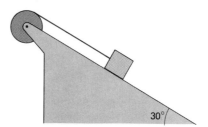

Fig. 12–55   Problem 20

21. A bicycle is under repair, and is upside-down with its 66-cm-diameter wheel spinning freely at 230 rpm. The mass of the wheel is 1.9 kg, and is concentrated mostly at the rim. The cyclist holds a wrench against the tire for 3.1 s, with a normal force of 2.7 N. If the coefficient of friction between the wrench and the tire is 0.46, what is the final angular speed of the wheel?

22. At the National Magnet Laboratory at MIT, energy is stored in huge solid flywheels of mass 7.7×10^4 kg and radius 2.4 m. If the flywheels ride on shafts 41 cm in diameter, and if the coefficient of friction between shafts and bearings is 0.018, how long would it take a flywheel to coast to a stop from its normal rotation rate of 360 rpm?

Section 12.3   *Energy and Rotational Motion*

23. The flywheel of the previous problem is used to generate a large amount of electric power for a short time. If the wheel's rotation rate is dropped from 390 to 300 rpm in 5.0 s, (a) what is the average power output and (b) what fraction of the rotational kinetic energy has been removed?

24. A 25-cm-diameter circular saw blade has a mass of 0.85 kg, distributed uniformly as in a disk. (a) What is its rotational kinetic energy at 3500 rpm? (b) What average power must be applied to bring the blade from rest to 3500 rpm in 3.2 s?

25. In the stellar collapse of Example 12–17, by what factor does the rotational kinetic energy of the star change?

26. An object of rotational inertia I is initially at rest. A torque is then applied to the object, causing it to begin rotating. The torque is applied for only one-quarter of a revolution, during which time its magnitude is given by $\tau = A\cos\theta$, where A is a constant and θ is the angle through which the object has rotated. What is the final angular speed of the object?

27. The rotational kinetic energy of a rolling automobile wheel is 40% of its translational kinetic energy. The wheel is then redesigned to have 10% lower rotational inertia and 20% less mass, while keeping its radius the same. By what percentage does its total kinetic energy at a given speed decrease?

28. Two solid cylinders of the same mass roll without slipping down an incline. If they start from the same height h, show that they reach the bottom with the same linear speed, even if their radii are not the same. What is the linear speed at the bottom?

29. A ball rolls without slipping down a slope of vertical height 35 cm, and reaches the bottom moving at 2.0 m/s. Is the ball hollow or solid?

30. A solid ball of mass M and radius R starts at rest at height h above the bottom of the path shown in Fig. 12–56. It rolls without slipping down the left side of the path. The right side of the path, starting at the bottom, is frictionless. To what height does the ball rise on the right?

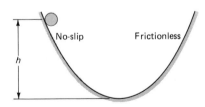

Fig. 12–56   Problem 30

31. A solid marble of radius r starts from rest and rolls without slipping on the loop-the-loop track shown in Fig. 12–57. Find the minimum starting height of the marble from which it will remain on the track through the loop.

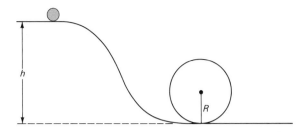

Fig. 12–57   Problem 31

32. A 100-g yo-yo consists of two disks joined by a narrower shaft. The rotational inertia of the yo-yo is 600 g·cm². A string 80 cm long is wound around the narrow shaft of the yo-yo. With the yo-yo initially 50 cm off the ground, the string is pulled vertically upward with a force of 0.98 N until the string comes entirely free of the yo-yo. The yo-yo then falls freely. What are its rotational and translational kinetic energies just before it hits the ground?

Section 12.4   *Angular Momentum*

33. What is the angular momentum of the earth about its own axis? About an axis through the center of the sun and perpendicular to earth's orbital plane? See Appendix F, and consider earth to be a uniform sphere.

34. A 150-g baseball has an angular momentum of 7.7 kg·m²/s about a vertical axis through the batter. If it is moving at 43 m/s without spinning, by how much does it miss the axis of the batter?

35. About 99.9% of the mass of the solar system lies in the sun. Referring to Appendix F, estimate what fraction of the angular momentum of the solar system about the sun's axis is associated with the sun. Where is most of the rest of the angular momentum?

36. Two ants, each of mass 0.012 g, are standing at rest on a motionless record turntable with rotational inertia 0.010 kg·m², one 5.0 cm from the center and the other 11 cm from the center. The ants start walking in the same direction around the turntable, at 0.90 cm/s relative to the turntable. How long will it take the turntable to complete one revolution?

37. A physics student is standing on an initially motionless, frictionless turntable with rotational inertia 0.31 kg·m². She is holding a wheel of rotational inertia 0.22 kg·m² spinning at 130 rpm about a vertical axis, as shown in Fig. 12–58. When she turns the wheel upside down, the student and turntable begin rotating at 70 rpm. (a) What is the student's mass, considering her to be a cylinder 30 cm in diameter? (b) How much work did it take to turn the wheel upside down?

38. A 3.0-m-diameter merry-go-round with a rotational inertia of 120 kg·m² is spinning freely at 0.50 rev/s. Four 25-kg children sit suddenly on the edge of the merry-go-round. Find the new angular speed and de-

Fig. 12–58   Problem 37

termine the total energy lost to friction between the children and the merry-go-round.

39. Derive Equation 12–38 for the case of a gyroscope whose angular momentum vector is not necessarily in a horizontal plane.

40. After being "spun up," the gyroscope of Example 12–12 is placed with one end on a frictionless stand, with its axis horizontal. The rotating ring of the gyroscope is 3.0 cm from the end, and the mass of the rest of the gyroscope is negligible compared with that of the ring. What is the precession rate of the gyroscope?

41. A gyroscope consists of a disk and shaft mounted on frictionless bearings in a circular frame of diameter d. Initially the gyroscope is spinning with angular speed ω, and is perfectly balanced so it is not precessing (Fig. 12–59). When a mass m is hung from the frame at one of the shaft bearings, the gyroscope precesses with angular speed Ω. Find the rotational inertia of the disk in terms of the quantities given.

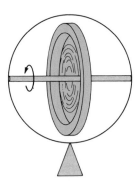

Fig. 12–59   Problem 41

Supplementary Problems

42. A 55-kg diver is spinning forward in a tuck position at 2.1 rev/s, as shown in Fig. 12–60a. In this position, the diver's rotational inertia is 3.9 kg·m². At the peak

of her trajectory, the diver suddenly straightens out in a horizontal position as shown in Fig. 12–60b. In this new position, the diver has an overall length of 2.2 m with center of mass midway along this length, and a rotational inertia of 21 kg·m² about the rotation axis. What is the minimum height of the diver above the water if she is to enter the water hands first? Neglect the diver's horizontal translational motion.

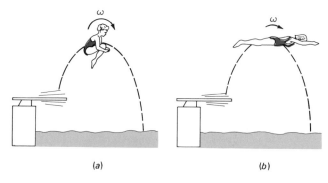

(a)   (b)

Fig. 12–60   Problem 42

43. A rod of length ℓ and mass m is suspended from a pivot, as shown in Fig. 12–61. The rod is struck midway along its length by a wad of putty of mass m moving horizontally at speed v. The putty sticks to the rod. What is the minimum speed v that will result in the rod's making a complete circle rather than swinging like a pendulum?

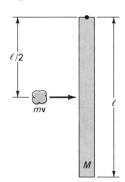

Fig. 12–61   Problem 43

44. A 30-cm-diameter phonograph record is dropped onto a turntable being driven at $33\frac{1}{3}$ rpm. If the coefficient of friction between record and turntable is 0.19, how far will the turntable rotate between the time when the record first contacts it and when the record is rotating at the full $33\frac{1}{3}$ rpm? Assume that the record is a homogeneous disk. *Hint:* You will need to do an integral to calculate the torque.

45. The contraption shown in Fig. 12–62 consists of two solid rubber wheels each of mass M and radius R that are mounted on an axle of negligible mass in a rigid square frame made of thin rods of mass m and length

ℓ. The axle bearings are frictionless, and the wheels are mounted symmetrically about the center line of the frame, just far enough apart that they don't touch. The whole contraption is floating freely in space, and the frame is not rotating. The wheels are rotating with angular speed ω in the same direction, as shown. A mechanism built into the frame moves the axles very slightly so that the wheels touch and frictional forces act between them. Describe quantitatively the motion of the system after the wheels have stopped slipping against each other.

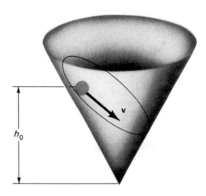

Fig. 12–63   Problem 46

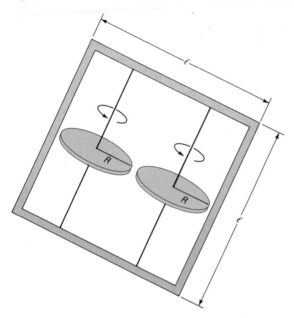

Fig. 12–62   Problem 45

46. A small mass m slides on the inside of a frictionless, cone-shaped surface as shown in Fig. 12–63. At its highest point, it is a distance h_0 above the vertex of the cone, and has speed v_0. What is the minimum height above the vertex that the mass reaches as it moves around the surface? *Hint:* What quantities are conserved?

47. In 1609, Johannes Kepler presented his famous equal-area law, stating that the planets move in such a way that a line from a planet to the sun sweeps out equal areas in equal times (Fig. 12–64). Later, Newton showed that the equal-area law is a consequence of his law of universal gravitation. In fact, the equal-area law holds for any *central* force—any force, like the gravitational force of the sun, that is always directed toward the same point. Show that a central force cannot exert any torque about the center of force, and that the angular momentum of an object moving under the influence of such a force cannot change. Use this result to prove the equal-area law. *Hint:* Consider the small area dA swept out in a small time dt, and show that the "areal velocity," dA/dt, is constant.

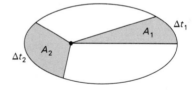

Fig. 12–64   (Problem 47) Kepler's equal-area law. If the time intervals Δt_1 and Δt_2 are equal, then so are the areas A_1 and A_2.

13

OSCILLATORY MOTION

oscillatory motion

When we disturb a system in stable equilibrium, forces within the system tend to restore its equilibrium. For example, if the ball of Fig. 13–1 is disturbed, it experiences a net force toward its equilibrium position at the bottom of the valley. This force accelerates the ball toward equilibrium, and if frictional energy losses are not too great it reaches the bottom with some kinetic energy. As a result, it overshoots its equilibrium, now moving up the opposite side of the valley. Again, a force develops that points back toward equilibrium. This force decelerates the ball, which eventually stops and then heads back toward the bottom. Again it may overshoot. The result is **oscillatory motion**—the movement of the ball back and forth about its equilibrium position. Were there no friction, this oscillation would continue forever. In a real system, of course, friction eventually dissipates the energy, and the system settles into equilibrium.

A wide variety of physical systems are in stable equilibrium, but subject to occasional disturbances. As a result, these systems frequently undergo oscillatory motion. Examples of oscillatory motion can therefore be found throughout the vast realm of physical phenomena. A uranium nu-

Fig. 13–1
(a) A ball in stable equilibrium at the bottom of a valley. *(b)* If the ball is disturbed from equilibrium, it oscillates about its equilibrium position.

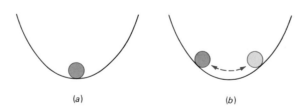

(a)   (b)

Fig. 13–2
The collapse of the Tacoma Narrows Bridge in Washington occurred because the bridge design did not anticipate large-scale oscillatory motion of the structure; see also Fig. 13–37.

cleus undergoes oscillations before it fissions. The oscillatory motion of the atoms in water molecules is responsible for cooking in a microwave oven. When your professor squeaks chalk on the blackboard, contact between chalk and board disturbs the equilibrium state of the chalk, setting up an oscillation that produces the horrible sound you hear. The timekeeping mechanism of a watch—whether an old-fashioned spring and balance wheel or a modern quartz crystal—is a carefully engineered system whose disturbance from equilibrium results in a very precise oscillatory motion. Musical instruments work because of oscillatory motion. Even large-scale structures like buildings and bridges undergo oscillatory motions about their equilibria; if these oscillations are not accounted for in the design of the structure, the result can be disaster (Fig. 13–2). Finally, many stars undergo oscillations about their equilibrium sizes (Fig. 13–3). Oscillatory motion is a universal phenomenon. In this chapter, we develop a detailed description of oscillatory motion and apply that description to a variety of physical systems.

13.1
PROPERTIES OF OSCILLATORY MOTION

amplitude
period
frequency

Oscillatory motion is characterized by the maximum size of the displacement from equilibrium—called the **amplitude** of oscillation—and by the time it takes for the motion to repeat itself—the **period** of oscillation. Closely related to the period is the **frequency** of oscillation, defined as the number of oscillation cycles per unit time. The frequency and period are simply inverses:

$$f = \frac{1}{T},\tag{13–1}$$

hertz

where f is the frequency and T the period. The unit of frequency is the cycle/second, which is given the name **hertz** (Hz), after the German physicist, engineer, and mathematician Heinrich Hertz (1857–1894), who was the first to produce and detect radio waves.

Fig. 13–3
The star δ Cephei exhibits periodic variations in brightness, corresponding to oscillatory motion of the star about its equilibrium configuration. Magnitude (vertical axis) is an astronomical measure of brightness, with lower magnitude representing higher brightness.

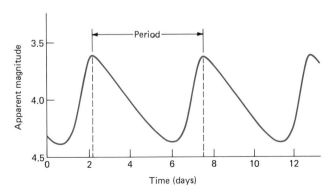

Example 13–1
Amplitude, Period, and Frequency

Bored by a physics lecture, a student holds one end of a flexible plastic ruler against a book and idly strikes the other end, setting it into oscillation (Fig. 13–4). The student notes that 28 complete cycles occur in 10 s, and that the end of the ruler moves a total distance of 8.0 cm, as shown in Fig. 13–4. What are the amplitude, period, and frequency of this oscillatory motion?

Solution

We have defined amplitude as the maximum displacement from equilibrium. Since the end of the ruler moves both left and right of equilibrium, the amplitude of the motion is 4.0 cm. (The full 8.0-cm motion is called the **peak-to-peak** amplitude.) With 28 cycles in 10 s, the time per cycle, or period, is

peak-to-peak

$$T = \frac{10 \text{ s}}{28} = 0.36 \text{ s.}$$

The frequency in Hz is the inverse of the period:

$$f = \frac{1}{T} = \frac{1}{0.36 \text{ s}} = 2.8 \text{ Hz.}$$

←8.0 cm→

Fig. 13–4
Top view of a ruler undergoing oscillatory motion (Example 13–1).

Amplitude and period or frequency do not completely characterize an arbitrary oscillatory motion, for they do not describe in detail how a system behaves over one cycle. The behavior of a ball bouncing between two rigid walls (Fig. 13–5a) is quite different from that of a ball rolling in a U-shaped valley (Fig. 13–5b). The difference lies in the force as a function of position. For the ball bouncing between rigid walls, the force acts only when the ball is in contact with the walls; between bounces, the ball moves with constant speed. For the ball in the valley, a force acts as soon as the ball moves from its equilibrium position, and the force increases as the ball moves up the ever-steepening sides of the valley. In the rest of this chapter, we examine the important case of oscillatory motion that results from a force that is proportional to displacement from equilibrium.

13.2
SIMPLE HARMONIC MOTION

For a wide variety of physical systems, the force that develops when a body is displaced from a stable equilibrium is directly proportional to the displacement. For many other systems, this proportionality is an approximation that becomes increasingly accurate for small displacements. The kind of motion that results from such a force is therefore common in physical systems, and is called **simple harmonic motion:**

simple harmonic motion

> Simple harmonic motion is the motion that results when an object is subject to a restoring force proportional to its displacement from equilibrium.

Mathematically, we can describe such a force by writing

$$F = -kx, \tag{13–2}$$

Fig. 13–2
The collapse of the Tacoma Narrows Bridge in Washington occurred because the bridge design did not anticipate large-scale oscillatory motion of the structure; see also Fig. 13–37.

cleus undergoes oscillations before it fissions. The oscillatory motion of the atoms in water molecules is responsible for cooking in a microwave oven. When your professor squeaks chalk on the blackboard, contact between chalk and board disturbs the equilibrium state of the chalk, setting up an oscillation that produces the horrible sound you hear. The timekeeping mechanism of a watch—whether an old-fashioned spring and balance wheel or a modern quartz crystal—is a carefully engineered system whose disturbance from equilibrium results in a very precise oscillatory motion. Musical instruments work because of oscillatory motion. Even large-scale structures like buildings and bridges undergo oscillatory motions about their equilibria; if these oscillations are not accounted for in the design of the structure, the result can be disaster (Fig. 13–2). Finally, many stars undergo oscillations about their equilibrium sizes (Fig. 13–3). Oscillatory motion is a universal phenomenon. In this chapter, we develop a detailed description of oscillatory motion and apply that description to a variety of physical systems.

13.1
PROPERTIES OF OSCILLATORY MOTION

amplitude
period
frequency

Oscillatory motion is characterized by the maximum size of the displacement from equilibrium—called the **amplitude** of oscillation—and by the time it takes for the motion to repeat itself—the **period** of oscillation. Closely related to the period is the **frequency** of oscillation, defined as the number of oscillation cycles per unit time. The frequency and period are simply inverses:

$$f = \frac{1}{T},\tag{13–1}$$

hertz

where f is the frequency and T the period. The unit of frequency is the cycle/second, which is given the name **hertz** (Hz), after the German physicist, engineer, and mathematician Heinrich Hertz (1857–1894), who was the first to produce and detect radio waves.

Fig. 13–3
The star δ Cephei exhibits periodic variations in brightness, corresponding to oscillatory motion of the star about its equilibrium configuration. Magnitude (vertical axis) is an astronomical measure of brightness, with lower magnitude representing higher brightness.

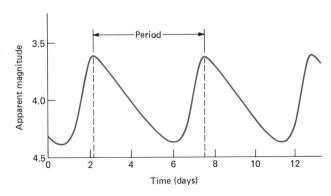

Example 13-1
Amplitude, Period, and Frequency

Bored by a physics lecture, a student holds one end of a flexible plastic ruler against a book and idly strikes the other end, setting it into oscillation (Fig. 13–4). The student notes that 28 complete cycles occur in 10 s, and that the end of the ruler moves a total distance of 8.0 cm, as shown in Fig. 13–4. What are the amplitude, period, and frequency of this oscillatory motion?

Solution

We have defined amplitude as the maximum displacement from equilibrium. Since the end of the ruler moves both left and right of equilibrium, the amplitude of the motion is 4.0 cm. (The full 8.0-cm motion is called the **peak-to-peak** amplitude.) With 28 cycles in 10 s, the time per cycle, or period, is

$$T = \frac{10 \text{ s}}{28} = 0.36 \text{ s}.$$

The frequency in Hz is the inverse of the period:

$$f = \frac{1}{T} = \frac{1}{0.36 \text{ s}} = 2.8 \text{ Hz}.$$

peak-to-peak

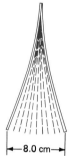

Fig. 13–4
Top view of a ruler undergoing oscillatory motion (Example 13–1).

Amplitude and period or frequency do not completely characterize an arbitrary oscillatory motion, for they do not describe in detail how a system behaves over one cycle. The behavior of a ball bouncing between two rigid walls (Fig. 13–5a) is quite different from that of a ball rolling in a U-shaped valley (Fig. 13–5b). The difference lies in the force as a function of position. For the ball bouncing between rigid walls, the force acts only when the ball is in contact with the walls; between bounces, the ball moves with constant speed. For the ball in the valley, a force acts as soon as the ball moves from its equilibrium position, and the force increases as the ball moves up the ever-steepening sides of the valley. In the rest of this chapter, we examine the important case of oscillatory motion that results from a force that is proportional to displacement from equilibrium.

13.2
SIMPLE HARMONIC MOTION

For a wide variety of physical systems, the force that develops when a body is displaced from a stable equilibrium is directly proportional to the displacement. For many other systems, this proportionality is an approximation that becomes increasingly accurate for small displacements. The kind of motion that results from such a force is therefore common in physical systems, and is called **simple harmonic motion:**

simple harmonic motion

> Simple harmonic motion is the motion that results when an object is subject to a restoring force proportional to its displacement from equilibrium.

Mathematically, we can describe such a force by writing

$$F = -kx, \tag{13–2}$$

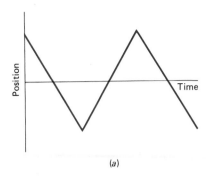

(a)

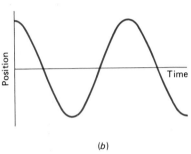

(b)

Fig. 13–5
Displacement as a function of time
(a) for a ball bouncing between rigid
walls and *(b)* for a ball rolling in a U-
shaped valley like that of Fig. 13–1.

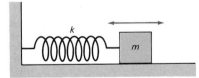

SIMPLE HARMONIC MOTION IS
THE MOTION THAT RESULTS FROM
A RESTORING FORCE THAT IS
PROPORTIONAL TO ITS DISPLACEMENT
FROM EQUILIBRIUM.

Fig. 13–6
Simple harmonic motion results from a
restoring force that is *proportional* to
the displacement from equilibrium.

differential equation
dependent variable
independent variable

Fig. 13–7
A mass attached to an ideal spring
undergoes simple harmonic motion.
The mass is on a frictionless surface,
and the spring exerts a force if it is
either stretched or compressed from its
equilibrium position.

where F is the force, x the displacement, and k a constant giving the proportionality between force and displacement. The minus sign in Equation 13–2 indicates a *restoring* force: if the object is displaced in some direction, the force is in the *opposite* direction, so that it tends to *restore* the equilibrium (Fig. 13–6).

Although we have phrased our definition of simple harmonic motion and Equation 13–2 in terms of force F and linear displacement x, simple harmonic motion also results if a body is subject to a restoring torque proportional to its angular displacement from equilibrium. More generally, any system whose tendency to return to equilibrium is proportional to how far it is displaced from equilibrium will exhibit simple harmonic motion.

Comparison with Equation 4–7 shows that Equation 13–2 is the equation for the force in an ideal spring, where k is the spring constant. For this reason, a system consisting of a mass on a spring (Fig. 13–7) undergoes a simple harmonic motion. Many other systems—even atoms and molecules—can often be understood by modelling them as miniature mass-spring systems (see Example 13–7 and Problem 5).

We have defined simple harmonic motion, but we have not described it. How does a body in simple harmonic motion actually move? We can find out by applying Newton's second law, $\mathbf{F} = m\mathbf{a}$, to the mass-spring system of Fig. 13–7. For this system, the force on the mass m is $-kx$ from Equation 13–2, so that Newton's law becomes

$$ma = -kx, \qquad (13\text{–}3)$$

where we take the x-axis to be along the direction of the motion with origin at the equilibrium position. To describe the motion in detail, we need to know how the position x depends on time. Although time does not appear explicitly in Equation 13–3, it is implicitly present because the acceleration a is the second time derivative of position. Thus Equation 13–3 may be written

$$m\frac{d^2x}{dt^2} = -kx. \qquad (13\text{–}4)$$

The unknown in Equation 13–4 is the displacement, x. But the solution to this equation is not a simple number like x = 5.0 m, but is rather a function of time. Equation 13–4 is a **differential equation**: the unknown (or **dependent**) variable—in this case x—appears in a derivative with respect to the known (or **independent**) variable—in this case t. Differential equations occur throughout physics and engineering, and finding their solutions often presents fascinating challenges to the theoretical physicist, applied mathematician, or computer scientist.

What does it mean to solve an equation? To solve the equation 2x = 6 means to find a value for x that, when substituted into the equation, yields an identity—in this case 6 = 6. Similarly, to solve Equation 13–4 means to find a function of time which, when substituted for x in the equation, results in an identity. What function might it be? We expect the system to undergo oscillatory motion. This *physical* insight suggests a *mathematical* solution that is periodic. Equation 13–4 tells us further that our solution x must be proportional to its own second derivative. You know two periodic functions with this property: sine and cosine. (Differentiate sine and you get cosine; differentiate cosine and you get −sine. Similarly, start with cosine and after two differentiations you have −cosine.) So we might expect our solution to Equation 13–4 to involve sine or cosine functions whose arguments depend on the time t. We know physically that we can have oscillatory motions with different amplitudes and frequencies; our solution should therefore be flexible enough to accommodate these differences. Just as we need an origin—the equilibrium position—from which to measure distances, so we also need to define an initial time $t=0$ from which our times are measured. A mathematical form that accommodates different amplitudes, frequencies, and initial times is

$$x = x_0 \cos(\omega t + \phi), \tag{13–5}$$

where x_0, ω, and ϕ are constants.

What is the physical significance of the constants in Equation 13–5? Since the cosine function itself varies between $+1$ and -1, the quantity x_0 multiplying the cosine is just the amplitude, or greatest displacement from equilibrium (Fig. 13–8).

The cosine—and thus the oscillatory motion that is described by Equation 13–5—undergoes a full cycle as its argument increases by 2π. In Equation 13–5, the argument of the cosine is $(\omega t + \phi)$. Since the time for one full cycle of the oscillatory motion is the period T, the argument of the cosine must increase by 2π as time increases by T:

$$\omega(t + T) + \phi = \omega t + \phi + 2\pi,$$

so that $\omega T = 2\pi$, or

$$T = \frac{2\pi}{\omega}. \tag{13–6}$$

The frequency of the motion is then

$$f = \frac{1}{T} = \frac{\omega}{2\pi}. \tag{13–7}$$

Equation 13–7 shows that ω is a measure of the frequency, although it differs from the frequency in Hz by a factor of 2π. ω is called the **angular frequency,** and is measured in rad/s or, since radians are dimensionless, simply in s^{-1}. The relation between angular frequency and frequency in Hz is the same as that between angular speed in rad/s and in rev/s that we encountered in the previous chapter. Here, as there, we introduce the quantity ω because it affords the simplest mathematical description of the motion we are studying—in this case simple harmonic motion. This similarity between the descriptions of simple harmonic motion and of rotational motion is no coincidence, as we will explore further in Section 13.4.

Finally, as Fig. 13–9 shows, the constant ϕ—called the **phase constant**—is related to our choice for $t=0$, or, equivalently, to the initial displacement of the oscillating mass. You might wonder why we chose a cosine instead of a sine in writing Equation 13–5. It really doesn't matter which we use; our freedom to choose the phase constant ϕ means that the special cases $\cos\omega t$ and $\sin\omega t$ are included in Equation 13–5. In the former case $\phi=0$, while in the latter $\phi=-\pi/2$, as you can see from the trigometric identity $\cos(\alpha - \pi/2) = \sin\alpha$.

angular frequency

phase constant

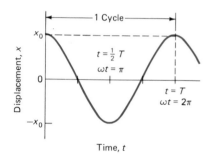

Fig. 13–8
The function $x_0 \cos\omega t$. This function varies between $\pm x_0$, and undergoes one full cycle as ωt increases by 2π.

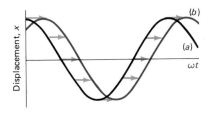

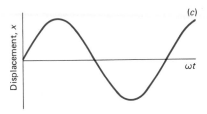

Fig. 13–9
The phase constant φ determines the displacement *x* at time *t* = 0. *(a)* φ = 0 corresponds to oscillation with maximum displacement at *t* = 0, while *(b)* φ = −π/4 shifts the cosine curve to the right by π/4. *(c)* φ = −π/2 corresponds to oscillation passing through equilibrium at time *t* = 0.

Merely writing the displacement x in the form 13–5 does not guarantee that we have a solution; we must now find out if the assumed solution 13–5 makes Equation 13–4 into an identity. Equation 13–4 contains both x and its second derivative. If x is given by Equation 13–5, then its first derivative is

$$\frac{dx}{dt} = \frac{d}{dt}[x_0 \cos(\omega t + \phi)] = -x_0\omega \sin(\omega t + \phi), \qquad (13\text{–}8)$$

where we have used the chain rule for differentiation (see Appendix B). Then the second time derivative is

$$\frac{d^2x}{dt^2} = \frac{d}{dt}\left(\frac{dx}{dt}\right) = \frac{d}{dt}[-x_0\omega \sin(\omega t + \phi)] = -x_0\omega^2 \cos(\omega t + \phi). \quad (13\text{–}9)$$

We can now try out the assumed solution for x (Equation 13–5), and its second time-derivative (Equation 13–9), in the differential equation 13–4. Substituting x and d^2x/dt^2 in the appropriate places gives

$$m[-x_0\omega^2 \cos(\omega t + \phi)] \overset{?}{=} -k[x_0 \cos(\omega t + \phi)], \qquad (13\text{–}10)$$

where the ? indicates that we are still trying to find out if this is indeed an equality. Now Equation 13–10 must be true *for all values of time t*. Why? Because Newton's law holds at all times, and Equation 13–10 is derived from Newton's law for the mass-spring system. Fortunately, the time-dependent term cos(ωt + φ) appears on both sides of the equation, so that we can cancel it. Had this not been possible, there would have been no way to make Equation 13–10 an identity for all times. We notice, too, that the constant x_0 and the minus sign cancel from the equation, leaving only

$$m\omega^2 = k. \qquad (13\text{–}11)$$

What is Equation 13–11 telling us? This equation follows directly from Newton's law, if we assume a solution of the form 13–5. If that solution is to be correct, then Equation 13–11 must be an identity. This is possible, but only if the angular frequency ω is given by

$$\omega = \sqrt{\frac{k}{m}}. \qquad (13\text{–}12)$$

So what does all this mean physically? Equation 13–5 *is* a possible description of the motion of a mass m on a spring of spring constant k, provided the angular frequency ω is given by Equation 13–12. Equivalently, using Equation 13–7, we can write the frequency in Hz as

$$f = \frac{\omega}{2\pi} = \frac{1}{2\pi}\sqrt{\frac{k}{m}}, \qquad (13\text{–}13)$$

so that the period of the motion is

$$T = \frac{1}{f} = \frac{2\pi}{\omega} = 2\pi\sqrt{\frac{m}{k}}. \qquad (13\text{–}14)$$

Does the relation of frequency to mass m and spring constant k make physical sense? If we increase the mass m, it becomes harder to accelerate, so that we expect slower oscillation. Indeed, this is reflected in Equations 13–12 and 13–13, where m appears in the denominator. Similarly, if we increase the spring constant k, the spring becomes stiffer, and able to provide larger accelerations. As a result, the frequency should increase—and it does, as shown by the presence of k in the numerator of Equations 13–12 and 13–13. Physical systems display a large range of m and k values and a correspondingly large range of oscillation frequencies. An atom, with its small mass and its "springiness" provided by electrical forces, may oscillate at 10^{13} Hz or more (see Problem 5 and Example 13–7). A massive skyscraper, whose "springiness" is in the inherent flexibility of its construction materials, may oscillate at frequencies around 0.1 Hz (see Example 13–2). You can verify that Equations 13–12 and 13–13 give the right units for frequency (see Problem 1).

What about the amplitude x_0? It cancelled from our equations, showing that Equation 13–5 is a solution of Equation 13–4 for *any* value of x_0. Physically, this means that we can have simple harmonic motion of any amplitude we choose, and that the frequency of that motion does not depend on amplitude. Independence of frequency and amplitude is one of the particularly simple features of simple harmonic motion, and arises because the restoring force is *directly proportional* to the displacement. When the restoring force does not take the mathematically simple form $F = -kx$, then the frequency does depend on amplitude and the analysis of oscillatory motion becomes much more complicated (see Problem 55). In particular, if a real spring is stretched too far, the force it develops is no longer directly proportional to displacement. For this reason, simple harmonic motion occurs in real springs and other systems usually only for relatively small-amplitude oscillations.

Just as the amplitude x_0 dropped out of our equations, so did the phase constant ϕ. Again, this shows that a solution of the form 13–5 satisfies Newton's law—Equation 13–4 for the mass-spring system—for any choice of phase constant. Thus of the three constants x_0, ω, and ϕ that appear in our solution, one—the angular frequency ω—is determined by the physical properties m and k of the system, while the other two can take on arbitrary values. What determines those values? They are determined by initial conditions—x_0 by how far the spring is stretched initially, and ϕ by the exact time when the motion starts.

You might wonder whether Equation 13–4 admits solutions other than Equation 13–5. The answer is no. In the study of differential equations, it is shown that an equation like 13–5 that contains a second derivative as its highest derivative has as its most general solution a function that, like Equation 13–5, contains two arbitrary constants. Specification of those constants in accordance with initial conditions then gives the particular solution valid in specific circumstances. So *all* oscillating mass-spring systems are described by Equation 13–5; specification of amplitude x_0 and phase constant ϕ then gives a description applicable in a specific situation.

Since Equation 13–5 is indeed a solution for the position of the oscillating mass as a function of time, the first and second derivatives of that

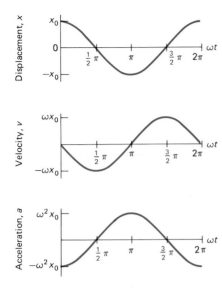

Fig. 13–10
Graphs of displacement, velocity, and acceleration in simple harmonic motion. The phase constant ϕ has been taken to be zero.

equation—Equations 13–8 and 13–9—must be the velocity and acceleration, respectively. From Equation 13–8, we see that the velocity is a sine function when the displacement is a cosine function. The velocity is therefore a maximum when the displacement is zero, and vice versa. As the mass moves through its equilibrium position, it has its maximum speed; at the extremes of its displacement, it is instantaneously stopped. Equation 13–8 also shows that the maximum velocity is given by $x_0\omega$. This makes sense because, for a fixed amplitude x_0, a higher-frequency oscillation requires a larger velocity in order to traverse the same distance in a shorter time. Similar analysis of Equation 13–9 shows that the displacement and acceleration reach their extremes at the same time, but with opposite signs. When the displacement has its greatest positive value, the acceleration has its greatest negative value, showing that the spring force tends to restore the system to equilibrium. This relation between acceleration and displacement makes good physical sense, for at maximum displacement the spring exerts its greatest force, and therefore provides the greatest acceleration. Figure 13–10 summarizes the relations among displacement, velocity, and acceleration over one cycle.

Example 13–2
A Skyscraper in Simple Harmonic Motion

The huge mass-spring system shown in Fig. 13–11 is mounted on the top floor of New York's 300-m (900-ft) high Citicorp Tower. It is engineered to oscillate at the same frequency as the building itself, but out of phase, thereby reducing the overall oscillation amplitude of the building. The 3.73×10^5-kg concrete block completes one oscillation in 6.80 s, and the oscillation amplitude in a high wind is 110 cm. Determine the spring constant and the maximum speed and acceleration experienced by the concrete block. Write an equation describing the position x of the block as a function of time, assuming it is in equilibrium and moving toward positive x at time $t = 0$.

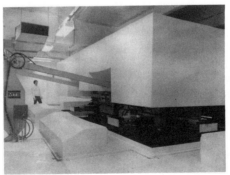

Fig. 13–11
(Left) New York's 59-floor Citicorp Tower. Like other skyscrapers, this building can be set into simple harmonic motion by the force of the wind. (Center) To reduce the building's tendency to oscillate, this huge mass-spring system was constructed on the top floor. The 410-ton mass and springs are engineered to have the same oscillation frequency as the building. The mass rides on a film of oil, and is set oscillating 180° out of phase with the building, thereby reducing the amplitude of the building's oscillation. (Right) A similar "tuned mass damper" in action on Boston's John Hancock Building. (See also color plate 2).

Solution

Since we know the frequency and period, we can solve Equation 13–12 for the spring constant k:

$$k = \frac{4\pi^2 m}{T^2} = \frac{(4\pi^2)(3.73 \times 10^5 \text{ kg})}{(6.80 \text{ s})^2} = 3.18 \times 10^5 \text{ N/m}.$$

Equations 13–8 and 13–9 show that the maximum speed and acceleration are $v_{max} = \omega x_0$ and $a_{max} = \omega x_0^2$. In terms of the period, the angular frequency may be obtained from Equation 13–6:

$$\omega = \frac{2\pi}{T} = \frac{2\pi}{6.80 \text{ s}} = 0.924 \text{ s}^{-1},$$

so that

$$v_{max} = \omega x_0 = (0.924 \text{ s}^{-1})(1.10 \text{ m}) = 1.02 \text{ m/s}$$

and

$$a_{max} = \omega^2 x_0 = (0.924 \text{ s}^{-1})^2 (1.10 \text{ m}) = 0.939 \text{ m/s}^2.$$

Finally, the position as a function of time is given by Equation 13–5, with our values $x_0 = 1.10$ m and $\omega = 0.924$ s^{-1}. What about the phase constant ϕ? We want a sinusoidal curve that has $x = 0$ at $t = 0$, with x increasing at that point. The appropriate form is $\cos(\omega t - \pi/2)$ or simply $\sin \omega t$. So we have

$$x = 1.10 \sin(0.924t),$$

with x in m and t in s.

13.3
APPLICATIONS OF SIMPLE HARMONIC MOTION

Simple harmonic motion occurs in many physical situations analogous to the mass-spring system. The basic criterion for simple harmonic motion is that the force or other physical quantity driving the system back toward equilibrium must be directly proportional to displacement from equilibrium. We now explore a number of instances in which this occurs. As you study each, consider how it meets the basic criterion for simple harmonic motion and yet differs from the mass-spring system we studied previously.

The Vertical Mass-Spring System

Shown in Fig. 13–12, this system is like the horizontal mass-spring system except that the force of gravity acts in addition to the spring force. What effect does gravity have? First, it changes the equilibrium position, since the spring must now exert an upward force to counteract the weight. If we take the origin at the end of the unstretched spring, and let the positive x-axis point downward, then in the new equilibrium the force-balance condition reads

$$mg - kx_1 = 0,$$

where x_1 is the new equilibrium position with the weight attached. Solving

Fig. 13–12
A vertical mass-spring system.

for x_1 gives

$$x_1 = \frac{mg}{k}.$$

The motion of the system is described by Newton's second law, which, with the gravitational force included, becomes

$$m\frac{d^2x}{dt^2} = -kx + mg, \tag{13–15}$$

where x is measured from the original, horizontal equilibrium. This equation resembles Equation 13–4 for the horizontal mass-spring system, except for the additional constant term mg. To solve the equation, we introduce a new position variable x', defined as the displacement from the new equilibrium position x_1:

$$x' = x - x_1 = x - \frac{mg}{k}.$$

Solving for x and substituting into Equation 13–15, we have

$$m\frac{d^2}{dt^2}\left(x' + \frac{mg}{k}\right) = -k\left(x' + \frac{mg}{k}\right) + mg.$$

The derivative of the constant mg/k is zero, so the left-hand side of this equation is just d^2x'/dt^2. Then, distributing the $-k$ on the right-hand side, we have

$$m\frac{d^2x'}{dt^2} = -kx' - k\left(\frac{mg}{k}\right) + mg,$$

or

$$m\frac{d^2x'}{dt^2} = -kx'. \tag{13–16}$$

Equation 13–16 is identical to Equation 13–4 except for the substitution of x' for x. The analysis of Equation 13–4 therefore applies here, and we conclude that the vertical mass-spring system oscillates about its equilibrium position in the same sinusoidal way and at the same frequency as its horizontal counterpart oscillates about its equilibrium. The effect of gravity is *only* to displace the equilibrium; it has no effect on the frequency or time dependence of the motion.

The Torsional Oscillator

A torsional oscillator is a system that is subject to a restoring torque when it is displaced from equilibrium. Common examples include the balance wheel in a mechanical watch (Fig. 13–13) and a child who twists up the ropes on a swing (Fig. 13–14). In these and all other torsional oscillators, the restoring torque sets up a back-and-forth rotational motion about the torque-free equilibrium state. When the restoring torque is directly proportional to the angular displacement from equilibrium, then the result is simple harmonic motion.

Fig. 13–13
The balance wheel in a watch is a torsional oscillator whose period is the basic time interval of the watch. The coiled hairspring provides the restoring torque.

Fig. 13–14
Twisting the swing ropes gives rise to a torque that tends to untwist them. The result is torsional oscillation about the untwisted equilibrium position.

A simple example of a torsional oscillator consists of an object of rotational inertia I suspended from a string, rope, or wire that develops a torque when it is twisted (Fig. 13–15). When the restoring torque is directly proportional to the angular displacement, we can write

$$\tau = -\kappa\theta, \tag{13–17}$$

where τ is the torque and θ the angular displacement. The constant κ is called the **torsional constant,** and is the rotational analog of the spring constant; it gives the torque developed per radian of twist.

The behavior of the torsional oscillator is described by the rotational analog of Newton's law (Equation 12–14):

$$\tau = I\alpha,$$

where α is the angular acceleration. Using Equation 13–17 for τ, and writing α as the second time derivative of angular displacement, the analog of Newton's law becomes

$$I\frac{d^2\theta}{dt^2} = -\kappa\theta. \tag{13–18}$$

This equation is identical to Equation 13–4 for the linear oscillator, except that I replaces m, θ replaces x, and κ replaces k. We need not repeat the mathematical analysis that led to the solution of Equation 13–4; rather, in direct analogy with the solution 13–15, we can write

$$\theta = \theta_0 \cos(\omega t + \phi), \tag{13–19}$$

where θ_0 and ϕ are the amplitude and phase constant as determined by the initial conditions, and where ω is given by analogy with Equation 13–12:

$$\omega = \sqrt{\frac{\kappa}{I}}. \tag{13–20}$$

Expressions for the period and frequency in Hz follow directly from Equations 13–6 and 13–7.

In addition to its timekeeping functions, the torsional oscillator provides an accurate way of measuring rotational inertia, as the example below illustrates.

torsional constant

Fig. 13–15
A simple torsional oscillator, consisting of a disk suspended by a wire.

Example 13–3

A Torsional Oscillator

An irregular object is suspended from a fiber of torsional constant $\kappa = 3.1$ N·m/rad. When the system is set into torsional oscillation, 68 complete oscillations occur in 2.0 minutes. What is the rotational inertia of the object about an axis lying along the fiber?

Solution

Solving Equation 13–20 for I, we have

$$I = \frac{\kappa}{\omega^2} = \frac{\kappa}{(2\pi f)^2} = \frac{3.1 \text{ N·m/rad}}{[(2\pi)(68 \text{ cycles}/120 \text{ s})]^2} = 0.24 \text{ kg·m}^2.$$

The Pendulum

simple pendulum

A **simple pendulum** consists of a point mass suspended from a massless cord that cannot stretch. This idealization is approximated by many physical systems in which a mass of small extent is suspended by a long structure of much lower mass.

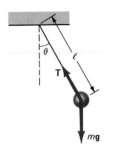

Fig. 13–16
A simple pendulum, showing the forces acting.

Figure 13–16 shows a pendulum of mass m and length ℓ displaced slightly from its equilibrium. We could analyze this situation either in terms of the forces on the mass m or, because the motion is a rotation about the suspension point, in terms of torque. We choose the latter because it is readily generalized to other types of pendulum. (Problem 20 analyzes the simple pendulum in terms of forces alone.) From Fig. 13–16, we see that tension force in the cord, acting along the line between the suspension point and the mass, exerts no torque. When the pendulum is displaced from equilibrium, however, the gravitational force does exert a torque whose magnitude is given by Equation 11–2:

$$\tau = mg\ell \sin\theta$$

(see Fig. 13–16). The direction of this torque is such that it tends to rotate the pendulum back toward equilibrium. Writing the rotational analog of Newton's law, $\tau = I\alpha$, we then have

$$I \frac{d^2\theta}{dt^2} = -mg\ell \sin\theta, \tag{13–21}$$

where the minus sign shows that the torque is opposite to the angular displacement, and where we have written the angular acceleration α as the second time derivative of angular displacement. Equation 13–21 looks somewhat analogous to Equation 13–4, with θ replacing x, I replacing m, and $mg\ell$ replacing k. But not quite—the right-hand side contains not θ but $\sin\theta$, so that the restoring torque is not *directly proportional* to the angular displacement. Because of this, Equation 13–21 does not, in general, describe simple harmonic motion. Although the motion of the system is usually oscillatory,* it does not have the simple sinusoidal time dependence of Equation 13–5, and the frequency is not truly independent of amplitude. The exact solution of Equation 13–21 is very difficult, and is usually done only in advanced mechanics courses. Problem 55 explores some aspects of the full pendulum problem.

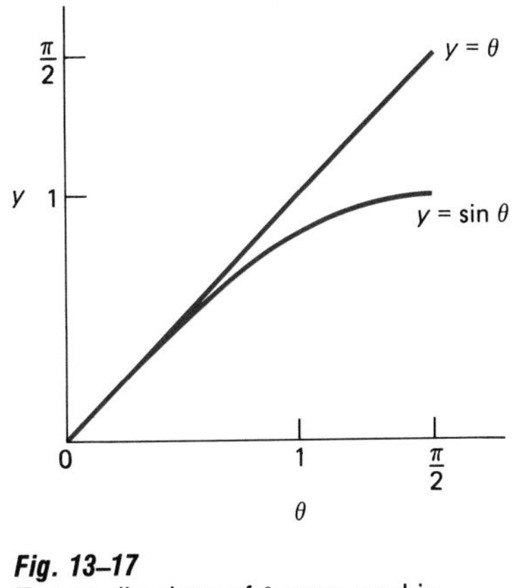

If, however, the amplitude of the pendulum motion is small, then it *approximates* simple harmonic motion. For small angles (much less than 1 radian), the sine of the angle is approximately equal to the radian measure of the angle itself (Fig. 13–17). For example, an angle of 0.11 rad has a sine equal to 0.1097—a difference of only 0.3 per cent—while even a 30° angle (0.52 rad) has a sine of 0.50, within 4 per cent of its radian measure. Thus if we restrict ourselves to small-amplitude oscillations, we can obtain an approximate equation by replacing $\sin\theta$ with θ itself:

$$I \frac{d^2\theta}{dt^2} = -mg\ell\theta.$$

Fig. 13–17
For small values of θ measured in radians, $\sin\theta$ and θ itself are nearly equal.

---

*It may not even be oscillatory. If given enough energy, the mass may go "over the top," exhibiting circular motion with nonuniform speed.

Comparing this approximate equation with Equation 13–4, we see immediately that the pendulum will exhibit simple harmonic motion. Since I plays the role of m, and $mg\ell$ plays the role of k, we can find the frequency by replacing m with I and k with $mg\ell$ in Equation 13–12:

$$\omega = \sqrt{\frac{mg\ell}{I}}. \tag{13-22}$$

For a *simple* pendulum, the rotational inertia I is that of a point mass m a distance ℓ from the rotational axis, or $I = m\ell^2$ (see Equation 12–15). Then we have

$$\omega = \sqrt{\frac{mg\ell}{m\ell^2}} = \sqrt{\frac{g}{\ell}}, \quad \text{(simple pendulum)} \tag{13-23}$$

or, from Equation 13–6,

$$T = \frac{2\pi}{\omega} = 2\pi\sqrt{\frac{\ell}{g}}. \quad \text{(simple pendulum)} \tag{13-24}$$

These equations show that the frequency and period of a simple pendulum are independent of its mass, depending only on length and gravitational acceleration. You can confirm this prediction experimentally by sitting on a swing next to another swing with a small child on it. Even though your mass is greater, both of you have nearly the same swing period.

Example 13–4
Tarzan and the Simple Pendulum

Tarzan stands on a branch, petrified with fear as a leopard approaches. Fortunately, Jane is on a branch of the same height in a nearby tree, 8.0 m away, and she is holding a 25-m-long vine of negligible mass attached directly above the point midway between her and Tarzan (Fig. 13–18). Seeing Tarzan in trouble, Jane grasps the vine and steps off her branch with negligible initial velocity. How soon does she reach Tarzan? When she does reach him, she grabs him and lets the vine take the two of them back to safety. How long does the return journey take, if Tarzan's mass is 50 per cent greater than Jane's? Use the approximation $\sin\theta \approx \theta$ throughout. How good is this approximation?

Solution

Jane and the vine constitute a pendulum that takes half a period to reach Tarzan. The period is given by Equation 13–24:

$$T = 2\pi\sqrt{\frac{\ell}{g}} = 2\pi\sqrt{\frac{25\text{ m}}{9.8\text{ m/s}^2}} = 10\text{ s},$$

so that Jane reaches Tarzan in 5.0 s. Since the period of a simple pendulum is independent of its mass, the return trip with Tarzan takes the same time.

How good is the approximation $\sin\theta \approx \theta$? From Fig. 13–18, we see that

$$\sin\theta_0 = \frac{4.0\text{ m}}{25\text{ m}} = 0.16.$$

Then $\theta_0 = \sin^{-1}(0.16) = 0.1607$ rad, for a difference of only 0.43 per cent at the greatest angular displacement. Remember to set your calculator to radians for this comparison of $\sin\theta$ with θ.

Fig. 13-18
Tarzan in trouble (Example 13-4).
Lengths are not to scale.

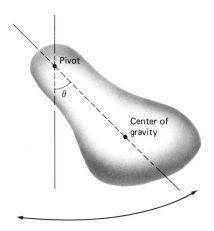

Fig. 13-19
A physical pendulum.

physical pendulum

A **physical pendulum** is an object of arbitrary shape that is free to swing in a plane (Fig. 13–19). The physical pendulum differs from the simple pendulum in that its mass is distributed over the whole length of the object, rather than being concentrated at the end. In our analysis of the simple pendulum, we used this distinction only at the very end, when we wrote $m\ell^2$ for the rotational inertia. Therefore, our analysis before that point applies to the physical pendulum as well. In particular, Equation 13–22 gives correctly the frequency of a physical pendulum executing small-amplitude oscillations, provided we interpret ℓ as the distance from the pivot to the center of gravity, rather than the overall length.

Example 13–5
A Physical Pendulum

A thin rod of mass m and length ℓ is suspended from a pivot at one end. Find the period of small-amplitude oscillations. What would be the length of a simple pendulum of the same period?

Solution

The center of gravity lies halfway along the rod, so we use $\ell/2$ in place of ℓ in Equation 13–22. Combining Equation 13–22 with Equation 13–6 then gives the period:

$$T = \frac{2\pi}{\omega} = 2\pi\sqrt{\frac{I}{mg\ell/2}}.$$

Table 12–2 shows that the rotational inertia of a rod about one end is $I = \frac{1}{3}m\ell^2$. Then we have

$$T = 2\pi\sqrt{\frac{m\ell^2}{3mg\ell/2}} = 2\pi\sqrt{\frac{2\ell}{3g}}.$$

Comparing this result with Equation 13–24, we see that this is the same period as that of a simple pendulum with length $\frac{2}{3}\ell$. Does this make sense? Yes: in the physical pendulum, much of the mass is concentrated more closely to the pivot, making the effective length shorter than the overall length.

13.4
SIMPLE HARMONIC MOTION IN TWO DIMENSIONS

So far, we have discussed systems in which motion is confined to a single one-dimensional path—either the straight line of a mass-spring system or the arc of a pendulum. But for many systems, oscillatory motion in several directions is possible at once. One example is a pendulum that is free to swing in any direction (Fig. 13–20). We could set this pendulum into one-dimensional simple harmonic motion by displacing it in some direction and then releasing it. But suppose we displace it in the x-direction, and as we release it give it an impulse in the y-direction. The result is a composite motion whose maximum x-displacement (and therefore zero x-velocity) occurs when the y-displacement is zero and the y-velocity a maximum. As the pendulum swings in toward $x=0$, it moves rapidly away from $y=0$. Since the periods of the x- and y-motions are the same (we are making the assumption of small amplitude oscillations), the pendulum reaches its maximum y-displacement as it passes through $x=0$. The result is a curved path—in fact, an ellipse (Fig. 13–21).

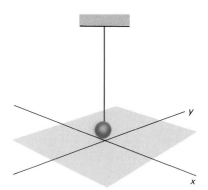

Fig. 13–20
The pendulum is free to swing in any direction. We can analyze its motion in terms of components along the x- and y-axes.

If the amplitudes of the x- and y-motions are equal, then the motion is circular. We can prove this by describing the motion vectorially. With maximum x-displacement at $t=0$, the x-motion is described by $x = r_0\cos\omega t$, where r_0 is the amplitude of both x- and y-components. With maximum velocity at $t=0$, the y-motion is described by $y = r_0\cos(\omega t - \pi/2) = r_0\sin\omega t$. A vector description of the motion is then given by

$$\mathbf{r} = x\hat{\mathbf{i}} + y\hat{\mathbf{j}} = r_0(\cos\omega t\,\hat{\mathbf{i}} + \sin\omega t\,\hat{\mathbf{j}}), \qquad (13\text{–}25)$$

where $\mathbf{r}$ is the displacement from equilibrium. The magnitude of $\mathbf{r}$ is then given by the Pythagorean theorem:

$$r^2 = x^2 + y^2 = r_0^2(\cos^2\omega t + \sin^2\omega t) = r_0^2,$$

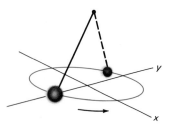

Fig. 13–21
When the motions are out of phase, the pendulum traces out an ellipse.

since $\cos^2\omega t + \sin^2\omega t = 1$ for any value of ωt.

We could have analyzed this circular pendulum motion without any reference to simple harmonic motion. In fact, we did so in Example 5–9, to illustrate uniform circular motion. In these two different approaches to the same physical situation, we can see the link between simple harmonic motion and uniform circular motion, and can understand why we speak of angular frequency in simple harmonic motion even when there are no physical angles involved. Our analysis of the circular pendulum in this chapter shows that its motion may be resolved into two simple harmonic motions, at right angles in space and $\pi/2$ rad out of phase. If we look at the pendulum from above, we see it trace out the circle—the composite motion. But if we peer in a horizontal plane at the pendulum bob (Fig. 13–22), we see only the projection of that circular motion onto one of the coordinate axes; it appears to us that the bob is executing simple harmonic motion. The period of the circular motion is the same as that of each simple harmonic motion component, and therefore the angular frequencies of the circular and harmonic motions are equal as well. The angular frequency of the circular motion has obvious physical meaning—it is the number of radians swept out in a unit time. We can think of the angular frequency of simple harmonic motion as the angular frequency of a circular motion of which that simple harmonic motion might be a component. The "angle" ωt then corresponds directly to a physical angle in the circular motion. Similarly, the phase constant ϕ corresponds to the initial angular displacement in the circular motion (Fig. 13–23).

We have considered the cases of two harmonic motions either in phase or $\pi/2$ rad out of phase. In the most general case, we can have two motions of different amplitudes with arbitrary phase difference. In that case, the motion is elliptical, but the axes of the ellipse do not generally correspond with the x- and y-axes. Problem 53 explores this situation further.

With the pendulum we have been discussing, the frequencies of the x and y harmonic motions are necessarily the same. But other two-dimensional oscillators may have different frequencies for the two perpendicular components of motion (Fig. 13–24). In this case the motion becomes interestingly complex. If, for example, the frequencies form a 2:1 ratio, the

Fig. 13–22
If we observe an object in the plane of its circular motion, we see only one component of that motion. This component is simple harmonic motion.

Fig. 13–23
Simple harmonic motion as a projection of uniform circular motion. The quantities ωt and ϕ in simple harmonic motion correspond to physical angles in the associated uniform circular motion.

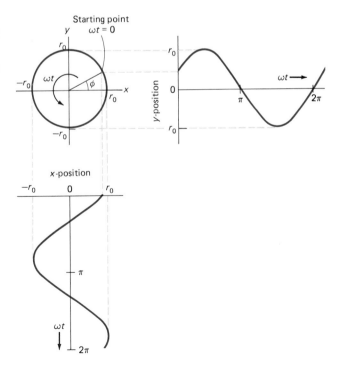

lissajous figures

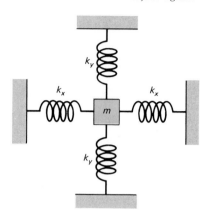

Fig. 13–24
A two-dimensional mass-spring system. If the spring constants differ in the x- and y-directions, the frequencies of the associated simple harmonic motions also differ.

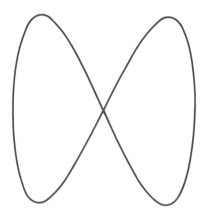

Fig. 13–25
A 2:1 frequency ratio gives rise to a figure-eight-shaped path.

oscillating object traverses a figure-eight-shaped path (Fig. 13–25). In general, such paths are called **lissajous figures.** They can be quite complex, striking, and even beautiful (Fig. 13–26). If the frequencies are not related as the ratio of a pair of integers, then the path is not closed but actually fills an entire region bounded by the amplitudes in both directions (Fig. 13–27).

13.5
ENERGY IN SIMPLE HARMONIC MOTION

Displacing an object from stable equilibrium requires energy, as we do work against the restoring force. In a mass-spring system, that work goes into increasing the potential energy of the spring. If we then release the mass, it accelerates toward its equilibrium position, gaining kinetic energy. Where does it get the kinetic energy? From the potential energy of the spring, which consequently decreases. When the mass moves through its equilibrium position, its potential energy is zero, and all its energy is now kinetic. As it moves away from equilibrium, the mass slows down as the spring stretches. Its kinetic energy decreases and potential energy again increases. If there is no energy loss or gain, this process repeats indefinitely. Oscillatory motion is therefore a process whereby energy is transferred back and forth between its kinetic and potential forms (Fig. 13–28).

In the case of a mass-spring system, the potential energy is given by Equation 7–3:

$$U = \tfrac{1}{2}kx^2,$$

where x is the displacement from equilibrium. Meanwhile, the kinetic energy is $K = \tfrac{1}{2}mv^2$. We can illustrate explicitly the interchange of kinetic and potential energy in simple harmonic motion by using x from Equation 13–5 and v (that is, dx/dt) from Equation 13–8 in the expressions for potential and kinetic energy. Then we have

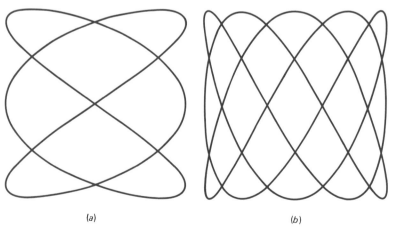

(a)   (b)

Fig. 13–26
Lissajous figure produced by perpendicular oscillations with (a) 2:3 and (b) 5:3 frequency ratios.

$$U = \tfrac{1}{2}kx^2 = \tfrac{1}{2}k(x_0\cos\omega t)^2 = \tfrac{1}{2}kx_0^2\cos^2\omega t \qquad (13\text{--}26a)$$

and

$$K = \tfrac{1}{2}mv^2 = \tfrac{1}{2}m(-x_0\omega\sin\omega t)^2 = \tfrac{1}{2}m\omega^2 x_0^2\sin^2\omega t = \tfrac{1}{2}kx_0^2\sin^2\omega t, \qquad (13\text{--}26b)$$

where we have used the fact that $\omega^2 = k/m$ in writing the final form of Equation 13–26. We set the phase constant to zero in Equation 13–26 to describe motion starting from maximum displacement. Comparing our expressions for potential and kinetic energy, we see that both have the same maximum value—$\tfrac{1}{2}kx_0^2$—equal to the initial potential energy of the stretched spring. Since $\sin\omega t = \cos(\omega t - \pi/2)$, the only difference between the two lies in their relative phase—the potential energy is a maximum when the kinetic energy is zero, and vice versa. What about the total energy? It is

$$E = U + K = \tfrac{1}{2}kx^2 + \tfrac{1}{2}mv^2 \qquad (13\text{--}27)$$

or

$$E = \tfrac{1}{2}kx_0^2\cos^2\omega t + \tfrac{1}{2}kx_0^2\sin^2\omega t = \tfrac{1}{2}kx_0^2,$$

where we have used the fact that $\sin^2\omega t + \cos^2\omega t = 1$. Although both kinetic and potential energy vary with time, their sum remains constant (Fig. 13–29).

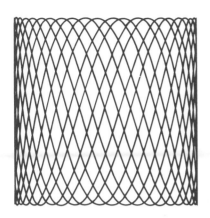

Fig. 13–27
In this case the frequency ratio is π, and the curve shown results after 7 cycles of the lower-frequency oscillator. Eventually the entire rectangular area would be traversed.

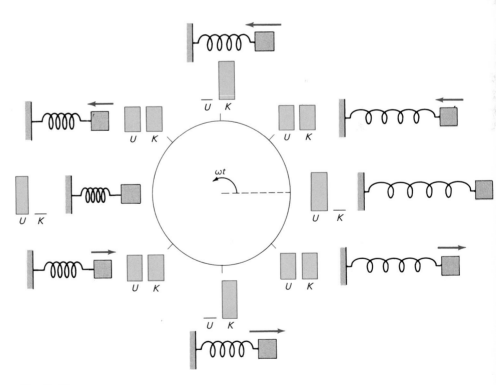

Fig. 13–28
Kinetic and potential energy in a simple harmonic oscillator. Figure shows the spring compression and velocity of oscillating mass, along with relative amounts of potential and kinetic energy, at eight points in the cycle.

Fig. 13–29
Fig. 13–29
The potential and kinetic energy of a simple harmonic oscillator vary with time, but their sum is constant.

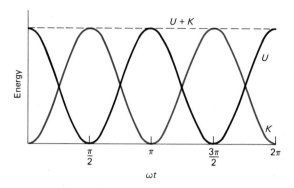

Example 13–6
Energy in Simple Harmonic Motion

A mass m is mounted on a spring of constant k, and the system is displaced a distance x_0 from equilibrium. Use energy conservation to find the maximum velocity attained by the mass once it is released, and show that the result is consistent with Equation 13–8. Find also the velocity when the potential and kinetic energies are equal.

Solution

The initial potential energy, which is the total energy, is $\frac{1}{2}kx_0^2$. At maximum velocity, all of this energy has been converted to kinetic energy, so that

$$\tfrac{1}{2}mv_{max}^2 = \tfrac{1}{2}kx_0^2,$$

or

$$v_{max} = \sqrt{\frac{k}{m}}\, x_0 = \omega x_0,$$

in agreement with the maximum velocity from Equation 13–8. A third way to obtain this maximum velocity would be to integrate the acceleration over time (see Problem 13).

When the kinetic and potential energies are equal, each must be equal to half the total energy. Therefore, for the kinetic energy,

$$\frac{1}{2}mv^2 = \frac{1}{2}\left(\frac{1}{2}mv_{max}^2\right),$$

so that

$$v = \frac{1}{\sqrt{2}}v_{max} = \frac{\sqrt{2}}{2}\omega x_0.$$

Recall that Equation 7–3 for the potential energy of a spring was derived by integrating the spring force, $-kx$, over distance. Since every simple harmonic oscillator has a restoring force or torque that is directly proportional to displacement, integration of that force or torque always results in a potential energy function that depends on the square of the displacement. Conversely, any system whose potential energy function is quadratic in the displacement exhibits simple harmonic motion. For many systems, it is easier to evaluate the potential energy than the force, so that this criterion is often a more useful one. Also, complicated potential energy func-

Fig. 13–30
Near its minima, which are points of stable equilibrium, the potential energy curve approximates a parabola.

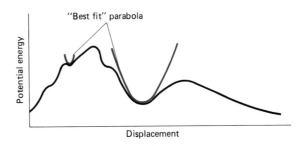

tions are often approximately quadratic near their stable equilibrium points (Fig. 13–30), so that small-amplitude oscillations about these points represent simple harmonic motion. As the example below illustrates, we can determine the properties of the motion by comparing the potential energy function with that of a linear or torsional oscillator, as appropriate.

Example 13–7

The Potential Energy Function for a Molecule

When a molecule of hydrochloric acid (HCl) is stretched or compressed from its equilibrium, its potential energy increases. For small displacements from equilibrium, the potential energy of the molecule is given by

$$U = \alpha x^2,$$

where $\alpha = 247$ J/m^2, and where we take the zero of potential energy at equilibrium. Determine the oscillation frequency of this molecule, under the approximation that the more massive chlorine remains essentially at rest while the hydrogen undergoes oscillatory motion.

Solution

The expression for the potential energy of HCl resembles that of a spring, $\frac{1}{2}kx^2$, with the quantity α playing the role of $\frac{1}{2}k$. So the effective "spring constant" of the molecule is 2α or 594 N/m. The mass of the hydrogen is essentially that of its single proton (see table inside the front cover), so that the oscillation frequency as given by Equation 13–13 is

$$f = \frac{1}{2\pi} \sqrt{\frac{k}{m}} = \frac{1}{2\pi} \sqrt{\frac{594 \text{ N/m}}{1.67 \times 10^{-27} \text{ kg}}} = 9.49 \times 10^{13} \text{ Hz}.$$

Like most molecular oscillations, this one occurs at a frequency associated with infrared radiation; for this reason the structure of molecules can be studied by measuring precisely the amounts of infrared radiation absorbed by a substance at different infrared frequencies, a process known as infrared spectroscopy.

13.6
DAMPED HARMONIC MOTION

In real oscillating systems, forces like friction are always present that dissipate the oscillation energy. Unless energy is somehow added, dissipation eventually brings the system to rest at equilibrium. The motion in this case is said to be **damped.** If the dissipation is sufficiently weak that only a small fraction of the system's energy is removed in each oscillation cycle,

damped harmonic motion

Fig. 13–31
Damped oscillatory motion.

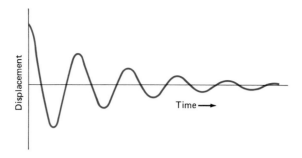

Displacement

Time →

then we expect that the system should behave essentially as in the un-damped case, except for a gradual decrease in the oscillation amplitude (Fig. 13–31).

Can we describe damped motion more quantitatively? Yes, but only if we have a mathematical expression for the damping force. In many systems—especially those involving friction associated with slow motion through a viscous fluid—the damping force is approximately proportional to the velocity, and in the opposite direction:

$$F_d = -bv = -b\frac{dx}{dt},$$

where b is a constant giving the strength of the damping. We can write Newton's law as before, now including the damping force along with the restoring force. For a mass-spring system, we have

$$m\frac{d^2x}{dt^2} = -kx - b\frac{dx}{dt},$$

or

$$m\frac{d^2x}{dt^2} + b\frac{dx}{dt} + kx = 0. \qquad (13-28)$$

We will not solve this equation, but simply state that its solution, provided the damping is not too large, is

$$x = x_0 e^{-bt/2m} \cos(\omega t + \phi). \qquad (13-29)$$

The solution 13–29 describes sinusoidal motion whose amplitude decreases exponentially with time. How fast the amplitude drops depends on the damping constant b and mass m: when $t = 2m/b$, the amplitude has dropped to $1/e$ of its original value (Fig. 13–32a). When the damping is so weak that only a small fraction of the total energy is lost in each cycle, then the frequency ω in Equation 13–29 is essentially equal to the undamped frequency $\sqrt{k/m}$. But with stronger damping, the damping force slows the motion, and the frequency becomes lower. At the same time, the oscillation amplitude decreases more rapidly with time (Fig. 13–32b). As long as any

underdamped

critical damping

oscillation occurs, the motion is said to be **underdamped.** For sufficiently strong damping, though, the effect of the damping force is as great as that of the spring force. Under this condition, called **critical damping,** the system returns to its equilibrium state without undergoing any oscillations (Fig. 13–32c). If the damping is made still stronger, the system becomes

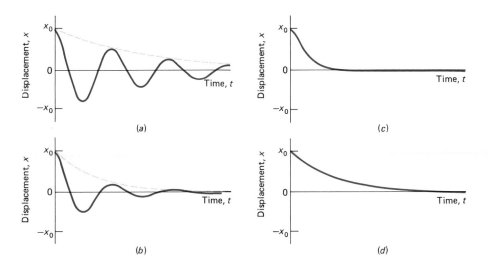

Fig. 13–32
(a) Weak damping, showing sinusoidal oscillations confined within the "envelope" of a decreasing exponential. The time in which the amplitude decreases to 1/e—about one-third—of its original value is considerably longer than the oscillation period, T. (b) Stronger damping, but still underdamped. (c) Critical damping. (d) Overdamping.

overdamped

Fig. 13–33
Dampers resting on piano strings cause the sound-producing oscillations to die out quickly.

overdamped. The damping force now dominates the motion, and as a result the system returns to equilibrium more slowly than with critical damping (Fig. 13–32d).

Many physical systems, ranging from atoms to the human leg, can be modelled as damped oscillators. Engineering systems are often designed with very specific amounts of damping. For example, automobile shock absorbers—the damping element in the automobile suspension—are designed along with the springs to give critical damping. This results in the most rapid return to equilibrium while successfully absorbing the energy imparted to the car by road bumps. When the shock absorbers wear out, the car's suspension becomes underdamped, and the whole car bounces up and down after hitting a bump. Similarly, the strings of a piano are damped—although not critically—immediately after a key is struck so that the note will die out quickly. Depressing the right foot pedal on the piano lifts the dampers from the strings, allowing them to vibrate much longer (Fig. 13–33).

Example 13–8
Damped Harmonic Motion

A damped mass-spring system has $m = 85$ g, $k = 22$ N/m, and $b = 0.016$ kg/s. How many oscillations will it make while its amplitude falls to half its original value?

Solution

The amplitude will be half its original value when the factor $e^{-bt/2m}$ in Equation 13–29 has the value one-half:

$$e^{-bt/2m} = \tfrac{1}{2}.$$

Taking the natural logarithm of both sides gives

$$\frac{-bt}{2m} = -\ln 2,$$

where we have used the facts that $\ln(x)$ and e^x are inverse functions and that $\ln(1/x) = -\ln(x)$. Then

$$t = \frac{2m}{b}\ln 2 = \frac{(2)(0.085 \text{ kg})}{0.016 \text{ kg/s}} \ln 2 = 7.4 \text{ s}$$

is the time for the amplitude to drop to half its original value. For small damping, the period is very close to the undamped period, which is

$$T = 2\pi\sqrt{\frac{m}{k}} = 2\pi\sqrt{\frac{0.085 \text{ kg}}{22 \text{ N/m}}} = 0.39 \text{ s}.$$

Then the number of cycles during the 7.4 s it takes the amplitude to drop in half is

$$\frac{7.4 \text{ s}}{0.39 \text{ s}} = 19.$$

That the number of oscillations is much greater than 1 tells us that the damping is fairly small, justifying our use of the undamped period.

13.7

DRIVEN OSCILLATIONS AND RESONANCE

The left photograph in Figure 13–34 shows a child pushing steadily on a tree with very little effect. Yet on the right we see the same child producing large-amplitude motion of the treetop—an experiment you may have performed in your own childhood. How is this possible? In the vibrating tree, we have a system analogous to a mass-spring system, one that has a natural frequency of oscillation. By pushing the tree at its own natural frequency,

Fig. 13–34
(Left) A child pushing steadily near the bottom of a tree produces very little displacement of the treetop. (Right) By applying small pushes at the tree's natural oscillation frequency, the same child produces large-amplitude motion of the treetop.

the child supplies a small amount of energy each cycle, and that energy accumulates to produce large-amplitude motion.

What would happen if the child pushed the tree very slowly—at a frequency much lower than the natural frequency? This would be very much like the steady push of Fig. 13–34 on the left, and amplitude of the motion would be very small. How about a very high frequency, much higher than the natural frequency? Try it! Vibrate a tree very fast near the bottom, and not much happens at the top. Only when you push at a frequency near the natural frequency can you build up energy in the oscillation.

When an oscillating system is pushed by an external force, we say that the system is **driven,** and the resulting motion is **driven harmonic motion.** We can understand this motion quantitatively by writing Newton's law for a driven system. Consider a mass-spring system, and suppose that the driving force is given by $F = F_0 \cos\omega_d t$, where ω_d is the **driving frequency,** not necessarily the natural frequency of the system. Then Newton's law is

$$m\frac{d^2x}{dt^2} = -kx - b\frac{dx}{dt} + F_0 \cos\omega_d t, \qquad (13\text{–}30)$$

where the first term on the right-hand side is the restoring force, the second the damping force, and the third the driving force. Since the system is being pushed at the driving frequency ω_d, we expect that it will undergo oscillatory motion at this frequency. (It might also oscillate at its natural frequency. But this undriven **transient solution** soon damps out, leaving only **steady-state** motion at the driving frequency.) So we guess that the solution to Equation 13–30 might have the form

$$x = x_0 \cos(\omega_d t + \phi). \qquad (13\text{–}31)$$

By substituting this expression and its derivatives into Equation 13–30 and using trig identities (see Problem 48), it is possible to satisfy Equation 13–30, provided that

$$x_0 = \frac{F_0}{m\sqrt{(\omega_d^2 - \omega_0^2)^2 + b^2\omega_d^2/m^2}} \qquad (13\text{–}32)$$

and

$$\phi = \tan^{-1}\left(\frac{\omega_d b}{m(\omega_0^2 - \omega_d^2)}\right), \qquad (13\text{–}33)$$

where ω_0 is the undamped natural frequency $\sqrt{k/m}$, as distinguished from the driving frequency ω_d.

Equations 13–32 and 13–33 show that the amplitude and phase of the motion vary with driving frequency for a fixed amplitude F_0 of the driving force. Figure 13–35 shows a **resonance curve,** which is just a plot of the amplitude 13–32 as a function of driving frequency, for several values of the damping. As long as the system is underdamped, the curve has a maximum at some nonzero frequency (see Problem 49), and for weak damping that maximum occurs at very nearly the natural frequency. The weaker the damping, the more sharply peaked is the resonance curve. Thus in a weakly damped system, it is possible to build up large-amplitude oscillations with relatively small driving forces. Such **resonance** can cause serious problems in physical systems. For example, we would not want a vary-

(margin notes) driven harmonic motion · driving frequency · transient solution · steady state · resonance curve · resonance

Fig. 13–35
Resonance curves for several damping strengths. ω_0 is the undamped natural frequency $\sqrt{k/m}$.

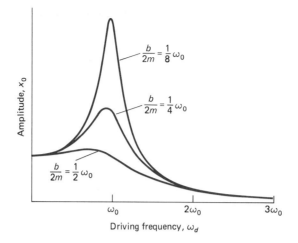

Amplitude, x_0

$\frac{b}{2m} = \frac{1}{8}\,\omega_0$

$\frac{b}{2m} = \frac{1}{4}\,\omega_0$

$\frac{b}{2m} = \frac{1}{2}\,\omega_0$

ω_0 $2\omega_0$ $3\omega_0$

Driving frequency, ω_d

Fig. 13–36
Resonant vibrations in the structure housing a large telescope would cause unacceptable blurring of astronomical images. This housing for the Multiple-Mirror Telescope on Mt. Hopkins in Arizona was intentionally made as small as possible, in part to limit such vibrations.

ing wind to set up vibrations in a skyscraper analogous to those of the tree in Fig. 13–34! (In fact, the device described in Example 13–2 is designed to minimize just such oscillations.) The building housing a large telescope (Fig. 13–36) must be free of resonant vibrations if the telescope is to produce clear images. A famous disaster involving resonance is the 1940 collapse of the Tacoma Narrows Bridge in Washington state (Fig. 13–37). Resonance is also important at the microscopic level. Microwave ovens work by pumping energy into food at a resonant frequency of the water molecule; the resulting oscillation energy is eventually dissipated as heat by damping forces. At the nuclear level, the process called nuclear magnetic resonance (NMR) uses resonant behavior of protons to probe the structure of matter and to produce images useful in medical diagnosis.

Fig. 13–37
(Left and center) Wind-driven resonant oscillations of the Tacoma Narrows Bridge caused its collapse in 1940 only 4 months after it was opened. (Right) The bridge has since been rebuilt, with additional bracing.

SUMMARY

1. **Oscillatory motion** occurs when a system is displaced from stable equilibrium. Restoring forces bring the system back toward equilibrium, and if friction is not great enough to prevent overshoot, the system oscillates back and forth about equilibrium.

2. **Simple harmonic motion** is a special case of oscillatory motion that occurs when the restoring force is directly proportional to displacement, a situation that holds approximately for many systems subject to small displacements from equilibrium. Simple harmonic motion is described by the equation

$$x = x_0 \cos(\omega t + \phi),$$

where x_0 is the maximum displacement, or **amplitude,** and ϕ the **phase constant,** which tells when the maximum amplitude occurs in relation to the time $t = 0$. ω is the **angular frequency** of oscillation, and is given by $\omega = \sqrt{k/m}$ for a mass m oscillating on a spring of constant k. Analogous expressions hold for other systems, including torsional oscillators and pendulums:

$$\omega = \sqrt{\frac{\kappa}{I}}, \quad \text{(torsional oscillator)}$$

$$\omega = \sqrt{\frac{g}{\ell}}, \quad \text{(simple pendulum)}$$

and

$$\omega = \sqrt{\frac{mg\ell}{I}}. \quad \text{(physical pendulum)}$$

The frequency in Hz (or cycles/second) and the period can be found from the angular frequency by using the equations

$$f = \frac{\omega}{2\pi}$$

and

$$T = \frac{1}{f} = \frac{2\pi}{\omega}.$$

3. When an object undergoes simple harmonic motion in two perpendicular directions, the resulting composite motion is, most generally, elliptical. For the special case when the two component motions have the same amplitude but differ in phase by $\pi/2$, the composite motion is circular. Conversely, simple harmonic motion may be considered a component of circular motion, with the quantities ωt and ϕ corresponding to real angles in the related circular motion.

4. In oscillatory motion, energy is transferred back and forth between potential and kinetic forms. In the absence of energy input or loss mechanisms, the total energy remains constant even as the relative amounts of potential and kinetic energy change. The special case of simple harmonic motion results when the potential energy is a quadratic function of displacement—a condition that is met approximately for many systems near stable equilibrium.

5. When energy loss occurs—for example, through friction—then the motion is **damped.** In an **underdamped** system, the damping force is relatively small, and the resulting damped motion is described by

$$x = x_0\, e^{-bt/2m} \cos(\omega t + \phi),$$

where b is a proportionality constant between damping force and velocity. When the damping is very small, the frequency ω is essentially the same as the undamped frequency $\sqrt{k/m}$; for larger damping, the frequency drops because of the slowing effect of friction. As the damping force is increased, the system eventually undergoes **critical damping,** in which the system reaches equilibrium without oscillation. For even greater damping, the system is **overdamped,** reaching equilibrium without oscillation, but more slowly than with critical damping.

6. When an oscillatory system is driven by an external force, it responds with motion at the driving frequency. The amplitude of motion is greatest for driving frequencies near the natural frequency of the system. Called **resonance,** this phenomenon is most prominent in weakly damped systems. The amplitude of resonant oscillations is given by

$$x_0 = \frac{F_0}{m\sqrt{(\omega_d^2 - \omega_0^2)^2 + b^2\omega_d^2/m^2}},$$

where F_0 is the amplitude of the driving force, ω_d its frequency, m the mass, b the damping constant, and $\omega_0 = \sqrt{k/m}$ the undamped natural frequency of the system.

QUESTIONS

1. Is a vertically bouncing ball an example of oscillatory motion? Of simple harmonic motion? Explain.
2. The vibration frequencies of atomic-sized systems are much higher than those of macroscopic mechanical systems. Why should this be?
3. What happens to the frequency of a simple harmonic oscillator when the spring constant is doubled? When the mass is doubled?
4. If the spring of a simple harmonic oscillator is cut in half, what happens to the frequency?

5. How does the frequency of a simple harmonic oscillator depend on its amplitude?

6. How would the frequency of a horizontal mass-spring system change if it were taken to the moon? Of a vertical mass-spring system? Of a simple pendulum?

7. In what ways is the motion of a damped pendulum not exactly periodic?

8. When is the acceleration of an undamped simple harmonic oscillator zero? When is the velocity zero?

9. Is the acceleration of a damped harmonic oscillator ever zero? Explain.

10. If the spring in a mass-spring system is not massless, how will this affect the frequency?

11. What will happen to the period of a mass-spring system if it is placed in a jetliner accelerating down the runway? What will happen to the period of a pendulum in the same situation?

12. Even if the springs in the two-dimensional oscillator of Fig. 13–24 are ideal, the system exhibits simple harmonic motion only for small-amplitude oscillations. Why?

13. Explain how simple harmonic motion might be used to determine the mass of objects in an orbiting spacecraft.

14. What is the phase constant of each of the simple harmonic motions shown in Fig. 13–38?

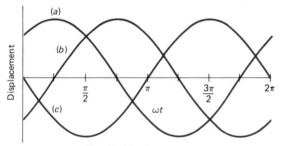

Fig. 13–38   Question 14

15. How does the solution of a differential equation differ from that of an algebraic equation?

16. One pendulum consists of a solid rod of mass m and length ℓ, another of a compact ball of the same mass m on the end of a massless string of the same length ℓ. Which has the greater period? Why?

17. Give an example of a system that can simultaneously undergo mass-spring simple harmonic motion, torsional simple harmonic motion, and pendulum motion.

18. Why doesn't the period of a simple pendulum depend on its mass?

19. When the amplitude of a pendulum's motion becomes large enough that the approximation $\sin\theta \approx \theta$ is no longer valid, the period lengthens with increasing amplitude. Why should this be?

20. Could a mass suspended from a string ever undergo non-oscillatory motion?

21. The needle of a sewing machine moves up and down, yet its driving force comes from an electric motor executing rotary motion. What sort of mechanism might convert rotary to up-and-down motion? *Hint:* What relevance would this conversion mechanism have for the present chapter?

22. The x- and y-components of motion of a body are both simple harmonic with the same frequency and amplitude. What shape is the path of the body if the component motions are (a) in phase; (b) $\pi/2$ out of phase; (c) $\pi/4$ out of phase?

23. List five oscillatory systems and identify the damping forces that act on each.

24. Why is critical damping desirable in many mechanical systems?

25. Explain why the frequency of a damped system is lower than that of the equivalent undamped system.

26. The quantity $2m/b$ in Equation 13–29 has the units of time. How should this time compare with the period of the motion if damping is to be considered small?

27. Opera singers have been known to break glasses with their voices. How?

28. Depressing the rightmost pedal on a piano causes notes to sound much longer than usual. What quantity changes in Equation 13–28 when the pedal is pressed?

29. Real physical systems often have more than one resonant frequency. How can this come about?

30. What does the collapse of the Tacoma Narrows Bridge (Fig. 13–37) say about the damping forces acting on the bridge?

PROBLEMS

Sections 13.1–13.2   Oscillations and Simple Harmonic Motion

1. Show that ω and f in Equations 13–12 and 13–13 have the units of inverse time.

2. Write expressions for simple harmonic motion (a) with amplitude 10 cm, frequency 5.0 Hz, and with maximum displacement at $t=0$ and (b) with amplitude 2.5 cm, angular frequency 5.0 s^{-1}, and with maximum velocity at $t=0$.

3. Determine the amplitude, angular frequency, and phase constant for each of the simple harmonic motions shown in Fig. 13–39.

4. A 200-g mass is attached to a spring of constant $k=5.6$ N/m, and set into oscillation with amplitude $x_0=25$ cm. Determine (a) the frequency in Hz, (b) the

period, (c) the maximum velocity, and (d) the maximum force in the spring.

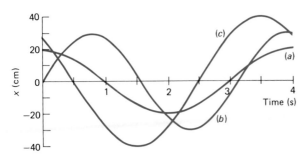

Fig. 13-39 Problem 3

5. A simple model of a carbon dioxide (CO_2) molecule consists of three mass points (the atoms) connected by two springs (electrical forces), as suggested in Fig. 13-40. One way this system can oscillate is if the carbon atom stays fixed and the two oxygens move symmetrically on either side of it. If the frequency of this oscillation is 4.0×10^{13} Hz, what is the effective spring constant? The mass of an oxygen atom is 16 u.

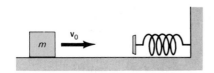

Fig. 13-40 Problem 5

6. Sketch the following simple harmonic motions on the same graph: (a) $x = (15 \text{ cm})[\cos(2.5t + \pi/2)]$; (b) motion with amplitude 30 cm, period 5.0 s, phase constant 0; (c) motion with amplitude 15 cm, frequency 0.40 Hz, phase constant 0.

7. Two identical mass-spring systems consist of 430-g masses on springs of constant $k = 2.2$ N/m. Both are displaced from equilibrium and the first released at time $t = 0$. How much later should the second be released so the two oscillations differ in phase by $\pi/2$?

8. The quartz crystal in a digital quartz watch executes simple harmonic motion at 32,768 Hz. (This frequency is chosen because it is 2^{15} Hz, so that after 15 circuits that divide frequency in half, an electrical signal at 1.00000 Hz is achieved.) If each face of the crystal undergoes a maximum displacement of 100 nm, find the maximum velocity and acceleration of the crystal faces.

9. A mass m slides along a frictionless horizontal surface at speed v_0. It strikes a spring of constant k attached to a rigid wall, as shown in Fig. 13-41. After a completely elastic encounter with the spring, the

Fig. 13-41 Problem 9

mass heads back in the direction it came from. In terms of k, m, and v_0, determine (a) how long the mass is in contact with the spring and (b) the maximum compression of the spring.

10. A 50-g mass is attached to a spring and undergoes simple harmonic motion. Its maximum acceleration is 15 m/s² and its maximum speed is 3.5 m/s. Determine (a) the angular frequency, (b) the spring constant, and (c) the amplitude of the motion.

11. Show by substitution that $x = x_0 \sin\omega t$ is a solution to Equation 13-4.

12. Show by substitution that $x = A\cos\omega t - B\sin\omega t$ is a solution to Equation 13-4, and that this form is equivalent to Equation 13-5 with

$$x_0 = \sqrt{A^2 + B^2} \quad \text{and} \quad \phi = \tan^{-1}(B/A).$$

13. Integrating the nonconstant acceleration of a harmonic oscillator over time from the time of maximum displacement to the time of zero displacement should give the velocity at zero displacement. Carry out this integration, using Equation 13-9 for the acceleration, and show that your answer is just the maximum velocity.

14. A 500-g block on a frictionless surface is connected to a rather limp spring of constant $k = 8.7$ N/m. A second block rests on the first, and the whole system executes simple harmonic motion with a period of 1.8 s. When the amplitude of the motion is increased to 35 cm, the upper block just begins to slip. What is the coefficient of static friction between the blocks?

Section 13.3 *Applications of Simple Harmonic Motion*

15. A 640-g hollow ball 21 cm in diameter is suspended by a wire and is undergoing torsional oscillations at a frequency of 0.78 Hz. What is the torsional constant of the wire?

16. A physics student, bored by a lecture on simple harmonic motion, idly picks up his pencil (mass 9.2 g, length 17 cm) by the tip with his frictionless fingers, and allows it to swing back and forth with small amplitude. If it completes 6279 full cycles during the lecture, how long does the lecture last? *Hint:* See Example 13-5.

17. A pendulum of length ℓ is mounted in a rocket. What is its period if the rocket is (a) at rest on its launch pad; (b) accelerating upward with acceleration $a = \frac{1}{2}g$; (c) accelerating downward with acceleration $a = \frac{1}{2}g$; (d) in free fall?

18. While waiting for your plane to leave the airport gate, you take out your keys and suspend them by a thread pulled from your coat. You set the resulting pendulum oscillating and note that it completes 90 full cycles in 1 minute. Finally the plane begins to take off. During the 60 s it takes to become airborne, your pendulum completes 91 cycles. Determine (a) the acceleration (assumed constant) of the plane and (b) its speed as it leaves the ground. *Hint:* Remember that the plane's acceleration and the force of gravity are at right angles.

19. A mass is attached to a vertical spring, which then goes into oscillation. At the high point of the oscillation, the spring is in the original unstretched equilibrium position it had before the mass was attached; the low point is 5.8 cm below this. What is the period of oscillation?

20. Derive the period of a simple pendulum by considering the horizontal displacement x and the force acting on the bob, rather than the angular displacement and torque.

21. A solid disk of radius R is suspended from a spring of linear spring constant k and torsional constant κ, as shown in Fig. 13–42. In terms of k and κ, what value of R will give the same period for the vertical and torsional oscillations of this system?

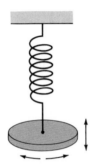

Fig. 13–42   Problem 21

22. A thin steel beam 8.0 m long is suspended from a crane and is undergoing torsional oscillations. Two 75-kg steelworkers leap onto opposite ends of the beam, as shown in Fig. 13–43. If the frequency of torsional oscillations diminishes by 20%, what is the mass of the beam?

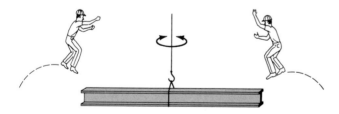

Fig. 13–43   Problem 22

23. Geologists use an instrument called a **gravimeter** to measure the local acceleration of gravity, thereby learning about the variations in density of rocks in earth's crust. A particular gravimeter uses the period of a 1-m-long pendulum to determine g. If g is to be measured to within 1 mgal (1 gal = 1 cm/s²) and if the period can be measured with arbitrary accuracy, how accurately must the length of the pendulum be known?

24. A pendulum consists of a 320-g solid ball 15.0 cm in diameter, suspended by an essentially massless string 80.0 cm long. Calculate the period of this pendulum, treating it first as a simple pendulum and then as a physical pendulum. How big is the error introduced by the simple pendulum approximation? *Hint:* Remember the parallel axis theorem.

25. A thin, uniform hoop of mass M and radius R is suspended from a thin horizontal rod and set oscillating with small amplitude, as shown in Fig. 13–44. Show that the period of the oscillations is $2\pi\sqrt{2R/g}$. *Hint:* You may find the parallel axis theorem useful.

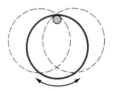

Fig. 13–44   Problem 25

26. A solid disk of mass M and radius R is mounted on a horizontal axle, as shown in Fig. 13–45. A spring is connected to the disk at a point $\frac{1}{2}R$ above the axle, and in equilibrium runs horizontally to a wall. If the disk is rotated slightly away from equilibrium, what is the angular frequency of the resulting small-amplitude oscillations? *Hint:* For small θ, $\sin\theta \approx \theta$ and $\cos\theta \approx 1$.

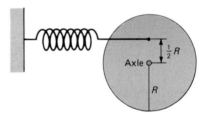

Fig. 13–45   Problem 26

27. A point mass m is attached to the rim of an otherwise uniform solid disk of mass M and radius R (Fig. 13–46). The disk is rolled slightly away from its equilibrium position and released. It rolls back and forth without slipping. Show that the period of this motion is given by

$$T = 2\pi\sqrt{\frac{3MR}{2mg}}.$$

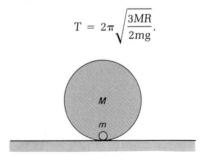

Fig. 13–46   Problem 27

28. Repeat the previous problem for the case when the disk does not contact the ground but is mounted on a frictionless horizontal axle through its center. Why is your answer different?

29. A cyclist turns her bicycle upside down to tinker with it. After she gets it upside down, she notices the front wheel executing a slow, small-amplitude back-and-forth rotational motion with a period of 12 s. Considering the wheel to be a thin ring of mass 600 g and radius 30 cm, whose only irregularity is the presence of the tire valve stem, determine the mass of the valve stem.

30. A mass m is mounted between two springs of constants k_1 and k_2, as shown in Fig. 13–47. Show that the angular frequency of oscillation is given by

$$\omega^2 = \frac{k_1 + k_2}{m}.$$

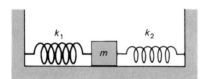

Fig. 13–47 Problem 30

31. Repeat the previous problem for the case when the springs are connected as in Fig. 13–48.

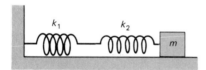

Fig. 13–48 Problem 31

Section 13.4 *Simple Harmonic Motion in Two Dimensions*

32. Differentiate Equation 13–25 to get the associated velocity vector, and show that the speed is constant.

33. The equation for an ellipse is

$$\frac{x^2}{a^2} + \frac{y^2}{b^2} = 1.$$

Show that two simple harmonic motions of different amplitudes that are at right angles in space and $\pi/2$ out of phase give rise to elliptical motion. How are the constants a and b related to the amplitudes?

34. What is the frequency ratio of the two simple harmonic motions that make the pattern shown in Fig. 13–49?

35. The x- and y-components of motion of a body are harmonic with frequency ratio 1.75:1. How many oscillations must each component undergo before the body returns to its initial position?

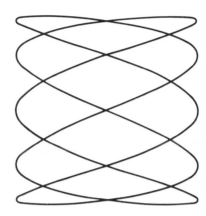

Fig. 13–49 Problem 34

Section 13.5 *Energy in Simple Harmonic Motion*

36. A 1400-kg car with poor shock absorbers is bouncing down the highway at 20 m/s, executing vertical harmonic motion at 0.67 Hz. If the amplitude of the oscillations is 18 cm, what is the total energy in the oscillations? What fraction of the car's total energy is this? Neglect rotational energy of the wheels, and the fact that not all the car's mass participates in the oscillation.

37. A 450-g mass on a spring is oscillating at 1.2 Hz. The total energy of the oscillation is 0.51 J. What is the amplitude of oscillation?

38. The motion of a particle is described by

$$x = (45 \text{ cm})[\sin(\pi t + \pi/6)].$$

At what time is the potential energy twice the kinetic energy? What is the position of the particle at this time?

39. A torsional oscillator of rotational inertia 1.6 kg·m^2 and torsional constant 3.4 N·m/rad has a total energy of 4.7 J. What are its maximum angular displacement and maximum angular speed?

40. Show that the potential energy of a simple pendulum is proportional to the square of the angular displacement in the small-amplitude limit.

41. Differentiate Equation 13–27 with respect to time and show that the result is Equation 13–4. *Hint:* Remember that $v = dx/dt$, and that total energy is constant.

42. A solid cylinder of mass M and radius R is mounted on an axle through its center. The axle is attached to a horizontal spring of constant k, and the cylinder rolls back and forth without slipping (Fig. 13–50).

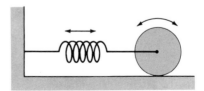

Fig. 13–50 Problem 42

Write the statement of energy conservation for this system, and differentiate it to obtain an equation analogous to Equation 13–4 (see previous problem). Comparing your result with Equation 13–4, determine the angular frequency of the motion.

43. A mass m is free to slide on a frictionless slope whose shape is given by $y = ax^2$, where a is a constant with the units of inverse length (Fig. 13–51). The mass is given an initial displacement from the bottom of the track and then released. Find the period of the resulting motion.

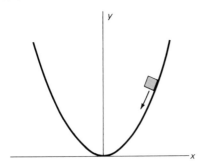

Fig. 13–51   Problem 43

Section 13.6   *Damped Harmonic Motion*

44. A 250-g mass is mounted on a spring of constant $k = 3.3$ N/m. The damping constant for this system is $b = 8.4 \times 10^{-3}$ kg/s. How many oscillations will the system undergo during the time the amplitude decays to $1/e$ of its original value?

45. The vibration of a piano string can be described by an equation analogous to Equation 13–29. If the quantity analogous to $b/2m$ in that equation has the value 2.8 s^{-1}, how long will it take the vibration amplitude to drop to half its original value?

Section 13.7   *Driven Oscillations and Resonance*

46. A mass-spring system has $b/m = \omega_0/5$, where b is the damping constant and ω_0 the natural frequency. How does its amplitude when driven at frequencies 10% above and below ω_0 compare with its amplitude at ω_0?

47. A car's front suspension has a natural frequency of 0.45 Hz. The car's front shock absorbers are worn out, so that they no longer provide critical damping. The car is driving on a bumpy road with bumps 40 m apart. At a certain speed, the driver notices that the car begins to shake violently. What speed?

48. Show by direct substitution that Equation 13–31 satisfies Equation 13–30 with x_0 and ϕ given by Equations 13–32 and 13–33.

49. A harmonic oscillator is underdamped provided that the damping constant b is less than $2m\omega_0$, where ω_0 is the natural frequency of undamped motion. Show that for an underdamped oscillator, Equation 13–32 has a maximum for a driving frequency less than ω_0.

Supplementary Problems

50. Repeat Probem 43 for a small solid ball of mass M and radius R that rolls without slipping on the parabolic track.

51. A child twirls around on a swing, twisting the swing ropes, as shown in Fig. 13–52. As a result, the child and swing rise slightly, with the rise, h, in cm equal to the square of the number of full turns of the swing. When the child stops twisting up the swing, it goes into torsional oscillation. What is the period of this oscillation, assuming that all the potential energy of the system is gravitational? The combined mass of the child and swing is 20 kg, and the rotational inertia of the pair about the appropriate vertical axis is 0.12 kg·m^2.

Fig. 13–52   Problem 51

52. A 1.2-kg block rests on a frictionless surface, and is attached to a horizontal spring of constant $k = 23$ N/m (Fig. 13–53). The block is oscillating with amplitude 10 cm and with phase constant $\phi = -\pi/2$ in Equation 13–4. A block of mass 0.80 kg is moving from the right at 1.7 m/s. It strikes the first block when the latter is at the rightmost point in its oscillation. The collision is completely inelastic, and the two blocks stick together. Determine the frequency, amplitude, and phase constant (relative to the *original* $t = 0$) of the resulting motion.

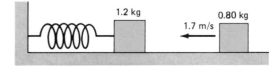

Fig. 13–53   Problem 52

53. The motion of a two-dimensional simple harmonic oscillator is described by

$$\mathbf{r} = A \sin\omega t \, \hat{\mathbf{i}} + A \sin(\omega t + \pi/4) \, \hat{\mathbf{j}}.$$

The path is an ellipse; find the orientation of its major axis. *Hint:* At what *time* is the magnitude of the displacement a maximum? What are the components of the displacement at this time? You will probably have to solve this problem numerically or graphically.

54. A small object of mass m slides without friction in a circular bowl of radius R. Derive an expression for small-amplitude oscillations about equilibrium, and compare with that of a simple pendulum.

55. A more exact expression than Equation 13–24 for the period of a simple pendulum is

$$T = T_0[1 + \tfrac{1}{4}\sin^2(\tfrac{1}{2}\theta_0) + \tfrac{9}{64}\sin^4(\tfrac{1}{2}\theta_0) + \cdots],$$

where $T_0 = 2\pi\sqrt{\ell/g}$ is the period in the limit of arbitrarily small amplitude, and θ_0 is the amplitude. The $\cdots$ indicates that additional terms (in fact, infinitely many more) are needed for an exact expression. For a pendulum with $T_0 = 1.00$ s, plot the period given above versus amplitude for amplitudes from 0 to 45°. By what percentage does the plotted period differ from T_0 for θ_0 of 30° and 45°?

56. A mass m is connected between two springs of length L, as shown in Fig. 13–54. At equilibrium, the tension force in each spring is F_0. Find the period of oscillations *perpendicular* to the springs, assuming sufficiently small amplitude that the magnitude of the spring tension is essentially unchanged.

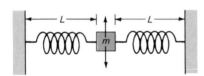

Fig. 13–54   Problem 56

57. A disk of radius R is suspended from a pivot somewhere between its center and edge (Fig. 13–55). For what pivot point will the period of this physical pendulum be a minimum?

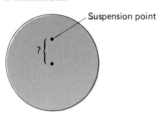

Suspension point

?{

Fig. 13–55   Problem 57

58. A uniform piece of wire is bent into a V-shape with angle θ between two legs of length ℓ. The wire is placed over a pivot, as shown in Fig. 13–56. Show that the angular frequency of small-amplitude oscillations about this equilibrium is given by

$$\omega = \sqrt{\frac{3g\,\cos(\theta/2)}{4\ell}}.$$

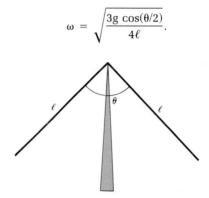

Fig. 13–56   Problem 58

59. An object in *unstable* equilibrium is subject to a force that increases linearly with displacement from equilibrium: $F = kx$. Set up Newton's law as a differential equation for this case, and show that one solution is

$$x = x_0\,e^{kt/m}.$$

Does this solution represent reasonable physical behavior? Can you find another solution? Is it physically very probable?

60. Imagine a small hole bored directly through the center of the earth and out to the other side. Show that an object dropped into this hole executes simple harmonic motion, and calculate the period of the motion. Assume that the earth has uniform density. *Hint:* Consult Section 8.5.

14

WAVE MOTION

Disturbing the equilibrium of the simple mass-spring system in Fig. 14–1a results in oscillatory motion. In the more complicated system of Fig. 14–1b, the motion of one mass is "felt" by the others—first by those nearby, then by those farther away. The result is a **wave**—a disturbance that moves or **propagates** through the system. Although the individual masses oscillate about their equilibrium positions, neither they nor the entire system move with the wave. What moves, instead, is energy, as evidenced by the temporary motions of the masses and stretching of the springs as the wave goes by. We clarify this point in defining a wave:

wave
propagates

definition of wave

A wave is a travelling disturbance that transports energy but not matter.

You are familiar with many examples of waves. Information in your professor's lectures is carried in the energy of sound waves that disturb the air and eventually your eardrums. But air from the professor's mouth does not move across the classroom. Similarly, ocean waves bring to shore the energy imparted to the water by distant winds. But the water itself does

Fig. 14–1
(a) Disturbance of the simple mass-spring system results in oscillatory motion about equilibrium. *(b)* The motion of one disturbed mass in the coupled mass-spring system is eventually communicated to adjacent masses. The result is a propagating wave—in this case, regions of compression and elongation that move along the system. The system is shown at two different times to illustrate the motion of the wave.

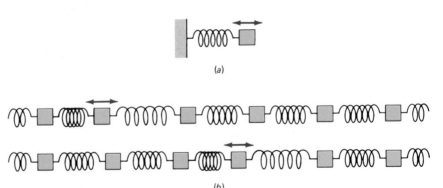

(a)

(b)

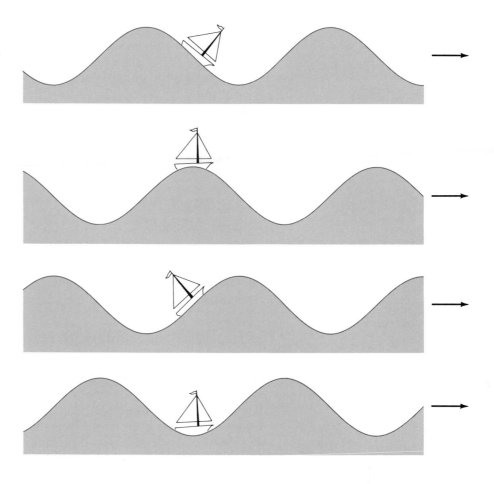

Fig. 14–2
A boat bobs up and down as a wave passes, but does not participate in the forward motion of the wave.

not move with the wave, as you can see by watching a boat bob up and down (Fig. 14–2). Ultrasound waves—sound waves of much higher pitch than the human ear can detect—are used to make images of the interior of the human body (Fig. 14–3). Earthquakes set up wave motion in the earth itself. Detection of these waves helps determine the location and strength of the quake (Fig. 14–4).

Fig. 14–3 (left)
Twin fetuses are evident in this image, produced by ultrasound waves reflecting within the human body.

Fig. 14–4 (right)
This seismograph at Kilauea Volcano, Hawaii, records not only tremors associated with the volcano but also waves generated by earthquakes thousands of kilometers distant.

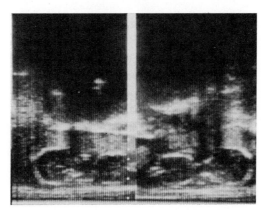

mechanical waves

The waves we have mentioned are **mechanical waves,** so called because they involve the disturbance of some mechanical medium, be it a mass-spring system, air, water, or the solid earth. In this chapter, we will develop a physical understanding of mechanical waves. Many properties of mechanical waves are also shared by electromagnetic waves, which include visible light, radio waves, x-rays, and others. We will explore electromagnetic waves further in Chapters 32 to 35.

14.1
PROPERTIES OF WAVES

Wave Amplitude

amplitude

A wave displaces a medium from its equilibrium state. The maximum displacement is the **amplitude** of the wave. Wave amplitude measures whatever physical quantity is affected by the wave—for example, the height of a water wave, or the pressure of a sound wave.

Longitudinal and Transverse Waves

longitudinal wave

transverse wave

We could set up a wave on the coupled mass-spring system of Fig. 14–1b by displacing a mass either along the direction of the spring or at right angles to it. In the first case we get a **longitudinal wave**—a travelling disturbance consisting of regions of stretched and compressed springs, with masses moving back and forth along the direction that the wave itself travels (Fig. 14–5). Sound waves provide another example of longitudinal waves (Fig. 14–6). Figure 14–7 illustrates a **transverse wave**—a travelling

Fig. 14–5
A longitudinal wave on a coupled mass-spring system. Large arrow indicates direction of wave propagation, small arrows the motions of the individual masses.

Fig. 14–6
A sound wave is a longitudinal disturbance characterized by changes in air pressure.

Fig. 14–7
This transverse wave on a coupled mass-spring system. Large arrow indicates direction of wave propagation, small arrows the motions of the individual masses.

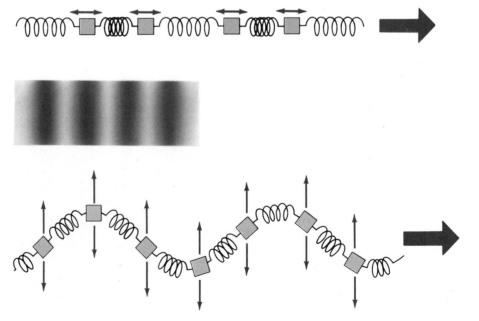

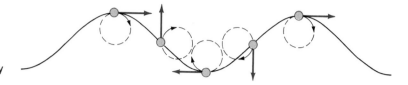

Fig. 14–8
Motion of the water in a water surface wave is approximately circular, giving the wave both transverse and longitudinal components. Colored arrows represent instantaneous velocity of the water: dotted circles are the paths followed by "parcels" of water as the wave passes. The wave is moving to the right.

disturbance in which masses move perpendicularly to the direction that the wave is travelling. Some waves—like those on the surface of water—are neither fully longitudinal nor transverse, but include components of each (Fig. 14–8).

Waveforms

Wave disturbances come in a variety of shapes, called waveforms. An isolated disturbance, travelling through an otherwise undisturbed medium, is a pulse (Fig. 14–9a). A pulse is produced when a medium is disturbed briefly. At the opposite extreme is a continuous wave (Fig. 14–9b), produced when a medium is disturbed at some point in a regular, periodic way. A wave train (Fig. 14–9c) occurs in the more realistic case when a periodic disturbance lasts for a finite time.

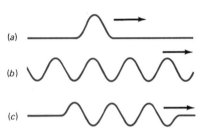

(a)

(b)

(c)

Fig. 14–9
(a) A pulse, (b) a continuous wave, and (c) a wave train.

Wavelength and Period

wavelength

period

A continuous wave is further characterized by two quantities that describe the variation of the wave disturbance in space and time. The **wavelength** λ is the distance between one point and the next point where the wave pattern begins to repeat (Fig. 14–10). In a continuous wave, each point in the medium undergoes oscillatory motion (Fig. 14–11). The wave **period** T is the period of this motion. Period is a *temporal* property of a wave, measured most directly by observing the wave at a single point in space. Wavelength, in contrast, is a *spatial* property, measured directly by observing the wave at a single instant of time.

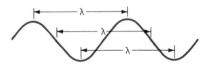

Fig. 14–10
The wavelength, λ, is the distance in which the wave pattern repeats.

Wave Speed

A wave travels at a characteristic speed through its medium. In air under typical conditions, the speed of sound waves is about 340 m/s. Small ripples on the surface of a pond move at only about 20 cm/s, while earthquake-generated waves in the earth's outer crust move at speeds on the order of 6 km/s. Wave speed, wavelength, and period are related quantities. In one wave period, an observer at a fixed point sees one complete wavelength go by (Fig. 14–12). Since the wave moves a full wavelength in one period, its speed is

$$v = \frac{\lambda}{T}.$$

(14–1)

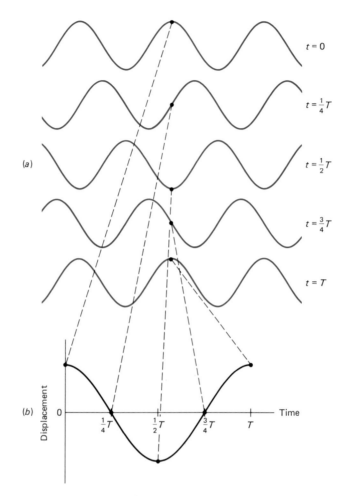

Fig. 14–11
(a) A continuous wave, seen at successive instants of time. A given point in the medium
undergoes oscillatory motion (b). The period of this motion is the wave period T.

For some kinds of waves, speed is independent of period and wavelength;
in that case the ratio λ/T is constant. For other waves, for example, those
on the surface of deep water, wave speed depends on wavelength, and the
ratio λ/T is not constant (see Problem 30).

Fig. 14–12
One full cycle—occupying a distance of
one wavelength—passes a given point
in one wave period. The wave speed is
therefore $v = \lambda/T$.

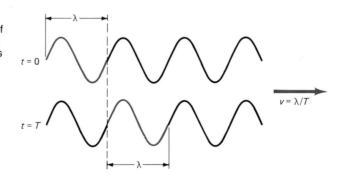

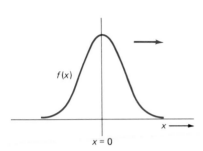

Fig. 14–13
A "snapshot" of a wave pulse at time
$t = 0$.

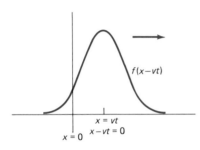

Fig. 14–14
At a later time t, the wave pulse looks
the same, but it has moved to the right.
It is described by the same function f,
but now the argument of the function is
$x - vt$, where v is the wave speed.

simple harmonic wave

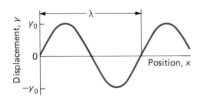

Fig. 14–15
"Snapshot" of the simple harmonic
wave described by Equation 14–2 at
time $t = 0$.

14.2
MATHEMATICAL DESCRIPTION OF WAVE MOTION

Figure 14–13 shows a "snapshot" of a wave pulse at an instant of time—say, $t = 0$. The pulse is moving along the x-axis, and we have chosen $x = 0$ to coincide with the position of its peak at time $t = 0$. We could describe the pulse at this instant by giving the displacement, y, as a function of position:

$$y = f(x),$$

where the peak of the function $f(x)$ occurs when its argument is zero. The displacement y could be any physical quantity disturbed by the wave. Assuming the wave pulse retains its shape,* Fig. 14–14 shows the situation at arbitrarily later time t. To describe the pulse now, we need a function that gives the same shape as $f(x)$, but moved to the right. How far to the right? The pulse is moving at speed v, and has been doing so for time t, so it has moved a distance vt. A function with the desired effect is $f(x - vt)$. Whatever shape $f(x)$ describes, $f(x - vt)$ describes the same shape. But where the peak of $f(x)$ occurred at $x = 0$, the peak of $f(x - vt)$ occurs when $x - vt = 0$, or when $x = vt$. As time increases, the value of x needed to give the peak of the pulse also increases, so that the function $f(x - vt)$ correctly represents the moving pulse.

Although our discussion focused on a single wave pulse, the same considerations apply as well to *any* functions $f(x)$, including continuous waves. Quite generally, any wave moving in the positive x-direction may be described by some function of the quantity $x - vt$, with v the wave speed. You can easily convince yourself that a wave moving in the negative x-direction is described by a function of $x + vt$.

A particularly important case is the **simple harmonic wave,** so called because the wave medium executes simple harmonic motion about its equilibrium. A simple harmonic wave is sinusoidal in shape, with one full cycle occurring in one wavelength λ. At time $t = 0$, we can therefore describe such a wave by

$$y = y_0 \cos\left(\frac{2\pi x}{\lambda}\right), \qquad \text{(at } t = 0) \qquad (14–2)$$

where y is the displacement from equilibrium and y_0 the amplitude or maximum displacement (Fig. 14–15). We could also have included a phase constant in this expression; for simplicity we assume a choice of origin $x = 0$ that makes the phase constant zero. We choose the argument of the cosine function so that it becomes 2π when $x = \lambda$, thereby ensuring that the function repeats itself in one wavelength. Equation 14–2 describes the wave at time $t = 0$; to get a description at all times, we simply replace x by $x \pm vt$, where v is the wave speed, and where our choice of sign depends on which direction the wave is moving:

---

*It may or may not, depending on the medium. See Section 14.5.

$$y = y_0 \cos\left[\frac{2\pi(x \pm vt)}{\lambda}\right] = y_0 \cos\left(\frac{2\pi x}{\lambda} \pm \frac{2\pi vt}{\lambda}\right). \qquad (14\text{--}3)$$

But Equation 14–1 shows that the wave speed v and wavelength λ are related to the period T by $v = \lambda/T$, so that $v/\lambda = 1/T$. We define the wave **frequency,** in cycles per second or Hz, to be the inverse of the period ($f = 1/T$), so that we have

frequency

$$y = y_0 \cos\left(\frac{2\pi x}{\lambda} \pm \frac{2\pi t}{T}\right) = y_0 \cos\left(\frac{2\pi x}{\lambda} \pm 2\pi ft\right). \qquad (14\text{--}4)$$

Equation 14–4 shows that if we sit at a fixed position—say $x = 0$—then the displacement varies in time as $y = y_0\cos(2\pi ft)$, which we recognize as simple harmonic motion with angular frequency $\omega = 2\pi f$ (Fig. 14–16). As in the previous chapter, it is usually more convenient mathematically to work with this angular frequency ω, defined by

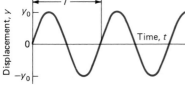

Fig. 14–16
Temporal variation of the wave displacement is simple harmonic motion with period T, frequency f, and angular frequency $\omega = 2\pi f$. Note the difference between this figure and the previous one; here we see the *temporal* variation at a *fixed position*, while there we saw the *spatial* variation at a *fixed time*.

$$\omega = 2\pi f = \frac{2\pi}{T}. \qquad (14\text{--}5)$$

Recall that the units of ω are rad/s or simply s^{-1}. Similarly, it is convenient to define the **wave number,** k:

wave number

$$k = \frac{2\pi}{\lambda}. \qquad (14\text{--}6)$$

The wave number is a spatial analog of the angular frequency; the latter describes the number of radians of wave cycle per unit *time*, the former the number of radians of wave cycle per unit *distance*. The definition of k shows that its units are m^{-1}. Using the angular frequency ω and wave number k, we can rewrite Equation 14–4:

$$y = y_0 \cos(kx \pm \omega t). \quad \text{(simple harmonic wave)} \qquad (14\text{--}7)$$

The simple way in which k and ω enter this description of the wave is the reason these quantities are so often used. If we solve Equations 14–5 and 14–6 for T and λ, respectively, and use the results in Equation 14–1, we can express the wave speed v in terms of ω and k:

$$v = \frac{\lambda}{T} = \frac{2\pi/k}{2\pi/\omega} = \frac{\omega}{k}. \qquad (14\text{--}8)$$

Example 14–1

A Simple Harmonic Wave

A sinusoidal water wave has a maximum height of 7.4 cm above the equilibrium water level, a distance of 55 cm between wave crests, and is propagating in the $-x$-direction at 93 cm/s. Express the wave in the form 14–7, assuming that a wave crest is at $x = 0$ at time $t = 0$.

Solution

To use the form 14–7, we need to know the amplitude y_0, the wave number k, and the angular frequency ω. The 7.4-cm height is the amplitude y_0, and the 55 cm between crests is the wavelength λ. Then the wave number is

$$k = \frac{2\pi}{\lambda} = \frac{2\pi}{55 \text{ cm}} = 0.11 \text{ cm}^{-1}.$$

The angular frequency can then be obtained from Equation 14–8:

$$\omega = vk = (93 \text{ cm/s})(0.11 \text{ cm}^{-1}) = 10 \text{ s}^{-1}.$$

With the wave moving in the $-x$-direction, we choose the plus sign in Equation 14–7, giving

$$y = 7.4 \cos(0.11x + 10t) \text{ cm}.$$

The phase constant is zero since a wave peak is at $x = 0$ at time $t = 0$.

14.3

WAVES ON A STRING: THE WAVE EQUATION

A Stretched String

What determines whether waves can propagate in a given medium, and if so how fast they go? The answer must lie in the physical properties of the medium. To find that answer, we apply Newton's law of motion to the medium. We now illustrate this procedure for transverse waves on a stretched string. Our results are directly applicable to strings and wires on musical instruments, as well as to other elongated structures (Fig. 14–17).

Our string has mass per unit length of μ kg/m, and is stretched to a tension force of F_0 newtons. In equilibrium, the string lies along the x-axis. Suppose the string is distorted slightly, displacing part of it in the y-direction. We want to show that wave motion results when the string is released, and to determine the wave speed.

To do so, we will apply Newton's law to a small mass element dm along the string. To apply Newton's law, we need the net force $d\mathbf{F}$ acting on the mass element. Suppose the string is distorted only slightly, so that the magnitude F_0 of the string tension does not change significantly. Figure 14–18 shows a magnified view of a small portion of the distorted string, including a mass element dm. Although we are assuming that the magnitude F_0 of the tension stays essentially constant, the *direction* of the tension force varies slightly along the mass element because the distorted string is curved. Therefore, the tension forces pulling at opposite ends of the mass element, although of the same magnitude, do not exactly cancel. Because the string is nearly horizontal, the y-components of tension differ more significantly than the x-components. Under our assumption that the distortion is small, then, we consider that the net force on the mass element dm is essentially in the y-direction. From Fig. 14–18, we see that the difference in the y-components of tension between the right and left ends of the mass element is

$$dF_y = F_0 \sin\theta_2 - F_0 \sin\theta_1.$$

But the string is very nearly horizontal, so that the sines of the angles are

Fig. 14–17
A wave pulse on a Slinky is an example of a transverse wave on an elongated structure.

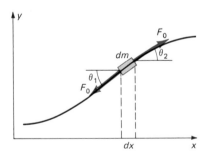

Fig. 14–18
A magnified view of a small portion of the stretched string. There is a nonzero net force on a small mass element dm because the string points in slightly different directions at opposite ends of the mass element. The vertical distortion is exaggerated for clarity: actually, the string remains very nearly horizontal.

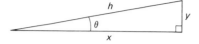

very nearly the same as their tangents (Fig. 14–19). But the tangents of the two angles are just the slopes of the string—or the derivatives dy/dx—at the ends of the mass element. Then the y-component of force on the mass element is approximately

$$dF_y \simeq F_0(\tan\theta_2 - \tan\theta_1) = F_0\left(\left.\frac{dy}{dx}\right|_{x+dx} - \left.\frac{dy}{dx}\right|_x\right).$$

The quantity in parentheses is just the change in the first derivative from one end of the interval dx to the other. Dividing that change by dx would give the rate of change of the first derivative—that is, the second derivative d^2y/dx^2. So we can write our equation more simply if we multiply the right-hand side by one in the form dx/dx, to get

$$dF_y = F_0 \frac{d^2y}{dx^2} dx. \tag{14–9a}$$

This equation gives the net force on the mass element dm. Before we use this force in Newton's law, we should be a little more careful in expressing derivatives. Figure 14–18, on which Equation 14–9a is based, is a picture of the string *at an instant of time*. Therefore, the derivative in Equation 14–9a is to be taken with *time fixed*. This distinction is important because the displacement of the string is a function not only of position but also of time, and changes if either of these variables changes. We encountered a similar situation in Section 7.5, where we developed the relation between force and potential energy when the latter varies in more than one dimension. Here, as in Chapter 7, we call a derivative taken with respect to one *partial derivative* variable while others are held constant a **partial derivative**, and write partial derivatives with the special symbol ∂ in place of the usual d. So Equation 14–9a should be written more correctly as

$$dF_y = F_0 \frac{\partial^2 y}{\partial x^2} dx. \tag{14–9b}$$

Now we can use Newton's law to equate the force to the mass times acceleration. The mass dm of the tiny piece of string we are considering is given by the mass per unit length, μ, times the length of the piece. Since the string is nearly horizontal, the length is nearly dx. Then

$$dm = \mu dx.$$

Using this mass and the force from Equation 14–9b in Newton's law $F = ma$, we have

$$F_0 \frac{\partial^2 y}{\partial x^2} dx = \mu \, dx \frac{\partial^2 y}{\partial t^2},$$

where we have written the acceleration as the second partial derivative of displacement with respect to time. Cancelling dx on both sides, we can write

$$\frac{\partial^2 y}{\partial x^2} = \frac{\mu}{F_0} \frac{\partial^2 y}{\partial t^2}. \tag{14–10}$$

What is this equation? It is just Newton's law applied to a mass element on our stretched string. Since there is nothing special about the particular mass element we considered, Equation 14–10 must apply along the entire string.

We developed Equation 14–10 to tell us about waves that might propagate on the string. To see what it has to say, let us assume we have a simple harmonic wave propagating on our string. In the previous section, we found that such a wave is described by

$$y = y_0 \cos(kx - \omega t), \tag{14–11}$$

where y is the displacement, y_0 the amplitude, k the wave number, and ω the angular frequency. If this mathematical form is consistent with Newton's law, then such waves are possible. To see if this is the case, we form the second partial derivatives of the wave displacement y in Equation 14–11 and substitute them into Equation 14–10 to see if we get an identity. The derivatives are

$$\frac{\partial^2 y}{\partial x^2} = -y_0 k^2 \cos(kx - \omega t)$$

and

$$\frac{\partial^2 y}{\partial t^2} = -y_0 \omega^2 \cos(kx - \omega t),$$

where in taking the partial derivative with respect to x we treat t as a constant and vice versa. Putting these derivatives into Equation 14–10 then gives

$$-y_0 k^2 \cos(kx - \omega t) = \left(\frac{\mu}{F_0}\right)[-y_0 \omega^2 \cos(kx - \omega t)].$$

Both the amplitude y_0 and the cosine term cancel, leaving

$$k^2 = \frac{\mu}{F_0}\omega^2,$$

or

$$\left(\frac{\omega}{k}\right)^2 = \frac{F_0}{\mu}. \tag{14–12}$$

What is Equation 14–12 telling us? Again, it follows directly from Newton's law applied to the stretched string, under the assumption that a wave described by Equation 14–11 is propagating along the string. So it tells us that such a wave can propagate, provided the wave properties ω and k are related to the properties F_0 and μ of the string in the way Equation 14–12 implies:

$$\frac{\omega}{k} = \sqrt{\frac{F_0}{\mu}}.$$

That is, only waves with ω and k related in this special way can propagate on the string. But ω/k is just the wave speed (see Equation 14–8), so that

$$v = \frac{\omega}{k} = \sqrt{\frac{F_0}{\mu}}. \tag{14–13}$$

Does this make sense? If we stretch the string tighter, increasing the tension F_0, then the restoring force is greater and the disturbance travels faster down the string. If the mass per unit length μ increases, the string's inertia is greater and it responds to the restoring force with less acceleration, resulting in a slower wave. Equation 14–13 also shows that the wave speed on the string is independent of wavelength.

The Wave Equation

Using Equation 14–13, we can write Equation 14–10 in the form

$$\frac{\partial^2 y}{\partial x^2} - \frac{1}{v^2} \frac{\partial^2 y}{\partial t^2} = 0. \tag{14–14}$$

wave equation

Equation 14–14 is known as the **wave equation.** Whenever analysis of a system results in an equation of the form 14–14, then we know that system supports waves propagating at speed v. The wave equation 14–14 usually holds only for small-amplitude disturbances, typified by our assumption of a stretched string that remains nearly horizontal. Large-amplitude disturbances usually result in a more complicated equation, and in a wave speed that depends on wavelength.

In analyzing waves on a string, we showed that a travelling simple harmonic wave of the form 14–11 is a solution to the wave equation. Is this the only possible kind of wave we can have? No: you can easily show that *any* function of the form $f(x \pm vt)$ also satisfies the wave equation (see Problem 16). Therefore, the shape of a wave that satisfies the wave equation is quite arbitrary; the actual wave shape that propagates depends on the way in which the string or other medium is initially displaced from equilibrium.

Example 14–2

The Wave Equation and Sound Waves

In analyzing the propagation of sound waves in a gas, the following equation arises:

$$\frac{\partial^2 P}{\partial x^2} - \frac{m}{\gamma kT} \frac{\partial^2 P}{\partial t^2} = 0, \tag{14–15}$$

where P is the gas pressure, x the position in the gas, t the time, T the gas temperature in kelvins (K), m the mass of a gas molecule in kg, γ a constant that depends on the molecular structure of the gas, and $k = 1.38 \times 10^{-23}$ J/K, a constant called Boltzmann's constant. We will encounter k in Chapter 17; note that it is unrelated to the wave number, whose symbol is also k. Obtain an expression for the speed of sound in a gas, and evaluate for air at room temperature (300 K), with $\gamma = 1.4$ and average molecular mass about 29 u.

Solution

Comparing the sound wave equation 14–15 with the general form 14–14, we see that the quantity multiplying the time derivative is the inverse of the wave speed squared:

$$\frac{1}{v^2} = \frac{m}{\gamma kT}.$$

Then the speed of sound is

$$v = \sqrt{\frac{\gamma kT}{m}}. \tag{14–16}$$

Evaluating this expression for air, we have

$$v = \left[\frac{(1.4)(1.38 \times 10^{-23} \text{ J/K})(300 \text{ K})}{(29 \text{ u})(1.7 \times 10^{-27} \text{ kg/u})} \right]^{1/2} = 340 \text{ m/s}.$$

This result amounts to about one-fifth of a mile per second, or about 1200 km/h. Problems 3 and 5 explore applications of the sound speed.

14.4
WAVE ENERGY AND INTENSITY

A wave propagates because motion in one part of a medium is communicated to adjacent parts. In the process, energy is passed through the medium.

For a simple harmonic wave on a stretched string, we can determine quantitatively the rate of energy transmission. To do so, we need to know both the kinetic and potential energy associated with a mass element dm on the string. With the displacement y of the string given by Equation 14–11, the speed u of a mass element dm at a point x and time t is

$$u = \frac{\partial y}{\partial t} = \omega y_0 \sin(kx - \omega t).$$

(We use partial derivatives because the displacement y is a function of both x and t, and we want its variation with t alone.) This is the speed of a mass element on the string, and *not* the wave speed v. For our transverse wave, the string motion is in fact at right angles to the wave motion. Since the string has mass μ per unit length, we have $dm = \mu dx$, where dx is the unstretched length of the mass element. Then the kinetic energy dK in the mass element dm becomes

$$dK = \tfrac{1}{2}dm u^2 = \tfrac{1}{2}\mu\omega^2 y_0^2 \sin^2(kx - \omega t)dx.$$

This expression shows that the kinetic energy oscillates between zero and a maximum of $\tfrac{1}{2}\mu\omega^2 y_0^2$ as the $\sin^2(kx - \omega t)$ term swings between 0 and 1. Over one full cycle, the average of $\sin^2$ is $\tfrac{1}{2}$, as shown in Fig. 14–20, so that the time-average kinetic energy, $\overline{dK}$, in our mass element is

$$\overline{dK} = \tfrac{1}{4}\mu\omega^2 y_0^2 dx.$$

What about the potential energy? In simple harmonic motion, the average kinetic energy and average potential energy are equal,* Therefore, the time-average energy $\overline{dE}$ in the mass element dm is twice its time-average kinetic energy, or

$$\overline{dE} = \tfrac{1}{2}\mu\omega^2 y_0^2 dx.$$

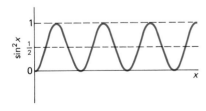

Fig. 14–20
The function $\sin^2$ swings symmetrically between 0 and 1, with average value $\tfrac{1}{2}$.

---

*Although at a given point in a propagating wave, both kinetic and potential energy peak at the same time, rather than being out of phase as in pure simple harmonic motion.

What happens to this energy? As the mass element moves, it does work on the *next* mass element, passing along its energy in the process. But over one cycle, it gains back the same amount of energy from the *previous* mass element. So on the average, the energy in the mass element remains constant (although it does fluctuate over one cycle, something our time-average derivation cannot reveal).

How long does it take the mass element to give up the energy dE? If the wave is moving at speed v, then the energy dE passes the mass element of length dx in time $dt = dx/v$. So the average rate of energy flow, or average power, is

$$\overline{P} = \frac{\overline{dE}}{dt} = \frac{\frac{1}{2}\mu\omega^2 y_0^2 \, dx}{dx/v} = \frac{1}{2}\mu\omega^2 y_0^2 v. \tag{14–17}$$

Although Equation 14–17 was derived for waves on a string, some aspects of the equation are common to all waves. For example, the power is always proportional to wave speed. Also, all waves share the property that the power is proportional to the *square* of the amplitude.

intensity

plane wave

The total power is a useful quantity in describing a wave confined to a narrow structure like a string for mechanical waves or a coaxial cable or optical fiber for electromagnetic waves. But when waves travel throughout a three-dimensional medium, it makes more sense to speak of the **intensity,** or power per unit area. In a **plane wave**—one whose wavefronts are planes—the intensity remains constant (Fig. 14–21a). With waves from a localized source, however, the wave energy spreads over an ever greater area. When waves emanate in all directions from a small, pointlike source, then the wavefronts are spherical and the area increases as the square of the distance from the source. As a result, intensity decreases as the inverse square of the distance (Fig. 14–21b). Similarly, for a linelike source, intensity decreases as the inverse of the distance (Fig. 14–21c). These decreases

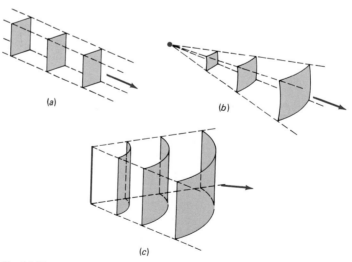

Fig. 14–21
(a) In a plane wave, wave energy remains spread over the same area as the wave propagates, so that the wave intensity remains constant. *(b)* As a wave spreads spherically from a point source, its intensity decreases as the inverse square of distance from the source. *(c)* Wave intensity decreases as the inverse of distance from a line source.

TABLE 14–1 ▬▬▬▬▬▬▬▬▬▬▬▬▬▬▬
Wave Intensities

Wave	Intensity, W/m²
Sound, 4 m from loud rock band	1
Sound, jet taking off, at 30 m	5
Sound, whisper, intensity at eardrum	10^{-10}
Light, sunlight intensity at earth's orbit	1368
Light, sunlight intensity at Jupiter's orbit	50
Light, 1 m from typical camera flash	4000
Light, at target of laser fusion experiment	10^{18}
TV signal, 5.0 km from 50-kW transmitter	1.6×10^{-4}
Microwaves, inside microwave oven	6000
Microwaves, maximum allowable outside microwave oven	50
Earthquake wave, 5 km from Richter 7.0 quake	4×10^{4}

do not occur because of any loss of wave energy, but only because the wave energy spreads over greater area. Table 14–1 lists the intensities of various waves. Even though different kinds of waves are listed, it still makes sense to compare intensity, which in each case measures power transmitted per unit area.

Example 14–3 ▬▬▬▬▬▬▬▬▬▬▬▬▬▬▬▬▬▬▬▬▬▬▬▬▬▬
Wave Intensity

Verify the entry in Table 14–1 for the 50-kW TV transmitter.

Solution

At a distance of 5.0 km, the 50-kW transmitter power is spread over the surface of a sphere 5.0 km in radius. The area of a sphere is $4\pi r^2$, so the intensity, or power per unit area, is

$$I = \frac{P}{A} = \frac{50 \times 10^3 \text{ W}}{(4\pi)(5.0 \times 10^3 \text{ m})^2} = 1.6 \times 10^{-4} \text{ W/m}^2.$$

14.5
THE SUPERPOSITION PRINCIPLE AND WAVE INTERFERENCE

The Superposition Principle

Any function of the quantity $x \pm vt$—a wave of any shape—satisfies the wave equation 14–14. Suppose we have two such functions, $f(x \pm vt)$ and $g(x \pm vt)$. Then the function $f(x \pm vt) + g(x \pm vt)$ is also a solution, as you can easily show by substituting it into the wave equation (see Problem 26). Physically, this means that if we have two waves propagating simultaneously in a medium, then the net displacement of the medium is just the sum of the displacements of the two waves. This property—the simple addition of waves to produce a composite wave—is called the **superposition principle.** Waves that obey the superposition principle include all those described by Equation 14–14, and others besides. Some waves—mostly

superposition principle

those in complicated media or those of very large amplitude—do not. We will be concerned primarily with waves that obey the superposition principle.

The superposition principle makes it possible to decompose a complicated wave shape into a sum of simple harmonic waves. Consider the two simple harmonic waves

$$y_1 = y_0 \sin(kx - \omega t)$$

and

$$y_2 = \tfrac{1}{3}y_0 \sin(2kx - 2\omega t)$$

shown in Fig. 14–22. Both have the same speed—the same ratio of angular frequency to wave number. But the second wave has twice the frequency and therefore half the wavelength of the first, along with one-third the amplitude. Their sum, also shown in Fig. 14–22, is a more complicated waveform. Other wave shapes can be built up out of sums of many simple harmonic waves. Figure 14–23 shows how a square wave is constructed from an infinite sum of simple harmonic waves of the form

$$y = \frac{4y_0}{\pi}\left[\frac{\sin(kx - \omega t)}{1} + \frac{\sin(3kx - 3\omega t)}{3} + \frac{\sin(5kx - 5\omega t)}{5} + \cdots\right].$$

Fig. 14–22
Two simple harmonic waves sum to make a more complex waveform.

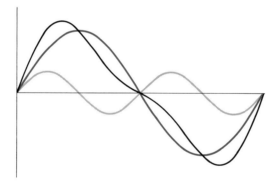

Fig. 14–23
A square wave built up as a sum of simple harmonic waves. Shown are the square wave, the first three harmonic waves in the series, and their sum (black).

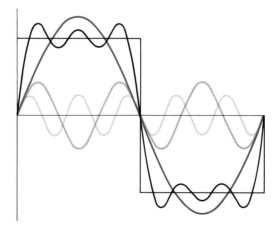

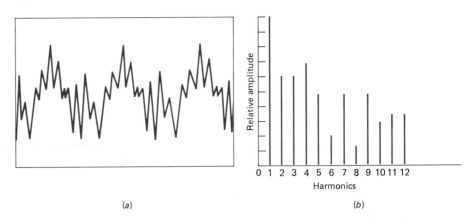

(a)

(b)

Fig. 14–24
(a) Waveform produced by a violin. (b) Fourier analysis shows the relative strength of the different harmonics whose sum is the complex waveform. To reproduce the waveform faithfully, an audio system must be capable of reproducing not only the fundamental frequency but also the significant harmonics.

Fourier analysis

This decomposition of an arbitrary wave shape into simple harmonic waves is called **Fourier analysis,** after the French mathematician Jean Baptiste Joseph Fourier (1786–1830). Fourier showed that *any* periodic function can be so decomposed. Fourier analysis has many applications, ranging from music to communications, because it allows us to understand how a complex wave shape behaves if we know how its composite simple harmonic waves behave (Fig. 14–24).

When wave speed is independent of wavelength, as it is for waves that satisfy Equation 14–14, then all simple harmonic waves making up a complex waveform travel at the same speed. As a result, the waveform maintains its shape. Other waves, obeying more complicated wave equations, often have speeds that depend on wavelength. Then, individual harmonic waves travel at different speeds, and a complex waveform changes shape

dispersion

as it moves. This phenomenon is called **dispersion,** and is illustrated in Fig. 14–25. Waves on the surface of deep water, for example, have speed given by

$$v = \sqrt{\frac{\lambda g}{2\pi}}, \tag{14–18}$$

where λ is the wavelength and g the acceleration of gravity. Since v depends on λ, the waves are dispersive. Long-wavelength waves from a storm at sea have the highest speeds and therefore reach shore well in advance of both the storm and the shorter-wavelength waves.

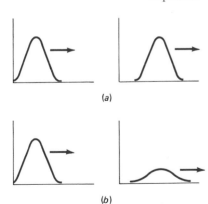

(a)

(b)

Fig. 14–25
(a) A wave pulse in a nondispersive medium holds its shape as it propagates. (b) In a dispersive medium, a pulse changes shape as it propagates.

Interference

Figure 14–26a shows two identical sinusoidal wave trains travelling in opposite directions on the same string. What happens when they meet? If the waves obey the superposition principle, then their amplitudes simply add at each instant. When they overlap crest-to-crest, that addition results instantaneously in a structure with twice the amplitude of either wave

Fig. 14–26
(a) Sinusoidal wavetrains propagating in opposite directions. *(b)* At this instant, the waves interfere constructively. *(c)* Half a cycle later, they interfere destructively. *(d)* The wavetrains continue on their way, unaffected by their encounter.

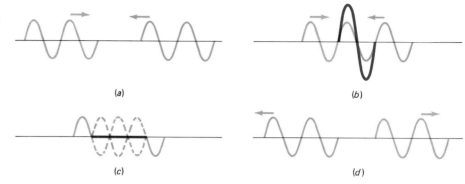

(a)

(b)

(c)

(d)

(Fig. 14–26*b*). Half a wave period later, crests line up with troughs, and the net amplitude is zero (Fig. 14–26*c*). While they are interacting, the two waves are said to **interfere.** If the composite amplitude is enhanced, the interference is **constructive;** if the amplitude is diminished, then it is **destructive.** Finally, the waves continue on their separate ways, completely unchanged by their interaction.

interference

constructive interference

destructive interference

An important example of wave interference is provided by waves from two point sources (Fig. 14–27). Along curves called **nodal lines,** crests and troughs meet, resulting in diminished amplitude. Between nodal lines are regions of constructive interference, where wave amplitude is enhanced. The same effect can be achieved by passing waves through two small holes or slits (Fig. 14–28). Two-slit interference experiments have considerable importance in optics (Chapter 35) and in modern physics, and are of historical interest because they were first used to demonstrate the wave nature of light (Fig. 14–29). Problem 32 explores the mathematics of two-slit interference.

nodal lines

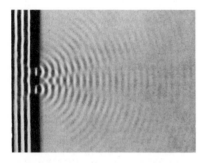

Fig. 14–27
Pattern produced when circular water waves from two point sources interfere. The nodal lines—regions of destructive interference—are clearly evident.

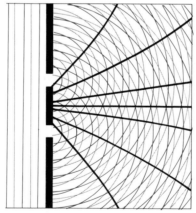

Fig. 14–28
Passing straight waves through two small slits also results in a two-source or double-slit interference pattern.

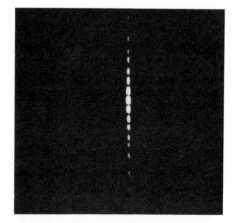

Fig. 14–29
Interference pattern produced by shining laser light through two narrow slits. The slits were produced by painting a microscope slide black, then scratching the paint with two razor blades held side-by-side.

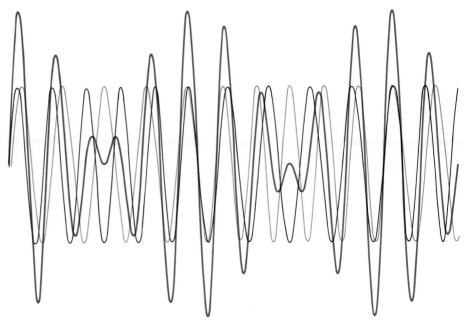

Fig. 14–30
Beats. Two waves of slightly different frequencies and their sum (color). Their sum shows a gradual variation in amplitude. The frequency of this variation is the difference in frequency of the individual waves.

Beats

When waves of two slightly different frequencies are superposed, whether the interference is constructive or destructive at a given point varies with time (Fig. 14–30). The result is a gradual variation in amplitude of the composite waveform at a frequency equal to the difference of the two individual frequencies (see Problem 56). For sound waves, the result is a *beats* periodic variation in sound intensity called **beats;** the closer the two frequencies, the longer the period of the beats. We hear beats whenever two sound sources operate at slightly different frequencies. In twin-engine propeller-driven aircraft, for example, pilots synchronize engine speeds by reducing the beat frequency toward zero. Beating of two nearby frequencies of electromagnetic waves forms the basis for some very sensitive measurement techniques.

14.6
WAVE REFLECTION

What happens to a wave when the properties of its medium change or the medium ends? Suppose, for example, that we clamp one end of a string tightly to a rigid wall. The wall is incapable of moving, so that none of the wave energy can pass into it. But the wave can't simply stop, for where *reflection* would its energy go? What happens is that the wave **reflects,** turning around and heading back down the string. For reflection at a rigid wall, the reflected wave is inverted, or turned 180° out of phase from the incident

Fig. 14–31 (left)
Reflection at a rigid wall. Incident and reflected pulses are 180° out of phase. During reflection, the string shape (shown in color) is the superposition of the incident and reflected waves.

Fig. 14–32 (right)
When the end of the string is free to move up and down, the wave is reflected with no change in phase. To maintain tension, the end is attached to a frictionless ring that is free to slide up and down a rod. The colored curve is actual string shape, a superposition of incident and reflected pulses.

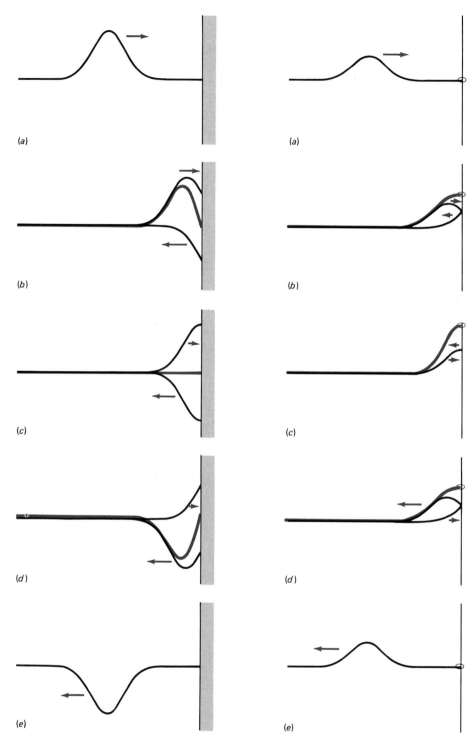

(a)

(b)

(c)

(d)

(e)

wave (Fig. 14–31). This must happen so the net displacement of the string—the superposition of incident and reflected waves—is zero at the clamped end. At the opposite extreme, our string might terminate in a way that allows its end to move freely up and down. Again, there is no place for the wave energy to go, so reflection must occur. Because the end is free to move, the displacement in this case must be a maximum at the end. Therefore, the incident and reflected waves interfere constructively, so that the two are in phase (Fig. 14–32).

Between the extremes of a rigid wall and a perfectly free end lies the case of a string connected to another string of different mass per unit length. In this case, we find that some wave energy is transmitted into the second string, and some is reflected back along the first (Fig. 14–33). In Problem 57, you can show that the amplitudes of the transmitted and reflected waves are given by

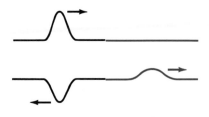

$$A_R = \frac{\sqrt{\mu_1} - \sqrt{\mu_2}}{\sqrt{\mu_1} + \sqrt{\mu_2}} A_I \qquad (14\text{--}19)$$

and

$$A_T = \frac{2\sqrt{\mu_1}}{\sqrt{\mu_1} + \sqrt{\mu_2}} A_I, \qquad (14\text{--}20)$$

Fig. 14–33
Partial reflection at a junction between two different strings.

where A_I, A_R, and A_T are the incident, reflected, and transmitted wave amplitudes, and μ_1 and μ_2 the mass per unit length of the two strings. Letting μ_2 become indefinitely large gives $A_R = -A_I$ and $A_T = 0$; this corresponds to total reflection at a rigid wall, with the minus sign signifying the phase change. Letting $\mu_2 = 0$ gives $A_R = A_I$, corresponding to reflection with the end of the string perfectly free. Finally, putting $\mu_1 = \mu_2$ gives $A_R = 0$ and $A_T = A_I$, showing that the wave simply continues on if the strings have the same mass per unit length. More generally, when $\mu_2 > \mu_1$, we get reflection with a phase change; otherwise there is no phase change.

The phenomenon of partial reflection and transmission at a junction of strings has its analog in the behavior of all sorts of waves at interfaces between two different media. For example, shallow water waves are partially reflected if the water depth changes suddenly. Light incident on even the clearest glass undergoes partial reflection because of the difference in the light-transmitting properties of air and glass. Partial reflection of ultrasound waves at the interfaces of body tissues with different densities makes ultrasound a valuable medical diagnostic (Fig. 14–34).

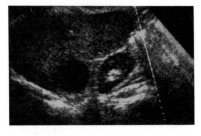

Fig. 14–34
A tumor is clearly visible in this image formed by reflected ultrasound waves. The waves reflect at the interface of the higher-density tumor tissue.

refraction

When waves strike the interface at an oblique angle, the phenomenon of **refraction**—changing of the direction of wave propagation—also occurs. We discuss refraction in Chapter 34.

14.7
STANDING WAVES

When a string is clamped rigidly at both ends, waves can propagate back and forth, reflecting off the ends. But now, only waves whose amplitude is always zero at the ends can meet the physical conditions imposed by clamping the string. How is this possible? The answer lies in the superposition of waves travelling in opposite directions.

Consider a string of length L lying along the x-axis with one end at $x = 0$ and the other at $x = L$. If a simple harmonic wave is propagating in the $+x$-direction, we may write

$$y_1 = y_0 \cos(kx - \omega t)$$

for the string displacement y_1. When this wave reflects from the rigidly clamped end at $x = L$, it gives rise to a wave propagating in the $-x$-direction:

$$y_2 = -y_0 \cos(kx + \omega t),$$

where the minus sign before the amplitude accounts for the phase change that occurs on reflection at a rigid wall. At the other end, this wave too is reflected. As a result, both waves exist simultaneously on the string, whose net displacement is therefore the superposition of the two waves:

$$y = y_1 + y_2 = y_0[\cos(kx - \omega t) - \cos(kx + \omega t)].$$

Using trigonometric identities for the cosine of a sum and difference of angles (see Appendix A), this becomes

$$
\begin{aligned}
y = {} & y_0\{\cos(kx)\cos(\omega t) + \sin(kx)\sin(\omega t) \\
& - [\cos(kx)\cos(\omega t) - \sin(kx)\sin(\omega t)]\} \quad\quad (14\text{--}21) \\
= {} & 2y_0\sin(kx)\sin(\omega t).
\end{aligned}
$$

What sort of wave does this equation describe? At each position x, the string oscillates up and down with frequency ω and amplitude $2y_0\sin(kx)$. The maximum amplitude always occurs at the same position x, so that the oscillating disturbance does not move along the string. This oscillation is a **standing wave**—a nonpropagating structure that nevertheless can be thought of as a superposition of two waves propagating in opposite directions.

standing wave

Because the ends of the string are clamped rigidly, they must correspond to points of zero displacement; such points are called **nodes.** In general, nodes lie half a wavelength apart. Therefore, a wave whose wavelength was twice the string length L could form a standing wave. So could a wave with wavelength equal to the string length; in addition to nodes at the clamped ends, this standing wave would have a node at the center of the string. More generally, if the string length L is any multiple of half a wavelength, then a standing wave can fit on the string (Fig. 14–35):

nodes

$$L = \frac{n\lambda}{2}, \quad n = 1, 2, 3, 4, \ldots,$$

or

$$\lambda = \frac{2L}{n}. \quad\quad (14\text{--}22)$$

Although this result is obvious from Fig. 14–35, we can also derive it using Equation 14–21 for the standing wave. The amplitude $2y_0\sin(kx)$ must be zero at the string ends. This requirement is obviously met at $x = 0$, and at the end $x = L$ it means that

$$2y_0\sin(kL) = 0.$$

This can be true for nonzero amplitude y_0 only when the sine is zero, which occurs at integer multiples of π radians:

$$kL = n\pi, \quad n = 1, 2, 3, 4, \ldots$$

But $k = 2\pi/\lambda$, so that this condition is

$$\frac{2\pi}{\lambda}L = n\pi,$$

or

$$\lambda = \frac{2L}{n},$$

which is Equation 14–22.

mode number
fundamental

The integer n in Equation 14–22 is called the **mode number.** The $n = 1$ mode is the **fundamental,** and represents the longest wavelength that can exist on the string. The higher modes are called the second harmonic, third harmonic, and so forth.

When a string is fixed rigidly at one end and free to move at the other, a similar analysis applies except that now the string length must be an odd multiple of a quarter wavelength so that the amplitude at the free end is a maximum (Fig. 14–36). In this case, the allowed wavelengths correspond

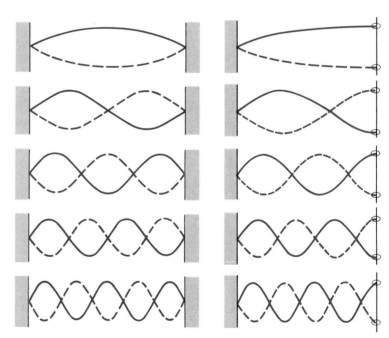

Fig. 14–35
Standing waves on a string. Clamped rigidly at the ends, the string can accommodate only an integer number of half wavelengths. Shown are the fundamental and next four harmonics. The nodes occur at the ends in all cases and in between for the harmonics. Solid line is the string at one instant, dashed line one-half cycle later.

Fig. 14–36
When one end of the string is fixed and the other free, the string can accommodate only an odd number of quarter wavelengths. (Here the string is clamped at the left end, but at the right end is attached to a ring that slides without friction along a vertical pole.)

Fig. 14–37

Waveforms of a clarinet (above) and oboe (below) playing the same note. Although the fundamental frequencies are the same, the waveforms and resulting sounds are quite different, reflecting the different mixtures of standing wave modes in the two instruments.

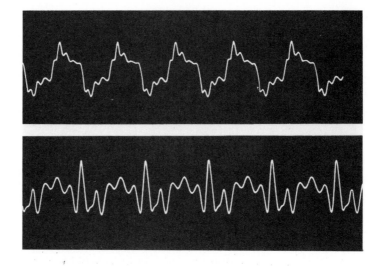

only to odd harmonics of the fundamental frequency, which is itself half the fundamental frequency of the previous case (see Problem 37).

Our analysis of standing waves on a string applies directly to stringed musical instruments like the violin and piano. Standing waves can also exist in other systems. An organ pipe or woodwind instrument, for example, supports standing sound waves whose allowed modes are determined by the structure of the instrument. In general, a mixture of different modes can exist at once, and it is the particular blend of harmonics that gives an instrument its distinctive sound (Fig. 14–37).

Other examples of standing waves include water waves in a confined space (Fig. 14–38), electromagnetic waves in a closed cavity, and standing waves in mechanical structures. In microwave ovens, a rotating piece of metal is sometimes used to reflect the waves in different directions. Otherwise, standing-wave patterns would be set up and parts of the food at the nodes would receive very little energy.

Fig. 14–38

Oscillations in the thermocline of Lake Ontario. Thermocline is the layer of rapidly changing water temperature separating warmer surface water from cool, deep water. Oscillations represent a standing wave of considerable amplitude, in which the entire lake participates.

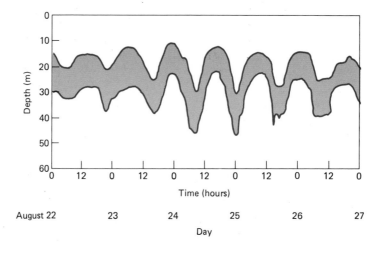

Standing waves can be started and their amplitude increased by supplying energy at exactly the right frequency—the frequency of an allowed mode. This is just like resonance in a simple harmonic oscillator, except that a standing-wave system can be driven to resonance at any of its many allowed frequencies. The oscillations of the Tacoma Narrows Bridge shown in Fig. 13–37 are actually torsional standing waves, driven in resonance by the wind.

In two- and three-dimensional systems, standing-wave patterns may be quite complex, with different sets of allowed wavelengths in different directions. Figures 14–39 and 14–40 show some standing-wave patterns on a drum head and guitar. Determination of the many resonant frequencies in complicated systems ranging from molecules to stars is an important task for physicists, engineers, and physical chemists (Fig. 14–41).

Fig. 14–39
Standing-wave patterns on a drum.

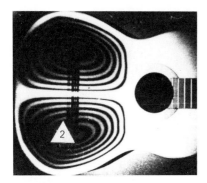

Fig. 14–40
Standing wave pattern on a guitar, imaged using holographic interference of laser light.

14.8
THE DOPPLER EFFECT

The intrinsic speed v of a wave is its speed measured relative to the medium through which the wave propagates. A point source that is stationary with respect to the medium produces waves that move outward uniformly in all directions (Fig. 14–42). But when the source moves with respect to the medium, then successive wavefronts originate from different locations (Fig. 14–43). As a result, wave crests moving in the direction of the source occur closer together than if the source were not moving—that is, the wavelength is shorter in this direction. In the opposite direction, source motion makes the wave crests farther apart, resulting in a greater wavelength. Both these effects are evident in Fig. 14–43. The change in wavelength due to relative motion between source and observer is called the **Doppler effect.**

To analyze the Doppler effect, let λ be the wavelength emitted by a stationary source, and λ' that emitted by the same source if it moves with speed u through a medium where the wave speed is v. At some instant, a wave crest is produced at the source. The next crest is emitted one wave period T later. How far does the source move in this time? Simply uT, since u is the source speed. How about the first wave crest? It moves one wavelength, so that the distance between successive crests in the direction of source motion is

$$\lambda' = \lambda - uT.$$

Writing $T = \lambda/v$, this expression becomes

$$\lambda' = \lambda - u\frac{\lambda}{v} = \lambda\left(1 - \frac{u}{v}\right). \quad \text{(source approaching)} \quad (14\text{–}23)$$

In the direction opposite the source motion, the situation is the same except that now the wavelength *increases* by the amount $\lambda u/v$, giving

$$\lambda' = \lambda\left(1 + \frac{u}{v}\right). \quad \text{(source receding)} \quad (14\text{–}24)$$

Doppler effect

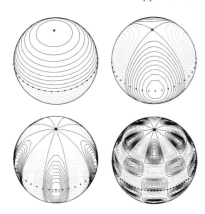

Fig. 14–41
Oscillations of the sun form complicated standing-wave patterns that provide a direct probe of conditions deep in the sun's interior. Shown here are computer simulations of four possible oscillation modes.

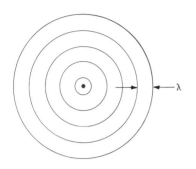

Fig. 14–42
Circular waves produced by a source at rest with respect to the medium.

Because the wave speed in the medium does not change, the relation $v = f\lambda$ still holds. When the wavelength decreases in front of a moving source, therefore, the frequency of the waves at a fixed point in the medium must increase. This is evident from Fig. 14–43, where an observer at point A will see more wave crests per unit time than if the source were not moving. Similarly, an observer at B will see fewer wave crests per unit time, and therefore a decreased frequency associated with the longer wavelength. Using Equations 14–23 and 14–24, and noting that $\lambda = v/f$, we are able to write

$$\frac{v}{f'} = \frac{v}{f}\left(1 \pm \frac{u}{v}\right)$$

or

$$f' = \frac{f}{1 \pm u/v}, \tag{14–25}$$

where f' is the Doppler-shifted frequency, and where the $+$ and $-$ signs correspond to receding and approaching sources, respectively. A Doppler shift also occurs when an observer moves relative to a source of waves. Problem 49 explores this situation.

You have probably experienced the Doppler effect for sound when standing near a highway. A loud truck approaching you makes a relatively high-pitched sound "aaaaaaaaaaa." As the truck passes you, the pitch drops abruptly, "aaaaaaaaaeiooooooooooooo," and stays low as the truck recedes. Practical uses of the Doppler effect are numerous. Measuring the Doppler shift in ultrasound reflected from moving body tissues allows measurement of blood flow, fetal heartbeat, and other physiological quantities. Although light and other electromagnetic waves do not require a material medium, they are nevertheless subject to the Doppler effect. Police radar works by measuring the Doppler shift of high-frequency radio waves reflected from

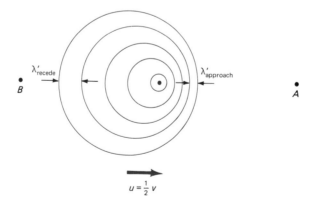

Fig. 14–43
Circular waves produced by a source moving with respect to the medium. The wavelength—the distance between wave crests—is shortened in the direction toward which the source is moving, and lengthened in the opposite direction. In the case shown, the source speed u is half the wave speed v.

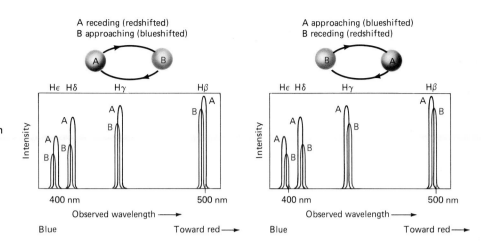

A receding (redshifted)
B approaching (blueshifted)

A approaching (blueshifted)
B receding (redshifted)

Fig. 14–44
The wavelengths of light emitted by hydrogen atoms in a double star system show clearly the presence of one star moving toward the observer, the other away. The pattern changes with time as the stars revolve around one another, so that the motion of the two can be studied in detail even though the individual stars cannot even be resolved in a photographic image.

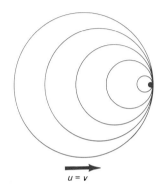

Fig. 14–45
When the source moves at the wave speed, waves pile on top of one another to create a shock wave.

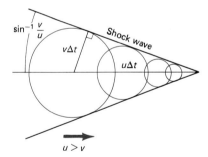

Fig. 14–46
Shock wave formed when source speed u exceeds wave speed v. In a time Δt from the generation of the largest circular wavefront shown, that wavefront moves a distance $v\Delta t$. But at the same time, the source moves a distance $u\Delta t$. As a result, the shock wave forms a cone of half-angle $\theta = \sin^{-1}(v/u)$.

shock wave

moving cars. The Doppler shift of starlight allows us to study stellar motion (Fig. 14–44). Finally, the Doppler shift of light from distant galaxies is evidence that our entire universe is expanding.

Example 14–4
The Doppler Effect

A car is roaring down the highway with its stereo blasting loudly. An observer with perfect pitch is standing by the roadside and, as the car approaches, notices that a musical note that should be G ($f = 392$ Hz) sounds like A ($f = 440$ Hz). How fast is the car moving? The speed of sound is about 340 m/s.

Solution

Solving Equation 14–25 for the source speed u, we have

$$u = v\left(1 - \frac{f}{f'}\right) = (340 \text{ m/s})\left(1 - \frac{392 \text{ Hz}}{440 \text{ Hz}}\right) = 37.1 \text{ m/s}.$$

14.9
SHOCK WAVES

Equation 14–23 suggests that wavelength goes to zero when a source approaches at the wave speed. What can this mean? When the source moves at exactly the wave speed, wave crests emitted in the forward direction cannot get away from the source, but instead pile up into a very large amplitude wave at the front of the source (Fig. 14–45). This large-amplitude wave is a **shock wave.** When the source moves faster than the wave speed, the waves pile up on a cone whose half-angle is given by $\sin\theta = v/u$, as shown in Fig. 14–46. Shock waves of this sort are formed by supersonic aircraft, and are called sonic booms (Fig. 14–47).

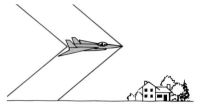

Fig. 14–47
Sonic booms produced by the nose and tail of a supersonic aircraft. If the shock waves are strong enough, they can break glass or do other damage to buildings in their path.

Fig. 14–48
The bow wave is a shock wave formed by a boat moving faster than the water wave speed. Note reflections of waves off channel walls.

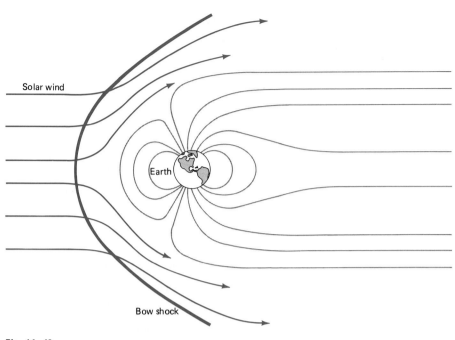

Fig. 14–49
(a) Earth's bow shock, formed as the solar wind encounters earth's magnetic field. Shock actually stands about 20 earth radii on the sunward side of the planet.

Shock waves occur in a wide variety of physical situations. The bow wave of a boat is a shock wave on the water surface (Fig. 14–48). On a much larger scale, a huge bow shock forms around the earth as the solar wind—a high-speed outflow of matter from the sun—encounters earth's magnetic field (Fig. 14–49). Other planets exhibit similar bow shocks, and in 1986 a bow shock was discovered sunward of Halley's Comet.

SUMMARY

---

1. A **wave** is a disturbance that moves through a medium, carrying energy but not matter. Waves have a number of distinguishing properties:
   a. The peak value of the disturbance is the wave **amplitude.**
   b. If the disturbance of the medium is in the direction of wave motion, the wave is **longitudinal.** If the disturbance is at right angles to the wave motion, then the wave is **transverse.** Some waves have both longitudinal and transverse aspects.
   c. A **pulse** is a single, isolated propagating disturbance. A **continuous wave** is a periodic disturbance that repeats indefinitely. A special case of a contin-

uous wave is a **simple harmonic wave,** whose shape is sinusoidal.
   d. A continuous wave is characterized by its **wavelength** λ—the distance between wave crests—and **period**—the time for a full wavelength to pass a given point. Since the wave travels one wavelength in one wave period, the wave speed is

$$v = \frac{\lambda}{T}.$$

   e. A wave propagating along the x-axis can be described by a function of the quantity $x \pm vt$, where the minus sign signifies propagation in the +x-direction, and vice versa. In the case of a simple harmonic wave, the function is sinusoidal, and may be written in the form

$$y = y_0\cos(kx \pm \omega t),$$

where y is the displacement of whatever quantity the wave disturbs and y_0 the amplitude. The quantities k and ω are the **wave number** and **angular frequency,** and are related to the wavelength λ and frequency f:

$$k = \frac{2\pi}{\lambda}$$

and

$$\omega = 2\pi f.$$

The wave speed is then given by

$$v = \frac{\omega}{k}.$$

2. Applying Newton's law for small-amplitude displacements of a stretched string results in the **wave equation:**

$$\frac{\partial^2 y}{\partial x^2} - \frac{1}{v^2}\frac{\partial^2 y}{\partial t^2} = 0,$$

where y is the displacement and v the wave speed. For waves on a stretched string, the speed is given by

$$v = \sqrt{F_0/\mu},$$

where F_0 is the string tension and μ the mass per unit length. The wave equation describes not only waves on a string, but also all small-amplitude waves whose speed is independent of wavelength.

3. Waves carry energy. The average rate of energy flow, or average power, is proportional to the wave speed and to the square of the wave amplitude. For waves propagating in three dimensions, a more useful measure is the wave **intensity,** or power per unit area. The intensity of a plane wave remains constant as the wave propagates. But when a wave spreads its energy over an ever-larger area, as in a spherical or circular wave, then the intensity drops with distance from the source. For a spherical wave, this spreading of the energy results in an intensity that is proportional to the inverse square of the distance.

4. Many waves obey the **superposition principle,** meaning that when two waves interact, their amplitudes simply add. The superposition principle allows complex wave shapes to be analyzed as sums of simple harmonic waves. When wave speed is independent of wavelength, the composite waves all travel at the same speed and the complex waveform retains its shape. But when wave speed varies with wavelength, a complex waveform changes shape as it propagates—a phenomenon called **dispersion.**

5. **Interference** occurs when two waves interact. For waves that obey the superposition principle, the overall wave shape is the sum of the displacements of the interfering waves. When the overall amplitude is enhanced, the interference is **constructive;** if the amplitude is decreased, it is **destructive.** The pattern produced by interfering waves may contain regions where destructive interference results in zero displacement of the medium from its equilibrium state; these regions are called **nodes** or **nodal lines.**

6. **Reflection** occurs when a wave encounters a change in the medium. If the new medium permits no displacement whatever—as when a string is clamped rigidly at one end—then the wave is totally reflected with a phase change of 180° or π radians. When the original medium simply ends—as when a string is completely free at one end—then the wave is totally reflected with no phase change. When the new medium differs from the old, but still permits wave propagation, then the wave is partially reflected and partially transmitted at the interface between media.

7. **Standing waves** occur when a medium is bounded at both ends, so that incident and reflected waves interfere to make a nonpropagating disturbance in which each point undergoes an oscillatory displacement from equilibrium. When the system is constrained so that there can be no displacement at either end, then its length must be an integer number of half wavelengths, or

$$L = \frac{n\lambda}{2}.$$

When there can be no displacement at one end of the medium, while the other end is completely free, then an odd integer number of quarter wavelengths must fit in between the ends of the medium. The longest wavelength—lowest frequency—standing wave is the **fundamental mode;** the higher frequencies are **harmonics.** Large-amplitude standing waves may be built up by applying a driving force at a frequency near that of a standing-wave mode—a phenomenon known as **resonance.**

8. The **Doppler effect** is the change in wavelength and/or frequency due to motion of the wave source or observer. For mechanical waves, the Doppler shifted wavelength from a moving source is

$$\lambda' = \lambda\left(1 \pm \frac{u}{v}\right),$$

where λ is the wavelength emitted by a source at rest, v the wave speed, and u the speed of a moving source. The $+$ sign applies to a source receding from the observer, the minus sign to an approaching source.

9. **Shock waves** are formed by sources moving through a medium at speeds greater than the wave speed. In a shock wave, the properties of the medium change abruptly because of the piling up of many wave crests all moving more slowly than the source. The shock wave from a moving source forms a cone of half-angle $\theta = \sin^{-1}(v/u)$, with v the wave speed and u the source speed.

QUESTIONS

1. What distinguishes a wave from an oscillation?
2. Red light has a longer wavelength than blue light. Compare their frequencies.
3. Consider a light wave and a sound wave that have the same wavelength. Which has the higher frequency?
4. A car stops suddenly on a highway. The subsequent stopping of cars behind it can be thought of as a "wave" of "stoppedness" propagating backward through the traffic. In what sense is this a wave? List several factors that determine its speed.
5. Must a wave be either transverse or longitudinal? Explain.
6. As a wave propagates along a stretched string, part of the string moves. Is the speed of this motion related to the wave speed? Explain.
7. The speed of sound in water is about 1500 m/s, compared with 340 m/s in air at room temperature. Yet water is about 1000 times more dense than air. Why might the sound speeds be so much closer than the densities?
8. If the intensity of waves from a point source decreases as the inverse square of the distance, how does the wave amplitude depend on distance?
9. How does the amplitude of circular waves on the surface of water depend on distance from the source? Assume the wave propagates without energy loss.
10. The intensity of light from a point source decreases as the inverse square of the distance from the source. Does this mean that light loses energy travelling great distances? Explain.
11. The wave number k is sometimes called the spatial frequency. Why is this name appropriate?
12. Small-amplitude waves on a stretched string propagate with speed that is independent of wavelength. Do you expect the speed of larger-amplitude waves to depend on wavelength? If so, should the speed increase or decrease with increasing wavelength? Give a physical reason for your answer.
13. An upward and a downward pulse, otherwise of identical shape, are travelling in opposite directions along a stretched string. As they pass through each other, there is an instant when destructive interference is complete and the displacement of the string is everywhere zero. How does this situation differ from true equilibrium? *Hint:* Where is the wave energy?
14. How do the tuning pegs on a guitar or violin work? Do they affect the wavelength of standing waves on the guitar string? The frequency? The wave speed?
15. A lightning flash produces a burst of electromagnetic waves that is heard in a nearby radio as a sudden crash of static. Waves from lightning flashes in the antarctic propagate through the ionosphere (an electrically conducting layer in earth's upper atmosphere) and arrive in the arctic, where a radio receiving them makes a long, whistling sound whose pitch decreases with time. What does this say about the speed of an electromagnetic wave in the ionosphere as a function of frequency? (Only in vacuum is the speed of electromagnetic waves necessarily independent of frequency.)
16. Do the nodal lines of Fig. 14–27 move as the waves propagate? Or does the pattern remain stationary?
17. How does the conservation of energy require that waves reflect when their medium ends abruptly?
18. If you drop a perfectly transparent piece of glass into perfectly clear water, you can still see the glass. Why?
19. Standing waves do not propagate. In what sense, then, are they waves?
20. Can standing waves satisfy the wave equation even though they don't propagate?
21. In the 1920s the astronomer Edwin Hubble observed distant galaxies and found that their light was always red-shifted relative to nearby galaxies. Assuming the red shift is due to the Doppler effect, what does this observation imply about the motion of distant galaxies relative to earth?
22. Why does a boat easily produce a shock wave on the water surface, while it takes very high speed aircraft to produce sonic booms?
23. In Fig. 14–33, which string has the greater mass per unit length? Explain.

PROBLEMS

Section 14.1 *Properties of Waves*

1. The speed of sound waves in air at room temperature is about 340 m/s. Find the wavelength, period, angular frequency, and wave number of a 1.0-kHz sound wave. (The human ear is most sensitive to sound at about this frequency.)
2. Calculate the wavelengths of (a) an AM radio wave (1.0 MHz); (b) a channel 9 TV signal (190 MHz); (c) a police radar (10 GHz; 1 GHz = 10^9 Hz); (d) infrared radiation emitted by a hot stove (4×10^{13} Hz); (e) green light (6.0×10^{14} Hz); and (f) x-rays (1.0×10^{18} Hz). All are electromagnetic waves that propagate at 3.0×10^8 m/s.
3. (a) Determine an approximate value for the normal speed of sound (340 m/s) in miles per second. (b) Suppose you see a lightning flash and, 10 s later, hear the thunder. How many miles from you did the flash occur? Neglect the travel time for the light (why?).
4. Detecting objects by reflecting waves off them is effective only for objects about one wavelength in size or larger. (a) What is the smallest object that can be seen

with visible light (maximum frequency about 8×10^{14} Hz)? (b) What is the smallest object that can be detected by a medical ultrasound device using waves at 5.0 MHz? The speed of sound in body tissue is about 1500 m/s.

5. Short races in a track meet are run over straight courses, and timers at the finish line start their watches when they see smoke from the starting gun, not when they hear the shot. Why? How much error would be introduced by timing a 100-m race at the sound of the shot? The speed of sound in air at normal temperatures is about 340 m/s.

6. Two boats are anchored off shore, and observers on each find themselves bobbing up and down at the rate of 6 complete up-down-up cycles each minute. When one boat is up, the other is down, and vice versa. If the waves causing the bobbing travel at 2.2 m/s, what is the minimum possible distance between the boats?

Section 14.2   *Mathematical Description of Wave Motion*

7. A simple harmonic wave of wavelength 16 cm and amplitude 2.5 cm is propagating along a string in the negative x-direction at 35 cm/s. Write a mathematical expression describing this wave.

8. Figure 14–50 shows a simple harmonic wave at time $t=0$ and later at $t=2.6$ s. Write a mathematical expression describing this wave.

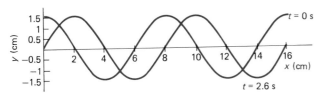

Fig. 14–50   Problem 8

9. What are the amplitude, the frequency in Hz, the wavelength, and the speed of a water wave whose displacement from the equilibrium water level is given by $y=0.25 \cos(0.52x - 2.3t)$, where y and x are in meters and t in seconds?

Section 14.3   *Waves on a String: The Wave Equation*

10. The main cables supporting the George Washington Bridge in New York City have a mass per unit length of 4100 kg/m, and are under tension of 2.5×10^8 N. At what speed would a transverse wave propagate along these cables?

11. A transverse simple harmonic wave 1.2 cm in amplitude and 85 cm in wavelength propagates along a string. The string is under 21 N tension, and has a mass of 15 g/m. Determine (a) the wave speed and (b) the maximum speed of a point on the string.

12. For a transverse simple harmonic wave on a stretched string, the requirement that the string be nearly horizontal used in deriving Equations 14–10 and 14–14 is met if the wave amplitude is much less than the

wavelength. Why? Show that, under this approximation, the maximum speed u of the string itself must be considerably less than the wave speed v. In particular, if the amplitude is not to exceed 1% of the wavelength, how large can the string speed u be in relation to the wave speed v?

13. A rope is stretched between supports 12 m apart; its tension is 35 N. If one end of the rope is tweaked, the resulting disturbance reaches the other end 0.45 s later. What is the total mass of the rope?

14. A 3.1-kg mass hangs from a 2.7-m-long string whose total mass is 62 g. What is the speed of transverse waves on the string? *Hint:* You can ignore the string mass in calculating the tension but not in calculating the mass per unit length. Why?

15. A uniform cable hangs vertically under its own weight. Show that the speed of waves on the cable is given by $v=\sqrt{yg}$, where y is the distance from the bottom of the cable. Why is this speed independent of the cable mass and length?

16. Show explicitly that any function of the form $f(x \pm vt)$ satisfies the wave equation (Equation 14–14). *Hint:* Use the chain rule to express the derivatives with respect to time and space in terms of derivatives with respect to $x \pm vt$.

17. In analyzing the behavior of shallow water, the following equation arises:

$$\frac{\partial^2 y}{\partial x^2} - \frac{1}{gh}\frac{\partial^2 y}{\partial t^2} = 0,$$

where h is the equilibrium depth and y the displacement from equilibrium. What is the speed of waves in shallow water? (Here shallow means water depth much less than wavelength.)

Section 14.4   *Wave Energy and Intensity*

18. A sound wave in air is a pressure disturbance. How does the amplitude of this pressure disturbance drop with distance from a point source of sound?

19. A loudspeaker emits energy at the rate of 50 W, spread uniformly in all directions. What is the intensity of sound 18 m from the speaker?

20. A large boulder drops from a cliff into the ocean, producing a circular wave. A small boat 18 m from the impact point measures the wave amplitude at 130 cm. At what distance will the wave amplitude be 50 cm?

21. Figure 14–51 shows a wave train consisting of two cycles of a sine wave propagating down a string. Obtain an expression for the total energy in this wave train, in terms of the string tension F_0, wave amplitude y_0, and wavelength λ.

Fig. 14–51   Problem 21; the wave amplitude shown is exaggerated.

22. Use the intensity of sunlight from Table 14–1 and radius of earth's orbit from Appendix F to calculate the total power output of the sun.

23. A steel wire has a mass of 5.0 g/m and is under 450 N tension. What is the maximum power that can be carried by transverse waves on this wire if the wave amplitude is not to exceed 10% of the wavelength? How would your answer change if the mass per unit length were doubled?

24. A 1-megaton nuclear explosion produces a shock wave whose amplitude, measured as excess pressure above normal atmospheric pressure, is 1.4×10^5 Pa (1 Pa = 1 N/m^2) at a distance of 1.3 km from the explosion. An excess pressure of 3.5×10^4 Pa will totally destroy a typical wood-frame house. At what distance from the explosion will such houses be destroyed? Compare with the size of a typical city. Assume that the wavefront is spherical, and ignore energy losses and reflection from the ground.

25. Use Table 14–1 to determine how close to a rock band you should stand for it to sound as loud as a jet plane at 100 m. Treat the band and the plane as point sources. Is this a reasonable approximation?

Section 14.5   *The Superposition Principle and Wave Interference*

26. Consider two functions $f(x \pm vt)$ and $g(x \pm vt)$ that both satisfy the wave equation (Equation 14–14). Show by direct substitution into the wave equation that any function of the form $Af(x \pm vt) + Bg(x \pm vt)$ also satisfies the wave equation, where A and B are constants.

27. Two pulses, each consisting of a single sinusoidal cycle, are launched from opposite ends of a taut string at time $t = 0$. The total length of the string is four wavelengths, and the pulses propagate at speed v. Sketch the appearance of the string at the following times: λ/v, $5\lambda/4v$, $3\lambda/2v$, and $7\lambda/4v$.

28. The triangular wave of Fig. 14–52 can be described by the following sum of simple harmonic waves:

$$y = \frac{8y_0}{\pi^2}\left[\frac{\sin kx}{1^2} - \frac{\sin 3kx}{3^2} + \frac{\sin 5kx}{5^2} - \cdots\right].$$

where n takes on all integer values. Plot the sum of the first three terms in this series at time $t = 0$ for $kx = 0$ to $kx = 2\pi$, and compare with Fig. 14–52. If you have access to a computer, you may wish to use more terms in your calculation.

29. Figure 14–53 shows a plot of angular frequency ω versus wave number k for waves in some medium. Locate (a) a wavelength range in which a composite wave will retain its shape, (b) a wavelength range in which higher-frequency waves propagate faster than lower-frequency waves, and (c) a wavelength range in which lower-frequency waves propagate faster than higher-frequency waves. The relation between ω and k plotted in Fig. 14–53 is called the dispersion relation for the medium.

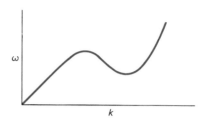

Fig. 14–53   Problem 29

30. Graph the dispersion relation (ω versus k, as explained in the previous problem) for surface waves in deep water (see Equation 14–18).

31. Two simple harmonic waves of equal frequency ω and amplitude y_0 differ in phase by ϕ, so that they may be described by

$$y_1 = y_0 \cos(kx - \omega t)$$

and

$$y_2 = y_0 \cos(kx - \omega t + \phi).$$

Show that their superposition is also a simple harmonic wave, and determine its amplitude y_s as a function of the phase difference ϕ.

32. Figure 14–54 shows two wave sources A and B, and a point P a long way from them. Suppose that P is on the innermost nodal line of the interference pattern produced by waves from the two sources. Show that the distance y on the diagram is given by $y = L\lambda/2d$, where λ is the wavelength, d the source spacing, and L the perpendicular distance from a line through the two sources to the point P. Assume that L is much greater than either the wavelength λ or the source spacing d. With this approximation, both lines AP and BP make very nearly the same small angle θ with

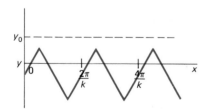

Fig. 14–52   A triangular wave (Problem 28).

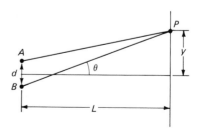

Fig. 14–54   Problem 32

the perpendicular bisector of the two sources. *Hint:* Study Figs. 14–28 and 14–54, and recognize that crests and troughs meet along a nodal line, so that one wave must travel half a wavelength farther than the other to reach the point *P*. *Another hint:* Draw a perpendicular from point *A* to the line *BP*, and think about the previous hint!

Section 14.6   *Wave Reflection*

33. A 3.3-m-long string of total mass 56 g is spliced to a 2.6-m cord of total mass 280 g. The two are under a uniform tension of 110 N. A 2.0-cm-high pulse is launched from the end of the lighter string at time $t = 0$. (*a*) At what time will the pulse reach the far end of the heavier cord? (*b*) What will its amplitude be in the heavier cord? (*c*) At what time will the pulse reflected off the junction return to its starting point? (*d*) What will be the amplitude of this reflected pulse?

34. A wave pulse carrying 0.12 J of energy is propagating down a string with $\mu = 14$ g/m, when it encounters a splice to a heavier string with $\mu = 22$ g/m. How much energy is carried down the heavier string?

35. A rope is made up of a number of identical strands twisted together. At one point, the rope becomes frayed so that only a single strand continues (Fig. 14–55). The rope is held under tension, and a 1.0-cm-high pulse is sent from the single strand. The first pulse reflected back along the single strand has an amplitude of 0.45 cm. How many strands are in the rope?

Fig. 14–55   Problem 35

Section 14.7   *Standing Waves*

36. The A-string (440 Hz) on a piano is 38.9 cm long, and is clamped tightly at both ends. If the string is under 667 N tension, what is its mass?

37. Show that only odd harmonics are allowed on a taut string with one end clamped and the other free.

38. A string is clamped at both ends and tensioned until its fundamental vibration frequency is 85 Hz. If the string is then held rigidly at its midpoint, what is the lowest frequency at which it will vibrate?

39. Show that the standing-wave condition shown in Equation 14–22 is equivalent to the requirement that the wave period be an integer multiple of the time it takes a wave to make one round trip between the ends of the medium.

40. The wave speed on a taut wire is 320 m/s. If the wire is 80 cm long, and is vibrating at 800 Hz, how far apart are the nodes?

41. "Vibrato" in a violin is produced by sliding the finger back and forth along the vibrating string. The G-string on a particular violin measures 30 cm between the bridge and its far end and is clamped tightly at both points. Its fundamental frequency is 297 Hz. (*a*) How far from the end should the violinist place a finger so that the G-string plays the note A (440 Hz)? (*b*) If the violinist executes vibrato by moving the finger 0.50 cm to either side of the position in part *a*, what range of frequencies is played?

42. Estimate the fundamental frequency of the human vocal tract, assuming it to be a cylinder 15 cm long that is closed at one end and open at the other. (These conditions make the tract analogous to a string fixed at one end and free at the other.) Assume that the speed of sound is 340 m/s.

43. A bathtub 1.7 m long contains 13 cm of water. By sloshing water back and forth with your hand, you can build up a large-amplitude oscillation. Determine the lowest frequency possible for such a resonant oscillation. *Hint:* At resonance in this case, the wave has a crest at one end and a trough at the other. The speed of shallow water waves is given by $v = \sqrt{gh}$.

44. Radio and TV antennas are often designed to be about a half wavelength long near the middle of the frequency band in which they operate. This way, weak radio or TV signals can build up relatively strong electrical oscillations in the antenna. What should be the approximate length of an antenna designed to receive FM radio signals ($f = 100$ MHz)? Compare with the length of FM antennas you or fellow students may have connected to FM receivers. The FM band lies near the middle of the channel 2–13 TV band. How does your computed FM antenna length compare with the size of a typical TV antenna?

45. The sun undergoes small-amplitude oscillations in size with a fundamental period of about 160 min. (*a*) Treating the sun like a string clamped at one end (the center) and free at the other end (the surface), estimate the average sound speed in the sun. (*b*) The sound speed in a gas is given by $v_s = \sqrt{\gamma kT/m}$, where T is the temperature in kelvins, γ a constant equal to $\frac{5}{3}$ for the gas composing the sun, k Boltzmann's constant (1.38×10^{-23} J/K), and m the mean molecular mass in kg (for the solar hydrogen/helium mixture, the mean molecular mass is 1.6 u). Use your result from (*a*) to estimate the average temperature of the sun. (The sun's temperature varies from 20×10^6 K at the core to 6000 K near the surface; your answer reflects this range of temperatures.) *Hint:* Consult Appendix F for the sun's radius.

Section 14.8   *The Doppler Effect*

46. A car horn emits 380-Hz sound waves. If the car moves at 18 m/s with its horn blasting, what frequency will a person standing in front of the car hear? The speed of sound is 340 m/s.

47. The Doppler effect in ultrasound is commonly used by obstetricians to detect fetal heartbeat. As the heart muscle pulsates in and out, ultrasound waves reflected off the moving heart are Doppler shifted from the frequency of the incident waves. If the ultrasound

transmitter sends out 5.0-Mhz waves, and the maximum frequency shift of the reflected waves is 100 Hz, how fast does the wall of the heart move? The speed of sound in body tissues is 1500 m/s.

48. Red light emitted by hydrogen atoms in the laboratory has a wavelength of 6563 Å (1 Å = 10^{-10} m). Light emitted in the same process on a distant galaxy is received at earth with a wavelength of 7057 Å. Describe the motion of the galaxy relative to earth.

49. Show that an observer moving toward a source of waves that is stationary with respect to the medium will not detect any shift in wavelength, but will measure a higher frequency given by $f' = f(1 + u/v)$, where f is the unshifted frequency, u the speed of the observer, and v the wave speed. Compare with Equation 14–25 and show that the two give nearly equal results when $u \ll v$.

Section 14.9 *Shock Waves*

50. Figure 14–56 shows a wind-tunnel test of a projectile; shock waves from the nose of the projectile are clearly evident. By making appropriate measurements, determine the speed of the projectile as compared with the sound speed.

Fig. 14–56 Problem 50

51. A supersonic airplane flies directly over you at 2.2 times the speed of sound. You hear its sonic boom 19 s later. What is the plane's altitude? Take the sound speed to be 340 m/s.

Supplementary Problems

52. A steel wire 0.85 mm in diameter propagates transverse waves at 270 m/s. What is the tension in the wire? The density of steel is 7.9 g/cm³.

53. A rectangular trough is 2.5 m long and is much deeper than its length, so that Equation 14–18 applies to waves propagating in it. Determine the wavelength and frequency of the longest and next-longest standing waves possible in this trough. Why isn't the higher frequency double the lower?

54. A supersonic airplane flies directly over you at an altitude of 6.5 km. 23 s later, you hear its sonic boom. What is the plane's speed? Take the sound speed to be 340 m/s.

55. An astronaut sets a rubber band whirling in a weightless environment, so that it forms a circle. If the rotating circle is disturbed to form a bump (Fig. 14–57), show that the bump remains fixed in space as the rubber band rotates.

Fig. 14–57 Problem 55

56. Beats: consider two waves of the same amplitude but of different frequencies ω_1 and ω_2. Consulting Appendix A for the appropriate trigonometric formulas, show that their superposition at position $x = 0$ may be written

$$y = 2y_0 \sin[\tfrac{1}{2}(\omega_1 + \omega_2)t] \cos[\tfrac{1}{2}(\omega_1 - \omega_2)t].$$

The second term in this expression represents the beats—an overall variation in amplitude at a frequency equal to the difference of the frequencies of the two superposed waves. On a graph, sketch about 10 cycles each of two waves whose frequencies differ by 10%. Add the two to form their superposition, and graph it. The low-frequency variation in amplitude should be quite evident.

57. A string of mass per unit length μ_1 is joined at $x = 0$ to a second string of mass per unit length μ_2. A wave (the incident wave) is propagating from the first string toward the second, and is characterized by a displacement y_I:

$$y_I = A_I \cos(k_1 x - \omega t).$$

At the junction between strings, the wave is partially reflected and partially transmitted. The reflected wave travels in the same medium as the incident wave, and therefore has the same wave number. It can be described by

$$y_R = A_R \cos(k_1 x + \omega t),$$

where the + sign shows that its direction is opposite to that of the incident wave. Finally, there is a transmitted wave in the second string, whose displacement is given by

$$y_T = A_T \cos(k_2 x - \omega t),$$

where k_2 is not the same as k_1 because the media and therefore wave speeds differ. If the string does not break, its total displacement approaching the junction from the left must equal that approaching it from the right, so that $y_I + y_R = y_T$. Furthermore, the derivatives dy/dx must be similarly continuous at the interface, or the acceleration at the junction would be infinite. Use these continuity conditions to derive Equations 14–19 and 14–20 for the reflected and transmitted amplitudes A_R and A_T.

15

FLUID MOTION

A tornado whirls out of a darkened sky, spreading destruction in its path. Over hundreds of millions of years, earth's continents drift about the planet, riding a viscous layer of liquid rock. The interplay of gravitational and magnetic forces shapes complex structures in the sun's outer atmosphere; occasionally the force balance fails, ejecting 10^{12} kg of solar material into space. A jetliner cruises through earth's atmosphere at 1000 km/h, propelled by hot gases spewing from its engines and supported by higher air pressure on the underside of its wings. A stream of gas leaves the surface of a giant star and forms a cosmic whirlpool before it plunges into the star's invisible companion, a black hole. You step on a car's brake pedal; the force of your foot is amplified by the pressure of liquid in the brake lines and cylinders. Your own body is sustained by air moving in and out of your lungs, and by flow of blood throughout your tissues. All these examples (Fig. 15–1) involve fluid motion.

Fig. 15–1
(Left to right) A tornado, a huge prominence in the sun's atmosphere, and the hydraulic brake system of a car all involve fluid motion.

fluid

Fluid is matter that cannot maintain its own shape and therefore flows readily under the influence of forces. Liquids and gases are fluids. Unlike solids, where strong intermolecular forces lock atoms into a rigid structure, fluids experience only weak intermolecular forces. In a liquid, these forces are strong enough to maintain molecules in close contact, but are too weak to keep them from moving readily past one another. In a gas, intermolecular forces are almost negligible, and the gas is held together essentially by external forces. These are provided by the walls of a container or, with astronomical-sized systems like stars or planetary atmospheres, by gravity. Plasmas—gases of electrically charged particles—are sometimes contained by magnetic forces.

Fluids are involved in most of the interesting happenings in our universe, whether on earth or in the most remote parts of the cosmos. Think for a minute about a world without fluids! The only motions would be those of rigid, macroscopic objects or the very slow changes in solids resulting from atomic diffusion and other subtle processes. The richness of activity to which we are accustomed would be drastically reduced; indeed, living things as we know them could not exist.

15.1
DESCRIBING FLUIDS

If we could observe a fluid on the molecular scale, we would find large numbers of molecules in continuous motion, colliding frequently with each other and with the walls of their containers. These collisions, and the responses of the individual molecules to external forces like gravity, are governed essentially by the laws of mechanics we developed earlier. In principle, therefore, we could study the behavior of fluids by applying the laws of mechanics to all the individual molecules in a fluid sample. But the number of molecules makes this a practical impossibility. Even a drop of water contains about 10^{21} molecules; to calculate the motions of all the molecules in the drop for one second would take the fastest computers many times the age of the universe!

Because the number of molecules in a typical fluid sample is so large, we approximate the fluid by considering it to be truly continuous rather than being composed ultimately of discrete particles. In this approximation, valid for fluid samples large compared with the spacing between molecules, we describe the fluid by specifying its density and pressure. We define these properties below; in general, each may be a function of both time and position.

Density

density

Density (symbol ρ, the Greek letter rho) measures the mass per unit volume of fluid; its SI units are kg/m^3. The density of water under normal conditions is about 1000 kg/m^3; that of air under normal conditions is about 1000 times smaller. Liquids are very nearly **incompressible,** meaning that their densities vary only slightly. Gases, in contrast, are **compressible,** meaning that their densities are easily altered.

incompressible
compressible

Pressure

pressure

Pressure (symbol P) measures the magnitude of the normal force per unit area that a fluid exerts either on the walls of its container or on an adjacent region of fluid (Fig. 15–2). Accordingly, the SI units of pressure

pascal

are N/m^2; this unit is given the special name **pascal** (Pa), after the French mathematician, scientist, and philosophical writer Blaise Pascal (1623–1662). (Pascal's mathematical contributions are honored through the popular computer language named for him.) Other common units of pressure

atmosphere

include the **atmosphere** (1 atm = 1.013×10^5 Pa, or the pressure of earth's

bar

atmosphere under normal conditions at sea level), the **bar** (1 bar = 10^5 Pa), the torr (1 torr = 1/760 atm, or the pressure that can support a column of mercury 1 mm high), and, in the English system, the lb/in^2 or psi.

Pressure as we have defined it is a scalar quantity. In a static fluid, pressure at a given point is exerted equally in all directions, so that it makes no sense to associate a directionality with pressure (Fig. 15–3). This is because a fluid, by definition, offers no resistance to deformation. Any imbalance in pressure over a small volume element of fluid would result in deformation of the volume element to equalize the pressure.

Fig. 15–2
Pressure describes force per unit area exerted by a fluid, either on its container walls or on adjacent fluid. (According to Newton's third law, container and adjacent fluid exert forces equal but opposite to those shown. These reaction forces are not shown.)

Example 15–1

Pressure in a Basketball

A basketball of radius 12 cm is inflated to a pressure of 5.5×10^4 Pa (8 psi) above atmospheric pressure. What is the force on one hemisphere of the basketball due to this pressure?

Solution

Figure 15–4a shows force vectors associated with the 5.5×10^4-Pa excess pressure inside the ball. Clearly the net force on the right half of the ball points horizontally to the right. To calculate this force, we sum the horizontal components of the force vectors arising from the excess internal pressure. Figure 15–4b shows a small patch of area da on the ball's surface. Calling ΔP the excess of internal over external pressure, and remembering that pressure is force per unit area, we can write the force dF on the area da as $dF = \Delta P\, da$. This force points radially outward; its horizontal component is $dF_x = dF \cos\theta$, where θ is defined in Fig. 15–4b. So we have

$$dF_x = \Delta P\, da \cos\theta.$$

But $da \cos\theta$ is just the projection of the infinitesimal area da onto the vertical plane.

Fig. 15–3
A fluid exerts pressure in all directions.

Fig. 15–4
(a) Force vectors associated with the excess pressure inside a basketball. The net pressure force on the right hemisphere of the ball points horizontally to the right. (b) The force on a small surface element of area da is $\Delta P\, da$, where ΔP is the excess internal pressure.

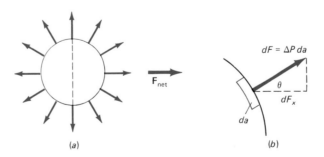

The sum—or integral—of all such projections is just the area πr^2 of a circle through the ball's center, so that the net force on the right half of the ball is

$$F_x = \int dF_x = \pi r^2 \Delta P = (\pi)(0.12 \text{ m})^2(5.5 \times 10^4 \text{ Pa}) = 2.5 \times 10^3 \text{ N},$$

or about 560 pounds! Since there is nothing special about the right hemisphere, this is the magnitude of the net force on any hemisphere of the basketball. (Of course, the net force on the entire ball is zero.) So what keeps the ball from flying apart? The ball itself stretches, and the tension within the rubber of the ball balances the force associated with excess pressure.

Example 15–2

Pressure in a Nuclear Explosion

One of the most destructive effects of a nuclear explosion is the generation of a shock wave consisting of a sudden increase in atmospheric pressure that propagates outward from the site of the explosion. At a distance of 8 km (5 miles) from a 1-megaton nuclear explosion, the shock wave exhibits an abrupt 25 per cent increase in atmospheric pressure. Estimate the net force exerted by the shock wave on a person standing and facing the blast.

Solution

As the wave passes, one side of the person is exposed to natural atmospheric pressure, the other to the excess pressure behind the shock wave. The force on the person is therefore $A\Delta P$, where ΔP is the excess pressure in the shock wave and A the person's area normal to the shock. Estimating that the person has approximately a rectangular cross section, 2 m high and 0.3 m wide, we have $A = 0.6 \text{ m}^2$. The pressure increase ΔP is 25 per cent of atmospheric pressure—that is, 0.25 atm or $2.5 \times 10^4 \text{ N/m}^2$. Then the net force on the person is

$$F = A\Delta P = (0.6 \text{ m}^2)(2.5 \times 10^4 \text{ N/m}^2) = 1.5 \times 10^4 \text{ N},$$

nearly 2 tons! Problem 24 of the preceding chapter also deals with the shock wave from a nuclear explosion.

15.2
FLUIDS AT REST

hydrostatic equilibrium

For a fluid to remain at rest, the net force on every part of the fluid must be zero. When this condition is met, the fluid is said to be in **hydrostatic equilibrium.** In the absence of gravity or other external forces that act throughout the volume of a fluid, hydrostatic equilibrium requires that the pressure be constant throughout the fluid. This is because, in hydrostatic equilibrium, the net force on *any* volume within the fluid must be zero. If the pressure varies with position, then there will be a net force on a fluid volume that extends from a region of lower to higher pressure (Fig. 15–5). As Fig. 15–5 suggests, it is not pressure itself but *pressure differences* that give rise to forces within fluids. In more advanced treatments of fluid behavior, the force per unit volume of fluid arising from pressure differences is shown to be equal to the negative of the pressure gradient, ∇P (we defined the gradient operator ∇ in Chapter 7). What happens at the boundaries of the fluid? The fluid exerts a pressure force on its container. By Newton's third law, the container must exert an equal but opposite force back

Fig. 15–5
(a) If the pressure varies with position, then there is a net pressure force on a volume of fluid. In the absence of other forces acting on the fluid volume, the fluid cannot be in hydrostatic equilibrium. (b) If the pressure is constant, the net pressure force on a fluid volume is zero, and the fluid is in hydrostatic equilibrium.

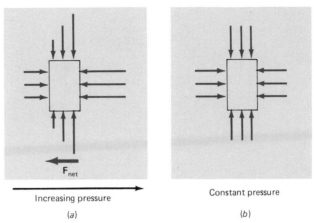

Increasing pressure

F_{net}

(a)

Constant pressure

(b)

Fig. 15–6
This tank burst because of excess pressure; it was propelled over 50 meters before coming to rest.

Pascal's principle

Fig. 15–7
The hydraulic cylinder actuating the power shovel is clearly visible in this photo.

on the fluid. This force is provided by tension associated ultimately with the stretching of interatomic bonds within the container. If the container is incapable of providing a force equal to the pressure force, it will burst (Fig. 15–6).

Suppose we increase the pressure in one part of a fluid—for example, by poking a finger into a balloon. In hydrostatic equilibrium, the pressure must be the same everywhere in the balloon. Therefore, the local pressure increase caused by poking the balloon at one point is transmitted equally throughout the fluid volume. This principle was first articulated by Pascal, and is now known as **Pascal's principle:** an increase in pressure at any point in a fluid is transmitted undiminished throughout the fluid volume and to the walls of the container. Pascal applied this principle in his invention of the hydraulic press; today hydraulic systems, based on Pascal's principle, are used in control of machinery ranging from automobile brake systems to aircraft wings to bulldozers, cranes, and robots (Fig. 15–7).

Application
A Hydraulic Braking System

Figure 15–8 shows a simplified diagram of an automobile braking system. The system includes a "master" cylinder containing a piston actuated by the brake pedal, and "slave" cylinders at each wheel. The cylinders are connected by metal and rubber hoses, and the whole system is filled with an incompressible liquid. When you step on the brake pedal, you force fluid out of the master cylinder. Since the fluid is incompressible, pistons in the slave cylinders move outward to keep the volume of the system constant. These slave pistons push friction pads against the rotating brake disks, thereby dissipating the kinetic energy of the car.

Typically the master cylinder has a much smaller cross-sectional area than the slave cylinders. Applying a given force to the master cylinder piston causes an increase in pressure equal to that force divided by the area of the piston. Pascal's principle tells us that this pressure increase is transmitted undiminished throughout the system. With the same pressure in the slave cylinders, the force—the pressure times the area—on the larger slave pistons is larger than the force on the master cylinder piston. The hydraulic system effectively amplifies the relatively weak force of your foot into the large force that ultimately stops the rotating disks and the wheels attached to them.

Isn't this force amplification giving us something for nothing? No—just as with a mechanical leverage system, the work we do on the master cylinder ends up as

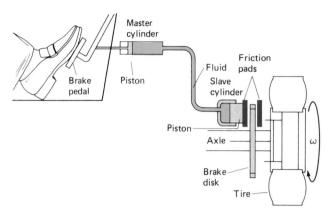

Fig. 15–8
Simplified diagram of an automobile braking system, showing master cylinder and one of the slave cylinders.

Fig. 15–9
Just before landing, the wing flaps on a jetliner are extended enough to expose the hydraulic pistons controlling the flaps.

an equal amount of work done by the slave cylinders. Work—the integral of force over the distance moved by the master piston—is equal to the sum of the four similar integrals for the work done by the slave pistons. Since each slave piston has larger area and therefore larger force, and since there are four of them for one master cylinder, the distance moved by the slave pistons is much less than the distance you move the master piston. The hydraulic system gives a large force applied over a small distance in exchange for a smaller force applied over a larger distance.

Our description gives the essential features of a modern automotive braking system. But a real system is more complicated. Today's cars include two independent hydraulic systems, so that a leak in one system will not leave the car without brakes. And in larger cars, an additional power assist is provided to allow further amplification of the force at the brake pedal.

Hydraulic systems similar to automotive braking systems are used in controlling machinery including aircraft wing and tail flaps (Fig. 15–9). Unlike mechanical linkages, hydraulic hoses may be routed conveniently anywhere, allowing easy control of remote pieces of a machine.

Hydrostatic Equilibrium with Gravity

For a fluid to be in hydrostatic equilibrium in the presence of gravity, the force associated with fluid pressure must counteract the gravitational force. Since a net pressure force arises only when pressure varies with position, we expect that hydrostatic pressure cannot be uniform when gravity is present.

Figure 15–10 shows a fluid element of horizontal cross-sectional area A and small vertical thickness dy in a uniform gravitational field. A gravitational force $g\,dm$ acts on this element, where dm is its mass and where we take the y-direction to be positive downward. In hydrostatic equilibrium this gravitational force must be balanced by an upward pressure force. Therefore, the pressure at the bottom of the fluid element must be greater than at the top. Suppose the pressure at the top of the fluid element is P, and the pressure at the bottom is $P + dP$, where dP is a small pressure increase. The downward force on the top of the fluid element is PA, and

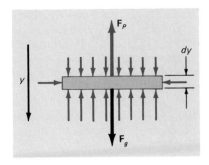

Fig. 15–10
Forces on a fluid element in hydrostatic equilibrium. The net pressure force must balance the gravitational force, requiring a higher pressure at the bottom of the fluid element.

the upward force on the bottom is $(P+dP)A$, so that the net pressure force on the fluid element is

$$dF = PA - (P+dP)A = -A\,dP,$$

where the minus sign indicates that the force is upward, or in the $-y$-direction. In hydrostatic equilibrium the net force—pressure force plus gravitational force—must be zero:

$$g\,dm - A\,dP = 0.$$

But the mass dm is just the density ρ of the fluid element multiplied by its volume, $A\,dy$, so that

$$g\rho A\,dy - A\,dP = 0,$$

or

$$\frac{dP}{dy} = \rho g. \qquad \text{(hydrostatic equilibrium)} \qquad (15\text{--}1)$$

This equation shows that dP/dy is always positive, indicating what we anticipated—that in the presence of a gravitational field pressure increases with depth. The exact form of that increase depends on the nature of the fluid. For a liquid, which is essentially incompressible, ρ is constant and we can multiply through by dy and integrate to get

$$\int_{P_0}^{P} dP = \int_{0}^{y} \rho g\,dy,$$

where P_0 is the pressure at the top of the fluid, where the depth y is zero. Carrying out the integration gives

$$P - P_0 = \rho g y,$$

or

$$P = P_0 + \rho g y. \qquad \text{(liquid)} \qquad (15\text{--}2)$$

Equation 15–2 shows pressure increasing linearly with depth in a liquid.

Example 15–3

Pressure in a Liquid

How far under water would you have to go for water pressure to be twice atmospheric pressure? What is the water pressure at the bottom of the Marianas trench, a 11.3-km-deep valley on the floor of the North Pacific Ocean? Assume in both cases that pressure at the top of the water is 1.0 atm, and that the density of the water is 1.0×10^3 kg/m^3.

Solution

When the water pressure is twice atmospheric pressure, we have $P = 2.0$ atm while $P_0 = 1.0$ atm. Solving Equation 15–2 for the depth y gives

$$y = \frac{P - P_0}{\rho g} = \frac{2.0 \times 10^5\ \text{Pa} - 1.0 \times 10^5\ \text{Pa}}{(1.0 \times 10^3\ \text{kg/m}^3)(9.8\ \text{m/s}^2)} = 10\ \text{m},$$

where we have converted pressures from atm to Pa for consistency with our other SI units. Our result shows that pressure on a diver would double at a depth of only 10 m or about 30 ft. Since pressure in a liquid increases linearly, water pressure continues to increase by 1 atm for every 10 m of depth. Figure 15–11 shows that

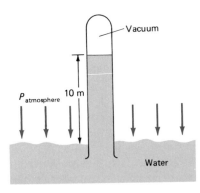

Fig. 15–11
The pressure at the bottom of a 10-m-high column of water is 1 atm higher than at the top. Therefore, normal atmospheric pressure can support a water column 10 m high with vacuum above it.

Fig. 15–12
Oceanographers bring deep ocean water to the surface in this titanium container that maintains the water and any inhabitants at their natural pressure.

barometer

manometer

gauge pressure
overpressure

our result also has implications for the design of water pumps, for it says that 1 atm of pressure can support a column of water 10 m high. Any water pump that works by suction—that is, by creating a vacuum and allowing air pressure to push water up—can raise water by at most 10 m. Pumping water from greater depths requires a submerged pump or one that somehow increases the pressure at the pumping depth.

To find the pressure in the Marianas Trench, we could solve Equation 15–2 for P. But we have already found that water pressure increases by 1 atm for every 10 m of depth. In the Marianas trench, we therefore have a total increase of

$$\Delta P = \frac{11.3 \times 10^3 \text{ m}}{10 \text{ m/atm}} = 1100 \text{ atm,}$$

or more than 8 tons per square inch! The 1 atm at the top of the ocean is a negligible addition to this huge pressure increase. Creatures living at these depths are in pressure equilibrium with their surroundings; to bring them to the surface for study, scientists must maintain their huge natural pressure or they will explode (Fig. 15–12). A similar plight affects scuba divers who foolishly hold their breath while ascending; air in the lungs expands, bursting the alveoli.

The variation in liquid pressure with depth provides a means of pressure measurement that is used in several practical instruments. Figure 15–13 shows a simple **barometer,** a device commonly used for measuring air pressure. The barometer consists of a tube sealed at its top end and containing the liquid metal mercury. The bottom end is open and immersed in a pool of mercury exposed to atmospheric pressure. Above the top of the mercury column is vacuum (pressure $P = 0$), so that the pressure at the surface of the mercury pool as given by Equation 15–2 is simply ρgh, where ρ is the density of mercury and h the height of the mercury column above the pool surface. In equilibrium, this pressure is exactly balanced by atmospheric pressure, so that the height of the mercury column is directly proportional to atmospheric pressure. Reading the height in mm or inches gives us two commonly used pressure units: torr (mm of mercury) and inches of mercury. Problem 1 explores the conversion of these units to the SI unit Pa, using the density of mercury. In principle, we could make a barometer using water, but then, as Example 15–3 shows, a 10-m-high water column would be needed.

Measurement of pressure differences is accomplished using a **manometer,** a simple U-shaped tube containing a liquid and open at both ends (Fig. 15–14). A difference in pressure between the ends of the tube results in a difference in the heights of the two liquid surfaces; Equation 15–2 shows that this height difference is directly proportional to the pressure difference. A manometer may use mercury, water, or other fluids. Very small pressure differences—associated, for example, with chimney drafts—are often measured in inches of water, a unit arising naturally with water-filled manometers. When one end of a manometer senses atmospheric pressure, the pressure it reads is called **gauge pressure** or **overpressure,** meaning the excess pressure above atmospheric pressure. Tire gauges and inflation specifications for tires, sports equipment, and the like are given in gauge pressure rather than absolute pressure. A tire inflated to 30 psig (pounds per square inch, gauge), for example, has an absolute pressure of

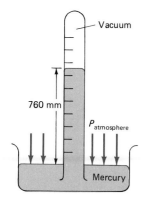

Fig. 15–13
A mercury barometer for measuring air pressure. Note the resemblance to Fig. 15–11, which is essentially a water barometer. Column height in a mercury barometer at 1.0 atm air pressure is 760 mm or 29.92 in.

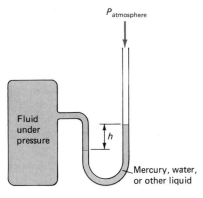

Fig. 15–14
A manometer being used to measure pressure difference between a closed container of fluid and the ambient atmosphere. Difference in height of the two liquid surfaces is a measure of the pressure difference across the ends of the tube.

scale height

about 45 psi because the 15 psi atmospheric pressure is not included in the 30 psig specification.

Although Equation 15–2 applies only to a liquid of constant density, Equation 15–1 is valid for any fluid. In a gas like a planetary atmosphere, it is convenient to measure height h upward from the surface rather than depth y downward. Taking $h = -y$, Equation 15–2 then becomes

$$\frac{dP}{dh} = -\rho g. \tag{15–3}$$

In an ideal gas of uniform temperature, pressure and density are directly proportional:

$$P = \rho \frac{kT}{m}, \tag{15–4}$$

where T is the temperature in kelvins, k a constant called Boltzmann's constant (equal to 1.38×10^{-23} J/K), and m the mass of a gas molecule in kg. (We will study this relation—called the ideal gas law—in Chapter 17.) Although the temperature of a planetary atmosphere is not exactly uniform, the approximation of uniform temperature allows us to make a rough calculation of the pressure and density structure of an atmosphere. Solving Equation 15–4 for the density ρ and using the result in Equation 15–3, we have

$$\frac{dP}{dh} = -P \frac{mg}{kT}.$$

Multiplying by dh and dividing by P gives

$$\frac{dP}{P} = -\frac{mg}{kT} dh.$$

To solve for the pressure P as a function of height h, we integrate both sides from the surface $h = 0$ where the pressure has the value P_0:

$$\int_{P_0}^{P} \frac{dP}{P} = -\frac{mg}{kT} \int_0^h dh.$$

The integral of dP/P is the natural logarithm of P, while the integral of dh is just h, so that

$$\ln P - \ln P_0 = -\frac{mg}{kT} h.$$

Recalling that $\ln P - \ln P_0 = \ln(P/P_0)$, and that $e^{\ln x} = x$, we can exponentiate both sides of this equation to get

$$P = P_0 e^{-mgh/kT}. \quad \text{(uniform-temperature atmosphere)} \tag{15–5}$$

This equation shows that the pressure in an atmosphere of uniform temperature decreases exponentially with height. The constant kT/mg is called the **scale height**, h_0, and is the height over which atmospheric pressure drops to $1/e$ of its ground-level value. In terms of scale height, Equation 15–5 may be written

$$P = P_0 e^{-h/h_0}. \tag{15–6}$$

In deriving Equations 15–5 and 15–6, we assumed that g does not vary significantly with height. For earth's relatively thin atmosphere, this is a good approximation. Problem 38 explores the case of the sun's atmosphere, which extends far enough from the surface that the variation in g is important.

Example 15–4

Pressure Variation in Earth's Atmosphere

Atmospheric pressure at sea level is 1.0 atm. Assuming a uniform temperature of 280 K and an average molecular mass of 28.8 u (4.8×10^{-26} kg; this reflects atmospheric composition of approximately 80 per cent N_2 and 20 per cent O_2), calculate the scale height in earth's atmosphere and from it the pressure atop Mt. Washington in New Hampshire (altitude 1900 m), atop Dhaulagiri in Nepal (altitude 8200 m), and in California's Death Valley, 86 m below sea level.

Solution

The scale height is

$$h_0 = \frac{kT}{mg} = \frac{(1.38 \times 10^{-23} \text{ J/K})(280 \text{ K})}{(4.8 \times 10^{-26} \text{ kg})(9.8 \text{ m/s})} = 8200 \text{ m}.$$

We then have

$$P_{\text{Washington}} = P_0 e^{-h/h_0} = (1.0 \text{ atm})(e^{-1900 \text{ m}/8200 \text{ m}}) = 0.79 \text{ atm}.$$

Similar calculations give $P_{\text{Dhaulagiri}} = 0.37$ atm and $P_{\text{Death Valley}} = 1.01$ atm, where in the last case the altitude h is negative, resulting in a pressure greater than that at sea level. Since gas density is proportional to pressure, these results show that much less air is available on a high mountain like Dhaulagiri, which is why most climbers carry oxygen tanks.

Archimedes' Principle and Buoyancy

Throw a rock into water and it sinks. Throw in a block of wood and it floats. Why the difference?

Consider an arbitrarily shaped volume within a fluid in hydrostatic equilibrium (Fig. 15–15a). We know that equilibrium is maintained by the upward pressure force that exactly balances the weight of the fluid volume. Now imagine replacing the fluid volume by a solid object of identical shape (Fig. 15–15b). The remaining fluid has not changed in any way, so it continues to exert an upward force on the object—a force whose magnitude equals the weight of the *original fluid volume*. This upward force is called the **buoyancy force;** as we have just shown, the buoyancy force on an object is always equal to the weight of the fluid displaced by that object. This principle—called **Archimedes' principle**—was first articulated in the third century B.C. by the Greek inventor and mathematician Archimedes. If the object weighs more than the original fluid volume, the net force on it— downward weight plus upward buoyancy force—is downward, and it sinks. If the object weighs less than the original fluid volume, the buoyancy force is greater than its weight, and it rises. Therefore, an object will float or sink, depending on whether its average density is greater or less than that of the fluid. An object whose density is equal to that of the fluid experiences zero net force, and is said to be in **neutral buoyancy.** Most fish

buoyancy force

Archimedes' principle

neutral buoyancy

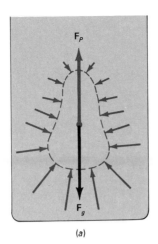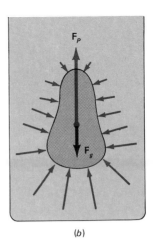

Fig. 15–15
(a) In hydrostatic equilibrium, there is an upward pressure force F_p equal to the weight F_g of any fluid volume. (b) If the fluid volume is replaced by a solid object of identical shape, the remaining fluid is not changed in any way, so that there remains an upward force equal to the weight of the fluid displaced by the object. In general, this upward buoyancy force may be more or less than the weight of the object.

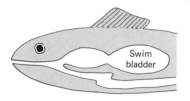

Fig. 15–16
Most fish contain a "swim bladder"—a gas-filled bladder whose volume adjusts to keep the fish in neutral buoyancy, with no tendency to float or sink.

Fig. 15–17
A hot-air balloon is in neutral buoyancy when it maintains a constant altitude.

contain a gas-filled bladder whose volume changes to maintain the fish in neutral buoyancy (Fig. 15–16). Submerged submarines, too, take on and discharge sea water as needed to maintain neutral buoyancy. A balloon cruising at constant altitude is another example of an object in neutral buoyancy (Fig. 15–17).

What happens when a buoyant object reaches the surface of the fluid? As the object breaks through the surface, only part of it remains immersed in the fluid. As always, the buoyant force is equal to the weight of the fluid displaced by the object. But now, the volume of fluid displaced is less than the volume of the entire object. So the buoyant force decreases. Eventually an equilibrium is reached in which the buoyant force balances the weight of the entire object. Since the buoyant force is equal to the weight of the fluid displaced by the submerged portion of the object, in equilibrium the object will float at such a level that its submerged portion displaces an amount of water whose weight equals the object's entire weight (Fig. 15–18).

Example 15–5
The Tip of the Iceberg

The average density of a typical arctic iceberg is about 0.86 that of sea water (iceberg density varies depending on the amount of air and rock entrained in the iceberg; for antarctic icebergs the typical density is closer to 0.80 that of sea water). What fraction of an arctic iceberg's volume is submerged?

Solution

In equilibrium, the entire weight of the iceberg is supported by the buoyant force, equal to the weight of the volume of water displaced by the submerged portion of the iceberg. If V_{sub} is the submerged volume, the weight of this volume of water—and therefore the magnitude of the buoyant force—is

$$F_b = W_{water} = m_{water}g = \rho_{water}V_{sub}g.$$

Fig. 15–18
Supertankers (left) empty and (right) fully loaded. In both cases the full weight of the supertanker and cargo is supported by the buoyant force, which is equal to the weight of the water displaced by the ship. The heavily loaded tanker must therefore displace more water, so it rides lower.

The weight of the iceberg is

$$W_{ice} = m_{ice}g = \rho_{ice}V_{ice}g,$$

where V_{ice} is the volume of the entire iceberg. Equating the iceberg weight to the magnitude of the buoyant force, we have

$$\rho_{ice}V_{ice}g = \rho_{water}V_{sub}g,$$

so that

$$\frac{V_{sub}}{V_{ice}} = \frac{\rho_{ice}}{\rho_{water}} = 0.86.$$

Roughly six-sevenths of an arctic iceberg is therefore submerged (Fig. 15–19).

Example 15–6
An Overloaded Boat

A flat-bottomed rowboat has length $\ell = 3.2$ m, width $w = 1.1$ m, and is 28 cm deep. Empty, it has a mass of 130 kg. Its maximum rated load is 450 kg. How much of the boat is submerged (a) when it is empty and (b) when it carries its maximum rated load? (c) What is the total mass it could carry without taking on water?

Fig. 15–19
Because the density of ice is close to that of water, only about one-seventh of an iceberg's volume is above water.

Solution

In equilibrium, the total weight of the boat and its load must be supported by the buoyant force of the water. By Archimedes' principle, this force is equal to the weight of water displaced by the boat. Letting y be the depth to which the boat is submerged, we have

$$\ell w y \rho_w g = Mg.$$

Here ρ_w is the density of water, and $\ell w y$ the submerged volume, so that the term on the left is the weight of displaced water, and is therefore equal to the magnitude of the buoyant force. M is the total mass of the boat plus its load, so that the term on the right is the weight supported by the buoyant force. Solving for y, we have

$$y = \frac{M}{\ell w \rho_w} = \frac{(130 \text{ kg})}{(3.2 \text{ m})(1.1 \text{ m})(1000 \text{ kg/m}^3)} = 3.7 \text{ cm}$$

for the empty boat, and

$$y = \frac{(130 \text{ kg} + 450 \text{ kg})}{(3.2 \text{ m})(1.1 \text{ m})(1000 \text{ kg/m}^3)} = 16 \text{ cm}$$

at maximum rated load. To find the absolute maximum load the boat can hold before water pours over its sides, we set y equal to the boat's 28-cm depth, and solve for M:

$$M = \ell w y \rho_w = (3.2 \text{ m})(1.1 \text{ m})(0.28 \text{ m})(1000 \text{ kg/m}^3) = 986 \text{ kg}.$$

This is the total mass; subtracting the 130-kg boat mass gives 856 kg for the absolute maximum load. In this example we assumed that the load is centered over the boat's center of mass, so that there is no net torque tending to tip the boat.

15.3
FLUID DYNAMICS

In the preceding section, we considered fluids in static equilibrium. We now turn our attention to moving fluids, which are described by giving the flow velocity at each point in the fluid and at each instant of time. Figure 15–20 shows some flow velocity vectors in a river. We can describe flow velocity either with individual vectors, as in Fig. 15–20, or by drawing continuous lines that connect the velocity vectors head-to-tail (Fig. 15–21). Called **streamlines,** these lines lie along the local direction of flow. Their spacing is a measure of flow speed, with closely spaced streamlines indicating high speed. Small particles, like smoke or dyes, are often introduced

streamlines

Fig. 15–20
Vectors describing flow velocity in a river. Note the higher velocity where the river is narrower.

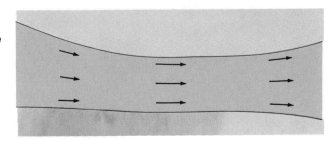

Fig. 15–21
Streamlines are an alternative way of representing the flow velocity. Note that the streamlines are closest together where the flow speed is highest.

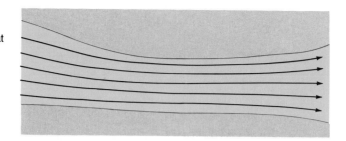

into moving fluids because they follow streamlines and therefore give a visual indication of the flow velocity pattern (Fig. 15–22).

steady flow

We distinguish two types of fluid motion. In **steady flow,** the pattern of fluid motion remains the same at each point, even though individual fluid elements are in continual motion. When you look at a river in steady flow, for example, it always looks the same, even though you are not seeing the same water each time you look. At a given point in the river, the water velocity, density, and pressure are always the same. **Unsteady flow,** in contrast, involves fluid motion that changes with time even at a fixed point.

unsteady flow

The flow of blood in your arteries is an example of unsteady flow; with each contraction of the heart ventricles, the pressure rises, and the flow velocity increases. Figure 15–23 shows other examples of steady and unsteady flow. We will restrict our quantitative description of fluid motion to steady flow. Like all other motion in classical physics, fluid motion is governed by Newton's laws. Indeed, it is possible to write Newton's second law in a form that involves explicitly the fluid velocity as a function of position and time. But the resulting equation is difficult to solve in any but the simplest cases. Instead of applying Newton's law directly, we will approach fluid dynamics from the point of view of energy conservation. In mechanics, we found that energy conservation gave us a "shortcut" in solving mechanics problems where the application of Newton's law would be

Fig. 15–22
In this aerodynamic test of an automobile, smoke particles introduced into the air follow streamlines and therefore allow airflow to be traced.

Fig. 15–23
(Left) Airflow above a burning cigarette is traced by the smoke. At first the flow is steady but soon becomes unsteady. (Right) Unsteady flows in Jupiter's atmosphere, photographed by a Voyager spacecraft. (See also color plate 2.)

difficult. The price we paid for the simplicity of the energy conservation method was a loss of detailed information about the behavior of a system as a function of time. Similarly, in fluid dynamics, use of energy conservation allows us to solve simply for many features of a moving fluid, but does not give all the details of the motion.

Conservation of Mass: The Continuity Equation

In mechanics problems, we had no trouble keeping track of the individual objects involved. But a fluid is continuous and deformable, so that it is not easy to follow an individual fluid element as it moves. Yet we know that fluid is conserved; as a fluid moves and deforms, new fluid is neither *continuity equation* created nor destroyed. The **continuity equation** is a mathematical statement of this fact that fluid (that is, matter itself) is conserved.

To develop the continuity equation, consider an arbitrary steady fluid flow represented by streamlines as shown, for example, in Fig. 15–24. Since streamlines are the paths followed by individual fluid elements, fluid flows along streamlines but never across them. In Fig. 15–24 we have *flow tube* shaded a **flow tube**—a small tube-like region bounded on its sides by a set of streamlines and on its ends by areas at right angles to the streamlines. We choose the flow tube to have sufficiently small cross-sectional area that fluid velocity and other fluid properties do not vary significantly over any cross section; however, fluid properties may vary along the flow tube. Although our flow tube has no physical boundaries, it nevertheless acts like a pipe of the same shape, for no fluid crosses the streamlines, just as no fluid crosses the walls of a pipe. With the direction of flow shown, fluid enters the flow tube at its left end and exits at the right. In steady flow, the rate at which fluid enters the flow tube must equal the rate at which it exits; otherwise, fluid would accumulate in or be displaced from the tube, and the fluid properties at each point could not be independent of time.

Figure 15–25 shows our flow tube in more detail, with a small fluid element just about to enter the tube, a process that will take a time Δt.

Fig. 15–24
Streamlines in steady flow. Shaded area outlines a flow tube. In steady flow, the rate at which fluid enters the flow tube equals the rate at which it leaves the tube.

Fig. 15–25
Fluid elements of equal mass entering and leaving the flow tube.

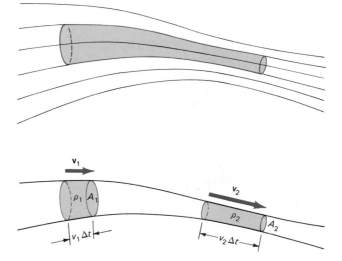

Another fluid element is shown just about to leave the flow tube; suppose that it contains the same amount of fluid as the entering fluid element. In steady flow, its exit from the tube must also take time Δt, so that there is no change in the total amount of fluid in the tube. The entering fluid element is moving at some speed v_1; since it takes time Δt for all of it to enter the tube, its length is $v_1\Delta t$. With cross-sectional area A_1, length $v_1\Delta t$, and density ρ_1, the mass of the entering fluid element is

$$m_1 = \rho_1 A_1 v_1 \Delta t.$$

Similarly, the exiting fluid element has length $v_2\Delta t$, and therefore its mass is

$$m_2 = \rho_2 A_2 v_2 \Delta t.$$

We have argued that, in steady flow, the mass entering in time Δt and that leaving in the same time must be equal. Therefore

$$\rho_1 v_1 A_1 = \rho_2 v_2 A_2. \tag{15–7}$$

Since the endpoints of the tube are arbitrary, we conclude that the quantity $\rho v A$ must have the same value anywhere along the flow tube:

$$\rho v A = \text{constant along a flow tube.} \quad \text{(any fluid)} \tag{15–8}$$

mass flow rate

What is this quantity $\rho v A$ that is constant along a flow tube? Its SI units are $(\text{kg/m}^{-3})(\text{m/s})(\text{m}^2)$, or simply kg/s. $\rho v A$ is therefore the **mass flow rate** or mass of fluid per unit time passing through the flow tube. Equations 15–7 and 15–8 are equivalent expressions of the continuity equation, both stating that the mass flow rate is constant along a flow tube in steady flow.

For a liquid, the density ρ is essentially constant, and the continuity equation 15–8 becomes simply

$$v A = \text{constant along a flow tube.} \quad \text{(liquid)} \tag{15–9}$$

volume flow rate

Now the constant quantity is just vA, with units of $(\text{m/s})(\text{m}^2)$, or m^3/s. vA is therefore the **volume flow rate**; in a fluid of constant density, constancy of mass flow rate (Equations 15–7 and 15–8) also implies constancy of volume flow rate, since a given mass of fluid does not change volume. In the form 15–9, the continuity equation makes obvious physical sense. Where the liquid has a large cross-sectional area, it can flow relatively slowly to transport a given mass of fluid per unit time. But where the area is more constricted, the flow must be faster to carry the same mass per unit time. With a gas, obeying Equation 15–8 but not necessarily 15–9, the situation is slightly more ambiguous, as density variations too play a role in the continuity equation. For flow speeds below the speed of sound in a gas, it turns out that lower area implies a higher flow speed just as for a liquid. But when gas flow speed exceeds the sound speed, density changes become so great that flow speed actually decreases with lower area.

Example 15–7

The Continuity Equation

In the lower part of its valley, the Ausable River in upstate New York is about 40 m wide. Under typical early summer conditions, it is 2.2 m deep and flows at 4.5 m/s. Just before it reaches Lake Champlain, the river enters Ausable Chasm, a deep gorge cut through rock (Fig. 15–26). At its narrowest, the gorge is only 3.7 m wide at the river surface. If the flow rate in the gorge is 6.0 m/s, how deep is the river at this point? Assume that the river has a rectangular cross section, with uniform flow speed over a cross section.

Solution

Writing the cross-sectional area as the product of width w and depth d, Equation 15–9 may be written

$$v_1 w_1 d_1 = v_2 w_2 d_2.$$

Solving for the gorge depth d_2 gives

$$d_2 = \frac{v_1 w_1 d_1}{v_2 w_2} = \frac{(4.5 \text{ m/s})(40 \text{ m})(2.2 \text{ m})}{(6.0 \text{ m/s})(3.7 \text{ m})} = 18 \text{ m},$$

or about 60 feet!

Fig. 15–26
The Ausable River in upstate New York cuts through a narrow chasm. To accommodate the flow, water depth in the chasm is much greater than elsewhere.

Conservation of Energy: Bernoulli's Equation

Having expressed conservation of matter through the continuity equation, we now turn to conservation of energy. Figure 15–27 again shows a flow tube with fluid elements entering and leaving. In steady flow, the properties of the fluid at a given point are always the same. So it doesn't matter whether we think of Fig. 15–27 as a "snapshot" showing two different fluid elements at the same time or the same fluid element at two different times. In deriving the continuity equation, we took the former viewpoint; here we will find the latter viewpoint more convenient. Our fluid element of mass m enters the flow tube with velocity v_1 and leaves with velocity v_2. Therefore, the change in kinetic energy of the fluid element as it traverses the flow tube is

$$\Delta K = \tfrac{1}{2}m(v_2^2 - v_1^2).$$

The work-energy theorem (Equation 6–19) tells us that the change in ki-

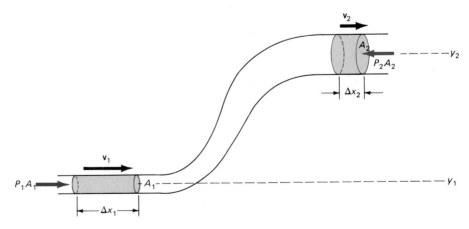

Fig. 15–27
A flow tube showing a fluid element entering and leaving. The work done by pressure and gravitational forces equals the change in the kinetic energy of the fluid element.

netic energy is equal to the net work done on the fluid element. As the fluid element enters the tube, it is subject to a pressure force P_1A_1 from the pressure of adjacent fluid outside the tube. This external force acts over the length Δx_1 of the fluid element as it enters the tube, so that it does work $W_1 = P_1A_1\Delta x_1$ on the fluid element. Similarly, as it leaves the tube, the fluid element is subject to a force P_2A_2 from the pressure of the fluid beyond the right-hand end of the tube. Since the direction of this force is opposite to the flow direction, it does negative work $W_2 = -P_2A_2\Delta x_2$ on the fluid element. Of course there are also forces on the fluid element arising from the pressure of adjacent fluid within the flow tube. But as the fluid element moves through the tube, it experiences such forces first on one end, then on the other, so that no net work is done by these forces internal to the flow tube. Forces from adjacent flow tubes act at right angles to the flow, so that they, too, do no work. Finally, the fluid element rises a distance $y_2 - y_1$ as it traverses the flow tube; the downward gravitational force therefore does negative work $W_g = -mg(y_2 - y_1)$ on the fluid element. Here y measures the vertical distance from an arbitrary reference height to the center of the flow tube; we assume that the flow tube is small enough that we can neglect variations in height and pressure across it. Summing the three contributions to the work done on the fluid element, and applying the work-energy theorem to equate that net work to the change in kinetic energy across the flow tube, we have

$$W_1 + W_2 + W_g = \Delta K,$$

or

$$P_1A_1\Delta x_1 - P_2A_2\Delta x_2 - mg(y_2 - y_1) = \tfrac{1}{2}m(v_2^2 - v_1^2).$$

The quantities $A_{1,2}\Delta x_{1,2}$ are the volumes of the fluid element as it enters and leaves; since density is mass per unit volume, we can write these volumes in the form $m/\rho_{1,2}$. Making this substitution, the mass m cancels from the equation. Rearranging the resulting equation so that all the terms associated with the entering fluid element are on the left side, and those associated with the departing fluid element are on the right, we have

$$\frac{P_1}{\rho_1} + \tfrac{1}{2}v_1^2 + gy_1 = \frac{P_2}{\rho_2} + \tfrac{1}{2}v_2^2 + gy_2. \tag{15–10}$$

What do the three terms on each side of this equation mean? The terms $\tfrac{1}{2}v^2$ and gy look like kinetic and potential energy, respectively, but divided by the mass. Therefore, they represent the kinetic energy per unit mass associated with the flow velocity v, and the potential energy per unit mass associated with the fluid height y. What about the terms P/ρ? They, too, have the units of energy per unit mass, and represent energy associated with the compression of the fluid. So Equation 15–10 expresses conservation of energy along the flow tube. Since the endpoints of the tube are arbitrary, we conclude that the quantity $P/\rho + \tfrac{1}{2}v^2 + gy$ is constant along a flow tube.

So far, we have assumed only that our flow tube is narrow and that there is no energy loss or gain from friction or from mechanical devices like pumps and turbines within the flow tube. If we further restrict consideration to liquids, which are essentially incompressible, the quantities ρ_1

and ρ_2 in Equation 15–10 become the constant density ρ, and we can re-write the equation in the form

$$P + \tfrac{1}{2}\rho v^2 + \rho gh = \text{constant along flow tube.} \quad \text{(liquid)} \quad (15\text{–}11)$$

Bernoulli's equation

Equation 15–11—the statement of energy conservation for an ideal incompressible fluid with no energy dissipation—is called **Bernoulli's equation,** after the Dutch-born Swiss mathematician Daniel Bernoulli (1700–1782). If we make the further restriction that the flow is irrotational—that a fluid element at any point has no angular momentum about that point—then it turns out that the constant in Bernoulli's equation has the same value for all flow tubes.

Applications of Fluid Dynamics

The laws of mass and energy conservation that we have just written for fluids can be used to explain a wide variety of natural and technological phenomena involving fluid motion.

Example 15–8

The Flow of Liquid from a Tank

A large, open tank is filled to a height h with liquid of density ρ. What is the speed of liquid emerging from a small hole at the base of the tank (Fig. 15–28)?

Solution

Assuming that the flow is steady and irrotational, we can apply Bernoulli's equation to the emerging fluid. Since the hole is open to the atmosphere, the fluid pressure at the hole must be atmospheric pressure, P_a. If we define the height at the tank bottom, where the hole is, to be $y=0$, then the quantity $P+\tfrac{1}{2}\rho v^2 + \rho gh$ evaluated at the hole has the value $P_a + \tfrac{1}{2}\rho v_{\text{hole}}^2$. To solve for v_{hole}, we need to know the value of this quantity at some other point. A suitable point is the top of the tank, where the pressure is also atmospheric pressure since the tank is open. What about the flow velocity? If the area of the tank is large compared with the hole area, the liquid level at the top drops only slowly, and we can take $v_{\text{top}} \approx 0$. Therefore, the quantity $P+\tfrac{1}{2}\rho v^2 + \rho gy$—the constant in Bernoulli's equation—has the value $P_a + \rho gh$ at the top of the tank. Bernoulli's equation states that $P+\tfrac{1}{2}\rho v^2 + \rho gy$ has the same value everywhere; equating our expressions for this quantity at the hole and at the tank top then gives

$$P_a + \tfrac{1}{2}\rho v_{\text{hole}}^2 = P_a + \rho gh.$$

Solving for the outflow velocity at the hole, we have

$$v_{\text{hole}} = \sqrt{2gh}.$$

Does this simple result make sense? We would get the same speed by dropping an object through a distance h—and for the same reason: conservation of energy. Draining a gram of water through the hole is energetically the same as removing a gram of water from the top of the tank and dropping it through the distance h. Just as the speed of a falling object is independent of its mass, so is the speed of the emerging liquid independent of its density, or mass per unit volume.

Our use of Bernoulli's equation—an equation derived for *steady* flow—is an approximation in this example, for the liquid level at the tank top must drop as liquid drains through the hole. The approximation is a reasonable one provided the tank area is large compared with the hole area. Problem 32 explores the long-term situation as the tank drains completely.

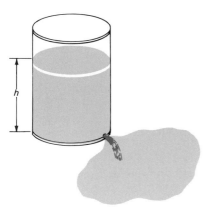

Fig. 15–28
Example 15–8

Example 15–9
Flow from a Faucet

Water leaves a faucet in steady, near-vertical flow (Fig. 15–29 left). Use the continuity equation and Bernoulli's equation to find the diameter of the falling water column as a function of the distance h measured downward from the faucet. Take D_0 to be the outlet diameter of the faucet and v_0 to be the flow velocity at the faucet.

Solution

Since the density of a liquid is constant, we use the continuity equation in the form 15–9:

$$\tfrac{1}{4}\pi D^2 v = \tfrac{1}{4}\pi D_0^2 v_0,$$

where D and v are the diameter and flow speed at any point, and where $\tfrac{1}{4}\pi D^2 = \pi r^2$ is the cross-sectional area of the water column. So the diameter and flow speed are related by

$$D^2 = \frac{D_0^2 v_0}{v}.$$

We want the diameter as a function of position, so we must express flow speed in terms of position. Bernoulli's equation provides the desired link. Since the entire water column is open to the atmosphere, the pressure anywhere is atmospheric pressure, P_a. (Here we neglect variation in atmospheric pressure over the height of the water stream.) So Bernoulli's equation reads

$$P_a + \tfrac{1}{2}\rho v^2 + \rho g y = P_a + \tfrac{1}{2}\rho v_0^2,$$

where the left-hand side applies to an arbitrary point and the right-hand side to the faucet outlet. We have taken $y=0$ at the faucet, and our y-axis is positive upward for consistency with our derivation of Bernoulli's equation. Then the downward distance from the faucet, h, is given by $h = -y$. Making this substitution and solving for v gives

$$v = \sqrt{v_0^2 + 2gh}.$$

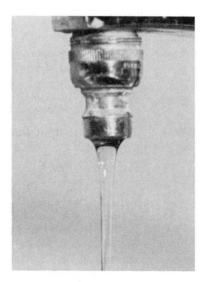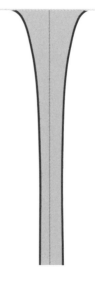

Fig. 15–29
(Left) Steady flow from a faucet. (Right) Theoretical result of Example 15–9 closely matches the observed behavior of the water.

This is just the result we would expect for an object with initial velocity v_0 falling a distance h. Using this result in our expression for the diameter D then gives

$$D = D_0 \left(\frac{v_0^2}{v_0^2 + 2gh} \right)^{1/4}.$$

Figure 15–29 (right) shows that this theoretical result matches closely the actual flow. Problem 30 explores this example further.

For horizontal flow, or when changes in gravitational potential energy density are small compared with kinetic energy density and with pressure, we can drop the $\rho g y$ terms in Bernoulli's equation to obtain

$$P + \tfrac{1}{2}\rho v^2 = \text{constant}. \tag{15–12}$$

This equation shows that flow speed is high where pressure is low, and vice versa. You might think that flow speed should be high where pressure is high. But conservation of energy requires the opposite. Both the pressure term P and the flow kinetic energy term $\tfrac{1}{2}\rho v^2$ represent energy per unit volume; if one increases, the other must decrease to conserve energy.

Equation 15–12, like Equation 15–11, applies strictly only to an incompressible fluid—a liquid. With a gas, density can vary with position and we must use Equation 15–10. If the density of a gas decreases sufficiently in a region where pressure increases, flow velocity can actually increase to satisfy energy conservation in the form 15–10. However, it is only with supersonic flows—flow speeds greater than the sound speed in the gas— that such effects occur. Indeed, density variations in a gas are generally insignificant when the flow speed is much less than the speed of sound. (Here we continue to assume that density variations due to gravity are negligible, as we are dealing only with small vertical distances.) For this reason we can treat slowly moving gas as an incompressible fluid, and apply Equation 15–12.

Nature and technology abound with applications of Equation 15–12 to liquids and to gases flowing at well below their sound speeds. Blow across the top of a piece of paper and it extends horizontally (Fig. 15–30). Why? Because you have increased the flow speed above the paper and therefore lowered the pressure. Higher pressure below the paper therefore results in a net upward force on the paper. Airplane wings work in the same way. They are shaped or angled so that air flows at a higher speed over the top surface, again resulting in a lower pressure above the wing and therefore a net upward force (Fig. 15–31). Think about a jumbo jet, and you will agree that this upward force can be substantial (Fig. 15–32). The dirt around a prairie dog's hole is mounded up in a shape that resembles the leading edge of an airplane wing (Fig. 15–33). When the wind blows, this design results in lower air pressure above the hole and therefore in an outflow of air from the burrow. Biologists speculate that prairie dogs have evolved this hole design to provide ventilation. The **Bernoulli effect,** as the occurrence of low pressure with high flow speed is called, can sometimes be strikingly counterintuitive. Figure 15–34 shows a Ping-Pong ball supported by a *downward* blowing column of air in an inverted funnel. Rapid divergence of the airflow in the expanded region of the funnel results in a

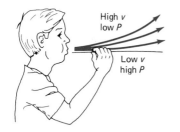

High v
low P

Low v
high P

Fig. 15–30
When you blow across the top of a piece of paper, higher flow speed at the top results in lower pressure, and therefore a net upward force.

Bernoulli effect

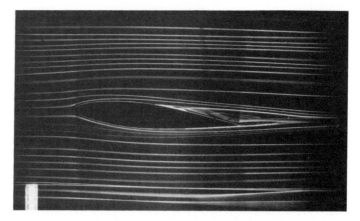

Fig. 15–31
A model wing being tested in a wind tunnel. Both wing shape and its inclination to the horizontal result in higher flow speed and therefore lower pressure over the top surface, giving a net upward pressure force. Note the slight downward deflection of airflow at upper right. This shows that the wing has exerted a downward force on the air, in reaction to the upward force of air on the wing.

Fig. 15–32
A fully loaded 747 jet can weigh nearly 4×10^6 N (over 400 tons). Its weight is supported by the force associated with the difference in air pressure across its wings.

lower flow speed and therefore higher pressure below the ball, whose weight is thereby supported. A new application of the Bernoulli effect is the so-called "Bernoulli box," a computer disk that combines the high speed of a hard disk with the ruggedness and low cost of floppy disks. In the Bernoulli box disk, pressure differences associated with airflow are used to maintain the otherwise flexible disk in a rigid state, allowing high-speed data access.

A constriction in a pipe carrying incompressible liquid requires that the flow speed increase in order to maintain constant mass flow. Such a constriction is called a **venturi.** Bernoulli's equation requires that the fluid

venturi

Fig. 15–33
Is the entrance of a prairie dog hole shaped to provide ventilation by lowering the pressure over the hole?

Fig. 15–34
A Ping-Pong ball supported by a downward-moving air stream.

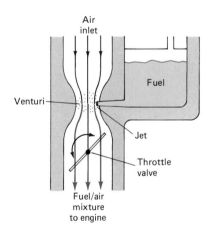

Fig. 15–35
Simplified view of a carburetor, showing the venturi and the jet where fuel enters.

pressure be lower in the venturi. The venturi is put to good use in an automobile carburetor. Air flows into the carburetor because of lower pressure created by the downward motion of the engine's pistons. In the venturi (also called the throat) of the carburetor, the flow speed increases and the pressure drops. A tiny hole, or jet, opens into the venturi and carries gasoline. The lower pressure in the venturi results in a flow of gasoline into the carburetor, where it mixes with the high-speed air stream to form the combustible mixture burned in the engine. A flat disk—the throttle valve—below the venturi is connected to the car's accelerator pedal. The throttle valve pivots to control the rate at which the fuel/air mixture reaches the engine. Figure 15–35 gives a simplified view of a carburetor.

Example 15–10

A Carburetor

With its throttle valve wide open, an automobile carburetor has a throat diameter of 2.4 cm. With each revolution, the car's 2.2-L, 4-cylinder engine sucks in 0.55 L of air through the carburetor. At an engine speed of 3000 revolutions per minute, what are (a) the volume flow rate, (b) the airflow speed, and (c) the difference between atmospheric pressure and air pressure in the carburetor throat? The density of air is 1.3 kg/m^3.

Solution:

With 0.55 L of air per revolution, at 3000 rpm the engine takes in air at the volume flow rate $\mathscr{F}$ given by

$$\mathscr{F} = (0.55 \text{ L/rev})(3000 \text{ rev/min}) = 1650 \text{ L/min} = 0.0275 \text{ m}^3/\text{s}.$$

But the volume flow rate is the product of flow speed and area, so that within the carburetor throat we have $\mathscr{F} = vA = \pi r^2 v$. Then

$$v = \frac{\mathscr{F}}{\pi r^2} = \frac{0.0275 \text{ m}^3/\text{s}}{(\pi)(0.012 \text{ m})^2} = 61 \text{ m/s}.$$

Since this is well below the sound speed (approximately 340 m/s), the air acts as though it were incompressible.

We can now use Bernoulli's equation to find the pressure difference, taking as our reference point the ambient atmosphere where the flow speed is essentially zero. Then Bernoulli's equation becomes

$$P_a + 0 = P_c + \tfrac{1}{2}\rho v^2,$$

where P_a is atmospheric pressure and P_c the pressure in the carburetor. Solving for the pressure difference $P_a - P_c$ gives

$$P_a - P_c = \tfrac{1}{2}\rho v^2 = \tfrac{1}{2}(1.3 \text{ kg/m}^3)(61 \text{ m/s})^2 = 2.4 \times 10^3 \text{ N/m}^2,$$

or about 2.4 per cent of atmospheric pressure. It is this pressure difference that drives gasoline through the jet and into the carburetor throat. Problem 33 explores the gasoline flow, which itself can be treated using the equations of fluid dynamics.

Figure 15–36 shows a venturi flowmeter, a device used to measure the rate of fluid flow. With no flow, the fluid is in static equilibrium and there is no pressure difference between the venturi and the main pipe. As the flow rate increases, the pressure in the venturi of the flowmeter decreases,

Fig. 15–36
A venturi flowmeter. According to Bernoulli's equation, the pressure difference across the manometer tube depends on the flow rate through the flowmeter.

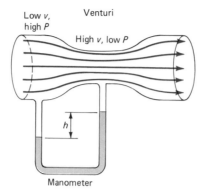

so that the manometer reading provides a measure of flow speed. Known values of the fluid density and venturi area can then be used to calculate the mass flow rate. Problem 29 explores the venturi flowmeter.

Natural examples of venturi flow range from airflow in caves to the exchange of matter in binary star systems (Fig. 15–37).

15.4
OTHER ASPECTS OF FLUID MOTION

viscosity

Our discussion of fluid motion has been limited to the special case of the steady flow of an ideal fluid. By ideal, we mean that fluid friction or **viscosity** is absent. The neglect of viscosity is often a valid approximation, especially in large-volume flows where not much of the fluid is near any boundary like a pipe wall. But when a fluid is confined to a narrow channel, and for any fluid near enough to a solid boundary, viscosity always plays an important role. The importance of viscosity depends not only on the dimensions of the flow but also on the fluid itself; water is relatively inviscid, while molasses is extremely viscous. Viscosity in a flowing fluid results in the conversion of flow kinetic energy into heat, and can also be

Fig. 15–37
This artist's conception of the binary star system Cygnus X-1 shows gas streaming off the large star and into a gaseous disk surrounding its companion black hole (left). A venturi-like constriction in the flow is provided by the effects of gravity and orbital motion.

Fig. 15–38
(a) In an inviscid fluid, flow velocity in a pipe would be uniform. (b) Viscosity reduces the flow velocity, especially near the pipe walls, resulting in a nonuniform velocity profile.

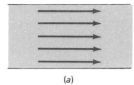

(a)

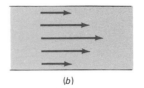

(b)

important in introducing angular momentum into the flow. At a solid surface bounding a viscous fluid, viscosity results in a drag force that slows the fluid, bringing it to a complete halt right at the wall. The result is a nonuniform flow velocity profile in enclosed structures like pipes (Fig. 15–38). Viscosity is the dominant influence on fluids confined to very small spaces, as, for example, the lubricating oils separating metal surfaces in machinery. And viscosity plays an important role in stabilizing flows that would otherwise become **turbulent,** or chaotically unsteady (Fig. 15–39).

turbulent

Since we have discussed only steady flows, we have not been able to consider the fascinating subject of waves in fluids. Physically, a slight buildup of pressure in an otherwise hydrostatic fluid will propagate through the fluid to form a sound wave. Mathematically, an analysis of this situation shows that common fluids obey, at least approximately, the wave equation we introduced in the last chapter. The sound speed revealed by such a wave equation depends on the properties of the fluid. Other types of waves are possible, too. Waves at the interface of two fluids—like waves on the surface of the ocean—are one example. In electrically conducting fluids, more complicated wave modes exist, often involving the interaction of electrical, magnetic, and pressure forces.

Fig. 15–39
A jet of water quickly becomes turbulent. In a more viscous fluid like molasses, viscosity would delay the onset of turbulence.

SUMMARY

1. **Fluid** is matter in a state that is capable of deforming and flowing in response to forces. Fluids are approximated as continuous distributions of matter, and are characterized by their density and pressure:
   a. **Density** (symbol ρ) measures mass per unit volume of fluid. Liquids are nearly incompressible, meaning that they are capable only of very small changes in density. Gases, on the other hand, can undergo large density changes.
   b. **Pressure** measures the force exerted normal to a unit area by the fluid.
2. In **hydrostatic equilibrium** there is no net force on any element of a fluid.
   a. In the absence of external forces like gravity that act throughout the volume of a fluid, hydrostatic equilibrium requires that pressure be constant through-

out the fluid. With constant pressure, there is no net pressure force on any fluid element; net pressure force arises only when pressure varies with position. An increase in pressure anywhere in the fluid is transmitted throughout the fluid to maintain hydrostatic equilibrium, a fact known as **Pascal's principle.**
   b. In hydrostatic equilibrium in the presence of gravity, pressure increases going downward, so that there is a net upward pressure force to balance the gravitational force on a fluid element. In a liquid, whose density is constant, pressure increases linearly with depth:

$$P = P_0 + \rho gy, \quad \text{(liquid)}$$

where P_0 is the pressure at the liquid surface and y

the depth. In a gas at constant temperature T, pressure decreases exponentially with height h:

$$P = P_0 e^{-h/h_0},$$

where $h_0 = mg/kT$ is the **scale height,** with m the mass of a gas molecule and k Boltzmann's constant $(1.38 \times 10^{-23}$ J/K).

c. **Buoyancy force** is an upward force on an object wholly or partly immersed in a fluid. The buoyancy force arises because the pressure force is greater at the bottom of the object than at the top. The buoyancy force is equal to the weight of the fluid displaced by the object—a fact known as **Archimedes' principle.** The net force on an object is the sum of its downward weight and the upward buoyancy force; therefore, objects float if they are less dense than the fluid and sink if they are more dense.

3. A moving fluid is characterized by giving its flow velocity at each point in space and at each instant of time. Flow velocity is conveniently represented by **streamlines,** curves traced out by individual fluid elements. Flow velocity is higher where streamlines are closer together and lower where they are farther apart. In **steady flow,** flow velocity is always the same at a given point; in **unsteady flow,** it varies with time as well as position. A simplified description of a moving fluid is provided by applying laws of conservation of mass and conservation of energy. Both laws are applied to a narrow **flow tube**—a volume bounded by a specific set of streamlines, so that fluid cannot cross the flow tube boundary. Conservation of mass results in the **continuity equation:**

$$\rho v A = \text{constant along a flow tube,}$$

where A is the tube area, which may vary with position along the tube. The quantity $\rho v A$ is the **mass flow rate,** with SI units kg/s. In a liquid, or in a gas flowing without significant density change, the continuity equation becomes

$$v A = \text{constant along a flow tube.}$$

The quantity vA is the **volume flow rate,** with SI units m³/s. With steady flow in the absence of fluid friction (viscosity) or energy addition by heat or mechanical means, energy conservation implies that the energy per unit mass of fluid is constant along a flow tube:

$$\frac{P}{\rho} + \tfrac{1}{2}v^2 + gy = \text{constant along a flow tube,}$$

where y is the vertical height of the flow tube, which may vary along the tube. For an incompressible fluid, the density ρ is constant, and the statement of energy conservation becomes **Bernoulli's equation:**

$$P + \tfrac{1}{2}\rho v^2 + \rho g h = \text{constant along a flow tube.}$$

Bernoulli's equation implies that a fluid in horizontal motion moves fastest where pressure is lowest, and vice versa. This **Bernoulli effect** is crucial in many technological and natural phenomena, including the flight of birds and aircraft.

4. Fluid friction, or **viscosity,** results in a loss of fluid energy to heat. Viscosity is especially significant in narrowly confined flows, or near flow boundaries. Viscosity exerts a stabilizing influence on flows that would otherwise become **turbulent,** or chaotically unsteady.

5. Disturbances in fluid properties can result in waves propagating through fluids at characteristic speeds that depend on the fluid.

QUESTIONS

1. What is the difference between hydrostatic equilibrium, steady flow, and unsteady flow?
2. Why do your ears "pop" when you drive up a mountain?
3. The cabins of commercial jet aircraft are usually pressurized to the pressure of the atmosphere at about 2 km (7000 ft) above sea level. Why do you not feel the lower pressure on your entire body?
4. Water pressure at the bottom of the ocean arises from the weight of the overlying water. Does this mean that the water exerts pressure only in the downward direction? Explain.
5. The three containers in Fig. 15–40 are filled to the same level and are open to the atmosphere. How do the pressures at the bottoms of the three containers compare?
6. Municipal water systems often include tanks or reservoirs mounted on hills or towers. Besides water storage, what function might these reservoirs have?

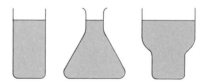

Fig. 15–40   Question 5

7. Why is it easier to float in the ocean than in fresh water?
8. Figure 15–41 shows a cork suspended from the bottom of a beaker of water. The beaker is mounted on a board that is rotating about a vertical axis through its far end. Explain the position of the cork as shown in the figure.
9. An ice cube is floating in a cup of water. Will the water level rise, fall, or remain the same when the cube melts?

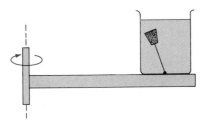

Fig. 15–41 Question 8

10. Figure 15–42 shows a hydraulic lift in which a person pushing with one finger lifts a large truck. Is this device really possible? Is it practical? How would the fluid pressure below the truck-supporting piston compare with that below the piston on which the person is pushing?

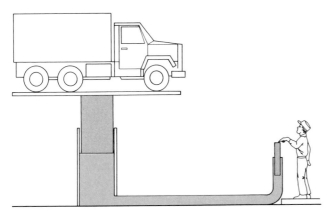

Fig. 15–42 Hydraulic lift (Question 10); the drawing is *not* to scale.

11. A mountain stream, frothy with entrained air bubbles, presents a serious hazard to hikers who fall into it, for they may sink in the stream where they would float in calm water. Why?
12. Why are dams thicker at the bottom than at the top?
13. It is not possible to breathe through a snorkel-like tube from a depth greater than a meter or so (Fig. 15–43). Why not?

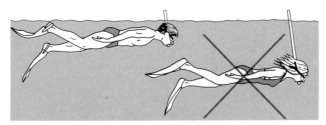

Fig. 15–43 It is not possible to breathe through a snorkel tube from significant depths (Question 13).

14. Most humans float naturally in fresh water. Yet the body of a drowning victim generally sinks, often rising several days later after bodily decomposition has set in. What might explain this sequence of floating, sinking, and floating again?
15. A helium-filled balloon stops rising long before it reaches the "top" of the atmosphere, while a cork released from the bottom of a lake rises all the way to the surface of the water. Explain the difference between these two behaviors.
16. A barge filled with steel beams overturns in a lake, spilling its cargo. Does the water level in the lake rise, fall, or remain the same?
17. Imagine a cylindrical tube filled with water and set rotating about its axis. If pieces of wood and stone are introduced into the cylinder, where will each end up? Assume the cylinder is in the weightless environment of an orbiting spacecraft.
18. When gas in steady, subsonic flow through a tube encounters a constriction, its flow speed increases. When it flows supersonically in the same situation, flow speed decreases in the constriction. What must be happening to the gas density at the constriction in the supersonic case?
19. A ball moves horizontally through the air without spinning. Where on the ball's surface is the air pressure greatest? Explain your answer in terms of Bernoulli's equation.
20. A ball supported by an upward-flowing air column (Fig. 15–44) is essentially in stable equilibrium; if the ball is displaced slightly in any direction, it returns to its original position. Explain. *Hint:* How do you expect the airflow velocity to vary with horizontal position from the center of the air column?

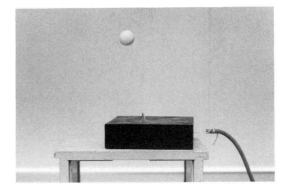

Fig. 15–44 Question 20

21. Under what conditions can a gas be treated as incompressible?
22. As you drive along a highway and are passed by a large truck, your car may experience a force toward the truck. Explain the origin of this force in terms of the Bernoulli effect.
23. A pump is submerged at the bottom of a 200-ft-deep well. Does it take more power to pump water to the surface when the well is full of water or nearly empty? Or doesn't it matter?

PROBLEMS

Section 15.1 *Describing Fluids*

1. Use the densities of water and mercury (1.00×10^3 kg/m^3 and 1.36×10^4 kg/m^3, respectively) to derive the conversion factors from inches of water, inches of mercury, and torr (mm of mercury) to the SI unit Pa.

2. A rubber suction cup 3.2 cm in diameter is used to suspend a hanging plant from a horizontal glass surface (Fig. 15–45). What is the maximum weight that can be so suspended (*a*) at sea level and (*b*) in Denver, where atmospheric pressure is about 0.80 atm?

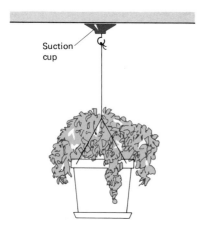

Fig. 15–45 Problem 2

3. 8.8 kg of compressed air is stored in a gas cylinder with a volume of 0.050 m^3. (*a*) What is the density of air in the cylinder? (*b*) How large a volume would the same gas occupy at standard atmospheric density of 1.3 kg/m^3?

4. The fuselage of a 747 jumbo jet can be considered roughly a cylinder 60 m long and 6 m in diameter. If the interior of the aircraft is pressurized at 0.75 atm, what is the force tending to tear any two half-cylinders of the aircraft apart when it is flying at an altitude of 10 km, where air pressure is about 0.25 atm? (The earliest commercial jets of the 1950's suffered structural failure from just such forces. Modern jets are better engineered to accommodate these forces.)

5. The piston of the hydraulic lift used in a garage has a diameter of 40 cm. If the piston supports a total mass of 3100 kg, what is the pressure in the hydraulic fluid supporting the piston? Assume normal atmospheric pressure of 1.0×10^5 Pa.

6. Exhaust hoods in the chemistry labs of the Middlebury College Science Center are kept on constantly for safety reasons. However, many of the ventilation fans supplying replacement air to the building have been shut down to conserve energy. The result is a lower pressure inside the building than outside. On a day when atmospheric pressure is 765.0 torr, a physics student leaving the building notices that a force of 150 N is required to open the door, which measures 90 cm × 210 cm. The door is hinged on one of its 210-cm vertical sides, and the student pulls a door handle at the opposite side. What is the pressure inside the building? *Hint*: What is the torque on the door?

7. Fully loaded, a model 240 Volvo station wagon has a mass of 1950 kg (4300 lb). If each of its four tires is inflated to a gauge pressure of 230 kPa (32 psig), and if each supports one-fourth of the car's weight, how much surface area of each tire is in contact with the road?

8. The emergency escape window over the wing of a DC-9 jetliner measures 50 cm × 90 cm. If the pressure inside the plane is 650 torr when it is flying at an altitude where atmospheric pressure is 320 torr, is there any danger that a passenger could open the emergency window while at that altitude? To answer this question, calculate the force that would be required to pull the escape window open (the window opens inward). For simplicity, assume that the window is pulled straight inward, rather than being pivoted about one edge.

9. The density of earth's atmosphere as a function of height h is given approximately by $\rho = \rho_0 e^{-h/h_0}$, where $\rho_0 = 1.3$ kg/m^3 is the density at the surface and $h_0 = 8.2$ km. Calculate the total mass of atmosphere within the first 150 km of the surface. *Hint*: Since the atmosphere is thin compared with the radius of the earth, you can treat it as a slab whose area is that of earth's surface and whose height is 150 km. But don't forget that the density varies with height! (See Problem 37 for more on the mass of the atmosphere.)

Section 15.2 *Fluids at Rest*

10. Figure 15–46 shows a schematic diagram of a hydraulic lift. The diameter of the large piston is 40 cm, and that of the small tube leaving the pump is 1.7 cm. A total load of 2800 kg (including the piston mass) is raised a distance of 2.3 m with the lift. (*a*) What volume of fluid passes through the pump? (*b*) What is

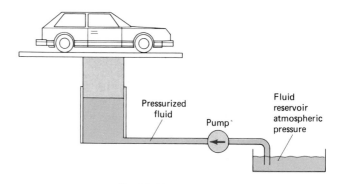

Fig. 15–46 Problem 10

the pressure at the pump outlet? (c) What is the force exerted by the pump on the fluid at its outlet? (d) How much work does the pump do? (e) If the lifting operation takes 40 s, what is the power output of the pump? In your calculations, neglect changes in pressure associated with varying depth of the hydraulic fluid. Neglect also the mass of the fluid raised in the cylinder.

11. In a properly functioning chimney, the pressure is slightly lower than the pressure in the building housing the fireplace, woodstove, or furnace connected to the chimney. This lower pressure is associated with the rise of hot, buoyant gases in the chimney, and ensures that these gases do not leak into the building. Chimney pressure, also called draft, can be measured with a water manometer, as shown in Fig. 15–47. If the difference in water levels on the two sides of the manometer tube is 0.04 in (typical of an oil furnace), by how much does the chimney pressure differ from atmospheric pressure in the building?

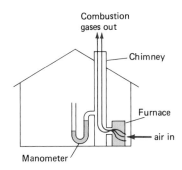

Fig. 15–47 Water manometer used on a chimney (Problem 11).

12. A research submarine can withstand external pressures of 7500 psi when its internal pressure is 1.0 atm. How deep can it dive?

13. Scuba diving equipment supplies the diver with air at the same pressure as the surrounding water, allowing the diver to breathe easily. But at air pressures greater than about 150 psi, nitrogen narcosis becomes a serious problem. (Nitrogen narcosis is an intoxicating effect of high-pressure nitrogen that can result in fatally abnormal behavior, like offering one's air supply to passing fish.) At what depth does nitrogen narcosis become a hazard?

14. The U-shaped tube shown in Fig. 15–48 initially contains only water, and then a quantity of oil occupying a 2.0-cm length of the tube is added, as shown. If the oil has a density 0.82 times that of water, what is the difference in heights of the liquid levels on the two sides of the tube? (Both ends of the tube are open to the atmosphere.)

15. A child attempts to drink water through a 100-cm-long straw, but finds that the water rises only 75 cm. By how much has the child reduced her mouth pressure below that of the atmosphere?

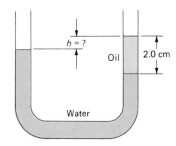

Fig. 15–48 Problem 14

16. The maximum cruising altitude of a DC-9 jetliner is 42,000 ft. Making the approximation that the atmosphere is at a uniform temperature of 280 K, what is the air pressure at this altitude (see Example 15–4)?

17. Barometric pressure in the eye of a typical hurricane is on the order of 0.91 atm (27.2 inches of mercury). How does the level of the ocean surface (assumed calm) under the eye compare with that under a distant fair-weather region where the pressure is 1.0 atm?

18. On land, the most massive concrete block you can carry has a mass of 25 kg. How massive a block could you carry under water? The density of concrete is 2.3×10^3 kg/m^3.

19. A partially full beer bottle with an interior diameter of 52 mm is floating upright in water, as shown in Fig. 15–49. A beer drinker takes a swig and replaces the bottle in the water, where it now floats 28 mm higher than before. How much beer did the beer drinker drink? Assume the density of beer is essentially that of water.

Fig. 15–49 Beer bottle before and after (Problem 19).

20. A glass beaker has a mass of 25 g and measures 10 cm high by 4.0 cm in diameter. Empty, it floats in water with one-third of its height submerged. How many 15-g rocks can be placed in the beaker before it sinks?

21. A typical supertanker has a mass of 2.0×10^8 kg and carries twice that amount of oil. If 9.0 m of the ship is submerged when it is empty, what is the minimum water depth needed for it to navigate when full? Assume the sides of the ship are vertical.

22. A balloon contains gas of density ρ_g, and is to lift a payload of mass M, not including the mass of the gas. Show that the minimum mass m of gas required is

$$m = \frac{M\rho_g}{\rho_a - \rho_g},$$

where ρ_a is the density of the atmosphere. Evaluate your result for a helium balloon carrying two people and a basket with total mass 230 kg. The density of helium is about one-seventh that of air.

23. A ship's hull has a V-shaped cross section, as shown in Fig. 15–50. The ship has vertical height h_0 from keel to deck, and a total length ℓ perpendicular to the plane of Fig. 15–50. Empty, the hull extends underwater to a depth h_1, as shown. Obtain an expression for the maximum load the ship can carry, in terms of the density ρ of water and the quantities h_0, h_1, ℓ, and θ shown in Fig. 15–50.

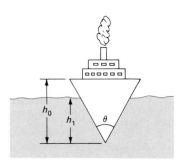

Fig. 15–50   Problem 23

Section 15.3   *Fluid Dynamics*

24. A fluid is flowing steadily, roughly from left to right. At the left it is flowing rapidly; then it slows down, and finally speeds up again. Its final speed at the right is not as fast as its initial speed at the left. Sketch some streamlines associated with this flow.

25. Show that pressure has the units of energy density and that the quantity P/ρ has the units of energy per unit mass.

26. A fire hose 10 cm in diameter delivers water at the rate of 15 kg/s. The hose terminates in a nozzle 2.5 cm in diameter. What are the flow speeds (a) in the hose and (b) in the nozzle?

27. A can of height h is full of water. At what distance y from the bottom of the can should a small hole be cut so that initially the water travels the same distance y horizontally as it does vertically? (See Fig. 15–51.)

28. The water in a garden hose is at a gauge pressure of 20 psig, and is moving at negligible speed. The hose terminates in a sprinkler consisting of many small holes. What is the maximum height reached by the jets of water emerging from these holes?

29. The venturi flowmeter shown in Fig. 15–52 is used to measure the flow rate of water in a solar collector system. The flowmeter is inserted in a pipe with internal diameter of $\frac{3}{4}$ in; at the venturi of the flowmeter, the diameter is reduced to $\frac{1}{4}$ in. The manometer tube contains oil with density 0.82 times that of water. If the difference in oil levels between the two sides of the

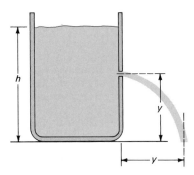

Fig. 15–51   Problem 27

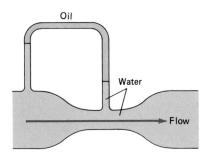

Fig. 15–52   Problem 29

manometer tube is 0.55 in, what is the flow rate in gallons/minute?

30. You turn on a bathroom sink faucet and notice that the water column has shrunk to about half its original diameter about 15 cm below the faucet outlet. What is the approximate outflow speed? *Hint*: See Example 15–9.

31. A drinking straw 20 cm long and 3.0 mm in diameter stands vertically in a cup of juice 8.0 cm in diameter. 6.5 cm of the straw extends above the surface of the juice. A child sucks on the straw, and the level of juice in the glass begins dropping at the rate of 0.2 cm/s. (a) By how much does the pressure in the child's mouth differ from atmospheric pressure? (b) What is the greatest distance above the liquid surface from which the child could drink, assuming the same mouth pressure? (Note: You are concerned here only with the *initial* rate, before the juice level has dropped appreciably, but after the straw has filled and approximately steady flow has been established.) The density of the juice is approximately that of water.

32. A can of height h and cross-sectional area A_0 is initially full of water. A small hole of area $A_1 \ll A_0$ is cut in the bottom of the can. Find an expression for the time it takes all the water to drain from the can. *Hint*: Call the water depth y, and use the continuity equation to write the rate dy/dt at which the depth decreases in terms of the outflow velocity at the hole. Then integrate.

33. While a car is driving at a steady 55 mph on a level road, the pressure inside the venturi of its carburetor is 4% below atmospheric pressure. (a) What is the speed of the air in the venturi? (b) Gasoline enters the venturi through a hole (the "jet") 0.90 mm in diameter, coming from a bowl within the carburetor that is exposed to atmospheric pressure and that is not significantly higher than the jet. How many miles per gallon of gasoline can the car drive under these conditions? The density of gasoline is 0.67 g/cm^3. See Example 15–10.

34. If the pressure just inside the tip of the firehose nozzle in Problem 26 is 10% greater than atmospheric pressure, what are the speed and diameter of the water column just outside the nozzle tip?

35. Figure 15–53 shows a simplified diagram of a Pitot tube, a device used for measuring flow speeds of fluids not confined in pipes. The Pitot tube arrangement shown here is used to measure airspeed on the underside of an aircraft wing. The device consists of two small tubes, A opening at right angles to the flow and B opening parallel to the flow. The other ends of both tubes are connected to an instrument—for instance, a manometer—that measures the pressure difference between them. Fluid at the entrance of tube A moves past with its normal velocity and pressure. But there is no flow through either tube, so that fluid stops at the entrance of tube B. Bernoulli's equation shows that the pressure in tube B is therefore higher. Show that the flow speed is given by

$$v = \sqrt{\frac{2\Delta P}{\rho}},$$

where ΔP is the pressure difference between the two tubes, and ρ is the density of the moving fluid.

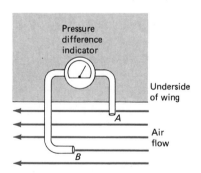

Fig. 15–53 Pitot tube mounted on the underside of an aircraft wing (Problem 35).

Supplementary Problems

36. How massive an object can be supported by a 2-in-diameter suction cup mounted on a vertical wall, if the coefficient of friction between the cup and wall is 0.72? Assume normal atmospheric pressure.

37. Use the information given in Problem 9 to calculate the height below which half of earth's atmosphere lies. Assume a uniform temperature of 280 K, as in Example 15–4. You can consider that the atmosphere ends at about 150 km, or that it extends forever. Why doesn't it make any significant difference which you assume?

38. The sun's atmosphere, or corona, extends far enough from the surface that the variation in gravity with height must be considered in accurate descriptions of the atmosphere. Show that the pressure of a uniform-temperature atmosphere surrounding a spherical star or planet of mass M and radius R is given by

$$P = P_0 \exp\left[\frac{GMm}{kT}\left(\frac{1}{r} - \frac{1}{R}\right)\right],$$

where G is the constant of universal gravitation, r the distance from the sun's center, m the mass of a gas molecule, k Boltzmann's constant (see Section 15.2), and P_0 the pressure at the base of the atmosphere. Using the approximation $e^x \approx 1 + x$ for $|x| \ll 1$, show that this result reduces to Equation 15–5 when $r - R \ll R$ (that is, close to the solar surface). Why does this make sense? (Note: exp(x) means e^x.)

39. A pencil is weighted at one end so that it floats vertically with length ℓ submerged. The pencil is pushed vertically downward (without being totally submerged), then released. Show that it undergoes simple harmonic motion with period $T = 2\pi\sqrt{\ell/g}$.

16

TEMPERATURE AND HEAT

In the next four chapters, we turn our attention to thermodynamics—the physics of heat, temperature, and related phenomena. Our own bodies provide a qualitative sense of what it means for things to be "hot" or "cold." In our study of thermodynamics, we will quantify this sense and explore the behavior of matter under different thermal conditions (Fig. 16–1). And we will learn something our sense of temperature could never tell us—namely, how thermal phenomena reflect the laws of mechanics operating on a microscopic scale.

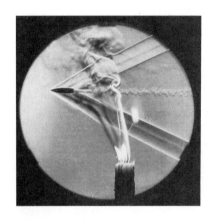

Fig. 16–1
A bullet travelling supersonically through the heated air above a candle flame. Flame, shock wave, and distortion of the shock by the heated air are all thermal effects.

16.1
MACROSCOPIC AND MICROSCOPIC DESCRIPTIONS

Some physical quantities—for example, mass—apply equally to microscopic objects like atoms and molecules and to macroscopic objects like cars, gas cylinders, and planets. But other quantities that we use to describe macroscopic objects—for example, temperature and pressure—have no meaning on a microscopic scale. We speak of the temperature and pressure of air, but it makes no sense to ask about the temperature or pressure of an individual air molecule. The study of temperature, heat, and related mac-

thermodynamics

roscopic quantities comprises the branch of physics called **thermodynamics.**

statistical mechanics

The thermodynamic behavior of matter is determined by the behavior of its constituent atoms and molecules in response to the laws of mechanics. **Statistical mechanics** is the branch of physics that relates the macroscopic description of matter to the underlying microscopic processes. Historically, thermodynamics was developed before the atomic theory of matter was well established. The subsequent explanation of thermodynamics in terms of statistical mechanics—the mechanics of atoms and molecules—was a triumph for classical physics. At the same time, the few ther-

404

modynamic phenomena that could not be explained successfully with classical physics helped point the way to the development of the quantum theory. In our study of thermal phenomena, we will interweave the macroscopic and microscopic descriptions to provide the fullest understanding of both viewpoints.

16.2
TEMPERATURE AND THERMODYNAMIC EQUILIBRIUM

Take a bottle of soda from the refrigerator and leave it on the kitchen counter. Eventually, it reaches room temperature. Come back much later and it's still at room temperature. No scientific definition of temperature is involved here. Our own sense of touch is enough to tell us that the bottle of soda is somehow different than when we first removed it from the refrigerator, and that after a long time it is no longer changing. We generalize from this simple experiment to say that when two systems—in this case the bottle and the surrounding air—are allowed to remain in contact until no further change occurs in any macroscopic property of either, then the two are in **thermodynamic equilibrium.**

thermodynamic equilibrium

The concept of thermodynamic equilibrium does not require a prior notion of temperature. To determine whether two systems are in equilibrium, we can consider *any* macroscopic properties. Typical properties include length, volume, pressure, electrical resistance, and fluid viscosity as well as temperature. If any macroscopic property changes when two systems are placed in contact, then they were not originally in equilibrium. If no macroscopic changes occur, then the systems were and are in equilibrium.

thermal contact

When we speak of placing systems in contact, we mean specifically **thermal contact.** Two systems are in thermal contact if heating one of them results in macroscopic changes in the other. For example, if we place two metal cups of water in physical contact, and heat one cup with a flame, the water in both cups gets hotter. In this case, the cups are in thermal contact. If we were to separate our cups by a layer of styrofoam, then heating one would have little effect on the other. In this case, the two systems are **thermally insulated.**

thermally insulated

Using the concept of thermodynamic equilibrium, we can now define quantitatively what we mean by temperature. We first state what it means for two systems to be at the same temperature:

systems at the same temperature

Two systems are at the same temperature if they are in thermodynamic equilibrium.

Conversely, two systems that are not in thermodynamic equilibrium are at different temperatures.

To define temperature more quantitatively, we associate numerical values of temperature with values of some macroscopic property of a system. We might associate temperature with the length of an iron rod so that increasing length corresponds to higher temperature. Or we could use the pressure of gas in a fixed-volume container, assigning increasing values of temperature to increasing pressures.

What makes such a temperature scale useful is a simple experimental fact. Consider two systems A and C that are each in thermal contact with a third system, B, but not with each other (Fig. 16–2a). We wait until system A is in thermodynamic equilibrium with B, and C with B. If we then bring systems A and C into thermal contact (Fig. 16–2b), we find that no further changes occur, showing that A and C had already reached thermodynamic equilibrium with each other. This basic fact—that two systems in thermodynamic equilibrium with a third system are also in equilibrium with each other—is often called the **zeroth law of thermodynamics.** Rephrased in terms of temperature, the zeroth law states that if system A is at the same temperature as system B, and B is at the same temperature as system C, then systems A and C are at the same temperature. The experimental fact embodied in the zeroth law therefore allows us to define temperature in a consistent way that applies equally to all systems.

zeroth law of thermodynamics

16.3
MEASURING TEMPERATURE

thermometer

A **thermometer** is a system with a conveniently observed macroscopic property that changes with temperature. In principle, any such system will do, and we can use any monotonic function of the macroscopic variable as our measure of temperature. We might choose an iron rod, whose length increases with temperature and thus serves as a measure of temperature. The length of a mercury column serves as the temperature indicator in a standard mercury thermometer. An approach often used in thermostats involves two different metals bonded together in a thin strip. When heated, the two metals expand at different rates, causing the strip to bend (Fig. 16–3). This bending can turn an indicating needle or open electrical contacts that, for example, control a furnace. Electrical resistance has long been used in scientific and engineering work to measure temperature, and, with the development of microelectronics, is rapidly replacing the mercury thermometer in medical and other applications (Fig. 16–4). Another, and very accurate, electrically based thermometer is the thermocouple, a junction of two different metals that generates a small electrical voltage that depends on the temperature of the junction.

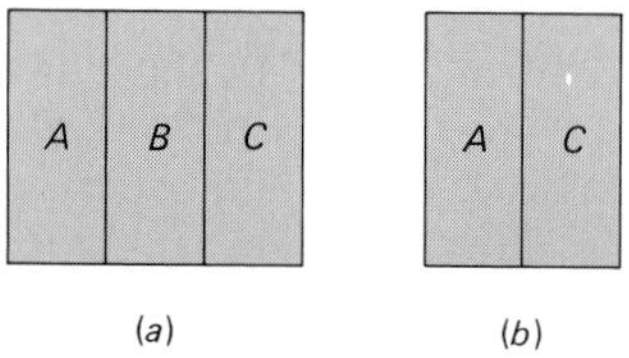

To use different thermometers quantitatively, we must make them all consistent with each other. We can get consistent results by choosing a standard thermometer that works over the widest possible range, and then relating actual values of temperature to physical processes that always occur at the same temperature. One class of thermometers—gas thermometers—fits our requirement at all but the most extreme temperatures. Gas thermometers use as their indication of temperature either (1) the volume of a gas kept at constant pressure, or (2) the pressure of a gas kept at constant volume. Figure 16–5 shows a constant-volume gas thermometer, widely used in scientific work or for calibrating other thermometers.

The temperature measured by a constant-volume gas thermometer is taken to be a linear function of the gas pressure. We need two points to define a linear function. The Kelvin scale, which is part of the SI system of

Fig. 16–2
(a) A and C are in thermal contact with B, but not with each other. (b) When they are brought together, A and C are found to be in thermodynamic equilibrium.

Fig. 16–3
The coil shown at the center of this thermostat is a bimetallic strip whose configuration changes with temperature, actuating a switch that controls a furnace.

measurement, defines the zero of temperature as zero gas pressure in a constant-volume gas thermometer. The second point is established by the **triple point** of water—the unique temperature at which water can exist with solid, liquid, and gas simultaneously in equilibrium.* At the triple point, the temperature is, by definition, 273.16 K. A plot of temperature versus gas pressure for a constant-volume gas thermometer then looks like Fig. 16–6. The linear relation shown in Fig. 16–6 is given algebraically by

$$T = 273.16 \frac{P}{P_3}, \tag{16–1}$$

where P_3 is the gas thermometer pressure at the triple point.[†]

The Kelvin temperature scale, with its zero point where a gas thermometer would have zero pressure, is a fundamental scale for dealing with basic thermodynamic phenomena. Since a gas cannot have negative pressure, the zero of the Kelvin scale represents an absolute lower limit of temperature, and is called absolute zero. For this reason the Kelvin scale is also called the absolute temperature scale. We will explore further this sense of "absoluteness," and the meaning of absolute zero, in Chapter 19. Figure 16–7 shows some important physical situations on the Kelvin scale.

Other scales in common use include the Celsius (°C), Fahrenheit (°F), and Rankine (°R) scales, although the latter two are now largely limited to the United States. One degree Celsius represents the same temperature difference as one kelvin (K), but the zero of the Celsius scale occurs at 273.15 K, so that

$$T_C = T - 273.15, \tag{16–2}$$

where T_C is the Celsius temperature and T the absolute temperature in kelvins. Historically, the Celsius scale was defined before the absolute scale, and was chosen so that the melting point of ice at standard atmospheric pressure—the ice point—is at exactly 0°C, while the boiling point of water at standard atmospheric pressure—the steam point—is at exactly 100°C. The triple point of water occurs at 0.01°C, which accounts for the 273.15 difference between the Kelvin and Celsius scales as well as for the difference of 0.01 in the constants of Equations 16–1 and 16–2.

The Fahrenheit scale, which is part of the British system of units, is defined with the ice point at 32°F and the steam point at 212°F. As a result, the relation between the Fahrenheit and Celsius scales is

$$T_F = \tfrac{9}{5}T_C + 32. \tag{16–3}$$

A fourth scale, often used in engineering work in the United States, is the Rankine scale. A Rankine degree is the same size as a Fahrenheit degree, but the zero of the Rankine scale coincides with the zero of the Kelvin scale, so that

$$T_R = \tfrac{9}{5}T = T_F + 459.67. \tag{16–4}$$

Figure 16–8 summarizes the four temperature scales in relation to some important physical processes.

triple point

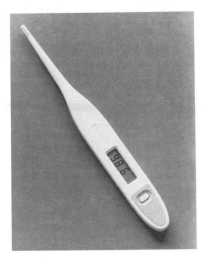

Fig. 16–4
An electronic digital fever thermometer.

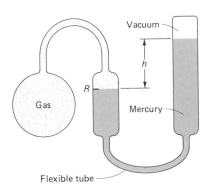

Fig. 16–5
A constant-volume gas thermometer. As the gas pressure increases with temperature, the closed tube at right is raised to keep the mercury at the reference point R, thus maintaining constant gas volume. The height difference h is then a measure of gas pressure. Temperature is taken to be directly proportional to h.

---

*The triple point is discussed in detail in the next chapter.
[†]This pressure depends on the particular thermometer, and should not be confused with the 610 Pa pressure of the water at its triple point.

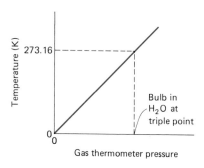

Fig. 16-6
Kelvin temperature scale defined using constant-volume gas thermometer.

Example 16-1

Temperature Scales

What is normal body temperature (98.6°F) on the Celsius, Kelvin, and Rankine scales? If you have a fever of 101.6°F, by how much has your temperature risen on each of these scales?

Solution

Solving Equation 16-3 for T_C gives

$$T_C = \tfrac{5}{9}(T_F - 32) = \tfrac{5}{9}(98.6 - 32.0) = 37.0°C.$$

Then Equation 16-2 gives the Kelvin temperature:

$$T = T_C + 273.2 = 310.2 \text{ K.}$$

Finally, from Equation 16-4,

$$T_R = T_F + 459.7 = 558.3°R.$$

A degree Rankine and a degree Fahrenheit are the same size, so that a fever of 101.6°F represents a 3.0° rise on either scale. A kelvin and a degree Celsius are the same size, and are both larger than a degree Fahrenheit by the factor $\tfrac{9}{5}$, so that a rise of 3.0°F is equivalent to a rise of

$$\frac{3.0}{9/5} = 1.7°C = 1.7 \text{ K.}$$

16.4

TEMPERATURE AND HEAT

A lighted match is hot enough to burn your finger, yet it would not provide much heat in a cold room. A large vat of hot water—although much cooler than the match—would do a better job of heating the room. This example illustrates our intuitive sense of the difference between temperature and heat: heat measures an *amount* of "something," while temperature measures the *intensity* of that "something."

caloric   What is that "something"? Before the early 1800's, heat itself was thought to be a substance, called **caloric,** that was transferred from a hot body to a cooler one in much the same way that fluid flows from a region of higher pressure to one of lower pressure. The theory of caloric was invoked to explain a wide variety of thermal phenomena, including heat

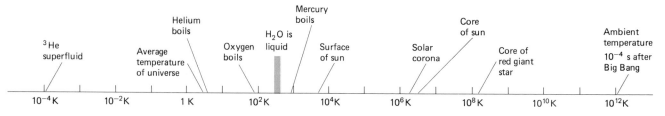

Fig. 16-7
Some physical processes on the Kelvin scale. Chart is logarithmic to allow display of a wide temperature range.

Fig. 16-8
Relations among the four temperature scales.

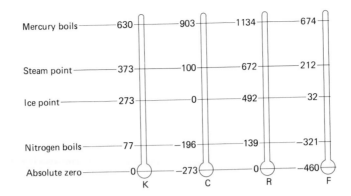

	K	C	R	F
Mercury boils	630	903	1134	674
Steam point	373	100	672	212
Ice point	273	0	492	32
Nitrogen boils	77	−196	139	−321
Absolute zero	0	−273	0	−460

transfer and the thermal expansion of materials. Temperature changes associated with mechanical friction were explained by saying that small pieces of material shaved off in the frictional process released large amounts of caloric.

In the late 1700's, Benjamin Thompson (1753–1814) made observations refuting the caloric theory. Thompson, an American who was hired to direct the Bavarian arsenal and who later became Count Rumford of Bavaria, supervised the boring of cannon. That heat was involved in this process was well known, and water was needed to cool both the cannon and the boring tool. But Thompson noticed that the heating was not associated with the production of metal chips that could release caloric, for a dull tool that did little drilling actually caused more heating. It appeared that caloric could be produced in endless quantities in such a frictional process. This observation was difficult to reconcile with the notion of caloric as a conserved physical substance. Thompson suggested instead that the heating was associated with the mechanical work done on the drill bit.

In the half-century following Thompson's observations, the caloric theory gradually faded in popularity as a series of experiments confirmed the association between heating and the expenditure of mechanical energy. These experiments culminated in the work of the British brewer and physicist James Joule (1818–1889), who explored the relation between heat and mechanical, electrical, and chemical energy. In 1843, Joule quantified the relation between heat and energy, bringing thermal phenomena under the powerful conservation-of-energy law. In recognition of this major synthesis in physics, the SI unit of energy is named after Joule. In Chapter 18 we will explore in detail the relation between heat and energy.

Our everyday experience of heat is nearly always associated with the movement of energy from one body to another. A hot stove burner heats a pot of soup, or a furnace heats water that, in turn, transfers heat to a house. We rarely make statements about the actual amount of "heat" in an object— we are concerned instead that the temperature of the object be appropriate. We want the furnace to transfer energy to our house in the winter, and are satisfied when the house reaches a certain temperature, not when it contains a certain amount of "heat." Our scientific definition of heat reflects this natural inclination to think of heat as energy in transit:

heat   **Heat** is energy that is transferred from one object to another because of temperature differences alone.

internal energy

Strictly speaking, the word heat refers only to energy in transit. Once heat has been transferred to an object, we say that the **internal energy** of the object has increased, but not that it contains more heat. This distinction reflects the fact that other processes than heating—such as transfer of mechanical or electrical energy—can also change the temperature of an object. As a result, heat alone does not uniquely determine the energy contained in a given object. In subsequent chapters, we will explore further this distinction between heat and internal energy.

calorie

We can determine the amount of heat transferred to an object by noting the change in its temperature. Before the equivalence of heat and energy was well established, heat was measured in **calories** (cal). One calorie was defined as the amount of heat needed to raise the temperature of one gram of water from 14.5°C to 15.5°C. Through the work of Joule, we now know that one calorie is equivalent to about 4.18 J of energy. Several slightly different definitions of the calorie have been established, based on different experimental conditions for determining the equivalence. In this book, we use the so-called thermochemical calorie, defined as exactly 4.184 J. The "calorie" used in describing the energy content of foods is actually a kilocalorie. In the British system of units, still widely used in engineering work in the United States, the unit of heat is the **British thermal unit** (Btu). One Btu is the amount of heat needed to raise the temperature of one pound of water from 63°F to 64°F, and is equal to 1055 J.

British thermal unit

Example 16–2
The Calorie

A typical fast-food hamburger contains "300 calories," which means 300 kcal. If the hamburger were consumed by a 50-kg person, and all its energy content released as heat, by how much would the person's body temperature rise? Assume that the person is made essentially of water, and is insulated so that heat from the hamburger cannot be released to the environment.

Solution

Each kcal can raise one kg of water by one degree Celsius. The hamburger releases 300 kcal, but these must be distributed among 50 kg, so that each kg gets 300/50 or 6.0 kcal, and therefore the temperature would rise by 6.0°C. (That this rise doesn't occur is a tribute to our bodies' ability to store excess energy as fat, as well as to the fact that we use food energy to do mechanical work and to keep our bodies warmer than our surroundings.)

16.5
HEAT CAPACITY AND SPECIFIC HEAT

specific heat

Matter usually gets hotter when we transfer heat to it. How much hotter? That depends on what kind of matter we have and on how much of it there is. We already know the answer for water: if we have 1.0 g of water, and transfer 4.184 J or 1.0 cal of heat to it, then its temperature will rise approximately 1.0°C. The amount of heat needed to produce a unit temperature rise in a unit mass of a substance is the **specific heat** of that substance.

Water's specific heat is 1.0 cal/g·°C, or 4184 J/kg·K. (Degrees Celsius and kelvins are interchangeable in describing specific heats, for we are concerned only with temperature *changes*.) More precisely, the specific heat of a substance is defined as the ratio of the heat ΔQ per unit mass required to produce a temperature change ΔT, in the limit of arbitrarily small ΔT:

$$c = \frac{1}{m} \lim_{\Delta T \to 0} \frac{\Delta Q}{\Delta T} = \frac{1}{m}\frac{dQ}{dT}. \tag{16-5}$$

The amount of heat needed to cause a unit temperature rise in a *particular* piece of matter of mass m is called the **heat capacity, C**:

heat capacity

$$C = \frac{dQ}{dT} = mc. \tag{16-6}$$

For many substances under normal conditions, specific heat is nearly independent of temperature, so we can find the total amount of heat needed to bring an object from temperature T_1 to T_2 by multiplying its heat capacity by the temperature difference:

$$Q = C(T_2 - T_1) = mc(T_2 - T_1). \tag{16-7}$$

When the specific heat itself depends on temperature, we write $c(T)$ and consider a series of infinitesimal temperature rises dT, with heat required for each given by $dQ = mc(T)dT$. To find the total heat required to raise the temperature from T_1 and T_2, we then integrate over the temperature interval:

$$Q = \int_{T_1}^{T_2} mc\, dT. \tag{16-8}$$

Table 16–1 lists the specific heats of common substances. Because the British thermal unit is defined as the amount of heat needed to raise one pound of water by one degree Fahrenheit, the specific heat in cal/g·°C is numerically equal to its value in Btu/lb·°F.

TABLE 16–1
Specific Heats (temperature range 0°C to 100°C except as noted)

Substance	Specific Heat (J/kg·K)	Specific Heat (cal/g·°C, kcal/kg·°C, Btu/lb·°F)
Aluminum	900	0.215
Copper	386	0.0923
Iron	447	0.107
Glass	753	0.18
Mercury	140	0.033
Steel	502	0.12
Stone (granite)	840	0.20
Water:		
Liquid	4184	1.00
Ice, −10°C	2050	0.49
Wood	1400	0.33

Example 16–3

Heat Capacity and Specific Heat

A water heater holds 150 kg of water. How much energy does it take to bring the water temperature from 18°C to 50°C? If the energy is supplied by a 5.0-kW electric heating element, how long must the electricity be on?

Solution

The heat required is given by Equation 16–7:

$$Q = mc(T_2 - T_1) = (150 \text{ kg})(4184 \text{ J/kg·K})(50°C - 18°C) = 2.0 \times 10^7 \text{ J}.$$

Note that we are justified in mixing K and °C here because a temperature *difference* is the same whether expressed in degrees Celsius or in kelvins.

The heating element supplies heat at the rate of 5.0 kW, or 5.0×10^3 J/s, so that it must be on for a time t given by

$$t = \frac{2.0 \times 10^7 \text{ J}}{5.0 \times 10^3 \text{ J/s}} = 4000 \text{ s},$$

or a little over an hour. (At 10¢/kWh, we would pay 56¢ for this energy.)

Example 16–4

A Temperature-Dependent Specific Heat

At very low temperatures, the specific heat of a solid is approximately proportional to the cube of the absolute temperature. For example, the specific heat of copper in the range from a few K to about 50 K is given very nearly by

$$c = 31 \left(\frac{T}{343 \text{ K}} \right)^3 \text{ J/g·K}.$$

How much heat is required to raise the temperature of a 40-g sample of copper from 10 K to 25 K?

Solution

Because the specific heat is temperature-dependent, we must use the integral in Equation 16–8 to calculate the heat:

$$Q = \int_{T_1}^{T_2} mc \, dT$$

$$= \frac{(40 \text{ g})(31 \text{ J/g·K})}{(343 \text{ K})^3} \int_{10\text{K}}^{25\text{K}} T^3 dT = (3.1 \times 10^{-5} \text{ J/K}^4) \left. \frac{T^4}{4} \right|_{10\text{K}}^{25\text{K}}$$

$$= (3.1 \times 10^{-5} \text{ J/K}^4) \frac{(25 \text{ K})^4 - (10 \text{ K})^4}{4} = 2.9 \text{ J}.$$

In contrast, it takes nearly 100 times as much energy—230 J—to raise the same 40-g piece of copper from 300 to 315 K, as you can confirm by applying Equation 16–7 and the data from Table 16–1.

When two objects at different temperatures are placed in thermal contact, energy is transferred from the hotter object to the cooler one until thermodynamic equilibrium is reached. If the two objects are thermally insulated from their surroundings, then all the energy leaving the hotter object is transferred to the cooler one. Mathematically, this statement may be written

$$m_1 c_1 \Delta T_1 + m_2 c_2 \Delta T_2 = 0, \qquad (16\text{-}9)$$

where mc is the heat capacity of an object of mass m and specific heat c, and ΔT is the temperature change (final temperature minus initial temperature) of that object. For the hotter object, ΔT is negative, so that the two terms in Equation 16-9 have opposite signs, and can therefore sum to zero. One term represents the outflow of heat from the hotter object, the other inflow into the cooler object.

Example 16-5
The Equilibrium Temperature

An aluminum frying pan with a mass of 1.5 kg is heated on a stove to 180°C, then plunged into a sink containing 8.0 kg of water at room temperature (20°C). Assuming that none of the water boils, and that no heat is lost to the surroundings, what is the equilibrium temperature of the water and pan?

Solution

We write Equation 16-9 in the form

$$m_p c_p (T - T_p) + m_w c_w (T - T_w) = 0,$$

where the subscripts p and w refer to the pan and the water, and where the equilibrium temperature T is the same for both. Solving for T and using the appropriate numerical values then gives

$$T = \frac{m_p c_p T_p + m_w c_w T_w}{m_p c_p + m_w c_w}$$

$$= \frac{(1.5 \text{ kg})(0.215 \text{ kcal/kg} \cdot {}^\circ\text{C})(180^\circ\text{C}) + (8.0 \text{ kg})(1.0 \text{ kcal/kg} \cdot {}^\circ\text{C})(20^\circ\text{C})}{(1.5 \text{ kg})(0.215 \text{ kcal/kg} \cdot {}^\circ\text{C}) + (8.0 \text{ kg})(1.0 \text{ kcal/kg} \cdot {}^\circ\text{C})}$$

$$= 26^\circ\text{C}.$$

16.6
HEAT TRANSFER*

So far we have said nothing about the mechanisms whereby energy is transferred between two objects. We have simply assumed that somehow we had the means either to facilitate or to inhibit this transfer. In engineering problems involving heat, we need to know precisely how to achieve a desired degree of energy transfer. In scientific work, knowledge of energy-transfer mechanisms helps us understand a system's thermal behavior.

Three heat-transfer mechanisms commonly occur. These are conduction, a process involving direct physical contact; convection, involving energy transfer by the bulk flow of a fluid; and radiation, or energy transfer by electromagnetic waves. In a given situation, one of the three may dominate, or we may have to take all three into account.

---

*Strictly speaking, the term "heat transfer" is redundant, for heat is energy being transferred from one object to another. But the term is in such widespread use that we will speak of heat transfer when we really mean energy transfer that occurs because of a temperature difference.

Conduction

conduction

Conduction is the transfer of heat through direct physical contact. Microscopically, conduction occurs because molecules in a hotter region transfer energy to those of an adjacent cooler region by colliding with them. The effect of these collisions in a given material is quantified in the material's **thermal conductivity,** k, whose SI units are W/m·K. Common materials exhibit a broad range of thermal conductivities, from about 400 W/m·K for copper—a good conductor of heat—to 0.029 W/m·K for styrofoam, a good thermal insulator.

thermal conductivity

Figure 16–9 shows a slab of material of thickness Δx and cross-sectional area A. Suppose one face of the material is held at temperature T and the other at $T + \Delta T$. Intuitively, we might expect the rate of heat flow through the slab to increase with increasing area A and to decrease with increasing thickness Δx. Of course, the heat-flow rate must also depend on the thermal conductivity of the material. We expect further that heat will flow from the hotter to the cooler face of the slab, with the heat-flow rate dependent also on the temperature difference between the two faces. Our intuition is borne out experimentally: for a rectangular slab of material, the heat-flow rate dQ/dt is given by

$$\frac{dQ}{dt} = -kA\frac{\Delta T}{\Delta x}.\qquad(16\text{–}10)$$

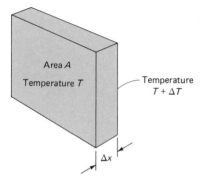

Fig. 16–9
Slab geometry for Equation 16–10.

Since dQ/dt is a rate of energy flow, its units are joules/second, or watts. The minus sign in Equation 16–10 shows that heat transfer is opposite to the direction of increasing temperature, that is, from hotter to cooler. Equation 16–10 is strictly correct only when the temperature varies uniformly from one surface to the other, and when the areas of the two surfaces are equal. Because the thermal conductivity k may vary with temperature, or because we may be interested in heat transfer in other geometries (such as through the insulation around a pipe), we often write Equation 16–10 in the limit of arbitrarily small Δx. In this limit we have

$$\frac{dQ}{dt} = -kA\frac{dT}{dx}.\quad\text{(heat conduction)}\qquad(16\text{–}11)$$

Many practical problems involve rectangular geometry and an insignificant variation of k with temperature; in those cases we can use Equation 16–10. Table 16–2 lists thermal conductivities of some common materials. Both SI and British units are listed, as the latter are commonly used in calculations involving heat loss in buildings.

Table 16–1 is remarkable for the range of thermal conductivities it contains. Metals are exceptionally good conductors of heat as well as of electricity: they contain free electrons that can move quickly, carrying internal energy and electric charge from one region to another. Gases, on the other hand, are poor conductors because the wide spacing of their molecules makes for infrequent transfers of energy. Good heat insulators, like fiberglass and styrofoam, owe their insulating properties not to their chemical composition but to their physical structure, which traps small volumes of air or other gas.

TABLE 16-2
Thermal Conductivities

Material	k (W/m·K)	k (Btu·in/h·ft²·°F)
Air	0.026	0.18
Aluminum	237	1644
Concrete (varies with mix)	1	7
Copper	401	2780
Fiberglass	0.042	0.29
Glass	0.7–0.9	5–6
Goose down	0.043	0.30
Helium	0.14	0.97
Iron	80.4	558
Steel	46	319
Styrofoam	0.029	0.20
Water	0.61	4.2
Wood (pine)	0.11	0.78

Example 16-6
Heat Conduction

A lake with a flat bottom and steep sides has a surface area of 1.5 km² and is 8.0 m deep. On a summer day, the surface water is at a temperature of 30°C, while the bottom water is at 4.0°C. What is the rate of heat conduction through the lake? Assume that the temperature declines uniformly from surface to bottom.

Solution

Our lake resembles the slab of Fig. 16–9, so that Equation 16–10 applies. Taking the thermal conductivity of water from Table 16–2, we have

$$\frac{dQ}{dt} = -kA\frac{\Delta T}{\Delta x} = -(0.61 \text{ W/m·K})(1.5 \times 10^6 \text{ m}^2)\frac{30°C - 4.0°C}{8.0 \text{ m}} = -3.0 \times 10^6 \text{ W}.$$

The minus sign indicates that heat flows in the direction of decreasing temperature, or downward into the lake. (Unless the surface and bottom temperatures are held fixed, this heat flow will eventually bring the entire lake to a single temperature.)

Example 16-7
Heat Conduction in Cylindrical Geometry

A 10-m-long copper pipe of diameter 2.0 cm carries hot water at 95°C. The pipe passes through an unheated space where the temperature is −20°C. The pipe is insulated with a 1.5-cm-thick layer of styrofoam. What is the rate of heat loss from the pipe?

Solution

Here we do not have rectangular geometry, so we cannot apply Equation 16–10. However, if we consider a very thin cylindrical layer within the insulation (Fig. 16–10), then the areas of the inner and outer surfaces of the layer must be nearly equal. Although the geometry is not rectangular, we could imagine unrolling the thin layer into a rectangular sheet, to make the situation resemble Fig. 16–9. Let dr be the thickness of the layer, and dT the temperature difference across it. The area of either side of the layer is approximately $2\pi rL$, where r is the distance to the layer from the center of the pipe, and L is the length of the pipe. Then Equation 16–11 becomes

$$\frac{dQ}{dt} = -kA\frac{dT}{dr} = -k2\pi rL\frac{dT}{dr}. \tag{16–12}$$

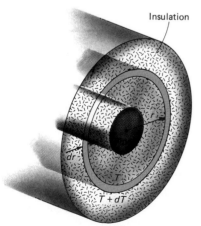

Insulation

Fig. 16–10
Cross section of cylindrical, insulated pipe, showing thin layer of insulation of thickness dr used in setting up integration.

We don't know what dQ/dt is—that's what we're trying to determine—but we do know that the heat flowing through our thin layer also flows through any other thin layer of the insulation, since there are no sources of heat in the insulation. Thus the rate at which heat leaves the copper pipe is equal to the rate at which heat flows across any layer of the insulation, and that in turn is equal to the rate of heat loss at the top surface of the insulation.

We can rewrite Equation 16–12 in the form

$$\frac{dQ}{dt}\frac{dr}{r} = -2\pi kL\, dT, \tag{16–13}$$

where our argument of the previous paragraph tells us that the heat-loss rate dQ/dt is a constant, independent of r. We need to relate this loss rate to the temperature difference across the insulation. We do so by integrating Equation 16–13 across the insulation. Letting R_1 be the inner radius of the insulation (equivalently, the pipe's radius), and R_2 the outer radius of the insulation, we have

$$\frac{dQ}{dt}\int_{R_1}^{R_2}\frac{dr}{r} = -2\pi kL\int_{T_1}^{T_2} dT.$$

Here we have pulled constants out of the integral, and have called the temperatures at the inner and outer edges of the insulation T_1 and T_2, respectively. Evaluating the integrals gives

$$\frac{dQ}{dt}\ln r\Big|_{R_1}^{R_2} = -2\pi kL\, T\Big|_{T_1}^{T_2},$$

or

$$\frac{dQ}{dt}\ln\left(\frac{R_2}{R_1}\right) = -2\pi kL(T_2 - T_1).$$

Solving for the heat-loss rate dQ/dt then gives

$$\frac{dQ}{dt} = \frac{2\pi kL(T_1 - T_2)}{\ln(R_2/R_1)}. \tag{16–14}$$

Our particular pipe has a radius of 1.0 cm and 1.5-cm-thick insulation, so that the outer radius of the insulation is $R_2 = 2.5$ cm. Obtaining k for styrofoam from Table 16–2, and inserting values for L, T_1, and T_2, we have

$$\frac{dQ}{dt} = \frac{2\pi(0.029\ \text{W/m·°C})(10\ \text{m})[95°C - (-20°C)]}{\ln(2.5/1.0)} = 230\ \text{W}.$$

Knowing the heat-loss rate, we could integrate Equation 16–13 from the inner edge of the insulation to an arbitrary position r within the insulation, and from this obtain an expression for the temperature as a function of position in the insulation (see Problem 35). Heat-loss calculations in spherical geometry are handled in a similar manner (see Problem 38).

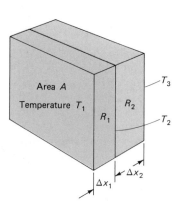

Fig. 16–11
A composite slab.

Frequently we have a situation in which heat flows through several different materials. A typical example is the wall of a building, which may contain wood, plaster, fiberglass insulation, and other materials. Figure 16–11 shows such a composite structure, involving two slabs of different materials with a temperature T_1 on one side of the composite and T_3 on the other. The two slabs may have different thicknesses Δx_1 and Δx_2, but they have the same area. The heat-flow rate dQ/dt through each slab must be the same, since energy does not accumulate or disappear at the interface between the two. We then have

$$\frac{dQ}{dt} = -k_1 A \frac{T_2 - T_1}{\Delta x_1} = -k_2 A \frac{T_3 - T_2}{\Delta x_2}, \tag{16–15}$$

where k_1 and k_2 are the thermal conductivities of the two materials, and T_2 is the temperature at the interface. We would like to express the heatflow rate in terms of the surface temperatures T_1 and T_3 alone, without having to worry about the intermediate temperature T_2. To do so, it is convenient to define the **thermal resistance,** R, of each slab:

thermal resistance

$$R = \frac{\Delta x}{kA}. \tag{16–16}$$

The SI units of R are K/W. Unlike the thermal conductivity, k, which is a property of a *material*, R is a property of a *particular piece* of material, reflecting both its conductivity and its geometry. In terms of thermal resistance, Equation 16–15 becomes

$$\frac{dQ}{dt} = -\frac{T_2 - T_1}{R_1} = -\frac{T_3 - T_2}{R_2},$$

so that

$$R_1 \frac{dQ}{dt} = T_1 - T_2$$

and

$$R_2 \frac{dQ}{dt} = T_2 - T_3.$$

Adding these two equations gives

$$(R_1 + R_2)\frac{dQ}{dt} = T_1 - T_2 + T_2 - T_3 = T_1 - T_3,$$

or

$$\frac{dQ}{dt} = \frac{T_1 - T_3}{R_1 + R_2}. \tag{16–17}$$

Equation 16–17 shows that the composite slab acts like a single slab whose thermal resistance is the sum of the resistances of the two slabs that compose it. We could easily extend this treatment to show that the thermal resistances of three or more slabs add when the slabs are arranged so that the same heat flows through all of them.

In the United States, the insulating properties of building materials are usually described in terms of the **R-factor,** $\mathcal{R}$, which is the thermal resistance per unit area:

R-factor

$$\mathcal{R} = \frac{R}{A} = \frac{\Delta x}{k}. \tag{16–18}$$

The units of $\mathcal{R}$, although rarely stated, are $\text{ft}^2 \cdot {}^\circ\text{F} \cdot \text{h/Btu}$. This means that $\mathcal{R}$-19 fiberglass insulation, now in common use as a wall insulation in the northern part of the United States, has a heat loss of $\frac{1}{19}$ Btu per hour for each square foot of insulation for each degree Fahrenheit temperature difference across the insulation.

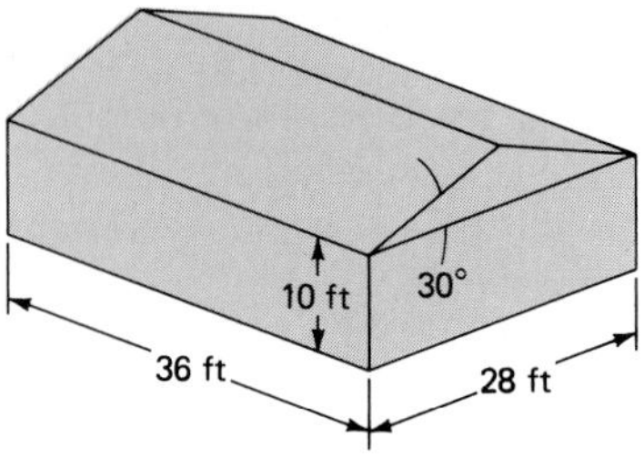

Fig. 16–12
House of Example 16–8.

Example 16–8

Building Insulation and Heat Loss

The rectangular house of Fig. 16–12 measures 36 ft long, 28 ft wide, and has 10-ft-high side walls. The peaked roof is pitched at a 30° angle, and consists of a $\frac{1}{2}$-in plaster ceiling ($k = 3.0$ Btu·in/h·ft²·°F for plaster), $9\frac{1}{2}$-in fiberglass insulation ($\mathcal{R} = 30$), $\frac{1}{2}$-in plywood ($\mathcal{R} = 0.65$), and $\frac{1}{2}$-in cedar shingles ($\mathcal{R} = 0.55$). The walls have similar construction, except that they contain only $3\frac{1}{2}$ in of fiberglass insulation ($\mathcal{R} = 11$). The average outdoor temperature during a winter month is 20°F, and the house is maintained at 70°F. If the house is heated by an oil furnace that produces 100,000 Btu of heat per gallon of oil, and if oil costs 83¢/gallon, how much does it cost to heat the house for the month? Ignore heat losses through the floor and windows (even though these may be substantial).

Solution

We must first calculate the R-factors for the composite walls and roof. We are given $\mathcal{R}$ for all but the plaster; from Equation 16–18 we have

$$\mathcal{R}_{\text{plaster}} = \frac{\Delta x}{k} = \frac{\frac{1}{2} \text{ in}}{3.0 \text{ Btu·in/h·ft}^2\text{·°F}} = 0.17.$$

Then the R-factors for the walls and roof are

$$\mathcal{R}_{\text{wall}} = \mathcal{R}_{\text{plaster}} + \mathcal{R}_{\text{fiberglass}} + \mathcal{R}_{\text{plywood}} + \mathcal{R}_{\text{shingles}}$$

$$= 0.17 + 11 + 0.65 + 0.55 = 12,$$

and

$$\mathcal{R}_{\text{roof}} = 0.17 + 30 + 0.65 + 0.55 = 31.$$

The total wall area is

$$A_{\text{wall}} = (28 \text{ ft} + 36 \text{ ft} + 28 \text{ ft} + 36 \text{ ft})(10 \text{ ft}) + (28 \text{ ft})(7 \text{ ft}) = 1480 \text{ ft}^2,$$

where the second term in the sum is the area of the two triangular portions of the side walls, whose height is (14 ft)(sin30°), or 7 ft. These $\mathcal{R}$-12 walls lose $\frac{1}{12}$ Btu/h/ft²/°F, and the temperature difference across them is 50°F, so that the total heat-loss rate through the walls is

$$\left(\frac{dQ}{dt}\right)_{\text{walls}} = (\tfrac{1}{12} \text{ Btu/h/ft}^2\text{/°F})(1480 \text{ ft}^2)(50\text{°F}) = 6170 \text{ Btu/h}.$$

The area of the pitched roof is increased over that of a flat roof by the factor 1/cos30°, so that the heat-loss rate through the roof is

$$\left(\frac{dQ}{dt}\right)_{\text{roof}} = (\tfrac{1}{30} \text{ Btu/h/ft}^2\text{/°F})\frac{(36 \text{ ft})(28 \text{ ft})}{\cos 30°}(50\text{°F}) = 1940 \text{ Btu/h}.$$

The total heat-loss rate is then

$$\left(\frac{dQ}{dt}\right) = 6170 \text{ Btu/h} + 1940 \text{ Btu/h} = 8110 \text{ Btu/h}.$$

In a month, this results in a total heat loss of

$$Q = (8110 \text{ Btu/h})(30 \text{ days/month})(24 \text{ h/day}) = 5.8 \times 10^6 \text{ Btu}.$$

With 10^5 Btu per gallon of oil burned, this requires 59 gallons of oil, at a cost of $(58)(\$0.83) = \$48$.

This estimate is low; losses through windows and doors are substantial, and cold air infiltration results in additional heat loss. A more accurate analysis would also consider heat lost to the ground and solar energy gained through the windows. Problem 31 provides a more realistic look at heat loss in this house.

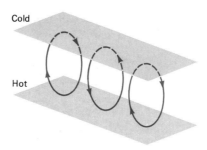

Fig. 16–13
Convection between two plates at
different temperatures.

Fig. 16–14
Top view of convection cells in a
laboratory experiment.

convection

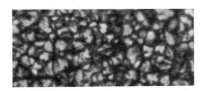

Fig. 16–15
Granulation of the sun's surface shown
in this photograph is caused by large-
scale convection cells that carry energy
from the solar interior to the surface.

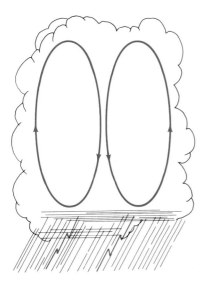

Fig. 16–16
Convection cells in a thunderstorm.
Violent up- and downdrafts may exceed
100 km/h.

Convection

Convection is the transfer of heat by the bulk motion of a fluid. Convection occurs in the presence of gravity because a fluid usually becomes less dense when heated, so that warmer fluid rises. Figure 16–13 shows two plates held at different temperatures, with fluid between them. If the lower plate is hotter, heat is transferred to the bottom fluid, which becomes less dense and rises. When the heated fluid reaches the top plate, it transfers heat to the plate, thus cooling and again sinking to the bottom. A steady fluid flow results, in which fluid rises, cools, and sinks again, in the process transferring heat from the lower to the upper plate, as suggested in Fig. 16–13. Often the flow pattern acquires a striking regularity. Figure 16–14 shows photographs of convection between two plates in the laboratory, while Fig. 16–15 shows remarkably similar patterns associated with convective motions in the sun.

Convection is an important heat-transfer mechanism in a wide range of technological and natural environments. When you heat water on a stove, convection carries heat from the bottom of the pan to the top. Houses usually rely on convection from heat sources near floor level to circulate warm air throughout a room. Insulating materials like fiberglass and goose down trap air and thereby inhibit convection that would otherwise cause excessive heat loss from our houses and our bodies. Convection associated with solar heating of the earth's surface drives the vast air movements that establish our overall climate. Violent convective movements, such as those in thunderstorms (Fig. 16–16), are associated with localized temperature differences. On a much longer time scale, convection in the earth's mantle, brought about because of earth's hot inner core, is responsible for continental drift (Fig. 16–17). Convection plays a crucial role in many astrophysical processes, including the generation of magnetic fields in stars and planets.

As with conduction, the convective heat-loss rate often is approximately proportional to the temperature difference. But the calculation of convective heat loss is complicated because of the need to understand the details of the associated fluid motion. The study of convection processes is an important research area in many fields of contemporary science and engineering.

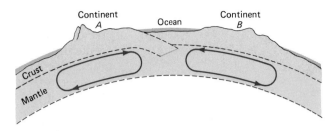

Fig. 16–17
Slow convection in the earth's mantle is responsible for continental drift. Continents A and B are on a collision course, while ocean between them is closing as crust from A is pulled into the mantle.

Radiation

When we turn an electric stove burner to "high," it glows a brilliant red-orange color. By virtue of its high temperature, the burner is emitting visible light. Set on "low," the burner does not appear to glow, but we can still tell it is hot by holding a hand near it. Again, the hot burner is emitting something that tells us it is hot. In both cases, that something is **radiation.** In later chapters, we will investigate the nature and origin of radiation in terms of electromagnetic phenomena and atomic theory; our "radiation" is really called "electromagnetic radiation." For now, we simply assert that heated objects lose energy by emitting radiation. Our stove-burner example suggests the amount of radiation emitted increases very rapidly with temperature. Experiment confirms this: the rate of energy loss by radiation—often called the **luminosity,** L—is given by the **Stefan-Boltzmann law:**

$$L = e\sigma AT^4, \tag{16-19}$$

where L is measured in watts, A is the surface area of the emitting material, T its absolute temperature, and σ a universal constant called the **Stefan-Boltzmann constant.** (In SI, $\sigma = 5.67 \times 10^{-8}$ W/m^2·K^4.)

The quantity e is called the **emissivity** of the material and measures its effectiveness at radiating energy. The value of e ranges between 0 and 1. Emissivity is closely related to a material's ability to absorb radiation; for this reason, materials with emissivity close to 1 appear dark, while light-colored materials have low emissivity. An object with $e = 1$ is called a **black body** because it absorbs all light and other radiation, and therefore appears black at normal temperatures. High emissivity and correspondingly high absorption is the reason why a dark-colored car gets hotter in bright sunlight, and why solar collector plates are painted black.

Materials not only emit radiation; they also absorb it. In thermodynamic equilibrium, the temperature of an object is the same as that of its surroundings. Therefore, the object must emit the same amount of radiation as it absorbs; otherwise it would gain or lose energy and its temperature would change. This statement must be true at *any* temperature, so in equilibrium the rate of absorption of radiation must also be given by Equation 16–19. When an object is not in equilibrium, there is a net transfer of energy between the object and its surroundings, given by the difference between its emission and absorption.

(margin terms)
radiation
luminosity
Stefan-Boltzmann law
Stefan-Boltzmann constant
emissivity
black body

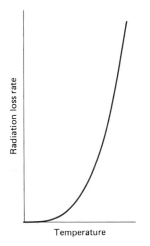

Fig. 16–18
Radiation loss as a function of temperature.

The rate of energy transfer by radiation depends on the fourth power of the temperature. Because of this very steep dependence on temperature (Fig. 16–18), radiation is generally unimportant at low temperatures, but dominates at high temperatures. Houses lose relatively little heat through radiation, while much of the energy transferred from a woodburning stove is through radiation. A hot stove burner with nothing on it glows red and loses energy predominantly by radiation; with a pan on the burner, it stays cooler and loses heat predominantly through conduction to the pan. (From the pan, heat may be lost through convection in a boiling liquid.) Very hot objects, like the surface of the sun or the filament of a light bulb, lose most of their energy through radiation. When an object is surrounded by vacuum, so that conduction and convection cannot occur, then all the energy loss is by radiation.

Example 16–9
The Surface Temperature of the Sun

The sun radiates energy at the rate $L = 3.9 \times 10^{26}$ W, and its radius is 7.0×10^{8} m. Assuming the sun to be a black body (emissivity = 1), what is the surface temperature of the sun?

Solution

The radiated power is given by the Stefan-Boltzmann law, Equation 16–19. Solving this equation for T and using $4\pi R^2$ for the surface area of the sun gives

$$T = \left(\frac{L}{4\pi R^2 \sigma} \right)^{1/4} = \left[\frac{3.9 \times 10^{26} \text{ W}}{4\pi (7.0 \times 10^8 \text{ m})^2 \, (5.7 \times 10^{-8} \text{ W/m}^2 \cdot \text{K}^4)} \right]^{1/4} = 5.8 \times 10^3 \text{ K},$$

in agreement with observational measurements.

16.7
THERMAL ENERGY BALANCE

thermal energy balance

When an object is in thermodynamic equilibrium with its surroundings, its temperature remains constant. But an object could also maintain a constant temperature, different from that of its surroundings, by balancing energy gain with energy loss. We call the latter situation **thermal energy balance.** A house in the winter provides a good example. If the house is unheated, it will come to equilibrium at the outdoor temperature. If the house is heated, it will lose heat because of the temperature difference between itself and its environment, but this loss will be balanced by energy input from a furnace, wood stove, solar collector, electricity, or other heat source.

The principle of thermal energy balance is a powerful one, used throughout physics and engineering to determine the temperatures of objects under different conditions. It is also used to calculate the output of heat sources needed to achieve desired temperatures. If we supply heat to an object at a constant rate, its temperature will at first rise. But as it rises, so does the heat-loss rate. Eventually the object reaches a temperature where the heat-loss rate is equal to the rate of heat input. At that point, the object is in thermal energy balance, and its temperature remains constant.

Example 16–10
Thermal Energy Balance

A poorly insulated electric water heater loses heat at the rate of 40 W for each degree Celsius difference between the water temperature and the ambient temperature outside the tank. The heater is located in a basement where the ambient temperature is 15°C. The tank is heated by a 2.5-kW electric heating element. Assuming that no water is drawn from the tank, and that the heating element operates continuously, what is the water temperature in the tank?

Solution

Heat is supplied to the water at the rate $(dQ/dt)_{gain} = 2.5$ kW; it is lost at a rate given by

$$\left(\frac{dQ}{dt}\right)_{loss} = (40 \text{ W/°C})(\Delta T),$$

where ΔT is the temperature difference between the water and its surroundings. In thermal energy balance, the heat loss and gain are equal, so that

$$(40 \text{ W/°C})(\Delta T) = 2.5 \text{ kW},$$

or

$$\Delta T = \frac{2.5 \text{ kW}}{40 \text{ W/°C}} = 63°C.$$

With the ambient temperature at 15°C, the water temperature is $15°C + 63°C = 78°C$.

Example 16–11
The Temperature of the Earth

The intensity, S, of sunlight at the earth's orbit is about 1.4 kW/m². If the earth had no atmosphere and behaved like a black body, what would be its average temperature? Assume that the entire planet is at the same temperature.*

Solution

Surrounded by the vacuum of space, earth exchanges energy with its environment through radiation.† From Fig. 16–19, we see that our planet presents an effective area of πR_e^2 to the nearly parallel rays of the incident sunlight. But the entire $4\pi R_e^2$ of earth's surface area radiates energy back into space. Thus the condition of energy balance becomes

$$\pi R_e^2 S = \sigma 4\pi R_e^2 T^4,$$

where we have set $e = 1$ for a black body. Then

$$T = \left(\frac{S}{4\sigma}\right)^{1/4} = \left(\frac{1.4 \times 10^3 \text{ W/m}^2}{(4)(5.7 \times 10^{-8} \text{ W/m}^2 \cdot \text{K}^4)}\right)^{1/4} = 280 \text{ K} = 7°C.$$

---

*The fairly rapid rotation of the earth, combined with the effects of atmospheric and oceanic circulation, moderates the extremes of temperature that would otherwise occur between the night and day sides of the planet.

†The earth gains a very small amount of energy from the tides—actually gravitational energy from the earth-moon system—as well as from the decay of radioactive elements in the planet's interior. And a very small fraction of the incident solar energy is not returned to space, but is stored in fossil fuels. Nevertheless, solar energy provides 99.98 per cent of the earth's energy input, and the earth is in nearly perfect energy balance. This nearly perfect balance accounts for our looming shortage of fossil fuels, as we release in a few decades energy that took hundreds of millions of years to accumulate.

Fig. 16–19
Earth presents an effective area πR_e^2 to incident solar energy, but radiates from its entire surface area.

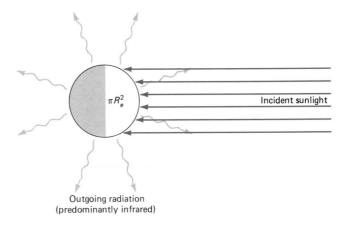

πR_e^2

Incident sunlight

Outgoing radiation
(predominantly infrared)

Although this result is only slightly below the 286 K average temperature of our planet, we have ignored several factors that have a substantial influence on that temperature. First, the atmosphere both reflects and absorbs some incident sunlight, so that the peak intensity at the surface is about 30 per cent lower than in space. Second, earth is not a black body. Third, carbon dioxide in the atmosphere, although transparent to the incident visible light, is relatively opaque to the infrared radiation emitted by the earth. The first effect tends to decrease the temperature, while the third increases it because a higher surface temperature is required to ensure that enough radiation escapes the atmosphere.

A change of only a few degrees in the average temperature might bring major changes in climate. The injection of particulate matter into the atmosphere could result in a decrease in sunlight intensity at the surface, thereby lowering the temperature. Air pollution from industrial processes and volcanic eruptions could have this effect. Substantial temperature changes might also occur in the aftermath of a nuclear war (Fig. 16–20). Although the original drastic predictions of a "nuclear winter" have been moderated by more sophisticated computer models, active research on this topic continues in several countries. At the other extreme, burning

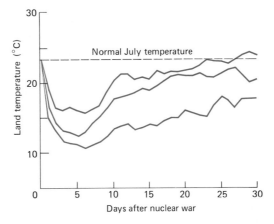

Fig. 16–20
The results of a 1986 study at the National Center for Atmospheric Research predict a maximum temperature drop of 12°C in the aftermath of a nuclear war. The three curves correspond to different amounts of smoke released into the atmosphere, including 18 billion kg (top), 54 billion kg (middle), and 160 billion kg (bottom).

fossil fuels release carbon dioxide that traps infrared radiation, leading to a temperature increase. Such climatic disasters are a major concern of atmospheric scientists, and are explored further in Problems 36 and 37.

Example 16–12
A Solar Greenhouse

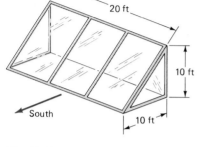

Fig. 16–21
Greenhouse for Example 16–12.

Figure 16–21 shows a solar-heated greenhouse. The south-facing diagonal wall is made entirely of double-pane glass ($\mathcal{R}=1.8$), while the other walls are opaque and have R-factors of 30. Although the glass walls are transparent to sunlight, they are essentially opaque to infrared radiation emitted by the warm greenhouse—just as the earth's atmosphere is transparent to visible light but somewhat opaque to infrared radiation. On a typical winter day, the intensity of sunlight on the glass wall amounts to 120 W/m² when averaged over the 24-hour period. If the greenhouse is maintained at 80°F, what is the lowest the daily average outdoor temperature can get before it is necessary to turn on a backup heat source? The greenhouse is built on a heavy concrete slab with enough heat capacity to store at least a day's energy, so that fluctuations in temperature and sunlight throughout the day are unimportant as long as appropriate daily averages are maintained. The slab is heavily insulated on the bottom, so that heat loss to the ground is negligible.

Solution

We first calculate the heat-loss rate for the greenhouse. The total area of insulated wall is

$$A_w = (10 \text{ ft})(20 \text{ ft}) + 2[\tfrac{1}{2}(10 \text{ ft})^2] = 300 \text{ ft}^2,$$

so that the heat-loss rate through these $\mathcal{R}$-30 walls is

$$\left(\frac{dQ}{dt}\right)_w = \frac{A_w \Delta T}{\mathcal{R}_w} = \frac{300 \Delta T}{30} = 10\Delta T \text{ Btu/h} \cdot {}^\circ\text{F}.$$

Similarly, the glass area is

$$A_g = (20 \text{ ft})(10\sqrt{2} \text{ ft}) = 280 \text{ ft}^2,$$

so that the heat-loss rate through the glass is

$$\left(\frac{dQ}{dt}\right)_g = \frac{A_g \Delta T}{\mathcal{R}_g} = \frac{280 \Delta T}{1.8} = 160\Delta T \text{ Btu/h} \cdot {}^\circ\text{F}.$$

If the greenhouse is to be in thermal energy balance, the total heat-loss rate of $170\Delta T$ Btu/h·°F must be balanced by the energy gained from the sun. The average solar input is $S = 120$ W/m², or

$$S = (120 \text{ W/m}^2)[3.4 \text{ (Btu/h)/W}](0.093 \text{ m}^2/\text{ft}^2) = 38 \text{ Btu/h} \cdot \text{ft}^2,$$

where we have obtained the conversion factors from Appendix D. Then the total solar gain on the 280-ft² glass is

$$(280 \text{ ft}^2)(38 \text{ Btu/h} \cdot \text{ft}^2) = 11 \times 10^3 \text{ Btu/h}.$$

Equating the heat loss to the solar gain gives

$$170\Delta T \text{ Btu/h} \cdot {}^\circ\text{F} = 11 \times 10^3 \text{ Btu/h},$$

so that

$$\Delta T = \frac{11 \times 10^3 \text{ Btu/h}}{170 \text{ Btu/h} \cdot {}^\circ\text{F}} = 65{}^\circ\text{F}.$$

With a greenhouse temperature of 80°F, the outdoor temperature can drop as low as 80°F − 65°F = 15°F before a backup heat source is needed.

___

What happens when a system is not in energy balance? Then it gains or loses heat at a rate given by the difference between the heat input and the heat output. A house, for example, is subject to varying outdoor temperatures, energy input from sunlight, and internal heat supply as the furnace cycles on and off. Energy input to the earth itself varies because of the ellipticity of the planet's orbit, changes in ice and cloud cover, changes in atmospheric and oceanic composition, and possibly very slight changes in the sun's power output. Although the analysis of such time-varying situations is conceptually straightforward, researchers are often hampered by imperfect knowledge of physical processes. For example, we simply do not know how much of the carbon dioxide released into the atmosphere ends up dissolved in the oceans, so that we cannot yet predict the climatic effects of carbon dioxide emission. Even when all factors are known, the complicated functional forms of the varying quantities mandate computer calculations in all but the simplest cases. Example 16–13 and Problems 39 to 41 explore such time-varying situations.

Example 16–13
The Approach to Equilibrium

In deep space, far from any star or planet, an astronaut releases a spherical container filled with 100 kg of water at 300 K. The container has a radius of 29 cm, negligible mass, and is painted black so that its emissivity is 1. Obtain an expression for the temperature of the water as a function of time while it remains in the liquid state, and determine how long it takes to reach the freezing point. The amount of radiation incident on the sphere in deep space is so low (corresponding to a temperature of only 3 K) that you can ignore it for the temperatures of interest here.

Solution

The sphere radiates energy at a rate given by the Stefan-Boltzmann law (Equation 16–19):

$$L = \sigma A T^4,$$

where we have set the emissivity to 1. This energy loss is associated with a temperature drop given by differentiating Equation 16–7:

$$L = -\frac{dQ}{dt} = -mc\,\frac{dT}{dt},$$

where the minus sign indicates a heat loss. Combining these two equations and solving for dT/dt gives

$$\frac{dT}{dt} = -\frac{\sigma A}{mc}T^4.$$

This is a differential equation for T; multiplying through by $T^{-4}dt$ gives

$$T^{-4}dT = -\frac{\sigma A}{mc}dt.$$

To find the temperature as a function of time, we integrate from the initial temperature T_0 and time $t_0 = 0$ to an arbitrary temperature T and time t:

$$\int_{T_0}^{T} T^{-4}dT = -\int_{0}^{t} \frac{\sigma A}{mc}dt,$$

or

$$\frac{T^{-3} - T_0^{-3}}{-3} = -\frac{\sigma A}{mc}t.$$

Solving for T gives our desired expression for temperature as a function of time:

$$T = \left(\frac{3\sigma A}{mc}t + T_0^{-3}\right)^{-1/3}.$$

To find when the water reaches the freezing point, we can solve instead for the time t:

$$t = \frac{mc}{3\sigma A}\left(\frac{1}{T^3} - \frac{1}{T_0^3}\right).$$

Using $m = 100$ kg, $c = 4184$ J/kg·K, and noting that $A = 4\pi r^2$, the drop from $T_0 = 300$ K to the freezing point at 273 K then takes 2.8×10^4 s, or about 8 hours.

SUMMARY

1. A system consisting of many atoms or molecules may be described from either a **microscopic** or a **macroscopic** viewpoint. A microscopic description is concerned with the properties of the individual particles—their masses, positions, and velocities. A macroscopic description involves general properties—like temperature, pressure, and volume—that are not associated directly with the individual particles. **Statistical mechanics** is the branch of physics dealing with the microscopic description of many-particle systems, while **thermodynamics** takes the macroscopic viewpoint.

2. Two systems are in **thermodynamic equilibrium** if, when they are brought into contact, none of the macroscopic properties of either system changes. Two systems in thermodynamic equilibrium have the same **temperature.**

3. The **zeroth law of thermodynamics** states that two systems in equilibrium with a third system are in equilibrium with each other. The zeroth law allows us to establish temperature scales.

4. A **thermometer** is a system one of whose macroscopic properties is used as an indicator of temperature. When a thermometer is in thermodynamic equilibrium with another system, then the temperature indicated by the thermometer is equal to the temperature of the other system. Although there are many practical thermometers and several different temperature scales in use, most laboratory standards are set using **gas thermometers,** either the constant-volume or constant-pressure type. The **kelvin,** or **absolute,** temperature scale is used in the SI system of units.

5. **Heat** is energy being transferred between two objects as a result of a temperature difference alone.

6. The **specific heat** of a material is the ratio of heat to temperature change for a unit mass of the material:

$$c = \frac{1}{m}\frac{dQ}{dT}.$$

The **heat capacity** of a given object is the ratio of heat to temperature change for that object:

$$C = mc = \frac{dQ}{dT}.$$

7. **Heat transfer** can occur through any of three distinct physical mechanisms:

   a. **Conduction** is the direct transfer of heat through physical contact between two objects. Microscopically, conduction involves energy exchange through molecular collisions. The rate of conduction depends on the **thermal conductivity** k of the material through which heat flows, on its thickness Δx and area A, and on the temperature difference ΔT across the material:

$$\frac{dQ}{dt} = -kA\frac{\Delta T}{\Delta x}.$$

   We often describe heat conduction through a particular piece of material in terms of its **thermal resis-**

tance, $R = \Delta x/kA$, or its **R-factor**, $\mathcal{R} = \Delta x/k$. The resistances or R-factors of a composite material are obtained by adding the individual R's or $\mathcal{R}$'s.

b. **Convection** is the transfer of heat through the bulk motion of material, as when heated air rises and transfers its energy to the ceiling of a room. Convection is difficult to describe quantitatively, but is an important process in the laboratory, in engineering situations, and in nature.

c. **Radiation** is electromagnetic energy that can be transferred across empty space. The net rate at which energy is lost from a hot object of area A through radiation is given by

$$L = e\sigma AT^4,$$

where T is the temperature of the object. The emissivity e is a number between 0 and 1 that describes the effectiveness of the object as an emitter of radiation, while σ is the universal **Stefan-Boltzmann constant,** $\sigma = 5.67 \times 10^{-8}$ W/m^2·K^4.

8. **Thermal energy balance** exists when an object gains and loses energy at the same rate. The temperature of an object in thermal energy balance remains constant as long as the rates of heat loss and gain are constant. If one of these changes, the temperature will rise or fall until heat loss again balances heat gain.

QUESTIONS

1. Two identical-looking physical systems are in the same macroscopic state. Must they be in the same microscopic state? Explain.

2. Two identical-looking physical systems are in the same microscopic state. Must they be in the same macroscopic state? Explain.

3. Given that there are three mechanisms of heat transfer, how would you construct a good insulator?

4. Does a thermometer measure its own temperature, or the temperature of its surroundings? Explain.

5. To get an accurate body temperature measurement, you must hold a standard glass-and-mercury fever thermometer under your tongue for about 3 minutes. Some of the new electronic fever thermometers require less than a minute. What might account for the difference?

6. Why is it better to define temperature scales in terms of a physical state, like the triple point of water, rather than by having an official, standard thermometer stored at the International Bureau of Weights and Measures?

7. Compare the relative sizes of the kelvin, the degree Celsius, the degree Fahrenheit, and the degree Rankine.

8. Does a vacuum have temperature? Explain.

9. If you put a thermometer in direct sunlight, what do you measure? The air temperature? The temperature of the sun? Some other temperature?

10. Why does the temperature in a stone building usually vary less than in a wooden building?

11. Why do large bodies of water exert a temperature-moderating effect on their surroundings?

12. A Thermos bottle (Fig. 16–22) consists of an evacuated, double-wall glass liner. The glass is coated with a thin layer of aluminum. How does a Thermos bottle work?

13. Stainless-steel cookware often has a layer of aluminum or copper embedded in the bottom. Why?

14. What method of energy transfer is involved in baking? In broiling?

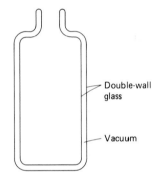

Fig. 16–22 Cross section of a Thermos bottle (Question 12).

15. Solar collectors often have a copper absorber surface coated with a black paint, whose emissivity is nearly 1 for visible and infrared radiation. Better results can be achieved, though, with a "selective surface," whose emissivity is high for visible light but low for the infrared radiation associated with lower temperatures than the sun's. Why is this?

16. Glass and fiberglass are made from the same material, yet have dramatically different thermal conductivities. How can this be?

17. A thin layer of glass does not offer much resistance to heat conduction. Yet double-glazed windows provide substantial energy savings. Why?

18. The insulating value of a double-glazed window actually decreases if the spacing is made too great. Can you think of a reason why this might be?

19. To keep your hands warm while skiing, you should wear mittens instead of gloves. Why?

20. Since earth is exposed to solar radiation, why is earth not at the same temperature as the sun?

21. On a clear night, a solar collector will often cool well below the ambient air temperature. Why? Why does this not happen on a cloudy night?

22. Is earth in perfect energy balance? If not, why not?

_____ **PROBLEMS** _____

Section 16.1   *Macroscopic and Microscopic Descriptions*

1. Consider a simple system consisting of three shelves, one at ground level, one 0.50 m above the ground, and one 1.0 m above the ground. The system also includes three masses A, B, and C. A and B are each 1.0 kg and C is 2.0 kg. A microscopic state of the system is specified by giving the location (shelf) of each mass. (There may be more than one mass on a shelf.) A macroscopic state of the system is specified by giving the total gravitational potential energy of the system. How many microscopic states correspond to the macroscopic state with energy 20 J? (Take $g = 10$ m/s^2.)

2. The macroscopic state of a carton capable of holding a half-dozen eggs is specified by giving the number of eggs in the carton. The microscopic state is specified by telling where each egg is in the carton. How many microscopic states correspond to the macroscopic state of a full carton?

Section 16.2   *Temperature and Thermodynamic Equilibrium*

3. Systems A and B are in thermodynamic equilibrium. The two systems are then separated by an insulating wall, and system C, not initially in equilibrium with B, is brought into contact with B and allowed to reach equilibrium. Can you reach any conclusion about the final temperatures of A and C?

4. If system A is not in thermodynamic equilibrium with system B, and B is not in equilibrium with C, can you draw any conclusions about the temperatures of the three systems?

Section 16.3   *Measuring Temperature*

5. A constant-volume gas thermometer is filled with air whose pressure is 1.01×10^5 Pa (1.00 atm) at the normal melting point of ice. Assuming ideal behavior, what would the pressure be at (a) the normal boiling point of water? (b) the normal boiling point of oxygen (90.2 K)? (c) the normal boiling point of mercury (630 K)?

6. A thermistor is a device whose electrical resistance, R, decreases exponentially with increasing temperature:

$$R = R_0 e^{-\alpha(T - T_0)},$$

where α is a constant and T_0 is some reference temperature at which the resistance is R_0. The unit of electrical resistance is the ohm (Ω). A particular thermistor has a resistance of 1.1×10^5 Ω at the melting point of ice and 9.0×10^3 Ω at the boiling point of water. What is its resistance at normal human body temperature?

7. At what temperature(s), if any, do the following temperature scales coincide? (a) Fahrenheit and Celsius (b) Celsius and Rankine, (c) Celsius and Kelvin, (d) Kelvin and Rankine, (e) Rankine and Fahrenheit.

8. The normal boiling point of nitrogen is 77 K. Express this temperature in Celsius, Fahrenheit, and Rankine, keeping only two significant figures in your answers.

9. The temperature at the center of the sun is about 16×10^6 K. Express this temperature in Celsius, Fahrenheit, and Rankine, keeping only two significant figures in your answers. Why is this problem easier than the preceding one?

10. The temperature of a constant-pressure gas thermometer is directly proportional to the gas volume. If the volume is 1.00 L at the triple point of water (0.01°C), what is it at the normal boiling point of water?

Sections 16.4–16.5   *Temperature and Heat, Heat Capacity and Specific Heat*

11. The average human diet contains about 2000 kcal per day. If all this food energy is used by the body in producing heat and mechanical energy, what is the average power output of the body?

12. A circular lake 1.0 km in diameter is 10 m deep. Solar energy is incident on the lake at an average rate of 200 W/m^2. Assuming that all this energy is absorbed by the lake, and that the lake does not exchange heat with its immediate surroundings (an unrealistic assumption!), how long will it take for the lake's temperature to rise from 10°C to 20°C?

13. 1700 J of heat is required to raise the temperature of 180 g of a substance by 15°C. (a) What is the heat capacity of this piece of the substance? (b) What is the specific heat of the substance?

14. How much heat is required to raise the temperature of an 800-g copper pan from 15°C to 90°C if (a) the pan is empty? (b) the pan contains 1.0 kg of water? (c) the pan contains 4.0 kg of mercury?

15. 100 g of water and 100 g of another substance from Table 16–1 are initially at 20°C. Both are placed over identical flames that transfer heat to the substances at the same rate. At the end of 1 minute, the water is at 32°C and the other substance at 76°C. (a) What is the other substance? (b) What is the heating rate?

16. Two neighbors return from Florida one winter to find their houses at a frigid 15°F. Each house has a furnace that produces heat at the rate of 100,000 Btu/h. One house is made of stone and weighs 75 tons. The other is made of wood and weighs 15 tons. How long does it take to bring each house to a comfortable 65°F? Neglect heat loss that occurs during the heating process, and assume that the houses are heavily insulated on the outside of the wall structure, so that the entire wall mass must reach 65°F.

17. A child complains that her cocoa is too hot. The temperature of the cocoa is 90°C. Her father pours 2 oz of milk at 3°C into the 6 oz of cocoa. Assuming that milk and cocoa have the same specific heat as water, what is the new temperature of the cocoa?

18. A piece of copper at 300°C is dropped into 1.0 kg of water at 20°C. If the equilibrium temperature is 25°C, what is the mass of the copper?

19. Two 1000-kg cars collide head-on at 90 km/h. If all their kinetic energy ends up as heat, what is the increase in temperature of the wrecks? The specific heat of the cars is essentially that of iron.

20. A 1500-kg car moving at 40 km/h is brought to a sudden stop. What is the increase in temperature of the car's four steel brake disks, each of mass 5.0 kg? Assume that all the kinetic energy of the car goes into heating the brake disks, and that no heat is lost to the environment or to other parts of the car.

21. A thermometer whose mass is 83.0 g is used to measure the temperature of a 150-g sample of water. The thermometer has a specific heat of 0.190 cal/g·°C, and it reads 20.0°C before being immersed in the water. The temperature of the water is 60.0°C. If the thermometer is perfectly accurate, what will it read after it comes to equilibrium with the water?

Section 16.6–16.7 *Heat Transfer and Thermal Energy Balance*

22. How thick a concrete wall would be required in order to give the same insulating value as $3\frac{1}{2}$ inches of fiberglass?

23. Compute the R-factors for 1-inch thicknesses of air, concrete, fiberglass, glass, styrofoam, and wood.

24. An iron rod is 50 cm long and 2.0 cm in diameter. One end of the rod is in contact with boiling water and the other end is in contact with ice water. If the rod is well insulated so that there is no heat lost out the sides, what is the rate at which the rod transfers heat from the boiling water to the ice water?

25. What is the R-factor for a wall consisting of $\frac{1}{4}$-in pine paneling, $\mathscr{R}$-11 fiberglass insulation, $\frac{3}{4}$-in pine sheathing, and 2.0-mm aluminum siding?

26. A house measures 28 ft × 28 ft × 9 ft high and has $\mathscr{R} = 30$ for its roof and $\mathscr{R} = 20$ for its walls. The house next door has the same height, the same wall and roof construction, and the same square shape, but twice the floor area. Compare the heating bills in the two houses, assuming they are maintained at the same temperature. Neglect heat loss through the floor.

27. A house is insulated so that its total heat loss is 700 Btu/h·°F. On an evening when the outdoor temperature is 55°F, the owner of the house throws a party and 40 people come. If the average power output of the human body is 100 W, and if there are no other heat sources on in the house, what will the house temperature be during the party?

28. The surface area of an electric clothes iron is 300 cm². If the surface temperature of the iron is 450 K and the room temperature is 300 K, at what rate is energy radiated by the iron? If electric power is supplied to the iron at the rate of 1000 W, what does your answer say about the importance of other energy transfer mechanisms in maintaining the iron at a steady temperature? Take the emissivity of the iron to be 0.95.

29. An electric stove burner has a total surface area of 325 cm². If the temperature of the burner is 900 K, and the electric power input to the burner is 1500 W, what fraction of the burner's heat loss is by radiation? The emissivity of the burner is 1.0, and room temperature is 300 K.

30. An electric current is passed through a metal strip that measures 0.50 cm × 5.0 cm × 0.10 mm. As a result, energy is supplied to the strip at the rate of 50 W. The strip is covered with a black coating that gives it an emissivity of 1, and is in a room where the temperature is 300 K. What is the temperature of the strip if (a) the strip is enclosed in a vacuum bottle that is transparent to all radiation? (b) the strip is enclosed in an insulating box with thermal resistance $R = 8.0$ K/W that blocks all radiation?

31. Rework Example 16–8, now assuming that the house has 10 single-glazed windows ($\mathscr{R} = 0.90$), each measuring 2.5 ft × 5.0 ft. Four of the windows are on the south, and supply the house with solar energy at an average rate of 30 Btu/h·ft². The rest of the windows contribute no solar gain but, of course, there is heat loss through *all* the windows. What is the total heating cost for the month? How much is the solar gain worth?

32. The average human body produces heat at the rate of 100 W. What is the coldest outdoor temperature in which a goose-down sleeping bag is usable, if the bag has a loft (thickness) of 5 cm? Make reasonable simplifying assumptions about the shape of the person and sleeping bag, and assume that normal body temperature (37°C) is to be maintained.

33. A steam pipe of 12-cm diameter carries steam at 550 K. How thick a layer of fiberglass insulation should be wrapped around this pipe if the heat loss per meter of pipe is to be 150 W? The air around the pipe is at 300 K.

34. A pipe of $\frac{3}{4}$-in diameter is surrounded by a $\frac{1}{2}$-in-thick layer of insulation. If the insulation thickness is doubled, how is the heat-loss rate changed?

35. Integrate Equation 16–13 from the inner edge of the insulation to an arbitrary point within the insulation, and use your result in conjunction with Equation 16–14 to show that the temperature as a function of position is given by

$$T = T_1 - (T_1 - T_2)\frac{\ln(r/R_1)}{\ln(R_2/R_1)},$$

where r is the position measured from the center of the pipe, and R_1 and R_2 are the inner and outer radii of the insulation. Show that this result reduces to the expected values at R_1 and R_2.

36. Combustion of fossil fuels like coal and oil puts large quantities of carbon dioxide into the atmosphere, effectively reducing the emissivity of the earth for the infrared radiation it emits. If we put enough CO_2 into the atmosphere to reduce the effective emissivity by 10%, what would happen to the average temperature of the earth?

37. In contrast to the previous problem, suppose that air

pollution in the form of soot reduced by 4% the amount of sunlight reaching earth's surface. What would be the average global temperature under these conditions? Use the assumptions of Example 16–11.

Supplementary Problems

38. A sphere of radius R_1 is maintained at a temperature T_1. The sphere is surrounded by insulation of thermal conductivity k and outer radius R_2. Show that the rate of heat loss from the sphere is

$$\frac{dQ}{dt} = \frac{4\pi k R_1 R_2 (T_1 - T_2)}{R_2 - R_1},$$

where T_1 and T_2 are the temperatures at the inner and outer edges of the insulation, respectively.

39. An object of heat capacity C is surrounded by insulation that results in a heat loss at the rate h per unit temperature difference between the object and its surroundings. If the object is at an initial temperature T_1, and its surroundings at a constant temperature T_0, and if there is no energy transferred to or from the object except by conduction through the insulation, show that the temperature of the object as a function of time is described by the equation

$$C\frac{dT}{dt} = -h(T - T_0).$$

Show that the solution to this equation is

$$T = T_0 + (T_1 - T_0)e^{-ht/C},$$

where t is the time. Show that this solution gives reasonable results when $t = 0$ and as $t \rightarrow \infty$.

40. A house is being maintained at 20°C on a winter day when the outdoor temperature is -15°C. Suddenly, the furnace fails. Determine how long it will take for the house temperature to reach the freezing point. The heat-loss rate for the house is 150 W/°C, and its heat capacity is 6.5×10^6 J/K. (See the result of the preceding problem.)

41. During a week in the summertime, the outdoor temperature fluctuates according to the equation

$$T_0 = 20°C - (5°C)\cos\left(\frac{\pi t}{12}\right),$$

where t is the time measured in hours since midnight. If there is no heat source in the house of the preceding problem, show that the house temperature also varies sinusoidally with time, and find the maximum and minimum house temperature. To do this, you can assume a sinusoidally varying house temperature, and see if it is a solution to the differential equation of Problem 39, when T_0 is varying sinusoidally. Does the house temperature peak at the same time as the outdoor temperature? Physically, why is this so?

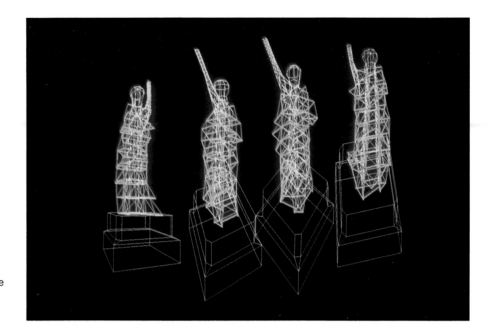

PLATE 1
A computer-aided set of drawings of the skeleton of the Statue of Liberty aided in the statue's reconstruction. [Swanke Hayden Connell Architects]

PLATE 2
The tuned-mass damper in action at the John Hancock Building in Boston. The system oscillates at the same frequency as the building, but out of phase, thus reducing building sway during high winds. (See Example 13−2.) [MTS Systems Corp.]

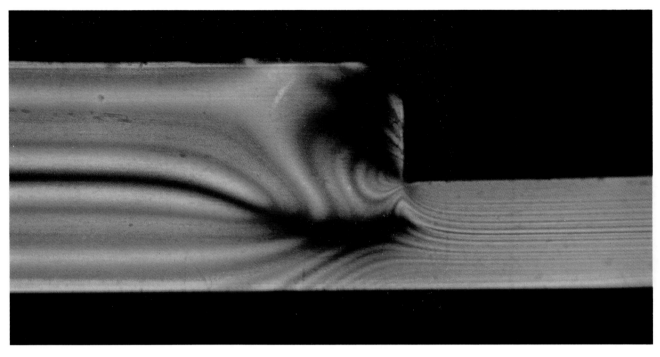

PLATE 3
Fluid flow at a step, shown by optical interferometry. [M. Horsmann, Ruhr University]

PLATE 4
Swirling patterns in Jupiter's atmosphere reveal complex fluid motion around the Great Red Spot. [NASA/Jet Propulsion Laboratory]

PLATE 5
Closeup of the Particle Beam Fusion Accelerator II at Sandia National Laboratories shows the ring of 36 accelerator modules converging on the target area. The cover of this text shows a test firing of this accelerator. On it, the PBFA is illuminated by electrical discharges that result from dielectric breakdown of air at the surface of the water that serves as dielectric in the accelerator's pulse-forming capacitors. [Sandia National Laboratories]

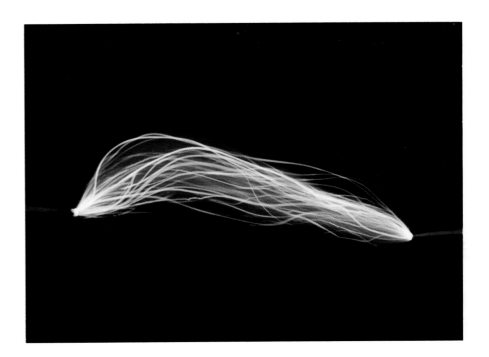

PLATE 6
Sparks between electrodes represent dielectric breakdown in air. [Jay M. Pasachoff]

PLATE 7
White light passed through a prism
breaks up into its component colors
as different wavelengths are refracted
through different angles. [Dennis
di Cicco]

PLATE 8
The Balmer series of hydrogen extends through the visible. H-alpha is the red spectral line at right.
[Alfred Leitner, Rensselaer Polytechnic Institute]

PLATE 9
The solar spectrum is a continuous spectrum crossed by dark absorption lines. H-alpha is at right in
the red. The sodium D lines are in the orange-yellow region near center. The magnesium b lines are
in the green, and H-beta lies in the blue-green region. [R. Giovanelli and H.R. Gillett, CSIRO
National Measurement Laboratory, Australia]

17

THE THERMAL BEHAVIOR OF MATTER

Matter responds in a variety of ways to the transfer of heat. It may get hotter, as we discussed in the preceding chapter, or it may undergo a change of phase, as when ice melts. It may experience changes in size or shape, or in pressure. In this chapter, we seek to understand the thermal behavior of matter. We start with a particularly simple state—the gaseous state—whose behavior can be accounted for by applying Newtonian mechanics at the molecular level. We then move to more complicated situations whose explanation is still grounded in the molecular properties of matter, but whose description is necessarily more empirical.

17.1
THE IDEAL GAS

A gas consists of matter in a rarefied state. Gas, like liquid, is a fluid, capable of deforming and flowing. But unlike liquid, gas is readily compressible, so that the volume of a given sample of gas is easily altered. The rarefied nature of a gas, resulting in weak interactions along its constituent particles, makes the thermal behavior of a gas particularly simple. In studying this behavior, we will develop a clear understanding of the relation between macroscopic properties—like temperature and pressure—and the underlying microscopic properties including the masses and velocities of the constituent particles.

The Equation of State of an Ideal Gas

At the macroscopic level, the state of a given sample of gas in thermodynamic equilibrium is determined completely by its temperature, pressure, and volume. A simple system for studying gas behavior consists of a cylinder sealed by a movable piston (Fig. 17–1). If we maintain the gas at a

Fig. 17–1
A piston-cylinder system.

431

constant temperature—by immersing the cylinder in a large reservoir of water or other substance at a fixed temperature—and push slowly on the piston to decrease the gas volume, we find that the pressure rises approximately in inverse proportion to the volume:

$$P \propto \frac{1}{V}. \quad \text{(at fixed } T)$$

Boyle's law

Known as **Boyle's law,** this relation was described by the English scientist Robert Boyle in 1660, a year before Isaac Newton entered college. If we keep the volume constant by holding the piston at a fixed position, and heat the gas, we find that the pressure is approximately proportional to the absolute temperature:

$$P \propto T. \quad \text{(at fixed } V)$$

Combining our two proportionalities, we have

$$P \propto \frac{T}{V}.$$

If we now hold T and V constant, but introduce more gas into the cylinder, we find that the pressure increases in proportion to the amount of gas. If N is the actual number of gas molecules, we can then write

$$P = \frac{NkT}{V},$$

or

$$PV = NkT, \quad (17\text{--}1)$$

where k is a constant.

ideal gas law

Equation 17–1 is called the **ideal gas law.** The constant k is found experimentally to have very nearly the same value for all gases:

$$k = 1.38 \times 10^{-23} \text{ J/K},$$

Boltzmann's constant

and is called **Boltzmann's constant,** after the Austrian physicist Ludwig Boltzmann (1844–1906), who was instrumental in developing the microscopic description of thermal phenomena.

Because the number of molecules, N, in a typical sample of gas is so large, we often express the ideal gas law in terms of the number of moles (mol) of gas. A mole of anything consists of Avogadro's number of that thing, where Avogadro's number is $N_A = 6.022 \times 10^{23}$. ($N_A$ is defined so that 12.0 g of ^{12}C atoms contains exactly one mole of those atoms.) If there are n moles of molecules in a gas, then $N = nN_A$ is the number of molecules, so that the ideal gas law can be written:

$$PV = nN_AkT = nRT, \quad (17\text{--}2)$$

universal gas constant

where the constant $R = N_Ak = 8.314$ J/K·mol is called the **universal gas constant.**

Example 17–1
The Ideal Gas Law

What is the volume occupied by 1.00 mol of an ideal gas at standard temperature and pressure (STP), where $T = 0°C$ and $P = 1.00$ atm?

Solution

Solving the ideal gas law for the volume V, and expressing all quantities in SI units, we have

$$V = \frac{nRT}{P} = \frac{(1.00 \text{ mol})(8.31 \text{ J/K·mol})(273 \text{ K})}{1.01 \times 10^5 \text{ Pa}} = 22.4 \times 10^{-3} \text{ m}^3 = 22.4 \text{ L.}$$

The ideal gas law is remarkably simple. It relates in a straightforward way the quantity of gas and its temperature, pressure, and volume. The form of the law and the numerical values of the constants k and R do not depend on such things as the substance making up the gas or the mass of the individual molecules. Yet the ideal gas law is obeyed very well by most real gases over a wide range of pressures. A plot of PV/nT versus pressure for an ideal gas (Fig. 17–2) gives a horizontal straight line whose vertical coordinate is the gas constant, R. As pressure is lowered, *all* gases converge to the ideal gas value of PV/nT and therefore exhibit ideal gas behavior. This nearly ideal behavior is what gives the gas thermometer its high precision over a wide temperature range.

Kinetic Theory of the Ideal Gas

Why do all gases obey such a simple relation among pressure, temperature, and volume? Studying the macroscopic properties themselves can never answer this question; instead, we must examine physical processes at the molecular level.

The gaseous state of matter is a rarefied state in which individual molecules are relatively far apart and interact only weakly. To understand how matter in this state might behave, we make a number of simplifying assumptions:

1. We assume that our gas consists of a very large number of identical point particles, having no size or internal structure, each of mass m. Of course, this assumption is false for real atoms or molecules, but if the gas is sufficiently rarefied, the size of the particles becomes negligible compared with the distance between them.
2. The particles do not interact with each other. This assumption rules out interactions—involving electrical, gravitational, or nuclear forces—that

Fig. 17–2
A plot of *PV/nT* versus pressure shows that real gases exhibit ideal behavior at low pressures. Even at 10 atm, oxygen deviates from the ideal gas law by less than 2 per cent.

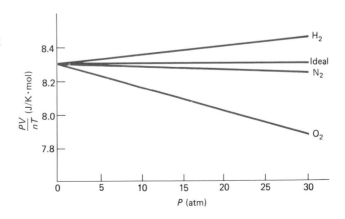

might cause the particles to attract or repel one another. Again, this assumption is not strictly correct for real particles, but normally, the average potential energy associated with such interactions is negligible compared with a particle's kinetic energy.

3. The particles are moving in random directions with a range of speeds. This assumption ensures that the gas has, on the average, no net momentum.

4. The particles undergo collisions with the walls of their container, and these collisions conserve momentum and energy. Here is where our gas model is tied to the laws of Newtonian mechanics. We also assume that the walls are perfectly smooth, so that forces exerted during these collisions are normal to the walls.

Consider a gas confined inside a rectangular container (Fig. 17–3). Each time a gas molecule collides with a container wall, it exerts a force on the wall. There are so many molecules that individual collisions are not evident on the macroscopic scale; instead the wall experiences a force that is essentially constant. The gas pressure is simply a measure of this force on a unit area.

To calculate the pressure, consider one molecule colliding with a wall (Fig. 17–4). Call the direction perpendicular to the wall the x-direction. Since the wall is smooth and the collision elastic, the component of the molecule's velocity in the y-direction remains unchanged during the collision, but the x-component of the velocity changes sign. Thus the molecule undergoes a change in momentum of magnitude $2mv_{xi}$, where v_{xi} is the magnitude of the x-velocity of this particular molecule. Now suppose that there are N_i molecules in the container whose x-components of velocity have the same magnitude v_{xi}. Because the molecules are moving in random directions, on the average half of them have x-components of velocity in the $+$x-direction and half in the $-$x-direction. Then in the time it takes one of these N_i molecules to cross the entire container, all $N_i/2$ of them that are moving toward the wall will hit it. The time in which these $N_i/2$ collisions take place is

$$\Delta t = \frac{\ell}{v_{xi}},$$

where ℓ is the length of the container in the x-direction. The change in momentum of all these molecules during this time is then

$$\Delta p_i = \frac{N_i}{2} 2mv_{xi} = N_i mv_{xi}.$$

By Newton's second law, the wall exerts a force whose average magnitude is equal to the average rate of change of the molecules' momentum:

$$F_i = \frac{\Delta p_i}{\Delta t} = \frac{N_i mv_{xi}}{\ell/v_{xi}} = \frac{N_i mv_{xi}^2}{\ell}.$$

By Newton's third law, this is also the magnitude of the average force exerted on the wall by the molecules whose x-component of velocity is v_{xi}. The total force on the wall is obtained by summing over all values of v_{xi}:

$$F = \sum F_i = \frac{\sum N_i mv_{xi}^2}{\ell}.$$

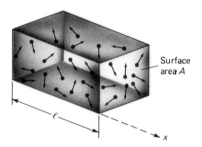

Fig. 17–3
Gas molecules confined to a rectangular box.

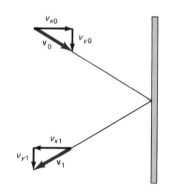

Fig. 17–4
A molecule undergoes an elastic collision with the container wall, reversing its x-component of velocity and transferring momentum $2mv_x$ to the wall. The y-component of velocity remains unchanged.

If the wall has area A, then the pressure, or force per unit area, is

$$P = \frac{F}{A} = \frac{\sum N_i m v_{xi}^2}{\ell A} = \frac{\sum N_i m v_{xi}^2}{V},$$

where $V = \ell A$ is the volume of the container. We can rewrite the pressure in a more meaningful way if we consider the average value of v_x^2 over all the molecules. This average, designated by $\overline{v_x^2}$, is obtained by summing the individual values v_{xi}^2, weighted by the number of molecules N_i with each value, and then dividing by N, the total number of molecules:

$$\overline{v_x^2} = \frac{\sum N_i v_{xi}^2}{N}.$$

Comparing this average with our expression for pressure shows that the pressure can be written

$$P = \frac{N m \overline{v_x^2}}{V},$$

so that

$$PV = N m \overline{v_x^2} = 2N(\tfrac{1}{2} m \overline{v_x^2}). \qquad (17-3)$$

We recognize the quantity $\tfrac{1}{2} m \overline{v_x^2}$ as the average kinetic energy per molecule associated with the motion of molecules in the x-direction. In Equation 17–3 we have the ideal gas law expressed in microscopic terms! Comparing with the macroscopic expression, $PV = NkT$, shows that

$$kT = 2\,(\tfrac{1}{2} m \overline{v_x^2}),$$

or

$$\tfrac{1}{2} m \overline{v_x^2} = \tfrac{1}{2} kT.$$

Because our molecules are moving in random directions, the average kinetic energy associated with motion in the x-direction is the same as in the y- and z-directions. The average total kinetic energy of a molecule is thus three times the average kinetic energy associated with its motion in any one direction, so that

$$\tfrac{1}{2} m \overline{v^2} = \tfrac{3}{2} kT. \qquad (17-4)$$

Our derivation of Equation 17–3 shows us why, in terms of simple Newtonian mechanics, we should expect a gas to obey the ideal gas law. In Equation 17–4 we get an added bonus: a microscopic understanding of the meaning of temperature. Temperature is simply a measure of the average kinetic energy associated with translational motion of the molecules.

Example 17–2

Temperature and Kinetic Energy

What is the average kinetic energy of a molecule in air at room temperature? What would be the speed of a nitrogen molecule with this energy?

Solution

The average kinetic energy is given by Equation 17–4, where room temperature is about 20°C (293 K):

$$\tfrac{1}{2}m\overline{v^2} = \tfrac{3}{2}kT = \tfrac{3}{2}(1.4 \times 10^{-23} \text{ J/K})(293 \text{ K}) = 6.2 \times 10^{-21} \text{ J}.$$

A nitrogen molecule consists of two nitrogen atoms (atomic mass 14 u; see Appendix E), so its mass is

$$m = 2(14 \text{ u})(1.7 \times 10^{-27} \text{ kg/u}) = 4.8 \times 10^{-26} \text{ kg}.$$

The kinetic energy is $K = \tfrac{1}{2}mv^2$, so that

$$v = \sqrt{\frac{2K}{m}} = \sqrt{\frac{2(6.2 \times 10^{-21} \text{ J})}{4.8 \times 10^{-26} \text{ kg}}} = 510 \text{ m/s}.$$

thermal speed

Not surprisingly, this **thermal speed** is of the same order of magnitude as the sound speed in air at room temperature. At the microscopic level, the speed of the individual molecules sets an approximate upper limit on the maximum rate at which information can be transmitted by disturbances—that is, sound waves—propagating through the gas.

The Distribution of Molecular Speeds

Our results 17–3 and 17–4 are fundamentally statistical. The macroscopic pressure is not associated with any one molecular collision, but only with huge numbers of collisions occurring so frequently, and each transferring so little momentum, that we sense only the average force and not the many individual impacts. Similarly, temperature does not tell us much about the energy of any individual molecule, but only about the average over all the molecules. In describing a gas—or any macroscopic system—in terms of temperature and pressure, we have averaged out the details of what is going on at the microscopic level. From this averaging, there emerges a simplicity at the macroscopic level that would be lost in the chaos of 10^{23} or so random molecular motions.

Frequently, though, we want to know more than just these average quantities. For example, is the range of molecular speeds limited to a narrow band about the thermal speed $v_{th} = \sqrt{3kT/m}$ obtained from Equation 17–4? Or are there many molecules moving much faster than the thermal speed? Such fast molecules would be important in determining, for example, the chemical reaction rates in a gas mixture, or the tendency of an atmosphere to escape the gravitational field of its planet.

In the 1860's, the Scottish physicist James Clerk Maxwell considered the sharing of energy that must result from collisions among the molecules of a gas. Although we have neglected such collisions, their presence changes none of our earlier conclusions as long as the collisions are elastic; in fact, collisions are responsible for bringing a gas into thermodynamic equilibrium. Maxwell showed that molecular collisions result in a speed distribution where the number of molecules in a very small speed range Δv about some speed v is given by

$$N(v)\,\Delta v = 4\pi N\left(\frac{m}{2\pi kT}\right)^{3/2} v^2 e^{-mv^2/2kT}\Delta v, \qquad (17-5)$$

where N is the total number of molecules in the gas, m the molecular mass, k Boltzmann's constant, and T the absolute temperature. The quantity $N(v)$ is the number of molecules per unit speed range; multiplying by a small

Fig. 17–5
Maxwell-Boltzmann distribution of molecular speeds for a sample of 10^{20} nitrogen molecules (N_2) at temperatures of 80 K and 300 K. Also marked is the thermal speed $v_{th} = \sqrt{3kT/m}$. Areas under both curves correspond to the total number of molecules, and are therefore the same.

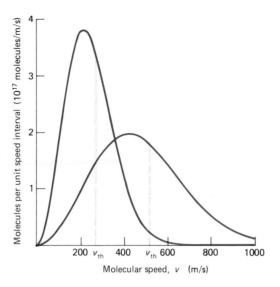

Maxwell-Boltzmann
distribution

speed range Δv gives $N(v)\Delta v$, or the actual number of molecules in that speed range. Equation 17–5 is known as the **Maxwell-Boltzmann distribution.** Figure 17–5 shows plots of this distribution for the same gas sample at two different temperatures. Each curve exhibits a single peak; this is the most probable value for a molecule's speed. Because the curve is not symmetric about the peak, the most probable speed lies below the mean thermal speed (see Problem 16). Figure 17–5 shows what Equation 17–4 already told us: that an increase in temperature corresponds to an increase in mean thermal speed. But it also shows something averages alone could not reveal: that an increase in temperature results in a broader speed distribution, with relatively fewer molecules near the most probable speed.

The Maxwell-Boltzmann distribution shows why the chemical reaction rates of gas mixtures increase dramatically with temperature; it is the excessively energetic molecules, of which there are more at higher temperatures, that are largely responsible for sustaining chemical reactions. A similar distribution holds for molecular speeds in liquids, and explains why a liquid evaporates at temperatures below its boiling point, and why evaporation results in cooling. Even though molecules with speeds near the thermal speed lack sufficient energy to escape the liquid, those in the high-speed "tail" of the Maxwell-Boltzmann distribution can escape. As they leave, the mean speed of the remaining molecules drops, and with it the temperature.

Real Gases

Though the ideal gas law is a good approximation to the behavior of real gases, the assumptions that led to our derivation of the ideal gas law are not entirely realistic. We should, therefore, expect a real gas to exhibit deviations from ideal behavior.

We assumed that our ideal gas molecules were point particles that underwent no intermolecular collisions or other interactions. But real mole-

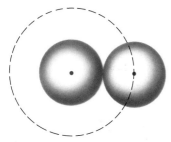

Fig. 17–6
At closest approach, the centers of two spherical molecules are separated by twice their radius. Each molecule therefore excludes the center of the other from a volume eight times its actual volume, as indicated by the dashed circle. We can regard either molecule as excluding the other, so that in the total sample half the molecules contribute to the excluded volume, which is therefore four times the molecular volume times the total number of molecules.

van der Waals force

van der Waals equation

cules have nonzero size, and therefore collide. During a collision, one molecule can only approach within two molecular radii of another (Fig. 17–6). Some of the total volume of the container is therefore excluded because of the finite molecular size, so that the volume V entering the gas law should be less than the total volume of the container. How much less? As Fig. 17–6 shows, a pair of spherical molecules has associated with it an excluded volume eight times the actual molecular volume. If b is this excluded volume summed over all the molecular pairs in one mole of gas, then we should write

$$P(V - nb) = nRT,$$

or

$$PV\left(1 - \frac{nb}{V}\right) = nRT,$$

where nb is the excluded volume in n moles of gas. In a rarefied gas, where the density of molecules n/V is small, we recover the ideal gas law.

Another deviation from ideal behavior occurs because each molecule exerts a weak attractive force on its neighbors. This force originates in electrical interactions among the particles making up the molecules; in Chapter 22 we will show how it arises.* Called the **van der Waals force,** it has a very short range and is therefore important only when molecules are relatively close. Physically, the van der Waals force reduces gas pressure because molecules slow down as they move apart, therefore lowering the force of molecular collisions with container walls. Mathematically, this results in an additional term in the equation describing gas behavior. Correcting for both finite molecular volume and the van der Waals force gives the **van der Waals equation:**

$$\left(P + \frac{n^2 a}{V^2}\right)(V - nb) = nRT, \tag{17–6}$$

where a and b are constants that depend on the particular gas. For low particle densities n/V, both correction terms in the van der Waals equation become negligible, showing that a rarefied gas closely follows the ideal gas law.

Example 17–3

The van der Waals Equation

The constants a and b in the van der Waals equation for nitrogen have the values $a = 0.14$ Pa·m^6/mol^2, $b = 3.91 \times 10^{-5}$ m^3/mol. If 1.000 mol of nitrogen is confined to a volume of 2.000 L, and is at a pressure of 10.00 atm, by how much does its temperature as predicted by the van der Waals equation differ from the ideal gas prediction?

Solution

Converting to SI units, we have $P = 1.013 \times 10^6$ Pa and $V = 2.000 \times 10^{-3}$ m^3. Then the ideal gas law predicts a temperature of

---

*Of course, there is also a gravitational attraction among the molecules. But this is entirely negligible.

$$T_{\text{ideal}} = \frac{PV}{nR} = \frac{(1.013 \times 10^6 \text{ Pa})(2.000 \times 10^{-3} \text{ m}^3)}{(1.000 \text{ mol})(8.314 \text{ J/K·mol})} = 244 \text{ K.}$$

The van der Waals equation predicts a temperature of

$$T_{\text{van}} = \frac{(P + n^2 a/V^2)(V - nb)}{nR}$$

$$= [1.013 \times 10^6 \text{ Pa} + (1.000 \text{ mol})^2 (0.14 \text{ Pa·m}^6/\text{mol}^2)/(2.000 \times 10^{-3} \text{ m}^3)^2]$$

$$\times \left(\frac{2.000 \times 10^{-3} \text{ m}^3 - (1.000 \text{ mol})(3.91 \times 10^{-5} \text{ m}^3/\text{mol})}{(1.000 \text{ mol})(8.314 \text{ J/K·mol})} \right) = 247 \text{ K.}$$

Even at 10 atm, nitrogen deviates from ideal gas behavior by only about 1 per cent.

17.2
PHASE CHANGES

For gases at high densities and pressures, the van der Waals equation provides a much better description of gas behavior than the ideal gas law. But even the van der Waals equation is only an approximation. At high enough pressure or low enough temperature, intermolecular forces become dominant, locking molecules so tightly together that the gas condenses into a liquid or even a solid. These different states of matter are called **phases.** A full theoretical description of phases and the **phase changes** that substances undergo is well beyond the scope of this book, but we can gain a qualitative understanding by examining the relation of pressure, volume, and temperature over a wide range of values.

phases
phase changes

PVT Diagrams

Suppose we have a gas at such pressure, temperature, and volume that it behaves very nearly like an ideal gas. Then if we hold the temperature constant but decrease the volume, the pressure rises inversely as the volume:

$$P \propto \frac{1}{V}.$$

The proportionality constant is, from the ideal gas law, just nRT. If we hold our gas at a different temperature, we get a similar inverse proportionality, but with a different proportionality constant. We can describe these different proportionalities by plotting curves of pressure versus volume for a variety of temperatures. Figure 17–7 shows such a PV diagram for an ideal gas. These constant-temperature curves are called **isotherms.**

iostherms

For a real gas, we expect van der Waals deviations from ideal behavior at very low volumes where intermolecular spacing becomes small. As we lower the gas temperature, these deviations become more pronounced. Eventually we reach a **critical temperature** T_C below which even the van der Waals description fails. Here the PV diagram exhibits a most unusual behavior: the pressure versus volume curves become flat over a finite range of volumes (Fig. 17–8).

critical temperature

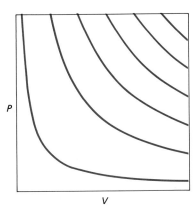

Fig. 17–7
PV diagram for an ideal gas, showing curves of constant temperature (isotherms). For an ideal gas, the isotherms are hyperbolas.

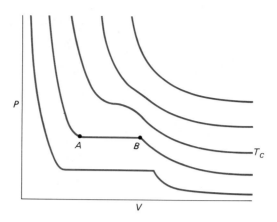

Fig. 17–8
Isotherms in the *PV* diagram for a substance that liquefies. T_C labels isotherm at the critical temperature. Moving along the horizontal line *AB* on the isotherm below T_C, the volume increases as the liquid changes to vapor. Since *AB* is part of an isotherm, temperature remains constant during this process.

What is going on here? One clue comes from looking on the left side of the *PV* curves for temperatures below T_C. Here the curves are almost vertical, indicating that any change of volume results in a huge pressure change. In other words, the substance is virtually incompressible: it has become a liquid. As we drop steeply down a curve of decreasing pressure, the volume increases only slightly until we reach the point where the *PV* curve suddenly becomes horizontal. For a range of volumes beyond this point, the pressure remains constant as the volume increases. In this region, the liquid and gas phases can coexist, so that as the volume increases, more and more of the liquid evaporates, greatly increasing the volume without any change in pressure. The pressure at which this occurs is called the **saturated vapor pressure**.

saturated vapor pressure

We entered the horizontal portion of the curve at point *A*, with our substance entirely in the liquid state. As we traverse from *A* to *B*, the liquid evaporates. Finally, at point *B*, the liquid is completely gone and the substance again exhibits the compressibility characteristic of a gas.

If we set a liquid at its vapor pressure and give it a large volume into which to expand, the liquid changes rapidly into a gas. This is just what happens when water boils. The saturated vapor pressure of water at 100°C is 1 atm, so that under these conditions water is on the horizontal portion of its *PV* curve. When exposed to the open air, so that its expansion is uninhibited, the liquid changes vigorously into gas. When we say that the boiling point of water is 100°C, we really mean that at this temperature, the vapor pressure equals atmospheric pressure. Water will boil at *any* temperature if placed in an environment where the pressure is equal to its vapor pressure at that temperature (Figure 17–9). The decrease in boiling point with decreasing pressure is a hazard to astronauts and high-altitude fliers, whose bodily fluids may boil abruptly if their air- or spacecraft lose pressure.

vapor

Figure 17–10 again shows the *PV* curves for a substance that liquefies, now with the boundaries between the liquid and gas phases clearly marked. The term **vapor** is sometimes applied to the gas phase at temperatures below T_c, although the term is sometimes restricted to mean gas in equilibrium with liquid. Notice that there is no really clear-cut distinction

Fig. 17–9
At the 3000-m altitude of Sacramento Peak Observatory in New Mexico, air pressure is sufficiently low that water boils at 92 °C.

between liquid and gas, except in the region of liquid-vapor equilibrium. If we start above the critical temperature and decrease the temperature while holding pressure constant, we move horizontally to the left in the *PV* diagram. As we do so, the volume drops and so the density of the substance rises, and we move smoothly from gas to liquid. In this region of the *PV* diagram, both gas and liquid are extremely dense fluids and are essentially

critical point   indistinguishable. It is only below the so-called **critical point,** where the *PV* curves begin to exhibit a horizontal portion, that there is a clear distinction between liquid and gas. Most of our experience with liquid-vapor transitions is below the critical point. Water, for example, has its critical point at the high temperature and pressure of 647 K and 218.3 atm. Figure 17–11 shows carbon dioxide in a sealed container being brought to its critical point. At that point the liquid-gas interface simply disappears as the two phases become indistinguishable.

The *PV* diagram of Fig. 17–10 is still incomplete, for it does not account for the third phase of matter—the solid phase. Modifying our *PV* diagram to include the solid phase would make it extremely complicated. Instead of trying to plot *PV* curves for each different temperature, we resort to a three-dimensional diagram with pressure, volume, and temperature along the three axes. The result is a *PVT* surface—a three-dimensional surface which, at each point, gives the relation between temperature, pressure,

Fig. 17–10
PV diagram showing boundaries between liquid and gas phases.

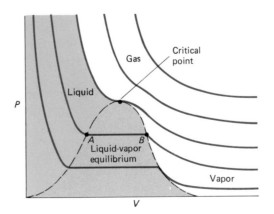

Fig. 17–11
The liquid-gas interface disappears as CO_2 in a glass container is brought to its critical point. Temperature and pressure at the critical point are 304 K and 73 atm.

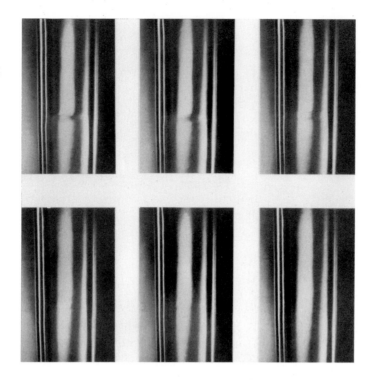

and volume. Figure 17–12 shows a PVT surface for a typical substance that expands when it melts.

The PVT surface describes fully the relation among the three variables P, V, and T. If we specify any two of these quantities, the third is determined by the distance to the PVT surface from the plane defined by the axes of the two known quantities. If we have a substance at some point on

Fig. 17–12
A PVT surface. Also shown are projections of the surface onto the PV plane—giving a PV diagram analogous to Fig. 17–10—and onto the PT plane, giving a PT diagram.

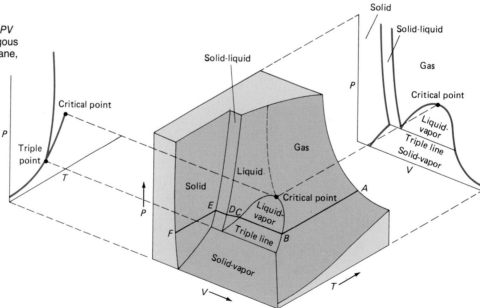

the *PVT* surface, we can perform processes that move it to other points. For example, we might enclose our substance in a tightly sealed cylinder with a movable piston. If we put the cylinder in a large bath of water at constant temperature, we can move along the isotherms—curves of constant temperature—in the *PVT* diagram by moving the piston and thus changing the volume of the substance. Or we might maintain a constant pressure, by exposing the piston to the atmosphere, and then move through the *PVT* surface by changing the temperature. Such a constant-pressure process is termed **isobaric,** and the resulting paths on the *PVT* diagram are **isobars.**

isobaric
isobars

Imagine moving along the isobar marked *ABCDEF* in Fig. 17–12. This path is an isobar because it is at constant height—that is, constant pressure—above the *VT* plane. At point *A* our substance is in the gas phase. As we lower the temperature, the volume drops proportionately, with slight deviations from a straight line because of van der Waals effects. At point *B* we reach the zone where liquid and vapor exist in equilibrium. As we try to cool the substance further, we find it remains at constant temperature while it liquefies. As it liquefies, the volume drops and we move along the liquid-vapor equilibrium curve, section *BC* of the isobar. When we reach point *C*, all the substance is in the liquid phase. Further cooling occurs with very little change in volume—the liquid is virtually incompressible—until we reach point *D*. Here the liquid begins to solidify, and again the substance experiences a significant change in volume without any change in temperature. Finally, at point *E*, the substance is entirely in the solid phase, and continues cooling with only slight changes in volume.

The particular isobar *ABCDEF* encompassed all three phases of the substance. Other isobars would have given different behaviors. At pressures greater than the critical pressure P_C we would have passed smoothly from gas to liquid without having noticed the distinction between the two, but would later have entered a region of solid-liquid equilibrium before the substance was entirely solid. At low enough pressure, we would have missed the liquid phase altogether, passing directly into a region of solid-vapor equilibrium. At the so-called triple-point pressure P_3, we would have moved through a region in which all three phases coexist in equilibrium. The unique values of pressure and temperature in this region define the **triple point** of the substance. We have already seen that the triple point of water is used in establishing the absolute temperature scale. To know that we are at the triple-point temperature, we need not make any measurements; it suffices to observe that all three phases are in equilibrium.

triple point

In our everyday experience, we associate heating a solid with eventual transitions to the liquid and then to the gas phase. The occurrence of these transitions, and the notion that the gas phase is always hotter than the liquid and solid phases, is partly an artifact arising because our everyday experience is confined to a pressure of 1 atm. For different pressures, or substances less common than those we deal with every day, rather different things may happen. For example, carbon dioxide at atmospheric pressure never enters the liquid phase—solid CO_2 (dry ice) goes directly from the solid to the vapor phase, giving it the name dry ice (Fig. 17–13). Only at much higher pressures does CO_2 liquefy. Fire extinguishers, for example, often contain liquid CO_2 under pressure. The interior of the earth is under

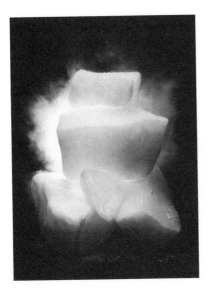

Fig. 17–13
Carbon dioxide subliming from solid to gas. The gray mist is actually a cloud of water droplets that condensed on contact with the cold CO_2 gas.

enormous pressure from the weight of the overlying layers, and is hot enough that some of it is liquid. But the central core of the planet, although the hottest part, is solid because the pressure is so high (Fig. 17–14).

The relations among the various phases are summarized in a **phase diagram,** which is the projection of the *PVT* surface onto the *PT* plane (Fig. 17–15). In plotting only this two-dimensional picture, we lose information about volume changes, but we show more clearly the regimes of temperature and pressure in which the different phases exist. The solid lines in the phase diagram show where the processes of melting (and its inverse, freezing or fusion), vaporization (and its inverse, liquefying), and sublimation (a change directly from solid to gas) occur. The special status of the triple and critical points shows clearly on the phase diagram.

Our *PVT* surface and phase diagram are for a particularly simple substance. Many real substances exhibit more complicated diagrams associated with subphases of liquid and solid that reflect changes in crystalline structure and molecular bonding.

phase diagram

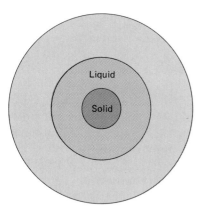

Fig. 17–14
Because of its extreme pressure, the inner core of the earth is solid, even though it is hotter than the surrounding liquid. Core temperatures range from a low around 4300 K in the outer core to a high around 6400 K in the solid inner core.

Heat and Phase Changes

Once we have heated a solid to its melting point, we can continue to add heat without any corresponding change in temperature until the solid has entirely liquefied. The same thing happens at the vaporization point. At a pressure of 1 atm, a pot of boiling water stays at 100°C until all the water is gone. Conversely, if we try to cool water below its freezing point, the temperature stays at 0°C until all the water has frozen. Only then does the temperature decrease again.

What is happening to the heat we add or remove during these phase changes? On a molecular level, the energy goes primarily into breaking the bonds that hold molecules in the tight configurations of solid or liquid. During a phase change, the potential energy of molecular separation changes, rather than the kinetic energy and therefore temperature. Macroscopically, we characterize phase changes by the energy required to melt or vaporize a given amount of material. We call this energy the **heat of fusion,** L_f, or **heat of vaporization,** L_v. Both are **heats of transformation.** If we determine that a mass m of a given substance requires heat Q to change completely from one phase to another with no temperature change, then the heat of transformation is given by:

heat of fusion
heat of vaporization
heats of transformation

$$L = \frac{Q}{m}. \tag{17–7}$$

To reverse the phase change—that is, to change water to ice—we must remove the energy implied by Equation 17–7. In addition to the heats of fusion and vaporization, we specify a heat of sublimation when a substance goes directly from the solid to the vapor state. Table 17–1 lists heats of transformation for some common materials at atmospheric pressure.

Heat transfers associated with phase changes are typically quite large. The heat of fusion of water, for example, is 334 kJ/kg or 80 cal/g—meaning that it takes as much energy to melt one gram of ice as it does to raise the resulting water from 0°C to 80°C.

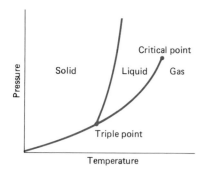

Fig. 17–15
A phase diagram is the projection of a *PVT* surface onto the *PT* plane.

TABLE 17–1 ▬▬▬▬
Heats of Transformation (at atmospheric pressure)

Substance	Melting Point, K	L_f, kJ/kg	Boiling Point, K	L_v, kJ/kg
Alcohol, ethyl	159	109	351	879
Copper	1356	205	2839	4726
Lead	600	24.7	2023	858
Mercury	234	11.3	630	296
Oxygen	54.4	13.8	90.2	213
Sulfur	388	38.5	718	287
Water	273	334	373	2257

Example 17–4
Heat of Transformation

A 500-g block of lead is at its melting point of 600 K. How much heat does it take to melt the lead?

Solution

Solving Equation 17–7 for Q gives the required heat:

$$Q = mL_f = (0.500 \text{ kg})(24.7 \text{ kJ/kg}) = 12.4 \text{ kJ}.$$

The heats of fusion and vaporization measure only the energy associated with phase changes, during which the temperature does not change. To calculate the energy required to change first the temperature and then the phase of a substance, we must use both the specific heat and the heat of transformation.

Example 17–5
Changes of Temperature and Phase

A 50-g ice cube is taken from a freezer at $-10°C$, put in a pan on a hot stove, and heated until it has all turned to steam. How much energy is required?

We break this problem into several parts, each dealing with either a temperature change or a phase change. Equation 16–7, along with the appropriate specific heat from Table 16–1, allows us to calculate the heat associated with temperature changes. Equation 17–7 and Table 17–1 allow us to calculate the heat of transformation, as in Example 17–4.

1. To bring 50 g of ice from $-10°C$ to its melting point of 0°C, we have

$$Q_1 = mc\Delta T = (0.050 \text{ kg})(2.1 \text{ kJ/kg·K})(10 \text{ K}) = 1.1 \text{ kJ}.$$

2. To melt the 50 g of ice takes

$$Q_2 = mL_f = (0.050 \text{ kg})(330 \text{ kJ/kg}) = 17 \text{ kJ}.$$

3. To raise the water from 0°C to its boiling point at 100°C takes

$$Q_3 = mc\Delta T = (0.050 \text{ kg})(4.2 \text{ kJ/kg·K})(100 \text{ K}) = 21 \text{ kJ}.$$

4. Finally, to boil the water completely takes

$$Q_4 = mL_v = (0.050 \text{ kg})(2.3 \times 10^3 \text{ kJ/kg}) = 120 \text{ kJ}.$$

Then the total energy required for the process is

$$Q = Q_1 + Q_2 + Q_3 + Q_4 = 160 \text{ kJ.}$$

Note that 75 per cent of this energy is associated with vaporization.

Example 17–6
Equilibrium Temperature Following a Phase Change

200 g of ice at $-10°C$ is added to 1.0 kg of water at 15°C, and the mixture kept in an insulated container. When thermodynamic equilibrium is reached, how much, if any, ice is left? What is the equilibrium temperature?

Solution

To solve this problem we must first find out whether the water is capable of melting all the ice. To bring the ice to its melting point and then melt all of it requires

$$Q_1 = mc\Delta T + mL_f$$

$$= (200 \text{ g})(0.49 \text{ cal/g·°C})(10°C) + (200 \text{ g})(80 \text{ cal/g})$$

$$= 980 \text{ cal} + 16 \times 10^3 \text{ cal} = 17 \text{ kcal.}$$

Here we have chosen to work in calories rather than joules because the specific heat of water is conveniently 1.0 cal/g·°C.

If we cool the water to 0°C, we can extract an amount of heat given by

$$Q_2 = mc\Delta T = (1.0 \times 10^3 \text{ g})(1.0 \text{ cal/g·°C})(15°C) = 15 \text{ kcal.}$$

This is not enough heat to melt all the ice. We know immediately that we will be left with at least some ice, and that the temperature cannot exceed 0°C. On the other hand, the heat we can extract from the water is far more than needed simply to raise the ice to 0°C, so that we will melt some of the ice and thus will have a mixture of ice and water in equilibrium at 0°C. Since our final temperature is 0°C, we will extract all 15 kcal from the water. From our calculation of Q_1, we know that 0.98 kcal is needed to raise the ice to 0°C. The remaining 15 kcal -0.98 kcal $= 14$ kcal goes into melting ice. We can solve Equation 17–7 for m to find how much ice melts:

$$m = \frac{Q}{L_f} = \frac{14 \text{ kcal}}{80 \text{ kcal/kg}} = 0.18 \text{ kg} = 180 \text{ g.}$$

Therefore, we are left with 20 g of ice in 1180 g of water, with the mixture at 0°C.

17.3
THERMAL EXPANSION

We have explored changes of temperature and phase that occur when matter is heated. But our *PVT* diagrams show that matter may also change in volume or pressure when heated. When an ideal gas is held at constant pressure, for example, then its volume is directly proportional to its temperature.

The volume and pressure relations in the liquid and solid phases are not quite so simple. Liquids and solids are far less compressible than gases, so that thermal expansion is less pronounced. This low compressibility is reflected in the steepness of the *PVT* surface in the regions of pure liquid or solid. A large change in pressure or temperature is required for even a small change in volume. On the microscopic level, the molecules in a liquid or solid are closely spaced; to move them requires that we do work against large (electrical) forces.

We can characterize the change in the volume of a substance with temperature in terms of its **coefficient of volume expansion,** β, defined as the fractional change in volume when the substance undergoes a small temperature change ΔT:

coefficient of volume expansion

$$\beta = \frac{\Delta V/V}{\Delta T}.$$

(17–8)

Because β varies with temperature, Equation 17–8 is strictly valid only in the limit as ΔT approaches 0, so that we should write

$$\beta = \lim_{\Delta T \to 0} \frac{\Delta V/V}{\Delta T} = \frac{1}{V}\frac{dV}{dT}.$$

(17–9)

In practice, we can often consider β constant over a finite temperature range, and thus use Equation 17–8. Our definition of β assumes that pressure remains constant. We could entirely inhibit thermal expansion with appropriate pressure increases.

Often we want to know how one linear dimension of a solid changes with temperature. This is especially true with long, rodlike structures where the absolute change is greatest along the long dimension (Fig. 17–16). We then speak of the **coefficient of linear expansion,** α, defined by

coefficient of linear expansion

$$\alpha = \lim_{\Delta T \to 0} \frac{\Delta L/L}{\Delta T} = \frac{1}{L}\frac{dL}{dT}.$$

(17–10)

The volume expansion coefficient and linear expansion coefficient α are related in a simple way:

$$\beta = 3\alpha,$$

(17–11)

Fig. 17–16
As its length increases on a hot summer day, a power line sags noticeably.

as you can show in Problem 40. This relation means that either of these coefficients fully characterizes the thermal expansion of a material. However, the linear expansion coefficient α is really only meaningful with solids, because liquids and gases deform readily and therefore do not generally expand proportionately in all directions (see Question 32). Table 17–2 lists expansion coefficients for some common substances.

TABLE 17–2
Expansion Coefficients*

Solids	α (K⁻¹)	*Liquids and Gases*	β (K⁻¹)
Aluminum	24×10^{-6}	Air	3.7×10^{-3}
Brass	19×10^{-6}	Alcohol, ethyl	75×10^{-5}
Copper	17×10^{-6}	Gasoline	95×10^{-5}
Glass (Pyrex)	3.2×10^{-6}	Mercury	18×10^{-5}
Ice	51×10^{-6}	Water, 1°C	-4.8×10^{-5}
Invar†	0.9×10^{-6}	Water, 20°C	20×10^{-5}
Steel	12×10^{-6}	Water, 50°C	50×10^{-5}

*At approximately room temperature unless noted.

†Invar, consisting of 64 per cent iron and 36 per cent nickel, is an alloy designed for use where thermal expansion must be minimized.

Example 17–7

Thermal Expansion

A steel girder is 15 m long at −20°C. It is used in the construction of a building where temperature extremes of −20°C to 40°C are expected. By how much does the girder length change between these extremes?

Solution

Table 17–2 gives the coefficient of linear expansion for steel: $\alpha = 12 \times 10^{-6}$ K^{-1}. Equation 17–10 then gives

$$\Delta L = \alpha L \Delta T = (12 \times 10^{-6}\ °C^{-1})(15\ m)(60°C) = 1.1\ cm.$$

What happens when a hollow object, like that shown in Fig. 17–17, expands? Do you expect that the exterior dimensions of the object would increase, while the hole shrinks in size as the material expands into it? In reality, *every* linear dimension in the solid expands in the same proportion, causing the hole to expand, too (Fig. 17–18).

Fig. 17–17
How does the size of the hole change when the solid expands?

Example 17–8

Thermal Expansion

A car's gasoline tank is made of steel and has internal dimensions of 60 cm × 60 cm × 20 cm at a temperature of 10°C. It is filled with gasoline at 10°C. If the temperature now increases to 30°C, by how much does the volume of the tank increase? How much gasoline spills out of the tank?

Solution

The initial volume of the tank is

$$V = (0.60\ m)(0.60\ m)(0.20\ m) = 0.072\ m^3.$$

Equation 17–8 then gives the volume increases:

$$\Delta V = \beta V \Delta T,$$

so that

$$\Delta V_{tank} = (3)(12 \times 10^{-6}\ °C^{-1})(0.072\ m^3)(20°C) = 5.2 \times 10^{-5}\ m^3,$$

and

$$\Delta V_{gas} = (95 \times 10^{-5}\ °C^{-1})(0.072\ m^3)(20°C) = 1.4 \times 10^{-3}\ m^3.$$

Since Table 17–2 gives the coefficient of linear expansion, α, for solids, we have used $\beta = 3\alpha$ in calculating the volume change of the tank.

The expansion of the steel tank is negligible compared with that of the gasoline, so that we lose $1.4 \times 10^{-3}\ m^3$, or 1.4 L of gasoline. Don't fill your gas tank to the very top!

Fig. 17–18
A hollow solid made of rectangular slabs. As the slabs expand, so does the interior hole.

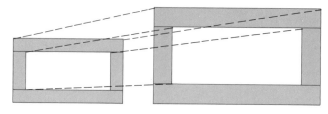

Fig. 17–19
Volume of one gram of water near its melting point.

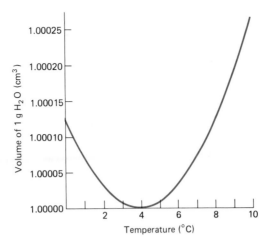

Thermal Expansion of Water

The entry for water at 1°C in Table 17–2 is remarkable, for the negative expansion coefficient shows that water at this temperature actually contracts on heating. Figure 17–19 shows the volume occupied by one gram of water as a function of temperature. From 0°C to 4.0°C, the volume decreases with increasing temperature, indicating a negative expansion coefficient. This curious behavior occurs only near the melting point, and is related to another anomalous property of water—that its solid state is less dense than its liquid state. This unusual situation occurs because the structure of the ice crystal prevents H_2O molecules in the solid state from coming as close as they do in the liquid state (Fig. 17–20). Just above the melting point, the same intermolecular forces that give the ice crystal its structure still influence the molecules strongly enough to cause the liquid to contract with increasing temperature. Finally, above 4.0°C, thermal motion of the molecules becomes more important than these intermolecular forces, and the liquid exhibits normal thermal expansion with increasing temperature.

The anomalous behavior of water has important consequences for aquatic life. Because water occupies the least volume when it is at 4.0°C, water at this temperature sinks to the bottom of freshwater lakes. For lakes deeper than a few meters, sunlight is insufficient to change this temperature, so that the deep water remains at 4.0°C year-round. In winter, the surface waters cool below 4.0°C and eventually freeze, forming an insulating layer of ice that, because of its lower density, floats on the surface. As a result, ice cover in temperate climates rarely exceeds a meter or so, and aquatic life can continue below the ice layer. If water behaved like most other liquids, the coldest water would be at the bottom and lakes would freeze from the bottom up, destroying most aquatic life.

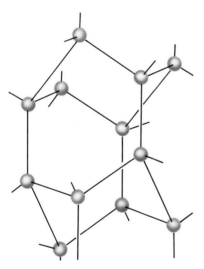

Fig. 17–20
The structure of an ice crystal makes the solid phase less dense than the liquid. Spheres represent individual H_2O molecules.

Example 17–9
Thermal Expansion of Water

The coefficient of volume expansion of liquid water in the temperature range from 0°C to about 20°C is given approximately by

$$\beta = a + bT + cT^2, \tag{17-12}$$

where T is the temperature in degrees Celsius, and where $a = -6.43 \times 10^{-5}\,°C^{-1}$, $b = 1.70 \times 10^{-5}\,°C^{-2}$, and $c = -2.02 \times 10^{-7}\,°C^{-3}$. If a sample of water occupies 1.00000 L at 0°C, what will its volume be at 12°C?

Since the expansion coefficient β is temperature-dependent, we must integrate, using Equation 17–9, to find the new volume. Multiplying that equation by dT, we have

$$\beta\, dT = \frac{dV}{V}.$$

We now integrate this expression from our initial temperature T_1 and volume V_1 to the final temperature T_2 and unknown final volume V_2, using Equation 17–12 for β:

$$\int_{T_1}^{T_2} (a + bT + cT^2)\,dT = \int_{V_1}^{V_2} \frac{dV}{V},$$

or

$$a(T_2 - T_1) + \tfrac{1}{2}b(T_2^2 - T_1^2) + \tfrac{1}{3}c(T_2^3 - T_1^3) = \ln\!\left(\frac{V_2}{V_1}\right).$$

Evaluating the left-hand side of this equation using the appropriate values of a, b, c, T_1, and T_2, we have

$$3.36 \times 10^{-4} = \ln\!\left(\frac{V_2}{V_1}\right),$$

so that

$$\frac{V_2}{V_1} = e^{3.36 \times 10^{-4}} = 1.00034.$$

With $V_1 = 1.00000$ L, the final volume V_2 is 1.00034 L.

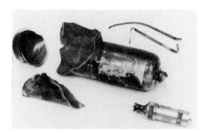

Fig. 17–21
This tank of torch fuel overheated and burst from the resulting pressure increase, injuring the operator of the torch.

Fig. 17–22
Thermal expansion distorted these railroad tracks, causing a derailment.

Thermal Stresses

By enclosing a liquid or gas in a rigid container, or by tightly clamping the ends of a solid in place, we can inhibit thermal expansion. Following its PVT diagram, the material must then experience a pressure increase. If a container is unable to withstand this increase, it will burst (Fig. 17–21). If a solid is restrained along only one dimension, it will deform and perhaps crack (Fig. 17–22). Similarly, when a solid is clamped at both ends and cooled, tension forces develop that may cause it to break. Finally, if a solid is heated or cooled unevenly, part of the material expands or contracts at a different rate from the surrounding material. Again, the resulting forces are often sufficient to destroy the material.

SUMMARY

1. An **ideal gas** is described by a the **ideal gas law,** a particularly simple relation among pressure, temperature, and volume:

$$PV = NkT.$$

Here N is the number of molecules in the gas and $k = 1.38 \times 10^{-23}$ J/K is **Boltzmann's constant.** The ideal gas law may also be written

$$PV = nRT,$$

where n is the number of moles of the gas and $R = 8.314$ J/K·mol is the **universal gas constant.**

2. **Kinetic theory** provides a description of the ideal gas in terms of the microscopic physics of large numbers of molecules. Kinetic theory allows us to interpret temperature as a measure of the average kinetic energy of the gas molecules:

$$\tfrac{1}{2}m\overline{v^2} = \tfrac{3}{2}kT.$$

3. **Real gases** deviate from ideal behavior at high densities, where the finite size of the molecules and the intermolecular forces become significant. Real gases are described approximately by the **van der Waals equation:**

$$\left(P + \frac{n^2a}{V^2}\right)(V - nb) = nRT,$$

where a and b are constants describing a particular gas.

4. Most substances can exist in any of several **phases,** including gas, liquid, and one or more solid phases. The relations among temperature, pressure, and volume in these phases are described by a **PVT surface.** Projection of the PVT surface onto the PT plane gives the **phase diagram** for the substance. Important points on this diagram are the **triple point,** where the three phases can coexist in equilibrium, and the **critical point,** where the boundary between liquid and gas phases ceases to be distinct.

5. During a phase change, energy is transferred to or from a substance without an accompanying temperature change. The energy associated with the phase change of a unit mass of the substance is called the **heat of transformation.** The heat of **fusion** describes the solid-liquid transition, the heat of **vaporization** describes the liquid-gas transition, and the heat of **sublimation** describes the solid-gas transition.

6. Most substances expand when heated at constant pressure. The change in volume of a substance is given by

$$\Delta V = \beta V \Delta T,$$

where the quantity β is called the **coefficient of volume expansion.** β is temperature-dependent, so that this equation holds only for relatively small temperature changes. The change in length of a solid is given by a similar expression:

$$\Delta L = \alpha L \Delta T,$$

where $\alpha = \beta/3$ is the **coefficient of linear expansion.** Water provides an important exception to the general rule that substances expand when heated. In the range from 0°C to 4°C, water contracts on heating, a phenomenon with significant implications for ecological systems.

7. If a substance is held rigidly, so that thermal expansion or contraction cannot occur, then **thermal stresses** develop that may destroy either the object or its container.

QUESTIONS

1. If the volume of an ideal gas is increased, must the pressure drop proportionately? Explain.
2. According to the ideal gas law, what should the volume of a gas be at absolute zero? Why is this result absurd?
3. Why are you supposed to check the pressure in a tire when the tire is cold?
4. Figure 17–23 shows a thin, flexible wire draped over an ice cube, with weights tied on either end of the wire. If the weights are heavy enough, the wire will gradually melt its way through the ice cube, nevertheless leaving the cube in one piece. Explain this unusual phenomenon in terms of the phase diagram for water (Fig. 17–24).

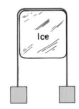

Fig. 17–23   Question 4

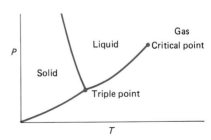

Fig. 17–24   Phase diagram for water.

5. Figure 17–25 shows a phase diagram for carbon dioxide. Could you perform the experiment in Fig. 17–23 on a dry ice cube at atmospheric pressure? At pressures above the triple-point pressure? See preceding question.

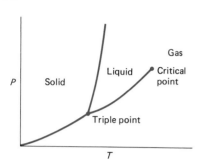

Fig. 17–25 Phase diagram for CO_2.

6. Suppose there are only two molecules in a container. Does it make sense to talk about the temperature and pressure of this gas?

7. The average *speed* of the molecules in a gas increases with increasing temperature. What about the average *velocity*?

8. Suppose you start running while holding a jar of air. Do you change the average speed of the air molecules? The average velocity? The temperature?

9. Gas thermometers containing different gases at relatively high pressures do not agree exactly at all temperatures, even if they were all calibrated at the triple point of water. Why not?

10. Do all molecules in a gas have the same speed? If so, how does this come about? If not, how can the notion of temperature be meaningful?

11. The speed of sound in a gas is closely related to the thermal speed of the molecules. Why?

12. Two different gases are at the same temperature, and both are at low enough densities that they behave like ideal gases. Do their molecules have the same thermal speeds? Explain.

13. The atmosphere of a small planet such as earth contains very little of the lighter gases like hydrogen and helium. The massive planet Jupiter has an atmosphere rich in light gases. Why might this be?

14. Is the van der Waals force attractive or repulsive? Justify your answer in terms of the sign of the term n^2a/V^2 appearing in the van der Waals equation, assuming that the parameter a is positive.

15. Some people think that ice and snow must always be at 0°C. Is this always true? Under what circumstances must it be true? Could you paraphrase the same arguments for another substance? Try steel, for example.

16. What is the temperature of water just under the ice layer of a frozen lake? At the bottom of the lake?

17. Deep lakes usually "turn over" twice each year. During the overturn, water from the lake bottom gets mixed with surface water, while at other times deep water and surface water do not mix. Under what conditions should such an overturn occur?

18. How is it possible to have liquid water at 0°C?

19. Some ice and water have been together in a glass for a long time. Is the water hotter than the ice?

20. Does it take more heat to melt a gram of ice at 0°C than to bring the resulting water to the boiling point? Once at the boiling point, does it take more heat to boil all the water than it did to bring it from the melting point to the boiling point?

21. Table 17–1 suggests that substances that are gases at room temperature and atmospheric pressure have relatively low heats of vaporization. Why might this be?

22. Why do we use the triple point of water for thermometer calibration? Why not just use the melting point or boiling point?

23. When the average temperature drops below freezing in winter, regions near large lakes remain warmer for some time. Why?

24. Must all substances exhibit solid, liquid, and gas phases? What might happen before a substance reaches the gas phase? Consider sugar, for example.

25. What would you do to air to liquefy it in such a way that the gas-liquid transition was obvious? In such a way that it was not obvious?

26. How is it possible to have boiling water at a temperature other than 100°C?

27. Why does the water in a car radiator boil explosively when you remove the radiator cap?

28. How does a pressure cooker work?

29. Why does a double boiler prevent food from burning?

30. Why can you boil water in a paper cup without burning the cup?

31. Suppose mercury and glass had the same coefficient of volume expansion. Could you build a mercury thermometer?

32. In calculating the change in length of the column in a mercury thermometer, should you use the coefficient of linear expansion or the coefficient of volume expansion? Explain.

33. Can you think of any circumstance in which you might heat a liquid and have it expand without deforming (that is, with no change in shape)?

34. Why are the coefficients of volume expansion generally greater for liquids than for solids?

35. A bimetallic strip consists of thin pieces of brass and steel bonded together (Fig. 17–26). Such strips are often used in thermostats. What will happen when the strip is heated? *Hint:* Consult Table 17–2.

Fig. 17–26 Bimetallic strip (Question 35).

36. Modern railroad tracks are often welded together, with no gaps for many miles. What do you suppose is done about thermal expansion of such tracks?

37. Why are power lines more likely to break in winter?

PROBLEMS

Section 17.1   *The Ideal Gas*

1. An ideal gas occupies a volume V at a temperature of 100°C. If the pressure of the gas is held constant, by what factor does the volume change when (a) the Celsius temperature is doubled? (b) the absolute (Kelvin) temperature is doubled?

2. 2.0 mol of an ideal gas are at an initial temperature of 250 K and pressure of 1.5 atm. (a) What is the volume occupied by the gas? (b) The pressure is now increased to 4.0 atm and the volume of the gas is observed to drop to half its original value. What is the new temperature of the gas?

3. The solar corona is an extended atmosphere of hot $(2 \times 10^6$ K) gas surrounding the cooler visible surface of the sun. The gas pressure in the solar corona is about 3×10^{-2} Pa. What is the number density (particles per cubic meter) in the solar corona? Compare with the number density of molecules in earth's atmosphere.

4. In a typical laboratory vacuum apparatus, a pressure of 1.0×10^{-10} Pa is readily achievable. If the residual air in this "vacuum" is at 0°C, how many air molecules are there in one liter?

5. A cubical metal box with thin walls measures 10 cm on each side. It is filled with 0.50 mol of an ideal gas at 180 K, and is surrounded by air at atmospheric pressure. What is the pressure force (magnitude and direction) on each side of the box?

6. A helium balloon occupies a volume of 8.0 L at 1.0 atm pressure and a temperature of 20°C. The balloon rises to an altitude where air pressure is 0.65 atm and where the temperature is −10°C. Assuming that the balloon stays in thermodynamic equilibrium with its surroundings, what is its volume at the new altitude?

7. A cylinder of compressed air stands 100 cm tall and has an internal diameter of 20.0 cm. At room temperature, the pressure in the cylinder is 180 atm. (a) How many moles of air are in the cylinder? (b) What volume would this air occupy at 1.0 atm pressure and room temperature?

8. A 3000-mL flask may be closed tightly by means of a valve. The valve is initially open while the flask is in a room containing air at 1.00 atm and 20°C. The valve is then closed, and the flask immersed in a bath of boiling water at 1.00 atm pressure. When the air in the flask has reached thermodynamic equilibrium, the valve is opened, air is allowed to escape, and the valve then closed. The flask is then cooled to room temperature. (a) What is the maximum pressure reached in the flask? (b) How many moles of gas escape when the air is let out of the flask? (c) What is the final pressure in the flask?

9. A student's dormitory room measures 3.0 m × 3.5 m × 2.6 m. (a) How many air molecules does it contain? (b) What is the total translational kinetic energy of these molecules? (c) How does this energy compare with the kinetic energy of the student's 1200-kg car going at 90 km/h?

10. At what temperature would the thermal speed of nitrogen molecules in air be one per cent of the speed of light? Why might it not be possible to achieve such a temperature in N_2 gas?

11. What is the thermal speed of hydrogen (H_2) molecules at a temperature of 800 K?

12. The principal gas components of earth's atmosphere are N_2, O_2, and Ar, with smaller amounts of CO_2 and H_2O and trace amounts of other gases. In air at 300 K, what are the thermal speeds of these gases?

13. The diameter of the hydrogen molecule (H_2) is on the order of 10^{-10} m. (a) What is the approximate value of the constant b in the van der Waals equation for hydrogen? *Hint:* Consult Fig. 17–6. (b) For H_2 gas at 273 K, at what pressure does the finite size of the H_2 molecule cause a deviation of one per cent from ideal gas behavior?

14. The van der Waals constants for helium gas (He) are $a = 0.0341$ L²·atm/mol² and $b = 0.0237$ L/mol. What is the temperature of 3.00 mol of helium at 90.0 atm pressure if the gas is confined to a volume of 0.800 L? How does this result differ from the ideal gas prediction?

15. Plot the Maxwell-Boltzmann distribution of molecular speeds for samples of monatomic helium gas (He) containing 10^{18} atoms at temperatures of 50 K and 250 K.

16. (a) Show that Equation 17–4 implies a thermal speed given by $\sqrt{3kT/m}$. (b) By differentiating the Maxwell-Boltzmann distribution (Equation 17–5), show that the most probable molecular speed is lower than this thermal speed by a factor of $\sqrt{2/3}$.

Section 17.2   *Phase Changes*

17. Carbon dioxide sublimes (changes from solid to gas) at 195 K. The heat of sublimation is 573 kJ/kg. How much heat must be extracted from 250 g of CO_2 vapor at 195 K in order to solidify it?

18. A 100-g block of ice is initially at −20°C. Heat is supplied to it at the rate of 500 W. (a) How long must the heat be applied in order to produce water at 50°C? (b) Make a graph showing temperature versus time from when the heating starts until the water reaches 50°C.

19. Repeat Example 17–6 for the case when the initial mass of ice is 50 g.

20. Water at 300 K is sprinkled onto 200 g of molten copper at its melting point of 1356 K. The water boils away completely, leaving the copper entirely solid but still at 1356 K. How much water was sprinkled on the copper? Assume that water vapor leaves the system and need not come to thermodynamic equilibrium with the copper, and that there is no heat transfer from the copper except to the liquid water.

21. A 50-g ice cube at $-10°C$ is placed in an equal mass of water. What must the initial temperature of the water be in order that there remain 50 g of ice and 50 g of liquid water?

22. The size of industrial air conditioning and refrigeration units is often given in tons, meaning the number of tons of water at its melting point that the unit could freeze in one day. (This unit is a holdover from the days when ice was cut from lakes in winter to use for cooling throughout the year.) What is the rate of heat extraction, in watts, of a 15-ton refrigeration unit?

23. 1.0 kg of ice at $-40°C$ is added to 1.0 kg of water at $5.0°C$. Describe the composition and temperature of the equilibrium mixture.

24. Repeat the preceding problem if the 1.0 kg of ice is initially at $-80°C$ and if there are only 200 g of water at $5.0°C$.

25. A 40-kg block of aluminum is at a temperature of $50°C$. A jet of steam at $100°C$ is directed at the aluminum. Assuming that all the steam hits the block and condenses, and that the resulting liquid water drops off immediately without transferring heat to the aluminum, how much steam must reach the aluminum to heat it to $100°C$?

26. A solar-heated house stores energy in 5.0 tons of Glauber salt ($Na_2SO_4 \cdot 10H_2O$), a substance that melts at $90°F$. The heat of fusion of Glauber salt is 104 Btu/lb, and the specific heats of the solid and liquid are, respectively, 0.46 Btu/lb·°F and 0.68 Btu/lb·°F. After a week of sunny weather, the storage medium is all liquid at $95°F$. Then a cool, cloudy period sets in, during which the house loses heat at a rate averaging 20,000 Btu/h. (a) How long is it before the temperature of the storage medium drops below $60°F$? (b) How much of this time is spent at $90°F$?

Section 17.3   *Thermal Expansion*

27. Show that the coefficient of volume expansion of an ideal gas is just the reciprocal of the absolute temperature.

28. Suppose that a single piece of welded steel railroad track stretched across the continental United States. If the track were free to expand, but not to bend sideways, estimate its length change in going from the coldest winter day ($-25°C$) to the hottest summer day ($+40°C$)?

29. A constant-volume gas thermometer like that of Fig. 16–5 is made from Pyrex glass. If the thermometer is calibrated at the triple point of water and then used to determine the boiling point of water, how much error will be introduced by ignoring the expansion of the glass?

30. A 2000-mL graduated cylinder is filled with liquid at 350 K. When the liquid is cooled to 300 K, the cylinder is full only to the 1925-mL mark. Using Table 17–2, identify the liquid.

31. A steel ball bearing 5.0 mm in diameter is encased in a Pyrex glass cube 1.0 cm on a side. At 330 K, the ball bearing fits tightly in its cavity. At what temperature will it have a clearance of 1.0 μm all around?

32. Referring to Example 17–9, calculate the change in volume of 1.00000 L of water in going from (a) $0°C$ to $4.0°C$, (b) $4.0°C$ to $15°C$. Compare your result in part (b) with the answer you would get assuming that the coefficient of volume expansion were constant and equal to its value at $0°C$.

33. Use Equation 17–12 to show that water has its greatest density at very nearly $4.0°C$.

34. A sample of water occupies 5.00000 L at $0°C$. At what other temperature would it have the same volume? (See Example 17–9.)

35. In Equation 17–12, the coefficients appropriate to carbon tetrachloride (CCl_4) are $a = 1.18 \times 10^{-3} °C^{-1}$, $b = 0.396 \times 10^{-6} °C^{-2}$, and $c = 0.390 \times 10^{-7} °C^{-3}$. If a sample of CCl_4 occupies 2.000 L at $10°C$, what is its volume at $50°C$? Compare with the result obtained by assuming that the coefficient of volume expansion is constant and is equal to its value at $0°C$.

36. A glass marble 1.000 cm in diameter is to be dropped through a hole in a steel plate. At room temperature the hole has a diameter of 0.997 cm. By how much must the temperature of the steel be raised so that the marble fits through the hole?

Supplementary Problems

37. Temperatures in the upper atmospheres of earth and Jupiter are about 1000 K and 300 K, respectively. Calculate the thermal speeds of hydrogen (H_2) and oxygen (O_2) at each planet, and compare with the escape speed for that planet. What implications do your results have for the atmospheric composition of the planets? Explain in terms of your numerical results and the shape of the Maxwell-Boltzmann distribution.

38. Ignoring air resistance, find the height from which you must drop an ice cube at $0°C$ so that it completely melts on impact. Assume no heat is exchanged with the environment.

39. The timekeeping of an old pendulum clock is regulated by a brass pendulum 20.0 cm long. If the clock is accurate at $20°C$, but is in a room at $18°C$, how long will it be before the clock is in error by one minute? Will it be too fast or too slow?

40. Prove the relation of Equation 17–11, showing that the volume expansion coefficient β and linear expansion coefficient α are related by $\beta = 3\alpha$. To do so, consider a cube of side L. Use the chain rule to write

$$\frac{dV}{dT} = \frac{dV}{dL}\frac{dL}{dT}.$$

Substitute this expression for dV/dT in Equation 17–9. Express the cube's volume V in terms of its side length L, and from this evaluate the quantity dV/dL. Show that your result then takes the form of Equation 17–10, with $\alpha = \frac{1}{3}\beta$.

18

HEAT, WORK, AND THE FIRST LAW OF THERMODYNAMICS

How do we change the temperature of a substance? One approach, described in detail in Chapter 16, involves heat transfer between substances at different temperatures. Another approach—which you have experienced if you ever pulled a rope too quickly through your hands, or touched a drill bit just after drilling a hole, or smelled your car's brakes burning as you went down a steep hill—involves mechanical energy. In this chapter, we will explore this second approach to temperature change, and in the process come to a deeper understanding of the relation between heat and mechanical energy.

18.1
THE FIRST LAW OF THERMODYNAMICS

We have defined heat as energy that is transferred from one object to another because of a temperature difference alone. As a result of heat transfer, a substance undergoes changes in its thermodynamic state—its temperature, pressure, and volume may change, and it may undergo a change of phase. But we can accomplish these same changes by the transfer of mechanical energy. As Fig. 18–1 suggests, we can raise the temperature of water by heating with a flame or by stirring violently with a spoon. For the flame to heat the water, the flame temperature must exceed the water temperature. But the temperature of the spoon is irrelevant, for it is the transfer of the mechanical energy associated with the spoon's motion that raises the water temperature. We use the term **work** to describe energy transfer that does not require a temperature difference. This sense is consistent with our earlier use of the word work to describe mechanical energy being transferred from one object to another. Sometimes that energy ends up as stored potential energy, but often it results in a temperature change—as when we do work against frictional forces. That both heat and mechanical work can

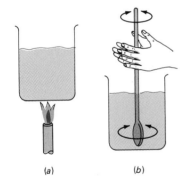

Fig. 18–1
Water temperature may be raised (a) with a flame or (b) by mechanical agitation.

work

have the same effect on a substance—namely, a temperature change—is what made possible Joule's quantitative identification of heat as a form of energy (Fig. 18–2).

What happens to the energy we transfer to a system, either through heat or through mechanical work? It ends up changing the **internal energy** of the system. We discussed internal energy earlier in connection with heat transfer alone. Now, recognizing that heat and mechanical energy can have the same effect on a system, we conclude that both are associated with internal energy changes. What, exactly, is internal energy? It is simply the sum of all the different kinds of energy associated with the individual particles making up a substance. For the ideal gas of the previous chapter, internal energy is the total kinetic energy of the gas molecules. In a more complicated substance, other forms of energy also contribute to the internal energy. In a gas of diatomic molecules, for example, we must consider also the energy of rotation or vibration of the molecules (Fig. 18–3). In a solid or liquid, with molecules close together, the potential energy associated with intermolecular forces also accounts for some internal energy. When energy is released in reactions that rearrange atoms or even atomic nuclei, then we must include the potential energy of atomic and nuclear configurations as internal energy. A fuel—like gasoline or uranium—has considerable internal potential energy.

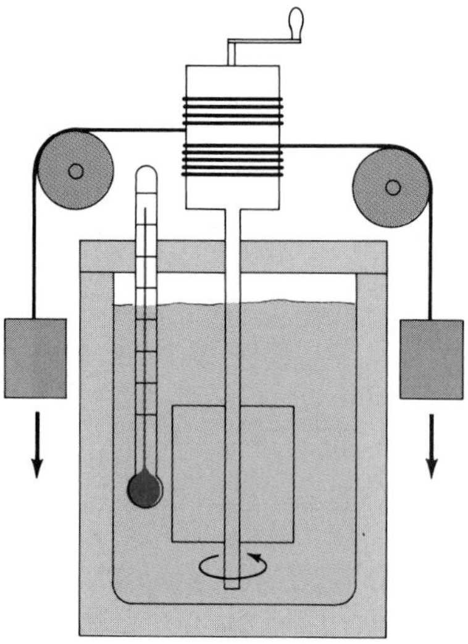

If we keep track of all energy entering and leaving a system, and monitor corresponding changes in internal energy, we find experimentally that energy is conserved. Given the extent to which we have stressed energy conservation in our study of mechanics, this is hardly surprising. But our statement of energy conservation is now broader, for it applies to heat transfer as well as to mechanical work. This broader statement of energy conservation is known as the **first law of thermodynamics:**

> The change in the internal energy of a system is equal to the heat added to the system minus the work done by the system.

Mathematically, we can state the first law by writing

$$\Delta U = Q - W, \tag{18–1}$$

where ΔU ($= U_{\text{final}} - U_{\text{initial}}$) is the change in internal energy of a system, Q the heat added *to* the system, and W the work done *by* the system. Why the minus sign in Equation 18–1? Had we defined W as the work done *on* the system, then positive values of both W and Q would represent energy input to the system, and both would enter the equation with plus signs. Historically, though, the first law was developed in connection with the study of engines, which typically take in energy from a heat source and give out energy in the form of mechanical work. Keeping track of energy transfers in an engine is made simpler by considering both the heat input and work output as positive quantities; their difference is then the change in internal energy.

The first law of thermodynamics shows that there are many ways to change the internal energy of a system. We could heat the system without doing any mechanical work on it. Or we could do only mechanical work, without allowing any heat transfer. Or we could apply some combination of heat and mechanical work. The change in internal energy does not de-

internal energy

Fig. 18–2
Schematic diagram of Joule's apparatus. Potential energy of the falling weights is converted to kinetic energy of the rotating paddle, which in turn becomes internal energy of the water.

first law of thermodynamics

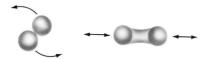

Fig. 18–3
Molecular rotation and vibration contribute to the internal energy of a diatomic gas.

pend on the details of the process whereby we change it—it depends only on the net amount of energy transferred to the system. Like temperature and pressure, internal energy is a **thermodynamic state variable**—a quantity whose value is independent of how the system got into a particular state.

thermodynamic
state variable

Frequently we are concerned with the rates of energy flow in a system. By differentiating the first law with respect to time, we obtain a statement about these rates:

$$\frac{dU}{dt} = \frac{dQ}{dt} - \frac{dW}{dt}. \qquad (18\text{–}2)$$

Here dU/dt is the rate of change of internal energy, dQ/dt the rate at which heat is supplied to the system, and dW/dt the rate at which the system does work in its surroundings.

Example 18–1

A Nuclear Power Plant

In a nuclear power plant, heat produced from nuclear fission boils water that, in turn, drives a steam turbine connected to an electric generator. The steam is then cooled and condensed through thermal contact with water from a river, lake, or ocean, and sent back to the reactor to be reheated and turned to steam again. The nuclear reactor in a certain power plant releases energy at the rate of 1500 MW. (This is the rate at which the internal energy of the nuclear fuel decreases.) The power plant supplies 500 MW of electric power. At what rate is heat transferred to the cooling water used to condense steam?

Solution

Solving Equation 18–2 for the rate of heat transfer gives

$$\frac{dQ}{dt} = \frac{dU}{dt} + \frac{dW}{dt}.$$

Here dU/dt is -1500 MW, since the internal energy is decreasing, while dW/dt, the work done *by* the power plant, is $+500$ MW, so that

$$\frac{dQ}{dt} = -1500 \text{ MW} + 500 \text{ MW} = -1000 \text{ MW}.$$

Since a positive Q represents heat transferred *to* a system, the minus sign in this answer shows that heat is being transferred *from* the power plant *to* the cooling water at the rate of 1000 MW. The numbers in this example are typical of a medium-sized power plant. We will see in the next chapter just why it is that so much of the energy released in the reactor ends up in the cooling water rather than as electricity.

18.2

THERMODYNAMIC PROCESSES

The first law of thermodynamics relates the two kinds of energy transfers—heat and work—that can change the thermodynamic state of a system. Although the law applies to *any* system, we can most readily illustrate its use in application to an ideal gas. The ideal gas law (Equation 17–1 or 17–2) relates the temperature, pressure, and volume of a given sample of gas:

$$PV = nRT.$$

If we know any two of these, the thermodynamic state of the gas is completely determined. One way to specify the thermodynamic state is to use a diagram whose two coordinate axes represent two of the thermodynamic state variables—for example, pressure P and volume V. Each point on such a diagram specifies a unique combination of pressure and volume, and therefore a thermodynamic state of the system. The utility of a PV diagram in specifying thermodynamic states is not limited to ideal gases, but applies to any substance whose thermodynamic state is uniquely specified by giving values of P and V.

Thermodynamic state is a meaningful concept only for a system in thermodynamic equilibrium, for only in equilibrium does the system have uniquely defined values of temperature and pressure. While we are changing the thermodynamic state of a system, we may temporarily find that the system has no uniquely defined state. Suppose we take a sealed container of gas in thermodynamic equilibrium at 100 K and plunge it suddenly into a vat of boiling sulfur at 718 K (Fig. 18–4). Before the gas comes to equilibrium with the sulfur, there will be a period of chaos when gas molecules nearest the container walls have much more energy than those near the center of the container. Only gradually, as the molecules share energy through collisions, will the system come to an equilibrium state in which the average molecular energy is the same throughout the gas and therefore in which temperature is well defined.

A very different approach to heating the same gas sample would be to put it in contact with a heat reservoir whose temperature we can control (Fig. 18–5). If we raise the reservoir temperature very slowly, its temperature will exceed that of our gas sample by only a very small amount, and the gas will be very nearly in equilibrium at all times. In the limit of an arbitrarily slow change in reservoir temperature, we can consider the gas sample to be exactly in equilibrium at all times. Such a slow change, in which a system remains in equilibrium, is called a **quasi-static process.** No real process quite achieves the quasi-static limit, but we can approach it by making the changes in a system occur very slowly. Because a system undergoing quasi-static change is always in thermodynamic equilibrium, it is always describable by a point in a thermodynamic state diagram like a PV diagram. The overall quasi-static change from one state to another is then described by a continuous set of points—a curve—in that diagram (Fig. 18–6).

In the quasi-static heating process we just described, our gas sample is always at very nearly the same temperature as its surroundings. If we make the surroundings infinitesimally cooler rather than warmer than the gas, the process will proceed in the opposite direction. For this reason, such a process is also called a **reversible process.** Once we have heated our gas sample quasi-statically to the boiling point of sulfur, for example, we could quasi-statically cool it to its original state. In the process, the gas would pass in the opposite direction through the same series of equilibrium states that it did while heating, thereby reversing the direction of its travel in the PV plane. A process like the sudden plunging of the gas sample into boiling sulfur is, in contrast, **irreversible.** A system undergoing irreversible change

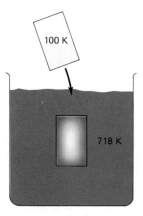

Fig. 18–4
An irreversible process. A gas sample at 100 K is plunged suddenly into boiling sulfur. A period of disequilibrium ensues, in which molecules near the container walls have more energy than those in the center of the sample. Eventually a new equilibrium is established as molecules share energy through collisions.

quasi-static process

reversible process

irreversible

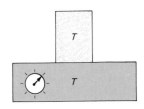

Fig. 18–5
A gas sample in contact with a heat reservoir whose temperature is adjustable. If the reservoir temperature is changed very slowly, the gas remains very nearly in equilibrium at all times.

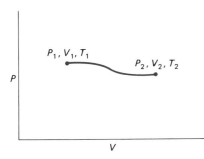

Fig. 18–6
Because a system undergoing a quasi-static change is always in equilibrium, the change may be described by a succession of states—a continuous path—in the PV diagram.

Fig. 18–7
A gas-cylinder system.

has no well-defined thermodynamic state until the process is complete, so it makes no sense to think of a path in the PV diagram. A process may be irreversible even though it returns a system to the same original state it started from. The distinction between reversible and irreversible refers not to the end states but to the process that takes a system between those states.

The terms reversible and irreversible refer to transfers of mechanical energy as well as of heat. For example, we hold a small balloon in our hands and suddenly, violently compress it. Gas molecules near the edges of the balloon are rapidly accelerated, while those at the center do not yet sense the compression. The system is temporarily out of equilibrium; we have performed an irreversible process. Eventually the system reaches equilibrium with constant temperature and pressure throughout, but during the compression these quantities have no well-defined values. If, on the other hand, we compress the balloon so slowly that the entire volume of gas can adjust continuously to the changing conditions, then we perform a reversible compression.

We can imagine many ways to change the thermodynamic state of a system—reversibly, irreversibly, while maintaining a constant pressure, while maintaining a constant temperature, changing both temperature and pressure simultaneously, and so on. We now consider several important special cases of such changes, as they apply to a simple ideal gas system. Although the system we use and the processes to which we subject it may seem artificial, our special cases in fact illustrate the physical principles behind a myriad of technological devices and natural phenomena, from the operation of a gasoline engine to the propagation of a sound wave to the oscillations of a variable star.

Our system consists of an ideal gas confined to a cylinder that is sealed with a movable piston (Fig. 18–7). The piston and the cylinder walls are perfectly insulating—they block all heat transfer—while the bottom of the cylinder is a perfect conductor of heat. By altering the gas temperature and/or the location of the piston, we can change the thermodynamic state of the gas. We will consider only reversible processes, which we can describe by paths in the PV diagram for our gas, and will use the first law of thermodynamics to relate the associated energy transfers by heat and work.

Before examining specific processes, we develop a general relation between volume change and work that holds for all processes. If A is the cross-sectional area of our piston and cylinder, and P the gas pressure, then $F = PA$ is the force exerted by the gas on the piston (Fig. 18–8). If the piston moves a small distance Δx, then the work done by the gas is

$$\Delta W = F\Delta x = PA\Delta x.$$

But $A\Delta x$ is the change in volume of the gas, so that

$$\Delta W = P\Delta V. \qquad (18\text{–}3)$$

The work ΔW is positive when the volume increases, indicating that the gas does work on the piston and its surroundings, while the work is negative if ΔV is negative, indicating that work is done on the gas. In general, the pressure may depend on the volume, so that Equation 18–3 is strictly valid only in the limit of arbitrarily small volume changes. To find the

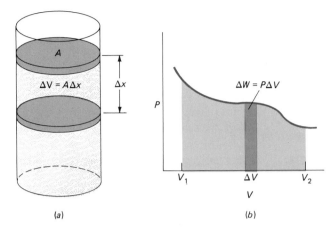

Fig. 18–8
(a) The force on piston of area A is PA, so that the work done in moving the piston a small distance Δx is PAΔx, or PΔV. (b) On the PV diagram, this work corresponds to the area under the PV curve on the volume interval ΔV associated with the change Δx in the position of the piston. The total work to take the system from V_1 to V_2 is the area under the PV curve between V_1 and V_2.

work associated with a larger volume change, we take the limit ΔV→0, and integrate over the volume change:

$$W = \int dW = \int_{V_1}^{V_2} P\,dV, \qquad (18\text{–}4)$$

where V_1 and V_2 are the initial and final volumes. Figure 18–8b shows the geometrical interpretation of this equation: the work done by the gas is simply the area under the PV curve.

Isothermal Processes

isothermal process

To perform a reversible **isothermal process,** we place the heat-conducting bottom of our gas cylinder in good thermal contact with a substance—a heat reservoir—whose temperature is held constant (Fig. 18–9). We then move the piston to change the volume of the system, doing so slowly enough that the gas remains in equilibrium with the heat reservoir. The system then moves from its initial state to its final state along a curve of constant temperature—an **isotherm**—in the PV diagram (Fig. 18–10). The work done in the process is given by Equation 18–4, and is equal to the area under the isotherm.

isotherm

For an ideal gas, we can relate pressure P and volume V through the ideal gas law:

$$P = \frac{nRT}{V}.$$

Then Equation 18–4 becomes

$$W = \int_{V_1}^{V_2} \frac{nRT}{V}\,dV.$$

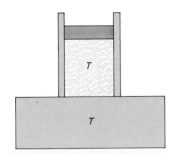

Fig. 18–9
During an isothermal process, the system is held in thermal contact with a heat reservoir at fixed temperature.

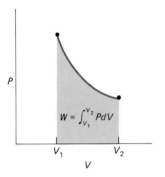

Fig. 18–10
PV diagram for an isothermal process. The work done during the process is the area under the curve.

For an isothermal process, the temperature T is constant, giving

$$W = nRT \int_{V_1}^{V_2} \frac{dV}{V} = nRT \ln V \Big|_{V_1}^{V_2} = nRT \ln\left(\frac{V_2}{V_1}\right). \qquad (18\text{–}5)$$

The internal energy of an ideal gas consists only of the kinetic energy of its molecules which, as we saw in the previous chapter, depends only on temperature. Therefore, in an isothermal process, there is no change in the internal energy of the ideal gas. The first law of thermodynamics then gives

$$\Delta U = 0 = Q - W,$$

so that

$$Q = W = nRT \ln\left(\frac{V_2}{V_1}\right). \qquad \text{(isothermal process)} \qquad (18\text{–}6)$$

What does this result $Q = W$ mean? Recall that Q is the heat transferred *to* the gas, while W is the work done *by* the gas. Therefore, our result says that when an ideal gas does work W on the external world with no change in temperature, it must absorb an equal amount of heat from outside. Similarly, if work is done on the gas, then the gas must reject an equal amount of heat to the outside if its temperature is not to change.

The result 18–6 applies only to an ideal gas, in which the internal energy depends only on the temperature. In a real gas, a change in volume results in a change in the potential energy associated with intermolecular forces. Because these forces are attractive, the potential energy increases as gas volume increases. In an isothermal expansion, in which the *kinetic* energy of the molecules remains constant, the *total* internal energy actually increases, and the gas must absorb an amount of heat Q that exceeds slightly the work W it does on the external world.

Example 18–2
A Diver Exhales

A scuba diver is swimming at a depth of 25 m, where the pressure is 3.5 atm (recall Example 15–3). The air she exhales forms bubbles 8.0 mm in radius. How much work is done by each bubble as it expands on rising to the surface? Assume that, because of their small size and relatively slow rise, the bubbles remain at the uniform 300 K temperature of the surrounding water.

Solution

Since the bubbles remain at constant temperature, the process is isothermal and Equation 18–6 applies. Just before they break the surface, the bubbles are at essentially 1 atm pressure, so that their pressure has decreased by a factor of 3.5. The ideal gas law, $PV = nRT$, shows that when temperature is constant, pressure and volume are inversely related. Therefore, the bubble volume has increased by a factor of 3.5. To apply Equation 18–6, we also need the quantity nRT for a bubble. We can get this from the ideal gas law using the given radius of the bubbles at 3.5 atm pressure:

$$nRT = PV = P(\tfrac{4}{3}\pi r^3),$$

so that Equation 18–6 becomes

$$W = nRT \ln\left(\frac{V_2}{V_1}\right) = \tfrac{4}{3}\pi r^3 P \ln\left(\frac{V_2}{V_1}\right)$$

$$= \tfrac{4}{3}\pi (8.0 \times 10^{-3}\ \text{m})^3 (3.5\ \text{atm})(1.01 \times 10^5\ \text{Pa/atm}) \ln(3.5)$$

$$= 0.95\ \text{J}.$$

Where does this energy go? As the bubble expands, it pushes water outward and, ultimately, upward. It therefore raises the gravitational potential energy of the ocean. (When the bubble breaks the surface, this excess potential energy becomes kinetic energy, appearing in the form of small waves on the water surface.)

Constant-Volume Processes and Specific Heat

constant-volume process

A **constant-volume process** (also called isometric, isochoric, or isovolumic), occurs in a rigid, closed container whose volume cannot change. In our piston-cylinder arrangement, we could tightly clamp the piston to achieve a constant-volume process. Because the piston does not move, the gas does no work, and the first law becomes simply

$$Q = \Delta U. \tag{18-7}$$

molar specific heat at constant volume

To express this result in terms of a temperature change ΔT, we introduce the **molar specific heat at constant volume,** C_V, defined by the equation

$$Q = nC_V\Delta T, \tag{18-8}$$

where n is the number of moles of gas in our system. This molar specific heat is like the specific heat we defined in Chapter 16, except that with a gas it is more convenient to consider the heat per mole rather than per unit mass. Introducing the definition of C_V into Equation 18-7 gives

$$nC_V\Delta T = \Delta U. \tag{18-9}$$

Solving for C_V, and taking the limit of very small temperature changes, gives

$$C_V = \frac{1}{n}\frac{dU}{dT}. \tag{18-10}$$

For an ideal gas, the internal energy is a function of temperature alone, so that dU/dT has the same value no matter what process the gas may be undergoing. Therefore, Equation 18-9, relating the temperature change ΔT and internal energy change ΔU, applies not only to a constant-volume process, but also to *any* ideal gas process. Why, then, have we been so careful to label C_V the specific heat *at constant volume?* Although the relation $nC_V\Delta T = \Delta U$ holds for any process, it is only when no work is done that the first law allows us to write $Q = \Delta U$, and therefore only for a constant-volume process that Equation 18-8 holds.

Isobaric Processes and Specific Heat

isobaric

Isobaric means constant pressure. In a reversible isobaric process, a system moves along an isobar, or curve of constant pressure, in its PV diagram (Fig. 18-11). The isobar is simply a horizontal straight line in the PV dia-

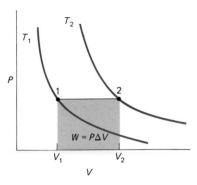

Fig. 18–11
PV diagram for an isobaric process.
Also shown are the isotherms for the
initial and final temperatures. Work
done is the shaded area under the
isobar.

gram, and the system moves from an initial state with temperature T_1 and volume V_1 to a final state (T_2, V_2) while its pressure P remains constant. The work done in the process is the area under the isobar, and is given by Equation 18–4:

$$W = \int_{V_1}^{V_2} P dV = P \int_{V_1}^{V_2} dV = P(V_2 - V_1) = P\Delta V,$$

since P is constant.

Solving the first law (Equation 18–1) for Q and using our expression for work gives

$$Q = \Delta U + W = \Delta U + P\Delta V.$$

But we found in the previous section that the change in internal energy of an ideal gas is given by $\Delta U = nC_V\Delta T$ for *any* process. Therefore

$$Q = nC_V\Delta T + P\Delta V \tag{18–11}$$

molar specific heat at
constant pressure

for an isobaric process. We define the **molar specific heat at constant pressure** as the heat required to raise one mole of gas through a unit temperature change when the gas is at constant pressure, so that

$$Q = nC_P\Delta T.$$

Equation 18–11 can then be written

$$nC_P\Delta T = nC_V\Delta T + P\Delta V. \quad \text{(isobaric process)} \tag{18–12}$$

Equation 18–12 is a useful form for calculating temperature changes in an isobaric process, if we know both specific heats C_P and C_V. However, we really need only one of these specific heats, for a simple relation holds between the two. The ideal gas law, $PV = nRT$, allows us to write

$$P\Delta V = nR\Delta T$$

for an isobaric process. Using this expression in Equation 18–12 gives

$$nC_P\Delta T = nC_V\Delta T + nR\Delta T,$$

so that

$$C_P = C_V + R. \tag{18–13}$$

Does this result make sense? Specific heat is a measure of the amount of heat needed to cause a given temperature change. In a constant-volume process, no work is done and all the heat goes into raising the internal energy and thus the temperature of an ideal gas. In a constant-pressure process, work is done and some of the added heat ends up as mechanical energy of the external world, leaving less energy available for raising the temperature. Therefore, in a constant-pressure process, *more* heat is needed to achieve a given temperature change, so that the specific heat at constant pressure is larger than that at constant volume, as reflected in Equation 18–13.

Why did we not distinguish specific heats at constant volume and constant pressure much earlier? Because we were concerned mostly with solids and liquids, whose coefficients of expansion are far lower than those of gases. As a result of its relatively small expansion, the work done by a solid or liquid is much less than that done by a gas. Since work is what gives rise to the difference between C_V and C_P, the distinction is less significant for solids and liquids. As a practical matter, measured specific heats are usually at constant pressure, because enormous pressures would be needed to prevent volume changes in solids or liquids.

Adiabatic Processes

adiabatic process   In an **adiabatic process,** we allow no heat transfer between a system and its environment. The easiest way to achieve this is to surround the system with a perfect insulator that prevents conduction, convection, and radiation. Because heat transfer takes time, a rapid process is often adiabatic simply because the time for the process to occur is short compared with the time needed to transfer significant amounts of heat into or out of the system. In a gasoline engine, for example, the compression of the gasoline-air mixture and subsequent expansion of the combustion products are nearly adiabatic because they occur so rapidly that little heat flows through the cylinder walls during the process.

In an adiabatic process, the heat Q is zero, so that the first law becomes

$$0 = \Delta U + W,$$

or

$$\Delta U = -W. \qquad \text{(adiabatic process)} \qquad (18\text{--}14)$$

This equation says that if a system does work on its surroundings, then its internal energy must decrease by the same amount. This makes sense because, if no heat is added, then any energy transferred out as work must come from the system's internal energy. Microscopically, the decrease in internal energy comes about as gas molecules bounce off the moving piston. Were the piston not moving, the molecules would return to the gas with the same speed and hence kinetic energy. But when the piston is moving, their speed is decreased by that of the piston, so that they return to the gas with less kinetic energy (Fig. 18–12). Conversely, if the piston moves inward, work is done on the gas, and its internal energy increases.

What is the path of a reversible adiabatic process in the PV diagram? Because the internal energy changes, so must the gas temperature. In an adiabatic expansion, for example, the temperature drops and the volume increases. The ideal gas law, $PV = nRT$, then requires that the pressure decrease as well—and by more than it would in an isothermal expansion between the same initial and final volumes. Figure 18–13 depicts such an adiabat adiabatic path, called an **adiabat.**

Fig. 18–12
Molecules bouncing off a stationary piston rebound with no loss of speed. But molecules bouncing off an outward-moving piston rebound with lower speed, having given energy to the piston.

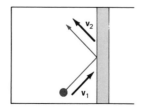
(a) Stationary piston

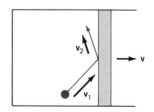
(b) Moving piston

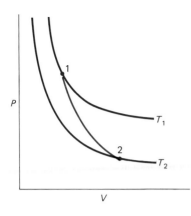

Fig. 18–13
PV diagram for an adiabatic process. Also shown are isotherms for the initial and final temperatures. The adiabatic path is steeper than the isotherms because the gas loses internal energy as it does work.

Because neither temperature nor pressure nor volume remains constant, we can determine the adiabatic path only by considering a sequence of infinitesimal adiabatic changes, and integrating to get a relation between the initial and final states. Earlier we found that for *any* process, the relation between temperature change and internal energy change is given by Equation 18–9, so that the internal energy change during an infinitesimal adiabatic process is

$$dU = nC_V dT.$$

The work done in the process is

$$dW = PdV.$$

Then the first law for an adiabatic process (Equation 18–14) becomes

$$nC_V dT = -PdV. \tag{18–15}$$

To determine the adiabatic path in the *PV* diagram, we must obtain a relation between *P* and *V* alone. To eliminate *dT* in Equation 18–15, we first differentiate the ideal gas law, allowing both *P* and *V* to change:

$$d(PV) = nRdT,$$

or

$$PdV + VdP = nRdT. \tag{18–16}$$

Solving Equation 18–16 for *ndT* and using the result in Equation 18–15 gives

$$C_V \frac{PdV + VdP}{R} = -PdV.$$

Multiplying through by *R* and bringing all terms to the left side of the equation results in

$$(C_V + R)PdV + C_V VdP = 0.$$

But we found earlier that C_P, the specific heat at constant pressure, is given by $C_P = C_V + R$, so that

$$C_P PdV + C_V VdP = 0.$$

Dividing this equation through by $C_V PV$ gives

$$\frac{C_P}{C_V} \frac{dV}{V} + \frac{dP}{P} = 0.$$

The specific heats are essentially constant (see Section 18.3), so that we can integrate this equation to obtain

$$\frac{C_P}{C_V} \ln V + \ln P = \ln(\text{constant}),$$

where we have chosen to call the constant of integration ln(constant). The ratio of specific heats, C_P/C_V, is commonly called γ, so that

$$\gamma \ln V + \ln P = \ln(\text{constant}).$$

Noting that $\gamma \ln V = \ln(V^\gamma)$, it follows by exponentiation that

$$PV^\gamma = \text{constant.} \quad \text{(adiabatic process)} \quad (18\text{--}17)$$

Equation 18–17 is the equation for an adiabatic path in the PV diagram of an ideal gas. For a particular adiabatic process starting from a given state (P_0, V_0), the constant in the equation is determined from the initial conditions:

$$\text{constant} = P_0 V_0^\gamma. \quad (18\text{--}18)$$

Because $C_P = C_V + R$, the ratio $\gamma = C_P/C_V$ is always greater than 1. Therefore, an adiabatic process with a given volume change results in a greater pressure change than would an isothermal process starting from the same initial conditions, for the isothermal process satisfies the condition $PV = \text{constant}$, while the adiabatic process satisfies $PV^\gamma = \text{constant}$. In the PV diagram, the adiabatic path is steeper than the isothermal path, as reflected in Fig. 18–13. Physically, the adiabatic path is steeper because the gas loses internal energy as it does work, so that its temperature drops.

It might seem that the equation $PV^\gamma = \text{constant}$ contradicts the ideal gas law, $PV = nRT$. But the ideal gas law does not imply $PV = \text{constant}$ except in an isothermal process, where T is constant. Since a reversible process cannot be both adiabatic (no heat addition) and isothermal (no temperature change), there is no contradiction. The adiabatic equation $PV^\gamma = \text{constant}$ was derived using the ideal gas law, and describes a situation in which the temperature changes. In fact, we can find the temperature change by using the ideal gas law to eliminate pressure from the adiabatic equation. We may write the adiabatic equation in the form

$$PV \cdot V^{\gamma-1} = \text{constant.}$$

But the ideal gas law gives $PV = nRT$, so that

$$nRTV^{\gamma-1} = \text{constant,}$$

or

$$TV^{\gamma-1} = \text{constant,} \quad (18\text{--}19)$$

where we have absorbed the constants n and R into a new definition of the constant on the right-hand side. Equation 18–19 shows explicitly that an adiabatic process is not isothermal, for the temperature must drop in an adiabatic expansion and rise in an adiabatic compression. We could have written the adiabatic equation in yet a third way, eliminating V in favor of T and P (see Problem 12).

Example 18–3

A Diesel Engine

Ignition of the fuel in a diesel engine occurs on contact with air heated by compression as the piston moves to the top of its stroke. (In this way a diesel differs from a gasoline engine, in which a spark ignites the fuel.) The compression occurs fast enough that very little heat flows out of the gas, so that the process may be considered adiabatic. If a temperature of 500°C is required for ignition, what must be the compression ratio (ratio of maximum to minimum cylinder volume) of the diesel engine? Air has a specific heat ratio γ of 1.4, and before compression its temperature is 20°C.

Solution

Equation 18–19 gives the relation between temperature and volume in an adiabatic process. Writing T_0 and V_0 for the temperature and volume at the bottom of the piston stroke, and T_1 and V_1 for the top of the stroke, Equation 18–19 becomes

$$T_1 V_1^{\gamma-1} = T_0 V_0^{\gamma-1}.$$

Solving for the compression ratio V_0/V_1 gives

$$\frac{V_0}{V_1} = \left(\frac{T_1}{T_0}\right)^{1/(\gamma-1)} = \left(\frac{773\ \text{K}}{293\ \text{K}}\right)^{1/0.4} = 11.$$

Practical diesel engines have considerably higher compression ratios to ensure reliable ignition. Conversely, compressional heating places an upper limit on the compression ratios of gasoline engines, where ignition by hot air would circumvent the carefully timed spark ignition system (see Problem 22).

Example 18–4
Adiabatic and Isothermal Expansion

Two identical gas-cylinder systems each contain 0.060 mol of ideal gas at 300 K and 2.0 atm pressure. The specific heat ratio γ of the gas is 1.4. The gas samples are allowed to expand, one adiabatically and one isothermally, until both are at 1.0 atm pressure. What are the final temperatures and volumes of each?

Solution

The initial volume V_0 of both samples may be obtained from the ideal gas law, $PV = nRT$. Solving for V gives

$$V_0 = \frac{nRT_0}{P_0} = \frac{(0.060\ \text{mol})(8.3\ \text{J/K·mol})(300\ \text{K})}{(2.0\ \text{atm})(1.0 \times 10^5\ \text{Pa/atm})}$$

$$= 7.5 \times 10^{-4}\ \text{m}^3 = 0.75\ \text{L}.$$

For the isothermal sample, T remains constant at 300 K. With constant temperature, $PV = $ constant, so that as the pressure drops in half, the volume doubles, becoming 1.5 L.

For the adiabatic expansion, we have

$$PV^\gamma = \text{constant} = P_0 V_0^\gamma.$$

Solving for V gives

$$V^\gamma = \frac{P_0 V_0^\gamma}{P},$$

or

$$V = \left(\frac{P_0}{P}\right)^{1/\gamma} V_0 = \left(\frac{2.0\ \text{atm}}{1.0\ \text{atm}}\right)^{1/1.4} (0.75\ \text{L}) = 1.2\ \text{L}.$$

We can obtain the temperature from Equation 18–19:

$$TV^{\gamma-1} = \text{constant} = T_0 V_0^{\gamma-1},$$

or

$$T = T_0 \left(\frac{V_0}{V}\right)^{\gamma-1} = (300\ \text{K})\left(\frac{0.75\ \text{L}}{1.2\ \text{L}}\right)^{0.4} = 250\ \text{K}.$$

Since we knew both P and V, we could equally well have determined the temperature from the ideal gas law.

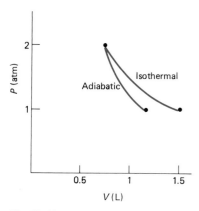

Fig. 18–14
PV diagram showing isothermal and adiabatic processes of Example 18–4.

In this example, both gas samples do work against the outside world as they expand. In the isothermal process, energy leaving the system as work is replaced by energy entering as heat, so that the temperature and internal energy remain constant. But heat cannot flow during the adiabatic process, so that the work done is at the expense of the internal energy of the gas. As a result, the gas temperature drops, and the volume does not increase as much. Figure 18–14 shows both processes on a PV diagram.

How much work is done during a reversible adiabatic process? For an adiabatic process, there is no heat transfer, so that $W = -U$. But earlier we found that $\Delta U = nC_V \Delta T$ for *any* process. Therefore, in an adiabatic process,

$$W = -nC_V \Delta T = nC_V(T_1 - T_2), \tag{18-20}$$

where T_1 and T_2 are the initial and final temperatures. We can express this result in terms of pressure and volume changes by solving the ideal gas law, $PV = nRT$, for temperature and using the result in Equation 18–20:

$$W = nC_V \frac{P_1 V_1 - P_2 V_2}{nR}.$$

But $R = C_P - C_V$, so

$$\frac{C_V}{R} = \frac{C_V}{C_P - C_V} = \frac{1}{(C_P/C_V) - 1} = \frac{1}{\gamma - 1}.$$

Then our expression for the adiabatic work becomes

$$W = \frac{P_1 V_1 - P_2 V_2}{\gamma - 1}. \tag{18-21}$$

We could also have obtained this result by direct integration of Equation 18–4 for an adiabatic process (see Problem 9).

Cyclic Processes

Many systems encountered in nature and in engineering applications undergo cyclic processes, in which the system returns periodically to the same thermodynamic state. Important examples include engines, refrigerators, and air compressors, whose mechanical construction ensures that a gas or fluid sample is eventually returned to the same state. Many natural oscillations, like those of a sound wave or a pulsating star, are very nearly cyclic.

Figure 18–15 shows a cyclic path $ABCDA$, in the PV diagram of an ideal gas. Starting from state A, we could traverse this path with our gas-cylinder system by

1. Heating the gas at constant pressure P_A until it expands to volume V_B. (We could maintain constant pressure by placing appropriate weights on the piston.)
2. Locking the piston in place and cooling the gas until its pressure drops to P_C.
3. Maintaining constant pressure P_C and continuing to cool until the gas volume drops to $V_D (= V_A)$.

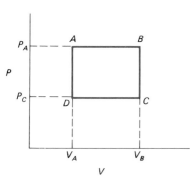

Fig. 18–15
A cyclic process.

4. Locking the piston and heating the system until its pressure again reaches P_A.

How much work does the gas do during the cyclic process $ABCDA$? During the constant-volume processes BC and DA, the piston is locked so that no work is done. During the isobaric expansion AB, the gas does work

$$W_{AB} = P\Delta V = P_A(V_B - V_A).$$

With a weighted piston, the work goes into raising the gravitational potential energy of the piston. In an engine, the work done by the gas would go into performing some useful task like lifting a weight, accelerating a vehicle, or overcoming friction. During the isobaric compression CD, work is done *on* the gas. (In our simple example, this work comes at the expense of the piston's gravitational energy.) Equivalently, the gas does negative work:

$$W_{CD} = P_C(V_D - V_C) = P_C(V_A - V_B),$$

where we have used the fact that $V_D = V_A$ and $V_C = V_B$. The net work done by the gas is then

$$W_{ABCDA} = W_{AB} + W_{CD} = P_A(V_B - V_A) + P_C(V_A - V_B) = (P_A - P_C)(V_B - V_A).$$

Geometrically, this result is just the area—measured not in cm^2 but in units of pressure times volume—enclosed by the rectangle $ABCD$. As we traverse the rectangle in the clockwise direction, the gas does this much work on its surroundings. Had we gone around in the counterclockwise direction, the area enclosed would have been the work done *on* the gas by the external world. Even when a cyclic path is not a simple rectangle, it remains true that the work done on or by the system is the area enclosed by the path in the PV diagram (Fig. 18–16).

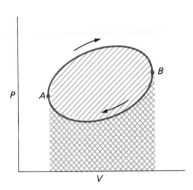

Fig. 18–16
An arbitrary cyclic path, traversed clockwise in the *PV* diagram. Area between the upper part of the curve and the *V*-axis (shaded with /////) is the work done *by* the gas as it goes from state *A* to state *B*. Area between the lower part of the curve and the *V*-axis (shaded also with \\\\\) is the work done *on* the gas as it returns from *B* to *A*. Net work done by the gas is the area enclosed by the path.

Example 18–5
A Cyclic Process

A sample of ideal gas occupies 4.0 L at 300 K and 1.0 atm pressure. The specific heat ratio of the gas is $\gamma = 1.4$. It is compressed adiabatically until its volume is 1.0 L, and then cooled at constant volume to 300 K. It is then allowed to expand isothermally to its original volume. How much work is done on the gas?

Solution

Figure 18–17 shows the cyclic path $ABCA$ in the PV diagram. The work done on the gas is the area enclosed by this curve, which may be calculated by considering the work done moving along each of the three sections AB, BC, and CA. The adiabatic work W_{AB} is given by Equation 18–21:

$$W_{AB} = \frac{P_A V_A - P_B V_B}{\gamma - 1}.$$

We know P_A, V_A, and V_B. To get P_B, we solve the adiabatic relation $P_B V_B{}^\gamma = P_A V_A{}^\gamma$ to obtain

$$P_B = P_A \left(\frac{V_A}{V_B}\right)^\gamma.$$

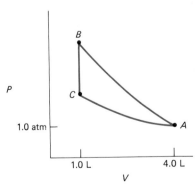

Fig. 18–17
The cyclic process *ABCA* of Example 18–5 includes adiabatic, constant-volume, and isothermal sections.

Then the adiabatic work is

$$W_{AB} = \frac{P_A[V_A - (V_A/V_B)^\gamma V_B]}{\gamma - 1}$$

$$= \frac{(1.0 \times 10^5 \text{ Pa})[4.0 \times 10^{-3} \text{ m}^3 - (4.0 \text{ L}/1.0 \text{ L})^{1.4}(1.0 \times 10^{-3} \text{ m}^3)]}{0.4}$$

$$= -7.4 \times 10^2 \text{ J}.$$

The minus sign indicates that work is done *on* the gas during this adiabatic compression.

No work is done during the constant-volume process BC, while the isothermal work W_{CA} is given by Equation 18–5:

$$W_{CA} = nRT \ln\left(\frac{V_A}{V_C}\right).$$

The quantity nRT is given by the ideal gas law:

$$nRT = PV,$$

where we can evaluate P and V at any point along the isotherm because it is a curve of constant temperature. We know these values at point A, so that the isothermal work becomes

$$W_{CA} = P_A V_A \ln\left(\frac{V_A}{V_C}\right) = (1.0 \times 10^5 \text{ Pa})(4.0 \times 10^{-3} \text{ m}^3) \ln\left(\frac{4.0 \text{ L}}{1.0 \text{ L}}\right) = 5.5 \times 10^2 \text{ J}.$$

The net work is then

$$W_{ABC} = W_{AB} + W_{BC} + W_{CA}$$

$$= -7.4 \times 10^2 \text{ J} + 0 \text{ J} + 5.5 \times 10^2 \text{ J} = -1.9 \times 10^2 \text{ J}.$$

The minus sign indicates that net work is done *on* the system.

Because the system returns to its original state, there can be no net change in its internal energy. Therefore, it must reject to its surroundings an amount of heat equal to the work done on it. When does this heat rejection occur? The compression AB is adiabatic, meaning that no heat flow takes place. During the isothermal expansion CA, the gas does work on its surroundings; since its temperature and internal energy do not change, it must therefore absorb heat from its surroundings. So the heat rejection occurs only during the constant-volume cooling period BC.

18.3

THE SPECIFIC HEATS OF AN IDEAL GAS

The first law of thermodynamics has given us a relation between the molar specific heats of an ideal gas at constant volume and constant presure:

$$C_P = C_V + R.$$

But what are the actual values of the specific heats? We need to know these in order to calculate the specific heat ratio, γ, that enters the adiabatic equations.

When we modeled an ideal gas in Chapter 17, we assumed that the gas was composed of structureless particles of zero size. The only energy associated with such particles is their kinetic energy of translation (linear motion), so that the internal energy U of the gas is simply the sum of the translational kinetic energies of all the particles. But we found that the average kinetic energy is directly proportional to the temperature:

$$\tfrac{1}{2}m\overline{v^2} = \tfrac{3}{2}kT.$$

If we have n moles of gas, the internal energy is then

$$U = nN_A(\tfrac{1}{2}m\overline{v^2}) = \tfrac{3}{2}nN_AkT,$$

where N_A is Avogadro's number. But $N_Ak = R$, the gas constant, so that

$$U = \tfrac{3}{2}nRT.$$

Using this result in Equation 18–10 for the molar specific heat gives

$$C_V = \frac{1}{n}\frac{dU}{dT} = \tfrac{3}{2}R. \qquad (18\text{--}22)$$

For this simple gas of structureless particles, the adiabatic exponent γ is then

$$\gamma = \frac{C_P}{C_V} = \frac{C_V + R}{C_V} = \frac{\tfrac{5}{2}R}{\tfrac{3}{2}R} = \frac{5}{3} = 1.67. \qquad (18\text{--}23)$$

Some gases, notably the inert gases helium (He), neon (Ne), argon (Ar), and others in the last column of the periodic table, behave as though their adiabatic exponents and specific heats were given correctly by Equations 18–23 and 18–22. But other gases do not. At room temperature, for example, hydrogen (H_2), oxygen (O_2), and nitrogen (N_2) obey adiabatic laws with γ very nearly $\tfrac{7}{5}$ ($=1.4$). Solving Equation 18–23 for C_V shows that for these gases

$$C_V = \frac{R}{\gamma - 1} = \frac{R}{\tfrac{7}{5} - 1} = \tfrac{5}{2}R.$$

On the other hand, sulphur dioxide (SO_2) and nitrogen dioxide (NO_2) have specific heat ratios close to 1.3 and therefore volume specific heats of about $3.4R$.

What is going on here? Has our ideal gas model failed? Not completely, for all these gases obey the relation $PV = nRT$ to a high degree of accuracy. But something is wrong in our model's assumptions about the internal energy of the gas. A clue lies in the structure of individual gas molecules, which is reflected in their chemical formulas. The inert gas molecules are monatomic—their molecules consist of single atoms. These atoms behave approximately like our model's structureless mass points.* Hydrogen, oxygen, and nitrogen molecules are diatomic. We can model these molecules by considering that they consist of two mass points at slightly different locations, as suggested in Fig. 18–18. Although a gas of such "dumbbell-shaped" molecules should still obey the ideal gas law $PV = nRT$, the molecules differ in one important respect from mass points. In addition to translational kinetic energy, they can have rotational energy as well. We broaden our notion of an ideal gas to include these molecules with internal

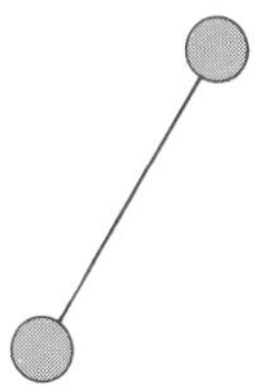

Fig. 18–18
"Dumbbell" model of a diatomic molecule.

---

*Of course, atoms do have structure. But that structure is not altered by collisions of gas molecules at normal temperatures, so that no change in internal energy is associated with atomic structure.

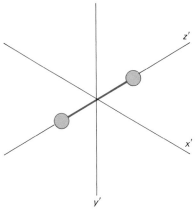

Fig. 18–19
A diatomic molecule can rotate about either of the two mutually perpendicular axes x' and y'.

structure. The "idealness" of the gas then lies not in the nature of the molecules but in the approximation that there is no potential energy of interaction among them.

How might a diatomic molecule rotate? Since the individual atoms behave approximately like mass points, rotation about the axis joining the atoms contributes little to the molecule's energy, for the rotational inertia about this axis is negligible. In a coordinate system x', y', z', whose z'-axis corresponds to the direction of the molecule axis at some instant, we can have energy associated with rotation about either of the other axes x' and y' (Fig. 18–19). Therefore, the total rotational energy of the molecule may be written

$$E_{rot} = \tfrac{1}{2}I_{x'}\omega_{x'}^2 + \tfrac{1}{2}I_{y'}\omega_{y'}^2,$$

where the I's are the rotational inertias about the two axes, and the ω's the corresponding angular velocities. Of course, the center of mass of the molecule may still be moving, so that there is also translational energy associated with each of the three possible directions of motion. Then the total energy of the molecule is

$$E = E_{trans} + E_{rot} = \tfrac{1}{2}mv_x^2 + \tfrac{1}{2}mv_y^2 + \tfrac{1}{2}mv_z^2 + \tfrac{1}{2}I_{x'}\omega_{x'}^2 + \tfrac{1}{2}I_{y'}\omega_{y'}^2.$$

The five terms in this equation correspond to the five independent kinds of motion our dumbbell molecule can have. The number of independent terms needed to describe the state of a system is called the number of *degrees of freedom* **degrees of freedom** of that system. A diatomic molecule has five degrees of freedom. A monatomic molecule, in contrast, has only three degrees of freedom, corresponding to the three independent components of translational motion.

The Equipartition Theorem

In Chapter 16, we showed that the average translational kinetic energy associated with the motion of a gas molecule in one direction is equal to $\tfrac{1}{2}kT$. We then argued that all three directions of motion are, on the average, equally likely, so that the total translational energy is, on the average, $\tfrac{3}{2}kT$. The argument from one direction to three is a statistical one, based on the notion that random collisions among the molecules will result in a "sharing" of energy among the three possible degrees of freedom. When a molecule can rotate as well as translate, we expect that collisions will result in energy sharing with the rotational motion as well. The more frequent and more violent the collisions, the more energy we will have in the rotation of molecules. We therefore expect rotational energy, like translational energy, to increase with temperature.

Just how much energy ends up in rotation? If we had a diatomic gas whose molecular motions were initially all translational, collisions would soon transfer energy to molecular rotations. If, on the other hand, the motions were initially nearly all rotational, collisions among the rotating molecules would transfer energy from rotation to translation. So we expect that in equilibrium the total energy will be shared among the various degrees of freedom. This qualitative expectation is expressed quantitatively in the *equipartition theorem* **equipartition theorem,** which states that

> When a system is in thermodynamic equilibrium, the average energy per molecule is $\frac{1}{2}kT$ for each degree of freedom.

We will not prove this theorem, which was originally derived by the Scottish physicist James Clerk Maxwell over 100 years ago, although our arguments above certainly suggest that it is reasonable. The equipartition theorem ensures that the internal energy of an ideal gas is still a function of temperature alone, even when the gas molecules have internal structure.

Using the equipartition theorem, we can now calculate the specific heat of a diatomic molecule as described by our dumbbell model. Each molecule has five degrees of freedom, three translational and two rotational. In equilibrium at temperature T, the average energy of a molecule is $5(\frac{1}{2}kT) = \frac{5}{2}kT$. If we have n moles of diatomic gas, the total internal energy is then

$$U = nN_A(\tfrac{5}{2}kT) = \tfrac{5}{2}nRT.$$

Equation 18–10 then gives the specific heat at constant volume:

$$C_V = \frac{1}{n}\frac{dU}{dT} = \tfrac{5}{2}R. \qquad \text{(diatomic molecule)} \qquad (18\text{–}24)$$

Our result $C_P = C_V + R$ still holds, since it was derived from the first law of thermodynamics without regard to molecular structure, so that we have

$$C_P = \tfrac{7}{2}R$$

and

$$\gamma = \frac{C_P}{C_V} = \frac{7}{5} = 1.4.$$

These results describe the observed behavior of diatomic gases like hydrogen, oxygen, and nitrogen at room temperature.

Example 18–6
The Equipartition Theorem

A gas mixture consists of 2.0 mol of oxygen (O_2) and 1.0 mol of argon (Ar). What is the volume specific heat of the mixture?

Solution

Being diatomic, each oxygen molecule has 5 independent components of motion. Then the equipartition theorem gives the total internal energy of all the oxygen molecules:

$$U_{O_2} = nN_A(\tfrac{5}{2}kT) = (2.0 \text{ mol})(\tfrac{5}{2}RT) = 5.0RT,$$

where we used $N_A k = R$. Similarly, the internal energy associated with all the monatomic argon molecules is

$$U_{Ar} = nN_A(\tfrac{3}{2}kT) = (1.0 \text{ mol})(\tfrac{3}{2}RT) = 1.5RT.$$

Then the total internal energy of the gas is

$$U = U_{O_2} + U_{Ar} = 6.5RT,$$

so that

$$C_V = \frac{1}{n}\frac{dU}{dT} = \frac{6.5R}{3.0 \text{ mol}} = 2.2R.$$

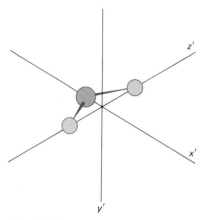

Fig. 18–20
A polyatomic molecule like NO₂ has three rotational degrees of freedom.

A polyatomic molecule like NO_2 (Fig. 18–20) has significant rotational inertias about any three perpendicular axes, and therefore exhibits three independent components of rotational motion. With the three independent components of translational motion, the molecule then has a total of six degrees of freedom. The equipartition theorem then tells us that the internal energy of n moles of such a gas is

$$U = (nN_A)(6)(\tfrac{1}{2}kT) = 3nRT,$$

so that

$$C_V = \frac{1}{n}\frac{dU}{dT} = 3R,$$

$$C_P = C_V + R = 4R,$$

and

$$\gamma = \frac{C_P}{C_V} = \frac{4}{3} = 1.33.$$

These values are reasonably close to the experimental values $\gamma = 1.29$ and $C_V = 3.47R$, although the agreement is noticeably poorer than for the monatomic and diatomic gases that we have considered.

Modifying the notion of an ideal gas to account for molecular structure with its additional degrees of freedom is an example of how remarkably successful the application of Newtonian mechanics can be. But at the molecular level, we are beginning to push the limits of Newtonian mechanics, and are approaching the realm where a quantum-mechanical description is necessary. We can see the breakdown of Newtonian mechanics in Fig. 18–21, which shows the volume specific heat of hydrogen (H_2) as a function of temperature. In the region from about 250 K to 750 K, the volume specific heat is very nearly equal to the value $\tfrac{5}{2}R$ predicted by our rotating dumbbell model. But between its melting point at 20 K and about 100 K, hydrogen shows a volume specific heat close to $\tfrac{3}{2}R$—like that of a mona-

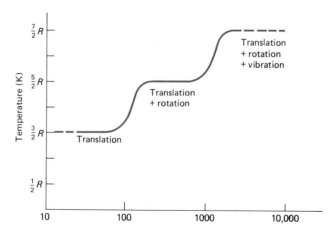

Fig. 18–21
Volume specific heat of H_2 as a function of temperature.

tomic gas. And above 3000 K* the specific heat is about $\frac{7}{2}R$, as though the molecules had somehow acquired two more degrees of freedom.

What is going on here? The molecules act as though they have different degrees of freedom, depending on temperature. In quantum mechanics, the energy associated with a periodic motion like a molecular rotation does not come in arbitrary amounts, but only in discrete multiples of some minimum amount. At low enough temperatures, the average thermal energy $\frac{1}{2}kT$ is less than this minimum, so that a molecule cannot rotate because it cannot acquire the minimum energy needed to do so. At these temperatures, it exhibits only the three translational degrees of freedom, and so the gas behaves as though it were composed of monatomic molecules. As temperature is increased, occasional molecules do acquire enough energy to rotate, and C_V increases. Eventually a plateau is reached, where the temperature and thermal energy are high enough that essentially all the molecules are rotating, and the gas obeys our rotating dumbbell model. That model assumed that the two atoms were joined by a rigid rod; in fact, they are joined by electrical forces and are free to vibrate like a mass-spring system along a line joining the two atoms (Fig. 18–22). At temperatures below about 800 K, the thermal energy is too low to excite these vibrations, but above that temperature more and more molecules are vibrating as well as rotating. There are two degrees of freedom associated with molecular vibrations—one with the kinetic energy of vibration, the other with the potential energy associated with the forces binding the atoms. As a result, each vibrating molecule has two more degrees of freedom. At high enough temperatures, essentially all the molecules are vibrating as well as rotating and translating, and the volume specific heat of $\frac{7}{2}R$ reflects the presence of all seven degrees of freedom. The presence of molecular vibrations is what made our calculation of γ and C_V for NO_2 rather imprecise.

Are you bothered by the strange restrictions quantum mechanics imposes on the rotation and vibration of molecules? You should be! Nothing in the physics you have studied until now, and nothing in your everyday experience, suggests that a rotating object cannot have any amount of energy you care to give it. But quantum mechanics deals with a realm of objects much smaller than those of our dialy experience. The quantization of energy levels is only one of many unusual things that occur in the quantum realm. We will explore more quantum phenomena in Chapter 36.

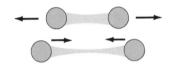

Fig. 18–22
The atoms of a diatomic molecule are free to vibrate along a line joining their centers.

―――――――
*At 3200 K the H_2 molecule dissociates into two atoms, so that the curve does not continue beyond this temperature.

―――――― **SUMMARY** ――――――

1. The **first law of thermodynamics** is a statement of the conservation of energy principle that explicitly includes the internal energy of a system. The first law says that the change in internal energy of a system is equal to the difference between the heat Q added to the system and the work W done by the system:

$$\Delta U = Q - W.$$

2. A **thermodynamic process** takes a system from one thermodynamic state to another. A **quasi-static process** occurs slowly enough that the system remains in thermodynamic equilibrium during the process. Consequently, a quasi-static process may be described by a path in a diagram relating thermodynamic variables of the system, such as pressure and volume. If the system can be made to retrace the same path, then the process

is also **reversible**. In a system like a gas in a cylinder, where work is done only by the expansion of the system, we can write

$$W = \int_{V_1}^{V_2} P\,dV \quad \text{(any process)}$$

for the work done by the system as it undergoes a thermodynamic process during which its volume changes from V_1 to V_2.

a. In an **isothermal process,** temperature does not change. When n moles of an ideal gas undergo a reversible isothermal process, the work done by the gas is

$$W = nRT \ln\left(\frac{V_2}{V_1}\right). \quad \text{(isothermal process)}$$

Since the internal energy of an ideal gas is a function of temperature alone, the internal energy does not change during an isothermal process, and therefore the system absorbs an amount of heat Q equal to the work it does on its environment.

b. In a **constant-volume process,** volume does not change, so the gas does no work, and therefore the change in internal energy is just equal to the heat added. For an ideal gas, the internal energy is a function of temperature alone, so the change in internal energy is

$$\Delta U = nC_V\Delta T, \quad \text{(any process)}$$

where C_V is the molar specific heat at constant volume. This relation between ΔU and ΔT is true for *any* process, whether constant-volume or not. In a constant-volume process, $\Delta U = Q$, so

$$Q = nC_V\Delta T. \quad \text{(constant-volume process)}$$

c. In an **isobaric process,** pressure does not change, and the work done is just

$$W = P\Delta V. \quad \text{(isobaric process)}$$

The heat Q gained by the system in an isobaric process may be calculated from the first law, $Q = \Delta U + W$, or, in terms of the molar specific heats at constant pressure and constant volume,

$$nC_P\Delta T = nC_V\Delta T + P\Delta V. \quad \text{(isobaric process)}$$

d. In an **adiabatic process,** no heat is transferred into or out of the system, so that any work done is at the expense of the system's internal energy:

$$\Delta U = -W. \quad \text{(adiabatic process)}$$

The pressure and volume of a gas undergoing a reversible adiabatic process are related by

$$PV^\gamma = \text{constant}, \quad \text{(adiabatic process)}$$

where $\gamma = C_P/C_V$ is the ratio of specific heats. The work done by the gas during the adiabatic process is

$$W = \frac{P_1V_1 - P_2V_2}{\gamma - 1}. \quad \text{(adiabatic process)}$$

e. In a **cyclic process,** the system returns to its original state, so that there is no change in its internal energy. The work done by the system is equal to the heat transferred to the system, and is given by the area enclosed by the cyclic path in the PV diagram.

3. The volume specific heat of an ideal gas is determined by the number of **degrees of freedom** available to the gas molecules. According to the **equipartition theorem,** the average energy associated with each degree of freedom in a gas at temperature T is $\frac{1}{2}kT$. In a gas of monatomic molecules, only the three degrees of freedom corresponding to the three directions of translational motion are present, and the volume specific heat is $C_V = \frac{3}{2}R$. For a gas of diatomic molecules undergoing rotation, there are two additional degrees of freedom associated with the two possible rotation axes, and the volume specific heat is $C_V = \frac{5}{2}R$. In general, the volume specific heat is $\frac{1}{2}R$ times the number of degrees of freedom of a gas molecule. It follows from the first law of thermodynamics that the specific heat at constant pressure, C_P, is given by

$$C_P = C_V + R.$$

4. At lower temperatures, the quantization of energy as described by quantum mechanics may prevent some degrees of freedom from absorbing any energy. As a result, the effective number of degrees of freedom, and with it the specific heats, may vary with temperature.

QUESTIONS

1. In Chapter 7, we wrote the conservation of energy principle in the form $K + U = \text{constant}$. In what way is the first law of thermodynamics a broader statement than that of Chapter 7?
2. The temperature of the water in a jar is raised by violently shaking the jar. Which of the terms Q and W in the first law of thermodynamics is involved in this case?
3. In Example 18–1, we considered the electric power leaving a power plant as part of the work done by the

power plant. Why did we include the electric power with the work rather than with the heat?
4. What is the difference between heat and internal energy?
5. Some water is tightly sealed in a perfectly insulated container. Is it possible to change the water temperature? Explain.
6. Is the internal energy of a van der Waals gas a function of temperature alone? Why or why not?
7. Are the initial and final equilibrium states of an irre-

versible process describable by points in a *PV* diagram? Explain.

8. Why can an irreversible process not be described by a path in a *PV* diagram?

9. Does the first law of thermodynamics apply to an irreversible process?

10. Is it possible to have a process that is isothermal but irreversible? Explain.

11. A quasi-static process begins and ends at the same temperature. Is the process necessarily isothermal?

12. Figure 18–23 shows two processes connecting the same initial and final states. For which process is more heat added to the system?

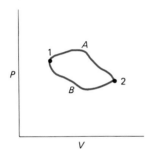

Fig. 18–23   Question 12

13. Two identical gas-cylinder systems are taken from the same initial state to the same final state, but by different processes. Is the work done in each case necessarily the same? The heat added? The change in internal energy?

14. When you let air out of a tire, the air seems cool. Why? What kind of process is occurring?

15. Blow on the back of your hand with your mouth wide open. Your breath will feel hot. Now tighten your lips into a small opening, and blow again. Now your breath feels cool. Why?

16. How is it posssible to have $PV^\gamma = $ constant in an adiabatic process and yet have $PV = nRT$?

17. Water is boiled in an open pan. Of which of the four specific processes we considered is this an example?

18. Is it possible to involve an ideal gas in a process that is simultaneously isothermal and isobaric? Is there any real substance that could undergo such a process? Give an example.

19. Three identical gas-cylinder systems expand from the same initial state to final states that have the same volume. One system expands isothermally, one adiabatically, and one isobarically. Which does the most work? Which does the least work?

20. If the relation $\Delta U = nC_V\Delta T$ holds for any ideal gas process, why do we call C_V the molar specific heat at constant volume?

21. Imagine a gas consisting of very complicated molecules, each with several hundred degrees of freedom. What, approximately, is the adiabatic exponent γ of this gas? Is it easy or hard to change the temperature of the gas?

22. In what sense can we consider a gas of diatomic molecules to be an ideal gas? After all, we must give up our assumption that gas particles are mass points in order to model such a gas successfully.

PROBLEMS

Section 18.1   *The First Law of Thermodynamics*

1. 1.0 kg of water in a closed but uninsulated container is shaken violently until its temperature rises by 2.0°C. The mechanical work required in the process is 14 kJ. How much heat is lost from the water during the process? How much mechanical work would have been required if the water were insulated perfectly in the container?

2. An 8.5-kg rock at 0°C is dropped into a well-insulated vat containing a mixture of ice and water at 0°C. When equilibrium is reached, it is found that there are 6.3 g less of ice in the vat. From what height was the rock dropped?

3. Water flows over Niagara Falls (height 50 m) at the rate of about 10^6 kg/s. Suppose that all the water is passed through a turbine connected to an electric generator, and that the generator supplies 400 MW of electric power. If the water has negligible kinetic energy after leaving the turbine, by how much does the temperature of the water increase between the top of the falls and the outlet of the turbine?

Section 18.2   *Thermodynamic Processes*

4. An ideal gas expands from the state (P_1,V_1) to the state (P_2,V_2), where $P_2 = 2P_1$ and $V_2 = 2V_1$. The expansion is quasi-static, and proceeds along a straight-line path, as shown in Fig. 18–24. How much work is done by the gas during this process?

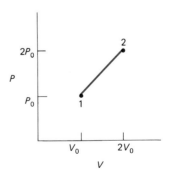

Fig. 18–24   Problem 4

5. Repeat the preceding problem for the case when the gas expands along a path given by

$$P = P_1\left[1 + \left(\frac{V - V_1}{V_1}\right)^2\right].$$

Sketch the path in the PV diagram, and calculate the work done.

6. What is the minimum amount of work needed to take the gas of Problem 4 between its initial and final states, if the gas pressure cannot be less than P_1 or more than P_2? What is the maximum amount of work that could be done under these same restrictions? Sketch the appropriate paths in the PV diagram.

7. A balloon contains 0.30 mol of helium at a temperature of 300 K and pressure of 1.0 atm. It is transported to a region where atmospheric pressure is 0.75 atm but where the temperature is still 300 K. How much work is done by the gas in the balloon? Assume that the process is isothermal.

8. Repeat the preceding problem, now assuming that the process is adiabatic. What should the final air temperature be if the gas is still to be in thermal equilibrium with its surroundings?

9. Show that the application of Equation 18–4 to an adiabatic process leads to Equation 18–21.

10. A gas sample expands isothermally from state A to state B, in the process absorbing 35 J of heat. It is then compressed isobarically to state C, for which its volume is the same as that of state A. During the compression, 22 J of work is done on the gas. The gas is then heated at constant volume until it returns to state A. (a) Draw a PV diagram showing this process. (b) How much work is done by or on the gas as it goes through the whole cyclic process? (c) How much heat is transferred to or from the gas as it goes from state B to state C and on to state A?

11. A piston-cylinder arrangement containing 0.30 mol of nitrogen at high pressure is in thermal equilibrium with an ice-water bath containing 200 g of ice. The pressure of the ambient air is 1.0 atm. The gas is allowed to expand isothermally until it is in pressure balance with its surroundings. After the process is complete, the bath contains 210 g of ice. What was the original pressure of the gas?

12. Derive an equation relating pressure and temperature for an adiabatic process.

13. 2.0 mol of an ideal gas with molar specific heat at constant volume $C_V = \frac{5}{2}R$ is initially at 300 K and 1.0 atm pressure. Determine the final temperature of the gas and the work done by the gas when 1.5 kJ of heat is added to the gas (a) isothermally, (b) isobarically, (c) at constant volume.

14. At any point in the PV diagram of an ideal gas, the slope of the adiabat passing through that point is γ times the slope of the isotherm passing through that point. Prove this.

15. Show that the work done by a van der Waals gas undergoing isothermal expansion from volume V_1 to

volume V_2 is given by

$$W = nRT \ln\left(\frac{V_2 - nb}{V_1 - nb}\right) + an^2\left(\frac{1}{V_2} - \frac{1}{V_1}\right).$$

Is the heat gained by the gas equal to the work done? *Hint:* Consult Section 17.1.

16. A bicycle pump consists of a cylinder whose internal dimensions are 20 cm long by 3.0 cm in diameter when the pump handle is all the way out. The pump contains air ($\gamma = 1.4$) at 20°C and 1.0 atm. If the outlet of the pump is blocked and the handle pushed quickly until the internal length of the pump cylinder is only 10 cm, by how much does the temperature of the air in the pump rise? Assume that no heat is lost during the compression, but that the process is quasi-static.

17. An ideal gas system is taken clockwise around a circular path in its PV diagram, eventually returning to its starting point. The minimum and maximum pressures during the process are 1.5 atm and 5.5 atm; the minimum and maximum volumes are 3.0 L and 11 L. How much work is done in the process?

18. If the system in the preceding problem consists of 1.3 mol of gas, what is the maximum temperature reached in the process? What are the pressure and volume at this hottest point?

19. A tightly sealed glass flask contains 5.0 L of air at 0°C and 1.0 atm pressure. How much heat is required to raise the temperature of the air to 20°C? The molar specific heat at constant volume for air is 2.5R.

20. A balloon contains 5.0 L of air at 0°C and 1.0 atm pressure. How much heat is required to raise the temperature of the air to 20°C, assuming that the gas stays in pressure equilibrium with its surroundings? The molar specific heat at constant volume for air is 2.5R. Ignore any force due to the stretching of the rubber balloon.

21. Figure 18–25 shows an apparatus used to study adiabatic free expansion, an irreversible process in which a gas is allowed to expand into a vacuum. The apparatus consists of two chambers of equal volume V connected by a valve. Both chambers are well insulated so that no heat flows into or out of the system. The left-hand chamber initially contains a gas at pressure P_0 and temperature T_0. The right-hand chamber is initially evacuated. When the valve is opened, the gas expands to fill both chambers. (a) Explain why this process is irreversible. (b) If the gas is ideal, what is its temperature after a new equilibrium has been

Fig. 18–25 Apparatus for adiabatic free expansion (Problem 21).

reached? (c) If the gas obeys the van der Waals equation, is the equilibrium temperature more or less than T_0?

22. A B21F Volvo engine has a compression ratio of 9.3 (ratio of maximum to minimum cylinder volume). If air at 320 K fills an engine cylinder when the piston is at the bottom of its stroke, what is the air temperature when the piston reaches the top of its stroke? The piston rises rapidly enough that there is negligible heat flow into the cylinder, so that the process is essentially adiabatic. The specific heat ratio γ for air is 1.4.

23. A piston-cylinder arrangement has an initial volume of 1.0 L and contains ideal gas at 1.0 atm pressure and a temperature of 300 K. The molar specific heat of the gas at constant volume is 2.5R. The piston is held fixed and the cylinder is placed in contact with a heat reservoir at 600 K. Once equilibrium is reached, the gas is allowed to expand isothermally. When the pressure reaches 1.0 atm, the gas is allowed to cool at constant pressure to 300 K. This cyclic process is repeated once a second. (a) Draw a PV diagram for the process. (b) What is the net rate at which work is done by the gas?

Section 18.3   *The Specific Heats of an Ideal Gas*

24. You have 2.0 mol of an ideal diatomic gas whose molecules can rotate but not vibrate. Suppose that somehow you arrange to give the gas 8.0 kJ of energy in such a way that it all goes initially into translational motion of the molecules. (a) When equilibrium is reached, what is the temperature of the gas? (b) Why is it not possible to ascribe a temperature to the state of the gas just after you added the energy?

25. A piston-cylinder system contains 0.50 mol of hydrogen (H_2) at 400 K and 3.0 atm pressure. An identical system contains 0.50 mol of helium (He) under the same conditions. Each gas undergoes an expansion that quadruples the system volume. Calculate the work done by each, if the expansion is (a) isothermal, (b) adiabatic.

26. A gas mixture contains 2.5 mol O_2 and 3.0 mol Ar at 350 K and 10 atm pressure. (a) What are the molar specific heats at constant volume and pressure for this mixture? (b) If the mixture expands adiabatically until its temperature is 300 K, what is its final volume?

27. An ideal gas in the apparatus of Problem 21 undergoes an irreversible free expansion that doubles its volume. Once equilibrium has been reached, the gas is compressed slowly back to its original volume. After this compression, the pressure is 1.3 times the original pressure before the free expansion. (a) Is the gas monatomic, diatomic, or polyatomic? (b) By what factor has the internal energy of the gas changed?

28. Figure 28–26 shows a simple model of a CO_2 molecule. Show that this molecule can undergo any of four independent vibrations. Determine the specific heat C_V under the assumption that all these vibrations, plus translation and rotation, are occurring.

Fig. 18–26 Model of a CO_2 molecule (Problem 28).

Supplementary Problems

29. A piston-cylinder arrangement is initially filled with 1.0 g of water at 50°C. The piston, which has negligible mass, is free to move and is exposed to atmospheric pressure (1.0 atm). The system is heated with a bunsen burner until the temperature inside the cylinder is 200°C. Assume that liquid water is incompressible and that steam is an ideal gas with volume specific heat 4.3R. How much work is done by the system, and how much heat is added to it?

30. A piston-cylinder system containing n moles of ideal gas is surrounded by air at fixed temperature T_0 and fixed pressure P_0. The piston has mass M and cross-sectional area A. If the piston is displaced slightly from its equilibrium position and then released, it will oscillate about its equilibrium position. Determine the angular frequency of oscillation, assumed to be of small amplitude, under the assumption (a) that the gas temperature remains constant at T_0; (b) that no heat flows into or out of the gas during the oscillation.

31. The piston of the preceding problem is connected to a massless spring whose spring constant is k. (a) Show that the angular frequency of adiabatic oscillations is now given by

$$\omega = \left(\frac{\gamma A^2 P_0^2 + knRT}{MnRT} \right)^{1/2}.$$

(b) Obtain an expression for the gas temperature as a function of time.

32. Under normal conditions, the temperature of earth's atmosphere near the ground decreases with altitude. The rate of temperature decrease is called the **lapse rate.** The decrease in pressure with altitude is such that a parcel of air that rises through the atmosphere and expands adiabatically cools at the rate of about 10°C per kilometer of rise. Show that if the lapse rate is less than 10°C/km, a rising parcel of air will end up less dense than its surroundings and thus will continue to rise. (Assume that the air rises adiabatically, so that it does not exchange heat with its surroundings.) In contrast, show that if the lapse rate is greater than 10°C/km, a rising parcel of air will be denser than its surroundings. In the former case, the air is unstable, in that a parcel of air that rises will continue to do so. In the latter case, the air is stable, and a parcel of air that is displaced will return to its original position. Unstable air results in vertical mixing of air, and the formation of typical "fair weather"

clouds. Stable conditions result in the trapping of air in layers, and can be especially serious in cities with significant air pollution.

33. A cylinder of cross-sectional area A is closed by a massless piston. The cylinder contains n mol of air ($\gamma = 1.4$), initially in temperature and pressure equilibrium with the surrounding air at T_0 and P_0. The piston is initially at height h_1 above the bottom of the cylinder. Sand is gradually sprinkled onto the piston, so that it moves downward quasi-statically, reaching a final height h_2. What is the total mass of the sand if (a) the process is isothermal? (b) the process is adiabatic?

19

THE SECOND LAW
OF THERMODYNAMICS

The first law of thermodynamics tells us that heat and other forms of energy are equivalent. Much of our world works because of this equivalence. A car runs because heat released by burning gasoline gets converted to mechanical work in the car's engine. Most of the electrical energy that operates our lights and motors and computers originates in heat that is released in the burning of coal, oil, or gas, or in the fissioning of uranium nuclei (Fig. 19–1). Our own bodily movements and even our thoughts make use of energy that was once released as heat deep in the core of the sun. But the first law does not tell the whole story. Heat and mechanical energy are not exactly the same thing, and it is the difference between them that makes the conversion of heat to work a more subtle task than the first law alone would imply.

Fig. 19–1
Strip mining coal. Much of our electrical energy comes from burning coal—itself a form of chemically stored solar energy.

19.1
REVERSIBILITY AND IRREVERSIBILITY

Suppose we make a movie of a ball bouncing off the floor. If we show the movie (Fig. 19–2a), we see the ball coming in from the left, bouncing, and going off to the right at the same angle and nearly the same speed as it came in. If we run the movie backward (Fig. 19–2b), it still makes good sense. Now let's film another simple physical event. Give a block of wood a push along a table. The block slows down and stops, and we find that both the block and the table top are slightly warmer (Fig. 19–3a). Show this movie in reverse (Fig. 19–3b) and it makes no sense. We never see a block at rest on a table suddenly start to move, while the block and table become cooler. Yet energy could be conserved in such a process, so the first law of thermodynamics would be satisfied.

Break an egg into a bowl and start beating; the yolk and white quickly blend into scrambled egg. You would be very surprised if, on reversing the

481

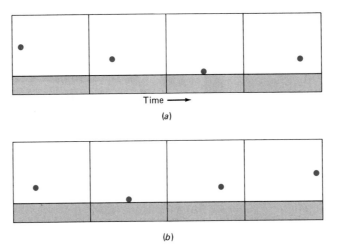

Fig. 19–2
A movie of a bouncing ball makes sense whether it is shown forward or backward.

direction of your beater, the scrambled yolk and white separated—even though there is nothing in the laws of mechanics to prevent it. Or put glasses of hot and cold water in thermal contact. The hot water gets cooler and the cold water gets hotter. The reverse never occurs—even though energy could be conserved if it did.

irreversible   Most happenings in the world are **irreversible,** in the sense of the block, the egg, and the water. Even the bouncing ball is not perfectly reversible, for it rebounds with slightly lower speed, leaving the ball and court warmer. What is the origin of this irreversibility? In each irreversible happening, we start with matter in an organized state. In our sliding block, all the molecules initially have velocity components in the same direction. The egg is initially organized so that all the yolk molecules are in one re-

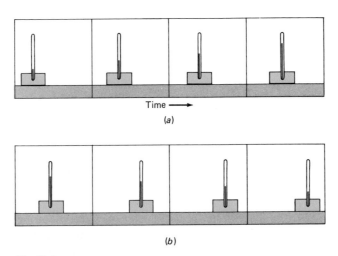

Fig. 19–3
Movie of a block warming as it dissipates kinetic energy through friction. Shown in reverse, it depicts an event that would not happen, even though it does not violate energy conservation.

gion, the white molecules in another. And a greater number of energetic molecules are initially in the hotter glass of water. Of all the possible states into which matter could arrange itself, these *organized* states are very special and relatively rare. There are many more *chaotic* states—for instance, all the possible arrangements of molecules in a scrambled egg—that display less overall organization. If we have a system in a relatively organized state, chances are that it will evolve to a less organized one, simply because there are far more such states available to it. Once a system is in a disorganized state, it is unlikely to assume spontaneously a more organized state. When we are dealing with macroscopic systems of many molecules, the number of less organized states is fantastically larger than the number of more organized states, so that the probability of spontaneously achieving a more organized state is impossibly low.

A key word in this discussion is spontaneous. We could restore some system to its original, organized state—for example, by putting one of our water glasses in the refrigerator and the other on the stove—but to do so we must carry out a rather deliberate and energy-consuming process.

The sense of irreversibility we have been discussing is fully consistent with the notion of an irreversible thermodynamic process that we introduced in the preceding chapter. In an irreversible process, we deliberately force a system into a state of thermodynamic disequilibrium. For example, we push suddenly on the piston of a gas cylinder, temporarily giving the molecules near the piston a common velocity component away from the piston. This temporary state shows more organization than the equilibrium state, for some gas molecules share a common motion. But molecular collisions soon randomize these motions, so that the gas ends up in a new equilibrium state. A spontaneous return to the original state is highly improbable, and in this sense the process is irreversible.

Irreversibility is a probabilistic notion. Events that could occur without violating the principles of Newtonian physics nevertheless do not occur because they are simply too improbable. We cannot make a car move forward by asking all its molecules suddenly to move in the same direction—even though there is more than enough energy in random thermal motion to give the car a substantial velocity. There is a phenomenal amount of energy in the thermal motion of water molecules in the oceans, but we cannot easily extract useful work by cooling the oceans (see Problem 3 and Section 19.3). The extreme improbability that molecules will spontaneously assume an ordered state makes a great deal of the world's energy unavailable to do work.

19.2
THE SECOND LAW OF THERMODYNAMICS

Heat Engines

heat engine   Although we cannot spontaneously convert into useful work *all* the random motion associated with the internal energy of a hot object, we can construct devices, called **heat engines,** that extract at least *some* useful work from internal energy. Examples include gasoline and diesel engines, fossil-fueled and nuclear power plants, and jet aircraft engines.

working fluid

second law of
thermodynamics,
Kelvin-Planck statement

To make a heat engine, we need a source of heat and a **working fluid**—a substance that undergoes changes of thermodynamic state and in the process does work on its surroundings. Conceptually, we can describe a heat engine by a diagram showing how heat flows into the engine and how work emerges. Figure 19–4a shows such an energy flow diagram for a "perfect" heat engine—one that extracts energy as heat from a reservoir and converts it all to work. Such an engine would do exactly what we argued against in the previous section—it would convert the random energy of thermal motion entirely into the ordered motion associated with mechanical work. In fact, a perfect heat engine is impossible, for the same reason that we cannot unscramble an egg or cause a wood block suddenly to accelerate from rest at the expense of its internal energy. This observational fact comprises one statement of the **second law of thermodynamics:**

> It is impossible to construct a heat engine operating in a cycle that extracts heat from a reservoir and delivers an equal amount of work.

This particular form of the second law was formulated by Lord Kelvin and Max Planck. The phase "in a cycle" refers to the fact that an engine must take on a series of mechanical and thermodynamic configurations that repeat themselves if it is to operate continuously. Typical engines have shafts that rotate, or pistons that go back and forth, so that the engine runs in a cyclic fashion.

One way to convert heat to work is to put a gas-cylinder system in contact with a heat reservoir. If the gas pressure is initially higher than the pressure associated with the piston weight and the external atmosphere, then the gas will expand and do work W on the piston. In this isothermal process, the gas simultaneously extracts heat $Q = W$ from the heat reservoir. Eventually the gas expands to reach pressure equilibrium, and the piston must then be returned to its original position if it is to do more work.

To return the piston, we could simply push it back. But this would take as much work as we extracted during the expansion, so that our engine would do no net work. Instead, we can cool the gas to reduce its volume, then complete the cycle by heating at constant volume until the gas is back to its initial state. But to cool the gas, we transfer heat to a reservoir at a lower temperature than our hot reservoir, so that some energy leaves the system as heat rather than work. The energy flow diagram for such a real heat engine is shown in Fig. 19–4b. A real engine extracts heat from a high-temperature reservoir and produces mechanical work. But the amount of work is always less than the heat extracted, and the difference between the two leaves the engine as heat rejected to a lower-temperature reservoir. The radiator in a car (Fig. 19–5) has the specific job of rejecting heat to the environment—heat that comes from burning gasoline but that does not contribute to the motion of the car. A large power plant may use a major river as its low-temperature reservoir, dumping enormous amounts of heat and possibly altering the ecology of the river (Fig. 19–6).

The second law of thermodynamics states that we cannot build a perfect heat engine. But how close can we come? Since we must pay for the fuel that heats our high-temperature reservoir, it is clearly advantageous to minimize the heat rejected to the low-temperature reservoir. We define the

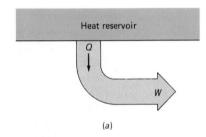

(a)

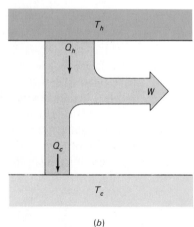

(b)

Fig. 19–4
(a) Energy flow diagram for a perfect heat engine. The engine extracts heat Q from a reservoir and delivers an equal amount of work. (b) Energy flow diagram for a real heat engine. Only a fraction of the heat Q_h extracted from the high-temperature reservoir is delivered as work: the remainder is rejected to the low-temperature reservoir.

Fig. 19–5
A car radiator's job is to reject heat to the environment.

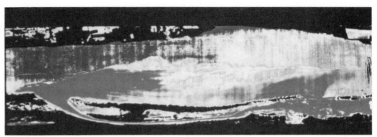

Fig. 19–6
Infrared aerial photo shows hot water discharge from a power plant.

efficiency

efficiency of an engine as the ratio of work, W, put out by the engine in one cycle to heat, Q_h, extracted from the high-temperature reservoir:

$$e = \frac{W}{Q_h}.$$

Since the process is cyclic, there is no net change in internal energy of the working fluid over one cycle, so that the first law of thermodynamics assures us that the work done is the difference between the heat Q_h extracted from the high-temperature reservoir and the heat Q_c rejected to the low-temperature reservoir. Then the efficiency can be written

$$e = \frac{Q_h - Q_c}{Q_h} = 1 - \frac{Q_c}{Q_h}. \tag{19–1}$$

Figure 19–7 shows a simple heat engine whose efficiency we can readily calculate. The engine consists of a closed cylinder containing an ideal gas. A movable piston is connected to a rod that drives a wheel. The wheel does mechanical work, such as lifting a weight or accelerating a vehicle. To operate the engine, we start with the piston in its most retracted position and place the cylinder in contact with the high-temperature reservoir at T_h. The gas expands isothermally along the path AB in its PV diagram (Fig. 19–8). In the process, the gas extracts heat Q_h from the reservoir and, because its internal energy does not change, delivers an equal amount of work to the wheel. At state B, we remove the cylinder from the reservoir so that the gas continues to expand, but now adiabatically. We design the engine so that, when the piston reaches its maximum displacement, the gas temperature has cooled to T_c, the temperature of the cool reservoir. At this

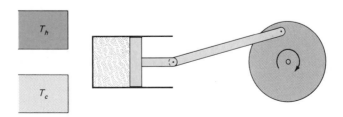

Fig. 19–7
A simple heat engine.

point the gas is in state C in its PV diagram. We then place the piston in contact with the cool reservoir. The inertia of the wheel causes it to continue turning, and the wheel does work on the gas, compressing it isothermally from state C to state D. This work ends up as heat rejected to the cool reservoir. Finally, at state D, we separate the cylinder from the reservoir and allow the compression to continue adiabatically until the gas temperature has risen to T_h and the piston is once again fully retracted.

The engine we have just described undergoes a cyclic process consisting of four reversible steps, two of them isothermal and two adiabatic. This cycle is called a **Carnot cycle,** and the engine a **Carnot engine,** after the French engineer Sadi Carnot (1796–1832), who explored the properties of such an engine even before the first law of thermodynamics was formulated. The particular configuration of our engine is not important, nor is the use of an ideal gas as working fluid necessary. What distinguishes the Carnot cycle from others is the sequence of thermodynamic processes and the fact that these processes are reversible. The Carnot engine is an example of a **reversible engine**—one in which thermodynamic equilibrium is maintained so that all steps could, in principle, be reversed.

What is the efficiency of a Carnot engine? In the preceding chapter we examined isothermal processes, and found that the heat absorbed during such a process is given by

$$Q = W = nRT \ln\left(\frac{V_B}{V_A}\right),$$

where V_A and V_B are the initial and final volumes. Then the heat Q_h absorbed during the isothermal expansion AB in the Carnot cycle of Fig. 19–8 is

$$Q_h = nRT_h \ln\left(\frac{V_B}{V_A}\right), \qquad (19\text{-}2)$$

while the heat Q_c rejected during the isothermal compression CD is

$$Q_c = -nRT_c \ln\left(\frac{V_D}{V_C}\right) = nRT_c \ln\left(\frac{V_C}{V_D}\right). \qquad (19\text{-}3)$$

We put the minus sign in Equation 19–3 because our statement of the first law describes Q as the heat *absorbed*, while Equation 19–1 for the engine efficiency requires that Q_c be the heat *rejected* by the engine. To calculate engine efficiency according to Equation 19–1, we need the ratio Q_c/Q_h:

$$\frac{Q_c}{Q_h} = \frac{T_c \ln(V_C/V_D)}{T_h \ln(V_B/V_A)}. \qquad (19\text{-}4)$$

This expression can be simplified by using Equation 18–19 that relates temperature and volume in an adiabatic process:

$$TV^{\gamma-1} = T_0 V_0{}^{\gamma-1}.$$

Applying this expression to the adiabatic processes BC and DA in the Carnot cycle gives

$$T_h V_B{}^{\gamma-1} = T_c V_C{}^{\gamma-1}$$

and

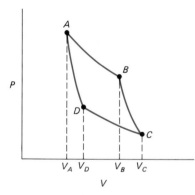

Fig. 19–8
PV diagram for the engine of Fig. 19–7.

$$T_h V_A{}^{\gamma-1} = T_c V_D{}^{\gamma-1}.$$

Dividing these two equations, we obtain

$$\left(\frac{V_B}{V_A}\right)^{\gamma-1} = \left(\frac{V_C}{V_D}\right)^{\gamma-1} \quad \text{or} \quad \frac{V_B}{V_A} = \frac{V_C}{V_D}$$

so that Equation 19–4 becomes simply

$$\frac{Q_c}{Q_h} = \frac{T_c}{T_h}.$$

The efficiency of the engine, given by Equation 19–1, is then

$$e = 1 - \frac{Q_c}{Q_h} = 1 - \frac{T_c}{T_h}, \tag{19–5}$$

where the temperatures are measured on an absolute scale (Kelvin or Rankine). Equation 19–5 tells us that the efficiency of a Carnot engine depends only on the highest and lowest temperatures of the working fluid. For a practical engine, the low temperature is usually the ambient temperature of the environment. Then to maximize the efficiency, we must make the high temperature as high as possible. Most real engines represent a compromise between efficiency and the ability of materials to withstand high temperatures and pressures.

Example 19–1
A Carnot Engine

A Carnot engine extracts 240 J of heat from a high-temperature reservoir during each cycle. It rejects 100 J of heat to a reservoir at 15°C. How much work does the engine do in one cycle? What is its efficiency? What is the temperature of the hot reservoir?

Solution

The first law of thermodynamics requires that energy not rejected as heat be delivered as work, so that the engine does

$$W = 240 \text{ J} - 100 \text{ J} = 140 \text{ J}$$

of work. The efficiency is defined by Equation 19–1 as the ratio of work to heat extracted from the hot reservoir:

$$e = \frac{W}{Q_h} = \frac{140 \text{ J}}{240 \text{ J}} = 0.583 = 58.3\%.$$

Knowing the efficiency, we can solve Equation 19–5 for the high temperature to get

$$T_h = \frac{T_c}{1-e} = \frac{288 \text{ K}}{1-0.583} = 691 \text{ K} = 418°C.$$

Note that in using Equation 19–5, we must work with absolute temperatures.

Engines, Refrigerators, and the Second Law

What is so special about the Carnot cycle? The efficiency of a Carnot engine is limited by the operating temperatures, but wouldn't it be possible for us to build another kind of engine that would operate at greater efficiency? The answer is no. The Carnot engine is reversible, in that all its thermodynamic steps are reversible. An engine that includes irreversible processes—whether they involve friction or deviations from thermodynamic equilibrium—will be less efficient because organized molecular motion will be lost through the irreversible processes. This idea is elaborated in **Carnot's theorem:**

Carnot's theorem

> All reversible engines operating between the same temperatures have the same efficiency (given by Equation 19–5), and no irreversible engine can be as efficient.

refrigerator

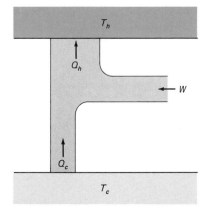

To prove Carnot's theorem, we introduce the notion of a **refrigerator.** A refrigerator is the opposite of an engine: it extracts heat from a cool reservoir and rejects it to a hotter reservoir, taking in work in the process. Figure 19–9 shows an energy flow diagram for a refrigerator. A refrigerator is an unusual device: it forces energy to flow in a way it doesn't spontaneously go. With a refrigerator, we can move heat from a glass of cold water into a glass of hot water, making the cold water colder and the hot water hotter. But to do so, we must do work—a household refrigerator uses electricity. Heat flow from colder to hotter does not happen spontaneously. This observational fact is reflected in another statement of the second law of thermodynamics, this one due to the German physicist Rudolph Clausius (1822–1888):

> It is impossible to make a refrigerator, operating in a cycle, whose sole effect is the transfer of heat from a cooler object to a hotter object.

Fig. 19–9
Energy flow diagram for a real refrigerator.

The Clausius statement rules out a "perfect" refrigerator, like that shown conceptually in Fig. 19–10.

How are the Clausius statement of the second law and the Kelvin-Planck statement related? The Clausius statement says that it is impossible to build a perfect refrigerator; the Kelvin-Planck statement says that it is impossible to build a perfect heat engine. Suppose the Clausius statement were false. Then we could build the device of Fig. 19–11, consisting of a real heat engine and a perfect refrigerator. In each cycle the engine would extract, say, 100 J from the hot reservoir, put out 60 J of useful work, and reject 40 J to the cool reservoir. We could arrange for the perfect refrigerator to transfer 40 J from the cool reservoir back to the hot reservoir. The net effect would be to extract 60 J of heat from the hot reservoir and convert it entirely into work. Conceptually, the composite device looks like Fig. 19–12, and would be a perfect heat engine. Therefore, if it is possible to build a perfect refrigerator, it is also possible to build a perfect heat engine. A similar argument can be made to show that if a perfect heat engine is possible, then so is a perfect refrigerator (see Problem 9). Thus the Clausius and Kelvin-Planck statements of the second law of thermodynamics are equivalent, in that if one is false then so is the other.

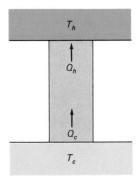

Fig. 19–10
Energy flow diagram for a perfect refrigerator. The Clausius statement of the second law rules out such a device.

How would we build a refrigerator? Because our Carnot engine is reversible, we could do work on the engine, traversing Fig. 19–8 in a counterclockwise direction. All thermodynamic processes would reverse, and the device would transfer heat from the cool reservoir to the hot reservoir.

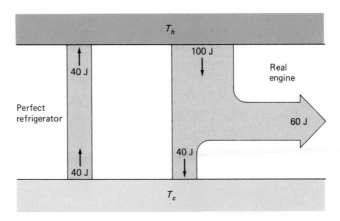

Fig. 19–11
Real engine combined with a perfect refrigerator.

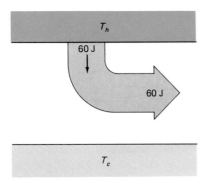

Fig. 19–12
The combination of Fig. 19–11 would be equivalent to a perfect heat engine.

Although real refrigerators are not designed exactly like engines, the two are, in principle, interchangeable.

With our two statements of the second law of thermodynamics, we are now ready to prove Carnot's assertion that the maximum efficiency of a heat engine is given by Equation 19–5. Consider again the reversible engine we showed in Fig. 19–11, which extracts 100 J from its hot reservoir and delivers 60 J of work. This engine is 60 per cent efficient. Since it is reversible, we could equally well run the engine as a refrigerator, supplying 60 J of work and transferring an additional 40 J of heat from the cool reservoir to the hot reservoir. Suppose we had another engine, this one with 70 per cent efficiency, operating between the same two reservoirs. Suppose we run the first engine as a refrigerator and the second as an engine, as suggested in Fig. 19–13. The net result is that we extract 10 J of heat from the cool reservoir and deliver 10 J of work, as shown in Fig. 19–14. The composite device acts like a perfect heat engine, violating the second law of thermodynamics. This same argument could be made for any combination of a

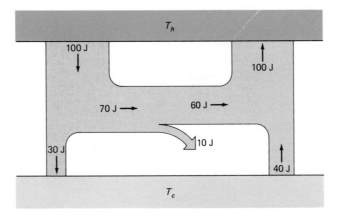

Fig. 19–13
60 per cent efficient reversible engine run as a refrigerator, along with a hypothetical engine of 70 per cent efficiency.

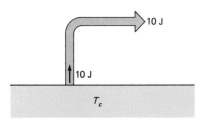

Fig. 19–14
The composite device of Fig. 19–13 would be equivalent to this perfect heat engine.

reversible engine and another, more efficient engine. Therefore, it is impossible to make a heat engine that is more efficient than any reversible heat engine. Since the Carnot engine is reversible, its efficiency is the maximum possible, and must be the same as that of any other reversible engine.

19.3
APPLICATIONS OF THE SECOND LAW OF THERMODYNAMICS

The world abounds with thermal energy, but the second law of thermodynamics imposes a fundamental limitation on our ability to turn that energy to our own uses. Any device we construct that involves the interchange of heat and work is a heat engine or refrigerator, and therefore subject to the second law.

Limitations on Heat Engines

Most of the electricity used in the United States is produced in large power plants that are basically heat engines driven by chemical reactions involving fossil fuels like coal, oil, or natural gas, or by nuclear reactions involving the fissioning of uranium nuclei (Fig. 19–15). Figure 19–16 shows a schematic diagram of such a power plant. The working fluid is water, heated in a boiler and converted to steam at high pressure. The steam expands adiabatically against the blades of a turbine—a fanlike device that spins when struck by the steam. The turbine is coupled mechanically to a generator that converts mechanical work into electrical energy (the principle behind such a generator is described in Chapter 29). Since the flow of electrical energy does not involve a temperature difference, the electrical energy leaving the power plant qualifies as work. Leaving the turbine, the steam is still in the gaseous state, and is usually hotter than

Fig. 19–15
A nuclear power plant showing two reactor containment vessels at left, and large cooling towers in the background.

Fig. 19–16
Schematic diagram of a typical electric power plant.

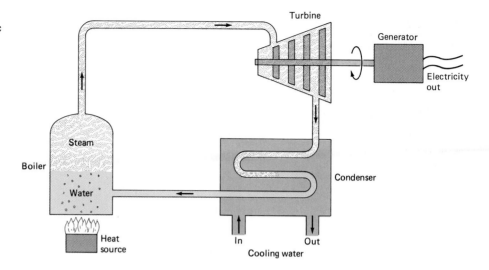

condenser

the water supplied to the boiler. Here is where the second law enters! Had the water returned to its original state, we would have extracted as work all the energy acquired in the boiler, in violation of the second law. To use the steam again, we run it through a **condenser**—a device where pipes carrying the steam are in contact with large volumes of cool water, typically from a river, lake, or ocean. The condensed steam, now cool water, is fed back into the boiler to repeat the cycle.

The maximum steam temperature in a power plant is limited by the materials used in its construction. For a fossil-fueled plant, current technology permits high temperatures around 650 K. Potential damage to nuclear fuel rods limits the temperature in a nuclear plant to around 570 K. Cooling-water temperature averages about 40°C (310 K), so that maximum possible efficiencies for these power plants, given by Equation 19–5, are

$$e_{fossil} = 1 - \frac{310 \text{ K}}{650 \text{ K}} = 0.52 = 52\%$$

and

$$e_{nuclear} = 1 - \frac{310 \text{ K}}{570 \text{ K}} = 0.46 = 46\%.$$

Any temperature difference between the exhausted steam and the cooling water, mechanical friction, and the need to divert energy for driving pumps and pollution control devices all reduce efficiency further, to about 40 per cent for fossil-fueled plants and 34 per cent for nuclear plants. This means that, when we make electricity, roughly two-thirds of the energy released from our expensive fuels ends up as waste heat.

Where does the waste energy go? Most of it is extracted by the condenser, whose cooling water constitutes the cool reservoir of the heat engine. A typical large power plant produces 1000 MW of electricity, so that another 2000 MW of waste heat is dumped into the cooling water. This rate of energy addition can cause a large temperature rise in even a major river,

Fig. 19–17
Fog forms above the cooling towers of this 1500-MW coal-fired power plant operated by Pennsylvania Power and Light Company.

and can lead to serious ecological problems. The need for power plant cooling water—imposed on us by the second law of thermodynamics—is so great that a substantial fraction of all rainwater falling on the United States eventually finds its way through the condensers of power plants (see Problem 23).

To reduce problems from this "thermal pollution," many modern power plants employ cooling towers—expensive devices in which waste heat is released to the air rather than to water. Environmental effects of the heated air are less serious than with heated water, although the increase in humidity can bring about unfortunate changes in local weather (Fig. 19–17).

Example 19–2
A Power Plant

The Vermont Yankee nuclear power plant at Vernon, Vermont, produces 540 MW of electric power. Within the plant's nuclear reactor, energy from nuclear fission is released as heat at the rate of 1590 MW. Steam produced in the reactor enters the turbine at a temperature of 556 K, and is discharged to the condenser at 313 K. Water from the Connecticut River is pumped through the condenser at the rate of 2.27×10^4 kg/s.

What is the maximum efficiency of the power plant, as limited by the second law of thermodynamics? What is the actual efficiency? How much does the temperature of the cooling water rise? If the minimum flow in the river is 3.40×10^4 kg/s, what is the maximum temperature rise of the entire river? How many houses like that of Example 16–8 could be heated with the waste heat from Vermont Yankee?

Solution

The second-law efficiency is given by Equation 19–5:

$$e = 1 - \frac{T_c}{T_h} = 1 - \frac{313 \text{ K}}{556 \text{ K}} = 0.437 = 43.7\%.$$

The actual efficiency is the ratio of electric power output to the rate of heat extraction at the high temperature:

$$e = \frac{dW/dt}{dQ_h/dt} = \frac{540 \text{ MW}}{1590 \text{ MW}} = 0.340 = 34.0\%.$$

Waste heat is discharged to the river at a rate given by the difference between the thermal power output and the electric power output:

$$\frac{dQ_c}{dt} = \frac{dQ_h}{dt} - \frac{dW}{dt} = 1590 \text{ MW} - 540 \text{ MW} = 1050 \text{ MW}.$$

We can relate the waste power output to the temperature rise of the cooling water by differentiating Equation 16–7 with respect to time:

$$\frac{dQ_c}{dt} = \frac{dm}{dt} c\Delta T,$$

where dm/dt is the mass flow rate and ΔT the fixed temperature rise. Solving for ΔT gives

$$\Delta T = \frac{dQ_c/dt}{c \, dm/dt} = \frac{1050 \times 10^6 \text{ W}}{(4184 \text{ J/kg·°C})(2.27 \times 10^4 \text{ kg/s})} = 11.1°C.$$

Eventually, the hot-water discharge mixes with the rest of the river. The result is the same as if the entire river had been used for cooling, so that

$$\Delta T_{\text{river}} = \frac{dQ_c/dt}{c\,(dm/dt)_{\text{river}}} = \frac{1050 \times 10^6 \text{ W}}{(4184 \text{ J/kg·°C})(3.40 \times 10^4 \text{ kg/s})} = 7.4\text{°C}.$$

The house of Example 16–8 uses 8110 Btu/h, or 2.38×10^{-3} MW. Then the waste heat from Vermont Yankee could heat $(1050 \text{ MW})/(2.38 \times 10^{-3} \text{ MW/house})$, or 440,000 houses. This example is unrealistic because real houses are considerably less energy-efficient than the house of Example 16–8. Nevertheless, waste heat from power plants is a potentially valuable energy source for heating buildings.

Gasoline and diesel engines used in vehicles provide another pervasive example of heat engines in our society. A typical automobile engine operates at a high temperature of 800 K, and rejects waste heat at the prevailing temperature of the environment, about 300 K. The maximum possible thermodynamic efficiency is then

$$e = 1 - \frac{300 \text{ K}}{800 \text{ K}} = 0.63 = 63\%,$$

but irreversible thermodynamic processes make the actual efficiency much lower. Mechanical friction consumes additional energy, so that less than 20 per cent of the energy released by burning gasoline is available at the driving wheels. When the car is not accelerating or climbing, all of this energy is then converted to heat through friction with air and road.

We would not be quite so concerned with the efficiency of engines if we did not have to pay for fuel. Even a small temperature difference between two regions can drive a heat engine. Surface temperatures in tropical oceans, for example, are about 25°C (298 K). Three hundred meters down, the ocean temperature has dropped to about 5°C (278 K). A heat engine operating between these temperatures would have an efficiency of only $1 - 278 \text{ K}/298 \text{ K} = 0.07$ or 7 per cent. Nevertheless, there are large amounts of warm water in the ocean, and its energy—ultimately solar in origin—is free. Although the engineering problems of such an **ocean thermal energy conversion** scheme (OTEC) are substantial, research on this method of power generation is currently underway. Figure 19–18 shows one proposal for an OTEC power plant.

Another approach to a fuel-less heat engine is shown in Fig. 19–19. This **solar thermal power plant** uses a large field of computer-controlled mirrors to focus sunlight on a central boiler mounted high on a tower. Pilot plants of this sort are currently operating in the desert southwest of the United States.

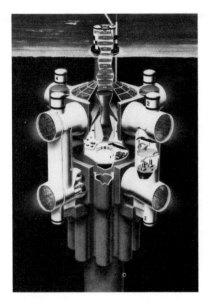

Fig. 19–18
Artist's conception of an OTEC power plant.

ocean thermal energy
conversion

solar thermal
power plant

Fig. 19–19
10 MW solar thermal power plant at Barstow, California.

Refrigerators and Heat Pumps

Reversing a heat engine gives us a refrigerator. We do mechanical work on the device, and as a result transfer energy from a cooler to a hotter reservoir (Fig. 19–20). Your home refrigerator uses electricity to transfer energy from cool things in the refrigerator to the kitchen environment. If you put your hand near the coils of the refrigerator, you can feel heat rejected to the air. By the first law of thermodynamics, this heat comprises both the energy removed from the refrigerator's contents as well as the energy that was supplied as work. In designing a refrigerator, we would like to minimize the amount of work needed to extract a given amount of heat.

We characterize a refrigerator by the ratio of total heat extracted at the lower temperature to work input to the refrigerator. This ratio is called the **coefficient of performance** (COP):

coefficient of performance

$$COP = \frac{Q_c}{W} = \frac{Q_c}{Q_h - Q_c}. \qquad (19\text{--}6)$$

For a reversible heat engine, we found in deriving Equation 19–5 that the heat ratio Q_c/Q_h is equal to the temperature ratio T_c/T_h. The best refrigerator we can build is a reversible one, so that for this refrigerator the heat and temperature ratios are also equal. Then Equation 19–6 becomes

$$COP = \frac{T_c}{T_h - T_c}. \qquad (19\text{--}7)$$

As the high and low temperatures encountered in a refrigerator become arbitrarily close, Equation 19–7 shows that the COP becomes large—meaning that the refrigerator requires relatively little work to do its job. But if we wish to effect a transfer of heat between two widely separated temperatures, then the COP drops and the amount of work we must do becomes considerable. In the limit of very large T_h, the COP approaches zero, indicating that we are simply converting mechanical work into a nearly equal amount of heat, and transferring very little additional heat from the cool reservoir. In operating a real refrigerator, we would like to minimize the work needed. But Equation 19–7—like Equation 19–5 for a heat engine—imposes a fundamental limitation on that minimum amount of work.

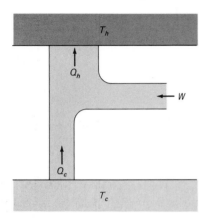

Fig. 19–20
Energy flow diagram for a refrigerator.

Example 19–3
A Refrigerator

A typical home freezer operates between a low temperature of 0°F (−18°C) and a high of 86°F (30°C). What is the maximum possible COP of this refrigerator? With this COP, how much electrical energy would the refrigerator require to freeze 500 g of water, initially at 0°C?

Solution

The COP is given by Equation 19–7:

$$COP = \frac{T_c}{T_h - T_c} = \frac{255 \text{ K}}{303 \text{ K} - 255 \text{ K}} = 5.3.$$

To produce 500 g of ice takes

$$Q_c = mL_f = (0.50 \text{ kg})(334 \text{ kJ/kg}) = 170 \text{ kJ},$$

where we obtained the heat of fusion from Table 17–1. This Q_c is the heat that must be removed from the low-temperature end of the refrigerator. The COP is the ratio of the heat removed to the work done, so that

$$W = \frac{Q_c}{COP} = \frac{170 \text{ kJ}}{5.3} = 32 \text{ kJ}.$$

In a real refrigerator, the COP would be lower, and the work correspondingly higher, because of irreversible processes in the refrigerator cycle.

Fig. 19–21
A heat pump. During the summer, the device acts as a refrigerator that cools the interior of the house and dumps heat into the environment. During the winter, its operation reverses, so that it effectively cools the outdoors and in the process transfers heat into the house. The work, *W*, is typically in the form of electricity.

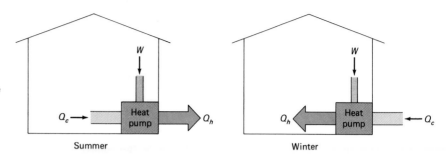

Refrigerators are not confined to kitchens. An air conditioner is a refrigerator designed to cool an entire room or building. A heat pump is a refrigerator that cools a building in the summer and heats it in the winter (Fig. 19–21). Heat pumps are widely used in the southern United States, where in winter they pump energy from the cool outside air to the warm house. A heat pump requires electricity, but it supplies more energy as heat than it uses in electricity. The excess energy comes from the outdoor environment, which becomes slightly cooler. Heat pumps operating from the constant 10°C temperatures found two meters below the ground have been used for years in the Scandinavian countries, and are becoming more popular in the northern United States.

Example 19–4
A Heat Pump

A heat pump extracts energy from the ground at 10°C and transfers it to water at 70°C. The water is circulated to heat a building. What is the COP of the heat pump? What is the electrical power consumption of the device if it supplies heat to the building at the rate of 20 kW? Would you be better off with a heat pump or an oil furnace if electricity costs 10 cents/kWh and oil costs 83¢/gallon? (At typical oil burner efficiencies, a gallon of oil releases about 100,000 Btu of useful heat when burned in a home furnace.)

Solution

The COP of the heat pump is given by Equation 19–7:

$$\text{COP} = \frac{T_c}{T_h - T_c} = \frac{283 \text{ K}}{343 \text{ K} - 283 \text{ K}} = 4.7.$$

For every joule of electricity used, the heat pump transfers an additional 4.7 joules from the ground to the house, giving a total of 5.7 joules to the house. If it supplies heat at the rate of 20 kW, the electrical power needed is then

$$P_{\text{electric}} = \frac{20 \text{ kW}}{5.7} = 3.5 \text{ kW}.$$

At 10¢/kWh, it costs 35¢ to run the heat pump for an hour. From Appendix D, we find that 20 kW is equivalent to 68,000 Btu/h, so that we must burn oil at the rate of

$$\frac{68,000 \text{ Btu/h}}{100,000 \text{ Btu/gal}} = 0.68 \text{ gal/h}.$$

At 83¢/gal, the oil costs 56¢/h.

Although the heat pump in this example is far less expensive, two factors make the comparison unrealistic. First, real heat pumps operate at a COP closer to half their theoretical maximum, approximately doubling the cost over that of this example. Second, the initial installation of a heat pump is very expensive, so that economic considerations may still favor oil. As heat pump technology advances, though, these devices will come into wider use.

If we are concerned about energy and not just cost, we should also consider the source of electricity to run the heat pump. If this is a thermal power plant operating at an efficiency of, say, ⅓, then three times as much energy is used at the power plant as is supplied to the heat pump. The overall efficiency of the heat pump and power plant may actually be less than that of the oil furnace (see Problem 22).

19.4
THE THERMODYNAMIC TEMPERATURE SCALE

When we discussed temperature measurement in Chapter 16, we encountered the problem that no single method works at all temperatures. We settled on the gas thermometer as the best laboratory standard, but even a helium gas thermometer becomes useless at temperatures below about 1 K.

The operation of a reversible heat engine provides a way of measuring temperature that is independent of the particular device used and that works at any temperature. We can take such an engine—or any thermodynamic system—through a complete cycle of reversible processes. Measuring the heats Q_c and Q_h, or one of these and the work output, allows us to calculate the efficiency and from it the temperature ratio. If one temperature is known, we can then determine the other. Although this may seem a rather obscure way to measure temperature, it is actually used in certain experiments at very low temperature. And it provides an absolute way of defining temperature that can, in principle, be used at any temperature. The temperature scale so defined is called the **thermodynamic temperature scale.**

thermodynamic temperature scale

absolute zero

The zero of the thermodynamic scale—**absolute zero**—is absolute in that it represents a state of maximum order. This is reflected in the fact that a heat engine rejecting heat at absolute zero would operate at 100 per cent efficiency, as suggested by Equation 19–5 with $T_c = 0$. Unfortunately, there are no heat reservoirs at absolute zero, and it can be shown that it is impossible to cool anything to this temperature in a finite number of steps. The statement that it is impossible to reach absolute zero in a finite number of steps is called the **third law of thermodynamics.**

third law of thermodynamics

19.5
ENTROPY AND THE QUALITY OF ENERGY

If offered a joule of energy, would you rather have that joule delivered in the form of mechanical work, heat transferred from an object at 1000 K, or heat transferred from an object at 300 K? Your answer might depend on what you wanted to do with the energy. If you wanted to lift or accelerate a mass, then you would be smart to take your energy as work. But if you want to keep warm, then heat from the 300 K object would be perfectly acceptable.

If, on the other hand, you weren't sure what you wanted to do with the energy, then which should you take? The second law of thermodynamics makes the answer clear: you should take the work. Why? Because you could use it directly to give one joule of mechanical energy—kinetic or potential—to a mass, or you could, through friction or other irreversible processes, use it to raise the temperature of something.

If you chose heat from the 300 K object as the form for your joule of energy, then you could supply a full joule only to objects cooler than 300 K. You could not do mechanical work unless you used your joule of 300 K heat to run a heat engine. With its maximum temperature only a little above ambient, your engine would be very inefficient, and you could only extract a small fraction of a joule of mechanical energy. Nor could you supply your energy to anything hotter than 300 K, unless you ran a refrigerator—and that would take additional work over and above your one joule. You would be better off with one joule of energy in the form of 1000 K heat, for you could transfer it to anything cooler than 1000 K, or could run a heat engine to produce a substantial fraction of a joule of mechanical energy.

Taking your one joule of energy in the form of work gives you the most options. Anything you can do with a joule of energy, you can do with the work. Energy taken as heat is less versatile, with the joule of 300 K energy the least useful of the three. We are not talking here about the quantity of energy—we have exactly one joule in each case—but about the **quality of energy,** indicated by the ability to do a variety of useful tasks. Schematically, we can describe energy quality on a diagram like Fig. 19–22. In Fig. 19–22, we list electricity with mechanical energy because electrical energy flows without requiring a temperature difference, and thus qualifies as work. We can readily convert an entire amount of energy from higher to lower quality, but the second law of thermodynamics prevents us from going in the opposite direction with 100 per cent efficiency.

The notion of energy quality has important implications for our efficient use of energy, suggesting that we try to match the quality of available energy sources to our energy needs. Figure 19–23 shows a breakdown of United States energy use by quality; it makes clear that much of our energy demand is for relatively low-quality heat. Although we can do anything we want with a high-quality energy source like flowing water or electricity, it makes sense to put our high-quality energy to high-quality uses, and use lower-quality sources to meet other needs. For example, we often heat water with electricity—a great convenience to the homeowner, but a thermodynamic folly, for in so doing we convert the highest-quality form of energy into low-grade heat. If our electricity comes from a thermal power plant, we already threw away two-thirds of the energy from fuel as waste heat. It makes little sense to run an elaborate and inefficient heat engine only to have its high-quality output of electricity used for low-temperature heating. On the other hand, if we heat our water by burning a fuel directly at the water heater, then we can, in principle, transfer all the fuel's energy to the water. Another approach to matching energy quality with energy needs is to combine the production of steam for heating with the generation of elec-

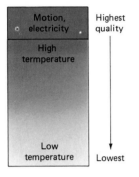

Fig. 19–22
Energy quality.

quality of energy

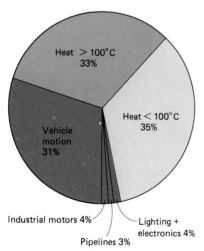

Fig. 19–23
Energy use in the United States, organized by energy quality.

cogeneration

tricity. Popular in Europe for many years, this idea of **cogeneration** is gaining popularity in the United States (Fig. 19–24).

Entropy

How can we make more precise the notion of energy quality? Consider the system shown in Fig. 19–25. It consists of two identical objects, initially at different temperatures and thermally insulated from each other and from the environment. The two objects are then brought into thermal contact, so that they eventually reach equilibrium at the same temperature.

What has changed between the initial and final states of our system? Not the energy, for we did no work on the system, and lost no heat, so that the total energy remained constant. But we have lost the ability to do useful work. In the initial state, we could have run a heat engine between the hot and cool objects. In the final state, there is no temperature difference and therefore no way to run a heat engine. The quality of the energy has decreased. Can we find some system property that reflects this change in energy quality?

Figure 19–26 shows another example of a system whose energy quality deteriorates without any loss of total energy. We start with a gas confined by a removable partition to one side of an insulated container; the other side is evacuated. We then remove the partition and allow the gas to expand freely to fill the whole container. Since the process is adiabatic, and no work is done, the total energy does not change. But we lose the ability to extract some of that energy as useful work. Starting with the initial state, we could have opened a small hole and used the escaping gas to run a turbine (Fig. 19–27). In the final state, there is no way to run the turbine, for there is no pressure difference across which the gas will flow. Again, can we find a quantity that describes this decrease in energy quality?

To find such a quantity, we consider an ideal gas undergoing a Carnot cycle (Fig. 19–28). In deriving the efficiency of such a cycle, we found that

$$\frac{Q_c}{Q_h} = \frac{T_c}{T_h}, \tag{19-8}$$

where Q_c is defined as the heat *rejected from* the system to the low-temperature reservoir at T_c, and Q_h is the heat *added to* the system from the high-temperature reservoir at T_h.

We now change our convention on the meaning of Q_c, taking it also to mean heat *added to* our system. Changing this convention changes the sign of Q_c, for positive heat rejected from our system is the same as negative heat added to it. With this change in convention, Equation 19–8 may be written

Fig. 19–24
This cogenerating unit at Middlebury College provides 600 kW of electricity from high-pressure steam produced in the college's heating plant. After passing through a turbine that drives the electric generator, the steam is used to heat buildings.

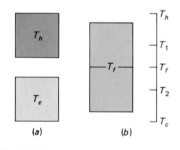

Fig. 19–25
Two identical objects, initially at different temperatures, are allowed to come to thermal equilibrium.

Fig. 19–26
Adiabatic free expansion. No energy leaves the system during the process, but energy quality deteriorates.

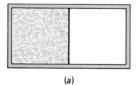
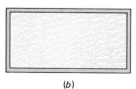

(a)   (b)

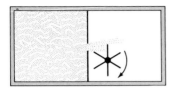

Fig. 19-27
A means of extracting useful work from the gas in the initial state of Fig. 19-26.

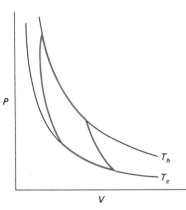

Fig. 19-28
Carnot cycle for an ideal gas.

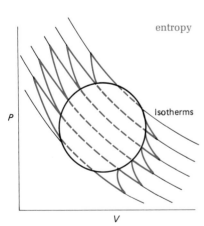

Fig. 19-29
An arbitrary cycle may be approximated by a sequence of adiabatic and isothermal steps.

$$\frac{Q_c}{T_c} + \frac{Q_h}{T_h} = 0,$$

or

$$\sum \frac{Q}{T} = 0. \tag{19-9}$$

Although this result follows specifically from consideration of a Carnot cycle consisting of two adiabatic and two isothermal steps, we can generalize it to any reversible cycle. Any such cycle can be approximated arbitrarily closely by a sequence of adiabatic and isothermal steps (Fig. 19-29). One way to move around our reversible cycle is to take in the entire sequence of Carnot cycles shown in Fig. 19-29. But Equation 19-9 holds for each of these Carnot cycles, so that we must have

$$\sum \frac{Q}{T} = 0 \tag{19-10}$$

over the entire set of Carnot cycles. Although the Q's in this expression are heats added in the isothermal sections of the individual Carnot cycles, those parts of a Carnot cycle that traverse the interior of our closed path are crossed twice, in opposite directions (Fig. 19-30), so that their heats cancel. Then Equation 19-10 applies equally well to the jagged approximation to our closed path. In the limit as we approximate the smooth curve ever closer by more and more adiabatic and isothermal segments, Equation 19-10 becomes

$$\oint \frac{dQ}{T} = 0, \tag{19-11}$$

where the circle on the integral sign indicates that we integrate over all the heat transfers dQ along a *closed* path.

Equation 19-11 tells us that there is some quantity—a small amount of which is given by dQ/T—that does not change when we traverse a closed path, bringing our system back to its original state. We call this quantity the **entropy**, S, and write

$$dS = \frac{dQ}{T} \tag{19-12}$$

and

$$S_2 - S_1 = \int_1^2 dS = \int_1^2 \frac{dQ}{T}, \tag{19-13}$$

where the limits on the integral refer to two thermodynamic states and S_1 and S_2 are the entropies of those states. Note that we are not concerned with the actual value of the entropy, but only with the *change* in entropy from one state to another. In this way, entropy is like potential energy, for which only changes are meaningful.*

---

*Actually, the zero of entropy is meaningful, in that zero entropy for any system corresponds to a temperature of absolute zero.

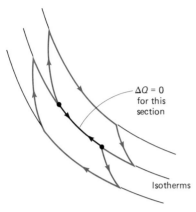

Fig. 19–30
Heat transfers cancel as two adjacent Carnot cycles are traversed across the interior of the original cycle.

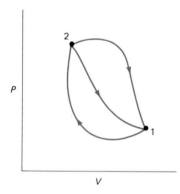

Fig. 19–31
Entropy change is zero when the system returns to its original state, so the entropy change from state 2 to state 1 must be independent of the path taken.

Example 19–5
Entropy

An object of mass m and specific heat c is taken from an initial temperature T_1 to a final temperature T_2. What is the change in its entropy?

Solution

The entropy change is given by Equation 19–13. To evaluate the integral in this equation, we must relate dQ and T. We can do so because we know the specific heat, so that

$$dQ = mcdT.$$

Then Equation 19–13 is

$$\Delta S = \int_{T_1}^{T_2} \frac{dQ}{T} = \int_{T_1}^{T_2} mc\, \frac{dT}{T} = mc\, \ln\left(\frac{T_2}{T_1}\right). \tag{19–14}$$

What good is this concept of entropy? Equation 19–11 shows that there is no entropy change when we take a system from some initial state and eventually return it to that state. Now suppose we take a system part way around a closed path, from state 1 to state 2 as shown in Fig. 19–31. The entropy change from state 2 back to state 1 must be exactly opposite the change from 1 to 2, so that the change going around the closed path is zero. But this must be true no matter how we get from state 2 to state 1, as suggested by the two possible paths in Fig. 19–31. That is, the entropy change given by Equation 19–13 is independent of path; it is a property only of the initial and final states themselves. Like pressure, temperature, and volume, entropy is a thermodynamic state variable—a quantity that characterizes a given state independently of how a system got into that state.

Equation 19–13 for calculating the entropy difference between two states is meaningful only for reversible processes, for the temperature T is defined only in thermodynamic equilibrium, while an irreversible process takes a system temporarily out of equilibrium. But because entropy is a property of a thermodynamic state, and not of how a system got into that state, entropy changes during an irreversible process can be calculated by finding a reversible process that takes a system between the same two states, then calculating the entropy change for that process using Equation 19–13.

At the beginning of this section, we considered some irreversible processes in which energy quality deteriorated although total energy remained constant. We can come to understand the meaning of entropy by calculating entropy changes for these processes.

Example 19–6
Adiabatic Free Expansion

An ideal gas, initially confined to one side of an insulated container, is allowed to expand freely to fill the entire container (Fig. 19–32). What is the entropy change for this process?

Solution

Since the free expansion is adiabatic and involves no work, the internal energy and temperature of the ideal gas are unchanged. But the gas volume doubles. To

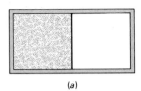
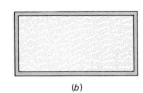

Fig. 19–32
Adiabatic free expansion.

(a)   (b)

find the entropy change for the irreversible free expansion, we can consider a reversible process taking the system between the same two states. Since the initial and final temperatures are the same, such a process is an isothermal expansion. The heat added during such an expansion is given by Equation 18–6:

$$Q = nRT \ln\left(\frac{V_2}{V_1}\right).$$

Since the temperature is constant during the reversible isothermal expansion, the entropy change of Equation 19–13 is given simply by

$$S_2 - S_1 = \frac{Q}{T} = nR \ln\left(\frac{V_2}{V_1}\right). \tag{19–15}$$

Since the final volume V_2 is larger than the initial volume, the entropy change is positive.

In calculating the entropy change, we used the heat Q that would be required during a reversible isothermal process. Of course, no heat is transferred during the actual adiabatic free expansion—but the entropy change is the same as for the reversible isothermal process, which is why we calculated the entropy change as though such a process had occurred.

Example 19–7
Irreversible Heat Transfer

An object at a high temperature T_h is placed in contact with another object at a lower temperature T_c. What is the change in the entropy of the system when the two have reached equilibrium?

Solution

Again, we have an irreversible process. But we can calculate the entropy change by considering any process that takes the two objects from their initial temperatures to the same final temperature T_f. For example, we cool the hot object by placing it in contact with a heat reservoir initially at T_h and gradually lowering its temperature to T_f. The entropy change is given by

$$\Delta S_h = \int_{T_h}^{T_f} \frac{dQ}{T}.$$

Although we could evaluate this integral, let us for the moment note only that there is some average temperature T_1 *that lies between* T_h and T_f, for which

$$\Delta S_h = \frac{Q_h}{T_1},$$

where Q_h is the heat absorbed by the hot object. Q_h is, in fact, negative since the hot object loses heat as it cools. Similarly, the entropy change of the cool object is

$$\Delta S_c = \int_{T_c}^{T_f} \frac{dQ}{T} = \frac{Q_c}{T_2},$$

where T_2 is some average temperature *that lies between* T_c and T_f, and Q_c is the heat absorbed by the cool object.

The total entropy change of the system is

$$\Delta S = \Delta S_c + \Delta S_h = \frac{Q_c}{T_2} + \frac{Q_h}{T_1}.$$

Because the two objects are insulated from the outside world, any heat lost by the hot object in the actual irreversible process is gained by the cool object. Therefore, $Q_h = -Q_c$, and the total entropy change becomes

$$\Delta S = \frac{Q_c}{T_2} - \frac{Q_c}{T_1} = Q_c\left(\frac{1}{T_2} - \frac{1}{T_1}\right).$$

Now whatever the value of T_1, it lies *above* the equilibrium temperature T_f. Similarly, T_2 lies *below* T_f, so that $T_2 < T_1$. Furthermore, Q_c is positive, since the cool object absorbs heat. Therefore, the first term in our expression for ΔS is greater than the second term, so that the entropy change is positive.

Problem 35 explores this example more quantitatively, but the significant conclusion remains: entropy increases during an irreversible heat transfer.

The systems in Examples 19–6 and 19–7 were closed systems, in that they did not exchange energy or matter with their environments. For each system, entropy increased as the system underwent an irreversible process. At the same time, energy that was originally available to do useful work became unavailable. This correlation between an increase in entropy and a decrease in energy quality can be made more quantitative. Consider the free expansion of Example 19–6. Had we let the gas expand isothermally but reversibly, it could have done an amount of work given by Equation 18–5:

$$W = nRT \ln\left(\frac{V_2}{V_1}\right). \tag{19–16}$$

After the irreversible free expansion, the gas could no longer do this work, even though its energy content was unchanged. During the expansion, its entropy increased by an amount given by Equation 19–15:

$$\Delta S = nR \ln\left(\frac{V_2}{V_1}\right).$$

Comparing this entropy change with the energy that became unavailable to do work (Equation 19–16), we have

$$E_{\text{unavailable}} = T\Delta S. \tag{19–17}$$

Equation 19–17 is an example of a more general relation between entropy and the quality or availability of energy:

entropy and the quality of energy

> During an irreversible process in which the entropy of a system increases by ΔS, an amount of energy $E = T_{\min}\Delta S$ becomes unavailable to do work. $T_{\min}$ is the temperature of the coolest reservoir available to the system.

This statement shows that entropy provides our desired measure of energy quality. Given two systems that are identical in composition and energy content, the system with the lowest entropy contains the highest-quality energy. An increase in entropy always corresponds to a degradation in energy quality, in that some energy becomes unavailable to do work. Problem 36 explores this energy degradation quantitatively.

Entropy and the Second Law

At the beginning of this chapter, we argued that processes in nature always occur irreversibly, going from ordered states to disordered states. It is this loss of order in thermodynamic systems—as when a hot and a cold object eventually reach the same temperature—that is responsible for energy becoming unavailable to do work. Entropy is a measure of this disorder. Given the tendency of systems to evolve toward disordered states, we can make a statement about entropy that is, in fact, a general statement of the second law of thermodynamics:

second law of
thermodynamics

> The entropy of a closed system can never decrease.

At best, the entropy of a closed system remains constant—and this happens only in an ideal, reversible process. If anything irreversible occurs—a slight amount of friction or a deviation from exact thermodynamic equilibrium—then entropy increases. There is no going back. As entropy increases, some energy becomes unavailable to do work, and nothing within the closed system can restore that energy to its original quality. This statement of the second law in terms of entropy is equivalent to our previous statements about the impossibility of perfect heat engines and refrigerators, for the operation of either device would require a decrease in entropy.

What about a system that is not closed? Can't we decrease its entropy? Yes—but only by supplying high-quality energy from outside. When we run a refrigerator, we decrease the entropy of the refrigerator's contents and its surroundings (Fig. 19–33). At the same time, however, we must do work on the refrigerator to effect this entropy decrease—to make heat flow in the direction it doesn't normally go. If we enlarge our closed system to include the power plant or whatever else supplies the refrigerator with work, we would find that the entropy of this new closed system does not decrease. The entropy decrease at the refrigerator is offset by entropy increases elsewhere in the system (Fig. 19–34). If irreversible processes occur anywhere in the system, then there is a net entropy increase.

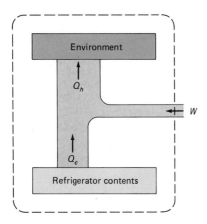

Fig. 19–33
The entropy decrease of the refrigerator's contents can occur only when work is done on the system from outside.

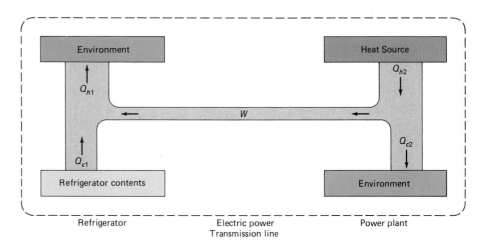

Fig. 19–34
When the source of work for the refrigerator is included in the system, then entropy of the entire system can at best remain constant.

Whenever we encounter a system whose entropy seems to decrease—that gets more rather than less organized—we do not have a closed system. If we enlarge the boundaries of our system to encompass anything that might exchange energy with it, then we will always find that the enlarged system undergoes an entropy increase. Ultimately, we can enlarge our system to include the entire universe. When we do, we have the ultimate statement of the second law of thermodynamics:

second law of
thermodynamics

> The entropy of the universe can never decrease.

As last examples of this broad statement, consider the growth of a living thing from the random mix of molecules in its environment, or the construction of a skyscraper from materials that were originally dispersed about the earth, or the appearance of ordered symbols on a printed page from what was a uniformly mixed bottle of ink. All these are processes in which matter goes from near chaos to a highly organized state—akin to separating yolk and white from a scrambled egg. They are certainly processes in which entropy decreases. But earth is not a closed system. It gets high-quality energy from the sun, energy that is ultimately responsible for life and all its actions. If we consider the earth-sun system, then the entropy decrease associated with life and civilization on earth is more than balanced by the entropy increase associated with nuclear fusion inside the sun. We living things represent a remarkable phenomenon—the organization of matter in a universe governed by a tendency toward disorder. But we do not escape the second law of thermodynamics. Our highly organized selves and society, and the entropy decreases they represent, come into being only at the expense of greater entropy increases elsewhere in the universe.

SUMMARY

1. Irreversible processes are those that result in a decrease in organization or order. Friction is an irreversible process that converts directed kinetic energy into the random kinetic energy of molecular motions. Thermodynamic processes that take a nonequilibrium state toward equilibrium are similarly irreversible. Because there are many more ways to arrange matter in a disorganized state than in an organized state, the probability that a disorganized state will become spontaneously organized is impossibly small.

2. The **second law of thermodynamics** comprises a set of equivalent statements to the effect that processes never take matter from less organized to more organized states.

   a. A **heat engine** is a device that extracts heat from a reservoir and converts some of the energy to useful work. Applied to heat engines, the second law states that it is impossible to build a perfect heat engine—one that converts heat to work with one hundred per cent efficiency. The most efficient engine is one that includes only reversible processes. The efficiency of such an engine—of which a **Carnot engine** is one example—is

   $$e = 1 - \frac{T_c}{T_h},$$

   where T_c and T_h are the temperatures of cool and hot reservoirs with which the engine exchanges energy. Because of the second law's limitation on efficiency, practical heat engines like automobile engines and electric power plants exhaust a great deal of waste heat to their environments.

   b. A **refrigerator** is a device that extracts heat from a cool object and transfers it to a warmer object. Applied to refrigerators, the second law states that it is impossible to build a perfect refrigerator—one that transfers heat without energy input from outside. Because of this second-law limitation, real refrigerators require a source of outside power. The efficiency of a refrigerator is measured by its **coefficient** of **performance,** defined as the ratio of heat extracted from the cool reservoir to the work input to the device:

$$\text{COP} = \frac{Q_c}{W}.$$

A refrigerator could be made by running a reversible engine backward, supplying work instead of producing it. Consideration of the second-law efficiency of such an engine then shows that the maximum COP of a refrigerator is

$$\text{COP} = \frac{T_c}{T_h - T_c}.$$

3. The second-law limit on the efficiency of a heat engine provides a definition of the **thermodynamic temperature scale.** This scale is independent of any particular type of thermometer or material, and coincides with the ideal gas temperature scale in the range where the latter is useful.
4. **Quality of energy** refers to the ability of a given amount of energy to be converted into work. The highest-quality energy is mechanical or electrical; these are associated with the organized motion or potential energy of particles in an object. Next in quality comes the energy associated with large temperature differences, followed by ever smaller temperature differences. The highest-quality energy can be readily converted to lower quality with one hundred per cent efficiency. The reverse process—conversion of lower-quality to higher-quality energy with one hundred per cent efficiency—is prohibited by the second law of thermodynamics.
5. **Entropy** provides a measure of the relative disorder of a system, so that increasing entropy is associated with decreasing energy quality. Entropy is a thermodynamic state variable, so that its value depends only on the state of a system and not on how the system got into that state. The entropy difference between two states may be calculated by evaluating the expression

$$\Delta S = \int_1^2 \frac{dQ}{T}$$

for any reversible process connecting those states.
6. In terms of entropy, the second law of thermodynamics becomes the statement that the entropy of the universe can never decrease. In this form, the second law is a universal statement about the tendency of systems to evolve toward states of higher disorder.

QUESTIONS

---

1. Which of the following processes is irreversible?
   a. Stirring sugar into coffee
   b. Building a house
   c. Demolishing a house with a wrecking crane
   d. Demolishing a house by taking it apart piece-by-piece
   e. Warming a bottle of milk by transferring it directly from a refrigerator to stove top
   f. Writing a sentence
   g. Harnessing the energy of falling water to drive machinery
2. Could you cool the kitchen by leaving the refrigerator door open? Explain.
3. Could you heat the kitchen by leaving the oven open? Explain.
4. Why don't we simply refrigerate the cooling water of a power plant before it goes to the plant, thereby increasing the plant's efficiency?
5. Should a car get better mileage in the summer or the winter? Explain.
6. Is there a limit to the maximum temperature that can be achieved by focusing sunlight with a lens? If so, what is it?
7. Name some irreversible processes that occur in a real engine.
8. A power company claims that electric heat is one hundred per cent efficient. Discuss this claim.
9. Steam leaves the turbine of a power plant at 150°C, and is condensed to water at the same temperature by contact with a river at 20°C. What temperature should be used as T_c in evaluating the thermodynamic efficiency?
10. A hydroelectric power plant, using the energy of falling water, can operate with an efficiency arbitrarily close to one hundred per cent. Why?
11. Viewed in isolation, a windmill is not a heat engine, for it converts mechanical energy of wind directly into electrical energy. But taking a broader view, the windmill is part of a natural heat engine. Explain.
12. To maximize the COP of a refrigerator, should you strive for a large or a small temperature difference? Explain.
13. The manufacturer of a heat pump claims that the device will heat your home using only energy already available in the ground. Is this true?
14. Proponents of ocean thermal energy conversion do not seem bothered by the extremely low efficiency of this process. Yet comparably low efficiencies would be intolerable in fossil-fueled or nuclear power plants. Why the difference?
15. Could anything useful be done with the waste heat from a power plant? Give two examples.
16. Does sunlight represent high- or low-quality energy? Explain. (See Question 6.)
17. The heat Q added during the process of adiabatic free expansion is zero. Why can't we then argue from

Equation 19–13 that the entropy change during this process is zero?

18. Energy is conserved, so why can't we recycle it like we do materials?

19. Name several systems that undergo entropy decreases. By considering larger systems of which they must be part, argue that the net entropy of the larger systems must increase.

PROBLEMS

Section 19.1   *Reversibility and Irreversibility*

1. An egg carton has places for one dozen eggs, as shown in Fig. 19–35. How many distinct ways are there to arrange six eggs in the carton? Of these, what fraction correspond to all six eggs being in the left half of the carton? Treat the eggs as distinguishable, so that an interchange of any two eggs gives rise to a new state of the system.

Fig. 19–35   Problem 1

2. Consider a gas consisting of only four molecules, all of which have the same speed v in the x-direction, but which can be moving either to the left or to the right. A microscopic state of the gas is specified by telling which molecules are moving in which direction. (a) How many possible microscopic states are there? (b) Of these, how many correspond to all the molecules moving in the same direction? If the system is in some random state, what are the chances that it will be a state with all molecules moving in the same direction? (c) Repeat parts a and b for the case of 10 molecules. (d) Repeat parts a and b for the case of 10^{23} molecules.

3. Estimate the energy that could be extracted by cooling all the oceans 0.1°C. Make reasonable assumptions about the volume of water in the oceans. How does this amount of energy compare with humanity's average yearly energy consumption of 2.5×10^{20} J?

Sections 19.2–19.3   *The Second Law and Applications*

4. What is the efficiency of a reversible engine operating between (a) the normal boiling and freezing points of water? (b) the surface temperature of the sun (5600 K) and the temperature of intergalactic space (3 K)? (c) the normal melting point of helium (1.76 K) and 1.0 K? (d) the temperature of surface water in the tropical oceans (80°F) and deep ocean water (39°F)? (e) a flame at 1000°C and room temperature?

5. The maximum temperature in a nuclear power plant is 570 K. The plant rejects heat to a river where the temperature is 0°C in the winter and 25°C in the summer. What are the maximum possible efficiencies of the power plant in the summer and winter? Why might the plant not achieve these efficiencies?

6. Consider a Carnot engine operating between a high temperature T_h and a low temperature T_c that is still above the ambient environmental temperature T_0. It should then be possible to operate a second Carnot engine between the output of the first engine and the environment, thereby producing more useful work. Show that the maximum overall efficiency of such a two-stage engine is the same as that of a single engine operating between T_h and T_0.

7. What is the COP of a reversible refrigerator operating between a minimum temperature of 0°C and a maximum temperature of 30°C?

8. 4.0 L of water at 9.0°C are put into a refrigerator. The refrigerator's 130-W motor then runs for 4.0 minutes to cool the water to the refrigerator's low temperature of 1.0°C. What is the COP of the refrigerator? How does this compare with the maximum possible COP if the refrigerator exhausts heat at 25°C?

9. In Section 19.2, we showed that the existence of a perfect refrigerator would allow us to construct a perfect heat engine, in violation of the Kelvin-Planck statement of the second law of thermodynamics. Make a similar argument, using appropriate energy flow diagrams, to show that the existence of a perfect heat engine would allow us to construct a perfect refrigerator, thus violating the Clausius statement of the second law.

10. An engine absorbs 900 J of heat each cycle and provides 350 J of work. (a) What is its efficiency? (b) How much heat is rejected each cycle? (c) If the engine is reversible and rejects heat to its environment at 10°C, what is its maximum temperature? (d) If the engine operates at 1000 cycles/minute, what is its mechanical power output?

11. Which would provide the greatest increase in the efficiency of a Carnot engine: an increase of 10°C in the maximum temperature or a decrease of 10°C in the minimum temperature?

12. Which would provide the greatest increase in the COP of a refrigerator: a decrease of 10°C in the maximum temperature or an increase of 10°C in the minimum temperature?

13. What is the maximum efficiency of an OTEC power plant operating between 20°C and 10°C?

14. 0.20 mol of ideal gas are taken through the Carnot cycle of Fig. 19–36. Calculate the heat Q_h absorbed, the heat Q_c rejected, and the work done. Find the maximum and minimum temperature, and show explicitly that the efficiency defined in Equation 19–1 is equal to the Carnot efficiency given by Equation 19–5.

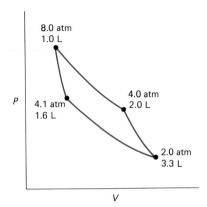

Fig. 19–36 Problem 14 (Diagram is not to scale.)

15. A reversible engine contains 0.20 mol of ideal monatomic gas, initially at 600 K and confined to 2.0 L. The gas is taken through a cycle consisting of the following steps:
    a. Isothermal expansion to a volume of 4.0 L
    b. Isovolumic cooling to 300 K
    c. Isothermal compression to 2.0 L
    d. Isovolumic heating to 600 K
    Calculate heat added during the cycle, and the work done. Determine the engine's efficiency, defined as the ratio of net work done to heat *absorbed* during the cycle. Compare with the efficiency of a Carnot engine operating between the same maximum and minimum temperatures. Why is your result not a violation of Carnot's statement that all reversible engines operating between the same reservoirs must have the same efficiency?

16. Repeat the preceding problem for the cycle of Fig. 19–37.

17. A power plant extracts energy from steam at 250°C and delivers 800 MW of electric power. It discharges waste heat to a river at 30°C. The overall efficiency of the power plant is 28%. (a) How does this efficiency compare with the maximum possible at these operating temperatures? (b) What is the rate of waste heat discharge to the river? (c) How many houses, each requiring 40,000 Btu/h of heat, could be heated with the waste heat from this power plant?

18. Show that it is impossible for two adiabats of any substance to intersect. *Hint:* Connect the two by a suitable isotherm, and show that the resulting cycle represents a heat engine operating at 100% efficiency.

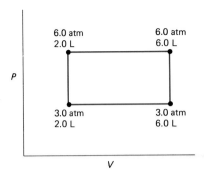

Fig. 19–37 Problem 16

19. Gasoline and diesel engines operate on the **Otto cycle,** shown in Fig. 19–38. In this idealization, we neglect the intake of fuel-air and exhaust of combustion products; the work involved in both is roughly equal but opposite. We also assume that all processes are reversible, which is not true for a real engine. In your calculations, assume that the gas in the cylinder has a fixed specific heat ratio γ. (a) What is the efficiency of this engine? (b) What is the maximum temperature, in terms of the minimum temperature T_c? (c) How does the actual efficiency compare with the efficiency of a Carnot engine operating between the same temperature extremes? (See Problem 15.)

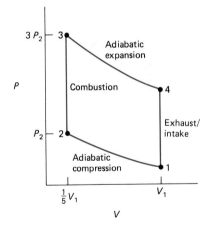

Fig. 19–38 Otto cycle (Problems 19, 20).

20. For the internal-combustion engine of the preceding problem, the **compression ratio** is defined as the ratio of maximum to minimum volume. In the engine of the preceding problem, the compression ratio is 5. What effect does an increase in compression ratio have on engine efficiency? Show by plotting efficiency versus compression ratio. Show also the efficiency of a Carnot engine operating between the same temperature extremes as your internal-combus-

tion engine. Assume that the pressure continues to triple during the combustion phase, as shown in Fig. 19–38.

21. A heat pump designed for use in southern climates extracts heat from the outside air and delivers hot air at 105°F to the inside of a house. (a) If the average outdoor temperature in winter is 40°F, what is the average COP of the heat pump? Assume that the heat pump operates in a reversible Carnot cycle. (b) Suppose the heat pump is used in a northern climate where the average winter temperature is 15°F. Now what is the COP? (c) Two *identical* houses, one in the north and one in the south, are heated by this pump. Both houses maintain indoor temperatures of 65°F. What is the ratio of electric power consumption in the northern house to that in the southern one? *Hint:* Think about heat loss as well as COP!

22. A house is heated by a heat pump with a COP of 2.3. The heat pump is run by electricity generated in a power plant whose efficiency is 28%. What is the overall efficiency of this process, defined as the ratio of heat delivered to the house to heat released from fuel at the power plant? Could the efficiency so defined ever exceed 100%?

23. The electric power output of all the thermal electric power plants in the United States is about 2×10^{11} W. These plants operate with an average efficiency around 33%. What is the rate at which all these plants use cooling water, if the average temperature rise in the water is 5°C? Compare with the average flow at the mouth of the Mississippi River $(1.8 \times 10^7 \text{ kg/s})$.

24. A solar "power tower" plant is to be built in a desert location where the only source of cooling water is a small creek with an average flow of 100 kg/s and an average temperature of 30°C. The plant is to cool itself by boiling away the entire creek. If the maximum temperature achieved in the plant is 500 K, what is the maximum electric power output it can sustain without running out of cooling water?

25. The McNeil generating station in Burlington, Vermont, came into operation in 1984 as the largest wood-burning power plant in the United States. Here are some facts about the McNeil power plant:
a. The electric power output of the plant is 59 MW.
b. Fuel consumption is 1.2×10^6 kg of wood per day. The wood has an energy content of 1.4×10^7 J/kg, and burns with 85% efficiency, meaning that 85% of this energy is transferred to water in the plant's boiler.
c. Steam enters the turbine of the power plant at 1000°F, and leaves at 200°F.
d. Cooling water circulates through the condenser at the rate of 1.8×10^5 kg/min. The temperature of the cooling water increases by 18°F as it extracts heat from steam leaving the turbine.
Calculate the efficiency of the plant in three ways: (a) Using only fact c above to get the maximum possible efficiency; (b) using only facts a and b; and (c) using only facts a and d.

Section 19.4   *The Thermodynamic Temperature Scale*

26. A small sample of material is taken through a Carnot cycle between a high-temperature reservoir of boiling helium ($T = 1.76$ K) and an unknown low temperature. During the process, 7.5 mJ of heat is absorbed from the boiling helium, and 0.44 mJ rejected to the low-temperature reservoir. What is the unknown low temperature?

27. A reversible engine operating between a vat of boiling sulfur and a bath of water at its triple point is observed to have an efficiency of 61.4%. What is the boiling point of sulfur?

Section 19.5   *Entropy and the Quality of Energy*

28. Calculate the entropy change associated with melting 1.0 kg of ice at 0°C.

29. The temperature of n moles of ideal gas is changed from T_1 to T_2. Show that the entropy change of the gas is

$$\Delta S = nC_V \ln\left(\frac{T_2}{T_1}\right).$$

30. 5.0 mol of an ideal diatomic gas is initially at 1.0 atm pressure and 300 K. What is the entropy change of the gas if it is heated to 500 K (a) isobarically, (b) isovolumically, and (c) adiabatically?

31. Since entropy and temperature are both state variables, we could describe thermodynamic processes on a TS diagram instead of a PV diagram. Sketch a graph of a Carnot cycle on such a diagram, with absolute temperature on the vertical axis and entropy on the horizontal axis.

32. A 500-g block of copper at 80°C is dropped into 1.0 kg of water at 10°C. (a) What is the final temperature? (b) What is the entropy change of the system?

33. n moles of an ideal monatomic gas are taken around the cycle shown in Fig. 19–39. The process BC is iso-

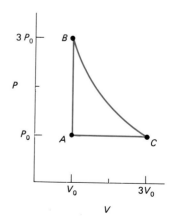

Fig. 19–39   Problem 33

thermal. Calculate the entropy change associated with each of the processes AB, BC, and CA, and show explicitly that there is no net change in entropy on traversing the whole cycle.

34. The interior of a house is maintained at 20°C while the outdoor temperature is −10°C. The house loses heat at the rate of 30 kW. At what rate does the entropy of the universe increase because of this irreversible heat flow?

35. Two identical objects of mass m and specific heat c at initial temperatures T_h and T_c are placed in an insulated box and allowed to come to equilibrium. Show that the entropy increase in the process is

$$\Delta S = mc \ln\frac{(T_h + T_c)^2}{4T_h T_c}.$$

Assume that the specific heat is independent of temperature.

36. You have available 50 kg of steam at 100°C, but no source of heat to maintain it in that condition. You also have a heat reservoir at 0°C. Suppose you operate a heat engine in this system, so that the steam condenses and gradually cools to 0°C as the engine runs. (a) Calculate the total entropy change of the steam and subsequent hot water. (b) Calculate the total entropy change of the reservoir. (c) What is the total amount of work that the engine can do? Hint: Consider the entropy change of the system. This change occurs as a result of a reversible process that supplies useful work. But if it had occurred irreversibly, how much energy would have become unavailable to do work?

37. Arguing by analogy with Example 19–7, show that a perfect refrigerator would reduce the entropy of the universe.

Supplementary Problems

38. In a steam engine, water is boiled under pressure so that its boiling point is above the normal 100°C. The resulting steam expands adiabatically and drives an engine that powers a railroad locomotive with a mass of 3.0×10^4 kg. The engine exhausts 100°C steam at the rate of 3.6 kg/s. If the engine can climb a 200-m-high hill in 45 s, what is the minimum possible temperature of the boiling water?

39. A Carnot engine extracts heat from a block of material of mass m and specific heat c that is initially at temperature T_{ho}. The engine rejects heat to a reservoir at T_c. The cool reservoir is large enough that its temperature never changes, but the hot material has no heat source, so that it cools as the engine runs. The engine is operated in such a way that the mechanical power

output is proportional to the temperature difference $T_h − T_c$:

$$P = P_0 \frac{T_h − T_c}{T_{ho} − T_c},$$

where T_h is the instantaneous temperature of the hot material and P_0 is the initial power output. Find an expression for T_h as a function of time, and determine how long it takes for the engine's power output to drop to zero. Your expression for T_h should suggest that T_h can drop below T_c. How is this possible?

40. Figure 19–40 shows a cyclic process in the PV diagram of an ideal diatomic gas. The process AB is isovolumic, BC isothermal, CD adiabatic, and DA isobaric. Fill in the blank spaces in the table below:

	P	V	T	$U − U_A$	$S − S_A$
A	P_0	V_0	T_0	0	0
B	$3.4P_0$	V_0			
C					
D	$3.0V_0$	P_0			

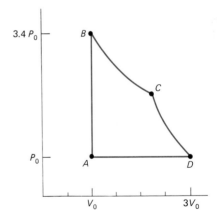

Fig. 19–40 Process for Problem 40. The location of point C is not quantitatively correct.

20

ELECTRIC CHARGE

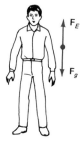

Fig. 20-1
Gravity exerts a downward force on your arms. Electrical forces acting at the atomic and molecular scale keep your arms from falling off.

As you sit at your desk gravity exerts a downward force on you. Why don't you fall right through the desk? Your arms hang down from your shoulders. Why don't they fall off (Fig. 20–1)? What holds tree branches up, or keeps mountains from spreading out into flat blobs of matter? You lift a glass of water. What holds the water in the glass? For that matter, what holds the individual water molecules together? You drive a car or bicycle around a curve. What keeps you from skidding off in a straight line? You blow up a balloon and feel the higher-pressure air pushing against the balloon. What keeps the air molecules in the balloon? How do they exert a pressure? Newton's laws tell us that in all these cases forces are involved. What is the nature of those forces?

Remarkably, all the forces other than gravity that we encounter in our daily lives, as well as in most scientific investigations, have a single origin. They are electromagnetic forces. In all our mechanics problems involving tension forces, friction, normal forces, and pressure forces, we were really talking about electromagnetic forces. Other, more subtle interactions are electromagnetic. Light hits the lens of your eye and refracts, allowing you to see clearly. The interaction of light with the eye is fundamentally electromagnetic. Sunlight hits a solar collector and is converted to heat. Radio waves excite vibrations in your radio receiver, bringing you music (Fig. 20–2). Again the interaction is electromagnetic.

In Chapter 4 we introduced the three forces—gravity, the electroweak force, and the strong nuclear force—that physicists currently regard as fundamental, and described ongoing efforts to explain these three as manifestations of a single underlying interaction. Even though the fundamental forces now appear to be related, this relationship is physically significant only at very high energies—those found in the very early universe or in the largest particle accelerators. Similarly, unification of electromagnetism with the weak nuclear force is significant only at high energies. In most applications, we can safely treat electromagnetism as a distinct force, operating independently of the others. Figure 20–3 reviews the relation of electromagnetism to the other forces of nature.

510

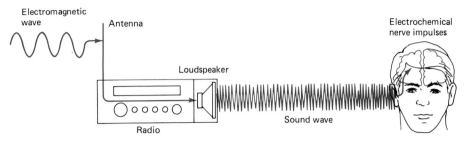

Fig. 20–2
Many different electromagnetic interactions are involved when you listen to the radio. The radio signal itself is an electromagnetic wave; electromagnetic forces drive electrons in the radio, ultimately exerting forces on the loudspeaker. Electrical forces among air molecules result in sound reaching your ear. Vibrations of your eardrum are converted to electrical signals carried to the brain by your auditory nerve.

20.1
THE STUDY OF ELECTROMAGNETISM

In the next twelve chapters, we will develop a thorough understanding of electromagnetism. That we devote so much time to this study is indicative of how all-encompassing the electromagnetic force is. Other than gravity it is the only force most of us will ever deal with.

Our study is motivated by three distinct themes, each indicating a different reason for understanding electromagnetism.

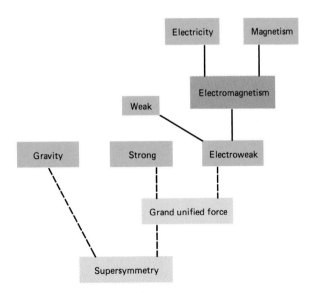

Fig. 20–3
The place of electromagnetism among the forces of nature. Electromagnetism comprises electricity and magnetism, once thought to be separate. Similarly, electromagnetism and the weak force are both aspects of the electroweak force. Physicists are close to a description of the electroweak and strong force as manifestations of a "grand unified force." Further off is a possible unification with gravitation; some current theories describing such a unified interaction are termed "supersymmetry" theories.

1. The electromagnetic force is solely responsible for the structure of matter from atomic up to roughly human size. Much of physics, all of chemistry, and much of biology deal in this realm. Only at still larger scales does gravity become significant.

   The wonderful diversity of chemical compounds making up our world is testimony to the rich possibilities contained in the electromagnetic interaction. Even life itself, and the DNA replicating mechanism at its heart, are manifestations of electromagnetism (Fig. 20–4). For students of physical and biological sciences, understanding electromagnetism means understanding the most fundamental basis of these disciplines.

2. We live in a technological world increasingly dominated by devices that operate on electromagnetic principles. Electric lights, motors, batteries, and generators have been essential to all twentieth-century technology. More recently the development of electronics and microelectronics has led to the proliferation of devices for conveying and processing information, for sensing and measuring, and for sophisticated control of industrial, scientific, medical, and even household systems.

3. Study of electromagnetism leads to an understanding of the nature of light and from there to the theory of relativity. Among humanity's most remarkable intellectual achievements, relativity profoundly alters our ideas of space and time—the very basis of physical reality.

We have been speaking of electromagnetism. Yet you are probably more familiar with electricity and with magnetism, not with some union of the two. We begin with separate studies of these two distinct phenomena, but eventually the connections between the two will come to dominate our considerations. Ultimately, we will understand electromagnetism as a single phenomenon. We will have been through the first step in a unification of the forces of nature, a step with numerous practical applications and with important consequences for our understanding of physical reality.

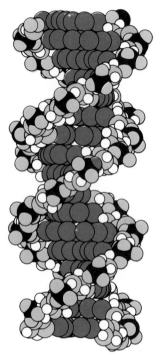

Fig. 20–4
Electrical forces govern the replication of DNA molecules.

20.2
ELECTRIC CHARGE

electric charge

Electric charge is a fundamental property of matter. Of the three familiar building blocks of ordinary matter—electrons, protons, and neutrons—two are electrically charged. Charge is not an incidental aspect of these particles, but a basic characteristic without which they would not be the same entities. What is electric charge? At the most fundamental level we cannot answer this question. We can only repeat that it is a basic property of matter and then proceed to describe the behavior of charged matter. As we come to understand electrical interactions we gain a familiarity with charge that is as close as we can come to knowing what charge "really" is.

If it bothers you that we cannot define charge, consider the question "what is mass?" We don't know the answer to that either, but we have spent all our lives pushing things and lifting them, so that we have an intuitive feel for how mass behaves—a feel that we have quantified through our study of physics.

Two Kinds of Charge

positive charge
negative charge

Electric charge comes in two varieties. Since the time when Benjamin Franklin suggested the names, we have called them **positive** and **negative,** but these names have no physical significance. There is nothing "missing" from negative charge. Positive and negative charge are complementary properties, not the presence and absence of something. The utility in the names is mathematical, since it is the algebraic sum of charges—described by positive and negative numbers—that has physical significance.

Quantities of Charge

Every electron carries the same amount of charge as every other electron. Every proton carries the same amount of charge as every other proton. Remarkably, the charge on the proton has *exactly* the same magnitude as the charge on the electron, although it has the opposite sign. Think about this! The electron and proton are very different particles—they differ in mass and many other characteristics—yet their electrical properties must be intimately related. Problem 1 shows how dramatically different our world would be if there were even a slight difference between the magnitudes of the electron and proton charge.

elementary charge
quantized

The magnitude of the charge on the electron or proton is called the **elementary charge,** e. Electric charge is **quantized**—that is, it comes only in discrete amounts. In a famous experiment performed in 1909, the American physicist R. A. Millikan used electric forces to suspend small charged oil drops, balancing the force of gravity (Fig. 20–5). From the electric force Millikan computed the charge. His results showed that the electric charge on the drops was always a multiple of a basic value that we now know as the elementary charge.

The theories of elementary particles that evolved during the 1970's suggest that the most basic unit of charge is actually $\frac{1}{3}e$. Such "fractional charges" are believed to reside on **quarks,** fundamental building blocks comprising protons, neutrons, and other particles. The force binding quarks together is probably such that the quarks can never be separated. As a result, we do not expect to encounter an isolated fractional charge. In combination, quarks appear always to join in such a way that the total charge is zero or an integer multiple of the full elementary charge.

quarks

coulomb

The SI unit of charge is the **coulomb** (C), named for the French physicist Charles Augustin de Coulomb (1736–1806). The coulomb is defined

Fig. 20–5
Millikan's oil-drop experiment. By balancing electrical and gravitational forces on oil drops, Millikan showed that electric charge is quantized.

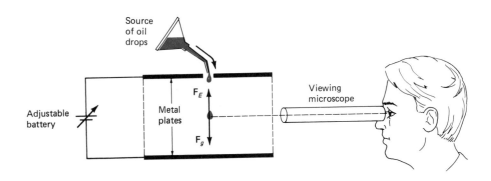

formally in terms of electric current, which we will study in Chapter 25. The coulomb was established before the discovery that charge comes in integer multiples of an elementary charge. Although not its official definition, it is convenient to think of a coulomb as simply a certain number of elementary charges. One coulomb is about 6.25×10^{18} elementary charges, so that one elementary charge e is 1.60×10^{-19} C.

Charge Conservation

net charge

Electric charge is a conserved quantity. If you confine charged matter to a closed region, the algebraic sum of the individual charges in that region remains constant no matter what. This sum is called the **net charge.** In centuries of experiments, from the simplest involving static electricity to the most complex elementary particle interactions in high-energy accelerators, no one has ever observed an isolated electric charge appearing or disappearing. Charged particles may be created or annihilated, but always in pairs of equal and opposite charge (Fig. 20–6). The net charge always remains the same.

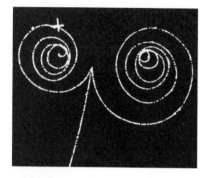

Fig. 20–6
This picture, taken in a bubble chamber attached to a high-energy accelerator, shows the creation of a pair of charged particles—an electron and a positron, or anti-electron. Their tracks are the two oppositely directed spirals that diverge from the common point where the particles were created. The net charge remains zero before and after the particle creation.

20.3
THE INTERACTION OF CHARGES: COULOMB'S LAW

We know about charge from observing the behavior of charged matter. Rub a balloon on your clothing and you can stick it to yourself. Saw a piece of styrofoam insulation and you end up with little beads of styrofoam stuck all over you. In each case a mechanical process has transferred charge from one object to another. Initially you and the balloon or styrofoam were electrically neutral, meaning that neither carried a net charge. After the charge transfer, you carry charge opposite that of the balloon or styrofoam. That you and the balloon or styrofoam stick to each other is a manifestation of a fundamental interaction of electric charges:

Charges of opposite sign attract each other.

If you charged two balloons and held them next to each other on the ends of strings (Fig. 20–7) you would find that they repel each other:

Charges of the same sign repel each other.

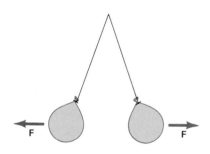

Fig. 20–7
The balloons carry similar electric charges, so they repel each other.

You have encountered numerous and often annoying instances of electrical interaction. Dust that sticks to phonograph records, nylon socks that cling to other clothing when they come out of the dryer, and shocks when you cross a carpet and then touch a doorknob are examples of the same underlying phenomenon. If you have ever had the dangerous experience of being on an exposed mountain ridge during a thunderstorm you may have felt your hair stand on end—a manifestation that charge is building up on you, with like charges in different strands of hair pushing away from each other. (Incidentally, this effect indicates that lightning may be about to strike you!)

These examples are situations in which you are directly aware of the electrical force. But electrical phenomena are not confined to those exam-

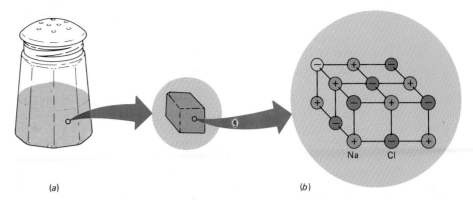

Fig. 20–8
(a) The salt shaker and even a single salt grain are electrically neutral, so that we are not directly aware of the role that electric forces play in these structures. *(b)* But if we study the salt crystal on the atomic scale, we see immediately that electric forces bind individual sodium and chlorine ions together and, in fact, are responsible for the cubical shape of the larger crystal.

ples. The study of electricity would be rather unimportant if the only significant electrical interactions were these obvious ones. In fact, all interactions of everyday matter—from the motion of a car on a road to the movement of a muscle to the growth of a tree—are dominated by the electric force. It is just that matter on the large scale is so very nearly perfectly neutral that electrical effects in bulk matter are not so obvious. When we study matter on the molecular or even cellular scale the appearance of individual charged particles makes the electrical nature of the interactions immediately obvious (Fig. 20–8).

The attraction and repulsion of electric charges implies that a force acts between charges. In a series of careful experiments performed in the late 1700's, the physicist Coulomb investigated this electric force. He found that the force between two charges acts along the line joining them, with a magnitude proportional to the product of the charges and inversely proportional to the square of the distance between them. We can summarize the magnitude and directional aspects of Coulomb's discovery by writing

$$\mathbf{F} \propto \frac{q_1 q_2}{r^2} \, \hat{\mathbf{r}}. \qquad (20\text{–}1)$$

Here $\mathbf{F}$ is the force that charge q_1 exerts on q_2, r the distance between the two charges, and $\hat{\mathbf{r}}$ a unit vector pointing from q_1 toward q_2 (Fig. 20–9). What about attractive and repulsive forces? If both charges have the same sign, the force on q_2 is in the direction of the unit vector, or away from q_1—that is, repulsive. If the charges have opposite signs, the quantity multiplying the unit vector is negative, and the force on q_2 is toward q_1, or attractive. Over the years, Equation 20–1 has been confirmed with extreme precision using methods far more accurate than the simple force measurements of Coulomb. We will describe some of these methods in succeeding chapters.

Equation 20–1 is known as **Coulomb's law.** Strictly speaking, the law applies only to **point charges**—charged objects of zero size. Only with point charges is the distance r defined unambiguously. However, we can use Coulomb's law whenever charged objects are very small compared with

Fig. 20–9
Quantities in Coulomb's law. The unit vector $\hat{\mathbf{r}}$ conveys directional information only, allowing the law to be written in vector notation.

Coulomb's law

point charges

the distance between them. When the sizes of charged objects become significant compared with their separation, Coulomb's law applies directly only to the interactions between each small part of each object. Later in this chapter we develop techniques for dealing with this situation.

Coulomb's law is similar to Newton's law of gravity. Given mass and charge, these laws with their inverse-square distance dependences are the simplest one could concoct—and the real universe has somehow opted for that simplicity. Because there are two kinds of charge, Coulomb's law admits both attractive and repulsive interactions. In contrast, no one has ever found "negative mass" (even antimatter has "positive mass"), so that the gravitational interaction described by Newton is only attractive.*

Equation 20–1 implies a proportionality constant between force and the quantity $q_1 q_2 / r^2$. In SI units this constant is called k and has the value 9.0×10^9 N·m²/C². We can then write

$$\mathbf{F} = \frac{k q_1 q_2}{r^2} \,\hat{\mathbf{r}}. \tag{20–2}$$

permittivity constant

As we proceed in our study of electromagnetism, we will find it convenient to express k in terms of an equivalent quantity, the **permittivity constant** ϵ_0. k and ϵ_0 are related by

$$k = \frac{1}{4\pi\epsilon_0}, \tag{20–3}$$

so that Coulomb's law may be written

$$\mathbf{F} = \frac{1}{4\pi\epsilon_0} \frac{q_1 q_2}{r^2} \hat{\mathbf{r}}. \tag{20–4}$$

Equations 20–2 and 20–4 are both expressions of Coulomb's law. We will use the form 20–2 as being more convenient in this chapter, but will often find the form 20–4 more appropriate in subsequent chapters.

Example 20–1
Coulomb's Law

A 1.0-μC charge is located at x = 1.0 cm, and a −1.5-μC charge at x = 3.0 cm. What is the force exerted by the positive charge on the negative charge? What would the force be if the distance between charges were tripled?

Solution

Coulomb's law gives

$$\mathbf{F} = \frac{k q_1 q_2}{r^2} \,\hat{\mathbf{r}}$$

$$= \frac{(9.0 \times 10^9 \text{ N·m}^2/\text{C}^2)(1.0 \times 10^{-6} \text{ C})(-1.5 \times 10^{-6} \text{ C})}{(0.020 \text{ m})^2} \,\hat{\mathbf{i}}$$

$$= -34\hat{\mathbf{i}} \text{ N},$$

---

*There is a more subtle distinction between Coulomb's law and Newton's law of gravity. We believe that Coulomb's law is an exact description of the electric force. But we know that Newton's law is only an approximation, with a more exact description of gravity given by Einstein's general theory of relativity.

where we have expressed the 2.0-cm distance between charges as 0.020 m, and where we write $\hat{\imath}$ for $\hat{r}$ because a unit vector pointing from the positive charge toward the negative charge is in the positive x-direction. The minus sign shows that the force is actually in the negative x-direction—that is, attractive.

This force varies as the inverse square of the distance, so that if the distance is tripled the force drops by a factor of $3^2 = 9$ to $-3.8\hat{\imath}$ N.

How would this example differ had we calculated the force that the negative charge exerts on the positive charge? The unit vector $\hat{r}$ would have reversed direction, but the magnitude would remain the same. Thus Newton's third law is explicitly satisfied, in that the two charges exert equal but opposite forces on each other.

This example suggests that charges in the microcoulomb range produce reasonably large forces at typical laboratory distances. In the realm of everyday experience, one coulomb would be an enormous charge! Could you hold two one-coulomb charges a meter apart?

Example 20–2 _____
Coulomb's Law and the Atom

The electron in a hydrogen atom has a kinetic energy of 13.6 eV (2.2×10^{-18} J; see Appendix D for conversion factors). Assume the electron is in circular orbit around the central proton, and use Coulomb's law to estimate the size of the atom.

Solution

The force on the electron is directed toward the proton and has magnitude given by Coulomb's law:

$$F = \frac{ke^2}{r^2},$$

with r the radius of the electron orbit and e the elementary charge. This electric force keeps the electron in uniform circular motion, so that

$$\frac{ke^2}{r^2} = \frac{mv^2}{r}.$$

Solving for r gives

$$r = \frac{ke^2}{mv^2}.$$

The quantity mv^2 in the denominator is just twice the kinetic energy of 2.2×10^{-18} J, so that

$$r = \frac{(9.0 \times 10^9 \ \text{N·m}^2/\text{C}^2)(1.6 \times 10^{-19} \ \text{C})^2}{(2)(2.2 \times 10^{-18} \ \text{J})} = 5.2 \times 10^{-11} \ \text{m}.$$

Thus the atomic diameter is about 10^{-10} m. Although our picture of the electron orbiting the proton like a planet orbiting the sun is not fully consistent with modern quantum mechanics, our Coulomb's law estimate of the atomic size is reasonable.

20.4
THE SUPERPOSITION PRINCIPLE

Coulomb's law describes the interaction between two charges. What if we have more than two charges? Again, the electric force is remarkably simple. To calculate the force on a charge q caused by two or more other charges, we simply calculate the individual forces between q and each of the other charges, then add them (vectorially, of course, because force is a vector). This sounds obvious, but nature need not have been this simple! In Einstein's general relativity, for example, the gravitational force is *not* just the sum of individual forces, a fact that greatly complicates the mathematical analysis of gravity. That the presence of a third charge has no influence on the interaction between two other charges is what makes the mathematical description of electromagnetism simpler than it otherwise would be.

superposition principle

The fact that electric forces add vectorially is known as the **superposition principle.** This principle allows us to solve relatively complicated problems in electromagnetism. With it we can break them into a series of simpler problems, and at the end take the sum of the results of these individual problems.

Example 20–3

Using the Superposition Principle

Two charges of the same magnitude q but opposite sign are located on the x-axis at $+a$ and $-a$, as shown in Fig. 20–10. What is the force on a positive charge Q located at the origin? On a positive charge Q on the y-axis at a distance a above the x-axis?

Solution

We use the superposition principle, finding individual forces between Q and each of the other charges. At the origin, Q experiences a force

$$\mathbf{F}_- = \frac{k(-q)Q}{a^2}(-\hat{\mathbf{i}}) = \frac{kqQ}{a^2}\hat{\mathbf{i}}$$

from the negative charge. The unit vector $\hat{\mathbf{r}}$ becomes $-\hat{\mathbf{i}}$ because it points from the negative charge toward the origin. The two minus signs multiply to a plus sign, showing that the force is actually in the $+x$-direction—that is, attractive. Similarly, the force on Q caused by the positive charge is just

$$\mathbf{F}_+ = \frac{kqQ}{a^2}\hat{\mathbf{i}}.$$

Both force vectors point in the $+x$-direction, so that the net force on Q is

$$\mathbf{F}_{\text{net}} = 2k\frac{qQ}{a^2}\hat{\mathbf{i}}.$$

With Q a distance a up the y-axis, its distance from each of the other charges becomes $\sqrt{2}a$. Then the force arising from the negative charge is

$$\mathbf{F}_- = \frac{k(-q)Q}{(\sqrt{2}a)^2}\hat{\mathbf{r}}_- = \frac{-kqQ}{2a^2}\hat{\mathbf{r}}_-.$$

Since $\hat{\mathbf{r}}_-$ points diagonally upward to the left, the minus sign shows that this force points diagonally downward to the right, as shown in Fig. 20–10. Similarly, the force from the positive charge is

$$\mathbf{F}_+ = \frac{kqQ}{(\sqrt{2}a)^2}\hat{\mathbf{r}}_+ = \frac{kqQ}{2a^2}\hat{\mathbf{r}}_+,$$

pointing diagonally upward to the right.

We must now add the two forces vectorially. Fig. 20–10 shows that the y-components of the vectors $\mathbf{F}_-$ and $\mathbf{F}_+$ cancel, while the x-components are equal. The

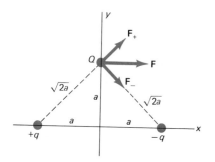

Fig. 20–10
The forces on Q from the individual charges $+q$ and $-q$ add vectorially.

angle between either force vector and the x-direction is 45°, so that either x-component is

$$F_{x+} = F_{x-} = \frac{kqQ}{2a^2}\cos 45° = \frac{kqQ}{2a^2}\frac{\sqrt{2}}{2}.$$

Adding the two equal x-components then gives the net force:

$$\mathbf{F}_{net} = \frac{\sqrt{2}kqQ}{2a^2}\hat{\imath}.$$

Since the quantity multiplying $\hat{\imath}$ is positive, this force points to the right, as Fig. 20–10 shows it must.

Example 20–4
More Superposition

A positive charge $+2q$ lies on the x-axis at $x = -a$. A charge $-q$ lies on the axis at $x = +a$, as shown in Fig. 20–11. Find a point where the electric force on a third charge would be zero.

Fig. 20–11
Example 20–4

Solution

If such a point exists, it must be on the x-axis. Off axis, the individual force vectors associated with the two charges could not point in opposite directions, and therefore their sum could not be zero. Could the point lie between the two charges? No, for in this region a positive charge would be pulled to the right by $-q$ and pushed to the right by $+2q$. Similarly, a negative charge would definitely experience a force to the left. Could the point lie to the left of $+2q$? A charge in this region is always closer to $+2q$ than to $-q$. The force arising from $+2q$ is always larger, partly because the third charge Q is closer to it and partly because of its greater charge. Could the point lie to the right of $-q$? Possibly; although a charge in this region is always closer to $-q$, the larger magnitude of $+2q$ may mean there is a point where its force is greater.

Let us then locate a third charge Q at some point x to the right of $-q$, so that $x > a$. At this point our charge Q is a distance $x - a$ from $-q$, so that the force on Q due to $-q$ is

$$\mathbf{F}_- = \frac{k(-q)Q}{(x-a)^2}\hat{\mathbf{r}} = \frac{-kqQ}{(x-a)^2}\hat{\imath},$$

where we have written $\hat{\imath}$ because a unit vector from $-q$ toward Q is in the $+x$-direction. The charge Q lies a distance $x - (-a) = x + a$ from $+2q$, so that the force on Q from $+2q$ is

$$\mathbf{F}_+ = \frac{k(2q)Q}{(x+a)^2}\hat{\imath}.$$

Again, the unit vector is chosen to point from $+2q$ toward Q. If the two forces are to cancel, we must have

$$\mathbf{F}_+ + \mathbf{F}_- = \frac{-kqQ}{(x-a)^2}\hat{\imath} + \frac{2kqQ}{(x+a)^2}\hat{\imath} = 0.$$

Note that k, q, Q, and $\hat{\imath}$ cancel from both terms. The fact that Q cancels shows that if we do find the desired point, it will be a point of no force whatever the sign or magnitude of the charge Q we put there. We then have

$$\frac{2}{(x+a)^2} = \frac{1}{(x-a)^2}.$$

Inverting both sides and taking square roots gives

$$\frac{x+a}{\sqrt{2}} = \pm(x-a).$$

We solve separately the two cases associated with the + and − signs in this equation. For the + sign, we have

$$x = a\frac{\sqrt{2}+1}{\sqrt{2}-1} = a\frac{(\sqrt{2}+1)^2}{(\sqrt{2}-1)(\sqrt{2}+1)} = (3+2\sqrt{2})a.$$

Since $x > a$, this point is indeed to the right of $-q$ and does give zero force. Physically, note that to the left of our solution point the force of $-q$ dominates (at least as long as we stay to the right of $-q$). Far to the right of our point we are nearly equal distances from both $+2q$ and $-q$, so that the greater magnitude of the positive charge causes its force to dominate. Somewhere in between the forces are equal but opposite; this is the point we have found.

What about the second equation that arose when we took the square root? You can verify that it gives a value for x that lies to the left of $-q$. In this region, our choice of the unit vector associated with $-q$ is incorrect, so that the mathematics does not give a meaningful answer.

Charge Distributions

charge distribution

In the examples above we were concerned with the force caused by a **charge distribution**—that is, by an arrangement of one or more charges in space. Any piece of matter containing electric charges—you, or a strand of DNA, or a molecule you are trying to synthesize, or an electrode in a TV tube—all these are charge distributions. To understand how these objects interact electrically requires knowledge of the net electric force arising from all the individual charges in the distribution. When a distribution contains very many charges, we often neglect the discrete nature of electric charge and consider that charge is continuously distributed over a volume, volume charge density an area, or a line. We then speak of the **volume charge density,** ρ, the surface charge density **surface charge density,** σ, or the **line charge density,** λ. These quantities line charge density describe the amount of charge per unit volume, per unit area, or per unit length, respectively. They have units of C/m^3, C/m^2, and C/m. When we calculate the force caused by a continuous distribution of charge, we consider the distribution to consist of infinitely many infinitesimal point charges dq. The vector sum of electric forces becomes an integral over the entire charge distribution. Example 20–5 shows how the force arising from a continuous distribution is calculated.

Example 20–5
A Continuous Charge Distribution

A thin rod of length ℓ is charged uniformly with line charge density λ. A charge Q is located a distance a from one end of the rod, as shown in Fig. 20–12. Calculate the electric force on the charge Q.

Solution

Let the y-axis lie along the rod, with the origin at the charge Q. Consider a small length dy of rod, containing a charge dq, as shown in Fig. 20–12. The mag-

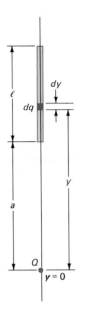

Fig. 20–12
The force on the charge Q is the sum of the forces associated with all the charge elements dq along the rod.

nitude dF of the force on Q arising from dq is given by Coulomb's law:

$$dF = \frac{kQdq}{y^2}, \tag{20–5}$$

since y is the distance from Q to dq. Because the problem involves only one dimension, we have not used the full vector notation. The force is attractive if Q and λ have opposite signs; otherwise it is repulsive. Now, our dq is one of only many tiny charge elements comprising the rod. The other dq's give rise to forces with the same form as Equation 20–5, but with values of y ranging from $y = a$ to $y = a + \ell$. To determine the total force F, we add the contributions of all the dq's along the rod. Since there are infinitely many infinitesimally small charge elements dq, this sum becomes an integral:

$$F = \int dF = \int_{y=a}^{y=a+\ell} \frac{kQdq}{y^2}. \tag{20–6}$$

To evaluate this integral, we must relate dq and y. How? The rod contains charge λ per unit length, and the charge dq occupies a length dy. Therefore, the charge dq is

$$dq = \lambda dy.$$

Using this expression in Equation 20–6 then gives

$$F = \int_a^{a+\ell} \frac{kQ\lambda dy}{y^2} = kQ\lambda \int_a^{a+\ell} \frac{dy}{y^2}$$

$$= kQ\lambda \left[-\frac{1}{y} \right]_a^{a+\ell} = kQ\lambda \left[\frac{1}{a} - \frac{1}{a+\ell} \right] \tag{20–7}$$

$$= \frac{kQ\lambda\ell}{a(a+\ell)} = \frac{kQq_{\text{rod}}}{a(a+\ell)},$$

where q_{rod}, the total charge on the rod, is just $\lambda\ell$ because the line charge density is constant. Does this result make sense? To see that it does, consider the special case when the charge Q is very far from the rod, so that $a \gg \ell$. Then our result reduces approximately to

$$F = \frac{kQq_{\text{rod}}}{a^2}.$$

This is just the force of one point charge q_{rod} on another point charge Q. In this special case, the length of the rod is negligible and it acts just like a single point charge. But as we move closer to the rod, so that the length becomes significant compared with the separation a, then the expression for the force becomes complicated, reflecting the fact that we are really dealing with a superposition of forces from numerous charges spread along the charged rod.

Having derived the result 20–7, we can again use the superposition principle to calculate the force arising from a combination of charged rods or rods and point charges. Problems 21 and 22 explore these cases.

In principle, superposition can always be used in conjunction with Coulomb's law to sum the forces from individual charges in a distribution. But with complicated charge distributions, the vector calculation is often extremely difficult. We can sometimes understand the behavior of a charge distribution by considering the forces associated with a simpler distribution that it resembles. In the next few chapters we develop powerful methods for exploring the electrical interactions of charge distributions. These methods remain rooted in the fundamental facts of Coulomb's law and superposition, but they assume a mathematical elegance that allows us to avoid horrendous vector sums and proceed directly to our answer.

SUMMARY

1. **Electromagnetism** is a manifestation of one of the three fundamental forces that are believed to govern all interactions in the universe. The electromagnetic interaction is particularly important over a distance scale ranging from atomic-sized up to macroscopic objects. Electromagnetism is responsible for most interactions studied in the natural sciences, and for the electrical and electronic technologies so pervasive in modern civilization. The study of electromagnetism leads directly to a confrontation with the Newtonian conceptions of absolute space and time, out of which we will develop a deeper understanding of physical reality through the theory of relativity.

2. **Electric charge** is a fundamental property of matter. Many of the so-called elementary particles carry electric charge. In particular, the electric charges on electrons and protons are responsible for most properties of ordinary matter.

    a. Electric charge comes in two types, arbitrarily called positive and negative.

    b. Electric charge is **quantized,** occurring only in discrete amounts. The SI unit of charge is the coulomb (C), consisting of about 6.25×10^{18} elementary charges. Thus the elementary charge is $e = 1.6 \times 10^{-19}$ C.

    c. Charge is **conserved.** That is, the algebraic sum of the charges in a closed region never changes.

3. Electric charges interact via the **electric force.** The force between like charges is repulsive; between opposite charges it is attractive. The force between two charges acts along the line joining them and is proportional to the product of their charges and inversely proportional to the distance between them, as described by **Coulomb's law:**

$$\mathbf{F} = \frac{kq_1q_2}{r^2}\hat{\mathbf{r}}.$$

4. The electric force on a charge in the presence of two or more other charges is simply the vector sum of the forces caused by the individual charges. This important property of the electric force is known as the **superposition principle.** It greatly simplifies the calculation of electric forces arising from various **charge distributions.**

QUESTIONS

1. Imagine a universe with only one kind of electric charge. How might it differ from our universe?

2. Discuss this statement: It is precisely because electric forces are so strong that the electrical nature of most everyday interactions is not obvious.

3. Verify that the units of the coulomb constant k are $N \cdot m^2/C^2$. What are the units of ϵ_0?

4. You are given two electric charges. Could you determine whether they had the same or opposite signs? Could you determine the signs of each?

5. In Example 20–4 we found a point where the electric force on any charge would be zero. If you placed a charge there, would it be in stable equilibrium?

6. Consider two charges confined to a straight line. They either repel or attract each other, so that the electric force alone cannot hold them in static equilibrium. If you put a third charge on the line, is there any choice for its sign, magnitude, and location that would put all three charges in stable equilibrium under the influence of the electric force alone?

7. In which of the following phenomena does electromagnetism play a dominant role?

    a. Gasoline burns in a car engine.
    b. The moon orbits the earth.
    c. A nerve impulse travels from the brain to a muscle.
    d. Protons and neutrons join to form an atomic nucleus.
    e. Sunlight falls on a solar collector and heats water.
    f. A chemist synthesizes a new polymer.
    g. Nutrients cross a cell membrane to keep the cell alive.
    h. You sit in a chair and the chair doesn't collapse.
    i. The atmosphere remains bound to the earth.

8. Outside the atomic nucleus, the neutron is not a stable particle, but undergoes radioactive decay with a half life of about ten minutes. One of the products of neutron decay is the proton. Must there be others? If so, what electrical properties must they have?

9. In Example 20–5, why couldn't we just apply Coulomb's law and write $F = kQq_{rod}/a^2$ for the force of the rod on the charge Q? $q_{rod} = \lambda \ell$ is the total charge on the rod and a the distance between Q and the end of the rod.

10. Two cubical blocks of wood, each 10 cm on a side and each carrying charge $q = 5.0 \ \mu C$, are placed with their centers 15 cm apart. To calculate the force acting between them, can you use Coulomb's law with $q_1 = q_2 = 5.0 \ \mu C$ and $r = 15$ cm? Why or why not? If not, how, in principle, could you calculate the force?

11. A spherical balloon is initially uncharged. If positive charge were spread uniformly over the surface of the balloon, would you expect it to expand slightly or to contract slightly? What would happen if negative charge were spread over the balloon?

12. In addition to the electric force between an electron and a proton, there is also a gravitational force acting between the two. What happens to the ratio of electric to gravitational force as the distance between the electron and proton is increased?

13. Why do we say that the superposition principle simplifies the mathematical description of the electric force? After all, the principle requires us to do vector sums, which can be quite complicated.

14. Suppose someone argued that the force we call gravity is really an electric force, caused by a net electric charge on the earth. How could you disprove this? *Hint:* Think about Galileo!

15. Why does the event in Fig. 20–6 not violate conservation of charge?

16. How is the superposition principle of this chapter similar to the superposition principle we introduced in our study of wave motion? Incidentally, it is precisely the superposition principle for electric forces that gives rise to the superposition principle for electromagnetic waves.

PROBLEMS

Section 20.2   *Electric Charge*

1. Suppose the electron and proton charges differed by one part in one billion. Estimate the net charge you would carry, assuming equal numbers (how many?) of electrons, protons, and neutrons in your body.

2. A typical lightning flash delivers about 25 C of negative charge from cloud to ground. How many electrons are involved in the flash?

3. Current theories of the so-called elementary particles suggest that many of these particles are made from more basic building blocks called **quarks.** The two most common quarks carry charges of $+\frac{1}{3}e$ and $-\frac{2}{3}e$, respectively, where e is the magnitude of the elementary charge found on electrons, protons, and many other elementary particles. In what ways could three such quarks combine to make protons ($+1$ elementary charge) and neutrons (neutral)?

4. A 2-g Ping-Pong ball rubbed against a wool jacket acquires a net positive charge of $1 \ \mu C$. Estimate the fraction of the ball's electrons that have been removed. Assume that roughly half the ball's mass is protons, with nearly all the rest neutrons.

Section 20.3   *The Interaction of Charges: Coulomb's Law*

5. If the charge imbalance of Problem 1 existed, what would be the approximate force between you and another person 10 m away? Compare with your weight. In this approximation, treat the people as point charges.

6. The strong nuclear force is an attractive force that acts between two protons (as well as between neutrons, or protons and neutrons). For two protons, the strong nuclear force overcomes the repulsive electric force only when the protons are closer than about 10^{-15} m; at 10^{-15} m the two forces have approximately equal magnitude. Estimate this magnitude.

7. Compare the gravitational force between an electron and proton with the electrical force between the two. At what distance(s) is your answer correct?

8. A 16-μC charge is in the x-y plane at the point (0 m, 2.0 m), while a 41-μC charge is at the point (1.0 m, 0 m). Express the electric force on the 41-μC charge in unit vector notation.

9. A 6.5-μC charge is held at rest, while a small charged sphere of mass 2.3 g is released 50 cm away. Immediately after its release, the sphere is observed to accelerate toward the charge with an acceleration of 340 m/s². What is the charge on the sphere? *Hint:* Ignore gravity. How do you know you are justified in doing so? Does the acceleration of the sphere remain constant?

10. How far apart should an electron and proton be so that the force of earth's gravity on the electron is equal to the electric force from the proton? Your answer shows why gravity is unimportant on the molecular scale!

11. A spring of spring constant 100 N/m is stretched 10 cm beyond its 90-cm equilibrium length. If you want to keep it stretched by attaching equal electric charges to the opposite ends, what magnitude of charge should you use? What would your answer be if the spring were stretched 20 cm?

12. The earth carries a net negative charge of 4.3×10^5 C. The force due to this charge is the same as if all the charge were concentrated at earth's center. How much charge would you have to place on a 1.0-g mass for it to be suspended just above the earth by the electric force? Consult Appendix F for the radius of the earth.

13. Two small spheres having the same mass m and the same charge q are suspended from massless strings of length ℓ, as shown in Fig. 20–13. Each string makes

an angle θ with the vertical. Show that the charge on each sphere is given by

$$q = \pm \sqrt{\frac{4mg\ell^2 \sin^3\theta}{k \cos\theta}}.$$

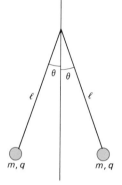

Fig. 20-13   Problem 13

14. Two identical small metal spheres initially carry charges q_1 and q_2, respectively. When the charges are 1.0 m apart they experience an attractive force of 2.5 N. The spheres are brought into contact and charge is transferred from one to the other until they have the same net charge. They are again placed 1.0 m apart, and now they repel each other with a force of 2.5 N. What were the original values of the charges on the two spheres?

Section 20.4   The Superposition Principle

15. Three charges are located in the x-y plane as shown in Fig. 20-14. Taking $q_1 = 68$ μC, $q_2 = -34$ μC, and $q_3 = 15$ μC, determine the force on q_3. Express your result both in unit vector notation and as a magnitude and direction.

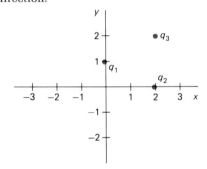

Fig. 20-14   Problem 15

16. Two charges q are located a distance a apart. A third charge Q is in line with the other two, and a distance x from a point midway between them, as shown in Fig. 20-15. Find expressions for the magnitude of the force on Q, for the two cases $|x| > a/2$ and $|x| < a/2$. Show that for $|x| \gg a$ your result is what you would expect for a point charge 2q acting on Q a distance x away. Is your result for $x = 0$ expected?

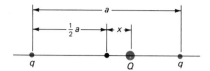

Fig. 20-15   Problem 16

17. Four identical charges q are placed at the corners of a square of side a, as shown in Fig. 20-16. Show that any charge placed at the center of the square will be in equilibrium. Is the equilibrium stable?

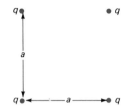

Fig. 20-16   Problems 17, 18

18. Suppose the bottom two charges of Fig. 20-16 are changed from q to −q. What is the force on a charge Q placed at the center of the square?

19. You are given two charges +4q and one charge −q. How would you space them along a line so that there is no net electric force on any of the three? Is this equilibrium stable?

20. A charge −q and a charge $\frac{4}{9}q$ are located a distance a apart, as shown in Fig. 20-17. Where would you place a third charge so that all three are in static equilibrium? What should be the sign and magnitude of the third charge? Is this equilibrium stable?

Fig. 20-17   Problem 20

21. A thin rod of length ℓ carries total charge q, distributed uniformly over its length. A point charge with the same charge q lies a distance a to the right of the rod end, as shown in Fig. 20-18. Where could you place a third charge so that it would experience zero net force?

Fig. 20-18   Problem 21

22. The thin rods shown in Fig. 20-19 are each 15 cm long and are oriented at 45° to the horizontal. Each rod carries a charge of 1.2 μC, distributed uniformly along its length. Lines drawn through the two rods

intersect 15 cm from their near ends. A small sphere carrying 5.2 μC is located at this intersection and is in static equilibrium, with the electric and gravitational forces in balance. What is its mass?

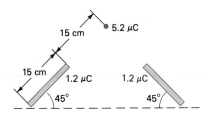

Fig. 20–19   Problem 22

Supplementary Problems

23. Repeat Problem 16 for the case where the charge Q lies a distance y along the perpendicular bisector of a line between the other two charges (Fig. 20–20). Show that the magnitude of the force on Q is now given by

$$F = \frac{8kqQy}{(a^2 + 4y^2)^{3/2}}.$$

Does this result make sense for $y = 0$ and $y \gg a$? What is the direction of the force?

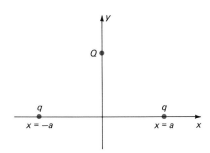

Fig. 20–20   Problem 23

24. A point charge carrying -2.5 μC is at rest 1.0 cm from another point charge carrying $+2.5$ μC. How much energy would be required to remove the negative charge to a great distance from the positive charge? *Hint:* Use the work-energy theorem; compare with the calculation of gravitational escape energy in Chapter 8.

25. A thin rod of length ℓ lies along the x-axis with its right end at $x = 0$, as shown in Fig. 20–21. The rod carries a nonuniform line charge density given by $\lambda = \lambda_0 x/\ell$, where λ_0 is a constant. Show that the magnitude of the force on a point charge Q at position $x = a$ along the positive x-axis is given by

$$F = \frac{kQ\lambda_0}{\ell}\left[\ln\left(\frac{a}{a+\ell}\right) + \left(1 - \frac{a}{a+\ell}\right)\right].$$

Fig. 20–21   Problem 25

26. Show that the result of the preceding problem reduces to

$$F = \frac{kQ\lambda_0\ell}{2a^2}$$

when $a \gg \ell$. By comparing this result for $a \gg \ell$ with Coulomb's law, show that the total charge on the rod is $\lambda_0\ell/2$. Confirm this by direct integration of $dq = \lambda dx$ over the rod. *Hint:* You will need to write the term $a/(a+\ell)$ as $(1 + \ell/a)^{-1}$, and use the binomial expansion and logarithmic expansion (see Appendix A) including the first three terms.

27. For the situation of Problem 23, find the point where the force has the greatest magnitude.

21

THE ELECTRIC FIELD AND GAUSS'S LAW

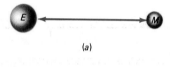

(a)

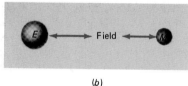

(b)

Fig. 21–1
(a) In the action-at-a-distance view, the earth directly influences a distant object, in this case the moon. *(b)* In the field view, earth gives rise to a field everywhere in space. The object, in turn, responds to the field in its immediate vicinity.

In Chapter 8, we introduced the concept of the gravitational field. Using this concept, we described the gravitational force on an object in earth's vicinity by saying that earth gives rise to a gravitational field at all points in space, and that the object responds to the field in its immediate vicinity (Fig. 21–1). This view contrasted sharply with our earlier "action-at-a-distance" picture of the earth somehow reaching out across empty space and pulling on the object.

Mathematically, we defined the gravitational field as a set of vectors describing the force per unit mass at all points (Fig. 21–2). Knowing the field at a point, we could calculate the force on *any* object at that point by multiplying the mass of the object by the gravitational field vector.

In our study of gravity, we did not use the field concept extensively. And we could always draw the same physical conclusions—the directions and magnitudes of forces on objects, or the parameters of a satellite's orbit—using either the field concept or the "action-at-a-distance" description. In light of our brief introduction of the gravitational field in Chapter 8, you are certainly justified in viewing the field as a purely mathematical construct we introduce for computational convenience or to satisfy some philosophical aversion to the notion of action-at-a-distance. But as we now advance into the study of electromagnetism, you will find that the field concept becomes virtually indispensable, and that fields become increasingly real to you. To the physicist, fields are every bit as real as matter.

21.1
THE ELECTRIC FIELD

electric field

The field concept applies to electrical as well as gravitational interactions. An **electric field** is a field that gives rise to forces on electric charges. Elec-

Fig. 21–2
Gravitational field vectors in the vicinity of earth. The vectors all point toward the center of the earth, and their magnitudes decrease with the inverse square of the distance from earth's center.

tric charge is the ultimate source of electric field, just as mass is of gravitational field. Instead of talking about forces between distant charges, we use the field concept to describe the force one charge experiences in the field of another. We say that one charge produces a field throughout space and that the second charge responds to the field at its particular location.

Just as the gravitational field is force per unit mass, so is the electric field force per unit charge. An electric field exists when charged objects experience forces proportional to their charges. We define the strength of the electric field as the magnitude of the force per unit charge. Thus in SI units, the electric field is measured in newtons/coulomb (N/C). We define the direction of the electric field to be the direction of the force on a positive charge.

To measure the electric field at a point, we put a known charge at that point and determine the force on it. We might, for example, use a spring scale to find the force required to hold the charge in place. Or we might release the charge and measure its acceleration, then use Newton's law to calculate the force. Since the magnitude of the field is defined as the force per unit charge, we then divide the measured force by the charge to determine the field strength. The field direction is that of the force if the charge is positive and opposite to the force if it is negative. This experimental procedure and the definition of electric field on which it rests can be summed up in one simple vector equation:

$$\mathbf{F} = q\mathbf{E}. \qquad (21\text{--}1)$$

Equation 21–1 relates the electric force on a charge to the electric field at the charge's location. As a vector equation, it expresses the relation between the direction of the field, the direction of the force, and the sign of the charge. If the charge q is positive, then the force $\mathbf{F}$ and the electric field $\mathbf{E}$ are in the same direction. But if q is negative, then the force is opposite the field direction.

Equation 21–1 may be regarded as a definition of $\mathbf{E}$, as a procedure for measuring $\mathbf{E}$, or, if $\mathbf{E}$ is known, as a formula for determining the force on a charge q placed in the field $\mathbf{E}$. Note that Equation 21–1 implies that the SI units of $\mathbf{E}$ are N/C.

There are two practical difficulties in using Equation 21–1 to measure electric fields. First, we must be sure that the force we measure is caused only by the electric field. If it is not, we must first subtract any other forces, such as gravity (see Problem 4). Second, the field we are trying to measure arises from one or more charges. If our test charge is sufficiently large, its own field may be strong enough to move the other charges, thereby changing the field we are trying to measure. For this reason we usually think of measuring electric fields with a very small test charge.

Example 21–1
The Electric Field

When placed at a certain point, a -5.0-μC charge experiences a force of 0.15 N in the $-x$-direction. What is the electric field at this point? What would be the force on a $+10$-μC charge at this point?

Solution

The electric field is a vector defined as the force per unit charge. This definition is embodied in Equation 21–1, which we solve for the field:

$$\mathbf{E} = \frac{\mathbf{F}}{q} = \frac{-0.15\hat{\imath} \text{ N}}{-5.0 \times 10^{-6} \text{ C}} = 3.0 \times 10^{4} \hat{\imath} \text{ N/C}.$$

Note that the minus sign on the charge results in a field in the $+x$-direction—opposite to the force on the negative charge.

The force on *any* charge placed in this field is given by Equation 21–1, so that the magnitude of the force on a $+10$-μC charge is simply

$$\mathbf{F} = q\mathbf{E} = (10 \times 10^{-6} \text{ C})(3.0 \times 10^{4} \hat{\imath} \text{ N/C}) = 0.30\hat{\imath} \text{ N}.$$

The force on the positive charge is in the same direction as the field.

21.2
CALCULATING THE ELECTRIC FIELD

Fig. 21–3
Electric fields in a TV tube accelerate electrons toward the front of the tube, where their energy is converted to the light we see. Shown here is the rear of the tube, where the acceleration takes place.

Knowing the electric field enables us to calculate the forces on electric charges and then, via Newton's law, the motion of those charges. Since all matter contains electrically charged particles, an understanding of electric fields helps us develop insight into the structure and behavior of matter. Furthermore, we can build devices in which electric fields accelerate particles in useful ways. Examples include television sets, x-ray machines, the fusion machine shown on the cover, and virtually every kind of electrical appliance (Fig. 21–3).

Given a distribution of charges, whether as simple as a single point charge or as complicated as a DNA molecule, we would like to know the associated electric field. In the case of a point charge, calculation of the electric field is particularly simple. We already know from Coulomb's law that if we place a point charge q_1 a distance r from another point charge q, the force on q_1 will be

$$\mathbf{F} = \frac{1}{4\pi\epsilon_0} \frac{qq_1}{r^2} \hat{\mathbf{r}}, \tag{21–2}$$

with $\hat{\mathbf{r}}$ a unit vector pointing from q toward the location of q_1. Since the electric field is defined as the force per unit charge, we divide the force in Equation 21–2 by the charge q_1 to obtain the field of q at the location of q_1:

$$\mathbf{E} = \frac{1}{4\pi\epsilon_0} \frac{q}{r^2} \hat{\mathbf{r}}. \quad \text{(field of a point charge)} \tag{21–3}$$

Since the location of q_1 is arbitrary, this equation gives the field arising from the charge q at any point a distance r from q. The equation contains no reference to the charge q_1, for the field of q exists independently of any other charge. The direction of $\mathbf{E}$ is away from q if q is positive and toward

Fig. 21–4
Field vectors arising from *(a)* a positive point charge and *(b)* a negative point charge. Both show that the field vectors point in the radial direction, directly away from or toward the charge, and that the field magnitude decreases with distance.

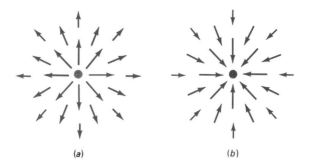

(a)   (b)

q if it is negative. Figure 21–4 shows some field vectors associated with positive and negative point charges.

What about the field of a more complicated charge distribution? Since the electric force obeys the superposition principle, so does the electric field—the force per unit charge. The field at a given point is therefore the vector sum of the fields of individual charges. Were this not true, the mathematical analysis of electromagnetic phenomena would be virtually impossible. Physically, the superposition principle is a reflection of the fact that the electric field, although emanating from charges, is itself uncharged. Think about this! Field and charge are different. Charge is the source of field, and field in turns acts on other charges. But the *field* of one charge does not act on the *field* of another. That is why the field of two or more charges can be obtained by calculating separately the fields of each charge as though no others existed, then adding the separate fields.

The superposition principle allows us to calculate the electric field of any charge distribution by summing the fields of individual point charges as given by Equation 21–3. Of course this sum is a vector sum, which may be written

$$E = E_1 + E_2 + E_3 + \cdots = \sum E_n = \sum \frac{1}{4\pi\epsilon_0} \frac{q_n}{r_n^2} \hat{r}_n, \qquad (21\text{–}4)$$

where the E_n's are the fields of the point charges q_n that are located at distances r_n from the point where we are evaluating the field. Equation 21–4 provides a prescription by which we can, in principle, calculate the field of any charge distribution. In practice, the process of summing the individual field vectors may be complicated unless our charge distribution contains relatively few charges organized in a symmetric way.

Example 21–2
Two Equal Point Charges

Consider two charges q lying on the x-axis a distance a on either side of the origin, as shown in Fig. 21–5. What is the electric field at a point on the y-axis?

Solution

Since any point on the y-axis is equidistant from each charge, the magnitude of the field from each is the same, and is given by

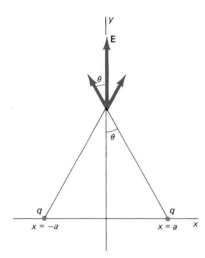

Fig. 21–5
At any point on the y-axis, the electric field vectors of two individual point charges sum to a net electric field pointing along the y-axis.

$$E_1 = E_2 = \frac{1}{4\pi\epsilon_0}\frac{q}{r^2} = \frac{1}{4\pi\epsilon_0}\frac{q}{a^2+y^2}.$$

But the field directions are different. Figure 21–5 shows that the x-components cancel, while the y-components are equal and add to give

$$E_y = E_{y1} + E_{y2} = 2\left(\frac{1}{4\pi\epsilon_0}\frac{q}{a^2+y^2}\cos\theta\right).$$

Figure 21–5 shows that

$$\cos\theta = \frac{y}{r} = \frac{y}{\sqrt{a^2+y^2}},$$

so that the electric field at P is given by

$$\mathbf{E} = \frac{1}{4\pi\epsilon_0}\frac{2qy}{(a^2+y^2)^{3/2}}\,\mathbf{\hat{j}}. \qquad (21\text{–}5)$$

Does this result make sense? Evaluating Equation 21–5 at $y=0$ gives zero field. Here, midway between the two charges q, a test charge would be pulled or pushed equally in opposite directions. There would be no net force, so the electric field should be zero. Evaluating Equation 21–5 at large distances, so large that a^2 can be neglected compared with y^2 in the denominator, we have

$$E = \frac{1}{4\pi\epsilon_0}\frac{2q}{y^2}. \qquad (y \gg a)$$

This is just the field of a point charge 2q located at the origin. Our example illustrates an important point. At a sufficiently great distance from any finite size distribution carrying a net charge, the detailed structure of the distribution becomes unimportant and the field of the distribution resembles that of a point charge, pointing in the radial direction and decreasing as the inverse square of the distance from the charge distribution.

Example 21–3 _____
An Electric Dipole

dipole

A **dipole** consists of two charges of equal magnitude but opposite sign that are held a fixed distance apart. The dipole configuration is important in practical devices such as radio antennas. Furthermore, many molecules can be regarded as dipoles, so that familiarity with the electric field of a dipole helps us to understand interactions among molecules.

Consider a dipole consisting of a positive charge $+q$ located at $x=a$ and a negative charge $-q$ located at $x=-a$, as shown in Fig. 21–6. What is the electric field at a point on the y-axis?

Solution

The situation is identical to that of Example 21–2, except for the direction of the field arising from the negative charge. As Fig. 21–6 indicates, the net electric field is now parallel to the x-axis. The field magnitudes are the same as in Example 21–2, but now the net field has only an x-component, twice the x-component of the field from either charge:

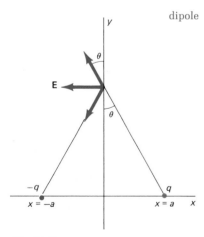

Fig. 21–6
An electric dipole.

$$E_x = E_{x1} + E_{x2} = -2\left(\frac{1}{4\pi\epsilon_0}\frac{q}{a^2+y^2}\sin\theta\right).$$

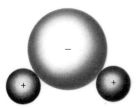

The minus sign appears because, as Fig. 21–6 indicates, the net field points in the negative x-direction. Also from Fig. 21–6, we see that $\sin\theta$ is given by

$$\sin\theta = \frac{a}{r} = \frac{a}{\sqrt{a^2 + y^2}},$$

so that

$$\mathbf{E} = -\frac{1}{4\pi\epsilon_0}\frac{2qa}{(a^2 + y^2)^{3/2}}\,\hat{\mathbf{i}}.$$

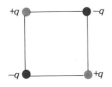

Fig. 21–7
A water molecule. Electrons spend more time in the vicinity of the single oxygen atom, giving rise to separate regions of positive and negative charge. Electrically, the molecule resembles a dipole. At distances large compared with its size, the molecule gives rise to an electric field just like that of the dipole of Fig. 21–6.

Does this result make sense? Midway between the charges, each contributes equally to the net electric field, pointing toward the negative charge. So we expect a resultant field twice that of either charge alone. Evaluating our field expression at $y = 0$ gives

$$\mathbf{E}(y=0) = -\frac{1}{4\pi\epsilon_0}\frac{2qa}{(a^2)^{3/2}}\,\hat{\mathbf{i}} = -\frac{1}{2\pi\epsilon_0}\frac{q}{a^2}\,\hat{\mathbf{i}},$$

which is indeed twice the field of either charge.

Frequently we are only interested in the field a great distance from the dipole. One needn't go far from a molecule for its size to be insignificant! Then a^2 can be neglected compared with y^2, giving

$$\mathbf{E} = -\frac{1}{4\pi\epsilon_0}\frac{2qa}{y^3}\,\hat{\mathbf{i}}. \qquad (21\text{--}6)$$

Fig. 21–8
An electric quadrupole, consisting of two oppositely directed dipoles, gives rise to a field that decreases as $1/r^4$.

The dipole field decreases as the inverse cube of the distance. Physically, this is because the dipole has no net charge. Its field arises entirely from the slight separation of two opposite charges. Because of this separation the dipole field is not exactly zero, but it is weaker and more localized than the field of a point charge, a fact reflected in its more rapid decrease with distance. Many more complicated charge distributions exhibit the essential property of a dipole—namely, they contain regions of equal but opposite charge separated from each other (Fig. 21–7). At great distances from these distributions, their more complicated structure becomes unimportant, and their fields approach that of Equation 21–6. Still other distributions, such as the oppositely directed dipoles—or **quadrupole**—of Fig. 21–8, have fields that decrease even faster than $1/r^3$.

quadrupole

The physical characteristics of the dipole that determine the field are the charge q and separation $2a$. At relatively great distances from the charges ($r \gg a$), only the product $2qa$ of the charge and separation is significant. We call this product the **dipole moment,** p. Its units are C·m. Using the dipole moment, Equation 21–6 may be written

dipole moment

$$\mathbf{E} = -\frac{p}{4\pi\epsilon_0 y^3}\,\hat{\mathbf{i}}. \qquad (21\text{--}7a)$$

It is useful to imagine a dipole whose charge separation shrinks to zero, while the magnitude of the charges grows indefinitely to maintain a constant dipole moment. The limit of zero size is an idealization called a **point dipole.** Since it has zero size, the field on the perpendicular bisector of the point dipole is given exactly by Equation 21–7a. Although the point dipole is an idealization, it is a useful approximation to a real dipole whose size is small compared with distances to other charge distributions.

point dipole

What about the field at points not on the y-axis? In Problem 8, the field along the x-axis is shown to be

$$\mathbf{E} = \frac{p}{2\pi\epsilon_0 x^3}\,\hat{\mathbf{i}}. \qquad (21\text{--}7b)$$

Problem 9 generalizes Equations 21–7a and 21–7b to obtain an expression for the field of a dipole at an arbitrary position **r**. Because the dipole is not spherically symmetric, its field is a function both of distance and of the angle between the position vector and the dipole axis. For this reason, it is often convenient to treat the **dipole moment** as a **vector** quantity. We define this quantity to be a vector of length p (the product of dipole charge and separation) that points from the negative charge toward the positive charge (Fig. 21–9).

dipole moment vector

Fig. 21–9
The dipole moment vector has a magnitude given by the charge times the separation, and points from the negative toward the positive charge.

In the preceding chapter, we introduced the notion of a continuous distribution of charge, characterized by its charge per unit volume (ρ), per unit area (σ), or per unit length (λ). Of course any charge distribution consists ultimately of discrete charges residing on elementary particles. But if we are interested in electric fields on scales much larger than, say, the spacing between atoms, then the idea of a continuous distribution introduces negligible error and greatly simplifies our calculations. To calculate the field of a continuous distribution, we break the charged region into very many small charge elements dq. Each dq produces a small electric field $d\mathbf{E}$ given by Equation 21–3:

$$d\mathbf{E} = \frac{1}{4\pi\epsilon_0} \frac{dq}{r^2} \hat{\mathbf{r}}. \qquad (21–8)$$

Then, in analogy with Equation 21–4, we form the vector sum of all the $d\mathbf{E}$'s. But now there is no limit to the number of charges dq, and the individual dq's and their fields are infinitesimally small, so that the sum becomes an integral:

$$\mathbf{E} = \int d\mathbf{E} = \int \frac{1}{4\pi\epsilon_0} \frac{dq}{r^2} \hat{\mathbf{r}}. \qquad (21–9)$$

The limits of this integral are chosen to include the entire charge distribution. Figure 21–10 illustrates schematically the meaning of Equation 21–9.

How are we to evaluate the integral in Equation 21–9? Because it is a vector integral—a sum over many vectors, which may not all point in the same direction—it is frequently necessary to break the integral into com-

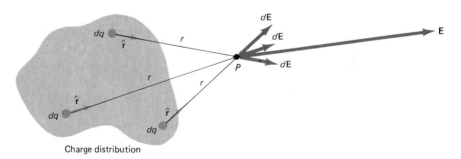

Charge distribution

Fig. 21–10
The electric field at point *P*, given by Equation 21–9, is the sum of the vectors *d*E arising from all the individual charge elements *dq* in the entire charge distribution, each calculated using the appropriate distance *r* and unit vector *r̂*, both of which may vary from one charge element to another.

ponents along appropriate coordinate axes. Furthermore, the integral contains the charge element dq and also varying quantities like r and any trigonometric functions arising from taking components. To evaluate the integral, we must express all these things in terms of a single integration variable. The charge can be related to the geometrical variables by writing it as the appropriate charge density multiplied by a small element of volume, area, or length:

$$dq = \rho dV,$$

$$dq = \sigma dA,$$

or

$$dq = \lambda dx.$$

Which of these we use depends on whether our charge is distributed over a volume, an area, or a line. Example 21–4 illustrates the application of Equation 21–9 to a line of charge.

Example 21–4
An Infinite Line of Charge

An infinite line of charge coincides with the x-axis and carries a line charge density λ C/m. What is the electric field at a point P on the y-axis?

Solution

Of course no such infinite line exists. But calculating the field of this relatively simple charge distribution gives insight into the fields of more complicated distributions. Furthermore, our result is a good approximation to the field of a finite line of charge at points that are much closer to the line than they are to either end.

Consider a small element of charge dq located a distance x to the right of the origin. Figure 21–11 shows the contribution that this element dq makes to the field at point P. To determine the net field at P, we could write this dE in terms of its x- and y-components, and then integrate each component over the entire line to get the field of all the charge elements dq along the line. But for each dq to the right of the origin, there is a corresponding dq the same distance to the left. The x-components of the fields from such a pair cancel, while the y-components are the same, as shown in Fig. 21–11. In fact, our pair of charge elements dq is just like the system of two point charges we examined in Example 21–2. There, for two charges lying a distance a on either side of the origin, we found an electric field on the y-axis given by Equation 21–5. Here our charges have size dq, are located a distance x on either side of the origin, and give rise to an infinitesimal field dE. So we replace a by x, q by dq, and $\mathbf{E}$ by $d\mathbf{E}$ in Equation 21–5 to get the field of our charge-element pair:

$$d\mathbf{E} = \frac{1}{4\pi\epsilon_0} \frac{2ydq}{(x^2+y^2)^{3/2}} \,\hat{\mathbf{j}}.$$

To evaluate the net field, we add—that is, integrate—the contributions of all charge element pairs along the line. Each pair contributes only a y-component to the field, so that

$$\mathbf{E} = \int d\mathbf{E} = \int_{x=0}^{\infty} \frac{1}{4\pi\epsilon_0} \frac{2ydq}{(x^2+y^2)^{3/2}} \,\hat{\mathbf{j}}.$$

Although the line extends from $-\infty$ to $+\infty$, we integrate over only half the line

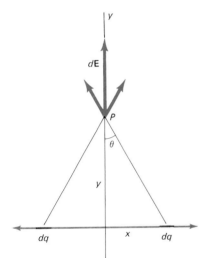

Fig. 21–11
A pair of charge elements dq on either side of the origin contributes a net field $d\mathbf{E}$ in the y-direction.

because the expression we are integrating is already the field of a charge pair, and accounts for charge elements on either side of the origin.

Many quantities under the integral sign are constant—they don't change as we range over the charge distribution and pick up contributions from all the charge elements dq. Certainly $2/4\pi\epsilon_0$ is constant. So is y, since the point P stays fixed on the y-axis. But x changes as we move along the line. The problem is to relate x and dq, so that we have only one variable under the integral sign. How? A small charge element dq takes up a small distance dx along the line. How much charge is contained in this distance dx? The line carries a charge λ per unit length, so that

$$dq = \lambda dx.$$

Substituting this expression for dq and bringing constants out of the integral gives

$$\mathbf{E} = \frac{\lambda y}{2\pi\epsilon_0} \int_0^\infty \frac{dx}{(x^2+y^2)^{3/2}} \,\hat{\mathbf{j}}. \tag{21-10}$$

The integral is a standard form that could be found in any table of integrals. Alternatively, we could make a change of variables to obtain an easily integrated expression. In this problem the geometry suggests such a substitution. The angle θ shown in Fig. 21–11 is just as good a measure of position along the line as the coordinate x. The two are related by

$$x = y\tan\theta$$

so that

$$dx = y\,\frac{d\theta}{\cos^2\theta},$$

where the derivative of $\tan\theta$ is given in Appendix A, or could be obtained by differentiating $\sin\theta/\cos\theta$. Then the integral in Equation 21–10 becomes

$$\int \frac{y}{(x^2+y^2)^{3/2}}\,\frac{d\theta}{\cos^2\theta} = \frac{1}{y^2}\int \left(\frac{y}{\sqrt{x^2+y^2}}\right)^3 \frac{d\theta}{\cos^2\theta} = \frac{1}{y^2}\int \cos\theta\,d\theta,$$

where we have used $y/\sqrt{x^2+y^2}=\cos\theta$. What about the limits of this integral? As x ranges from 0 to infinity, θ ranges from 0 to $\pi/2$. Then we have

$$\frac{1}{y^2}\int_0^{\pi/2} \cos\theta\,d\theta = \frac{1}{y^2}\sin\theta\Big|_0^{\pi/2} = \frac{1}{y^2}(1-0) = \frac{1}{y^2}.$$

Substituting this result for the integral in Equation 21–10 then gives the field:

$$\mathbf{E} = \frac{\lambda y}{2\pi\epsilon_0}\frac{1}{y^2}\,\hat{\mathbf{j}} = \frac{\lambda}{2\pi\epsilon_0 y}\,\hat{\mathbf{j}}. \tag{21-11}$$

This field decreases as the inverse of the distance, rather than as the inverse square. This happens even for great distances because the charge distribution is of infinite extent, so that it never looks like a point charge. Although Equation 21–11 describes the field along our particular y-axis, the infinite length of line and its cylindrical symmetry mean that all directions perpendicular to the line are equivalent. We could have placed our y-axis anywhere perpendicular to the line, so that our result describes the field at any point a distance y from the line. The field vectors point radially outward from the line and their magnitude decreases inversely with distance (Fig. 21–12).

This example required a fairly involved calculation. As charge distributions get more complicated, calculating electric fields by summing the fields of individual charges remains simple in principle but becomes difficult in practice. For some charge distributions, including the infinite line of charge, there is a much easier and more elegant way to determine the electric field, which we will now develop.

Fig. 21–12
The electric field associated with an infinite line of charge points radially outward from the line, and its magnitude decreases inversely with distance.

21.3
ELECTRIC FIELD LINES

electric field lines

The electric field of any charge distribution extends throughout space. We can picture the entire field using continuous lines called **electric field lines.** To draw a single field line, start at some point and determine the field there. Move a very small distance in the direction of the field and evaluate the field at the new point. Repeating this procedure traces out a path which is the electric field line. One can also move backward from the initial point, tracing out the rest of the field line. The resulting field line is a path whose direction at any point is the direction of the electric field vector at that point. More mathematically, the tangent to the field line gives the direction of the electric field at the tangent point. Drawing many such field lines gives a sense of the overall structure of the electric field.

For the field of a single point charge, tracing the field lines is particularly simple. For a positive charge, the field at any point is directed away from the charge. Move a little distance farther away and the field still points directly away from the charge. So the field lines are straight lines pointing radially outward from the point charge. The lines start on the charge and extend outward indefinitely. Figure 21–13 shows some field lines for an isolated point charge.

A field line picture shows the direction of the electric field, but what about its magnitude? In Fig. 21–13 the field lines spread apart as they extend farther from the point charge. Coulomb's law tells us that the field gets weaker farther from the charge. So in Fig. 21–13 the field is stronger where the field lines are closer together and weaker where they are farther apart. This qualitative statement is always true of electric field lines, and allows us to infer relative field strength as well as direction from field line pictures.

A quantitative relation between field lines and field strengths holds too. Defining the density of field lines as the number of lines that cross a unit area oriented at right angles to the lines, we find that field line density is directly proportional to field strength. We can easily see this for the field of a point charge. Consider the two spherical shells shown surrounding the point charge in Fig. 21–14. The same number of field lines crosses each shell, and the lines are perpendicular to the shells. The larger shell has twice the radius of the smaller one, so that its surface area is four times greater. Then the field-line density is one-fourth as great for the larger sphere. This is just the decrease in electric field strength given by the $1/r^2$ dependence in Coulomb's law. It is precisely because the inverse-square law holds for the electric field of a single point charge that we can represent the field quantitatively with field lines.

You might argue that "number of field lines" is a vague term since we can always draw as many field lines as we want. To make the field-line picture useful, however, we associate a fixed number of field lines with a charge of given magnitude. Thus in Fig. 21–15 we adopt the convention that eight field lines emanate from a charge q, as shown in Fig. 21–15a. Then a charge $2q$ is represented by 16 field lines, as in Fig. 21–15b. A negative charge is represented by field lines pointing toward the charge, as in Fig. 21–15c. In two-dimensional drawings like Fig. 21–15, we must re-

Fig. 21–13
Field lines of an isolated point charge.

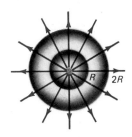

Fig. 21–14
The outer shell has twice the radius and four times the area of the inner shell. Because the same number of field lines crosses each, the field line density decreases as $1/r^2$, and is therefore proportional to the field magnitude.

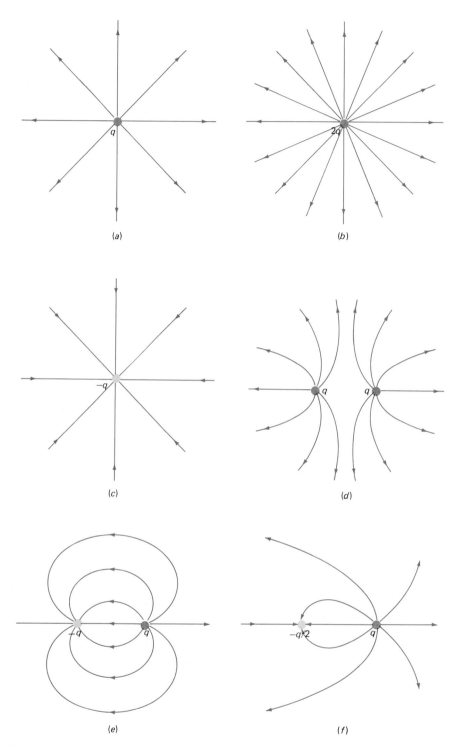

Fig. 21–15
Field lines for (a) a positive charge q, (b) 2q, (c) −q, (d) the two equal charges of Example 21–2, (e) the dipole of Example 21–3, (f) unequal charges +q and −½q. In all six drawings, eight field lines represent a charge q, with different charges represented by proportionately different numbers of lines.

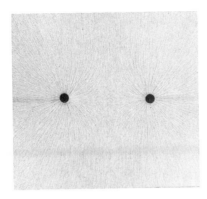 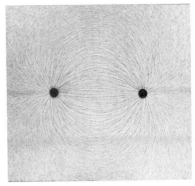

Fig. 21–16
Field line patterns associated with charged objects are evident in these photographs by the alignment of small fibers floating on a liquid. (Left) Two equal charges. (Right) Two opposite charges. Compare with Fig. 21–15*d* and *e*, respectively.

member that it is the number of field lines per unit *area* that is strictly proportional to field strength. For the point charges in Fig. 21–15, the spacing of field lines on the two-dimensional page decreases as $1/r$, but their density in three dimensions decreases as $1/r^2$.

In drawing the field lines of more complicated distributions, we must add vectorially the fields of individual charges to get the net field that gives the field line direction. Figure 21–15*d–f* show the fields for the charge distributions of Examples 21–2 and 21–3, and for two opposite but unequal charges. In Fig. 21–15*d* and *f*, note how the field far from the charges begins to resemble that of a single point charge.

It is possible to display electric field line patterns experimentally by floating short fibers on a liquid in which charged objects are placed (Fig. 21–16). Similarly, hairs on the head of a highly charged person align themselves with the electric field (Fig. 21–17).

Fig. 21–17
Hairs of a highly charged person align themselves with the electric field.

Counting Field Lines

Figure 21–18 shows the same charge distributions as Fig. 21–15. For each distribution a number of surfaces are indicated by gray lines. (The figure shows only the two-dimensional cross section of each surface.) Each surface is closed; it is impossible to get from inside to outside without crossing the surface. We now ask a very simple question: How many field lines cross each surface?

Consider Fig. 21–18a. For surfaces 1 and 2 the answer is obvious: eight. Surface 3 is a little ambiguous, but is clarified if we adopt the convention that a field line crossing from inside to outside counts as positive and a crossing from outside to inside as negative. Then the one field line that crosses surface 3 three times does so twice going out and once going in, for a net gain of one crossing. So eight field lines also cross surface 3. In fact, any closed surface you might draw that enclosed the charge q would have eight field lines crossing it. Surface 4 is different. Two lines cross going in and two going out, making zero net crossings. Although surface 4 lies in the field of q, q is not inside the surface. By drawing other surfaces you can convince yourself that any surface that does not enclose q has zero net field line crossings.

The situation for Fig. 21–18b is identical except that now surfaces enclosing the charge have sixteen field line crossings, reflecting the double charge $2q$ in this figure. Surfaces that do not enclose the charge still have zero net crossings. Figure 21–18c is also like Fig. 21–18a except that its charge is $-q$. According to our sign convention, -8 field lines cross any surface enclosing the charge $-q$.

In Fig. 21–18d, surfaces 1 and 2 each enclose one of the charges q. These surfaces, and any others you draw that enclose only one of the charges, have eight field line crossings. But how many field lines cross surface 3? Sixteen. What is the total charge enclosed by surface 3? $2q$. Meanwhile surface 4, which is in the field of both charges but encloses no charge, has zero net field line crossings.

On to Fig. 21–18e, the dipole. Surface 1 encloses charge q and has eight field line crossings. Surface 2 encloses $-q$ and has -8 field line crossings. Surface 3 is interesting. It encloses both charges q and $-q$, for a net enclosed charge of zero. Count the field lines. As many go out as come in, so the total number of field line crossings is zero.

Figure 21–18f has two opposite but unequal charges q and $-q/2$. Eight field lines cross surface 1, which encloses only the charge q. Surface 2 encloses only $-q/2$, and has -4 field line crossings. So would any other surface enclosing only $-q/2$. Surface 3 encloses both charges, for a net enclosed charge of $q/2$. Four field lines cross surface 3—the same number of lines we would expect from a point charge of magnitude $q/2$, given our convention that eight field lines represent a charge q. Finally surface 4, which is in the field of both q and $-q/2$, but encloses neither, has zero net field line crossings.

The exercise of counting field line crossings in Fig. 21–18 is deceptively simple. It is probably one of the easier things you've done in your study of physics, yet from it has emerged a simple but profound statement about the electric field:

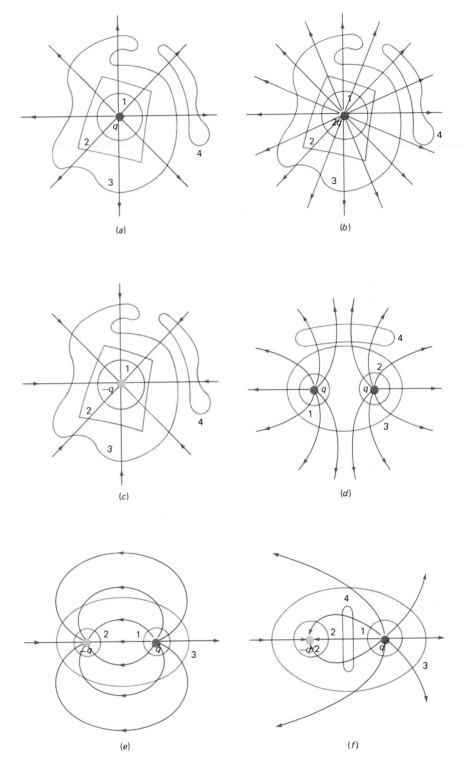

Fig. 21–18
In all cases, the number of field lines emerging from a closed surface is proportional to the net charge enclosed.

> The number of field lines crossing a closed surface is proportional to the net charge enclosed by that surface.

You may have reached this conclusion before you finished considering the charge distributions of Fig. 21–18. Notice how general the statement is. It does not matter what shape the surface is, or whether the enclosed charge is a single point or a lot of charges adding to the same net charge. Nor does it matter how those charges are arranged as long as they are enclosed by the surface. The number of field lines crossing a closed surface depends *only* on the net charge enclosed by that surface. Our statement is at once simple, elegant, and extraordinarily powerful. We will now rephrase it in a more mathematically rigorous way, obtaining one of the four fundamental laws that govern the behavior of electromagnetic fields throughout the universe. As we do so, defining new terms and writing integrals, remember that we are merely rewriting the simple truth obvious in Fig. 21–18. You should refer to that figure and accompanying discussion if you begin to lose the physical significance amid the mathematics.

Electric Flux

To describe more rigorously what we mean by "number of field lines crossing a surface," consider Fig. 21–19, which shows a flat sheet of area A placed in a uniform electric field $\mathbf{E}$. How many field lines cross this

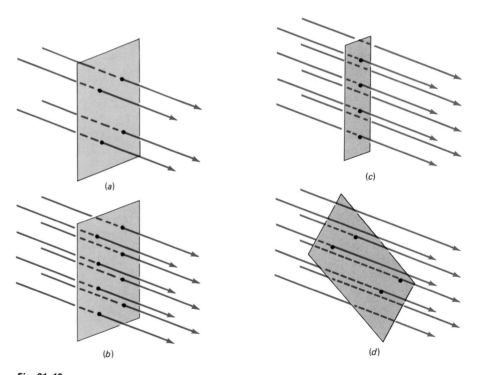

Fig. 21–19
The number of field lines crossing a flat surface depends on the field strength, the surface area, and the orientation of the surface relative to the field direction.

surface? For Fig. 21–19a the answer is four. In Fig. 21–19b, the field strength is twice as great as in Fig. 21–19a (how can you tell?), and there are eight field line crossings. In Fig. 21–19c the field is the same as in Fig. 21–19b, but the area is halved and so is the number of field line crossings. In Fig. 21–19d the larger area is oriented so it is no longer perpendicular to the field, and the number of field line crossings is reduced. We see that the number of field line crossings depends on three things: field strength, surface area, and orientation of surface relative to field.

How can we express the orientation of the surface? For a flat surface, we use the direction of the perpendicular to specify the surface orientation (Fig. 21–20). If θ is the angle between the electric field vector and this perpendicular, as indicated in Fig. 21–21, then the number of field line crossings range from a maximum when θ is zero down to zero when θ is $\pi/2$ or 90°. Given this behavior, you can probably guess what trigonometric function of θ is involved. By examining Fig. 21–21, you should convince yourself that the number of field lines is proportional to the *cosine* of θ.

Putting together the three quantities on which the number of field lines depends gives

$$\text{Number of field lines} \propto E\,A\,\cos\theta. \qquad (21\text{–}12)$$

Now "number of field lines" is still a vague term, since we can draw as many field lines as we like. But the quantity on the right side of this equation has a definite value, and it captures the spirit of what we mean by "number of field lines crossing a surface." We call this quantity the **electric flux** ϕ_E through the surface. For open surfaces like those of Figs. 21–19 and 21–21, there is an ambiguity in the sign of the flux, since we could choose either direction along the perpendicular in determining θ. But for closed surfaces, we unambiguously define the direction to use in determining θ as the direction of the *outward*-pointing perpendicular. If we define a vector **A** whose direction is that of the perpendicular to the surface, and whose magnitude is the surface area, then we can express Equation 21–12 more compactly as

$$\phi = \mathbf{E}\cdot\mathbf{A}, \qquad (21\text{–}13)$$

where $\cos\theta$ is now included in the dot product.

Although the flux ϕ is related to the electric field **E**, the two quantities are quite distinct. The field is a vector defined at every point. Flux is a global property of the field, meaning that it depends not on the field at a single point but rather on many field vectors taken over an area. Unlike field, flux is a scalar quantity. Electric flux is simply a quantification of the notion "number of field lines crossing a surface."

What if a surface is curved, and/or the electric field varies from one point to another? The vectors **E** and **A** in Equation 21–13 are then ambiguous. In this case, we divide the surface into many small patches, each small enough that it is essentially flat. With very small patches, the field will be nearly uniform over each patch (Fig. 21–22). Equation 21–13 can then be used to evaluate the flux through the patch:

$$d\phi = \mathbf{E}\cdot d\mathbf{A},$$

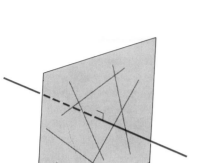

Fig. 21–20
Many lines of different directions lie within the surface; only the perpendicular to the surface uniquely specifies the surface orientation.

electric flux

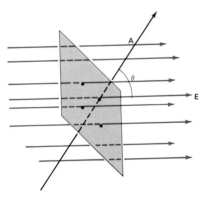

Fig. 21–21
The number of field lines crossing the area is proportional to $\cos\theta$, with θ the angle between the field and the perpendicular to the surface.

Fig. 21–22
Even though the surface is curved and the field varies, a small enough patch of surface is nearly flat and the field nearly uniform over the patch.

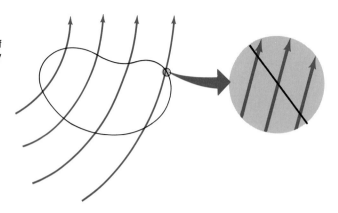

where **E** is the electric field at the patch, $d\mathbf{A}$ a vector specifying the orientation and area of the small patch, and $d\phi$ the resulting flux. To find the total flux through the surface, we add the fluxes through all the patches. Letting each patch become infinitesimally small and the number of such patches arbitrarily large, the sum becomes an integral:

$$\phi = \int_{\text{surface}} \mathbf{E}\cdot d\mathbf{A}. \quad \text{(electric flux)} \qquad (21\text{--}14)$$

The limits of this integral are chosen to range over the entire surface, picking contributions from all the patches $d\mathbf{A}$. Although the integral can be difficult to evaluate, we will find it most useful in cases where its evaluation is almost trivial. Since **E** is measured in N/C, the units of flux are $\text{N}\cdot\text{m}^2/\text{C}$.

21.4
GAUSS'S LAW

In Section 21.3, we found the simple result that the number of electric field lines crossing a closed surface is proportional to the net charge enclosed. Now that we have developed flux as a concept expressing more rigorously the idea of "number of field lines," we can rephrase this statement:

The electric flux through any closed surface is proportional to the net charge enclosed by that surface.

Writing the same thing in mathematical symbols gives

$$\phi = \oint \mathbf{E}\cdot d\mathbf{A} \propto q_{\text{enclosed}}. \qquad (21\text{--}15)$$

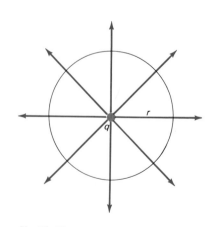

Fig. 21–23
The electric field of a point charge has the same magnitude over a spherical surface centered on the charge, and is everywhere perpendicular to the surface.

Here the circle on the integral sign is a reminder that we are to integrate over a *closed* surface.

To evaluate the proportionality constant in Equation 21–15, consider a positive point charge q and a spherical surface of radius r centered on q (Fig. 21–23). The flux through this surface is given by Equation 21–14:

$$\phi = \oint \mathbf{E} \cdot d\mathbf{A} = \oint E \, dA \cos\theta,$$

where θ is the angle between the electric field and the outward perpendicular to the sphere, and where the integral is over the entire surface of the sphere. The magnitude of the electric field is given by Equation 21–3:

$$E = \frac{1}{4\pi\epsilon_0}\frac{q}{r^2}.$$

The field points radially outward, so that it is everywhere parallel to the outward perpendicular to our spherical surface. Then $\theta = 0$, so that $\cos\theta = 1$, and the flux becomes

$$\phi = \oint_{\text{sphere}} \frac{1}{4\pi\epsilon_0}\frac{q}{r^2}\, dA = \frac{1}{4\pi\epsilon_0}\frac{q}{r^2}\oint_{\text{sphere}} dA,$$

where we have taken the expression for the field magnitude outside the integral because it has the same value everywhere on our spherical surface. The remaining integral is just the sum of the areas of all the infinitesimal patches on the surface of the sphere—in other words, the surface area of the sphere, $4\pi r^2$. Then the flux becomes

$$\phi = \frac{1}{4\pi\epsilon_0}\frac{q}{r^2}\, 4\pi r^2 = \frac{q}{\epsilon_0}.$$

Comparing this result with Equation 21–15 shows that the proportionality constant between the flux through a closed surface and the charge enclosed by that surface is $1/\epsilon_0$. So Equation 21–15 becomes

$$\oint \mathbf{E} \cdot d\mathbf{A} = \frac{q_{\text{enclosed}}}{\epsilon_0}, \qquad \text{(Gauss's law)} \qquad (21\text{–}16)$$

where the integral is taken over *any* closed surface that encloses the net charge q.

Although we used a specific field—that of a point charge—and a specific surface in evaluating the proportionality constant, once that constant is known Equation 21–16 becomes a general statement about the behavior of electric fields. Known as **Gauss's law,** it is one of four fundamental relations that govern all electromagnetic phenomena throughout the universe. Whether you journey into the interior of a star in some remote galaxy, down among the strands of a DNA molecule, or into the generator of an electric power plant, you will find that the flux of the electric field through any closed surface you care to consider will depend only on the net charge enclosed by that surface. In over a century of experiments, no electric field has ever been observed to violate Gauss's law.

We stress again that Gauss's law, although clothed in the mathematical finery of a surface integral, is simply a more rigorous way of saying that the number of field lines crossing a closed surface depends only on the enclosed charge. This in turn is true because electric field lines do not begin or end in empty space, but always on charges, a fact that reflects the inverse-square nature of the electric force between point charges.

Gauss's law

21.5
USING GAUSS'S LAW

gaussian surface

Gauss's law is true for *any* surface enclosing *any* charge distribution. (We call such a surface a **gaussian surface.**) However, evaluation of the surface integral is often difficult when the field strength or field direction relative to the surface vary. But when the charge distribution has sufficient symmetry, evaluation of the integral becomes simple. Then Gauss's law allows us to calculate the electric field far more easily than we could using Coulomb's law. We now illustrate the use of Gauss's law for three important symmetries.

Spherical Symmetry

center of symmetry

A charge distribution is spherically symmetric if the charge density depends only on the distance from some point, called the **center of symmetry.** A point charge, a uniformly charged sphere, and a spherical surface carrying a uniform surface charge density are all spherically symmetric charge distributions. Spherical symmetry of a charge distribution implies that the magnitude of the electric field also depends only on the distance r from the center of symmetry. What about the field direction? The only possible direction consistent with the symmetry is the radial direction—inward for a negative charge and outward for a positive charge (Fig. 21–24).

Gauss's law applies to *any* surface we might draw around the spherically symmetric charge distribution. But a useful surface is a sphere of arbitrary radius r centered on the center of symmetry. Then the magnitude E is the same at all points on this gaussian surface. Furthermore, **E** is everywhere in the same direction as the perpendicular to the surface, so that $\cos\theta = 1$. (Here we assume a net positive charge; for a negative charge, the field points radially inward, and $\cos\theta = -1$.) Thus, although the direction of **E** and of the surface vary, the product $E\cos\theta$ remains constant and is simply the field magnitude E. Then the flux through our gaussian sphere becomes

$$\phi = \oint \mathbf{E}\cdot d\mathbf{A} = \oint E\cos\theta \, dA = E\oint dA = 4\pi r^2 E, \qquad (21\text{--}17)$$

where the last step follows because $\oint dA$ is just the surface area of the sphere, $4\pi r^2$. This expression for the flux does not depend on the details of the charge distribution, so long as it is spherically symmetric.

Gauss's law says that the flux through the sphere is given by q/ϵ_0, where q is the net charge *enclosed* by the sphere. Suppose our spherically symmetric charge distribution carries total charge Q and has radius R. That is, whatever the particular distribution of charge for $r \leq R$, there is no charge at $r > R$. For any gaussian sphere with $r > R$, like surface a in Fig. 21–25, the enclosed charge is the total charge Q. Equating the flux in Equation 21–17 to Q/ϵ_0 through Gauss's law gives

$$4\pi r^2 E = \frac{Q}{\epsilon_0},$$

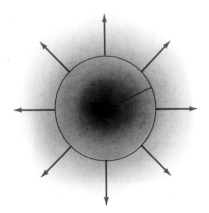

Fig. 21–24
For a spherically symmetric charge distribution, the field vectors at a given radius all have the same magnitude and point in the radial direction.

so that

$$E = \frac{1}{4\pi\epsilon_0}\frac{Q}{r^2}. \qquad (r>R) \qquad\qquad (21\text{–}18)$$

This is just the field of a point charge! Equation 21–18 is a remarkably simple result. It says that *outside* any spherically symmetric distribution of charge, the field is identical to that of a point charge located at the center of symmetry. This is not just an approximation valid at great distances from the charge—it is exactly true right up to the surface r=R! Imagine how hard it would have been to calculate this field from Coulomb's law! Yet somehow all the charge elements throughout the spherically symmetric distribution produce d**E**'s that add vectorially to give the same field as a single point charge. Like Gauss's law itself, this result is a consequence of the inverse-square force law. (In Chapter 8, we proved the analogous result for the gravitational field, which also obeys an inverse-square force law. There, since we didn't know about Gauss's law, our calculation involved a complicated integration.)

The field inside the charge distribution depends on how charge is distributed. This is because a gaussian sphere with r<R, such as surface *b* of Fig. 21–25, does not enclose the entire charge Q. How much it encloses depends on the charge distribution. In the examples below, we consider two important special cases.

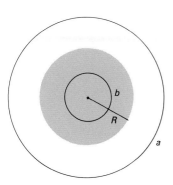

Fig. 21–25
The surface *a* encloses the entire charge *Q*, while surface *b* encloses only some of the charge. The quantity *q* in Gauss's law represents only the *enclosed* charge.

Example 21–5
A Uniformly Charged Sphere

A total charge Q is spread uniformly throughout a sphere of radius R. What is the electric field at all points in space?

Solution

This charge distribution is spherically symmetric, so that the field for r>R is like that of a point charge, and is given by Equation 21–18.

Inside the sphere, Equation 21–17 for the flux still holds, but now the charge enclosed is some fraction of Q. What fraction? The volume of the sphere is $\frac{4}{3}\pi R^3$, and it contains a total charge Q. Since charge is spread uniformly throughout the sphere, the volume charge density is constant, and is given by

$$\rho = \frac{Q}{V} = \frac{Q}{\frac{4}{3}\pi R^3}.$$

The charge enclosed by a sphere of radius r is just the volume of that sphere multiplied by the volume charge density:

$$q_{enclosed} = V\rho = \frac{4}{3}\pi r^3 \frac{Q}{\frac{4}{3}\pi R^3} = Q\frac{r^3}{R^3}.$$

Equating the flux of Equation 21–17 to $q_{enclosed}/\epsilon_0$, we have

$$4\pi r^2 E = \frac{Qr^3}{\epsilon_0 R^3},$$

so that

$$E = \frac{1}{4\pi\epsilon_0}\frac{Qr}{R^3}. \qquad (r<R) \qquad\qquad (21\text{–}19)$$

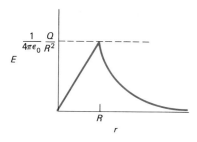

Fig. 21–26
Field strength versus radial distance for a uniformly charged sphere of radius R. For $r > R$ the field has the inverse-square dependence of a point-charge field.

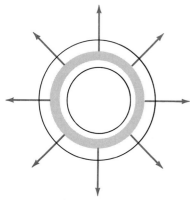

Fig. 21–27
A charged spherical shell. Any gaussian sphere outside the shell encloses the entire charge Q, and the field outside is that of a point charge. Any gaussian sphere inside the shell encloses zero net charge, and the field inside is zero.

The field *inside* this charge distribution increases linearly with distance from the center. This result is entirely consistent with the inverse-square law for point charges. Although the field of each charge element decreases as $1/r^2$, in this case the amount of charge enclosed increases more rapidly—as r^3—resulting in a field that increases linearly with r. Figure 21–26 shows the combined results for the fields both inside and outside the sphere. The field direction is, of course, radial, pointing outward if Q is positive and inward if Q is negative.

Example 21–6

A Thin Spherical Shell

A thin spherical shell of radius R carries a total charge Q distributed uniformly over its surface. What is the electric field inside and outside the shell?

Solution

Since this distribution is spherically symmetric, we already know that the field outside is just the point-charge field of Equation 21–18.

For any gaussian sphere inside the shell, the enclosed charge is zero (Fig. 21–27). Equating the flux from Equation 21–17 to this zero enclosed charge then gives

$$4\pi r^2 E = 0,$$

so that the field is zero everywhere inside the shell! How can this be? Again, it is a manifestation of the inverse-square law. At any point inside the shell, the larger fields of nearby portions of the shell are exactly cancelled by the weaker fields of more distant parts of the shell. This cancellation occurs because the distant charges are more abundant in just the right amount to compensate for their weaker fields (Fig. 21–28).

Our conclusion that the field is zero inside the shell did not follow only from the fact of zero enclosed charge. We also required the spherical symmetry that led to Equation 21–17. In nonsymmetric situations, it is quite possible to have zero net charge enclosed by a surface and still have electric fields within and on that surface (Fig. 21–29). Gauss's law alone requires only that the flux—the net number of field line crossings—be zero for a closed surface containing zero net charge.

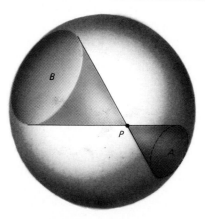

Fig. 21–28
At the arbitrary point P inside a charged shell, the field arising from the relatively few but nearby charges in region A is exactly cancelled by the field arising from the more numerous but more distant charges in the region B.

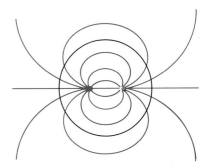

Fig. 21–29
The net charge enclosed by the sphere is zero, but the field within the sphere is not zero. Here the charge distribution—a dipole—is not spherically symmetric, so that Equation 21–17 is not a valid expression for the flux.

This section has illustrated the steps needed to calculate the electric field using Gauss's law. To summarize:

1. Study the symmetry to see if you can construct a gaussian surface on which the field magnitude and its direction relative to the surface are constant. If this is impossible, then Gauss's law, although true, will not provide a simple calculation of the field. Although we used a spherical gaussian surface in this section, the gaussian surface can be any size or shape as long as it is closed. It need not coincide with any real surfaces.

2. Evaluate the flux. This should be easy because your choice of a gaussian surface makes the term $E\cos\theta$ constant, so that this term can come outside of the flux integral, leaving an integral equal to the surface area.

3. Evaluate the *enclosed* charge. This is not the same as the total charge if your gaussian surface is within the charge distribution. Trust Gauss! The flux really does depend only on the *enclosed* charge. Whenever you doubt this, go back to Fig. 21–18 for a simple reaffirmation.

4. Equate the flux to $q_{enclosed}/\epsilon_0$ and solve for E. The direction of **E** should be clear from the symmetry.

Line Symmetry

symmetry axis

A charge distribution has line symmetry when it is infinitely long and has a charge density that depends only on the perpendicular distance from a line, or **symmetry axis** (Fig. 21–30). The symmetry of this situation requires that the electric field point radially outward from the axis, and that the field magnitude depend only on perpendicular distance from the axis. (Here we assume positive charge; for negative charge the field points inward.) We need a surface on which the magnitude of **E** and its orientation relative to the surface remain constant. Figure 21–31 shows a cylindrical surface concentric with the symmetry axis. The surface has arbitrary length ℓ and radius r. The electric field is everywhere perpendicular to the curved

Fig. 21–30
(a) A charge distribution with line symmetry. The charge density depends only on the distance r from the symmetry axis, and the distribution extends in both directions. *(b)* End view.

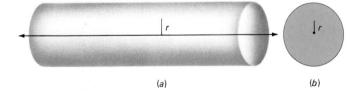

(a)   (b)

Fig. 21–31
A cylindrical gaussian surface surrounding a charge distribution with line symmetry. The flux through the ends is zero because field lines do not cross the ends. The flux through the curved portion is $2\pi r\ell E$. **E** is radial and has the same magnitude over the gaussian cylinder.

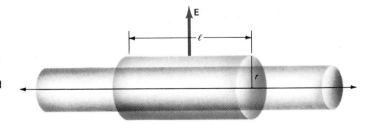

Fig. 21–32
Unrolling the cylinder makes a flat sheet of area $2\pi r\ell$.

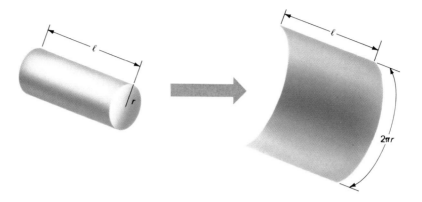

part of this surface, so that $\cos\theta = 1$ (-1 for negative charge) in the expression for flux. Then the flux through the curved portion of the surface is

$$\phi = \int \mathbf{E}\cdot d\mathbf{A} = \int E\,dA = E\int dA = 2\pi r\ell E. \qquad (21\text{–}20)$$

You can understand the last step in this expression by imagining unrolling the cylinder to make a flat sheet of width ℓ and length $2\pi r$ (Fig. 21–32). What about the flux through the ends of the cylinder? Here the field lies along the surface, so that no field lines cross the surface, giving zero flux through the ends. Mathematically, the vectors $\mathbf{E}$ and $d\mathbf{A}$ are perpendicular, so that $\cos\theta = 0$ in the dot product $\mathbf{E}\cdot d\mathbf{A}$. Therefore, the only flux is through the curved part of the cylinder, as given by Equation 21–20. Gauss's law tells us that the flux is proportional to the enclosed charge:

$$2\pi r\ell E = \frac{q_{\text{enclosed}}}{\epsilon_0},$$

so that

$$E = \frac{q_{\text{enclosed}}}{2\pi\epsilon_0 r\ell}. \qquad (21\text{–}21)$$

Here q is the charge within our cylindrical gaussian surface of length ℓ and radius r. We must still determine that charge; how we do so depends on the details of the charge distribution, as the examples below illustrate.

Example 21–7

An Infinite Line of Charge

Use Gauss's law to calculate the electric field of an infinite line of charge carrying line charge density λ.

Solution

This is the same problem we solved in Example 21–4 through a tedious Coulomb's law calculation. If the line charge density is λ, then the charge enclosed by a gaussian cylinder of length ℓ is $\lambda\ell$. Using this expression for q_{enclosed} in Equation 21–21 gives

$$E = \frac{\lambda\ell}{2\pi\epsilon_0 r\ell} = \frac{\lambda}{2\pi\epsilon_0 r},$$

the same result we found in Example 21–4. Notice how much simpler is the Gauss's law calculation! The symmetry of the situation and an intelligent choice of gaussian surface helped us bypass the entire integration of Example 21–4.

Although this example dealt with an infinitesimally thin line of charge, you can easily convince yourself that the same result must hold *outside* any charge distribution that exhibits line symmetry.

Example 21–8

A Hollow Pipe

A thin-walled copper pipe 3.0 m long and 2 cm in diameter carries a net charge $q = 5.8 \ \mu C$, distributed uniformly. What is the electric field 8.0 mm from the pipe axis? 8.0 cm from the axis? Assume in both cases that the point where you are evaluating the field is not too close to the ends of the pipe.

Solution

Here we do not have a truly infinite line. But for points close to the pipe and sufficiently far from the ends, the contribution to the field from distant charges becomes very small, so the field becomes approximately that of an infinite line.

A point 8.0 mm from the axis lies inside the 2-cm-diameter pipe. A gaussian cylinder entirely inside the pipe encloses zero net charge; therefore, Equation 21–21 shows that the field is zero everywhere inside the pipe.

For a point outside the pipe our gaussian cylinder might as well enclose the entire pipe, so the length ℓ in Equation 21–21 is 3.0 m and the enclosed charge q is 5.8 μC.* Evaluated at $r = 8.0$ cm, Equation 21–21 then gives

$$E = \frac{q_{enclosed}}{2\pi\epsilon_0 r\ell} = \frac{5.8 \times 10^{-6} \ C}{(2\pi)(8.9 \times 10^{-12} \ C^2/N \cdot m^2)(8.0 \times 10^{-2} \ m)(3.0 \ m)} = 4.3 \times 10^5 \ N/C.$$

Note that the radius r used in Equation 21–21 is measured from the pipe axis, not from the wall of the pipe.

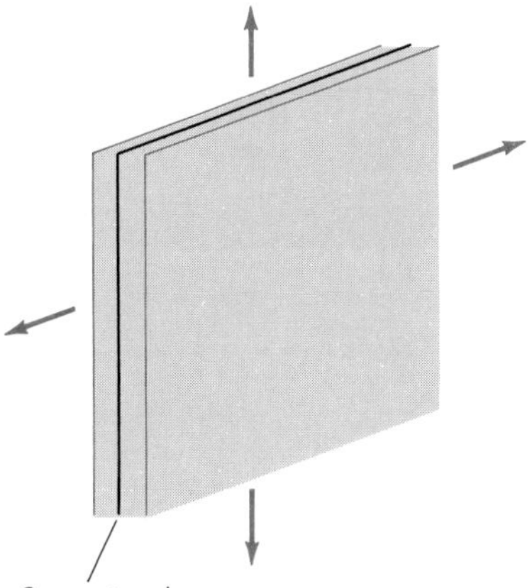

Other examples involving line symmetry are handled in a similar way—all we have to do is to evaluate the charge enclosed by a gaussian cylinder at the radius where we want to evaluate the field (see Problem 28).

Plane Symmetry

A charge distribution has plane symmetry when the charge density depends only on the perpendicular distance from a plane (Fig. 21–33). Nothing varies in any direction parallel to this symmetry plane. Therefore, the electric field must be perpendicular to that plane; any other orientation would favor one direction parallel to the plane.

We can evaluate the flux integral in Gauss's law by constructing a gaussian surface whose sides are perpendicular to the symmetry plane and whose ends are parallel to it (Fig. 21–34). Our gaussian surface straddles the symmetry plane, extending an equal distance on either side, as Fig. 21–34 indicates. Since no field lines cross the sides, the flux through the sides

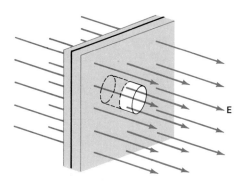

Fig. 21–33
A charge distribution exhibiting plane symmetry. The charge density depends only on the perpendicular distance from the plane of symmetry. The distribution extends infinitely in both directions parallel to the symmetry plane.

Fig. 21–34
A charge distribution with plane symmetry, showing the electric field and a gaussian surface.

---

*Since the pipe is finite, there is flux emerging from the ends. Ignoring this relatively small flux is equivalent to the approximation that the pipe is infinite.

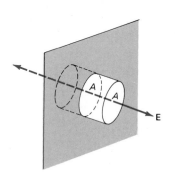

Fig. 21–35
The area of the sheet enclosed by the gaussian surface is the same as the area A of its ends.

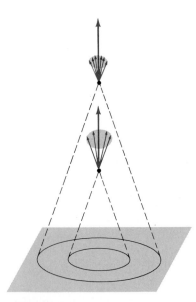

Fig. 21–36
As we rise above an infinite sheet, the amount of charge within a given angular region increases and just compensates for the inverse-square decrease in the fields of individual charges.

is zero. But field lines cross perpendicular to the ends, so that **E** and the area element vectors $d\mathbf{A}$ on the ends are parallel. Then $\cos\theta$ in the product **E**·$d\mathbf{A}$ is 1 over both ends (-1 if the charge is negative). Since the electric field cannot depend on position parallel to the symmetry plane, the magnitude E is the same and constant over both ends. Since the flux through the sides is zero, the total flux through our gaussian surface then becomes

$$\phi = \int_{\substack{\text{both}\\\text{ends}}} E\,dA = 2EA,$$

where A is the cross-sectional area of each end. The factor of 2 arises because there are two ends, and from symmetry **E** must point outward from each end. Then Gauss's law gives

$$2EA = \frac{q_{enclosed}}{\epsilon_0},$$

so that

$$E = \frac{q_{enclosed}}{2\epsilon_0 A}. \qquad (21\text{–}22)$$

Equation 21–22 holds for any charge distribution with plane symmetry; the particular distribution of charge perpendicular to the symmetry plane determines how we evaluate the enclosed charge.

Example 21–9
An Infinite Sheet of Charge

An infinite sheet of charge carries a uniform surface charge density σ. What is the electric field arising from this sheet?

Solution

Since the surface charge density is uniform, our distribution has plane symmetry. Figure 21–35 shows the sheet and an appropriate gaussian surface, to which Equation 21–22 applies. Because the sides of the gaussian surface are perpendicular to the charge sheet, the area of the sheet enclosed by the cylinder is the same as the area A of the two ends. The surface charge density—charge per unit area—is σ, so the enclosed charge is just σA. Then Equation 21–22 becomes

$$E = \frac{q_{enclosed}}{2\epsilon_0 A} = \frac{\sigma A}{2\epsilon_0 A} = \frac{\sigma}{2\epsilon_0}. \qquad (21\text{–}23)$$

What a simple result! The field points outward from the sheet with magnitude that doesn't depend on distance from the sheet. How can this be? By symmetry, the field must be perpendicular to the sheet. There is no charge anywhere except on the sheet, so that is the only place where field lines can begin or end. Therefore, the density of field lines—a measure of field strength—is constant everywhere off the sheet. What about Coulomb's law? How can the field not depend on distance? Figure 21–36 shows what happens as we rise above an infinite plane. The influence of charge elements directly below us diminishes as the inverse square of the distance. But within any given angular region, more and more total charge appears, increasing as the square of our distance. The net effect is a field whose magnitude is independent of position. We have a similar situation with the acceleration of gravity, g, near earth's surface. When we treat g as constant, we are approximating the earth as an infinite plane with a uniform gravitational field. Problems 11 and 44 explore the field of an infinite sheet by direct integration of the fields of all charge elements on the sheet.

Although this example treated only an infinitesimally thin sheet of charge, we would find a uniform electric field *outside* any charge distribution exhibiting plane symmetry. The field *inside* such a distribution would depend on how the charge density varies in the direction perpendicular to the symmetry plane (see Problem 38).

Example 21–10
A Distribution with Plane Symmetry

A flat sheet 50 cm square carries a uniform surface charge density. An electron 1.5 cm from a point near the center of the sheet experiences a force of 1.8×10^{-12} N, directed away from the sheet. Find the total charge on the sheet.

Solution

1.5 cm from the sheet and far from its edges, the sheet looks effectively infinite, so its field is given by Equation 21–23. From the definition of electric field, Equation 21–1, the force on the electron is just $-eE$, with e the elementary charge, and where the minus sign designates a force away from the sheet. Meanwhile the charge density σ on the sheet is the total charge, q, divided by the sheet area A. Then we can write

$$F = -eE = -e\frac{\sigma}{2\epsilon_0} = \frac{-eq}{2\epsilon_0 A},$$

so that

$$q = \frac{2\epsilon_0 AF}{-e} = \frac{2(8.9 \times 10^{-12} \text{ C}^2/\text{N} \cdot \text{m}^2)(0.50 \text{ m})^2(1.8 \times 10^{-12} \text{ N})}{-1.6 \times 10^{-19} \text{ C}} = -50 \; \mu\text{C}.$$

The minus sign is expected because the electron is repelled by the sheet, which must therefore carry a negative charge. Notice that the 1.5-cm distance from the sheet is irrelevant in the calculation; we used it only to establish that we had approximately the field of an infinite sheet, which does not depend on distance. Could we have made the same approximation if the electron were 1.5 m from the sheet? How about 15 m? (See Problem 40.)

21.6
FIELDS OF ARBITRARY CHARGE DISTRIBUTIONS

The examples of Section 21.5 show how easy Gauss's law can sometimes make problems that would be difficult to solve using Coulomb's law. In each case the symmetry allowed us to construct a gaussian surface on which $E\cos\theta$ was constant. Only then could we take E outside the integral and solve for it. The methods we used are applicable to *any* charge distributions with the appropriate symmetries. But many situations do not possess the symmetry needed to apply Gauss's law in calculating the field. Try, for example, to calculate the field of a dipole using Gauss's law. The attempt fails because it is impossible to draw an appropriate surface.

We can gain a qualitative understanding of the fields associated with more complicated charge distributions by considering the fields of simpler distributions that we have already calculated using either Coulomb's law or Gauss's law. Figure 21–37 summarizes the fields of a dipole, a point charge, a uniformly charged line, and a uniformly charged plane. For the last three, note the simple relation between the number of dimensions in the charge distribution and the way the field strength depends on distance. The plane has two dimensions and its field strength is independent of distance. The line has one dimension and its field falls as $1/r$. The point has no dimensions and its field falls as $1/r^2$. In a sense, the dipole continues this progression, for it consists of two opposite point charges whose effects very nearly cancel. No wonder its field falls even faster, as $1/r^3$. In fact, one can construct a hierarchy of charge distributions whose fields fall off ever faster as dipoles nearly cancel dipoles, and so on. Such **multipole expansions** are useful in the mathematical analysis of complicated charge distributions such as complex molecules or radio antennas.

multipole expansions

Fig. 21–37
The fields of a dipole, a point charge, a charged line, and a charged plane.

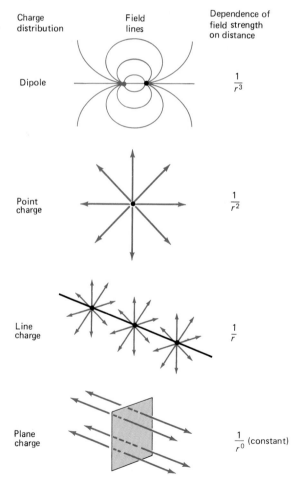

Charge distribution	Field lines	Dependence of field strength on distance
Dipole		$\dfrac{1}{r^3}$
Point charge		$\dfrac{1}{r^2}$
Line charge		$\dfrac{1}{r}$
Plane charge		$\dfrac{1}{r^0}$ (constant)

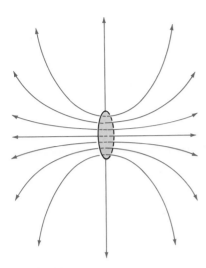

Fig. 21–38
The field of a uniformly charged disk.

Frequently we have a charge distribution that lacks the symmetry required to make Gauss's law useful and for which a Coulomb's law calculation would be impossibly difficult. Often we would like to know the approximate shape of the field lines without investing the effort needed to solve the problem exactly. Such an approximation often gives more valuable insights than the result of a detailed calculation. Good thinking coupled with knowledge of simpler charge distributions can go a long way toward providing a reasonable approximation to the actual field. Consider, for example, the uniformly charged disk shown in Fig. 21–38. For points much closer to the disk surface than to the edge, the disk looks almost like an infinite flat plane of charge. For these points the field is approximately that of an infinite plane—a field that points directly away from the plane and does not fall off with distance. (Here we assume that the disk is positively charged. The field would point toward a negatively charged disk.) Sufficiently far from the disk, meanwhile, the disk's exact size and shape are unimportant. Its field closely resembles that of a single charged point: far from the disk the field points radially outward in all directions and falls off as the inverse square of the distance from the disk. How far from the disk is "sufficiently far"? That depends on how good an approximation we want. Many disk diameters away, the disk will begin to resemble a point, and the approximation will be a good one. The field at intermediate distances is harder to determine. But somehow the infinite plane field lines close to the disk must connect smoothly to the point charge field lines far away. If we sketch these in, as in Fig. 21–38, we have a rough picture of the field everywhere. Don't underestimate the value of a simple approximation like this one! It can often tell all we need to know about a situation and may provide a much clearer understanding than a detailed calculation would.

SUMMARY

1. The **electric field** is a set of vectors given at each point by the electric force per unit charge at that point:

$$\mathbf{F} = q\mathbf{E}.$$

The field concept allows us to describe interactions locally, rather than in terms of action at a distance. Charges give rise to fields that extend throughout space. Other charges respond to the fields at their particular locations. The electric field of a point charge q is given by

$$\mathbf{E} = \frac{1}{4\pi\epsilon_0} \frac{q}{r^2} \hat{\mathbf{r}},$$

where $\hat{\mathbf{r}}$ is a unit vector pointing radially outward from the point charge.

2. The **superposition principle** states that the electric field due to a distribution of charges is the vector sum of the fields of the individual charges making up the distribution:

$$\mathbf{E} = \sum_n \frac{1}{4\pi\epsilon_0} \frac{q_n}{r_n^2} \hat{\mathbf{r}}_n.$$

With continuous distributions of charge, the sum becomes an integral over the entire charge distribution:

$$\mathbf{E} = \int \frac{1}{4\pi\epsilon_0} \frac{dq}{r^2} \hat{\mathbf{r}}.$$

3. **Electric field lines** are a visual way of representing an electric field. At a given point a field line has the same direction as the electric field vector at that point. The number of field lines crossing a unit area perpendicular to the field is a measure of field strength, so that where the field is strong the lines are close together. Electric field lines always begin or end on electric charges.

4. The concept of **electric flux** quantifies the notion "number of field lines crossing a surface." Flux is de-

fined as the surface integral of the electric field over a surface:

$$\phi = \int \mathbf{E} \cdot d\mathbf{A},$$

where $d\mathbf{A}$ is an infinitesimal vector whose direction at any point is perpendicular to the surface at that point.

5. **Gauss's law** is a fundamental relation governing the behavior of electric fields throughout the universe. Loosely, Gauss's law states that the number of field lines crossing any closed surface depends only on the net charge enclosed by that surface. More rigorously, Gauss's law is the statement that the electric flux through any closed surface is proportional to the charge enclosed by that surface:

$$\int \mathbf{E} \cdot d\mathbf{A} = \frac{q_{enclosed}}{\epsilon_0}.$$

Gauss's law is a reflection of the inverse-square nature of the force between electric point charges. We believe that it is always true. Gauss's law is particularly useful in calculating the fields of charge distributions exhibiting sufficient symmetry. To make such a calculation, we draw an imaginary **gaussian surface** including regions on which the magnitude of the electric field and its orientation relative to the surface are constant. Then we can evaluate the flux integral in terms of the unknown field magnitude E, determine the charge enclosed by our gaussian surface, and then solve for E.

6. Important charge distributions whose fields we calculated using Gauss's law include those with spherical, line, and planar symmetry. Using the superposition principle, we earlier calculated the field of an electric dipole. Summarized in Fig. 21–37, the fields of these various distributions can be used to gain an approximate sense of the fields of more complicated distributions.

QUESTIONS

1. Is the electric force on a charge always in the direction of the electric field?
2. Can electric field lines ever cross? Why or why not?
3. The electric flux through a certain surface is zero. Must the field strength be zero on that surface? If not, give an example.
4. Why should the test charge used to measure an electric field be small?
5. Why does a dipole produce any electric field at all? After all, the dipole consists of two opposite charges and therefore has no net charge.
6. Why does Gauss's law imply that electric field lines begin or end only on charges?
7. If you determined that the flux of the gravitational field through a certain closed surface was zero, what could you conclude about the region interior to that surface?
8. A spherical shell carries a nonuniform surface charge density. Can you conclude that the electric field inside the shell is zero? Why or why not?
9. Give several examples of charge distributions whose fields do not vary as the inverse square of the distance from the distribution. How is this not a violation of Coulomb's law?
10. A point charge is located a fixed distance from the center of a uniformly charged sphere, outside the sphere. If the sphere shrinks in size without losing any charge, what happens to the force on the point charge?
11. In applying Equation 21–18 for the field of a spherically symmetric charge distribution, is r the distance from the edge of the distribution or from its center? Explain.
12. In a certain region there is an electric field that points straight to the right and whose magnitude decreases as one moves to the right. Does the region contain net positive charge, net negative charge, or no net charge? *Hint:* Draw a picture, including an appropriate gaussian surface.
13. Under what conditions can the electric flux through a surface be written as EA, where A is the area of the surface?
14. A charge distribution consists of a charge $+3q$ and another of $-5q$, separated by a small distance. Describe the electric field very far from this distribution. What is meant by "very far"?
15. Why can't you use Gauss's law to determine the field of a uniformly charged cube? Wouldn't a cubical gaussian surface work in this case?
16. You are sitting inside an uncharged hollow spherical shell. Suddenly someone dumps a billion coulombs of charge uniformly over the shell. What happens to the electric field at your location?
17. No matter how far you get from an infinite line of charge, its field never resembles that of a point charge. Why doesn't this violate our statement that the field of a localized distribution carrying a net charge resembles a point charge field at sufficiently large distances?
18. Why is it that the field within a uniformly charged sphere actually increases with distance from the center? How is this consistent with Coulomb's law?
19. Consider a flat surface in a uniform electric field, as in Fig. 21–19. For most orientations of the surface the flux through this surface is not zero, and yet there is no net charge in the region. Is this a violation of Gauss's law? Explain.
20. Does Gauss's law apply to a spherical surface *not* centered on a point charge, as shown in Fig. 21–39? If you didn't know the field of the point charge,

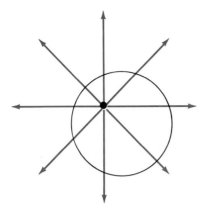

Fig. 21–39   Question 20

would this be a useful gaussian surface for calculating it?

21. An infinite plane carries a surface charge density that varies with position on the plane. Is the resulting electric field necessarily independent of distance from the plane? Could you use Gauss's law to calculate the field?

22. We have suggested in this chapter that the ability to represent electric fields using field lines is a consequence of the inverse-square dependence of the point-charge field. Show why field lines would not be a satisfactory representation of the electric field if the field of a point charge did not fall off as $1/r^2$. Would Gauss's law hold in this case?

PROBLEMS

Section 21.2   *Calculating the Electric Field*

1. What is the magnitude of the force on a 2.0-μC charge in an electric field of 100 N/C?

2. A 1.0-g mass carries a 5.0-μC charge. Placed in a uniform electric field, it experiences an acceleration of 2.0 m/s². What is the electric field strength?

3. The electron in a hydrogen atom is about 0.51×10^{-10} m from the proton. (*a*) What is the electric field strength at this distance? (*b*) What is the magnitude of the electric force on the electron?

4. A 3.8-g particle carries a charge of 4.0 μC. At a certain point, it experiences a downward force of 0.24 N. What is the electric field at that point? Assume the particle is near the surface of the earth, and do not ignore gravity.

5. A 1.0-μC charge experiences a force of 10 N in a certain electric field. What would be the force on a proton in the same field?

6. Two opposite charges, one +2.0 μC and the other −2.0 μC, are separated by 5.0 cm, as shown in Fig. 21–40. Find the electric field (*a*) 10 cm directly above the point midway between the charges; (*b*) 10 cm directly to the right of the point midway between the charges; (*c*) midway between the charges.

Fig. 21–40   Problem 6

7. A 1.0-μC charge and a 2.0-μC charge are located 10 cm apart, as shown in Fig. 21–41. Find a point where the electric field due to these two charges is zero.

Fig. 21–41   Problem 7

8. Calculate the electric field on the x-axis for the dipole of Example 21–3, and show that for large distances ($x \gg a$) the field can be written $E = p/2\pi\epsilon_0 x^3$, with $p = 2qa$ the dipole moment.

9. Consider a point P at a distance r and angle θ to a point dipole **p**, as shown in Fig. 21–42. By resolving the dipole moment vector into components perpendicular and parallel to a line from the dipole to the point P, and using the results of Example 21–3 and the preceding problem, show that the electric field at P is

$$\mathbf{E} = \frac{p}{4\pi\epsilon_0 r^3}(\hat{\mathbf{i}}\sin\theta + 2\hat{\mathbf{j}}\cos\theta),$$

where the x- and y-axes are oriented as in Fig. 21–42.

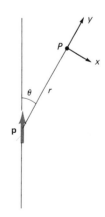

Fig. 21–42   Problem 9

10. A thin ring of radius a carries a total charge q distributed uniformly around the ring, as shown in Fig. 21–43. Show that the field on the axis of the ring points away from the ring and has magnitude

$$E = \frac{1}{4\pi\epsilon_0} \frac{qx}{(a^2 + x^2)^{3/2}}.$$

Does your result make sense when $x \gg a$?

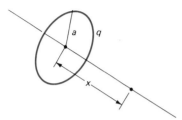

Fig. 21–43   Problem 10

11. An infinite plane carries a uniform surface charge density σ. Imagine the plane divided into thin rings of radius r and width dr, as suggested in Fig. 21–44. (a) Show that the area of such a ring is very nearly $2\pi r dr$, so that the charge dq on the ring is $dq = 2\pi\sigma r dr$. (b) Use the result of the preceding problem to write the infinitesimal electric field dE of this charged ring. (c) Integrate over all such rings (that is, from $r = 0$ to ∞), showing that the electric field is independent of the distance x from the plane and is given by Equation 21–23.

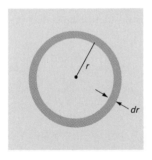

Fig. 21–44   Problem 11

12. A thin, uniformly charged rod of length ℓ carries total charge q. What is the electric field a distance x from the end of the rod? *Hint:* See Example 20–5.
13. A semicircular loop of radius a carries positive charge q. Facing it is another loop carrying charge $-q$, as shown in Fig. 21–45. Find the electric field at the center of the circle formed by these loops. Assume that charge is distributed uniformly over the loops.
14. Two parallel wires of negligible diameter are 2.5 m long and 1.5 cm apart. They carry equal but opposite charges, distributed uniformly over their lengths. An

electron midway between the wires experiences an acceleration of 1.2×10^{18} m/s². What is the charge on the wires?

Fig. 21–45   Problem 13

Section 21.3   *Electric Field Lines*

15. What is the net charge shown in Fig. 21–46? The magnitude of the middle charge is 3.0 μC.

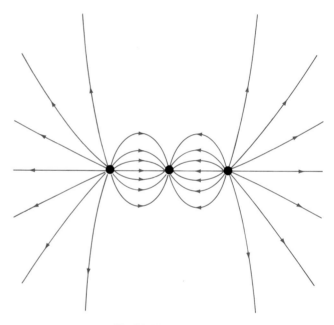

Fig. 21–46   Problem 15

16. Two charges $+q$ and one charge $-q$ are located at the vertices of an equilateral triangle. Sketch approximately the electric field lines associated with this configuration.
17. A sphere carries a total charge $+3q$ spread uniformly over its surface. A point charge $-q$ is located nearby, as shown in Fig. 21–47. Sketch approximately the field lines of this configuration, assuming that the presence of the point charge does not distort the charge distribution on the sphere.
18. A point charge q is at the center of a spherical shell carrying charge $-2q$. Sketch the field lines associated with this configuration.

Fig. 21–47 Problem 17

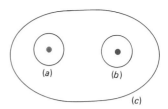

Fig. 21–50 Problem 24

19. What is the flux through the hemispherical *open* surface shown in Fig. 21–48? The strength of the uniform field is E and the radius of the hemisphere is R. *Hint:* Don't do a messy integral!

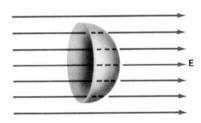

Fig. 21–48 Problem 19

20. Consider a flat surface with area 2.0 m^2 placed in a uniform electric field of 800 N/C. What is the electric flux through the surface when it is (*a*) at right angles to the field? (*b*) at 45° to the field? (*c*) parallel to the field?

Section 21.4   Gauss's Law

21. What is the electric flux through each of the closed surfaces shown in Fig. 21–49?

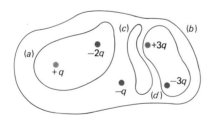

Fig. 21–49 Problem 21

22. A 2.6-μC charge is at the center of a cube 7.5 cm on a side. What is the electric flux through one face of the cube? *Hint:* Think about symmetry, and don't do an integral!

23. If the charge in the preceding problem is now inside the cube but not at the center, what is the flux through the entire cube? Could you still calculate the flux through one face without doing an integral?

24. A dipole consists of two charges $\pm 6.1 \times 10^{-7}$ C located 1.2 cm apart. What is the electric flux through each surface shown in Fig. 21–50?

25. The electric field in a certain region is parallel to the x-axis with its x-component given by $E_x = 40x$ N/C, with x in meters. What is the volume charge density in the region? *Hint:* Apply Gauss's law to a cube 1 meter on a side.

26. Figure 21–51 shows a rectangular box of sides $2a$ and length ℓ surrounding a line of charge with uniform line charge density λ. The line passes through the center of the box faces. To show directly that Gauss's law is satisfied for this box, divide one of the faces into strips of width dx, as shown, and write an expression for the flux $\mathbf{E} \cdot d\mathbf{A}$ over this strip. Integrate your expression over the face, and multiply by 4 to get the flux through the entire box. Show that your result is consistent with Gauss's law.

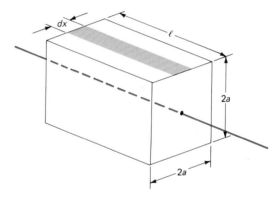

Fig. 21–51 Problem 26

Section 21.5   Using Gauss's Law

27. During fair weather, an electric field of about 100 N/C points vertically downward in earth's atmosphere. Assuming that this field arises from charge distributed in a spherically symmetric manner over the earth, determine the net charge on the earth.

28. An infinite rod of radius a carries a uniform volume charge density ρ. Obtain expressions for the electric field inside and outside the rod.

29. A sphere of radius $2a$ contains a concentric spherical hole of radius a, as shown in Fig. 21–52. The solid part of the sphere carries a uniform volume charge density ρ. Determine the electric field (*a*) inside the hole; (*b*) inside the solid part of the sphere; (*c*) outside the sphere as functions of the distance r from the center of the sphere.

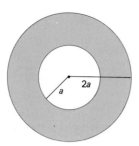

Fig. 21–52 Problem 29

30. How should the charge density within a solid sphere vary in order that the magnitude of the electric field within the sphere may be constant?

31. Two infinite sheets of charge are parallel and carry the same surface charge density σ. Use the superposition principle to determine the electric field between the sheets as well as beyond the sheets.

32. Repeat the preceding problem for the case when the two sheets carry surface charge densities σ and $-\sigma$.

33. A charged sphere 10 m in diameter carries a spherically symmetric charge distribution. A small spring scale is placed on the surface of the sphere and used to "weigh" a 5.0-mm-diameter styrofoam ball carrying a 9.4-μC charge. The "weight" (spring scale reading) is 850 N. What is the charge on the large sphere?

34. A spherical shell 30 cm in diameter carries a total charge of 85 μC distributed uniformly over its surface. A 1.0-μC point charge is located at the center of the shell. What is the electric field at a point (a) 5.0 cm from the center and (b) 45 cm from the center? How would your answers change if the charge on the shell were doubled?

35. A long, thin wire carries a uniform line charge density $\lambda = -6.8$ μC/m. It is surrounded by a thick concentric cylindrical shell of inner radius 2.5 cm and outer radius 3.5 cm. What uniform volume charge density within the shell will result in zero electric field outside the shell?

36. A sphere of radius R carries a uniform surface charge density σ. Consider a patch of the sphere's surface so small that its curvature is insignificant. (a) Apply Gauss's law to the gaussian surface shown in Fig. 21–53, and show that the electric field at the surface of the sphere is given by $E = \sigma/\epsilon_0$. Hint: What is the field inside the sphere? (b) Show that your result is the same as that obtained by applying Gauss's law to a spherical surface just outside the surface of the charged sphere. Why is your result in part (a) twice the result obtained for a uniform sheet of charge? After all, a small patch of the sphere looks like a charged sheet from close by. Hint: Is the field **E** in Gauss's law produced only by the charges within the gaussian surface?

37. A proton is released from rest 1.0 cm from a large sheet carrying a surface charge density of -2.4×10^{-8} C/m². How much later does it strike the sheet?

Fig. 21–53 Problem 36

38. A slab of charge extends infinitely in two dimensions, and has thickness d in the third dimension, as shown in Fig. 21–54. The slab carries a uniform volume charge density ρ. What is the electric field at all points, both inside and outside the slab? Express your answer in terms of the distance x from the center plane of the slab.

Fig. 21–54 Edge-on view of an infinite slab of charge (Problem 38).

Section 21.6 *Fields of Arbitrary Charge Distributions*

39. A rod 0.5 m long with radius 1.0 cm carries a charge of 2.0 μC distributed uniformly along it. What is the approximate magnitude of the electric field (a) 4.0 mm from the surface of the rod and not near either end and (b) 23 m from the rod?

40. What is the electric field 15 m from the charged plate of Example 21–10?

41. Two circular plates are 10 cm in diameter and 2.0 mm apart. They carry equal but opposite charges ± 0.50 μC, uniformly distributed over their surfaces. What is the electric field (a) between the plates but

not near either edge? (b) 2.5 m from the plates on a plane passing midway between the plates and parallel to them? *Hint on (b):* See Example 21–3. What is the field of either plate alone at large distances?

Supplementary Problems

42. Two thin, uniformly charged rods of length ℓ carrying charges $-q$ and $+q$ are arranged end-to-end, as shown in Fig. 21–55. (a) What is the electric field a distance x to the right of the point where the rods join? (b) Obtain an approximate expression for the electric field when $x \gg \ell$, and from it show that the configuration of rods behaves like a dipole of dipole moment $q\ell$. *Hint:* See Problem 8.

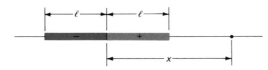

Fig. 21–55 Problem 42

43. A 1.2-g particle carrying a charge $+0.26$ μC is attached to the center of a square plate 50 cm on a side by means of a thread 1.0 cm long (Fig. 21–56). When disturbed slightly from its equilibrium, the particle oscillates with a period of 1.4×10^{-2} s. What is the charge on the plate?

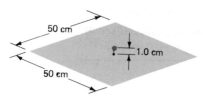

Fig. 21–56 Problem 43

44. Calculate the field of a uniformly charged infinite sheet by considering the sheet to consist of infinitely many line charges lying side-by-side, and integrating the fields of those line charges. Your result should agree with Equation 21–23.

45. A thick spherical shell of inner radius a and outer radius b carries a charge density given by

$$\rho = \frac{ce^{-r/a}}{r^2},$$

where c is a constant and r the distance from the center of the shell. Find the electric field strength at all points.

46. Two thin rods of length ℓ carry the same line charge density λ. The rods lie close together but at a slight angle, as shown in Fig. 21–57. Their separation is a at the closer end, b at the farther end. Assuming that $b \ll \ell$, and ignoring the fringing fields at the ends of the rods, show that the force on the upper rod due to the lower rod has magnitude

$$F = \frac{\ell}{b-a} \frac{\lambda^2}{2\pi\epsilon_0} \ln(b/a).$$

What is the direction of this force?

Fig. 21–57 Problem 46

22
MATTER IN ELECTRIC FIELDS

We previously defined the electric field at a point as the force per unit charge at that point. Even this definition should serve as a reminder that electric fields are important because of the effect they have on charged matter. Indeed, the behavior of matter from the atomic scale through the size of human beings or even larger objects is primarily determined by the response of electric charges to electric fields (Fig. 22–1).

In the previous chapter, we considered properties of the electric field and learned how to calculate the field of a given charge distribution. Here we consider the opposite problem: given an electric field, how do charges respond when placed in that field? How, for example, do electrons move in a TV tube? Why do gas molecules experience a weak attractive force? We begin with a discussion of individual, isolated charges. However, matter on the macroscopic scale consists of many charges, often closely interacting, and to understand fully the behavior of matter in electric fields we must consider also the combined response of these many charges to a field. How, for example, do all the electrons in the lens of your eye behave in the electric field of a light wave? The answer explains the optical properties of your eye.

Fig. 22–1
Sparks fly when an externally applied electric field overcomes the internal fields of air molecules. This phenomenon is a reminder that electric forces govern the behavior of matter on the molecular scale.

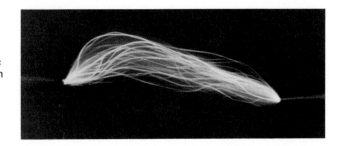

22.1
A SINGLE CHARGE IN AN ELECTRIC FIELD

The motion of a single charge in an electric field is governed both by the definition of electric field (Equation 21–1), which can be written

$$\mathbf{F} = q\mathbf{E},$$

and by Newton's law:

$$\mathbf{F} = m\mathbf{a}.$$

Combining these two equations gives the acceleration of a charge in an electric field:

$$\mathbf{a} = \frac{q}{m}\mathbf{E}. \tag{22--1}$$

This equation shows that it is the ratio of charge to mass that determines a particle's acceleration in a given electric field. This makes sense because a greater charge experiences a greater electric force, while a greater mass responds with less acceleration to that force. This explains why electrons, about 2000 times less massive than protons but carrying the same charge, are readily accelerated in electric fields. Many practical devices, from x-ray machines to television tubes, make use of the high accelerations possible with electrons even in modest electric fields.

Of course, Equation 22–1 correctly describes only a situation in which the electric field is alone responsible for all the force on a particle. Other forces, such as gravity, may also need to be considered in calculating the acceleration of a charged particle. But for atomic- and molecular-sized particles, the force of gravity is usually insignificant compared with the electric force. Even in the relatively weak 100 N/C electric field in earth's atmosphere, for example, the electric force on an electron is 10^{12} times the gravitational force (see Problem 1). On the other hand, Millikan's famous oil-drop experiment (see Section 20.2) required electric forces equal in magnitude to the gravitational force (see Problem 2).

When the electric force is not exactly balanced by some other force, then a charged object in an electric field will follow Newton's law and accelerate in response to the net force. If the electric field is uniform and independent of time, the result is particularly simple. Here Equation 22–1 shows that the particle undergoes constant acceleration given by $\mathbf{a} = q\mathbf{E}/m$. In Chapters 2 and 3 we developed a thorough description of motion under constant acceleration, which we applied most frequently using the acceleration of gravity in the uniform gravitational field near earth's surface. The results of Chapters 2 and 3 apply as well to acceleration in a uniform electric field, although there is a subtle difference between the constant acceleration caused by gravity and that of a uniform electric field. For gravity the acceleration is the same for all objects, no matter what their mass. But the acceleration in an electric field varies from object to object, depending on the ratio of charge to mass. This is a useful phenomenon, for it allows us to separate charged objects (such as ions) according to their charge-to-mass ratios.

Example 22–1

Acceleration in a Uniform Electric Field

An electric field of 2.0 N/C points vertically upward. An electron is released from rest in this field. How far and in what direction does it move in 1.0 μs?

Solution

Finding the distance traveled is a standard problem in accelerated motion, to which Equation 2–11 applies:

$$y = \tfrac{1}{2}at^2.$$

With the acceleration given by Equation 22–1, this is

$$y = \frac{1}{2}\frac{qE_y}{m}t^2 = \frac{(-1.6\times10^{-19}\ \text{C})(2.0\ \text{N/C})}{(2)(9.1\times10^{-31}\ \text{kg})}(1.0\times10^{-6}\ \text{s})^2 = -0.18\ \text{m}.$$

The minus sign indicates motion downward, opposite to the field direction because the electron carries a negative charge. A proton in the same situation would move upward, but not nearly as far because its acceleration is much less as a result of its much higher mass. You should convince yourself that we were justified in neglecting gravity in this example.

Application

The Electrostatic Precipitator

electrostatic precipitator

The most effective way of reducing air pollution by small particles such as smoke from oil- and coal-fired electric power plants is to use an **electrostatic precipitator,** a device in which a strong electrostatic field removes these particles from combustion gases on their way to the smokestack. A typical precipitator consists of parallel metal plates with a thin wire between them (Fig. 22–2). Stack gases flow in the region between the plates. Application of a high voltage between the wire and plates sets up a strong electric field. Near the wire the field is so strong that some gas molecules are ionized. These ions drift through the gas, attaching themselves to small pollutant particles.

Fig. 22–2
Cutaway diagram of an electrostatic precipitator.

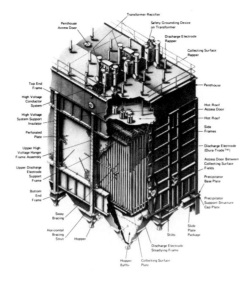

The particles, now carrying net charges, are driven to the collecting plates by the electric field. There they accumulate. Every few minutes a mechanical vibrator taps the plates and the particles fall into a hopper, where they can be trucked away to use for fill or for making cinder blocks and similar products. Precipitators are engineered so that the electric force on typical particles is many thousands of times greater than the forces associated with gravity or with the upward motion of the hot gases.

Two-dimensional motion in a uniform electric field is analogous to projectile motion in the uniform gravitational field near earth's surface. The constant acceleration formulas of Chapter 3 apply, and a charged object describes a parabolic path. A charged particle moving initially in a straight line may be deflected by passing through a uniform electric field. Such electrostatic deflection is used to "steer" electron beams in a number of practical devices including oscilloscopes.

Example 22–2

Deflecting an Electron Beam

An electron is moving horizontally to the right at a speed of 4.0×10^6 m/s. It enters a region in which there is an electric field of 1.0×10^3 N/C pointing downward. The region containing the electric field extends horizontally for a distance of 2.0 cm, as shown in Fig. 22–3. (We could make such a uniform-field region using two uniformly charged plates. If the plates are closely spaced, the transition from no field to a nearly uniform field will be quite abrupt.) By how much and in what direction is the electron deflected when it leaves the electric field? Describe its subsequent motion.

Solution

Because the field points vertically downward, it does not affect the horizontal component of the electron's velocity. Thus the electron spends a time

$$t = \frac{\Delta x}{v_x} = \frac{2.0 \times 10^{-2} \text{ m}}{4.0 \times 10^6 \text{ m/s}} = 0.50 \times 10^{-8} \text{ s}$$

in the electric field region. During this time it experiences a vertical acceleration qE/m, and undergoes a vertical deflection

$$y = \tfrac{1}{2}at^2 = \frac{1}{2}\frac{qE_y}{m}t^2$$

$$= \frac{(-1.6 \times 10^{-19} \text{ C})(-1.0 \times 10^3 \text{ N/C})}{(2)(9.1 \times 10^{-31} \text{ kg})}(0.50 \times 10^{-8} \text{ s})^2 = 2.2 \text{ mm}.$$

Since we described the downward-pointing electric field as being in the $-y$-direction, this positive answer means that the electron is deflected upward. In the field region, it follows an upward-curving parabola, as shown in Fig. 22–3. Once it leaves the field region, the electron again moves in a straight line. Its x-component of

Fig. 22–3
The electron is deflected from its straight-line path by a uniform electric field.

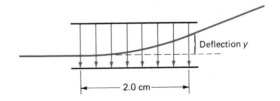

Deflection y

2.0 cm

velocity is unchanged, but it has gained a y-component given by

$$v_y = at = \frac{q}{m}E_y\, t$$

$$= \frac{(-1.6\times10^{-19}\text{ C})}{(9.1\times10^{-31}\text{ kg})}(-1.0\times10^3\text{ N/C})(0.50\times10^{-8}\text{ s})$$

$$= 8.8\times10^5\text{ m/s}.$$

Thus the electron leaves the region with a speed of

$$v = \sqrt{v_x^2+v_y^2} = \sqrt{(4.0\times10^6\text{ m/s})^2+(8.8\times10^5\text{ m/s})^2} = 4.1\times10^6\text{ m/s}$$

at an angle of

$$\theta = \tan^{-1}\!\left(\frac{v_y}{v_x}\right) = \tan^{-1}\!\left(\frac{8.8\times10^5\text{ m/s}}{4.0\times10^6\text{ m/s}}\right) = 12°$$

above the horizontal.

When the electric field is not uniform, it is generally more difficult to calculate particle trajectories. An important exception, however, is the case of a particle moving always at right angles to a field that emanates from a central point or line. Such a particle is subject to a force that changes only the direction of its motion, so that it undergoes uniform circular motion. A point charge, a line charge, and other distributions with spherical or line symmetry provide fields in which uniform circular motion is possible (Fig. 22–4). An atom, for example, may be thought of as a miniature "solar system" with its electrons held in circular orbit by electric forces. Although this simple classical picture is not fully consistent with modern quantum mechanics, it does provide some insight into the nature of the atom. On a larger scale, use of charge distributions with the appropriate symmetry allows us to "steer" charged particles into circular paths, as the example below illustrates.

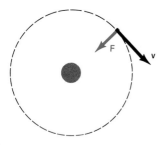

Fig. 22–4
Given the right initial speed, a particle may undergo uniform circular motion in a centrally directed field.

Example 22–3

An Electrostatic Analyzer

An electrostatic analyzer is used to select charged particles with a particular speed from a beam of particles with random speeds. The purpose may be to prepare a beam whose particles have very nearly the same speed, or to measure the unknown distribution of speeds in a gas of charged particles. The device consists of a pair of closely spaced curved plates that carry uniform surface charge densities. Neglecting the slight outward bulging of the field at the ends, the field between such plates approximates that between two concentric cylinders, and therefore decreases inversely with distance from the center of curvature (Fig. 22–5; see Section 21.5). With the inner plate negative, the field points radially inward, so that a positively charged particle with the right initial velocity will move in a circular arc parallel to the plates.

A particular analyzer consists of an inner plate of radius 5.0 cm with a spacing of 4.0 mm between the plates. The electric field between the plates is given by

$$E = E_0 \frac{b}{r},$$

where $E_0 = 2.4\times10^4$ N/C, $b = 5.0$ cm, and r is the distance from the center of curvature. A beam of protons is incident on the device, as shown in Fig. 22–5. Collima-

Fig. 22–5
An electrostatic analyzer, showing the trajectories of protons in the radial electric field. Only the protons with the correct speed get through the analyzer.

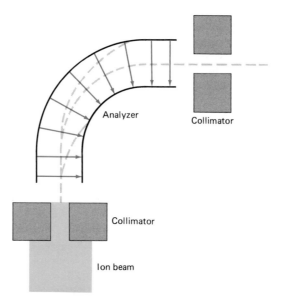

tors—widely spaced plates with small holes—ensure that particles enter the field region at right angles to the field, and that only those particles leaving at right angles remain in the outgoing beam. What is the speed of protons in the outgoing beam?

Solution

In the field region, the protons undergo uniform circular motion, with their acceleration provided by the electric field:

$$\frac{v^2}{r} = \frac{e}{m}E = \frac{e}{m}E_0\frac{b}{r}.$$

Then

$$v = \sqrt{\frac{eE_0b}{m}} = \sqrt{\frac{(1.6 \times 10^{-19}\ \text{C})(2.4 \times 10^4\ \text{N/C})(0.050\ \text{m})}{1.7 \times 10^{-27}\ \text{kg}}} = 3.4 \times 10^5\ \text{m/s}.$$

Notice that it does not matter where the protons enter the analyzer: the $1/r$ decrease in the field strength matches the $1/r$ dependence of the acceleration, so that all the protons emerge with the same speed. The speed distribution of the incident beam may be determined by varying the field between the plates, thereby selecting different speeds, and recording the strength of the outgoing beam. A device of this sort has been used on spacecraft to analyze charged particles in interplanetary space.

22.2
CONDUCTORS

Frequently we are interested in the behavior of macroscopic pieces of matter in electric fields. By "macroscopic" we mean "large enough to contain a great many atoms." In principle, all we need to know is the way an individual charge interacts with the electric field; we could determine the behavior of a piece of matter by calculating the individual motions of all its constituent electrons and protons. But with 10^{23} particles (typical of a cubic centimeter of matter or so) such a calculation is a practical impossibility! Besides, we don't really care what each individual particle does. Rather, we want some sort of overall picture, in which the individual motions on the atomic scale are averaged to give the large-scale behavior we observe.

In considering the response of bulk matter to electric fields, we classify matter into two categories—conductors and insulators. A **conductor** is a material in which some charges are relatively free to move throughout the material. The most familiar example of a conductor is a metal, in which some electrons are not bound to individual atoms but are free to move throughout the metal. In an **insulator,** by contrast, all charges are bound to individual atoms or molecules and cannot move through the material. Glass, plastic, rubber, dry wood, air, and many other common materials are insulators. Intermediate between conductors and insulators are a few materials known as **semiconductors.** These materials permit limited but highly controllable movement of charge, using a complex mechanism to be discussed in Chapter 25. Semiconductors—of which silicon is the most common example—lie at the heart of the ongoing revolution in microelectronics.

conductor

insulator

semiconductors

Electrostatic Equilibrium

Imagine a piece of electrically conducting material (Fig. 22–6*a*) placed in a uniform electric field (Fig. 22–6*b*). We can make a remarkably simple statement about this and any other situation in which a conductor is placed in a static electric field. Because charges in the conductor are free to move, they will do so in response to the electric force $q\mathbf{E}$. If the conductor is a metal, the negative electrons will move to the left—against the field direction. As negative charge accumulates at the left of the conductor, an excess of positive charge remains on the right (Fig. 22–6*c*). This separation of charge within the material gives rise to an electric field, this one pointing to the left and thus opposite to the original applied field. This internal field makes it increasingly difficult for more electrons to move to the left.

How long, then, will electrons continue to move? Eventually, the internal electric field developed by the charge separation becomes equal in magnitude but still opposite in direction to the original field. The net electric field within the material is then zero and there is no net force on the free charges in the material. The conductor is then said to be in **electrostatic equilibrium** (Fig. 22–6*d*).

In equilibrium, there is no more net motion of charge, although individual charges continue to move about in random thermal motion. Electrostatic equilibrium in a conductor always requires that there be *no electric field within the conductor*. Were this not true, charges within the conductor would experience electric forces, and would move until those forces disappeared—until the electric field within the conductor was zero. In electrostatic equilibrium, the electric field in a conductor is zero regardless of the shape of the conductor, the magnitude and direction of the applied field, or even the nature of the conducting material. This ability of a conductor to cancel electric fields is the basis for **shielding**—the use of conductive enclosures to keep out unwanted electric fields (Fig. 22–7).

electrostatic equilibrium

shielding

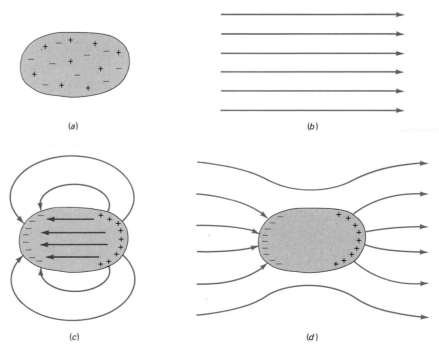

Fig. 22–6
(a) A piece of conducting material. *(b)* A uniform electric field. *(c)* In response to the electric field, charges within the conductor move apart, resulting in an internal field that cancels the original field within the conductor. *(d)* The net electric field is the vector sum of the original uniform field and the field resulting from the redistribution of charge in the conductor.

This discussion of conductors in electrostatic equilibrium is a macroscopic one. It considers only overall average fields within the material. Of course there are still electric fields, some quite strong, near individual electrons and protons. These fields are not eliminated by charge redistribution because they occur in spaces as small as the distance between individual electrons and protons. But the *average* field, encompassing distances many times the spacing between individual particles, must be zero in electrostatic equilibrium. As you think about the simple statements we

Fig. 22–7
The shielded cable connecting a turntable to a stereo amplifier keeps out stray electric fields that would otherwise introduce noise into the amplifier. Although the situation is not strictly static, the fields change slowly enough that charges in the shield can move to cancel them.

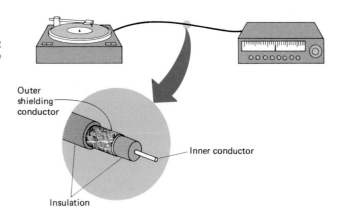

Outer shielding conductor

Inner conductor

Insulation

can make about conductors in electrostatic equilibrium, do not look at the conductor with too fine a microscope! We have here one of many cases in physics where a many-particle system obeys very simple laws while its constituent particles may be exhibiting complicated behavior. If you start thinking about interactions between individual particles, some of our statements may no longer be true and may seem confusing.

Although free charges do exist in conducting matter, most conductors remain electrically neutral; they contain the same number of electrons and protons. Even when charge separation occurs, the whole piece of material remains neutral (although different regions of it may have excesses of one kind of charge). But what happens when we have a charged conductor—one with a nonzero net charge? Imagine putting excess charge into the interior of a conductor—for example, electrons into a metal. There is a mutual repulsion among the electrons, but because these are excess electrons there is no compensating attraction from protons. So we might expect the excess electrons to move as far from one another as possible—to the surface of the conductor.* When the surface has an elongated shape, we might expect more charge to accumulate near the distant ends (Fig. 22–8). We will now use Gauss's law to prove rigorously our expectation that excess charge should move to the surface of a conductor. In the next chapter, we will show further that the charge density is related to the curvature of the conductor surface.

Fig. 22–8
Excess charge accumulates at the ends of an elongated conductor.

Gauss's Law and Conductors

Figure 22–9 shows a piece of conducting material, with a gaussian surface drawn just below the material surface. (Recall that a gaussian surface is any surface we draw for convenience in applying Gauss's law. A gaussian surface need not correspond to any physical surface.) The gaussian surface may be arbitrarily close to the material surface, as long as it all lies below that surface. In equilibrium there can be no electric field inside the conductor, so that there is no electric field anywhere on the gaussian surface. But then there is no flux through the gaussian surface. (Mathematically, **E** in the flux integral is everywhere zero; physically, there are no field lines crossing the gaussian surface). Gauss's law states that the flux through any closed surface is proportional to the net charge enclosed. Our gaussian surface has zero flux, and therefore encloses no net charge. Therefore, any excess charge must lie outside the gaussian surface—that is, on the physical surface of the conductor.

The fact that excess charge resides only on the surface of a conductor can be used as a very sensitive test of Gauss's law, and hence of the inverse-square law for the electric field. An experiment of this sort is shown schematically in Fig. 22–10. Basically, we place a charged conductor inside an uncharged hollow conductor. When the two conductors touch, all charge moves to the surface of the hollow conductor, leaving the originally charged conductor with no net charge. In practice, the experiment is often

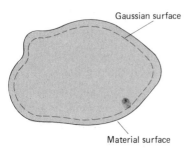

Gaussian surface

Material surface

Fig. 22–9
Since the electric field inside the conductor is zero, there is no charge enclosed by a gaussian surface inside the conductor surface.

*If the excess charge is huge, electrons may actually leave the surface, a process known as *field emission*. But at the surface, there are strong unbalanced electric forces from charges within the conductor. These forces are normally sufficient to prevent excess charge from leaving the conductor surface.

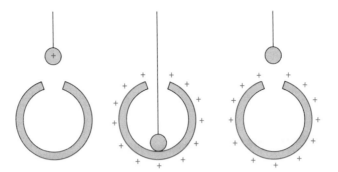

Fig. 22–10
Experimental text of Gauss's law. When the small charged conductor is placed in contact with the interior of the uncharged hollow conductor, all charge moves to the outside of the hollow conductor.

done in reverse. An uncharged conductor is placed inside the hollow outer conductor. The outer conductor is then charged, and sensitive instruments are used to detect any movement of charge to the inner conductor. If no such motion is detected, the inverse-square law is confirmed. Recent experiments of this type show that the exponent 2 appearing in the inverse-square law is indeed 2 to within 3×10^{-16}. Such tests are far more sensitive than direct measurements of how the force between charges varies with distance.

Example 22–4
Gauss's Law and Conductors

An irregularly shaped conductor has a hollow cavity in it, as shown in Fig. 22–11. The conductor carries a net charge of 1.0 μC. A small charged metal object carrying 2.0 μC is inside the cavity, and is not touching the conductor. Use Gauss's law to determine the net charge on the outer surface of the conductor.

Solution

The electric field is zero everywhere within the conductor, in particular on the gaussian surface shown in Fig. 22–11. So the flux $\int \mathbf{E} \cdot d\mathbf{A}$ through this surface is zero. Gauss's law then assures us that the surface encloses zero net charge. How is this possible? Since the cavity contains 2.0 μC, the inner wall of the cavity must carry -2.0 μC. This surface charge is attracted to the cavity wall by the presence of the positive charge within the cavity. But the conductor carries a net charge of 1.0 μC. With -2.0 μC on the cavity wall, the outer surface of the conductor must then carry 3.0 μC.

What would happen if we touched the charged object in the cavity to the cavity wall? You should convince yourself that we would still find 3.0 μC on the outside of the conductor.

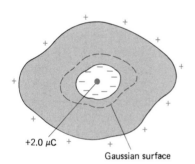

Fig. 22–11
A conductor with a hollow cavity in it. A gaussian surface is shown surrounding the cavity.

+2.0 μC

Gaussian surface

The fact that neither electric field nor excess charge exists inside a conductor in electrostatic equilibrium allows us to calculate the electric field near the surface of a charged conductor. At the surface, the electric field cannot have a component parallel to the surface. If it did, charges would move along the surface and we would not have equilibrium. So the electric field must be everywhere perpendicular to the surface.

Consider the very small cylindrical gaussian surface shown in Fig. 22–12. We make the gaussian surface so small that we can neglect any curvature of the conductor. The gaussian surface is oriented so that its sides are

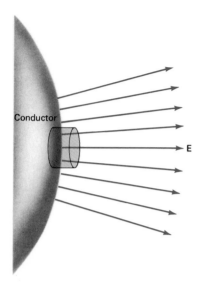

Fig. 22–12
A gaussian surface straddling the surface of a charged conductor.

perpendicular to and its top and bottom ends parallel to the conductor's physical surface. Because the electric field is perpendicular to the conductor surface, there is no flux through the sides of the gaussian surface. Because there is no electric field inside the conductor, the flux through the bottom end of the gaussian surface is zero. Flux emerges only through the top end of the gaussian surface, where, because the sides are arbitrarily short, the field is essentially that at the surface.* Because the gaussian surface is so small, the field magnitude is essentially constant over its top, so that the flux integral becomes

$$\phi = \oint \mathbf{E} \cdot d\mathbf{A} = EA,$$

where A is the area of the top of the gaussian surface and where $\cos\theta = 1$ because $\mathbf{E}$ is perpendicular to the surface (i.e., parallel to $\mathbf{A}$). If the conductor carries a surface charge density σ, then the gaussian surface encloses charge σA. Relating the flux and enclosed charge through Gauss's law gives

$$EA = \frac{\sigma A}{\epsilon_0},$$

or

$$E = \frac{\sigma}{\epsilon_0}. \quad \text{(field at conductor surface)} \qquad (22\text{--}2)$$

This result applies to any conductor in electrostatic equilibrium, and suggests that we find large electric fields where the charge density on a conductor is high. Engineers designing electrical devices must be careful to avoid high charge densities, which can result in dangerous sparks, arcs, and breakdown of electrical insulation.

Example 22–5

A Charged Metal Plate

One side of a large, flat conducting plate carries a uniform surface charge density σ. What is the electric field at the surface of the plate?

Solution

Applying Equation 22–2, we have

$$E = \frac{\sigma}{\epsilon_0}.$$

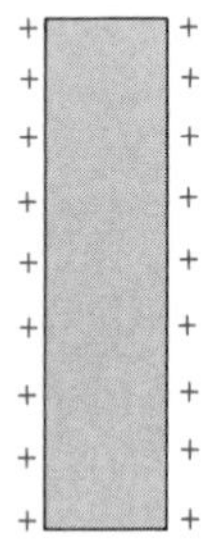

Fig. 22–13
Edge-on view of an isolated conducting plate. Free charges move to equalize the surface charge densities on both sides.

But wait! Doesn't this contradict Equation 21–23 for the field of a charged sheet? There, we found a field $E = \sigma/2\epsilon_0$. Here we have twice that result. But there is no contradiction. If our conducting plate is isolated, then symmetry and the fact that charge is free to move through the plate require that its other side carry the same charge density σ. We have essentially two charge sheets, and our result $E = \sigma/\epsilon_0$ includes the fields of both (Fig. 22–13). Inside the plate, the fields of the two sheets cancel; outside they add to make twice the field of one sheet alone.

---

*How do we know the field doesn't suddenly develop a component parallel to the surface? Although such discontinuous behavior may seem unreasonable, Gauss's law alone does not rule it out. But with a static field, it cannot happen. In the next chapter we will develop further understanding of the electrostatic field that will show why its component parallel to the conductor surface cannot change discontinuously.

Fig. 22–14
Two oppositely charged conducting plates. Charge accumulates on the surfaces facing each other.

What if our sheet is not isolated, so that symmetry does not require equal charge densities on both sides? Our result $E = \sigma/\epsilon_0$ comes right from Gauss's law, and must still apply. How is this possible? Our charge distribution might be asymmetric if, for example, our plate were in the vicinity of another plate with opposite charge (Fig. 22–14). Now we have two charged sheets, one on each plate. Each gives rise to the field $E = \sigma/2\epsilon_0$ of a single charge sheet. But between the plates these fields add, again giving the net field $E = \sigma/\epsilon_0$ of Equation 22–2. Problem 12 explores the case where the two plates carry unequal charge densities.

Remember that the results we have just considered—the absence of electric field or excess charge inside a conductor, and the presence of a perpendicular electric field given by Equation 22–2 at the surface of a conductor—are valid only in electrostatic equilibrium. When we relax the requirement of equilibrium, we will find situations in which conductors contain electric fields. Even in these cases the conductor strives to reach equilibrium through charge motion that tends to cancel any applied electric field. The better the conductor, the more quickly it approaches and the closer it gets to a condition of true equilibrium. In a truly perfect conductor—physically realizable in certain materials at very low temperature—there can be no electric field even when the conductor is not at equilibrium.

22.3
A DIPOLE IN AN ELECTRIC FIELD

In the preceding chapter (Example 21–3) we considered the field of the electric dipole, an object consisting of two equal but opposite charges held a small distance apart. Because dipoles contain electric charges they not only produce electric fields but also respond to electric fields. Many simple molecules can be described as electric dipoles, so that understanding the response of dipoles to electric fields provides insight into molecular behavior.

A Dipole in a Uniform Electric Field

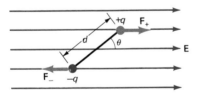

Fig. 22–15
A dipole in a uniform electric field experiences a net torque but no net force.

Consider first a dipole placed in a uniform electric field **E,** as shown in Fig. 22–15. The dipole consists of charges $+q$ and $-q$ separated by a distance d. The dipole moment vector **p** has magnitude qd and points from the negative to the positive charge. The positive charge experiences a force of magnitude qE in the direction of the field. Because the dipole charges are equal in magnitude but opposite in sign, the negative charge experiences a force of the same magnitude but in the opposite direction. Thus there is no net force on the dipole.

However, unless the dipole happens to be aligned exactly with the field, it does experience a net torque. This is readily seen in Fig. 22–15, where the rightward force on the positive charge and leftward force on the negative charge produce a torque that tends to align the dipole with the electric field. How strong is this torque? Recall from Chapter 11 that torque depends on the force, the distance from the pivot point to the point where the force is applied, and the angle between the force vector and a vector from the pivot to the force application point. This relationship is conveniently summarized by the vector equation 11–3:

$$\boldsymbol{\tau} = \mathbf{r} \times \mathbf{F}.$$

Consider the net torque about the center of the dipole. For either charge the distance from the force point to this pivot point is half the dipole separation, or $d/2$. The direction of a vector from pivot to force point at the positive charge is the same as the direction of the dipole moment $\mathbf{p}$. The force on the positive charge is qE, so that the torque on the positive charge has magnitude

$$\tau_+ = rF\sin\theta = \tfrac{1}{2}d \, qE\sin\theta.$$

From Fig. 22–15, it is clear that the negative charge contributes equally to the net torque, which is then

$$\tau = qdE\sin\theta = pE\sin\theta, \qquad (22\text{–}3)$$

where $p = qd$ is the dipole moment. The direction of this torque is such as to twist the dipole in a clockwise direction. Applying the right-hand rule shows that the torque vector points into the page. This is the same direction as the cross product of the dipole moment vector with the electric field. Furthermore, the torque magnitude of Equation 22–3 is the same as the magnitude of that cross product. So we can write

$$\boldsymbol{\tau} = \mathbf{p} \times \mathbf{E}. \qquad \text{(torque on a dipole)} \qquad (22\text{–}4)$$

This result is quite general. Any dipole in an electric field experiences a torque whose magnitude and direction are given by Equation 22–4.

Imagine a dipole that is aligned with a uniform electric field. To twist this dipole out of line with the field, we must do work against the torque that develops. Conversely, a dipole at some angle to an electric field has potential energy associated with its tendency to align with the field. How much work or energy is involved? In Chapter 12, we found that the work required to rotate an object from some initial orientation θ_0 to a final orientation θ against a torque is given by Equation 12–28:

$$W = \int_{\theta_0}^{\theta} \tau \, d\theta.$$

The integral is required if the torque changes with angle, as Equation 22–3 shows it does in the dipole case. Using Equation 22–3, the work integral becomes

$$W = \int_{\theta_0}^{\theta} pE\sin\theta \, d\theta,$$

where θ refers to the angle between the dipole moment $\mathbf{p}$ and electric field $\mathbf{E}$. Both the magnitudes p and E are constants, so that

$$W = pE\int_{\theta_0}^{\theta} \sin\theta \, d\theta = pE\left(-\cos\theta\right)\Big|_{\theta_0}^{\theta} = pE(\cos\theta_0 - \cos\theta). \qquad (22\text{–}5)$$

The work-energy theorem (Section 6.5) assures us that this result is also the change in potential energy of the dipole. It is convenient to choose the zero of potential energy when the dipole is at right angles to the field. We are free to do this because only differences in potential energy are physically meaningful. Then the reference angle θ_0 becomes $\pi/2$, and using Equation 22–5 we can write the potential energy of the dipole as

$$U = -pE\cos\theta,$$

or, more compactly in vector form,

$$U = -\mathbf{p}\cdot\mathbf{E}. \tag{22–6}$$

Our choice of reference angle was made in part so that this simple relation would hold. With this choice, the potential energy is negative if the dipole is within $\pi/2$ of the field direction and positive if it is not.

Example 22–6

A Dipole in a Uniform Electric Field

An isolated water molecule has dipole moment $p = 6.2 \times 10^{-30}$ C·m. Such a molecule is oriented with its dipole moment at right angles to a uniform electric field, and experiences a torque of 3.1×10^{-28} N·m. What is the strength of the electric field? How much energy would be released if the molecule were allowed to align with the field?

Solution

Solving Equation 22–3 for E gives

$$E = \frac{\tau}{p\sin\theta} = \frac{3.1 \times 10^{-28}\ \text{N·m}}{(6.2 \times 10^{-30}\ \text{C·m})(\sin 90°)} = 50\ \text{N/C}.$$

When the molecule is aligned with the field, $\cos\theta = 1$, and Equation 22–6 gives the potential energy:

$$U = -\mathbf{p}\cdot\mathbf{E} = -pE = -(6.2 \times 10^{-30}\ \text{C·m})(50\ \text{N/C}) = -3.1 \times 10^{-28}\ \text{J}.$$

Since its original orientation was that of zero potential energy according to Equation 22–6, its potential energy has decreased by 3.1×10^{-28} J.

As with an object falling in a gravitational field, this decrease of potential energy is associated with an increase in some other form of energy. For a truly isolated molecule, the potential energy becomes rotational kinetic energy as the molecule swings through equilibrium, then again potential energy as it swings toward an orientation opposite its original state. The molecule swings back and forth in a torsional oscillation, continually interchanging potential and kinetic energy (Problem 27 treats such an oscillation quantitatively). If other molecules are nearby, this oscillation is quickly damped through collisions, and the energy ends up heating the entire mass of water.* If energy is supplied continuously by periodically reversing the electric field, then electrical energy can be converted continuously to heat within a material containing water molecules. This is how cooking occurs in a microwave oven, and why paper plates—which contain no water—can be used in such ovens.

---

*Even the oscillations of an isolated dipole eventually damp out, as the system loses energy in the form of electromagnetic radiation.

A Dipole in a Nonuniform Electric Field

Our conclusion that a dipole in a uniform electric field experiences only a torque but no net force depends critically on the field being uniform. When the field is not uniform—when its magnitude and/or direction differs from one end of the dipole to the other—then the forces on the two charges comprising the dipole are not equal and opposite. For this reason a dipole in a nonuniform field generally experiences a net force that depends not on the field itself but rather on how the field varies with position.

Figure 22–16 shows two dipoles in a nonuniform field. (How can you tell that the picture represents a nonuniform field?) Our previous arguments about torque still hold, so that dipole A experiences a torque that tends to twist it into line with the field. Dipole B is already aligned with the field, so it experiences no torque. But the field near the positive end of dipole B is stronger (how can you tell?) so that the rightward force on the positive charge is greater than the leftward force on the negative charge. Thus the dipole experiences a net force to the right. Suppose the electric field at the negative charge is $\mathbf{E}$, and at the positive charge $\mathbf{E}+\Delta\mathbf{E}$. Then the net force on the dipole is

$$\mathbf{F} = \mathbf{F}_+ + \mathbf{F}_- = q(\mathbf{E}+\Delta\mathbf{E}) + (-q)\mathbf{E} = q\Delta\mathbf{E},$$

where q is the magnitude of either dipole charge. If we call the dipole separation Δx, so that the dipole moment is $p=q\Delta x$, we may write $q=p/\Delta x$. Then the net force becomes

$$\mathbf{F} = p\,\frac{\Delta\mathbf{E}}{\Delta x}.$$

In the limit of a point dipole (see Example 21–3), Δx becomes arbitrarily small and we have

$$\mathbf{F} = p\,\frac{d\mathbf{E}}{dx}. \tag{22–7}$$

This result holds when both the dipole moment vector and the electric field are in the x-direction. In the more general case of arbitrary dipole and field orientations, as with dipole A in Fig. 22–16, it is readily shown that the net force on the dipole is

$$\mathbf{F} = \mathbf{p}\cdot\nabla\mathbf{E}, \tag{22–8}$$

where ∇ is the gradient operator introduced in Section 7.5. Equations 22–7 and 22–8 show clearly that the net force on a dipole depends on the rate of change of field with position, rather than on the field itself.

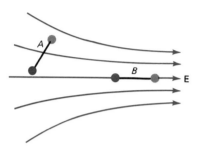

Fig. 22–16
Two dipoles in a nonuniform field. Dipole *A* experiences both a torque and a force; dipole *B* a net force.

Example 22–7

A Dipole in a Nonuniform Field

Find the net force on a water molecule 15 Å from a proton. Assume that the molecule's dipole moment vector is aligned with the electric field (Fig. 22–17), and that there are no other charges in the vicinity. Approximate the water molecule by a point dipole of dipole moment $p=6.2\times10^{-30}$ C·m.

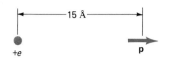

Fig. 22–17
A water molecule, treated as a point dipole, in the electric field of a proton.

Solution

Our molecule is essentially in the same configuration as dipole B of Fig. 22–16, lying along a straight field line, so that Equation 22–7 applies. Calling the direction of the dipole moment vector the x-direction, and taking the origin at the proton, Equation 21–3 for the field of the proton becomes

$$\mathbf{E} = \frac{1}{4\pi\epsilon_0}\frac{e}{x^2}\hat{\mathbf{i}}.$$

Then Equation 22–7 gives the force on the molecule:

$$\mathbf{F} = p\frac{d\mathbf{E}}{dx} = p\frac{d}{dx}\left(\frac{1}{4\pi\epsilon_0}\frac{e}{x^2}\hat{\mathbf{i}}\right) = \frac{p}{4\pi\epsilon_0}\left(-\frac{2e}{x^3}\right)\hat{\mathbf{i}}$$

$$= -(6.2\times10^{-30}\text{ C·m})(9.0\times10^9\text{ N·m}^2/\text{C}^2)\frac{(2)(1.6\times10^{-19}\text{ C})}{(15\times10^{-10}\text{ m})^3}\hat{\mathbf{i}}$$

$$= -5.3\times10^{-12}\,\hat{\mathbf{i}}\text{ N},$$

where $1/4\pi\epsilon_0 = 9.0\times10^9$ N·m^2/C^2, the constant k we introduced in Chapter 20. The minus sign indicates that this force is toward the proton, which makes sense because the negative end of the molecule is closer to the proton and consequently the attractive force on the negative end is greater than the repulsive force on the positive end. You can verify that the dipole exerts an equal but opposite force on the proton (see Problem 20).

Application
Liquid Crystals

liquid crystal display

Fig. 22–18
A liquid crystal display.

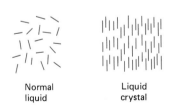

Normal   Liquid
liquid   crystal

Fig. 22–19
Alignment of dipole-like molecules in a liquid crystal.

Glance at your watch or calculator and you will probably see black numbers against a gray background (Fig. 22–18). This **liquid crystal display** contains matter in a unique state—the liquid crystal—that exhibits characteristics of both solids and liquids. The liquid crystal responds readily to an electric field, making it ideally suited for displaying information from electronic devices like watches and calculators.

A liquid crystal consists of long molecules whose chemical structure gives rise to excess charge on one end. Interaction between the charges causes most of the molecules in a given sample to align with each other, as suggested in Fig. 22–19. It is this alignment, or regular order, that makes the substance a crystal. But in response to an electric field, the liquidity of the material allows the orientation of the molecules to change. In a liquid crystal display, small cells of liquid crystal are fabricated into a structure consisting of seven distinct segments (Fig. 22–20). Optical devices called polarizers are placed on either side of the liquid crystals. Normally each segment appears transparent, reflecting a silvery-gray color. But when an electric field is applied to a segment, the liquid crystal molecules rotate, and the polarizing properties of the material change. The segment then appears black. By selecting which segments are so activated, numbers and letters can be formed in the seven-segment display.

Liquid crystal displays have two distinct advantages. First, they require very little electric power. Second, since they do not produce their own light, but only reflect ambient light, they can be seen even in the brightest sunshine. A corresponding disadvantage is that they are not visible in the dark, although a small flashlight bulb can be used to illuminate the display when it is dark.

Liquid crystals show great promise for more complex information displays, including flat-screen television and computer displays.

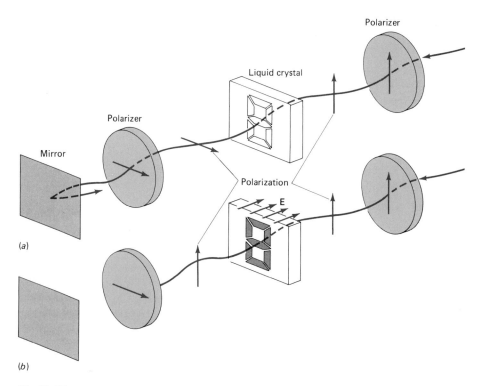

Fig. 22–20
A seven-segment display consists of seven small bands of liquid crystal sandwiched between optical polarizers. *(a)* Electric field off. *(b)* Electric field on.

Other kinds of liquid crystals respond to temperature changes, altering the degree of molecular alignment and changing color. You have probably seen digital thermometers using such liquid crystals.

Dipole-Dipole Interactions and the van der Waals Force

In Chapter 17, we introduced the van der Waals force—a weak, short-range force between gas molecules that causes deviations from ideal gas behavior. Here we show how the van der Waals force arises from electrical interactions among dipole-like molecules.

Consider two dipoles of moment p_1 and p_2, lying on the x-axis as shown in Fig. 22–21. We could calculate the force between the dipoles directly by applying Coulomb's law for the forces between pairs of charges (see Problem 24). Or, if the dipoles are widely separated, we can use Equation 21–7b for the electric field on the axis of a point dipole. Applied to the field of dipole A at the location x of dipole B, Equation 21–7b gives

$$\mathbf{E} = \frac{p_1}{2\pi\epsilon_0 x^3}\, \hat{\mathbf{i}}.$$

Since dipole B is aligned with a straight field line from dipole A, Equation 22–7 gives the force on B:

Fig. 22–21
Two dipoles with their dipole moments aligned.

Fig. 22-22
The induced dipole moment of molecule *B* arises from the field of dipole *A*, and is aligned with that field.

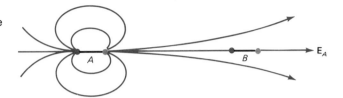

$$\mathbf{F} = p\,\frac{d\mathbf{E}}{dx} = p_2\,\frac{d}{dx}\,\frac{p_1}{2\pi\epsilon_0 x^3}\,\hat{\mathbf{i}} = -\,\frac{3p_1p_2}{2\pi\epsilon_0 x^4}\,\hat{\mathbf{i}}. \tag{22-9}$$

The minus sign indicates that the force is attractive, while the x^4 in the denominator shows that the force drops very rapidly with the distance between dipoles.

Most common gases consist of nonpolar molecules—molecules that do not normally exhibit any charge separation and therefore have no dipole moment. But random fluctuations in the configuration of a molecule can lead to charge separation, so that the molecule temporarily develops a dipole moment. This dipole moment gives rise to an electric field that, in turn, leads to charge separation in nearby molecules, giving them dipole moments as well. These **induced dipole moments** come into existence aligned with the field of the first dipole, since that field is what causes charge to separate in nearby molecules (Fig. 22–22). For a second molecule located along the axis of the first, the configuration is identical to Fig. 22–21, so that there is an attractive force between the molecules given by Equation 22–9. But now the dipole moment p_2 of the second molecule is itself proportional to the field of the first molecule, which decreases as $1/x^3$. The result in Equation 22–9 is a moment p_2 decreasing as $1/x^3$, multiplying a term decreasing as $1/x^4$, for a force that decreases as $1/x^7$. This very short-range attractive force is the **van der Waals force** that we considered in Chapter 17. Its short range means that it is significant only when molecules are relatively close—that is, in dense gases. The van der Waals force is a subtle manifestation of a much simpler phenomenon—the electric force between point charges that is described by Coulomb's law.

induced dipole moments

van der Waals force

22.4
DIELECTRICS

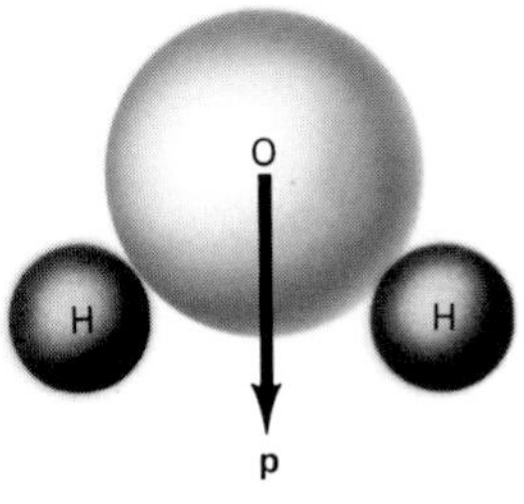

Some materials consist of molecules that have intrinsic electric dipole moments. The structure of these **polar molecules** is such that electrons spend more of their time nearer one end of the molecule, giving rise to the charge separation needed to make a dipole. Water (H_2O) is a polar molecule, whose dipole moment results from the affinity of its single oxygen atom for electrons (Fig. 22–23). Figure 22–24 shows another common substance containing intrinsic electric dipoles. When an electric field is applied to a material containing polar molecules, the molecular dipoles experience torques that tend to align them with the field. Random thermal rotation of the molecules prevents complete alignment but, nevertheless, the effect of the electric field is to produce some order in what was initially a random orienta-

Fig. 22-23
Electrons in a water molecule spend most of their time in the vicinity of the oxygen, giving the molecule an electric dipole moment.

tion of the molecular dipoles. The degree of alignment of the molecular dipoles increases with increasing electric field strength.

nonpolar molecule   Even a **nonpolar molecule**—one with no intrinsic dipole moment—will develop a dipole moment under the influence of an electric field (Fig. 22–25). Although the centers of positive and negative charge initially coincide, the electric field induces a slight charge separation, giving rise to a dipole moment. In a material consisting of such molecules, the molecular dipoles come into existence necessarily aligned with the field. The size of the individual dipole moments depends on the magnitude of the applied electric field. For most materials, in modest electric fields, the induced dipole moment is proportional to the field strength.

dielectrics   Materials whose molecules have either intrinsic or induced dipole moments are termed **dielectrics.** Most insulators have dielectric properties. When a dielectric is placed in an electric field, the molecular dipoles align with the field and produce an overall polarization of the material. What effect does this have on the electric field? Each dipole gives rise to its own electric field. This field is strongest between the two charges of the dipole, and points from positive charge to negative. As Fig. 22–26 shows, the molecular dipoles line up so that their positive charges are in the direction of the applied field. As a result, the internal fields of the molecular dipoles oppose the applied field, causing a reduction in the net average electric field within the dielectric.

The reduction in field strength in a dielectric depends on the strength (for nonpolar molecules) or degree of alignment (for polar molecules) of the individual molecular dipoles. The reduction also depends on the density of dipoles in the material as well as on variables like temperature. When the individual dipole moments or their degree of alignment are themselves proportional to the field strength, then the field is reduced by a constant factor. The inverse of this factor is called the **dielectric constant, κ,** of the material. The applied field and the reduced internal field of the dielectric are related by

Fig. 22–24
Soaps and detergents owe their cleaning power to highly polar molecules that attach themselves to particles of dirt and grease.

dielectric constant

$$\mathbf{E}_{\text{internal}} = \frac{\mathbf{E}_{\text{applied}}}{\kappa},$$ (22–10)

where κ is the dielectric constant.

Fig. 22–25
(a) A nonpolar molecule with no external electric field. (b) An externally applied field gives rise to a dipole moment.

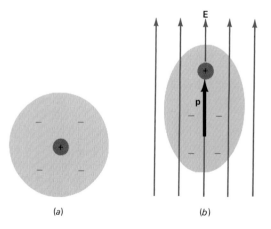

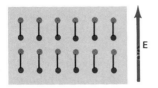

Fig. 22–26
The alignment of molecular dipoles in a dielectric results in a reduction in the net electric field within the dielectric.

dielectric breakdown

Fig. 22–27
Dielectric breakdown in a solid produced this tree-like pattern.

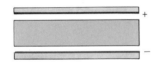

Fig. 22–28
Two metal plates separated by a dielectric (Example 22–8).

The dielectric constant determines the effect of insulating materials on electric fields and influences, for example, the speed at which electrical signals can be transmitted down cables. As we will see in Chapter 32, the dielectric constant is intimately related to the index of refraction that describes the propagation of light through transparent materials. Because the dielectric constant is directly related to the properties of individual molecules, measuring the dielectric constant provides information about the molecular structure of a material.

If the electric field applied to a dielectric becomes too great, individual charges are ripped free from the material, which begins to act more like a conductor. Called **dielectric breakdown,** this phenomenon can cause severe damage in electrical equipment (Fig. 22–27). Lightning is the result of large-scale dielectric breakdown in air. We will discuss dielectrics and dielectric breakdown further in Chapter 24, where we include a table of dielectric properties.

Example 22–8
The Dielectric Constant and Dielectric Breakdown

Pyrex glass has a dielectric constant of 5.6, and undergoes dielectric breakdown when the field in the glass reaches 1.4×10^7 N/C. If a piece of Pyrex glass is used to separate two large, flat, closely spaced metal plates (Fig. 22–28), what is the maximum charge density on either plate? Assume that the plates carry equal but opposite charges with uniform surface charge density.

Solution

In Example 22–5, we found that the field strength between two closely spaced metal plates carrying opposite charges is $E = \sigma/\epsilon_0$, with σ the surface charge density on either plate. With a dielectric between the plates, the field is reduced according to Equation 22–10, so that

$$E = \frac{\sigma}{\kappa \epsilon_0}.$$

Solving for σ and using for E the field at dielectric breakdown, we have

$$\sigma = \kappa \epsilon_0 E = (5.6)(8.9 \times 10^{-12} \ \text{C}^2/\text{N} \cdot \text{m}^2)(1.4 \times 10^7 \ \text{N/C}) = 7.0 \times 10^{-4} \ \text{C/m}^2.$$

Note that κ is dimensionless, since it is just a ratio of electric fields.

SUMMARY

1. The force on a charged particle in an electric field follows directly from the definition of the electric field as the force per unit charge. The force is simply the product of the charge and the electric field:

$$\mathbf{F} = q\mathbf{E}.$$

When no other forces act on the particle, the resulting acceleration, given by Newton's law, is just

$$\mathbf{a} = \frac{q}{m} \mathbf{E}.$$

Thus the response of a charged particle to an electric

field depends on the charge to mass ratio, q/m, of the particle.

2. A **conductor** is a material in which charges are free to move. **Electrostatic equilibrium** is defined as a state in which there is no net charge motion. Because electric fields give rise to forces on charges, it follows that the average electric field inside a conductor in electrostatic equilibrium is zero. Gauss's law assures us that, in consequence, there can be no net charge inside a conductor in electrostatic equilibrium. Any excess charge on a conductor resides on the surface of the conductor, where the electric field is perpendicular to the surface

and is given by

$$E = \frac{\sigma}{\epsilon_0}.$$

3. Because a **dipole** consists of two equal but opposite charges, it experiences no net force in a uniform electric field. But the opposite forces on the charges making up a dipole result in a **torque** on the dipole. This torque is given by the cross product of the dipole moment and the electric field:

$$\boldsymbol{\tau} = \mathbf{p} \times \mathbf{E}.$$

As a result, it takes work to twist a dipole out of alignment with an electric field. The potential energy associated with this work is given by

$$U = -\mathbf{p} \cdot \mathbf{E},$$

where the zero of potential energy occurs when the dipole is at right angles to the field.

4. A dipole in a nonuniform electric field generally experiences different forces at its two ends, giving rise to a net force on the dipole. This force depends on the rate of change of the field with position, and on the orientation of the dipole relative to the field:

$$\mathbf{F} = \mathbf{p} \cdot \nabla \mathbf{E}.$$

The force between dipolar gas molecules is responsible for the weak van der Waals forces that cause deviations from ideal gas behavior.

5. A **dielectric** is a material containing molecular dipoles. Orientation of these individual dipoles with an applied electric field causes a reduction of the field within the material. The ratio of the applied field to the reduced field within the dielectric is called the **dielectric constant**. The dielectric constant is an easily measured property that reflects the molecular structure of the material.

QUESTIONS

1. An electric field is established that suspends a certain charged particle in earth's gravitational field. What characteristic must a second charged particle have if it, too, is to be suspended by the same electric field?
2. A deuteron (heavy hydrogen nucleus) has twice the mass but the same charge as a normal hydrogen nucleus (a proton). A deuteron and a proton are released from rest in a uniform electric field. Compare the distances each travels in the same time.
3. Under what circumstances is the path of a charged particle in an electric field a parabola? A circle?
4. A charged particle is released from rest in an electric field. Under what conditions will the particle's subsequent trajectory coincide with a field line?
5. In electrostatic equilibrium, the electric field at the surface of a conductor points at right angles to the surface. Give a simple argument to show that, if the field is not perpendicular to the surface, the conductor cannot be in electrostatic equilibrium.
6. In electrostatic equilibrium, the electric field at the surface of an insulator need not be perpendicular to the surface. Why not?
7. In Section 22.2, we argued that excess charge in an elongated conductor would concentrate at the ends. Where near this conductor would you expect to find the strongest electric field?
8. The electric field of a flat sheet of charge is $\sigma/2\epsilon_0$. Yet the field of a flat conducting sheet—even a thin one like a piece of aluminum foil—is σ/ϵ_0. Explain this apparent discrepancy.
9. A metal is an electrical conductor because it contains free electrons not bound to individual atoms. Does Gauss's law require that all these free electrons be on the surface of the conductor? Explain.
10. In our discussion of metallic conductors, we assumed that electrons are free to move within the metal but not to leave it altogether. What prevents them from leaving the metal?
11. A dipole is placed in the electric field of a point charge. Is there an orientation for which there is no net torque on the dipole? If so, describe it.
12. A dipole is placed in the electric field of a point charge. Is there an orientation for which there is no net force on the dipole? If so, describe it.
13. How could a point dipole experience a force in a nonuniform electric field? After all, the point dipole has zero size. *Hint:* Consider the limiting process that led to the concept of a point dipole (see Example 21–3).
14. Why should there be a force between two dipoles? Each dipole, after all, carries zero net charge.
15. Is there any way to arrange two dipoles so that there is no net force between them? Are there torques in this arrangement?
16. Explain why a nonuniform field is required for a net force on a dipole.
17. If dipole A of Fig. 22–16 is released from rest, describe qualitatively its subsequent behavior. Repeat for dipole B.
18. Figure 22–29 shows a dipole in a nonuniform electric field. The field strength is the same at both ends of the dipole. Is there a net force on the dipole? If not, why not? If so, what is its direction?
19. Why does the van der Waals force fall more rapidly with distance than the force between dipoles with permanent dipole moments?
20. Could the dielectric constant of a polar material be less than 1? Why or why not?
21. How would Equation 22–10 be modified if the induced dipole moment of a nonpolar molecule were proportional not to the magnitude of the applied field but to its square root?

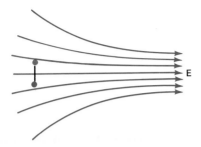

Fig. 22–29 The field strength is the same at both ends of the dipole (Question 18).

22. In a polar material, molecular rotation associated with thermal energy of the material prevents perfect alignment of molecular dipoles with an applied electric field. Would you expect the dielectric constant of such a material to increase or decrease with increasing temperature?

PROBLEMS

Section 22.1   *A Single Charge in an Electric Field*

1. Compare the gravitational force on an electron with the electric force in the 100 N/C field of the earth.

2. In his famous 1909 experiment that demonstrated quantization of electric charge, R. A. Millikan suspended small oil drops in an electric field. With a field strength of 2.0×10^7 N/C, what mass drop can be suspended when the drop carries a net charge of ten elementary charges?

3. A uniform electric field **E** is set up in a region between two metal plates of length ℓ, as shown in Fig. 22–30. The plates are a distance d apart. An electron enters the region midway between the plates and moving at right angles to the field with speed v, as shown. What is the minimum speed the electron must have when it enters the region between the plates in order to get through the region without hitting either plate? Neglect gravity.

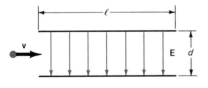

Fig. 22–30 Problem 3

4. Electrons in a TV tube move down the tube at one-tenth the speed of light. If the electrons are accelerated from rest in a region 5.0 cm long, how strong an electric field is needed?

5. An electron is moving in a circular path around a long, uniformly charged wire carrying 2.5 nC/m. (1 nC = 10^{-9} C.) (a) What is the electron's speed? (b) Describe the electron's path if it also had a velocity component parallel to the wire.

6. Two parallel metal plates measure 10 cm on a side and are 8.0 mm apart. They carry equal but opposite uniform surface charge densities. An electron is launched from the edge of the positive plate, moving at a 10° angle to the plate at 9.0×10^6 m/s (Fig. 22–

31). (a) What is the minimum value of the charge density for which the electron could leave the space between the plates without striking either plate? (b) What is the maximum value of the charge density for the electron to leave the space between the plates?

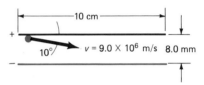

Fig. 22–31 Problem 6

7. Figure 22–32 shows a device that its inventor claims will separate a desired isotope of a particular element. (Isotopes of the same element have nuclei with the same charge but different masses.) Atoms of the element are first stripped completely of their electrons, then accelerated from rest through an electric field chosen to give the desired isotope exactly the right velocity to pass through the electrostatic analyzer (see Example 22–3). Prove that the device will not work—that it will not separate the isotopes.

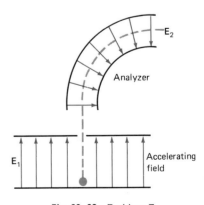

Fig. 22–32 Problem 7

8. A small object of mass m and charge Q is attached by a massless, uncharged thread of length ℓ to a large metal plate carrying a uniform surface charge density σ of the same sign as Q (Fig. 22–33). If the object is displaced slightly from its equilibrium, show that it undergoes simple harmonic motion with period

$$T = 2\pi\sqrt{\frac{\epsilon_0 m\ell}{Q\sigma}}.$$

Assume that gravity is negligible.

Fig. 22-33 Problem 8

Section 22.2 Conductors

9. A net charge of 5.0 μC is applied on one side of a solid metal sphere 2.0 cm in diameter. After electrostatic equilibrium is reached, what are the (a) the volume charge density inside the sphere and (b) the surface charge density on the sphere's surface? Assume that no other charges or conductors are nearby. (c) Which of your answers depends on this assumption, and why?

10. A conductor carries a surface charge density of -0.60 C/m^2. Calculate the acceleration an electron would experience when placed at the surface of this conductor.

11. An irregular conductor containing an irregular, empty cavity carries a net charge Q (Fig. 22–34). (a) Show that the electric field inside the cavity is zero. (b) If you put a point charge inside the cavity, what value should it have in order to make the surface charge density on the outer surface of the conductor everywhere zero? (c) Does it matter where in the cavity you put the point charge?

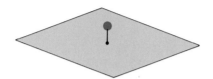

Fig. 22-34 Problem 11

12. Two flat, parallel metal plates of area 0.080 m^2 carry net charges of -2.1 μC and $+3.8$ μC. What are the surface charge densities on the inner and outer faces of each plate?

13. A coaxial cable consists of an inner wire and a concentric cylindrical outer conductor (Fig. 22–35). If the conductors carry equal but opposite charges, show that there is no net surface charge density on the *outside* of the outer conductor.

Fig. 22-35 Problem 13

14. A neutral dime is placed in a uniform electric field of 6.2×10^5 N/C, with its faces perpendicular to the field. (a) What is the charge density on the faces of the dime? (b) What is the total charge on each face? Neglect edge effects.

15. A point charge is placed at the center of a neutral, hollow spherical conducting shell of inner radius 2.5 cm and outer radius 4.0 cm (Fig. 22–36). The outer surface of the shell acquires a surface charge density of 7.1×10^{-2} μC/cm^2. Find (a) the value of the point charge within the cavity and (b) the surface charge density on the inner wall of the cavity.

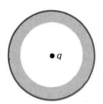

Fig. 22-36 Problem 15

Section 22.3 A Dipole in an Electric Field

16. In Fig. 22–16 a dipole in a nonuniform field experiences a force in the direction of the field. Sketch a situation in which the force on a dipole is opposite to the field direction.

17. A dipole with dipole moment 1.5×10^{-9} C·m is oriented at 30° to an electric field of 4.0×10^6 N/C. (a) What is the magnitude of the torque on the dipole? (b) How much work is required to turn the dipole so that it is antiparallel to the field?

18. A molecule has its dipole moment aligned with an electric field of 1.2×10^3 N/C. If it takes 3.1×10^{-27} J to reverse the molecule's orientation, what is its dipole moment?

19. A water molecule (dipole moment $p = 6.2 \times 10^{-30}$ C·m) is between two parallel metal plates carrying surface charge densities of $\pm 2.2 \times 10^{-9}$ C/m^2. If the molecule's dipole moment vector is parallel to the plates, what is the torque on the molecule?

20. Calculate the force exerted on the proton by the water

molecule in Example 22–7, and show explicitly that Newton's third law is satisfied.

21. A dipole of moment $p = 1.4 \times 10^{-8}$ C·m is 2.8 cm from the axis of a 3.0-m-long wire carrying a total charge of $+12$ μC. What are the force and torque on the dipole when **p** is (a) parallel to the wire and (b) perpendicular to the wire? (c) How much work is done on the dipole when it goes from (a) to (b)?

22. Two dipoles of moment $p = 8.3 \times 10^{-12}$ C·m are 12 cm apart and are oriented as shown in Fig. 22–37. What is the torque on dipole B? (See Example 21–3 for dipole fields; see Problem 32 for more on this situation.)

Fig. 22–37 Problems 22, 32

23. A dipole and a point charge are oriented as shown in Fig. 22–38. Use Equation 22–8 to calculate the force on the dipole. Check your result by calculating the force on the point charge (see Example 21–3 for dipole field), then applying Newton's third law.

Fig. 22–38 Problem 23

24. Consider two dipoles, each made up of charges $\pm q$ separated a small distance a, as shown in Fig. 22–39. By considering forces between pairs of charges in opposite dipoles, calculate the net force between the dipoles. Show that your result reduces to Equation 22–9 in the case $x \gg a$.

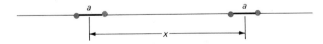

Fig. 22–39 Problem 24

Section 22.4 *Dielectrics*

25. The dielectric constant of water is 78. How strong an electric field must be applied outside a water sample to obtain a field of 100 N/C within the water?

26. A slab of dielectric material with dielectric constant κ is placed in a uniform electric field **E,** as in Fig. 22–26. As dipoles in the material align with the field, opposite ends of individual dipoles are adjacent within the material, and there is no net charge associated with the interior dipoles. But the positive ends of the dipoles at one surface, and the negative ends at the other surface, have no adjacent dipoles to cancel them. This results in a net surface charge density, called the **polarization surface charge density.** The electric field at all points is the vector sum of the original field and the fields of these two polarization charge sheets. Show that Equation 22–10 requires that the polarization charge density σ' is

$$\sigma' = \epsilon_0 E\left(1 - \frac{1}{\kappa}\right).$$

Supplementary Problems

27. A dipole consists of two small objects each of mass m that carry charges $\pm q$. The objects are separated by a massless rod of length ℓ. If the dipole is displaced slightly from alignment with a uniform electric field **E,** it undergoes simple harmonic oscillations. Show that the period of these small-amplitude oscillations is

$$T = \pi \sqrt{\frac{2m\ell}{qE}}.$$

28. Consider the electric field **E** at a point on the surface of the irregular hollow conductor shown in Fig. 22–40. In Section 22.2, we showed that the electric field is perpendicular to the conductor surface and has magnitude σ/ϵ_0, where σ is the local surface charge density. Now imagine removing a small patch of the conductor (Fig. 22–40b), leaving a small hole. Show that the electric field at this hole is now $\sigma/2\epsilon_0$, pointing outward. To do so, consider the removed part as a single flat charge sheet. Evaluate the field of this charge sheet, and show that when it is back in place the rest of the conductor must contribute a field $\sigma/2\epsilon_0$ to give zero field inside the conductor and σ/ϵ_0 just outside. Your result shows that half the field at the

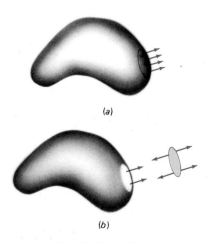

(a)

(b)

Fig. 22–40 Problem 28.

surface of a conductor in electrostatic equilibrium is contributed by charges in the immediate vicinity, the other half by all the rest of the charges on the conductor surface. Charges must distribute themselves over the surface to make this true.

29. Use the result of the preceding problem to show that the pressure (force per unit area) arising from the electric force on a uniformly charged sphere is $\sigma^2/2\epsilon_0$. *Hint:* What is the electric force on a small patch in the field of all the rest of the sphere?

30. At a temperature of 500 K, the rotational states of water molecules in the gaseous state are fully excited (see Section 18.3). From the viewpoint of classical physics, the molecules are simply rotating end-over-end. Each water molecule has a dipole moment $p = 6.2 \times 10^{-30}$ C·m. How strong an electric field must be applied to change the rotation into an oscillation for a molecule whose rotational energy is the average for this temperature?

31. Although we wrote Equation 22–6 for a uniform field, the result holds for a nonuniform field as well. Recalling from Chapter 7 that force is the gradient of the potential energy, take the gradient of the potential energy given by Equation 22–6, and show that Equation 22–8 results.

32. Calculate the torque on dipole *A* of Problem 22. Compare with the result of that problem for the torque on dipole *B*, and comment on the apparent violation of the strong form of Newton's third law (see Chapter 12, Section 2, especially Fig. 12–16).

23

ELECTRIC POTENTIAL

23.1
WORK AND THE ELECTRIC FIELD

You must do work to lift an object in earth's gravitational field. Lowering the object releases the energy that was stored as potential energy when it was lifted. In Chapter 7, we appplied the term **conservative** to the gravitational force because it "gives back" all the stored energy; similarly, we say that the gravitational field is a **conservative field.** An important property of conservative forces and fields that we developed in Chapter 7 is path-independence: the work required to move an object from one point to another does not depend on the path taken, but only on the endpoints of the path. You could lift a book straight from the floor to your desktop or you could raise it a meter above the desktop, whirl it around in circles, and finally lower it to the desktop. In either case, provided you accounted for both positive and negative energy changes, you would find that it took the same amount of work to get the book to the desktop (Fig. 23–1).

It is an experimental fact that the static electric field is also conservative. You must do work to move a charge against the electric force, and you can recover the associated energy by allowing that force to do work on the charge. Because the field is conservative, the work involved in moving a charge from one point to another in an electric field does not depend on the path taken, but only on the endpoints of the path.

Consider a positive charge q being moved between two points A and B in a uniform electric field **E,** as shown in Fig. 23–2. How much work is required to move the charge at constant velocity between these points if they are a distance ℓ apart? In Chapter 6, we found that the work done on an object *by* a force **F** is $\int \mathbf{F} \cdot d\boldsymbol{\ell}$, where the integral is taken over the path travelled by the object. The work *we* must do is the opposite of the work done by the force, so that

$$W_{AB} = -\int_A^B \mathbf{F} \cdot d\boldsymbol{\ell} = -\int_A^B q\mathbf{E} \cdot d\boldsymbol{\ell}, \qquad (23\text{–}1)$$

where we have written the electric force on q as $q\mathbf{E}$. Because the electric

conservative force

conservative field

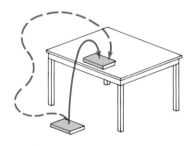

Fig. 23–1
The gravitational field is conservative, so the work required to move between two points is independent of the path taken.

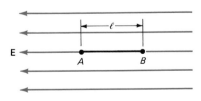

Fig. 23–2
The work required to move a charge q from A to B in a *uniform* electric field is $qE\ell$.

field is conservative, the work is the same for any path beginning at A and ending at B. We are free to choose a path that will make our calculation simple. In this case, the simplest path is a straight line between points A and B. Since $\mathbf{E}$ is uniform, it is constant along this path, and the angle between the field and the vectors $d\boldsymbol{\ell}$ is always 180° (we are going *opposite* to the field direction). Then for this simple uniform-field case, the work becomes

$$W_{AB} = -\int_A^B qE\cos 180° \, d\ell = qE\int_A^B d\ell = qE\ell, \tag{23-2}$$

where the minus sign in front of the integral has cancelled with $\cos 180° = -1$. We didn't really have to do an integral to get this result, since we know from Chapter 6 that the work done moving directly against a constant force is simply the force times the distance moved. But we will often encounter nonuniform electric fields, for which we must use the full integral form to calculate the work.

23.2
POTENTIAL DIFFERENCE

Equations 23–1 and 23–2 give the work required to move a charge q between points A and B. A charge $2q$ would require twice as much work; $\frac{1}{10}q$ would require one-tenth the work. Because the work is directly proportional to the charge, it is convenient to speak of the work per unit charge required to move between points A and B. We define this work per unit charge as the **electric potential difference** between the two points A and B.

electric potential
difference

Always think of electric potential in terms of two points. This is ultimately a very practical matter. If you forget that potential difference is a property of two points, you will not be able to hook up a voltmeter properly or to attach jumper cables safely to your car battery! Later we will speak of "the potential at a point." This is a shorthand way of talking, for we always have in mind some other point that serves as a reference, and we always mean the potential difference between our point and that reference point.

To obtain the potential difference $V_B - V_A$ between two points A and B, we simply divide the work W_{AB} of Equation 23–1 by the charge q:

$$V_B - V_A = \frac{W_{AB}}{q} = -\int_A^B \mathbf{E} \cdot d\boldsymbol{\ell}. \qquad \text{(potential difference)} \tag{23-3}$$

For the uniform-field case of Fig. 23–2, we can divide Equation 23–2 by q to get

$$V_B - V_A = E\ell. \qquad \text{(moving against uniform field)} \tag{23-4}$$

Potential difference can be positive or negative, depending on whether our path goes with or against the field. Moving a positive charge through a positive potential difference is like going uphill: we must do work on the charge. Moving a positive charge through a negative potential difference is like going downhill: we do negative work or, equivalently, the charge does

volt

work on us. From the definition of potential difference, it follows directly that its units are joules/coulomb. Potential difference is so important that one J/C is given its own special name—the **volt** (V). To say that a car has a 12-volt battery, for example, means that it takes 12 joules of work to move one coulomb of charge from the negative to the positive terminal of the battery. Alternately, one coulomb would gain 12 joules if allowed to move from the positive to the negative terminal. Note that this practical discussion of a car battery continues to stress the fundamental meaning of potential difference: that it is a property of two points, being the work per unit charge associated with charge moving between those points.

voltage

We will sometimes use the term **voltage** to mean potential difference, especially in describing electric circuits. Strictly speaking the two terms are not synonymous, since voltage is used even in nonconservative situations that arise when fields change with time. But in common usage this subtle distinction is usually not bothersome.

electron volt

In working with systems on molecular or smaller scales, it is often convenient to measure energy in **electron volts** (eV), defined as the energy gained by a particle with one elementary charge as it moves through a potential difference of one volt. Since one elementary charge is 1.6×10^{-19} C, one eV is 1.6×10^{-19} J. Although the energy in eV is easy to calculate for an object whose charge is given in elementary charges, the electron volt is *not* a standard SI unit and should first be converted to joules if it is to be used in calculating other quantities.

Example 23–1

Work and Potential Difference

Suppose that we must do 3.2×10^{-19} J of work on a proton to move it between two points. What is the potential difference between those points? How much energy would an electron gain if it moved freely between the same two points? What would be the electron's final speed if it started from rest? Assume that only the electric force is acting.

Solution

The potential difference is the work required to move a unit charge between the two points. The proton carries one elementary charge, or 1.6×10^{-19} C, so that the potential difference in this case is

$$V = \frac{W}{q} = \frac{3.2 \times 10^{-19} \text{ J}}{1.6 \times 10^{-19} \text{ C}} = 2.0 \text{ J/C} = 2.0 \text{ V}.$$

This potential difference is a property of the two points, not of the charge being moved, so that the same 2.0 V applies to an electron. The electron carries a negative elementary charge, so it would gain 2.0 eV or 3.2×10^{-19} J of kinetic energy in moving between the same two points. The electron's final speed is then obtained from

$$K = \tfrac{1}{2}mv^2,$$

or

$$v = \sqrt{\frac{2K}{m}} = \sqrt{\frac{(2)(3.2 \times 10^{-19} \text{ J})}{9.1 \times 10^{-31} \text{ kg}}} = 8.4 \times 10^5 \text{ m/s}.$$

Note that although it was easiest to calculate and to state the electron's energy in eV, we needed to use joules in our speed calculation for consistency with other SI units.

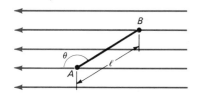

Fig. 23–3
For a path at an angle to a uniform field, the potential difference is $-E\ell\cos\theta$. Note that $\cos\theta$ is negative in this case, so the potential difference is positive.

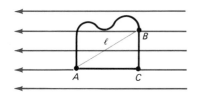

Fig. 23–4
The potential difference between points C and B is zero, so that $V_B - V_A = V_C - V_A$. The potential difference for the curved path is also $V_B - V_A$.

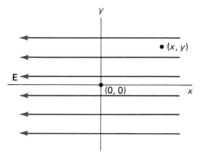

Fig. 23–5
Example 23–2

To calculate the potential difference along a straight path at some angle to a uniform field (Fig. 23–3), we must use the angle in evaluating the dot product in Equation 23–3. Here E and $\cos\theta$ are both constants, so that Equation 23–3 gives

$$V_B - V_A = -\int_A^B \mathbf{E}\cdot d\boldsymbol{\ell} = -E\cos\theta \int_A^B d\ell = -E\ell\cos\theta.$$

If we define a vector $\boldsymbol{\ell}$ that coincides with our straight path, we can write this expression more compactly in the form

$$V_B - V_A = -\mathbf{E}\cdot\boldsymbol{\ell}. \tag{23–5}$$

In Fig. 23–3, $\theta > 90°$, so that $\cos\theta$ is negative, making the potential difference positive. In particular, if θ were 180°, we would recover our earlier result for a path along the field. More generally, the quantity $-\ell\cos\theta$ is a component of the path *along* the field. Just as the work mgh needed to lift an object in earth's gravitational field depends only on the *height h*, so the electric potential difference depends only on the distance moved *along* the field direction. You can see this explicitly by considering the alternate path ACB shown in Fig. 23–4. The potential difference between points C and B is zero, since it takes no work to move at right angles to the electric force. So the potential difference $V_B - V_A$ is the same as the difference $V_C - V_A$ associated with a path of length $\ell\cos\theta$ parallel to the field. It would be difficult to calculate the potential difference using the curved path shown in Fig. 23–4, but the conservative nature of the electric field assures us that the answer is still the same $V_B - V_A$ that we get using either of the other paths.

Example 23–2
Potential in a Uniform Field

A uniform electric field $\mathbf{E}$ points in the $-x$ direction, as shown in Fig. 23–5. What is the potential difference between the origin and any arbitrary point (x,y)?

Solution

To get from the origin to an arbitrary point (x,y), we can move first along the x-axis and then parallel to the y-axis. It takes no work to move in the y-direction—perpendicular to the field—so that the potential difference depends only on x. Moving in the positive x direction, we are going against a uniform electric field, so that the potential difference is Ex from Equation 23–4. Moving in the negative x-direction, the potential difference is negative. But this is also accounted for by the expression Ex, which is negative for negative values of x. So the potential difference between the origin and any point (x,y) is given by

$$V(x,y) - V(0,0) = Ex.$$

We could have obtained this result more formally by writing $\mathbf{E} = -E\hat{\mathbf{i}}$ and using Equation 23–3.

This uniform-field situation is analogous to a hill with constant slope in earth's gravitational field. If we move upslope from a given point, the work required is proportional to how far we go. If we move downslope, we gain energy in proportion to our distance downslope from the given point.

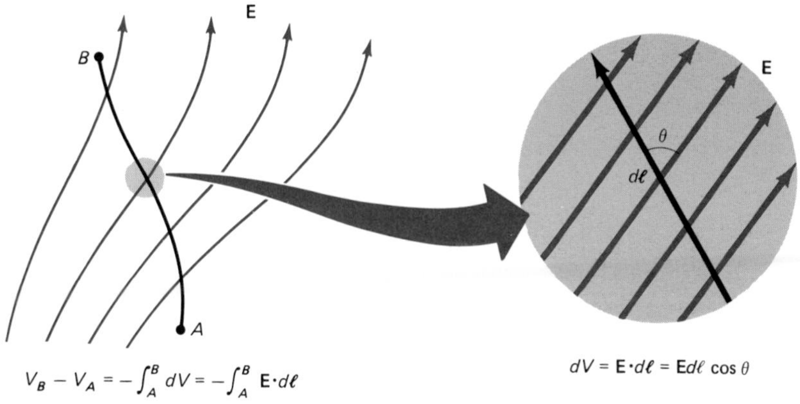

$$V_B - V_A = -\int_A^B dV = -\int_A^B \mathbf{E} \cdot d\boldsymbol{\ell}$$

$$dV = \mathbf{E} \cdot d\boldsymbol{\ell} = E d\ell \cos\theta$$

Fig. 23–6
The line integral of Equation 23–3 is the sum of infinitely many infinitesimally small potential differences.

If the field is not uniform or the path is not straight, then we must use the integral form 23–3 to calculate the potential difference because the magnitude of **E** and/or the angle between **E** and the path is changing. Figure 23–6 shows the meaning of Equation 23–3 in this case. If we look at a sufficiently small part of the curve in Fig. 23–6, so small that the field is essentially uniform and the path essentially straight, then Equation 23–5 for a straight path in a uniform field should apply. Describing our short segment of path by a small vector $d\boldsymbol{\ell}$, we can write Equation 23–5 in the form

$$dV = -\mathbf{E} \cdot d\boldsymbol{\ell},$$

where dV is the potential difference between the ends of the vector $d\boldsymbol{\ell}$. (We do not write $d\mathbf{E}$ because the field **E** has a value at each point. **E** does not become smaller as we consider ever smaller regions.) The integral in Equation 23–3 is simply the sum of all the dV's over the path:

$$V_B - V_A = -\int_A^B dV = -\int_A^B \mathbf{E} \cdot d\boldsymbol{\ell}.$$

line integral   Like the work integral we introduced in Chapter 6, this **line integral** is simply a sum of scalar quantities—dot products of the vectors **E** and $d\boldsymbol{\ell}$—over some path. The limits of the integral are the endpoints of the path. Because the electrostatic field is conservative, it is not necessary to specify which of the infinitely many paths between these endpoints is to be taken; all such paths give the same result.

Example 23–3 _____
Evaluating the Line Integral

In Example 23–2 we evaluated the potential difference between the origin and an arbitrary point in the uniform electric field $\mathbf{E} = -E\hat{\imath}$. Here we repeat this calculation using a curved path to illustrate application of the line integral in Equation 23–3. In this particular case, of course, the method of Example 23–2 is far easier, but often the situation does not admit such an easy solution.

Fig. 23–7
Example 23–3

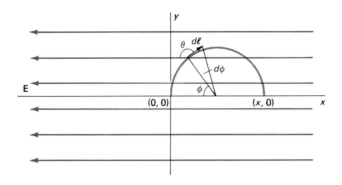

Solution

Figure 23–7 shows a semicircular path between the origin and the point (x,0). The field vector **E** and path vector $d\ell$ are shown for one point along this path. As we move along the path, the angle θ between the field and the path ranges from $\pi/2$ (90°) at the point (0,0) to $3\pi/2$ (270°) at (x,0). The line integral for the potential is then

$$V(x,0) - V(0,0) = -\int_{\theta=\pi/2}^{\theta=3\pi/2} \mathbf{E} \cdot d\ell = -\int_{\pi/2}^{3\pi/2} E \cos\theta \, d\ell.$$

To evaluate this integral, we must somehow relate $d\ell$ and θ so that we have only one variable under the integral sign. Because $d\ell$ is very small and tangent to the semicircular path, its length is essentially that of the arc subtended by the angle ϕ shown in Fig. 23–7. So we can write

$$d\ell = \tfrac{1}{2}x\,d\phi,$$

where $\tfrac{1}{2}x$ is the radius of the semicircle. How are θ and ϕ related? As θ ranges from $\pi/2$ to $3\pi/2$, ϕ ranges from 0 to π. Our angles θ and ϕ differ only by the constant $\pi/2$, so that $d\phi = d\theta$. Then $d\ell = \tfrac{1}{2}x\,d\theta$, and the potential becomes

$$V(x,0) - V(0,0) = -\int_{\pi/2}^{3\pi/2} \tfrac{1}{2}Ex \cos\theta \, d\theta = -\tfrac{1}{2}Ex \sin\theta \Big|_{\pi/2}^{3\pi/2} = -\tfrac{1}{2}Ex(-1-1) = Ex,$$

in agreement with Example 23–2. Not only does this more difficult calculation illustrate the use of the line integral, but it also suggests that the potential difference is truly path-independent.

In this example an integral is needed even though the electric field is uniform, because the orientation between path and field is changing. Often we will encounter situations in which we can choose a path lying along a field line, so that the orientation between field and path does not change. But if the field changes in magnitude in these situations, it is still necessary to use an integral. Only when both the field magnitude and the orientation between the field and path are constant does the integral expression for potential difference reduce to a simple product.

We have stressed that potential difference is a property of two points and that it makes no sense to speak of the potential at a point. Often, however, we choose a reference point at which we say that the potential is zero. We then speak of "the potential at a point," meaning the potential difference between that point and our reference point. The choice of a reference point has no physical significance, although a sensible choice may greatly simplify mathematical calculations or make an electrical situation easier to understand. Changing the reference point may change the values of potential everywhere, but it cannot change potential differences, which are the only quantities of direct physical significance.

Choice of the reference point where potential is zero is often suggested by a situation. In electric power systems, for example, the earth—called "ground"—is taken as the zero of potential. Of course, the earth is not a single point, but it is a good enough conductor that it does not sustain large electric fields. Thus the potential difference between two nearby points on earth is small, and it makes sense to assign the earth a fixed value for potential. In a car, whose rubber tires insulate it from earth, the car's metal body is usually considered the zero of potential. The point of zero potential in an electric circuit is often called "ground" even though it may be different from the physical ground of earth.

When we deal with isolated charges, it is convenient to take the zero of potential to be infinitely far from all charges under consideration. This has the advantage of providing a reference point that is equally distant from all the charges. As we will soon see by evaluating Equation 23–3 for the field of a point charge, it does not take an infinite amount of work to move from infinity to the vicinity of a point charge.

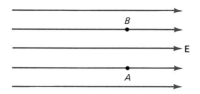

Fig. 23–8
The potential difference is zero between two points aligned perpendicular to the field.

equipotential surfaces
equipotentials

23.3
EQUIPOTENTIALS AND THE ELECTRIC FIELD

Figure 23–8 shows a uniform electric field and two points aligned perpendicular to the field. The potential difference between these points is zero because it requires no work to move a charge at right angles to the electric force. If the field is uniform in three dimensions, then the potential difference between any two points on any plane perpendicular to the field is in fact zero. Such planes are called **equipotential surfaces** or simply **equipotentials.** By definition, the potential difference between any two points on an equipotential surface is zero. It takes no work to move a charge around on an equipotential surface.

How do we locate equipotentials? The only way a charge can move in a nonzero field without work being involved is to move at right angles to the field. Therefore, the equipotential surfaces are always perpendicular to the electric field. This is true even when the field lines are not straight; in that case the equipotentials are curved surfaces. Figure 23–9 shows some equipotential surfaces in both uniform and nonuniform electric fields.

Fig. 23–9
(a) Equipotential surfaces in a uniform electric field are planes perpendicular to the field. (b) In a nonuniform field, equipotential surfaces are curved but still perpendicular to the field.

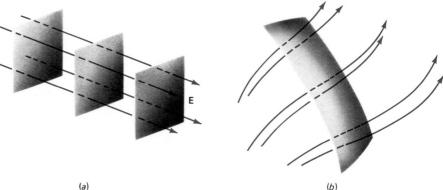

(a)   (b)

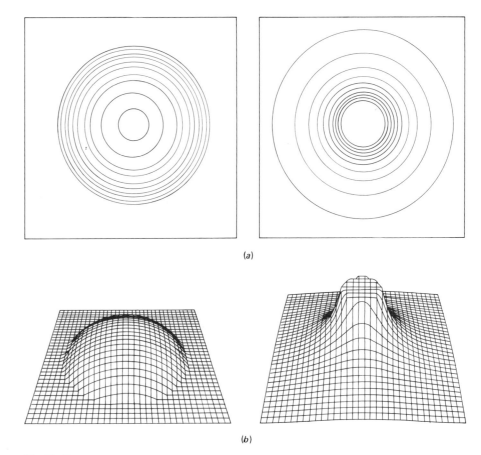

Fig. 23–10
(a) A contour map of a hill and a set of equipotentials for a spherical shell of charge. *(b)* A physical diagram of the hill and a plot of potential as a function of position in the horizontal plane.

Equipotentials are like contour lines used on a map to show land elevation. A contour line is a line of constant elevation: it takes no work to move along a contour line in earth's gravitational field. The contour lines lie at right angles to the uphill or downhill direction. Contour lines are usually spaced at equal increments of elevation. Where lines are close together, there is a large change in elevation over a small distance, and the land is steep. Similarly, an equipotential diagram usually shows curves or surfaces spaced at equal increments of potential. Where equipotentials are close together, a relatively large amount of work must be done to move a charge at right angles to the equipotentials, and the electric field is therefore strong.

Figure 23–10*a* compares a contour map of a hill with a set of equipotential curves for a spherical shell of charge. (The equipotential surfaces themselves are spherical, but the figure shows only a plane cross section of these surfaces.) From the contour map, we see that the hill is steeper at the bottom and more rounded at the top. Analogously, the equipotential diagram reflects the inverse-square decrease in field strength, with the strongest field adjacent to the charged shell. The interior of the shell contains no electric field and therefore shows up as a potential "plateau" in the diagram. Figure 23–10*b* shows three-dimensional diagrams of the hill and of the electric potential that are represented in two dimensions in Fig. 23–10*a*. Of course, there is no physical hill in the potential case; here the third coordinate signifies the value of the potential as a function of position along the other two coordinates.

Just as a contour map tells us all there is to know about the shape of the land, so an equipotential diagram contains full information on the configuration of the electric field. Given the equipotentials, we can draw field lines simply by moving at right angles to the equipotentials (why?). Figure 23–11*a* shows some field lines and equipotentials for a dipole. Figure 23–11*b* is a three-dimensional graph of the dipole potential. You can see how difficult it would be to push a positive charge "up" the huge "potential hill" of the dipole's positive end, and how readily it would "fall down" the potential "hole" of the negative end.

Given field lines, we can construct equipotentials. Equivalently, we can calculate the potential from the field using Equation 23–3. Conversely, given equipotentials, we could construct field lines. Is there a mathematical relation that is the inverse of Equation 23–3, allowing us to calculate the field from the potential? Because Equation 23–3 expresses the potential as an integral over this field, we might expect the inverse relation to involve a derivative.

We have seen that field strength is related to the spacing of the equipotentials, or to the rate of change of potential with distance. If we move in a particular direction, the component of the field in that direction is related to the rate of change of potential in that direction. More precisely,

$$E_\ell = -\frac{dV}{d\ell},\qquad (23\text{–}6)$$

where E_ℓ is the field component in some direction and $dV/d\ell$ the rate of

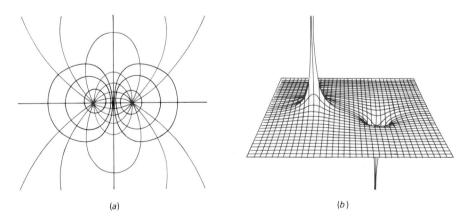

(a)   (b)

Fig. 23–11
(a) Equipotentials (black) and field lines (color) for a dipole. The two sets of curves are everywhere perpendicular. *(b)* A three-dimensional diagram showing the dipole potential as a function of position in the *x-y* plane.

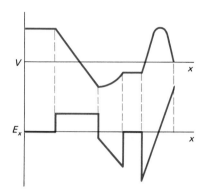

Fig. 23–12
The *x*-component of the electric field is the negative rate of change of the potential with respect to *x*.

change of potential in that direction. The minus sign in Equation 23–6 is the same one that appears in Equation 23–3. It shows that if we move toward a region of higher potential, then the electric field is opposite to our direction of motion. If we move along an equipotential, $dV/d\ell = 0$, so that there is no component of field along the equipotential. If we move at right angles to an equipotential, then Equation 23–6 gives the full magnitude of the field. Figure 23–12 and Example 23–4 illustrate the relation between potential and field.

Example 23–4
Potential and Electric Field

Show that the potential $V(x,y) - V(0,0) = Ex$ of Example 23–2 gives rise to the electric field $\mathbf{E} = -E\hat{\imath}$ of that example.

Solution

Equation 23–6 shows that the x-component of $\mathbf{E}$ is the negative of the derivative of the potential with respect to x:

$$E_x = -\frac{dV}{dx} = -\frac{d(Ex)}{dx} = -E.$$

The y- and z-components are zero because the field depends only on x, so that the derivatives of V with respect to y and z are zero. Therefore, the field is entirely in the x-direction:

$$\mathbf{E} = E_x\hat{\imath} = -E\hat{\imath}.$$

Through Equations 23–3 and 23–6 or, equivalently, field line and equipotential diagrams, we see that specifying either the potential or the electric field supplies all the information needed to determine the other. In practical situations this means that we are free to calculate either the potential or the field, whichever is easier, and then determine the other as needed. Calculating the potential has a substantial advantage: only one number is required to specify the potential at a point; potential is a scalar. But electric field is a vector, requiring three numbers at each location for its specification. Adding the contributions of individual charges to determine the net electric field is often complicated because of different orientations of the field vectors. This complication does not arise in calculating potential.

Incidentally, Equation 23–6 shows that the units of electric field can be expressed as volts/meter (see Problem 1).

23.4
THE POTENTIAL OF A CHARGE DISTRIBUTION

The Potential of a Point Charge

The electric field of a point charge q is given by Equation 21–3:

$$\mathbf{E} = \frac{1}{4\pi\epsilon_0} \frac{q}{r^2} \hat{\mathbf{r}},$$

where $\hat{\mathbf{r}}$ is a unit vector from the charge toward the point where the field

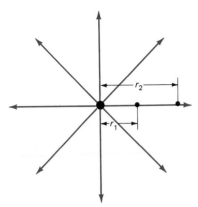

Fig. 23–13
The potential difference between points r_1 and r_2 is obtained by integrating the field over a path between the two points.

is being evaluated. Consider two points at distances r_1 and r_2 from a positive point charge, as shown in Fig. 23–13. What is the potential difference between these points? The simplest path to use in calculating potential difference is a straight line between the points. The distance between the points is $r_2 - r_1$, but we cannot simply multiply this distance by the electric field because the field magnitude changes continuously between r_1 and r_2. Instead, we integrate according to Equation 23–3:

$$V_2 - V_1 = -\int_{r_1}^{r_2} \mathbf{E} \cdot d\boldsymbol{\ell} = -\int_{r_1}^{r_2} \frac{1}{4\pi\epsilon_0} \frac{q}{r^2} \hat{\mathbf{r}} \cdot d\boldsymbol{\ell}.$$

As we move from r_1 toward r_2, our path element vectors $d\boldsymbol{\ell}$ correspond to small increments dr in radial position, and these path element vectors point in the radial, or $\hat{\mathbf{r}}$, direction. Therefore, $d\boldsymbol{\ell} = \hat{\mathbf{r}}\,dr$, and the potential becomes

$$V_2 - V_1 = -\int_{r_1}^{r_2} \frac{1}{4\pi\epsilon_0} \frac{q}{r^2} \hat{\mathbf{r}} \cdot \hat{\mathbf{r}}\,dr = -\frac{q}{4\pi\epsilon_0} \int_{r_1}^{r_2} r^{-2}\,dr,$$

since the dot product of the unit vector $\hat{\mathbf{r}}$ with itself is simply 1. Evaluating the integral then gives

$$V_2 - V_1 = -\frac{q}{4\pi\epsilon_0}\left(-\frac{1}{r}\right)\Bigg|_{r_1}^{r_2} = \frac{q}{4\pi\epsilon_0}\left(\frac{1}{r_2} - \frac{1}{r_1}\right). \qquad (23\text{–}7)$$

Does this result make sense? For $r_2 > r_1$, the potential difference is negative, showing that the potential is lower at the outer point. A positive test charge at r_1 would "fall down" the potential "hill" toward r_2, accelerated by the repulsive electric field of the charge q. If $r_1 > r_2$, the potential difference is positive—we must do work to push a positive test charge "up" the potential "hill" toward the charge q. Although we considered a positive point charge q in deriving Equation 23–7, our result holds as well for $q < 0$, in which case the equation gives a potential difference of the opposite sign.

Although we derived Equation 23–7 for two points on the same radial line, the equation holds for any two points at radii r_1 and r_2. This is clear from Fig. 23–14, where two points are located at different radii and different angular positions. Since the electrostatic field is conservative, we can evaluate the potential difference over any path between the two points—in particular, over a path that is partly along a radius and partly along a circular arc centered on the point charge. No work is required to move along the circular arc, since this section of path is at right angles to the field. So the only contribution to the potential difference comes from the radial section of the path, where the potential difference is given by Equation 23–7.

Equation 23–7 shows why it is convenient to choose the zero of potential at infinity when dealing with isolated charges. If we let r_1 become arbitrarily large, and drop the subscript from r_2 because it can be any radial distance r, Equation 23–7 becomes

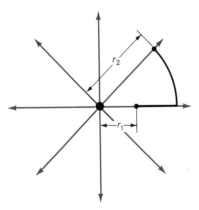

Fig. 23–14
Equation 23–7 gives the potential difference even if the two points do not lie along the same radius. This is so because one path joining them consists of a purely radial segment to which Equation 23–7 applies, and a circular section along which no work is done against the radially directed field.

$$V(r) - V(\infty) = V(r) = \frac{1}{4\pi\epsilon_0} \frac{q}{r}. \qquad \text{(point charge potential)} \qquad (23\text{–}8)$$

We often call this expression $V(r)$ "the potential of a point charge." What we really mean is that the expression gives the potential *difference* between a point a distance r from the point charge and a point very far from the charge. Because the field outside any spherically symmetric charge distribution is the same as that of a point charge, Equation 23–8 also gives the potential outside a spherically symmetric charge distribution (see Problem 19).

It might bother you that the potential difference can be finite over an infinite distance. How can you push a charge an infinite distance against a force and not do infinite work? The answer lies in the inverse-square law. The field drops so rapidly with distance that the sum of all the infinitesimal potential differences $\mathbf{E} \cdot d\boldsymbol{\ell}$ remains finite. We encountered an analogous result in Chapter 8 for the inverse-square force of gravity, when we found that it took only a finite amount of energy to escape completely the gravitational attraction of a finite mass distribution. For fields that drop only as $1/r$ or more slowly, it would be useless to put the zero of potential at infinity because for these fields the potential integral is truly infinite. Whenever we deal with a charge distribution of finite size, however, the field always drops at least as fast as $1/r^2$ when we are sufficiently far from the charge distribution (why?). For such finite-sized charge distributions, placing the zero of potential at infinity is always a possible choice.

The Potential of an Arbitrary Charge Distribution

Given an arbitrary charge distribution, we could always calculate the field using Coulomb's law or Gauss's law, and then use Equation 23–3 to calculate the potential. If we already have a simple expression for the field, as is the case with symmetric charge distributions, then using Equation 23–3 is often the easiest way to calculate the potential.

Example 23–5
Potential Difference in the Field of a Line Charge

Consider two points located distances r_1 and r_2 from an infinite line of charge, as shown in Fig. 23–15. What is the potential difference between these two points?

Solution

In Chapter 21 we used Gauss's law to obtain the result

$$\mathbf{E} = \frac{\lambda}{2\pi\epsilon_0 r}\,\hat{\mathbf{r}},$$

where λ is the charge line density. As we move from r_1 toward r_2, we can again write $d\boldsymbol{\ell} = \hat{\mathbf{r}}dr$, just as we did in evaluating the point-charge potential. Then Equation 23–3 becomes

$$V_2 - V_1 = -\int_{r_1}^{r_2} \mathbf{E} \cdot d\boldsymbol{\ell} = -\int_{r_1}^{r_2} \frac{\lambda}{2\pi\epsilon_0 r}\,\hat{\mathbf{r}} \cdot \hat{\mathbf{r}}\,dr$$

$$= -\frac{\lambda}{2\pi\epsilon_0}\int_{r_1}^{r_2} \frac{1}{r}\,dr = -\frac{\lambda}{2\pi\epsilon_0}\ln r\,\bigg|_{r_1}^{r_2} \tag{23–9}$$

$$= -\frac{\lambda}{2\pi\epsilon_0}[\ln r_2 - \ln r_1] = \frac{\lambda}{2\pi\epsilon_0}\ln\left(\frac{r_1}{r_2}\right).$$

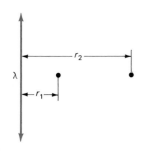

Fig. 23–15
Two points in the field of a line charge.

The sign of the potential difference in Equation 23–9 depends both on the sign of the charge density λ and on whether we move inward or outward. Since the logarithm of a number less than 1 is negative, the equation shows that for positive λ, the potential decreases as we move outward. Conversely, the potential increases going inward, as we must do work on a positive test charge against the repulsive force of the positive line charge. These conclusions are reversed if the line is negatively charged.

Note that in this case we cannot let r_1 go to infinity, for this would give an infinite potential difference. Physically, this reflects the fact that our charge distribution is itself of infinite extent. Mathematically, it reflects the slow $1/r$ decrease in field strength.

Although we derived Equation 23–9 for an infinitesimally thin line of charge, considerations of Section 21.5 show that this result holds outside *any* charge distribution with line symmetry.

When the field of our charge distribution is not simple and symmetric, then calculating the field and subsequently the potential is more difficult. In such cases it is usually better to calculate the potential first, and from it the field. The superposition principle makes this an easy task. Consider a charge distribution consisting of some isolated point charges. As shown symbolically in Fig. 23–16, the work required to bring a test charge from infinity to a point P in the vicinity of the charge distribution is simply the sum of the amounts of work needed to bring it to that point in the field of each of the charges individually. This is a restatement of the superposition principle, which says that the force between two charges is unaffected by a third charge. In terms of potential, this means that we can calculate the potential at point P by adding the potentials at P of the individual point charges, each taken with respect to the same reference point, most conveniently a point infinitely far from all the charges. Because potential is a scalar, this is a simple addition requiring no consideration of angles or vector components.

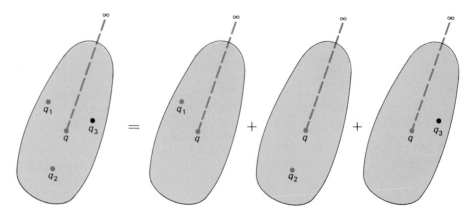

Fig. 23–16
The work required to bring a test charge from infinity to a point P in the vicinity of several charges is the sum of the amounts of work required to bring it to P in the field of each charge alone.

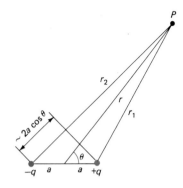

Fig. 23–17
A dipole and a point *P* where its potential is to be evaluated. When *P* is far from the dipole (compared with the dipole separation), then the distance r_2 from *P* to the negative charge is longer than the distance r_1 by approximately $2a\cos\theta$. (The angle whose vertex is $-q$ is very nearly θ.)

Example 23–6
The Potential of a Dipole

The dipole of Fig. 23–17 consists of two point charges $\pm q$ separated by a distance $2a$. What is the potential of this dipole at an arbitrary point *P*? Take the zero of potential at infinity.

Solution

To calculate the dipole potential, we need only determine how far the point *P* is from each charge in the dipole, and then add the point-charge potentials given by Equation 23–8. If r_1 and r_2 are the distances from the point *P* to the two charges, then this addition of point-charge potentials gives

$$V(P) = \frac{q}{4\pi\epsilon_0 r_1} + \frac{-q}{4\pi\epsilon_0 r_2} = \frac{q}{4\pi\epsilon_0}\left(\frac{1}{r_1} - \frac{1}{r_2}\right) = \frac{q(r_2 - r_1)}{4\pi\epsilon_0 r_1 r_2}.$$

Of course, we do not yet have expressions for r_1 and r_2 in terms of the coordinates of the point *P*. We have seen that in practical situations we are usually interested in the potential a great distance from a dipole. If r is the distance from the center of the dipole to the point *P*, and θ the angle between the dipole axis and a line from the dipole center to *P*, then if $r \gg a$ the following approximate expressions become highly accurate:

$$r_1 r_2 = r^2$$

and

$$r_2 - r_1 = 2a\cos\theta,$$

as you can see by examining the small triangle in Fig. 23–17, which becomes very nearly a right triangle for $r \gg a$. Then the dipole potential becomes

$$V(r,\theta) = \frac{q}{4\pi\epsilon_0}\frac{2a\cos\theta}{r^2} = \frac{p\cos\theta}{4\pi\epsilon_0 r^2}, \tag{23–10}$$

with $p = 2aq$ the dipole moment.

Note that this dipole potential drops more rapidly with distance than the point-charge potential, just as the dipole field drops more rapidly than the point-charge field. Along the perpendicular bisector of the dipole, Equation 23–10 gives $V = 0$. This makes sense because, as we saw in Chapter 21, the field is at right angles to the bisector so that it takes no work to move along the bisector. Another way of looking at this is to note that points on the perpendicular bisector are equidistant from both charges. Then the work required to move a positive test charge against the field of the positive charge is exactly balanced by the energy gained moving toward the negative charge. Equation 23–10 was used in producing the dipole potential diagrams of Fig. 23–11. In Fig. 23–11 you can see that the perpendicular bisector of the dipole is a flat line amidst otherwise rugged terrain, so that a charge moving in along this line is moving along an equipotential.

We can calculate the dipole field from the potential using Equation 23–6. Consider the field at points on the perpendicular bisector of the dipole. There is no change in potential along the bisector, and therefore no field component along the bisector. What about the field component at right angles to the bisector? If we consider an infinitesimal distance $d\ell$ perpendicular to the bisector, we can write $d\ell = r\,d\theta$ (Fig. 23–18), where $d\theta$ is the infinitesimal change in angle over the distance $d\ell$. Over this distance, the potential changes from zero to a very small value

$$dV = \frac{p}{4\pi\epsilon_0 r^2}\,d(\cos\theta),$$

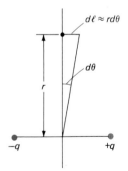

Fig. 23–18
A small distance at right angles to the dipole bisector is very nearly $r\,d\theta$.

where $d(\cos\theta)$ is the change in $\cos\theta$ corresponding to the change $d\theta$ in θ. Then Equation 23–6, which states that the electric field is the negative rate of change of potential with distance, becomes

$$E = -\frac{dV}{d\ell} = -\frac{p}{4\pi\epsilon_0 r^2}\frac{d(\cos\theta)}{rd\theta} = -\frac{p}{4\pi\epsilon_0 r^3}\frac{d(\cos\theta)}{d\theta}.$$

The derivative of cosine is $-$sine, and at the bisector our angle θ relative to the dipole axis is $\pi/2$, so that

$$E = \frac{p}{4\pi\epsilon_0 r^3}\sin(\pi/2) = \frac{p}{4\pi\epsilon_0 r^3}.$$

This is exactly the result we found in Chapter 21 (Equation 21–7a) by adding vectorially the field contributions of the individual charges. When a charge distribution becomes much more complicated than a simple dipole, calculating the field from the potential as we did here is almost always simpler than calculating the field directly using vector addition. Problem 45 explores the calculation of the dipole field from the potential for points not on the dipole bisector or axis.

When a charge distribution consists not of discrete charges but of continuously distributed charge, it is still possible to calculate the potential by considering the distribution to be made up of many infinitesimal charge elements dq. Each charge element acts like a point charge and contributes to the potential an amount dV given by

$$dV = \frac{dq}{4\pi\epsilon_0 r},$$

where the zero of potential is taken to be at infinity. The potential at some point P is the sum—in this case an integral—of the contributions dV from all the charge elements:

$$V = \int\frac{dq}{4\pi\epsilon_0 r} = \frac{1}{4\pi\epsilon_0}\int\frac{dq}{r}, \tag{23–11}$$

where the limits on the integral are chosen to include all dq's in the charge distribution. Note again that vector addition is not necessary to calculate the potential.

Example 23–7
A Ring of Charge

A total charge Q is distributed uniformly around a ring of radius a, as shown in Fig. 23–19. What is the potential at points along the axis of this charged ring?

Solution

Let x be the distance from the center of the ring to some arbitrary point on the axis. The distance from each point on the ring to a point on the axis is the same, and is given by

$$r = \sqrt{x^2 + a^2}.$$

The potential on the axis is obtained by summing the potentials of all the charge elements dq around the ring, as described by Equation 23–11:

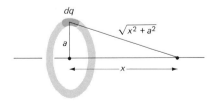

Fig. 23–19
Example 23–7

$$V = \frac{1}{4\pi\epsilon_0} \int_{ring} \frac{dq}{r} = \frac{1}{4\pi\epsilon_0 \sqrt{x^2+a^2}} \int_{ring} dq,$$

where we have taken $r = \sqrt{x^2+a^2}$ outside the integral because it is the same for all charge elements as long as our point lies on the ring axis. The remaining integral means the sum of all the dq's over the charged ring, so that its value is simply the total charge Q. Thus we have

$$V = \frac{1}{4\pi\epsilon_0} \frac{Q}{\sqrt{x^2+a^2}}. \qquad (23\text{--}12)$$

Does this result make sense? At great distances from the ring ($x \gg a$), we may ignore a^2 in the denominator so our result becomes

$$V = \frac{1}{4\pi\epsilon_0} \frac{Q}{x}.$$

This is just the potential a distance x from a point charge Q. Far from the ring, so far that its finite extent is no longer important, its potential is essentially the same as that of a point charge at the origin. What about the potential at the center of the ring? Equation 23–12 gives

$$V = \frac{1}{4\pi\epsilon_0} \frac{Q}{a}$$

for this case. At the center we are a distance a from all elements of the charged ring. Because potential is a scalar, the directions to those elements do not matter, and the potential is just what we would expect a distance a from a point charge Q.

Example 23–8
The Potential of a Charged Disk

A charged disk of radius a carries a total charge Q distributed uniformly over its surface. What is the potential on the disk axis, a distance x from the disk? From the potential, determine the electric field on the axis.

Solution

To use Equation 23–11 for the potential of a continuous charge distribution, we must divide the disk into charge elements dq. In the preceding example, we calculated the potential of a charged ring. That result, and the symmetry of our disk, suggest that we divide the disk into thin rings (Fig. 23–20) and then integrate the result of the preceding example over all the rings comprising the disk. If a ring-shaped charge element has radius r and carries charge dq, then Equation 23–12 gives

$$dV = \frac{1}{4\pi\epsilon_0} \frac{dq}{\sqrt{x^2+r^2}}$$

for the potential at x arising from the ring of radius r. The potential from the entire disk is then

$$V = \frac{1}{4\pi\epsilon_0} \int_{r=0}^{r=a} \frac{dq}{\sqrt{x^2+r^2}}.$$

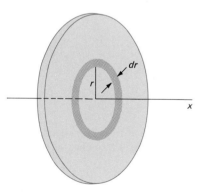

Fig. 23–20
A charged disk, showing a ring-shaped charge element dq of radius r and width dr.

To evaluate the integral, we must relate the variable r to the charge element dq. If we "unwind" the thin ring (Fig. 23–21), we have a strip of area $2\pi r\, dr$. Because the disk is uniformly charged, its surface charge density σ is just the total charge divided by the disk area:

$$\sigma = \frac{Q}{\pi a^2}.$$

Fig. 23–21
Unwinding the thin ring gives a strip of width dr and length $2\pi r$.

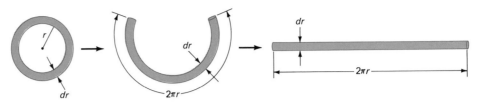

Then the charge dq on our infinitesimal strip of area $2\pi r\,dr$ is

$$dq = \sigma\,2\pi r\,dr = \frac{Q}{\pi a^2}\,2\pi r\,dr = \frac{2Q}{a^2}\,r\,dr.$$

Using this result in our integral for the potential, we have

$$V = \frac{1}{4\pi\epsilon_0}\int_0^a \frac{2Q}{a^2}\frac{r\,dr}{\sqrt{x^2+r^2}} = \frac{Q}{2\pi\epsilon_0 a^2}\int_0^a \frac{r\,dr}{\sqrt{x^2+r^2}}.$$

To evaluate the integral, we note that

$$r\,dr = \tfrac{1}{2}d(r^2) = \tfrac{1}{2}d(x^2+r^2)$$

since x is constant with respect to the integration. For convenience, we give the quantity x^2+r^2 a new name, z. As r ranges from 0 to a, z ranges from x^2 to x^2+a^2, so that

$$V = \frac{Q}{2\pi\epsilon_0 a^2}\int_{x^2}^{x^2+a^2} z^{-1/2}(\tfrac{1}{2}dz) = \frac{Q}{2\pi\epsilon_0 a^2}\left.\frac{\tfrac{1}{2}z^{1/2}}{\tfrac{1}{2}}\right|_{x^2}^{x^2+a^2} \tag{23–13}$$

$$= \frac{Q}{2\pi\epsilon_0 a^2}(\sqrt{x^2+a^2} - x).$$

Problem 40 shows that this complicated-looking result reduces to the potential of a point charge for $x \gg a$. To find the electric field, we use Equation 23–6 with the derivative in the x-direction, which symmetry assures us is the field direction on the disk axis:

$$E_x = -\frac{dV}{dx} = -\frac{Q}{2\pi\epsilon_0 a^2}\frac{d}{dx}[(x^2+a^2)^{1/2} - x)]$$

$$= -\frac{Q}{2\pi\epsilon_0 a^2}[\tfrac{1}{2}(x^2+a^2)^{-1/2}(2x) - 1]$$

$$= \frac{Q}{2\pi\epsilon_0 a^2}\frac{\sqrt{x^2+a^2} - x}{\sqrt{x^2+a^2}}.$$

How are we to trust such a complicated-looking result? We can see whether it reduces to the simpler results we expect in special cases. For example, very close to the disk we can neglect x compared with a, giving

$$E = \frac{Q}{2\pi\epsilon_0 a^2}\frac{a}{a} = \frac{Q}{2\pi\epsilon_0 a^2} = \frac{\sigma}{2\epsilon_0},$$

where $\sigma = Q/\pi a^2$ is the surface charge density. But this result is just the field of an infinite sheet of charge (Equation 21–23), and that is exactly what we should expect very close to the disk surface. You can also verify that our general result reduces to the point-charge field for $x \gg a$ (see Problem 40).

23.5

THE POTENTIAL OF AN ISOLATED CONDUCTOR

The fact that there is no electric field in a conductor in electrostatic equilibrium means that it takes no work to move a small test charge within such a conductor. Also in electrostatic equilibrium, any field at the conductor surface must be perpendicular to the surface; otherwise charge would move along the surface, and we would not have equilibrium. Therefore, the potential difference between any two points on or in the conductor is zero. Every conductor in electrostatic equilibrium is an equipotential.

A solid conducting sphere provides a particularly simple example of an isolated conductor. We know from Gauss's law that outside the sphere its field is that of a point charge. We need not recalculate the potential outside the sphere, since it must also be that of a point charge. What about the potential at the sphere? We know that there is no potential difference between points on or inside the sphere. But this does *not* require the potential of the sphere to be zero: the zero field only means that the potential does not *change* from point to point over the sphere. It still takes work to move a test charge from infinity to the surface of the sphere. Once on the sphere, it takes no more work to move about the surface or into the interior. Thus the potential of any point on or inside the sphere is just the point-charge potential (Equation 23–8) evaluated at the surface of the sphere:

$$V_{\text{sphere}} = \frac{q}{4\pi\epsilon_0 R}, \tag{23–14}$$

where R is the radius of the sphere. Equation 23–14 shows that the potential of a spherical conductor increases with increasing charge on the conductor. For a fixed charge, the equation also shows that the potential of the conductor increases with decreasing radius. This makes sense because the surface of a small spherical conductor is closer to the center of charge than is the surface of a large conductor, so that stronger fields are encountered in approaching the smaller conductor.

Another way of viewing Equation 23–14 is to consider a number of widely separated spheres of different sizes all held at the same potential. You could achieve this situation by connecting the spheres with thin wires (Fig. 23–22). We require that the spheres be widely separated so that their charge distributions remain spherically symmetric; then Equation 23–14 for an isolated sphere applies to each. Equation 23–14 shows that in order

Fig. 23–22
Two conducting spheres held at the same potential by means of a conducting wire. The smaller sphere has a smaller charge in proportion to its radius, but a larger surface charge density and therefore a larger surface electric field.

for the spheres to have the same potential, they must have the same ratio of charge to radius. Therefore, the smaller sphere carries less charge in proportion to its smaller radius. However, the *area* of the smaller sphere is less by the *square* of its radius. Therefore, the surface charge density, which is the ratio of total charge to surface area, is *larger* for the smaller sphere. Indeed, the surface charge density is inversely proportional to the radius (see Problem 32).

Our understanding of spherical conductors provides a qualitative description of nonspherical conductors as well. All parts of an irregularly shaped conductor must be at the same potential. Where the conductor surface curves sharply, it is like a small sphere and generally carries a higher charge density. In the preceding chapter, we showed that the field at the surface of a conductor is proportional to the surface charge density: $E = \sigma/\epsilon_0$. Therefore, the field at the surface of an irregular conductor is highest where the charge density is highest—which is usually where the surface curves most sharply.*

Knowing that a conductor is an equipotential and that the field is stronger where the conductor curves more sharply, we can easily sketch qualitatively what the fields and equipotentials of charged conductors must look like. Figure 23–23 shows an irregular charged conductor. Because the conductor surface is an equipotential, and because the electric field is perpendicular to the surface, equipotentials just above the surface cannot be shaped too differently from the surface. Near the surface the field lines must be closer together where the surface curves sharply. On the other hand, far from the conductor it must look like a point charge, with evenly spaced radial field lines and spherical equipotentials. With these limiting cases in mind, we can sketch the approximate form of the field lines and equipotentials everywhere, as in Fig. 23–23.

We stress that the conclusions of this section apply only to *isolated* conductors—those far from any other charges. The field of a nearby charge will modify the charge distribution on a conductor, altering the surface charge distribution (Fig. 23–24).

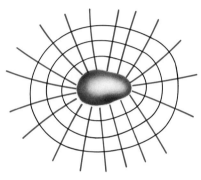

Fig. 23–23
An irregular, charged conductor. The field is strongest, and therefore the equipotentials more closely spaced, where the conductor is most sharply curved.

---

*This association of strong field and sharp curvature is only approximate, and in some unusual configurations may not hold at all. See "The Lightning Rod Fallacy," by R. H. Price and R. J. Crowley, *American Journal of Physics*, vol. 53, p. 843 (September 1985).

Fig. 23–24
(a) An isolated conducting sphere carries a uniform surface charge density and has a spherically symmetric electric field. (b) The presence of a nearby charge distorts the surface charge distribution and therefore the electric field.

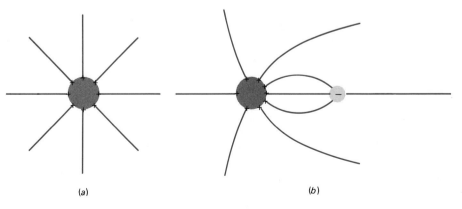

(a)   (b)

Application
Corona Discharge

The uneven distribution of charge on a nonspherical conductor makes it possible to develop large electric fields simply by charging conductors that have very sharp edges. This effect can be useful or annoying, depending on where it occurs. Fields above about 3×10^6 V/m are strong enough to ionize air (that is, to strip electrons from air molecules), making it a conductor. When this occurs, electric charge (and electrical energy) can be lost from a conductor. The electrical breakdown of air is often evidenced by a blue glow (Fig. 23–25) in the vicinity of the high fields. Called **corona discharge,** this glow results from the recombination of free electrons with ions and the subsequent settling of the newly recombined atoms into their lowest-energy states. To avoid corona discharge in high-voltage equipment it is necessary to eliminate sharp corners. Despite careful engineering, high-voltage power lines often display corona discharge at points where wires meet or are suspended from poles. Corona discharge is especially likely in humid weather when the breakdown field—the field strength at which air ionizes—is reduced.

Corona discharge is put to good use in the electrostatic precipitator discussed in the preceding chapter. The central wire of the precipitator has so small a radius that corona discharge occurs in its vicinity. This produces ions, which attach to pollutant particles in the gas stream and allow the electric field to force those particles to the collecting plates.

Corona discharge also occurs at the sharp points of lightning rods placed on buildings. The discharge reduces buildup of excess charge, thereby helping to prevent damage from lightning. If lightning does strike the rod, its charge is harmlessly conducted to the earth by thick conducting wires.*

corona discharge

Fig. 23–25
Corona discharge on a 208 kV power line.

Example 23–9
Isolated Conductors

Two metal spheres of diameters 5.0 cm and 1.0 cm are 2.0 m apart and are connected by a wire of negligible thickness. If a 1000 V/m electric field is desired at the surface of the smaller sphere, how much charge should be placed on the system?

Solution

Let Q_1 and Q_2, and R_1 and R_2, be the charges and radii of the larger and smaller spheres, respectively. The surface area of the smaller sphere is $4\pi R_2^2$, so that the field at its surface, as given by Equation 22–2, is

$$E_2 = \frac{\sigma}{\epsilon_0} = \frac{Q_2}{4\pi\epsilon_0 R_2^2}$$

(we could also write this result directly since the field outside the sphere is a point-charge field; in either case we use the fact that the spheres are widely separated to

---

*Although lightning rods were invented by Benjamin Franklin, there is still controversy about exactly how they work. Our conclusion that the field is strongest where the conductor is sharpest is tempered by the effects of a "sheath" of charge that may gather around the conductor in a medium like ionized air. This effect may actually make a rounded lightning rod more effective; see "Lightning Rods: Franklin Had It Wrong," by John Noble Wilford, *New York Times,* June 14, 1983, p. 15.

conclude that the charge distribution is uniform). Then the charge on the smaller sphere is

$$Q_2 = 4\pi\epsilon_0 R_2^2 E_2$$
$$= 4\pi(8.9 \times 10^{-12} \text{ N·m}^2/\text{C}^2)(0.50 \times 10^{-2} \text{ m})^2(1000 \text{ V/m}) = 2.8 \text{ pC}.$$

The wire forces the two spheres to the same potential, so that Equation 23–14 gives

$$\frac{1}{4\pi\epsilon_0}\frac{Q_1}{R_1} = \frac{1}{4\pi\epsilon_0}\frac{Q_2}{R_2},$$

or

$$Q_1 = \frac{R_1}{R_2}Q_2 = \frac{2.5 \text{ cm}}{0.50 \text{ cm}}(2.8 \text{ pC}) = 14 \text{ pC}.$$

Then the total charge is $Q = Q_1 + Q_2 = 17$ pC.

SUMMARY

1. The **electric potential difference between two points** is the work per unit charge required to move charge between those points. Often called **potential difference** or simply **potential,** electric potential difference is inherently a property of two points. The conservative nature of the static electric field ensures that the work done moving between two points is independent of the path taken, so that the potential difference between two points has a unique value. The potential may be written as a **line integral** over any path joining the two points:

$$V_B - V_A = -\int_A^B \mathbf{E}\cdot d\boldsymbol{\ell}.$$

2. Given the potential difference V between two points, the work required to move a charge q between these points follows immediately from the definition of potential difference as the work per unit charge:

$$W = qV.$$

If W is positive, some agent must do work on the charge to move it between the points. If W is negative, the charge does work on the agent or, if no agent is present, the charge gains kinetic energy as it moves between the two points.

3. When dealing with isolated charges it is often useful to choose one of the two points used in establishing potential difference to be at infinity. Then we speak of the **potential** at a point, although we actually mean the potential difference between that point and infinity.

4. **Equipotential surfaces** or **equipotentials** are surfaces on which the potential (measured with respect to some fixed reference point) is constant. It requires no work to move charges around on equipotential surfaces. Equipotential surfaces are everywhere perpendicular to the electric field. Where equipotentials are closely spaced, the electric field is strong; where they are widely spaced, the field is weak. Mathematically, the field component in a given direction is the negative of the rate of change of potential with position in that direction:

$$E_\ell = -\frac{dV}{d\ell}.$$

5. The potential of a point charge (with respect to a reference point at infinity) follows from evaluating the line integral defining potential difference:

$$V(r) = \frac{1}{4\pi\epsilon_0}\frac{q}{r}. \quad \text{(point charge)}$$

Because the electric field obeys the superposition principle, we can add or integrate the potentials of point charges to obtain the potential of a more complicated charge distribution. Because potential is a scalar, such a calculation is simpler than the corresponding calculation for the electric field vector. We can then determine the field by differentiating the potential along three mutually perpendicular directions to obtain the field components in those directions.

6. Since there is no electric field inside or parallel to the surface of a conductor in electrostatic equilibrium, such a conductor is an equipotential. The surface charge density and therefore the electric field at the surface of a conductor are usually greatest where the conductor curves most sharply. Very strong electric fields occur at sharp bends in conductors; if strong enough, these fields may result in **corona discharge**—the stripping of free charge from the conductor.

QUESTIONS
---

1. You have probably seen a bird or squirrel sitting on a power line and wondered why it didn't get electrocuted. Given the definition of potential, and the fact that the wire is an excellent conductor, why is this so?

2. One proton is accelerated from rest by a uniform electric field, another by a nonuniform field. If they move through the same potential difference, how do their final velocities compare?

3. Would a free electron move toward a region of higher or lower potential?

4. The potential difference between points A and B in Fig. 23–26 is zero, since they are equidistant from the point charge Q. How can this be, when a test charge moved along the path shown clearly experiences a force that is not everywhere perpendicular to its path?

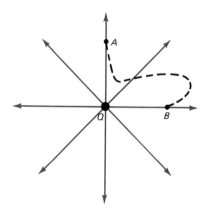

Fig. 23–26   Question 4

5. A proton and a positron (a positively charged particle with the mass of the electron) are accelerated from rest through potential differences of the same magnitude. How do their final kinetic energies compare?

6. The electric field in the center of a uniformly charged ring is, by symmetry, obviously zero. Yet the result of Example 23–7 shows that the potential in the center of the ring is not zero. How is this possible?

7. Must the potential be zero at any point where the electric field is zero? Explain.

8. Must the electric field be zero at any point where the potential is zero? Explain.

9. The potential is constant throughout an entire volume. What must be true of the electric field within this volume?

10. In considering the potential of an infinite flat sheet of charge, why would it not be useful to set the zero of potential at infinity?

11. Trucks equipped with "cherry-picker" lifts for working in trees and power lines often carry signs warning that touching the truck could result in electrocution

(Fig. 23–27). Explain how this hazard might arise and why it would be less of a hazard to a worker on the truck than to someone on the ground who contacts the truck.

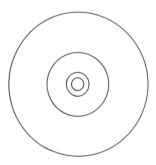

Fig. 23–27   A cherry-picker truck (above) carries a sign (below) warning of electrocution hazard should the truck contact power lines (Question 11).

12. In a certain region, the equipotentials are spheres whose spacing is *directly* proportional to their radii (Fig. 23–28). What can you conclude about the distribution of charge in the region?

Fig. 23–28   Question 12

13. In a certain region, the equipotentials are spheres whose spacing is independent of radius (Fig. 23–29).

What can you conclude about the charge density in the region? *Hint:* See Example 21–5.

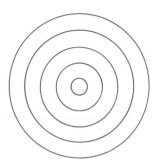

Fig. 23–29   Question 13

14. Is it possible for equipotential surfaces to intersect?
15. Why does our conclusion about the charge density on an isolated spherical conductor being inversely proportional to the sphere's radius require that the sphere be isolated?
16. The conductor shown in Fig. 23–30 carries a net positive charge. One student argues that the surface charge density on the concave portion should be negative because at that point the surface has "negative curvature." Another student argues that a negative surface charge density would not be consistent with the conductor's being an equipotential. Which is right? *Hint:* Try drawing field lines, and remember that potential difference is path-independent.

Fig. 23–30   Question 16

PROBLEMS

Section 23.2   *Potential Difference*

1. Show that 1 V/m is the same as 1 N/C.
2. Figure 23–31 shows a uniform electric field of magnitude E. Calculate the potential differences $V_B - V_A$ and $V_C - V_B$ for the points shown. Both pairs of points have the same spacing d, but the line from B to C is at 45° to the field. Use your result to calculate the potential difference $V_C - V_A$.

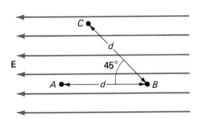

Fig. 23–31   Problem 2

3. The huge particle accelerator at Fermilab near Chicago is 4 miles in circumference and produces a beam of 800 GeV protons. (*a*) What is the proton energy in joules? (*b*) The accelerator operates in a "pulsed" mode, producing a beam of 10^{13} protons every minute. What is the average power required to accelerate the particles?
4. A proton, an alpha particle (a helium nucleus with no electrons), and a singly ionized helium atom (a helium atom missing one of its two electrons) are all accelerated through a potential difference of 100 V. Calculate the energy gained by each.
5. Electrons in a color TV tube are accelerated through a 25-kV potential difference before they hit the inside of the TV screen. What is their speed when they hit the screen?
6. A 12-V car battery stores 2.8 MJ of energy. How much charge can move between the battery terminals before it is totally discharged? Assume that the potential difference remains at 12 V, an assumption that is not very realistic.
7. A 5.0-g object carries a net charge of 3.8 μC. It acquires a speed v when accelerated from rest through a potential difference V. A 2.0-g object acquires twice the speed under the same circumstances. What is its charge?
8. Two large, flat, parallel metal plates are a distance d apart, where d is small compared with the plate size. If the plates carry equal but opposite charge densities σ, show that the potential difference between them is given by $V = \sigma d/\epsilon_0$.
9. The electric field in a region is given by

$$\mathbf{E} = \frac{E_0 x}{x_0} \,\hat{\mathbf{i}},$$

where E_0 and x_0 are constants. (*a*) Find the potential as a function of position, taking the zero of potential at $x = 0$. (*b*) Find an expression for the work that would need to be done on a proton to move it in a circular arc from the point $(x_0, 0)$ to the point $(-3x_0, 2x_0)$.
10. A proton and a positron (a positively charged particle with the mass of an electron) are accelerated through a potential difference of 440 V. What are the final kinetic energies and velocities of each?

11. Figure 23–32 shows three paths leading from infinity to a point P on the perpendicular bisector of a dipole. For each path, how much work is required to bring a positive test charge q from infinity to the point P?

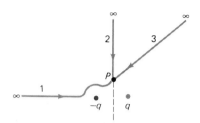

Fig. 23–32   Problem 11

12. The potential difference between the two contacts in a standard electrical outlet is 120 V. (a) How much work is done by the electric field on 1.0 C of charge as it moves between the two contacts? (b) If 0.83 C move each second through a light bulb that is plugged into the outlet, what is the rate at which the electric field does work? (c) What is the source of the energy supplied to the charge that moves through the light bulb?

Section 23.3   Equipotentials and the Electric Field

13. In a uniform electric field, equipotential planes that differ by 1.0 V are 2.5 cm apart. What is the electric field strength?
14. Figure 23–33 shows a plot of potential versus position along the x-axis. Make a plot of the x-component of the electric field for this situation.

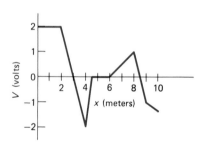

Fig. 23–33   Problem 14

15. Sketch some equipotentials and field lines for a distribution consisting of two equal point charges.
16. Sketch some equipotentials and field lines for a distribution consisting of a point charge $-q$ and a point charge $+2q$.
17. The potential in a certain region is given by $V = axy$, where a is a constant. The zero of potential is taken to be at the origin. (a) Determine the electric field in the region. (b) What are the units of a? (c) Sketch some equipotentials and field lines.

Section 23.4   The Potential of a Charge Distribution

18. In the view of classical physics, a hydrogen atom consists of an electron in circular orbit around a proton. The orbital radius is 0.51 Å (1 Å $= 10^{-10}$ m). What is the electric potential at the electron orbit? Take the zero of potential to be at infinity. Compare the numerical value of your answer with the energy required to ionize a hydrogen atom (13.6 eV), and comment on your comparison. *Hint:* See Chapter 8, Section 8.4.
19. A spherically symmetric charge distribution has radius R and total charge Q. Show that the potential at its surface is

$$V = \frac{1}{4\pi\epsilon_0}\frac{Q}{R},$$

where the zero of potential is taken at infinity.
20. A proton is released from rest at the surface of a 3.5-cm-diameter metal sphere carrying a total charge of 0.86 μC. Use the result of the preceding problem to find the proton's velocity when it is far from the sphere.
21. A solid sphere of radius R carries a uniform volume charge density ρ. Find (a) the potential at points outside the sphere, taking the zero of potential at infinity; (b) the potential difference $V(0) - V(R)$ between the sphere's surface and its center; and (c) the potential at any point inside the sphere, taking the zero of potential at infinity. *Hint:* See Example 21–5.
22. A thin, spherical shell of charge has radius R and total charge Q distributed uniformly over its surface. What is the potential at its center? *Hint:* See Problem 19.
23. A thin ring of radius R carries a charge 3Q distributed uniformly over three-quarters of its circumference, and $-Q$ over the remainder. What is the potential at the center of the ring?
24. Nuclear fusion—the joining of two atomic nuclei to make a heavier nucleus, accompanied by the release of energy—occurs only when nuclei approach within about 10^{-15} m. At that point the strong nuclear force overcomes the electrical repulsion, allowing the nuclei to fuse. Consider a deuteron (a heavy hydrogen nucleus, or combination of a proton and a neutron) that is held at a fixed position. Treat the deuteron as a point charge and a point mass. (a) What is the electric potential 10^{-15} m from the deuteron? (b) How much total energy must a second deuteron have if it is to get this close to the first? (c) If the second deuteron is part of a hot gas, with the energy of part (b) being the mean thermal energy, how hot is the gas? *Hint:* Consult Section 17.1. Your answer is higher by a factor of 100 than the actual temperatures anticipated in fusion reactors. Nevertheless, your huge answer suggests one reason why it has proven so difficult to harness fusion as a controlled energy source.
25. A charge $+2q$ and another of $-q$ are located a distance d apart, as shown in Fig. 23–34. Find points on

the line passing through the two charges where (a) the electric field is zero and (b) the potential is zero. Sketch some field lines and equipotentials.

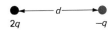

2q   d   −q

Fig. 23–34 Problem 25

26. A coaxial cable consists of a 2.0-mm-diameter inner conductor and an outer conductor whose diameter is 1.6 cm (Fig. 23–35). If the conductors carry equal but opposite line charge densities of 0.56 μC/m, what is the potential difference between them? *Hint:* See Example 23–5.

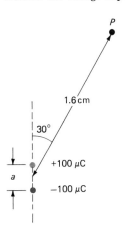

2.0 mm

1.6 cm

Fig. 23–35 Problem 26

27. The potential at point P in Fig. 23–36 is 37 mV. The charge separation a is much smaller than the 1.6-cm distance to P. What is the charge separation?

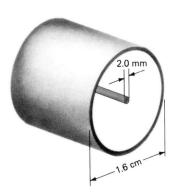

P

1.6 cm

30°

+100 μC

a

−100 μC

Fig. 23–36 Problem 27

28. Three equal charges q form an equilateral triangle of side ℓ. What is the potential at the center of the triangle?

29. A thin rod of length ℓ carries a charge Q distributed uniformly over its length. (a) Show that the potential in the plane that perpendicularly bisects the rod is given by

$$V = \frac{Q}{2\pi\epsilon_0\ell} \ln\left[\frac{\ell}{2R} + \sqrt{1 + \frac{\ell^2}{4R^2}} \right],$$

where R is the distance from the rod center and where the zero of potential is taken at infinity. (b) Show that this expression reduces to an expected result when $R \gg \ell$. *Hint:* See Appendix A for a series expansion of the logarithm.

Section 23.5 *The Potential of an Isolated Conductor*

30. (a) How much charge can be placed on a metal sphere 1.0 cm in diameter before corona discharge occurs to the surrounding air? (b) What is the sphere's potential at that point?

31. A spark plug in an automobile engine has a center electrode in the form of a wire about 2 mm in diameter. If the end of the wire has worn to a hemispherical shape, what is the minimum potential difference that will cause the plug to spark in air? Assume that the field above the center electrode is essentially that of a sphere, despite the influence of the other electrode.

32. Two isolated conducting spheres of radii R_1 and R_2 are connected by a thin wire, forcing them to have the same potential. If the potential is V, calculate the surface charge density σ on each sphere, and show explicitly that the surface charge densities are inversely proportional to the radii of the spheres.

33. Sketch some equipotentials and field lines for the conductor shown in Fig. 23–30 (with Question 16). Assume the conductor is positively charged.

34. Two conducting spheres are each 5.0 cm in diameter and each carries 0.12 μC of charge. The spheres are 8.0 m apart. Determine (a) the potential on each sphere, referenced to infinity; (b) the electric field strength at the surface of each sphere; (c) the potential midway between the spheres; (d) the electric field strength midway between the spheres; (e) the potential difference between the spheres.

35. Two small metal spheres are located 2.0 m apart. One sphere has a radius of 0.50 cm and carries a charge of +0.20 μC. The other has a radius of 1.0 cm and carries +0.080 μC. (a) What is the potential difference between the spheres? (b) If the spheres were connected by a thin wire, how much charge would move along it and in what direction? *Hint:* The two spheres must end up at the same potential.

Supplementary Problems

36. The earth carries a net charge of about -4×10^5 C. (a) If no other charges were present, what would be the potential at the earth's surface, with respect to infinity? (b) Actually, this net negative charge is balanced by a positive charge layer in the earth's ionosphere,

an electrically conducting layer about 60 km above the earth. Estimate the potential difference between earth and ionosphere.

37. A long, thin *conducting* rod of length ℓ carries a total charge Q. Show that the potential on the plane that perpendicularly bisects the rod is given by

$$V = \frac{1}{4\pi\epsilon_0} \frac{Q}{\sqrt{R^2 + \ell^2/4}},$$

where R is the distance from the rod center, and where V is taken to be zero at infinity. Hint: How does charge distribute itself along the rod?

38. A power transmission line consists of two parallel steel wires 3.0 cm in diameter spaced 2.0 m apart. If the potential difference between the wires is 4000 V, what is the charge per unit length on each wire? Assume the wires carry equal but opposite charges.

39. A steel sphere 1.0 cm in diameter is insulated with a thin coating of polyethylene (dielectric constant $\kappa = 2.3$, breakdown field 50 kV/mm). What is the maximum potential the sphere can have (measured relative to infinity) without dielectric breakdown occurring? Assume that the insulating layer is so thin that the potential difference across it is negligible.

40. Show that the potential and electric field of the charged disk in Example 23–8 reduce to the appropriate point-charge expressions when $x \gg a$.

41. An electric power transmission line is at a potential 350 kV higher than that of the ground, 20 m below. Assuming that the ground is uncharged, and that its presence does not significantly alter the electric field in the immediate vicinity of the wire, determine the minimum wire diameter possible without dielectric breakdown of the air (dielectric breakdown in air occurs at a field strength of about 3×10^6 V/m). *Hint: You may need to do an iterative calculation, guessing a value for the answer and seeing how close it comes to satisfying your equation.*

42. Calculate the electric field in the perpendicular bisecting plane for the charged rod of Problem 29, and show that your result reduces to a point-charge field for $R \gg \ell$.

43. Let the two charges of Problem 25 lie in the x-y plane with $-q$ at the origin and $2q$ at $x = -d$. Find an expression (y as a function of x) for points in the x-y plane that lie on the equipotential whose potential is zero. Sketch this curve.

44. A thin rod of length ℓ lies on the y-axis with its center at the origin. It carries a line charge density given by $\lambda = 4\lambda_0 y^2/\ell^2$, where λ_0 is a constant. Calculate the potential at an arbitrary distance x from the rod in its perpendicular bisecting plane.

45. A point dipole of dipole moment p lies at the origin with its dipole moment vector along the y-axis. Use Equations 23–10 and 23–6 to show that the electric field at an arbitrary point (x,y) may be written

$$\mathbf{E} = \frac{1}{4\pi\epsilon_0} \left[\frac{3pxy}{(x^2+y^2)^{5/2}} \hat{\imath} + \frac{(2y^2-x^2)p}{(x^2+y^2)^{5/2}} \hat{\jmath} \right].$$

24

ELECTROSTATIC ENERGY AND CAPACITORS

24.1
THE ENERGY OF A CHARGE DISTRIBUTION

Imagine holding two positive charges in your outstretched arms (Fig. 24–1*a*). Now bring the charges together. To do so takes work, as you move each charge against the electric field of the other. With the charges close together, you must exert considerable force to keep them from flying apart (Fig. 24–1*b*). Clearly there is potential energy in the situation, potential energy that would be converted to kinetic energy if you released the charges. This potential energy is associated with the new distribution of charge you created by moving two widely separated charges closer together. Because the static electric field is conservative, the change in potential energy from the original, widely spaced charge distribution to the new, closely spaced distribution is equal to the work you did in moving the charges together.

Fig. 24–1
(*a*) Widely separated charges exert little force on each other. Moving them together takes work, and (*b*) holding them close together requires considerable force. Note how the bending of field lines suggests the repulsive force.

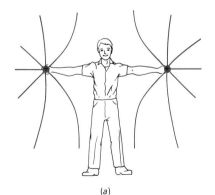

(*a*)

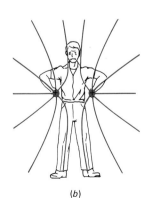

(*b*)

In the preceding chapter we defined electric potential difference as the work per unit charge needed to move charge between two points. Then the work needed to move an arbitrary charge between those points is given by the charge multiplied by the potential difference. In the simple case of two point charges, suppose that charge q_1 is initially an infinite distance from q_2. You then assemble the new charge distribution by moving q_1 to a distance r from q_2. Equation 23–8 gives the electric potential difference between infinity and any point a distance r from q_2:

$$V = \frac{1}{4\pi\epsilon_0}\frac{q_2}{r}.$$

Multiplying by the charge q_1 gives the work required to assemble the new charge distribution:

$$W = \frac{1}{4\pi\epsilon_0}\frac{q_1 q_2}{r}. \tag{24–1}$$

electrostatic potential energy

If we define the zero of potential energy when the charges are infinitely far apart, then we can call the quantity in Equation 24–1 the **electrostatic potential energy** of the charge distribution. Equation 24–1 shows that the potential energy of two point charges is positive if the charges have the same sign and negative if they have opposite signs. In the latter case, it would take positive work to separate the charges. Some thought should convince you that we would have obtained the same potential energy if we had moved charge q_2 in the field of q_1. That is, the potential energy of the charge distribution is independent of how it is assembled.

These considerations hold for any charge distribution. In general, it takes work to assemble a charge distribution and that work can be considered stored as potential energy. The potential energy can be positive or negative, depending on whether the net work required to assemble the charge distribution is positive or negative. Because the electric force obeys the superposition principle, the potential energy is independent of how the charge distribution is assembled: the total energy is simply the sum of the potential energies of every charge pair making up the distribution.

Example 24–1 _____

The Energy of a Charge Distribution

Three point charges each carrying $+q$ and a fourth carrying $-q/2$ are initially infinitely far apart. They are brought together to form the square charge distribution shown in Fig. 24–2. What is the electrostatic potential energy of this charge distribution?

Solution

We can assemble the charge distribution in any order without altering the potential energy. Assume that the positive charge at the upper left is brought in first. This takes no work because no other charge is in place. Next the positive charge q at the upper right is brought to a distance a from the first positive charge. Equation 24–1 tells us that the work required is

$$W_2 = \frac{1}{4\pi\epsilon_0}\frac{q^2}{a}.$$

Now the third positive charge is brought to its place at the lower left. This point is

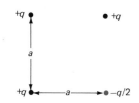

Fig. 24–2
Example 24–1

a distance a from the first charge q and $\sqrt{2}a$ from the second charge, so that the work required is

$$W_3 = \frac{1}{4\pi\epsilon_0}\left(\frac{q^2}{a} + \frac{q^2}{\sqrt{2}a}\right).$$

Finally, the negative charge $-q/2$ is brought to a point a distance a from the second and third charges and $\sqrt{2}a$ from the first charge. The work required is again the sum of the potential differences multiplied by the charge $-q/2$:

$$W_4 = \frac{1}{4\pi\epsilon_0}\left(-\frac{q^2}{2a} - \frac{q^2}{2a} - \frac{q^2}{2\sqrt{2}a}\right).$$

This work is negative, indicating that an applied force would have to do positive work to remove the negative charge from the distribution. Adding the work required to bring in the second, third, and fourth charges gives the electrostatic potential energy of the charge distribution:

$$W = W_2 + W_3 + W_4$$

$$= \frac{1}{4\pi\epsilon_0}\left(\frac{q^2}{a} + \frac{q^2}{a} + \frac{q^2}{\sqrt{2}a} - \frac{q^2}{2a} - \frac{q^2}{2a} - \frac{q^2}{2\sqrt{2}a}\right)$$

$$= \frac{q^2(2\sqrt{2}+1)}{8\pi\sqrt{2}\epsilon_0 a}.$$

That this is a positive quantity indicates that the work needed to assemble the three positive charges is greater than the energy gained bringing in the negative charge. Problem 4 explores this example for a different value of the fourth charge.

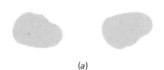

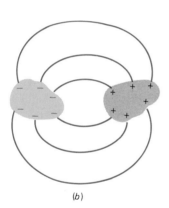

(a)

(b)

Fig. 24–3
A pair of isolated conductors.
(a) Initially they are uncharged, and there is no potential difference between them. *(b)* When they carry opposite charges, there is an electric field associated with the charged conductors, and consequently a potential difference between them.

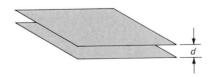

Fig. 24–4
A pair of closely spaced conducting plates.

24.2
TWO ISOLATED CONDUCTORS

An important practical charge distribution consists of two isolated conductors on which we place equal but opposite charges. Figure 24–3 shows two such conductors, each initially carrying no net charge. With no net charge on either conductor, there is no electric field, and no potential difference between the conductors. But each conductor is made up of equal numbers of positive and negative elementary charges. Imagine moving a small quantity of charge from one conductor to the other, giving rise to a net positive charge on one conductor and a net negative charge on the other. This results in an electric field and therefore in a potential difference between the conductors. If we try to transfer more charge between the conductors, we must do work traversing this potential difference. The more charge we move, the harder it gets to transfer additional charge. The work it takes to transfer the charge is stored as potential energy of the charge distribution.

It is generally difficult to calculate the stored potential energy for a pair of irregularly shaped conductors like those of Fig. 24–3. An important practical case for which the potential energy may be calculated is a pair of identical, flat, parallel conducting plates whose separation is small compared with their width (Fig. 24–4). We start with the plates uncharged, and then transfer charge q from one plate to the other. (In practice, we would accomplish this by connecting the plates to the terminals of a battery.

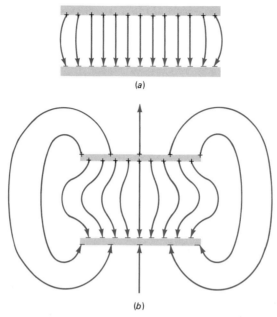

Fig. 24–5
The electric field of a pair of parallel conducting plates. *(a)* When the plate spacing is small, all the charge resides on the inner edges of the plates and the surface charge density is nearly uniform, giving rise to a uniform electric field. In this situation we can neglect the "fringing field" at the edges. *(b)* Neglecting the "fringing field" is not appropriate for widely-spaced plates, whose field and charge distribution are not uniform.

Charge moves off one plate, through the battery, and onto the other plate.) Charging the plates results in an electric field between them. For closely spaced plates, this field is essentially uniform except very near the edges (Fig. 24–5), and we may neglect this nonuniform "fringing field."

What is the electric field strength between the plates? In Chapter 22 we used Gauss's law to show that the field near the surface of a conductor carrying surface charge density σ is given by

$$E = \frac{\sigma}{\epsilon_0}.$$

With our closely spaced plates, every point between the plates is close to the surface of a conductor, so that this expression applies. What is the charge density on the plates? If the plates were isolated, charge would distribute itself symmetrically, with half the total charge covering each surface of a plate. But the presence of the oppositely charged plate causes all the excess charge to appear on the inner surface of each plate, as shown in Fig. 24–5a. Then the magnitude of the surface charge density σ on the inner surface of either plate is just

$$\sigma = \frac{q}{A},$$

where A is the plate area. Using this surface charge density in our expression for the electric field gives

$$E = \frac{\sigma}{\epsilon_0} = \frac{q}{\epsilon_0 A}. \tag{24-2}$$

You might think that the field strength in Equation 24–2 should be doubled, because there are two plates each carrying surface charge density of magnitude $\sigma = q/A$. But we already took into account the presence of both plates when we argued that all charge would reside on the inner surfaces of the plates, thereby doubling the surface charge density over the value it would have had if only one plate had been present. Equation 24–2 was derived strictly in accordance with Gauss's law. To double the field would violate Gauss's law by saying that the number of field lines emerging from a closed surface depends on something other than the net charge enclosed! We discussed this point in Example 22–5.

The presence of the electric field means that there is a potential difference between the plates, so that it takes work to move more charge between them. Because the field is uniform, the potential difference is a simple product of the field strength by the distance between the plates:

$$V = Ed = \frac{qd}{\epsilon_0 A}. \tag{24-3}$$

Here q is the magnitude of the charge on either plate, so the potential difference is directly proportional to the charge.

Now imagine moving an additional very small positive charge dq from the negative to the positive plate. (Although the negative plate has a *net* negative charge, it still consists of great numbers of positive and negative elementary charges.) How much work does this take? The work depends on the potential difference between the plates, which, through Equation 24–3, depends on how much charge has already been transferred. Because potential difference is the work per unit charge required to move between two points, the work dW required to move the charge dq between the plates is just

$$dW = V dq = \frac{qd}{\epsilon_0 A} dq. \tag{24-4}$$

Suppose that we start with zero net charge on either plate and gradually transfer a total charge Q from one plate to the other. Each dq that we move requires an amount of work given by Equation 24–4, so that the total work required is the sum of all the dW's for all the small pieces of charge dq that make up Q. In the limit of infinitesimally many infinitesimal charges dq, this sum becomes an integral, and we have

$$W = \int_0^Q dW = \int_0^Q \frac{qd}{\epsilon_0 A} dq = \frac{d}{\epsilon_0 A} \int_0^Q q\, dq.$$

That the variable q remains under the integral sign reflects the physical fact that the continually increasing charge on the plates results in an increasing potential difference and therefore makes it harder to move each additional charge dq. Continuing the integration, we have

$$W = \frac{d}{\epsilon_0 A} \frac{q^2}{2} \Big|_0^Q = \frac{d}{2\epsilon_0 A} Q^2. \tag{24-5}$$

Thus the work required to charge the plates increases as the square of the charge Q. The work done in charging the plates ends up as stored potential energy of the final charge distribution, so that the stored energy is

$$U = W = \frac{d}{2\epsilon_0 A} Q^2. \tag{24-6}$$

The quadratic dependence of the stored energy on charge suggests that a pair of parallel plates is an excellent device for storing electrostatic energy.

24.3
ENERGY AND THE ELECTRIC FIELD

We have seen that the work required to assemble a distribution of electric charge ends up as electrostatic potential energy. Just where is this energy stored? To answer this question, we must find something that changes as the charge distribution is assembled. In Example 24–1, we considered the assembly of four point charges to form a square. Surely the point charges themselves did not change; we only moved them closer together. Similarly, when we charged our pair of metal plates we did not alter the individual charges; we only moved them from one plate to another. What has changed in both these cases? The electric field has changed. In the first case we started with four isolated point charges and an electric field that looked like four isolated point-charge fields. We ended with a new charge distribution whose field did not look at all like a point-charge field.

In the case of the parallel plates, we started with uncharged plates and no electric field. As soon as we began transferring charge from one plate to the other, an electric field appeared between the plates, and this field grew in strength as more charge was transferred.

So where is the energy stored? It is stored in the electric field. As we create or alter a charge distribution, we do work and an altered electric field configuration develops. The work we do in moving the charges ultimately goes into the alteration of the field. If the work done by the applied force is positive, we have added energy to the field. If the work is negative, we have removed energy from the field.

You might argue that it makes no sense to speak of energy stored in an electric field because the field is just a mathematical construct describing electric forces on charges. But to a physicist, the electric field is just as real as matter. One of the strongest reasons for believing this is that the field represents stored energy that can be transported and extracted to perform useful work. Next time you are sunning yourself on a beach, consider the physicist's description of sunlight as energy that is coming to you partly in the form of electric fields that originated in charge distributions on the sun!

Every electric field represents stored energy. If the field is altered, energy is either accumulated or released, depending on whether the work done is positive or negative. If the field could be made to disappear entirely, all its stored energy would be released. Because electric forces are primarily responsible for the behavior of everyday matter, many seemingly different forms of energy storage really involve electric field energy. When you stretch a spring, for example, you alter the relative orientation of atoms, storing energy in the altered electric field between the charged particles making up the atoms. Similarly, when you burn gasoline you are rearranging electric charges into a configuration whose field contains less energy.

If energy is stored in an electric field, then it makes sense to suppose that the amount of energy stored in a given region depends on the strength of the field in that region. In general the electric field may vary continuously throughout space, so that we would like to know the **energy density,** or energy per unit volume, at each point where an electric field exists. We can readily determine this energy density for the case of two parallel plates. The uniform electric field between the plates is given by Equation 24–2, which may be solved for the charge to give

energy density

$$Q = \epsilon_0 A E.$$

Putting this expression for Q into Equation 24–6 for the stored energy gives

$$U = \frac{d}{2\epsilon_0 A} Q^2 = \frac{d}{2\epsilon_0 A} (\epsilon_0 A E)^2 = \tfrac{1}{2}\epsilon_0 E^2 \, Ad. \qquad (24\text{–}7)$$

Our assumption that the plates are very close together allowed us to conclude that the field is very nearly uniform between the plates and essentially zero outside the plates. Therefore, the stored energy given by Equation 24–7 is stored in the region between the plates, and is distributed uniformly because the field is uniform. The volume between the plates is just the plate area times the separation, or Ad, so that the energy per unit volume, or energy density u, is given by $u = U/Ad$, or

$$u = \tfrac{1}{2}\epsilon_0 E^2. \quad \text{(electric energy density)} \qquad (24\text{–}8)$$

Although we derived this expression for the uniform field between two parallel plates, it is in fact a universal expression that holds for any electric field. At any point in the universe where an electric field exists, there is stored energy whose density, in J/m^3, is given by Equation 24–8.

Note that the electric field energy density of Equation 24–8 depends on the square of the field. This means that we cannot calculate the energy density in the field of a charge distribution by simply adding the energy densities associated with the fields of the individual charges. Instead, we must first calculate the net electric field, then use Equation 24–8 to get the energy density. Mathematically, this is because the square of a sum is not equal to the sum of the squares $[2^2 + 3^2 \neq (2+3)^2]$. Physically, it would not make sense to sum the separate energy densities of individual charges because this sum would not change as we altered the positions of the charges in the distribution, and therefore would not reflect the work done in moving the charges.

The deepest significance of Equation 24–8 lies in its statement that every electric field represents stored energy. As we observe a variety of physical phenomena, from everyday happenings on earth to events in distant galaxies, we can understand that the driving energy for many of these phenomena comes from the release of energy stored in electric fields.

Example 24–2
Electrical Energy of a Thunderstorm

Electric fields inside a thunderstorm have typical values of 10^5 V/m and get even higher just before electrical energy is unleashed as lightning (Fig. 24–6). The origin of these fields and hence of the energy stored in them is still not fully understood, but is generally believed to be associated with charge transfer to rising and falling water droplets or ice crystals in the intense updrafts and downdrafts of the thunderstorm. Consider a typical thundercloud that rises to an altitude of 10 km and has a diameter of 20 km. Assuming an average field strength of 10^5 V/m, estimate the total electrostatic energy stored in the cloud. How many gallons of gasoline would you have to burn to release the same amount of energy?

Solution

The energy density is given by Equation 24–8:

$$u = \tfrac{1}{2}\epsilon_0 E^2 = \tfrac{1}{2}(8.9\times10^{-12}\ \text{C}^2/\text{N·m}^2)(10^5\ \text{V/m})^2 = 4.5\times10^{-2}\ \text{J/m}^3.$$

(You should verify that the units work out!) We are assuming that this energy density is the same throughout the storm, so that we find the total energy by multiplying the energy density by the volume. The storm is roughly cylindrical in shape, so its volume is

$$v = \pi r^2 h = \pi(10\ \text{km})^2(10\ \text{km}) = 3100\ \text{km}^3 = 3.1\times10^{12}\ \text{m}^3.$$

Then the total stored energy is

$$U = uv = (4.5\times10^{-2}\ \text{J/m}^3)(3.1\times10^{12}\ \text{m}^3) = 1.4\times10^{11}\ \text{J}.$$

A gallon of gasoline contains about 10^8 J (see Appendix D), so that the electrical energy stored in a thunderstorm at any given instant is equivalent to about 1000 gallons or 4000 L of gasoline. This comparison is not quite fair to the thunderstorm, though, because its electrical energy is continually dissipated in lightning strikes and at the same time renewed by the violent motion of the air. Problem 46 explores thunderstorm energetics in more detail.

Fig. 24–6
Lightning is the sudden release of energy stored in atmospheric electric fields.

When the electric field is uniform, as in our thunderstorm example, we can find the stored energy simply by multiplying the energy density by the volume. But when the field changes with position we must resort to calculus. Consider a small volume element dv, so small that the electric field is essentially uniform over this volume. The stored energy dU in the volume element is just the energy density times the volume, or

$$dU = u\,dv = \tfrac{1}{2}\epsilon_0 E^2 dv.$$

The total energy U is then the sum of all the dU's. In the limit of infinitesimally small volumes dv and energies dU, this sum becomes an integral:

$$U = \tfrac{1}{2}\epsilon_0 \int E^2 dv, \qquad (24\text{--}9)$$

where the limits on the integral are chosen to cover the entire region in which the electric field of interest exists.

We derived Equation 24–9 for the electric field energy using our previously determined expression for the work needed to assemble a simple charge distribution. We can also reverse that process, using the electric field of a charge distribution to calculate the energy density and from it the stored energy and therefore the work needed to assemble the distribution. Example 24–3 illustrates this procedure for a case when the energy density varies with position.

Example 24–3

A Shrinking Sphere

A sphere of radius R_1 carries a total charge Q distributed evenly over its surface (Fig. 24–7). How much work does it take to shrink the sphere to a smaller radius R_2? Practical applications in which this question might prove important include the behavior of cell membranes, charged bubbles, and raindrops in thunderclouds.

Solution

Shrinking the sphere moves all the charge elements on its surface closer together, so we expect the shrinking to require positive work. It would be difficult to calculate this work by considering individually the inward movement of each charge element, for moving individual charge elements one at a time would destroy the spherical symmetry. But we do know that the work is equal to the change in stored electric field energy given by Equation 24–9. With all charge distributed evenly over the sphere's surface, Gauss's law tells us that there is no electric field within the sphere. Because of the spherical symmetry, Gauss's law also tells us that the field outside the sphere is identical to that of a point charge Q at the sphere's center (recall Section 21.5). This means that the field at and beyond the original radius R_1 does not change as we shrink the sphere. What does change is the field between R_1 and R_2 (Fig. 24–8). Originally this field was zero. After the sphere has shrunk, this region, too, is filled with a point-charge field. This newly created field is the site of the additional energy stored in shrinking the charge distribution.

Because the point-charge field between R_1 and R_2 changes with position, we must use the integral form 24–9 to calculate the stored energy. The electric field outside the sphere is that of a point charge:

$$E = \frac{1}{4\pi\epsilon_0}\frac{Q}{r^2},$$

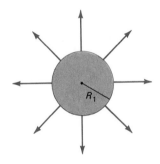

Fig. 24–7
Charged sphere and its electric field before shrinking.

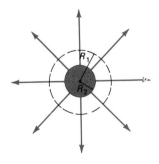

Fig. 24–8
Shrinking the sphere creates a new field in the region $R_2 < r < R_1$.

so that, from Equation 24–8, the energy density as a function of r is

$$u(r) = \tfrac{1}{2}\epsilon_0 E^2 = \tfrac{1}{2}\epsilon_0 \left(\frac{1}{4\pi\epsilon_0} \frac{Q}{r^2} \right)^2 = \frac{Q^2}{32\pi^2\epsilon_0 r^4}.$$

To determine the total stored energy we integrate this energy density over the volume between R_2 and R_1. Because of the spherical symmetry we consider volume elements made of thin spherical shells of thickness dr (Fig. 24–9). What is the volume dv of such a shell? The inner and outer surfaces of the shell are spherical, with area $4\pi r^2$ essentially the same for both because the shell is so thin. Then the volume of the shell is very nearly $dv = 4\pi r^2 dr$, so that Equation 24–9 becomes

$$U = \int_{R_2}^{R_1} u\, dv = \int_{R_2}^{R_1} \frac{Q^2}{32\pi^2\epsilon_0 r^4} 4\pi r^2 dr$$

$$= \frac{Q^2}{8\pi\epsilon_0} \int_{R_2}^{R_1} r^{-2} dr = \frac{Q^2}{8\pi\epsilon_0} \left(-\frac{1}{r} \right) \Big|_{R_2}^{R_1}$$

$$= \frac{Q^2}{8\pi\epsilon_0} \left(\frac{1}{R_2} - \frac{1}{R_1} \right).$$

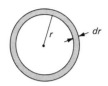

Fig. 24–9
A very thin spherical shell of thickness dr and radius r has volume $4\pi r^2 dr$.

This is the total energy stored in the new electric field between R_2 and R_1, and is therefore also the work done in shrinking the sphere from R_1 to R_2. Were the sphere allowed to expand back to its original radius R_1, this amount of energy would be released. If we let R_1 go to infinity, our result becomes the work required to assemble a sphere of radius R_2 carrying surface charge Q. Equivalently, the result gives the energy stored in the field of the sphere. Because the stored energy becomes infinite as R_2 approaches zero, our result suggests that the notion of a point charge is an impossible idealization. Problem 49 explores some implications of this result in the theory of elementary particles, especially the electron.

24.4
CAPACITORS

capacitor

In electrical and electronic equipment, the storage of electrical energy is often accomplished using a pair of charged conductors separated by an insulator. Such a device is called a **capacitor.** (An older term, still in occasional use, is "condenser.") Capacitors are typically used for short-term energy storage in situations where it is necessary to store or release electrical energy relatively quickly. Most practical electronic devices, including radio, TV, computers, and audio equipment, would be virtually impossible to construct without capacitors. When you tune a radio, you are adjusting a capacitor. Failure of the simple, inexpensive capacitor in your car's ignition system could leave you stranded on the highway. And many high-energy experiments in physics and engineering use so much power that, were it not for capacitors, they could not be done without disrupting the supply of electric power to the rest of the world!

When a capacitor is uncharged, both its conductors are neutral so that there is no electric field in the capacitor, and no energy is stored. A capacitor is charged by transferring charge (usually electrons) from one conductor to the other. The work required is generally supplied by some other source of electrical energy that is connected to the capacitor through wires leading to its two conductors. An electric field develops as a result of the charge separation, and the stored energy resides in this field. Although we speak of a capacitor as being charged, the capacitor remains overall electrically neutral, in that the net charge on the whole capacitor is zero. But the two individual conductors making up a charged capacitor are not neutral: one carries a charge $+q$, the other $-q$. When we say that the charge on a capacitor is q, we really mean that q is the magnitude of the charge on either conductor, but with one conductor positive and the other negative.

Once a capacitor is charged there is an electric field and therefore a potential difference between its two conductors. As the charge is increased the potential difference increases proportionately. Conversely, imposing a potential difference on a capacitor (by connecting it to a battery, for example) causes the capacitor to become charged in proportion to the potential difference imposed. The ratio of charge to potential difference is characteristic of a given capacitor, and is called its **capacitance:**

capacitance

$$C = \frac{q}{V}.$$

(24–10)

Here q is the magnitude of the charge on either conductor and V the potential difference (or voltage) between the conductors. Clearly the units of capacitance are coulombs/volt. One coulomb/volt is given the name **farad** (F), in honor of the nineteenth century scientist Michael Faraday. One farad is so huge a capacitance that the smaller units microfarad (10^{-6} F; abbreviated μF) and picofarad (10^{-12} F; abbreviated pF and often pronounced "puff") are widely used.

farad

Capacitance is a property of the physical construction of a capacitor—the shapes of its two conductors, their separation, and the choice of insulating material between them. Although q and V enter the defining relation 24–10, capacitance does not depend on either q or V. The capacitance C is a constant that gives the ratio of charge to potential difference for a particular capacitor. If V is increased, q increases proportionately, maintaining the constant ratio C that characterizes the capacitor.

Any arrangement of two insulated conductors constitutes a capacitor. Practical capacitors are manufactured in a variety of configurations. Often they are made from two long strips of aluminum foil separated by a thin layer of plastic or paper. This foil "sandwich" is then rolled into a compact cylinder, wires are attached, and the whole assembly is covered with a protective coating. Another common arrangement is the variable capacitor, whose configuration can be altered to change its capacitance. When you tune a radio, you are adjusting the capacitance of a variable capacitor. Figure 24–10 shows several typical capacitors.

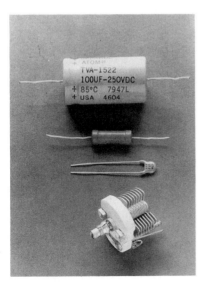

Fig. 24–10
Typical capacitors. (Top to bottom) 100-μF, 250-V electrolytic; 0.1-μF, 400-V mylar; 47 pF, 1000-V ceramic; air-insulated variable.

Calculating Capacitance

Capacitance is defined through Equation 24–10 as the ratio of charge to potential difference. To calculate the capacitance of a particular configuration of two conductors, we assume there is a charge q on the capacitor and calculate the corresponding potential difference. Because the capacitance depends only on the physical configuration of the capacitor, it doesn't matter what value we assume for the charge—that will cancel when we take the ratio of charge to potential difference. For two irregularly shaped conductors we are back to the problem of determining the potential difference between the conductors from the distribution of charge on their surfaces. When the capacitor design includes sufficient symmetry, this calculation becomes straightforward.

By far the most important capacitor design is the parallel-plate configuration. In Section 24.2 we examined such a capacitor in some detail, although at that time we did not call the configuration a capacitor. There we found that for closely spaced plates, the uniform electric field in the capacitor is given by

$$E = \frac{q}{\epsilon_0 A},$$

where q is the magnitude of the charge on either plate and A is the plate area. Corresponding to this uniform field is a potential difference

$$V = Ed = \frac{qd}{\epsilon_0 A},$$

with d the plate spacing. Solving for the ratio $C = q/V$ gives

$$C = \frac{q}{V} = \frac{\epsilon_0 A}{d}. \qquad \text{(parallel-plate capacitor)} \qquad (24\text{–}11)$$

Equation 24–11 gives the capacitance of a parallel-plate capacitor in terms of the universal constant ϵ_0 and factors that describe the physical configuration of the capacitor. (Strictly speaking, this expression holds only for capacitors insulated by vacuum. Later we will modify Equation 24–11 to account for other insulating materials.) Note that neither charge nor potential difference enters the final expression for capacitance, showing that the capacitance is indeed a constant. Equation 24–11 suggests that the way to make a capacitor with large capacitance is to use two plates of large area but small separation. Incidentally, Equation 24–11 shows that the units of ϵ_0 may be expressed as farads/meter (F/m); see Problem 19.

Example 24–4

A Parallel-Plate Capacitor

A capacitor consists of two circular metal plates of radius 10 cm separated by an air gap of 5.0 mm. What is its capacitance? When a 12-volt battery is connected to the capacitor, how much charge appears on the plates?

Solution

Since the plate spacing is much smaller than the plate size, Equation 24–11 holds and the capacitance is

$$C = \frac{\epsilon_0 A}{d} = \frac{\epsilon_0 \pi r^2}{d} = \frac{(8.85 \times 10^{-12} \text{ F/m})(\pi)(0.10 \text{ m})^2}{5.0 \times 10^{-3} \text{ m}} = 5.6 \times 10^{-11} \text{ F} = 56 \text{ pF}.$$

Equation 24–10 defines capacitance as the ratio of charge to potential difference. We can rewrite this defining relation to solve for the charge:

$$q = CV = (56 \text{ pF})(12 \text{ V}) = 670 \text{ pC}.$$

What this really means, of course, is that the positive plate carries 670 pC and the negative plate -670 pC. Overall, the capacitor remains neutral. Note that by working with the capacitance in pF, the charge automatically comes out in pC:

Example 24–5
A Cylindrical Capacitor

A capacitor consists of two long concentric pieces of cylindrical metal tubing of length L, as shown in Fig. 24–11. The inner cylinder has radius a and the outer cylinder has radius b. What is the capacitance of this cylindrical capacitor?

Solution

Equation 24–11 does not apply to this configuration because the electric field between the cylinders is not the uniform field of a parallel-plate capacitor. To calculate the capacitance, we need a relation between charge and potential difference for this new configuration. In Example 23–5, we found that the potential difference between two points outside a charge distribution with line symmetry is given by

$$V_2 - V_1 = \frac{\lambda}{2\pi\epsilon_0} \ln\left(\frac{r_1}{r_2}\right),$$

where λ is the line charge density. Because our capacitor is long compared with its diameter, we can neglect the fringing fields at its ends so that this expression is a good approximation to the potential difference due to the inner conductor. What about the outer conductor? Recall (Example 21–8) that the electric field inside an empty, hollow pipe is zero; therefore, the outer conductor contributes nothing to the electric field or potential difference between the conductors. If the magnitude of the charge on either conductor is q, then the line charge density λ is q/L, and our expression for the potential difference can be written

$$V = V_a - V_b = \frac{q}{2\pi\epsilon_0 L} \ln\left(\frac{b}{a}\right),$$

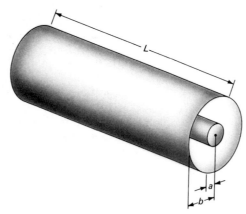

Fig. 24–11
A cylindrical capacitor (Example 24–5).

where we have written $V_a - V_b$ rather than $V_b - V_a$ because we want a positive potential difference to use in calculating the capacitance, which is intrinsically positive. Capacitance is the ratio of charge to potential difference, so we have

$$C = \frac{q}{V} = \frac{q}{(q/2\pi\epsilon_0 L)\ln(b/a)} = \frac{2\pi\epsilon_0 L}{\ln(b/a)}. \qquad (24\text{-}12)$$

Does this result make sense? We already found that the capacitance of a parallel-plate capacitor increases with increasing plate area or with decreasing plate separation. With the cylindrical capacitor we can increase the area of both conductors by increasing the length L of the capacitor, and indeed Equation 24-12 shows the capacitance increasing proportionately. We can decrease the spacing of the conductors by making the radii a and b more nearly equal. This makes b/a closer to one, and $\ln(b/a)$ closer to zero, again increasing the capacitance. Although the geometries of the cylindrical and parallel-plate capacitors are quite different, the same physical considerations apply to both: a large capacitance is achieved with large conductor areas and small separation. When the separation is very small, the curvature of the cylindrical capacitor cannot matter, and Equation 24-12 should reduce to Equation 24-11 for the parallel-plate capacitor. You can show this using a series expansion for the logarithm in Equation 24-12 (see Problem 39).

Energy Storage in Capacitors

In Section 24.3, we found that any electric field represents stored energy. The example that guided us to that conclusion was a parallel-plate capacitor. For that configuration, we found a stored energy U given by Equation 24-7:

$$U = \frac{d}{2\epsilon_0 A} Q^2.$$

Since $\epsilon_0 A/d$ is the capacitance of the parallel-plate capacitor, this stored energy may be written

$$U = \frac{Q^2}{2C}. \qquad (24\text{-}13)$$

It is usually easier to measure voltage than charge. To express the stored energy in terms of voltage, we can solve the equation defining capacitance, $C = Q/V$, for Q and use the result in Equation 24-13:

$$U = \frac{Q^2}{2C} = \frac{(CV)^2}{2C} = \tfrac{1}{2}CV^2. \qquad \text{(energy in capacitor)} \qquad (24\text{-}14)$$

Although Equations 24-13 and 24-14 were derived for a parallel-plate capacitor, they hold for any capacitor regardless of its configuration (see Problem 38).

The dependence of the stored energy on the square of the potential difference in Equation 24-14 has important implications for the efficient storage of energy in capacitors, for it suggests that more energy can be stored in a small capacitor at high voltage than in a large capacitor at a lower voltage. Practically, difficulties of handling high voltages weaken this conclusion somewhat, but the fact remains that stored energy in a capacitor increases very rapidly with increasing voltage across its plates.

Energy storage in capacitors has important safety implications for people working on electronic equipment. Typical devices such as stereo amplifiers or TV sets usually contain one or more large capacitors used in producing steady direct current power for the device (Fig. 24–12). Even after the device has been unplugged, these may store enough energy to give a dangerous shock. The first thing a trained repair person does is to locate these capacitors and discharge them by holding a large screwdriver with an insulated handle across the capacitor terminals. The resulting spark and loud noise are testimony to the amount of stored energy released! The pitted appearance of a screwdriver used for this purpose (Fig. 24–13) attests to the fact that the energy stored in a capacitor can vaporize steel!

Fig. 24–12
The large cylindrical objects are 4700-μF capacitors in the power supply of a stereo amplifier; they can give a dangerous shock even after the amplifier has been unplugged.

Example 24–6
Energy Stored in Capacitors

A 1.0-μF capacitor is charged to 100 V, and a 100-μF capacitor is charged to 1.0 V. How much energy does each store?

Solution
Applying Equation 24–14, we find

$$U_{1.0\,\mu F} = \tfrac{1}{2}CV^2 = \tfrac{1}{2}(1.0\ \mu F)(100\ V)^2 = 5.0 \times 10^3\ \mu J$$

and

$$U_{100\,\mu F} = \tfrac{1}{2}CV^2 = \tfrac{1}{2}(100\ \mu F)(1.0\ V)^2 = 50\ \mu J.$$

In this example the 1.0-μF capacitor stores 100 times as much energy as the 100-μF capacitor. Will this always be true? Of course not. If we put equal voltages on the two capacitors, which would store more energy? It is only because of the larger voltage, and the dependence of the stored energy on the square of the voltage, that the smaller capacitor in this example stores more energy.

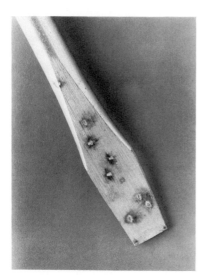

Fig. 24–13
This screwdriver has been used to discharge a large capacitor.

Application
Camera Flashes, Toilet Flushes, and Laser Fusion

You have probably used a camera equipped with an electronic flash. The flash unit contains a special tube filled with xenon gas. When a large potential difference is applied across two metal electrodes in the tube, dielectric breakdown occurs and the xenon suddenly ionizes. Recombination of electrons with xenon ions then results in a bright pulse of white light. After a flash picture has been taken the photographer must wait a while—typically around 10 seconds—before the flash is ready again. Why is this? The flash is powered by a small battery. Although the total energy used by the flash is small, during the short interval that the xenon is being ionized the rate of energy use far exceeds the maximum power output of the battery (see Problem 40). So the battery is used to charge a capacitor at a slow rate. Once the capacitor is charged to the voltage needed to fire the flashtube, the capacitor's stored energy can be dumped suddenly into the flashtube, providing the short burst of high power needed to give the intense light. The flashtube cannot be used again until the capacitor is recharged.

This situation is exactly analogous to one you encounter every day in household plumbing. Flushing a toilet requires a great amount of water in a short time—far more than typical household plumbing could supply in that time. So the water is accumulated gradually in the toilet tank, then suddenly dumped when needed

Fig. 24–14
(Left) The NOVA pulsed laser system at Lawrence Livermore National Laboratory can deliver a 15-kJ pulse lasting 0.1 ns or a 150-kJ pulse lasting 3 ns. (Right) Huge banks of capacitors are used to store the pulse energy for laser fusion; this capacitor bank at the Los Alamos National Laboratory has a capacitance of 2.3 F.

for flushing. In this analogy the household plumbing, with its relatively narrow pipes, corresponds to the small battery of the camera flash. The toilet tank, which gradually accumulates water, corresponds to the capacitor, which gradually accumulates electrical energy. Of course you need to wait between flushes for the toilet tank to fill, just as you need to wait between flash pictures for the capacitor to charge.

Professional photographers needing to take flash pictures in rapid succession often carry around a large, heavy battery pack that is capable of supplying the flash power directly. Similarly, institutional buildings with large, high-pressure water pipes often have toilets without tanks that can be flushed in rapid succession.

The simple example of the camera flash, scaled up many times in size, shows how very large amounts of power may be obtained briefly for industrial or scientific applications. Indeed, some experiments involving high-power pulsed lasers for nuclear fusion and ballistic missile defense research require more power than all the world's electric generating stations produce. The required energy is accumulated in huge banks of capacitors, which are suddenly discharged to provide energy to the laser. Think here about the difference between energy and power! The pulsed laser is only on for about 10^{-9} seconds or less, so although it consumes energy at an enormous rate while on, it does not use all that much total energy (see Problem 41). The laser is not fired very often (at least in today's test configurations), so that there is plenty of time to charge the capacitors. The *average* power consumption of the experiment is modest. Figure 24–14 shows the NOVA laser for the Lawrence Livermore Laboratory laser fusion experiment and the capacitor banks for a laser fusion experiment at the Los Alamos Scientific Laboratory.

Connecting Capacitors Together

The most important consideration in using a capacitor is whether its capacitance is the right value for the particular application. But there is another practical consideration that must be taken into account if the capacitor is to function safely and effectively. In Chapter 22 we introduced the phenomenon of dielectric breakdown, in which strong electric fields rip charge from a dielectric material, destroying its insulating properties. To avoid dielectric breakdown of their insulating materials, capacitors are

working voltage

rated according to their **working voltage,** meaning the maximum voltage that should normally be applied across the two conductors of the capacitor.

Large capacitances are most easily achieved using small plate separations, as suggested by Equation 24–11. But a small plate separation implies a rapid change in potential with position, or a large electric field between the plates. Thus in practical capacitors there is a trade-off between capacitance and working voltage. High working voltage and high capacitance together require large plate separation to keep the electric field small and avoid dielectric breakdown, while at the same time requiring large plate area to keep the capacitance up. Thus large, high-voltage capacitors are physically bulky and expensive to build. Often economics as well as physics dictates the final design of a circuit involving capacitors.

When a single capacitor with a desired combination of capacitance and working voltage is not available, we can often obtain the desired combination by connecting two or more capacitors together. There are only two ways in which we can connect two capacitors (and indeed any electronic components with only two wires coming from them). Called **parallel** and **series,** these two possible connections are shown in Fig. 24–15. We would like to know the equivalent capacitance of each combination.

parallel
series

Consider first the parallel combination of Fig. 24–15a. If we impose a potential difference V across the two wires coming from the combination, what will be the potential difference across each capacitor? The key to answering this question is the recognition that the wires connecting the capacitors are conductors, so that in electrostatic equilibrium there can be no potential difference between any points connected to the same wire.* Thus all points connected directly together—including the top plates of each capacitor—are at the same potential. Similarly, the bottom plates and the wires connecting them are all at the same potential. Therefore, the potential differences across the two capacitors are equal. This is a very important point in the practical understanding of electric circuits, and applies to any two circuit components that are connected in parallel:

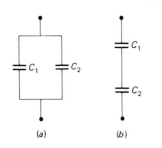

Fig. 24–15
Connecting capacitors together.
(a) Parallel; *(b)* series. $\perp$ is the standard circuit symbol for a capacitor.

parallel circuit
components

> The potential differences across two circuit components in parallel are equal.

Recognizing this simple fact is essential in developing your understanding of electric circuits!

What is the equivalent capacitance of the two parallel capacitors? The equivalent capacitance is the ratio of the total charge on both capacitors to the voltage across the parallel combination. Solving the defining relation $C = q/V$ for charge, we can write the charge on capacitor C_1 as

$$q_1 = C_1 V$$

and that on C_2 *as*

$$q_2 = C_2 V.$$

---

*Even when we relax the equilibrium assumption, this conclusion will still hold in the approximation that the wires are perfect conductors.

The potential difference V is the *same* in both cases because the capacitors are connected in *parallel*. Thus the total charge is

$$q = q_1 + q_2 = C_1V + C_2V = (C_1 + C_2)V.$$

Taking the ratio of total charge q to the voltage V across the parallel combination gives the equivalent capacitance:

$$C = C_1 + C_2. \qquad \text{(parallel capacitors)} \qquad (24\text{--}15)$$

Equation 24–15 is frequently expressed by the statement that "capacitors in parallel add." You can understand this result physically by considering two parallel-plate capacitors with equal spacing. Connecting them in parallel amounts to adding their plate areas, giving a larger capacitance. Although we derived Equation 24–15 for two parallel capacitors, the result that parallel capacitances add is easily extended to three or more capacitors (see Problem 30).

What about the working voltage of the parallel combination? Both capacitors experience the full potential difference V, so that the working voltage of the combination is that of whichever capacitor has the lower working voltage.

Example 24–7
Parallel Capacitors

A 1.0-μF capacitor with a maximum working voltage of 100 V is connected in parallel with a 2.0-μF capacitor whose maximum working voltage is 50 V. What is the equivalent capacitance of the combination? What is the maximum voltage that can be applied safely to the combination? If a 12-volt battery is connected across the combination, how much charge appears on each capacitor?

Solution

From Equation 24–15, the equivalent capacitance is just the sum of the individual capacitances, or 3.0 μF. Because each capacitor experiences the full applied voltage, the maximum safe voltage is 50 volts, as determined by the capacitor with the smallest voltage rating. When the parallel combination is connected to a 12-volt battery, a 12-volt potential difference appears across each capacitor, so that

$$q_1 = C_1V = (1.0\ \mu\text{F})(12\ \text{V}) = 12\ \mu\text{C}$$

and

$$q_2 = C_2V = (2.0\ \mu\text{F})(12\ \text{V}) = 24\ \mu\text{C}.$$

This example shows that while the voltage on parallel capacitors is the same, the charge need not be. Our total charge, 36 μC, is just what we would expect on a 3.0-μF capacitor charged to a potential difference of 12 V, so that the parallel combination does indeed behave like a 3-μF capacitor.

Now consider two capacitors connected in series, as in Fig. 24–15b. There are no wires connecting the middle two plates to the external circuitry, so that the net charge on these plates together is zero and cannot change. If one of these plates is charged, the other must carry an opposite charge of equal magnitude. Now suppose we charge the whole system, placing $+q$ on the upper plate of the top capacitor and $-q$ on the lower plate of the bottom capacitor. What happens to the middle plates? Negative charge $-q$ is pulled to the bottom plate of the upper capacitor by the positive charge on the top plate. This leaves $+q$ on the upper plate of the lower capacitor. As a result, the two capacitors in *series* carry *equal charges*. We can solve the defining relation $C = q/V$ for the potential difference across each capacitor:

$$V_1 = \frac{q}{C_1}$$

and

$$V_2 = \frac{q}{C_2}.$$

Here there is no need to label the q's because the charge on each of the two series capacitors is the same. But now the potential differences need not be the same. Since the electric fields in both capacitors point the same way, the potential difference across the series combination is just

$$V = V_1 + V_2.$$

Inserting our expressions for the individual potential difference gives

$$V = \frac{q}{C_1} + \frac{q}{C_2} = q\left(\frac{1}{C_1} + \frac{1}{C_2}\right).$$

Dividing by q gives an expression for V/q, which the defining relation $C = q/V$ shows is the reciprocal of the equivalent capacitance:

$$\frac{1}{C} = \frac{1}{C_1} + \frac{1}{C_2}. \quad \text{(series capacitors)} \qquad (24\text{--}16)$$

This result is frequently described by saying that "capacitors in series add reciprocally." The result is easily extended to three or more capacitors (see Problem 30). When there are just two capacitors, the reciprocals may be combined over a common denominator to give

$$C = \frac{C_1 C_2}{C_1 + C_2}. \qquad (24\text{--}17)$$

Equations 24–16 and 24–17 show that the combined capacitance of two series capacitors is less than the capacitance of either. You can make physical sense of this by considering parallel-plate capacitors with equal plate areas. Putting them in series effectively adds the plate separations of the two capacitors, yielding a smaller overall capacitance.

What about the voltage rating of the series combination? The full applied voltage V is the sum of the voltages across each capacitor, so that each can be rated for less than the full applied voltage. The fraction of the applied voltage that appears across each capacitor depends on the relative capacitances. This situation is explored further in Problems 32 and 33.

Example 24–8
Series Capacitors

The capacitors of Example 24–7 are now connected in series. What is the equivalent capacitance? If a 12-volt battery is connected across the combination, what are the charge and voltage on each capacitor?

Solution

Equation 24–17 gives the series capacitance:

$$C = \frac{C_1 C_2}{C_1 + C_2} = \frac{(1.0 \ \mu F)(2.0 \ \mu F)}{1.0 \ \mu F + 2.0 \ \mu F} = 0.67 \ \mu F.$$

The charge on each capacitor is the same as that on the series combination, and is given by

$$q = CV = (0.67 \ \mu F)(12 \ V) = 8.0 \ \mu C.$$

Then the voltages across the two capacitors are

$$V_1 = \frac{q}{C_1} = \frac{8.0 \ \mu C}{1.0 \ \mu F} = 8.0 \ V$$

and

$$V_2 = \frac{q}{C_2} = \frac{8.0 \ \mu C}{2.0 \ \mu F} = 4.0 \ V.$$

Note that V_2 could also have been calculated once V_1 was known by subtracting V_1 from the total applied voltage.

More complicated combinations of capacitors frequently occur. Often these may be broken down into a number of series or parallel combinations, which are handled as in the examples above (see Fig. 24–16 and Problems 24 and 34).

Capacitors and Dielectrics

The insulating material between the plates of a capacitor serves several purposes. It maintains physical separation of the plates, and minimizes charge leakage between the plates. The molecular properties of the insulation also influence the capacitance. In Chapter 22 we described the dielectric properties of insulating materials, and showed how the alignment of individual molecular dipoles with an applied electric field acts to reduce the field in the material. We described this behavior using the dielectric constant κ, which is the ratio of the externally applied field to the reduced field in the material.

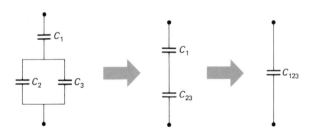

Fig. 24–16
The combination of three capacitors may be analyzed by first calculating the equivalent capacitance C_{23} of the parallel capacitors C_2 and C_3. Then the equivalent capacitance of the series combination consisting of C_{23} and C_1 may be found; this is the equivalent capacitance C_{123} of the whole combination.

Consider a parallel-plate capacitor with vacuum (or air, whose dielectric constant is very nearly 1) between its plates. If the potential difference across the capacitor is V_0 and the charge on its plates is q, then Equation 24–11 gives the capacitance:

$$C_0 = \frac{q}{V_0}.$$

Suppose the capacitor is not connected to any external circuit, so there is no way for charge to move on or off its plates. What happens if a dielectric material is inserted into the space between the plates? Alignment of molecular dipoles reduces the field between the capacitor plates from its original value E_0 to a new value E_0/κ. It takes less work to move a charge between the plates against this reduced field, so the potential difference is reduced by the same factor. However, the charge cannot change. Therefore, the capacitance—the ratio of charge to potential difference—must increase. The new capacitance is given by

$$C = \frac{q}{V} = \frac{q}{(V_0/\kappa)} = \kappa \frac{q}{V_0} = \kappa C_0. \qquad (24\text{--}18)$$

The increase in capacitance associated with a dielectric helps us build physically smaller capacitors of a given capacitance. Capacitors are among the most difficult electronic components to miniaturize, so that the ongoing revolution in microelectronics has spurred a search for suitable dielectrics with large dielectric constants, good insulating characteristics, and the ability to withstand dielectric breakdown. Tantalum oxide has become widely used in recent years because its dielectric constant of 26 permits substantial space savings. In a few unusual applications, water, with its dielectric constant of 78, is used as a dielectric. In high-energy experiments where it is necessary to store a large amount of energy for only a short time, the advantage of water's large dielectric constant outweighs the disadvantages of poor insulating quality (see color plate 5). Table 24–1 lists dielectric constants and breakdown fields of common materials.

In addition to helping build better capacitors, the relation between capacitance and dielectric constant serves as a useful probe of the structure

TABLE 24–1
Properties of Some Common Dielectrics

Dielectric Material	Dielectric Constant	Breakdown Field (kV/mm)
Air	1.0006	3
Aluminum oxide	8.4	670
Glass (Pyrex)	5.6	14
Mica	5.4	100
Neoprene	6.9	12
Paper	3.5	14
Plexiglass	3.4	40
Polyethylene	2.3	50
Polystyrene	2.6	25
Quartz	3.8	8
Tantalum oxide	26	500
Teflon	2.1	60
Water	78	—

of matter. Introducing a dielectric material between capacitor plates lowers the potential difference by the reciprocal of the dielectric constant. We can measure the change in potential difference and therefore calculate the dielectric constant. This, in turn, gives us information about the density and structure of the individual molecular dipoles. Conversely, we can use the measured dielectric constant to help identify an unknown material.

Example 24–9
Capacitors and Dielectrics

An air-insulated capacitor is charged by connecting it to a 12-V battery. The battery is then disconnected and replaced with a voltmeter that indicates the potential difference across the plates. When the space between capacitor plates is filled with an unknown plastic, the voltage drops to 3.8 V. What is the unknown material? If the plate spacing is 0.10 mm, how much voltage can the capacitor withstand with this material between its plates?

Solution

The voltage has dropped by a factor $\kappa = 10/3.8 = 2.6$. From Table 24–1, we see that a plastic with this dielectric constant is polystyrene. With a dielectric breakdown field of 25 kV/mm, the 0.1-mm-thick piece of polystyrene can withstand 2.5 kV. The rated working voltage of the capacitor would actually be lower, to allow a margin of safety.

What happens to the energy stored in a capacitor when a dielectric is inserted between its plates? Insertion of the dielectric increases the capacitance by a factor of the dielectric constant κ, but it also decreases the potential difference by the same factor. If the energy is initially $U_0 = \frac{1}{2}C_0V_0^2$, then after the dielectric is inserted it becomes

$$U = \tfrac{1}{2}(\kappa C_0)\left(\frac{V_0}{\kappa}\right)^2 = \frac{1}{2\kappa}C_0V_0^2 = \frac{U_0}{\kappa}. \qquad (24\text{–}19)$$

Since $\kappa > 1$, the energy has decreased. Where has it gone? As the dielectric slab moves into the capacitor, the electric field causes charge separation within the molecules, giving rise to internal dipoles aligned with the field. (In the case of polar molecules—those with intrinsic dipole moments—the molecular dipoles tend to align with the field, giving the same effect.) In the absence of any damping mechanism, the dipoles would oscillate, and the "missing" energy would reside in the energy of these oscillations. In practice, the dipoles in a solid interact strongly, and the energy is quickly dissipated as heat (see Problem 52). By writing Equation 24–19 explicitly for a parallel-plate capacitor, you can show that the energy density in the presence of a dielectric is not given by Equation 24–8, but rather by

$$u = \tfrac{1}{2}\kappa\epsilon_0 E^2 \qquad (24\text{–}20)$$

(see Problem 45). Equation 24–20 is a general relation that gives the energy density anywhere inside a dielectric. The electric field E that appears here is the field averaged over many molecules ($E = V/d$ for a parallel-plate capacitor). The factor κ that distinguishes Equation 24–20 from Equation 24–8 reflects the presence of electric fields on the microscopic scale that

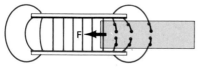

Fig. 24–17
The nonuniform fringing field outside the capacitor is stronger nearer the plates, and therefore results in a net force on the molecular dipoles in a dielectric slab.

are associated with the stretching (for nonpolar molecules) or reorientation (for polar molecules) of individual molecules. When no molecules are present, $\kappa = 1$ and the two equations are identical.

That the energy of a capacitor is lower with a dielectric inserted suggests that a force acts to propel the dielectric slab into the capacitor. What is the physical origin of this force? Figure 24–17 shows that it originates in something we have heretofore intentionally ignored—the nonuniform fringing field beyond the plates of the capacitor. This nonuniform field acts on the dipoles in the dielectric to produce a net force toward the interior of the capacitor (see Problem 43).

SUMMARY

1. The work required to assemble a charge distribution is stored as the **electrostatic potential energy** of the distribution. Electrostatic potential energy resides in the electric field. Whenever an electric field is altered, energy is added to or removed from the field.

   a. The **electric field energy density** is given by
   $$u = \tfrac{1}{2}\epsilon_0 E^2,$$
   where the SI units of u are J/m^3. In a dielectric material, the energy density is increased by a factor of the dielectric constant κ.

   b. The electrostatic potential energy of a charge distribution may be determined either by computing the work required to assemble the individual charges of the distribution or, knowing the electric field of the distribution, by integrating the energy density over the volume containing the field:
   $$U = \int u\, dv = \tfrac{1}{2}\epsilon_0 \int E^2 dv.$$

2. A **capacitor** is an arrangement of two conductors separated by an insulator. Transferring charge from one conductor to the other results in an electric field in the region between the conductors. The capacitor is an energy storage device, storing energy in the electric field between its conductors.

   a. The **capacitance** of a capacitor is the ratio of the magnitude of charge on either conductor to the potential difference between the conductors:
   $$C = \frac{q}{V}.$$

   The capacitance of a parallel-plate capacitor is given by
   $$C = \frac{\epsilon_0 A}{d},$$
   where A is the plate area and d the spacing. The capacitance of other configurations may be determined by assuming a charge, computing the associated potential difference, and taking the ratio $C = q/V$.

   b. The energy stored in a capacitor depends on the capacitance and on the square of the potential difference:
   $$U = \tfrac{1}{2}CV^2.$$

   c. Capacitors may be connected in **series** or **parallel.** The capacitances of parallel capacitors add:
   $$C = C_1 + C_2 + C_3 + \cdots . \quad \text{(parallel capacitors)}$$

   The capacitances of series capacitors add reciprocally:
   $$\frac{1}{C} = \frac{1}{C_1} + \frac{1}{C_2} + \frac{1}{C_3} + \cdots . \quad \text{(series capacitors)}$$

   d. The **working voltage** of a capacitor is the maximum potential difference that can be applied across the capacitor without risk of dielectric breakdown in the insulating material.

   e. The dielectric constant of the insulating material used in a capacitor affects the capacitance, increasing it by a factor of the dielectric constant.

QUESTIONS

1. Two positive point charges are initially infinitely far apart. Is it possible, using only a finite amount of work, to move them until they are located a small distance d apart?

2. Two positive point charges are initially a small distance d apart. Is it possible, using a finite amount of work, to move them together until there is no separation between them?

3. The work required to assemble a certain charge distribution is exactly zero. Does this mean the assemblage of charges is in static equilibrium under the influence of the electric force alone? Explain.

4. How does the energy density a certain distance from a negative point charge compare with the energy density the same distance from a positive point charge of equal magnitude?

5. A dipole consists of two equal but opposite charges. Is the total energy stored in the field of the dipole zero? Why or why not?

6. Charge is spread over the surface of a balloon. The balloon is then allowed to expand. What happens to the energy of the electric field? If it is reduced, where does it go? If it is increased, where does the extra energy come from?

7. Why doesn't the superposition principle hold for electric field energy densities? That is, if you double the field strength at some point, why don't you simply double the energy density as well?

8. A student argues that the total energy associated with the electric field of a charged sphere must be infinite because its field extends throughout an infinite volume. Criticize this argument.

9. A capacitor is said to carry a charge Q. What is the net charge on the entire capacitor?

10. Does the capacitance of a capacitor describe the maximum amount of charge it can hold, in the same way that the capacity of a bucket describes the maximum amount of water it can hold? Explain and compare the meanings of capacitance and capacity.

11. A cylinder for storing compressed gas has a fixed volume, yet knowing this volume does not tell us how much gas the cylinder can hold. Why not? How is the cylinder like a capacitor? Form analogies between quantities used in describing a capacitor and the amount of gas in the cylinder, the cylinder pressure, and the maximum pressure the cylinder can withstand.

12. A capacitor of capacitance C is charged to a potential difference V and carries charge Q. Why isn't the stored energy given simply by CV^2? After all, the work required to move a charge Q against a potential difference V is QV, and $C = Q/V$ so that $Q = CV$.

13. Is a force needed to hold the plates of a charged capacitor in place? Explain.

14. Why do we say that capacitance depends only on the physical configuration of conductors making up a capacitor, not on the charge or potential difference, and yet we define capacitance as $C = Q/V$?

15. Why can't useful capacitors of arbitrarily large capacitance be made by simply reducing the spacing between parallel plates?

16. A solid conducting slab is inserted between the plates of a capacitor, as shown in Fig. 24–18. Does the capacitance increase, decrease, or remain constant?

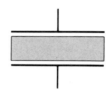

Fig. 24–18   Question 16

17. Why is a capacitor needed to store energy from the battery in a camera flash? After all, the battery supplies the energy to charge the capacitor, so it is surely capable of supplying the energy needed to light the flashtube.

18. Two capacitors are connected in series. Is the equivalent capacitance more or less than that of either one?

19. Two capacitors are connected in series. Could the maximum working voltage of the combination be as great as the sum of the working voltages of both capacitors? Could it be lower than this sum? Could it be lower than the working voltage of either capacitor?

20. Explain why the potential differences across parallel capacitors must be the same.

21. Two capacitors are storing equal amounts of energy, yet one has twice the capacitance of the other. How do their voltages compare?

22. Explain why inserting a dielectric between capacitor plates increases the capacitance.

23. An air-insulated parallel-plate capacitor is connected to a battery that imposes a potential difference V across the capacitor. If a dielectric slab is inserted between the capacitor plates, what happens to (a) the potential difference; (b) the capacitor charge; and (c) the capacitance?

24. An ideal dielectric would be a material whose internal dipoles did not dissipate energy as heat as they respond to an electric field. If a slab of ideal dielectric is placed part way between the plates of a capacitor, what will be its subsequent motion? Assume that the capacitor is charged but is not connected to a battery or other external circuitry.

25. A capacitor is charged and left connected to the charging battery. If you insert a dielectric slab between the capacitor plates, do you do work on the slab, or does it do work on you? Explain.

PROBLEMS

Section 24.1   *The Energy of a Charge Distribution*

1. Three point charges each of charge $+q$ are moved from infinity to the corners of an equilateral triangle of side a. How much work is required?

2. Repeat the preceding problem for the case of two charges $+q$ and one charge $-q$.

3. Suppose that two of the charges of Problem 1 are held in place by some agent, while the third is allowed to move freely under the influence of the electric force. If this third charge has mass m, what will be its speed far from the other two charges?

4. Repeat Example 24–1 for the case when the negative charge is $-q$ instead of $-q/2$.

5. To a very crude approximation, a water molecule can be considered to consist of a negatively charged oxygen atom (net charge $-2e$) and two "bare" protons (charge $+e$ on each), arranged as in Fig. 24–19. Calculate the electrostatic potential energy of this configuration and therefore the energy that must have been released in forming the molecule from its constituent atoms, initially widely separated. *Note:* Your answer is an overestimate, because electrons are actually "shared" between the oxygen and the two hydrogens, spending more time in the vicinity of the oxygen.

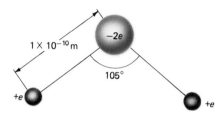

Fig. 24–19 A water molecule (Problem 5).

Section 24.2   *Two Isolated Conductors*

6. Two square conducting plates each measure 5.0 cm on a side. They are parallel and are separated by a distance of 1.2 mm. (*a*) If the plates are initially uncharged, how much work is required to transfer 6.2 μC from one plate to the other? (*b*) After the final state in part (*a*) is reached, how much work is required to transfer another 6.2 μC between the plates?

7. A conducting sphere of radius a is inside a concentric spherical conducting shell of radius b. Initially, there is zero net charge on either conductor. How much work does it take to transfer charge from one conductor to the other until they have opposite charges of equal magnitude Q?

8. Consider two conducting spheres each of radius a separated by a far greater distance, so that the influence of each one's electric field on the other is negligible. (*a*) If one sphere carries charge q and the other $-q$, show that the potential difference between the spheres is $q/2\pi\epsilon_0 a$. (*b*) Using this result, write an expression for the work dW involved in moving an infinitesimal positive charge dq from the negative to the positive sphere. (*c*) Assuming the spheres start out uncharged, and that charge is transferred until the magnitude of the charge on each is Q, integrate the result of part (*b*) to show that the total work involved in charging this system of two conductors is $Q^2/4\pi\epsilon_0 a$.

9. Equation 24–6 gives the energy stored between a pair of parallel conducting plates. By differentiating this equation with respect to the plate separation, show that there is an attractive force

$$F = \frac{Q^2}{2\epsilon_0 A}$$

acting between the plates. Compare with the answer you would get by multiplying the field due to one

plate by the charge of the other. Which answer is right? Why?

Section 24.3   *Energy and the Electric Field*

10. The energy density in a uniform electric field is 3.0 J/m^3. What is the field strength?

11. A 12-V car battery stores about 1 kWh of energy. If all this energy were used to create a uniform electric field of 2000 V/m, how much volume would the field occupy?

12. Air undergoes dielectric breakdown at a field strength of 3×10^6 V/m. Could you store energy in a uniform electric field in air with the same energy density as that of liquid gasoline? (See Appendix D.)

13. The electric field strength as a function of position x in a certain region is given by

$$E = E_0\frac{x}{x_0},$$

where $E_0 = 2.4\times10^4$ V/m and $x_0 = 6.0$ m. The field strength does not depend on position in the y- or z-direction. How much energy is stored in a cube one meter on a side, located between $x=0$ and $x=1.0$ m?

14. A sphere of radius a carries a charge Q distributed uniformly over its surface. Show that the total energy stored in its electric field is $U = Q^2/8\pi\epsilon_0 a$.

15. A sphere of radius a carries a charge Q distributed uniformly throughout its volume. How much more energy is stored in its electric field than in that of the preceding problem, where the charge was all on the sphere's surface? *Hint:* See Example 21–5.

16. A nucleus of uranium-235 contains 92 protons and 143 neutrons, and has a diameter of 6.6×10^{-15} m. Assume that the positive charge of the protons is distributed uniformly throughout the nucleus, and use the results of the two previous problems to calculate the electrostatic energy stored in the field of this nucleus. You can ignore the neutrons in this calculation because they don't experience the electric force. What keeps the nucleus from flying apart?

17. Two 4.0-mm-diameter drops of water each carry 1.5×10^{-8} C of charge. They are initially separated by a great distance. How much work is required to bring them together so they form a single spherical drop? Assume that all charge resides on the surfaces of the drops.

Section 24.4   *Capacitors*

18. A certain capacitor stores 0.040 J of energy when charged to 100 V. (*a*) How much energy would it store when charged to 25 V? (*b*) What is its capacitance?

19. Show explicitly that the units of ϵ_0 may be written as F/m.

20. A commercially produced 0.020-μF capacitor is constructed from two long strips of aluminum foil each 4.0 cm wide and separated by a piece of paper (dielectric constant $\kappa = 3.5$) 0.050 mm thick. How long are the foil strips?

21. A parallel-plate capacitor consists of two foil plates

of area 75 cm² separated by a 0.020-mm-thick piece of polyethylene. (*a*) What is its capacitance? (*b*) What is the maximum voltage that can be applied before dielectric breakdown occurs?

22. A 0.01-μF, 300-V capacitor costs 25¢, a 0.1-μF, 100-V capacitor costs 35¢, and a 30-μF, 5-V capacitor costs 88¢. (*a*) Which can store the most charge? (*b*) Which can store the most energy? (*c*) Which is the most cost-effective energy storage device, as measured by energy stored per unit cost?

23. A 5.0-μF capacitor is charged to 50 V, and a 2.0-μF capacitor is charged to 100 V. The two are disconnected from their charging batteries, and connected in parallel with no connections to any other circuitry. What are the voltages across each after they are connected? *Hint: Charge is conserved!* (See Problem 50 for more on this situation.)

24. Four identical capacitors of capacitance C are connected as shown in Fig. 24–20. What is the equivalent capacitance of the combination when it is charged through points A and B?

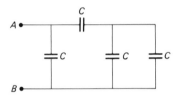

Fig. 24–20 Problem 24

25. A 10-μF, air-insulated parallel-plate capacitor is charged to 2500 V. The capacitor is disconnected from its charging battery, and its plate separation doubled. By how much does the energy stored in the capacitor change? Where does the extra energy come from?

26. Repeat the preceding problem, except now assume that the capacitor remains connected to a 2500-V battery while the separation is being increased.

27. What is the capacitance of the two widely separated spheres of Problem 8?

28. A variable "trimmer" capacitor used to make fine adjustments in circuits has a capacitance range from 5.0 pF to 15 pF. If the trimmer capacitor is placed in parallel with a capacitor of about 0.001 μF, over what percentage range can the capacitance of the combination be varied?

29. Capacitors are usually marked with a value for capacitance and a tolerance range within which the actual capacitance lies. For example, a 1-μF, ±20% capacitor may have an actual capacitance anywhere from 0.8 μF to 1.2 μF. If you connect a 0.01-μF, ±20% capacitor in series with a 0.02-μF, ±30% capacitor, in what capacitance range will the equivalent capacitance lie? Express your answer also as a capacitance and its associated tolerance percentage.

30. Repeat the derivations leading to Equations 24–15 and 24–16 for parallel and series capacitors, now assuming combinations of three capacitors.

31. You are given three capacitors: 1.0 μF, 2.0 μF, and 3.0 μF. What possible values of equivalent capacitance can you obtain using all three capacitors?

32. Two capacitors C_1 and C_2 are in series, with a voltage V across the series combination. Show that the voltages V_1 and V_2 across C_1 and C_2, respectively, are

$$V_1 = \frac{C_2}{C_1 + C_2} V \quad \text{and} \quad V_2 = \frac{C_1}{C_1 + C_2} V.$$

33. A 0.10-μF capacitor rated at 50 V is in series with a 0.20-μF capacitor rated at 200 V. What is the maximum voltage that should be applied across the series combination? *Hint: See the preceding problem.*

34. (*a*) What is the equivalent capacitance of the combination shown in Fig. 24–21? (*b*) Determine the voltages across each capacitor. (*c*) Determine the charges on each capacitor.

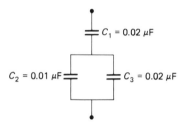

Fig. 24–21 Problem 34

35. A capacitor consists of a conducting sphere of radius a surrounded by a conducting shell of radius b. Show that its capacitance is given by

$$C = 4\pi\epsilon_0 \frac{ab}{b - a}.$$

36. Show that the result of the preceding problem reduces to Equation 24–11 for a parallel-plate capacitor when the separation $b - a$ between the spheres becomes small compared with the radius a. *Hint: Use the binomial theorem (Appendix A).*

37. We live inside a giant capacitor! Its two conductors are earth's surface and the ionosphere, a conducting layer of the atmosphere beginning at an altitude of about 60 km. (*a*) What is the capacitance of this system? *Hint: You can treat it as either a spherical capacitor or a parallel-plate capacitor. Why?* (*b*) The potential difference between earth and ionosphere is about 300 kV. How much energy is stored in the electric field between earth and ionosphere?

38. The cylindrical capacitor of Example 24–5 is charged to a voltage V. Obtain an expression for the energy density as a function of radial position in the capacitor, and integrate over the volume of the capacitor to show explicitly that the stored energy is $\frac{1}{2}CV^2$.

39. By using an approximation for the logarithm (see Appendix A) show that Equation 24–12 for the capacitance of a cylindrical capacitor reduces to Equation 24–11 for a parallel-plate capacitor when the two cylinders are closely spaced.

40. A camera flashtube requires 5.0 J of energy per flash. The duration of the flash is 1.0 ms. (a) What is the average power used by the flashtube *while it is actually flashing*? (b) If the flashtube operates at 200 V, what size capacitor is required to supply the flash energy? Assume that all the capacitor energy can be utilized, although in practice this is not possible. (c) If the flashtube is fired once every 10 s, what is the average power drain on the battery? Compare with your answer to part (a).

41. The NOVA laser fusion experiment at Lawrence Livermore Laboratory in California delivers 1.5×10^{14} W (roughly 100 times the output of all the world's electric power plants) when its lasers are on. But the short-pulse lasers are on for only about 10^{-10} s. (a) How much energy is delivered in one laser pulse? (b) The capacitor bank supplying this energy has a total capacitance of 0.26 F. Only about 0.03 per cent of the energy in the capacitor bank actually appears in the laser pulse. To what voltage must the capacitor bank be charged?

42. A solid conducting slab is inserted between the plates of an air-insulated parallel-plate capacitor. The thickness of the slab is 80% of the plate spacing, the slab area is that of the plates, and the slab is inserted parallel to the plates. (a) What happens to the capacitance? (b) Does it matter whether or not the slab is centered between the plates? (c) Does it matter if the slab touches one of the plates?

43. An air-insulated parallel-plate capacitor of capacitance C_0 is charged to a voltage V_0 and the charging battery is then disconnected. A slab of dielectric constant κ, whose thickness is essentially equal to the capacitor spacing, is then inserted halfway into the capacitor (Fig. 24-22). The capacitor has length ℓ in the direction the slab is being inserted. Determine (a) the capacitance; (b) the stored energy; and (c) the force on the slab in terms of C_0, V_0, ℓ, and κ.

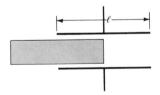

Fig. 24-22 Problems 43 and 44

44. Repeat the preceding problem, now assuming that the battery remains connected while the slab is inserted.

45. Apply Equation 24-19 explicitly to a parallel-plate capacitor containing a dielectric, and show that the energy density between the plates is given by Equation 24-20.

Supplementary Problems

46. A typical lightning flash transfers 30 C across a potential difference of 3×10^7 V. Assuming that such flashes occur once every 5 seconds in the thunderstorm of Example 24-2, roughly how long could the thunderstorm continue if its electrical energy were not replenished?

47. Use the fact that the static electric field is conservative to show that there *must* be a fringing field at the edges of a parallel-plate capacitor. *Hint:* Suppose there were no fringing field, and consider the work needed to move a charge along the two paths shown in Fig. 24-23. Remember that the plates are equipotentials. By sketching the fringing field, and using conservative field arguments, show also that the fringing field must be weaker than the field between the plates.

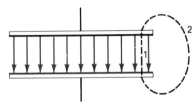

Fig. 24-23 Problem 47: What can you say about the work needed to move a charge along the paths (1) and (2) shown?

48. A long, thin nonconducting rod of length ℓ carries a charge Q distributed uniformly throughout its volume. Show that the electrostatic energy stored *within* the rod itself is independent of its diameter and is given by

$$U = \frac{Q^2}{16\pi\epsilon_0 \ell}.$$

Assume that the diameter is much less than the rod length.

49. Einstein's theory of relativity shows that a mass m at rest is equivalent to an amount of energy mc^2, where c is the speed of light (see Chapter 7, Section 7). One way to think about an electron from the viewpoint of classical physics—physics that does not take into account the quantum theory—is to consider the electron a purely electrical entity, with its mass arising entirely from the energy associated with its electric field. Assume that the electron is a sphere of radius R whose charge e is distributed uniformly over its surface. Calculate the energy stored in the electric field of the electron, and by setting that energy equal to the Einstein rest energy mc^2, show that

$$R = \frac{e^2}{4\pi\epsilon_0 mc^2} = 2.8 \times 10^{-15} \text{ m}.$$

R is called the **classical electron radius**. Although it is a useful quantity for describing the electron's behavior in some electromagnetic interactions, the "size of an electron" is a meaningless concept in quantum theory, so one should not interpret the classical electron radius too literally.

50. Compare the stored energy before and after the capac-

itors in Problem 23 are connected. Speculate on the discrepancy.

51. A TV antenna cable consists of two 0.50-mm-diameter wires spaced 12 mm apart. Estimate the capacitance per unit length of this cable, neglecting dielectric effects of the insulation.

52. An air-insulated parallel-plate capacitor has a plate area of 100 cm^2 and a spacing of 0.50 cm. The capac-

itor is charged to a certain voltage and then disconnected from the charging battery. A thin-walled, nonconducting box of the same dimensions as the capacitor is filled with water at 20.0°C. The box is released at the edge of the capacitor and moves without friction into the capacitor. When it has reached equilibrium, the water temperature is 21.5°C. What was the original voltage on the capacitor?

25

ELECTRIC CURRENT

So far our discussion of electrical phenomena has been based on the assumption of electrostatic equilibrium. We now relax that assumption, and focus on situations in which charges are moving. Such motion usually occurs only in materials containing free charges, so our discussion will emphasize electrical conductors. Occasionally we will also deal with charges moving in an otherwise empty region—for example, electrons in a TV picture tube.

25.1
ELECTRIC CURRENT

electric current

A net motion of electric charge constitutes an **electric current.** We describe current in terms of the amount of charge passing through a conductor in a given time. Accordingly, the units of current are coulombs per second (C/s). Electric current is so important that it is given its own special unit, the ampere (A). Named after the French physicist André Marie Ampère (Fig. 25–1), an ampere is one coulomb per second. When the current is steady we can write simply

$$I = \frac{\Delta q}{\Delta t},$$ (25–1)

ampere

where Δq is the net charge moving past a given point in time Δt. When current is not steady, we must consider the ratio of charge to time for very small time intervals, giving an instantaneous current I that may vary with time:

$$I(t) = \lim_{\Delta t \to 0} \frac{\Delta q}{\Delta t} = \frac{dq}{dt}. \quad \text{(electric current)}$$ (25–2)

639

Fig. 25–1
André Marie Ampère (1775–1836), whose pioneering work with electric currents is honored in the SI unit of current.

We define the direction of current as the direction in which positive charge flows. If the moving charge is negative, as with electrons in a metal, then the current is opposite to the flow of the actual charges. You can blame this confusing situation on Benjamin Franklin! It was Franklin who assigned the names "negative" and "positive" to the two kinds of electric charge. His choice was based on electrostatic experiments. Had Franklin known that the free charges in metals are electrons, he might well have reversed his nomenclature.

An electric current may consist of only one sign of charge in motion, or it may involve both positive and negative charge. In the latter case the current is determined by the *net* charge motion—that is, by the algebraic sum of the currents associated with both kinds of charges. You, for example, consist of equally large numbers of positive and negative charges. As you walk along, these charges move. Do you constitute a current? No, because your positive and negative charges both move in the same direction. The positive charges comprise a current in the direction you are moving, but the negative charges comprise an equal current in the opposite direction. The net current is zero. On the other hand, if you were to move only positive charge in a material, or only negative charge, or positive and negative charge in opposite directions, there would be a net current.

To determine the electric current—the rate at which charge passes a fixed point—we must know the density of charge carriers, their velocity, and the charge carried by each. Consider a conductor containing n free charges per unit volume, each carrying charge q and moving to the right with speed v_d, as shown in Fig. 25–2. The quantity v_d is called the **drift velocity.** In some cases—a beam of electrons in vacuum, for example—the drift velocity is the actual particle velocity. More commonly, v_d represents the time-average velocity of charge carriers, averaging out the effects of random thermal motions and collisions. If the cross-sectional area of the conductor is A, then a length ℓ of the conductor contains $nA\ell$ moving charges, or a total moving charge $\Delta Q = nA\ell q$ coulombs. With drift velocity v_d, this length ℓ of charge moves past a fixed point in time $\Delta t = \ell/v_d$, so that the current is

drift velocity

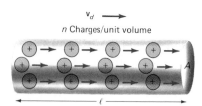

Fig. 25–2
The electric current in a conductor is the number of charge carriers per unit length times their drift speed. The number of charge carriers per unit length is the product of the number per unit volume with the cross-sectional area of the conductor.

$$I = \frac{\Delta Q}{\Delta t} = \frac{nA\ell q}{\ell/v_d} = nAqv_d. \qquad (25\text{–}3)$$

Example 25–1

The Current in a Wire

A piece of copper wire 1.0 mm in diameter carries a current of 5.0 A. The charge carriers in copper are electrons, and there are 8.4×10^{28} free electrons per cubic meter. What is the drift velocity of the electrons?

Solution

Solving Equation 25–3 for v_d gives

$$v_d = \frac{I}{nAq}.$$

Here $A = \pi r^2$ for the cylindrical wire and q is the elementary charge, so that

$$v_d = \frac{5.0 \text{ A}}{(8.4 \times 10^{28} \text{ m}^{-3})(\pi)(0.5 \times 10^{-3} \text{ m})^2(1.6 \times 10^{-19} \text{ C})} = 4.7 \times 10^{-4} \text{ m/s}.$$

(Check that the units work out.) Because the electrons are negative the direction of the drift velocity is opposite to that of the current.

Example 25–2

The Current in a Solution

Charge carriers in salt water are positive sodium ions and negative chlorine ions, each carrying plus or minus one elementary charge. A certain solution contains 6.0×10^{26} of each type of ion per cubic meter. The solution is contained in a long glass tube whose cross-sectional area is 3.0×10^{-4} m². The ions drift parallel to the long dimension of the tube at 2.6×10^{-5} m/s, with positive and negative ions moving in opposite directions. What is the current in the solution?

Solution

The current due to the positive ions is given by Equation 25–3:

$$I = nAqv_d$$
$$= (6.0 \times 10^{26} \text{ m}^{-3})(3.0 \times 10^{-4} \text{ m}^2)(1.6 \times 10^{-19} \text{ C})(2.6 \times 10^{-5} \text{ m/s})$$
$$= 0.75 \text{ C/s} = 0.75 \text{ A}.$$

The negative ions drift in the opposite direction, but they also carry the opposite charge, so that they provide a current of the same magnitude and in the same direction as the positive ions. The total current is then 1.5 A.

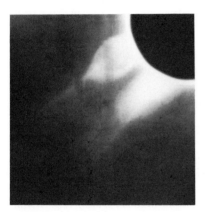

Fig. 25–3
Strong electric currents are flowing near the bright regions in this image of the solar corona, taken by the Solar Maximum Mission satellite. The artificially eclipsed sun is at upper right.

current density

There are many cases where electric currents are not so neatly confined as in a wire. Examples include currents in the oceans and solid earth, in the atmosphere, in chemical solutions, in the ionized gases of nuclear fusion experiments, and in the ionized gases that make up the stars and indeed much of the matter in the universe (Fig. 25–3). In all these situations the current is spread over a rather ill-defined region, and its magnitude and direction may vary from point to point as the density and/or velocity of the charge carriers vary. In analyzing such cases, we are concerned not only with the total current, but with the **current density,** defined as the current per unit area at a given point. Because the current may vary in both direction and magnitude, current density is a vector, given the symbol **J**. In deriving Equation 25–3 we explicitly multiplied by the area of our wire to determine the total current in the wire. Dividing by the area gives the magnitude of the current density:

$$J = \frac{I}{A} = nqv_d. \tag{25–4}$$

We can include the direction of **J** by accounting for the direction of the drift velocity, so that

$$\mathbf{J} = nq\mathbf{v}_d. \tag{25–5}$$

Equation 25–5 shows that the current density is in the same direction as the drift velocity for positive charges and in the opposite direction for negative charges. Although we obtained Equations 25–4 and 25–5 from Equation 25–3, which described a situation of uniform charge-carrier density and

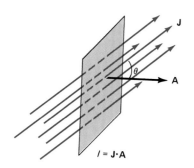

Fig. 25–4
The current through a surface of area *A* whose perpendicular makes an angle θ with the current density vector is *JA* cosθ, or **J·A**.

drift velocity, our results describe the local current density in more general situations.

If current density is uniform, as in a wire, we can find the total current by multiplying the current density by the area of the wire. If the current density is not at right angles to the area, then we consider only the component of **J** perpendicular to the area. If we define a vector **A** whose magnitude is the cross-sectional area and whose direction is perpendicular to the area, then a uniform current density **J** gives rise to a total current $I = $ **J·A** (Fig. 25–4). When the current density and/or surface orientation vary with position, we can do the same thing for many small areas, and then sum the results (Fig. 25–5). The current through a small area d**A** is **J·**d**A**, so that the total current through a surface is

$$I = \int \mathbf{J} \cdot d\mathbf{A}, \tag{25–6}$$

where the limits of the integral are chosen to cover the entire surface. Equation 25–6 and the discussion leading to it should remind you of our definition of electric flux in Chapter 21. (Compare Equation 25–6 with Equation 21–14.) Indeed, the electric current through a surface is the flux of the current density through that surface.

Example 25–3 _____

Current Density in a Wire

Twelve-gauge copper wire used in house wiring has a diameter of 0.21 cm. The maximum safe current density in such wire is 5.8×10^6 A/m². What is the maximum safe current? (Above this current there is a danger that heat developed in the wire might start a fire.) If the density of free electrons in the wire is 8.4×10^{28} m^{-3}, what is the drift velocity of the electrons at the maximum current?

Solution

Assuming that the current density is uniform over the wire, the maximum safe current is

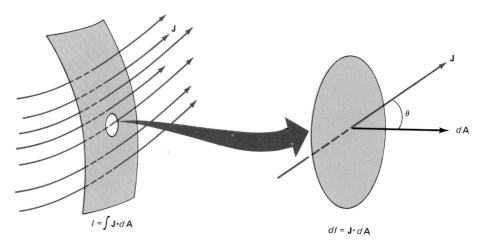

$I = \int \mathbf{J} \cdot d\mathbf{A}$ $dI = \mathbf{J} \cdot d\mathbf{A}$

Fig. 25–5
When the current density and/or surface are nonuniform, the total current may be written as an integral over the entire surface.

25

ELECTRIC CURRENT

So far our discussion of electrical phenomena has been based on the assumption of electrostatic equilibrium. We now relax that assumption, and focus on situations in which charges are moving. Such motion usually occurs only in materials containing free charges, so our discussion will emphasize electrical conductors. Occasionally we will also deal with charges moving in an otherwise empty region—for example, electrons in a TV picture tube.

25.1
ELECTRIC CURRENT

electric current

A net motion of electric charge constitutes an **electric current.** We describe current in terms of the amount of charge passing through a conductor in a given time. Accordingly, the units of current are coulombs per second (C/s). Electric current is so important that it is given its own special unit, the **ampere** (A). Named after the French physicist André Marie Ampère (Fig. 25–1), an ampere is one coulomb per second. When the current is steady we can write simply

ampere

$$I = \frac{\Delta q}{\Delta t},$$ (25–1)

where Δq is the net charge moving past a given point in time Δt. When current is not steady, we must consider the ratio of charge to time for very small time intervals, giving an instantaneous current I that may vary with time:

$$I(t) = \lim_{\Delta t \to 0} \frac{\Delta q}{\Delta t} = \frac{dq}{dt}. \quad \text{(electric current)}$$ (25–2)

Fig. 25–1
André Marie Ampère (1775–1836), whose pioneering work with electric currents is honored in the SI unit of current.

drift velocity

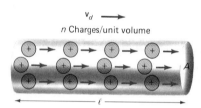

Fig. 25–2
The electric current in a conductor is the number of charge carriers per unit length times their drift speed. The number of charge carriers per unit length is the product of the number per unit volume with the cross-sectional area of the conductor.

We define the direction of current as the direction in which positive charge flows. If the moving charge is negative, as with electrons in a metal, then the current is opposite to the flow of the actual charges. You can blame this confusing situation on Benjamin Franklin! It was Franklin who assigned the names "negative" and "positive" to the two kinds of electric charge. His choice was based on electrostatic experiments. Had Franklin known that the free charges in metals are electrons, he might well have reversed his nomenclature.

An electric current may consist of only one sign of charge in motion, or it may involve both positive and negative charge. In the latter case the current is determined by the *net* charge motion—that is, by the algebraic sum of the currents associated with both kinds of charges. You, for example, consist of equally large numbers of positive and negative charges. As you walk along, these charges move. Do you constitute a current? No, because your positive and negative charges both move in the same direction. The positive charges comprise a current in the direction you are moving, but the negative charges comprise an equal current in the opposite direction. The net current is zero. On the other hand, if you were to move only positive charge in a material, or only negative charge, or positive and negative charge in opposite directions, there would be a net current.

To determine the electric current—the rate at which charge passes a fixed point—we must know the density of charge carriers, their velocity, and the charge carried by each. Consider a conductor containing n free charges per unit volume, each carrying charge q and moving to the right with speed v_d, as shown in Fig. 25–2. The quantity v_d is called the **drift velocity**. In some cases—a beam of electrons in vacuum, for example—the drift velocity is the actual particle velocity. More commonly, v_d represents the time-average velocity of charge carriers, averaging out the effects of random thermal motions and collisions. If the cross-sectional area of the conductor is A, then a length ℓ of the conductor contains $nA\ell$ moving charges, or a total moving charge $\Delta Q = nA\ell q$ coulombs. With drift velocity v_d, this length ℓ of charge moves past a fixed point in time $\Delta t = \ell/v_d$, so that the current is

$$I = \frac{\Delta Q}{\Delta t} = \frac{nA\ell q}{\ell/v_d} = nAqv_d. \qquad (25–3)$$

Example 25–1
The Current in a Wire

A piece of copper wire 1.0 mm in diameter carries a current of 5.0 A. The charge carriers in copper are electrons, and there are 8.4×10^{28} free electrons per cubic meter. What is the drift velocity of the electrons?

Solution

Solving Equation 25–3 for v_d gives

$$v_d = \frac{I}{nAq}.$$

Here $A = \pi r^2$ for the cylindrical wire and q is the elementary charge, so that

$$v_d = \frac{5.0 \text{ A}}{(8.4 \times 10^{28} \text{ m}^{-3})(\pi)(0.5 \times 10^{-3} \text{ m})^2(1.6 \times 10^{-19} \text{ C})} = 4.7 \times 10^{-4} \text{ m/s}.$$

(Check that the units work out.) Because the electrons are negative the direction of the drift velocity is opposite to that of the current.

Example 25-2

The Current in a Solution

Charge carriers in salt water are positive sodium ions and negative chlorine ions, each carrying plus or minus one elementary charge. A certain solution contains 6.0×10^{26} of each type of ion per cubic meter. The solution is contained in a long glass tube whose cross-sectional area is 3.0×10^{-4} m^2. The ions drift parallel to the long dimension of the tube at 2.6×10^{-5} m/s, with positive and negative ions moving in opposite directions. What is the current in the solution?

Solution

The current due to the positive ions is given by Equation 25-3:

$$I = nAqv_d$$
$$= (6.0 \times 10^{26} \text{ m}^{-3})(3.0 \times 10^{-4} \text{ m}^2)(1.6 \times 10^{-19} \text{ C})(2.6 \times 10^{-5} \text{ m/s})$$
$$= 0.75 \text{ C/s} = 0.75 \text{ A}.$$

The negative ions drift in the opposite direction, but they also carry the opposite charge, so that they provide a current of the same magnitude and in the same direction as the positive ions. The total current is then 1.5 A.

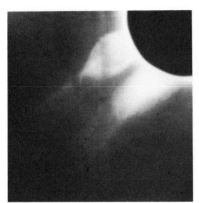

Fig. 25-3
Strong electric currents are flowing near the bright regions in this image of the solar corona, taken by the Solar Maximum Mission satellite. The artificially eclipsed sun is at upper right.

current density

There are many cases where electric currents are not so neatly confined as in a wire. Examples include currents in the oceans and solid earth, in the atmosphere, in chemical solutions, in the ionized gases of nuclear fusion experiments, and in the ionized gases that make up the stars and indeed much of the matter in the universe (Fig. 25-3). In all these situations the current is spread over a rather ill-defined region, and its magnitude and direction may vary from point to point as the density and/or velocity of the charge carriers vary. In analyzing such cases, we are concerned not only with the total current, but with the **current density,** defined as the current per unit area at a given point. Because the current may vary in both direction and magnitude, current density is a vector, given the symbol **J.** In deriving Equation 25-3 we explicitly multiplied by the area of our wire to determine the total current in the wire. Dividing by the area gives the magnitude of the current density:

$$J = \frac{I}{A} = nqv_d. \tag{25-4}$$

We can include the direction of **J** by accounting for the direction of the drift velocity, so that

$$\mathbf{J} = nq\mathbf{v}_d. \tag{25-5}$$

Equation 25-5 shows that the current density is in the same direction as the drift velocity for positive charges and in the opposite direction for negative charges. Although we obtained Equations 25-4 and 25-5 from Equation 25-3, which described a situation of uniform charge-carrier density and drift velocity, our results describe the local current density in more general situations.

If current density is uniform, as in a wire, we can find the total current by multiplying the current density by the area of the wire. If the current density is not at right angles to the area, then we consider only the component of **J** perpendicular to the area. If we define a vector **A** whose magnitude is the cross-sectional area and whose direction is perpendicular to the area, then a uniform current density **J** gives rise to a total current $I = $ **J·A** (Fig. 25–4). When the current density and/or surface orientation vary with position, we can do the same thing for many small areas, and then sum the results (Fig. 25–5). The current through a small area d**A** is **J·**d**A,** so that the total current through a surface is

$$I = \int \mathbf{J} \cdot d\mathbf{A}, \qquad (25\text{–}6)$$

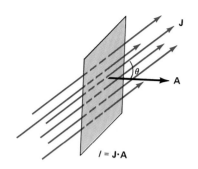

Fig. 25–4
The current through a surface of area A whose perpendicular makes an angle θ with the current density vector is $JA\cos\theta$, or **J·A.**

where the limits of the integral are chosen to cover the entire surface. Equation 25–6 and the discussion leading to it should remind you of our definition of electric flux in Chapter 21. (Compare Equation 25–6 with Equation 21–14.) Indeed, the electric current through a surface is the flux of the current density through that surface.

Example 25–3

Current Density in a Wire

Twelve-gauge copper wire used in house wiring has a diameter of 0.21 cm. The maximum safe current density in such wire is 5.8×10^6 A/m^2. What is the maximum safe current? (Above this current there is a danger that heat developed in the wire might start a fire.) If the density of free electrons in the wire is 8.4×10^{28} m^{-3}, what is the drift velocity of the electrons at the maximum current?

Solution

Assuming that the current density is uniform over the wire, the maximum safe current is

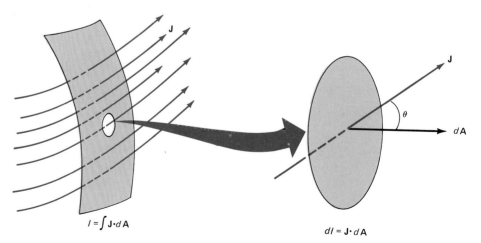

Fig. 25–5
When the current density and/or surface are nonuniform, the total current may be written as an integral over the entire surface.

$$I_{max} = J_{max}A = J_{max}\pi(\tfrac{1}{2}d)^2$$

$$= (5.8\times10^6 \text{ A/m}^2)(\pi)(0.105\times10^{-2} \text{ m})^2 = 20 \text{ A}.$$

Solving Equation 25–4 for v_d gives

$$V_d = \frac{J}{nq} = \frac{5.8\times10^6 \text{ A/m}^2}{(8.4\times10^{28} \text{ m}^{-3})(1.6\times10^{-19} \text{ C})} = 0.43 \text{ mm/s}.$$

This remarkably small value is typical of drift velocities in conductors.

How can the drift velocity be so small? When you turn on a light switch, the light comes on almost immediately, not several thousand seconds later as this answer might suggest. Here it is important to distinguish between the speed at which the electrons move and the speed at which electrical signals move. Once electrons start moving at one end of a wire, their electric fields almost immediately influence electrons farther down the wire, so that the current starts up everywhere almost instantaneously. The individual electrons at the switch do not have to travel all the way to the bulb before it lights. The same thing happens when you turn on a garden hose. If the hose is full of water (and a wire is always full of free electrons), then water comes out the far end almost as soon as you turn on the faucet, even though the water at the faucet has not had time to travel to the far end of the hose. In an electric circuit, the electrical impulse travels at nearly the speed of light, many orders of magnitude faster than the drift speed of the charge carriers.

Example 25–4

A Nonuniform Current Density

Evaporation of surface water results in an increase in salt content of the ocean near the surface. As a result, the water is a better conductor nearer its surface, and can more easily sustain electric currents. In a certain region of the ocean, an electric current is flowing horizontally. Because of the nonuniform salt content, the current density varies with depth, being given by

$$\mathbf{J} = J_0 e^{-y/y_0}\,\hat{\mathbf{k}},$$

where $J_0 = 5.0$ mA/m^2, $y_0 = 185$ m, and y is the depth measured from the surface. The geometry of the situation is shown in Fig. 25–6. What is the total current flowing through a square 1.0 km on a side, oriented parallel to the x-y plane and with its top at the water surface?

Solution

Since $\mathbf{J}$ varies with position, we must integrate to get the total current. J does not vary in the x-direction, so we can divide the area into narrow strips of width $\ell = 1.0$ km and height dy, as shown in Fig. 25–7. The area of each element is then $da = \ell\,dy$. Since the current density is at right angles to our area elements (that is, in the same direction as the vectors $d\mathbf{A}$ perpendicular to the area elements), the dot product $\mathbf{J}\cdot d\mathbf{A}$ in Equation 25–6 becomes simply $J\,dA$, and we have

$$I = \int \mathbf{J}\cdot d\mathbf{A} = \int J\,dA = \int_0^{1.0 \text{ km}} J_0 e^{-y/y_0}\,\ell\,dy$$

$$= J_0\ell \int_0^{1.0 \text{ km}} e^{-y/y_0}dy = J_0\ell \left.\frac{e^{-y/y_0}}{-1/y_0}\right|_0^{1.0 \text{ km}} = J_0\ell y_0 \left. e^{-y/y_0}\right|_{1.0 \text{ km}}^{0}$$

$$= (5.0 \text{ mA/m}^2)(1000 \text{ m})(185 \text{ m})(e^{-0} - e^{(-1000 \text{ m})/(185 \text{ m})}) = 920 \text{ A}.$$

Because of the exponential decrease in current density, this result is far below the 5000-A current arising from a uniform 5.0 mA/m^2 current density flowing through 1.0 km^2.

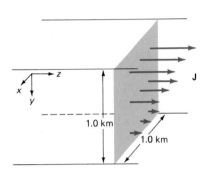

Fig. 25–6
What is the total current through the shaded area? (Example 25–4)

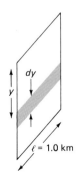

Fig. 25–7
An area element of area ℓdy.

25.2
MECHANISMS OF CONDUCTION

What causes electric current? If free charges in a conductor are initially at rest, or in purely random motion, we know of only one thing that can give them all a component of motion in the same direction—an electric field. So it is not surprising that electric current comes about when an electric field is applied to a conductor. You might think that it would suffice to apply a field briefly to get the charges moving. You could then remove the field and expect the current to continue. In most conductors, however, the charges cannot move unimpeded. They bump into things—usually ions— and quickly lose the directed motion that makes for a net current. To sustain a net current, it is necessary to accelerate the charge carriers in the same direction during the intervals between collisions. We can accomplish this by applying a steady electric field within the conductor. Having such a field does not violate our earlier conclusion that there can be no electric field in a conductor because we are no longer assuming electrostatic equilibrium; we are now explicitly allowing charges to move.

Although our picture of an electric field giving rise to a current is quite general, the details of how the current and field are related varies significantly with the type of conductor. In addition to metals, important conductors include ionic solutions, plasmas (ionized gases), semiconductors, and superconductors, each of which we will discuss later in this section. The response of each type of conductor to an electric field depends on the nature of the charge carriers in the conductor as well as on the distribution of fixed charge and neutral matter. The response may also be influenced by factors like temperature or the presence of impurities in the material.

We can characterize a given conductor by the relation between current density and applied electric field. This relation may be written

$$\mathbf{J} = \sigma\mathbf{E}, \tag{25-7}$$

conductivity

where σ is called the **conductivity** of the material. Equation 25–7 shows that conductivity is the ratio of the magnitude of the current density to the magnitude of the applied electric field. For many common conductors, this

ohmic

ratio is independent of the applied field; such materials are called **ohmic**. In ohmic materials, Equation 25–7 shows that the current density is directly proportional to the electric field. For some other materials, the con-

nonohmic

ductivity may itself depend on the electric field. In these **nonohmic** materials, the relation between current density and field is not linear.

ohm

Equation 25–7 shows that the units of conductivity are $(A/m^2)/(V/m)$, or $A/V{\cdot}m$. One V/A is given the name **ohm** (symbol Ω), after the German physicist Georg Ohm (1789–1854), whose experiments clarified the relation between current and voltage in electric circuits. The SI unit of conductivity

resistivity

can therefore be written $(\Omega{\cdot}m)^{-1}$. We frequently speak of the **resistivity** ρ of a material, defined simply as the inverse of the conductivity:

$$\rho = \frac{1}{\sigma}, \tag{25-8}$$

so that Equation 25–7 may also be written

TABLE 25–1 ▬▬▬▬▬▬▬▬▬▬▬▬▬▬▬▬▬▬▬
Resistivities at 20°C

Material	*Resistivity ($\Omega \cdot m$)*
Metallic conductors:	
Aluminum	2.8×10^{-8}
Copper	1.7×10^{-8}
Iron	10×10^{-8}
Silver	1.6×10^{-8}
Mercury	98×10^{-8}
Ionic solutions (in water at 18°C):	
1-molar copper sulfate ($CuSO_4$)	3.9×10^{-4}
1-molar hydrochloric acid (HCl)	1.7×10^{-2}
1-molar sodium chloride (NaCl)	1.4×10^{-4}
Water, pure (H_2O)	2.6×10^{5}
Sea water (typical; varies considerably)	0.22
Semiconductors (pure):	
Germanium	0.45
Silicon	640
Insulators:	
Ceramics	10^{11}–10^{14}
Glass	10^{10}–10^{14}
Polystyrene	10^{15}–10^{17}
Rubber	10^{13}–10^{16}
Wood (dry)	10^{8}–10^{14}

$$\mathbf{J} = \frac{\mathbf{E}}{\rho}. \tag{25-9}$$

The units of resistivity are $\Omega \cdot m$.

Conductivity and resistivity vary dramatically from one material to another; indeed, the range of measurable conductivities is one of the broadest ranges of any physical quantity, spanning over 24 orders of magnitude. Table 25–1 lists some typical resistivities.

Conduction in Metals

In metals at normal temperatures, conduction occurs because of the response of individual free electrons to an applied electric field. In the absence of an applied field, the free electrons in a metal are in random thermal motion at rather high speeds. When an electric field is applied, these electrons experience an acceleration in a direction opposite to that of the field, so that they acquire a velocity component antiparallel to the field. Were there nothing in the metal but electrons, this component of velocity would increase because of the steady electric force. However, the conduction electrons are embedded in a crystal lattice made up of positive ions. Electrons collide frequently with the ions, and when they do two things happen (Fig. 25–8). First, the direction of the electron's velocity is altered in a random fashion, so the component of velocity that contributes to the net current is usually destroyed. Second, the electron gives some of its energy—including energy it gained from the electric field—to the ion. This energy is gradually shared with neighboring ions. The net result is that directed motion imparted to the electron by the electric field is transformed to random motion, or internal energy, of the entire piece of metal.

Fig. 25–8
The path of an electron moving through a metal. The motion is almost completely random, but in the presence of an electric field there is a net drift antiparallel to the field. Curvature of the path segments due to electric field acceleration, and the tendency to a net drift antiparallel to the field, are both highly exaggerated in this diagram.

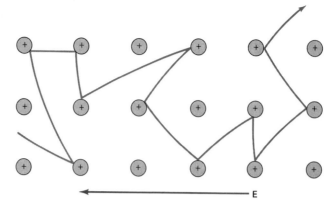

What happens to the electron after collision with an ion? It moves off in some new and quite random direction. However, it still experiences the applied electric field, so it continues to accelerate, again gaining a velocity component antiparallel to the field. Sooner or later it encounters another ion, and again gives up some energy and loses the directed motion that contributes to the net current. This situation is repeated over and over again for every free electron in the metal. Although the electrons are continually accelerated by the field, collisions prevent their field-aligned components of velocity from becoming large. The overall velocity distribution remains quite random (Fig. 25–9), but superimposed on each random velocity is a very small, time-average velocity v_d that is antiparallel to the field. No single electron moves steadily with this drift velocity, but the effect—averaged over many electron-ion collisions—is the same as if all electrons did. Small though it is, the drift velocity is entirely responsible for the electric current in the metal.

The motion of a conduction electron in a metal is roughly analogous to that of a car in heavy city traffic. When the car is moving, chances are that

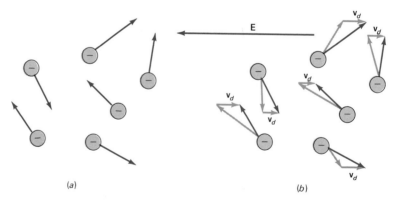

Fig. 25–9
(a) In the absence of an applied field, the velocities of the individual electrons are completely random. (b) When a field is applied, the electrons acquire a time-average velocity component antiparallel to the field. This drift velocity is what gives rise to the current. Drift velocity is highly exaggerated in this figure; in a metal at room temperature it is actually orders of magnitude smaller than the random thermal speeds.

it is accelerating after a recent stop. Soon it stops again, transferring energy gained from burning gasoline into heat energy in its brakes and ultimately in the surrounding air. Although the car is accelerating much of the time, it never gets going very fast. Its progress across the city is anything but steady, and yet could be characterized by a single average velocity obtained by dividing the total distance by the total time. The electron's situation is more complicated because of its random thermal motion and because it moves in three dimensions. However, both car and electron share the net effect of a constant average velocity arising from acceleration in the presence of repeated disruption.

Application
Noise in Electronic Equipment

Although the time-average current associated with random thermal motion of charge carriers is zero, at any given instant random fluctuations may result in more charge carriers moving in one direction than in another. The effect is a very small current, whose sign and magnitude fluctuate randomly with time. Called thermal noise, this current may disturb and even overwhelm currents of interest in sensitive electronic instruments. Good instrument design strives to minimize thermal noise, but it cannot be eliminated altogether. In very sensitive instruments such as radio telescopes, it is often necessary to cool amplifiers to liquid helium temperatures of about 4 K to bring thermal noise down to acceptable levels. Ultimately thermal noise limits our ability to detect and study very weak electrical signals.

Can we evaluate more quantitatively the relation between field and current for a particular metal? The conductivity must depend on such factors as the electron charge, the density of electrons, and the electron mass. If we double the density of electrons, then in a given electric field twice as much charge will flow, and the current density will double. So the conductivity should be directly proportional to the electron density. If we could double the mass of an electron, we would halve the acceleration in a given electric field, and therefore the drift velocity and hence current density should decrease proportionately. So the conductivity should be inversely proportional to the electron mass. If we could double the electron charge, two things would happen: first, every charge carrier would have twice as much charge. All else being equal, this factor alone would double the current density. Second, a doubled charge would result in a doubled electric force and therefore acceleration. As a result, the drift velocity and therefore current density would increase proportionately. The presence of both these effects suggests that the conductivity should depend on the square of the electron charge. Is there any other factor that influences conductivity? Yes: remember that the occurrence of collisions is what gives the electrons a time-average velocity in the presence of a steady electric field. This time-average velocity depends on how long the electric field acts on the electrons before they collide and lose their newly gained component of motion antiparallel to the field. The longer the time between collisions, the greater the time-average electron velocity—the drift velocity—and therefore the greater the current density. Therefore we expect the conductivity to depend collision time   also on the **collision time,** τ, defined as the average time an electron spends between collisions. Putting together all the factors influencing conductivity, we have

$$\sigma = \frac{ne^2\tau}{m}, \qquad (25\text{–}10)$$

where n is the number of electrons per unit volume, e the electron charge, τ the collision time, and m the electron mass.

Can we calculate the collision time that appears in Equation 25–10? We might expect it to depend on things like the spacing of ions in the metal and on the mean thermal speed of the electrons. The greater the thermal speed, the more frequent the collisions, so that the collision time τ and conductivity σ should decrease with increasing thermal speed—that is, with increasing temperature. Newtonian physics suggests that the thermal speed is proportional to the square root of the temperature, since temperature is a measure of mean thermal energy (see Section 17.1). But experimentally, we do not measure a conductivity that decreases with the square root of temperature. The detailed description of conductivity in metals is one place where classical Newtonian physics becomes inappropriate. Instead, the quantum-mechanical interaction of the electron with the metallic crystal lattice must be considered. The more difficult quantum calculation gives results in agreement with observation (Fig. 25–10).

Should the time between collisions depend also on the drift speed? In principle, yes. But in metallic conductors at normal temperatures, the drift speed is many orders of magnitude smaller than the thermal speed. A change in the drift speed—in response to a change in the electric field—has a negligible effect on the time between collisions. As a result, the conductivity is to a very good approximation independent of the electric field. For this reason Equations 25–7 and 25–9 give linear relationships between current density and electric field in metals.

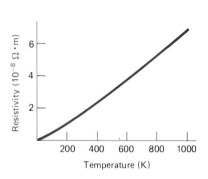

Fig. 25–10
Resistivity—the inverse of conductivity—as a function of temperature for copper. Except at very low temperatures, the resistivity increases approximately linearly with temperature. Classical physics would predict a dependence on $\sqrt{T}$ rather than the observed linear relation.

Example 25–5
Conductivity and Collision Time

Copper at room temperature contains 8.4×10^{28} free electrons per cubic meter. For these free electrons, what is the mean time between collisions with the copper ions?

Solution

Solving Equation 25–10 for the collision time gives

$$\tau = \frac{m\sigma}{ne^2}.$$

Writing $\sigma = 1/\rho$, and obtaining ρ from Table 25–1, we have

$$\tau = \frac{m}{ne^2\rho} = \frac{9.1 \times 10^{-31} \text{ kg}}{(8.4 \times 10^{28} \text{ m}^{-3})(1.6 \times 10^{-19} \text{ C})^2(1.7 \times 10^{-8}\ \Omega \cdot \text{m})}$$
$$= 2.5 \times 10^{-14} \text{ s}.$$

Problem 12 uses this result for the collision time to estimate the mean thermal speed of the electrons.

Conduction in Ionic Solutions

An ionic solution, such as salt water, contains positive and negative ions from a dissolved substance. When an electric field is applied, the different ions move in opposite directions, both contributing to the net current. The current in an ionic solution is limited by collisions between the ions and neutral atoms. In addition to heating the solution, some of the energy of these collisions may go into chemical reactions that result in energy storage. When you charge a car battery, for example, you drive a current through the battery's acid solution. Some electrical energy goes into reversing the chemical reactions by which the battery supplies energy. Conduction in ionic solutions also plays an important role in the corrosion of metals, for example, those exposed to salt solutions. And it is the presence of an ionic solution—sweat—on the skin that makes the typical human a good enough conductor to justify substantial worry about electrical safety! Table 25–1 includes resistivities of some ionic solutions. Note that these solutions are poorer conductors than metals.

Conduction in Plasmas

A plasma is an ionized gas that conducts electric current because it contains free electrons and positive ions. It takes substantial energy to ionize atoms, so that plasmas usually occur only in high-temperature environments. The few examples of plasmas on earth occur in fluorescent lamps, neon signs (Fig. 25–11), devices for nuclear fusion research, the ionosphere, flames, and lightning flashes. The earth is cool enough that plasmas are not common here. Yet most of the matter in the universe is probably in the plasma state. In particular, the stars are almost entirely plasma.

The electrical properties of plasma make it so different from ordinary gas that plasma is often called "the fourth state of matter." In many cases, especially in astrophysical plasmas, the gas is so diffuse that collisions between particles are rare. Then there is nothing to prevent large currents from developing in response to an electric field, so that these "collisionless" plasmas exhibit large conductivity. So high is the conductivity of many plasmas that as a first approximation it is often assumed infinite!

Fig. 25–11
A neon sign contains ionized neon gas that sustains an electric current.

Semiconductors

No insulator is perfect. Even the best insulators contain a few free charges that can carry minuscule currents. One reason for this is that random thermal motions occasionally dislodge an outer electron from an atom in the insulating material. If this electron becomes separated from its atom, it may wander through the material until it finds another atom that has lost an electron. It then "falls" into the "hole" left by the missing electron and settles down to its normal role as a bound electron. But while it is free, the electron may respond to an electric field and therefore carry a current.

In most insulators the number of electrons that separate from their atoms is insignificant at room temperature. At higher temperatures the number increases, though most materials decompose or melt before their

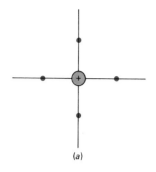

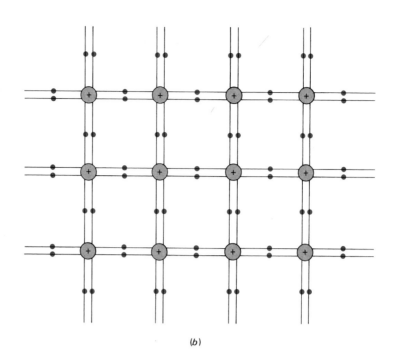

Fig. 25-12
(a) A single silicon atom, showing its four valence electrons. (b) In a silicon crystal, each atom is bonded to each neighbor by a pair of electrons.

semiconductors

conductivity becomes significant. But a few materials—notably the element silicon—exhibit significant conductivity at room temperature. The electrical properties of these **semiconductors** make possible the microelectronic technology that is an increasingly dominant influence on our civilization.

Figure 25-12 shows a highly simplified diagram of a silicon crystal. Each silicon atom contains four outermost electrons that participate in bonding the atom to its neighbors. Suppose one of these bonds is broken, and an electron freed. (The electron might be freed by random thermal motion, or by the influence of light, as in a photovoltaic cell used for solar energy conversion.) The electron leaves behind a "hole" in the crystal structure. If an electric field is applied, the electron will move in a direction opposite to the field. But another thing happens too. The bound electron nearest the "hole" is bound more weakly than its neighbors because of the disruption of the crystal structure. The electric force may nudge it over until it "falls" into the hole and takes the place of the original free electron, thus opening a hole at this adjacent position. The net result is that the hole has moved in the direction of the field. Now another electron can fall into the hole, so its motion continues. As a result there are two kinds of charge carriers in a semiconductor—electrons and holes. A hole is nothing but the absence of an electron, but it nevertheless acts like a positive charge. Figure 25-13 shows conduction by holes and electrons in a semiconductor.

As suggested in Table 25-1, a pure semiconductor has rather low conductivity, which makes it useful only in a few limited applications. The key to semiconductor technology lies in the careful control of semiconductor conductivity by adding small amounts of impurities—a process called **doping**. Doping radically alters the electrical properties of the material. Figure 25-14 shows a piece of silicon to which a very few atoms of arsenic

doping

Fig. 25–13
Conduction by electrons and holes in a semiconductor. An electron moves antiparallel to the electric field, the hole along the field. The movement of the hole is really the jumping of an adjacent bound electron into the hole, creating a new hole at the electron's original site.

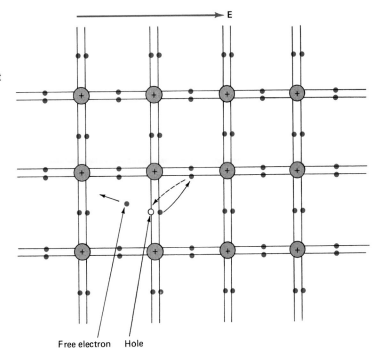

Free electron   Hole

Fig. 25–14
(a) An arsenic atom, showing its five valence electrons. (b) Incorporated into a silicon crystal, the arsenic atom does its best to fit into the crystal structure. One of its electrons cannot participate in the bonding, but moves throughout the crystal lattice as a free conduction electron. Because its free charge carriers are predominantly electrons, the arsenic-doped silicon is an *N*-type semiconductor.

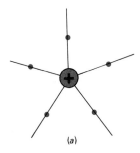

(a)

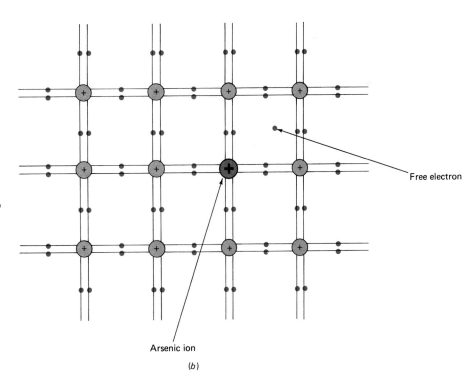

Free electron

Arsenic ion

(b)

(a)

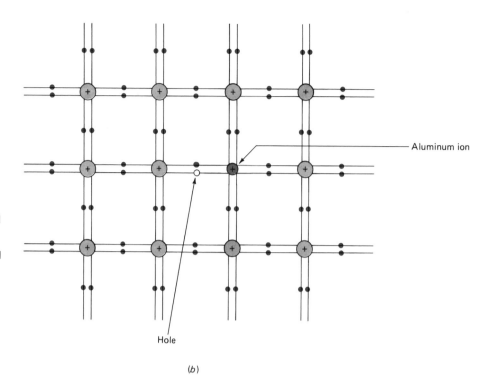

Aluminum ion

Hole

(b)

Fig. 25–15
(a) An aluminum atom, showing its three valence electrons. (b) Incorporated into a silicon crystal, the aluminum cannot complete the normal bonds. The result is a hole that can move throughout the material, giving rise to a current. The aluminum-doped silicon is a *P*-type semiconductor.

have been added. With their five outermost electrons, the arsenic atoms don't quite fit into the crystal lattice, but they do the best they can, bonding to their four silicon neighbors. One electron remains, unable to participate in the bonding. This electron wanders through the material and is free to carry electric current. Even at doping levels as low as one arsenic atom per 10^8 silicon atoms, the conductivity of the material is determined almost entirely by these excess electrons. Holes play essentially no role in the conduction. Because its charge carriers are predominantly negative (they are electrons), the material of Fig. 25–14 is called **N-type** semiconductor. However, the material does *not* carry a net negative charge, since excess electrons are balanced by positive charge in the arsenic ions.

By doping with aluminum or other atoms having only three outermost electrons to participate in bonding, it is possible to make **P-type** semiconductor in which positive charge carriers—holes—are the dominant charge carriers. Figure 25–15 shows such a material.

Our description of semiconductors is necessarily qualitative. A quantitative description would require that we consider the quantum-mechanical interactions of ions, electrons, and holes in the crystal lattice. Nevertheless the classical description is sufficient for a basic understanding of practical semiconductor devices.

N-type semiconductor

P-type semiconductor

Application

Transistors and Integrated Circuit Chips

With the exception of the nuclear bomb, no other single invention of the twentieth century has the transistor's potential for reshaping human civilization. Transistors are small, inexpensive, and powerful. What is a transistor? How does it

Fig. 25–16
(Left) ENIAC, one of the first computers, was built in 1946, occupied an entire room, and broke down frequently. (Right) A personal computer of the 1980's is more powerful than ENIAC, costs orders of magnitude less, and is extraordinarily reliable.

work? How can hundreds of thousands of them be placed on a single integrated circuit chip?

A transistor is a semiconductor device that allows a weaker electrical signal to control a stronger one. For example, the transistors in a stereo amplifier use a very small potential difference—developed by a phonograph cartridge, a radio signal, an audio or video cassette player—to control much more powerful sources of electrical energy which feed loudspeakers in order to produce sound. In a computer or calculator, transistors manipulate electrical signals that represent information coded as a sequence of relatively high or low potential differences. The transistor shares its control function with older devices like vacuum tubes. But whereas vacuum tubes are large, fragile, expensive, unreliable, hot, slow, and consume large amounts of power, transistors are small, rugged, cheap, reliable, cool, fast, and consume little power. The difference between tubes and transistors is dramatized by the contrast between a modern personal computer and the first computers of the 1950's (Fig. 25–16).

field effect transistor

One widely used type of transistor is the **field effect transistor** (FET). Figure 25–17 shows a simplified cross section of an FET. This particular type of FET is a metal-oxide-semiconductor FET or MOSFET, a type in common use in stereo systems, calculators, and microcomputers. It consists of a slab of P-type semiconductor with two separate regions of N-type material at the top. These regions are called the drain and source. Part of the structure is coated with a layer of silicon dioxide (SiO_2), an excellent insulator. A layer of metal, called the gate, is coated on top of the SiO_2. Metallic connections are made to the drain and source.

Now, a junction between N- and P-type semiconductors has the property that current flow from N to P is strongly inhibited, while current flows readily from P to N. The reason is that an electric field pointing from N to P at the junction drives charge carriers on either side away from the junction, making the junction a very poor conductor. A field pointing from P to N, on the other hand, drives charge carriers to the junction region, making it a good conductor (Fig. 25–18).

What happens if we apply a potential difference between the two N-type regions of the transistor? Note that there are two PN junctions separating the N-type regions, and that they face in opposite directions. Whichever way we apply our potential difference, one of those junctions is in the orientation of Fig. 25–18b, so that very little current flows. Suppose now that we place positive charge on the gate. Although negative charge carriers—electrons—are scarce in the P-type slab,

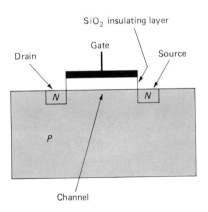

Fig. 25–17
A metal-oxide-semiconductor field effect transistor, or MOSFET.

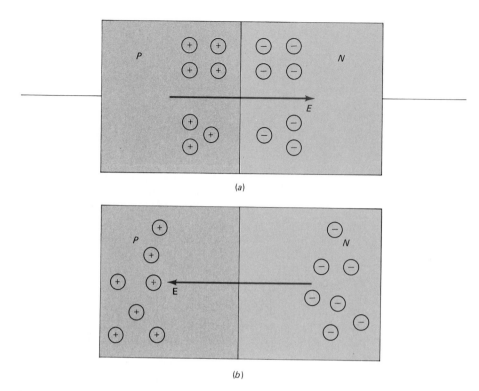

Fig. 25–18
A junction between *P*- and *N*-type semiconductors conducts in one direction but not the other. (*a*) An electric field pointing from *P* to *N* brings charge carriers into the junction region, since electrons move against the field and holes move in the field direction. This makes the junction a good conductor. (*b*) Pointing from *N* to *P*, the electric field depletes the junction of charge carriers, making it a poor conductor.

there are some of them. The presence of positive charge on the gate attracts electrons from throughout the slab to the channel between the two N-type regions. Although this channel is made of P-type material, the influx of negative charge makes it temporarily N-type, thereby permitting conduction between source and drain. The conductivity of the channel is strongly influenced by the amount of charge on the gate, so that by altering the gate charge we control the current between the N-type source and drain. In practice, the gate charge is established by applying a potential difference between the gate and the rest of the transistor. In a well-designed transistor, very small changes in gate potential can control large currents. If the gate potential is set by a weak electrical signal from, say, a microphone, then the current through the FET will be a stronger representation of the same signal. The transistor is then an amplifier, increasing the strength of electrical signals.

Construction of an FET may sound very difficult, with all its layers and different regions. Actually, though, the entire transistor is made from a single piece of silicon. By exposing different regions to impurity chemicals, the N, P, and insulating regions are easily formed, and metal is coated where needed.

The same process is used to produce entire circuits containing hundreds of thousands of transistors on a single piece of silicon, smaller than a single vacuum tube or transistor (Fig. 25–19). These **integrated circuits**—often called **chips**—lie at the heart of modern electronic devices. Entire computers—called **microprocessors**—now come on single chips (Fig. 25–20; compare with the ENIAC vacuum-

integrated circuits
chips
microprocessors

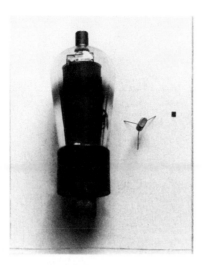

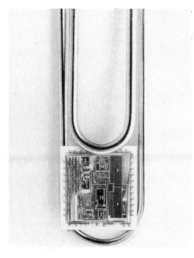

Fig. 25–19
A vacuum tube, transistor, and
integrated circuit chip. The chip
contains about 100,000 transistors.

Fig. 25–20
A microprocessor chip contains an
entire computer. It is shown here
against a paper clip. Compare with the
ENIAC computer of Fig. 25–16.

tube computer shown in Fig. 25–16). Microprocessors are used increasingly for sophisticated control of everything from automobile engines to household energy systems to video games to cash registers and automatic bank tellers, and they are at the heart of every personal computer.

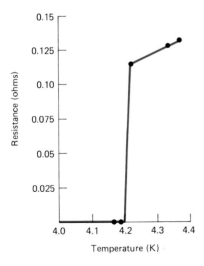

Fig. 25–21
The resistance of a mercury sample at low temperatures, showing superconductivity—a state of zero resistance—below 4.2 K.

Superconductors

In 1911, the Dutch physicist H. Kamerlingh Onnes, studying the electrical properties of mercury at very low temperatures, found a sudden drop in resistivity at a temperature of 4.2 K (Fig. 25–21). The resistivity below this temperature proved immeasurably low. Subsequent research has shown that a number of metals exhibit this phenomenon of **superconductivity** at temperatures near absolute zero. Currents in superconductors persist for years without any measurable decrease in magnitude, suggesting that the resistivity of a superconductor is truly zero.

Although superconductivity was discovered in 1911, a satisfactory explanation of this unusual phenomenon was not given until 1957. In that year John Bardeen, Leon Cooper, and J. Robert Schrieffer published a theory showing that superconductivity arises from the quantum-mechanical interaction of the conduction electrons in the metal. In a way that has no analog in classical physics, all the electrons behave in a coherent fashion, moving through the crystal lattice of the metal with no energy loss. This can occur only at very low temperatures because otherwise random thermal motion destroys the coherence of the electron behavior. For their explanation of superconductivity, Bardeen, Cooper, and Schrieffer received the 1972 Nobel Prize in physics.

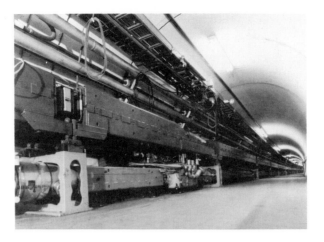

Fig. 25–22
Superconducting electromagnets (large rectangular structures at bottom) at Fermilab near Chicago steer high-energy protons around a circular path 2 km in diameter.

Superconductivity may be an esoteric quantum-mechanical phenomenon, but it has great practical significance, for it offers the possibility of loss-free transmission of electric power. At present, 5 to 8 per cent of commercially generated electric power is lost in transmission to users, so the development of superconducting transmission lines would result in significant energy savings. It would also permit the transmission of electric power at lower voltages, making underground power lines possible. Superconducting electromagnets offer substantial energy savings in large particle accelerators for high-energy physics research (Fig. 25–22), and a successful nuclear fusion reactor will undoubtedly employ superconducting magnets. Finally, a superconducting control device called the Josephson junction may eventually supplant the transistor as the basic element in computers. A superconducting computer would be orders of magnitude smaller and faster than today's computers.

The problem with superconductors, of course, is that they must be kept cold. It takes energy to run the refrigeration units that keep superconductors near absolute zero. In some cases, the energy savings are still substantial. But at present, economic and energy considerations make superconductors impractical for many potential applications. A major technological breakthrough would be the development of a room-temperature superconductor. Although this goal presently appears a long way off, and may well be impossible, intensive research efforts have led to compounds that exhibit superconductivity at temperatures as high as 20 K.

25.3
RESISTANCE AND OHM'S LAW

So far we have discussed conduction of electric current in terms of conductivity and resistivity—properties of a material that relate the current density and electric field within the material. We now consider the electrical properties of a specific piece of material. Figure 25–23 shows a cylindrical piece of conducting material—for example, a wire. Suppose there is

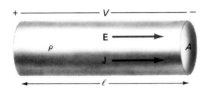

Fig. 25–23
A piece of wire made from a material with resistivity ρ. A uniform electric field **E** drives current along the wire. The electric field is associated with a potential difference V across this piece of wire.

a uniform electric field **E** within the wire, parallel to its axis. Then there must be a uniform current density given by Equation 25–9:

$$\mathbf{J} = \frac{\mathbf{E}}{\rho}.$$

Since **J** is also parallel to the wire, the total current of Equation 25–6 is simply the product of the current density and the cross-sectional area of the wire:

$$I = JA = \frac{EA}{\rho}$$

(recall that current density is current per unit area). Now, if the wire has length ℓ, then there is a potential difference, or voltage, between its two ends given by Equation 23–4:

$$V = E\ell,$$

where we write a simple product because the field is uniform over the wire. How are the voltage and current related? Taking the ratio of our expressions for voltage and current, we have

$$\frac{V}{I} = \frac{E\ell}{(EA/\rho)} = \frac{\rho\ell}{A}. \tag{25–11}$$

This equation shows that for an ohmic material, in which resistivity is independent of electric field, the ratio of voltage to current in a particular piece of material depends only on the resistivity of the material and on the dimensions of the particular piece. This ratio of voltage to current is called the electrical **resistance.** Clearly the units of resistance are volts/ampere, or ohms (Ω). From Equation 25–11 we see that the resistance of a piece of material with uniform cross-sectional area is just

resistance

$$R = \frac{\rho\ell}{A}. \tag{25–12}$$

Using this expression, the ratio of voltage to current may be written

$$\frac{V}{I} = R. \tag{25–13}$$

Although we derived this result using Equation 25–12, which applies only to a material of uniform resistivity and cross section, Equation 25–13 remains true even when resistance cannot be calculated using Equation 25–12 (see Problems 30 and 37).

As applied to ohmic materials, for which R is constant, Equation 25–13 is known as **Ohm's law.** Unlike Gauss's law, which holds universally, Ohm's law is an empirical statement about the electrical behavior of particular materials. When a material is not ohmic, so that its resistivity depends on electric field, Equation 25–13 may be considered the definition of resistance. In this case, however, resistance varies with voltage, so the ratio of voltage to current is not constant.

Ohm's law

Ohm's law is often written in the equivalent forms

$$V = IR$$

and

$$I = \frac{V}{R}.$$

open circuit

short circuit

This last form makes good sense, for it shows that a given voltage (energy per charge) can push more current through a lower resistance. Two extreme cases are worth noting. An ideal **open circuit**—a set of wires and electrical components with a nonconducting gap somewhere—has infinite resistance. No matter what the voltage across an open circuit, there can be no current. An ideal **short circuit,** in contrast, has zero resistance. In a short circuit, arbitrarily large currents can flow without requiring any potential differences to drive them. All real situations, with the exception of superconductors, lie between these two extremes.

Example 25–6 _____
Resistance and Ohm's Law

A copper wire 0.50 cm in diameter and 70 cm long is used to connect a car battery to the starter motor of the car. What is the resistance of the wire? If the starter motor draws a current of 170 A, what is the potential difference across the wire?

Solution

From Table 25–1 we see that the resistivity of copper is 1.7×10^{-8} Ω·m. Putting this value into Equation 25–12 along with the dimensions of the wire gives

$$R = \frac{\rho\ell}{A} = \frac{(1.7 \times 10^{-8}\ \Omega \cdot m)(0.70\ m)}{(\pi)(0.25 \times 10^{-2}\ m)^2} = 6.1 \times 10^{-4}\ \Omega.$$

A wire with this low resistance is required because the starter motor draws such a large current from the battery.

We now apply Ohm's law to find the voltage across the wire:

$$V = IR = (170\ A)(6.1 \times 10^{-4}\ \Omega) = 0.10\ V.$$

This voltage is small compared with the 12 volts available from a car battery, showing that the wire size is well chosen for this application. A larger voltage difference—which would occur with a narrower wire—would mean a significant reduction in power delivered to the motor.

25.4
ELECTRIC POWER

In Section 25.2, we described conduction in a metal in terms of collisions between electrons and fixed ions. In these collisions, energy gained from the electric field is transformed into heat. We can describe this transformation quantitatively if we recall that voltage is work or energy per unit charge and that current is the rate of charge flow.

Suppose, for example, a current of 2 A is flowing in a conductor, and that there is a voltage difference of 5 V across the conductor. If allowed to move freely from one end of the conductor to the other, each coulomb of charge would gain 5 joules of energy (a volt is a joule/coulomb). In fact each coulomb does gain this much energy, but gives it to the conductor as heat because of collisions between electrons and ions. So 5 joules are transferred from electrical energy to heat for each coulomb that passes through the wire. But there are 2 A, or 2 coulombs/second, passing through the wire. Therefore 5 joules/coulomb × 2 coulombs/second, or 10 joules/second, are transferred from electrical to heat energy. One joule/second is one watt, so electrical energy is being converted to heat at the rate of 10 watts. In general, we can write

$$P = IV \quad \text{(electric power)} \tag{25–14}$$

for the power in an electrical conductor. Although we developed Equation 25–14 for the case when electrical energy is converted into heat, the equation holds any time electrical energy is converted to another form. If, for example, we measure 5 V across an electric motor and 2 A of current through the motor, we can conclude that electrical energy is being converted into mechanical energy at the rate of 10 watts. (In a real motor some of the power goes into heating the motor as well.)

If we solve Ohm's law ($R = V/I$; Equation 25–13) for V and put the result into Equation 25–14, we obtain:

$$P = I^2R. \tag{25–15}$$

Solving Ohm's law instead for I gives

$$P = \frac{V^2}{R}. \tag{25–16}$$

If I is in amperes, V in volts, and R in ohms, the power P in these two equations will be in watts. Equation 25–15 is useful when we know the current but not the voltage, and Equation 25–16 in the opposite case. Although the two equations seem to exhibit very different dependences of power on R, they are really equivalent because both come from Ohm's law and Equation 25–14. (See Question 17 for more on this point.)

Example 25–7
Electric Power in a Light Bulb

The voltage in typical household wiring is about 120 V. How much current does a 100-W light bulb draw? What is the resistance of the light bulb under these conditions?

Solution

Solving Equation 25–14 for I gives

$$I = \frac{P}{V} = \frac{100 \text{ W}}{120 \text{ V}} = 0.830 \text{ A}.$$

Because we now know the current, we can determine the resistance directly from Ohm's law:

$$R = \frac{V}{I} = \frac{120 \text{ V}}{0.830 \text{ A}} = 144 \ \Omega.$$

We could have bypassed the calculation of current and obtained R directly from Equation 25–16:

$$R = \frac{V^2}{P} = \frac{(120 \text{ V})^2}{100 \text{ W}} = 144 \ \Omega.$$

Finally, had we known the current but not the voltage, we could have used Equation 25–15:

$$R = \frac{P}{I^2} = \frac{100 \text{ W}}{(0.830 \text{ A})^2} = 144 \ \Omega.$$

The three methods are equivalent. Use of Equations 25–15 and 25–16 merely amounts to solving symbolically for I or V when the other is known.

Because a light bulb filament undergoes a huge temperature change when turned on, its resistance is not independent of voltage and current. Our value of 144 Ω holds when the light is on. When it is off, and cool, its resistance is much lower.

SUMMARY

1. **Electric current** is a net flow of electric charge. Current is defined as the charge per unit time passing a given point, or

$$I = \frac{dq}{dt}.$$

a. If a material contains n free charges per unit volume, each carrying charge q, and moving with average velocity v_d (called the **drift velocity**), then the current through an area A at right angles to the drift velocity is

$$I = nAqv_d.$$

b. **Current density** (symbol **J**) is a vector specifying the current per unit area and the direction in which the current is flowing:

$$\mathbf{J} = nq\mathbf{v_d}.$$

The total current through a surface is the flux of the current density over that surface:

$$I = \int \mathbf{J} \cdot d\mathbf{A}.$$

2. **Conductivity** (symbol σ) is a property of a given material that describes the ratio of electric field to current density in the material:

$$\mathbf{J} = \sigma \mathbf{E}.$$

Resistivity (symbol ρ) is the inverse of conductivity.

3. Mechanisms of conduction vary with different materials and states of matter, but all are characterized by the presence of free electric charge in the material.
   a. **Ohmic conductors** are those in which conductivity and resistivity are independent of electric field, so that there is a linear relation between current density and field. Metals are ohmic conductors in which the combined effects of acceleration of free electrons in an applied electric field and collisions between electrons and metal ions result in a time-average drift velocity, and therefore current density, proportional to the applied field.
   b. **Ionic solutions** are conductors because dissolved substances separate into positive and negative ions that can move throughout the solution.
   c. **Plasmas** are ionized gases, in which atoms are stripped of one or more electrons. The presence of free electrons and the infrequency of collisions, especially in diffuse plasmas, make most plasmas excellent conductors. Plasmas are rare on earth, but comprise much of the matter in the universe.
   d. **Semiconductors** are substances that conduct only poorly in their pure state, but whose electrical properties may be altered radically by the addition of very small amounts of impurities. P-type semiconductors, containing positive charge carriers (holes) and N-type, containing negative charge carriers (electrons) are both available. Semiconductor technology is the basis of modern microelectronics.

e. **Superconductors** are substances that have zero resistivity. Currently known materials exhibit superconductivity only at extremely low temperatures (less than 20 K).

4. **Resistance** is a property of a particular piece of material describing the ratio of voltage across the material to current through the material:

$$R = \frac{V}{I}.$$

The resistance of a piece of material depends on its resistivity and physical dimensions. For an object of length ℓ, uniform cross-sectional area A, and resistivity ρ, the resistance is

$$R = \frac{\rho\ell}{A}.$$

5. **Ohm's law** states that the resistance of ohmic materials is independent of voltage, so that the ratio of voltage to current in these materials is constant:

$$R = \frac{V}{I}.$$

Ohm's law is an empirical law, which only holds for certain materials including metals and other commonly used conductors.

6. **Electrical energy conversion** occurs in an electrical device when there is a potential difference across the device and a current through it. The rate at which electrical energy is converted to another form is given by the product of voltage and current:

$$P = IV.$$

In an object whose only electrical property is resistance, the energy is converted entirely into heat, and the power can be written in the two equivalent forms

$$P = I^2R$$

and

$$P = \frac{V^2}{R}.$$

QUESTIONS

1. If you physically move a neutral conductor, does this constitute an electric current?
2. In the previous chapters, we stressed that there could be no electric field inside a conductor in electrostatic equilibrium. Why in this chapter do we so readily discuss electric fields inside conductors?
3. A wire carries a steady current. If the wire diameter decreases in the direction of the current, what happens to the current density?
4. When you talk on the telephone, your voice is heard essentially immediately by the party at the other end. Yet the drift velocity of electrons in the telephone wires is on the order of mm/s. Explain the apparent contradiction.
5. What is the difference between current and current density? Why might an electrical engineer be more concerned with the former and a physicist with the latter?
6. How can a steady electric field give rise to a constant drift velocity? This sounds like a throwback to Aristotelian ideas, in which a force was thought necessary to keep an object in motion. Is there a contradiction with Newton's law?
7. Why does the conductivity of a metal depend on the square of the electron charge?
8. When caught in the open in a lightning storm, it is advisable to crouch low with the feet close together. If lightning strikes nearby, why is this a safer position than lying flat on the ground?
9. A simplifying approximation often used in plasma physics is that there can be no electric field in a plasma. How does this approximation follow from the fact that the conductivity of a plasma is very large?
10. What are P-type and N-type semiconductors? Does either type carry a net electric charge?
11. In our discussion of semiconductors, we noted that conduction in a doped semiconductor occurs almost entirely through charge carriers of one sign contributed by the impurity atoms used in doping. Yet there remain a few "minority" charge carriers contributed by the semiconductor itself. Explain why these are vital to the operation of a field-effect transistor.
12. Does an electric stove burner draw more current when it is first turned on or when it is fully hot? Assume that the voltage applied to the burner remains constant.
13. Macroscopic electric fields cannot exist inside a superconductor. Why not?
14. In developing Ohm's law for metallic conductors we assumed that the drift speed was very small compared with the thermal speed of the electrons. Why was this assumption necessary to our conclusion that the resistance of a metal is independent of voltage?
15. Two devices having different resistances are designed to operate at 120 V. Which consumes more power, the lower or the higher resistance?
16. A power line with small but nonzero resistance is used to transmit 350 MW of power from a nuclear

power plant to a city. Is this power transmission best accomplished using a high voltage and low current, or a low voltage and high current? What factors might influence an engineer's choice of voltage and current?

17. Equation 25–15 suggests that no power can be dissipated in a superconductor ($R = 0$). Yet Equation 25–16 suggests that the power should be infinite. Resolve this dilemma. *Hint:* Think about Ohm's law and how the two equations relate to it. Also think about the electric field inside the superconductor.

18. An electric motor made with superconducting wires and perfectly frictionless bearings is rotating at con-

stant angular speed. The motor is doing no mechanical work. There is a steady voltage across the motor. Give an argument suggesting that there can be no current through the motor. What happens if a mechanical load is attatched to the motor shaft, so that it begins doing work—for example, lifting a weight?

19. The resistivity of a pure semiconductor decreases approximately exponentially with increasing temperature. Speculate on what might happen if a fixed potential difference were applied across a piece of such material.

PROBLEMS

Section 25.1   *Electric Current*

1. The electron beam that "paints" the picture on a particular TV screen contains 5.0×10^6 electrons per cm of its length. If the electrons move down the tube at 6.0×10^7 m/s, how much current does the electron beam carry? What is the direction of this current?

2. The filament of the light bulb in Example 25–7 has a diameter of 0.050 mm. What is the current density in the filament? Compare with the current density in a 12-gauge wire (diameter 0.21 cm) supplying current to the light bulb.

3. A car battery is rated at 80 ampere-hours, meaning that the battery can supply 80 amperes of current for one hour before becoming discharged. If the battery is discharged by connecting it to a light bulb, how much total charge moves through the bulb?

4. A metal bar has a rectangular cross section 5.0 cm by 10 cm, as shown in Fig. 25–24. The bar has a non-uniform conductivity, ranging from zero at the bottom to a maximum at the top. As a result, the current density increases linearly from zero at the bottom to 0.10 A/cm² at the top. What is the total current in the bar?

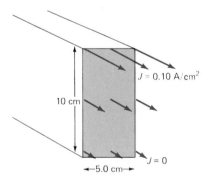

Fig. 25–24   Problem 4

5. At the Stanford Linear Accelerator, electrons are accelerated to very nearly the speed of light. These high-energy electrons are produced in pulses containing 5×10^{11} electrons each, and lasting 1.6 μs.

(a) Assuming an electron speed approximately that of light, what is the physical length of each pulse? (b) What is the peak current (i.e., the rate of charge flow past a point when a pulse of electrons is going by)? (c) If the accelerator produces 180 pulses/second, what is the average current?

6. Helium gas is confined in a rigid container at a density of 5.0×10^{18} atoms/m³. While still confined at this density, all the helium is completely ionized (stripped of its free electrons). Under the influence of an electric field, the electrons in the plasma drift in one direction at 40 m/s, while the protons drift in the opposite direction at 6.5 m/s. (a) What is the current density in the plasma? (b) What fraction of the current is carried by electrons?

7. A piece of copper wire joins a piece of aluminum wire whose diameter is twice that of the copper wire. The same current flows through both wires. The density of conduction electrons in copper is 8.4×10^{28} m⁻³; in aluminum it is 18×10^{28} m⁻³. Compare the drift velocities in the two conductors.

8. In Fig. 25–25, a 100-mA current flows through a copper wire 0.10 mm in diameter, a 1.0-cm-diameter glass tube containing a salt solution, and a vacuum tube where the current is carried by an electron beam 1.0 mm in diameter (the complete circuit is not shown). The density of conduction electrons in copper is 8.4×10^{28} m⁻³. In the solution, the current (assumed uniform over the tube cross section) is carried equally by positive and negative ions, each with two elementary charges; the density of each ion type is 6.1×10^{23} m⁻³. The density of electrons in the beam is 2.2×10^{16} m⁻³. What is the drift velocity in each conducting medium?

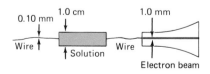

Fig. 25–25   Problem 8

Section 25.2   *Mechanisms of Conduction*

9. There is a potential difference of 2.5 V between opposite ends of a 6.0-m-long iron wire. (a) Assuming a uniform electric field in the wire, what is the current density? (b) If the wire diameter is 1.0 mm, what is the total current?

10. The maximum safe current in a 12-gauge (2.1-mm-diameter) copper wire is 20 A. What are the maximum safe current density and electric field in this wire?

11. The density of conduction electrons in aluminum is 18×10^{28} m^{-3}. What is the mean time between electron-ion collisions in aluminum?

12. Metallic copper (atomic weight 64, density 8.9 g/cm^3) forms a crystal structure with copper atoms located at the corners of cubes. (a) Using the density and atomic weight given, calculate the distance between adjacent copper atoms in this structure. (b) Use your result, and the collision time of Example 25–5, to estimate the mean thermal speed of electrons in copper. Compare with a typical drift speed, e.g., that of Example 25–3.

13. A pure silicon crystal contains 4.9×10^{28} silicon atoms per cubic meter. At room temperature, the density of electron-hole pairs is 1×10^{16} m^{-3}. In what concentration (Al atoms per Si atom) must aluminum be added to give a conductivity 1000 times that of pure silicon? Assume that the conductivity is directly proportional to the density of free charge carriers.

Section 25.3   *Resistance and Ohm's Law*

14. The "third rail" of a subway track is made from an iron bar measuring 10×15 cm. This rail carries current that runs the motors on the subway train. What is the resistance of a 5.0-km piece of rail?

15. How must the diameters of copper and aluminum wire be related if they are to have the same resistance per unit length?

16. 18-gauge copper wire (diameter 1.0 mm) is often used in extension cords. (a) What is the resistance per unit length of this wire? (b) An electric saw that draws 7.0 A is operated at the end of an 8.0-m-long extension cord made of 18-gauge copper wire. What is the potential difference between the point where the cord leaves the wall outlet and where it reaches the saw?

17. A high-voltage transmission line is to have a resistance per unit length of 50 mΩ/km. What wire diameter is required if the line is made of (a) copper or (b) aluminum? (c) If the costs of copper and aluminum wire are \$1.53/kg and \$1.34/kg, which material is more economical? Calculate the cost per unit length. The densities of copper and aluminum are 8.9 g/cm^3 and 2.7 g/cm^3, respectively.

18. The presence of a few ions in air makes the atmosphere a conductor, although a rather poor one. If the total resistance between the ionosphere and earth is 200 Ω, how much current flows as a result of the 300-kV potential difference between earth and ionosphere?

19. A solid, rectangular iron bar measures 0.50 cm by 1.0 cm by 20 cm. What is the resistance measured between each of the three pairs of opposing faces? Assume that the faces in question are equipotentials when the resistance between them is measured.

20. Corrosion at the junction of metallic conductors often gives rise to high resistance in an otherwise low-resistance circuit. This phenomenon is a frequent cause of hard starting in cars. In an effort to find out why a car won't start, a mechanic measures the voltage between the battery terminal and the wire carrying current to the starter motor. While the motor is cranking, this voltage is 4.2 V. If the motor draws 125 A, what is the resistance of the connection at the battery terminal?

21. Two parallel metal plates are placed in a plastic tank of water, as shown in Fig. 25–26. The plates measure 15 cm wide by 5.0 cm high, and are 2.5 cm apart. If a potential difference of 60 V is applied between them, how much current flows when the water is (a) pure H_2O; (b) sea water? Neglect edge effects (that is, assume that the current flows only between the plates and that the current density is uniform). See Table 25–1.

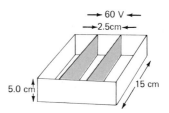

Fig. 25–26   Problem 21

Section 25.4   *Electric Power*

22. An electric stove burner has a resistance of 6.3 Ω and consumes 1.5 kW of power. At what voltage does it operate?

23. If the electrons of Problem 1 are accelerated through a potential difference of 20 kV, how much power must be supplied to produce the electron beam?

24. An immersion heater can bring a cup (about 250 mL) of water from 10°C to boiling in 4 minutes. The heater operates at 120 V, and is about 80% efficient in an open, uninsulated cup. (a) How much current does it draw? (b) What is its resistance?

25. The motion of Jupiter's satellite Io through the Jovian magnetic field results in a potential difference of about a million volts across the satellite (in Chapter 29 you will learn how this comes about). The environment near Jupiter consists of diffuse plasma, and a current of about a million A is driven through this plasma by the voltage across Io. Most of the associated power is dissipated in Jupiter's ionosphere, where it gives rise, among other things, to radio emission detected at earth. How much power is involved in the Io current system? How does this compare with electric power use on earth?

26. During a "brown-out" the power line voltage drops from 120 V to 105 V. By how much does the thermal output of a 1500-W stove burner drop, assuming that its resistance remains constant?

27. Two cylindrical resistors are made from the same material and have the same length. When connected across the same battery, one dissipates twice as much heat as the other. How do their diameters compare?

28. Your author's house uses approximately 90 kWh of electrical energy each week. If this energy is supplied at 240 V, what is the average electrical resistance that the house presents to the power lines?

29. A 2000-hp electric railroad locomotive gets its power from overhead wires with a potential difference of 10 kV. (a) How much current does the locomotive draw? (b) If the wires have a resistance of 0.20 Ω/km, how far from the power plant can the train go before 1% of the energy it uses is lost in the wires?

Supplementary Problems

30. A circular pan of radius b has a plastic bottom and metallic sides of height h. It is filled with a solution of resistivity ρ. A metal disk of radius a and height h is placed at the center of the pan, as shown in Fig. 25–27. The sides and the disk may be assumed perfect conductors compared with the solution. Show that the electrical resistance measured between the electrodes is $\rho \ln(b/a)/2\pi h$.

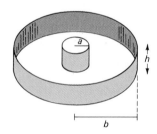

Fig. 25–27 Problem 30

31. By considering the power dissipation in a cylindrical resistor, show that the power dissipation per unit volume is given by $J^2\rho$.

32. A thermally insulated container of sea water carries a uniform current density of 75 mA/cm². How long does it take to raise its temperature from 15°C to 20°C? Use the result of the preceding problem, and assume that the specific heat of sea water is the same as that of pure water. Consult Table 25–1, and assume that the resistivity is approximately temperature-independent over this limited temperature range (actually it varies by about 10%).

33. Use the results of Section 17.1 to estimate the mean thermal energy of electrons in the "gas" of conduction electrons in copper at room temperature. Your result is, of course, based entirely on classical physics. Compare with the result of Problem 12, where the mean thermal velocity is determined from empirical measurements.

34. A 100-Ω resistor of negligible mass is mounted inside a calorimeter, with wires leading through the calorimeter insulation. When the wires are connected across a 12-V battery for 5.0 minutes, the temperature inside the calorimeter rises 26°C. What is the heat capacity of the calorimeter contents?

35. An ideal electric motor is lifting a 15-N weight vertically at the rate of 25 cm/s. If the motor is connected to a 6.0-V battery, how much current is flowing?

36. A parallel-plate capacitor has plates of area 10 cm² separated by a 1.0-mm layer of glass insulation (resistivity $\rho = 1.2 \times 10^{13}$ Ω·m, dielectric constant $\kappa = 5.6$). The capacitor is charged to 100 V and the charging battery disconnected. (a) What is the initial rate of discharge (that is, the initial current through the insulation)? (b) At this rate, how long would it take for the capacitor to discharge completely? *Note:* The capacitor does not actually discharge at a steady rate; see Section 26.6 for details.

37. A thick-walled pipe of inner radius a and outer radius b has a length ℓ much greater than its outer diameter (Fig. 25–28). The material of the pipe has conductivity σ. If a voltage V is applied between the inner and outer surfaces of the pipe, show that the resulting current is $I = 2\pi\sigma V\ell/\ln(b/a)$.

Fig. 25–28 Problem 37

38. A 12-gauge copper wire (diameter 0.21 cm) is surrounded by a 2.0-mm-thick layer of insulation whose thermal conductivity is 4.5 W/m·K. If the outside of the insulation is maintained at 25°C, what is the maximum current if the wire temperature is not to exceed 40°C? *Hint:* See Example 16–7.

26

ELECTRIC CIRCUITS

26.1
CIRCUITS AND SYMBOLS

electric circuit
circuit elements

An **electric circuit** is a collection of electrical devices, called **circuit elements,** connected by conductors. A circuit usually contains a source of electrical energy, and is designed to do something useful with that energy. You are most familiar with human-made electrical circuits, but important circuits also exist in nature. Examples include nervous systems in living organisms, circuits in astrophysical objects, and the earth's global atmospheric circuit (Fig. 26–1), in which thunderstorms are the batteries and the atmosphere a resistor. Your study of electric circuits should prove immensely practical, for it will help you to understand and to use effectively and safely the growing myriad of electrical and electronic devices you encounter. Basic circuit knowledge can even help you design new devices and troubleshoot old ones.

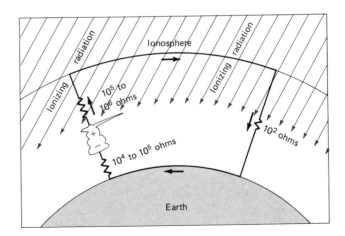

Fig. 26–1
In earth's atmospheric circuit, thunderstorms are the batteries and the atmosphere a resistor.

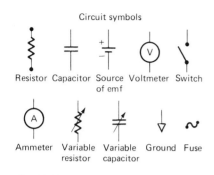

Circuit symbols

Resistor Capacitor Source Voltmeter Switch
                     of emf

Ammeter Variable Variable Ground Fuse
         resistor capacitor

Fig. 26–2
Commonly used circuit symbols.

It is often helpful to represent circuits symbolically. We do so using standard symbols for circuit elements, with lines to represent the wires connecting them. We usually assume that the wires are perfect conductors, so that all points connected by wires alone are at the same potential; such points are electrically equivalent. Realizing this will greatly facilitate your interpretation of circuit diagrams! Figure 26–2 shows some common circuit symbols.

26.2
ELECTROMOTIVE FORCE

How do we sustain current in a conductor? In the preceding chapter, we found that an electric field is necessary to drive a current in any conductor that has nonzero resistivity. But if we simply apply an electric field, say by putting excess charge on one end of the conductor, the charge will quickly redistribute itself until electrostatic equilibrium is reached and the electric field disappears. Somehow we must maintain the electric field, and with it the current, despite the tendency toward equilibrium. This requires that we compensate for the energy lost through the collisions that give the material its resistivity.

What we need is a device that can maintain charge separation by converting energy from some other form into electrical energy. We call such a device a source of **electromotive force** or emf. (The name has historical origins; emf is not actually a force.) Most sources of emf have two electrical contacts, or **terminals,** for connection to other circuit elements. Energy conversion processes within the source move an excess of positive charge to one terminal, negative to the other. The circuit symbol for a source of emf is shown in Fig. 26–2. The most familiar example is a battery, in which chemical energy drives electric charge to the two terminals. Other examples include electric generators, which convert mechanical energy to electrical energy; photovoltaic cells, which convert sunlight; fuel cells, which "burn" hydrogen to produce electrical energy and water; and biological cell membranes, which develop charge separation to control the movement of ions into the cell.

Electromotive force is quantified by the work per unit charge done by a source as it separates positive and negative charge to its two terminals. The units of emf are therefore volts, and the emf of a source is often called, loosely, its voltage. We speak, for example, of a 1.5-volt flashlight battery or a 12-volt car battery. An **ideal source of emf**—one with no internal energy losses—maintains the same voltage across its terminals under all conditions. Real sources of emf always have internal energy losses, so the terminal voltage may not equal the rated emf. We discuss this situation in the next section.

When a source of emf is not connected to any external circuit, no work is needed to maintain its terminal voltage. But current flows when the source is connected to an external circuit. This current would quickly deplete the charge at the terminals were it not for work done inside the source to separate more charge. The simple analogy of Fig. 26–3 illustrates the operation of a source of emf. The energy conversion mechanism in the

electromotive force

terminals

ideal source of emf

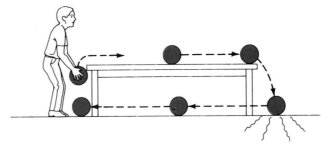

Fig. 26–3
A gravitational analog of a source of emf. The person lifting the bowling balls represents the energy conversion mechanism in the source, which does work on charges as it moves them against the electric field. The bowling balls fall off the table, dissipating their energy in collisions with the floor. Similarly, charges "fall" through an external circuit, giving up energy they gained from the source.

source is like a person lifting bowling balls to a table top. Once the balls are all on the table, no more work need be done. But if the balls begin to roll off the table, they dissipate their energy in collisions with the floor. To maintain a supply of bowling balls on the table, the person must do work against the gravitational field, continually lifting the balls after they have fallen. The source of emf does the same thing: it "lifts" charge against the internal electric field that points from its positive to its negative terminal. The charge then "falls" through the external circuit, dissipating energy in the circuit resistance. When the charge returns to the source it is again "lifted" and the process continues.

26.3
SIMPLE CIRCUITS: SERIES AND PARALLEL RESISTORS

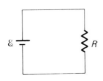

Fig. 26–4
A circuit containing a battery and a resistor.

In the circuit of Fig. 26–4 an ideal battery of emf $\mathscr{E}$ drives a current through the resistor R. How much current? The voltage across the ideal battery is its emf $\mathscr{E}$. The battery is connected to the resistor by ideal wires. Because they have zero resistance, there is no potential difference across either wire. Therefore, the voltages across the resistor and battery are the same. In terms of energy, we can say that each coulomb of charge passing through the battery gains $\mathscr{E}$ joules of energy. There is no energy loss as charge passes through the ideal wires, and then all the energy gained from the battery is dissipated as heat in the resistor.

The resistance R is simply the ratio of voltage across the resistor to current through the resistor, so that

$$R = \frac{\mathscr{E}}{I},$$

or

$$I = \frac{\mathscr{E}}{R}.$$

It may bother you that when charge gets to the bottom of the resistor it has lost all the energy it gained from the battery, for then how does it get back to the battery? Remember that the connection between resistor and battery is assumed to be an ideal wire. Because its resistance is zero, it takes no voltage to drive current through the wire.* The current in this circuit is determined entirely by the battery voltage and the resistance. The wires simply pass whatever current those circuit elements determine. If you try to use Ohm's law to calculate currents and voltages in ideal wires, you are needlessly complicating things!

Series Resistors and the Loop Law

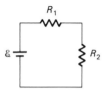

Fig. 26–5
A circuit containing two resistors in series.

Figure 26–5 shows a circuit containing two resistors in series. What is the current through these resistors? What is the voltage across each? Note that neither resistor is connected directly across the battery, so we cannot argue that the voltage across either is the battery voltage. We cannot divide the battery voltage by the resistance of one resistor to get the current through that resistor. Think about this! Ohm's law relates the voltage *across a resistor* to the current *through that resistor*. It does not relate arbitrary voltages and currents anywhere in a circuit. Just because there is a battery of emf $\mathcal{E}$ in a circuit does not mean that $\mathcal{E}$ volts appear across every resistor! However, in Fig. 26–5 the full battery voltage does appear across the series combination of two resistors, so that if we knew the equivalent resistance of this combination, we could solve for the current. What current? The current in both resistors—in fact, in the whole circuit. This simple circuit has only one loop through which current can pass. Charge flows steadily from the battery through the wires and resistors and back to the battery. Once current leaves one circuit element it has nowhere to go but into the next element. If there is to be no buildup of charge in the circuit, the current must be the same everywhere. (See Problem 51 for more on this point.) This situation is characteristic of circuit elements in series:

circuit elements in series

> The current in all elements of a series circuit is the same.

Imagine charge moving around the circuit. It starts at the positive terminal of the battery with $\mathcal{E}$ joules per coulomb. After it gets through the resistor R_1 it has lost some of this energy. How much? If the current in the circuit is I, then a voltage

$$V_1 = IR_1$$

must appear across R_1 to drive the current through R_1. So IR_1 joules per coulomb are lost in resistor R_1. Charge then passes through R_2, where a voltage

$$V_2 = IR_2$$

is needed to drive the current. Thus IR_2 joules per coulomb are lost in R_2.

---

*In practice, wires do have some resistance. Then the resistor voltage is slightly less than the battery voltage. A small voltage appears across the wires, and this voltage drives the current through them.

We used the same current in both these calculations because the resistors are in series so that the same current flows through both. The total energy per coulomb lost in the resistors is equal to that gained in the battery, so we can write

$$\mathscr{E} = IR_1 + IR_2$$

or

$$\mathscr{E} - IR_1 - IR_2 = 0. \qquad (26\text{–}1)$$

Equation 26–1 is an example of a general law that is true for any electric circuit loop, even if that loop is embedded in a more complicated circuit:

> The sum of the voltage differences around a loop is zero.

loop law, or Kirchhoff's voltage law

Known as the **loop law** or **Kirchhoff's voltage law** (after the German physicist Gustav Kirchhoff, 1824–1887), this is really just a statement of energy conservation for electric circuits. In applying the loop law, we must account carefully for the signs of the voltage differences. If we follow current through a battery from the negative to the positive terminal, the voltage increases as charges gain energy. As we move with the current through a resistor, the voltage decreases as energy is dissipated in the resistor. Figure 26–6 illustrates the loop law for the circuit of Fig. 26–5.

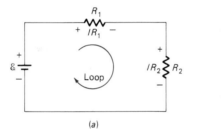

(a)

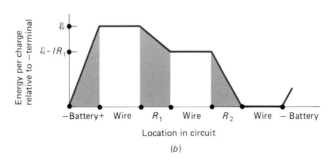

(b)

Fig. 26–6
(a) The series resistor circuit, showing the voltages across each element. (b) Energy per unit charge as a function of position in the circuit. The negative terminal of the battery has been taken as the zero of potential energy.

We can now solve Equation 26–1 for the current:

$$I = \frac{\mathcal{E}}{R_1 + R_2}.$$

(26–2)

Comparison of this expression with Ohm's law ($I = V/R$) shows that the two resistors in series behave like an equivalent resistance equal to the sum of their resistances. In an obvious generalization to more resistors, we have

$$R_{\text{series}} = R_1 + R_2 + R_3 + \cdots.$$

(26–3)

In other words, resistors in series add.

Finally, we multiply the current in Equation 26–2 by the individual resistances to get the voltages (often called voltage drops) across the two resistors, obtaining

$$V_1 = \frac{R_1}{R_1 + R_2}\mathcal{E}$$

(26–4a)

and

$$V_2 = \frac{R_2}{R_1 + R_2}\mathcal{E}.$$

(26–4b)

These expressions show that the battery voltage divides between the two resistors in proportion to their resistance. A group of resistors in series is often called a **voltage divider.**

Fig. 26–7
A 1.5-V calculator battery and a 1.5-V D-cell flashlight battery have the same terminal voltage, but the internal resistance of the calculator battery is higher.

voltage divider

Real Batteries

Figure 26–7 shows two batteries—a 1.5-volt calculator battery and a 1.5-volt size D flashlight battery. What is the difference between these two? If they were both ideal sources of emf there would be no electrical distinction because both would maintain 1.5 volts across their terminals no matter how much current was flowing. But these are real batteries. Both the resistance of the battery materials and the rate of internal chemical reactions limit the current that each can supply. Not surprisingly, the larger battery can supply more current. We characterize a real battery by giving its **internal resistance** as well as its voltage. We imagine the battery to be made of an ideal source of emf in series with this internal resistance, as shown in Fig. 26–8. Of course this is not how batteries are manufactured, since no manufacturer can make an ideal source! The internal resistance is intrinsic to the battery, and there is no way to circumvent it. The more powerful battery is the one with the lower internal resistance. It approaches more closely the ideal, which would have zero internal resistance, and as a result it can supply more current, hence more power.

internal resistance

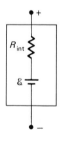

Fig. 26–8
A real battery can be modelled approximately as an ideal source of emf in series with a resistor called the internal resistance.

We can understand the effect of internal resistance by considering the simple circuit of Fig. 26–9. This circuit is just the series circuit of Fig. 26–5, with R_1 the internal resistance R_{int} and R_2 the external resistance R_L. R_L is called the load resistor because it is the thing to which we wish to deliver electric power; it is the electrical load on the battery. From Equation 26–4b we see that if R_{int} is small compared with R_L, then the voltage across the load resistance will be very close to that of the battery's internal

We used the same current in both these calculations because the resistors are in series so that the same current flows through both. The total energy per coulomb lost in the resistors is equal to that gained in the battery, so we can write

$$\mathcal{E} = IR_1 + IR_2$$

or

$$\mathcal{E} - IR_1 - IR_2 = 0. \qquad (26-1)$$

Equation 26–1 is an example of a general law that is true for any electric circuit loop, even if that loop is embedded in a more complicated circuit:

> The sum of the voltage differences around a loop is zero.

loop law, or Kirchhoff's voltage law

Known as the **loop law** or **Kirchhoff's voltage law** (after the German physicist Gustav Kirchhoff, 1824–1887), this is really just a statement of energy conservation for electric circuits. In applying the loop law, we must account carefully for the signs of the voltage differences. If we follow current through a battery from the negative to the positive terminal, the voltage increases as charges gain energy. As we move with the current through a resistor, the voltage decreases as energy is dissipated in the resistor. Figure 26–6 illustrates the loop law for the circuit of Fig. 26–5.

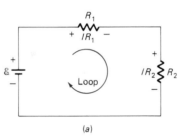

(a)

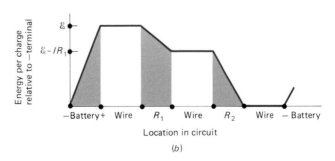

(b)

Fig. 26–6
(a) The series resistor circuit, showing the voltages across each element. (b) Energy per unit charge as a function of position in the circuit. The negative terminal of the battery has been taken as the zero of potential energy.

We can now solve Equation 26–1 for the current:

$$I = \frac{\mathcal{E}}{R_1 + R_2}.$$ (26–2)

Comparison of this expression with Ohm's law ($I = V/R$) shows that the two resistors in series behave like an equivalent resistance equal to the sum of their resistances. In an obvious generalization to more resistors, we have

$$R_{series} = R_1 + R_2 + R_3 + \cdots.$$ (26–3)

In other words, resistors in series add.

Finally, we multiply the current in Equation 26–2 by the individual resistances to get the voltages (often called voltage drops) across the two resistors, obtaining

$$V_1 = \frac{R_1}{R_1 + R_2}\mathcal{E}$$ (26–4a)

and

$$V_2 = \frac{R_2}{R_1 + R_2}\mathcal{E}.$$ (26–4b)

Fig. 26–7
A 1.5-V calculator battery and a 1.5-V D-cell flashlight battery have the same terminal voltage, but the internal resistance of the calculator battery is higher.

voltage divider

These expressions show that the battery voltage divides between the two resistors in proportion to their resistance. A group of resistors in series is often called a **voltage divider.**

Real Batteries

Figure 26–7 shows two batteries—a 1.5-volt calculator battery and a 1.5-volt size D flashlight battery. What is the difference between these two? If they were both ideal sources of emf there would be no electrical distinction because both would maintain 1.5 volts across their terminals no matter how much current was flowing. But these are real batteries. Both the resistance of the battery materials and the rate of internal chemical reactions limit the current that each can supply. Not surprisingly, the larger battery can supply more current. We characterize a real battery by giving its **internal resistance** as well as its voltage. We imagine the battery to be made of an ideal source of emf in series with this internal resistance, as shown in Fig. 26–8. Of course this is not how batteries are manufactured, since no manufacturer can make an ideal source! The internal resistance is intrinsic to the battery, and there is no way to circumvent it. The more powerful battery is the one with the lower internal resistance. It approaches more closely the ideal, which would have zero internal resistance, and as a result it can supply more current, hence more power.

internal resistance

We can understand the effect of internal resistance by considering the simple circuit of Fig. 26–9. This circuit is just the series circuit of Fig. 26–5, with R_1 the internal resistance R_{int} and R_2 the external resistance R_L. R_L is called the load resistor because it is the thing to which we wish to deliver electric power; it is the electrical load on the battery. From Equation 26–4b we see that if R_{int} is small compared with R_L, then the voltage across the load resistance will be very close to that of the battery's internal

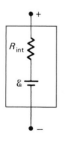

Fig. 26–8
A real battery can be modelled approximately as an ideal source of emf in series with a resistor called the internal resistance.

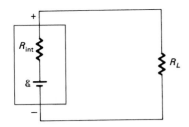

Fig. 26–9
When a real battery is connected to an external circuit, some voltage drops across the internal resistance. The voltage across the battery terminals—and across the external circuit—is therefore less than the battery's rated voltage.

source of emf. In this case the battery's behavior is nearly ideal, since it is maintaining essentially $\mathscr{E}$ volts across its terminals. But if we lower R_L, so that it becomes comparable with R_{int}, then the voltage across R_L decreases and our battery no longer seems ideal. As we lower R_L we draw more current from the battery. It takes a higher voltage to drive this current through the fixed resistance R_{int}, so that more voltage drops across R_{int}, leaving less across R_L. Even if we short-circuit the battery terminals (which is not good for the battery!) we will not get infinite current—in fact, we will simply have

$$I = \frac{\mathscr{E}}{R_{int}}. \qquad \text{(battery short-circuited)}$$

We conclude that a battery or other source of emf behaves more or less ideally depending on the size of its load resistance relative to its internal resistance. A calculator, for example, has a very high resistance and draws little current. It is quite happy with a small battery whose internal resistance, while relatively high, is still small compared with the calculator's resistance. A car starter motor, on the other hand, draws a large current and thus requires a battery of very low internal resistance.

Example 26–1
Internal Resistance

Your car's starter motor draws 125 A. The car has a 12-V battery, but while the starter motor is running the voltage across the battery terminals measures only 9.5 V. What is the internal resistance of the battery?

Solution

This circuit is just that of Fig. 26–9, with the starter being the load. With 9.5 V across the starter, there must be 2.5 V left across the internal resistance to make a total of 12 V. The current is the same throughout this series circuit, so that 125 A is the current through R_{int}. Knowing current and voltage, we apply Ohm's law:

$$R = \frac{V}{I} = \frac{2.5 \text{ V}}{125 \text{ A}} = 0.020 \ \Omega.$$

A battery voltage between 9 and 11 volts is typical of a car being started. A battery voltage much below 9 volts usually indicates a weak battery, a defective starter motor, or very cold weather!

Parallel Resistors and the Node Law

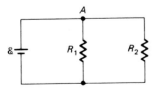

Fig. 26–10
Parallel resistors connected across a battery.

Figure 26–10 shows two resistors in parallel, connected across an ideal battery. What are the current through and voltage across each? How much current does the battery supply? As we found when discussing parallel capacitors in Section 24.4, the voltage across two circuit elements in parallel is the same. (Here we're assuming ideal wires that are always equipotentials, so that the positive battery terminal and the positive [top] ends of both resistors are all at the same voltage. So are the negative terminal and the negative [bottom] ends of both resistors.) So the voltage across each resistor is the battery emf $\mathscr{E}$, and we can use Ohm's law to calculate the currents:

$$I_1 = \frac{\mathcal{E}}{R_1} \tag{26-5a}$$

and

$$I_2 = \frac{\mathcal{E}}{R_2}. \tag{26-5b}$$

node

The point marked A in the Fig. 26–10 is a junction, or **node,** where three wires meet. Current flows into this node from the battery and out to the two resistors. If charge is not to accumulate at the node, then the rate at which charge flows into the node must equal the rate at which it flows out. If I is the battery current, then we have

$$I = I_1 + I_2. \tag{26-6}$$

If we think of a current flowing into the node as positive, and a current flowing out as negative, then it makes sense to write Equation 26–6 in the equivalent form

$$I + (-I_1) + (-I_2) = 0. \tag{26-7}$$

node law, or
Kirchhoff's current law

Equation 26–7, describing conservation of charge at the node, is an expression of the **node law,** also called **Kirchhoff's current law:**

> The algebraic sum of currents at a node is zero.

Using our currents from Equation 26–5 in Equation 26–6 gives

$$I = \frac{\mathcal{E}}{R_1} + \frac{\mathcal{E}}{R_2} = \mathcal{E}\left(\frac{1}{R_1} + \frac{1}{R_2}\right).$$

Comparison with Ohm's law in the form $I = V/R$ shows that the equivalent resistance of the parallel combination is given by

$$\frac{1}{R_{\text{parallel}}} = \frac{1}{R_1} + \frac{1}{R_2}. \tag{26-8}$$

This result is readily generalized to more than two parallel resistors:

$$\frac{1}{R_{\text{parallel}}} = \frac{1}{R_1} + \frac{1}{R_2} + \frac{1}{R_3} + \cdots. \tag{26-9}$$

In other words, resistors in parallel add reciprocally. Equation 26–9 shows that the resistance of a parallel combination is always lower than that of the lowest resistance in the combination. When there are only two parallel resistors, we can rewrite Equation 26–8 using a common denominator to obtain

$$R_{\text{parallel}} = \frac{R_1 R_2}{R_1 + R_2}. \tag{26-10}$$

Note that *parallel* resistors combine in the same way that *series* capacitors do, and vice versa.

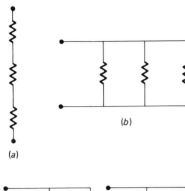

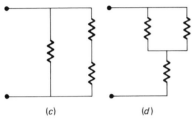

Fig. 26–11
The four possible combinations of three equal resistors.

Example 26–2

Parallel and Series Resistors

You have available three 2.0-Ω resistors. What different resistances can you make by combining all three resistors?

Solution

Figure 26–11 shows the four possible combinations. Resistors in series add, so that combination (*a*) has 6.0 Ω. Resistors in parallel reciprocally add, so that combination (*b*) has

$$\frac{1}{R} = \frac{1}{2.0 \ \Omega} + \frac{1}{2.0 \ \Omega} + \frac{1}{2.0 \ \Omega} = 1.5 \ \Omega^{-1},$$

for a resistance of 0.67 Ω. Combination (*c*) has two resistors in series, giving 4.0 Ω. This 4.0-Ω combination is in parallel with 2.0 Ω, so that Equation 26–10 gives

$$R = \frac{(2.0 \ \Omega)(4.0 \ \Omega)}{2.0 \ \Omega + 4.0 \ \Omega} = 1.3 \ \Omega.$$

Finally, combination (*d*) has two resistors in parallel, giving 1.0 Ω. This combination is in series with 2.0 Ω, for a total of 3.0 Ω. Thus you can make combinations ranging from 0.67 Ω to 6.0 Ω with these three equal resistors.

26.4
ANALYZING MORE COMPLEX CIRCUITS

We analyzed series and parallel resistor combinations using two basic physical principles—conservation of energy and conservation of charge. For electric circuits these principles take the form of the loop law and the node law. Any electric circuit containing batteries and resistors, and indeed other components, can be analyzed completely using these two laws.

In many circuits we can recognize series and parallel combinations. Because we have already applied the loop and node laws to these combinations we can often analyze a circuit without rewriting the loop and node laws. Instead we simplify the circuit by treating series and parallel combinations as one resistor.

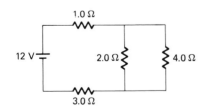

Fig. 26–12
Example 26–3

Example 26–3

Analyzing a Circuit

In the circuit of Fig. 26–12, what is the current through the 2-Ω resistor?

Solution

We approach this problem by simplifying the circuit until we can solve for something—in this case the total current. Then we reverse the process, analyzing the circuit details until we can solve for the quantity we want. Figure 26–13 shows the steps we take in simplifying the circuit. We get from the original circuit, Fig. 26–13*a*, to Fig. 26–13*b* by calculating the resistance of the parallel combination of 2.0 Ω and 4.0 Ω:

$$R_{\parallel} = \frac{(2.0 \ \Omega)(4.0 \ \Omega)}{2.0 \ \Omega + 4.0 \ \Omega} = 1.3 \ \Omega.$$

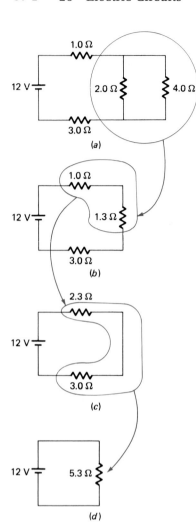

Fig. 26-13
Simplifying the circuit allows us to solve for the total current.

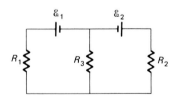

Fig. 26-14
This circuit cannot be analyzed using series and parallel combinations.

Then we move to Fig. 26–13c by evaluating the series combination of $R_\parallel$ with the 1-Ω resistor. Finally, we move to the simplest circuit by forming the series combination of the 3-Ω resistor with the 2.3-Ω combination of the other three, giving an equivalent resistance of 5.3 Ω. From Fig. 26–13d we can calculate the total current:

$$I = \frac{\mathcal{E}}{R} = \frac{12 \text{ V}}{5.3 \text{ }\Omega} = 2.3 \text{ A}.$$

Where does this current flow? It flows from the battery through the 3-Ω resistor, then on through the parallel combination of the 2-Ω and 4-Ω resistors, then through the 1-Ω resistor and back to the battery. It does *not* all flow through the 2-Ω resistor because there are two paths the current can take when it gets to the parallel combination. However, it does all flow through the parallel combination. We already found that this combination has a resistance of 1.3 Ω, and now we know that 2.3 A flows through the combination. So the voltage across the combination is

$$V = IR = (2.3 \text{ A})(1.3 \text{ }\Omega) = 3.0 \text{ V}.$$

This same voltage appears across each of the two resistors making up the parallel combination (why?) so that the current through the 2-Ω resistor is

$$I = \frac{V}{R} = \frac{3.0 \text{ V}}{2.0 \text{ }\Omega} = 1.5 \text{ A}.$$

In solving for the current in the 2-Ω resistor we effectively reversed our original simplification of the circuit, first considering Fig. 26–13b to get the voltage across the parallel combination, and then going to the full circuit to get our answer. At each stage we applied Ohm's law to solve for either a voltage or a current as needed.

Sometimes we encounter circuits where simplification by series and parallel combinations is impossible. This often happens when there is more than one source of emf in the circuit or when circuit elements are connected in complicated ways. Figure 26–14 shows a circuit containing two batteries and three resistors. Are resistors R_1 and R_3 in parallel? Not quite, because of the battery $\mathcal{E}_1$ between them. Are they in series? Not quite, because current flowing out the bottom of R_1 can go two ways when it gets to R_3. Circuit elements are in series when all the current from one has nowhere else to go but into the next element. They are in parallel when they are connected together at both ends. In Fig. 26–14 neither of these situations describes R_1 and R_3, so we cannot simplify the circuit using series or parallel combinations. The same argument holds for R_2 and R_3. Figure 26–15 shows another circuit where we cannot use series and parallel combinations. Are R_1 and R_2 in series? Parallel? How about R_1 and R_3? Again, there are no strictly series or parallel combinations.

We can always solve situations like those of Figs. 26–14 and 26–15 by considering conservation of energy and conservation of charge, which for electric circuits become the loop and node laws. In Fig. 26–14 we can identify three distinct loops and two nodes, as shown in Fig. 26–16. We would like to solve for the currents in all three resistors. We certainly don't know the values of these currents, and it is not even obvious which way they are flowing. We nevertheless label them with arrows to indicate direction. If the answers come out negative, this simply means the actual direction is opposite to our arrows.

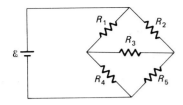

Fig. 26–15
Another circuit where series-parallel analysis is useless.

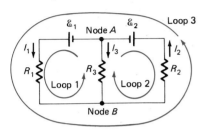

Fig. 26–16
This circuit includes three loops and two nodes. Current directions chosen are arbitrary: the algebraic signs in the solution will determine the actual directions.

Now consider traversing loop 1 in a counterclockwise direction, starting at node A. Voltage increases across battery $\mathscr{E}_1$, then decreases by $V = I_1 R_1$ across R_1. What happens across R_3? We are going *against* the assumed direction of current to complete the loop, so that we go from the low-voltage end of the resistor to the higher end, for a voltage *increase* of $I_3 R_3$. (The voltage may not actually increase; if I_3 turns out to be negative, our "increase" will actually be a decrease. But we need not worry about this—the algebra will take care of the signs!) The loop law states that the sum of the voltage differences around a loop is zero, so that

$$\mathscr{E}_1 - I_1 R_1 + I_3 R_3 = 0. \tag{26–11}$$

We can write a similar expression for loop 2:

$$\mathscr{E}_2 + I_2 R_2 + I_3 R_3 = 0. \tag{26–12}$$

Here we have chosen to go clockwise around the loop, so that each term enters with a positive sign. That the three terms add to zero tells us immediately that one of the currents cannot be in the direction we have assumed. You can probably guess that I_3, at least, does not flow the way we have indicated. Rather than change our conventions now, though, we will wait for the algebra to confirm our guess.

What about loop 3? Physically, it consists of parts of loops 1 and 2. Mathematically, we could obtain the equation for loop 3 by taking the difference of our previous two loop equations. In other words, the equation for loop 3 contains no new information and we need not bother with it. In general, in an N-loop circuit, it suffices to write any $N-1$ of the loop equations.

Conservation of charge tells us that the algebraic sum of currents into a node is zero. At node B this means that

$$I_1 + I_3 - I_2 = 0. \tag{26–13}$$

We could write a similar equation for node A, but we need not, for we now have three independent equations and that is enough to solve for the three unknown currents.

Example 26–4

Using Loop and Node Laws

The components in Fig. 26–16 have the following values: $\mathscr{E}_1 = 6.0$ V, $\mathscr{E}_2 = 9.0$ V, $R_1 = 2.0 \ \Omega$, $R_2 = 4.0 \ \Omega$, $R_3 = 1.0 \ \Omega$. What are the currents in the three resistors?

Solution

We could approach this problem by solving Equations 26–11 to 26–13 symbolically for the currents and then using our specific values, or we could write the equations with our values and solve them. Problem 21 illustrates the former approach; here we take the more specific approach.

With our values, Equations 26–11 to 26–13 become

$$6 - 2I_1 + I_3 = 0, \tag{26–14}$$

$$9 + 4I_2 + I_3 = 0, \tag{26–15}$$

and

$$I_1 + I_3 - I_2 = 0.$$

(Here we temporarily drop units and significant figures for clarity; we will recover them in our final answers.) Solving this last equation for I_3 gives

$$I_3 = I_2 - I_1. \tag{26-16}$$

Using this expression in Equations 26–14 and 26–15 gives

$$6 - 3I_1 + I_2 = 0, \tag{26-17}$$

$$9 - I_1 + 5I_2 = 0. \tag{26-18}$$

Solving Equation 26–17 for I_2 gives

$$I_2 = 3I_1 - 6. \tag{26-19}$$

Using this result in Equation 26–18 gives

$$14I_1 - 21 = 0,$$

so that $I_1 = 1.5$ A. This value can now be used in Equation 26–19 to obtain $I_2 = -1.5$ A. Equation 26–16 then gives $I_3 = -3.0$ A.

Do these numbers make sense? Clearly, I_3 must be negative because, as defined in Fig. 26–16, a positive I_3 would imply current flow out of the negative terminals of both batteries. The signs of I_1 and I_3 also seem consistent with the circuit diagram. However, a change in the voltage of either battery could alter the signs of these two currents (see Problem 22).

26.5
ELECTRICAL MEASURING INSTRUMENTS

Voltmeters

voltmeter

A **voltmeter** is a device that indicates the potential difference or voltage across its two terminals.* The result may be indicated by the position of a needle, by a digital readout, or by the deflection of an electron beam (Fig. 26–17). How do we use a voltmeter? Potential difference, or voltage, always refers to *two points* (it is the work per unit charge needed to move charge between those points). So to measure the potential difference between two points in a circuit, we connect the two terminals of the voltmeter to those two points. To measure the voltage across resistor R_2 in the circuit of Fig. 26–18, we connect the two terminals of the voltmeter to the two ends of the resistor, as shown in Fig. 26–19a. We do *not* break the circuit and insert the meter, as in Fig. 26–19b, for then we are not measuring the potential difference *across* R_2—in fact, we are altering radically the circuit we are trying to measure.

How good is our voltage measurement? There are two considerations here. First, how accurately does the meter indicate the voltage across its terminals? For digital meters this is usually expressed as the number of significant figures in the digital display, while for analog (moving-needle) meters, accuracy is given as a percentage of the full-scale reading. But there

---

*Strictly speaking, this definition of a voltmeter is meaningful only when the field is static. After you study Faraday's law in Chapter 29, you might enjoy reading "What Do 'Voltmeters' Measure?" [R. H. Romer, *American Journal of Physics*, vol. 50, no. 12 (1982).]

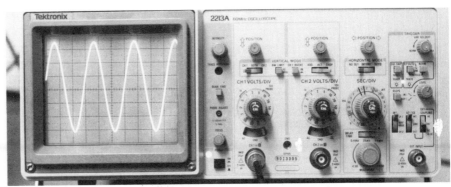

Fig. 26–17
(Left) An analog voltmeter. (Right) An oscilloscope is essentially a pair of voltmeters in which two different voltages are indicated simultaneously by deflection of an electron beam in the vertical and horizontal directions. By changing the horizontal deflection voltage linearly with time, the voltage causing vertical deflection is plotted as a function of time on the oscilloscope screen.

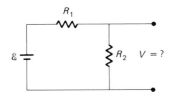

Fig. 26–18
How do you measure the voltage across R_2?

is a more subtle question of accuracy. Even if the meter reads exactly the voltage across its terminals, can we be sure that voltage is the same as it was before the meter was connected? The simple circuit of Fig. 26–20 provides the answer. What is the voltage between points A and B in Fig. 26–20a? There is an open circuit between these points, so that no current flows in the resistor R. With no current through the resistor, there is no voltage across the resistor, so that the voltage between points A and B is the same as the battery voltage. Think about this! A resistor resists current, not voltage. It takes voltage to drive current through a resistor. If there is no current *through* a resistor, then there can be no voltage *across* it. We could have arrived at this result more formally using our voltage divider equation 26–4b, with R_2 set to infinity in order to represent the open circuit.

Now let us connect a voltmeter between points A and B. Suppose the voltmeter has a resistance R_m between its terminals. Called the meter's internal resistance or input resistance, R_m is a property of the meter itself. Now the circuit looks like Fig. 26–20b. We have a complete circuit, with current flowing from the battery through R, through the meter, and back to the battery. A voltage drop across R is required to drive the current, so that the voltage between points A and B is now less than it was before the meter

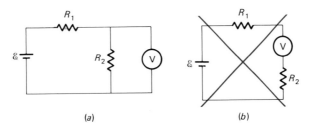

(a)   (b)

Fig. 26–19
(a) Correct and (b) incorrect ways to measure the voltage across R_2.

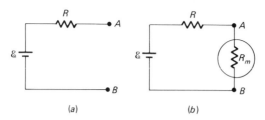

Fig. 26–20
(a) Since no current is flowing, there is no voltage drop across the resistor, so that the voltage between points *A* and *B* is the battery voltage. *(b)* If we connect a voltmeter with finite resistance between *A* and *B*, then some current flows and the measured voltage is less than the battery voltage.

Fig. 26–21
In a moving-coil meter, the indicating needle is attached to a coil that pivots between the poles of a magnet. Here the coil and concave pole pieces are clearly visible at the bottom.

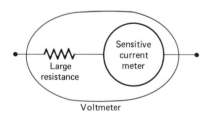

Fig. 26–22
A voltmeter is made by putting a large resistance in series with a moving-coil meter that is sensitive to current.

was connected. The voltmeter reading is not the same as the voltage in the absence of the meter. This discrepancy occurs not because the meter is inherently inaccurate, but rather because it draws current from the circuit it is measuring. How far off is the reading? That depends on the meter resistance relative to the rest of the circuit. If the meter resistance is high, it will draw little current, and its effect on the circuit will be small. Our circuit with the meter included is identical to the voltage divider of Fig. 26–5, with R_1 replaced by R and R_2 by R_m. The meter voltage is then given by Equation 24–4*b*:

$$V_m = \frac{R_m}{R + R_m}\mathscr{E}.$$

As R_m becomes large compared with R, the fraction $R_m/(R + R_m)$ approaches 1 and the meter voltage becomes essentially the open-circuit voltage $\mathscr{E}$. But if R_m is not large compared with R, then the meter reading will be substantially lower. We conclude that a voltmeter should have a substantially higher resistance than typical resistances in the circuit being measured.

How much higher? That depends on how accurate a reading we require. For 1 per cent accuracy, a rough rule of thumb is that the meter resistance should be 100 times the circuit resistance. If we are troubleshooting a car's electrical system, where currents are large and resistances low, we can get away with a fairly low meter resistance. But if we want to measure the voltage developed by the electrode in a chemist's pH meter, we must use a very high resistance indeed, for the pH electrode looks like a nonideal source of emf with internal resistance as high as 10^{14} Ω.

An ideal voltmeter should draw no current and so should have infinite resistance. How do we approach this ideal? Electrically, the simplest meter is the analog moving-needle meter. This meter uses a small coil of wire delicately suspended between the poles of a magnet (Fig. 26–21). When current flows through the coil a torque develops (we will see the reason for this in the next chapter) and the coil turns, taking the needle with it. This type of meter inherently requires a current to operate. We make it into a voltmeter by connecting a large resistance in series with the coil, as shown in Fig. 26–22. A voltage across the series combination then drives a small current. The more sensitive the meter, as measured by the smallness of the current needed to drive it to full scale, the higher the resistance we can use and so the more ideal the voltmeter.

But there is a practical and economic limit to the sensitivity we can manufacture into a moving-coil meter. For many scientific applications, the resistance of even the best moving-coil meter is too low. We can circumvent this problem by using the voltage we are measuring not to drive the meter directly but to control another source of electric power. To do this, we use a transistor amplifier whose output drives either a moving-needle meter or a digital display. Such a compound meter requires an internal power source, but with it input resistances of 10^7 to 10^8 Ω are typical, and many orders of magnitude higher can be achieved. These figures compare with 10^5 to 10^6 Ω for the best moving-needle meters driven directly by the circuit being measured. Integrated circuit technology has advanced to the point where transistor amplifier, digital converter, and digital display are manufactured as a single package. Such integrated digital meters are now the most economical variety, and provide greater inherent accuracy and higher input resistance than analog meters.

Example 26–5
Voltmeters

You wish to measure the voltage across the 40-Ω resistor of Fig. 26–23. What reading would an ideal voltmeter give? A voltmeter with a resistance of 1000 Ω?

Solution

An ideal voltmeter has infinite resistance and therefore would not alter the circuit, which is a simple voltage divider. Applying Equation 26–4b to this divider circuit gives the voltage across the 40-Ω resistor:

$$V_{40} = \frac{(40\ \Omega)(12\ \text{V})}{80\ \Omega + 40\ \Omega} = 4.0\ \text{V}.$$

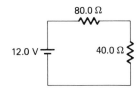

Fig. 26–23
Example 26–5

Because the meter is connected in parallel with the resistor, this is also the voltage read by the meter.

With the nonideal voltmeter in place, the circuit becomes that of Fig. 26–24. The meter and 40-Ω resistor form a parallel combination whose resistance is given by Equation 26–10:

$$R_{\parallel} = \frac{(40\ \Omega)(1000\ \Omega)}{40\ \Omega + 1000\ \Omega} = 38.5\ \Omega.$$

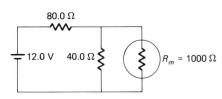

Fig. 26–24
A nonideal voltmeter alters the circuit, lowering the overall resistance.

The circuit now looks like a voltage divider with $R_{\parallel}$ the lower resistor. Applying Equation 26–4b to this circuit gives

$$V_{\parallel} = \frac{(38.5\ \Omega)(12\ \text{V})}{80\ \Omega + 38.5\ \Omega} = 3.9\ \text{V}.$$

This $V_{\parallel}$ is the voltage across the parallel combination consisting of the meter and 40-Ω resistor. Since the voltage across parallel circuit elements is the same, $V_{\parallel}$ is both the meter reading and the voltage across the 40-Ω resistor. And this value is 0.10 V—about 2.5 per cent—lower than the value indicated by an ideal voltmeter.

Ammeters

ammeter

An **ammeter** is a device that indicates current flowing through itself. Like a voltmeter, an ammeter has two terminals. But there the similarity ends. A voltmeter measures the potential difference *across* its terminals.

An ammeter measures the current *through* itself. We connect a voltmeter in parallel with the circuit element whose voltage we wish to measure. Conversely, we connect an ammeter in series with the element whose current we wish to measure. Why? Because only then will all the current through the circuit element go also through the ammeter. If you get used to voltage appearing *across* circuit elements and current flowing *through* them, you will have no trouble with meters. But if you insist on talking about "the voltage through" something you will be unable to hook up a meter accurately or safely. The way to hook up meters, and the words "across" for voltage and "through" for current, go right back to the definitions of potential difference as a property of two points and of current as a flow.

Figure 26–25 shows a simple circuit in which we wish to measure the current through the resistor. To do so we must break the circuit to insert our ammeter in series with the resistor, giving the circuit of Fig. 26–26a. We could equally well have connected the meter on the other side of the resistor, as in Fig. 26–26b (why?). But to connect the meter across the resistor would be incorrect, for then the current through the resistor would not flow through the meter (Fig. 26–26c).

What electrical properties must the ammeter have if it is not to alter the circuit it is measuring? If the meter had any resistance, then the total resistance in the circuit would increase with the meter connected. This in turn would decrease the current, resulting in an incorrect reading. We conclude that an ideal ammeter should have zero resistance. In practice, ammeter resistance should be much smaller than typical resistances in the circuit being measured.

Fig. 26–25
How do you measure the current through *R*?

Ohmmeters and Multimeters

Often we would like to measure the resistance of a particular circuit element. We can do this by connecting a source of known voltage in series with an ammeter and the unknown resistance, as in Fig. 26–27. Knowing the voltage and measuring the current then allows us to calculate the unknown resistance. A meter used for this purpose can be calibrated directly in ohms even though it is really measuring current; it is then called an ohmmeter   **ohmmeter.**

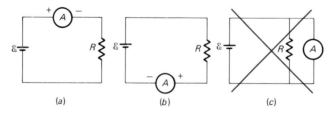

(a)   (b)   (c)

Fig. 26–26
To measure the current through the resistor, we break the circuit and insert an ammeter so that all the current through the resistor also flows through the meter. It doesn't matter if we put the meter (*a*) before or (*b*) after the resistor. (*c*) But we must not put the meter across the resistor, for then the current through the resistor does not go through the meter. In fact, this connection would probably destroy the ammeter!

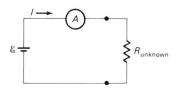

Fig. 26–28
Digital (above) and analog (below) multimeters.

The functions of voltmeter, ammeter, and ohmmeter are often combined in a single instrument called a **multimeter.** Multimeters include switches for selecting the quantity and range to be measured, and may be either analog or digital. An analog multimeter with no amplifiers in it is often called a VOM, for volt-ohm-milliammeter (the ampere itself is too large a unit to be convenient for most purposes where a multimeter would be used). Figure 26–28 shows typical multimeters.

26.6
CIRCUITS WITH CAPACITORS

So far we have considered only circuits in which current and voltage are steady in all components. When you turn on a flashlight, for example, current starts to flow almost immediately through the bulb, batteries, and connecting metal parts. The current continues to flow steadily until you turn off the switch.

When we put a capacitor into a circuit, this picture changes. A capacitor introduces time dependence. Circuit quantities change more gradually because of the capacitor. Why is this? Recall that a capacitor is a pair of insulated conductors that stores electrical energy when opposite charges are put on the conductors. A capacitor is characterized by its capacitance (Equation 24–10 of Chapter 24),

$$C = \frac{q}{V}, \qquad (26-20)$$

where q is the magnitude of the charge on either conductor and V the voltage between the conductors. (See Section 24.4 for a review of capacitors.) Because charge and voltage are proportional in a capacitor, it is not possible to change the voltage across a capacitor without changing the charge. In a circuit, we change the capacitor charge by moving charge on or off the capacitor plates through wires connecting them to the rest of the circuit. This charge movement constitutes a current. The magnitude of the current—in amperes, or coulombs/second—is the rate at which charge is entering or leaving the capacitor. As long as the current is finite, as it is in any real circuit, the charge on the capacitor cannot change instantaneously. Because charge and voltage are proportional in a capacitor, we conclude that

> The voltage across a capacitor cannot change instantaneously.

This simple statement is the key to understanding circuits containing capacitors.

Consider the circuit of Fig. 26–29. We arbitrarily assign the zero of potential to the negative terminal of the battery, which is at the same potential as the lower capacitor plate. (Remember that we are always free to choose the location of zero potential because only differences in potential are really meaningful.) Assume that the capacitor is initially uncharged, so that the voltage across it is zero. What happens when we close the switch? The voltage at the left end of the resistor becomes the battery voltage $\mathcal{E}$.

capacitor voltage

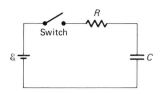

Fig. 26–29
An RC circuit. The switch is closed at time $t = 0$.

Fig. 26–30

Relationships between circuit quantities in a charging *RC* circuit.

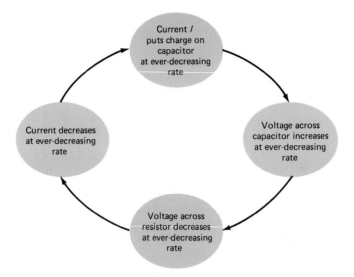

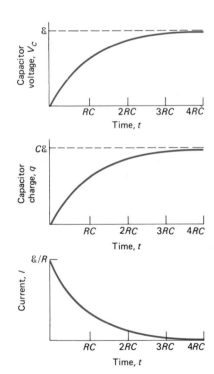

Fig. 26–31

Time dependence of capacitor voltage, capacitor charge, and current in a charging *RC* circuit.

The right end of the resistor is at the same voltage as the upper capacitor plate. But the voltage across the capacitor cannot change instantaneously, and therefore remains zero just after the switch is closed. With zero volts across the capacitor, the full battery voltage $\mathscr{E}$ appears across the resistor. Because its voltage is zero and cannot change instantaneously, the uncharged capacitor acts instantaneously like a short circuit. With $\mathscr{E}$ volts across the resistor, there must be a current $I = \mathscr{E}/R$ through the resistor. This current cannot flow "through" the capacitor, but serves instead to pile positive charge on the upper plate, negative charge on the lower. The same current I flows everywhere in the circuit except in the insulated gap between the capacitor plates.

Now that current is flowing, charge accumulates on the capacitor, and the capacitor voltage increases in proportion to this charge. As the capacitor voltage rises, the resistor voltage must fall because the voltage across the series combination of resistor and capacitor is just the battery voltage $\mathscr{E}$. But the current through the resistor is proportional to the resistor voltage, so that the resistor current falls as well. This in turn decreases the rate at which charge accumulates on the capacitor plates, lowering the rate at which the capacitor voltage increases. The voltage across the capacitor continues to increase, and the current through the resistor to decrease, but at an ever slower rate.

What happens if we wait a long time? As the capacitor voltage approaches the battery voltage, the voltage across the resistor, hence the current through the resistor, and therefore the rate of charge buildup on the capacitor, all become very small. The whole system tends more and more slowly toward a final state in which the capacitor is charged to the full battery voltage and the current in the circuit is zero. Figure 26–30 summarizes the interplay among current, charge, and voltage, while Fig. 26–31 shows the time dependence of these quantities.

We can perform a more quantitative analysis of this circuit using the loop law. Going clockwise around the loop, we first encounter a voltage increase of $\mathcal{E}$ volts across the battery, then a drop of IR volts across the resistor, then a drop V_C from the upper to lower capacitor plate. But Equation 26–20, the definition of capacitance, gives $V_C = q/C$, so that the loop equation becomes

$$\mathcal{E} - IR - \frac{q}{C} = 0. \tag{26–21}$$

This equation contains the two unknown quantities I and q. Can we relate them? Yes—but not through a single proportionality or other algebraic equation. Rather, the current is the rate at which charge is accumulating on the capacitor, or

$$I = \frac{dq}{dt}.$$

To use this relation, we take the time derivative of the loop equation 26–21:

$$-R\frac{dI}{dt} - \frac{1}{C}\frac{dq}{dt} = 0.$$

The battery voltage $\mathcal{E}$ does not appear in this differentiated equation because it is constant and thus its derivative is zero. Using $I = dq/dt$ and rearranging the equation slightly gives

$$\frac{dI}{dt} = -\frac{I}{RC}. \tag{26–22}$$

This equation shows that the rate of change of current is proportional to the current itself, expressing mathematically what Fig. 26–30 shows schematically. Equations like this arise whenever the rate of change of a quantity is proportional to the quantity itself. Population growth, the increase of money in a bank account, and the decay of a radioactive element are all described by similar equations.

Like the equation for simple harmonic motion we encountered in Chapter 13, Equation 26–22 is a differential equation, so called because the unknown quantity I occurs in a derivative. The solution to a differential equation is not a single number but rather a function expressing the relation between the unknown quantity—in this case current—and the independent variable—in this case time. We can solve this particular differential equation by multiplying both sides by dt/I, in order to collect all terms involving I on one side of the equation. This gives

$$\frac{dI}{I} = -\frac{dt}{RC}.$$

We then integrate both sides, noting that RC is constant:

$$\int \frac{dI}{I} = -\frac{1}{RC}\int dt.$$

The integral on the left is just the natural logarithm of I, $\ln I$, and that on the right is just the time t. We must also add a constant of integration, which we choose to write as $\ln I_0$. Then our integrated equation is

$$\ln I = -\frac{t}{RC} + \ln I_0.$$

We now exponentiate both sides:

$$e^{\ln I} = e^{-t/RC + \ln I_0} = e^{\ln I_0} e^{-t/RC},$$

where $e = 2.71827\ldots$ is the base of natural logarithms. But exponentiation and logarithm are inverse operations ($e^{\ln x} = x$), so that

$$I = I_0 e^{-t/RC}.$$

Now $e^0 = 1$, so that at time $t = 0$—the instant we close the switch—we have $I = I_0$. As we argued earlier, the capacitor voltage cannot change instantaneously, so there is at this instant no voltage across the capacitor and the full battery voltage $\mathcal{E}$ across the resistor. Then the current at $t = 0$ is

$$I = I_0 = \mathcal{E}/R,$$

so that the current at any time may be written

$$I = \frac{\mathcal{E}}{R} e^{-t/RC}. \tag{26-23}$$

This equation shows that the current decreases exponentially from its initial value $\mathcal{E}/R$ as time increases—just as our qualitative analysis suggested.

What about the capacitor voltage? The capacitor and resistor voltages must add to the battery voltage $\mathcal{E}$, and the resistor voltage is easy to calculate: it is just $V_R = IR$, or

$$V_R = \mathcal{E} e^{-t/RC}.$$

Thus the capacitor voltage is

$$V_C = \mathcal{E} - V_R = \mathcal{E}(1 - e^{-t/RC}). \tag{26-24}$$

Equation 26-24 shows the capacitor voltage starting at zero, and rising rapidly at first but with its rate of rise ever slowing, as it gradually approaches the battery voltage $\mathcal{E}$. Figure 26-31, shown at the end of our qualitative analysis, was in fact plotted using Equations 26-23 and 26-24.

When is the capacitor fully charged? Never, according to our equations! But the rate at which it approaches full charge is determined by the quantity RC that appears in Equations 26-23 and 26-24. (Problem 35 will convince you that this quantity has the units of time.) Called the **time constant,** RC is a characteristic time for changes to occur in a circuit containing a capacitor. Equation 26-24 shows that in one time constant, the voltage rises to $\mathcal{E}(1 - 1/e)$, or to about two-thirds of the battery voltage. A practical rule of thumb says that in five time constants ($t = 5RC$) a capacitor is 99 per cent charged (see Problem 37). The RC time constant clarifies our statement that the voltage across a capacitor cannot change instantaneously. We can now say that in times small compared with the time constant, the voltage across a capacitor cannot change appreciably. On the other hand, if we wait a long time—many time constants—we will find essentially no current flowing to the capacitor.

RC circuits with appropriate time constants are used in electronic timing applications covering microseconds to hours. In other circuits where

time constant

we want voltages to change rapidly, the time constant can be annoyingly long. For example, capacitance in audio equipment can limit high-frequency response, decreasing the quality of music reproduction. You intentionally alter the response of a stereo system by adjusting bass and treble controls, which are simply variable resistors in *RC* circuits.

What if we connect a charged capacitor across a resistor, as in Fig. 26–32? The voltage across the capacitor drives a current through the resistor. As charge leaves the capacitor, this voltage decreases, and with it the current. Mathematical analysis of this situation is very similar to that of the charging capacitor, and leads to exponential decay with the same time constant *RC* (see Problem 47).

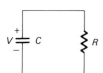

Fig. 26–32
A discharging *RC* circuit.

Example 26–6
An RC Circuit

In Chapter 24, we considered an electronic camera flash using a capacitor to store energy. A particular camera flashtube obtains its energy from a 150-μF capacitor and requires 170 V to fire. If the capacitor is charged by a 200-V battery* through a 30-kΩ resistor, how long must the photographer wait between flashes? What is the peak power drawn from the battery? Assume that the capacitor is fully discharged during a flash.

Solution

The time between flashes is the time it takes the capacitor voltage to reach 170 V. To find this time, we solve Equation 26–24 for the exponential term that contains the time:

$$e^{-t/RC} = 1 - \frac{V_C}{\mathscr{E}}.$$

We then take the natural logarithm of both sides, recalling that $\ln(e^x) = x$, so that

$$-\frac{t}{RC} = \ln\left(1 - \frac{V_C}{\mathscr{E}}\right).$$

Solving for *t* and setting $V_C = 170$ V, $\mathscr{E} = 200$ V, $R = 30$ kΩ, and $C = 150$ μF gives

$$t = -RC \ln\left(1 - \frac{V_C}{\mathscr{E}}\right)$$

$$= -(30 \times 10^3 \ \Omega)(150 \times 10^{-6} \ \text{F}) \ln\left(1 - \frac{170 \ \text{V}}{200 \ \text{V}}\right) = 8.5 \ \text{s}.$$

The maximum current drain on the battery occurs when the capacitor voltage is zero, at which point the full battery voltage occurs across the resistor, giving a current $I = 200$ V/30 k$\Omega = 6.7$ mA. Problem 48 explores the interesting question of power in this charging *RC* circuit, showing that energy from the battery cannot all end up stored in the capacitor.

Example 26–7
Long- and Short-Time Behavior of an RC Circuit

In Fig. 26–33*a* the capacitor is initially uncharged. What is the current through R_1 the instant after the switch is closed? A long time after the switch has been closed?

---

*Actually, much lower voltage batteries are used. But their voltage is increased using transistors and a transformer, working on principles described in Chapter 29.

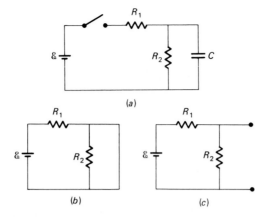

Fig. 26–33
(a) A circuit containing two resistors and a capacitor. (b) Just after the switch is closed, the voltage across the capacitor is still zero, so the capacitor is effectively a short circuit.
(c) Long after the switch is closed, the capacitor is fully charged. No more current flows into it, so it acts like an open circuit.

Solution

The capacitor voltage cannot change instantaneously, so that just after the switch is closed there can be no voltage across the capacitor and therefore none across R_1. Then the full battery voltage is across R_1, so that the current in R_1 is $\mathscr{E}/R_1$. After a very long time the capacitor will be fully charged (to what voltage?— see Question 25 and Problem 45), and no current will flow into it. The capacitor then acts like an open circuit, and we simply have two resistors in series. The current in each is $\mathscr{E}/(R_1 + R_2)$.

How simple this example is! The uncharged capacitor has no voltage across it, so it acts instantaneously like a short circuit. The fully charged capacitor has no current into it, so it acts like an open circuit. To solve the problem we could simply redraw the circuit, once with the capacitor replaced by a wire (Fig. 26–33b), the second time with the capacitor simply erased from the circuit diagram (Fig. 26–33c). Only if we wanted to know what was happening at intermediate times would we have to resort to the full solution of the differential equation describing the circuit.

26.7
ELECTRICAL SAFETY

Whether you find yourself in a laboratory hooking up electronic equipment, or in a hospital connecting instrumentation to a patient, or designing electrical devices, or simply at home plugging in appliances and tools, you should be concerned with electrical safety.

Everyone knows enough to be wary of "high voltage." People with a little more electrical sophistication are fond of saying "it isn't the voltage but the current that kills you." In fact, both points of view are partially correct. Current through the body is dangerous, but as with any resistor it takes voltage to drive that current.

TABLE 26–1 ▬▬▬

Effects of Externally Applied Current on the Human Organism

Current Range	Effect
0.5–2 mA	Threshold of sensation
10–15 mA	Involuntary muscle contractions; can't let go
15–100 mA	Severe shock; muscle control lost; breathing difficult
100–200 mA	Fibrillation of heart; death within minutes
>200 mA	Cardiac arrest; breathing stops; severe burns

Table 26–1 shows typical effects of electric currents introduced into the body through skin contact. Currents below the lethal 100 mA may result in involuntary muscle contraction that unfortunately keeps the victim in contact with the circuit. A primary danger is disturbance of the biologically generated electrical signals that pace heartbeat; this is reflected in the lethal zone of 100 to 200 mA at which currents the heart is thrown into fibrillation—uncontrolled spasms of the cardiac muscle. With electrical signals applied internally to local regions of the body, much smaller currents can be lethal. Doctors performing cardiac catheterization, for example, must worry about currents at the microampere level.

Above 200 mA of externally applied current, complete cardiac arrest may occur, breathing may stop, and there may be severe burns both internally and at the points of skin contact. Sometimes such high currents are useful: when a heart is fibrillating, doctors or emergency technicians briefly apply a high enough current to stop the heart. The heart often restarts, beating normally. The figures of Table 26–1 are rough averages, and vary widely from person to person as well as with duration of the shock and whether alternating or direct current is involved. In particular, very young children and people with heart conditions are at higher risk.

Under dry conditions, the typical human being has a resistance from one point to another on unbroken skin of about 10^5 Ω. What voltages are dangerous to such a person? At 10^5 Ω it takes

$$V = IR = (0.1 \text{ A})(10^5 \text{ Ω}) = 10,000 \text{ V}$$

to drive the fatal 100 mA. But a person who is wet or sweaty has a much lower resistance and may be electrocuted by 120-V household electricity. People have been electrocuted at voltages as low as 30 V, although such cases are rare.

It takes current to harm a person, but it takes voltage to drive that current. To be dangerous, an electric circuit must have high voltage *and* be capable of driving sufficient current. For example, a car battery can deliver 300 A, but it cannot electrocute you because its 12 volts will not drive much current through you (although you could be hurt by the energy released if you accidentally short-circuit such a battery). On the other hand the 20,000 V that runs your car's spark plugs will not electrocute you either, since the internal resistance of this high-voltage circuit is so high that it cannot deliver more than a few mA.

Because potential difference is a property of two points, receiving an electric shock requires that two parts of the body be in contact with conductors at different potentials. In typical 120-V wiring used throughout North America, one of the two wires is connected physically to the earth. This ground connection is to prevent the wiring from reaching arbitrarily high potentials with respect to the earth, as might otherwise happen in a thunderstorm or if a short circuit occurred in a power line. At the same time it means that an individual contacting the "hot" side of the circuit and any grounded conductor such as the ground, a water pipe, or a bathtub (which is grounded through its plumbing) will receive a shock.

A potentially dangerous situation occurs when power tools, instruments, or appliances are used by an operator who is likely to be in contact with a grounded conductor. Examples include working outdoors with an

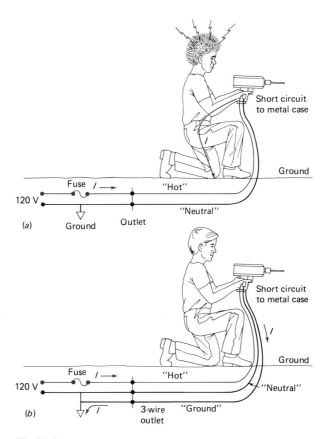

Fig. 26–34
(*a*) A short circuit in an ungrounded tool could result in a lethal shock. (*b*) With a grounded tool, the short circuit causes a blown fuse or circuit breaker, thereby protecting the operator.

electric drill, in a kitchen with an electric mixer, or in a laboratory with an oscilloscope. Suppose, for example, you are using a power tool that is plugged in through a regular two-wire cord. Normally the metal case of the tool is not connected to either wire. Now suppose something goes wrong in the tool and a wire short-circuits to the metal case. If it happens to be the wire that is plugged into the grounded side of the power line there is no problem, but if it is the other wire the case is suddenly 120 V above ground. If you are standing on the ground, or in a damp basement, or are leaning on the kitchen sink, you will receive a potentially lethal shock (Fig. 26–34*a*).

To avoid this danger many electrical devices are equipped with three-wire cords (Fig. 26–35). The third wire runs from exposed metal parts of the device to a grounded wire in the outlet, and normally carries no current. If a short circuit occurs the third wire provides a very low resistance path to ground (Fig. 26–34*b*). Large currents will flow and will blow the circuit breaker or fuse, shutting off the current. Held at ground potential by the ground wire, the operator of the device will be safe.

Because many older homes are not wired with grounded outlets, some manufacturers produce two-wire tools and other devices that are "double-

Fig. 26–35
Plugs. (Left) Grounded; (right) polarized.

insulated" to provide an extra margin of safety. Newer appliances are sometimes equipped with "polarized plugs," which can only be plugged in one way (Fig. 26–35), ensuring that exposed metal parts are most likely to end up at ground potential. Finally, electronic devices called ground-fault interrupters (GFI's; Fig. 26–36) are used in kitchen, bathroom, basement, and other hazardous circuits in new homes. These devices sense a slight imbalance—5 mA or less—in current along the two wires of a circuit. When such an imbalance is detected the interrupter shuts off the current in less than a millisecond, on the assumption that the excess current is leaking to ground—perhaps through a person. (Do ground-fault interrupters know about the node law?)

Fig. 26–36
A ground-fault interrupter protects against shock by sensing small currents leaking to ground. Shown is a portable unit; GFI's are also available for permanent installation in circuit-breaker panels.

SUMMARY

1. A source of **electromotive force** is a device that maintains a separation of positive and negative charge between its two terminals. The source of emf is an energy-conversion device, changing chemical, mechanical, or other forms of energy into the electrical energy associated with charge separation. An ideal source of emf maintains the same voltage across its terminals under all conditions, but energy losses in a real source result in a terminal voltage that decreases with increasing current.

2. Conservation of energy implies that the sum of potential differences around a closed loop in an electric circuit is zero. An important consequence of this **loop law** is that resistances in series add.

3. Conservation of charge implies that as much current flows out of a circuit node, or junction, as flows in. An important consequence of this **node law** is that resistances in parallel add reciprocally.

4. Electric circuits may often be analyzed using series and parallel resistor combinations. Where this approach fails, the loop and node laws may always be used to write simultaneous equations for the unknown circuit quantities.

5. Instruments are used to measure quantities in electric circuits, but care must be taken that the instrument does not itself change the quantity being measured.

   a. A **voltmeter** indicates the voltage between its two terminals. An ideal voltmeter draws no current and has infinite resistance. In practice, a voltmeter's resistance must be much higher than typical resistances in the circuit being measured.

   b. An **ammeter** indicates the current through itself. An ideal ammeter has no voltage drop across it and thus has zero resistance. In practice, an ammeter's resistance must be much lower than typical resistances in the circuit being measured.

   c. An **ohmmeter** compares current and voltage in an unknown resistor to determine its resistance.

   d. A **multimeter** combines the functions of voltmeter, ammeter, and ohmmeter.

6. It takes time to move charge on and off the plates of a capacitor. As a result the voltage across a capacitor cannot change instantaneously. Capacitors therefore introduce time-dependent behavior into electric circuits. Voltage and current in a resistor-capacitor (RC) circuit change with **time constant** given by the product of capacitance and resistance. While an RC circuit is charging from a battery of emf $\mathscr{E}$, for example, the voltage and current are

$$V_C = \mathscr{E}(1 - e^{-t/RC})$$

$$I = \frac{\mathscr{E}}{R} e^{-t/RC}$$

For a discharging capacitor, both current and voltage decrease exponentially with time constant RC.

7. **Electrical safety** is a serious problem. The danger of electric shock depends on the current at which biological damage occurs, the resistance of the organism, and the potential difference available to drive current through the organism.

QUESTIONS

1. In each of the circuits of Fig. 26–37, which, if any, of the resistors are in series? In parallel?

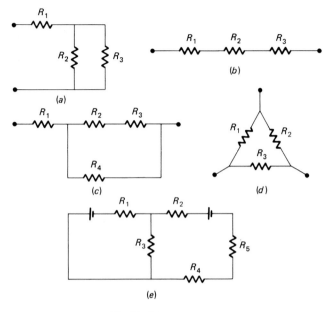

Fig. 26–37   Question 1

2. Are the electrical outlets in a home connected in series or parallel? How do you know?

3. In which of the circuits of Fig. 26–38 does the battery supply the same current? Assume that all the resistors have the same resistance.

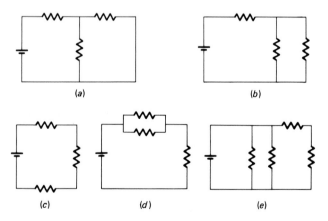

Fig. 26–38   Question 3

4. Can the voltage across a battery's terminals differ from the rated voltage of the battery? Explain.

5. Can the voltage across a battery's terminals be higher than the rated voltage of the battery? Explain.

6. In some cities, streetlights are wired in such a way that when one light burns out, they all go out. Are the lights in series or parallel?

7. If you know the battery voltage in Fig. 26–39, can you determine the voltage between points B and C without knowing the resistance R?

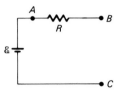

Fig. 26–39   Questions 7 and 17

8. Must we assign zero volts to the negative terminal of a battery?

9. When the switch in Fig. 26–40 is open, what is the voltage across the resistor? Across the switch?

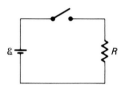

Fig. 26–40   Question 9

10. Two identical resistors in series dissipate equal power. How can this be, when electric charge loses energy in flowing through the first resistor?

11. What is the current through resistor R_2 in Fig. 26–41? Assume all the wires are ideal.

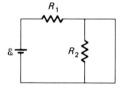

Fig. 26–41   Question 11

12. The resistors in Fig. 26–42 all have the same resistance. If an ideal voltmeter is connected between points A and B, what will it read?

13. When a large electrical load such as a washing machine, oven, or oil burner comes on, lights throughout a house often dim. Why is this? *Hint:* Think about real wires.

14. If the node law were not obeyed at some node in an electric circuit, what would happen to the voltage at that node?

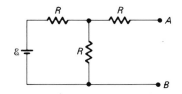

Fig. 26–42 Question 12

15. How would you connect a pair of equal resistors across an ideal battery in order to get the most power dissipation in the resistors?

16. You have a battery whose voltage and internal resistance are unknown. Using an ideal voltmeter and an ideal ammeter, how would you determine both these battery characteristics?

17. An ideal voltmeter is used to measure the voltages between points A and B and between C and B in the circuit of Fig. 26–39. How do the measurements compare?

18. You wish to measure the resistance R_2 in Fig. 26–43 with an ohmmeter. Can you do so while R_2 is in the circuit? Why or why not?

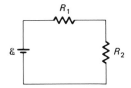

Fig. 26–43 Question 18

19. A student who is confused about voltage and current hooks a nearly ideal ammeter across a car battery. What happens?

20. A student who is confused about voltage and current tries to measure the voltage across a light bulb by inserting an ideal voltmeter in series with the bulb. Does the bulb stay lit?

21. Our equations describing charging of a capacitor suggest that the voltage and charge on a capacitor can have arbitrary values and change smoothly between values. If we look at the situation on a microscopic scale, this simple picture must be modified. Why?

22. What does it mean for a capacitor to be "fully charged"?

23. Is the current into a charging capacitor in an RC circuit greatest when the capacitor voltage is greatest or when it is smallest?

24. If it takes forever to charge a capacitor fully, why is the RC time constant of any significance in describing the charging?

25. The two resistors in Fig. 26–44 have equal resistance. If the circuit has been connected for a long time, what is the voltage across the capacitor?

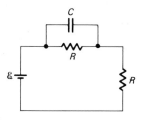

Fig. 26–44 Question 25

26. In one time constant, a charging capacitor reaches approximately $\frac{2}{3}$ of full charge. In one time constant, the voltage across a discharging capacitor falls to approximately $\frac{1}{3}$ of its original value. What is the origin of the approximate numerical factors $\frac{2}{3}$ and $\frac{1}{3}$ in these statements?

27. What's wrong with this news report: "A power-line worker was seriously injured when 4000 volts passed through his body"?

PROBLEMS

Section 26.1   *Circuits and Symbols*

1. Sketch a circuit diagram for a circuit that includes a resistor R_1 connected to the positive terminal of a battery, a pair of parallel resistors R_2 and R_3 connected to the lower-voltage end of R_1, then returned to the battery's negative terminal, and a capacitor across R_2.

2. A circuit consists of two batteries, a resistor, and a capacitor, all in series. Sketch this circuit. Does the description allow any flexibility in how you draw the circuit?

3. Resistors R_1 and R_2 are connected in series, and this series combination is in parallel with R_3. This parallel combination is connected across a battery whose internal resistance is R_{int}. Draw a diagram representing this circuit.

Section 26.2   *Electromotive Force*

4. If you accidentally leave your car headlights (current drain 5 A) on for an hour, how much of the 12-V battery's chemical energy is used up?

5. A battery stores 50 W·h of chemical energy. If it uses up this energy moving 3.0×10^4 C through a circuit, what is its voltage?

Section 26.3   *Simple Circuits: Series and Parallel Resistors*

6. A defective starter motor in a car draws 300 A from the car's 12-V battery, dropping the battery terminal voltage to only 6 V. A good starter motor should draw only 100 A. What will the battery terminal voltage be with a good starter?

7. What is the internal resistance of the battery in the preceding problem?

8. Three 1.5-V batteries, with internal resistances of 0.01 Ω, 0.1 Ω, and 1 Ω, each have 1-Ω resistors connected across their terminals. To three significant figures, what is the voltage across each resistor?

9. When a 9-V battery is temporarily short-circuited, a 200-mA current flows. What is the internal resistance of the battery?

10. What possible resistance combinations can you form using three resistors whose values are 1.0 Ω, 2.0 Ω, and 3.0 Ω?

11. A partially discharged car battery can be modelled as a 9-V source of emf in series with an internal resistance of 0.08 Ω. Jumper cables are used to connect this battery to a fully charged battery, modelled as a 12-V source of emf in series with a 0.02-Ω internal resistance. How much current flows into the discharged battery?

12. You have a number of 50-Ω resistors, each capable of dissipating 0.50 W without overheating. How many resistors would you need, and how would you connect them, so as to make a 50-Ω combination that could be connected safely across a 12-V battery?

13. A 50-Ω resistor is connected across a battery, and a current of 26 mA flows through the resistor. When the 50-Ω resistor is replaced with a 26-Ω resistor, a 47-mA current flows. What are the voltage and internal resistance of the battery?

14. A 6.0-V battery has an internal resistance of 2.5 Ω. If the battery is short-circuited, what is the rate of energy dissipation in its internal resistance?

15. How many 100-W light bulbs can be connected in parallel before they blow a 20-A circuit breaker? Assume that the power-line voltage is 120 V.

16. In the circuit of Fig. 26–45, all resistors are 2.0 Ω and all batteries are 6 V. What is the current through the resistor marked R_0? *Hint:* You do not need to do a complicated analysis! Why not?

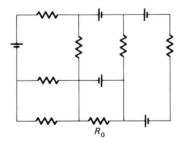

Fig. 26–45   Problem 16

17. A load resistance R is connected across a battery whose internal resistance is R_{int}. Show that the maximum power dissipation in R occurs when $R = R_{int}$. *Note:* This is not the way to treat a battery! But it is the basis for matching loads in amplifiers and other devices. If your stereo amplifier is rated at, say, 20 watts into an 8-Ω loudspeaker, this suggests that the internal resistance of the amplifier is 8 Ω. Connecting a 4-Ω or 16-Ω speaker will result in less power delivered to the speaker.

Section 26.4   *Analyzing More Complex Circuits*

18. A voltage divider consists of two 1-kΩ resistors connected in series across a 160-V battery. If a 10-kΩ resistor is connected across one of the 1-kΩ resistors, what is the voltage across it?

19. In the circuit of Fig. 26–46, R_1 is a variable resistor and the other two resistors have equal resistances R. Obtain an expression for the voltage across R_1, and sketch a graph of this voltage versus R_1 as R_1 varies from zero to 10R. What would happen if R_1 were increased to infinity?

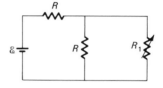

Fig. 26–46   Problem 19

20. In the circuit of Fig. 26–47, how much power is being dissipated in the 4-Ω resistor?

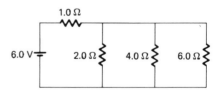

Fig. 26–47   Problem 20

21. Obtain an algebraic equation for the current I_2 in Fig. 26–16, in terms of the quantities $\mathscr{E}_1$, $\mathscr{E}_2$, R_1, R_2, R_3.

22. Rework Example 26–4 with $\mathscr{E}_2$ changed to 2 V. Discuss the significance of your value for I_2. What would happen if $\mathscr{E}_2$ were less than 2 V?

23. What is the equivalent resistance as measured between points A and B in each of the circuits of Fig. 26–48? *Hint:* In (c), think about symmetry and the current that would flow through R_2.

24. How much current flows through the ammeter in Fig. 26–49? Assume that the ammeter has zero resistance.

25. In the circuit of Fig. 26–50, it makes no difference whether the switch is open or closed. What is $\mathscr{E}_3$?

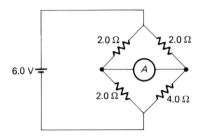

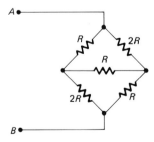

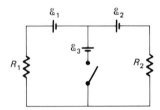

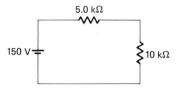

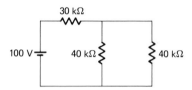

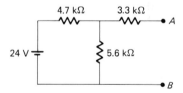

(a)   *(b)*   *(c)*

Fig. 26–48   Problem 23

Fig. 26–49   Problem 24

Fig. 26–50   Problem 25

26. In Fig. 26–51, what is the equivalent resistance measured between points A and B?

Fig. 26–51   Problem 26

Section 26.5   *Electrical Measuring Instruments*

27. A voltmeter with a resistance of 200,000 Ω is used to measure the voltage across the 10-kΩ resistor in the circuit of Fig. 26–52. By what percentage is the mea-

surement in error because of the finite meter resistance?

Fig. 26–52   Problem 27

28. The voltage across the 30-kΩ resistor in Fig. 26–53 is measured with (a) a VOM with 50-kΩ resistance; (b) a VOM with 250-kΩ resistance; (c) a digital multimeter with 10-MΩ resistance. To two significant figures, what does each meter indicate?

Fig. 26–53   Problem 28

29. What are the readings on the following instruments when connected between points A and B of Fig. 26–54? (a) An ideal voltmeter; (b) an ideal ammeter; (c) a voltmeter with 50-kΩ internal resistance; (d) an ammeter with 150-Ω internal resistance.

Fig. 26–54   Problem 29

30. A resistor draws 1.00 A from an ideal 12.0-V battery. (a) If an ammeter with resistance 0.10 Ω is inserted in the circuit, what will it read? (b) If this current is used to calculate the resistance, how will the calculated value compare with the actual value?

31. A neophyte mechanic foolishly connects an ammeter with resistance 0.1 Ω directly across a 12-V car battery whose internal resistance is 0.01 Ω. How much power is dissipated in the meter? No wonder it gets destroyed!

32. A moving-coil meter with a full-scale reading of 100 μA and 50-Ω resistance is to be made into a 10-V full-scale voltmeter. What resistance should you connect and how should you connect it?

33. The meter of the preceding problem is to be made

into a 100-mA full-scale ammeter. What resistance should you connect and how should you connect it?

34. Two identical resistors are connected in series across a battery, as shown in Fig. 26–55a. To determine the resistance, a voltmeter and an ammeter are connected to measure simultaneously the voltage across and current through one resistor. The resistance is then calculated as the ratio of voltage to current. There are two possible ways to connect the meters, as shown in Figs. 26–55b and 26–55c. If the ammeter resistance is 0.010R, and the voltmeter resistance 100R, how do the calculated resistances for each configuration compare with the actual resistance R?

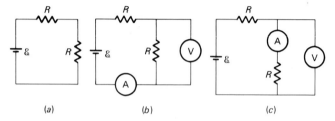

(a)   (b)   (c)

Fig. 26–55 Problem 34

Section 26.6   *Circuits with Capacitors*

35. Show that the quantity RC has the units of time (seconds).

36. If capacitance is measured in μF, what will be the units of the RC time constant when resistance is measured in (a) Ω; (b) $k\Omega$: (c) $M\Omega$? Your answers eliminate the need for tedious power-of-10 conversions when calculating time constants.

37. Show that a capacitor is charged to approximately 99% of the applied voltage in about 5 time constants.

38. A 1.0-μF capacitor is initially charged to 10.0 V. It is then connected across a 500-kΩ resistor. How long does it take (a) for the capacitor voltage to reach 5.0 V? (b) For the energy stored in the capacitor to decrease to half its initial value?

39. In the circuit of Fig. 26–56, the switch is initially open and the capacitors initially uncharged. What is the current through each resistor (a) just after the switch is closed and (b) a long time after the switch is closed? Assume that the resistors have the same resistance, R, and express your answers in terms of $\mathcal{E}$ and R. (c) Sketch qualitatively the current in each resistor as a function of time.

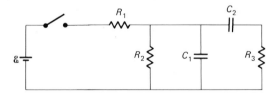

Fig. 26–56 Problem 39

40. A capacitor is charged until it contains 5.0 J of energy. It is then connected across a 10-kΩ resistor. In 8.6 ms, the resistor dissipates 2.0 J. What is the capacitance of the capacitor?

41. An uncharged 10-μF capacitor and a 470-kΩ resistor are connected in series, and 250 volts is applied across the combination. How long does it take the capacitor voltage to reach 200 V?

42. To protect service personnel against shock, large capacitors in the power supplies of electronic equipment are usually connected in parallel with "bleeder resistors" whose job is to discharge ("bleed") the capacitors after the equipment is turned off. If a 200-μF capacitor operating at 350 V is in parallel with a 56-kΩ bleeder resistor, (a) how long will it take the capacitor voltage to reach a safe 30 V after the equipment is turned off? (b) How much power does the bleeder resistor dissipate while the equipment is on?

43. A 2.0-μF capacitor is charged to 150 V. It is then connected to an uncharged 1.0-μF capacitor through a 2.2-kΩ resistor, as shown in Fig. 26–57. (a) What is the rate of power dissipation in the resistor the instant the circuit is connected? (b) What is the total energy dissipated in the resistor while the circuit comes to equilibrium? *Hint:* No integration needed! Think about charge conservation!

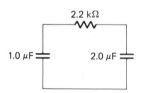

Fig. 26–57 Problem 43

44. The voltage across a charging capacitor in a simple RC circuit rises to $1 - 1/e$ times the battery voltage in 5.0 ms. (a) How long will it take to reach $1 - 1/e^3$ times the battery voltage? (b) If the capacitor is charging through a 22-kΩ resistor, what is its capacitance?

45. For the circuit of Fig. 26–33a (in Example 26–7), take $\mathcal{E} = 100$ V, $R_1 = 4.0$ kΩ, and $R_2 = 6.0$ kΩ. What are the currents through both resistors and the voltage across the capacitor (a) just after the switch is closed? (b) a long time after the switch is closed? A long time after the switch is closed, it is again opened. What are the currents through both resistors and the voltage across the capacitor (c) just after the switch is opened? (d) a long time after the switch is opened?

46. Obtain an expression for the rate (dV/dt) at which the voltage across a charging capacitor increases. Evaluate your result at time $t = 0$, and show that if the capacitor continued to charge at this rate (which it does *not*), then it would be fully charged in exactly one time constant.

47. Set up the equation describing a capacitor charged initially to voltage V_0 and then connected across a resistor R, as in Fig. 26–58. By relating charge and cur-

rent, obtain a differential equation and solve it to show that charge, current, and voltage all decrease exponentially with time constant RC.

Fig. 26–58   Problem 47

48. Of the total energy drawn from a battery in charging an RC circuit, show that only half ends up as stored energy in the capacitor. *Hint:* What happens to the rest of it? You will need to integrate.

Section 26.7   *Electrical Safety*

49. The threshold for feeling electric shock is about 1 mA. At what voltage will a 100-kΩ person feel shock?

50. A 20-kV power supply can deliver a current of 150 mA. It is to be used, however, to power a piece of equipment that needs a maximum current of only 2.5 mA. To make the equipment safer for people working on it, a resistor is placed between the power supply and the equipment, as shown in Fig. 26–59, so that a person contacting high-voltage points within the equipment cannot receive a fatal shock. (a) What is the maximum resistance for which the voltage on the equipment side of the resistor will still be at least 18.5 kV? (b) With this resistance, what is the maximum current that could flow through a person contacting the equipment? Assume a worst case in which the person's resistance is negligible.

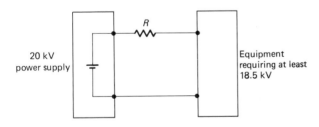

Fig. 26–59   Problem 50

Supplementary Problems

51. Suppose the currents into and out of a circuit node differed by 1 μA. If the node consisted of a small metal sphere of diameter 1 mm, how long would it take for the electric field around the node to reach the breakdown field in air (3×10^6 V/m)?

52. A student measures the voltage across a charged 26-μF capacitor that is not connected to any circuitry, using a voltmeter with a resistance of 250 kΩ. If the student notes the meter reading 1 second after connecting the meter to the capacitor, by what percent-

age will the reading differ from the initial capacitor voltage?

53. In Fig. 26–60, what is the current in the 4-Ω resistor in each of the following situations: (a) when an ideal ammeter is connected between points A and B? (b) when an ideal voltmeter is connected between points A and B? (c) when another 4-Ω resistor is connected between points A and B? (d) just after an uncharged capacitor is connected between points A and B? (e) a long time after an initially uncharged capacitor is connected between points A and B? (f) when an ideal 12-V battery is connected between points A and B, with its positive terminal at A? (g) just after a capacitor, charged to 12 V, is connected between points A and B (the positive capacitor plate is connected to point A)? (h) a long time after the capacitor above is connected?

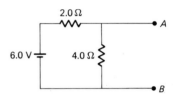

Fig. 26–60   Problem 53

54. The circuit shown in Fig. 26–61 extends forever to the right, and all the resistors have the same value R. Show that the equivalent resistance measured across the two terminals at the left is $\frac{1}{2}R(1 + \sqrt{5})$. *Hint:* You do not need to sum an infinite series!

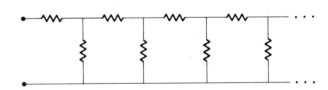

Fig. 26–61   Problem 54

55. Write the loop and node laws for the circuit of Fig. 26–62, and show that the time constant for the circuit is $R_1R_2C/(R_1 + R_2)$.

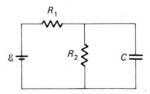

Fig. 26–62   Problem 55

56. When the relation between voltage and current in all circuit elements is linear (as it is with resistors, capacitors, and sources of emf), then the superposition

principle can be used to solve complex circuits. To solve a circuit containing several sources of emf, for example, you can set all but one of the sources to zero (that is, replace all but one of the sources by wires) and solve the resulting circuit for the unknown currents. Repeat this procedure for each other source. Then the currents for the complete circuit with all sources in place are given by the sums of the corresponding currents you obtained for the individual circuits each containing only one source. Figure 26–63 is a symbolic description of this procedure. Apply this procedure to the circuit of Example 26–4, and show that it gives the same answers obtained there.

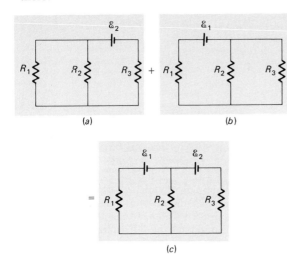

Fig. 26–63 The currents in the complex circuit *(c)* are the sums of the currents in individual circuits *(a)* and *(b)* obtained by setting all but one of the emf's to zero (Problem 56).

57. **Potentiometric voltage measurement** is a means of measuring voltages by comparing them with a known voltage. Figure 26–64 shows a simple device for such measurement. A precision, calibrated, adjustable voltage divider is connected across a battery whose voltage is precisely known. The output of this divider is connected through a sensitive meter to the unknown voltage, as shown. The voltage divider is adjusted until the meter reads zero. At this point the unknown voltage is equal to the output of the calibrated voltage divider, and the circuit is said to be nulled. The distinct advantage of this measurement technique is that at null, there is no current drawn from the circuit being measured, and therefore the measurement does not affect the voltage being measured. (In practice, the nulling procedure is often automated either electromechanically or electronically. Such **negative feedback** circuits form the basis for powerful measurement and control devices.) Suppose that a 1.5-V unknown is being measured potentiometrically, using a null-indicating microammeter that can be read to 0.050 μA. If the operator misjudges the exact null by this amount of current, show that the potentiometric measurement disturbs the measured circuit as would a voltmeter with 3.0×10^7 Ω resistance.

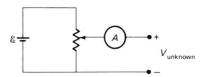

Fig. 26–64 Potentiometric voltage measurement (Problem 57).

27

THE MAGNETIC FIELD

Fig. 27–1
An audio cassette, a video cassette, a computer disk, and credit cards all use magnetism to store information.

Most people are fascinated with magnets. Magnetism—the seemingly mysterious force you feel when you try to push two magnets together in a way they don't want to go—is always intriguing. Some uses of magnets—holding notes on refrigerators, or screws on screwdrivers—are mundane. But others—holding gas at a temperature of one hundred million K in a nuclear fusion reactor, or converting electrical into mechanical energy in the motors of a railroad locomotive—are more impressive. And magnetism, like electricity, is at the heart of many natural phenomena and technological devices that are essential features of our world. Video and audio tape recorders, electric motors, TV picture tubes, computer disks, and electric power plants would be impossible without magnetism (Fig. 27–1). Earth's magnetism helps us find our way around, provides historical evidence for the evolution of our planet, and protects us from harmful radiation. The recent discovery of magnetic bacteria (Fig. 27–2) and of magnetic navigation in birds shows that biological systems have evolved to make use of

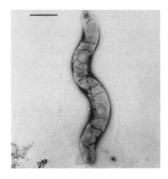 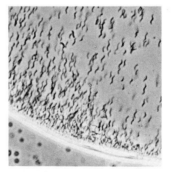

Fig. 27–2
(Left) A magnetic bacterium. The dark objects forming a chain down the center of the cell are particles of magnetite (Fe_3O_4). (Right) In the northern hemisphere, magnetic bacteria swim northward and downward to find food in sediments at the bottom of shallow water.

magnetism. Without magnetism we would not even see, for light itself originates in an interaction between magnetism and electricity. As with electricity, we often do not recognize the magnetic character of everyday phenomena.

In this chapter we discuss various aspects of magnetism and its relation to electricity. In subsequent chapters we will say more about the fundamental relation between magnetism and electricity, and eventually we will come to understand how electricity and magnetism are manifestations of the same underlying phenomenon.

The occurrence of natural magnets in a rock called "magnetite" led the Greeks to discover magnetism over 2000 years ago. With natural magnets for compass needles, sailors have used magnetic navigation for at least eight centuries. By the thirteenth century, it was known that all magnets have two different poles, and that like poles repel and unlike poles attract. Quantitative experiments with magnetism began in the 1700's, and by the mid-1800's much of the basis for our present understanding of magnetism and its relation to electricity had been developed.

What is a magnet? For the present, we can say that a magnet is a piece of material—usually containing iron—that attracts other pieces of iron. We will later replace this rather unsatisfactory description—a description that seems to emphasize an unusual property specific to the element iron—with a deeper understanding based on universal principles of electromagnetism.

magnetic field

The interaction between two magnets may be described in terms of a **magnetic field** (symbol **B**). We say that one magnet gives rise to a magnetic field at all points in space and that a second magnet responds to the field in its immediate vicinity. As with electricity, describing magnetism in terms of fields eliminates the awkward "action-at-a-distance" picture, in which one magnet somehow reaches across empty space to influence another. We represent magnetic fields by drawing field lines, whose properties are analogous to those of electric field lines. The direction of a field line at a given point is the local direction of the magnetic field vector, and the number of lines at right angles to a unit area is a measure of the field strength.

How does a magnet interact with a magnetic field? We can answer this question for a bar magnet, which is a particularly simple magnetic structure. The ends of the bar magnet are its two **poles,** which have opposite magnetic polarity. We arbitrarily call these poles north and south, in reference to earth's magnetism. The interaction of magnetic poles is very similar to the interaction of electric charges: like poles repel and unlike poles attract. We visualize magnetic field lines emerging from north poles and going into south poles, just as electric field lines emerge from positive charge and go into negative charge. A magnetic north pole experiences a force in the direction of the field, while a south pole experiences a force opposite to the field direction. The resulting effect on a bar magnet is a torque that tends to align the magnet with the magnetic field (Fig. 27–3).

poles

By exploiting this tendency of a bar magnet to align itself with the field we can trace out magnetic field lines. We place a small bar magnet on a pivot (a compass is just such a device) and allow it to swing into line with the field. Once we have determined the field direction at one point, we

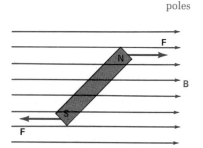

Fig. 27–3
The poles of a bar magnet experience opposite forces in a uniform magnetic field, giving rise to a torque that tends to align the magnet with the field.

Fig. 27–4
(Left) Field lines of a bar magnet.
(Right) The field-line pattern is indicated
by the alignment of iron filings.

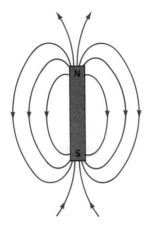

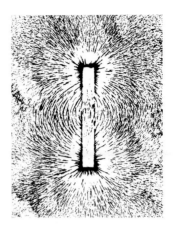

then move our small test magnet in the direction of the field and repeat the procedure, thus tracing out a field line. We can then start somewhere else in the field and trace out additional field lines. When we carry out this procedure for the field of a bar magnet, we obtain the field-line pattern shown on the left in Fig. 27–4. Or we can see the magnetic field all at once by dropping iron filings near the magnet (Fig. 27–4).

27.1
MAGNETIC DIPOLES AND MONOPOLES

Compare the magnetic field of Fig. 27–4 with the electric field of the electric dipole shown in Fig. 27–5. The fields look remarkably similar. Furthermore, both the bar magnet and the electric dipole show the same tendency to align with their respective fields. A bar magnet is, in fact, a magnetic dipole. The magnetic field it produces and its behavior in another magnetic field are analogous to the electric field and behavior of an electric dipole. Our knowledge of electric dipoles applies by analogy to magnetic dipoles. For example, a magnetic dipole in a uniform magnetic field experiences a torque, while in a nonuniform field the dipole experiences a force as well. The dipole itself gives rise to a magnetic field that falls as the inverse cube of distance from the dipole.

Fig. 27–5
The electric field of an electric dipole
resembles the magnetic field of the bar
magnet shown in Fig. 27–4.

magnetic monopoles

Why do we not introduce magnetic charges that behave analogously to electric charges? Why do we speak first of the more complicated dipole configuration? While electric charges play the simplest and most fundamental role in electrical phenomena, the same seems not to be true for magnetism. **Magnetic monopoles,** or isolated north and south magnetic poles, may not exist and, in any case, appear to play little or no role in our everyday world. For decades, physicists have debated the existence of such monopoles. Fundamental symmetry between electric and magnetic phenomena suggests that these magnetic analogs of electric charge might exist. Theoretical arguments put forth by P.A.M. Dirac in 1931 link the quantization of electric charge with the existence of magnetic monopoles. And a major class of contemporary elementary-particle theories suggests that magnetic monopoles were created and played an important role in the early

Fig. 27–6
Stanford University physicist Blas Cabrera with his monopole-detection apparatus.

history of the universe, just after the Big Bang event some 15 billion years ago. According to these theories, if monopoles exist, then they are the most massive of elementary particles—each with the mass of a small virus!

The possibility that monopoles exist has inspired many experiments to search for them. Looking through pieces of iron, where monopoles might become lodged, or in cosmic rays, or in lunar materials, investigators to date have found no conclusive evidence for a monopole. A 1982 experiment at Stanford University gave tantalizing evidence that a monopole passed through the experimental apparatus (Fig. 27–6). However, only one other similar event has been detected since.

27.2
ELECTRIC CHARGE AND THE MAGNETIC FIELD

How can we have magnetic dipoles, such as bar magnets, if we do not have magnetic monopoles? After all, an electric dipole is made up of two point charges, each of which is nothing but an electric monopole. Are not magnetic dipoles similarly made up of two opposite monopoles? No: attempts to take apart magnetic dipoles have never succeeded in producing two monopoles. If you cut a bar magnet in half you have not separated the north and south pole. Each piece remains a full dipole, complete with north and south pole (Fig. 27–7). You can repeat this procedure right down to the atomic scale, but you will never separate a bar magnet into two monopoles.

If magnetic monopoles play no part in the magnetic dipoles with which we are familiar, then what does? What else can respond to the magnetic field? In the answer to this question lies a profound link between electric

Fig. 27–7
Cutting a bar magnet into pieces only produces ever-smaller bar magnets—never isolated north and south magnetic poles.

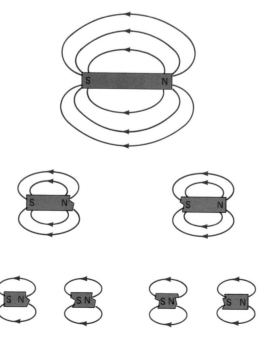

and magnetic phenomena, a link that we will ultimately develop into a unified picture of electromagnetism. In fact magnetic fields influence not only the hypothetical magnetic monopoles, but also *electric* charges.

The interaction of an electric charge with a magnetic field is more complicated than its interaction with an electric field. A charged particle experiences a magnetic force that depends not only on the magnetic field and the particle's charge, but also on its velocity. Furthermore, the magnetic force is not in the direction of the magnetic field but is at right angles both to the field direction and to the velocity (Fig. 27–8). A charge moving along the field direction experiences no magnetic force, while the charge experiences the greatest magnetic force when it moves at right angles to the magnetic field direction. At intermediate orientations of the velocity relative to the field, the magnitude of the force varies as the sine of the angle between velocity and magnetic field. Putting all these observed facts together gives the magnitude of the magnetic force on a moving charged particle:

$$F = qvB \sin\theta, \tag{27–1}$$

where q is the charge, v its speed, B the magnetic field strength, and θ the angle between field and velocity. What about the direction of this force? We have already noted that it is at right angles to both the velocity and the field. Recall (Chapter 11, Section 11.1) that the vector cross product of two vectors **A** and **B** gives a third vector whose magnitude is $AB \sin\theta$ and whose direction is perpendicular to both **A** and **B.** Using the cross product, we can write the magnetic force on a moving charged particle in terms of the cross product of the charge velocity and the magnetic field:

$$\mathbf{F} = q\mathbf{v} \times \mathbf{B}. \quad \text{(magnetic force)} \tag{27–2}$$

The direction of this force is obtained from the right-hand rule, as shown in Fig. 27–9.

Equations 27–1 and 27–2 show that the magnetic field B has the units of N·s/C·m. This unit is given the name **tesla** (T), after the Yugoslav-American inventor Nikola Tesla (1865–1943). One tesla is a strong magnetic field, so that a smaller unit called the **gauss** (G), equal to 10^{-4} T, is often used. The earth's magnetic field, for example, is a little under 1 G,

tesla

gauss

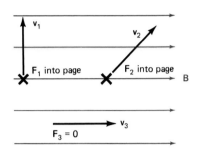

Fig. 27–8
The magnetic force on an electrically charged particle is at right angles to the particle's velocity and to the magnetic field.

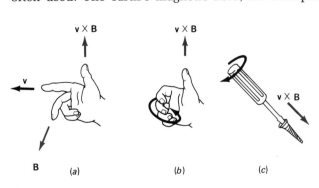

Fig. 27–9
Three ways to remember the right-hand rule. In *(b)* and *(c)* the curved arrow represents a motion taking the vector **v** onto the vector **B.**

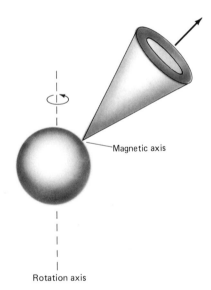

Fig. 27–11
A pulsar. Charged particles falling onto the magnetic poles of a rapidly spinning neutron star give rise to a beam of radio waves that spins like a searchlight beam, giving rise to radio pulses received at earth.

while the field at the poles of a toy magnet may be 100 G. In geophysics and space physics a still smaller unit, the **gamma** (γ), is used. One gamma is equal to 10^{-5} G or 10^{-9} T. Most fields of interest in laboratory apparatus are in the range from 1 G to 1 T. Fields as high as 5 to 10 T are produced routinely with superconducting magnets, and the world's record for the strongest steady magnetic field—33 T—is held by the National Magnet Laboratory in Cambridge, Massachusetts (Fig. 27–10). Elsewhere in the universe, rapidly rotating neutron stars—the stellar remains of supernova explosions—have fields of 10^8 T. In these huge fields the behavior of charged particles is completely dominated by the magnetic force. We detect these stars by the radio radiation we receive from them; this radiation comes in pulses, so astronomers call the neutron stars "pulsars" (Fig. 27–11). Analyzing the radio radiation gives a direct measure of a pulsar's magnetic field.

Example 27–1
The Magnetic Force

In a certain region a magnetic field of 0.10 T points vertically upward. (This field strength is typical of magnets used to "steer" charged particles in low-energy nuclear physics experiments, for example). Three protons enter the region, two horizontally and one vertically, as shown in Fig. 27–12. All three are moving at 2.0×10^3 m/s. What is the magnetic force on each proton?

Solution

Proton 2 is moving vertically, or parallel to the field. Therefore, it experiences no magnetic force. Protons 1 and 3 are moving at right angles to the field, so $\sin\theta = 1$ in Equation 27–1, and thus the forces on these two protons have the same magnitude:

$$F = qvB = (1.6 \times 10^{-19}\text{ C})(2.0 \times 10^3\text{ m/s})(0.10\text{ T}) = 3.2 \times 10^{-17}\text{ N}.$$

Since the protons carry positive charge, the direction of each force is the direction of the product $\mathbf{v} \times \mathbf{B}$. For proton 1, moving to the right, $\mathbf{v} \times \mathbf{B}$ is out of the page. For proton 3, moving to the left, the force is into the page. This example shows clearly that the magnetic field alone does not determine the magnetic force. Identical particles in the same field may experience different forces if their velocities are not identical. Had our particles been electrons, the negative sign of the electron charge would have indicated a magnetic force opposite to the direction of $\mathbf{v} \times \mathbf{B}$.

Example 27–2
The Magnetic Force

A particle carrying a 1.0-μC charge is moving at 20 m/s in a region where the magnetic field strength is 1.3 T. The particle experiences a force of 1.4×10^{-5} N. At what angle to the field is the particle moving?

Solution

Solving Equation 27–1 for $\sin\theta$ gives

$$\sin\theta = \frac{F}{qvB} = \frac{1.4 \times 10^{-5}\text{ N}}{(1.0 \times 10^{-6}\text{ C})(20\text{ m/s})(1.3\text{ T})} = 0.54,$$

so that

$$\theta = \sin^{-1}(0.54) = 33°.$$

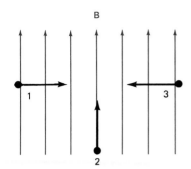

Fig. 27–12
Example 27–1

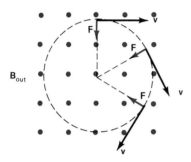

Fig. 27–13
A charged particle moving at right angles to a uniform magnetic field describes a circular path.

27.3
THE MOTION OF CHARGED PARTICLES IN MAGNETIC FIELDS

Because it is always at right angles to a particle's velocity, the magnetic force does no work on the particle. (Recall from Chapter 6 that power—the rate at which work is done—is the dot product of force with velocity; $P = \mathbf{F} \cdot \mathbf{v}$.) Because no work is done, the particle's kinetic energy cannot change—both the speed v and kinetic energy $\frac{1}{2}mv^2$ remain constant. Therefore the magnetic force changes only the direction of the particle's motion, but not its speed.

What sort of motion results from the magnetic force? Consider the specific case of a particle of charge q moving at right angles to a uniform magnetic field $\mathbf{B}$, as shown in Fig. 27–13. At some instant, the velocity $\mathbf{v}$ points to the right, so with the field out of the page, the cross product $\mathbf{v} \times \mathbf{B}$ points downward. If the particle is positive, it therefore experiences a downward force. This force changes the direction of the particle's motion, but not its speed. A little while later, the particle is moving downward and to the right. Now the cross product points downward and to the left. The particle speed v still has its original value, and the velocity is still at right angles to the field, so the magnitude of the force remains the same. The particle describes a path in which the force is always at right angles to its motion and always has the same magnitude. The rate of deviation from straight-line motion is the same everywhere, resulting in the simplest possible curved path—a circle. The magnetic force acts exactly like the tension in a string when you tie a mass to the string and whirl it around in a circle. The tension has constant magnitude and is always at right angles to the motion, and therefore changes only the direction of the motion but not the speed.

How big is the circle, and how long does it take to get around? We can answer these questions using Newton's second law. In its circular path, the charged particle undergoes an acceleration v^2/r, directed toward the center of the circle. What causes this acceleration? The magnetic force! With the field and velocity at right angles, the magnitude of the magnetic force is just

$$F = qvB,$$

and the force points toward the center of the circle. Writing Newton's law $\mathbf{F} = m\mathbf{a}$ then gives

$$qvB = m\frac{v^2}{r},$$

so that

$$r = \frac{mv}{qB}. \tag{27–3}$$

This result makes sense: the larger the particle's momentum mv, the harder it is for the magnetic force to bend it out of a straight line, so the larger the radius of the orbit. On the other hand, if we make the field or charge larger, then the magnetic force increases, giving a tighter orbit.

The circumference of the orbit is $2\pi r$, so that the period—the time it takes the particle to complete one full orbit—is

$$T = \frac{2\pi r}{v}.$$

Using Equation 27–3 for the radius r gives

$$T = \frac{2\pi r}{v} = \frac{2\pi}{v}\frac{mv}{qB} = \frac{2\pi m}{qB}. \qquad (27\text{–}4)$$

This is a remarkable result, for it shows that the period of the circular motion is independent of the particle's speed and the size of the orbit. It depends only on the magnetic field and the charge-to-mass ratio of the particle. The frequency in revolutions per second is simply $1/T$, or

$$f = \frac{qB}{2\pi m}. \qquad (27\text{–}5)$$

cyclotron frequency

This quantity is often called the **cyclotron frequency**, for it is the frequency at which charged particles circulate in a cyclotron particle accelerator. In astrophysics, the cyclotron frequency provides a simple way to measure the magnetic fields of distant objects (Fig. 27–14). Electrons circulating in magnetic fields emit electromagnetic radiation at the cyclotron frequency and its harmonics (in Chapter 32 we will explore the origin of this radiation). Because the electron charge-to-mass ratio (e/m) is the same everywhere, measurement of the cyclotron frequency immediately yields the magnetic field strength.

Fig. 27–14
The Crab Nebula, remnant of a supernova explosion observed nearly 1000 years ago. Intense radio emission occurs as electrons undergo circular motion in the magnetic fields of the nebula. Analysis of radio frequencies observed at earth allows radio astronomers to determine the magnetic field strength.

Example 27–3
A TV Picture Tube

In a TV picture tube, deflection of the electron beam that "paints" the TV picture is accomplished using magnetic fields. The geometry of a certain tube requires that the electron beam be bent in a circular arc with a minimum radius of 4.5 cm (Fig. 27–15). If the electrons are accelerated from rest through a 25-kV potential difference before they enter the magnetic field region, what is the magnetic field strength required? What direction should the field point to accomplish the deflection shown in Fig. 27–15? Assume that the field is uniform over the deflecting region, and zero elsewhere.

Solution

Solving Equation 27–3 for the magnetic field strength B gives

$$B = \frac{mv}{er},$$

with e the elementary charge. "Falling" through a potential difference $V = 25$ kV, the electron of charge e acquires a kinetic energy $\frac{1}{2}mv^2 = Ve$, so that its speed is

$$v = \sqrt{\frac{2Ve}{m}}.$$

Our expression for the field then becomes

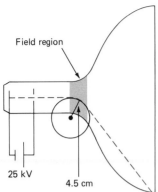

Fig. 27–15
A TV picture tube requiring a 4.5-cm bending radius for maximum deflection of the electron beam.

$$B = \frac{m}{er}\sqrt{\frac{2Ve}{m}} = \frac{1}{r}\sqrt{\frac{2mV}{e}}$$

$$= \frac{1}{0.045 \text{ m}}\left(\frac{(2)(9.1\times10^{-31} \text{ kg})(25\times10^{3} \text{ V})}{1.6\times10^{-19} \text{ C}}\right)^{1/2}$$

$$= 0.012 \text{ T} = 120 \text{ G}.$$

To achieve an initially downward deflection, the force $q\mathbf{v}\times\mathbf{B}$ must be initially downward. But the electron charge is negative, so that $\mathbf{v}\times\mathbf{B}$ must be upward. Application of the right-hand rule to the rightward-moving electrons shows that the magnetic field $\mathbf{B}$ must be into the page in Fig. 27–15.

What if a particle's motion is not initially perpendicular to the field? We can then resolve the velocity into two vectors, $\mathbf{v}_\perp$ perpendicular to the field and $\mathbf{v}_\parallel$ along the field. Then Equation 27–2 can be written

$$\mathbf{F} = q(\mathbf{v}_\perp + \mathbf{v}_\parallel)\times\mathbf{B} = q\mathbf{v}_\perp\times\mathbf{B} + q\mathbf{v}_\parallel\times\mathbf{B}.$$

The second term on the right-hand side of this equation is zero, since it is the cross product of parallel vectors, and therefore the force is simply

$$\mathbf{F} = q\mathbf{v}_\perp\times\mathbf{B}. \tag{27–6}$$

Since it is perpendicular to the magnetic field $\mathbf{B}$, this force influences the particle's motion only in the plane perpendicular to the field. Our earlier considerations of motion strictly in that plane apply here, so that the motion perpendicular to the magnetic field is circular, with frequency given by Equation 27–5. What about motion along the field? Equation 27–6 shows that there is no component of force along the magnetic field, and therefore this component of motion is unaffected by the field. In the absence of other forces, the particle moves with a uniform velocity $\mathbf{v}_\parallel$ along the magnetic field, even as it executes circular motion with velocity $\mathbf{v}_\perp$ perpendicular to the field. The resulting path is a helix, as shown in Fig. 27–16.

The absence of magnetic forces associated with motion along the magnetic field means that particles move readily in the field direction. But if you try to push them at right angles to the field they simply move in larger circles around the field. As a result, charged particles are often described as being "frozen" to the field lines. The particles act somewhat like beads strung on a wire, with the wire being the magnetic field (Fig. 27–17). Nonuniform fields and collisions between particles make this "freezing" of par-

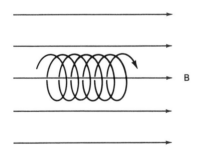

Fig. 27–16
A particle in a uniform magnetic field describes a helical path of constant radius and pitch.

Fig. 27–17
(a) Charged particles undergoing helical motion about magnetic field lines are like *(b)* beads that are free to move along a wire but not at right angles to it.

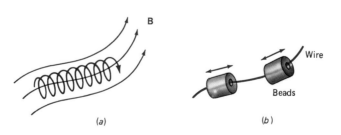

(a)   *(b)*

Fig. 27–18
Magnetic fields shape the delicate structures shown in this image of the solar corona, taken by the Solar Maximum Mission satellite. The black circle covers the sun itself, providing an artificial eclipse.

ticles and field less than perfect, but in many cases the density of charged particles is sufficiently low that the "frozen" assumption is an excellent approximation. Unlike our relatively cool planet earth, most of the universe consists of free electrons and protons, not bound together into neutral atoms. As a result, magnetic fields are a dominant influence on the motion of matter throughout most of the universe (Fig. 27–18).

Application

Earth's Magnetosphere

The earth possesses a magnetic field whose origin we will consider in the next chapter. Near earth, the field is approximately that of a dipole, and provides us a means of finding our way around the planet using magnetic compasses. Were there no external influences, earth's field would exhibit the dipole form with its $1/r^3$ drop-off out to large distances. But the earth and the other planets are immersed in a wind of charged particles emanating from the sun and sweeping through the solar system at speeds of typically 400 km/s. The origin of this wind probably involves magnetic interactions in the sun's atmosphere. Because charged particles are deflected into circular paths when they move at right angles to a magnetic field, the solar wind cannot readily penetrate earth's magnetic field. The field acts like an obstacle in the path of the wind. A huge shock wave, analogous to an airplane's sonic boom, forms about 65,000 km above the sunward side of earth. After crossing the shock, most of the solar wind is deflected around earth's magnetic field. In the process the field is compressed on the sunward side and drawn out into a long "magnetotail" on earth's night side (Fig. 27–19).

magnetosphere

The region in which the field is confined by the solar wind is termed the **magnetosphere,** and is the region in which earth's field is the dominant influence on charged particles. Occasionally high-energy particles from the sun do penetrate into the field, where they become trapped on field lines. They move back and forth between northern and southern hemispheres, entering earth's atmosphere near the poles where the field lines converge. Spiralling down the field lines, these high-

Fig. 27–19
The interaction of earth's magnetic field with the solar wind gives rise to a bow shock wave on the sunward (daytime) side of earth, and a magnetotail stretching beyond the nighttime side. The magnetopause marks the boundary of the magnetosphere—the region where earth's magnetic field is the dominant influence on charged particles. The actual magnetosphere is less symmetric than shown here, because earth's magnetic dipole and rotation axis are not aligned.

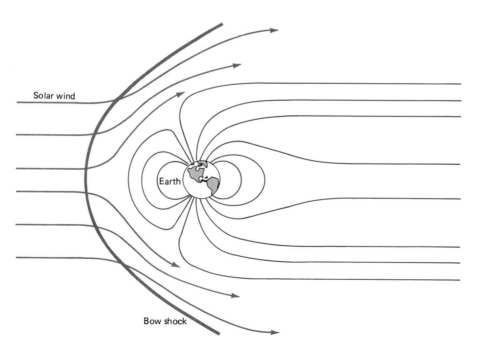

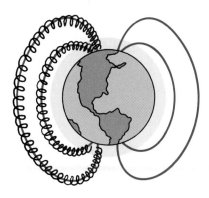

Fig. 27–20
The aurora arises from the interaction of high-energy particles trapped in earth's magnetic field with the upper atmosphere. (Left) Schematic diagram showing particles trapped on earth's magnetic field lines. (Center) The aurora can be seen almost every night from extreme northern latitudes, but only rarely from as far south as New York City. (Right) From high in space, the Dynamics Explorer I spacecraft took this ultraviolet image of the aurora surrounding earth's north magnetic pole.

energy particles interact with atmospheric atoms to produce auroras (Fig. 27–20). Variations in the solar wind, caused by the temporal and spatial variation of magnetic fields on the sun, buffet earth's magnetic field, giving rise to "magnetic storms" that can severely disrupt communications, electric power transmission, and even compass needles.

Studies of "fossil" magnetism in rocks indicate that earth's magnetic field reverses every half million years or so, with north and south poles interchanging. Acting like a giant tape recorder, newly formed rocks "remember" the field direction at their formation, providing geophysicists with evidence for the spreading of ocean floors over millions of years (Fig. 27–21). During the reversal the field is

Fig. 27–21
The direction of earth's magnetic field is "frozen" into the rocks formed as the Atlantic Ocean floor widens at the mid-Atlantic ridge. Reversals of earth's field result in strips of rock with opposite magnetic polarity. This figure shows a region adjacent to the mid-Atlantic ridge off Iceland; colored bands are rocks whose magnetization is in the same direction as earth's present field. White zones between colored bands represent rock whose magnetization is opposite earth's present field. Lightest colored rocks are the most recently formed, with darker coloring indicating older rocks.

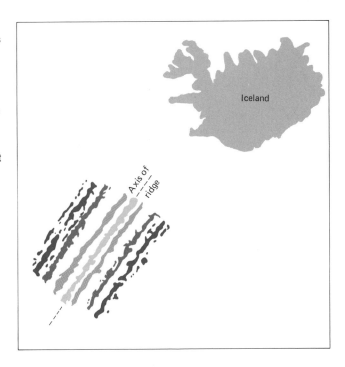

weakened, allowing more high-energy particles to reach earth's surface. Some scientists believe that these particles result in high rates of mutation and species extinctions. Of more immediate interest is the possibility that short-term changes in weather are influenced by solar particles and their interaction with earth's magnetic field. Much research effort is being devoted to the sun-earth connection, in which the magnetic field plays a central role.

Application

The Cyclotron

Physicists studying the basic structure of matter need tools that can probe the atomic nucleus and its constituent particles. The only probes sufficiently small are subatomic particles themselves, accelerated to high enough energies that they can disrupt the strong nuclear force. How is this acceleration accomplished? One way is to accelerate particles of charge q through a large potential difference V, giving each particle energy qV. But there are practical problems in the generation and handling of potential differences much over a million volts (see Application: Corona Discharge, in Chapter 23). To achieve higher energies, devices are used that circumvent the need for a single large potential difference. One of the earliest and most successful such devices is the **cyclotron** (Fig. 27–22). The device consists of an evacuated chamber between the poles of a magnet. At the center of the chamber is a source of the particles to be accelerated, usually protons or light ions. The ions undergo circular motion in the magnetic field.

Initially the radii of the orbits are small, since the ion energy is small. But also within the evacuated chamber are two hollow conducting structures each shaped like the letter D (Figs. 27–22 and 27–23). A modest potential difference is applied across these two "dees," and this potential difference is made to alternate in polarity with the same frequency as the circular motion of the ions. Recall that this frequency depends only on the magnetic field strength and the charge-to-mass ratio of the ions, but not on their energy. As the ions circle around in the cyclotron, they are accelerated across the gap between the dees by the strong electric field associated with the potential difference. Because each dee is a hollow, nearly closed conducting structure, there is no electric field within the dees.

Once inside a dee, the ions simply follow a circle in the magnetic field. Halfway round they again cross the gap between dees. If the electric field were steady, the ions would be *decelerated* at this crossing. But the electric field changes direction in step with the ions' circular motion, so that each time the ions cross the dee gap they are accelerated and gain more energy. They move faster and in ever larger circles, but always with the same orbital period. Eventually the ion orbits become nearly the size of the machine. At this point an electrostatic field provided by a high-voltage electrode deflects the ions out of the magnetic field and toward a target, where their interactions with target nuclei cause nuclear reactions.

In addition to providing experimental data on nuclear structure, cyclotrons are valuable in producing short-lived radioactive isotopes for a variety of purposes, particularly medical research and diagnosis. A number of large hospitals have their own cyclotrons, which produce radioactive carbon, oxygen, and nitrogen used in clinical and research procedures (Fig. 27–24).

At very high energies, the theory of relativity comes into play, and alters our conclusion that the cyclotron frequency is independent of particle energy. As a result, the cyclotron cannot be used to achieve these relativistic energies. An alternate accelerator design is the **synchrotron**, in which both the magnetic field and frequency of an alternating electric field are varied to account for increasing particle energy, while the orbital radius is held constant. Figure 27–25 shows the huge—2-km-diameter—proton synchrotron at Fermilab near Chicago.

cyclotron

synchrotron

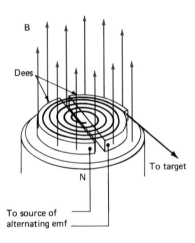

Fig. 27–22
A cyclotron, showing one magnet pole, dees, and a typical ion trajectory. Not shown are the vacuum chamber surrounding the dees, and frame supporting pole pieces and electromagnetic windings.

Fig. 27–23
A dee from the first cyclotron, developed in 1934 by E. O. Lawrence and M. S. Livingston. Lawrence was awarded the 1939 Nobel Prize in physics for his work on the cyclotron.

Example 27–4

Designing a Cyclotron

A cyclotron is to accelerate protons to a kinetic energy of 5.0 MeV. If the magnetic field in the cyclotron is 2.0 T, what must be the radius of the cyclotron and the frequency at which the dee voltage is alternated?

Solution

The cyclotron frequency is given by Equation 27–5:

$$f = \frac{qB}{2\pi m} = \frac{(1.6 \times 10^{-19}\ \text{C})(2.0\ \text{T})}{(2\pi)(1.7 \times 10^{-27}\ \text{kg})} = 3.0 \times 10^{7}\ \text{Hz}.$$

This is the frequency required to accelerate protons at each crossing of the dee gap; incidentally, it is about the frequency of a citizens band (CB) radio transmitter.

An energy of 5.0 MeV is equal to

$$(5.0 \times 10^{6}\ \text{eV})(1.6 \times 10^{-19}\ \text{J/eV}) = 8.0 \times 10^{-13}\ \text{J},$$

so the proton kinetic energy is

$$K = \tfrac{1}{2}mv^{2} = 8.0 \times 10^{-13}\ \text{J}.$$

Solving for the speed v gives

$$v = \sqrt{\frac{2K}{m}} = \left[\frac{(2)(8.0 \times 10^{-13}\ \text{J})}{1.7 \times 10^{-27}\ \text{kg}}\right]^{1/2} = 3.1 \times 10^{7}\ \text{m/s}.$$

Equation 27–3 then gives the radius needed to accommodate 5-MeV protons:

$$r = \frac{mv}{qB} = \frac{(1.7 \times 10^{-27}\ \text{kg})(3.1 \times 10^{7}\ \text{m/s})}{(1.6 \times 10^{-19}\ \text{C})(2.0\ \text{T})} = 0.16\ \text{m}.$$

To ensure a uniform magnetic field over the particle trajectories, the radii of the magnet pole pieces would have to be somewhat larger than this value.

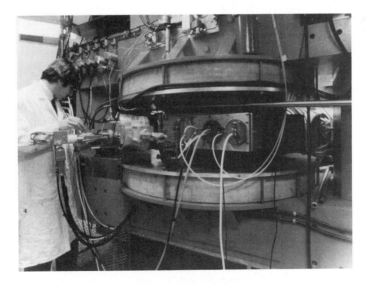

Fig. 27–24
Cyclotron at Massachusetts General Hospital accelerates deuterons (heavy hydrogen nuclei) to 6 MeV. The deuterons strike target materials, producing short-lived isotopes of oxygen, nitrogen, and carbon for medical studies.

Fig. 27–25
The 2-km-diameter proton synchrotron at Fermilab accelerates protons to 1 Tev (10^{12} eV). The main accelerator ring is in the background; in the foreground is a Fermilab office building and cooling pond.

Application

Fusion Reactors

One of the more promising energy sources for the future is nuclear fusion, the same process that powers the sun and stars. Fusion occurs when light nuclei join or fuse to make heavier nuclei. Unlike nuclear fission, which requires heavy elements such as uranium, nuclear fusion reactors could use hydrogen from sea water as fuel. Proposed fusion reactors would use only deuterium—the heavy isotope of hydrogen that makes up only about one ten-thousandth of naturally occurring hydrogen. Nevertheless, deuterium fusion releases so much energy that the deuterium in a gallon of sea water contains the energy equivalent of 400 gallons of gasoline!

The problem with fusion is that nuclei must be brought very close together before the strong nuclear force becomes effective in binding them. But nuclei are positively charged. To get them close together requires work to overcome the very strong electric field. One way to accomplish this is to heat an ionized gas to very high temperatures. Then some of the positive nuclei in the gas will have enough energy that when they collide they may overcome the electric repulsion and get close enough to fuse. For a practical fusion reactor, temperatures around 100 million kelvins are required!

How are we to contain a gas at such temperatures? A material container would either melt or, through collisions of gas particles with the walls, cool the gas to the point where fusion was not possible.

The sun readily provides containment with its huge gravity, but this option is not available to us. Because the hot gas consists of electrically charged particles, it interacts with magnetic fields. Some promising approaches to controlled fusion involve containment in "magnetic bottles." A uniform magnetic field would keep par-

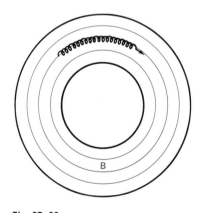

Fig. 27–26
Charged particles spiral about the circular field lines in a simplified fusion reactor. With its toroidal shape, the machine has no "ends" from which magnetic field lines emerge.

ticles from moving sideways to the field, but they could still escape out the "ends" of the field. The "bottle" can be "plugged" by making the field stronger near the ends, but there is still some leakage. Another approach is to eliminate the ends altogether, bending the magnetic field lines into a circular configuration (Fig. 27–26). Most fusion machines take this approach, giving a reaction chamber which is toroidal (doughnut-like) in shape. A common toroidal design is the **tokamak,** a name reflecting its Russian origin (Fig. 27–27). Current fusion research centers on field configurations in which the ionized gas is in stable equilibrium.

tokamak

Fig. 27–27
Tokamak Fusion Test Reactor at Princeton University, shown under construction.

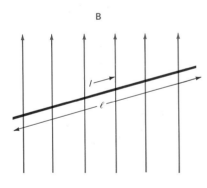

Fig. 27–28
A straight wire carrying current through a magnetic field.

27.4
THE MAGNETIC FORCE ON A CURRENT

So far we have considered the magnetic force on individual charged particles. An electric current is simply a group of charged particles sharing a common motion, so we should expect a current to interact with a magnetic field.

The Force on a Straight Wire

Imagine a long straight wire carrying a current I through a magnetic field **B,** as shown in Fig. 27–28. If each charge carrier has a drift velocity $\mathbf{v}_d$ along the wire, and if each carries charge q, then the magnetic force on each charge carrier is given by Equation 27–2:

$$\mathbf{F}_q = q\mathbf{v}_d \times \mathbf{B}.$$

(Of course the charge carriers also have random thermal velocities. But these average to zero and therefore make no contribution to the magnetic force.) If the wire has length ℓ and cross-sectional area A, and contains n charge carriers per unit volume, then the net force on all the charge carriers in the wire is

$$\mathbf{F} = nA\ell q\mathbf{v}_d \times \mathbf{B}.$$

The product $nAqv_d$ is just the current I, as we found in deriving Equation 25–3 two chapters ago. If we define a vector $\boldsymbol{\ell}$ whose magnitude is the length ℓ of the wire and whose direction is along the current, then we can write

$$\mathbf{F} = I\boldsymbol{\ell} \times \mathbf{B}. \qquad \text{(magnetic force on a current)} \qquad (27\text{–}7)$$

The direction of this force is at right angles to both the current and the magnetic field, or out of the page in Fig. 27–28. For a given direction of the current, the direction of the force does not depend on the sign of the charge carriers. If the current is to the right, then positive charges move to the right and the force on each is out of the page. If the charges are negative, they move to the left and the force is still out of the page because both the sign of the velocity and the sign of the charge are reversed, giving the same sign for the force $q\mathbf{v} \times \mathbf{B}$.

Strictly speaking, Equation 27–7 gives the net magnetic force only on the charge carriers in the wire. But the motion of the charge carriers—typ-

Fig. 27–29
Electrons moving to the left are deflected upward by the magnetic force, resulting in charge separation and an upward electric force on the fixed ions. As a result the entire wire is ultimately influenced by the magnetic force on the moving electrons.

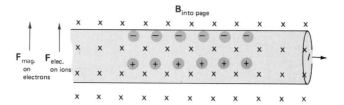

ically electrons—under the influence of the magnetic force causes charge separation in the wire, and the resulting electric field exerts a force on the fixed charges—typically ions—in the wire (Fig. 27–29). Thus the entire wire experiences the force. Although its origin is not entirely magnetic, we loosely call the force given by Equation 27–7 the magnetic force on the wire.

The magnetic force on a current-carrying wire is the basis for many practical devices, including electric motors that start cars and run refrigerators, clocks, record turntables, subway trains, pumps, food processors, power tools, and myriad other useful instruments of modern society.

Example 27–5

The Force on a Straight Wire

A power line runs along earth's equator, where earth's magnetic field points horizontally from south to north and has a strength of about 0.5 G. The current in the power line is 500 A, flowing from west to east. What are the magnitude and direction of the magnetic force on 1 km of the power line?

Solution

Let eastward be the x-direction, northward the y-direction, and upward the z-direction. After we convert gauss to tesla, Equation 27–7 gives

$$\mathbf{F} = I\boldsymbol{\ell} \times \mathbf{B} = (500 \text{ A})(1000\hat{\mathbf{i}} \text{ m}) \times (0.5 \times 10^{-4}\,\hat{\mathbf{j}} \text{ T}) = 25\hat{\mathbf{k}} \text{ N}.$$

This 25-N upward force is negligible compared with the weight—on the order of 2×10^4 N—of 1 km of power line.

Application

Magnetic Levitation

Problems with wheels on tracks limit the speeds of conventional trains to a maximum of about 300 kilometers per hour. These problems disappear if the train is suspended above its rails, which can be done using magnetic force. Engineers in a number of countries are actively developing such magnetically suspended vehicles (Fig. 27–30). Riding only centimeters above a conducting rail, a magnetically levitated vehicle should be capable of 500-kilometer-per-hour speeds. The magnetic fields—on the order of 2 T—needed to suspend the vehicle would be provided by superconducting electromagnets. These magnets would be part of the vehicle, and would interact with currents in the conducting rails below. Those currents themselves would be generated by the motion of the train, as we will see in subsequent chapters. Magnetic-levitation vehicles should prove especially effective for travel between cities in densely populated regions.

Fig. 27–30
This magnetically levitated train is being developed by the Japanese National Railway.

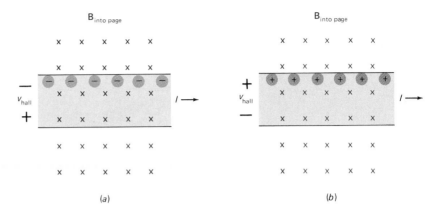

Fig. 27–31
Although the magnetic force on a wire is independent of the sign of the charge carriers, the direction of the electric field and potential difference arising from the charge separation does depend on the sign of the charge carriers. *(a)* Negative and *(b)* positive charge carriers.

Application

The Hall Effect

We have seen that the force on a current is independent of the sign of the charge carriers. In Fig. 27–29, for example, electrons moving to the left (and therefore carrying a current to the right) are forced upward by the magnetic field. To carry the same current, positive charge carriers would move to the right and the product $q\mathbf{v} \times \mathbf{B}$ would still be upward (Fig. 27–31). But something is different in the two cases. The top of the conductor becomes negatively charged for negative charge carriers and positively charged if the charge carriers are positive. This phenomenon is known as the **Hall effect.** The electric field associated with this effect gives rise to a potential difference between top and bottom of the conductor. The sign of this **Hall potential** depends on the sign of the charge carriers. Measurement of the Hall potential gives information about the nature of the charge carriers and therefore tells us about conduction processes in different materials. Furthermore, in a given conductor carrying a known current, the Hall potential is directly proportional to the magnetic field strength, so that the Hall effect provides a simple way to measure the magnetic field (Fig. 27–32). Problem 46 explores the Hall effect more quantitatively. The Hall effect turns out to be quantized—it occurs in discrete steps. The discovery of the quantized Hall effect won Klaus von Klitzing the 1985 Nobel Prize in Physics.

Hall effect

Hall potential

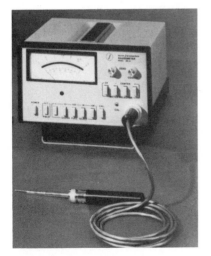

Fig. 27–32
This magnetometer uses the Hall potential developed in its probe to measure magnetic fields.

Nonuniform Fields and Curved Conductors

Equation 27–7 applies to a straight conductor at some arbitrary angle to a uniform magnetic field. What if the conductor bends, so that its orientation relative to the field changes? Or what if the field changes in magnitude or direction over the length of the conductor? We can still make use of Equation 27–7 if we apply it only to a very small segment of the conductor, so small that it is essentially straight and the field essentially constant over its length. If the length of this small segment is $d\boldsymbol{\ell}$, then Equation 27–7 gives

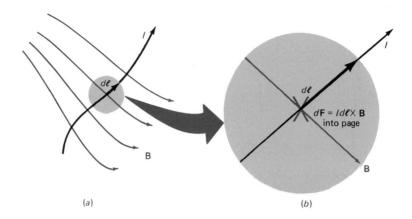

Fig. 27–33
(a) A curved conductor in a nonuniform magnetic field. (b) A small enough segment of the conductor can be treated as a straight wire in a uniform field.

$$d\mathbf{F} = I \, d\boldsymbol{\ell} \times \mathbf{B} \tag{27–8}$$

for the small force on the segment (Fig. 27–33). To find the total force on our conductor, we sum the forces on all such segments. In the limit of very small segments, this sum becomes an integral, and we have

$$\mathbf{F} = \int d\mathbf{F} = \int I \, d\boldsymbol{\ell} \times \mathbf{B}. \tag{27–9}$$

The integration is taken over the entire section of conductor on which we are calculating the force.

Example 27–6
The Force on a Curved Conductor

A semicircular wire connects two points C and D a horizontal distance $2R$ apart. The wire carries a current I from C to D, and is in a uniform magnetic field **B** pointing upward, as in Fig. 27–34. Show that the magnetic force on this semicircular wire is the same as the force that a straight wire from C to D would experience if it carried the same current I.

Solution

Since the orientation of the wire relative to the field varies, we must use the integral equation 27–9 to calculate the force. In Fig. 27–35a, we show a small segment $d\ell$ of the wire, and the angle θ that specifies its position on the semicircular arc. We make the segment infinitesimally small, so that it is essentially straight. This infinitesimal segment subtends an infinitesimal angle $d\theta$, as shown in Fig. 27–35a. A blown-up view of our segment (Fig. 27–35b) shows that the angle between the segment $d\ell$ and the magnetic field **B** is also θ. Then the force on the segment is given by Equation 27–8:

$$d\mathbf{F} = I \, d\boldsymbol{\ell} \times \mathbf{B}.$$

The magnitude of this force element $d\mathbf{F}$ is then

$$dF = I \, d\ell \, B \sin\theta.$$

Application of the right-hand rule shows that the direction of the force element is out of the page for all values of θ on our semicircle—that is, for $0 \le \theta \le \pi$. Therefore,

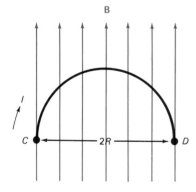

Fig. 27–34
Example 27–6

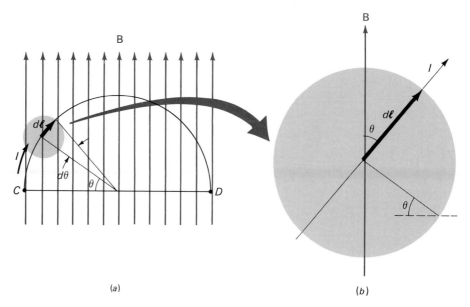

(a) (b)

Fig. 27–35
(a) The position of an infinitesimal segment $d\ell$ is specified by the angle θ. The segment subtends the infinitesimal angle $d\theta$. *(b)* A blow-up of the segment shows that the angle θ is also the angle between $d\ell$ and the magnetic field **B**.

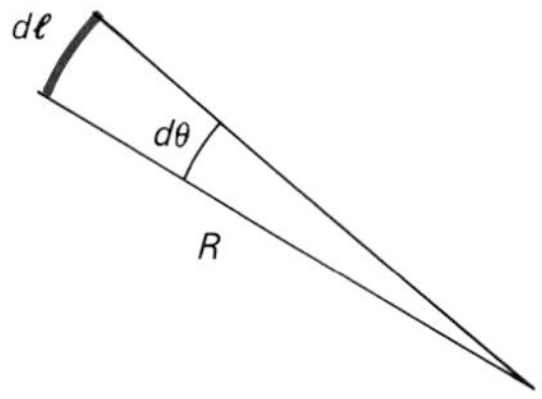

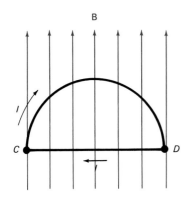

Fig. 27–36
An angle in radians is defined as the ratio of subtended arc to radius. Here the angle $d\theta$ is so small that the difference between a circular arc and the straight segment of length $d\ell$ is insignificant, so that $d\ell = Rd\theta$.

the integral in Equation 27–9 describes a sum of vectors all of which point in the same direction, so that the magnitude of the net force is

$$F = \int dF = \int_0^\pi IB \sin\theta \; d\ell.$$

But $d\ell$ subtends the angle $d\theta$, so that $d\ell = Rd\theta$ (Fig. 27–36). Then, taking the constants I, B, and R outside the integral, we have

$$F = IBR \int_0^\pi \sin\theta d\theta = IBR \left. (-\cos\theta) \right|_0^\pi$$
$$= IBR \left[-(-1 - 1) \right] = IB(2R).$$

This is just the force we would get from Equation 27–7 for a straight wire of length $2R$ carrying current I perpendicular to a magnetic field **B**.

The result of this example implies that the net force on a closed current loop like that of Fig. 27–37 is exactly zero. This result is in fact true regardless of the loop shape, as long as it is in a *uniform* magnetic field (see Problem 30).

Fig. 27–37
A closed current loop made from a semicircle and straight wire. Since the semicircle and straight wire carry currents in opposite directions and yet experience forces of the same magnitude, the net force on the closed loop must be zero.

27.5
A CURRENT LOOP IN A MAGNETIC FIELD

Consider a closed, current-carrying loop in a magnetic field. We can approximate such a closed loop with a nearly closed structure fed by two parallel and closely spaced wires as shown in Fig. 27–38. With only a small gap where the wires join, the effect of a magnetic field on the open loop

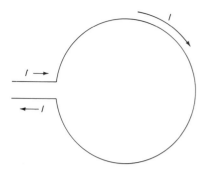

Fig. 27–38
A closed loop may be approximated by a nearly closed loop fed by closely spaced wires.

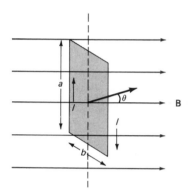

Fig. 27–39
A rectangular loop in a uniform magnetic field.

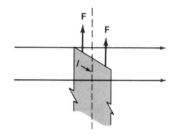

Fig. 27–40
Torques due to the forces on the loop top cancel in pairs, giving no net torque on the top of the loop alone.

will be nearly the same as on a closed loop. The wires feeding the loop are adjacent and carry current in opposite directions, so that any magnetic forces or torques associated with these feed wires essentially cancel. We could eliminate the feed wires by using a completely closed superconducting loop, in which a current will persist indefinitely. Or, at the microscopic level, an electron circling an atomic nucleus constitutes a miniature current loop.

For simplicity we consider a rectangular loop in a uniform magnetic field, as shown in Fig. 27–39. The normal to the plane of the loop makes an angle θ with the magnetic field. The net force on the loop is the sum of the forces on its four sides. Since the current flows in opposite directions on opposite sides, and since the field is uniform, the forces on the top/bottom and right/left sides of the loop cancel in pairs, and there is no net magnetic force on the loop. However, there can still be a net torque. In Chapter 11 we showed that when there is no net force on an object, then the torque is independent of the choice of pivot. For convenience, we can therefore choose our pivot to be a vertical axis through the center of the loop.

Consider first the top of the loop. The current runs in the same direction all across the top, so the magnetic force on all elements of the top has the same direction—namely, upward. Therefore, elements on opposite sides of the pivot contribute torques in opposite directions, so there is no net magnetic torque on the top of the loop (Fig. 27–40). Similar considerations hold for the bottom of the loop.

What about the vertical sides? In Fig. 27–41 we look down on the loop, and show the magnetic forces resulting from the upward- and downward-flowing currents in the vertical sides. From the figure, it is clear that these forces give rise to a net torque tending to twist the loop clockwise. Each vertical side is a straight wire of length a carrying current I at right angles to the horizontally directed magnetic field **B**, so the magnitude of the force on each side is simply

$$F_{\text{side}} = IaB.$$

The distance from the central axis pivot to the sides where this force acts is half the loop width, or $b/2$. The geometry of Fig. 27–41 shows that the angle used in calculating the torque is the same as the angle θ between the loop normal and the magnetic field. Therefore, the torque due to the force on each side is

$$\tau_{\text{side}} = F\frac{b}{2}\sin\theta = Ia\frac{b}{2}B\sin\theta.$$

Accounting for the contributions from both sides gives the magnitude of the net torque on the loop:

$$\tau = IabB\sin\theta.$$

We can express the torque in vector notation if we define a vector **A** whose magnitude is the area ab of the loop and which is perpendicular to the loop. We choose the direction of **A** by the right-hand rule: wrap your fingers around the loop in the direction of the current and your thumb points in the direction of **A.** Then we can write

$$\boldsymbol{\tau} = I\mathbf{A} \times \mathbf{B}. \qquad (27\text{–}10)$$

Although we derived this equation for a rectangular loop, it holds in fact for any current loop. The torque on a current loop depends on the current, the loop area, the magnetic field, and the orientation between loop and field.

Equation 27–10 should remind you of Equation 22–4 for the torque on an electric dipole in an electric field. There we had

$$\boldsymbol{\tau} = \mathbf{p} \times \mathbf{E},$$

with **p** the electric dipole moment and **E** the electric field. Comparison with Equation 27–10 suggests that a current loop in a magnetic field behaves analogously to an electric dipole in an electric field. Recognizing this analogy, we call the quantity I**A** the **magnetic dipole moment** $\boldsymbol{\mu}$ of the current loop:

$$\boldsymbol{\mu} = I\mathbf{A}. \qquad \text{(single-turn loop)}$$

More generally, if a loop consists of N turns of conducting wire, each turn contributes I**A** to the magnetic moment, which becomes

$$\boldsymbol{\mu} = NI\mathbf{A}. \qquad \text{(magnetic dipole moment)} \qquad (27\text{–}11)$$

Clearly the units of magnetic moment are A·m². Using the magnetic moment vector, Equation 27–10 then becomes

$$\boldsymbol{\tau} = \boldsymbol{\mu} \times \mathbf{B}, \qquad \text{(torque on a current loop)} \qquad (27\text{–}12)$$

in analogy with the electric case. The magnetic moment vector of a current loop is perpendicular to the plane of the loop, and the torque given in Equation 27–12 tends to align the magnetic moment with the field (Fig. 27–42). It takes work to twist the loop out of alignment with the field. In exact analogy with the electric-dipole case summarized in Equation 22–6 of Chapter 22, we express the associated potential energy as

$$U_{\text{magnetic}} = -\boldsymbol{\mu} \cdot \mathbf{B}. \qquad (27\text{–}13)$$

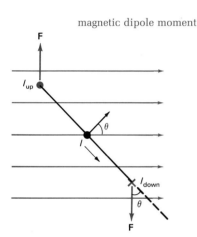

magnetic dipole moment

Fig. 27–41
Top view of loop, showing that magnetic forces on the vertical sides give rise to a net torque. The angles used in the torque calculation are the same as the angle θ between the loop normal and the magnetic field.

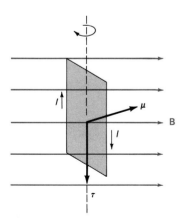

Fig. 27–42
The torque on a current loop tends to align the loop's magnetic-moment vector with the magnetic field.

Example 27–7
A Current Loop

A circular loop of radius 5.0 cm consists of 10 turns of wire. A current of 3.0 A flows in the wire. What is the magnitude of the loop's magnetic moment? Initially the magnetic moment is aligned with a uniform magnetic field of 100 G. The loop is then turned 90° from its original orientation. How much work does this take? How much torque is required to hold the loop in its new orientation?

Solution

The magnetic moment is the product of the loop area, current, and number of turns. As described by Equation 27–11:

$$\mu = NIA = (10)(3.0 \text{ A})(\pi)(0.050 \text{ m})^2 = 0.24 \text{ A·m}^2.$$

The loop's initial potential energy U_1 is given by Equation 27–13, with the angle θ being zero in the dot product, so that

$$U_1 = -\boldsymbol{\mu} \cdot \mathbf{B} = (-0.24 \text{ A·m}^2)(0.010 \text{ T})(\cos 0°) = -2.4 \times 10^{-3} \text{ J}.$$

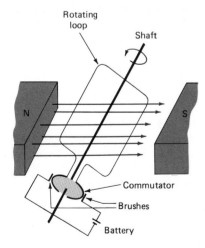

Fig. 27–43
Electric motors are made in a variety of sizes and power outputs. The large motor at bottom drives the world's largest fully steerable radio telescope.

Fig. 27–44
A simple electric motor.

In the new orientation, with magnetic moment and field at right angles, the potential energy U_2 is zero. The work involved in twisting the loop is equal to the change in potential energy, or

$$U_2 - U_1 = 0 \text{ J} - (-2.4 \times 10^{-3} \text{ J}) = 2.4 \times 10^{-3} \text{ J}.$$

This work is positive, as it should be because we are twisting the loop away from its lowest-energy orientation.

The magnitude of the torque needed to hold the new orientation is given by the magnitude of the cross product in Equation 27–12:

$$\tau = \mu B \sin\theta = (0.24 \text{ A·m}^2)(0.010 \text{ T})(\sin 90°) = 2.4 \times 10^{-3} \text{ N·m}.$$

Application

Electric Motors

Electric motors are so much a part of our lives that we hardly think of them. But sewing machines, car starters, refrigerators, vacuum cleaners, power saws, subway trains, food processors, tape recorders, washing machines, fans, hair dryers, water pumps, oil burners, and most industrial machinery would be difficult or impossible to build without electric motors (Fig. 27–43).

At the heart of every electric motor is a current loop in a magnetic field. But instead of a steady current, the loop carries a current that reverses periodically. In direct-current (DC) motors, this reversal is achieved through the electrical contacts that provide current to the loop. Figure 27–44 shows a simple motor consisting of a loop in the field of a permanent magnet. Current from an external battery reaches the loop through a set of stationary "brushes," which make contact with a pair of semicircular conductors called the "commutator." The commutator is attached rigidly to the loop, which rotates to align its dipole moment with the field. Just as it reaches alignment, however, the brushes cross the gaps in the commutator. This crossing reverses the connections of loop to battery, resulting in a reversal of the loop current. This in turn reverses the magnetic moment of the loop, so that it is no longer aligned with the field. The loop then rotates another 180° to align its new magnetic moment vector with the field. Just as it reaches alignment, the current again reverses. This process repeats, resulting in continuous rotation of the loop. A rigid shaft through loop and commutator delivers useful mechanical work to the device powered by the motor. The source of this work is the battery or whatever supplies electrical energy to the motor. The motor itself is a device that converts electrical to mechanical energy; the magnetic field is an intermediary in this conversion.

What determines the rotational speed of the motor? Why doesn't it undergo a constant rotational acceleration, reaching ever greater speeds? The answer to these questions lies in another deep interaction between electricity and magnetism, an interaction that we will explore in Chapter 29.

27.6
MAGNETIC MATTER

We began this chapter with a discussion of something familiar—the interaction of magnets. We quickly moved on to consider the behavior of moving electric charges in magnetic fields, and found that a current loop, consisting of moving charges, acts like a magnetic dipole. How are we to reconcile our new understandings of electric charges and magnetism with our older knowledge of magnets?

In fact the two are one and the same phenomenon. When an electron orbits an atomic nucleus, its circular motion and charge constitute a miniature current loop, giving rise to a magnetic dipole moment. In addition the electron possesses an intrinsic magnetic dipole moment associated with its "spin," or intrinsic angular momentum. Interactions among these magnetic moments determine the magnetic properties of individual atoms and of bulk matter. Although an accurate description of magnetism in matter necessarily involves quantum mechanics, we can nevertheless use our knowledge of magnetic dipoles to gain a qualitative understanding of magnetic matter.

Ferromagnetism

Experimentally, we observe three types of magnetic behavior in macroscopic matter. **Ferromagnetism**—the magnetic behavior you are most familiar with—is actually limited to a few substances, notably the elements iron, nickel, cobalt, and a few compounds of other elements. In a ferromagnetic material, a quantum-mechanical interaction among nearby atomic magnetic moments results in regions—called **magnetic domains**—in which all the atomic magnetic moments point in the same direction. A typical domain contains 10^{17} to 10^{21} atoms, and occupies a volume on the order of 10^{-12} to 10^{-8} m^3 (Fig. 27–45). The magnetic moment of a single domain can be large, since it is the sum of many atomic moment vectors all pointing in the same direction. A typical piece of ferromagnetic material, however, contains many domains with their moments in random directions, and therefore exhibits no net magnetic moment. But when the material is placed in an external magnetic field, a net magnetic moment develops. This

ferromagnetism

magnetic domains

Fig. 27–45
Pattern of magnetic domains in a cobalt-samarium casting. Domains with different magnetization directions are distinguished by their effects on polarized light. Although strong magnets can be made from cobalt-samarium, how can you tell that this sample is not a magnet?

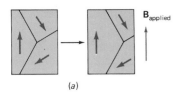

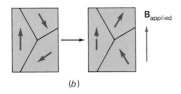

Fig. 27–46
Domain changes resulting in a net magnetic moment in a ferromagnetic material occur through *(a)* domain growth and *(b)* domain realignment.

Curie temperature

ferrimagnetism

paramagnetism

occurs because domains that already happen to be aligned with the field can grow by realignment of individual atomic moments in adjacent domains (Fig. 27–46a). In addition, the magnetic moments of entire domains can rotate (Fig. 27–46b). Since any dipole experiences a force in a nonuniform field, a piece of ferromagnetic material in a nonuniform magnetic field experiences a net force; this is why pieces of iron and other ferromagnetic materials are attracted to magnets even though they may not themselves be magnets.

Removal of the applied magnetic field does not entirely destroy the overall alignment that gives rise to a net magnetic moment in a ferromagnetic substance. This remanent magnetization is what allows permanent magnets to exist. In so-called hard ferromagnetic materials, the remanent magnetism is strong; these materials are used specifically for making permanent magnets. Remanent magnetism is relatively weak in soft ferromagnetic materials; these materials are used in applications like heads for tape recorders and computer disk drives where permanent magnetism is undesirable (the tapes and disks themselves would use harder materials to facilitate permanent information storage).

Random thermal motions tend to destroy the alignment of individual magnetic moments in a material. As a result, ferromagnetic effects weaken with increasing temperature. Above the so-called **Curie temperature,** ferromagnetism ceases altogether. Curie temperatures for the common ferromagnetic elements nickel, iron, and cobalt are 631 K, 1043 K, and 1395 K, respectively. The rarer ferromagnetic elements dysprosium and gadolinium have much lower Curie temperatures of 85 K and 289 K. The disappearance of ferromagnetism at the Curie temperature is an instance of a phase transition, analogous to the solid/liquid/gas transitions we studied in Chapter 17.

When a normally ferromagnetic element like iron is in a compound with other elements, the magnetic properties of the other elements may modify the magnetic behavior of the material. The result is **ferrimagnetism**—behavior resembling ferromagnetism but with more complex temperature dependence. The naturally occurring magnetic material magnetite (Fe_3O_4; lodestone) is ferrimagnetic. Ferrimagnetic materials are often used when magnetic materials are needed in electronic components because their higher resistivity results in lower power dissipation.

Paramagnetism

Many substances that are not ferromagnetic nevertheless consist of atoms or molecules with permanent magnetic moments. What distinguishes these **paramagnetic** materials from ferromagnetic materials is the absence of a strong interaction that tends to align nearby moments. As a result, individual atomic moments are not organized into domains, but point in random directions. Nevertheless, an externally applied field brings the atomic magnetic moments into some degree of alignment. At all but the coldest temperatures, this alignment is far less complete than in a ferromagnetic material, so that a paramagnetic material exhibits far weaker magnetic interactions. For example, paramagnetic substances are attracted only weakly to permanent magnets (Fig. 27–47).

Diamagnetism

Even materials with no intrinsic atomic magnetic moments can have magnetic moments induced when a magnet approaches or in other instances when the magnetic field in the material changes. Such materials are termed **diamagnetic.** In contrast to paramagnetic and ferromagnetic materials, diamagnetic materials are weakly repelled by magnets. We will explore the origins of diamagnetism, and the reason for this repulsive interaction, in Chapter 29.

Actually, the effects responsible for diamagnetism occur in all materials, but in paramagnetic and ferromagnetic materials they are generally overwhelmed by the influence of permanent atomic magnetic dipoles. But diamagnetism, unlike paramagnetism and ferromagnetism, does not weaken with increasing temperature, so that diamagnetic effects become relatively more prominent as temperature increases.

diamagnetism

Fig. 27–47
Liquid oxygen (O_2), one of the more strongly paramagnetic materials, is attracted to the poles of a magnet.

Magnetic Susceptibility

We can characterize quantitatively the effect of atomic magnetic dipole moments just as we did the effect of electric dipole moments in Chapter 22. There, we found that atomic electric dipoles align with an electric field to reduce the field within the material; we introduced the dielectric constant κ as the factor by which the field is reduced.

In a magnetic material, alignment of atomic magnetic moments has the opposite effect: the field within the material is *increased*. Figure 27–48 shows the reason for this. An electric dipole consists of two separated point charges. The strongest field associated with this dipole is the internal field pointing from the positive to the negative charge. When the dipole aligns with an applied electric field, this internal field is *opposite* to the applied field, and therefore serves to reduce it. A magnetic dipole, in contrast, arises from a current loop. There are no magnetic charges—monopoles—on which the field lines begin and end. Instead, as we will state more quantitatively in the next chapter, the magnetic field lines form closed loops.

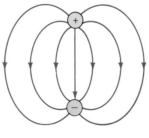

Electric dipole

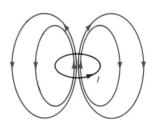

Magnetic dipole

Fig. 27–48
Although the electric and magnetic fields far from electric and magnetic dipoles have the same form, the fields *within* the atomic structure giving rise to the dipole moment have opposite directions. This is because electric dipoles ultimately arise from two separated point charges, while the magnetic dipoles arise from current loops.

When the loop's magnetic moment vector aligns with an applied magnetic field, the strong magnetic field within the loop is in the *same* direction as the applied field, and therefore the two fields reinforce.

relative permeability

In analogy with dielectric constant, we introduce the quantity κ_M, called the **relative permeability,** as the factor by which the magnetic field within a material increases because of the alignment of magnetic dipoles within the material. For paramagnetic materials, κ_M is slightly greater than 1; for ferromagnetic materials, it is much greater than 1. In diamagnetic materials, the dipoles align anti-parallel to the field; this makes $\kappa_M < 1$ for these materials. Because the relative permeabilities of diamagnetic and paramagnetic materials are very close to 1, it is more convenient to work with

magnetic susceptibility

the **magnetic susceptibility,** defined by

$$\chi_M = \kappa_M - 1.$$

The internal field in a magnetic material is then given by

$$B_{\text{int}} = \kappa_M B_{\text{applied}} = (\chi_M + 1)B_{\text{applied}}. \tag{27--14}$$

hysteresis

Equation 27–14 is most useful for paramagnetic and diamagnetic substances. In a ferromagnetic material, the relative permeability and susceptibility themselves depend on the applied field, and in fact on the past history of the material's exposure to magnetic fields (Fig. 27–49). This phenomenon of **hysteresis,** in which a ferromagnetic material "remembers" past fields, is what makes permanent magnets possible. Table 27–1 lists some magnetic susceptibilities for all three types of magnetic materials. Our definition of susceptibility shows that it is a dimensionless quantity. Note the entry for superconductors, which are perfectly diamagnetic. Their susceptibility of -1 shows that $\kappa_M = 0$, so that a superconductor completely excludes magnetic fields from its interior. We will explore the reasons for this in Chapter 29.

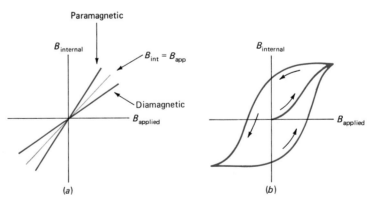

Fig. 27–49
Internal versus applied magnetic field for the different types of magnetic materials. *(a)* In diamagnetic and paramagnetic materials, the relationship is linear. *(b)* In ferromagnetic materials, the relationship depends on the strength of the applied field and on the past history of the material. In *(b)*, field strengths along the vertical axis are much greater than along the horizontal axis. Arrows indicate the direction in which fields are changed.

TABLE 27-1
Susceptibilities of Magnetic Materials*

Material	Magnetic Susceptibility, χ_M
Diamagnetic materials:	
Copper	-9.6×10^{-6}
Lead	-1.6×10^{-5}
Mercury	-2.8×10^{-5}
Nitrogen (gas, 293 K)	-6.7×10^{-9}
Sodium chloride	-1.4×10^{-5}
Any superconductor	-1.0
Water	-9.1×10^{-6}
Paramagnetic materials:	
Aluminum	2.1×10^{-5}
Chromium	3.1×10^{-4}
Oxygen (gas, 293 K)	1.9×10^{-6}
Oxygen (liquid, 90 K)	3.5×10^{-3}
Sodium	8.5×10^{-6}
Ferromagnetic materials (field- and history-dependent; maximum value listed):	
Iron (annealed)	5.5×10^3
Permalloy (55% Fe, 45% Ni)	2.5×10^4
Supermalloy (15.7% Fe, 79% Ni, 5.0% Mo; 0.30% Mn)	8.0×10^5
μ-metal (77% Ni, 16% Fe, 5% Cu, 2% Cr)	1.0×10^5

*At 300 K unless noted.

SUMMARY

1. **Magnetism** is a fundamental interaction that may be described in terms of **magnetic fields.**
   a. Magnetic **monopoles**—hypothetical particles analogous to electric charges—would interact with magnetic fields in exactly the same way that electric charges interact with electric fields. However, monopoles have never been detected and, if they do exist, are not known to play any role in the present-day universe.
   b. Instead, the most basic magnetic entities we know of are **magnetic dipoles,** whose behavior in magnetic fields is analogous to that of electric dipoles in electric fields.
2. The fundamental magnetic interaction we do observe involves **magnetic fields** and **moving electric charges.** A moving charged particle experiences a force that depends on its charge, its velocity, and the magnetic field. This force is at right angles to both velocity and field, and is given by

$$\mathbf{F} = q\mathbf{v} \times \mathbf{B}.$$

3. An electric current is made up of moving electric charges, so that there is a net magnetic force on an electric current. The force on small current elements of length $d\ell$ is

$$d\mathbf{F} = I\,d\ell \times \mathbf{B},$$

where I is the current, $\mathbf{B}$ the magnetic field, and $d\ell$ an infinitesimal vector pointing in the local direction of the current. For a straight wire of length ℓ in a magnetic field, this equation becomes

$$\mathbf{F} = I\ell \times \mathbf{B};$$

for other cases the infinitesimal force $d\mathbf{F}$ can be integrated over a current to obtain the force on the entire current.

4. A particularly important case of a current in a magnetic field is a closed current loop. Such a loop behaves like a **magnetic dipole** with magnetic dipole moment

$$\boldsymbol{\mu} = NI\mathbf{A},$$

where N is the number of turns in the loop, I the loop current, and $\mathbf{A}$ a vector perpendicular to the plane of the loop and whose magnitude is the loop area.
   a. A current loop in a magnetic field experiences a torque given by

$$\boldsymbol{\tau} = \boldsymbol{\mu} \times \mathbf{B}.$$

If the field is nonuniform, the loop experiences a net force as well.
   b. The potential energy associated with a current loop in a magnetic field may be written

$$U_M = -\boldsymbol{\mu}\cdot\mathbf{B},$$

where the zero of potential energy is taken when the loop's magnetic moment vector is perpendicular to the field.

5. Individual elementary particles as well as orbiting atomic electrons constitute miniature current loops, which are responsible for the magnetic effects we observe in macroscopic matter. We characterize magnetic materials by their **relative permeability**, κ_M, describing the ratio of the internal field in the material to the applied field, or equivalently, by the **magnetic susceptibility** given by $\chi_M = \kappa_M - 1$.

  a. In **ferromagnetic** materials, quantum-mechanical interactions result in strong alignment of atomic magnetic moments with applied magnetic fields. As a result, the field within a ferromagnetic mate-

rial is greatly increased. Ferromagnetic materials retain a remnant magnetization even when the applied field is removed; this phenomenon accounts for permanent magnets. $\chi_M \gg 0$ for ferromagnetic materials, although the value of χ_M depends on both the applied field and past history.

  b. In **paramagnetic** materials, individual atomic moments become partially aligned with an applied field, but there are no cooperative interactions among the individual moments, so that paramagnetism is a much weaker phenomenon than ferromagnetism. $\chi_M > 0$ for paramagnetic materials.

  c. **Diamagnetic** materials have no intrinsic magnetic moments. When they experience an applied field that changes with time, magnetic moments are induced that result in a repulsive interaction. $\chi_M < 0$ for diamagnetic materials.

QUESTIONS

1. Would a magnetic monopole gain energy while moving in a uniform magnetic field? Would an electric charge?

2. Why do we say that a bar magnet is a magnetic dipole?

3. An electron moving with velocity **v** through a magnetic field **B** experiences a magnetic force **F**. Which of the vectors **F**, **v**, and **B** *must* be at right angles?

4. Figure 27–50 shows the creation of matter from energy in a particle accelerator. A high-energy gamma ray moving in from the top has spontaneously given rise to an electron and its positively charged antiparticle, a positron. A magnetic field points out of the plane of the photograph, and the electron and positron spiral in this field. Which path belongs to which particle? Why might the paths be spirals rather than circles of constant radius?

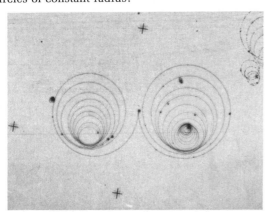

Fig. 27–50 Creation of an electron-positron pair, photographed in the bubble chamber of a high-energy particle accelerator (Question 4).

5. Any piece of matter consists of electrically charged particles. But if you move most pieces of matter through a magnetic field, they experience no force. Why is this?

6. Suppose you move a piece of electrically neutral conducting material (assumed nonmagnetic) through a magnetic field. If there is no current in the conductor, is there a net force on it? Does anything happen to the charges within the conductor?

7. A magnetic field points out of this page. Will a positively charged particle circle clockwise or counterclockwise as viewed from above?

8. An electron moves through a region in a straight line at constant speed. Can you conclude that there is no magnetic field in the region? Could you so conclude if you knew that there were no electric or gravitational forces on the electron? Explain.

9. High-resolution color TV monitors sometimes have a built-in circuit that compensates for changes in the orientation of earth's magnetic field relative to the TV picture tube as the monitor is moved from one place to another. Why is this necessary?

10. What is meant by the statement that charged particles can be "trapped" on magnetic field lines?

11. An electron beam comes straight to the center of a TV screen, where it makes a spot of light. If you hold the north pole of a bar magnet on the left side of the picture tube, which way will the spot move?

12. Do particles in a cyclotron gain energy from the electric field, the magnetic field, or both? Explain.

13. A cyclotron is designed to accelerate either hydrogen or deuterium nuclei. (Deuterium, a heavy isotope of hydrogen, has one proton and one neutron in its nucleus.) If the magnetic field is unchanged, how must the frequency of the alternating dee voltage be changed in order to switch from hydrogen to deuterium?

14. An electron and a proton moving at the same speed enter a region containing a uniform magnetic field. Which is deflected more from its original path?

15. An electron and a proton with the same kinetic energy enter a region containing a uniform magnetic field. Which is deflected more from its original path?

16. For what orientation of electric and magnetic fields could the net force on a particle be zero?

17. What does magnetism have to do with the fact that auroras are seen near the earth's poles?

18. In a certain region uniform electric and magnetic fields are at right angles to one another. A positively charged particle is released from rest in this region. Describe qualitatively its subsequent motion.

19. How do the period and radius of an electron's orbit in a magnetic field depend on its velocity? Assume that the electron is moving at right angles to the field.

20. Current in a certain ionic solution is carried equally by positive and negative ions. Would you expect the Hall effect to occur in this solution?

21. Two identical particles carrying equal charge are moving in opposite directions along a magnetic field, when they collide elastically head-on. Describe their subsequent motion.

22. Repeat the above question for the case when the two particles are moving instantaneously perpendicular to the field when they collide.

23. Under what conditions will a current loop in a magnetic field experience zero force? Zero torque?

24. When we approximate a closed current loop by an open loop fed by two wires, why is it important that the wires be close to each other?

25. A magnetic dipole is twisted out of alignment with a uniform magnetic field, and then released. Describe its subsequent motion.

26. If the dipole of the preceding question also has angular momentum parallel to its magnetic moment vector, what will be its motion when released? (This situation describes, for example, atomic nuclei. The behavior of such nuclei in a magnetic field forms the basis for the technique of nuclear magnetic resonance, widely used in determining the chemical structure of molecules and increasingly as a medical diagnostic tool.)

27. What would happen to a motor with no commutator? Assume instead that the current fed to the moving coil is always in the same direction.

28. An unmagnetized piece of iron normally has no net magnetic dipole moment. Yet it is attracted to either pole of a bar magnet. Why is this?

29. How would you determine experimentally whether a substance was paramagnetic or diamagnetic?

30. Would permanent magnets be possible if the relative permeability of ferromagnetic materials were strictly constant? Explain.

31. Why do paramagnetic and ferromagnetic effects weaken with increasing temperature?

PROBLEMS

Section 27.2   Electric Charge and the Magnetic Field

1. (a) What is the minimum magnetic field needed to exert a 5.4×10^{-15}-N force on an electron moving at 2.1×10^7 m/s? (b) What magnetic field strength would be required if the field were at 45° to the electron's velocity?

2. An electron moving at right angles to a 0.10-T magnetic field experiences an acceleration of 6.0×10^{15} m/s². (a) What is the electron's speed? (b) By how much does its *speed* change in 1 ns ($= 10^{-9}$ s)?

3. What is the magnitude of the magnetic force on a proton moving at 2.5×10^5 m/s (a) at right angles; (b) at 30°; (c) parallel to a magnetic field of 0.50 T?

4. A magnetic field of 0.10 T points in the x-direction. A charged particle carrying 1.0 μC enters the field region moving at 20 m/s. What are the magnitude and direction of the force on the particle when it first enters the field region if it does so moving (a) along the x-axis; (b) along the y-axis; (c) along the z-axis; (d) at 45° to both x- and y-axes?

5. A particle carrying a 50-μC charge moves with velocity $\mathbf{v} = 5.0\hat{\imath} + 3.2\hat{k}$ m/s through a uniform magnetic field $\mathbf{B} = 9.4\hat{\imath} + 6.7\hat{\jmath}$ T. (a) What is the force on the par-

ticle? (b) Form the dot products $\mathbf{F} \cdot \mathbf{v}$ and $\mathbf{F} \cdot \mathbf{B}$ to show explicitly that the force is perpendicular to both $\mathbf{v}$ and $\mathbf{B}$.

6. Moving in the x-direction, a particle carrying 1.0 μC experiences no force. Moving with speed v at 30° to the x-axis, the particle experiences a force of 2.0 N. What is the magnitude of the force it would experience if it moved along the y-axis with speed v?

7. A proton moving with velocity $\mathbf{v}_1 = 3.6 \times 10^4 \hat{\jmath}$ m/s experiences a force of $7.4 \times 10^{-16} \hat{\imath}$ N. A second proton moving on the x-axis experiences a magnetic force of $2.8 \times 10^{-16} \hat{\jmath}$ N. Find the magnitude and direction of the magnetic field, and the velocity of the second proton.

8. The magnitude of earth's magnetic field is a little less than 1 G near earth's surface. What is the maximum possible magnetic force on an electron with kinetic energy of 1 keV? Compare with the gravitational force on the same electron.

9. A certain region contains an electric field $\mathbf{E}$ and a magnetic field $\mathbf{B}$ at right angles to each other. A charged particle enters the region, moving at right angles to both fields. Show that it will be undeflected only if its speed is E/B, where E and B are the field magnitudes. Does the sign or magnitude of the charge

matter? *Note:* The field structure of this problem constitutes a velocity selector, used to prepare charged-particle beams whose constituent particles all have the same velocity.

Section 27.3 *The Motion of Charged Particles in Magnetic Fields*

10. A beam of electrons moving horizontally at 8.7×10^6 m/s enters a region where a uniform magnetic field of 180 G points vertically. (*a*) How far into the field region does the beam penetrate? (*b*) Describe the subsequent motion of the beam.

11. Radio astronomers detect electromagnetic radiation at a fundamental frequency of 42 MHz from an interstellar gas cloud. If this radiation is caused by electrons spiralling in a magnetic field, what is the field strength in the gas cloud?

12. Electrons and protons with the same kinetic energy are moving at right angles to a uniform magnetic field. How do their orbital radii compare?

13. A proton describes a helical path about a magnetic field of 1500 γ. The proton has a 200-km/s velocity component along the field and a 40-km/s velocity component at right angles to the field. (*a*) How far does the proton move along the field direction during the time it takes to complete one orbit around the field line? (*b*) How would your answer change if the velocity component perpendicular to the field were doubled?

14. In 2.0 μs, an electron moves 15 cm in the direction of a 1000-G magnetic field. If the electron's velocity components perpendicular and parallel to the field are equal, (*a*) what is the length of its actual helical trajectory and (*b*) how many orbits about the field direction does it complete?

15. Show that the orbital radius of a charged particle moving at right angles to a magnetic field B can be written

$$r = \frac{\sqrt{2Km}}{qB},$$

where K is the kinetic energy in joules, m the particle mass, and q its charge.

16. The Van Allen belts are regions in space where high-energy charged particles are trapped, spiralling in earth's magnetic field. If the field strength at the Van Allen belts is 0.10 G, what are the period and radius of the helical path described (*a*) by a proton with a 1.0-MeV kinetic energy? (*b*) by a 10-MeV proton?

17. Typical particle energies in a nuclear fusion reactor are on the order of 10 keV. If the smallest dimension of the reactor is on the order of 1 m, estimate the minimum magnetic field strength needed to ensure that protons have orbits smaller than the size of the reactor. Will this field be sufficient for electrons of the same energy?

18. Two protons, moving at 3.8×10^4 m/s in a plane perpendicular to a uniform magnetic field of 500 G, un-

dergo an elastic head-on collision. How much time elapses before they collide again?

19. Repeat the preceding problem for the case of a proton and an antiproton colliding head-on. (An antiproton has the same mass as a proton, but carries the opposite charge.)

20. A cyclotron is designed to accelerate deuterium nuclei. (Deuterium, a heavy isotope of hydrogen, has one proton and one neutron in its nucleus.) (*a*) If the cyclotron uses a 2.0-T magnetic field, at what frequency should the dee voltage be alternated? (*b*) If the vacuum chamber has a diameter of 0.90 m, what is the maximum kinetic energy of the deuterons? (*c*) If the magnitude of the potential difference between the dees is 1500 V, how many orbits do the deuterons complete before achieving the energy of part (*b*)?

21. Without changing the magnetic field, how could the cyclotron of the preceding problem be modified to accelerate (*a*) protons and (*b*) alpha particles (helium nuclei, consisting of two protons and two neutrons)? What would be the maximum energy achievable with (*c*) protons and (*d*) alpha particles? Assume that the mass of a nucleus is just the sum of the masses of its constituent particles. (This assumption is only approximately correct because of the Einstein mass-energy equivalence applied to the potential energy associated with electrical and nuclear forces.)

22. Figure 27–51 shows a simple mass spectrometer, designed to analyze and separate atomic and molecular ions with different charge-to-mass ratios. In the design shown, ions formed in an oven or by contact

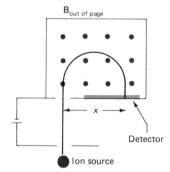

Fig. 27–51 A mass spectrometer (Problem 22).

with a hot wire are accelerated through a potential difference *V*, after which they enter a region containing a uniform magnetic field. They describe semicircular paths in the magnetic field, and land on a photographic film or other detector a lateral distance *x* from where they entered the field region, as shown. Show that *x* is given by

$$x = \frac{2}{B}\sqrt{\frac{2V}{(q/m)}},$$

where B is the magnetic field strength, V the accelerating potential, and q/m the charge-to-mass ratio of the ion. By counting the number of ions accumulated at different positions x, one can determine the relative abundances of different atomic or molecular species in a sample.

23. A mass spectrometer like that of the preceding problem has $V = 2000$ V and $B = 1000$ G. It is used to analyze a gas sample suspected of containing Ne, O_2, CO, SO_2, and NO_2. Ions are detected at distances of 2.6 cm, 3.6 cm, and 5.9 cm from the entrance to the field region. Which gases are actually present? Assume that all molecules are singly ionized. *Hint:* Consult Appendix E.

24. A mass spectrometer is used to separate the fissionable uranium isotope U-235 from the much more abundant isotope U-238. To within what percentage must the magnetic field be held constant if there is to be no overlap of these two isotopes? Both isotopes appear as constituents of uranium hexafluoride gas (UF_6), and the gas molecules are all singly ionized.

Section 27.4   *The Magnetic Force on a Current*

25. What is the force on a 50-cm-long wire carrying 15 A at right angles to a 500-G magnetic field?

26. A 10-cm-long wire with a mass of 2.2 g is suspended from two massless, conducting springs as shown in Fig. 27–52. A uniform magnetic field of 500 G points out of the page. A current flows down one spring, through the wire, and up the other spring. What current is required to keep the springs from stretching at all? Which way should this current flow?

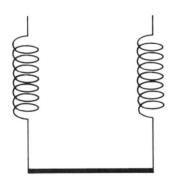

Fig. 27–52 Problem 26

27. A piece of wire whose mass is 75 grams per meter of length runs horizontally at right angles to a horizontally oriented magnetic field. A 6.2-A current in the wire results in its being suspended against gravity by the magnetic force. What is the magnetic field strength?

28. A nonuniform magnetic field points in the y-direction and increases in strength at the rate of 20 G/cm as one moves in the x-direction. A square wire loop 15 cm on a side lies in the x-y plane. A 2.5-A current flows

in the loop, going counterclockwise as viewed from the $+z$-direction (Fig. 27–53). What are the magnitude and direction of the net force on the loop?

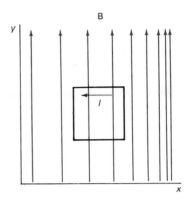

Fig. 27–53 Problem 28

29. Magnetic field lines in a certain region form concentric circles, with field strength given by $B = A/r$, where A is a constant and r the distance from the center of the circular field lines. A piece of wire carrying a current I extends in the radial direction from r_1 to r_2, as shown in Fig. 27–54. (The current is supplied by wires running perpendicular to the plane of the page.) Calculate the magnetic force on this piece of wire shown in the figure.

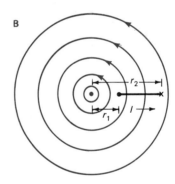

Fig. 27–54 Problem 29

30. Apply Equation 27–9 to a closed current loop of arbitrary shape located in a *uniform* magnetic field, and show that the net force on such a loop is zero. *Hint:* Both I and B are constants. What is the vector integral you are left with?

Section 27.5   *A Current Loop in a Magnetic Field*

31. Show that the units of magnetic moment ($A \cdot m^2$) can also be expressed as J/T.

32. A single-turn wire loop 10 cm in diameter carries a current of 12 A. It experiences a torque of 0.015 N·m

when the normal to the loop plane makes a 25° angle with a uniform magnetic field. What is the magnetic field strength?

33. A simple electric motor like that of Fig. 27–44 consists of a 100-turn coil of wire 3.0 cm in diameter. The coil is mounted between the poles of a permanent magnet that produces a 0.12-T field in the vicinity of the coil. When a current of 5.0 A flows in the coil, (a) what is its magnetic dipole moment and (b) what is the maximum torque developed by the motor?

34. A 10-turn wire loop measuring 8.0 cm by 16 cm and carrying 2.0 A lies in a horizontal plane but is free to pivot about an axis through the center of its long dimension. A 50-g mass hangs from one side of the loop, as shown in Fig. 27–55. A uniform magnetic field points horizontally, perpendicular to the rotation axis of the loop. What magnetic field strength is required to hold the loop in its horizontal orientation?

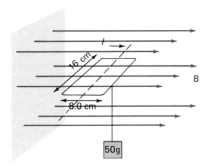

Fig. 27–55   Problem 34

35. Some smaller spacecraft use magnetic torquing to control their orientation in space. Three mutually perpendicular coils are mounted in the spacecraft. When current is passed through a coil, its magnetic moment vector rotates toward alignment with earth's magnetic field. Since electricity to run the coils is supplied from solar panels, there is no fuel to run out. A satellite with rotational inertia 20 kg·m^2 is in orbit at a height where earth's magnetic field strength is 0.18 G. Its magnetic torquing system uses 1000-turn coils 30 cm in diameter. During the torquing operation, the satellite is to achieve an angular acceleration of 0.0015 s^{-2}. What should be the current in the coils?

36. A square loop of side a is free to pivot about a horizontal rod passing through the centers of two sides of the loop. The loop carries a current I and is in a uniform magnetic field **B** pointing upward, as shown in Fig. 27–56. A mass m hangs from one side of the loop, as shown. Find the angle between the loop plane and the horizontal for which this system will be in equilibrium.

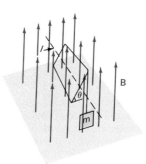

Fig. 27–56   Problem 36

37. Nuclear magnetic resonance is a widely used technique for performing chemical analysis. In NMR, different chemical structures are identified through sensitive measurements of the energy needed to flip atomic nuclei upside-down in a given magnetic field. This energy depends on the configuration of electrons surrounding the nucleus, which in turn depends on the molecular structure of which the nucleus is a part. If a magnetic field of 0.20 T is used in a certain NMR apparatus, how much energy is required to flip a proton (hydrogen nucleus) from a parallel to an antiparallel orientation relative to the field? The magnetic moment of the proton is 1.41×10^{-26} A·m^2.

38. A disk of radius R carries a uniform surface charge density σ, and is rotating with angular frequency ω. Show that the magnetic dipole moment of this disk is $\frac{1}{4}\pi\sigma\omega R^4$. *Hint:* Divide the disk into infinitesimal current loops, and integrate the magnetic moments of these loops.

39. A wire of length ℓ carries a current I. Find an expression for the magnetic dipole moment that results when the wire is wound into an N-turn circular coil, and show that this moment is maximum for $N=1$.

40. A closed current loop is made from two semicircular wire arcs of radius R, joined at right angles as shown in Fig. 27–57. The loop carries a current I and is oriented so that the plane of one semicircle is parallel to a uniform magnetic field **B,** as shown. Calculate (a) the magnetic dipole moment of this nonplanar loop and (b) the torque on the loop. *Hint:* Think of the loop as being made of two separate semicircular loops, whose magnetic moment vectors add. You can close the loops with a line between the two junction points without affecting anything (why?).

Section 27.6   *Magnetic Matter*

41. A container of liquid oxygen at 90 K is placed in a 500.0-G magnetic field. What is the field strength within the liquid oxygen?

42. Fig. 27–58 shows a hysteresis curve for the ferromagnetic alloy Alnico V, commonly used in making permanent magnets. (a) What is the maximum inter-

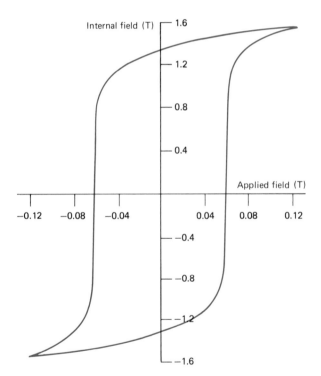

Fig. 27–57 Problem 40

nal field strength obtainable inside this material, in the absence of an externally applied field? (b) What is the maximum field strength obtainable inside this material, in the presence of an externally applied field?

Fig. 27–58 Internal versus external magnetic field for Alnico V (Problem 42).

Supplementary Problems

43. A certain region contains both an electric field and a magnetic field. An electron moving at 2.0×10^4 m/s in the y-direction experiences a force of $2.2 \times 10^{-15}\hat{\imath} + 4.8 \times 10^{-15}\hat{\jmath}$ N. Moving in the x-direction with the same speed, an electron experiences a force of $2.6 \times 10^{-15}\hat{\jmath} + 2.2 \times 10^{-15}\hat{k}$ N. Moving in the z-direction with a speed of 1.0×10^4 m/s, an electron experiences a force of $-1.1 \times 10^{-15}\hat{\imath} + 4.8 \times 10^{-15}\hat{\jmath}$ N. Find a combination of uniform electric and magnetic fields that could be responsible for these forces.

44. Electrons with a kinetic energy of 30 keV are moving through a TV picture tube, heading for the exact center of the screen. They traverse a region 40 cm long. What is the maximum deflection from screen center that could be caused by earth's 0.50-G magnetic field?

45. A circular wire loop of mass m and radius R carries a current I. The loop is hanging below a bar magnet, suspended by the magnetic force, as shown in Fig. 27–59. If the field lines at the location of the loop make an angle θ with the vertical, show that the magnetic field strength at this location is given by $B = mg/2\pi RI\sin\theta$. Which direction does the current circulate around the loop?

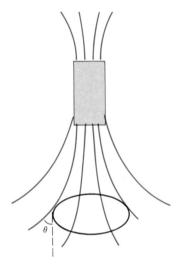

Fig. 27–59 Problem 45

46. The Hall effect: A metallic conductor is shaped like a flat bar with a rectangular cross section, as shown in Fig. 27–60. The conductor carries a current I. A uniform magnetic field **B** points at right angles to the current and is parallel to the long dimension of the conductor's rectangular cross section. Assume that

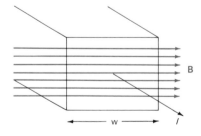

Fig. 27–60 Problem 46

charge separation due to the magnetic force results in all excess charge residing on the top and bottom of the conductor. Show that the resulting Hall potential is given by $V_H = IB/nqw$, where n is the number density of charge carriers in the conductor, q the charge on each, and w the width of the conductor.

47. A square wire loop of mass m carries a current I. It is initially in equilibrium, with its magnetic moment vector aligned with a uniform magnetic field **B.** If the loop is twisted slightly away from this equilibrium orientation and then released, show that it executes simple harmonic motion with period $T = 2\pi\sqrt{m/6IB}$.

48. Although it is not consistent with the quantum-mechanical picture of the atom, a calculation based on the classical notion of an electron orbiting an atomic nucleus nevertheless gives an accurate value for the magnetic dipole moment associated with the electron's orbital motion. Assuming the electron orbit in a hydrogen atom has a radius of 0.053 nm, show that the associated magnetic dipole moment is 9.3×10^{-24} A·m^2. *Hint*: In one orbital period, the full electron charge passes any given point in the orbit. Use this fact to calculate the average current represented by the electron's orbital motion.

28

SOURCES OF THE MAGNETIC FIELD

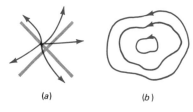

Fig. 28–1
In the absence of magnetic monopoles, the magnetic flux through a closed surface must be zero. (a) This means there can be no point where magnetic field lines begin or end, for a closed surface surrounding such a point would have nonzero net flux. (b) Instead, magnetic field lines generally form closed loops.

In the preceding chapter we introduced the magnetic field and discussed its effect on matter. We now consider the opposite question: how does matter give rise to magnetic fields?

28.1
GAUSS'S LAW FOR MAGNETISM

If magnetic charges—monopoles—existed, they would give rise to point-charge magnetic fields like the electric fields of electric point charges. These fields—and those of magnetic charge distributions—would be described by laws analogous to Coulomb's and Gauss's laws. In particular, we would find that the flux of the magnetic field through any closed surface would depend only on the enclosed magnetic charge. But the very existence of magnetic monopoles is uncertain. And even if they do exist, they seem to play no significant role in our world. In the absence of magnetic monopoles, the magnetic flux through any closed surface must be zero. We state this mathematically as **Gauss's law for magnetism:**

Gauss's law for
magnetism

$$\oint \mathbf{B} \cdot d\mathbf{A} = 0. \tag{28–1}$$

Like the analogous law for electric fields, this statement is one of the four fundamental equations describing fully the behavior of electric and magnetic fields. A consequence of Gauss's law for magnetism is that magnetic field lines can never begin or end (Fig. 28–1). Despite the zero on its right-hand side, Equation 28–1 is not an empty statement. As Fig. 28–1 shows, it places definite constraints on the behavior of magnetic fields. And it definitely does not say that there cannot be magnetic fields: rather, it says that if magnetic fields exist then they look different from the fields of point charges.

28.2

THE BIOT-SAVART LAW

Fig. 28–2
Oersted's experiment linking electricity and magnetism is commemorated on the Oersted medal, awarded by the American Association of Physics Teachers to honor "notable contributions to the teaching of physics."

Biot-Savart law

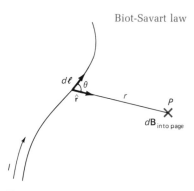

Fig. 28–3
The Biot-Savart law gives the magnetic field at the point P arising from the current I flowing along the infinitesimal vector $d\ell$. The unit vector $\hat{r}$ points from this source element $d\ell$ to the field point P.

If there are no magnetic monopoles, how can there be any magnetic fields? What else but a magnetic charge gives rise to a magnetic field? We asked a similar question in the preceding chapter when we wondered what other than a magnetic monopole might interact with a magnetic field. Our answer—a moving *electric* charge—hinted at a deep relation between electricity and magnetism. That relation is in fact a two-way street: not only do moving electric charges experience forces in magnetic fields, but they also give rise to magnetic fields.

When we investigated the behavior of a current loop in a magnetic field, we found that the loop behaves analogously to an electric dipole in an electric field. On the basis of this analogy we identified the current loop as a magnetic dipole. Now an *electric* dipole not only interacts with an electric field, it also gives rise to its own electric field. Similarly, a magnetic dipole gives rise to a magnetic field with properties analogous to the electric dipole field. Since all the magnetic dipoles we have encountered originate ultimately in electric current loops, we can argue that electric current—the motion of electric charge—is the source of the magnetic field.

That a current loop gives rise to a dipole magnetic field is an instance of the more general fact that moving electric charges produce magnetic fields. How can we calculate these fields? Can we show that a current loop has the magnetic field of a dipole? Interest in questions like these developed immediately after the discovery by Hans Christian Oersted in 1820 that a compass needle is deflected by an electric current—historically, the first hint of a link between electricity and magnetism (Fig. 28–2). A mere month after Oersted's discovery became known in Paris, the French scientists Jean Baptiste Biot and Félix Savart had experimentally determined the form of the magnetic field arising from a steady current. Known as the **Biot-Savart law,** their result gives the magnetic field at a point due to a small element of current.

Figure 28–3 shows part of a wire carrying a *steady* current I, and a point P—called the field point—where we wish to calculate the magnetic field. The Biot-Savart law gives the contribution $d\mathbf{B}$ to the field at P arising from the current I flowing along a vector $d\ell$ that coincides with an infinitesimal length $d\ell$ of the wire. The distance from this small section of wire to the point P is r, and we indicate the direction to point P by a unit vector $\hat{r}$ pointing from $d\ell$ toward P. Then the Biot-Savart law says that the magnetic field at P due to the current in the small segment of wire $d\ell$ is

$$d\mathbf{B} = \frac{\mu_0}{4\pi}\frac{I\,d\ell \times \hat{r}}{r^2},\qquad(28\text{–}2)$$

where μ_0 is a constant whose value is exactly $4\pi \times 10^{-7}$ N/A^2.* We emphasize that Equation 28–2 holds strictly only for *steady* currents—currents that *never* change.

---

*That this constant has an exact value is a consequence of the definition of the ampere in terms of magnetic forces. See the end of Section 28.2.

Compare the Biot-Savart law with Coulomb's law. Coulomb's law gives the electric field of a point charge in terms of the charge and the distance from the charge to the field point. The electric field of a point charge decreases as the inverse square of the distance, and the field direction lies along the line joining the charge with the field point. Analogously, the Biot-Savart law gives the magnetic field at a given point in terms of the current element that is its source and the distance to the field point. Like the electric field of a point charge, the magnetic field of an isolated current element falls as the inverse square of the distance. But here the similarity ends. Unlike the electric charge in Coulomb's law, the current element $I d\ell$ has associated with it a direction as well as a magnitude. As a result the magnetic field of the current element is not symmetric about the element, but depends on the direction of the field point relative to the direction of the current element. This directional character is expressed by the cross product in Equation 28–2. As shown in Fig. 28–3, the magnetic field is at right angles to both the current element and the vector from the current element to the field point.

Does the Biot-Savart law make sense? We should certainly not be surprised that the field depends on the current, and that it decreases with distance from the current. But what about its direction? Wouldn't it be easier if the field pointed radially outward from the current element? Yes—but that cannot be. The current element is *not* a magnetic charge. Field lines cannot begin or end on it. The cross product in Equation 28–2 ensures instead that the field lines *encircle* the current (Fig. 28–4).

There is another distinction between the Biot-Savart law and Coulomb's law. Both describe the fields of localized structures—current elements and point charges—that are the basic sources of the fields. It makes sense to talk about the electric field of an isolated point charge. But can we have an isolated current element? Not in a steady-state situation. The current flowing into our current element must be the same as the current flowing out; otherwise, charge would accumulate. Thus any Biot-Savart calculation necessarily involves the fields produced by many small current elements from an entire circuit. Experimentally, we find that the magnetic field obeys the superposition principle, so that the net field is the vector sum—or integral—of the fields of individual current elements:

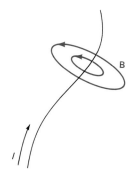

Fig. 28–4
Magnetic field lines generally encircle a current.

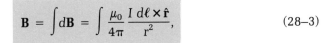

$$\mathbf{B} = \int d\mathbf{B} = \int \frac{\mu_0}{4\pi} \frac{I \, d\ell \times \hat{\mathbf{r}}}{r^2},$$
(28–3)

where the integration is taken over the entire circuit in which the current I flows.

Example 28–1 _____

The Field of a Current Loop

We found in the preceding chapter that a current loop behaves like a magnetic dipole when placed in a magnetic field. In the present chapter we asserted that the loop also produces a field with the characteristic dipole configuration. Can we use the Biot-Savart law to show this?

Solution

Figure 28–5*a* shows a circular loop of radius *a* carrying current *I*. We will calculate the field at all points on the axis of this loop, which we take to be the *x*-axis, with $x=0$ at the loop center. Consider a small segment *dℓ* at the top of the loop, as shown in Fig. 28–5*b*. The current element *I dℓ* produces a magnetic field *d***B** given by Equation 28–2. The direction of this field is that of the cross product *dℓ* × **r̂**, where **r̂** is a unit vector from *dℓ* toward the point on the axis where we wish to evaluate the field. Therefore *d***B** is at right angles to both *dℓ* and **r̂**, or upward and to the right as shown in Fig. 28–5*b*. For any point on the axis, the vectors *dℓ* and **r̂** are at right angles. Therefore, the magnitude of the cross product *dℓ* × **r̂** is just *dℓ*, since **r̂** is a *unit* vector [$|d\boldsymbol{\ell} \times \hat{\mathbf{r}}| = |d\boldsymbol{\ell}||\hat{\mathbf{r}}|\sin 90° = (d\ell)(1)(1) = d\ell$]. Furthermore, the distance *r* from *dℓ* to our field point on the *x*-axis is given by

$$r^2 = x^2 + a^2,$$

with *a* the loop radius. Then Equation 28–2 gives the magnitude *dB* of the field at *x* arising from the segment *dℓ*:

$$dB = \frac{\mu_0 I}{4\pi} \frac{|d\boldsymbol{\ell} \times \hat{\mathbf{r}}|}{r^2} = \frac{\mu_0 I}{4\pi} \frac{d\ell}{x^2 + a^2}. \tag{28–4}$$

What about segments *dℓ* at other positions around the loop? All are the same distance from the field point at *x*, and all are at right angles to the **r̂** vectors pointing toward the axis. Therefore all the *dℓ*'s give rise to field elements *d***B** whose magnitudes are given by Equation 28–4. But the directions of the individual *d***B**'s

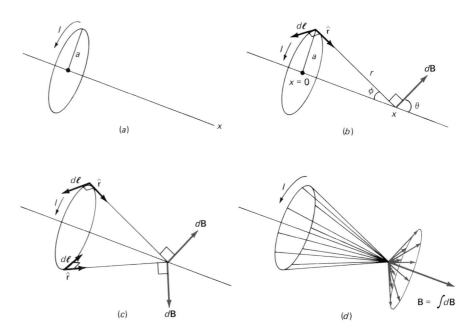

Fig. 28–5
(a) A current loop whose axis is the *x*-axis. *(b)* The field element *d***B** arising from the current element *Idℓ* at the top of the loop. *(c)* Another field element arises from the current element at the bottom of the loop. *(d)* The field elements from the entire loop lie on a cone about the axis, giving a net field along the axis.

differ. Consider a $d\ell$ at the bottom of the loop, as shown in Fig. 28–5c. Its field contribution dB points downward and to the right. The component of this field element at right angles to the x-axis cancels the corresponding component from the top of the loop. However, the components along the x-direction add. The same is true as we go around the loop. The field contributions in the x-direction all add, while those at right angles to the x-direction cancel in pairs. Therefore the net magnetic field on the axis is in the x-direction (Fig. 28–5d).

The x-component of any of the dB's is

$$dB_x = dB \sin\theta = \frac{\mu_0 I}{4\pi} \frac{d\ell}{x^2 + a^2} \cos\theta,$$

where θ is the angle between the magnetic field element dB and the x-axis, as shown in Fig. 28-5b. From the geometry of that figure, we see that $\theta = 90° - \phi$, so that $\cos\theta = \sin\phi$. But Fig. 28-5b also shows that

$$\sin\phi = \frac{a}{r},$$

so we have

$$\cos\theta = \sin\phi = \frac{a}{r} = \frac{a}{\sqrt{x^2 + a^2}},$$

and therefore

$$dB_x = \frac{\mu_0 I}{4\pi} \frac{d\ell}{x^2 + a^2} \frac{a}{\sqrt{x^2 + a^2}} = \frac{\mu_0 I}{4\pi} \frac{a\,d\ell}{(x^2 + a^2)^{3/2}}.$$

The magnitude of the net field is obtained by integrating this expression around the loop, summing the contributions from all the $d\ell$'s:

$$B = B_x = \int_{\text{loop}} \frac{\mu_0 I}{4\pi} \frac{a\,d\ell}{(x^2 + a^2)^{3/2}}.$$

How are we to evaluate this integral? μ_0 and I are constants. So are the loop radius a and the distance x along the axis. None of these quantities changes as we move around the loop. So we can write

$$B = \frac{\mu_0 I}{4\pi} \frac{a}{(x^2 + a^2)^{3/2}} \int_{\text{loop}} d\ell.$$

The integral in this expression means the sum of the lengths of small segments $d\ell$ all around the loop, or the loop circumference $2\pi a$. So we have

$$B = \frac{\mu_0 I}{4\pi} \frac{a}{(x^2 + a^2)^{3/2}} 2\pi a = \frac{\mu_0 I a^2}{2(x^2 + a^2)^{3/2}}. \qquad (28\text{–}5)$$

Does this current-loop field resemble that of a dipole? When we calculated the electric field associated with two equal but opposite point charges, we found that the field exhibited the $1/r^3$ drop-off characteristic of a dipole only when we were far from the charges. How far? A long distance compared with their spacing. Similarly, we can evaluate the magnetic field of our current loop for distances x much greater than our loop radius a. Taking $x \gg a$ in our loop-field equation 28–5 gives

$$B = \frac{\mu_0 I a^2}{2(x^2)^{3/2}} = \frac{\mu_0 I a^2}{2x^3}.$$

This equation shows clearly the inverse-cube fall-off characteristic of a dipole field. We can rewrite the equation in the form

$$B = \frac{\mu_0}{2\pi} \frac{I\pi a^2}{x^3}.$$

But πa^2 is the loop area, and in the preceding chapter we identified the product of current and loop area as the magnetic dipole moment μ. Thus we can write

$$B = \frac{\mu_0}{2\pi} \frac{\mu}{x^3}. \tag{28–6}$$

Although we derived it for a circular loop, Equation 28–6 holds at large distances regardless of the loop shape. Of course we have shown only that the field of the loop falls as the inverse cube of the distance for points on the loop axis. A much more difficult calculation would be required to find the field off the axis. Such a calculation would confirm that the inverse-cube field dependence holds everywhere, and that the field has the angular dependence typical of a dipole. Indeed, the full expression for the field of an electric dipole (Problem 45 of Chapter 23) holds for a magnetic dipole if we replace $1/4\pi\epsilon_0$ by $\mu_0/4\pi$, **E** by **B**, and **p** by **μ**. Problem 34 illustrates a Biot-Savart calculation of the magnetic field in the plane of the current loop.

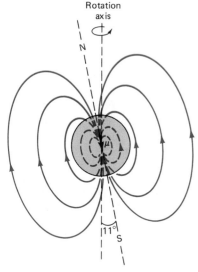

Fig. 28–6
Earth's magnetic field closely approximates that of a magnetic dipole located at the center of the earth and inclined at 11° to the rotation axis. Interior field lines are dashed to show that the field does not really arise from a single point dipole at the center, but rather from electric current flowing in the liquid outer core. Note from field direction that earth's "north" magnetic pole is really magnetic south pole.

Application

The Magnetic Fields of Earth and Sun

Earth, sun, and many other astronomical objects possess magnetic fields. As we saw in the preceding chapter, our planet's field shields us from direct impact of high-energy particles. Reasonably close to earth, where the distorting effects of the solar wind are not significant, the field approximates that of a magnetic dipole of magnetic moment $\mu = 8.0 \times 10^{22}$ A·m² (Fig. 28–6). The direction of the dipole moment vector differs by about 11° from that of earth's rotation axis; this accounts for the difference between magnetic and true north. Locally, the field often deviates significantly from a pure dipole form, and these deviations provide geologists with clues to the detailed structure of the planet. Furthermore the field is not constant. The local field direction may vary significantly on times as short as a few years, and the overall field reverses about every half million years or so (Fig. 28–7).

What causes earth's magnetic field? We know that magnetic fields arise from electric currents. Deep inside the earth are a liquid outer core and a solid inner core, both rich in iron. Through an interaction that we do not yet fully understand, the rotation of the planet combined with convective motions due to internal heat flow gives rise to electric currents in the liquid core. For reasons even less well understood, those currents and the resulting dipole field undergo reversals on a time scale of approximately a million years. Problems 7 and 9 deal with earth's dipole field and its origin.

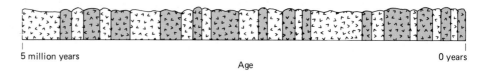

5 million years   0 years

Age

Fig. 28–7
Changing magnetic polarity of lava flows provides evidence for reversals of earth's magnetic field. Here, colored bands represent lava whose magnetization is in same direction as earth's present field; white bands have opposite magnetization. (See also Fig. 27–21.)

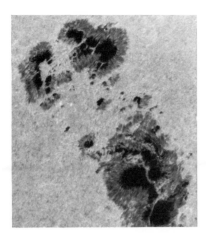

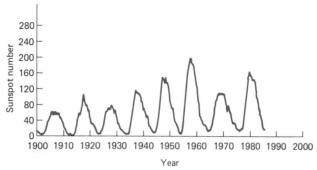

Fig. 28–8
(Left) A group of sunspots, each a region of intense solar magnetic field. (Right) The solar activity cycle is clearly evident in this plot of the number of sunspots observed each year.

The magnetic field of the sun probably arises in the same way, although the gaseous nature of the star and the intense energy flow associated with nuclear fusion within the sun's core make the behavior of the sun's magnetic field much more dynamic. The sun's field reverses every 11 years, giving rise to the solar activity cycle whose best-known indicator is the count of sunspots—regions of intense magnetic fields at the solar surface (Fig. 28–8). Between reversals the field is much more complex than that of a dipole, and even when the field is dipolar it is distorted by the outflow of the solar wind (Fig. 28–9).

The complex behavior of astrophysical magnetic fields is believed to be governed entirely by Newton's laws and the laws of electromagnetism. That we do not yet fully understand these fields is testimony to the rich variety of phenomena subsumed under those laws.

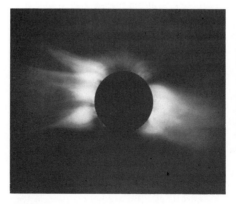

Fig. 28–9
Photographs of the solar corona taken during solar eclipses. Bright regions serve to trace approximately the sun's magnetic field lines. (Left) During the June 30, 1973 eclipse, near minimum activity in the solar cycle, the field resembles that of a dipole distorted by the outflowing solar wind. (Right) During the Feb. 16, 1980 eclipse, near maximum solar activity, the solar field is much more complicated, without clear evidence of dipolar structure.

Example 28-2

The Field of a Straight Wire

What is the magnetic field produced by a long, straight wire carrying a steady current I? We let the wire coincide with the x-axis, and we seek the field at an arbitrary point P a distance y from the wire. Because there is cylindrical symmetry about the wire, our result will hold at any angular position.

Solution

Consider a current element $I\,d\boldsymbol{\ell}$ as shown in Fig. 28-10. At point P the field element $d\mathbf{B}$ produced by this current element is perpendicular to both the wire and the vector from the wire to P—that is, $d\mathbf{B}$ points out of the page. Clearly this is true no matter where $d\boldsymbol{\ell}$ is along the wire, so that at point P all the $d\mathbf{B}$'s from all the current elements $d\boldsymbol{\ell}$ point in the same direction.

What about the magnitude of $d\mathbf{B}$? The distance $d\ell$ in the Biot-Savart law is just dx because the current flows along the x-axis. However, the angle between the vectors $d\boldsymbol{\ell}$ and $\hat{\mathbf{r}}$ is not $90°$ as in the current-loop case, but is the angle ϕ shown in Fig. 28-10. Thus the magnitude of $d\mathbf{B}$, as given by the Biot-Savart law (Equation 28-2), is

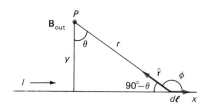

Fig. 28-10
Geometry for calculating the field at P of a straight wire carrying steady current I along the x-axis.

$$dB = \frac{\mu_0 I}{4\pi}\frac{dx\,\sin\phi}{r^2}, \tag{28-7}$$

where r is the distance from dx to P. To get the net field at P we must integrate this expression over the whole wire. As we move along the wire several quantities in the expression change, including dx, r, and ϕ. We must somehow relate these quantities to a single variable that depends on position along the wire. One obvious choice might be the position x of the current element. However, the integration is much easier if we express everything in terms of the angle θ shown in Fig. 28-10. From the figure we see that $\phi + (90° - \theta) = 180°$, so that $\theta = \phi - 90°$. Therefore $\sin\phi = \cos\theta$. We now have $\sin\phi$ in terms of θ. What about dx? Figure 28-11 shows that an infinitesimal change $d\theta$ in θ corresponds to a change dx in x such that $dx\cos\theta = r\,d\theta$. Then the infinitesimal change dx is

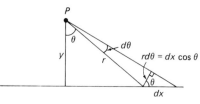

Fig. 28-11
For an infinitesimal change $d\theta$ in θ, the arc of length $r\,d\theta$ subtended by $d\theta$ is very nearly straight, and is related to dx by $r\,d\theta = dx\cos\theta$.

$$dx = \frac{r\,d\theta}{\cos\theta}.$$

Putting our expressions for dx and $\sin\phi$ into Equation 28-7 gives

$$dB = \frac{\mu_0 I}{4\pi}\frac{r\,d\theta}{\cos\theta}\frac{\cos\theta}{r^2} = \frac{\mu_0 I}{4\pi}\frac{d\theta}{r}.$$

From Fig. 28-10 we see that $r\cos\theta = y$, so that $r = y/\cos\theta$ and we have

$$dB = \frac{\mu_0 I}{4\pi y}\cos\theta\,d\theta.$$

To obtain the field at P we now integrate over the wire:

$$B = \int dB = \int_{\theta_1}^{\theta_2}\frac{\mu_0 I}{4\pi y}\cos\theta\,d\theta,$$

where θ_1 and θ_2 are the angles corresponding to the ends of the wire, as shown in Fig. 28-12. Now I, y, and $\mu_0/4\pi$ are constants—they don't change as we move along the wire—so we have

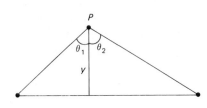

Fig. 28-12
The angles θ_1 and θ_2 are measured from the y-axis to the ends of the wire.

$$B = \frac{\mu_0 I}{4\pi y}\int_{\theta_1}^{\theta_2}\cos\theta\,d\theta = \frac{\mu_0 I}{4\pi y}\sin\theta\Big|_{\theta_1}^{\theta_2} = \frac{\mu_0 I}{4\pi y}(\sin\theta_2 - \sin\theta_1). \tag{28-8}$$

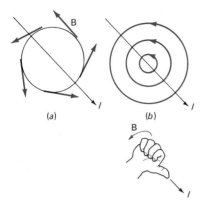

Fig. 28–13
(a) Some magnetic field vectors associated with a straight wire carrying steady current. (b) The corresponding magnetic field lines are circles, concentric with the wire. The right-hand rule gives their direction.

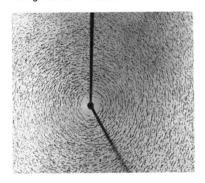

Fig. 28–14
Iron filings trace out the circular field lines surrounding a wire.

What is the direction of the field whose magnitude is given by Equation 28–8? In Fig. 28–10 we saw that the field is at right angles to both the wire and the line joining the point P with the wire. Since there is symmetry around the wire, the magnetic field lines must be circles, as shown in Fig. 28–13. The right-hand rule gives the sense in which these lines encircle the wire. What else could the field lines look like? The pattern must be symmetric about the wire, and in the absence of magnetic monopoles Gauss's law for magnetism forbids the field lines from beginning or ending anywhere. The Biot-Savart requirement that the field be perpendicular to the current rules out helical field lines, so that for the straight wire only circular field lines can satisfy the laws governing the magnetic field. Figure 28–14 shows the field of a long wire as traced out by iron filings.

Although Equation 28–8 was derived without approximations, it nevertheless represents an idealization because it is impossible to have an isolated segment of wire carrying a steady current. Other wires must connect to the segment to complete a circuit, and these will contribute magnetic fields of their own. Nevertheless, Equation 28–8 remains a useful approximation for long wires at points far from any connecting wires, and is also valuable for calculating the fields of structures that contain straight sections of wire (see Problems 5 and 6).

An important special case of Equation 28–8 is the field of an infinitely long wire. Of course no such wire exists, but when a wire is long compared with our distance from it, the approximation that it is infinitely long is often useful. For an infinitely long wire the angles θ_1 and θ_2 become $-90°$ and $+90°$, respectively, so that Equation 28–8 is simply

$$B = \frac{\mu_0 I}{4\pi y}[1 - (-1)] = \frac{\mu_0 I}{2\pi y}. \qquad \text{(infinite wire)} \qquad (28–9)$$

The field given by Equation 28–9 falls as the inverse of the distance from the wire. This should not be surprising: we found the same dependence on distance for the electric field of an infinitely long charged wire. This $1/r$ dependence is characteristic of static electric and magnetic fields associated with an elongated source. The difference between the electric and magnetic cases lies in the field-line pattern. The electric field lines of a long charged wire point radially outward, reflecting the presence of a net electric charge on the wire. The magnetic field lines of a long current-carrying wire encircle the wire, reflecting the absence of magnetic charge and the presence of electric current in the wire.

The Force between Two Conductors

In the preceding chapter we considered the force on a straight wire of length ℓ carrying a current I through a magnetic field **B**:

$$\mathbf{F} = I\boldsymbol{\ell} \times \mathbf{B},$$

where $\boldsymbol{\ell}$ is a vector whose magnitude is the wire length ℓ and whose direction is that of the current along the wire. Through Equation 28–9 we now know the magnetic field produced by a long wire. If we have two long, parallel wires carrying current in the same direction, as shown in Fig. 28–15, then each will experience a force arising from the field of the other. What are the magnitudes and directions of these forces?

If d is the distance between the wires, then at wire 2 the field B_1 due to the current I_1 is, from Equation 28–9,

$$B_1 = \frac{\mu_0 I_1}{2\pi d}.$$

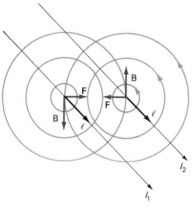

Fig. 28–15
The magnetic force between two parallel wires carrying current in the same direction is attractive.

This field is at right angles to wire 2, so the magnitude of the force on a length ℓ of wire 2 is just

$$F_2 = I_2\ell B_1 = \frac{\mu_0 I_1 I_2 \ell}{2\pi d}. \tag{28-10}$$

Calculating the force on a length ℓ of wire 1 would amount to interchanging the subscripts 1 and 2 in this expression, giving a force of the same magnitude. What is the direction of this force between the wires? As shown in Fig. 28–15, the force on wire 2 is toward wire 1. Similarly, the force on wire 1 is toward wire 2. The magnetic force between two parallel wires carrying current in the same direction is attractive. If the currents are in opposite directions, the force has the same magnitude but is repulsive.

The force between nearby conductors must be considered in the construction of electrical devices carrying large currents. In electromagnets, particularly, adjacent conductors must be given enough physical support that magnetic forces do not crush the device. The hum you often hear around electrical equipment comes from the mechanical vibration of tightly wound conductors in transformers and other electrical components. This vibration results from the changing magnetic force associated with the 60-Hz alternating current.

The force between conductors forms the basis for the definition of the ampere and consequently the coulomb. One ampere is defined as the current in each of the two very long parallel conductors carrying equal currents, when those conductors are 1 meter apart and experience a force of 2×10^{-7} N per meter of length. This definition provides a standard of current that is reproducible in any laboratory. The coulomb is then defined as the amount of charge delivered in one second by a current of one ampere.

28.3
AMPÈRE'S LAW

In Chapter 21 we did several relatively cumbersome calculations of electric fields using Coulomb's law. For example, we calculated the field of an infinite line of charge by integrating the fields of all the infinitesimal charge elements along the wire. We then made some simple observations about the number of field lines emerging from closed surfaces, and formulated Gauss's law—a simple but elegant statement that made electric field calculations much easier in situations with sufficient symmetry. Calculating the electric field of a line charge using Gauss's law took far less mathematical effort than did the Coulomb's law calculation.

Can we make a statement analogous to Gauss's law that would enable us to calculate the magnetic fields of currents with comparable ease? We have already stated Gauss's law for magnetism: the number of magnetic field lines emerging from a closed surface is proportional to the magnetic charge enclosed. Because we do not observe magnetic monopoles, all our closed surfaces enclose zero magnetic charge. Gauss's law for magnetism would be useful in calculating the fields of monopoles, but it does not go far enough to be of much help with the magnetic fields arising from electric currents.

Gauss's law for the electric field relates the amount of charge enclosed by a surface to the number of field lines emerging from that surface. We quantify "number of field lines" with the concept of flux. Is there an analogous concept that would prove useful in describing the magnetic field of a current? There is—and a clue to its nature comes from the fact that magnetic field lines are, generally, closed loops surrounding a current.

Consider the magnetic field of a long, current-carrying wire, as shown in Fig. 28–16. The magnitude of this field decreases inversely with distance from the wire, as given by Equation 28–9. Imagine drawing an arbitrary closed loop around the wire. The quantity that will prove useful is the length of this loop, weighted at each point by the field component in the direction of the loop. We call this quantity the **circulation** around the loop. It is easy to calculate the circulation for a circular loop that coincides with a field line. At all points on the circle labelled 1 in Fig. 28–16, for example, the field strength has the value B_1 given by Equation 28–9:

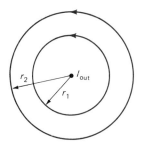

circulation

$$B_1 = \frac{\mu_0 I}{2\pi r_1},$$

where r_1 is the radius of this circle. Since it coincides with a field line, our circular loop is always in the same direction as the field, and the circulation becomes simply the loop circumference times the field strength:

$$\text{Circulation}_{\text{line 1}} = 2\pi r_1 B_1 = 2\pi r_1 \frac{\mu_0 I}{2\pi r_1} = \mu_0 I. \qquad (28\text{–}11)$$

Fig. 28–16
Magnetic field lines surrounding a current-carrying wire.

This circulation does not depend on the loop radius r_1! What about the circulation around the circular loop labelled 2, that also coincides with a field line? We would calculate it in the same way, just replacing r_1 with r_2. Our result would again be given by Equation 28–11. We would obtain this result no matter what field line we used. The circulation around any field line in the field of a long straight wire is the same.

On what does this circulation depend? The simple result in Equation 28–11 shows that the circulation around any loop coinciding with a field line is proportional to the current. What current? The current in the wire, which is encircled by all the field lines.

What if our closed loop does not coincide with a field line? Figure 28–17 shows two field lines of a current-carrying wire and a closed loop that encircles the wire but does not coincide with a single field line. To calculate the circulation around this loop, we move around the loop, taking the product of the distance moved with the field component in the direction of motion. When we move at right angles to the field, we get no contribution to the circulation because there is no field component in the direction of motion. Therefore the straight sections ca and bd of the loop in Fig. 28–17 contribute nothing to the circulation around the loop. What about the contribution to the circulation from the arc ab that lies on the inner field line? It is the same as the contribution we would get if we moved along the arc cd that is *not* part of our loop. Why? Because (1) the circulation around the inner and outer field lines is the same, as Equation 28–11 shows, and (2) the arc ab occupies the same fraction of the inner field line as the arc cd does of the outer field line. So the circulation around our closed loop is just what we would get going around the outer field line

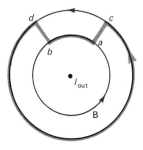

Fig. 28–17
A closed loop that does not coincide with a single field line. The circulation around this loop is the same as the circulation around a field line.

only, and therefore, according to Equation 28–11, has the value $\mu_0 I$ that we would get going around *any* field line.

We can imagine approximating an arbitrary loop by a series of concentric circular arcs that coincide with field lines, joined by radial segments that are perpendicular to the field lines (Fig. 28–18). Applying the same arguments we used for Fig. 28–17, we conclude that the circulation around such a loop is also the same as that around any field line, or $\mu_0 I$. We have in these observations a simple statement whose truth in fact extends to any situation involving steady current:

> The circulation around any closed loop encircling a steady current is proportional to the current encircled by that loop, and is given by $\mu_0 I$.

Ampère's law

This statement is a simplified version of **Ampère's law,** one of the four fundamental laws of electromagnetism. Ampère's law says that whatever arrangement of currents we might have, and however complicated the resulting magnetic field, the field will have a form that makes the circulation around any closed loop proportional to the current encircled by that loop. Compare this with Gauss's law, which says that whatever arrangement of charges we might have, and however complicated the resulting electric field, the field will have a form that makes the flux emerging from any closed surface proportional to the net charge enclosed.

Our statement of Ampère's law is true for any closed loop whatsoever, as long as the encircled current is steady (never changing in time). It does not matter whether the current is in a single wire or in a number of wires or distributed throughout space—we simply add all the currents to obtain the net current encircled by our loop. If there are currents flowing in opposite directions, then we must give opposite signs to opposite directions of current. It is this net current—the algebraic sum of currents encircled by the loop—that alone determines the circulation around the loop.

In talking about circulation we must specify the sense—clockwise or counterclockwise—in which we traverse the loop. Going around the loop in opposite directions gives opposite values for the circulation. We adopt the convention that circulation is positive if, when we curl the fingers of our right hand around the loop, our right thumb points in the general direction of the net current encircled by the loop.

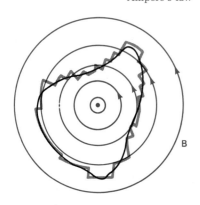

Fig. 28–18
An irregular loop may be approximated as a series of arcs and radial segments. The circulation around the loop is $\mu_0 I$, as it is for any closed loop.

Example 28–3

Ampère's Law

Figure 28–19 shows two equal currents I, one flowing into the page and one flowing out. What is the circulation going counterclockwise around each loop?

Solution

Loop 1 encloses the current I that points out of the page. Curl the fingers of your right hand around loop 1 in the counterclockwise direction and your thumb points out—the same direction as the current. Therefore this circulation is positive. Since the total current encircled is I, the circulation is just $\mu_0 I$. Loop 2 encircles the single current I that goes into the page, so that its circulation is just $-\mu_0 I$. Loop

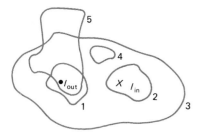

Fig. 28–19
Example 28–3

3 encircles both currents. Since the currents flow in opposite directions, the net current is zero, so that the circulation around loop 3 is zero. Note that this is true even though the magnetic field is not zero on loop 3. Loop 4, also, encloses no current and therefore has zero circulation. Finally loop 5, although different in shape and size from loop 1, encloses the same current and therefore has the same circulation.

We can quantify our statement of Ampère's law by quantifying the notion of circulation. To evaluate the circulation around a loop, we examine each small section of the loop and consider both the magnetic field strength and the extent to which the loop lies in the same direction as the field. Figure 28–20a shows part of an irregular loop L in a magnetic field. In Fig. 28–20b we examine a small part of the loop—so small that it is essentially straight and small enough that the magnetic field is essentially constant in magnitude and direction over it. We represent the small segment by a vector $d\boldsymbol{\ell}$, whose magnitude $d\ell$ is the length of the segment and whose direction is the local direction of the loop. Then the contribution dC from $d\boldsymbol{\ell}$ to the circulation C around the loop is the length of $d\boldsymbol{\ell}$ weighted by the component of the magnetic field in the direction of $d\boldsymbol{\ell}$, or

$$dC = d\ell\, B\cos\theta = \mathbf{B}\cdot d\boldsymbol{\ell},$$

where θ is the angle between the field and the direction of the loop segment $d\boldsymbol{\ell}$.

The circulation around the loop is just the sum of the contributions from all the segments $d\boldsymbol{\ell}$. As the segments get arbitrarily small, this sum becomes an integral for the total circulation:

$$C = \oint_{\text{loop}} \mathbf{B}\cdot d\boldsymbol{\ell}.$$

We encountered *line integrals* similar to this one in Chapter 6 when we considered the work done by a force, and again in Chapter 23 when we

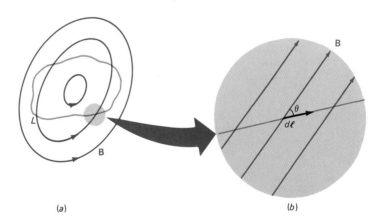

(a)   (b)

Fig. 28–20
(a) An irregular loop in a magnetic field. (b) A magnified view shows that the contribution to the circulation from an infinitesimal segment $d\ell$ of the loop is just $\mathbf{B}\cdot d\boldsymbol{\ell}$.

defined electric potential. The meaning of the line integral is straightforward: it is just a sum of many dot products of the field with segments $d\ell$ of the loop. The circle on the integral sign is a reminder that we are dealing with a closed loop.

Having quantified circulation in terms of a line integral, we can now express Ampère's law mathematically:

$$\oint_{\text{loop}} \mathbf{B} \cdot d\ell = \mu_0 I_{\text{encircled}}. \qquad \text{(Ampère's law, steady currents)} \qquad (28\text{--}12)$$

This statement is true for any arbitrary loop provided the current I on the right-hand side is steady and is the *net current encircled by the loop*. Equation 28–12—Ampère's law for steady currents—is a universal statement describing the relation between the magnetic field and steady electric currents. It holds in electromagnetic devices we build, in atomic and molecular systems, in the interaction of fluid motion and electric charge that gives rise to earth's magnetic field, and in distant astrophysical objects. Although it is difficult to show mathematically, the Biot-Savart law follows logically from Ampère's law in the same way that Coulomb's law follows from Gauss's.

28.4
USING AMPÈRE'S LAW

For charge distributions with sufficient symmetry we used Gauss's law to solve for the electric field in a simple and elegant way. Similarly, for current distributions with sufficient symmetry we can use Ampère's law to solve for the magnetic field. To do so we must find closed loops—called **amperian loops**—over which we can evaluate the circulation integral.

amperian loops

The Field of a Straight Wire

Here we use Ampère's law to calculate the field of an infinitely long, straight wire. This is a special case of the same calculation we did with great difficulty in Example 28–2, using the Biot-Savart law.

Figure 28–21 shows an end view of the wire, which carries a current I out of the page. The wire is cylindrically symmetric, so that the magnitude of B cannot depend on angular position around the wire. The wire is infinitely long, so that all points along it are equivalent. Furthermore we know that the field lines must be closed loops—they cannot begin on the wire and go radially outward because there are no magnetic monopoles on the wire. The only field lines that are both closed loops and exhibit cylindrical symmetry are circles concentric with the wire. By the right-hand rule these field lines circle counterclockwise around the current, as shown in Fig. 28–21.

We have used symmetry arguments to show what the field must look like. Now we must find an amperian loop around which we can evaluate the line integral in Ampère's law. In this example field lines themselves are appropriate loops, for symmetry makes the field magnitude constant on

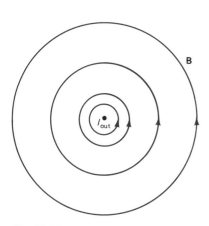

Fig. 28–21
Circular field lines surrounding an infinitely long wire.

the circular field lines. When our amperian loop coincides with a field line, the field is everywhere in the same direction as the loop, so the dot product $\mathbf{B} \cdot d\boldsymbol{\ell}$ in Ampère's law becomes simply $B \, d\ell$. Then the line integral in Ampère's law is

$$\oint_{\text{loop}} \mathbf{B} \cdot d\boldsymbol{\ell} = \oint_{\text{loop}} B \, d\ell = B \int_{\text{loop}} d\ell,$$

where we have taken the field magnitude B outside the integral because it is constant on the circular amperian loop. The remaining integral is just the loop circumference $2\pi r$. Thus the left-hand side of Ampère's law—the circulation around the loop—is $2\pi r B$. Ampère's law says that this circulation is μ_0 times the encircled current, so that

$$2\pi r B = \mu_0 I,$$

or

$$B = \frac{\mu_0 I}{2\pi r}. \tag{28-13}$$

This result is the same as Equation 28–9, and was arrived at with much less difficulty.

Our use of Ampère's law to derive the field of an infinitely long wire depends crucially on the symmetry of the situation. We cannot arbitrarily pull B outside the line integral unless we know—as we do here from symmetry—that it is constant in magnitude and in direction relative to our amperian loop. Clearly symmetry is crucial in allowing us to do this calculation. It may be less obvious why the wire has to be infinitely long for us to apply Ampère's law. Question 17 at the end of the chapter pursues this point.

In this calculation we made no assumptions about the diameter of our wire. Our result therefore holds for any long wire, thick or thin, as long as we restrict ourselves to points *outside* the wire, so that our amperian loop always encloses the entire current. In fact, you can easily convince yourself that our result must hold *outside* any current distribution with line symmetry. To calculate the field *inside* a current distribution, however, we must be careful to use only the actual current encircled by our amperian loop, as shown in the following example.

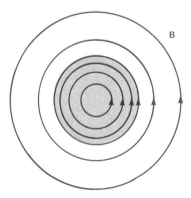

Fig. 28–22
Cross-sectional view of a long, cylindrical wire carrying current out of the page. Symmetry requires that the magnetic field lines be circular inside as well as outside the wire.

Example 28–4
The Field inside a Wire

A long, straight wire of radius R carries a current I distributed uniformly over its cross-sectional area. What is the magnetic field as a function of position within the wire?

Solution

All the symmetry arguments we used to find the field outside a wire still apply here: in particular, the field lines must be circles concentric with the wire axis, and these lines themselves constitute appropriate amperian loops (Fig. 28–22). The field magnitude is constant on such loops, and therefore the circulation around a loop of radius r is $2\pi r B$, just as we found in evaluating the field outside the wire. What is different is the encircled current. An amperian loop within the wire does not encircle the entire current I. How much current does it encircle? With current distributed uniformly over the wire, our amperian loop encircles a fraction of the total current that is proportional to the area encircled by the loop:

$$I_{encircled} = I \left(\frac{\text{amperian loop area}}{\text{cross-sectional area of wire}} \right) = I \frac{\pi r^2}{\pi R^2} = I \frac{r^2}{R^2}.$$

Equating the circulation $2\pi r B$ to μ_0 times the encircled current gives

$$2\pi r B = \mu_0 I \frac{r^2}{R^2},$$

so that

$$B = \frac{\mu_0 I r}{2\pi R^2}. \qquad (28\text{--}14)$$

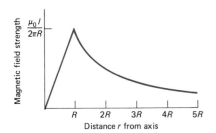

Fig. 28–23
The magnetic field strength inside and outside a wire of radius R carrying a uniform current I.

This field increases linearly with distance from the wire axis. We found a similar dependence when we evaluated the electric field within a cylindrically symmetric charge distribution. The increase occurs because we encircle more and more current—with $I_{encircled}$ growing as r^2—as long as we are inside the wire. Once we reach the surface, of course, the encircled current remains constant and the field begins to decrease inversely with distance, as described by Equation 28–13. Figure 28–23 plots the field strengths both inside and outside a wire.

A Current Sheet

A current sheet is a flat, wide distribution of current. Applying a potential difference across the ends of a piece of aluminum foil, for example, would give a current sheet. Or we could approximate a current sheet by placing many parallel wires side by side. We encounter current sheets less

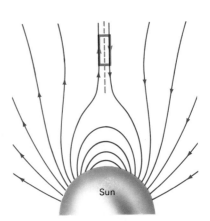

Fig. 28–24
(Left) A coronal streamer in the sun's outer atmosphere contains oppositely directed magnetic fields in close proximity. (Right) A model calculation of the coronal magnetic field. Since $\int \mathbf{B} \cdot d\boldsymbol{\ell}$ is clearly nonzero around the loop shown, there must be a current flowing through the loop. In this instance the current forms a thin, flat sheet that separates the regions of oppositely directed field.

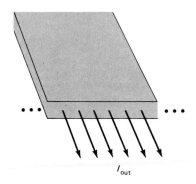

Fig. 28-25
Cross section of a wide current sheet.

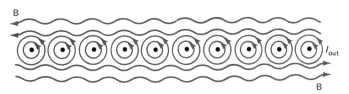

Fig. 28-26
A current sheet may be thought of as consisting of many parallel wires. The magnetic field of the sheet is the vector superposition of the fields of the individual wires.

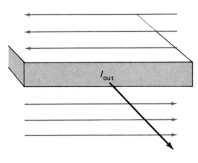

Fig. 28-27
The magnetic field of a current sheet.

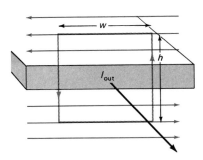

Fig. 28-28
Rectangular amperian loop for evaluating the magnetic field of a current sheet.

frequently than we do wires in our everyday experience with electric currents. But in many situations in nature the formation of current sheets is a natural consequence of the interaction of charged particles and magnetic fields. Whenever magnetic fields of opposite direction get close together, we usually find current sheets, and these sheets are often the sites of dynamic activity (Fig. 28-24).

Figure 28-25 shows a cross section of a current sheet, assumed to be infinitely wide and long, with current flowing out of the page. Because the sheet is infinitely wide, it makes more sense to talk about the current *i* per meter of width than about the total current.

What could the magnetic field of this current sheet look like? One way to see this is to think of the sheet as made up of many parallel wires, each with its own circular field lines, as shown in Fig. 28-26. Above the sheet, the fields of the individual wires add to produce a net field pointing to the left. Below the sheet, the net field is to the right. Field components perpendicular to the sheet cancel. In the limit of infinitely many, infinitely close wires, the field lines must be straight and parallel to the sheet (Fig. 28-27). Thinking of the field in terms of individual wires suggests that one way to calculate the field would be to sum the fields of the individual wires. Problem 35 explores this approach.

Symmetry considerations and the fact that the field lines cannot begin or end could also guide us to the picture of straight, parallel lines shown in Fig. 28-27. That the sheet is infinitely wide means that the field must look the same at all points a given distance from the sheet. Symmetry even tells us that the field must point opposite ways on opposite sides of the sheet. The easiest way to see this is to turn your book upside down. Doing so does not change the direction of the current, which is still out of the page. So the field must look just as it did before—and this can only happen if the field directions are opposite on opposite sides of the sheet.

Having established the general field configuration, we are ready to apply Ampère's law. Figure 28-28 shows an appropriate amperian loop, consisting of a rectangle of width *w* and height *h* centered on the current sheet. We can easily calculate the circulation around this loop, for symmetry requires the same field magnitude at all points on the two sections of width *w* that coincide with the field lines. The contribution to the circulation from each of these sections is just *Bw*. As we go counterclockwise around the loop, we move with the field on both these sections, so that their contributions to the circulation add, giving 2*Bw*. The vertical segments of the loop are perpendicular to the field, so they contribute nothing to the net circulation.

How much current is encircled by the loop? The loop encloses a width w of the current sheet. The sheet carries i amperes per meter, so that the encircled current is iw. Relating the circulation and encircled current through Ampère's law gives

$$2Bw = \mu_0 iw,$$

so that

$$B = \tfrac{1}{2}\mu_0 i. \tag{28-15}$$

This result shows that the magnetic field of an infinite current sheet does not drop off with distance from the sheet. We found a similar result for the electric field of an infinite, uniformly charged plane.

The Fields of Simple Current Distributions

We have calculated the magnetic fields of two simple, symmetric current distributions. Although the resulting magnetic fields look quite different from the electric fields of correspondingly symmetric charge distributions, they exhibit the same dependences on distance. Table 28–1 summarizes the electric and magnetic fields of several simple charge and current distributions. (Why is there no spherically symmetric case listed for the magnetic field?)

Why are we interested in the fields of such impossible things as infinitely long wires or infinite current sheets? First, because they are easy to calculate and give us valuable insight into the behavior of magnetic fields. Second, as in the electric case, we can often approximate real situations in terms of these idealized cases. When we are very close to any wire, for example, it looks long and straight and its magnetic field is given approximately by Equation 28–13. Problem 32 applies this approximation to the field very close to a current loop. For those situations in which we have nothing like the symmetry needed to apply Ampère's law, we can always calculate the magnetic field of a steady current using the Biot-Savart law, just as we can always calculate the electric field of an arbitrary charge distribution using Coulomb's law.

28.5
SOLENOIDS AND TOROIDS

In laboratory apparatus and other devices, we are often interested in producing a uniform magnetic field. We have found that we can produce a uniform electric field between the two closely spaced, charged conducting plates of a capacitor. Is there an analogous device that will produce a uniform magnetic field?

Figure 28–29 shows a hollow coil of wire carrying a current I. Close to any part of the wire are magnetic field lines encircling the wire. We show these field lines at the top and bottom of the coil, where the wires cross the plane of the page. The net field anywhere is, of course, the vector sum of the fields of the individual parts of the loop. You can see that inside the coil, the fields from elements of wire at the top and bottom all have a component to the right, and so tend to reinforce. Above the top of the coil,

TABLE 28–1

Fields of Some Simple Charge and Current Distributions

Distribution	Field Dependence on Distance*	Electric or Magnetic Field
Electric dipole	$\dfrac{1}{r^3}$	
Magnetic dipole	$\dfrac{1}{r^3}$	
Spherically symmetric charge distribution	$\dfrac{1}{r^2}$	
Spherically symmetric current distribution	Impossible for steady current	
Charge distribution with line symmetry	$\dfrac{1}{r}$	
Current distribution with line symmetry	$\dfrac{1}{r}$	
Infinite flat sheet of charge	Uniform field; no variation	
Current sheet	Uniform field; no variation	

*For field *outside* distribution.

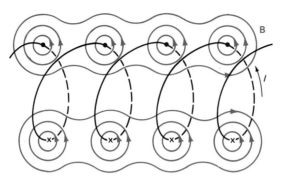

Fig. 28–29
A loosely wound coil of wire. The magnetic field arising from a current in the wire is strongest within the coil. Field is shown only in the plane of the page; dots are points where current emerges from that plane, crosses where current goes into plane of page.

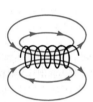

Fig. 28–30
A longer, more tightly wound solenoid.

solenoid

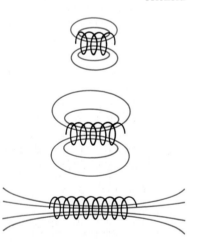

Fig. 28–31
As the solenoid gets longer, the interior field stays nearly constant, but the exterior field lines spread farther apart, so that the exterior field strength decreases.

however, the fields arising from elements at the top all have a component to the left, while fields from elements at the bottom all have a component to the right, thereby weakening the net field. A similar weakening occurs below the bottom of the coil. The net field is strong and points to the right within the coil, and is weaker and points to the left outside the coil, as shown in the figure.

Without doing a detailed calculation we cannot say much more about the loosely wound coil of Fig. 28–29. A calculation using the Biot-Savart law would be very difficult because of all the current elements at different positions and orientations making up the coil. In this short coil with widely spaced turns, there is not sufficient symmetry for us to make effective use of Ampère's law. But now imagine winding the coil more tightly, and making it much longer than its diameter, as suggested in Fig. 28–30. We call such a long, tightly wound coil a **solenoid.** The field is still strong inside the coil of the solenoid, and as the individual turns get arbitrarily close, the irregularities in the field disappear, giving straight field lines inside the solenoid.

What about the field lines outside the solenoid? Because field lines cannot begin or end, these exterior field lines must connect the field lines emerging from the right of the solenoid to those going into the left. Field lines very close to the solenoid axis bend very gradually and so spread far from the solenoid before they return to the other end.

Now imagine making the solenoid longer and longer, as shown in Fig. 28–31. As the solenoid lengthens, the exterior field strength decreases, as evidenced by the decreasing field line density in Fig. 28–31. In the limit of an infinitely long solenoid, all the field lines are confined to the interior and the exterior field is exactly zero. The situation is similar to that of a parallel-plate capacitor, where the interior field becomes uniform and the exterior field zero as the capacitor plates become large compared with their spacing (Fig. 28–32). For solenoids whose length is much greater than their diameter we can make the approximation that the interior field is parallel to the solenoid axis and that the exterior field is zero. As long as we stay away from the ends this approximation is a good one. Figure 28–33 shows the field of a solenoid as traced by iron filings.

Fig. 28–32
(a) A short solenoid is like a widely spaced capacitor, with significant exterior field and nonuniform interior field. *(b)* A long solenoid is like a closely spaced capacitor, with nearly uniform interior field and almost zero exterior field.

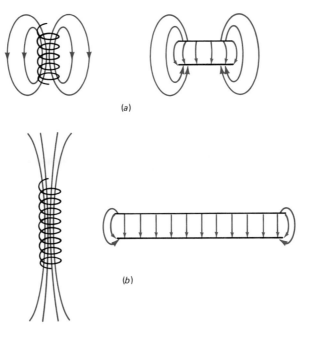

(a)

(b)

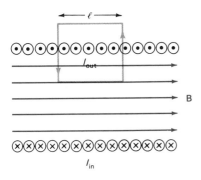

Fig. 28–33
In this photograph, iron filings trace the magnetic field of a solenoid.

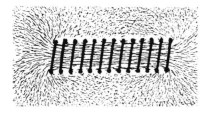

Fig. 28–34
Cross section of a long solenoid, showing a rectangular amperian loop.

Using this long-solenoid approximation, we can apply Ampère's law to calculate the field within the solenoid. Figure 28–34 shows a cross section of a long solenoid. With interior field parallel to the long direction of the solenoid, and zero exterior field, an appropriate amperian loop is the rectangle shown in the figure. With zero exterior field, there is no contribution to the circulation $\int \mathbf{B} \cdot d\ell$ from the top of the loop. The two vertical segments are at right angles to the field, so they too contribute nothing to the net circulation around the loop. In the long-solenoid approximation, the field magnitude B is constant along the interior segment of the amperian loop. As we traverse the loop in a counterclockwise direction we go with the field, so that the contribution to the circulation from this interior segment is just $B\ell$, where ℓ is the length of the amperian rectangle. In fact, this is all the circulation because the other three segments contribute zero.

How much current is encircled by the amperian loop? If there are n turns per unit length of solenoid, the amperian loop encircles $n\ell$ turns. The same current I flows through each turn (why?), so that the total current encircled by the amperian loop is $n\ell I$. Using Ampère's law to relate the circulation around the amperian loop to the encircled current gives

$$B\ell = \mu_0 n\ell I,$$

or

$$B = \mu_0 nI. \quad \text{(solenoid)} \quad (28\text{–}16)$$

Since the vertical dimension of our amperian rectangle never entered the calculation, this field magnitude does not depend on position relative to the solenoid axis. The magnetic field inside the solenoid is therefore truly uniform (see Problem 15).

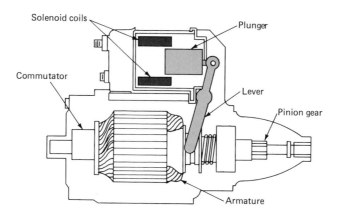

Fig. 28–35
Cross-section of a car starter motor, showing solenoid and mechanical linkage that engages starter pinion gear to start engine.

Solenoids of all sizes are used in a wide variety of experimental and practical devices. If you have ever heard the word "solenoid" before, it may have been in connection with cars. What is a solenoid doing in a car? Because a solenoid is hollow, an iron rod will be pulled into the solenoid by the nonuniform magnetic field near the solenoid ends. When you turn the key to start your car, you apply voltage across a solenoid mounted on the car's starter motor. Current flows in the solenoid, a magnetic field develops, and an iron rod is pulled inside the solenoid. Movement of the rod does two things. First, it joins a set of electrical contacts allowing current to flow to the starter motor. The starter draws 100 A or so, and it would be difficult and costly to switch this large current directly from inside the car (why?). At the same time, the rod moves a small gear to the end of the motor shaft, engaging a gear on the engine's flywheel so that the starter motor can turn the engine (Fig. 28–35).

Whenever you wash a load of clothes in a washing machine you also use solenoids. The valves that control the flow of water into the machine are solenoid valves, opened and closed by the action of an iron rod moving in a solenoid (Fig. 28–36). Solenoids are used in many other mechanical devices where limited motion in a straight line is needed.

Fig. 28–36
A solenoid valve from a washing machine, showing plunger retracted (left) and extended (right).

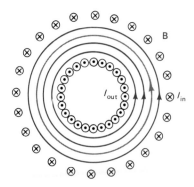

Fig. 28–37
A toroid. The coiled wire is shown only as it pierces the plane of the figure. Symmetry requires that the field lines be circular. Also shown is an amperian loop (gray) for use in calculating the field.

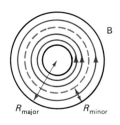

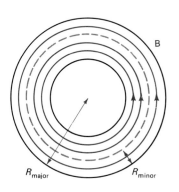

Fig. 28–38
The magnetic field of a toroid can be made more uniform by making the minor radius much smaller than the major radius.

Example 28–5
A Solenoid

A long piece of insulated copper wire is wound into a solenoid. The wire is 0.20 mm in diameter, including a negligibly thick layer of enamel insulation. If the wire is wound as tightly as possible in a single layer, what current must flow in order to produce a 250-G magnetic field in the solenoid?

Solution

With 0.20-mm wire tightly spaced, we have five turns per mm, or $n = 5000$ turns/meter. Solving Equation 28–16 for I then gives

$$I = \frac{B}{\mu_0 n} = \frac{0.025 \text{ T}}{(4\pi \times 10^{-7} \text{ N/A}^2)(5000 \text{ m}^{-1})} = 4.0 \text{ A}.$$

A useful configuration related to the solenoid is the toroid, which is just a solenoid bent into a doughnut shape (Fig. 28–37). The circular symmetry of the toroid requires that the field lines be circular, concentric with the center of the toroid, and with constant field magnitude along any field line. Choosing as our amperian loop a circle of radius r that coincides with a field line, we can readily calculate the circulation around this loop:

$$\oint_{\text{loop}} \mathbf{B} \cdot d\boldsymbol{\ell} = 2\pi r B.$$

Here we could evaluate the line integral because B is constant on the amperian loop and because the loop coincides with a field line. As a result the integral is just the field strength times the circumference $2\pi r$ of the loop. How much current is encircled? If the toroid consists of N turns, and carries a current I, then an amperian loop inside the toroid coil encircles a total current NI. Using Ampère's law to relate this current to the circulation gives

$$2\pi r B = \mu_0 N I,$$

so that

$$B = \frac{\mu_0 N I}{2\pi r}. \qquad (28\text{–}17)$$

This result holds when our amperian loop is within the toroid itself. On the other hand, if our amperian circle is inside the inner edge of the toroidal coils, then there is no current encircled, and the magnetic field is zero. If the amperian circle is outside the outer edge of the coils, it encircles equal but opposite currents, again giving zero field.

The toroidal field given by Equation 28–17 is not uniform, but exhibits a $1/r$ decrease. This nonuniform field causes problems with plasma confinement in fusion reactors, for it results in a drift of particles perpendicular to the field and therefore to the walls of the machine. The nonuniformity may be minimized by building a toroid whose minor radius is much smaller than its major radius, as shown in Fig. 28–38 (see Problem 30).

SUMMARY
---

1. **Gauss's law for magnetism** relates the number of magnetic field lines emerging from a closed surface to the net magnetic monopole charge enclosed. In the observed absence of magnetic monopoles, Gauss's law for magnetism says that the magnetic flux through any closed surface is zero:

$$\oint \mathbf{B} \cdot d\mathbf{A} = 0.$$

Equivalently, Gauss's law for magnetism says that magnetic field lines have no beginning or end. This gives the magnetic field a closed-loop structure very different from the electric field associated with electric charges.

2. Magnetic fields arise from **moving electric charges.** The field associated with a steady electric current is described by the **Biot-Savat law:**

$$d\mathbf{B} = \frac{\mu_0}{4\pi} \frac{I\, d\boldsymbol{\ell} \times \hat{\mathbf{r}}}{r^2},$$

where $d\mathbf{B}$ is the contribution to the field from a current I flowing along an infinitesimal vector $d\boldsymbol{\ell}$ a distance r from where the field is being evaluated. μ_0 is a constant whose value is $4\pi \times 10^{-7}$ N/A^2, and $\hat{\mathbf{r}}$ is a unit vector from the current element $I d\boldsymbol{\ell}$ toward the point where the field is being evaluated. The field of an entire current distribution is obtained by integrating over the fields of the current elements comprising the distribution.

An important special case of a current distribution is a current loop, which gives rise to a dipole magnetic field whose magnitude decreases as the inverse cube of the distance at distances large compared with the loop size.

3. **Ampère's law** provides a more elegant description of the relation between electric current and the magnetic field. Ampère's law relates the circulation, or line integral, of the field around an arbitrary closed loop to the current encircled by the loop:

$$\oint_{\text{loop}} \mathbf{B} \cdot d\boldsymbol{\ell} = \mu_0 I_{\text{encircled}}.$$

This form of Ampère's law is true for all steady currents. It may be used to calculate the magnetic field in situations with sufficient symmetry. These include:
a. configurations with line symmetry (e.g., straight wires)
b. current sheets and other configurations with plane symmetry
c. long solenoids
d. toroids
The Biot-Savart law follows logically from Ampère's law.

QUESTIONS
---

1. Ampère's law states that the circulation around any closed loop is proportional to the current encircled by the loop. Does this mean that if the circulation around a loop is zero, the magnetic field on the loop must be zero?

2. If magnetic monopoles exist, then a net motion of monopoles would constitute a magnetic current. Speculate on what sorts of fields might arise from such a current.

3. The field lines of a magnetic dipole field and an electric dipole field look the same. How, experimentally, could you tell which field was which?

4. Speculate on how a magnetic monopole would interact with an electric field.

5. The field of a long, straight wire consists of circular field lines. Does Ampère's law hold for a square loop surrounding this wire? For a circle that is not concentric with the wire? See Fig. 28–39.

6. Since magnetic monopoles are not observed, why do we even bother to write Gauss's law for magnetism? In the absence of monopoles, does this law have any physical significance?

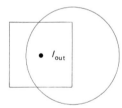

Fig. 28–39   Question 5

7. Figure 28–40 shows four different time-independent fields. Which are electric and which magnetic? Assume no magnetic monopoles are present.

8. Explain why the magnetic field of a current element, as given by the Biot-Savart law, cannot point radially outward from the current element.

9. Why is it advantageous to define the ampere in terms of magnetic forces rather than in terms of a standard ammeter from which all other meters are calibrated?

10. The Biot-Savart law shows that the field of a single current element falls as the inverse square of the dis-

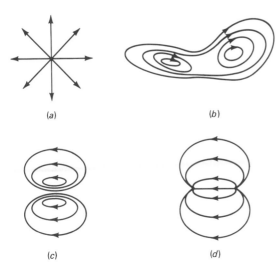

(a)

(b)

(c)

(d)

Fig. 28–40   Question 7

tance from the element. Could you put together a complete circuit whose field falls as the inverse square of the distance from the circuit? Explain.

11. What is the net circulation around each loop shown in Fig. 28–41? Assume each loop is traversed counterclockwise.

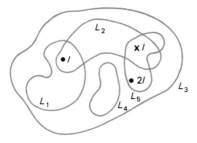

Fig. 28–41   Question 11

12. What must be going on inside the earth in order that it may have a magnetic field?

13. A piece of very flexible wire is wound into a spiral shape, as shown in Fig. 28–42. When current is passed through the wire (through connecting wires coming out of the plane of the page) what happens to the spiral?

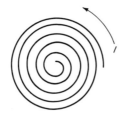

Fig. 28–42   Question 13

14. Does Equation 28–16 apply to the short solenoid of Fig. 28–43? What might be a better approximation to the field on the axis of this coil?

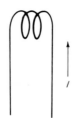

Fig. 28–43   Question 14

15. A solid cylinder and a hollow pipe of the same outer diameter carry the same current along their long dimensions. How do the magnetic field strengths at their surfaces compare?

16. Figure 28–44 shows some magnetic field lines associated with two parallel wires carrying equal currents perpendicular to the page. Are the currents in the same or opposite directions? How can you tell?

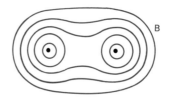

Fig. 28–44   Question 16

17. A short, straight length of wire certainly has cylindrical symmetry. Why can we not use Ampère's law for steady currents to calculate the magnetic field in this situation? *Hint:* Think about steady currents and a short piece of wire.

18. In Section 28.2 we developed an expression for the force between two parallel, current-carrying wires. Suppose two wires cross at right angles, without quite touching. Do they exert forces on each other? What about torques?

19. What would happen to the magnetic field inside a solenoid if you (a) doubled the solenoid length without changing the number of turns per unit length or (b) doubled the length without changing the total number of turns?

20. One way to heat a plasma (ionized gas) is to start a strong current flowing in one direction through the plasma. In response, the plasma compresses or "pinches" in the direction perpendicular to the current. This sudden, adiabatic compression heats the plasma. Explain, in terms of magnetic forces, why this compression occurs.

PROBLEMS

Section 28.1 Gauss's Law for Magnetism

1. Figure 28–45 shows several closed surfaces surrounding an electric dipole and several closed loops surrounding a current loop. What is the electric flux through surfaces 1, 2, and 3, and the magnetic flux through loops 4, 5, and 6?

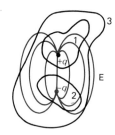

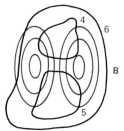

Fig. 28–45 Problem 1

2. Electric charge is distributed uniformly over the surface of a spherical balloon. The balloon is then inflated so it expands; as a result electric current flows radially outward everywhere on the expanding surface. Use Gauss's law for magnetism to argue that the magnetic field resulting from this spherical current distribution must be zero.

Section 28.2 The Biot-Savart Law

3. Part of a long piece of wire is bent into a semicircle of radius a, as shown in Fig. 28–46. A current I flows from left to right through the wire. Use the Biot-Savart law to calculate the magnetic field at the center of the semicircle (point P in the figure).

Fig. 28–46 Problem 3

4. An electron is moving at 3.1×10^6 m/s parallel to a 1.0-mm-diameter wire carrying 20 A. If the electron is 2.0 mm from the center of the wire, with its velocity in the same direction as the current, what are the magnitude and direction of the force it experiences?

5. A square loop of side a carries a current I. Determine the magnetic field (a) at the center of the loop and (b) on the loop axis, a distance $x \gg a$ from the loop.

6. A piece of wire is bent into an isosceles right triangle whose shorter sides have length a. If the wire carries a current I, what is the magnetic field at the point P shown in Fig. 28–47?

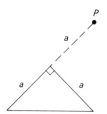

Fig. 28–47 Problem 6

7. The magnetic dipole moment of the earth is 8.0×10^{22} A·m². What is the magnetic field strength at earth's surface at either magnetic pole?

8. A power line carries a 500-A current toward magnetic north, and is suspended 10 m above the ground. The horizontal component of earth's magnetic field at the power line's latitude is 0.24 G. If a magnetic compass is placed on the ground directly below the power line, in what direction will it point?

9. Suppose the earth's field arises from a single loop of current flowing at the outer edge of the planet's liquid core (core radius 3000 km). How large must the current be to give the observed magnetic dipole moment of 8.0×10^{22} A·m²?

10. A single piece of wire is bent so that it includes a circular loop of radius a, as shown in Fig. 28–48. If a current I flows in the direction shown, what is the magnetic field at the center of the loop?

Fig. 28–48 Problem 10

11. The ampere is defined as the current flowing in two long straight wires when they are one meter apart and experience a force of 2×10^{-7} N per meter of length. It would take a rather large apparatus to determine the ampere in this way. Suppose you wished to use smaller wires, each 50 cm long and separated by only 2.0 cm. What force would correspond to a current of one ampere?

12. A long, straight wire carries a 20-A current. A 5.0-cm

by 10-cm rectangular wire loop carrying 500 mA is located 2.0 cm from the wire, as shown in Fig. 28–49. What is the net magnetic force on the loop?

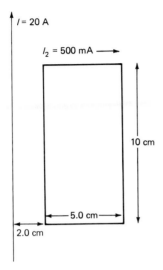

Fig. 28–49   Problem 12

13. A long, flat conducting ribbon of width w is parallel to a long straight wire, with its near edge a distance a from the wire (Fig. 28–50). Wire and ribbon carry the same current I; this current is distributed uniformly over the ribbon. Show that wire and ribbon each experience a force per unit length of magnitude

$$\frac{\mu_0 I^2}{2\pi w}\ln\left(\frac{a+w}{a}\right).$$

What is the direction of each force?

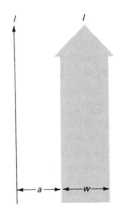

Fig. 28–50   Problem 13

Section 28.3   Ampère's Law

14. In Fig. 28–51, $I_1 = 2.0$ A flowing out of the page. $I_2 = 1.0$ A, also out of the page, and $I_3 = 2.0$ A, into the

page. What is the circulation around each loop shown?

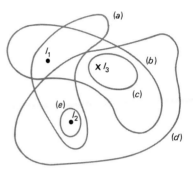

Fig. 28–51   Problem 14

15. Suppose that magnetic field lines in an empty region (the interior of a solenoid, for example) are perfectly straight. Use Ampère's law to show that the field strength cannot vary in the direction perpendicular to the field lines. How could you show that it cannot vary along the field lines?

16. The solar wind draws earth's magnetic field out into a long "magnetotail" on the night side of the planet, as discussed in the preceding chapter and shown in Fig. 27–19 of that chapter. Note in that figure that oppositely directed field lines are adjacent at the central axis of the magnetotail. (a) Use Ampère's law to argue that there must be a current sheet in the magnetotail. Which way is the current flowing in Fig. 27–19? (b) If the field lines are straight and antiparallel on opposite sides of the magnetotail axis, and if the field strength is 2×10^{-4} G, what is the current per unit length along the magnetotail axis?

17. We often draw situations like that of Fig. 28–52, in which the magnetic field changes abruptly from zero to some uniform field strength. Use Ampère's law to

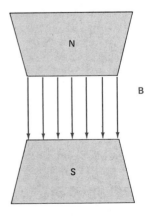

Fig. 28–52   Show that this abrupt change in field is impossible (Problem 17).

show that such an abrupt change is impossible in a current-free region. What do you suppose the actual field in Fig. 28–52 should look like? How does your proposed field circumvent the Ampère's-law objection to the field shown?

Section 28.4 *Using Ampère's Law*

18. (a) What is the magnetic field strength 0.10 mm from the axis of a 1.0-mm-diameter wire carrying a current of 5.0 A distributed uniformly over its cross section? (b) What is the field strength at the surface of the wire?

19. A solid wire 2.1 mm in diameter carries a 10-A current with uniform current density. What is the magnetic field strength (a) at the axis of the wire; (b) 0.20 mm from the axis; (c) at the surface of the wire; (d) 4.0 mm from the axis?

20. A long, hollow conducting pipe of radius R carries a uniform current I along the direction of the pipe, as shown in Fig. 28–53. Use Ampère's law to calculate the magnetic field both inside and outside the pipe.

Fig. 28–53   Problem 20

21. A long, hollow conducting pipe of radius R and length ℓ carries a uniform current flowing around the pipe, as shown in Fig. 28–54. What is the magnetic field both inside and outside the pipe? *Hint:* Does the situation resemble any other configuration you have seen?

Fig. 28–54   Problem 21

22. Two large, flat conducting plates lie parallel to the x-y plane. They carry equal currents, one in the +x-direction, the other in the −x-direction. In each plate there are i amperes per meter of width traversed in the y-direction. What is the magnetic field strength (a) between the plates and (b) outside the plates?

23. Repeat the preceding problem for the case when one current flows in the +x-direction and the other in the +y-direction.

24. A coaxial cable consists of a 1.0-mm-diameter inner conductor and an outer conductor of inner diameter 1.0 cm and 0.20 mm thickness (Fig. 28–55). A 100-mA current flows down the center conductor and

back along the outer conductor. Plot the magnetic field as a function of the distance r from the cable axis, including values of r from zero to beyond the outer conductor.

Fig. 28–55   Problem 24

25. A conducting slab extends infinitely in the x- and y-directions and has thickness h in the z-direction. It carries a uniform current density $\mathbf{J} = J\hat{\imath}$. Calculate the magnetic field strength (a) inside and (b) outside the slab, as functions of distance z from the center plane of the slab.

26. A hollow conducting pipe of inner radius a and outer radius b carries a current I parallel to the pipe axis and distributed uniformly throughout the pipe material (Fig. 28–56). Obtain expressions for the magnetic field (a) for r<a; (b) for a<r<b; (c) for r>b, where r is the radial distance from the pipe axis.

Fig. 28–56   Problem 26

Section 28.5 *Solenoids and Toroids*

27. A solenoid used in a plasma physics experiment is 10 cm in diameter, 1.0 m long, and carries a current of 35 A to produce a magnetic field of 1000 G. (a) How many turns are in the solenoid? (b) If the resistance of the solenoid is 2.7 Ω, how much power is dissipated in the solenoid? (c) How much would be dissipated if the solenoid were a superconductor?

28. You have 10 m of 0.50-mm-diameter copper wire and a battery capable of passing 15 A through this wire. Calculate the magnetic fields you would obtain (a) in-

side a 2.0-cm-diameter solenoid wound with the wire as closely spaced as possible and (b) at the center of a single circular loop made from the entire wire.

29. A toroidal coil of inner radius 15 cm and outer radius 17 cm is wound from 1200 turns of wire. What are (a) the minimum and (b) the maximum magnetic field strengths within the toroid when a 10-A current flows?

30. A toroidal fusion reactor requires a magnetic field that varies by no more than 10% from its central value of 1.5 T. If the minor radius of the toroidal coil producing this field is 30 cm, what is the minimum value for the major radius of the device?

31. Derive Equation 28–16 for the magnetic field of a solenoid by considering the solenoid to be made of a large number of adjacent current loops. Use Equation 28–5 for the field on the axis of a current loop, and integrate over all current loops.

Supplementary Problems

32. A circular wire loop of radius 15 cm and negligible thickness carries a current of 2.0 A. What is the magnetic field arising from this loop (a) in the plane of the loop, 1.0 mm outside the loop? (b) on the axis of the loop, 3.0 m from the loop?

33. A long, flat conductor of finite width w carries a total current I along its long dimension, as shown in Fig. 28–57. The current is distributed uniformly over the cross-sectional area of the conductor. Use suitable approximations to obtain expressions for the magnitude of the magnetic field (a) near the conductor but not near either edge (distance $r \ll w$), and (b) very far from the conductor (distance $r \gg w$). Sketch the magnetic field lines associated with this conductor.

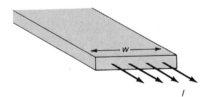

Fig. 28–57   Problem 33

34. Calculate the magnetic field a distance $x \gg a$ from the center of a square current loop of side a, along a line bisecting two of the sides. The loop carries a current I. Show that your result is consistent with the result of Problem 45 of Chapter 23 for the field of a dipole in its perpendicular bisecting plane.

35. Consider an infinite sheet of current to be made up of many closely spaced wires, as suggested in Fig. 28–26. Calculate the field of the current sheet by integrating the fields of the individual wires.

36. A wide, flat spring of spring constant $k = 20$ N/m consists of two 6.0-cm diameter turns of wire, as shown in Fig. 28–58. In its equilibrium position, the coils are nearly touching. A 10-g mass is hung from the spring, and at the same time a current I is passed

through it. The spring stretches 2.0 mm. Find the value of the current I. Assume that the spring coils remain close enough together that they can be treated approximately as parallel straight wires.

Fig. 28–58   Problem 36

37. A solid conducting wire of radius R runs parallel to the z-axis and carries a current density given by $\mathbf{J} = J_0 (1 - r/R)\hat{\mathbf{k}}$, where J_0 is a constant and r the radial distance from the wire axis. (a) What is the total current in the wire? (b) What is the magnetic field strength for points outside the wire ($r > R$)? (c) What is the magnetic field strength for points inside the wire ($r < R$)?

38. A current flows in the x-direction, with current density given by $\mathbf{J} = J_0 e^{-z/z_0}\hat{\mathbf{i}}$, where J_0 and z_0 are constants. Determine the magnetic field at all points.

39. A long, cylindrical conductor of radius R contains a cylindrical cavity of radius a whose axis is parallel to the conductor axis and a distance h from it (Fig. 28–59). The solid part of the conductor carries a uniform current density $\mathbf{J}$ parallel to its long dimension. Use the superposition principle to show that there is a uniform magnetic field within the cavity whose magnitude is $\frac{1}{2}\mu_0 J h$ and that points perpendicular both to the conductor axis and to a line connecting the conductor axis to the cavity axis.

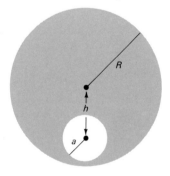

Fig. 28–59   Cross-sectional view of a long, cylindrical conductor containing a cylindrical cavity (Problem 39).

29

ELECTROMAGNETIC INDUCTION

All the electric and magnetic fields we encountered in the previous chapters had their ultimate origins in electric charge, either stationary or moving, or possibly in magnetic monopoles. We stressed a relation between electricity and magnetism, whereby electric charge gives rise to and interacts with both the electric field and the magnetic field. The remainder of our study of electromagnetism is devoted to a much more intimate relation between electricity and magnetism, a relation in which the fields themselves interact directly. This interaction forms the basis of new electromagnetic technologies, leads toward an understanding of the nature of light, and points the way to the theory of relativity.

29.1
INDUCED CURRENTS

In 1831, the English scientist Michael Faraday and the American Joseph Henry independently carried out experiments in which electric currents arose in circuits subjected to changing magnetic fields. Figure 29–1 shows one such experiment, in which a magnet is moved in the presence of a circuit consisting of a loop of wire and an ammeter. There is no battery or other obvious source of emf in the circuit. As long as the magnet is held still, there is no current. But while the magnet is moving, the ammeter

Fig. 29–1
When a magnet is moved near a closed circuit, a current flows in the circuit.

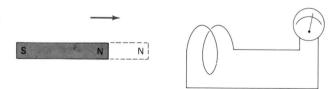

Fig. 29–2
An induced current arises when the magnet is replaced by a current-carrying circuit.

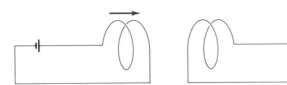

induced current

registers a current—an **induced current.** If we modify the experiment, holding the magnet still but moving the coil, we again observe an induced current. Apparently only the relative motion matters. If we replace the magnet with another circuit in which a battery drives a current, and move the two circuits relative to each other, we get an induced current in the circuit without the battery (Fig. 29–2). If we hold the two circuits still we get no induced current. But now if we close—or open—a switch in the circuit with the battery, we find that the ammeter indicates a momentary induced current (Fig. 29–3).

What is going on here? What do all these experiments have in common? Can we explain all these observations with what we already know of electromagnetism? The answer to this last question is no. We are observing a new phenomenon—**electromagnetic induction.** The one common element in these induction experiments is a *changing magnetic field.* It does not matter whether that field changes because a magnet is moved, or because a circuit is moved near a magnet, or because the current giving rise to the field changes. In all cases, an induced current appears in a circuit subjected to a changing magnetic field.

electromagnetic induction

29.2
FARADAY'S LAW

induced emf

What does it take to drive a current in a circuit? In Chapters 25 and 26 we explored this question, and found that we needed a source of electromotive force—a device like a battery that supplies energy to the circuit. When an induced current flows in a circuit, an emf is similarly present. This **induced emf** is not generally localized at one point in the circuit, as in a battery, but may be spread throughout the conductors making up the circuit.

Experimentally, we find that the induced emf in a circuit depends on the rate of change of magnetic flux through that circuit. Before quantifying this relationship, we explore the notion of magnetic flux and show how to calculate it.

Fig. 29–3
We can induce a current without any motion by changing the current in an adjacent circuit.

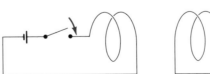
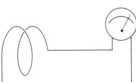

Fig. 29–4
A closed conducting loop encircles a
number of field lines from the bar
magnet. Magnetic flux is a quantitative
measure of the field lines encircled.

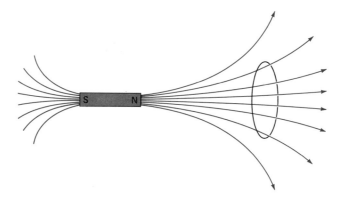

Fig. 29–4
A closed conducting loop encircles a
number of field lines from the bar
magnet. Magnetic flux is a quantitative
measure of the field lines encircled.

Magnetic Flux

A circuit in a magnetic field may encircle, or link, a number of field lines, as shown in Fig. 29–4. We quantify this "number of field lines" with the concept of **magnetic flux.** In analogy with our definition of electric flux in Chapter 21, magnetic flux, ϕ_B, is the integral of the field over a surface:

magnetic flux

$$\phi_B = \int \mathbf{B} \cdot d\mathbf{A}. \tag{29–1}$$

With a flat surface at right angles to the field, this equation reduces simply to a product of the magnetic field strength and surface area. But more generally, when the surface is not flat and/or the field varies with position, then Equation 29–1 is a prescription that says to divide our surface into infinitesimal patches described by vectors $d\mathbf{A}$ perpendicular to the patches and then to sum the infinitesimal flux contributions $d\phi = \mathbf{B} \cdot d\mathbf{A}$ over the entire surface. For a *closed* surface, the result of this integration is zero in the absence of monopoles, but for an *open* surface it need not be zero, as suggested in Fig. 29–5. In calculating the flux linked by a circuit, we can use any open surface bounded by our circuit. Imagine dipping the circuit into a soap solution and beginning to blow a bubble; any soap surface you produce is a suitable surface for evaluating the flux as long as it remains stuck to the circuit. Figure 29–5 shows some appropriate flux surfaces for a simple loop circuit.

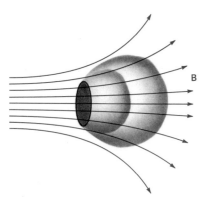

Fig. 29–5
In calculating the flux through a circuit, we can use *any* surface bounded by our circuit. Three such surfaces are shown here; they include the flat circular surface in the plane of the loop-shaped circuit, and two bubble-like surfaces bounded by the loop.

Example 29–1

Magnetic Flux

A solenoid of circular cross section has radius R, consists of n turns per unit length, and carries a current I. What is the magnetic flux through each turn of the solenoid?

Solution

Away from the solenoid ends, the field is uniform and parallel to the solenoid axis, with magnitude given by Equation 28–16:

$$B = \mu_0 n I.$$

A flat surface bounded by one turn of the solenoid lies at right angles to this uniform field, so that the flux is simply the product of the magnetic field and the area:

$$\phi_B = \int \mathbf{B} \cdot d\mathbf{A} = BA = \mu_0 n I \pi R^2.$$

We are being a little loose here in thinking of a single turn of the solenoid as a closed loop to bound a flux surface, but if the solenoid is tightly wound this is an excellent approximation. We will be concerned frequently with the flux through a multi-turn coil, and in calculating this flux it is convenient to view each turn as an individual loop. The flux through an N-turn coil in a uniform magnetic field is just N times the flux through each turn.

Example 29–2
Magnetic Flux

A long, straight wire carries a current I. A rectangular wire loop of dimensions ℓ by w lies with its closest edge a distance a from the wire, as shown in Fig. 29–6. What is the magnetic flux through the loop?

Solution

The magnetic field of the wire is given by Equation 28–13:

$$B = \frac{\mu_0 I}{2\pi r},$$

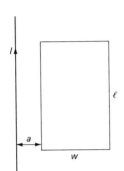

Fig. 29–6
Example 29–2

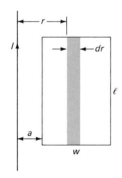

Fig. 29–7
Area elements for the flux calculation are strips of length ℓ and infinitesimal width dr.

where r is the distance from the wire. At the site of the loop, this field points straight into the page, perpendicular to the plane of the loop. However, the field varies with distance from the straight wire, so we cannot simply multiply the field by the loop area to get the flux. Instead, we divide the loop into thin strips of width dr and area $dA = \ell dr$, as shown in Fig. 29–7. With the field at right angles to each strip, $\mathbf{B} \cdot d\mathbf{A} = BdA$, and the flux through any strip is

$$d\phi = BdA = B\ell dr = \frac{\mu_0 I}{2\pi r}\ell dr.$$

Then the total flux through the loop is the integral over all such strips contained within the loop, that is, over all strips between $r = a$ and $r = a + w$:

$$\phi = \int_a^{a+w} \frac{\mu_0 I}{2\pi r}\ell dr = \frac{\mu_0 I \ell}{2\pi}\int_a^{a+w}\frac{dr}{r} = \frac{\mu_0 I \ell}{2\pi}\ln r \Big|_a^{a+w} = \frac{\mu_0 I \ell}{2\pi}\ln\left(\frac{a+w}{w}\right).$$

Flux and Induced emf

Having quantified the notion of magnetic flux, we are now ready to state rigorously the experimental fact that changing magnetic flux induces an emf in a circuit. Our statement is a special case of **Faraday's law of induction,** which constitutes another of the four basic laws of electromagnetism:

Faraday's law of induction

> The induced emf in a circuit is proportional to the rate of change of magnetic flux through any surface bounded by that circuit.

We stress that this statement is a special case; we will later broaden its scope to include situations where no circuit is present. In SI units the proportionality constant between emf and rate of change of flux is just -1, so that mathematically Faraday's law is

$$\mathcal{E} = -\frac{d\phi_B}{dt}, \tag{29-2}$$

where $\mathcal{E}$ is the emf induced in a circuit and ϕ_B the magnetic flux through any surface bounded by that circuit. Problem 1 will help convince you that the units of the rate of change of magnetic flux are indeed volts, the same as the units of emf. The minus sign in Equation 29–2 is essential, and we will have a great deal more to say about it subsequently. We stress that Faraday's law, Equation 29–2, applies whenever the magnetic flux through a circuit changes. We could change that flux by moving a magnet near the circuit, or the circuit near a magnet. We could also change it by changing the strength of the magnetic field of a stationary electromagnet. As the dot product in the definition $\phi = \int \mathbf{B} \cdot d\mathbf{A}$ shows, flux depends not only on field but also on area and on the orientation of area relative to the field. Thus we can change the flux, and hence induce an emf, by changing the area of a circuit or its orientation relative to a constant field.

Example 29–3
A Changing Magnetic Field

A wire loop of radius 10 cm has a resistance of 2.0 Ω. The loop is at right angles to a uniform magnetic field **B**, as shown in Fig. 29–8. The field strength is changing at the rate of 0.10 tesla/second. What is the magnitude of the induced current in the loop?

Solution

To find the induced current, we need to know the induced emf. Faraday's law tells us that the induced emf is related to the rate of change of magnetic flux through the circuit. With a uniform field at right angles to the loop, the flux is just the field strength times the loop area, or

$$\phi_B = \int \mathbf{B} \cdot d\mathbf{A} = \pi r^2 B.$$

Even though the magnetic field is changing with *time*, at any given instant it is uniform in *space*, which is why we did not need to do an integral to calculate the flux.

We don't know B, but this doesn't matter because we are really interested in the *rate of change* of the flux, not in the flux itself. With the loop area constant, the rate of change of flux is

$$\frac{d\phi_B}{dt} = \pi r^2 \frac{dB}{dt} = (\pi)(0.10 \text{ m})^2(0.10 \text{ T/s}) = 3.1 \times 10^{-3} \text{ V}.$$

By Faraday's law, this is the magnitude of the induced emf. We then calculate the current using Ohm's law, as we would for any emf:

$$I = \frac{\mathcal{E}}{R} = \frac{3.1 \times 10^{-3} \text{ V}}{2.0 \text{ Ω}} = 1.5 \times 10^{-3} \text{ A} = 1.5 \text{ mA}.$$

Example 29–4
A Changing Area

A circuit consists of two parallel conducting rails a distance ℓ apart connected at one end by a resistance R. A conducting bar slides along the rails. The whole circuit is in a constant, uniform magnetic field **B** at right angles to the plane of the

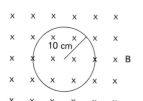

Fig. 29–8
A circular conducting loop at right angles to a uniform magnetic field (Example 29–3).

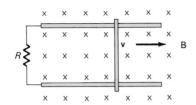

Fig. 29–9
When the bar is pulled to the right, the area of the circuit increases, thereby increasing the magnetic flux and giving rise to an induced emf that drives a current (Example 29–4).

circuit, as shown in Fig. 29–9. The bar is pulled to the right with constant speed v. What is the current in the circuit? Assume the bar and rails are ideal conductors, so that the total circuit resistance is R.

Solution

Here the current is driven by an induced emf arising from the change in magnetic flux that occurs as the circuit area increases. We determine the emf using Faraday's law, $\mathscr{E} = -d\phi_B/dt$. The circuit area is the rail spacing ℓ times the distance x from resistor to bar. With a uniform field perpendicular to the circuit, the flux integral $\int \mathbf{B} \cdot d\mathbf{A}$ reduces to the product of field strength with the area:

$$\phi_B = \int \mathbf{B} \cdot d\mathbf{A} = BA = B\ell x.$$

Never mind that we don't know x—we do know its rate of change and that is all we need. With the field and the rail spacing constant, the rate of change of flux is

$$\frac{d\phi_B}{dt} = B\ell \frac{dx}{dt} = B\ell v,$$

since dx/dt is just the bar velocity v. Faraday's law tells us that the magnitude of the induced emf is equal to the rate of change of flux, so that

$$|\mathscr{E}| = B\ell v.$$

This emf drives a current I around the circuit:

$$|I| = \frac{|\mathscr{E}|}{R} = \frac{B\ell v}{R}.$$

You might wonder about the direction of the induced current, since we concerned ourselves only with the magnitude of the induced emf. Although we could determine the direction of current mathematically from the minus sign in Faraday's law, we will find in Section 29.3 a physical justification for the minus sign that will make it clear how to determine the direction of the current. For now, we solve only for the magnitude of the current.

Example 29–5
A Changing Orientation

A circular wire loop of radius a and resistance R is initially perpendicular to a constant, uniform magnetic field B. The loop rotates with angular velocity ω about an axis through a diameter, as shown in Fig. 29–10. What is the current in the loop?

Solution

Again, we must find the rate of change of magnetic flux, from which we can get the induced emf and then the current. Here the field is uniform over the area, but the orientation of the field relative to the area is changing. The definition of magnetic flux contains a dot product to account for this orientation. Evaluating the flux, we have

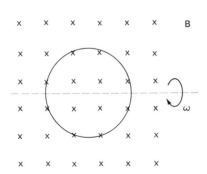

Fig. 29–10
A wire loop rotating in a uniform magnetic field. The flux changes because of the changing orientation between loop area and field, and as a result an emf is induced that drives a current through the loop (Example 29–5).

$$\phi_B = \int \mathbf{B} \cdot d\mathbf{A} = \int B \, dA \cos\theta = B \cos\theta \int dA = \pi a^2 B \cos\theta,$$

where θ is the angle between the field and a perpendicular to the loop area. Here we could take the field magnitude B outside the integral because the field is uniform over the loop. And even though the orientation between loop and field changes with *time*, it does not change with *position* at a given instant, so that we could also take $\cos\theta$ outside of the integral.

The changing orientation is described by giving θ as a function of time. Since the loop rotates with constant angular velocity ω, we can write simply

$$\theta = \omega t,$$

where we take the zero of time when $\theta = 0$. Then the flux is

$$\phi_B = \pi a^2 B \cos\omega t,$$

so that

$$\frac{d\phi_B}{dt} = \pi a^2 B \frac{d}{dt}(\cos\omega t) = -\pi a^2 B\omega \sin\omega t.$$

By Faraday's law, the emf is then

$$\mathcal{E} = -\frac{d\phi_B}{dt} = \pi a^2 B\omega \sin\omega t,$$

giving a current

$$I = \frac{\mathcal{E}}{R} = \frac{\pi a^2 B\omega}{R} \sin\omega t.$$

(Check the units!) Unlike the current in the previous two examples, this one changes with time. Its sinusoidal time dependence is in fact just like that of standard alternating current used for electric power—and with good reason: our rotating loop constitutes a simple alternating-current generator. Sinusoidally varying emf's and currents arise whenever conducting loops are rotated in uniform magnetic fields.

29.3
INDUCTION AND THE CONSERVATION OF ENERGY

When an induced current flows through a resistance, electrical energy is dissipated as heat. Where does this energy come from? The induced current is caused by a changing magnetic flux, so that ultimately whatever agent changes the flux must do work to supply the energy.

Consider the simple case of a bar magnet moved near a loop of wire, as shown in Fig. 29–11. If the loop were not present, it would take no work to move the magnet horizontally. But with the loop present, it must take work to move the magnet. Otherwise, we would get something for nothing: doing no work on the magnet, we would nonetheless produce heat in the loop. As we push the magnet toward the coil, we are doing work against a force. What force? A magnetic force, arising from the interaction of the bar magnet with the magnetic field of the induced current. Think about this! In the absence of the loop, the only magnetic field is that of the bar magnet, so there are no forces on the magnet and it takes no work to move it. With the loop present, the motion of the magnet induces a current in the loop. Like any other current, this induced current gives rise to its own magnetic field. The magnet experiences a force in this field, and we must do work to move the magnet against the force. With the magnetic fields as intermediaries, the work we do in moving the magnet ends up as heat in the loop.

As we push the magnet toward the coil, the force on it must be repulsive so that we do work, rather than having work done on us. In Fig. 29–11, we are moving the north pole of the magnet toward the loop. With

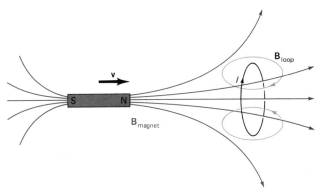

Fig. 29–11
As the bar magnet moves toward the conducting loop, the changing magnetic flux induces an emf that drives a current around the loop. Conservation of energy requires that the magnetic field arising from the induced current *oppose* the motion of the magnet, so that the agent moving the magnet must do work that ends up as heat in the conducting loop.

its induced current, the loop behaves like a magnetic dipole, giving rise to a field similar to the bar magnet's. For a repulsive force on the magnet, the north pole of the loop dipole must be toward the approaching magnet. Applying the right-hand rule by wrapping the right fingers around the current loop in the direction that makes the thumb point left, we see that the induced current must flow in the direction shown in the figure.

Our analysis leading to the direction of the induced current was based on one simple principle—conservation of energy. For electromagnetic induction, this universal principle always requires that

> The direction of the induced current is such as to oppose the change giving rise to it.

Lenz's law   This statement, called **Lenz's law,** is reflected mathematically in the minus sign on the right-hand side of Faraday's law (Equation 29–2).

What happens, for example, when we pull our magnet away from the loop? Conservation of energy requires that we work against a force to do this. The loop must then present the magnet with a south pole, resulting in an attractive force opposing the withdrawal of the magnet. As a result, the loop current is in the opposite direction (Fig. 29–12).

Fig. 29–12
When the magnet is pulled away from the conducting loop, the direction of the induced current is such that the magnetic force from the loop opposes the withdrawal of the magnet.

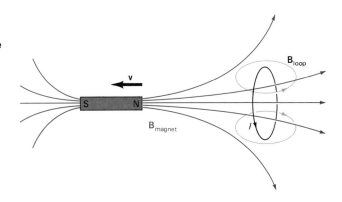

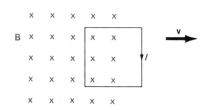

Example 29–6 _____
Conservation of Energy

A square wire loop of side ℓ and electrical resistance R is being pulled at constant speed v out of a uniform magnetic field **B** that points into the page, as shown in Fig. 29–13. Show that the rate of joule heating in the loop is equal to the rate at which the agent pulling the loop does mechanical work, and determine the direction of the induced current.

Solution

The agent pulling the loop must do work in order to supply the energy that ends up as heat in the loop resistance. Therefore, there must be a force on the loop that opposes the loop's motion. This force results from the interaction of the induced current with the magnetic field. Equation 27–7 gives the force on each side of the loop:

$$\mathbf{F} = I\boldsymbol{\ell} \times \mathbf{B},$$

where I is the current and $\boldsymbol{\ell}$ is a vector of length ℓ whose direction is that of the current. The right-hand side of the loop lies outside the field, and therefore experiences no force. The cross product shows that the forces on the top and bottom of the loop are in opposite directions (Fig. 29–14), and therefore contribute nothing to the net force. Then the magnetic force on the loop is that on the left-hand side alone. For this force to oppose the loop's rightward motion, the right-hand rule shows that the current must be upward, so that the current in the entire loop is clockwise.

We could equally well determine the current direction from magnetic flux considerations. As the loop leaves the field region, the flux through it decreases. The direction of the induced current is such as to oppose this decrease in flux. Therefore the magnetic field of the induced current points into the page, as the induced current tries to maintain the flux. By the right-hand rule, a field within the loop and into the page requires that the induced current flow in the clockwise direction.

To calculate the current, we must find the induced emf, which in turn is related to the rate of change of magnetic flux through the loop. With the field and loop perpendicular, and with the field uniform in the region where it is nonzero, the magnetic flux integral becomes a simple product of the magnetic field strength and the loop area that lies within the field:

$$\phi_B = B\ell x,$$

where x is the distance between the left edge of the loop and the right edge of the magnetic field region. The magnetic field remains constant, but as the loop moves the distance x decreases at the rate $dx/dt = -v$ (the minus sign indicates a decrease). Then the rate of change of flux is

$$\frac{d\phi_B}{dt} = \frac{d(B\ell x)}{dt} = B\ell\frac{dx}{dt} = -B\ell v,$$

so that Faraday's law gives

$$\mathcal{E} = -\frac{d\phi_B}{dt} = B\ell v.$$

This induced emf drives a current I around the loop, where

$$I = \frac{\mathcal{E}}{R} = \frac{B\ell v}{R}.$$

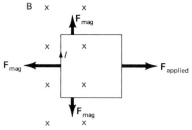

Fig. 29–13
The magnetic flux through the loop decreases as it is withdrawn from the field region. The induced current acts to oppose this decrease, and therefore the magnetic field arising from the induced current tends to reinforce the existing field. The direction of the induced current may also be determined by requiring that the magnetic force on the part of the loop within the field region be opposite to the direction of the loop's motion, so that the agent pulling the loop must do work.

Fig. 29–14
Forces on the loop. The net magnetic force is that on the left-hand side, and the agent pulling the loop must exert an equal but opposite force to maintain constant velocity.

The rate of energy dissipation in the loop is the product of the emf and the current (Equation 25–14):

$$P = I\mathscr{E} = \frac{B\ell v}{R} B\ell v = \frac{B^2\ell^2 v^2}{R}.$$

Since we know the current, we can also evaluate the magnetic force and from it the power supplied by the agent pulling the loop. We have seen that the total magnetic force on the loop is just that on the left-hand side, and that the agent pulling the loop must exert an equal but opposite force to maintain constant velocity (Fig. 29–14). With the current and magnetic field at right angles, the magnitude of this force is

$$F = |I\boldsymbol{\ell} \times \mathbf{B}| = I\ell B \sin 90° = I\ell B = \frac{B\ell v}{R}\ell B = \frac{B^2\ell^2 v}{R}.$$

The rate at which the agent pulling the loop does work is $P = \mathbf{F} \cdot \mathbf{v}$ (Equation 6–24). Here the force and velocity are in the same direction, so we have simply

$$P = \mathbf{F} \cdot \mathbf{v} = Fv = \frac{B^2\ell^2 v}{R}v = \frac{B^2\ell^2 v^2}{R},$$

in agreement with our expression for the electrical power dissipation in the loop. We have shown explicitly that energy is conserved, for all the work done by the agent pulling the loop ends up dissipated as heat in the loop resistance.

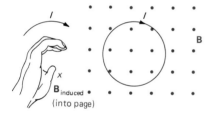

Fig. 29–15
A wire loop perpendicular to a uniform magnetic field pointing out of the page. Although the field is uniform in space, its strength is increasing with time. Within the loop, the field of the induced current opposes this increase, so that the current flows clockwise.

Lenz's law—conservation of energy applied to electromagnetic induction—determines the direction of the current even when no motion is involved. Figure 29–15 shows a wire loop in a magnetic field that points out of the page. Suppose the field is increasing in strength. Then the direction of the induced current must be such as to oppose this strengthening. The loop field in the interior of the loop must point into the page, reducing the net field strength. From the right-hand rule, the loop current is clockwise, as shown. If, on the other hand, the field is decreasing in strength, then a counterclockwise induced current will flow, giving rise to a field that reinforces the decreasing field, opposing its decrease.

Example 29–7
Lenz's Law

Two coils are arranged as shown in Fig. 29–16. If the resistance of the variable resistor is being increased, what is the direction of the induced current in the fixed resistor R?

Solution

Applying the right-hand rule, we find that the magnetic field of coil 1 emerges from the right side of the coil, pointing toward coil 2. As the resistance increases, the current in coil 1 decreases, and with it the strength of coil 1's magnetic field. This results in a decrease in the magnetic flux linked by coil 2. The induced current in coil 2 acts to oppose this decrease in flux, so that the magnetic field arising from the induced current reinforces the field from coil 1. Thus the field of coil 2 emerges from the right end of the coil and enters on the left end. By the right-hand rule, this requires a current from right to left in the fixed resistor.

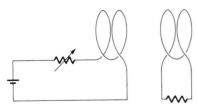

Fig. 29–16
As the current in the first coil decreases, an induced current appears in the second coil. The direction of this current is such that its associated magnetic field reinforces the field of the first coil, thereby opposing the decrease in magnetic flux.

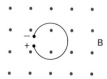

Fig. 29–17
A loop with a gap. A uniform magnetic field points out of the page, and is increasing in strength. Although no current flows in the loop, there is an induced emf that results in charge buildup at the gap.

Sometimes we have a situation involving an open circuit, in which no induced current can flow. Figure 29–17, for example, shows a wire loop with a small gap in it. The loop is at right angles to a uniform magnetic field pointing out of the page and increasing in strength. Although the loop is not quite closed, Faraday's law can still be used to determine the induced emf provided that the gap is very small. In response to this emf, charges pile up on either side of the gap, giving rise to a voltage across the gap. Once the gap voltage is equal to the induced emf, the electric field associated with charge separation counters the induced emf's tendency to move charge, and a steady state is reached. In the open-circuit case, the entire emf given by Equation 29–2 is available at the terminals. The polarity can be determined by thinking about what would happen if current did flow. In the closed circuit of Fig. 29–15, the induced current is clockwise. In the open circuit, this means that positive charge will pile up at the bottom of the gap, and negative charge will pile up at the top, as shown in Fig. 29–17.

Example 29–8

Induction in an Open Circuit

The left side of the loop of Example 29–6 is cut to produce a very small gap. Once a steady state is reached, what are the magnitude and polarity of the emf that appears at this gap? What is the current in the loop? The rate of joule heating in the loop? The rate at which the agent pulling the loop does work?

Solution

With a very small gap, the emf has the same value it did when the circuit was closed:

$$\mathcal{E} = B\ell v.$$

If the loop were closed, current would flow in the clockwise direction (see Example 29–6), so that positive charge piles up on the bottom of the gap, and negative on the top.

During the brief interval when charge is separating, there is a momentary current and the agent pulling the loop does work to separate the charge. But once the steady state is achieved, no current flows, so there is no power dissipation, no magnetic force, and no work done by the agent pulling the loop.

Electromagnetic induction is the principle behind many important technological devices. Induction permits us to transform mechanical into electrical energy, and provides great flexibility in the handling of electric power.

Application

The Electric Generator

Probably the most important technological application of induction occurs in the electric generator. The world currently uses electrical energy at the phenomenal rate of about 10^{13} watts—roughly equal to the power output of one hundred billion human bodies—and virtually all of this power comes from generators. A generator

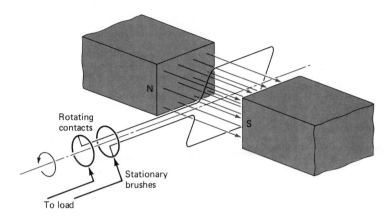

Fig. 29–18
Simplified diagram of an electric generator. As the loop rotates in the magnetic field, the changing flux results in an induced emf. Current flows through the rotating contacts and stationary brushes and on to an electrical load.

is nothing but a system of conductors in a magnetic field (Fig. 29–18). Mechanical energy is supplied to rotate the conductors, resulting in a changing magnetic flux. An emf is induced and current flows through the generator and on to whatever electrical loads are connected to it. Any source of mechanical energy can power the generator, but the most common sources are falling water and steam from burning fossil fuels or from nuclear fission (Fig. 29–19). On a smaller scale, electrical energy may be generated from the kinetic energy of wind (Fig. 29–20). A small electric generator, often called an alternator, is used to recharge the battery in an automobile while the engine is running.

Lenz's law, the conservation of energy in electromagnetic induction, is very much applicable to electric generators. Were it not for Lenz's law, which requires that induced currents *oppose* the changes giving rise to them, generators would turn on their own and happily supply electricity without the need for coal, oil, or ura-

Fig. 29–19
1000-MW steam turbine and generator at a New York City power plant.

Fig. 29–20
Wind-driven electric generators at a "wind farm" in Hawaii.

Fig. 29–21
A 110-car trainload of coal being unloaded at a 1300-MW power plant in Kansas. 14 such trains arrive at the plant each week.

nium. The voluminous quantities of fuel (Fig. 29–21) consumed by power plants are dramatic testimony to the minus sign appearing on the right-hand side of Equation 29–2!

An instructive introduction to Lenz's law comes about if you have access to a hand-cranked electric generator. Without any electrical load across the generator, it is easy to turn. But as you switch on increasingly heavy loads—by *lowering* the electrical load resistance—the generator gets harder to turn (Fig. 29–22). Most people find they can just sustain a 100-W light bulb with a hand generator. Think about this next time you leave a light on! You also experience Lenz's law when you turn on the headlights of a car that is idling slowly. You can hear the engine speed drop, and the car may even stall, as the car's generator gets harder for the engine to turn.

Electric generators in miniature are used extensively in the conversion of information into electrical signals. When you play a conventional record, for example, the grooves in the record wiggle the needle at the end of the record player's tone arm. This needle is connected mechanically to a tiny magnet, mounted inside a

Fig. 29–22
(Above) With no electrical load, a hand-cranked generator is easy to turn. (Below) With a 200-W load of light bulbs connected to the generator, it is much more difficult to turn. Voltmeter (left-hand meter) reads the same in both cases, showing that the generator is being turned at the same rate. But only in the lower photograph is the ammeter reading nonzero, giving nonzero power $(P = IV)$ and therefore requiring work to turn the generator.

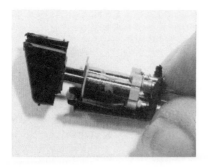

Fig. 29–23
A phonograph cartridge is a miniature electric generator.

Fig. 29–24
A tape head in a cassette player is a small coil in which a changing flux is induced by the changing magnetic field of the moving tape.

stationary coil (Fig. 29–23). As the magnet wiggles, an emf is induced that corresponds to the sounds recorded in the record grooves. After amplification, this signal drives loudspeakers which themselves use magnetic forces to convert electrical signals into sound. And the original recording was probably made using a microphone in which a ferromagnetic diaphragm, wiggling in response to sound waves, gave rise to a time-varying emf corresponding to the sound.

When you play a tape, the magnetized tape moves in front of a coil called the tape head (Fig. 29–24). The magnetic flux through the head changes with the changing magnetic field of the tape, inducing an emf that corresponds to the information—audio, video, or computer data—recorded on the tape. Magnetic disks used for information storage in computers work on the same principle.

Example 29–9

Designing a Generator

An electric generator like that shown in Fig. 29–18 consists of a 10-turn square wire loop 50 cm on a side. The loop is turned at 60 revolutions per second, to produce standard 60-Hz alternating current like that used throughout the United States and Canada. How strong must the magnetic field be for the peak output voltage of the generator to be 170 V? (This is actually the peak voltage of standard 120-V household wiring; 120 V is an appropriate average value.)

Solution

We need to evaluate the induced emf as a function of magnetic field strength, and use our desired 170-V emf to determine the necessary field strength. Faraday's law tells us that the emf is related to the rate of change of magnetic flux through the loop. With a uniform magnetic field, the flux through one turn of the loop is $\int \mathbf{B} \cdot d\mathbf{A} = BA\cos\theta$, where θ is the angle between the field and the normal to the loop. But the loop rotates with angular frequency $\omega = 2\pi f$, so that $\theta = 2\pi ft$. The loop area A is s^2, with s the length of the loop side, so the flux through N turns of the loop is

$$\phi_B = NBs^2\cos(2\pi ft).$$

To find the induced emf, we take the time rate of change of this flux:

$$\mathscr{E} = -\frac{d\phi_B}{dt} = -NBs^2 \left[-2\pi f \sin(2\pi ft) \right] = 2\pi NfBs^2 \sin(2\pi ft).$$

The peak emf is the quantity multiplying the sine; we want this to be 170 V. Solving for the unknown magnetic field B then gives

$$B = \frac{\mathscr{E}_{peak}}{2\pi Nfs^2} = \frac{170 \text{ V}}{(2\pi)(10)(60 \text{ Hz})(0.50 \text{ m})^2} = 0.18 \text{ T}.$$

This is a typical field strength near the poles of a strong permanent magnet.

Application

Eddy Currents

Our discussion of induced currents has centered on conducting loops. But induced currents also appear in solid conductors through which magnetic flux is changing. The resistance of a solid piece of conductor is low, so that the induced currents are large, resulting in substantial energy dissipation. The presence of these induced currents, called **eddy currents,** can make it difficult to move a conductor

eddy currents

Fig. 29–25
An aluminum pendulum bob swings between the poles of a magnet. This strobe photo shows that magnetic forces associated with eddy currents in the aluminum abruptly stop the pendulum.

rapidly through a magnetic field. For example, if you try to push a piece of metal—it need not be a ferrous metal like iron—between the poles of a magnet, you will find yourself working against a magnetic force.

A common demonstration of eddy currents consists of a pendulum with a metal bob that swings between the poles of a magnet (Fig. 29–25). As it swings toward the magnet, the pendulum experiences an increasing magnetic flux that induces eddy currents. The energy dissipated by these currents comes ultimately from the kinetic energy of the pendulum, which therefore stops abruptly between the magnet poles.

Eddy currents provide an alternative to friction brakes for stopping moving machinery. Rapidly rotating saw blades, for example, can be stopped abruptly by an electromagnet activated next to the blade. Similarly, eddy-current brakes are sometimes used on trains and in other applications involving rotating conductors.

In some instances eddy currents are a nuisance, acting just like friction in reducing the efficiency of machinery. To solve this problem, slots are often cut into moving conductors to make the current paths longer, thus increasing the electrical resistance and reducing the eddy currents. For example, if the solid pendulum bob in Fig. 29–25 is replaced by a slotted piece, it then swings more freely through the magnet.

29.4
INDUCED ELECTRIC FIELDS

Our discussion of electromagnetic induction has focused on induced currents and the emf's that drive them. What do we really mean by an induced emf? In a circuit containing a battery, the notion of emf is clear—the emf arises in a specific device where chemical energy is converted to electrical energy associated with separated charge. This charge separation sets up an electric field that drives current in a circuit external to the battery.

When we have an induced current associated with a moving conductor, we can also understand the origin of the current in terms of charge separation. Figure 29–26 shows a pair of conducting rails in a magnetic field. At one end the rails are connected by a resistor. A conducting bar rests across the rails and is free to move along them. If we pull this bar to the right, magnetic force on the charge carriers results in charge separation in the bar. Applying the right-hand rule, we see that positive charge is forced to the top of the bar, negative charge to the bottom. (In a metal, of course, only the negative charge moves, but the result is the same: opposite ends of the bar take on opposite charges.) The bar acts like a battery, its charge separation forcing a current through the rails and the resistor. The emf of this "battery" comes from the force on the charge carriers, and the ultimate source of electrical energy is the agent pulling the bar (see Problem 22 and Example 29–6). The emf induced by the motion of a conductor through a magnetic field is called **motional emf.**

motional emf

But now consider a current induced, in a simple conducting loop, by a magnetic field that changes with time (Fig. 29–27). No physical motion is involved, and there is no obvious force to move the charge carriers. Yet something must drive the current. That something is an **induced electric field.** In the simple case of a circular loop of wire, this induced electric field exists everywhere in the wire and points in the direction of the induced current. The field lines go along the wire loop, themselves forming closed circles, as shown in Fig. 29–27. By the induced emf, we simply mean the energy imparted per unit charge by the induced electric field. This is the same thing we mean by the emf of a battery, except that with the battery the energy is imparted chemically and the process is localized to a particular part of the circuit.

induced electric field

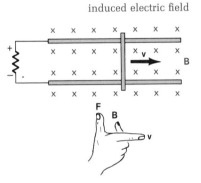

Fig. 29–26
As the bar is pulled to the right, the magnetic force drives positive charge carriers toward the top of the bar, negative toward the bottom.

Induced electric fields arise whenever magnetic flux changes with time. Electric circuits are not necessary for the existence of induced electric fields. When circuits are subject to changing magnetic flux, then the induced electric fields drive currents. But the induced fields—not the induced currents—are fundamental. If, for example, we remove the conducting ring in Fig. 29–27 there will still be an induced electric field. If we put a single, stationary electron in the changing magnetic field, the electron will experience an electric force arising from the induced electric field.

Induced electric fields give rise to forces on charged particles just as do the electric fields of other charges. But the field-line configuration of an induced electric field is typically quite different from that of an electric field originating from electric charges. Because induced fields are not associated with electric charges, their field lines have no beginnings or ends, but typically form closed loops much like the magnetic fields arising from electric currents. (In a region containing both electric charges and a changing magnetic field, the net electric field is the superposition of the induced field and the field arising from charges.)

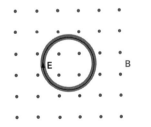

Fig. 29–27
A magnetic field that changes with time gives rise to an induced electric field (black) that is the source of the emf in the conducting loop (gray). Here the magnetic field (color) points out of the page and is increasing in strength.

We have seen that static electric fields—those arising from stationary charge distributions—are conservative, meaning that the work required to move a charge between two points is independent of the path taken. In particular, it takes no work to move around any closed path in an electrostatic field. Mathematically, we can express this by writing

$$\oint \mathbf{E} \cdot d\boldsymbol{\ell} = 0. \quad \text{(electrostatic field)}$$

But this cannot be true for an induced electric field! If we move clockwise around the loop in Fig. 29–27, we gain energy all the way around. But if we move counterclockwise, we must do work against the induced electric field. The amount of work done in moving around a closed path is not zero, and in fact depends on which path we choose. *An induced electric field is not conservative.*

For Further Thought
Nonconservative Electric Fields

The nonconservative nature of the induced electric field is strikingly demonstrated if you attempt to measure potential differences in nonconservative fields. In Chapter 23, we defined potential difference as the work required to move a unit charge between two points, and stressed that this work is independent of path for a conservative field. But when the field is nonconservative, the work is not independent of path, and the concept of potential becomes ambiguous.

Figure 29–28 shows an end view of a long solenoid surrounded by three resistors bent into circular arcs. Each resistor has the same resistance R. If the solenoid current is increasing, an induced electric field appears in the resistors, and drives a current I in the counterclockwise direction.

Because they have the same resistance and carry the same current, the potential difference across each resistor should be the same. We could try to measure the potential difference across one resistor, for example, by connecting a voltmeter as shown in Fig. 29–28b. With current I flowing through the resistance R, this meter will read IR. Since the current flows counterclockwise, we must connect the positive voltmeter terminal to point B.

But now let us try to measure the potential difference across the other two resistors together. To do so, we connect a voltmeter between points A and B as shown in Fig. 29–28c. With the current I flowing through the total resistance $2R$, the meter now reads $2IR$. Furthermore, with the current counterclockwise, the positive terminal of the voltmeter must be connected to point A. We have two voltmeters with their terminals connected to the same points, and yet they indicate different voltages. Not only are the magnitudes of the voltages different, but even their polarities differ. How can this be? We are seeing a manifestation of the non-

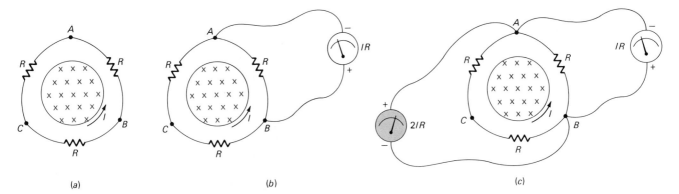

Fig. 29–28
(a) End view of a long solenoid surrounded by three resistors in series. A changing magnetic field in the solenoid induces an emf in the resistors, and the same current flows through each. *(b)* A voltmeter connected between points A and B as shown indicates the voltage IR across one resistor. *(c)* A second voltmeter connected to the *same points* indicates voltage $2IR$ of the opposite sign.

conservative nature of the induced electric field. The two voltmeters are positioned differently with respect to the changing magnetic flux, so that they sample different regions of the induced electric field. The work per unit charge—which is the line integral of the electric field—is not path-independent. Even though the meters are connected to the same points, they measure the line integral of the induced electric field over *different* paths, and so they do not read the same voltage.*

We have been thinking of Faraday's law—the law of electromagnetic induction—as a relation between the emf induced in a circuit and the rate of change of magnetic flux linked by that circuit. We now know that induced electric fields are the fundamental manifestation of changing magnetic flux, and that these fields arise whether or not circuits are present. We need to reformulate our statement of Faraday's law to describe induced electric fields without reference to circuits. The induced emf $\mathcal{E}$ that we have been writing on the left-hand side of Faraday's law means simply the work per unit charge gained by a test charge that is moved around a circuit. Since work is the line integral of force over distance, and electric field is the work per unit charge, we can write the emf as the line integral of the electric field. Then Faraday's law becomes

$$\oint \mathbf{E} \cdot d\boldsymbol{\ell} = -\frac{d\phi}{dt} = -\frac{d}{dt} \int \mathbf{B} \cdot d\mathbf{A}. \quad \text{(Faraday's law)} \qquad (29\text{--}3)$$

Here the line integral on the left-hand side is taken over *any* closed loop, which need not coincide with a circuit. The flux on the right-hand side is the surface integral of the magnetic field over *any* open surface bounded by the loop on the left-hand side.

Faraday's law in the form 29–3 makes no reference to wires or other circuits. It simply describes induced electric fields, which arise whenever there are changing magnetic fields. If electric circuits are present, then induced currents arise as well—but it is the induced electric fields that are fundamental. We can state Faraday's law loosely but powerfully by saying that

> A changing magnetic field gives rise to an electric field.

This direct interaction between the fields is the basis for numerous practical devices and, as we shall see in Chapter 32, is essential to the existence of light.

Note the similarity between Faraday's law describing the induced electric field and Ampère's law (Equation 28–12) describing the magnetic field of a steady current. Faraday's law gives the line integral of the electric field around a closed loop in terms of the rate of change of magnetic flux through a surface bounded by that loop. Ampère's law gives the line integral of the magnetic field around a closed loop in terms of the steady electric current through a surface bounded by that loop. In symmetric situations, we can

---

*For a fascinating discussion of voltage measurement in induced fields, see R. Romer, "What do 'Voltmeters' Measure?: Faraday's Law in a Multiply Connected Region," *American Journal of Physics*, vol. 50, no. 2, pp. 1089–1091 (December 1982).

evaluate the induced electric field with the same mathematical procedure we used for magnetic fields described by Ampère's law. Example 29–10 illustrates this procedure.

Example 29–10
An Induced Electric Field

Within a region of circular cross section, a uniform magnetic field points into the page, as shown in Fig. 29–29. Outside the region, whose radius is R, the field is zero. The field magnitude is increasing with time, at a rate given by

$$B = bt,$$

where b is a constant. (This field might be produced by a long solenoid, in which the current is increased at a constant rate.) What is the electric field within the region of nonzero magnetic field? If a stationary electron were placed at the point P halfway from the center to the edge of the region, what force would it experience?

Solution

The electric field in this example is produced by the changing magnetic field, so the electric field lines can have no beginning or end. Symmetry requires that the field lines be circles, as suggested in Fig. 29–29, and that the magnitude of the electric field depends only on the distance r from the symmetry axis.

Faraday's law (Equation 29–3) says that the line integral of the electric field around any closed loop is proportional to the rate of change of magnetic flux through a surface bounded by that loop. An appropriate loop for this example is a field line itself. If we traverse the field line in the direction of the field, the vector $d\ell$ is everywhere parallel to the field, and the field magnitude is constant, so that

$$\oint \mathbf{E} \cdot d\ell = \oint E d\ell = E \oint d\ell = 2\pi r E,$$

since $\int d\ell$ is just the loop circumference. Faraday's law relates this quantity to the rate of change of magnetic flux through any surface bounded by the integration loop. Here the magnetic field is uniform and perpendicular to the loop, so the magnetic flux through the surface is

$$\phi_B = \int \mathbf{B} \cdot d\mathbf{A} = B\pi r^2 = bt\pi r^2.$$

Then the rate of change of the flux is

$$\frac{d\phi_B}{dt} = \frac{d}{dt}(\pi r^2 b t) = \pi r^2 b.$$

Equating the negative of this rate of change of flux to the line integral of the electric field gives

$$2\pi r E = -\pi r^2 b,$$

so that

$$E = -\frac{br}{2}$$

for points inside the magnetic field region ($r < R$). (Compare this result with that of Example 28–4 for the magnetic field inside a current-carrying wire; both show fields increasing linearly with distance from the symmetry axis.)

The minus sign in our expression describes, through the right-hand rule, the direction of the induced electric field. However, the easiest way to determine this direction is to imagine what would happen if there were a conducting loop carrying

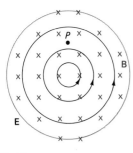

Fig. 29–29
A uniform magnetic field confined to a region of circular cross section. If the field increases with time, it gives rise to an induced electric field. Symmetry requires that the electric field lines be concentric circles.

an induced current. Such a current would be counterclockwise to oppose the change in magnetic field, so that the induced electric field must be counterclockwise.

Just as it would in any other electric field, a charged particle would experience a force $\mathbf{F} = q\mathbf{E}$ in this induced electric field. For an electron (charge $-e$), the force at point P (where $r = \frac{1}{2}R$) would then have magnitude $eE = \frac{1}{4}bR$. Since the electron is negative, this force points opposite the field direction—that is, to the right at point P.

Is there an electric field outside the region of changing magnetic field? Yes—because even though there is no magnetic field for $r > R$, a circular loop still encloses changing magnetic flux. Problem 25 explores the question of this exterior electric field.

Application

The Tokamak

Attempts to harness nuclear fusion—the energy source of the sun and of the hydrogen bomb—hinge on our ability to confine matter at temperatures near 100 million K long enough for significant numbers of fusion reactions to take place. These high temperatures are required to give individual nuclei the kinetic energy needed to overcome their electrical repulsion, coming close enough for the strong but short-range nuclear force to take over. At these temperatures, the reactants [typically the hydrogen isotopes deuterium (^{2}H) and tritium (^{3}H)] are fully ionized, forming a plasma consisting of positively charged nuclei and negative electrons.

Many fusion schemes make use of magnetic fields for confining the hot plasma. The most successful magnetic confinement device available today is the tokamak, a Russian invention now under study in the United States, Japan, and Europe as well as in the Soviet Union (Fig. 29–30).

The dual challenge to any fusion device is first, to heat the plasma to the extraordinary temperatures needed for fusion reactions and, second, to confine the hot plasma long enough that a sufficient number of fusion reactions take place. In

Fig. 29–30
The Tokamak Fusion Test Reactor (TFTR) at Princeton University's Plasma Physics Laboratory has reached temperatures of 200 million K—over 10 times that of the sun's core. It is soon expected to achieve scientific breakeven—the production of as much fusion energy as it takes to heat the plasma.

a tokamak, the plasma is confined inside a toroidal chamber by magnetic field lines that circle around the toroid. These field lines arise from currents flowing in coils wrapped around the toroid. An additional vertical field component is produced in order to oppose the tendency of the plasma to expand when current flows through it. Finally, the magnetic field is given a component around the minor radius of the toroid to inhibit instabilities that would throw plasma to the walls of the chamber. The net magnetic field spirals around the toroidal chamber, as shown in Fig. 29–31.

The tokamak is distinguished from other toroidal fusion reactors in that the field component around the minor radius arises from current flowing in the plasma itself. This current is driven by an induced electric field, which itself arises from a changing magnetic flux in the "hole" in the toroidal "doughnut." The plasma current not only provides a magnetic field component that stabilizes the plasma, but it also heats the plasma by joule heating. As in any conductor, the electric field accelerates individual charged particles, which then collide and share their energy, resulting in an increased plasma temperature. Unfortunately, the resistance of a plasma drops at high temperatures, making it difficult to reach fusion temperatures with joule heating alone. Nevertheless, the induced electric field brings the plasma close to the so-called ignition temperature, and at the same time results in a more stable plasma. Because the tokamak's induced electric field requires a changing magnetic field, the tokamak cannot be run continuously, but must be pulsed as large capacitors are suddenly discharged through coils to make the rapidly changing magnetic field.

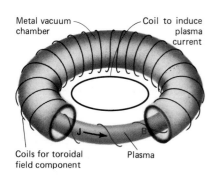

Fig. 29–31
The magnetic field of a tokamak spirals around the toroidal plasma chamber. **J** represents current flowing in the plasma, giving the field its twist. Plasma current is induced by changing flux associated with coils located in the "hole" of the toroidal vacuum chamber.

29.5
DIAMAGNETISM

In Section 27.6 we discussed paramagnetic and ferromagnetic materials, in which atomic magnetic dipoles align with an applied magnetic field, giving rise to an attractive interaction between the material and a magnet. We also mentioned diamagnetic materials, in which induced magnetic dipoles align antiparallel to the applied field, giving rise to a repulsive force. We are now ready to understand diamagnetism as a manifestation of Faraday's law at the microscopic level.

In a purely diamagnetic material, current loops associated with pairs of atomic electrons exactly cancel, leaving atoms with no intrinsic magnetic moments. Fig. 29–32a shows a simplified model to describe such an atom. The picture should not be taken too literally, for it uses classical physics to describe a phenomenon properly within the domain of quantum mechanics. Nevertheless, it shows qualitatively how diamagnetism is an electromagnetic induction effect.

The electron orbiting clockwise in Fig. 29–32a has a magnetic moment pointing out of the page (apply the right-hand rule, and remember that the electron carries a negative charge). In the absence of an applied magnetic field, this moment is exactly cancelled by the equal but oppositely directed moment of the other electron. But now suppose a magnetic field is applied, for example by moving a magnet toward the material. Orbiting atomic electrons experience a changing magnetic field as the magnet approaches. This changing field gives rise to an induced electric field that acts on the electrons. The direction of this field is such that the electrons respond in a way

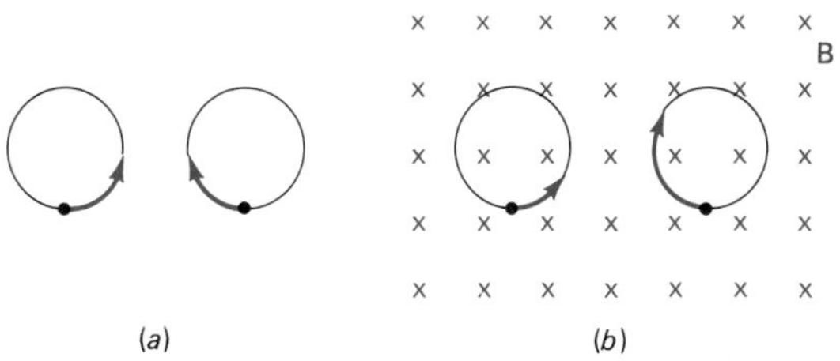

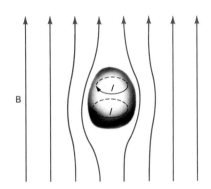

Fig. 29–32
Simplified, classical model for a diamagnetic atom. *(a)* In the absence of an applied field, magnetic moments associated with a pair of electrons exactly cancel. *(b)* As a magnetic field is applied, the changing flux induces an electric field that speeds up one electron and slows down the other. As a result, the atom acquires a net magnetic moment directed opposite to the applied field. (In applying the right-hand rule, remember that electrons are negative.)

that opposes the change in magnetic field. Since the field points into the page and is increasing, the electron magnetic moment that points out of the page must increase to oppose the increase in magnetic field. Conversely, the moment pointing into the page must decrease. Physically, these changes occur because the induced electric field increases the electron speed in one case and decreases it in the other (Fig. 29–32b).

Once the field is applied, the orbital speeds of the two electrons are no longer identical, and the atom represented by these electrons now has a net magnetic moment that points out of the page in Fig. 29–32b, opposing the field of the approaching magnet and resulting in a repulsive force. We could repeat our analysis for an approaching south pole, and would find again that the induced magnetic moment results in a repulsive force. This repulsion from a magnet is the distinguishing characteristic of diamagnetism. We listed a number of diamagnetic materials in Table 27–1.

An important special case of a diamagnetic material is a superconductor. With magnetic susceptibility −1, the superconductor totally excludes magnetic fields from its interior. Because the superconductor has zero resistance, an induced electric field of finite strength would give rise to an infinite current density. This impossibility is prevented by induced currents whose associated magnetic fields exactly cancel the applied field. With zero electrical resistance, the induced currents persist, so that the net magnetic field inside a superconductor always remains exactly zero (Fig. 29–33).

Fig. 29–33
Induced currents in a superconductor completely cancel an applied magnetic field within the superconductor. Net result is that a magnetic field cannot penetrate the superconductor.

SUMMARY

1. **Electromagnetic induction** is a fundamental phenomenon linking magnetism and electricity. Induction is described by **Faraday's law,** which states that a **changing magnetic field** gives rise to an **induced electric field.** Unlike the conservative electrostatic field of an electric charge, this induced field is **nonconservative,** meaning that it can do work on charges as they move around a closed loop. Faraday's law relates the line integral of this nonconservative electric field around an arbitrary loop to the rate of change of magnetic flux through a surface bounded by that same loop:

$$\oint \mathbf{E}\cdot d\boldsymbol{\ell} = -\frac{d\phi_B}{dt} = -\frac{d}{dt}\int \mathbf{B}\cdot d\mathbf{A}.$$

2. In order for energy to be conserved, the induced electric field is in such a direction as to oppose the change in flux that gives rise to it. This energy-conserving aspect of Faraday's law is called **Lenz's law,** and is reflected mathematically in the minus sign on the right-hand side of Faraday's law.

3. When a conductor is present, the nonconservative electric field manifests itself as an **induced emf:**

$$\mathcal{E} = -\frac{d\phi_B}{dt}.$$

This emf drives an **induced current** in any circuit with finite resistance. It does not matter whether the magnetic flux is changed by moving a conductor in a magnetic field, or by moving a magnetic field near a con-

ductor, or by altering the shape or orientation of the conductor. The generation of electric power by moving conducting loops in magnetic fields is an important technological example of induced currents.

4. **Diamagnetism** is a manifestation of electromagnetic induction on the atomic scale. Application of a magnetic field to a diamagnetic material results in induced atomic magnetic moments that cause the material to be repelled from a magnet.

QUESTIONS

1. A copper penny falls on a vertical path that takes it between the poles of a magnet. Does it hit the ground faster or slower than if no magnet were present?

2. A bar magnet is moved toward a conducting ring, as shown in Fig. 29–34. What is the direction of the induced current in the ring?

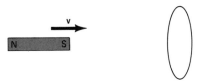

Fig. 29–34   Question 2

3. If you run along the earth's magnetic equator, you move at right angles to the lines of earth's magnetic field. Could you run fast enough to electrocute yourself? Assume no limit to how fast you can run.

4. Figure 29–35 shows two concentric conducting loops, the outer connected to a battery and a switch. The switch is initially open. It is then closed, left closed for a while, then opened again. Describe the currents in the inner loop during the entire procedure.

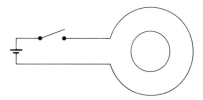

Fig. 29–35   Question 4

5. An electric generator is being turned at constant speed. A load resistor R is connected across the generator terminals. If the electrical resistance of the load is lowered, does the generator get easier or harder to turn?

6. Service manuals for cars often tell you to set the idle speed of the engine with the headlights on. Why? What does this have to do with electromagnetic induction?

7. Figure 29–36 shows two square wire loops, the first containing a battery and variable resistor. The resistor is initially at the midpoint of its resistance range. Should its resistance be lowered or raised in order to induce a clockwise current in the second loop?

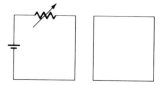

Fig. 29–36   Question 7

8. In Chapter 27 we discussed electric motors. Consider the simple motor shown in Fig. 27–44 of that chapter. What happens if you connect a resistor across the motor terminals, and turn the motor by hand? What is the difference between a motor and a generator?

9. Figure 29–37 shows an open loop of wire and a magnetic field pointing into the page. The field is decreasing in strength. If you connect an ideal voltmeter across the loop ends A and B, will it give a nonzero reading? If so, to which loop end should you connect the positive terminal of the voltmeter? If not, why not? Assume that the two voltmeter leads are so closely spaced that they enclose essentially no magnetic flux.

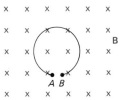

Fig. 29–37   Question 9

10. A constant, uniform magnetic field points into the page. A flexible, circular conducting ring lies in the plane of the page. If the ring is stretched, maintaining its circular shape, what is the direction of the induced current?

11. A student argues that it takes work to stretch the ring in the preceding question, and that therefore the ring should release all the work as energy if allowed to shrink. Is this right? Why or why not?

12. When a magnet is moved near a superconductor, the magnetic field lines never enter the superconductor. Why not?

13. Is it possible to produce an induced current that never changes? How or why not? Could you produce an induced current that was steady for some finite time? How or why not?

14. When you push a bar magnet into a conducting loop, you do work. What happens when you pull it out the other side?

15. You are turning a generator in such a way that the current it delivers remains constant. As you lower the load resistance across the generator, does the generator get easier or harder to turn?

16. Devise a way of measuring a magnetic field using Faraday's law.

17. In Fig. 29–38, a copper ring was originally resting on the wooden structure, surrounding the coil. When a rapidly changing current was applied to the coil, the ring was ejected into the air. Explain this phenomenon.

Fig. 29–38   Question 17

18. Fluctuations in earth's magnetic field due to changing solar activity can wreak havoc with communications, even those using underground cables. How is this possible?

19. Why is it not possible to run a tokamak on a continuous basis?

20. Conventional brakes on a car need large surface areas to dissipate the heat of friction when the brakes are applied. Would eddy-current brakes have the same problem?

21. In Chapter 27, we pointed out that a static magnetic field cannot change the energy of a charged particle. Is this true of a changing magnetic field? Discuss.

22. A long solenoid of circular cross section is oriented so that its magnetic field points out of the page, as shown in Fig. 29–39. The solenoid current is increasing. (a) What is the direction of the induced electric field at points A and B in the figure? (b) What is the magnitude of the induced electric field in the center of the solenoid? (Don't calculate! Argue from symmetry.)

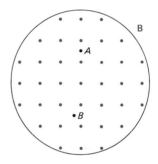

Fig. 29–39   Question 22

23. Is the concept of electric potential (Chapter 23) useful in a nonconservative electric field? Give an example to substantiate your answer.

24. Can an induced electric field exist in the absence of a conductor?

25. Could you tell whether a given electric field arises from electric charge or from a changing magnetic field? How or why not?

26. Does a diamagnetic material experience a force in a uniform magnetic field?

═══════════════════ **PROBLEMS** ═══════════════════

Section 29.2   *Faraday's Law*

1. Show that the volt is the correct SI unit for the rate of change of magnetic flux, so that Faraday's law is dimensionally correct.

2. A bar magnet is moved steadily through a wire loop (Fig. 29–40). Sketch qualitatively the current and power dissipation in the loop as functions of time. Indicate the position of the magnet on your time axes.

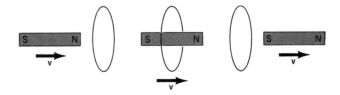

Fig. 29–40   Problem 2

3. A 2.0-m-long solenoid of circular cross section has a diameter of 15 cm and consists of 2000 turns of wire. The current in the solenoid is increasing at the rate of 100 A/s. A wire loop of diameter 10 cm is inside the solenoid and concentric with the solenoid axis. The loop has a resistance of 5.0 Ω. (a) What is the current in the loop? (b) If the loop had a diameter of 25 cm, so that the loop was outside the solenoid, would your answer to part (a) change? If so, would the loop current increase, decrease, or go exactly to zero?

4. A solenoid 2.0 m in length and 30 cm in diameter consists of 5000 turns of wire. A 5-turn wire loop is wrapped around the solenoid, and connected to a 180-Ω resistor, as shown in Fig. 29–41. The direction of the current in the solenoid is such that the solenoid's magnetic field points to the right in Fig. 29–41. The solenoid current is reduced steadily from 40 A to zero over a period of 10 ms. During this period, what are the direction and magnitude of the current in the resistor?

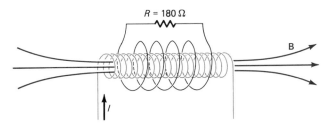

Fig. 29–41   Problems 4, 9

5. A small coil 1.0 cm in diameter consists of 5 closely spaced turns of wire. The coil is inserted into a uniform magnetic field and rotated at 10 revolutions per second about an axis perpendicular to the field. A voltmeter is connected to the coil through rotating contacts. If the peak reading on the voltmeter is 380 μV, what is the magnetic field strength?

6. The wingspan of a 747 jetliner is 60 m. If the plane is flying at 960 km/hour in a region where the vertical component of earth's magnetic field is 0.20 G, what emf develops between the plane's wingtips?

7. A square wire loop 3.0 m on a side lies at right angles to a uniform magnetic field of 2.0 T. A 6-V light bulb is in series with the loop, as shown in Fig. 29–42. The magnetic field is decreased steadily to zero over a time interval Δt. How long must Δt be if the light bulb is to shine at full brightness during this time?

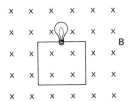

Fig. 29–42   Problem 7

8. A square conducting loop of side s and electrical resistance R moves to the right with speed v. At time t=0, its rightmost edge enters a region containing a uniform magnetic field **B** pointing into the page, as shown in Fig. 29–43. The magnetic field covers a region of width w. Take s=0.50 m, w=0.75 m, v=0.25 m/s, R=5.0 Ω, and B=1.0 T. (a) Plot the magnetic flux in the loop as a function of time. Take a flux into the page as positive. (b) Plot the current in the loop as a function of time. Take a clockwise current as positive. (c) Plot the power dissipation in the loop. (d) The top segment of the square is broken and an ideal voltmeter inserted in it, with its positive terminal to the right. Plot the voltmeter reading as a function of time. (e) With the voltmeter still in the loop, plot the power dissipation in the loop as a function of time.

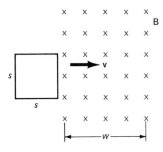

Fig. 29–43   Problem 8

9. A small coil is wrapped around a solenoid, and a resistor connected across it, as in Fig. 29–41. The current in the solenoid is varied as shown in Fig. 29–44. Make a qualitative plot of the current in the small coil and resistor as a function of time.

Fig. 29–44   Problem 9

10. Repeat the preceding problem if the solenoid current is described by Fig. 29–45. The curves in that figure are segments of parabolas.

Fig. 29–45   Problem 10

11. Figure 29–46 shows a pair of parallel conducting rails in a uniform magnetic field **B** pointing into the page. The rails are a distance ℓ apart, and a resistance R is connected across them, as shown. A conducting bar is free to move without friction along the rails. The bar is being pulled to the right with a constant speed v. (a) What is the direction of the current in the resistor? (b) What is the magnitude of the current in the resistor? (c) At what rate must work be done by the agent pulling the bar?

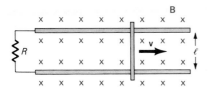

Fig. 29–46   Problem 11

12. The resistor in the preceding problem is replaced by an ideal voltmeter. (a) To which rail should the positive terminal of the meter be connected? (b) What does the voltmeter read? (c) At what rate must work be done by the agent pulling the bar?

13. The voltmeter of the preceding problem is replaced by a battery of emf $\mathscr{E}$ in series with a resistor R. The bar is initially at rest, and now no agent pulls it. (a) Describe the subsequent motion of the bar. (b) The bar eventually reaches a constant speed. Why? (c) What is the final speed of the bar? Express your answer in terms of the magnetic field strength, the battery emf, and the rail spacing ℓ. Does the resistance affect the final speed? If not, what effect does it have?

14. A car alternator consists of a 250-turn wire loop of diameter 10 cm, as shown in Fig. 29–47. The magnetic field in the alternator is 0.10 T. (This field arises from stationary coils supplied with direct current from the car battery.) If the alternator is turning at 1000 revolutions per minute, what is its peak output voltage?

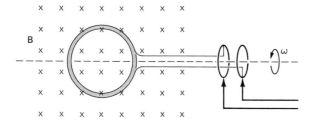

Fig. 29–47   Problem 14

15. A pair of parallel conducting rails lies at right angles to a uniform magnetic field **B** of magnitude 2.0 T, as shown in Fig. 29–48. The rails are 10 cm apart. A 5.0-Ω resistor and a 10-Ω resistor lie across the rails and are free to slide without friction along them. (a) The 5.0-Ω resistor is held fixed and the 10-Ω resis-

tor is pulled to the right at 50 cm/s. What are the direction and magnitude of the induced current? (b) Now the 10-Ω resistor is held fixed and the 5.0-Ω resistor is pulled to the left at 50 cm/s. What are the direction and magnitude of the induced current? (c) What is the rate of joule heating in the 10-Ω resistor in both cases above?

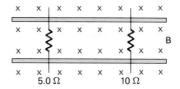

Fig. 29–48   Problem 15

16. Figure 29–49 shows two parallel conducting rails a distance ℓ apart lying perpendicular to a uniform magnetic field. A resistor R is inserted in one of the rails. Two identical conducting bars are free to slide without friction along the rails. (a) The right-hand bar is suddenly accelerated to speed v and is pulled to the right with this constant speed. Describe the subsequent motion of the left-hand bar, and give an expression for its terminal speed, if any. (b) A constant force F to the right is applied to the right-hand bar. Describe the subsequent motion of both bars.

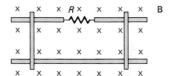

Fig. 29–49   Problems 16, 17

17. In Fig. 29–49, a battery of emf $\mathscr{E}$ is inserted in series with the resistor R. The positive terminal of the battery is to the right. (a) What is the final speed reached by each bar? (b) At what rate is work being done by the battery when the bars have reached their final speeds?

18. A pair of conducting rails stands in earth's gravitational field, making an angle θ to the vertical, as shown in Fig. 29–50. The rails are a distance ℓ apart, and are connected at the top by a resistor R. A uniform magnetic field points horizontally, into the page. A conducting bar of mass m is free to slide down the rails without friction. When released, the bar soon reaches a constant velocity. Obtain an expression for this velocity.

19. A magnetic field points into the page, and its magnitude increases to the right, as shown in Fig. 29–51. This increase is linear, so that the field is given by $B = B_0 + bx$, where b is a constant. A square conducting loop of side a and resistance R moves to the right

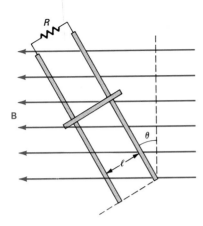

Fig. 29–50   Problem 18

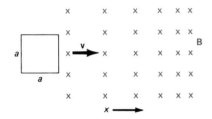

Fig. 29–51   Problem 19

with constant speed v. At time $t=0$, the left edge of the loop is at $x=0$. (a) What is the flux through the loop as a function of the position of the left edge of the loop? (b) What is the direction of the induced current in the loop? (c) What is the magnitude of the induced current? (d) At what rate does the agent pulling the loop do work on the loop?

20. A square wire loop 80 cm on a side is placed at right angles to a uniform magnetic field of 2.0 T, as shown in Fig. 29–52. At time $t=0$, all four sides of the loop begin to lengthen at the same rate. Obtain an expression for the length of a side as a function of time, if the induced emf in the loop is to be a constant 100 mV.

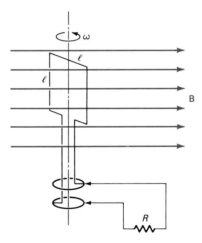

Fig. 29–52   Problem 20

Section 29.3   *Induction and the Conservation of Energy*

21. A bar magnet of mass m falls vertically at constant speed v down a hollow copper pipe. What is the rate of joule heating in the pipe? Neglect air resistance.

22. A pair of parallel conducting rails 10 cm apart lies at right angles to a uniform magnetic field of 0.50 T, as shown in Fig. 29–53. A 4.0-Ω resistor is connected across the rails. The bar is being pulled to the right at 2.0 m/s. (a) What is the current in the resistor? (b) What is the magnetic force on the bar? (c) What is the rate of joule heating in the resistor? (d) What is the rate at which the agent pulling the bar does work on it? Compare with your answer to part (c).

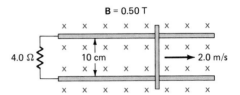

Fig. 29–53   Problem 22

23. A generator consists of a single-turn square loop of side ℓ, rotating at angular velocity ω in a uniform magnetic field **B**, as shown in Fig. 29–54. At time $t=0$ the loop plane is perpendicular to the magnetic field. A load resistor R is connected across the loop by means of rotating contacts. (a) Obtain an expression for the current in the resistor as a function of time. (b) Obtain an expression for the power dissipation in the resistor as a function of time. (c) Obtain an expression for the torque on the loop as a function of time. (d) Use your expression for torque to show that the power supplied by the agent turning the loop is equal to the power dissipated in the resistor.

Fig. 29–54   Problem 23

24. A circular conducting ring of radius r_0 lies at right angles to a uniform magnetic field **B** that points into the page. At time $t=0$, the ring begins to expand, so that its radius as a function of time is given by $r=r_0+vt$, where v is constant. At time $t=t_1$, the ring stops expanding. (a) What are the magnitude and direction of the induced current as a function of time

while the ring is expanding? (b) Integrate the power dissipation in the ring over time to calculate the total amount of joule heating while the ring is expanding. (c) By considering the magnetic force on the ring and integrating the rate at which the agent expanding the ring does work, show that the total mechanical work done in expanding the ring is equal to the joule heating.

Section 29.4 *Induced Electric Fields*

25. For the situation of Example 29–10, obtain an expression for the electric field outside the magnetic field region. How do points outside the region know anything about the changing magnetic field?

26. A uniform magnetic field points into the page, as shown in Fig. 29–55. In the same region, an electric field points straight up, but increases in strength at the rate of 10 volts/meter^2 as you move to the right. By applying Faraday's law to a rectangular loop, show that the magnetic field must be changing with time, and calculate its rate of change.

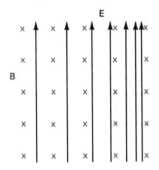

Fig. 29–55 Problem 26

27. A uniform magnetic field **B** is confined to a circular region 60 cm in diameter, as shown in Fig. 29–56. The field strength is decreasing at the rate of 2.0 T/s. (a) Use Faraday's law to calculate the line integral of the induced electric field around a circular loop halfway between the center and outer edge of the magnetic field region. (b) An electron circulates once around the loop, moving opposite to the direction of the electric field. How much energy does it gain? (c) What is the magnitude of the electric field at the position of the loop?

28. Consider the preceding problem, but now let the magnetic field fill a square region 60 cm on a side. (a) Use Faraday's law to calculate the line integral of the induced electric field around a square loop whose sides are halfway between the center and the edge of the magnetic field region. (b) An electron circulates once around the square loop, moving opposite to the direction of the electric field. How much energy does it gain? (c) Is it possible to specify the magnitude of the electric field on the loop? Why or why not?

29. The plasma in a tokamak is heated in part by the current driven by the induced electric field. Figure 29–57 shows a top view of a tokamak. The magnetic

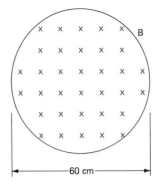

Fig. 29–56 Problem 27

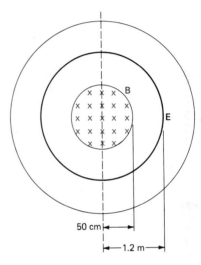

Fig. 29–57 Top view of a tokamak (Problem 29).

field **B** in the center of the tokamak is provided by coils that are not shown. This field is confined to a circular area of radius 50 cm. (a) If the magnetic field changes at the rate of 5100 T/s, what is the induced electric field 1.2 m from the center of the tokamak? (b) If a proton were to circle the tokamak twice during the time the field was changing, how much energy would it gain? The proton is 1.2 m from the center.

Supplementary Problems

30. At times prior to $t=0$, there is no current in either the solenoid or the small coil of Problem 4. Subsequently, the current in the small coil is observed to increase at the rate of 10 μA/s. What is the current in the solenoid as a function of time? Neglect any effects the changing current in the small coil might have on the solenoid.

31. Repeat Problem 18, assuming now that the magnetic field points vertically upward.

32. A long, straight wire carries an alternating current given by $I=I_0 \sin\omega t$. A conducting loop of length ℓ,

width w, and resistance R is located with its closest side a distance a from the wire, as in Fig. 29–58. (a) Write an expression for the magnetic field of the wire, as a function of time and of distance r from the wire. Assume that the field is given by Ampère's law for steady currents, even though the current is not steady in this case. (This assumption is reasonable provided that ω is not too big.) (b) What is the magnetic flux through the loop as a function of time? (c) What is the current in the loop as a function of time? (d) If the resistance of the loop were lowered, would this have any effect on the agent supplying current to the long wire?

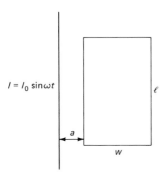

Fig. 29–58 Problem 32

33. Clever farmers whose lands are crossed by large power lines have been known to steal power from the power company by stringing wire near the power line, and making use of the induced current. The scene of a particular crime is shown in Fig. 29–59. The power line carries 60-Hz alternating current, and its peak voltage is 36 kV above ground. At the time the farmer plots the dirty deed, the peak current in the power line is 10^4 A. (a) If the farmer wants 170 V peak AC voltage, what should be the length of the loop shown in Fig. 29–61? (170 V is the peak of stan-

dard 120-V AC power.) (b) If all the equipment the farmer connects to the loop has an equivalent resistance of 5 Ω, what fraction of the power being transmitted down the line is the farmer stealing? (c) If the power company charges 10¢ per kilowatt hour, how much money does the theft amount to each day? (d) Without examining the farmer's lands, how, in principle, can the power company know that a crime is being committed?

34. A circular wire loop of resistance R and radius a lies at right angles to a uniform magnetic field. The strength of the field changes from an initial value B_1 to a final value B_2. Show, by integrating the loop current over time, that the total charge that moves around the ring is

$$q = \frac{\pi a^2}{R}(B_2 - B_1).$$

Note that this result is independent of how the field changes with time.

35. A disk of radius a and thickness h is made from a material whose resistivity is ρ. The disk is inside a solenoid of circular cross section, and the disk axis is coincident with the solenoid axis. The magnetic field of the solenoid is changing at the rate dB/dt. (a) What is the current density in the disk as a function of distance r from the disk center? (b) What is the rate of power dissipation by joule heating in the entire disk? Hint: Divide the loop into small conducting loops, and integrate the power dissipation over all the loops. (c) Obtain a numerical answer for part (b), assuming that the disk is made of aluminum (see Table 25–1), and that $a = 10$ cm, $h = 0.50$ cm, and $dB/dt = 1.0$ T/s.

36. A pendulum consists of a mass m suspended from two identical copper wires of negligible mass. At equilibrium, the mass is a vertical distance ℓ below the supports, and the wires make 45° angles with the vertical, as shown in Fig. 29–60. A uniform magnetic field **B** points into the page. The pendulum is displaced from the plane of the page by a small angle θ_0, then released. Obtain an expression for the voltmeter reading as a function of time.

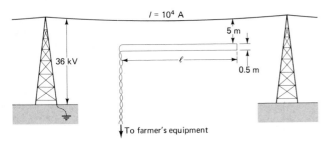

Fig. 29–59 Problem 33

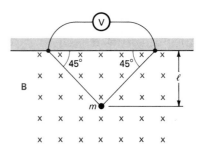

Fig. 29–60 Problem 36

30

INDUCTANCE AND MAGNETIC ENERGY

Faraday's law implies that a changing magnetic flux through a circuit gives rise to an induced emf in the circuit. In this chapter we consider the special case when that changing flux is itself caused by a changing current in an *inductance* electric circuit. We then speak of the **inductance** of the circuit.

30.1
MUTUAL INDUCTANCE

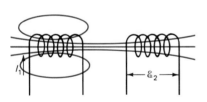

Fig. 30–1
The mutual inductance between two coils is the ratio of the magnetic flux through one coil to the current in the other coil. A changing current in one coil induces an emf in the other.

Consider two coils arranged so that some of the magnetic flux associated with current in one coil also passes through the second coil, as in Fig. 30–1. If we change the current I_1 in the first coil, an induced emf $\mathscr{E}_2$ appears in the second. As we discussed in the preceding chapter, $\mathscr{E}_2$ depends on the rate of change of magnetic flux through the second coil. The magnetic flux depends, in turn, on the current in the first coil and on the geometrical arrangement of the two coils that determines how much flux from the first coil actually links the second. We characterize this geometrical arrangement by the ratio of the total magnetic flux through the second coil to the current in the first coil. This ratio defines the **mutual inductance,** M, of the two coils:

$$M = \frac{\phi_2}{I_1}. \tag{30–1}$$

Solving Equation 30–1 for ϕ_2 and differentiating, we obtain

$$\frac{d\phi_2}{dt} = M\frac{dI_1}{dt}.$$

The left-hand side of this equation is just the rate of change of the total flux ϕ_2 through the second coil, so that Faraday's law gives

789

$$\mathcal{E}_2 = -M \frac{dI_1}{dt}, \tag{30-2}$$

where the minus sign describes the polarity of the induced emf.

We could equally well have considered the case where a changing current in the second coil induces an emf in the first. Although it is not at all obvious, the same value of mutual inductance applies in this case even when the arrangement of the two coils is far from symmetric.

From Equation 30–2, we see that the unit of mutual inductance is the volt-second/ampere. This unit is given the name henry (H) in honor of the American scientist Joseph Henry (1797–1878), who was also the first secretary of the Smithsonian Institution. Mutual inductances found in common electronic circuits usually range from microhenrys (μH) on up to several henrys.

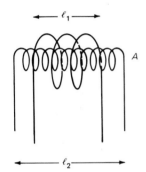

Fig. 30–2
Example 30–1 (not to scale).

Example 30–1

Mutual Inductance

A short solenoid of length ℓ_2 is wound around a much longer solenoid of length ℓ_1 and area A, as shown in Fig. 30–2. Both solenoids have n turns per meter. What is the mutual inductance of this arrangement?

Solution

Equation 28–16 gives the magnetic field produced by the long solenoid:

$$B = \mu_0 nI.$$

This field is uniform, is at right angles to the plane of the solenoid coils, and is confined to the interior of the longer solenoid. Therefore, the flux linked by each turn of the shorter solenoid is the product of this field strength with the area of the longer solenoid:

$$\phi_{1\ \text{turn}} = \int \mathbf{B} \cdot d\mathbf{A} = BA.$$

With n turns per unit length, there are $n\ell_2$ turns in the shorter solenoid, so the total flux through the shorter solenoid is $n\ell_2$ times the flux through a single turn. Then Equation 30–1 gives

$$M = \frac{\phi_2}{I} = \frac{n\ell_2 BA}{I} = \frac{n\ell_2 \mu_0 nIA}{I} = \mu_0 n^2 \ell_2 A.$$

This calculation was simple because we considered the flux from the longer solenoid that was linked by the shorter one. Had we done the opposite, we would have faced a difficult calculation because of the diverging field lines at the ends of the short solenoid. Yet, remarkably, such a calculation—best done with the aid of a computer—would have yielded exactly the same result.

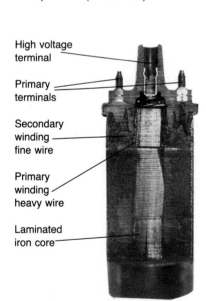

High voltage
terminal

Primary
terminals

Secondary
winding
fine wire

Primary
winding
heavy wire

Laminated
iron core

Fig. 30–3
An automobile ignition coil uses electromagnetic induction to produce high voltage that drives sparks to ignite the gasoline.

Example 30–2

An Ignition Coil

Gasoline in your car's engine is ignited when a high voltage applied to a spark plug causes a spark to jump between two conductors of the plug. This high voltage, in turn, is provided by the ignition coil—an arrangement of two coils wound tightly one on top of the other (Fig. 30–3). Current from the car's 12-volt battery flows through the coil with fewer turns. This current is interrupted periodically by a mechanical or electronic switch in the car's distributor (the so-called distributor points). The sudden change in current induces a large emf in the coil with more turns, and this emf drives the spark. Interruption of the current is carefully timed so that the spark occurs at exactly the right point in the engine cycle. An important part of a "tune-up" is the precise adjustment of this ignition timing.

A typical ignition coil draws a maximum current of 3.0 A, and supplies 20 kV to the spark plugs. If the distributor points interrupt the current in 0.10 ms, what is the mutual inductance of the ignition coil?

Solution

The rate of change of current is

$$\frac{dI}{dt} = \frac{3.0 \text{ A}}{1.0 \times 10^{-4} \text{ s}} = 3.0 \times 10^4 \text{ A/s}.$$

Solving Equation 30–2 for M then gives

$$M = \frac{|\mathscr{E}|}{|dI/dt|} = \frac{20 \times 10^3 \text{ V}}{3.0 \times 10^4 \text{ A/s}} = 0.67 \text{ H}.$$

30.2
SELF-INDUCTANCE

In the preceding chapter, and in our discussion of mutual inductance, we considered emf's and currents induced by changes—such as a moving magnet or a varying current in another circuit—that were external to our circuit. But the changing magnetic field associated with the changing current in a circuit also affects that same circuit.

Consider a circular loop carrying a current I, as shown in Fig. 30–4. Magnetic field lines arising from this current loop pass through the loop, as suggested in the figure, so there is a magnetic flux through the loop. If the current is steady this flux is constant in time, and there is no induced electric field. But if we change the loop current, then the flux changes, and an induced electric field arises. In order to conserve energy this field opposes the change that causes it—in this case the change in loop current. If the current is counterclockwise and we try to increase its strength, an induced electric field will appear in the clockwise direction to oppose the current increase. If we try to decrease the current, the induced electric field will have the opposite sense, now trying to maintain the current. The induced electric field makes it difficult to change the current in the circuit, as suggested in Fig. 30–5.

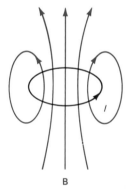

Fig. 30–4
Magnetic field lines of a circular current loop give rise to a magnetic flux through that loop.

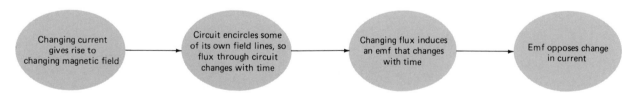

Fig. 30–5
Self-inductance occurs when a circuit encircles some of its own magnetic flux.

self-inductance

This property of a circuit whereby its own magnetic field opposes changes in the circuit current is termed **self-inductance.** All circuits possess self-inductance, but this inductance is important only when the circuit encircles a great many of its own magnetic field lines or when current changes very rapidly. A simple piece of wire exhibits very little opposition to current changes in the 60-Hz alternating current used in electric power systems. But in a TV set or computer, where currents change on time scales of billionths of a second, self-inductance of the wires themselves must be taken into account.

inductor

An **inductor** is a device designed specifically to exhibit self-inductance. A typical inductor consists of a coil of wire, constructed so that a great deal of its own magnetic flux is encircled. Some inductors are wound on iron cores to promote flux concentration (Fig. 30–6).

As long as there is a steady current in an inductor, the inductor acts just like a piece of wire. But when we try to change the current, the changing magnetic flux induces an emf that opposes the changing current. The more rapidly we change the current, the greater the rate of change of flux and so the greater the emf. We characterize the inductor by its self-inductance, L, defined as the ratio of magnetic flux through the inductor to current in the inductor:

$$\phi_B = LI. \tag{30–3}$$

The unit of self-inductance, like that of mutual inductance, is the henry. Inductance is a constant determined by the physical design of an inductor. In principle, we can calculate the inductance of any inductor, but in practice this is difficult unless the geometry is particularly simple. Inductors for use in electronic circuits are available commercially in a wide range of inductance values.

Fig. 30–6
Typical inductors.

Example 30–3

The Inductance of a Solenoid

A long solenoid of cross-sectional area A has length ℓ and consists of N turns of wire. What is the inductance of this solenoid?

Solution

To find the inductance, we must relate the current in the solenoid to the magnetic flux through it. In Section 28.5, we used Ampère's law to find the magnetic field of a long solenoid:

$$B = \mu_0 nI.$$

Here n is the number of turns per unit length and I is the solenoid current. For our N-turn solenoid of length ℓ, the number of turns per unit length is $n = N/\ell$, so that the field is

$$B = \frac{\mu_0 NI}{\ell}.$$

The total flux through all N turns of the solenoid is then

$$\phi = N \int_{1 \text{ turn}} \mathbf{B} \cdot d\mathbf{A} = NAB = \frac{\mu_0 N^2 IA}{\ell},$$

where we wrote the flux integral as a simple product because the solenoid field is uniform and perpendicular to the individual turns. We now have an expression relating the flux ϕ and current I. According to Equation 30–3, the inductance L is just the ratio of these quantities, or

$$L = \frac{\phi}{I} = \frac{\mu_0 N^2 A}{\ell}. \qquad (30\text{–}4)$$

Does this result make sense? As we increase either the number of turns or the solenoid area, we increase the flux and therefore the inductance. If we lengthen the solenoid without increasing the number of turns, we reduce the magnetic field strength and hence the flux. Equation 30–4 correctly reflects these trends. Can you see why the inductance should be proportional to the *square* of the number of turns?

The induced emf in an inductor is determined by Faraday's law, which relates the emf to the rate of change of magnetic flux:

$$\mathscr{E} = -\frac{d\phi_B}{dt}.$$

Differentiating Equation 30–3, the definition of inductance, gives

$$\frac{d\phi_B}{dt} = L\frac{dI}{dt}.$$

Then Faraday's law becomes

$$\mathscr{E} = -L\frac{dI}{dt}. \qquad (30\text{–}5)$$

This equation gives the emf $\mathscr{E}$ induced in an inductor L when the current in the inductor is changing at the rate dI/dt. The minus sign again tells us that the emf opposes the change in current. For this reason the inductor **back emf**   emf is often called a **back emf;** it works against changes brought about by externally applied emf's.

Although the physical interpretation of the signs in Equation 30–5 is straightforward, their algebraic interpretation requires some clarification. Figure 30–7a shows sign conventions for an inductor. If moving through the inductor in the direction of the current takes us from the negative to the positive end, then we define the emf to be positive. But if we move with the current and encounter the positive end of the inductor first, then the inductor emf is, by definition, negative. Because of the minus sign in Equation 30–5, a negative emf arises when dI/dt is positive—that is, when the current is increasing. Figure 30–7b shows the actual polarity for this case. If, on the other hand, current is decreasing, then the actual polarity of the inductor emf will be like that in Fig. 30–7a, as the inductor tries to keep the current going.

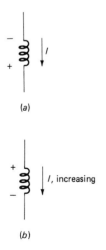

Fig. 30–7
Sign conventions for an inductor. *(a)* If moving with the current takes us from the negative to the positive end of the inductor, then the emf is defined to be positive. *(b)* Actual polarity in an inductor when current is increasing. Since we encounter the positive end first when moving with the current, this emf is by definition negative; it is a back emf that opposes the increase in current.

When the current in an inductor is steady, so that $dI/dt = 0$, then there is no emf in the inductor. In this case, the inductor acts just like a piece of wire. But when we change the current the inductor responds by producing a back emf that opposes the change in current. Now the inductor acts very much like a battery, with the magnitude of its emf dependent on how fast the current changes. If we try to start or stop current instantaneously, dI/dt is very large and a very large back emf appears. This is not merely mathematics! Rapid switching of inductive devices such as solenoids, solenoid valves, or motors can result in destruction of delicate electronic devices by induced currents. And people have been killed opening switches in circuits containing large inductors.

Example 30–4

The emf of an Inductor

A current of 5.0 A is flowing in a 2.0-H inductor. The current is reduced steadily to zero in 1.0 ms. What is the magnitude of the emf in the inductor while the current is being turned off?

Solution

Because the current changes steadily, its time rate of change has magnitude

$$\frac{dI}{dt} = \frac{5.0 \text{ A}}{1.0 \text{ ms}} = 5000 \text{ A/s},$$

so that

$$|\mathscr{E}| = L\frac{dI}{dt} = (2.0 \text{ H})(5000 \text{ A/s}) = 10{,}000 \text{ V},$$

enough to produce a lethal shock. Note that this voltage is quite unrelated to the voltage of the battery or whatever else was supplying the inductor current. We could have a 6-volt battery and still be electrocuted trying to open the circuit rapidly when a large inductance is present.

30.3
INDUCTORS IN CIRCUITS

The emf in an inductor depends on the rate of change of the inductor current, and acts to oppose that change in current. Because an infinite emf is impossible, so is an instantaneous change in inductor current:

> The current through an inductor cannot change instantaneously.

The more rapidly we try to change the current in an inductor, the greater the emf that in turn opposes the change in current.

The effect an inductor has on current is analogous to the effect a capacitor has on voltage. In Chapter 26, we found that the voltage across a capacitor cannot change instantaneously. Much of our understanding of capacitors applies to inductors if we simply interchange the words voltage and current.

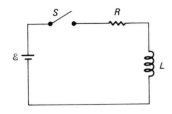

Fig. 30–8
An *LR* circuit.

Consider the circuit of Fig. 30–8, which is like the *RC* circuit of Chapter 26 but with the capacitor replaced by an inductor. What happens when we close the switch? We can understand qualitatively the behavior of this circuit in much the same way as we did the *RC* circuit, basing our analysis on the fact that the current through the inductor cannot change instantaneously. Just before we close the switch, the current through the inductor is zero. The current cannot change instantaneously, so it is zero just after we close the switch. How can this be? Better to ask "how could it not be?" If the current took some nonzero value just after the switch was closed, then the rate of change of current would be infinite, and an infinite back emf would be generated in the inductor. Not only is this physically impossible, but any emf larger than the battery emf would result in a current flowing counterclockwise around the circuit. Just after the switch is closed, in fact, the inductor emf must be exactly equal to the battery emf, so that no current flows in the resistor or anywhere else in the circuit. The inductor acts like an instantaneous open circuit.

The presence of the back emf means that, although the current is zero, the rate of change of current is not zero. So the current in the inductor begins to rise. This current flows through the resistor as well, so that a voltage develops across the resistor. This resistor voltage is the difference between the battery and inductor emf's, so that the inductor emf, initially equal to the battery emf, drops as the current increases. But the inductor emf is proportional to the rate of change of current through the inductor, so the rate of change of current decreases as the inductor emf decreases. We have a cycle in which increasing current implies decreasing inductor emf. But this decreasing inductor emf means that the rate of increase of current goes down, so the whole process tends toward a steady state in which nothing is changing. This sequence of events is summarized in Fig. 30–9, which is analogous to Fig. 26–30 for the *RC* circuit. The resulting

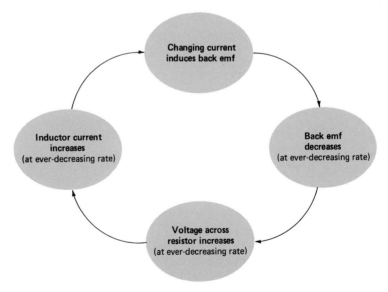

Fig. 30–9
Relation between circuit quantities in an *LR* circuit.

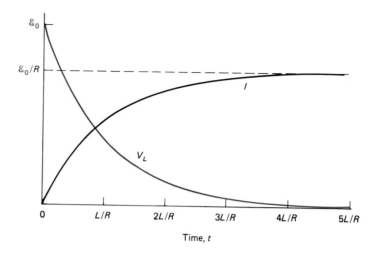

time dependence of the inductor current and emf are shown in Fig. 30–10. If we wait long enough, the current reaches a steady value. At this time nothing is changing, so that the inductor emf is zero and the inductor acts like a short circuit. The circuit current then is simply the battery emf divided by the resistance.

To analyze the *LR* circuit quantitatively, we apply the loop law. Traversing the circuit clockwise, starting at the negative terminal of the battery, we first encounter a voltage increase $\mathscr{E}_0$ due to the battery. Then the voltage decreases by *IR* in the resistor. Finally, there is a voltage change $\mathscr{E}_L$ in the inductor. This change is actually a decrease, because the inductor emf opposes the increase in current. However, we will simply call the inductor emf $\mathscr{E}_L$ and let Equation 30–5 take care of the signs. Then the loop law becomes

$$\mathscr{E}_0 - IR + \mathscr{E}_L = 0.$$

If we differentiate this equation with respect to time, the battery emf $\mathscr{E}_0$ drops out because it is constant, and we can write

$$\frac{d\mathscr{E}_L}{dt} = R\frac{dI}{dt}.$$

But the inductor emf $\mathscr{E}$ and rate of change of current *dI/dt* are related according to Equation 30–5, so

$$\frac{dI}{dt} = -\frac{\mathscr{E}_L}{L},$$

and our differentiated loop equation becomes

$$\frac{d\mathscr{E}_L}{dt} = -R\frac{\mathscr{E}_L}{L}. \tag{30–6}$$

The differential equation 30–6 describes a quantity—$\mathscr{E}_L$—whose rate of change is proportional to itself. We discussed such equations in Chapter 26 when we considered the *RC* circuit. Equation 30–6 is similar to Equation 26–22, but with current *I* replaced by the inductor emf $\mathscr{E}_L$ and capacitance

C by $1/L$. The solution to Equation 30–6 is like that of Equation 26–22, provided we make the appropriate substitutions for I and C:

$$\mathcal{E}_L = \mathcal{E}_0 e^{-Rt/L}. \tag{30–7}$$

This equation shows that the inductor emf decays exponentially to zero, starting from an initial value equal to the battery emf. What about the inductor current? In this series circuit, the inductor and resistor currents are the same. This current is determined by the resistance R and the voltage V_R across the resistor:

$$I = \frac{V_R}{R} = \frac{\mathcal{E}_0 - \mathcal{E}_L}{R} = \frac{\mathcal{E}_0 - \mathcal{E}_0 e^{-Rt/L}}{R} = \frac{\mathcal{E}_0}{R}(1 - e^{-Rt/L}). \tag{30–8}$$

inductive time constant With a capacitor, we characterized the exponentially changing quantities in terms of the capacitive time constant RC. With an inductor, we have an **inductive time constant** L/R. Significant changes in current cannot occur on time scales much shorter than L/R. On the other hand, an LR circuit will approach a steady state, with zero $\mathcal{E}_L$, only after many inductive time constants.

Why is the inductive time constant a quotient of L and R rather than a product, as in the capacitor case? In Problem 13 you will convince yourself mathematically that L/R does indeed have the units of seconds. But you can also understand this physically. The larger L, the larger the back emf and the longer it takes the current to build up. The larger R, the smaller the final current and so, all else being equal, the smaller the rate of change of current, and therefore the smaller the inductive effects. In problems involving very short or very long times we need not use Equations 30–7 and 30–8 to analyze an LR circuit. It is only when we are concerned with times comparable with L/R that we must calculate the exponentials.

Example 30–5
An LR Circuit

A 12-V battery, 1000-Ω resistor, and 5.0-mH inductor are connected in series. What is the steady-state current after many time constants have passed? How long does it take the current to reach half of its final value?

Solution

In the steady state nothing is changing, so there is no emf in the inductor and we can treat it like a short circuit. Then the current is simply the battery voltage divided by the resistance:

$$I = \frac{12\ \text{V}}{1000\ \Omega} = 12\ \text{mA}.$$

The current gradually increases to this steady-state value according to Equation 30–8. When the current has half its final value $\mathcal{E}_0/R$, Equation 30–8 becomes

$$\tfrac{1}{2} = 1 - e^{-Rt/L}.$$

Moving the exponential to the left-hand side and the $\tfrac{1}{2}$ to the right gives

$$e^{-Rt/L} = \tfrac{1}{2}.$$

Taking the natural logarithm of both sides [recall that $\ln(e^x) = x$]:

$$-Rt/L = \ln(\tfrac{1}{2}) = -\ln 2,$$

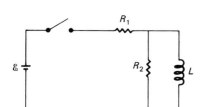

(a)

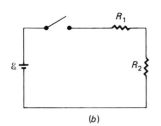

(b)

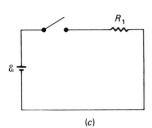

(c)

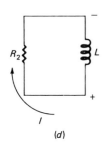

(d)

Fig. 30–11
(a) Circuit for Example 30–6. (b) Just after the switch is closed, the inductor acts like an open circuit. (c) Long after the switch is closed, nothing is changing and the inductor acts like a short circuit. (d) Current in the inductor cannot change instantaneously, so when the switch is again opened, current flows through R_2 in the direction shown.

so that

$$t = \frac{L \ln 2}{R} = \frac{(5.0 \times 10^{-3} \text{ H})(0.69)}{1000 \ \Omega} = 3.5 \ \mu s.$$

Although short, this time would be significant in a computer, radio, or other device involving high-speed electrical activity.

Example 30–6
Another LR Circuit

In the circuit of Fig. 30–11a, the switch is initially open. What is the current in resistor R_2 immediately after the switch is closed? A long time after the switch is closed? Long after the switch is closed, it is again opened. What is the current in R_2 just after it is opened? A long time after?

Solution

In this example, we are concerned only with times long or short compared with the inductive time constant. Consequently, we need not calculate the mathematical details describing the circuit at all times. Instead, we apply directly the fundamental property of the inductor: the current through the inductor cannot change instantaneously.

Just before we close the switch, the current in the inductor is zero. The current cannot change instantaneously, so it remains zero just after the switch is closed. At this instant the inductor might as well be an open circuit, giving the circuit shown in Fig. 30–11b. Then all the current from R_1 flows through R_2, giving a simple series circuit in which the current is

$$I = \frac{\mathscr{E}}{R_1 + R_2}.$$

If we wait long enough, the circuit will reach a steady state in which nothing is changing. With $dI/dt = 0$, there is no inductor emf, and the inductor acts like a short circuit. We can redraw the circuit as Fig. 30–11c, from which we see that all the current from R_1 goes through L, and none through R_2. The resulting current in R_1 and L is just $\mathscr{E}/R_1$.

Now the switch is opened again. Current in R_1 stops abruptly, since there is no way charge can get through the open switch. But the current through the inductor, which was $\mathscr{E}/R_1$ just before the switch was opened, remains $\mathscr{E}/R_1$ the instant after the switch is opened. There is only one place this current can go—through R_2, from bottom to top (Fig. 30–11d). So just after the switch is opened, the current in R_2 is $\mathscr{E}/R_1$. Notice that the value of R_2 has no effect on this current, which is determined entirely by the battery emf and the resistance R_1.

What about the voltage across R_2? This is given by Ohm's law:

$$V_2 = I_2 R_2 = \frac{\mathscr{E}R_2}{R_1}.$$

The larger R_2, the larger the voltage that appears when the switch is opened. If R_2 has infinite resistance, or is not in the circuit, the voltage will be arbitrarily large as the inductor seeks at all cost to keep the current flowing. This dangerous situation can result in ionization of the air, leading to arcing and vaporization of circuit conductors, and even to electric shock (Fig. 30–12). In circuits with large inductance, resistors are often placed in parallel with inductors in order to alleviate these dangers.

Fig. 30–12
These switch contacts show damage associated with arcing when the switch was used to interrupt current to a large inductor.

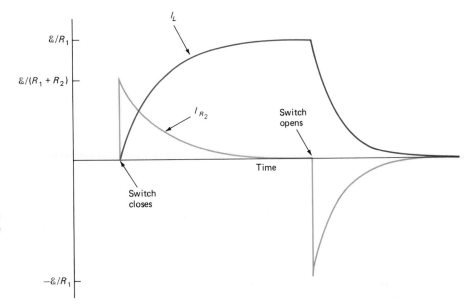

Fig. 30–13
Currents in R_2 and L for the circuit of Example 30–6. The different time constants before and after the switch opens reflect the fact that only R_2 is in the circuit after the switch opens.

Finally, when the switch has been open for a long time, the current decays to zero. Plots of the currents in R_2 and L as functions of time are shown in Fig. 30–13.

30.4
MAGNETIC FIELD ENERGY

In Example 30–6 we found that when we open the switch in Fig. 30–11a, current continues to flow in the inductor and in the resistor R_2. Heat energy must therefore appear in the resistor. Where does this energy come from?

We can gain insight into this question by noting what else happens in the circuit. As soon as the switch is opened, a gradual decrease in current begins, as Fig. 30–13 shows. As a result, the magnetic field of the inductor decreases. It is this decrease in field that, through Faraday's law, produces the inductor emf that drives the current.

When we first open the switch, there is a significant magnetic field associated with the inductor. As the current and magnetic field decay toward zero, heat energy appears in the resistor. Finally we reach a state in which there is no more current and no more magnetic field. What has happened between our initial and final states? Simply this—a magnetic field has disappeared and energy has appeared in the resistor. So where did the energy come from? It came from the magnetic field.

Like the electric field, the magnetic field, too, contains stored energy. Our decaying LR circuit is analogous to a discharging capacitor, in which

Fig. 30–14
This eruption of a prominence from the sun's surface involves the release of energy stored in magnetic fields.

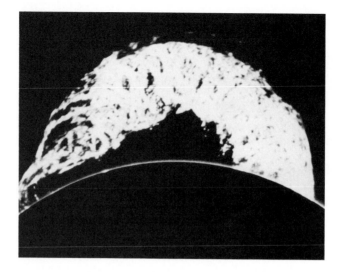

the electric field of the capacitor disappears and heat appears in a resistor. As in the electric case, the energy of a magnetic field is not limited to circuits. *Any* magnetic field represents stored energy. This energy may be converted to other forms, including heat, mechanical energy, or electrical energy. Conversely, energy initially in some other form may be stored in a magnetic field. The release of stored magnetic energy makes possible a number of practical devices, and also powers many violent events throughout the universe (Fig. 30–14).

Knowing that the magnetic field contains energy, we can reinterpret Example 30–6 in terms of energy transfers. When we close the switch in the circuit of Fig. 30–11a, current flows through R_1 and the inductor. Some energy from the battery is dissipated in R_1, but some goes into building up the magnetic field of the inductor. When we open the switch again, the battery is eliminated from the circuit. Now energy is dissipated in R_2, energy that comes from the decaying magnetic field of the inductor. The energy that heats R_2 during this decaying phase originally came from the battery, but was stored temporarily in the magnetic field of the inductor. Figure 30–15 outlines the energy transfers in this circuit.

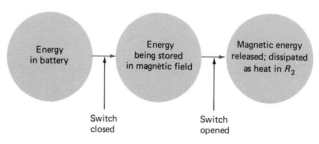

Fig. 30–15
Energy transfers in the circuit of Example 30–7.

Magnetic Energy in an Inductor

How much energy is stored in the magnetic field? In Chapter 24 we answered the analogous question for the electric field by considering the work required to charge a parallel-plate capacitor, thus creating an electric field.

We create a magnetic field every time we start a current flowing. We can calculate the energy stored in this field by considering the work required to build up current in an inductor. Earlier, we found that an inductor develops an emf that depends on the rate of change of current through the inductor (Equation 30–5):

$$\mathcal{E}_L = -L\frac{dI}{dt},$$

where I is the inductor current. From Section 25.4, we know that the rate at which the inductor consumes electrical energy is the magnitude of this emf times the current through the inductor:

$$P = I_L|\mathcal{E}_L| = LI\frac{dI}{dt}. \tag{30–9}$$

Unlike a resistor, the inductor does not dissipate this energy as heat. Rather, it stores the energy in its magnetic field. Equation 30–9 gives the rate at which this energy is being stored.

Suppose we have an inductor with no current, and therefore no magnetic field and no stored energy. We connect the inductor in a simple LR circuit like that of Section 30.3, and let current build up to some value I_0. Although the current is continuously changing as it rises from zero to I_0, at any instant the rate of energy storage in the magnetic field is given by Equation 30–9. In some small time interval dt, the amount of energy stored is the rate of energy storage times the time, or

$$dU = P\,dt = LI\frac{dI}{dt}\,dt = LI\,dI.$$

To find the total energy stored, we integrate over all the dU's as the current goes from zero to its final value I_0:

$$U = \int_0^{I_0} dU = \int_0^{I_0} LI\,dI = L\int_0^{I_0} I\,dI = \tfrac{1}{2}LI_0^2. \tag{30–10}$$

Equation 30–10 gives the total energy stored in the magnetic field of an inductor L when the inductor carries a current I_0. If the magnetic field is allowed to decay, as in Example 30–6, then this stored energy is released.

Example 30–7

The Energy in an Inductor

When the switch in Example 30–6 is opened, the inductor current flows through R_2, and the energy stored in the magnetic field is gradually dissipated as heat. Show explicitly, by integrating the resistor power over time, that all the magnetic energy gets dissipated in the resistor.

Solution

Once the switch is open, the current in the inductor and resistor decays exponentially:

$$I = I_0 e^{-Rt/L},$$

where I_0 is the inductor current just as the switch is opened, and I the instantaneous value of the current as it decays. The rate of energy dissipation in the resistor is I^2R, or

$$P = I^2R = (I_0 e^{-Rt/L})^2 R = I_0^2 R e^{-2Rt/L}.$$

Because it takes forever for the current to decay completely to zero, we find the total energy dissipated by integrating the resistor power from time $t=0$ to $t = \infty$:

$$U = \int_0^\infty I_0^2 R e^{-2Rt/L}\, dt = I_0^2 R \frac{-L}{2R} e^{-2Rt/L} \Big|_0^\infty = -\frac{I_0^2 L}{2}(0-1) = \tfrac{1}{2}LI_0^2.$$

Because $\tfrac{1}{2}LI_0^2$ is the magnetic energy initially stored in the inductor, we have shown that all the magnetic energy is eventually dissipated as heat in the resistor.

Magnetic Energy Density

A long solenoid is a particularly simple inductor in which the magnetic field is essentially uniform. We can exploit the simplicity of the solenoid to evaluate the energy density in the magnetic field, just as we used the parallel-plate capacitor to evaluate the electric field energy density. In Section 28.5, we showed that the magnetic field of a long solenoid is given by

$$B = \mu_0 n I,$$

where n is the number of turns per unit length. If our solenoid has length ℓ and a total of N turns, then $n = N/\ell$, so the field becomes

$$B = \frac{\mu_0 N I}{\ell}. \tag{30–11}$$

In Example 30–3, we found the inductance of a solenoid to be

$$L = \frac{\mu_0 N^2 A}{\ell},$$

with A the solenoid area. If we put this expression into Equation 30–10 for the energy stored in an inductor, we obtain

$$U = \tfrac{1}{2}LI^2 = \frac{\mu_0 N^2 A}{2\ell} I^2.$$

We would like to express this energy in terms of the magnetic field. To do so, we multiply numerator and denominator by $\mu_0 \ell$, to get

$$U = \frac{\mu_0 N^2 A}{2\ell} I^2 \frac{\mu_0 \ell}{\mu_0 \ell} = \frac{\mu_0^2 N^2 I^2}{2\ell^2} \frac{A\ell}{\mu_0}.$$

Comparison with Equation 30–11 shows that the quantity $\mu_0^2 N^2 I^2/\ell^2$ is just the square of the magnetic field, so that

$$U = \frac{B^2}{2\mu_0} A\ell.$$

This is the total energy stored in the magnetic field of the solenoid. For a long solenoid, the field is appreciable only in the interior of the solenoid, whose volume is $A\ell$. Then the energy density, or energy per unit volume, in the magnetic field is

$$u_B = \frac{U}{A\ell} = \frac{B^2}{2\mu_0}. \tag{30–12}$$

Although we derived this equation for the field of a solenoid, it is in fact a universal expression for the local magnetic energy density. Wherever there is a magnetic field, there is stored energy.

Compare Equation 30–12 with Equation 24–8 for the energy stored in an electric field:

$$u_E = \tfrac{1}{2}\epsilon_0 E^2.$$

The expressions for electric and magnetic energy densities are similar. Each is proportional to the *square* of the field strength, and each contains the appropriate constant. That the constant appears in the numerator in the electric case but in the denominator in the magnetic case is of no significance; it is merely a consequence of our definition of units in the SI system.

Example 30–8
Magnetic and Electric Energy

An electric field of 10^6 V/m is established between the plates of a parallel-plate capacitor. (This field strength is typical of capacitors in common applications.) The volume between the plates is 50 cm³. What magnetic field in the same volume would contain the same amount of energy as this electric field?

Solution

Because we are considering equal volumes, we can simply equate the electric and magnetic energy densities:

$$\frac{B^2}{2\mu_0} = \tfrac{1}{2}\epsilon_0 E^2.$$

Solving for the magnetic field gives

$$B = \sqrt{\mu_0 \epsilon_0 E^2}$$
$$= [(4\pi \times 10^{-7}\text{ H/m})(8.9 \times 10^{-12}\text{ F/m})(1.0 \times 10^6\text{ V/m})^2]^{1/2}$$
$$= 3.3 \times 10^{-3}\text{ T}.$$

(Check the units; see Problem 27.) This is a modest field, about 100 times the earth's field, and readily producible in the laboratory. This example suggests that electric and magnetic fields with typical laboratory values are roughly comparable energy storage media.

Example 30–9
Magnetic Energy in a Toroid

A toroidal coil of square cross section has inner radius R and side s, as shown in Fig. 30–16. The coil consists of N turns, and carries a current I. What is the total magnetic energy stored in the toroid?

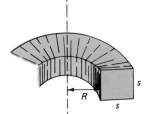

Fig. 30–16
Cross section of a toroidal coil (Example 30–9).

Solution

In Section 28.5, we found that the magnetic field in a toroid is given by

$$B = \frac{\mu_0 N I}{2\pi r},$$

with r the distance from the central axis of the toroid. In deriving this result, we considered a toroid of circular cross section. However, our derivation would give the same result for a toroid of any uniform cross section. The magnetic energy density in our toroid is then

$$u = \frac{B^2}{2\mu_0} = \frac{\mu_0 N^2 I^2}{8\pi^2 r^2}.$$

To calculate the total energy, consider a thin ring of thickness dr and width s located at a distance r from the axis of the toroid, as shown in Fig. 30–17. The volume of this ring is

$$dV = 2\pi r s \, dr,$$

so that the magnetic energy in the ring is

$$dU = u \, dV = \frac{\mu_0 N^2 I^2}{8\pi^2 r^2} 2\pi r s \, dr = \frac{\mu_0 N^2 I^2 s}{4\pi r} \, dr.$$

To find the total energy, we integrate over all such rings within the toroid; that is, from $r=R$ to $r=R+s$:

$$U = \int_R^{R+s} dU = \frac{\mu_0 N^2 I^2 s}{4\pi} \int_R^{R+s} \frac{dr}{r} = \frac{\mu_0 N^2 I^2 s}{4\pi} \ln\left(\frac{R+s}{R}\right).$$

Comparing this result with the expression $U = \frac{1}{2}LI^2$ would allow us to calculate the inductance of the toroid (see Problem 34).

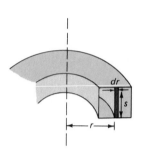

Fig. 30–17
Volume element for integration of the magnetic energy density in the toroid is a ring of thickness dr, radius r, and width s.

Example 30–10 _____

Energy in Earth's Magnetic Field

The strength of the earth's magnetic field is about 0.5 G at earth's surface. Estimate the magnetic energy stored in this field in the first 100 km above the surface. How many gallons of gasoline would you have to burn to get the equivalent amount of energy?

Solution

Although earth's dipole field drops as $1/r^3$, 100 km is such a small distance compared with earth's radius that the field remains nearly constant over this distance, so the energy density is essentially constant:

$$u = \frac{B^2}{2\mu_0} = \frac{(0.5 \times 10^{-4} \text{ T})^2}{(2)(4\pi \times 10^{-7} \text{ H/m})} = 10^{-3} \text{ J/m}^3.$$

With constant energy density, we can calculate the total energy as a simple product of the energy density and volume. Again, because 100 km is small compared with earth's radius, we can neglect the curvature of the earth and simply multiply earth's surface area by 100 km to get the volume:

$$V = (4\pi R_e^2)(10^5 \text{ m}) = (4\pi)(6 \times 10^6 \text{ m})^2(10^5 \text{ m}) = 5 \times 10^{19} \text{ m}^3,$$

where in estimating, we have rounded all quantities to one significant figure. Then the total magnetic energy is

$$U = \int u \, dV = uV = (10^{-3} \text{ J/m}^3)(5 \times 10^{19} \text{ m}^3) = 5 \times 10^{16} \text{ J}.$$

A gallon of gasoline contains about 10^8 J (see Appendix D), so this magnetic energy is equivalent to about

$$\frac{5 \times 10^{16} \text{ J}}{10^8 \text{ J/gallon}} = 5 \times 10^8 \text{ gallons}$$

of gasoline. This represents roughly one day's gasoline consumption in the United States. Our answer in this example is only a rough estimate, as we have neglected a factor-of-2 variation in field strength from equator to poles.

SUMMARY

1. **Mutual induction** describes the electromagnetic interaction of a pair of coils or other conductors placed so that the magnetic flux from one coil links the other. A changing current in the first coil then induces an emf in the second coil. The **mutual inductance** of such a system is defined as the ratio of the total flux in the second coil to the current in the first:

$$M = \frac{\phi_2}{I_1}.$$

By Faraday's law, the emf in the second coil is proportional to the rate of change of current in the first:

$$\mathscr{E}_2 = -M \frac{dI_1}{dt}.$$

The same mutual inductance M describes the emf developed in the first coil as a result of changing current in the second coil.

2. A changing current in a coil or circuit gives rise to a changing magnetic flux through that same circuit. This changing flux, in turn, gives rise to an induced electric field that opposes the original change in current. A device constructed especially to exploit this property of **self-inductance** is called an **inductor.** Usually a tightly wound coil of wire, an inductor opposes instantaneous changes in current, but has no effect on steady current. The self-inductance L of an inductor is the ratio of magnetic flux to current:

$$L = \frac{\phi}{I}.$$

Faraday's law relates the emf in an inductor to the rate of change of current:

$$\mathscr{E} = -L \frac{dI}{dt}.$$

The direction of the emf is such as to oppose changes in the inductor current. Self-inductance prevents the inductor current from changing instantaneously.

3. In a circuit containing a resistor R and inductor L, changes occur exponentially with an **inductive time constant** L/R. The rising current in a series LR circuit is given by

$$I = \frac{\mathscr{E}_0}{R} (1 - e^{-Rt/L}),$$

where $\mathscr{E}_0$ is the battery emf and t the time measured from when the circuit is first connected. If this current is subsequently allowed to decay, it goes exponentially to zero:

$$I = \frac{\mathscr{E}_0}{R} e^{-Rt/L},$$

where now t is measured from the start of the decay.

4. To build up current and therefore magnetic field in an inductor takes work, which ends up as stored magnetic energy in the inductor. An inductor of inductance L, carrying current I, contains stored energy

$$U = \tfrac{1}{2}LI^2.$$

5. All magnetic fields, not only those of inductors, contain stored energy, with the local **magnetic energy density** given by

$$u_B = \frac{B^2}{2\mu_0}.$$

QUESTIONS

1. Figure 30–18 shows two pairs of identical coils in different geometrical arrangements. For which arrangement is the mutual inductance greatest?
2. A car battery has an emf of 12 V, yet energy from the battery provides the 20,000-V spark that ignites the gasoline. How is this possible? Describe the energy transfers from battery to spark.
3. When two coils are connected in series but physically

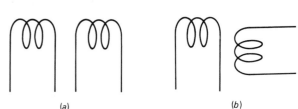

(a)   (b)

Fig. 30–18 Question 1

far apart, they behave as a single inductor whose inductance is the sum of the individual inductances. Why might this not be true if they are close together?

4. You have a fixed length of wire to wind into an inductor. Will you get more inductance if you wind a short coil with large diameter, or a long coil with small diameter?

5. You have a fixed length of wire of resistance R. You want to wind the wire into a small space and use it as a resistor. How would you wind it so as to minimize its self-inductance?

6. In wiring circuits that operate at high frequencies, like TV sets or computers, it is important to avoid extraneous loops in wires. Why? Why is this not a problem at lower frequencies, where voltages and currents change more slowly?

7. In a popular demonstration of induced emf, a light bulb is connected across a large inductor in an LR circuit, as shown in Fig. 30–19. When the switch is opened, the light bulb flashes brightly and burns out. Why?

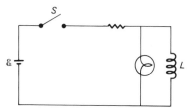

Fig. 30–19   Question 7

8. In an LR circuit like Fig. 30–8, what is the maximum emf that can appear in the inductor when the switch is initially closed? When the switch is later opened?

9. Does it take more or less than one time constant for current in an LR circuit to build up to half its steady-state value?

10. If you increase the resistance in an LR circuit, what effect does this have on the inductive time constant?

11. Speculate on what would happen if you connected an ideal battery directly across an ideal inductor, with no resistance anywhere in the circuit.

12. How could you modify the simple LR circuit of Fig. 30–8 to prevent dangerous voltages from developing when the switch is opened?

13. List some similarities and differences between inductors and capacitors.

14. A 1-H inductor carries 10 A, and a 10-H inductor carries 1 A. Which inductor contains more stored energy?

15. Does the energy density in a magnetic field depend on the direction of the field?

16. The field of a magnetic dipole extends to infinity. Is there an infinite amount of energy stored in the dipole field? Why or why not?

17. In Example 30–10, is our estimate of the magnetic field energy in the 100 km above earth too high or too low? Assume that the 0.5-gauss field strength applies at the poles.

18. It takes work to push two bar magnets together with like poles facing each other. Where does this energy go? *Hint:* Consult Chapter 24, Section 24.1.

PROBLEMS

Section 30.1   *Mutual Inductance*

1. Two coils have a mutual inductance of 2.0 H. If current in the first coil is changing at the rate of 60 A/s, what is the emf in the second coil?

2. In the circuit of Fig. 30–20, the variable resistor R_1 is being adjusted to change the current supplied by the battery. The mutual inductance of the two coils is 250 mH. If a steady current of 50 mA flows in the 100-Ω resistor R_2, what is the rate at which the current in R_1 is changing?

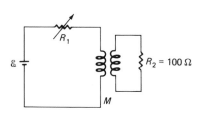

Fig. 30–20   Problem 2

3. An alternating current given by $I = I_0 \sin(2\pi ft)$ is applied to one of two coils whose mutual inductance is M. (a) Obtain an expression for the emf in the second coil as a function of time. (b) Take $I_0 = 1.0$ A, $f = 60$ Hz, and assume that the peak emf in the second coil is 50 V. What is the mutual inductance?

4. A short coil of radius a contains N turns. The coil is placed inside a long solenoid of radius b ($b > a$) consisting of n turns per meter. The axes of the two coils coincide. This arrangement is shown in Fig. 30–21. (a) Show that the mutual inductance of this arrangement is given by $M = \pi \mu_0 n N a^2$. (b) How would M change if the small coil were moved off the solenoid axis, but not so far that it hit the solenoid coils? (c) How would M change if the small coil were turned so that its axis was perpendicular to the solenoid axis? Assume that the small coil remains inside the solenoid.

5. A rectangular loop of length ℓ and width w is located with its nearest side a distance a from a long, straight wire, as shown in Fig. 30–22. What is the mutual inductance of this arrangement?

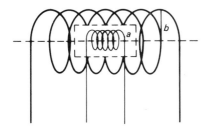

Fig. 30–21   Problem 4

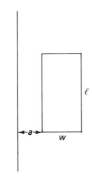

Fig. 30–22   Problem 5

Section 30.2   *Self-Inductance*

6. A solenoid 50 cm long and 4.0 cm in diameter contains 1000 turns of wire. What is the inductance of the solenoid?

7. The current in an inductor is changing at the rate of 100 A/s, and an emf of 40 volts appears in the inductor. What is its self-inductance?

8. A current of 60 mA is flowing in a 100-mH inductor. Over a period of 1.0 ms, the current in the inductor is reversed, going steadily from 60 mA down to zero and on to 60 mA in the opposite direction. What emf appears in the inductor during this reversal?

9. The current flowing in a 2.0-H inductor is given by $I = 3t^2 + 15t + 8$, where t is in seconds and I in amperes. Obtain an expression for the inductor emf as a function of time.

10. The emf in a 50-mH inductor has magnitude given by $|\mathscr{E}| = 0.020t$, with t in seconds and $\mathscr{E}$ in volts. At time $t = 0$ the current in the inductor is 300 mA. (a) If the current is increasing in strength, what is its value at $t = 3.0$ s? (b) Repeat for the case when the current is decreasing in strength. (c) Draw a circuit symbol for the inductor, showing the direction of the current and the polarity of the inductor emf for cases (a) and (b).

11. A coaxial cable consists of an inner conductor of radius a surrounded by a hollow cylindrical conducting shell of radius b, as shown in Fig. 30–23. Assuming that the two conductors of this cable carry equal currents in opposite directions, show that the inductance per unit length of the cable is

$$\frac{\mu_0}{2\pi} \ln\frac{b}{a}.$$

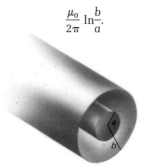

Fig. 30–23   Problem 11

12. Two long, straight wires are parallel to each other a distance d apart. The wires themselves have radius a, where $a \ll d$. They carry equal but opposite currents. Calculate the inductance per unit length of the parallel wires. Assume that the radius a of the wires is so small that you can neglect the magnetic flux of one wire that actually penetrates the interior of the other wire.

Section 30.3   *Inductors in Circuits*

13. Show that the inductive time constant has the units of seconds.

14. In the LR circuit of Fig. 30–8, the resistance R is 2.5 kΩ and the battery emf is 50 V. When the switch is closed, the current through the inductor rises to 10 mA in 30 μs. (a) If you wait a very long time, what will the inductor current be? (b) What is the inductance L?

15. For the circuit of Fig. 30–8, take $R = 100 \ \Omega$, $L = 2.0$ H, and $\mathscr{E} = 12$ V. 20 ms after the switch is closed, what are (a) the current in the circuit? (b) the voltage across the inductor? (c) the voltage across the resistor? (d) the rate of change of current in the circuit? (e) the rate of energy dissipation in the resistor?

16. Resistor R_2 is placed across the inductor in the circuit of Fig. 30–24 to limit the emf that develops when the switch is opened. What should be the value of R_2 in order that the inductor emf will not exceed 100 V?

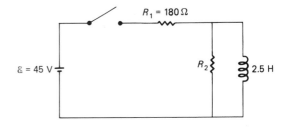

Fig. 30–24   Problem 16

17. Because real inductors are made of wire, they have

resistance as well as inductance. A certain 10-H inductor has a resistance of 2.0 Ω. If this inductor is connected across an ideal 12-V battery, how long will it take the current to reach 99% of its steady-state value?

18. A current of 5.0 A is flowing through a nonideal inductor with inductance $L = 500$ mH. If the inductor is suddenly short circuited, the inductor current drops to 2.5 A in 6.9 ms. What is the resistance of the inductor?

19. In the circuit of Fig. 30–25, take $\mathscr{E} = 12$ V, $R_1 = 4.0\ \Omega$, $R_2 = 8.0\ \Omega$, $R_3 = 2.0\ \Omega$, and $L = 2.0$ H. Assume initially that the switch has been open for a very long time. (a) Redraw the circuit as it looks immediately after the switch is closed (see Example 30–7), and determine the currents in R_1, R_2, and R_3 at this time. (b) Repeat part (a) for the case of a long time after the switch is closed. (c) After the circuit has reached a steady state with the switch closed, the switch is again opened. Now what are the currents in the three resistors? (d) A long time after the switch is opened, what are the currents in the three resistors?

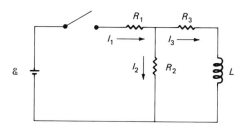

Fig. 30–25 Problem 19

20. In the circuit of Fig. 30–26, take $\mathscr{E} = 20$ V, $R_1 = 10\ \Omega$, $R_2 = 5.0\ \Omega$, and $L = 1.0$ H. Assume initially that the switch has been open for a long time. (a) What is the current in the inductor immediately after the switch is closed? (b) What is the current in the inductor a long time after the switch is closed? (c) Sketch qualitatively the behavior of the current as a function of time after the switch is closed.

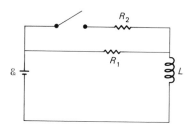

Fig. 30–26 Problem 20

Section 30.4   Magnetic Field Energy

21. A 5.0-H inductor carries a current of 35 A. How much energy is stored in the inductor?

22. What is the current in a 10-mH inductor when the stored energy is 50 μJ?

23. A nonideal inductor has an inductance of 3.0 H and a resistance of 0.50 Ω. The inductor is used to store 1.0 J of energy. (a) What is the current in the inductor? (b) In the steady state, how much time elapses before the energy dissipated as heat in the resistance of the inductor is equal to the stored energy?

24. A circuit contains a battery, switch, resistor, and inductor, all connected in series. When the switch is closed, the current rises to half its steady-state value in 1.0 ms. How long does it take for the stored energy in the inductor to rise to half its steady-state value?

25. The current in a 2.0-H inductor is decreased from 5.0 A to zero over a period of 10 ms. What is the average rate at which energy is being extracted from the inductor during this time? If the current decreases linearly with time, is this power constant?

26. The current in an ideal 2.0-H inductor is increasing with time. At some instant, the current is 3.0 A and the emf in the inductor is 5.0 V. At what rate is the stored energy increasing at this instant?

27. Show that the quantity $B^2/2\mu_0$ has the units of energy density (that is, J/m^3).

28. The ALCATOR fusion experiment at MIT has a magnetic field of 50 T. What is the magnetic energy density in ALCATOR?

29. A single-turn loop of radius R carries a current I. How does the magnetic energy density at the center of this loop compare with that at the center of a long solenoid with the same radius, carrying the same current, but consisting of n turns per unit length?

30. A sunspot is a region of intense magnetic field near the solar surface. A typical sunspot has a magnetic field of 0.2 T, is 50,000 km across, and may extend roughly 30,000 km into the sun. What is the total magnetic energy in this sunspot?

31. The magnetic field at the surface of Jupiter is about 10 G. (a) Assuming that the field strength drops as $1/r^3$, obtain an order-of-magnitude estimate of the total magnetic energy stored in the magnetic field external to Jupiter. Ignore variations in the field with latitude, and see Appendix F for relevant data. (b) In fact, Jupiter's magnetic field is confined by the solar wind to a magnetosphere whose radius is about 100 Jupiter radii. Does this confinement affect your estimate significantly?

32. A wire of radius R carries a current I, distributed uniformly over the cross section of the wire. What is the total magnetic energy per unit length stored *within* the wire? Is this equal to the total energy per unit length in the entire magnetic field of the wire?

33. For the coaxial cable described in Problem 11, (a) what is the magnetic energy density as a function of distance r from the axis of the cable? (b) Use the result of part (a) to find the total energy in a thin cylindrical shell of radius r, thickness dr, and length ℓ, as shown in Fig. 30–27. (c) Integrate your result from part (b) over the interior of the cable to show that the total energy in a length ℓ of the cable is

$$U = \frac{\mu_0}{4\pi} I^2 \ell \ln \frac{b}{a}.$$

(d) Show that this same result may also be obtained using the result of Problem 8 for the inductance per unit length of the cable, and applying the expression $U = \frac{1}{2}LI^2$.

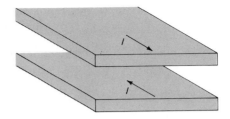

Fig. 30–28   Problem 35

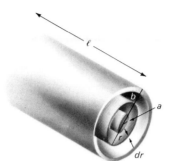

Fig. 30–27   Problem 33

Supplementary Problems

34. Use the result of Example 30–9 for the magnetic energy in a toroid to calculate the inductance of the toroid. Show that, when the major dimension of the toroid is much larger than the minor dimension, your expression reduces to that for the inductance of a solenoid (see Example 30–3). Does this make sense?

35. Two long, flat, parallel bars of width w and spacing d carry equal but opposite currents I, as shown in Fig. 30–28. (a) Assuming that the bars are close enough to ignore fringing fields, use Ampère's law to calculate the magnetic field between the bars. (b) Use the result of part (a) to determine the magnetic energy stored per unit length of the bars. (c) Compare your result in (b) with the expression $U = \frac{1}{2}LI^2$ to show that the inductance per unit length of the bars is given by $\mu_0 d/w$.

36. An electric field of magnitude E and a magnetic field of magnitude B have the same energy density. Obtain an expression for the ratio E/B and evaluate this ratio numerically. What are the units of this ratio? Does your answer look like any of the fundamental constants listed inside the front cover?

37. The earth's magnetic field does not extend forever, but is stopped in the sunward direction by the pressure of the solar wind, a high-speed flow of particles from the sun. The point at which the earth's field stops is determined approximately by the condition that the magnetic energy density equal the density of kinetic energy in the solar wind. Typically, the solar wind contains about 5 electrons and 5 protons per cubic centimeter, and its speed is about 400 km/s. Earth's magnetic field is approximately that of a dipole with dipole moment $\mu = 8.0 \times 10^{22}$ J/T. (a) Show that the density of kinetic energy in matter with density ρ moving at speed v is $u_K = \frac{1}{2}\rho v^2$, and use this result to calculate the kinetic energy density in the solar wind. (b) Use appropriate expressions for the dipole field above the equator to write an equation describing the energy density in earth's magnetic field as a function of radius. (c) Equate your expressions for magnetic energy density of earth's magnetic field and kinetic energy density of the solar wind, and solve for the radius r at which they are equal. This is roughly the distance at which earth's magnetic field ends abruptly. Approximately how many earth radii is this?

31

ALTERNATING-CURRENT CIRCUITS

In Chapter 26 we considered direct-current (DC) circuits, in which the source of electrical energy is a battery or other device whose emf does not change with time. When we turn on a circuit containing only resistors and a DC emf, current starts to flow immediately and remains steady until the circuit is turned off. Even when we add capacitance, as in Section 26.6, or inductance, as in Section 30.3, all currents and voltages eventually reach steady values.

We now turn our attention to alternating current (AC) circuits, in which sources of electrical energy vary with time. A familiar AC circuit is standard household wiring. Alternating current with a frequency of 60 Hz is used almost universally for electric power generation and transmission, for reasons we will discuss in Section 31.6. Devices such as stereos, TV's, radios, and microwave ovens involve more rapidly varying alternating currents.

31.1

ALTERNATING CURRENT

Although electrical quantities can exhibit many different types of time variation, we will consider only sinusoidal variations. More complicated variations can be analyzed as superpositions of sinusoidally varying quantities, as we described in Section 14.5. Figure 31–1 graphs one full cycle of an AC voltage as a function of time; this pattern repeats indefinitely in either direction. The AC signal is characterized by its amplitude, frequency, and phase—the same quantities we developed in Chapter 13 to describe simple harmonic motion. We describe amplitude in terms of either

root mean square   the peak value V_0 or the **root mean square** (rms) value, V_{rms}. The rms value is an average obtained by squaring the signal, taking the time average of the square over one cycle, then taking the square root. This procedure is used

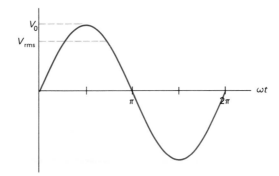

Fig. 31–1
A sinusoidally-varying AC voltage, showing peak voltage V_0 and rms voltage.

because the simple average of an AC signal is zero whatever its amplitude, since the AC signal spends as much time below zero as above. Use of rms values also permits simple calculation of the average power in AC circuits. For a sine wave, the rms value is the peak value divided by $\sqrt{2}$ (see Problem 4). When we say that the voltage in standard household wiring is 120 V, we are referring to the rms value.

In practical and engineering situations we usually describe frequency f in cycles per second, or hertz (Hz). In mathematical analysis of alternating current, it is usually more convenient to use the angular frequency ω, measured in radians per second or, equivalently, inverse seconds (s^{-1}). The relation between the two,

$$\omega = 2\pi f, \tag{31–1}$$

is the same as for rotational and simple harmonic motion, and arises because a full cycle contains 2π radians.

Sometimes we are interested in the phase constant ϕ of an AC signal—that is, when the sine curve "starts," crossing zero as it rises. Signal B in Fig. 31–2, for example, is $\pi/2$ or 90° ahead of signal A, while signals C and A are 180° out of phase.

A full mathematical description of a sinusoidal AC signal—for example, a voltage—includes amplitude V_0, frequency ω, and phase constant ϕ:

$$V = V_0 \sin(\omega t + \phi). \tag{31–2}$$

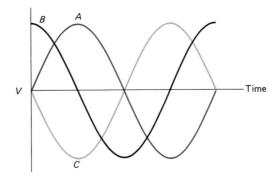

Fig. 31–2
Because its rising portion crosses zero a quarter-cycle earlier, signal B is $\pi/2$ or 90° ahead of signal A. Signals A and C are 180° out of phase.

Example 31–1
Alternating Current

Standard household wiring supplies 120 V rms at a frequency of 60 Hz. Express this voltage mathematically in the form of Equation 31–2. Assume that at time $t = 0$ the voltage is at its positive peak.

Solution

The rms voltage is the peak voltage divided by $\sqrt{2}$, so that

$$V_0 = \sqrt{2}V_{rms} = (\sqrt{2})(120 \text{ V}) = 170 \text{ V}.$$

The angular frequency ω is 2π times the frequency in Hz, so that

$$\omega = 2\pi f = (2\pi)(60 \text{ Hz}) = 380 \text{ s}^{-1}.$$

Our signal peaks when a sine wave of zero phase constant is just passing through zero, so the phase constant is 90° or $\pi/2$ radians. Then Equation 31–2 becomes

$$V = 170 \sin(380t + \pi/2) = 170 \cos(380t).$$

31.2
CIRCUIT ELEMENTS IN AC CIRCUITS

Here we examine individually the AC behavior of resistors, capacitors, and inductors so we can subsequently understand what happens when we combine these elements in AC circuits.

Resistors

An ideal resistor is a device whose current and voltage are always strictly proportional:

$$I = \frac{V}{R}.$$

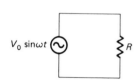

$V_0 \sin\omega t$ R

Fig. 31–3
A resistor connected across an AC generator.

Figure 31–3 shows a resistor R connected across an AC generator, so that the voltage across the resistor is equal to the generator voltage. The generator voltage may be described by Equation 31–2, so the current is

$$I = \frac{V}{R} = \frac{V_0\sin(\omega t + \phi)}{R} = \frac{V_0}{R}\sin(\omega t + \phi).$$

The current has the same frequency and phase as the voltage, and the maximum current is simply the maximum voltage divided by the resistance. Because voltage and current are both sinusoidal, the rms current is given by V_{rms}/R.

Capacitors

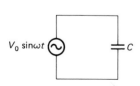

$V_0 \sin\omega t$ C

Fig. 31–4
A capacitor across an AC generator.

Figure 31–4 shows a capacitor connected across an AC generator. In Chapter 25, we defined a capacitor as a device in which voltage and charge are directly proportional:

$$q = CV. \tag{31–3}$$

If we differentiate this relation, we obtain

$$\frac{dq}{dt} = C\frac{dV}{dt}.$$

But dq/dt is the capacitor current I, so that

$$I = C\frac{dV}{dt}. \tag{31–4}$$

The generator voltage $V_0\sin\omega t$ appears directly across the capacitor, so the current is

$$I = C\frac{d}{dt}(V_0\sin\omega t)$$

$$= \omega CV_0\cos\omega t = \omega CV_0\sin(\omega t + \pi/2). \tag{31–5}$$

This equation shows clearly the phase and amplitude relations between current and voltage in a capacitor. Because the cosine curve is just a sine curve shifted left by $\pi/2$ or 90°, Equation 31–5 tells us that:

phase relation in a capacitor

> The current in a capacitor leads the voltage by 90°.

Figure 31–5 shows graphically this relation between current and voltage in a capacitor.

The term ωCV_0 multiplying the cosine in Equation 31–5 is the amplitude of the current, so we can write

$$I_0 = \omega CV_0,$$

or, in a form resembling Ohm's law,

$$I_0 = \frac{V_0}{1/(\omega C)}. \tag{31–6}$$

Equation 31–6 shows that the capacitor acts somewhat like a resistor of resistance $1/\omega C$. But not quite! This "resistance" does give the relation between the peak voltage and peak current, but it does not tell the whole story. The capacitor also introduces a phase difference between voltage and current. This phase difference reflects a fundamental physical difference between resistors and capacitors. A resistor dissipates electrical energy into

Fig. 31–5
In an AC circuit, current in a capacitor leads the voltage by 90°.

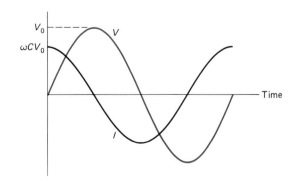

Fig. 31–6
(a) An *RC* circuit. (b) When the switch is closed, current flows to the capacitor before voltage develops across it, showing that current leads voltage in a capacitor.

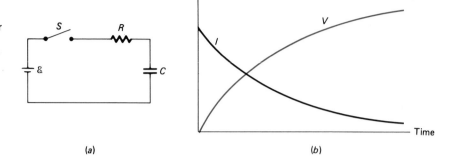

(a)   (b)

heat. A capacitor stores and releases electrical energy. Over a complete cycle, the agent turning the generator in Fig. 31–4 does no net work, while the agent turning the generator with the resistive load of Fig. 31–3 continually does work that gets dissipated as heat in the resistor. Because the quantity $1/\omega C$ in Equation 31–6 does not act quite like a resistance, we give

capacitive reactance

it the special name **capacitive reactance,** X_C. Like resistance, reactance is measured in ohms (Ω).

Does it make sense that X_C depends on frequency? Yes—as frequency goes to zero, X_C goes to infinity. At zero frequency nothing is changing, there is no need to move charge on or off the plates, so no current flows, and the capacitor might as well be an open circuit. As frequency increases, larger currents flow to move charge on and off the capacitor in ever shorter times, so the capacitor looks increasingly like a short circuit. We often summarize this behavior qualitatively by saying that, at low frequencies, a capacitor acts like an open circuit, while at high frequencies it acts like a short circuit. The terms "low frequency" and "high frequency" are relative. What we really mean by "low frequency" is a frequency so low that the capacitive reactance $1/\omega C$ is much larger than any resistance in the circuit. At "high frequency," the reactance is much smaller than any circuit resistance.

If you have trouble remembering how capacitive reactance depends on frequency or how capacitor current and voltage are related in phase, just think of the simple *RC* circuit we considered in Chapter 26 (Fig. 31–6). When you first close the switch, there is no voltage across the capacitor, but current begins to flow. So the current in a capacitor *leads* the voltage. If you wait long enough, everything is steady and there is no current flowing, so the capacitor acts like a *DC open circuit.* Thinking about the extreme cases of high and low frequency will greatly facilitate your understanding of capacitors.

Inductors

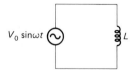

Figure 31–7 shows an inductor connected across an AC generator. The loop law for this circuit is

$$V_0 \sin\omega t + \mathscr{E}_L = 0.$$

Fig. 31–7
An inductor across an AC generator.

From the preceding chapter, we know that the inductor emf is given by

$$\mathscr{E}_L = -L\frac{dI}{dt},$$

so the loop law becomes

$$V_0\sin\omega t = L\frac{dI}{dt}. \qquad (31-7)$$

To obtain a relation involving the current I rather than its derivative, we integrate Equation 31–7:

$$\int V_0\sin\omega t \; dt = \int L\frac{dI}{dt}\,dt,$$

or

$$-\frac{V_0}{\omega}\cos\omega t = LI.$$

(We have done an indefinite integral—no limits—because we are interested not in a number but in a relation between two functions. A nonzero constant of integration would represent a steady current superimposed on the AC; here we have only AC so that the constant is zero.) Solving for I, we obtain

$$I = -\frac{V_0}{\omega L}\cos\omega t. \qquad (31-8)$$

Equation 31–8 shows that the voltage applied to the inductor and the current in the inductor are 90° out of phase, with the voltage leading:

phase relation in an inductor

> **The voltage in an inductor leads the current by 90°.**

This phase relation is shown in Fig. 31–8.

We also see from Equation 31–8 that the peak current is

$$I_0 = \frac{V_0}{\omega L}. \qquad (31-9)$$

Again, this equation resembles Ohm's law, with a "resistance" of ωL. But as with the capacitor, no power is dissipated in the inductor. Instead, energy is alternately stored and released as the inductor's magnetic field

Fig. 31–8
Voltage in an inductor leads current by 90° or $\pi/2$.

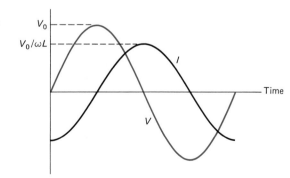

TABLE 31–1 ▄▄▄▄▄▄▄▄▄▄▄▄▄▄▄▄▄▄▄▄▄▄▄▄▄▄▄▄▄▄▄▄▄▄▄▄▄▄▄
Phase and Amplitude Relations in Circuit Elements

Circuit Element	Peak Current/Voltage	Phase Relation
Resistor	$I_0 = \dfrac{V_0}{R}$	V, I in phase
Capacitor	$I_0 = \dfrac{V_0}{X_C} = \dfrac{V_0}{1/\omega C}$	I leads V by 90°
Inductor	$I_0 = \dfrac{V_0}{X_L} = \dfrac{V_0}{\omega L}$	V leads I by 90°

inductive reactance

builds up, then decays. To distinguish it from dissipative resistance, we call the quantity ωL the **inductive reactance,** X_L. Inductive reactance, too, is measured in ohms.

Does it make sense that the inductive reactance increases with increasing ω and increasing L? An inductor is a device that, through its induced back emf, opposes changes in current. The greater the inductance, the greater the opposition to changing current. And the more rapidly the current is changing, the more vigorously the inductor opposes the change, so that the inductive reactance increases at high frequencies. In the extreme case of very high frequencies, an inductor looks like an open circuit. But at very low frequencies it looks more and more like a short circuit, until with direct current (zero frequency), an inductor exhibits no reactance because there is no change in current.

Table 31–1 summarizes the phase and amplitude relations in resistors, capacitors, and inductors.

Example 31–2 _____
Inductors and Capacitors

A capacitor is connected across the 60-Hz, 120-V rms power line, and an rms current of 200 mA flows. What is the capacitance? What inductance would have to be connected across the power line for the same current to flow? Would there be anything different about the circuit containing the inductor?

Solution

The peak current and voltage are related through Equation 31–6:

$$I_0 = \frac{V_0}{1/\omega C},$$

so that

$$C = \frac{I_0}{\omega V_0}.$$

We are given the rms voltage and current, but since only the ratio of these quantities appears in our equation, it does not matter whether we use rms or peak values. With $f = 60$ Hz, $\omega = 2\pi f$ or 380 s^{-1}, so that

$$C = \frac{I}{\omega V} = \frac{0.20 \text{ A}}{(380 \text{ s}^{-1})(120 \text{ V})} = 4.4 \ \mu\text{F}.$$

An inductor that passes the same current must have the same reactance, so that

$$\omega L = \frac{1}{\omega C},$$

or

$$L = \frac{1}{\omega^2 C} = \frac{1}{(380 \text{ s}^{-1})^2(4.4 \times 10^{-6} \text{ F})} = 1.6 \text{ H}.$$

Although the currents are the same, the two situations are different in that current leads voltage by 90° in the capacitor and lags by 90° in the inductor.

Phasor Diagrams

phasor diagrams

Phase and amplitude relations in AC circuits may be summarized graphically in **phasor diagrams.** A phasor is an arrow whose fixed length represents the amplitude of an AC voltage or current. One end of the phasor is at the origin, and we imagine the whole arrow to rotate about the origin with the angular frequency ω of the AC quantity. The component of the phasor on the vertical axis then represents the sinusoidally varying AC signal. Figure 31–9a shows phasors for the current and voltage in a resistor. The lengths of the phasors are related by Ohm's law, $V_0 = I_0 R$. The current and voltage phasors always point in the same direction, showing that current and voltage in the resistor are in phase. Figures 31–9b and 31–9c show phasor diagrams for a capacitor and an inductor. In each, the lengths of the phasors are related by the appropriate reactance, so that $V_0 = I_0 X$. As the phasors rotate, they remain at right angles, indicating the phase relation between current and voltage in these reactive circuit elements. You should convince yourself that all the relationships of Table 31–1 are correctly described by the phasor diagrams of Fig. 31–9. Although phasor diagrams do not add much to our understanding of AC circuits containing only one circuit element, they will greatly simplify our analysis of more complicated circuits.

Fig. 31–9
Phasor diagrams showing current and voltage in (a) a resistor, (b) a capacitor, and (c) an inductor. Length of phasors corresponds to peak values I_0 and V_0, while projection on vertical axis corresponds to instantaneous values $I(t)$ and $V(t)$.

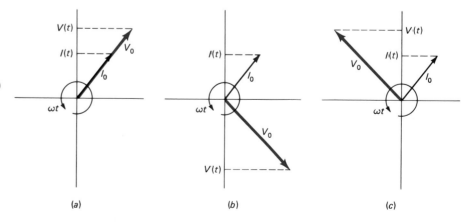

(a)   (b)   (c)

TABLE 31–2
Capacitors and Inductors

	Capacitor	Inductor
Defining relation	$C = \dfrac{q}{V}$	$L = \dfrac{\phi_B}{I}$
Defining relation, differential form	$I = C\dfrac{dV}{dt}$	$\mathscr{E} = -L\dfrac{dI}{dt}$
Opposes changes in	Voltage	Current
Energy storage	In electric field $U = \frac{1}{2}CV^2$	In magnetic field $U = \frac{1}{2}LI^2$
Behavior in low-frequency limit	Open circuit	Short circuit
Behavior in high-frequency limit	Short circuit	Open circuit
Reactance	$X_C = 1/\omega C$	$X_L = \omega L$
Phase	Current leads by 90°	Voltage leads by 90°

Capacitors and Inductors: A Comparison

Here and in previous chapters, we have considered in detail the separate behavior of capacitors and inductors. Many of the properties of these devices are analogous. A capacitor opposes instantaneous changes in voltage, while an inductor opposes instantaneous changes in current. In an RC circuit with a DC emf, voltage builds up exponentially across the capacitor, with time constant RC. In the analogous RL circuit, current builds up exponentially in the inductor, with time constant L/R. A capacitor stores electrical energy given by $\frac{1}{2}CV^2$. An inductor stores magnetic energy given by $\frac{1}{2}LI^2$. A capacitor acts like an open circuit at low frequencies, an inductor like a short circuit at low frequencies. Each exhibits the opposite behavior at high frequencies.

Capacitors and inductors are complementary devices, reflecting a deeper complementarity between electric and magnetic fields. Any verbal description of a capacitor applies to an inductor if we replace the words "capacitor" with "inductor," "electric" with "magnetic," and "voltage" with "current." Table 31–2 details many complementary aspects of capacitors and inductors.

31.3
LC *CIRCUITS*

In this section we consider circuits containing both inductors and capacitors. The properties of these circuits reflect directly the complementary nature of the two devices.

LC *Oscillations*

Figure 31–10 shows a simple circuit containing only a capacitor C and inductor L. Suppose the capacitor is initially charged to some voltage V_0 and corresponding charge q_0, then connected to the inductor. What happens?

Fig. 31–10
An *LC* circuit.

Fig. 31–11
Oscillation in an *LC* circuit, showing energy transfer between electric and magnetic fields.

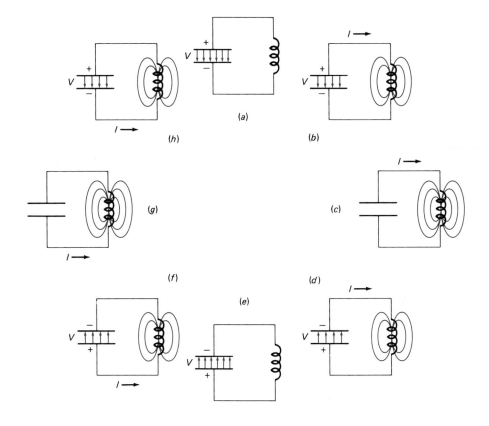

Initially, the capacitor is fully charged, while the inductor current is zero. There is electrical energy stored in the capacitor, but no energy in the inductor. This initial state is shown in Fig. 31–11a. Then the capacitor begins to discharge through the inductor. It cannot do so all at once, because the inductor opposes changes in current. So current in the inductor rises gradually, and with it the magnetic energy stored in the inductor. At the same time the capacitor voltage, charge, and stored electrical energy decrease. Some time later, the initial energy is divided equally between the capacitor and inductor, as in Fig. 31–11b. But the capacitor keeps discharging, eventually reaching zero charge, as in Fig. 31–11c. Now there is no voltage across the capacitor, and no stored electrical energy. All the energy that was initially in the electric field of the capacitor is in the magnetic field of the inductor.

Does everything stop at this point? No—at this instant that the capacitor is fully discharged, a substantial current is flowing in the inductor. The current in the inductor cannot change abruptly, so that now positive charge begins to pile up on the bottom plate of the capacitor. Stored electrical energy increases as the capacitor charges, and the inductor current and stored magnetic energy decrease. Eventually (Fig. 31–11e), the capacitor is fully charged in the opposite direction from its initial state. Again all the energy is in the capacitor, and none in the inductor. Again the capacitor begins to discharge, and the process repeats, now with a counterclockwise current. All the energy is transferred to the inductor (Fig. 31–11g), and then

TABLE 31–3 ▰▰▰▰▰▰
LC Circuits and Mass-Spring Systems

LC Circuit	Mass-Spring
Charge q	Displacement x
Current $I = dq/dt$	Velocity $v = dx/dt$
Inductance L	Mass m
Capacitance C	$1/k$ ($k =$ spring constant)
Magnetic energy $U_B = \frac{1}{2}LI^2$	Kinetic energy $U_K = \frac{1}{2}mv^2$
Electric energy $U_E = \frac{1}{2}(1/C)\,q^2$	Potential energy $U = \frac{1}{2}kx^2$
Frequency $\omega = 1/\sqrt{LC}$	Frequency $\omega = \sqrt{k/m}$
Resistance	Friction

back to the capacitor, which again attains its initial state (Fig. 31–11a). Provided there is no energy loss, the oscillation repeats indefinitely.

This LC oscillation should remind you of the mass-spring system we studied in Chapter 13. There, energy was transferred back and forth between kinetic energy of the mass and potential energy of the spring. Here, energy is transferred back and forth between magnetic energy of the inductor and electrical energy of the capacitor. The mass-spring system exhibits simple harmonic (that is, sinusoidal) motion with frequency determined by the mass m and spring constant k. Similarly, the LC circuit oscillates sinusoidally with frequency determined by the inductance L and capacitance C. Table 31–3 shows some analogies between mass-spring systems and LC circuits. We will develop these analogies more rigorously in the next section.

Analogies with LC circuits are so useful that engineers sometimes simulate complicated systems, such as bridges, automobile suspensions, or world energy usage, with networks of LC circuits (Fig. 31–12). Such a network is called an analog computer, because its behavior is analogous to that of the system under study.

Analyzing the LC Circuit

We described the LC circuit qualitatively in terms of transfer between electric and magnetic energy. This description suggests a way to analyze the circuit quantitatively. The total energy in the circuit is the sum of the magnetic and electric energy:

$$U = U_B + U_E = \frac{1}{2}LI^2 + \frac{1}{2}\frac{q^2}{C}. \tag{31–10}$$

The time derivative of this equation is

$$\frac{dU}{dt} = \frac{d}{dt}\left(\frac{1}{2}LI^2 + \frac{1}{2}\frac{q^2}{C}\right). \tag{31–11}$$

Fig. 31–12
This analog computer is a network of LC circuits used to simulate systems from cameras and copiers to aircraft and rockets.

But since the total energy does not change, $dU/dt = 0$. Carrying out the differentiations of the right-hand side of Equation 31–11, we then have

$$LI \frac{dI}{dt} + \frac{q}{C} \frac{dq}{dt} = 0. \tag{31–12}$$

Substituting $I = dq/dt$ and $dI/dt = d^2q/dt^2$, we obtain

$$L \frac{d^2q}{dt^2} + \frac{1}{C} q = 0. \tag{31–13}$$

Equation 31–13 is a differential equation describing the capacitor charge as a function of time. We encountered a similar equation in Chapter 13 when we studied the mass-spring system:

$$m \frac{d^2x}{dt^2} + kx = 0. \tag{31–14}$$

We found that Equation 31–14 could be satisfied by a sinusoidal function of time, with any amplitude and phase constant, and with frequency given by

$$\omega = \sqrt{k/m}. \tag{31–15}$$

Equation 31–13 is identical to Equation 31–14 except that q replaces x, L replaces m, and $1/C$ replaces k. Therefore the solution of Equation 31–13 is a sinusoidal oscillation of arbitrary amplitude and phase whose frequency is given by Equation 31–15 with L replacing m and $1/C$ replacing k:

$$q = q_0 \cos(\omega t + \phi), \tag{31–16}$$

where

$$\omega = \frac{1}{\sqrt{LC}}. \tag{31–17}$$

Differentiating this expression gives the current in the circuit:

$$I = \frac{dq}{dt} = \frac{d}{dt} [q_0 \cos(\omega t + \phi)] = -\omega q_0 \sin(\omega t + \phi). \tag{31–18}$$

(Our choice of cosine rather than sine to describe q is arbitrary; we can always turn a sine into a cosine with proper choice of phase constant ϕ.)

All other circuit quantities are readily determined from Equations 31–17 and 31–18. For example, the capacitor voltage, obtained from the capacitor defining relation $q = CV$, is

$$V_C = \frac{q}{C} = \frac{q_0}{C} \cos\omega t,$$

where we choose the phase constant to correspond to the initially charged capacitor of Fig. 31–11. The electrical energy stored in the capacitor is

$$U_E = \frac{q^2}{2C} = \frac{(q_0 \cos\omega t)^2}{2C} = \frac{q_0^2}{2C} \cos^2\omega t, \tag{31–19}$$

while the magnetic energy stored in the inductor is

$$U_B = \tfrac{1}{2}LI^2 = \tfrac{1}{2}L(-\omega q_0 \sin\omega t)^2 = \tfrac{1}{2}L\omega^2 q_0^2 \sin^2\omega t. \tag{31–20}$$

Fig. 31–13
Electric and magnetic energies in an *LC*
circuit. Their sum is constant.

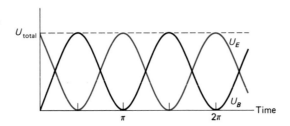

We can verify that our solution does conserve energy by adding the electric and magnetic energies:

$$U_{total} = U_E + U_B = \frac{q_0^2}{2C} \cos^2\omega t + \tfrac{1}{2}L\omega^2 q_0^2 \sin^2\omega t.$$

But Equation 31–17 tells us that $\omega^2 = 1/LC$, so we have

$$U_{total} = \frac{q_0^2}{2C} \cos^2\omega t + \frac{1}{2}\frac{L}{LC} q_0^2 \sin^2\omega t$$

$$= \frac{q_0^2}{2C}(\cos^2\omega t + \sin^2\omega t) = \frac{q_0^2}{2C},$$

since $\cos^2\omega t + \sin^2\omega t = 1$. Thus the total energy is indeed independent of time, and is equal to the initial energy stored in the capacitor. Figure 31–13 is a plot of the electric and magnetic energies as functions of time, showing that the two always sum to a constant.

Example 31–3

LC Oscillations

You wish to make an *LC* circuit oscillate at 440 Hz (A above middle C) to assist in tuning pianos. You have available a 2.0-H inductor. What value of capacitance should you use? If you initially charge the capacitor to 5.0 V, what will be the peak current and peak charge on the capacitor?

Solution

The oscillation frequency is given by Equation 31–17. Solving for *C* gives

$$C = \frac{1}{\omega^2 L} = \frac{1}{(2\pi f)^2 L} = \frac{1}{[(2\pi)(440 \text{ Hz})]^2(2.0 \text{ H})} = 0.065 \ \mu\text{F}.$$

The capacitor charge and current are related through the capacitor definition $C = q/V$, so that

$$q_0 = CV_0 = (0.065 \ \mu\text{F})(5.0 \text{ V}) = 0.33 \ \mu\text{C}.$$

The charge varies sinusoidally with time:

$$q = q_0 \cos\omega t,$$

so the current is

$$I = \frac{dq}{dt} = -\omega q_0 \sin\omega t.$$

This equation shows that the peak current I_0 is just ωq_0, or

$$I_0 = 2\pi f q_0 = (2\pi)(440 \text{ Hz})(0.33 \text{ }\mu\text{C}) = 910 \text{ }\mu\text{A} = 0.91 \text{ mA}.$$

Could you have calculated the current using energy considerations rather than by differentiating the charge? (See Problem 14.)

Resistance in LC Circuits—Damping

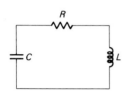

Fig. 31–14
An *RLC* circuit.

Real inductors, capacitors, and wires have resistance. Both this intrinsic resistance and any external resistance are represented by the resistor R in Fig. 31–14. What happens if we initially charge the capacitor in such a resistive *LC* circuit?

Provided the resistance is small—small enough that only a small fraction of the energy is lost in one cycle—then our analysis of the preceding section applies. The circuit oscillates at a frequency given very nearly by Equation 31–17. But as current flows back and forth through the resistor, energy is dissipated as heat. On each successive cycle, the total energy decreases. Consequently the amplitude of the oscillations—the peak charge and current—decreases with time.

We can analyze this *RLC* circuit by starting with Equation 31–11, but now we set dU/dt equal to the rate of energy dissipation in the resistor:

$$LI\frac{dI}{dt} + \frac{q}{C}\frac{dq}{dt} = -I^2R,$$

where the minus sign indicates that energy is *lost* in the resistor. Simplifying the left-hand side of this expression as we did in the preceding section leads to

$$L\frac{d^2q}{dt^2} + R\frac{dq}{dt} + \frac{q}{C} = 0.$$

This equation is mathematically identical to Equation 13–28 for damped simple harmonic motion, showing that our analogies of Table 31–3 continue to hold when resistance is present. Using Equation 13–29, which is the solution to Equation 13–28, and the appropriate analogies from Table 31–3, we can construct the solution for our decaying *RLC* circuit:

$$q = q_0 e^{-Rt/2L}\cos\omega t. \tag{31–21}$$

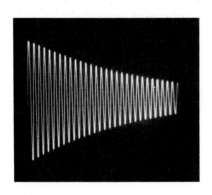

Fig. 31–15
An oscilloscope displays the capacitor voltage in a damped *RLC* circuit.

Other quantities show similar behavior, with oscillation amplitude decaying exponentially with time constant $2L/R$. Figure 31–15 shows an oscilloscope trace of the capacitor voltage in a circuit undergoing damped oscillations.

Equations 13–29 and 31–21 are correct only when the energy dissipation is small. As the electrical resistance increases, the oscillations decay more rapidly and the frequency of oscillation decreases. Finally, when the exponential time constant $2L/R$ equals the inverse of the natural frequency, much of the energy is lost in the time of one undamped oscillation period. **critical damping** This situation is termed **critical damping,** and at this value of R circuit quantities decay exponentially to zero, in analogy with a critically damped mechanical system (Section 13.6). For greater values of R, the circuit is **overdamped,** and also exhibits no oscillation.

overdamped

31.4
DRIVEN RLC CIRCUITS AND RESONANCE

In the LC and RLC circuits just considered, we supplied an initial energy, then let the circuit undergo oscillations. What happens if, instead, we connect an RLC circuit to an AC generator, as in Fig. 31–16? Because the RLC circuit is analogous to a mass-spring system with friction, we might expect the circuit to exhibit resonant behavior analogous to the mechanical resonance we discussed in Section 13.7.

Imagine that we can vary the frequency of the generator in Fig. 31–16, but that its peak voltage V_0 does not change. Let us explore the response of the circuit—specifically, the current that flows—for different frequencies of the applied voltage.

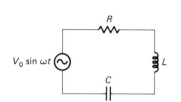

Fig. 31–16
(Above) A series RLC circuit driven by a sinusoidally varying voltage. (Below) The RLC circuit is analogous to a swing or other mechanical system, in that its greatest response occurs when the driving frequency is near the natural frequency of the system.

A Qualitative Analysis

We can begin to understand the driven RLC circuit by considering the two extreme cases of very low and very high frequency. In the limit of zero frequency—direct current—we reach a steady state in which nothing is changing. No charge moves on or off the capacitor plates, so there is no current anywhere in the series circuit. It is the capacitor that blocks low-frequency current, so at low frequencies the behavior of the capacitor dominates the circuit.

In the limit of very high frequencies, any current in the circuit is changing very rapidly. But the inductor opposes changes in current—the more rapid the changes, the more vigorously the inductor opposes them. So in the high-frequency limit, the current also goes to zero. In this limit the inductor, by blocking the current, dominates the circuit.

Unless the current is zero at all frequencies, it must have at least one maximum somewhere between the high- and low-frequency limits. You can probably guess that a maximum occurs at the natural frequency

Fig. 31–17
Peak current versus frequency in a driven series RLC circuit. The current is a maximum at the resonant frequency ω_r.

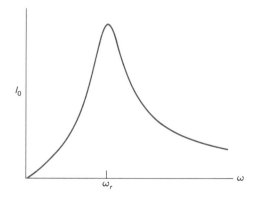

Fig. 31–18
Phase of current relative to driving voltage in a driven series *RLC* circuit.

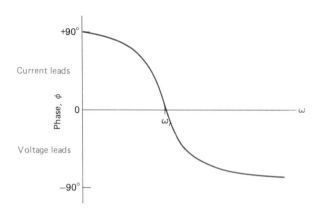

$1/\sqrt{LC}$. Driving an *RLC* circuit is just like pushing someone on a swing: the swing responds with the greatest amplitude of oscillation when you push at its natural frequency (Fig. 31–16). Like mechanical resonance, the peaking of the electric current at the natural frequency of the *LC* combination is

resonance

termed **resonance.** The fact that there is only one special frequency suggests that the current should have only one maximum. Putting together the circuit behavior at frequency extremes and at resonance, we can sketch qualitatively the dependence of current on frequency, shown in Fig. 31–17.

What about the phase of current relative to the applied voltage? We know that, in a capacitor, current leads voltage by 90°. The circuit behavior is dominated by the capacitor at very low frequencies, so we expect the current to lead at low frequencies. At high frequencies, on the other hand, the inductor dominates the circuit. Applied voltage leads current in an inductor, so at high frequencies the circuit current lags the applied voltage. Somewhere in between, the current and voltage must be in phase. Where? At resonance! Figure 31–18 summarizes our qualitative analysis of the phase relation between current and applied voltage in the *RLC* circuit.

We can understand the resonant behavior of the *RLC* circuit in more physical terms if we consider in detail just the *LC* part of the circuit. The inductor and capacitor are in *series,* so the current flowing in each must be the *same.* But the capacitor voltage lags the current by 90°, and the inductor voltage leads the current by 90°, so the inductor and capacitor voltages must be 180° out of phase. Figure 31–19 shows the phase relations

Fig. 31–19
Capacitor voltage lags current by 90°, while inductor voltage leads current by 90°. When an inductor and capacitor are in series, so that the current through each is the same, then the capacitor and inductor voltages are 180° out of phase.

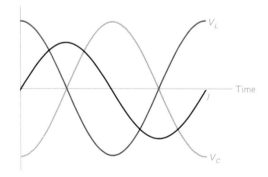

between current—which is the same in both inductor and capacitor—and the voltages across each of these circuit elements.

The phase relations of Fig. 31–19 hold at any frequency. From the figure, we see that the net voltage across the series combination of inductor and capacitor is actually less than the voltage across either. This surprising situation occurs because the two voltages are 180° out of phase—they have opposite polarity—so that they subtract rather than adding.

What are the relative amplitudes of these two voltages? At very low frequencies, the capacitor acts nearly like an open circuit. Its reactance $1/\omega C$ is much greater than either the resistance R or the inductive reactance ωL, so that most of the applied voltage drops across the capacitor, as suggested in Fig. 31–20a.

At high frequencies, in contrast, the capacitive reactance goes to zero while the inductive reactance becomes large. Now the inductor blocks the current, and therefore sustains most of the voltage drop (Fig. 31–20b).

At resonance, the reactances are equal and so are the magnitudes of the voltage drops V_C and V_L. But because the inductor and capacitor voltages are 180° out of phase, the voltage across the series LC combination at resonance is precisely zero (Fig. 31–20c). At resonance, the inductor and capacitor together have absolutely no effect on the circuit current, which is then determined entirely by the resistor. At resonance—but only at resonance—

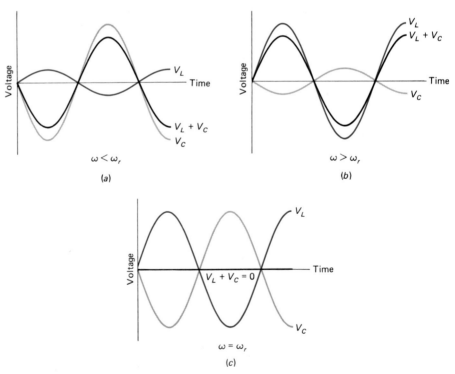

Fig. 31–20
(a) At low frequencies, most of the applied voltage appears across the capacitor. *(b)* At high frequencies, most of the applied voltage appears across the inductor. *(c)* At resonance, capacitor and inductor voltages have equal amplitude, but since they are 180° out of phase, they precisely cancel.

the series *LC* combination could be replaced by a wire! At resonance, the complementarity of the inductor and capacitor—reflecting the complementarity of magnetic and electric fields—exactly cancels the effects of these two circuit elements. At any other frequency, one of the two elements dominates, reducing the circuit current.

A Quantitative Analysis

We have argued that the current in our series *RLC* circuit tends to zero at high and low frequencies. We can also determine the current at resonance, where the effects of the inductor and capacitor exactly cancel, leaving only the resistor to limit the current:

$$I_{\text{res}} = \frac{V_0}{R}.$$

To calculate the current for all frequencies, we must find a general relation between current and applied voltage. We can do this by noting that the *same* current flows through all three elements in the series circuit. In the phasor diagram of Fig. 31–21a, we represent this current by a single phasor of length I_0. The angle of this phasor is arbitrary, as the figure is a "snapshot" of a situation in which the phasor rotates with angular frequency ω about the origin. The resistor voltage is a phasor in the same direction as the current. But because they are 90° out of phase with the current, the capacitor and inductor voltages are represented by phasors at right angles to the current, as shown.

Applying the loop law to Fig. 31–16, we see that the applied voltage equals the sum of the voltages across the three elements. In terms of the phasor diagram, the applied voltage phasor is the vector sum of the voltage phasors for the resistor, capacitor, and inductor. The component of this phasor along the vertical axis is the instantaneous value of the applied voltage, and is equal to the sum of the components of the three voltage phasors. Figure 31–21b shows the result of this phasor summation. Applying the

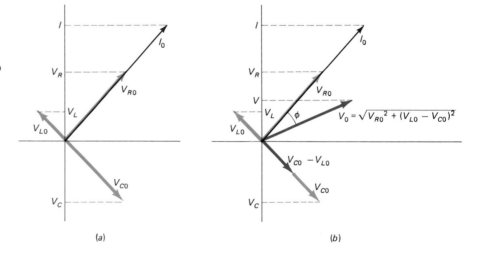

Fig. 31–21
Phasor diagrams for the driven *RLC* series circuit. *(a)* The current in the three circuit elements is the same, but the voltages are out of phase. *(b)* The three voltage phasors sum vectorially to the applied voltage.

(a)   (b)

Pythagorean theorem, we see that the magnitude of the applied voltage phasor is given by

$$V_0^2 = V_R^2 + (V_C - V_L)^2.$$

Expressing this in terms of the current and the resistance and reactance gives

$$V_0^2 = I_0^2 R^2 + (I_0 X_C - I_0 X_L)^2,$$

or

$$I_0 = \frac{V_0}{\sqrt{R^2 + (X_C - X_L)^2}}. \qquad (31\text{--}22)$$

Equation 31–22 has the form of Ohm's law, with the quantity $\sqrt{R^2 + (X_C - X_L)^2}$ playing the role of resistance. We call this quantity the impedance **impedance,** Z, of the circuit. Impedance is a generalization of resistance to include the frequency-dependent effects of capacitance and inductance. Putting in our expressions for the reactances gives

$$Z = \sqrt{R^2 + \left(\frac{1}{\omega C} - \omega L\right)^2}. \qquad (31\text{--}23)$$

In agreement with our qualitative analysis of the driven *RLC* circuit, this equation shows that the circuit impedance becomes very large at high and low frequencies, and has its lowest value, R, at resonance.

Figure 31–22 is a plot of Equation 31–22, showing peak current versus frequency for several values of resistance. As we lower the resistance, the peak current at resonance rises. Although the current at other frequencies rises, too, it does so to a much lower extent than at resonance. This is because the impedance at resonance depends only on the resistance, but includes reactive effects at other frequencies. As a result, the resonance curve becomes more sharply peaked as the resistance drops. For a circuit with very low resistance, the current at resonance is dramatically different from that at even a slightly different frequency. Such a circuit, called a high-Q **high-Q** (for high-quality) circuit, does a good job of distinguishing its resonant frequency from nearby frequencies. A low-Q circuit, in contrast, has a broad resonance curve and does not do a very good job selecting out the

Fig. 31–22
Resonance curves for an *RLC* circuit
with $L = 5.0$ mH, $C = 0.22$ μF, and
several resistance values.

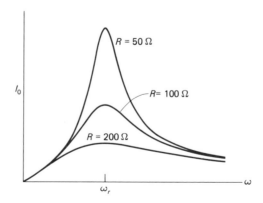

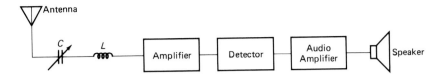

Fig. 31–23
Simplified diagram of a radio.

resonant frequency. A rigorous definition of Q can be given in terms of the width of the resonance peak (see Problem 38).

Application

Tuning a Radio

When you tune a radio, you are adjusting a variable capacitor to change the resonant frequency of an LC circuit. Figure 31–23 shows a simplified diagram of a radio. The antenna (symbol ∇) is subjected to radio signals of different frequencies from various stations. Currents induced by these signals must flow through the LC circuit. The impedance of this circuit is very low at resonance, and high at all other frequencies, so that only signals at the resonant frequency produce significant currents. In this way the hodgepodge of signals reaching the antenna is eliminated in favor of the one signal from the station that broadcasts at the selected frequency. Once the desired signal gets past the LC circuit, it is amplified, then converted to an electrical signal in the audio-frequency range, amplified again, and finally sent to a loudspeaker, which converts it to sound.

Equation 31–22 and the notion of impedance give us the relation between peak applied voltage and peak current in an RLC circuit. But this amplitude relation does not tell the whole story. What about the relative phases of current and voltage? From Fig. 31–21b, we see that the phase constant ϕ is given by

$$\tan\phi = \frac{V_{C0} - V_{L0}}{V_{R0}}.$$

Because the peak voltages are proportional to the reactances and resistance, this expression may be written

$$\tan\phi = \frac{X_C - X_L}{R}, \tag{31–24a}$$

or

$$\tan\phi = \frac{1/\omega C - \omega L}{R}. \tag{31–24b}$$

Equations 31–24a and 31–24b are consistent with our earlier qualitative discussion of phases in RLC circuits. At resonance, $X_C = X_L$, and the phase difference is zero. For frequencies below resonance, the capacitive reactance is larger and the phase difference is positive, showing that current leads voltage. Above resonance, the inductive reactance dominates, so that voltage leads current. At the high- and low-frequency extremes, $\tan\phi$ becomes arbitrarily large in magnitude, and the phase differences approach

Fig. 31–24
Phase relations for the *RLC* circuits
whose resonance curves are shown in
Fig. 31–22.

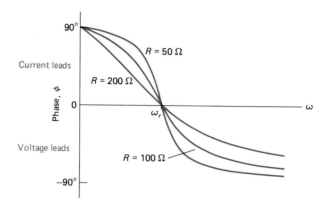

Fig. 31–24
Phase relations for the *RLC* circuits
whose resonance curves are shown in
Fig. 31–22.

90°. The sharpness of the transition from one extreme to the other depends on the Q of the circuit, as shown in Fig. 31–24.

Example 31–4

A Driven RLC Circuit

A driven *RLC* circuit consists of a 250-pF capacitor and a 0.10-mH inductor. What is the resonant frequency? What should the circuit resistance be if the current at a frequency 10 per cent below resonance is to be half that at resonance?

Solution

The resonant frequency is

$$\omega = \frac{1}{\sqrt{LC}} = \frac{1}{\sqrt{(0.10\times10^{-3}\text{ H})(250\times10^{-12}\text{ F})}} = 6.3\times10^{6}\text{ s}^{-1},$$

or

$$f = \frac{\omega}{2\pi} = 1.0\times10^{6}\text{ Hz}.$$

This frequency is in the middle of the AM radio band.

The peak current is given by Equation 31–22. At resonance, the denominator in this equation is just R. To cut the current in half, we must double this denominator, or quadruple the quantity under the square root sign. Designating by ω_1 the frequency at which we want half the maximum current, we can write

$$R^2 + \left(\frac{1}{\omega_1 C} - \omega_1 L\right)^2 = 4R^2,$$

or

$$R = \frac{1}{\sqrt{3}}\left(\frac{1}{\omega_1 C} - \omega_1 L\right). \tag{31–25}$$

Our frequency at half maximum current is 10 percent below resonance, so that

$$\omega_1 = 0.90\omega = (0.90)(6.3\times10^{6}\text{ s}^{-1}) = 5.7\times10^{6}\text{ s}^{-1}.$$

Using this value in Equation 31–25 then gives $R = 76\ \Omega$.

Near a large city, the AM radio dial is so crowded with stations that our criterion of half maximum current at 10 per cent below resonance would not be sufficient to separate individual stations. A lower resistance in the *LC* circuit would be required.

31.5
POWER IN AC CIRCUITS

In Section 31.1, we noted that average power dissipation over one cycle is zero in a circuit containing only a capacitor or an inductor. We can understand this physically because the reactive element alternately stores and releases energy rather than dissipating it as heat. Mathematically, we can see this from Fig. 31–25, which shows the current, voltage, and instantaneous power in a capacitor. The power is the product of the current and voltage. Because these two are out of phase, the power is positive half the time, and negative half the time. When the power is positive, the capacitor is absorbing energy from the source of emf that drives the current. When the power is negative, the capacitor is returning energy to the driving source. The net energy transferred to the capacitor over one cycle is $\int P\,dt$, or the area under the power versus time curve, and is zero in this case. Figure 31–26, in contrast, shows current, voltage, and instantaneous power in a resistor. Since current and voltage are always in phase, the power is always positive, and the resistor always takes energy from the source. Comparison of Fig. 31–25 and Fig. 31–26 suggests that the phase difference between current and voltage is important in determining the average power consumption of an AC circuit. We can see this more clearly if we imagine slipping the current and voltage just slightly out of phase, as in Fig. 31–27. Then the power is mostly positive, but there are narrow regions where it is negative, so the average power over one cycle is slightly less than in the resistor case. As the phase difference increases, so does the amount of time

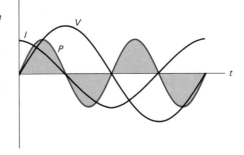

Fig. 31–25
Current, voltage, and their product—the instantaneous power consumption in a capacitor. Positive power means the capacitor is absorbing energy; negative power means it is releasing energy. Total energy transferred to the capacitor is $\int P\,dt$, or the area under the power versus time curve. With voltage and current 90° out of phase, the energy transferred over one cycle is zero.

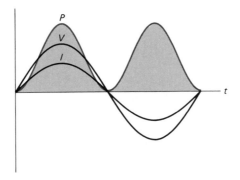

Fig. 31–26
Since voltage and current in a resistor are always in phase, the instantaneous power is never negative, so the resistor always absorbs energy.

Fig. 31–27
As the current and voltage go out of phase, the average power is reduced.

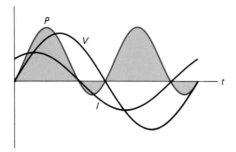

that the power is negative, until at 90° phase difference, the time-average power is zero.

We can develop a general expression for power in AC circuits by considering the time-average product of voltage and current with arbitrary phase difference ϕ:

$$<P> = <(I_0 \sin\omega t)[V_0 \sin(\omega t + \phi)]>,$$

where $<>$ indicates a time average over one cycle. Expanding the voltage term using a trig identity (see Appendix A) gives

$$<P> = I_0 V_0 <(\sin^2\omega t)(\cos\phi) + (\sin\omega t)(\cos\omega t)(\sin\phi)>.$$

The average of $(\sin\omega t)(\cos\omega t)$ is zero, as we argued previously for the case when two signals are 90° out of phase. The quantity $\sin^2\omega t$ swings from 0 to 1, and is symmetric about $\frac{1}{2}$, so its average value is $\frac{1}{2}$ (Fig. 31–28). Then we have

$$<P> = \tfrac{1}{2}I_0 V_0 \cos\phi.$$

Writing the peak values as $\sqrt{2}$ times the rms values gives

$$<P> = \tfrac{1}{2}\sqrt{2}I_{rms} \sqrt{2}V_{rms} \cos\phi = I_{rms}V_{rms}\cos\phi. \tag{31–26}$$

This equation confirms our earlier graphical arguments. When the voltage and current are in phase, the average power is just the product $I_{rms}V_{rms}$. (This, in fact, is a principal reason for using rms values: with them, the expression for average power is the same as in the DC case.) But with current and voltage out of phase, the average power is smaller; at 90° phase difference it is zero.

Fig. 31–28
The function $\sin^2\omega t$ swings between 0 and 1, and is symmetric about $\frac{1}{2}$, so its average value is $\frac{1}{2}$.

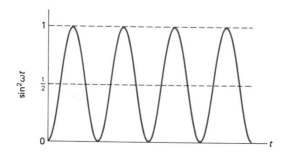

power factor

The factor cosϕ in Equation 31–26 is called the **power factor.** It is a measure of the resistive versus reactive qualities of a circuit. A power factor of 1 means a purely resistive circuit. A circuit containing only resistors, as well as an RLC circuit at resonance, both have power factors of 1. A circuit containing just an inductor or capacitor has a power factor of zero. Most circuits have power factors somewhere between zero and one. For the series RLC circuit, for example, we can see from Fig. 31–21b that the cosine of the phase angle ϕ is

$$\cos\phi = \frac{V_{R0}}{V_0}.$$

But $V_{R0} = I_0 R$ and $V_0 = I_0 Z$, so that

$$\cos\phi = \frac{R}{Z}, \tag{31–27}$$

where Z is the impedance given in Equation 31–23. At resonance, $Z = R$, and the power factor is indeed 1. At other frequencies, $Z > R$, and the power factor decreases.

31.6
TRANSFORMERS AND POWER SUPPLIES

Transformers

In our introduction to Faraday's law, we noted a simple experiment in which two circuits, one with a battery and one without, were placed in close proximity. When we changed the current in the circuit with the battery, we observed an induced current in the second circuit. Through electromagnetic induction, electrical energy was transferred from the battery to the second circuit even though there was no direct electrical connection between the two.

transformer

Our two circuits constitute a crude **transformer.** A more practical transformer consists of two coils of wire arranged so that almost all the magnetic flux from one coil is encircled by the second coil. This is often achieved by winding the two coils on an iron core, which increases and concentrates the magnetic flux. Figure 31–29 shows a simplified diagram of a typical transformer, along with the corresponding circuit symbol. Real transformers come in all shapes and sizes, depending on their applications (Fig. 31–30).

To use a transformer, we connect one coil, called the primary, to a time-varying voltage, and the other coil, called the secondary, to an electrical load to which we wish to supply power. The changing current in the primary gives rise to a changing magnetic field. The changing flux from this field passes through the secondary coil, producing an induced emf which drives current through the load.

Why not just connect the load directly to the original voltage source? This would be fine if our load were designed to run at the available voltage. But often we have devices that must run at higher or lower voltages. Most transistor circuits, for example, cannot handle more than a few tens of volts; the 120-V AC power line is too much for them. On the other hand,

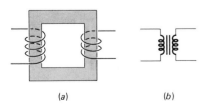

Fig. 31–29
(a) A simple transformer, consisting of two coils wound on an iron core.
(b) Transformer circuit symbol.

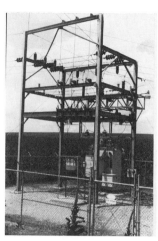

Fig. 31–30
Transformers. (Left to right) (1) Low-power audio-frequency transformers in a stereo amplifier. (2) Transformer in an AC adapter reduces line voltage to a level suitable for battery-powered equipment. (3) Transformer on pole converts 4000-V line voltage to 120 and 240 V for household use. (4) Large transformer at a power substation.

TV picture tubes and neon signs require much higher voltages than the AC line can provide. A transformer transforms a time-varying voltage like the 120-V AC line from one level to another.

The transformer of Fig. 24–29*a* has four turns of wire in its primary coil and two turns in its secondary. Since the same changing magnetic flux passes through both coils, the induced emf in the secondary is exactly half that of the primary. Had we put eight turns in the secondary, we would have doubled the primary voltage. By setting the ratio of turns from primary to secondary, we can make a **step-up** or **step-down** transformer that transforms a given AC voltage to any level we want.

step-up transformer, step-down transformer

Aren't we getting something for nothing with a step-up transformer? No—a step-up transformer increases voltage, but not power. In an ideal transformer, all the power in the primary circuit is transferred to the secondary circuit. The product of voltage and current in the primary is equal to the product of voltage and current in the secondary. If we increase the voltage in the secondary circuit, we decrease the current by the same ratio. Real transformers have losses associated with the resistance of their windings and heating of their iron cores, but with good design these losses amount to only a few per cent of the total power transferred.

A transformer only works with time-varying currents. We cannot increase the voltage of a battery using a transformer, unless we somehow interrupt the current from the battery. One of the main reasons that alternating current is used for electric power is the ease with which AC power is transformed from one voltage level to another. Generators in large power stations typically produce power at about 20 kV. This voltage is stepped up to hundreds of kV for transmission over long distances. Because loss in the power-line resistance is I^2R, it is best to transmit power at high voltage

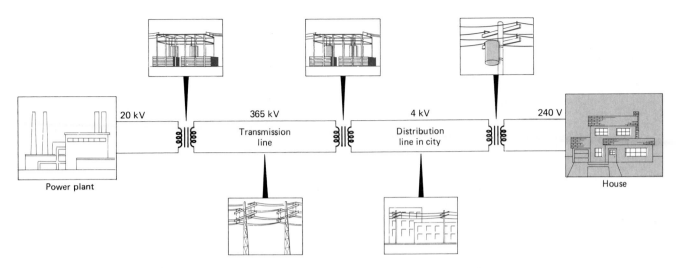

Fig. 31–31
Transformers are used throughout the power distribution network.

and low current to minimize power loss. At a city or town, the voltage is stepped down to several thousand volts for distribution to individual buildings. Transformers near each building reduce the voltage further—for example, to 120 and 240 V for household use. Within a building, individual electrical devices employ transformers to meet particular voltage requirements. Figure 31–31 outlines the voltage transformations in power transmission.

DC Power Supplies

Fig. 31–32
Diode symbol, with preferred current direction indicated.

Devices like motors, light bulbs, and electric heaters often work equally well on AC or DC. But electronic circuits almost always require DC power. How do we provide this power from the AC power line?

diode

In our discussion of transistors in Chapter 25, we considered the *PN* junction—a junction between *P*-type and *N*-type semiconductors with the property that current flows readily in one direction but not the other. A *PN* junction designed specifically to be such a "one-way valve" for electric current is called a **diode.** Figure 31–32 shows the circuit symbol for a diode. An ideal diode acts like a short circuit to current flowing in this preferred direction, and like an open circuit in the opposite direction.

Figure 31–33*a* shows a diode and load resistor connected to a source of AC power. The diode only passes current in one direction, giving the resistor current shown in Fig. 31–33*b*. Frequently we want not only direct current, but also some voltage different from the AC line voltage, so we put a transformer between the power line and our diode (Fig. 31–34). The ar-

power supply

rangement of Fig. 31–34 constitutes a crude **power supply,** designed to transform AC power into a form useful to a particular device. The diode in

rectifier

the power supply acts as a **rectifier,** changing the current from two-way to one-way flow.

Although current from our power supply flows in one direction, it still varies dramatically with time. If we sent this current to a stereo, we would

Fig. 31–33
(a) A diode and load resistor connected to an AC power source. The diode passes current in only one direction, cutting off the negative half of each cycle and giving the resistor current the time dependence shown in (b).

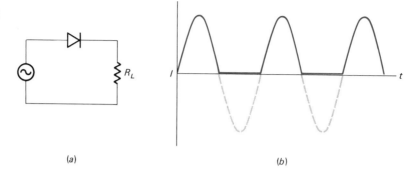

(a)

(b)

filter

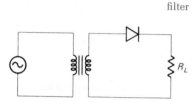

Fig. 31–34
A transformer is used when the desired voltage differs from that of the AC power line.

ripple

hear a loud hum superimposed on the music. We must somehow smooth, or **filter,** the power to produce steady direct current.

The simplest filter is a capacitor connected across the load resistor (Fig. 31–35a). As the voltage on the left-hand side of the diode rises, the capacitor voltage rises rapidly because of the short time constant associated with the low forward resistance of the diode (Fig. 31–35b). But then the AC voltage on the left of the diode begins to fall. The capacitor cannot discharge through the reverse direction of the diode, so its only discharge path is through the load resistor. If the RC time constant of the capacitor and load resistor is long enough, the capacitor voltage drops only slightly as the AC voltage on the left side of the diode falls to zero and then goes negative. When the AC voltage again rises above the capacitor voltage, the diode passes current and the capacitor voltage follows the AC to its peak. The process then repeats, with the resistor voltage and current varying much less than the AC voltage. The capacitor stores charge during times when the AC voltage goes low, maintaining a more steady voltage across the load.

There is still some time variation, or **ripple,** in the voltage and current to the load resistor. But by making the capacitor large—so large that the RC time constant of capacitor and load is huge compared with the period ($\frac{1}{60}$ s) of the AC line—this ripple can be minimized. We define the ripple

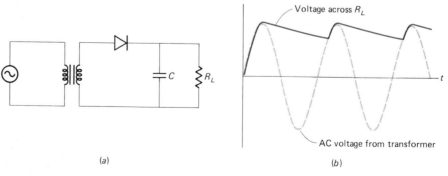

(a)

(b)

Fig. 31–35
(a) A capacitor is added to reduce variations in the power-supply voltage and current.
(b) Resistor voltage as a function of time.

Fig. 31–36
A full-wave bridge rectifier and its
output waveform.

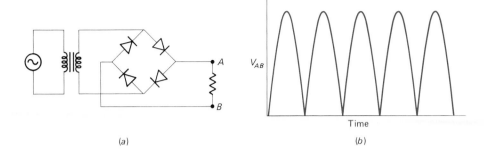

V_{AB}

Time

(a)   (b)

quantitatively as the ratio of the fluctuations in the DC voltage to the DC voltage itself. High-quality power supplies for electronic instruments may achieve ripple factors of 10^{-5} or less. Such supplies often use more complicated filters, involving additional capacitors, inductors, and even transistors. More complicated rectifier circuits, too, can be used to produce a full-wave signal **full-wave signal** that requires less filtering (Fig. 31–36).

SUMMARY

1. **Alternating current** is electric current that varies with time. The simplest and most common alternating current exhibits sinusoidal time variation. A sinusoidal AC signal is characterized by its amplitude, frequency, and phase:

$$I = I_0 \sin(\omega t + \phi).$$

2. In a resistor, the ratio of voltage to current is always constant, so that

$$I_0 = \frac{V_0}{R},$$

and the current and voltage are in phase.

3. In a capacitor the ratio of peak voltage and current is determined by the **capacitive reactance:**

$$I_0 = \frac{V_0}{X_C},$$

where

$$X_C = \frac{1}{\omega C}.$$

The current in the capacitor **leads** the voltage by 90°.

4. In an inductor the ratio of peak voltage and current is determined by the **inductive reactance:**

$$I_0 = \frac{V_0}{X_L},$$

where

$$X_L = \omega L.$$

The current in the inductor **lags** the voltage by 90°.

5. A **phasor diagram** uses vector-like arrows to signify the amplitude and phase of AC signals, and is useful in analyzing AC circuits containing more than one circuit element.

6. In an undriven LC circuit, energy oscillates between electric and magnetic forms at the **resonant frequency** $\omega = 1/\sqrt{LC}$. The amplitude of the oscillation decays exponentially as energy is dissipated in the circuit resistance.

7. In a series RLC circuit, the effects of inductance and capacitance exactly cancel at the resonant frequency. At this frequency the circuit exhibits the minimum **impedance,** and therefore passes the maximum current. At resonance the current and voltage are in phase. At lower frequencies the capacitor dominates and current leads voltage, while at higher frequencies the inductor dominates and voltage leads current. The impedance of a series RLC circuit is

$$Z = \sqrt{R^2 + (X_C - X_L)^2},$$

while the phase difference between current and voltage is given by

$$\cos\phi = \frac{R}{Z}.$$

RLC circuits can be used to select one among many frequencies, as in tuning a radio.

8. The power dissipated in an AC circuit depends on the relative effects of resistance and reactance. In a purely reactive circuit, current and voltage are 90° out of phase, and no power is dissipated. In a purely resistive circuit, the average power dissipation is just the product $I_{rms} V_{rms}$. When both resistance and reac-

tance are present, the power dissipation is determined by the **power factor,** $\cos\phi$, where ϕ is the phase angle between current and voltage. In general, the time-average power dissipated in an AC circuit is

$$P = I_{rms}V_{rms}\cos\phi.$$

9. A **transformer** is an inductive device in which two separate coils encircle the same magnetic flux. A changing current in one coil induces an emf in the other. Transformers are used to change voltage levels in AC circuits.

10. A DC **power supply** uses **diodes** to change alternating to direct current. Time variations are then smoothed with a **filter,** consisting of a capacitor or more complicated networks of capacitors and inductors. Power supplies provide the DC power required by transistor circuitry in electronic devices.

QUESTIONS

1. Two AC signals have the same amplitude but different frequencies. Are their rms amplitudes the same?

2. Does it make sense to talk about the phase difference between two AC signals of different frequencies? Sketch a diagram to confirm your answer.

3. What is meant by the statement "a capacitor is a DC open circuit"?

4. How can current keep flowing in an AC circuit containing a capacitor? After all, a capacitor contains a gap between two conductors, and no charge can cross this gap.

5. Why does it make sense that inductive reactance increases with frequency?

6. The same AC voltage appears across a capacitor and a resistor, and the same rms current flows into each. Is the power dissipation the same in each?

7. Two AC signals have voltages that are exactly opposite at every instant (that is, when one is at $+V$, the other is at $-V$). What is the phase difference between the two signals?

8. When a particular inductor and capacitor are connected across the same AC voltage, the current in the inductor is larger than in the capacitor. Will this be true at all frequencies?

9. An inductor and capacitor are connected in series across an AC generator, and the rms voltage across the inductor is found to be larger than that across the capacitor. Is the generator frequency above or below resonance?

10. When the capacitor voltage in an undriven LC circuit reaches zero, why don't the oscillations stop?

11. Why is the quantity ωL not called the resistance of an inductor?

12. Does an ideal inductor have resistance? Does a real inductor?

13. Why is Equation 31–6 not a full description of the relation between voltage and current in a capacitor? What equation does give the full relation?

14. If you double both the capacitance and inductance in an LC circuit, what effect does this have on the resonant frequency?

15. The capacitance of the variable capacitor in a radio varies approximately linearly with the angular position of the capacitor's rotating shaft. Why, then, are the frequency markings on the AM radio dial not evenly spaced? The AM band covers about 540 kHz to 1600 kHz. The FM band markings, covering about 88 MHz to 108 MHz, are more evenly spaced. Why?

16. In a series RLC circuit, the applied voltage lags the current. Is the frequency above or below resonance?

17. In a series RLC circuit, the applied voltage leads the current. Is the peak voltage greater across the capacitor or the inductor?

18. At a certain frequency, the impedance of a series RLC circuit is twice the resistance of the circuit. Can you tell whether the frequency is above or below resonance? Which is it, or why can't you tell?

19. We say that the capacitor in a driven RLC circuit dominates at low frequencies. What does this mean?

20. The voltage across two circuit elements in series is zero. Is it possible that the voltages across the individual elements are nevertheless not zero? Give an example.

21. If you measure the rms voltages across the resistor, capacitor, and inductor in a series RLC circuit, will they add to the rms value of the generator voltage? Reconcile your answer with the loop law. (See also Problem 27.)

22. In a fluorescent light fixture an inductor, called the ballast, is used to limit current to the lamp. Why is an inductor preferable to a resistor?

23. In a series RLC circuit, the power factor is 1. How does the frequency of the applied voltage compare with the resonant frequency?

24. What is the power factor in a circuit containing only a resistor? How does this power factor change with frequency?

25. To save electrical energy, should you strive for a large or small power factor?

26. When an AC motor runs with no significant mechanical load, its power factor is very nearly zero. What happens when the motor begins doing mechanical work?

27. Two DC power supplies are identical except that one has a larger filter capacitor. Which can supply the most current before the ripple reaches 1%?

28. Why is it easier to filter the output of a full-wave rectifier than of a half-wave rectifier?

29. A step-up transformer increases voltage, or energy per unit charge. How is this possible without violating conservation of energy?

30. The iron cores of transformers are often made by laminating together thin sheets of iron separated by an insulating substance. Why is this preferable to making the cores out of solid pieces of iron? *Hint:* Think about eddy currents.

31. A battery charger runs off the 120-V AC power line. It supplies up to 30 A to recharge a 12-V car battery, yet it can be plugged into a 15-A circuit without blowing the circuit breaker. How is this possible?

32. Manuals for electronic instruments that run off the AC power line, including stereo amplifiers, often caution against connecting the instrument to DC power sources. Why? *Hint:* What would happen if a DC voltage were imposed across the transformer in such an instrument?

PROBLEMS

Section 31.1 *Alternating Current*

1. What are the angular frequency and peak voltage of the 120-V rms, 60-Hz AC power line?

2. A 10-V rms AC signal with a frequency of 1000 kHz is 90° ahead of another signal whose frequency is the same but whose peak voltage is 7.1 V. On a single graph, plot voltage versus time for both these signals. Indicate values of voltage and time on your axes.

3. To show the effect of phase constant on an AC signal, plot on a single graph the functions

$$f(\omega t) = \sin(\omega t + \phi)$$

for values of ϕ including 0, 30°, 60°, 90°, and 120°. Let ωt range from 0 to 360° for each plot.

4. The rms amplitude of a time-dependent signal is defined as the square root of the average of the square of the signal. For a periodic signal the time average is given by the integral of the signal over one period, divided by the period. Consider the signal

$$V = V_0 \sin\omega t,$$

and show that $V_{rms} = V_0/\sqrt{2}$.

5. Figure 31–37 shows a square wave. How are the rms and peak voltages related for this signal? (See Problem 4.)

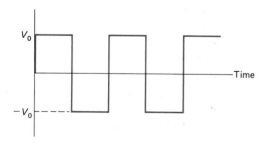

Fig. 31–37 A square wave (Problem 5).

6. Figure 31–38 shows a triangle wave. How are the rms and peak voltages related for this signal? (See Problem 4.)

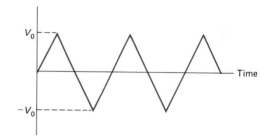

Fig. 31–38 A triangle wave (Problem 6).

7. The most general expression for a sinusoidally oscillating current may be written either

$$I = I_1 \sin\omega t + I_2 \cos\omega t,$$

or

$$I = I_0 \sin(\omega t + \phi).$$

Find relations between I_0, ϕ, I_1, and I_2 that make these two expressions equivalent. (See Appendix A for trig identities.)

Section 31.2 *Circuit Elements in AC Circuits*

8. Use the fundamental definitions of inductance and capacitance to show that inductive and capacitive reactance have the units of ohms.

9. A 1.0-μF capacitor is connected across the 120-V rms, 60-Hz AC line. What is the rms current?

10. A 5.0-mH inductor is connected across a 10-V rms AC generator, and an rms current of 2.0 mA flows. What is the frequency of the AC voltage?

11. A 2.0-μF capacitor has a capacitive reactance of 1000 Ω. (a) What is the frequency of the AC voltage applied to the capacitor? (b) What value of inductance would be needed to get the same inductive reactance at this frequency? (c) If the frequency were doubled, how would the reactances of the capacitor and inductor then compare?

12. A 0.75-H inductor is used in series with a fluorescent lamp to limit the current. The series combination is connected across the 120-V, 60-Hz AC power line. If

the voltage drop across the lamp itself is 30 V rms, what is the rms current in the lamp?

Section 31.3   LC Circuits

13. You have a 2.0-mH inductor and wish to make an *LC* circuit whose resonant frequency can be tuned across the AM radio band (550 kHz to 1600 kHz). What range of capacitance should your variable capacitor cover?

14. In Example 31–3, we obtained the maximum current by differentiating the capacitor charge. Show that the same result is obtained by equating the maximum energy in the capacitor to the maximum energy in the inductor.

15. The FM radio band covers the frequency range from 88 MHz to 108 MHz. The variable capacitor in a certain FM receiver ranges from 10.9 pF to 16.4 pF. What value of inductance is used with the capacitor to make an *LC* circuit whose resonant frequency covers the entire FM band?

16. In an *LC* circuit, what fraction of a cycle must pass before the energy in the capacitor falls from its peak value to one-fourth of its peak value?

17. The 2000-μF capacitor in the circuit of Fig. 31–39 is initially charged to 200 V. Describe how you would manipulate the switches *A* and *B* in order to charge the 500-μF capacitor to 400 V. Include in your description the times at which you would open and close the switches.

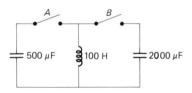

Fig. 31–39   Problem 17

18. One-eighth of a cycle after the capacitor in an *LC* circuit is fully charged, what are each of the following as fractions of their peak values: (a) the charge on the capacitor; (b) the energy in the capacitor; (c) the current in the inductor; (d) the energy in the inductor?

19. An *LC* circuit contains a 20-μF capacitor and has a period of 5.0 ms. The peak current in the circuit is 25 mA. (a) What is the value of the inductance? (b) What is the peak voltage across the capacitor?

20. A damped *LC* circuit consists of a 0.15-μF capacitor and a 20-mH inductor. The resistance of the inductor is 1.6 Ω. For how many cycles will the circuit oscillate before the peak voltage on the capacitor drops to half its initial value?

21. A damped *RLC* circuit includes a 5.0-Ω resistor and a 100-mH inductor. If half the initial energy in the circuit is lost after 15 cycles, what is the capacitance of the circuit?

Section 31.4   Driven RLC Circuits and Resonance

22. A series *RLC* circuit consists of a 500-pF capacitor, a 1.0-mH inductor, and a 2.0-kΩ resistor. (a) Obtain an expression for the impedance of this circuit as a function of frequency. (b) Assume that the *RLC* circuit is connected across a 10-V rms AC generator. Plot the rms current in the circuit as a function of generator frequency. Over what frequency range does the current have more than half the value it has at resonance?

23. If the *RLC* circuit of Example 31–4 is driven at resonance by a 24-V peak AC generator, what is the peak voltage across the capacitor?

24. A series *RLC* circuit has $R = 75$ Ω, $L = 20$ mH, and a resonant frequency of 4.0 kHz. (a) What is the capacitance? (b) What is the impedance of the circuit at resonance? (c) What is the impedance of the circuit at 3.0 kHz?

25. TV channel 2 occupies the frequency range from 54 MHz to 60 MHz. A series *RLC* tuning circuit in a certain TV contains an 18-pF capacitor. (a) When the TV is tuned to channel 2, the *RLC* circuit resonates in the middle of the channel 2 band. What, then, is the inductance? (b) To let the whole TV signal in, the resonance curve of the *RLC* circuit must be broad enough to let signals in the entire band through without much attenuation. Suppose that the current at the edges of the band is to be no less than 70% of the current at resonance. What constraints does this place on the circuit resistance?

26. An *RLC* circuit consists of a 10-Ω resistor, 1.5-μF capacitor, and 50-mH inductor. The circuit is driven by an AC source whose peak output voltage is 100 V. The capacitor is rated for a maximum voltage of 1500 V. (a) What peak voltage would appear across the capacitor at resonance? (b) Make a graph of the peak capacitor voltage as a function of frequency, and from it determine the frequency range(s) over which the circuit may be operated safely with the 100-V peak applied voltage.

27. A 2.0-H inductor and 3.5-μF capacitor are connected in series with a 50-Ω resistor and the combination is connected to an AC generator supplying 24 V peak at 60 Hz. (a) At the instant that the applied voltage is at its peak, what is the instantaneous voltage across each of the circuit elements? Show explicitly that these voltages sum to the applied voltage. (b) If rms-reading voltmeters are connected across each of the three circuit elements, what will each read? Do their readings sum to the rms value of the applied voltage? Does this contradict the loop law?

28. Use the expression for the resonant frequency of an *LC* circuit to show that the impedance of an *RLC* circuit can be written

$$Z = \sqrt{R^2 + \omega^2 L^2 (1 - \omega_r^2/\omega^2)^2},$$

where ω_r is the resonant frequency.

29. Figure 31–40 shows the phasor diagram for an *RLC* circuit. (*a*) Is the circuit being driven above or below resonance? (*b*) Complete the diagram by adding a phasor for the applied voltage, and from your completed diagram determine the phase difference between the applied voltage and the current.

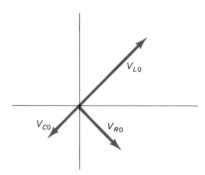

Fig. 31–40 Problem 29

30. The resistance of a series *RLC* circuit is such that the peak current at half the resonant frequency is half the peak current at resonance. (*a*) Show that the peak current at twice the resonant frequency is also half that at resonance. (*b*) Sketch phasor diagrams showing the current, voltages across the resistor, capacitor, and inductor, and applied voltage for the resonant frequency and for half and twice the resonant frequency.

Section 31.5 *Power in AC Circuits*

31. A series *RLC* circuit has a resistance of 100 Ω and an impedance of 300 Ω. (*a*) What is the power factor? (*b*) If the rms current is 200 mA, how much power is dissipated in the circuit?
32. A series *RLC* circuit consists of a 10-Ω resistor, 2.0-μF capacitor, and 500-mH inductor. It is connected to an AC source supplying 80 V rms. (*a*) What is the power factor when the applied voltage alternates at half the resonant frequency of the *RLC* circuit? (*b*) At twice the resonant frequency? (*c*) How much power does the circuit dissipate at half the resonant frequency? (*d*) At the resonant frequency? (*e*) At twice the resonant frequency?
33. A power plant supplies 60-Hz power at 365 kV rms and 200 A rms to a small city. The resistance of the transmission line leading to the city is 100 Ω. (*a*) What per cent of the power is lost in transmission if the power factor of the city is 1.0? (*b*) What per cent of the power is lost in transmission if the power factor of the city is 0.60? (*c*) Is it more economical for the power plant operators if their plant supplies a load with a large power factor or with a small one? Why?
34. A series *RLC* circuit has a power factor of 0.80 at 60 Hz. The impedance of the circuit is 100 Ω. (*a*) What is the resistance of the circuit? (*b*) If the inductance is 0.10 H, what is the resonant frequency?

Section 31.6 *Transformers and Power Supplies*

35. A car battery charger supplies 10 A at 14 V to recharge a 12-V car battery. The charger runs off the 120-V AC power line. Assume that there is no power loss in the charger itself. (*a*) How much power does the charger draw from the AC line? (*b*) How much current does the charger draw from the AC line? (*c*) If the battery stores 80% of the electrical energy coming into it, how much energy does the battery gain if the charger is connected to it for 10 hours? (*d*) If electricity costs 9¢ per kilowatt-hour, how much does it cost to run the charger for 10 hours?
36. For the power supply of Fig. 31–35*a*, the output of the transformer is 12 V rms at 60 Hz, and the filter capacitance is 50 μF. (*a*) With no load across the power supply, what is the DC output voltage? (*b*) What is the minimum load resistance if the ripple is not to exceed 1%?
37. The power supply shown in Fig. 31–41 delivers 28 V DC at 100 mA to its load. (*a*) What is the load resistance? (*b*) What must be the value of the capacitance if the DC voltage across the load is to vary by less than 1%? Assume that the AC voltage from the transformer is a 28-V peak, 60-Hz sine wave.

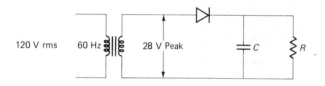

Fig. 31–41 Problem 37

Supplementary Problems

38. For *RLC* circuits in which the resistance is not too large, the *Q* factor may be defined as the ratio of the resonant frequency to the difference between the two frequencies where the power dissipated in the circuit is half that dissipated at resonance. Show, using suitable approximations, that this definition leads to the expression

$$Q = \frac{\omega_r}{\Delta\omega} = \frac{\omega_r L}{R},$$

where ω_r is the resonant frequency.

39. Figure 31–42 shows a parallel *LC* circuit. (*a*) Make

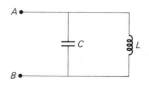

Fig. 31–42 A parallel *LC* circuit (Problem 39).

simple arguments to show that the impedance measured between points A and B should be zero in the limits of very low and very high frequency. (b) Using the fact that the voltage across parallel circuit elements is the same, show that the impedance is infinite at resonance.

40. Figure 31–43 shows a parallel LC circuit containing a resistance R. The circuit is at resonance when the current in the LR branch is equal in magnitude to the capacitor current but 180° out of phase. (a) Construct a phasor diagram showing the relevant currents, and from it show that the impedance measured between points A and B is given by

$$\frac{1}{Z} = \frac{1}{X_C} + \frac{1}{\sqrt{R^2+X_L^2}}.$$

(b) This circuit is at resonance when the current supplied by the driving voltage is at a minimum. Show that resonance occurs at a frequency given by

$$\omega_r^2 = \frac{1}{LC} - \frac{R^2}{L^2}.$$

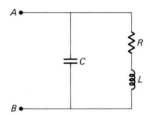

Fig. 31–43   Problem 40

41. Consider a series circuit containing an AC generator, a resistor, and a capacitor. Construct a phasor diagram, and derive expressions for the circuit impedance and the phase angle between the applied voltage and the current. Show that the current always leads the voltage.

42. Consider a series circuit containing an AC generator, a resistor, and an inductor. Construct a phasor diagram, and derive expressions for the circuit impedance and the phase angle between the applied voltage and the current. Show that the voltage always leads the current.

43. You wish to make a "black box" with two input connections and two output connections, as shown in Fig. 31–44. When you put a 12-V rms, 60-Hz signal across the input, a 6.0-V rms, 60-Hz signal should appear at the output, and the output voltage should lead the input voltage by 45°. Design a circuit that could be used in the "black box."

Fig. 31–44   Problem 43

44. A 2.5-H inductor is connected across a 1500-μF capacitor. A 5.0-kg mass is connected to a spring. What should the spring constant be if the mass-spring system and LC circuit are to have the same resonant frequency?

45. Problem 35 of Chapter 30 gives the inductance per unit length of a pair of long, flat, parallel conductors. (a) Calculate the capacitance per unit length of such a system. *Hint:* Consult Chapter 24. (b) Use your results to determine the resonant frequency of a system of such conductors, each 1.0 cm wide, spaced 1.0 mm apart, and 1.0 m long.

46. A piece of coaxial cable has an inner conductor of diameter 1.2 mm and an outer conductor of diameter 1.0 cm. The conductors are separated by polyethylene insulation. (a) Calculate the resonant frequency of a 5.0-m-long piece of this cable. *Hint:* Consult Example 24–5, Table 24–1, and Problem 30–11. (b) Could resonance in such a cable cause any problems? If so, what?

47. A bar magnet of mass m is suspended at rest inside a solenoid of inductance L by a spring of spring constant k, as shown in Fig. 31–45. The solenoid is connected through a switch to a charged capacitor C. (a) What relation must hold between L, C, m, and k for the mechanical and electrical systems to have the same resonant frequency? (b) Suppose that the relation of part (a) is met. Speculate on what will happen when the switch is closed.

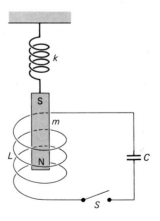

Fig. 31–45   Problem 47

32

MAXWELL'S EQUATIONS AND ELECTROMAGNETIC WAVES

At this point you have seen all four of the fundamental laws of electromagnetism. You understand that Gauss's law for electricity, Gauss's law for magnetism, Ampère's law, and Faraday's law together govern the behavior of electromagnetic fields throughout the universe. You have seen how these four laws determine the electric and magnetic interactions that make matter act as it does. You have examined many practical devices that exploit the laws of electromagnetism.

32.1
THE FOUR LAWS OF ELECTROMAGNETISM

Table 32–1 summarizes the four laws as you now understand them. As the title of the table implies, these laws do not yet provide a full description of electromagnetic phenomena, for our statement of Ampère's law is still valid only for steady currents.

TABLE 32–1
Four Laws of Electromagnetism (still incomplete)

Law	Mathematical Statement	What It Says
Gauss for **E**	$\oint \mathbf{E} \cdot d\mathbf{A} = \dfrac{q}{\epsilon_0}$	How charges give rise to electric field; field lines begin and end on charges
Gauss for **B**	$\oint \mathbf{B} \cdot d\mathbf{A} = 0$	No magnetic charge; magnetic field lines do not begin or end
Faraday	$\oint \mathbf{E} \cdot d\boldsymbol{\ell} = -\dfrac{d\phi_B}{dt}$	Changing magnetic field gives rise to electric field
Ampère (steady currents only)	$\oint \mathbf{B} \cdot d\boldsymbol{\ell} = \mu_0 I$	Electric current gives rise to magnetic field

As you look at these four laws together, you can't help but notice some striking similarities. On the left-hand sides of the equations, the two laws of Gauss are identical but for the interchanging of **E** and **B**. Similarly, the laws of Ampère and Faraday have left-hand sides that differ only in the interchange of **E** and **B.**

On the right-hand sides, things are rather more different. Gauss's law for electricity involves the charge enclosed by the surface of integration, while Gauss's law for magnetism has zero on the right-hand side. Actually, though, these laws are similar. Since we have no conclusive evidence for the existence of isolated magnetic charge, the enclosed magnetic charge on the right-hand side of Gauss's law for magnetism is zero. If and when magnetic monopoles are discovered, then the right-hand side of Gauss's law for magnetism would be nonzero for any surface enclosing net magnetic charge.

The right-hand sides of Ampère's and Faraday's laws are distinctly different. In Ampère's law we find the current—the flow of electric charge—as a source of the magnetic field. We can understand the absence of a similar term in Faraday's law because we have never observed a flow of magnetic monopoles. If we had such a flow, then we would expect this magnetic current to produce an electric field encircling the magnetic current.

Two of the differences among the four laws of electromagnetism would be resolved if we knew for sure that magnetic monopoles exist. That current theories of elementary particles suggest the existence of monopoles is a tantalizing hint that there may be a fuller symmetry between electric and magnetic phenomena. The search for symmetry, based not on logic or experimental evidence but on an intuitive sense that nature should be simple, has motivated some of the most important discoveries in physics.

32.2
AMBIGUITY IN AMPÈRE'S LAW

There remains one difference between the equations of electricity and magnetism that would not be resolved by the discovery of magnetic monopoles. On the right-hand side of Faraday's law we find the term $d\phi_B/dt$ that describes changing magnetic field as a source of electric field. We find no comparable term in Ampère's law. Are we missing something? The near symmetry of the equations suggests that perhaps we are. Is it possible that a changing electric field produces a magnetic field? So far, we have described no experimental evidence for such a conjecture. It is suggested only by our sense that the near symmetry between electricity and magnetism is not a coincidence, but a reflection of an underlying physical relation. If a changing electric field did produce a magnetic field, just as a changing magnetic field produces an electric field, then we would expect a term $d\phi_E/dt$ on the right-hand side of Ampère's law.

In our statement of Ampère's law in Chapter 28, we emphasized that the law applied only to steady currents—those that never change. The reason for this restriction is suggested in Fig. 32–1, which shows a simple RC circuit. While the capacitor charges, there is a current in the circuit that decreases with time. We expect that this current will set up a magnetic

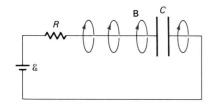

Fig. 32–1
A charging RC circuit, showing some magnetic field lines surrounding the current-carrying wires.

field, as suggested in the figure. Let us apply Ampère's law to calculate this field.

If the wire between resistor and capacitor is long and straight, then the magnetic field lines should be circles around the wire. Ampère's law tells us that the line integral around a field line (or, for that matter, around any closed loop) is proportional to the encircled current:

$$\oint \mathbf{B}\cdot d\boldsymbol{\ell} = \mu_0 I. \qquad (32\text{--}1)$$

By the encircled current, we mean the current through *any* surface bounded by the loop. Figure 32–2 shows four such surfaces. The same current flows through surfaces 1, 2, and 4. But there is no current through surface 3, which passes between the capacitor plates. So surface 3 does not give the same answer for the magnetic field. Ampère's law is ambiguous here; it does not give us a definitive answer for the field.

This problem does not arise with steady currents, for then capacitors are neither charging nor discharging, and there is no current in any wire in series with a capacitor. In particular, the only steady current possible in the circuit of Fig. 32–2 is zero current.* Then the current through all four surfaces is the same, and there is no ambiguity in the application of Ampère's law. But when currents are changing there may be situations like that of Fig. 32–2 where Ampère's law becomes ambiguous. That is why the form of Ampère's law we have used until now is valid only for steady currents.

Can we salvage Ampère's law? Is there any way to extend the law to unsteady currents without altering its validity in the steady case? Symmetry between Ampère's and Faraday's laws has already suggested that a changing electric flux might give rise to a magnetic field. Between the plates of our charging capacitor, we have a changing electric field, as shown in Fig. 32–3. This changing field means that there is a changing electric flux through surface 3 of Fig. 32–2.

It was the Scottish physicist James Clerk Maxwell (Fig. 32–4) who, about 1860, suggested that a changing electric flux should give rise to a magnetic field. Since that time many experiments, including direct measurement of the magnetic field associated with a huge capacitor, have confirmed Maxwell's remarkable insight. To quantify his idea, Maxwell introduced a new term into Ampère's law, writing

$$\oint \mathbf{B}\cdot d\boldsymbol{\ell} = \mu_0 I + \mu_0 \epsilon_0 \frac{d\phi_E}{dt}. \qquad (32\text{--}2)$$

Now there is no ambiguity. The integral is taken around any loop, I is the current through *any* surface bounded by the loop, and ϕ_E is the electric flux through that surface. With our charging capacitor, Equation 32–2 gives the same magnetic field no matter which surface we choose. For surfaces

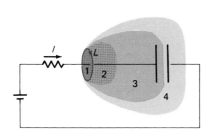

Fig. 32–2
Ampère's law relates the line integral around the loop *L* to the current through any surface bounded by the loop. There is no current through surface 3, so that Ampère's law is ambiguous.

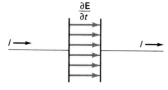

Fig. 32–3
There is a changing electric field in the charging capacitor, and therefore a changing electric flux through surface 3 of Fig. 32–2.

---

*You might argue that we could produce a steady current to a capacitor by steadily increasing the applied emf. Indeed we could—but we could not do so *forever* without requiring infinite energy. That is the reason why we have stressed that by steady current we mean current that *never* changes. Then a current that is steady for only a finite time does not strictly count as a steady current.

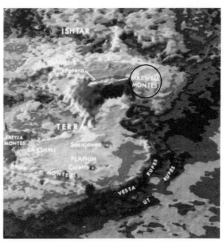

Fig. 32–4
(Left) James Clerk Maxwell. (Right) Maxwell's contributions to electromagnetic theory were honored recently by naming a mountain on Venus for him; this image showing the mountain was prepared from radar data, obtained by reflection of electromagnetic waves beamed from a satellite in Venus orbit to the Venusian surface and back.

1, 2, and 4 of Fig. 32–2, the current I makes all the contribution to the right-hand side of the equation (here we assume that the electric field outside the capacitor is zero). For surface 3, through which no current flows, the right-hand side of Equation 32–2 comes entirely from the changing electric flux. You can readily verify that the term $\epsilon_0 d\phi_E/dt$ has the units of current, and that, for the charging capacitor, this term is numerically equal to the current I (see Problem 3). Although the changing electric flux is not an electric current, it has the same effect as a current in producing a magnetic field. For this reason Maxwell called the term $\epsilon_0 d\phi_E/dt$ the **displacement current.** The word "displacement" has historical roots that do not provide much physical insight. But the word "current" is meaningful in that the effect of displacement current is indistinguishable from that of real current in producing magnetic fields.

displacement current

Example 32–1
Displacement Current

A parallel-plate capacitor with circular plates of radius R and spacing d is fed by long, straight wires as shown in Fig. 32–5. If the potential difference between the plates is increasing at the rate dV/dt, what is the magnetic field as a function of position between the plates?

Solution

With long, straight feed wires, the situation has cylindrical symmetry, so the magnetic field can depend only on the distance r from the symmetry axis. The only magnetic field lines that exhibit this symmetry and satisfy Gauss's law for magnetism are circles, as shown in Fig. 32–6. A magnetic field line within the capacitor encircles no current—no flow of electric charge—but it does encircle a changing electric field, and therefore there is a displacement current through a flat circular surface bounded by the field line. If the field line has radius r, the electric flux encircled is given by

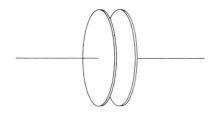

Fig. 32–5
Example 32–1

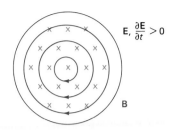

Fig. 32–6
Electric and magnetic field lines between the circular capacitor plates.

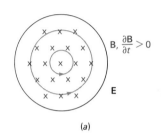

(a)

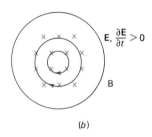

(b)

Fig. 32–7
(a) Applying Faraday's law to an electric field arising from a changing magnetic field gives the field direction shown. (b) Since the changing flux term—the displacement current—enters Ampère's law with a positive sign, the magnetic field induced by a changing electric flux has the opposite direction.

Maxwell's equations

$$\phi_E = \int \mathbf{E} \cdot d\mathbf{A} = \pi r^2 E = \pi r^2 \frac{V}{d},$$

where the uniformity of the field allows us to calculate the field as the ratio of potential difference to plate spacing, and the flux as a simple product of field and area. Then the displacement current is

$$I_D = \mu_0 \epsilon_0 \frac{d\phi_E}{dt} = \frac{\mu_0 \epsilon_0 \pi r^2}{d} \frac{dV}{dt}.$$

With cylindrical symmetry, the line integral on the left-hand side of Ampère's law becomes

$$\oint \mathbf{B} \cdot d\boldsymbol{\ell} = 2\pi r B.$$

Equating this quantity to μ_0 times the encircled displacement current gives

$$2\pi r B = \frac{\mu_0 \epsilon_0 \pi r^2}{d} \frac{dV}{dt},$$

so that

$$B = \frac{\mu_0 \epsilon_0 r}{2d} \frac{dV}{dt}.$$

This field, with its magnitude increasing linearly with r, should remind you of the magnetic field inside a cylindrical wire (see Example 28–4). Problem 2 extends this calculation to the field outside the capacitor.

What about the direction of the induced magnetic field? When we determined the direction of induced *electric* fields, we emphasized the use of energy conservation, reflected in the minus sign on the right-hand side of Faraday's law. In Ampère's law, the displacement current term appears with a positive sign, so that our induced magnetic field has direction opposite to that of an induced electric field that would arise in the analogous situation with a changing magnetic field (Fig. 32–7). Does this violate energy conservation? No—because the induced magnetic field does no work on electrically charged particles, so the induced field need not oppose the effect giving rise to it.

32.3
MAXWELL'S EQUATIONS

It was Maxwell's genius to recognize that Ampère's law should be modified to reflect the symmetry suggested by Faraday's law. The physical consequences of Maxwell's discovery go far beyond anything he could have imagined. To honor Maxwell, the four complete laws of electromagnetism are given the collective name **Maxwell's equations.** This full and complete set of equations, first published by Maxwell in 1864, governs the behavior of electric and magnetic fields everywhere. Table 32–2 summarizes Maxwell's equations.

These four simple, compact statements are all it takes to describe electromagnetic phenomena. Everything electric or magnetic that we have considered and will consider—from polar molecules to electric current; resistors, capacitors, inductors, and transistors; solar flares and cell membranes; electric generators and thunderstorms; computers and TV sets; the northern lights and fusion reactors—all these can be described using Maxwell's

TABLE 32–2 ■
Maxwell's Equations

Law	Mathematical Statement	What It Says	Equation Number
Gauss for **E**	$\oint \mathbf{E} \cdot d\mathbf{A} = \dfrac{q}{\epsilon_0}$	How charges give rise to electric field; field lines begin and end on charges	(32–3)
Gauss for **B**	$\oint \mathbf{B} \cdot d\mathbf{A} = 0$	No magnetic charge; magnetic field lines do not begin or end	(32–4)
Faraday	$\oint \mathbf{E} \cdot d\boldsymbol{\ell} = -\dfrac{d\phi_B}{dt}$	Changing magnetic field gives rise to electric field	(32–5)
Ampère	$\oint \mathbf{B} \cdot d\boldsymbol{\ell} = \mu_0 I + \epsilon_0\mu_0\dfrac{d\phi_E}{dt}$	Electric current and changing electric field give rise to magnetic field	(32–6)

equations. And despite this wealth of phenomena, we have yet to discuss one of the most important manifestations of electromagnetic fields.

Maxwell's Equations in Vacuum

Consider Maxwell's equations in a region free of any matter—in vacuum. We have learned enough about electromagnetism to anticipate that the fields themselves will still be able to interact, change, and carry energy even in the absence of matter. To express Maxwell's equations in vacuum, we simply remove all reference to matter—that is, to electric charge:

$$\oint \mathbf{E} \cdot d\mathbf{A} = 0 \tag{32–7}$$

$$\oint \mathbf{B} \cdot d\mathbf{A} = 0 \tag{32–8}$$

$$\oint \mathbf{E} \cdot d\boldsymbol{\ell} = -\frac{d\phi_B}{dt} \tag{32–9}$$

$$\oint \mathbf{B} \cdot d\boldsymbol{\ell} = \mu_0\epsilon_0\frac{d\phi_E}{dt}. \tag{32–10}$$

In vacuum the symmetry is complete, in that electric and magnetic fields enter on an equal footing.* The only source of each field is a change in the other field.

32.4
ELECTROMAGNETIC WAVES

Faraday's law—Equation 32–9—shows that a changing magnetic field gives rise to an electric field. In general, this induced electric field is itself changing with time. Ampère's law—Equation 32–10—shows that a changing

---

*The appearance of the constants ϵ_0 and μ_0 in Ampère's law but not in Faraday's law is an accident of our choice of units. That Faraday's law contains a minus sign while Ampère's does not is actually a symmetry, reflecting the complementary way in which electric and magnetic fields give rise to each other.

electric field gives rise to a magnetic field, which itself may be changing with time. Together, the two laws suggest the possibility of self-sustaining electromagnetic fields, in which a change in the electric field continually gives rise to the magnetic field, and vice versa. We will now show, directly from Maxwell's equations, that such fields are indeed possible, and that they are among the most important phenomena in our universe. When we are finished, we will have grasped the nature of light and provided a starting point for studying the theory of relativity.

In Chapter 14, we considered waves moving along strings, and showed that they could be described by the **wave equation:**

wave equation

$$\frac{\partial^2 y}{\partial x^2} - \frac{1}{v^2}\frac{\partial^2 y}{\partial t^2} = 0, \tag{32–11}$$

where y is the displacement of the string, x the position along the string, and t the time. The symbol ∂ indicates a partial derivative—a derivative taken with respect to one of the independent variables (x,t) while the other is held constant. v is the wave speed, which depends on the string tension and mass per unit length.

We found that one solution to the wave equation has the form

$$y = y_0\sin(kx - \omega t), \tag{32–12}$$

where ω is the angular frequency and k the wave number. These are related to the more familiar frequency f and wavelength λ by the relations

$$\omega = 2\pi f \tag{32–13}$$

and

$$k = \frac{2\pi}{\lambda}. \tag{32–14}$$

We will now show that electric and magnetic fields can satisfy the wave equation. We will do this by demonstrating that fields similar in form to Equation 32–12 satisfy Maxwell's equations. Although our demonstration is highly mathematical, its significance is profound, for it implies the existence of electromagnetic waves—structures consisting of electric and magnetic fields that travel freely through empty space.

Our electromagnetic wave consists of electric and magnetic fields oriented at right angles to each other and to the direction of wave motion, as suggested in Fig. 32–8. The electric field points in the y-direction, the magnetic field points in the z-direction, and the wave travels in the x-direction. In Fig. 32–8 the field lines are pictured only on the surface of a rectangular slab. We must imagine these lines continuing straight forever in the y- and z-directions, giving a wave whose properties do not vary with y or z. Such a wave, whose properties are independent of position in planes perpendicular to the propagation direction, is called a **plane wave.** Mathematically, our plane wave fields are described by

plane wave

$$\mathbf{E} = E_0\sin(kx - \omega t)\,\hat{\mathbf{j}} \tag{32–15}$$

and

$$\mathbf{B} = B_0\sin(kx - \omega t)\,\hat{\mathbf{k}}, \tag{32–16}$$

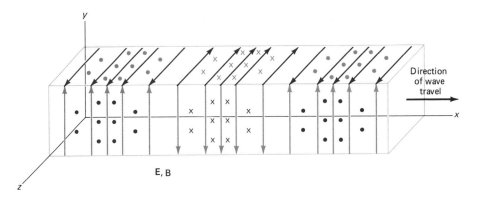

Fig. 32–8
Rectangular slab showing fields of a plane electromagnetic wave. The electric field (gray) is induced by the changing magnetic field (color), and vice versa.

where $\hat{\jmath}$ and $\hat{\mathbf{k}}$ are unit vectors in the y- and z-directions. In Fig. 32–8 the variation in spacing of the field lines, and their reversal from one region of densely spaced lines to another, reflect the sinusoidal dependence of the wave fields on position in the x-direction. We chose a sinusoidal wave shape because we are familiar with such waves from Chapter 14 and because a specific mathematical form will make our derivation more concrete. But we emphasize that here, as in Chapter 14, the wave equation admits as solutions *any* functions of $x \pm vt$ or, equivalently, $kx \pm \omega t$ (see Problem 6 in this chapter and Problem 16 in Chapter 14).

We now show that the electric and magnetic fields pictured in Fig. 32–8 and described by Equations 32–15 and 32–16 do indeed satisfy Maxwell's equations. Note first that our field lines continue forever, with no beginnings or ends; therefore Gauss's laws for electricity and magnetism in vacuum (Equations 32–7 and 32–8) are satisfied.

To see that Faraday's law is satisfied, consider an observer looking directly toward the x-y plane in Fig. 32–8. Such an observer would see electric field lines going up and down and magnetic field lines coming straight in and out, as shown in Fig. 32–9. Consider the small rectangular loop of height h and infinitesimal width dx shown in the figure. Evaluating the line

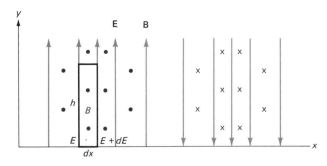

Fig. 32–9
Cross section of the wave slab of Fig. 32–8, showing the x-y plane.

integral of the electric field **E** around this loop, we get no contribution from the short ends because they are at right angles to the field. Going around counterclockwise, we get a contribution $-Eh$ as we go down the left side against the field direction. Then we get a positive contribution going up the right side. Because of the variation in field strength with position, the field strength on the right side of the loop is different from that on the left. Let the change in field be dE, so that the field on the right side of the loop is $E+dE$, giving a contribution of $(E+dE)h$ to the line integral. Then the line integral of **E** around the loop is

$$\oint \mathbf{E}\cdot d\ell \ = \ -Eh \ + \ (E+dE)h \ = \ h\,dE. \qquad (32\text{--}17)$$

Physically, this nonzero line integral means that we are dealing with an induced electric field. Induced by what? By a changing magnetic flux through the loop. The electric field of the wave arises because of the changing magnetic field of the wave. The area of the loop is $h\,dx$, and the magnetic field **B** is at right angles to this area, so that the magnetic flux through the loop is just

$$\phi_B \ = \ Bh\,dx.$$

The rate of change of flux through the loop is then

$$\frac{d\phi_B}{dt} \ = \ h\,dx\,\frac{dB}{dt}.$$

Faraday's law relates the line integral of the electric field to the rate of change of flux:

$$\oint \mathbf{E}\cdot d\ell \ = \ -\frac{d\phi_B}{dt},$$

or, using our expressions for the line integral and the rate of change of flux,

$$h\,dE \ = \ -h\,dx\,\frac{dB}{dt}.$$

Dividing through by $h\,dx$, we have

$$\frac{dE}{dx} \ = \ -\frac{dB}{dt}. \qquad (32\text{--}18a)$$

In deriving this equation, we considered changes in E with position at a fixed instant of time, as pictured in Fig. 32–9, so that our derivative dE/dx means the rate of change of E with position while time is held fixed. Similarly, in evaluating the derivative of magnetic flux, we were concerned only with the time rate of change at the fixed position of our loop. Both our derivatives represent rates of change with respect to one variable while the other variable is held fixed, and are therefore partial derivatives. Equation 32–18a should then be written more properly:

$$\frac{\partial E}{\partial x} \ = \ -\frac{\partial B}{\partial t}. \qquad (32\text{--}18b)$$

Equation 32–18—which is just Faraday's law applied to our electromagnetic wave—tells us that the rate at which the electric field changes with

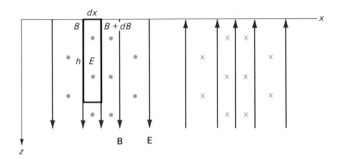

Fig. 32–10
Cross section of the wave slab of Fig. 32–8, showing the *x-z* plane.

position is related to the rate at which the magnetic field changes with time.

Now imagine an observer looking down on Fig. 32–8 from above. This observer sees the magnetic field lines lying in the x-z plane, and electric field lines emerging perpendicular to the x-z plane as shown in Fig. 32–10. We can apply Ampère's law (Equation 32–10) to the infinitesimal rectangle shown, just as we applied Faraday's law to a similar rectangle in the x-y plane. Going counterclockwise around the rectangle, we get no contribution to the line integral of the magnetic field on the short sides, since they lie perpendicular to the field. Going down the left side, we get a contribution Bh to the line integral. Going up the right side, against the field, we get a negative contribution $-(B+dB)h$, where dB is the change in B from one side of the rectangle to the other. So the line integral in Ampère's law is

$$\oint \mathbf{B}\cdot d\ell = Bh - (B+dB)h = -h\,dB.$$

The electric flux through the rectangle is simply $Eh\,dx$, so the rate of change of electric flux is

$$\frac{d\phi_E}{dt} = h\,dx\,\frac{dE}{dt}.$$

Ampère's law relates the line integral of the magnetic field to this time derivative of the electric flux, giving

$$-h\,dB = \epsilon_0\mu_0 h\,dx\,\frac{dE}{dt}.$$

Dividing through by $h\,dx$ and noting again that we are really dealing with partial derivatives, we have

$$\frac{\partial B}{\partial x} = -\epsilon_0\mu_0\,\frac{\partial E}{\partial t}. \tag{32–19}$$

Equations 32–18 and 32–19—derived from Faraday's and Ampère's laws—express fully the requirements that Maxwell's universal laws of electromagnetism pose on the field structure postulated in Fig. 32–8. The two equations are remarkable in that each describes an induced field that arises from the changing of the other field. That other field, in turn, arises from the changing of the first field. Thus we have a self-perpetuating electromagnetic structure, whose fields exist and change without the need for charged matter. If Equations 32–15 and 32–16, which describe the fields in Fig. 32–8, can be made consistent with Equations 32–18 and 32–19, then we will have demonstrated that our electromagnetic wave does indeed satisfy Maxwell's equations and is thus a possible configuration of electric and magnetic fields.*

To see that Equation 32–18 can be satisfied, we differentiate the electric field of Equation 32–15 with respect to x, and the magnetic field of Equation 32–16 with respect to t:

$$\frac{\partial E}{\partial x} = \frac{\partial}{\partial x} [E_0 \sin(kx - \omega t)] = kE_0 \cos(kx - \omega t)$$

and

$$\frac{\partial B}{\partial t} = \frac{\partial}{\partial t} [B_0 \sin(kx - \omega t)] = -\omega B_0 \cos(kx - \omega t).$$

Putting these expressions in for the derivatives in Equation 32–18 gives

$$kE_0 \cos(kx - \omega t) = -[-\omega B_0 \cos(kx - \omega t)].$$

The cosine term cancels from this equation, showing that the equation holds provided that

$$kE_0 = \omega B_0. \tag{32–20}$$

To see that Equation 32–19 also holds, we now differentiate the magnetic field with respect to position and the electric field with respect to time, obtaining

$$\frac{\partial B}{\partial x} = kB_0 \cos(kx - \omega t)$$

and

$$\frac{\partial E}{\partial t} = -\omega E_0 \cos(kx - \omega t).$$

Using these expressions in Equation 32–19 then gives

$$kB_0 \cos(kx - \omega t) = -\epsilon_0 \mu_0 [-\omega E_0 \cos(kx - \omega t)].$$

Again, the cosine term cancels, showing that this equation is satisfied if

$$kB_0 = \epsilon_0 \mu_0 \omega E_0. \tag{32–21}$$

Our analysis has shown that electromagnetic waves whose form is given by Fig. 32–8 and Equations 32–15 and 32–16 can exist, provided that the peak electric and magnetic fields, and the frequency ω and wave number k, are related by Equations 32–20 and 32–21. Physically, the existence of these waves is possible because a change in either kind of field—electric

---

*A more general approach would be to show that Equations 32–18 and 32–19 imply the wave equation. This approach is explored in Problem 7.

or magnetic—induces the other kind of field, giving rise to a self-perpetu-ating electromagnetic field structure. Maxwell's theory thus leads to the prediction of an entirely new phenomenon—the electromagnetic wave. We will now explore some properties of these waves.

32.5
THE SPEED OF ELECTROMAGNETIC WAVES

In Chapter 14, we found that the speed of a sinusoidal wave is given by the ratio of the angular frequency and wave number:

$$\text{wave speed} = \frac{\omega}{k}. \tag{32-22}$$

To determine the speed of our electromagnetic wave, we solve Equation 32–20 for E_0:

$$E_0 = \frac{\omega B_0}{k},$$

and use this expression in Equation 32–21:

$$kB_0 = \epsilon_0 \mu_0 \omega E_0 = \frac{\epsilon_0 \mu_0 \omega^2 B_0}{k}.$$

Solving for the wave speed ω/k then gives

$$\text{wave speed} = \frac{\omega}{k} = \frac{1}{\sqrt{\epsilon_0 \mu_0}}. \tag{32-23}$$

This remarkably simple result shows that the speed of an electromagnetic wave depends only on the electric and magnetic constants ϵ_0 and μ_0. All electromagnetic waves, regardless of frequency or amplitude, share this speed when travelling through empty space. Although we derived this re-sult for sinusoidal waves, superposition considerations of Section 14.5 show that it holds for any wave shape.

We can easily calculate the speed given in Equation 32–23:

$$\frac{1}{\sqrt{\epsilon_0 \mu_0}} = \frac{1}{[(8.85 \times 10^{-12} \text{ F/m})(4\pi \times 10^{-7} \text{ H/m})]^{1/2}} = 3.00 \times 10^8 \text{ m/s}.$$

But this is precisely the speed of light! As early as 1600, Galileo had tried to measure the speed of light by uncovering lanterns on different mountain tops. He was able to conclude only that "If not instantaneous, it is extraor-dinarily rapid." In 1728, James Bradley, in England, used changes in the apparent positions of stars resulting from earth's orbital motion to calculate a value of 2.95×10^8 m/s for the speed of light. Bradley's result differs by less than 2 per cent from the value 2.99792458×10^8 m/s used in the 1983 definition of the meter in terms of the speed of light. Furthermore, the Dutch physicist Christian Huygens had suggested in 1678—again, about 200 years before Maxwell—that light is a wave. A substantial body of op-tical experiments had confirmed Huygens' theory, although neither theory nor experiment could say what sort of wave light might be. Now, in the 1860's, came Maxwell. Using a theory developed from laboratory experiments involving electricity and magnetism, with no reference whatever to optics or light, Maxwell showed how the interplay of electric and magnetic fields could result in electromagnetic waves. The speed of those waves—calculated from the quantities ϵ_0 and μ_0 that were determined in laboratory experiments having nothing to do with light—was precisely the known speed of light. Maxwell was led inescapably to the conclusion that light is an electromagnetic wave.

Maxwell's identification of light as an electromagnetic phenomenon is a classic example of the unification of knowledge toward which science is ever striving. With one simple calculation, Maxwell brought the entire science of optics under the umbrella of electromagnetism. Maxwell's work stands as one of the crowning intellectual triumphs of humanity, an achievement whose implications are still expanding our view of the universe.

32.6
PROPERTIES OF ELECTROMAGNETIC WAVES

Our demonstration that electromagnetic waves of the form given by Equations 32–15 and 32–16 and Fig. 32–8 can satisfy Maxwell's equations places definite constraints on the properties of those waves. The wave frequency ω and wave number k are not both arbitrary, but must be related through

$$\frac{\omega}{k} = c, \tag{32-24a}$$

where $c = 1/\sqrt{\epsilon_0 \mu_0}$ is the speed of light. Equivalently, Equations 32–13 and 32–14 allow us to write

$$f\lambda = c. \tag{32-24b}$$

Furthermore, Equation 32–20 shows that

$$E = \frac{\omega}{k} B = cB. \tag{32-25}$$

Thus the field magnitudes in the wave are not arbitrary, but are in the ratio of the speed of light. Equations 32–15 and 32–16 for our assumed form of the wave show that the electric and magnetic fields are in phase. (This is why we wrote E and B in Equation 32–25 even though Equation 32–20 relates only the peak values E_0 and B_0.) Finally, Fig. 32–8 requires that the electric and magnetic fields be perpendicular to each other and to the direction of wave propagation. Although it is not clear that we had to start with a wave of the form we assumed, it is in fact the case that only waves with **E** and **B** in phase and with **E, B,** and the propagation direction all perpendicular to each other can satisfy Maxwell's equations in vacuum (see Problem 8).

Example 32–2

An Electromagnetic Wave

Light from a laser is propagating in the $+z$-direction. If the amplitude of the electric field in the light wave is 6.0×10^3 V/m, and if the electric field points in the x-direction, what are the direction and amplitude of the magnetic field?

Solution

If we imagine reorienting the wave of Fig. 32–8 so it is propagating along the z-direction, then rotate the wave about the z-direction so the electric field points in the x-direction, we find that the magnetic field points in the y-direction. The magnetic field amplitude is obtained by solving Equation 32–25 for B_0:

$$B_0 = \frac{E_0}{c} = \frac{6.0 \times 10^3 \text{ V/m}}{3.0 \times 10^8 \text{ m/s}} = 2.0 \times 10^{-5} \text{ T}.$$

The Electromagnetic Spectrum

Although an electromagnetic wave must have its frequency and wavelength related by Equation 32–24b, we are still free to choose one of these quantities. This means that we can have electromagnetic waves of any frequency, or, equivalently, any wavelength. Direct measurement shows that visible light occupies a wavelength range from about 400 nm to 700 nm (1 nm $= 10^{-9}$ m), corresponding to frequencies from 7.5×10^{14} Hz to 4.3×10^{14} Hz. The different wavelengths or frequencies correspond to different colors, with red at the long-wavelength, low-frequency end of the visible region and blue at the short-wavelength, high-frequency end (see color plates 7–9).

The range of frequencies occupied by visible light is rather limited. What about electromagnetic waves whose frequencies lie above and below the visible range? Such invisible electromagnetic waves were unknown in Maxwell's time. A brilliant confirmation of Maxwell's theory came in 1888, when the German physicist Heinrich Hertz succeeded in generating and detecting electromagnetic waves of much lower frequency than visible light (Fig. 32–11). Hertz intended his work only to verify Maxwell's modification of Ampère's law, but the practical consequences have proven enormous. In 1896, the Italian scientist Guglielmo Marconi demonstrated that he could generate and detect the so-called "Hertzian waves." In 1901, he transmitted

Fig. 32–11
Apparatus used by Heinrich Hertz to produce and detect electromagnetic waves. Top row shows detectors; bottom row is the transmitter.

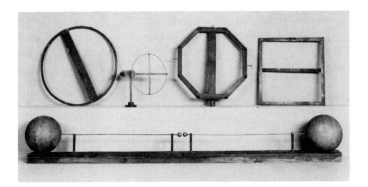

Fig. 32–12
On the Cape Cod National Seashore, we can still find ruins of the transatlantic wireless site operated by Guglielmo Marconi between 1901 and 1917. The first formal communication was between President Theodore Roosevelt and King Edward VII on January 19, 1903. Most of the station has been destroyed by erosion.

electromagnetic waves across the Atlantic Ocean, creating a public sensation (Fig. 32–12). From the pioneering work of Hertz and Marconi, spurred by the theoretical efforts of Maxwell, came the entire technology of radio, television, and microwaves that so dominates modern society. We now consider all electromagnetic waves in the frequency range from a few Hz to about 3×10^{11} Hz as radio waves, with ordinary AM radio at about 10^6 Hz, FM radio at 10^8 Hz, and microwaves used for radar, cooking, and satellite communications at 10^9 Hz and above.

Between radio waves and visible light lies the infrared frequency range. Electromagnetic waves in this region are emitted by warm objects, even when they are not hot enough to glow visibly. For this reason, infrared detectors are used to determine subtle body temperature differences in medical diagnosis, to examine buildings for heat loss, and to study the birth of stars in clouds of interstellar gas and dust (Fig. 32–13).

Beyond the visible region, we encounter the ultraviolet rays responsible for sunburn, then the highly penetrating x-rays, and finally the gamma rays

Fig. 32–13
Regions of significant heat loss—especially chimneys and windows—show as bright regions on this infrared photo.

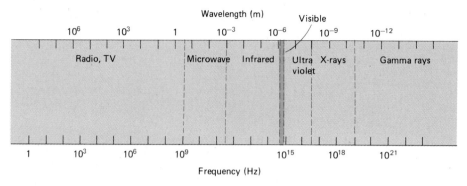

Fig. 32–14
The electromagnetic spectrum ranges from the lowest-frequency radio waves to the highest-frequency gamma rays. Note the logarithmic scale, in which equal intervals on the diagram correspond to factor-of-10 changes in frequency and wavelength.

whose primary terrestrial source is radioactive decay. All these phenomena, from radio to gamma rays, are essentially the same: they are all electromagnetic waves, differing only in frequency and wavelength. All travel with speed c, and all consist of electric and magnetic fields that arise from each other through the induction process described by Faraday's and Ampère's laws. The distinction between different kinds of electromagnetic waves is a mere convenience, since there are no gaps in the possible frequencies. Figure 32–14 shows the range of electromagnetic waves, or the **electromagnetic spectrum,** displayed in a single diagram.

electromagnetic spectrum

Application

The New Astronomy

A glance at Fig. 32–14 shows that visible light occupies a small fraction of the electromagnetic spectrum. For centuries our only information about the universe beyond earth—except for an occasional meteorite—came from visible light. Processes like those occurring on the visible surface of our sun and many other stars produce predominantly visible light. Optical astronomy, utilizing visible light, gave

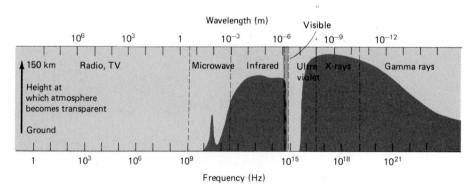

Fig. 32–15
Earth's atmosphere is opaque to most electromagnetic waves, although "windows" of transparency exist for several wavelength ranges. Bottom of diagram represents earth's surface, with top of gray curve the height to which waves of a given wavelength penetrate.

Fig. 32–16
(Left) The Very Large Array, a group of 27 radio telescopes near Socorro, New Mexico. By superposing signals from the different telescopes, it is possible to form high-resolution images of celestial radio sources. (Right) VLA image of the radio source Cygnus A shows two giant lobes of radio-emitting gas on either side of the center.

a good picture of the universe to the extent that it consists of objects not too different from the visible part of the sun. The restriction to optical astronomy was in part imposed by earth's atmosphere. Transparent to visible light, the atmosphere is largely opaque to other forms of electromagnetic radiation, although "windows" of relative transparency exist in parts of the radio and infrared bands (Fig. 32–15). The discovery by Bell Telephone Laboratories electrical engineer Karl Jansky in 1931 that radio waves from outer space can be detected on earth led to the development of radio astronomy. For decades, radio astronomy has given a picture of the universe that complements the optical view, showing phenomena that are simply not detectable by optical means (Fig. 32–16).

The onset of the space age in the late 1950's finally opened the entire electromagnetic spectrum to astronomers. Before this time there were surprisingly few suggestions that anything interesting might be found beyond the visible range. But satellites carrying infrared, ultraviolet, x-ray, and gamma-ray detectors have literally revolutionized our view of the universe (Fig. 32–17). Exotic objects like neutron stars—with the mass of the sun crushed to a diameter of a few kilometers—and

Fig. 32–17
(Left to right) (1) The Infrared Astronomical Satellite (IRAS), sent aloft in 1983, was operated jointly by the United States, the Netherlands, and Great Britain. (2) The arrow marks a newborn star in this IRAS image. (3) The second High Energy Astronomy Observatory satellite, also called the Einstein Observatory. (4) The Crab Nebula, imaged in x-rays by the Einstein Observatory, has a bright pulsar in its midst.

Fig. 32–18
Artist's conception of a black hole orbiting a giant star. Our understanding of such systems comes from observations of visible light from the star and of x-rays from the hot gas as it falls into the black hole, which is in the middle of the disk at right.

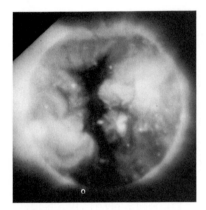

Fig. 32–19
An image of the sun taken in x-rays shows the structure of the hot, tenuous solar corona.

black holes, whose gravity is so strong that not even light can escape—are now objects of astronomical study (Fig. 32–18). The opening of the entire electromagnetic spectrum has brought a new richness to astronomy, showing that our universe contains some of the most unusual objects that the laws of physics permit. Phenomena that were once bizarre conjectures of theoreticians are now observed regularly. Closer to home, observations of the sun with ultraviolet and x-ray instruments have brought new understandings of the sun's hot, outer corona (Fig. 32–19). By turning space-borne infrared detectors toward earth, we have learned much about the structure and resources of our own planet (Fig. 32–20).

Fig. 32–20
Night-time infrared image of the region around Buffalo, New York (center), showing also the temperature structure in Lake Erie (bottom) and Lake Ontario (top). Image was made by a Landsat satellite.

Generating Electromagnetic Waves

So far we have shown that electromagnetic waves can exist, and have described some of their properties. But how do electromagnetic waves originate?

To initiate an electromagnetic wave, we must create a changing electric or magnetic field. Once we have done so, the self-regenerating process described by Ampère's and Faraday's laws occurs, and the wave propagates on its own. Ultimately, we create changing fields when we alter the motion of electric charge—that is, whenever we accelerate charge. Accelerated electric charge is the source of electromagnetic waves. In a radio transmitter, the accelerated charges are electrons moving back and forth in an antenna, driven by an alternating voltage from an *LC* circuit (Fig. 32–21). The emission of visible light is associated with electrons jumping between different atomic orbits. X-rays are produced when electrons abruptly decelerate as they collide with a massive target in an x-ray tube. Electrons spiralling in strong magnetic fields generate electromagnetic waves whose frequency depends on the magnetic field strength.

Polarization

An electromagnetic wave is characterized by its amplitude and frequency. Because the electric and magnetic fields making up electromagnetic waves are vectors, we must also specify the direction of the fields in order to describe a wave fully. The term **polarization** is used to describe the direction of the electric field. If this and the direction of wave propagation are both known, then the direction of the magnetic field is determined as well, since the two fields must be at right angles to each other and to the propagation direction.

Electromagnetic waves used in radio, TV, and radar are usually generated by sending alternating current through metal structures called antennas. The orientation of the antenna sets a preferred direction for the electric field, so that the waves are **polarized** (Fig. 32–22). On the other hand, visible light from hot objects like the sun or a light bulb is **unpolarized.** Such light originates in many individual atoms and consists of a mixture of electromagnetic wave pulses with random orientations and hence no definite polarization direction.

Unpolarized light may be polarized either by reflection off surfaces or when it passes through substances whose internal structure has a preferred direction, called the transmission axis. The plastic Polaroid, invented by

polarization

polarized
unpolarized

Fig. 32–21
Simplified diagram of a radio transmitter, consisting of an oscillating *LC* circuit connected to an antenna. The changing electric field associated with alternating current in the antenna gives rise to an electromagnetic wave.

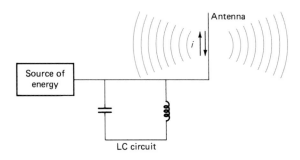

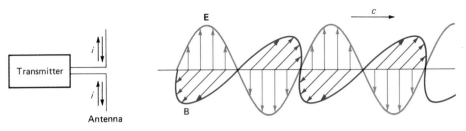

Fig. 32–22
Because its current-carrying elements are vertical, this radio antenna produces vertically polarized electromagnetic waves, meaning that the electric field vectors are vertical. The magnetic field vectors, in contrast, are horizontal.

Fig. 32–23
A pair of Polaroid sheets with their transmission axes at right angles. Where they overlap, no light can get through.

E. H. Land, consists of long molecules all aligned in the same direction. An electromagnetic wave passes easily through the material if its electric field is perpendicular to the direction of the molecules. But if the field is parallel to that direction, the wave is blocked. If unpolarized light is passed first through one piece of Polaroid it becomes polarized. If a second sheet is oriented with its polarizing direction at right angles to the first, none of the polarized light will get through (Fig. 32–23). Sunlight reflecting off a road or the hood of a car becomes partially polarized in the horizontal direction. Polaroid sunglasses, with their polarization direction vertical, eliminate this glare without excessively reducing the total light intensity.

We can calculate the effect a polarizing material has on light intensity by considering that the material passes unattenuated that component of the electric field that lies along the preferred direction of polarization. If θ is the angle between the field and the preferred direction, then the field component along the preferred direction is $E\cos\theta$. As we will show in the next section, the intensity S of an electromagnetic wave is proportional to the square of the field strength, so that when a polarized beam of light of intensity S_0 encounters a polarizing material, the transmitted intensity is

$$S = S_0\cos^2\theta. \qquad (32\text{–}26)$$

law of Malus

This relation is called the **law of Malus,** after its discoverer, an engineer in Napoleon's army. When unpolarized light is incident on a polarizing material, the transmitted intensity is half the incident intensity. This follows because the unpolarized light consists of many waves whose polarization directions make random angles θ with the transmission axis of the material. Averaging $\cos^2\theta$ over all angles gives the factor $\frac{1}{2}$.

Polarization can tell us much about the sources of electromagnetic waves or about materials through which the waves travel. Many astrophysical processes give rise to polarized electromagnetic waves. Measuring polarization then provides a clue to the mechanisms operating in a particular star or other object (Fig. 32–24). Polarization resulting from transmission of light through crystals identifies minerals and helps geologists understand the formation of rocks (Fig. 32–25). Polarized light helps engineers locate stresses in materials (Fig. 32–26).

linear polarization
circular polarization

The type of polarization we have been discussing, in which the electric field lies in a definite plane, is called **linear polarization.** Some processes result in **circular polarization,** in which the electric and magnetic fields

Fig. 32–24
Light from the Crab Nebula is highly polarized, as shown by the difference between these two images taken at different polarizations. Arrows indicate directions of the electric field vector.

Fig. 32–25
Photomicrograph of a thin section of rock placed between crossed polarizers. Individual mineral crystals within the rock rotate the electric field of the light, allowing different amounts of light to be transmitted.

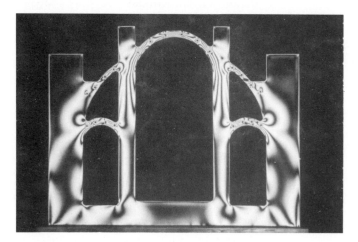

Fig. 32–26
A plastic model of a Gothic cathedral, photographed between polarizing sheets. The resulting fringe patterns reveal stresses and help architects understand the response of the building to wind and to the weight of structural members.

Fig. 32–27
In a circularly polarized electromagnetic wave, the direction of the electric field vector rotates as the wave propagates.

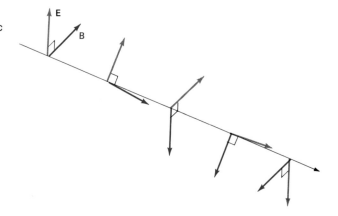

rotate about the propagation direction as the wave advances (Fig. 32–27; see Problem 37). Electrons spiralling in magnetic fields, for example, give rise to circularly polarized electromagnetic waves. Analysis of such waves provides astrophysicists a means of measuring the magnetic field strength in distant objects.

Energy in Electromagnetic Waves

In previous chapters, we showed that both electric and magnetic fields contain energy. Here we have considered electromagnetic field structures— waves—in which a self-sustaining combination of electric and magnetic fields travels through empty space. As the wave moves, it transports the energy of the electric and magnetic fields.

intensity

We use the term **intensity** to characterize the energy transport in an electromagnetic wave. Intensity is the rate at which a wave transports energy across a unit area. Accordingly, its units are joules/second/square meter, or W/m^2. To calculate the wave intensity, consider a rectangular slab of length dx and cross-sectional area A as shown in Fig. 32–28. Within this slab are the wave fields **E** and **B.** The energy densities of these fields are given by Equations 24–8 and 30–12:

$$u_E = \tfrac{1}{2}\epsilon_0 E^2$$

$$u_B = \frac{B^2}{2\mu_0}.$$

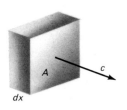

Fig. 32–28
A slab of length dx and cross-sectional area A at right angles to the propagation of an electromagnetic wave.

If dx is sufficiently small, the fields don't change much over our slab, so the total energy in the slab is just the sum of the electric and magnetic energy densities multiplied by the slab volume $A\,dx$:

$$dU = (u_E + u_B)\,A\,dx = \frac{1}{2}\left(\epsilon_0 E^2 + \frac{B^2}{\mu_0}\right) A\,dx.$$

This wave energy moves with speed c, so that all the energy contained in the slab of length dx moves out of the slab in a time

$$dt = \frac{dx}{c}.$$

The rate at which energy moves through the cross-sectional area A is then

$$\frac{dU}{dt} = \frac{1}{2}\left(\epsilon_0 E^2 + \frac{B^2}{\mu_0}\right)\frac{A\,dx}{dx/c} = \frac{c}{2}\left(\epsilon_0 E^2 + \frac{B^2}{\mu_0}\right)A.$$

Then the intensity S, or rate of energy flow per unit area, is then

$$S = \frac{c}{2}\left(\epsilon_0 E^2 + \frac{B^2}{\mu_0}\right). \tag{32–27}$$

We can recast this equation in simpler form by noting that, for an electromagnetic wave, $E = cB$ and $B = E/c$. Using these expressions to replace one of the E's in the term E^2 with B and similarly one of the B's in the term B^2 with E, we have

$$S = \frac{c}{2}\left(\epsilon_0 cEB + \frac{EB}{\mu_0 c}\right) = \frac{1}{2\mu_0}(\epsilon_0 \mu_0 c^2 + 1)\,EB.$$

But $c = 1/\sqrt{\epsilon_0 \mu_0}$, so that $\epsilon_0 \mu_0 c^2 = 1$, giving

$$S = \frac{EB}{\mu_0}. \tag{32–28}$$

Although we derived Equation 32–28 for an electromagnetic wave, it is in fact a special case of the more general result that nonparallel electric and magnetic fields are accompanied by a flow of electromagnetic energy. In general, the rate of energy flow per unit area is given by

$$\boxed{\mathbf{S} = \frac{\mathbf{E} \times \mathbf{B}}{\mu_0}.} \tag{32–29}$$

Poynting vector

Here a vector $\mathbf{S}$ is used to signify not only the magnitude of the energy flow, but also its direction. For an electromagnetic wave in vacuum, in which $\mathbf{E}$ and $\mathbf{B}$ must be at right angles, Equation 32–29 reduces to Equation 32–28, with the direction of energy flow the same as the direction of wave travel. The vector intensity $\mathbf{S}$ is called the **Poynting vector** after the English physicist J. H. Poynting, who suggested it in 1884. Poynting's name is especially fortuitous, for the Poynting vector points in the direction of energy flow. Problem 40 explores an important application of the Poynting vector to fields that do not constitute an electromagnetic wave.

Equations 32–28 and 32–29 give the rate of energy flow at the instant when the fields have magnitudes E and B. In an electromagnetic wave, these fields oscillate, and so does the wave intensity. We are usually not interested in this rapid oscillation. For example, when we consider the intensity of sunlight, it doesn't really matter that the intensity is fluctuating at about 10^{14} times per second. In such cases we are interested in the *average* intensity, $\overline{S}$. Because the instantaneous intensity of Equation 32–28 contains a product of sinusoidally varying terms, the average intensity is just the product of the peak intensities times the average of $\sin^2$ over one cycle. As we found in the preceding chapter, this average is just one-half (look at the curve for $\sin^2$; it varies between 0 and $+1$, and is symmetric about $\frac{1}{2}$), so that the average intensity is

$$\overline{S} = \frac{\overline{EB}}{\mu_0} = \frac{E_0 B_0}{2\mu_0}. \tag{32–30}$$

Although we have written the instantaneous and average intensities in terms of both E and B, we can use the wave condition $E = cB$ to eliminate either field in terms of the other. Equation 32–30, for example, can be written

$$\overline{S} = \frac{E_0^2}{2\mu_0 c}. \tag{32–31}$$

or

$$\overline{S} = \frac{c}{2\mu_0} B_0^2. \tag{32–32}$$

Example 32–3
The Intensity of Sunlight

The average intensity of sunlight on a clear day at noon is about 1 kW/m^2. What are the electric and magnetic field strengths in sunlight? How many solar collectors would you need to produce power equivalent to a 5-kW electric water heater in noonday sun? Assume the collectors have area 2.0 m^2, are 50 per cent efficient, and are perpendicular to the sun's rays. Assume that sunlight consists of sinusoidal electromagnetic waves.

Solution

Solving Equation 32–31 for the electric field gives

$E_0 = \sqrt{2\mu_0 c \overline{S}} = [(2)(4\pi \times 10^{-7} \text{ H/m})(3.0 \times 10^8 \text{ m/s})(1 \times 10^3 \text{ W/m}^2)]^{1/2} = 900 \text{ V/m}.$

The peak magnetic field is then given by $B_0 = E_0/c$, so that

$$B_0 = \frac{E_0}{c} = \frac{900 \text{ V/m}}{3.0 \times 10^8 \text{ m/s}} = 3 \times 10^{-6} \text{ T}.$$

At 1 kW/m^2, we would need 5 m^2 of collector area at 100 per cent efficiency to produce 5 kW. Our collectors are only 50 per cent efficient, so we need 10 m^2, for a total of 5 collectors of 2 m^2 each.

As an electromagnetic wave travels through empty space, the total energy carried by the wave does not change. With the plane waves we have been discussing, the intensity—the energy flow per unit area—is also the same everywhere. But when the wave originates in a localized source like an atom, a radio transmitting antenna, or the sun, its wavefronts are not planes but expanding spheres (Fig. 32–29). Since no energy is lost from the wave, the rate of energy flow must be the same through *any* closed surface surrounding the source. Put more mathematically, we can say that the flux of the Poynting vector through any closed surface depends only on the power of the source enclosed. (Sound familiar? Think of "field lines" associated with the Poynting vector, and go back to the discussion of Gauss's law in Chapter 21.) In particular, if we consider a spherical surface surrounding a point source of waves, then the area of the surface increases as the square of the radius (Fig. 32–30). But the rate of energy flow, or power, through the whole surface must be constant. Then the intensity—the power per unit area—varies as the inverse square of the distance from the point source:

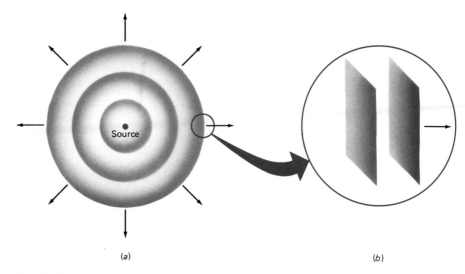

Fig. 32–29
(a) Spherical waves originate from a localized source. (b) A small region of the expanding spherical wave approximates a plane wave.

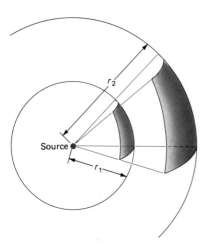

Fig. 32–30
Wave energy passes through the two patches at the same rate. The outer patch is larger than the inner one by the ratio of their radii squared, so that the wave intensity is lower by the same factor.

$$S = \frac{P}{4\pi r^2}. \qquad (32\text{--}33)$$

Here P is the total power emitted by the source, and r the distance from the source. This inverse-square decrease occurs not because the waves "weaken" and lose energy as they travel, but because the same energy gets spread more thinly as the wave expands. Because the intensity of an electromagnetic wave is proportional to the square of the field strengths (see Equations 32–31 and 32–32), Equation 32–33 implies that the electric and magnetic fields of a spherical wave themselves decrease as $1/r$.

Example 32–4
The Sun's Power

The intensity of sunlight just above earth's atmosphere is about 1.37 kW/m². (About one-third of the sunlight is reflected or absorbed by the atmosphere.) What is the intensity of sunlight at the planet Mercury? What is the total power emitted by the sun? To how many 100-W light bulbs is the sun equivalent?

Solution

From Appendix F, we find that earth's mean distance from the sun is 1.5×10^{11} m, while Mercury's is 0.58×10^{11} m. Because sunlight intensity falls as the inverse square of the distance, the ratio of intensities is the inverse of the square of the ratio of distances from the sun, so that

$$S_{\text{Mercury}} = \left(\frac{r_{\text{earth}}}{r_{\text{Mercury}}}\right)^2 S_{\text{earth}}$$

$$= \left(\frac{1.5}{0.58}\right)^2 (1.37 \times 10^3 \text{ W/m}^2) = 9.2 \times 10^3 \text{ W/m}^2,$$

about 6.7 times the intensity at earth.

At earth, the power falling on 1 square meter is 1370 W. The total power passing through a sphere at earth's distance from the sun is this number times the area of the sphere, or

$$P = 4\pi r^2 S = (4\pi)(1.5 \times 10^{11} \text{ m})^2(1370 \text{ W/m}^2) = 3.9 \times 10^{26} \text{ W}.$$

Since all energy leaving the sun passes through surfaces surrounding it, this is the sun's total power output. The sun equals 3.9×10^{24} 100-W light bulbs!

Momentum in Electromagnetic Waves

In your study of mechanics you found that moving objects carry both kinetic energy and momentum. The same is true for electromagnetic waves. Maxwell showed that an electromagnetic wave carrying energy at the rate of S W/m^2 also transports momentum at the rate S/c kg/m·s^2 (check that these are the right units for momentum per time per area). If an electromagnetic wave is absorbed by an object, this is the rate at which momentum is transferred to a unit area at right angles to the wave propagation. By Newton's law, the rate of change of momentum of an object is equal to the force on the object, so that the quantity S/c is the force per unit area, or **radiation pressure,** exerted on an object that absorbs electromagnetic waves. If the wave is reflected, then the radiation pressure is twice as great. (Consider an analogy: you're standing on ice skates and you catch a ball. You absorb the ball's momentum, and start moving backward. If you then throw the ball back, you acquire additional momentum. For the same reason, reflecting an electromagnetic wave doubles the momentum transfer.) Equivalently, a part of a light wave carrying energy U also carries momentum

radiation pressure

$$p = \frac{U}{c}. \tag{32–34}$$

The very small force exerted by ordinary light was first measured by Dartmouth College physicists E. F. Nichols and G. F. Hull in 1903. With high-energy laser light, or when ordinary light impinges on objects of low mass but large surface area, the force of electromagnetic radiation becomes appreciable. Lasers can exert enough light pressure to levitate small particles (Fig. 32–31; see Problem 32). The force of sunlight has been suggested as a means of driving interplanetary "sailing ships" (Fig. 32–32; see Problem 33). Finally, the idea that electromagnetic waves carry momentum played a crucial role in Einstein's development of his famous equation $E = mc^2$ relating energy and mass.

Fig. 32–31
The star-like image is a 20-micron particle levitated by a laser beam reflected upward by the prism shown at the bottom. The star-like rays are due to diffraction (see Chapter 35) inside the camera.

Fig. 32–32
Artist's conception of a spacecraft driven by the radiation pressure of sunlight. This particular design was proposed for a 1986 U.S. mission to Halley's Comet before the U.S. effort was cancelled for budgetary reasons. The blade-like sails are shown only partially extended; ultimately they would have extended 7.4 km.

SUMMARY

1. Maxwell's modification of Ampère's law adds a **displacement current** term $\epsilon_0 d\phi_E/dt$ to the right-hand side of Ampère's law. The modified Ampère's law states that a changing electric field gives rise to a magnetic field, just as Faraday's law states that a changing magnetic field gives rise to an electric field.

2. Maxwell's modification of Ampère's law completes the set of electromagnetic field equations. Called **Maxwell's equations,** the complete set (Table 32–2) describes fully the behavior of electromagnetic fields throughout the universe.

3. The interplay of electric and magnetic fields described by Ampère's and Faraday's laws gives rise to **electromagnetic waves**—structures consisting of changing electric and magnetic fields that travel through empty space, with each changing field giving rise to the other. Maxwell's equations show that electromagnetic waves can exist, and that they have a number of specific properties:

   a. All electromagnetic waves travel at the same speed $c = 3.0 \times 10^8$ m/s. This speed can be calculated from laboratory measurements of the electric and magnetic constants ϵ_0 and μ_0:

$$c = \frac{1}{\sqrt{\epsilon_0 \mu_0}}.$$

   b. The electric and magnetic fields of the wave are at right angles to each other and to the direction of wave travel.

   c. The magnitudes of the two fields are not arbitrary, but are related by $E = cB$.

   d. An electromagnetic wave can have any frequency or wavelength, but the two must be related through the equivalent relations $c = f\lambda$ and $c = \omega/k$.
      Radio waves, television, microwaves, infrared, visible light, ultraviolet, x-rays, and gamma rays are all forms of electromagnetic radiation. They differ only in frequency and wavelength, and together constitute the **electromagnetic spectrum.**

   e. Electromagnetic waves may be characterized by their **polarization,** which describes the direction of the electric field vector. In a **linearly polarized** wave, the electric vector oscillates in magnitude along a fixed direction. In a **circularly polarized** wave, the electric vector describes a circle as the wave propagates. Study of polarized electromagnetic waves provides information about the processes occurring where the waves originate, and about media through which the waves propagate.

   f. Electromagnetic waves carry energy. The energy transported across a unit area per unit time is the wave **intensity.** The **Poynting vector,**

$$\mathbf{S} = \frac{\mathbf{E} \times \mathbf{B}}{\mu_0},$$

   describes this energy transport for any configuration of electromagnetic fields, including the fields of a wave. In a sinusoidal electromagnetic wave, the time-average rate of energy transport is half the peak value, or

$$\bar{S} = \frac{E_0 B_0}{2\mu_0}.$$

   g. The intensity of electromagnetic waves emitted by a localized source falls off as the inverse square of distance from the source, as the energy is spread out over increasingly larger areas:

$$S = \frac{P}{4\pi r^2}.$$

   h. Electromagnetic waves carry momentum at the rate S/c per unit area. As a result, electromagnetic waves exert a **radiation pressure,** S/c, when they are absorbed by a surface at right angles to their propagation direction. If waves are reflected, the radiation pressure is doubled.

QUESTIONS

1. If Maxwell's modification of Ampère's law were not correct, so that a changing electric field did not give rise to a magnetic field, would electromagnetic waves be possible? Why or why not?

2. The presence of magnetic monopoles would require a modification in Gauss's law for magnetism (Equation 32–4). Would any other modification of Maxwell's equations be required?

3. When a capacitor is charging, we say that displacement current flows between its plates. Yet no charge actually moves between the plates. Why, then, do we speak of a current?

4. Is there displacement current in an electromagnetic wave? If so, in what direction(s) does it flow?

5. In what ways are electromagnetic waves similar to sound waves? In what ways are they different?

6. What feature of the electromagnetic wave considered in Section 32.4 ensures that Gauss's laws for electricity and for magnetism are satisfied?

7. Could an electromagnetic wave exist in vacuum if its electric and magnetic fields were parallel rather than perpendicular? Explain.

8. The speed of an electromagnetic wave is given by $c = \omega/k$ or $c = \lambda f$. How does the wave speed depend on frequency? On wavelength?

9. Some stars end their lives in violent events called supernova explosions. During such an event the star explodes in a matter of hours. When we observe a su-

pernova explosion in a distant galaxy, we see a rapid rise in visible light and other forms of electromagnetic waves emitted by the supernova. Why is this good evidence that the speed of light is independent of frequency?

10. If you turn a TV antenna so its metal rods point vertically, you may change the quality of your TV reception. Why? Think about polarization.

11. Unpolarized light is incident on two pieces of Polaroid with their polarization directions at right angles, and no light gets through. A third sheet is inserted between the two, and now some light gets through all three. How can this be?

12. Why is it not possible to define exactly where the visible portion of the electromagnetic spectrum ends?

13. Most of the energy output of the sun is in the form of electromagnetic waves in the visible part of the spectrum, with the most energy in the yellow-green region. We are well-equipped creatures, for our eyes happen to be sensitive to the sun's dominant emission, and are indeed most sensitive right in the yellow-green region. Do you think this is a coincidence? Explain.

14. Suppose your eyes were sensitive to radio waves rather than visible light. What would you see as you look around? What things would be especially bright?

15. If you double the electric field strength in an electromagnetic wave, how does the wave intensity change?

16. An *LC* circuit is made from ideal components connected by perfect wires with no resistance. The capacitor is charged, and the circuit starts oscillating. Will it oscillate forever? If not, how does it lose energy?

17. The intensity of light falls off as the inverse square of the distance from a localized source. Does this mean that light loses energy travelling large distances? If so, where does the energy go? If not, why do distant stars appear faint?

18. You are standing up having your picture taken. When the flash bulb goes off, a significant amount of electromagnetic wave energy strikes you. Why aren't you knocked over by the momentum transferred by the waves?

PROBLEMS

Section 32.1   *The Four Laws of Electromagnetism*

1. For each of the four pre-Maxwellian laws listed in Table 32–1, describe a situation that exhibits cylindrical symmetry and in which the law can be used to solve for the electric or magnetic field. Why is one case trivial? What is the variation in field strength with distance from the symmetry axis in the other three cases?

Section 32.2   *Ambiguity in Ampère's Law*

2. A parallel-plate capacitor has circular plates of radius 1.0 m and spacing 1.0 mm. A uniform electric field is established between the plates. If the field changes at the rate of 1.0×10^6 V/m·s, what is the magnetic field between the plates (*a*) on the symmetry axis of the plates? (*b*) 15 cm from the symmetry axis? (*c*) 150 cm from the symmetry axis? Assume that the capacitor is fed by long wires lying along the axis of the plates.

3. A parallel-plate capacitor of plate area *A* and spacing *d* is charging through a resistor *R* as shown in Fig. 32–33. Show that the displacement current between the plates is equal to the conduction current in the rest of the circuit.

4. An electric field points into the page and occupies a circular region of radius 1.0 m, as shown in Fig. 32–34. There are no electric charges in the region, but there is a magnetic field whose field lines form closed circles pointing counterclockwise. The magnetic field strength 50 cm from the center of the region is 2.0 μT. (*a*) What is the rate of change of the electric field? (*b*) Is the electric field increasing or decreasing?

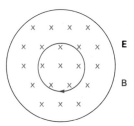

Fig. 32–34   Problem 4

Section 32.4   *Electromagnetic Waves*

5. At a particular point, the instantaneous electric field of an electromagnetic wave points in the $+y$-direction, while the magnetic field points in the $-z$-direction. In what direction is the wave propagating?

6. Show by direct substitution that Equations 32–18*b* and 32–19 can be satisfied by *any* function of the form $f(kx \pm \omega t)$.

7. Show that Equations 32–18*b* and 32–19 may be combined to yield a wave equation like Equation 32–11. What term in your equation corresponds to the v^2

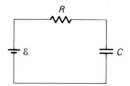

Fig. 32–33   Problem 3

term in Equation 32–11? *Hint:* Take the partial derivative of one equation with respect to x, and of the other with respect to t, and use the fact that

$$\frac{\partial^2 f}{\partial x \partial t} = \frac{\partial^2 f}{\partial t \partial x},$$

for any well-behaved function $f(x,t)$.

8. Show that it is impossible for an electromagnetic wave in vacuum to have a time-varying component of its electric field in the direction of its magnetic field. *Hint:* Assume **E** does have such a component, and show you cannot satisfy both Gauss and Faraday.

Section 32.5 *The Speed of Electromagnetic Waves*

9. A light-minute is defined as the distance light travels in one minute. Show that the sun is about 8 light-minutes from earth.

10. Roughly how long does it take light to go one foot? Some computers are so fast that this time is significant, so that the speed of the computer is limited by the travel times of electrical signals over the wires inside the computer.

11. If you speak by radio to an astronaut on the moon, how long is it before you can get a reply?

12. The rms magnetic field in a light wave is 1.0 G. What is the rms electric field?

13. "Ghosts" on a TV screen occur when part of the TV signal goes directly from transmitter to receiver, while part of it takes a longer route, reflecting off buildings, water, mountains, or other things. The electron beam in a TV tube "paints" the picture on the screen at such a rate that the beam scans from left to right across the screen in about 10^{-4} s. If a "ghost" image appears displaced about 1 cm from the main image, what is the difference in the path lengths taken by the direct and indirect signals? Assume that the TV screen has a horizontal width of 50 cm.

14. In A.D. 1054, Chinese astronomers and Native Americans noted a spectacular stellar explosion that produced an object brighter than the planet Venus (Fig. 32–35). The remains of this supernova explosion are visible today as the Crab Nebula. If the Crab Nebula is 6500 light-years from earth, when did the explosion actually occur?

Fig. 32–35 Native American petroglyph from Chaco Canyon, New Mexico, shows the Crab supernova explosion seen on earth in A.D. 1054 (Problem 14).

Section 32.6 *Properties of Electromagnetic Waves*

15. An AM radio wave has a frequency of 1.0 MHz and an average intensity of 2.0×10^{-4} W/m². What are its (a) wavelength, (b) angular frequency, (c) wave number, (d) peak electric field, and (e) peak magnetic field?

16. What are the wavelengths of each of the following: (a) a 100-MHz FM radio wave; (b) a 3.0-GHz radar wave (1 GHz = 10^9 Hz); (c) a 6.0×10^{14}-Hz light wave; and (d) a 1.0×10^{18}-Hz x-ray?

17. 60-Hz power lines emit small amounts of electromagnetic radiation. What is the wavelength of this radiation?

18. Antennas for transmitting and receiving electromagnetic radiation usually have typical dimensions on the order of half a wavelength. Look at a TV antenna and estimate the wavelength and frequency of a TV signal.

19. An electromagnetic wave is propagating in the z-direction. What is its polarization direction, if its magnetic field is in the y-direction?

20. Unpolarized light of intensity S_0 is incident on a "sandwich" of three polarizers. The outer two have their transmission axes perpendicular to each other, while the axis of the middle one lies at 45° to both the others. What is the intensity of light emerging from this "sandwich"?

21. Polarized light of average intensity S_0 passes through a sheet of polarizing material. The polarizing material is rotating at 10 revolutions/second. Plot the intensity (time-averaged over the electromagnetic wave cycle) of the transmitted light as a function of time.

22. Polarized light is incident on a piece of polarizing material, and only 20% of the light gets through. What is the angle between the electric field of the light and the transmission axis of the material?

23. 1000 V/m is a typical electric field encountered in the laboratory or in electronic equipment. What is the average intensity of an electromagnetic wave with this value as its rms electric field strength?

24. Show that the energy densities of the electric and magnetic fields in an electromagnetic wave are the same.

25. A 1000-W radio transmitter broadcasts its signal uniformly in all directions. What is the intensity of the radio signal at a distance of 5.0 km from the transmitting antenna?

26. At what rate does earth receive energy from the sun? How does this compare with the roughly 10^{13}-W rate at which the human race consumes energy?

27. Studies of the topography of the planet Venus, whose surface is invisible under a thick cloud layer, have been done by bouncing radar signals off Venus. The round-trip travel time for the waves provides a measure of the exact distance to the particular part of the planet's surface under study. Assume that Venus is 50 million km from earth during such a study, and that a radar transmitter on earth with a power output of 450 kW is used. (a) What is the intensity of the radar signals that reach Venus? (b) Assuming the signals are reflected in all directions and with no energy

loss, what is the intensity of the signal reflected normally from a 1.0-km-square patch of Venus' surface as received back at earth? (c) This radar technique can distinguish features that differ in altitude by 100 m on the surface of Venus. How accurately must the travel time of the signals be measured? Express as an actual time and as a fraction of the round-trip travel time.

28. Much astronomical debate centers on the nature of quasars, mysterious objects that are very distant from earth. Although quasars do not appear particularly bright in visible light, their observed brightness combined with their distance suggests an enormous energy output, so large that astrophysicists have had trouble imagining energy generation mechanisms for quasars. Suppose a quasar that is 10 billion light-years from earth appears the same brightness as a sunlike star 50,000 light-years from earth. How does the power output of the quasar compare with that of the sun?

29. An electromagnetic wave is emitted from a localized source. How do the magnitudes of the electric and magnetic fields in the wave depend on distance from the source? Compare with the corresponding dependence for the field of a stationary point charge.

30. A typical fluorescent lamp is 4 feet long and a few cm in diameter. (a) If you are near the lamp, but not near its ends, how do you expect the intensity of the light to depend on distance from the lamp? (b) What if you are very far from the lamp?

31. A 5.0-mW laser produces a beam whose diameter is 2.0 mm. (a) What is the intensity of the laser beam? (b) What is the rms electric field strength in the laser beam? (c) What is the rms magnetic field strength in the laser beam?

32. The laser of Problem 31 shines vertically upward onto a flat piece of aluminum foil whose diameter is the same as the laser beam's. The piece of foil has a mass of 5.0 mg. Is the laser beam powerful enough to support the foil against its weight? If not, how powerful should the beam be?

33. Serious proposals have been made to "sail" spacecraft to the outer parts of the solar system using the pressure of sunlight. How much sail area must a 1000-kg spacecraft have if its acceleration at earth's orbit is to be 1 m/s²? *Hint*: Can you ignore the sun's gravity?

34. You are floating in empty space with a 1.0-W flashlight. If you shine the flashlight in some direction, how long will it take you to accelerate to a speed of 10 m/s?

35. What is the force of sunlight on a 2.0-m² solar collector? The intensity of the sunlight is 1.0 kW/m².

36. A possible propulsion mechanism for interplanetary or interstellar spacecraft is the "photon rocket," a rocket that emits a powerful beam of light instead of the hot gases of an ordinary rocket. (a) How powerful a light source would be needed to develop a thrust equal to that of the space shuttle (3.5×10^7 N)? (b) Compare your answer with the total electric power consumption on earth (about 10^{13} W). Is the photon rocket a practical way to leave earth? What advantages does the photon rocket have for long-distance space travel?

Supplementary Problems

37. A circularly polarized wave is one in which the field vectors do not oscillate in magnitude, but rather rotate in direction. A circularly polarized wave may be thought of as a superposition of two linearly polarized waves whose polarization directions are perpendicular and whose oscillations are 90° out of phase. Show that the combination

$$\mathbf{E} = E_0 [\sin(kx - \omega t)\hat{\mathbf{j}} + \cos(kx - \omega t)\hat{\mathbf{k}}],$$

$$\mathbf{B} = B_0 [\sin(kx - \omega t)\hat{\mathbf{k}} - \cos(kx - \omega t)\hat{\mathbf{j}}]$$

does indeed represent such a circularly polarized wave (that is, show that the field magnitudes stay constant but that the field vectors rotate with angular frequency ω).

38. As you saw in Chapter 22, alignment of electric dipoles in a dielectric material reduces the electric field strength by a factor of $1/\kappa$, where κ is the dielectric constant. By appropriately modifying our proof that electromagnetic waves satisfy Maxwell's equations, show that the speed of light in a dielectric medium is $v = c/\sqrt{\kappa}$. This is the reason for our hint in Chapter 22 that the optical properties of a dielectric material are closely related to its dielectric properties, for the expression above shows that $\sqrt{\kappa}$ is just the index of refraction of the material. Things are not quite so simple, though, for the dielectric constant depends on frequency, and at the high frequencies of visible light, it may differ substantially from its low-frequency value of Table 24–1.

39. Studies of the origin of the solar system suggest that sufficiently small particles might be blown out of the solar system by the force of sunlight. To see how small such particles must be, compare the force of sunlight with the force of gravity and solve for the particle radius at which the two are equal. Assume the particles are spherical and have a density of 2 g/cm³. Why do you not need to worry about where in the solar system the particles are?

40. A cylindrical resistor of length ℓ and radius a has resistance R. The resistor carries a current I. Assume that the electric field is uniform throughout the wire, including right at its surface. Calculate the electric and magnetic fields at the surface of the wire, and show that they are at right angles to one another. Use Equation 32–29 to calculate the Poynting vector, and show that it points into the resistor. Calculate the flux of the Poynting vector (that is, $\int \mathbf{S} \cdot d\mathbf{A}$) over the surface of the resistor to get the total rate of energy flow into the resistor. Show that this power is just equal to the Joule heating $I^2 R$. Your result shows that the energy heating the resistor comes from the fields surrounding it. These fields are sustained by the source of electric power driving the current.

33

THE THEORY OF RELATIVITY

By the last quarter of the nineteenth century, the basic laws of electromagnetism had been formulated. Maxwell's theory had demonstrated the electromagnetic nature of light, and in the same era the work of Samuel Morse (telegraph), Alexander Graham Bell (telephone), Hertz and Marconi (radio), Thomas A. Edison (electric light, phonograph), and many others laid the foundation of electromagnetic technology. Yet at the same time the insights of Maxwell led to baffling questions and contradictions that shook the roots of physical understanding and even of common sense.

From the resolution of these contradictions arose the theory of relativity, a theory that radically altered the philosophical basis for our understanding of the physical world, and whose influence spilled over into all areas of twentieth-century thought. The theory of relativity stands as a monument to human intellect and imagination. It transcends the everyday world of common sense, and shows us a universe whose richness is almost beyond imagination.

33.1
SPEED c RELATIVE TO WHAT?

Maxwell's equations show that electromagnetic waves can exist, and that all such waves travel with speed c. Speed c relative to what? When we described the mechanics of a taut string in Chapter 14, we encountered a wave equation showing that waves could propagate along the string with a certain speed. A certain speed relative to what? Clearly, to the string. Similarly, sound waves propagate through the air with a certain speed relative to the air. If you move through the air, sound from sources in front of you approaches you at a speed greater than the speed of sound waves relative to the air. Sound from sources behind you approaches you at a lower speed (Fig. 33–1). You could even move so fast—in a supersonic airplane—that sound coming from behind could not catch up with you.

Fig. 33–1
Sound originating ahead of the plane moves, *relative to the plane,* at a speed greater than the speed of sound in air. Sound from behind the plane moves more slowly relative to the plane.

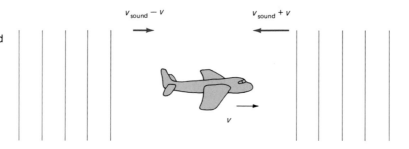

The Ether Concept

Each type of wave seems to have a characteristic speed *relative to the medium in which it propagates.* What about light? What is the medium through which it propagates? Light reaches us from the most distant galaxies, travelling through seemingly empty space. Yet all our other experience with waves suggests that there should be a medium—and that the speed c in Maxwell's equations should be the speed relative to that medium. Nineteenth-century scientists, understandably supposing that light waves were like mechanical waves, believed that light propagated through a tenuous substance called the **ether.** The ether permeated the entire universe, filling the smallest voids and permitting light to go anywhere. Electric and magnetic fields were visualized as stresses and strains in the ether. This mechanical view—that electromagnetic phenomena, including light, were disturbances of some substance—was deeply ingrained in the nineteenth-century scientists because of their previous experience that Newton's laws explained all known physical phenomena.

ether

The ether had to have some unusual properties. First, it must be tenuous and without significant viscosity, or it could not creep into every corner of the universe. And it must offer no resistance to the motion of material bodies, or the planets would soon lose their energy and spiral into the sun. At the same time the ether must be very stiff, for the speed of light is large. (If you make a spring stiffer, waves travel more quickly along it.) Indeed, the constants ϵ_0 and μ_0 must describe the mechanical properties of the ether that account for the high speed of light. These and other mechanical requirements make the ether a rather improbable substance, but without the ether it seemed there could be no waves, and the question "speed c relative to what?" would leave us floundering for an answer.

The existence of electromagnetic waves travelling at speed c follows from Maxwell's equations. But this result could be true only in a frame of reference fixed with respect to the ether, for if we move relative to the ether we should expect light to travel at a different speed relative to us. Therefore Maxwell's equations—that is, our description of electromagnetism—were presumably correct only in the ether's frame of reference.

This situation put electromagnetism in a rather different position from mechanics. In mechanics, the concept of absolute motion is meaningless. You can eat your dinner, or throw a ball, or do any mechanical experiment, as well on an airplane moving steadily at 1000 km/h as you can when the

Galilean relativity

airplane is standing still on the ground. You need not take the uniform motion of the plane into account. This is the principle of **Galilean relativity,** which states that the laws of mechanics are valid in all frames of reference in uniform motion (see Section 3.5). But the laws of electromagnetism could only be valid in the ether's frame of reference, for it seemed that only in this frame could the prediction of electromagnetic waves moving at speed c be correct.

33.2
MATTER, MOTION, AND THE ETHER

Given the existence of the ether, it is natural to ask about earth's motion relative to it. If earth is moving through the ether, we should expect light to travel faster relative to us when it comes from the direction toward which earth is moving. On the other hand, earth might be at rest relative to the ether. Because other planets, stars, and galaxies move with respect to earth, it is hard to imagine that ether is everywhere fixed with respect to earth alone, for this violates the Copernican view that earth does not occupy a privileged spot in the universe. But maybe earth drags with it the ether in its immediate vicinity. If this "ether drag" occurs, then the speed of light must be independent of direction, but if ether drag does not occur then the speed of light measured on earth must depend on direction. Through observation and experiment, nineteenth-century physicists sought to resolve the question of earth's motion through the ether.

Aberration of Starlight

Imagine standing in a rainstorm in which rain is falling vertically. To keep dry, you hold your umbrella with its shaft pointing straight up, as shown in Fig. 33–2a. But if you run, as in Fig. 33–2b, you will keep driest if you tilt your umbrella toward the direction in which you are running. Why? Because then the direction of rainfall *relative to you* is not straight down but at an angle, as shown in Fig. 33–2c. This argument presupposes that you do not drag with you a large volume of the air around you. If such an "air drag" occurred, raindrops entering the region around you would be accelerated quickly in the horizontal direction by the air moving with you, so that they would now fall vertically relative to you, as in Fig. 33–2d. No matter which way you ran, as long as you dragged air with you, you would point your umbrella vertically upward to stay dry.

This simple umbrella example is exactly analogous to the observation of light from stars, with the rain being starlight and the umbrella our telescope. If earth does not drag ether with it, the direction from which starlight comes will depend on the motion of earth relative to the ether. But if "ether drag" occurs in analogy with Fig. 33–2d, then light from a particular star will always come from the same direction.

In fact we do observe a change in the direction of starlight. As earth swings around in its orbit, we must first point our telescope one way to see a particular star. Then, six months later, earth's orbital motion is in exactly

Fig. 33–2
(a) Standing still in a vertically falling rain, you hold your umbrella overhead to keep driest. *(b)* Running through the rain, you tilt your umbrella to compensate for the rain's motion relative to you. *(c)* The situation of *(b)* seen from the runner's frame of reference. *(d)* If you dragged a volume of air with you, you would hold the umbrella overhead whatever your state of motion.

aberration of starlight

the opposite direction, and we must point our telescope in a slightly different direction. This phenomenon, called **aberration of starlight,** was discovered by James Bradley in England in 1725. Although the aberration angles are small, they are readily measurable with astronomical photography (see Problem 3). The aberration of starlight shows that *earth does not drag the ether.*

The Michelson-Morley Experiment

Aberration of starlight tells us that ether is not dragged by the earth. If we reject the pre-Copernican notion that earth alone is at rest relative to the ether, then we must conclude that earth moves through the ether. Furthermore, the relative velocity of that motion must change throughout the year, as earth orbits the sun.

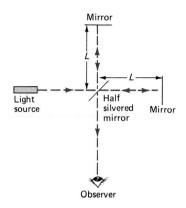

Fig. 33–3
Simplified diagram of the Michelson-Morley experiment.

In a series of experiments done in 1881–1887, the American scientists Albert A. Michelson and Edward W. Morley attempted to measure earth's velocity relative to the ether. Figure 33–3 shows a diagram of the apparatus Michelson and Morley used. Light from a source travels to a partially silvered mirror set at 45° to the light path. Half the light goes right through, and half is reflected through 90°. Travelling in mutually perpendicular directions, the two light beams are then reflected by ordinary mirrors and returned to the half-silvered mirror. Again, some of each beam is transmitted and some reflected, so that light from the two beams is again combined. The combined light is detected by an observer positioned at 90° relative to the light source. Together, the two beams form an interference pattern reflecting the relative phase of the light waves. If the paths travelled by the two beams have exactly the same length, and if the speed of light is the same in both directions, the waves will arrive exactly in phase. It is actually not important that the path lengths be the same. If they are not, the observer will see an interference pattern reflecting the phase difference due to difference in travel time over the different paths. If the apparatus is now rotated through 90°, and if the speed of light changes with direction, then there will be a change in that interference pattern.

Now suppose that earth moves at speed v relative to the ether. Then from the viewpoint of an observer on earth, there is an "ether wind" blowing past earth. Suppose that the Michelson-Morley apparatus is oriented with one light path parallel to the wind, the other perpendicular. Consider a light beam moving the distance L at right angles to the ether wind. The beam must be aimed slightly upwind, in order that it will actually move perpendicular to the wind. The light moves in this direction at speed c relative to the ether, but the ether wind sweeps it back so its path in the Michelson-Morley apparatus is at right angles to the wind. From Fig. 33–4, we see that its speed relative to the apparatus is

$$u = \sqrt{c^2 - v^2},$$

so that the round-trip travel time is

$$t_{\text{perpendicular}} = \frac{2L}{u} = \frac{2L}{\sqrt{c^2 - v^2}}. \qquad (33\text{--}1)$$

Fig. 33–4
Vector diagram showing resultant velocity **u** of light moving at right angles to the ether wind. v_{wind} is speed of ether wind and c that of light relative to the ether.

Light sent a distance L "upstream"—against the ether wind—travels at speed c relative to the ether but at speed $c - v$ relative to earth. It therefore takes a time

$$t_{\text{upstream}} = \frac{L}{c - v}.$$

Returning, the light moves at $c + v$ relative to earth, taking

$$t_{\text{downstream}} = \frac{L}{c + v}$$

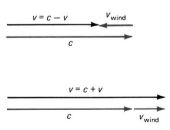

Fig. 33–5
Vector diagrams showing resultant velocities for light moving with and against ether wind. v_{wind} is the ether wind speed, c the speed of light relative to the ether, and u the resultant speed of the light.

(Fig. 33–5). So the round-trip time parallel to the wind is

$$t_{\text{parallel}} = \frac{L}{c - v} + \frac{L}{c + v} = \frac{2cL}{c^2 - v^2}. \qquad (33\text{--}2)$$

The two round-trip travel times differ, with the trip parallel to the ether wind always taking longer (see Problems 1, 2, and 4). Light on the parallel trip is slowed when it moves against the ether wind, then speeds up when it moves with the wind. But slowing always dominates, because the light spends more time moving against the wind than with it.

The Michelson-Morley experiment of 1887 was sensitive enough to detect differences in the speed of light at least an order of magnitude smaller than earth's orbital speed. The experiment was repeated with the apparatus oriented in different directions, and at different times throughout the year, and the same simple but striking result always emerged: there was never any difference in the travel times for the two light beams. In terms of the ether concept, the Michelson-Morley experiment showed that *earth does not move relative to the ether.*

A Contradiction in Physics

Aberration of starlight shows that earth does not drag ether with it. Earth must therefore move relative to the ether. But the Michelson-Morley experiment shows that it does not. This contradiction is a deep one, rooted in the fundamental laws of electromagnetism and in the analogy between mechanical waves and electromagnetic waves. The contradiction arises directly in trying to answer the simple question "with respect to what does light move at speed c?"

Physicists at the end of the nineteenth century made many ingenious attempts to resolve the dilemma of light and the ether, but their explanations were either inconsistent with experiment or lacked sound conceptual bases.

Fig. 33–6
The Einstein statue in Washington, D.C.

33.3
SPECIAL RELATIVITY

In 1905, at the age of 26, Albert Einstein (Fig. 33–6) proposed a theory that resolved the dilemma and at the same time altered the very foundation of physical thought. Einstein declared simply that the ether is a fiction. But then with respect to what does light move at speed c? With respect, Einstein declared, to anyone who cares to observe it. This statement is at once simple, radical, and conservative. Simple, because its meaning is clear and obvious. Anyone who measures the speed of light will get the value $c = 3.0 \times 10^8$ m/s. Radical, because it alters our common-sense notions of space and time. Conservative, because it asserts for electromagnetism what had long been true in mechanics: that the laws of physics do not depend on the motion of the observer. Einstein summarized his new ideas in the *special theory of relativity,* which is expressed in the simple sentence:

> The laws of physics are the same in all frames of reference in uniform motion.

This statement encompasses *all* laws of physics, including the laws of electromagnetism. The prediction that electromagnetic waves move at speed c

must, in the absence of an ether, be a universal prediction that holds in *all* frames of reference in uniform motion. The *special* theory of relativity is special because it is valid only for the special case of *uniform motion*. Later we will discuss the general theory of relativity, in which this restriction is removed.

Einstein's relativity readily explains the result of the Michelson-Morley experiment,* for no matter what the speed of earth relative to anything, an observer on earth should measure the same speed for light in all directions. But at the same time, we will see that relativity flagrantly violates our common-sense notions of space and time.

Space and Time in Relativity

Consider a car driving past a person standing by the roadside (Fig. 33–7). The driver of the car and the person by the roadside each measure the speed of the light from a blinking traffic signal. Relativity says they will get the same answer, $c = 3.0 \times 10^8$ m/s, even though the car is moving toward the source of light. How can this be? Consider how each observer might make the measurement. Let each be equipped with a meter stick and an accurate, high-speed electronic stopwatch. Suppose that a light pulse passes the front ends of both meter sticks just as they coincide. Each observer measures the time it takes the light pulse to cross the meter stick, then divides the distance (one meter) by the measured time to get the speed of light. Since the stick on the car is moving toward the light source, common sense suggests that the light will pass the far end of this stick first, and therefore that the time on the car's stopwatch will be shorter. But this violates relativity, for if both observers use the same path length for their speed-of-light measurement, and if they get different times, then they will have measured different speeds for light. In fact, both stopwatches will read the same time, even though common sense tells us that the light passes the end of the "moving" meter stick "earlier."

How can this be? It follows logically from the statement of special relativity, which in turn is consistent with physical experiments. But how can it be? Something must be "wrong" with someone's meter stick or stopwatch. Maybe the motion of the car somehow affects the stopwatch on the car. But no: this suggestion violates the spirit of relativity, which says that steady, uniform motion is undetectable—that it makes no more sense to say that the car is moving and the person by the road is at rest than to say the opposite. That is the whole point of relativity—any frame of reference in uniform motion is as good as any other for doing physics. So there can be nothing wrong with the clocks and meter sticks.

The only things left to go "wrong" are time and space. Time and space—the seemingly passive, universal backgrounds in which all physical events take place—must themselves depend on the observer. Two observers in different frames of reference, moving uniformly relative to each other, are measuring different quantities when they use clocks to record the

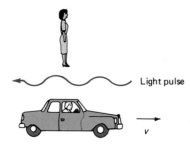

Fig. 33–7
As the car passes the person by the roadside, a light pulse goes by. Both measure the same speed *c* for the light, even though they are in relative motion.

Light pulse

v

---

*Nevertheless, there is controversy about whether Einstein was fully aware of that experiment. See *Subtle Is the Lord: The Science and Life of Albert Einstein*, by Abraham Pais (New York: Oxford University Press, 1982), for a discussion of the historical development of Einstein's thought.

passage of time and meter sticks to determine distances in space. In relativity it is the laws of physics—not measures of time and space—that must be the same for everyone. Time and space are altered in ways that allow the laws of physics, including Maxwell's theory and its prediction that light waves travel at speed c, to be the same for all observers in uniform motion.

Time Dilation

To see how time is altered, consider the simple device shown in Fig. 33–8. It consists of a box of length L with a light source at one end and a mirror at the other. Let a flash of light leave the source, travel to the mirror, and return to the source. We want to know the time interval between two distinct events: the light flash leaving the source and the flash returning to the source.

In Fig. 33–8a we consider the experiment in a frame of reference S' at rest with respect to the box. The light travels a distance $2L$ in this frame, giving a round-trip travel time of

$$\Delta t' = \frac{2L}{c}. \tag{33–3}$$

Now consider the *same* experiment, viewed from a frame of reference S in which the box is moving to the right with speed v. In this frame there are two clocks along the path of the box. These clocks are synchronized, and are in the same frame of reference, so they both measure the same quantity. The box passes the first clock just as the light flash goes off, and at this instant the clock reads zero. Just as the light flash returns to the source, the box passes the second clock. The time this clock reads at the

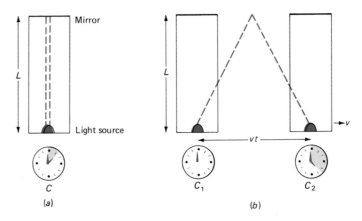

Fig. 33–8
A "light box," consisting of a box in which a light pulse leaves a source at one end, reflects from a mirror at the other end, and returns to its source. Dashed line is light path. The clock C is attached to the box, while the box moves between clocks C_1 and C_2 at speed v. (a) In a frame of reference at rest with respect to the box, the light travels a distance $2L$ at the speed of light c. (b) In a frame of reference in which the box is moving, the light travels a greater distance. But since its speed must still be c, the time interval between the light's leaving the source and its return must be longer. Part (b) shows the box both when the light is emitted and again when it returns to its source.

instant the box passes is the time, measured in frame S, between the emission and return of the light flash.

We can calculate this time interval Δt in frame S, just as we did in S', by figuring the total distance travelled by the light and dividing by its speed. Figure 33–8b shows the situation. In frame S, the box moves to the right a distance $v\Delta t$ in the time Δt between emission and return of the light flash. Meanwhile the light takes a diagonal path up to the mirror of the moving box, then back down to the source. The total length of this path is twice the diagonal from source to mirror, or, using the Pythagorean theorem,

$$\text{light path} = 2\sqrt{L^2 + (v\Delta t/2)^2}.$$

The time required for light to go this distance is just the distance divided by the speed of light, or

$$\Delta t = \frac{2\sqrt{L^2 + (v\Delta t/2)^2}}{c}. \tag{33–4}$$

Notice that we explicitly used the theory of relativity in writing Equation 33–4. We did not vectorially add the horizontal speed of the box to the vertical speed of light to get a new speed of light in frame S, for relativity says that the speed of light is the same in all uniformly moving frames of reference. Had we altered the speed, we would have had an increased path length in S, but an increased speed of light as well, and would have found that the time intervals in both frames were the same. But no! Relativity requires that we use the same speed c in both frames, even though the path lengths differ. That is why we get different answers for the time.

Our unknown time Δt appears on both sides of Equation 33–4. Multiplying through by c and squaring gives

$$c^2(\Delta t)^2 = 4L^2 + v^2(\Delta t)^2.$$

We then solve for $(\Delta t)^2$ to get

$$(\Delta t)^2 = \frac{4L^2}{c^2 - v^2} = \frac{4L^2}{c^2}\left(\frac{1}{1 - v^2/c^2}\right).$$

Taking the square root of both sides, and noting that $2L/c$ is just the time $\Delta t'$ measured in the frame S' at rest with respect to the box, we have

$$\Delta t = \frac{\Delta t'}{\sqrt{1 - v^2/c^2}} \quad \text{or} \quad \Delta t' = \Delta t\sqrt{1 - v^2/c^2}. \tag{33–5}$$

time dilation

Equation 33–5 describes the phenomenon of **time dilation,** in which the time between two events is always shortest in a frame of reference in which the two events occur at the same place. In our example, the two events are the emission and return of the light flash, and they occur at the same place—the bottom of the box—in the box frame S', but at different places in S. Time dilation is sometimes characterized by saying that "moving clocks run slow," although this statement is not strictly correct because relativity rules out our saying that one clock is moving and another not. What the statement means is that the time interval between two events will

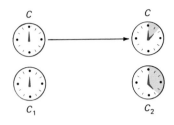

Fig. 33–9
Clock C moves between clocks C_1 and C_2, which are at rest relative to each other and synchronized in their rest frame. Time between the event of C passing C_1 and the event of C passing C_2 is shorter in C's frame of reference than in the frame of C_1 and C_2.

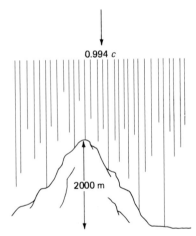

Fig. 33–10
The rate at which high-speed muons are incident is measured on a mountaintop and again at sea level. The muons decay at such a rate that only a small fraction should survive to reach sea level. But experimentally, the number reaching sea level is much larger—showing the effect of time dilation on the muon lifetimes.

be shortest in a frame of reference where the two events occur at the same place (Fig. 33–9).

We have illustrated time dilation with a very special device—a "light box." But the phenomenon would occur with any other timing device, for it is not that something unusual happens to our clock, but to time itself. If we take away the light box in Fig. 33–8, giving Fig. 33–9, the clocks will still show the same discrepancy. There is no use searching for a physical mechanism that slows things down. All manifestations of time—the oscillations of the quartz crystal in a digital watch, the swing of a pendulum clock, the period of vibration of atoms in an atomic clock, biological rhythms, and human lifetimes—all are affected in the same way.

Why don't we notice time dilation as we travel about in our everyday lives? Because the factor v^2/c^2 in Equation 33–5 is so small for any velocities we have relative to earth. Even in a jet airplane, we are moving at 1000 km/h or only about $10^{-6}c$. Then time in the airplane is different from that on earth by only about 1 part in $(10^6)^2$, which amounts to a few milliseconds per century. This illustrates an important point: any results predicted by relativity should agree with our common-sense, Newtonian ideas when relative velocities are small compared with the speed of light. Only at substantial fractions of c do relativistic effects become obvious. Since our intuitions and common sense are built on experience at low relative velocities, it is not surprising that effects at high velocities seem counter to common sense.

Example 33–1
Confirming Time Dilation

Time dilation is clearly illustrated in experiments with subatomic particles moving relative to us at speeds near that of light. A classic experiment, first performed by Rossi and Hall in 1941, uses the lifetimes of unstable particles called muons to measure time dilation.* Muons are created by the interaction of cosmic rays with earth's upper atmosphere. The experiment consists in counting the number of muons incident each hour on a mountain—Mt. Washington in New Hampshire, about 2000 m above sea level. The measurement is then repeated at sea level (Fig. 33–10). Using a detector arrangement that records only those muons moving at about $0.994c$ at the mountaintop altitude, it is found that on the average muons with this speed are incident on the mountaintop at the rate of about 560 per hour. The muons are unstable, meaning that they eventually decay into other particles. Their decay rate is such that, given 1000 muons, about 740 would remain 0.73 μs later, while only about 45 would remain 6.7 μs later. This statement is true in a frame of reference at rest with the muons.

Given 560 muons incident at $0.994c$ on the top of Mt. Washington, (a) how long would it take, as measured by an observer at rest with respect to the mountain, for the muons to reach sea level, 2000 m below? (b) If time for the muons ran at the same rate as for the observer at rest with respect to the mountain, how many muons would survive to reach sea level? (c) Time dilation suggests that the 2000-m trip may take less time in the muon frame of reference. Calculate the time in the muon

*This experiment is summarized by D. H. Frisch and J. H. Smith in the *American Journal of Physics*, vol. 31, pp. 342–355 (1963), and by the same authors in the film *Time Dilation—an Experiment with mu-Mesons*, produced by the Educational Development Center of Newton, MA, in 1963.

Fig. 33–11
At departure, the twins are the same age.

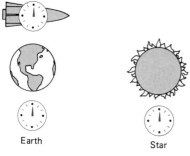

Earth   Star

(a)

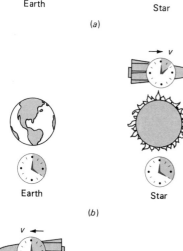

Earth   Star

(b)

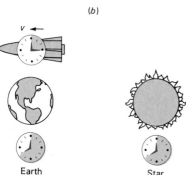

Earth   Star

(c)

frame, and from it infer how many muons should reach sea level if time dilation really occurs.

Solution

Moving at 0.994c, the muons take a time

$$\Delta t = \frac{2000 \text{ m}}{(0.994)(3.0 \times 10^8 \text{ m/s})} = 6.7 \ \mu s$$

to reach sea level. This is just the time it takes 955 out of 1000 muons to decay, leaving only 45. Out of 560 muons incident each hour on the mountaintop, then, we should expect

$$(560)\left(\frac{45}{1000}\right) = 25$$

muons to reach sea level each hour if time in the muon frame of reference is the same as in the mountain frame of reference. But if time dilation occurs, then the elapsed time in the muon frame is given by Equation 33–5:

$$\Delta t' = \Delta t\sqrt{1 - v^2/c^2} = (6.7 \ \mu s)\sqrt{1 - 0.994^2} = 0.73 \ \mu s.$$

According to our information on decay rates, we should expect 740 out of 1000 muons to survive for this time. Out of 560 original muons, then, we should find

$$(560)\left(\frac{740}{1000}\right) = 414$$

muons surviving to reach sea level.

So what happens? When the experiment is done, muon counts of just over 400 per hour are recorded at sea level. This is no subtle effect! The difference between 25 and 414 is dramatic. At 0.994c, the non-relativistic description is hopelessly inadequate. At this speed, the factor $\sqrt{1 - v^2/c^2}$ is one-ninth, and time dilation is obvious.

The Twin Paradox

The phenomenon of time dilation has startling consequences, for it allows us to travel into the future! The famous "twin paradox" illustrates this possibility. One of two twins boards a fast spaceship for a journey to a distant star (Fig. 33–11). The other stays behind on earth. Imagine that there are clocks at earth and star, like the two clocks in frame S of our light-box experiment (Fig. 33–12a). There is a clock on the spaceship, like the one clock in our light-box frame S′. When the ship arrives at the distant star, less time will have elapsed on the ship clock than on the earth and star clocks (Fig. 33–12b). Now the ship turns around and comes home. Again, the situation is identical to our light-box example, so that again less time elapses on the ship clock, and the travelling twin arrives home younger than the earthbound twin (Fig. 33–13)! Depending on how far and how fast the travelling twin goes, the difference in ages could be arbitrarily

Fig. 33–12
The travelling twin journeys to a distant star, then returns. Earth and star are at rest with respect to each other, and their clocks are synchronized. The figure is drawn from the earth-star frame of reference. (a) Ship and earth clocks agree as the ship leaves earth. (b) When the ship reaches the star, less time has elapsed on the ship than on earth and star. (c) When the ship returns to earth, its clock and earth's clock no longer agree.

Fig. 33–13
It is not only clock readings, but all manifestations of time that differ. The travelling twin really has aged less!

large. The travelling twin could even return to earth millions of years in the future, even though only hours had elapsed on the ship. But this is a one-way trip to the future! If the traveller doesn't like what he or she finds a million years in the future, there is no going back! Example 33–2 and Problem 10 explore the twin paradox more quantitatively.

Example 33–2
The Twin Paradox

Earth and a star are 10 light-years (ly) apart, measured in a frame at rest with respect to earth and star (here we neglect any slight relative motion between earth and star). Twin A boards a space ship and travels at 0.80c to the star, then immediately turns around and returns to earth at 0.80c. Twin B remains behind. Determine the round-trip travel time in the earth-star frame of reference and in the ship frame. By how much will the twins' ages differ when they get back together?

Solution

At 0.80c, the time to go 10 ly in the earth-star frame is just

$$\Delta t = \frac{10 \text{ ly}}{0.80 \text{ ly/y}} = 12.5 \text{ y},$$

where in working with light-years and years, we use $c = 1$ ly/y. The round-trip time is then 25 y. Equation 33–5 for $\Delta t'$ then gives the one-way travel time in the ship frame:

$$\Delta t' = \Delta t \sqrt{1 - v^2/c^2} = (12.5 \text{ y})(\sqrt{1 - 0.80^2} = 7.5 \text{ y}.$$

Then the round-trip time in the ship frame is 15 y, so the twins' ages differ by 10 y when twin A returns.

The paradox in the twin example is not just that something strange happens to time—we already expect that of special relativity. But now look at things from the spaceship's frame of reference. Doesn't the spaceship see the earth recede into the distance, turn around, and come back? And then shouldn't the earthbound twin be younger? This is the paradox. It is resolved by considering what is *special* about the special theory of relativity. The special theory applies only to frames of reference in *uniform* motion. The travelling twin must accelerate in order to return to earth, and is therefore not always in uniform motion. Absolute motion has no meaning in special relativity, but absolute acceleration does. The travelling twin feels inertial forces when the ship turns around, but the earthbound twin does not feel any comparable jolt. Although we cannot say that one twin is moving and the other is not, we can say that one twin's motion changes and the other's does not. The situation is not symmetric, and that is why the travelling twin really does return younger.

Could it really happen? It could, and it has. Atomic clocks are now so accurate that experiments have been done to detect the minuscule time difference between a clock flown around the earth on an airplane and one left behind.*

---

*See "Around the World Atomic Clocks: Predicted Relativistic Time Gains," and "Around the World Atomic Clocks: Observed Relativistic Time Gains," both by J. C. Hafele and R. E. Keating, *Science*, vol. 177, pp. 166–170 (July 1972).

What if the travelling twin did not turn around? We could still argue that the ship clock runs slower than clocks on the earth and star. But then isn't the situation symmetric? Couldn't the travelling twin argue that a clock on earth should run slower than clocks on the ship? Yes—and we would find this to be true if we set up a series of clocks in the ship frame and measured time intervals on the earth clock as earth passed first one, then another of the ship clocks. But there is no contradiction. Unless the two twins get together again, there is no way they can directly compare their clocks or their ages at the same place. Instead, they must compare one clock with a sequence of clocks that are all synchronized. And as we will soon see, clocks that are synchronized—all reading noon at the same instant—in one frame of reference are not synchronized in another frame! Only if the twins get back together can they compare just two clocks without having to worry about synchronization of distant clocks. And they can get back together only if at least one of them accelerates.

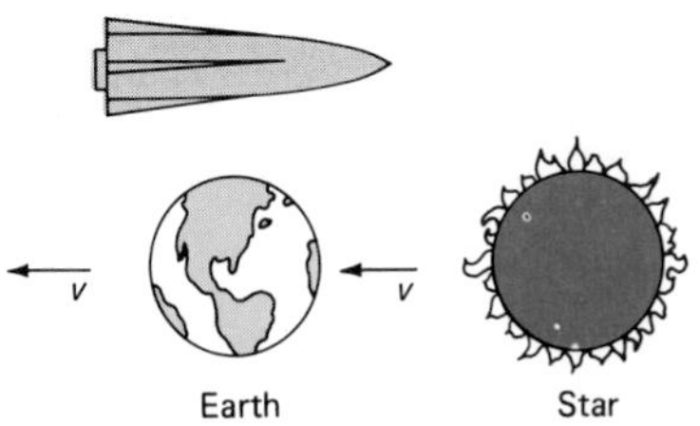

The Lorentz Contraction

Fig. 33–14
Earth-star trip viewed from the spaceship's frame of reference. Note that earth-star distance is contracted relative to Fig. 33–12, which is drawn from the earth-star frame.

In Example 33–2 the spaceship moved 10 ly in 12.5 y at speed 0.80c. These quantities are related by the simple expression $\Delta x = v\Delta t$, where Δx is the distance between earth and star *measured in the earth-star frame of reference S*. Now an observer in the ship frame of reference S' sees earth and star moving past at speed v. First earth passes the ship, then the star passes (Fig. 33–14). We found that the time interval between these two events, measured in the ship frame, is $\Delta t' = \Delta t\sqrt{1 - v^2/c^2} = 7.5$ y. Since earth and star are moving past at $v = 0.80\ c$, the distance between earth and star as measured in the ship frame must be

Lorentz contraction

$$\Delta x' = v\Delta t' = v\Delta t\sqrt{1 - v^2/c^2} = \Delta x\sqrt{1 - v^2/c^2}, \qquad (33\text{–}6)$$

or 6.0 ly in our example. This equation shows that the distance between two points is always greatest in a frame (the earth-star frame S, in this example) fixed with respect to those points. In any other frame the distance is smaller. This phenomenon is called the **Lorentz contraction,** or the Lorentz-Fitzgerald contraction, after the Dutch physicist H. A. Lorentz and the Irish physicist George F. Fitzgerald, who, in the 1890's, independently proposed it as an ad hoc way of explaining the Michelson-Morley experiment. Only through Einstein's theory did the contraction acquire a solid conceptual basis.

Although we developed the Lorentz contraction using the distance between separate objects—earth and a distant star—the effect occurs for any observer moving with respect to two points that are fixed with respect to each other. In particular, a rigid object like a meter stick or spaceship is shorter when measured by an observer with respect to whom it is moving. *Is* shorter? Shouldn't we say "seems shorter"? Isn't a meter stick "really" always one meter long? No! To claim otherwise is to violate the spirit of relativity. All observers in uniform motion have equal claim on "reality," in that the laws of physics are equally valid for all of them. To say that an object is "really" one meter long is to give special status to a particular state of motion, and that is precisely what relativity prohibits. What we can say is that a meter stick will measure one meter long for an observer at rest with respect to it. For any other observer, it is shortened in the direction of its relative motion.

As with time dilation, do not look for some physical mechanism that squashes moving objects. Rather, it is space itself that is different for different observers. In order to accept the simple fact that absolute motion is meaningless, we must alter our common-sense notions of time and space. Lorentz contraction and time dilation are manifestations of that alteration.

Example 33–3
Time Dilation and Lorentz Contraction

At the Stanford Linear Accelerator Center (SLAC), subatomic particles are accelerated to high energies over a straight path 3.2 km long (measured in a frame at rest with respect to SLAC; Fig. 33–15). During a particular experiment, electrons are accelerated to 0.999 999 5 of the speed of light. In the SLAC frame, how long would it take electrons with this speed to travel the full length of the device? How long would the trip take in the rest frame of the electrons? How long would the accelerator be in the rest frame of the electrons?

Solution

The electron speed is so close to that of light that the travel time is, to a very good approximation,

$$\Delta t = \frac{\Delta x}{c} = \frac{3.2 \times 10^3 \text{ m}}{3.0 \times 10^8 \text{ m/s}} = 1.1 \times 10^{-5} \text{ s}.$$

In the rest frame of the electrons, the time to traverse the accelerator is given by Equation 33–5, or

$$\Delta t' = \Delta t \sqrt{1 - v^2/c^2} = (1.1 \times 10^{-5} \text{ s})\sqrt{1 - 0.999\,999\,5^2}$$

$$= (1.1 \times 10^{-5} \text{ s})(1.0 \times 10^{-3}) = 1.1 \times 10^{-8} \text{ s}.$$

Our time calculation shows that the relativistic factor $\sqrt{1 - v^2/c^2}$ is 10^{-3}, so in the electron frame of reference the accelerator length is

$$\Delta x' = \Delta x \sqrt{1 - v^2/c^2} = 3.2 \times 10^{-3} \text{ km} = 3.2 \text{ m}.$$

Fig. 33–15
The Stanford Linear Accelerator is 3.2 km (2 miles) long in earth's frame of reference. But for electrons moving along the accelerator at 0.9999995c, the accelerator is only 3.2 m long.

Example 33–4

Time Dilation and Lorentz Contraction

A physics student from New York flies to San Francisco to do an experiment at SLAC. She travels a distance of 4800 km on a plane going at 1000 km/h. How long does the trip take in a frame at rest with respect to earth? How long does it take according to the student's watch? How far is it from New York to San Francisco in the airplane's frame of reference?

Solution

At 1000 km/h, the 4800-km trip takes

$$\Delta t = \frac{\Delta x}{v} = \frac{4800 \text{ km}}{1000 \text{ km/h}} = 4.8 \text{ hours.}$$

In the frame of the moving airplane, time and distance are altered by the relativistic factor $\sqrt{1-v^2/c^2}$. The speed of the plane is 1000 km/h, or 280 m/s, so that

$$\sqrt{1-v^2/c^2} = \left[1 - \frac{(280 \text{ m/s})^2}{(3.0 \times 10^8 \text{ m/s})^2}\right]^{1/2} \simeq 1 - (\tfrac{1}{2})(8.7 \times 10^{-13}) = 0.999\,999\,999\,999\,56.$$

Here we used the binomial theorem, $(1+x)^n \simeq 1 + nx$ for $|nx| \ll 1$ because most calculators do not carry enough significant figures to distinguish the result from 1 (see Appendix A). We really need not carry the calculation further. The time on the student's watch is the same as the time on the ground to about five parts in 10^{13}, and the distance in the plane's frame is the same as the ground distance to within this factor. The student need take no account of time dilation and Lorentz contraction, except in a physics exam or high-energy experiment!

Examples 33–3 and 33–4 show that relativistic effects are significant only at high relative velocities—so high that the quantity v^2/c^2 is comparable with 1. In our daily lives we have no experience with such velocities. It is for this reason that relativity so offends our common-sense notions of space and time. Those notions are built on a groundwork of limited experience that does not include high velocities. If we did move regularly with respect to our surroundings at speeds near that of light, the relativity of space and time would be as obvious as our common-sense notions now seem.

For physicists working with high-energy elementary particles, relativistic effects *are* obvious. Unstable particles moving through the laboratory at high speeds live longer than they would at rest in the lab. And high-energy particle accelerators would not work if their design did not take relativity into account.

Events and Simultaneity

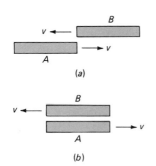

Fig. 33–16
(a) In the frame *S*, both sticks have the same speed and both are contracted by the same amount. *(b)* Therefore their opposite ends coincide at the same time.

Consider two identical sticks *A* and *B*, each of length *L* when measured in its rest frame. Suppose these sticks are moving toward each other. For a frame of reference *S* in which both sticks are moving at the same speed *v*, the situation is shown in Fig. 33–16. Both sticks are Lorentz contracted, but since both are moving at the same speed *v* relative to the frame *S*, both are contracted by the same amount and therefore have the same length. What happens as the sticks pass each other? First, the right end of stick *A* and the left end of stick *B* pass (Fig. 33–16*a*). A little while later, the right end of *A* passes the right end of *B*. At the same time, *because the sticks have the same length in S*, the left end of *A* passes the left end of *B* (Fig. 33–16b). The passing of the two right ends of the sticks is an event that we designate E_1. Similarly, the passing of the two left ends we designate E_2. We have shown that, in the frame S, the two events E_1 and E_2 are **simultaneous**—they occur at the same time.

Now let us look at the situation from a frame of reference S′ in which stick *A* is at rest. In this frame, stick *B* moves toward stick *A*. Since we are at rest relative to stick *A*, it has its full length L. But stick *B* is contracted more than in frame S because of its higher relative velocity. The situation is shown in Fig. 33–17. As the figure indicates, the event E_1 occurs before E_2; the two events are not simultaneous in the frame S′. What happens in a frame S″ at rest with stick *B*? As Fig. 33–18 shows, the events E_1 and E_2 are again not simultaneous, and this time event E_2 occurs first.

Isn't this all just an illusion arising from the apparent length differences due to motion of the sticks? Isn't the picture in frame S (Fig. 33–16) "really" the right one? No! Relativity theory assures us that all uniformly moving frames—including the frames S, S′, and S″ of Figs. 33–16 through 33–18—are equally valid frames of reference for describing physical reality. The length differences and the changes in time ordering of events E_1 and E_2 are not "apparent" and they are not "illusions." They arise from valid descriptions in different frames of reference, and each has equal claim to "reality." If you insist that one of the frames—say S—somehow has more validity, then you are reasserting the nineteenth-century notion that there is one favored reference frame in which alone the laws of physics are valid.

But how can observers disagree on the time order of events? Doesn't this violate causality? After all, if one event is a cause of another, we certainly expect the cause to precede the effect. It would be disturbing if some observer, with equal claim on "reality," found that cause and effect occurred in the reverse order. But there is no violation of causality. As we will soon show, the only events that can have their time order reversed are those that are so far apart in space, and so close in time, that not even light can travel fast enough to be at both events. There is no way that such events can influence each other, and therefore they cannot be causally related. In a very real sense it does not matter which event occurs first, and indeed different observers will disagree on their relative time order. For example, an event on earth now and another occurring five minutes from now on the sun cannot be causally related, for it takes light from the sun eight minutes to reach earth. For observers moving at high enough speeds relative to earth and sun, the solar event occurs first.

Only when events are close enough in space and separated enough in time so that light can travel from one to the other can the two be causally related. In that case, all observers will agree about their time order, although they may disagree about the actual time interval between the events. For example, an event on earth now and another occurring fifteen

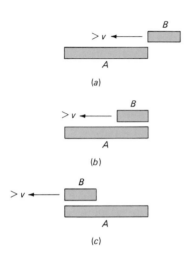

Fig. 33–17
The passing sticks described in a frame of reference S′ at rest with respect to stick *A*. Relative speed is greater than the speed *v* of each stick with respect to frame S. *(a)* First the left end of stick *B* passes the right end of *A*. *(b)* A while later, the right end of *B* passes the right end of *A*. This is event E_1. *(c)* Still later, the left end of *B* passes the left end of *A*. This is event E_2. The events E_1 and E_2 are not simultaneous in the frame S′.

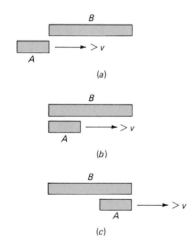

Fig. 33–18
The passing sticks described in a frame of reference S″ at rest with respect to stick *B*. *(a)* First the right end of stick *A* passes the left end of stick *B*. *(b)* A while later, the left end of *A* passes the left end of *B*. This is event E_2. *(c)* Still later, the right end of *A* passes the right end of *B*. This is event E_1. The events E_1 and E_2 are not simultaneous in the frame S″, and their time order is opposite what it was in frame S′.

minutes from now on the sun could be causally related, and therefore all observers will agree that the terrestrial event occurs first. We will explore these notions more quantitatively in the next section.

The Lorentz Transformations

Lorentz transformations

Our demonstration that the time order of events may be relative deals implicitly with the coordinates—position and time—of specific events, and suggests that those coordinates may differ for different observers. Similarly, time dilation and Lorentz contraction arise as specific instances of the way positions and times in one frame of reference are related to their values in another frame. We now seek more general expressions—called **Lorentz transformations**—relating the space and time coordinates of an event in two frames of reference in relative motion. Consider coordinate axes in a frame of reference S and in another frame S' moving in the x-direction with speed v relative to S. Suppose that the origins of the two coordinate systems coincide at time $t = t' = 0$. If an event has coordinates x, y, z, and t in S, what are its coordinates x', y', z', and t' in S'? Were it not for relativity, we would expect the coordinates y, z, and the time t to remain unchanged from one coordinate system to the other. For an event occurring at time $t=0$, when the two origins coincide, we would expect the x-coordinates to be the same too. But as S' moves in the positive x-direction relative to S, a given x-value in S would correspond to a value $x' = x - vt$ in S' (Fig. 33–19).

How does relativity alter this coordinate transformation? First, there can be no change in the coordinates y and z at right angles to the relative motion, for if there were then in one frame distances along the y- and z-axes would be unambiguously shorter, and observers in both frames would agree about this (Fig. 33–20). But then there must be something special about the frame with the shorter y or z distances, and it is just such specialness that relativity prohibits.

Because of the Lorentz contraction of distances along the direction of relative motion, we expect that the simple expression $x'=x-vt$ will need modification to be consistent with relativity. Any new expression we develop, however, must reduce to the nonrelativistic expression $x'=x-vt$ in the limit when $v \ll c$. A simple form that could have this property is

$$x' = \gamma(x - vt), \tag{33–7}$$

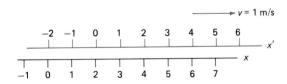

Fig. 33–19
Nonrelativistic picture of two coordinate axes in relative motion. The x-axis of S' moves to the right with speed v relative to the x-axis of S. At time $t=0$ the two axes coincided. Were it not for relativity, the x-coordinate of an event in S' would be its coordinate in S minus the distance the origin of S' has moved, or $x'=x-vt$. Here the relative speed is 1 m/s, so an event occurring at $x=5$ m and $t=2$ s occurs in S' at $x'=3$ m.

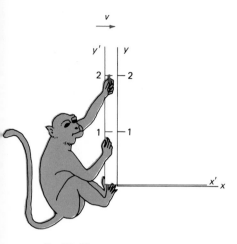

Fig. 33–20
An observer holding a pen at $y'=2$ in S' marks the y-axis in S as the two axes pass. Observers in both frames agree unambiguously about the location of the mark. Unless it is at $y=2$, there must be something special about the frame S' that makes y distances shorter. Relativity precludes such "specialness."

where γ is a number that depends on the relative velocity v but not on the coordinates x and t. We have simply guessed at the form 33–7, using our knowledge of how the transformation must look for very low velocities. There is at this point no guarantee that our guess is correct. But if we can find a value of γ that makes Equation 33–7 consistent with relativity, and with the property that $\gamma\to1$ for $v\ll c$, then we will have the relativistic transformation we seek.

Equation 33–7 is our guess for the transformation from coordinates x and t in the frame of reference S to x' in S'. But we could also transform the other way. The only difference is the direction of relative motion; frame S' is moving in the positive x-direction relative to S, while S is moving in the negative x-direction relative to S'. Therefore, the transformation from x' and t' to x must look just like Equation 33–7 except with v replaced by $-v$:

$$x = \gamma(x'+vt'). \qquad (33\text{--}8)$$

To see if we can make Equations 33–7 and 33–8 consistent with relativity, we impose the requirement that the speed of light be the same in both frames of reference. Specifically, suppose that a light flash goes off at the origin $x=0$ at time $t=0$. Since the origins of our frames coincide at this time, and since clocks at the origin in S and S' both read zero when the origins coincide, the light flash also occurs at $x'=0$ and $t'=0$ in frame S'. Let us call this event—the emission of the light flash—event E_1. Some time later, an observer at some position x in S observes the light flash. Let us call this event E_2. When does E_2 occur? Since light travels at speed c, we must have $x=ct$. Now in frame S' event E_2 has some coordinates x' and t'. But relativity requires that the speed of light in S' also be c, so we must have $x'=ct'$. Substituting $x=ct$ and $x'=ct'$ into Equations 33–7 and 33–8 gives

$$ct' = \gamma(ct-vt) = \gamma t(c-v)$$

and

$$ct = \gamma(ct'+vt') = \gamma t'(c+v).$$

Multiplying together the left-hand sides of these equations, and then the right-hand sides, and equating the results gives

$$c^2t't = \gamma^2tt'(c-v)(c+v),$$

so that

$$c^2 = \gamma^2(c^2-v^2),$$

or

$$\gamma = \frac{1}{\sqrt{1 - v^2/c^2}}. \qquad (33\text{--}9)$$

That we could find a value of γ depending on the relative velocity but not on the coordinates shows that our guess for the form of the transformation equations was correct. Equations 33–7 and 33–8, with γ given by Equation 33–9, are the relativistically correct transformations for the coordinates x and x'. Taking $v\ll c$ in Equation 33–9 shows that $\gamma\to1$ in this limit, so our transformation equations correctly reduce to the nonrelativistic result at low relative velocities.

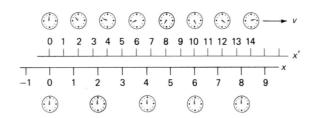

Fig. 33–21
As measured in frame S, lengths in S' are contracted and the clocks in S' are not synchronized.

Solving Equations 33–7 and 33–8 simultaneously for t' and t (see Problem 14) gives the transformation equations for time:

$$t' = \gamma\left(t - \frac{vx}{c^2}\right) \qquad (33\text{–}10)$$

and

$$t = \gamma\left(t' + \frac{vx'}{c^2}\right). \qquad (33\text{–}11)$$

Because we know about time dilation, we should not be too surprised to find that measures of time differ between the two frames of reference. But why should the time in one frame depend not only on the time in the other frame but also on location in space? Because, as we have found, events that are simultaneous in one frame of reference are not simultaneous in a frame of reference moving relative to the first. In this case, our simultaneous events are the pointing of all clock hands to the same time. For the clocks in S', these events are simultaneous in S'. But they are not simultaneous in S. In fact, as Equation 33–10 shows, clocks that are farther to the right in S' read successively earlier times (Fig. 33–21). The term vx/c^2 in Equation 33–10 and its analog in Equation 33–11 account for this non-synchronism of clocks in one frame as measured from the other frame.

Our earlier qualitative discussion of simultaneity can be made quantitative using the Lorentz transformations, as Example 33–5 and Problems 11 to 13 illustrate. Similarly, applying the Lorentz transformations to the coordinates describing the emission and return of the light flash in our lightbox example results in a derivation of time dilation (see Problem 15). Table 33–1 summarizes the Lorentz transformations between coordinates in frames S and S', where S' is moving at speed v in the positive x-direction relative to S.

TABLE 33–1

The Lorentz Transformations

S to S'	S' to S
$y' = y$	$y = y'$
$z' = z$	$z = z'$
$x' = \gamma(x - vt)$	$x = \gamma(x' + vt')$
$t' = \gamma(t - vx/c^2)$	$t = \gamma(t' + vx'/c^2)$

where $\gamma = \dfrac{1}{\sqrt{1 - v^2/c^2}}$

Example 33–5

The Lorentz Transformations

Our Milky Way galaxy and the Andromeda galaxy are approximately at rest with respect to each other, and are 2.0×10^6 light-years apart. At time $t = 0$ in the frame of reference of these two galaxies, supernova explosions occur in both galaxies. Are the explosions simultaneous to the pilot of a spacecraft travelling at 0.80c from the Milky Way toward Andromeda? How far apart in both distance and time does the pilot judge the explosions to be?

Solution

Let the origin of the galaxy frame S be at the supernova in our Milky Way, and let the x-axes of the galaxy frame S and spacecraft frame S' lie on the line connecting the two supernovae. Let the two frames coincide at time $t = t' = 0$. Then the space and time coordinates in S of the two supernova explosions are $x_1 = 0$, $t_1 = 0$, $x_2 = 2.0 \times 10^6$ ly, $t_2 = 0$. Similarly, the coordinates of the Milky Way explosion in the spacecraft frame S' are $x_1' = 0$, $t_1' = 0$. We seek the coordinates x_2' and t_2' of the Andromeda explosion in S'.

Referring to Table 33–1, we first calculate the relativistic factor γ:

$$\gamma = \frac{1}{\sqrt{1 - v^2/c^2}} = \frac{1}{\sqrt{1 - 0.80^2}} = 1.67.$$

(Since a light year is the distance light travels in one year, the speed of light is simply 1 ly/y, so that $v/c = 0.80$.) Then using the Lorentz transformations themselves, we have

$$x_2' = \gamma(x_2 - vt_2) = (1.67)[2.0 \times 10^6 \text{ ly} - (0.80 \text{ ly/y})(0 \text{ ly})]$$

$$= 3.3 \times 10^6 \text{ ly} = 3.3 \text{ Mly}.$$

and

$$t_2' = \gamma(t_2 - vx_2/c^2)$$

$$= (1.67)\left[0 \text{ y} - \frac{(0.80 \text{ ly/y})(2.0 \times 10^6 \text{ ly})}{(1 \text{ ly/y})^2}\right] = -2.7 \text{ My}.$$

Do these results make sense? In the spacecraft frame, the Andromeda supernova occurs nearly three million years before the Milky Way supernova! (The minus sign tells us that t_2' is earlier than the time $t_1' = 0$ of the Milky Way supernova.) Here is an example of events that are simultaneous in one frame but not in another. To make matter worse, consider an observer moving at 0.80c from Andromeda toward the Milky Way. For an observer in this frame, we reverse the sign of v in the transformation equations, obtaining a time of $+2.7$ million years. For this observer, the Milky Way supernova occurs first! How can this be? This is no contradiction, and no violation of cause and effect. In the spaceship's frame S' the two supernova events occur 3.3 Mly apart in space but only 2.7 My apart in time. Light from the "earlier" event cannot travel to the "later" event, so there can be no causal influence between the two. It really doesn't matter which occurs first, and indeed which does depends on the observer.

Had we considered supernova events occurring not simultaneously but a long time apart—longer than 2 million years—in the galaxy frame of reference, we would find that all observers would agree on the time order of the two events, although not necessarily on the actual value of the time interval (see Problems 11 to 13).

Relativistic Velocity Addition

If you are in an airplane moving at 1000 km/h relative to the ground, and if you walk toward the front of the plane at 5 km/h, common sense suggests that you move at 1005 km/h relative to the ground. But relativity implies that measures of time and distance vary among frames of reference in relative motion. For this reason the velocity of an object with respect to one frame does not simply add to the relative velocity between frames to give the object's velocity with respect to another frame. In our airplane example, this means that your velocity with respect to the ground is a little less than 1005 km/h as you stroll down the aisle of the plane, though the difference is insignificant at such a low speed.

relativistic velocity addition

We can derive the correct expression for **relativistic velocity addition** from the Lorentz transformations. Consider a frame of reference S and another frame S′ moving in the positive x-direction with speed v relative to S. Let their origins coincide at time $t = t′ = 0$, so the Lorentz transformations of Table 33–1 apply. Suppose an object moves with velocity $u′$ along the x′-axis in S′. In our airplane example, S′ would be the airplane frame of reference, $u′$ the velocity at which you walk through the plane, and v the velocity of the plane relative to the ground, or S frame. We seek the velocity u of the object relative to the frame S (that is, your velocity relative to the ground as you walk in the plane).

In either frame, velocity is the ratio of change in position to change in time, or

$$u = \frac{\Delta x}{\Delta t}.$$

Designating the beginning of the interval Δt by the subscript 1 and the end by 2, we can use Equations 33–8 and 33–11 to write

$$\Delta x = x_2 - x_1 = \gamma[(x_2′ - x_1′) + v(t_2′ - t_1′)] = \gamma(\Delta x′ + v\Delta t′)$$

and

$$\Delta t = t_2 - t_1 = \gamma[(t_2′ - t_1′) + v(x_2′ - x_1′)/c^2] = \gamma(\Delta t′ + v\Delta x′/c^2).$$

Forming the ratio of these quantities, we have

$$\frac{\Delta x}{\Delta t} = \frac{\Delta x′ + v\Delta t′}{\Delta t′ + v\Delta x′/c^2} = \frac{(\Delta x′/\Delta t′) + v}{1 + v(\Delta x′/\Delta t′)/c^2}.$$

But $\Delta x′/\Delta t′$ is the velocity $u′$ of the object in frame S′, and $\Delta x/\Delta t$ is the velocity u, so that

$$u = \frac{u′ + v}{1 + u′v/c^2}. \tag{33–12}$$

The numerator of this expression is just what we would expect from common sense. But this simple sum of two velocities is altered by the second term in the denominator, which is significant only when both the object's velocity $u′$ and the relative velocity v between frames are comparable with c. Solving Equation 33–12 for $u′$ in terms of u, v, and c gives the inverse transformation:

$$u′ = \frac{u - v}{1 - uv/c^2}. \tag{33–13}$$

Example 33–6
Relativistic Velocity Addition

Two spacecraft approach earth from opposite directions, each moving at 0.80c relative to earth, as shown in Fig. 33–22. How fast do the spacecraft move relative to each other?

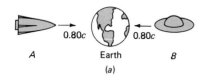

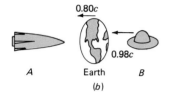

Fig. 33–22
(a) Two spaceships approaching earth at 0.80c. *(b)* The situation in the frame of reference of the left-hand spaceship. Note changes in lengths of the ships and earth.

Solution

Call the earth frame of reference S', and let S be the frame of spacecraft B. Then S' is moving at speed $v = 0.80c$ relative to S, while spacecraft A is moving at $u' = 0.80c$ relative to S'. Then the velocity of A relative to B is given by Equation 33–12:

$$u = \frac{u' + v}{1 + u'v/c^2} = \frac{0.80c + 0.80c}{1 + (0.80c)(0.80c)/c^2} = \frac{1.6c}{1.64} = 0.98c.$$

The relative speed remains less than the speed of light. This result is quite general: Equations 33–12 and 33–13 imply that as long as an object moves at a speed $v < c$ relative to some frame of reference, its speed relative to any other frame of reference will also be less than c (see Problem 20).

Example 33–7
Relativistic Velocity Addition

A light wave moves past earth at the speed of light c. You try to chase the light wave by hopping a fast spacecraft, moving at $0.95c$ relative to earth. What is the speed of the light relative to the spacecraft?

Solution

Call the earth frame S, so that $u = c$, and the spacecraft frame S', so that $v = 0.95c$. Then u', the speed of light relative to the spacecraft, is given by Equation 33–13:

$$u' = \frac{u - v}{1 - uv/c^2} = \frac{c - 0.95c}{1 - 0.95c^2/c^2} = \frac{0.05c}{0.05} = c.$$

We really didn't need to calculate this result, since a fundamental premise of relativity is that the speed of light is the same for all observers. The equations of relativistic velocity addition reflect this basic fact. No matter what the relative velocity v between two frames, light moving at c in one frame moves at c in any other frame. You cannot even begin to catch up with light!

Mass and Momentum in Relativity

A cornerstone of Newtonian mechanics is the law of conservation of linear momentum. We use this law to analyze collisions in which no external forces act. We find that the sum of the momenta before and after the collision remains the same. Observers in different frames of reference will calculate different values for the total momentum, but each will find that value is conserved in the collision. But in a collision at high speeds, we must take account of relativistic velocity addition. Because velocities do not simply add as we go from one frame to another, the quantity $m\mathbf{u}$ cannot be conserved in all frames. Must we discard the law of conservation of momentum?

No: the problem lies not with conservation of momentum but with our expression for momentum itself. Momentum is conserved in all frames of reference, but the quantity $m\mathbf{u}$ is not the conserved quantity. $m\mathbf{u}$ is an approximation valid only at speeds u much less than c. The measure of momentum that is appropriate at any speed is

$$\mathbf{p} = \frac{m\mathbf{u}}{\sqrt{1 - u^2/c^2}} = \gamma m \mathbf{u}, \qquad (33\text{--}14)$$

where γ is the familiar relativistic factor. At low velocities this relativistic momentum reduces to the familiar form $\mathbf{p} = m\mathbf{u}$.

In Example 33–6, we found that two spaceships approaching earth at $0.8c$ nevertheless approached each other at less than c. This example suggests something you have undoubtedly heard about relativity—that it is impossible to accelerate a material object to the speed of light. Why not? Can't you just keep pushing an object and have it go faster and faster? Equation 33–14 suggests an answer to this dilemma. We can interpret Equation 33–14 as saying that relativistic momentum is the product of the particle velocity $\mathbf{u}$ with its **relativistic mass:**

relativistic mass

$$m = \frac{m_0}{\sqrt{1 - u^2/c^2}}. \qquad (33\text{--}15)$$

rest mass

In this equation, and from now on, we use m_0 to signify the **rest mass,** or mass measured in a frame of reference at rest with respect to the object. At low speeds m_0 and the relativistic mass m are indistinguishable, but as the particle speed u approaches c, the particle behaves as though it were much more massive.

Relativistic mass increase makes it clear why we cannot accelerate a particle to the speed of light. As a particle's speed approaches c, its mass grows without bound, and a given force produces ever smaller acceleration. No matter how close to c the particle is going, it would always require an infinite force to push its speed to c.

Relativistic mass increase is insignificant as we accelerate macroscopic objects like baseballs, airplanes, and even spacecraft. But elementary particles are readily accelerated to speeds near that of light, and these particles show dramatically the mass increase given by Equation 33–15. Data taken as early as 1909 clearly show the relativistic mass increase for electrons (Fig. 33–23), and we now regularly accelerate more massive particles to speeds where the mass increases many-fold.

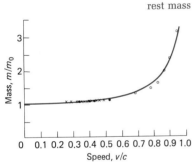

Fig. 33–23
Relativistic mass increase for electrons. Circles and crosses are data points from three different experiments, while solid line is a plot of Equation 33–15.

Example 33–8
Relativistic Mass Increase

In Example 33–3, we considered electrons moving at $0.9999995c$ in the Stanford Linear Accelerator. If the rest mass of an electron is 9.1×10^{-31} kg, what is the relativistic mass of the electrons in SLAC?

Solution

In Example 33–3, we found that the relativistic factor $\gamma = 1/\sqrt{1 - u^2/c^2}$ is 1000 for particles moving at $0.9999995c$, so Equation 3–15 gives

$$m = \frac{m_0}{\sqrt{1 - u^2/c^2}} = 1000m_0 = 9.1 \times 10^{-28} \text{ kg.}$$

Energy in Relativity

As we accelerate an object, we continually do work, thereby increasing the object's energy. But as the speed approaches c, we find that less and less of that energy goes into increasing the speed. Where does the rest go? According to Equation 33–15, much of that energy must somehow result in increasing the object's mass. This suggests that mass and energy may be closely related.

The most widely known result of the special theory of relativity makes this suggestion precise. Einstein's equation

$$E = mc^2 \qquad (33\text{–}16)$$

states that mass and energy are in fact equivalent.

Einstein arrived at Equation 33–15 through a simple "thought experiment." He imagined a closed box of mass M and length L. A light flash is emitted from one end of the box. We found in the preceding chapter that light carrying energy E also carries momentum $p = E/c$. Initially Einstein's box is at rest with zero momentum. When the light flash leaves one end, it carries momentum E/c. To conserve momentum, the box recoils in the opposite direction (Fig. 33–24). If the box is massive, its recoil velocity u will be small compared with c, so that momentum conservation may be written

$$Mu = \frac{E}{c},$$

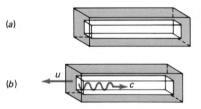

(a)

(b)

Fig. 33–24
Since light carries momentum, the box recoils when the light pulse is emitted.

or

$$u = \frac{E}{Mc}.$$

The light then moves down the box, taking a time

$$\Delta t = \frac{L}{c},$$

where again we assume that the box speed u is much less than c so that the distance travelled by the light is approximately L. In this time the box moves a very small distance Δx, given by

$$\Delta x = u\Delta t = \frac{EL}{Mc^2}.$$

Then the light flash hits the end of the box, transferring just enough momentum to bring the box to a stop.

But now the box is in a new position. It looks as if its center of mass has moved, and yet the box is an isolated system whose center of mass cannot move! To escape this dilemma, Einstein assumed that the light carried not only energy and momentum, but mass as well. If m is the mass carried by the light, we must have

$$mL = M\Delta x,$$

in order that center of mass of the system (box + light) will not move. Using our expression for Δx and solving for m gives

$$m = \frac{M\Delta x}{L} = \frac{M}{L}\frac{EL}{Mc^2} = \frac{E}{c^2},$$

or

$$E = mc^2,$$

where E is the energy of the light and m its equivalent mass.

Although we derived this expression for light energy, it is in fact a universal statement of the equivalence of mass and energy. Energy, like mass, exhibits inertia. A hot object is slightly harder to accelerate than a colder one because of the inertia of its thermal energy. When an object loses energy, it loses mass as well. The sun, for example, radiates energy at the rate of 3.9×10^{26} watts. Equivalently, the sun loses mass at the rate

$$\frac{dm}{dt} = \frac{1}{c^2}\frac{dE}{dt} = \frac{3.9 \times 10^{26} \text{ W}}{(3.0 \times 10^8 \text{ m/s})^2} = 4.3 \times 10^9 \text{ kg/s}.$$

To the general public, the equation $E = mc^2$ is synonymous with nuclear energy. The equation does describe mass changes that occur in nuclear reactions, but it applies equally well to chemical reactions and all other occurrences in which energy enters or leaves a system. If you weigh a nuclear power plant just after it has been refueled, then weigh it again a month later, you will find it weighs slightly less. If you weigh a coal-burning power plant, and all the coal and oxygen that go into it for a month, and all the carbon dioxide and other combustion products that come out, you will find a discrepancy between the mass of what goes in and what comes out. If both plants produce the same amount of energy, the mass discrepancy will be the same for both. This discrepancy represents the equivalent mass of the energy sent down the power lines, as well as of the waste heat released. The only difference between the two power plants lies in the amount of mass released as energy in each individual reaction. The fissioning of a single uranium nucleus involves about 10 million times as much energy, and therefore mass, as the reaction of a single carbon atom with oxygen to make carbon dioxide (Fig. 33–25). That is why many trainloads

Fig. 33–25
Energy release in chemical and nuclear reactions. A mass change occurs in both cases, but is approximately 10^7 times greater in the nuclear reaction.

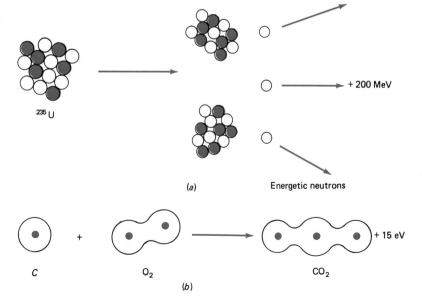

of coal are needed for the coal-burning plant to produce as much energy as the nuclear plant does on a single refueling.

So far we have discussed mass-energy equivalence associated with the kinetic and potential energy involved in rearranging atoms or atomic nuclei. In all these cases individual elementary particles—electrons, protons, and neutrons—retain their identities. But even the mass associated with the very existence of an object—the object's rest mass—is equivalent to energy. At the elementary particle level, the interchange of matter and energy is complete. A particle and its antimatter opposite, an antiparticle, can come together and annihilate totally. The particles disappear and in their place appear bursts of electromagnetic radiation. The opposite occurs, too. Pure electromagnetic radiation, travelling at speed c, can suddenly disappear, while a particle and its antiparticle materialize (Fig. 33–26). Because the energy $E = mc^2$ required to produce even the lightest particles is large, **pair creation** such **pair creation** does not play a significant role in our everyday lives. But in high-energy particle accelerators, in supernova explosions, and in the high-temperature conditions that prevailed in the early universe, the interchange of particles and electromagnetic radiation is an important phenomenon.

rest energy   Equation 33–16 describes both the **rest energy** $m_0 c^2$ associated with the rest mass of an object, as well as kinetic energy. If the object is moving relative to us, its mass increases according to Equation 33–15. The corresponding increase in its energy is reflected by using the relativistic mass in Equation 33–16:

$$E = mc^2 = \frac{m_0 c^2}{\sqrt{1 - u^2/c^2}}.$$

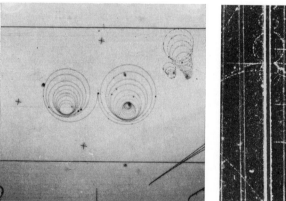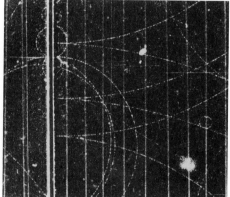

Fig. 33–26
Pair creation events produced in particle accelerators. Energy of electromagnetic radiation (high-energy gamma rays) has been converted into electrons and their antiparticles, positrons. On the left, electron and positron tracks spiral in opposite directions in a strong magnetic field perpendicular to the page. On the right, three pair creation events occur in a weaker magnetic field.

The difference between this total energy and the rest energy must be the kinetic energy, K:

$$K = \frac{m_0 c^2}{\sqrt{1 - u^2/c^2}} - m_0 c^2 = m_0 c^2 \left[(1 - u^2/c^2)^{-1/2} - 1\right].$$

To compare this with our familiar Newtonian expression, $K = \frac{1}{2}mu^2$, we can expand the quantity $(1 - u^2/c^2)^{-1/2}$ using the binomial theorem (Appendix A). This gives

$$(1 - u^2/c^2)^{-1/2} \simeq 1 + \frac{1}{2}\frac{u^2}{c^2},$$

so the kinetic energy becomes approximately

$$K \simeq m_0 c^2 \left(1 + \frac{1}{2}\frac{u^2}{c^2} - 1\right) = \frac{1}{2}mu^2.$$

Thus the Newtonian expression for kinetic energy is an approximation valid only at low velocities. At the other extreme, when an object's velocity differs only slightly from c, then its kinetic energy far exceeds its rest energy and is very nearly equal to its total energy mc^2 (see Problem 23).

33.4
WHAT IS NOT RELATIVE

In relativity, quantities like space, time, and mass are not absolute but depend on the reference frame of the observer. But the speed of light c remains the same for all observers. Are there other such **relativistic invariants**? Yes—and these invariants, not the shifting measures of space and time, are at the basis of relativity's objective description of physical relativity.

relativistic invariants

One invariant is electric charge. Although an electron's mass increases with speed, its charge does not; all observers measure the same value for the charge. Other invariants may be formed from combinations of quantities that are themselves not invariant. In Newtonian physics the distance between two points is the same no matter who measures it. In relativity, distance depends on the observer. So does time. But there is a quantity, analogous to distance but incorporating time as well, that remains invariant. Called the **spacetime interval,** it is a kind of "four-dimensional distance," not between two points in space but between two events in space and time. The spacetime interval Δs is given by an expression that looks like a modified Pythagorean theorem:

spacetime interval

$$(\Delta s)^2 = c^2(\Delta t)^2 - [(\Delta x)^2 + (\Delta y)^2 + (\Delta z)^2], \tag{33-17}$$

where the Δ quantities describe the differences between the space and time coordinates of two events. Equation 33–17 follows directly from the Lorentz transformations (see Problem 27), and its use is illustrated by the following example.

Example 33–9

The Spacetime Interval

A spaceship travels at 0.80c from earth to a star 10 light-years distant, as measured in the earth frame. In earth frame and ship frame, how far apart are the events of the ship leaving earth and arriving at the star? How long does the journey take? Show that, although the times and distances differ, the spacetime interval between the events is the same in both frames.

Solution

In the earth frame the distance is 10 ly, so the travel time is

$$\Delta t = \frac{10 \text{ ly}}{0.80 \text{ ly/y}} = 12.5 \text{ y}.$$

The ship is analogous to our light box of Section 33.3, so time dilation occurs, giving a ship time

$$\Delta t' = \Delta t \sqrt{1 - v^2/c^2} = (12.5 \text{ y})\sqrt{1 - 0.80^2} = 7.5 \text{ y}.$$

Let the origins of earth and ship coordinate systems coincide just as the ship leaves earth, at which event $x = x' = 0$ and $t = t' = 0$. Then in the earth frame, the coordinates of the arrival event are $x = 10$ ly, $t = 12.5$ y. The spacetime interval is then given by

$$(\Delta s)^2 = c^2(\Delta t)^2 - (\Delta x)^2 = (1 \text{ ly/y})^2(12.5 \text{ y})^2 - (10 \text{ ly})^2 = 56 \text{ ly}^2,$$

where in working with light-years and years, we use $c = 1$ ly/y.

In the ship frame, the time coordinate of the arrival is just $t' = 7.5$ years. What about the difference between the x'-coordinates of the two events? This is *not* the distance between earth and star, even when the Lorentz contraction is calculated. Rather, it means the distance *relative to the ship* that an observer has to travel to get between earth and star. Since the ship itself travels between earth and star, a passenger on the ship need not move *relative to the ship*. So $x' = 0$, and the spacetime interval is

$$(\Delta s)^2 = c^2(\Delta t)^2 - (\Delta x)^2 = (1 \text{ ly/y})^2(7.5 \text{ y})^2 - 0 \text{ ly}^2 = 56 \text{ ly}^2,$$

the same as in the earth frame.

The invariance of the spacetime interval suggests that something absolute remains behind the shifting sands of relativistic space and time. That absolute is **spacetime**—a four-dimensional framework linking space and time. The points in spacetime are events, specified by four coordinates. The time interval or space interval between two events depends on the particular frame of reference of the observer, but the spacetime interval—a four-dimensional "distance" that takes all four coordinates into account—is the same for all observers.

spacetime

In more advanced treatments of relativity, it is convenient to consider four-dimensional vectors called **four-vectors.** The displacement between two events in spacetime—specified by the four quantities Δx, Δy, Δz, and Δt—is a four-vector, with "length" given by Equation 33–17. With two- and three-dimensional vectors in nonrelativistic physics, it is possible to break a vector into components in many different ways. Although the values of the individual components depend on your choice of coordinate system, the length of the vector does not (Fig. 33–27). Similarly, the individual

four-vectors

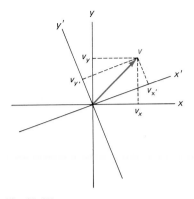

Fig. 33–27
Although the x- and y-components of an ordinary vector depend on the choice of coordinate axes, the length of the vector does not.

space and time components of a four-vector depend on your choice of reference frame—that is, on your velocity. But the spacetime interval does not.

Other quantities that relativity has stripped of their absoluteness may be combined to make four-vectors whose length is invariant. Energy and the three space components of momentum, for example, combine to form a single four-vector (see Problem 30). So do electric charge density and the three components of electric current density. Use of four-vectors in four-dimensional spacetime is an elegant way of saying that relativity really does deal with the absolute and unchanging aspects of physical reality. It is only when we limit our perspective to a single frame of reference that we lose sight of those features of our world that are truly universal.

33.5
ELECTROMAGNETISM AND RELATIVITY

Historically, relativity arose in Einstein's mind from deep questions presented by Maxwell's equations with regard to electromagnetic waves. We have seen that relativity profoundly alters the basic concepts of space and time that stand at the foundation of Newtonian mechanics. As a result, fundamental ideas like mass, momentum, and energy must be altered for relativistic consistency. What analogous changes does relativity require of Maxwell's electromagnetic theory? The answer is simple: none. Maxwell's theory culminated in the prediction of light waves travelling through empty space at speed c. Relativity requires that the laws of physics be the same in all frames of reference in uniform motion. But that is exactly what Maxwell's equations suggest—that a light wave in one frame should be a light wave in any other frame, and that such a wave should have speed c with respect to any observer. Even the simple fact that electromagnetic induction occurs equally well when you move a magnet near a conductor, or a conductor near a magnet, suggests that only relative motion should be important in electromagnetism. Indeed, Einstein thought a great deal about induction, and mentioned it at the beginning of his 1905 paper introducing special relativity. Even the title of that famous paper—"On the Electrodynamics of Moving Bodies"—shows how intimately related are electromagnetism and relativity. Maxwell's equations are relativistically correct, and require no modification.

Although electric and magnetic fields in any frame of reference obey the same Maxwell equations, this does not require the fields themselves to be independent of frame. If, for example, you sit in the rest frame of a point charge, you see a spherically symmetric point-charge field. If you move relative to the charge, you see a magnetic field as well, associated with the moving charge. Relativity accounts naturally for such a field transformation. In relativity, the electric and magnetic fields are not absolutes, but are considered manifestations of a more fundamental electromagnetic field. To one observer, this electromagnetic field breaks up in a certain way into electric and magnetic parts, while to another observer the individual electric and magnetic fields are different (see Problem 31).

Fig. 33–28
A current-carrying wire with equal
densities of positive and negative
charge moving in opposite directions. A
magnetic field surrounds the wire, so a
charge *q* moving parallel to the wire
experiences a magnetic force.

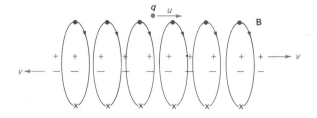

We can illustrate the deep relationship that relativity imposes between electricity and magnetism by considering the force on a charged particle in the vicinity of a current-carrying wire. For simplicity, we consider a wire containing equal line charge densities of positive and negative charge, moving in opposite directions at the same speed v relative to the wire (Fig. 33–28).

Consider a particle of positive charge q a distance r from the wire. Since the densities of positive and negative charge are equal, the wire is neutral and has no electric field. If the particle is at rest with respect to the wire, it experiences no force. But there is a magnetic field associated with the current in the wire. The magnetic field lines encircle the wire, and the right-hand rule shows that they point out of the page at the location of our charged particle. Now suppose the particle is moving to the right with velocity **u** relative to the wire. It experiences a magnetic force $\mathbf{F} = q\mathbf{u} \times \mathbf{B}$; the right-hand rule shows that this force is toward the wire.

So this is the situation from the frame of reference of the wire: the wire is electrically neutral and therefore produces no electric field. But it does carry a current, and therefore produces a magnetic field. If a positively charged particle moves to the right, it experiences a magnetic force directed toward the wire. To describe the situation we needed to know about electric charges, about Ampère's law for magnetism, and about the magnetic force $q\mathbf{u} \times \mathbf{B}$—in short, about a variety of phenomena that were discovered independently during the nineteenth century.

Now let's look at the situation in the frame of reference of the charged particle. Moving to the right, its speed relative to the positive charges is lower than its speed relative to the negative charges. As measured by the particle, distances between negative charges are Lorentz contracted—and by a greater amount than the distances between positive charges. But charge is invariant, so that charge density—the charge per unit length—is *greater* for the negative charges. So in the frame of the charged particle, *there is a net negative charge on the wire!* The negatively charged wire gives rise to an electric field pointing toward the wire. As a result, our positively charged particle experiences an electric force toward the wire (Fig. 33–29). Of course there is still a magnetic field as well, but since the particle is at rest in its own frame of reference, it experiences no magnetic force. The force it does experience is entirely electric. What appeared as a magnetic phenomenon in the wire frame of reference—the existence of a force directed toward the wire—is explained entirely as an electric phenomenon in the particle's frame of reference.

Although we will not do so, it is possible through judicious application of relativistic velocity addition formulas and Lorentz contraction to show

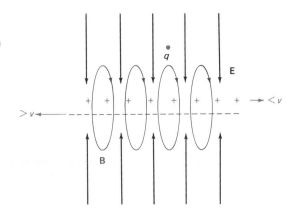

Fig. 33–29
The situation in the rest frame of the charged particle. The distance between negative charges is Lorentz contracted more than the distance between positive charges. There is therefore a net negative charge on the wire, giving rise to an electric field and an electric force on the charged particle.

that the electric force arising from our particle-frame description is the same as the magnetic force calculated from Ampère's law and $\mathbf{F} = q\mathbf{u} \times \mathbf{B}$ in the wire frame of reference.*

We have given two quite different descriptions of the force on a charged particle moving near a current-carrying wire. Our second description—from the frame of reference of the charged particle—required no knowledge of magnetism whatsoever. We only needed Coulomb's law for electric charge and the principle of relativity. And this illustrates a profound point: electricity and magnetism are not separate phenomena that happen to be related through Ampère's and Faraday's laws. Rather, they are two aspects of a single phenomenon—electromagnetism. In a universe obeying the principle of relativity, it is not even logically possible to have electricity without magnetism, or vice versa. Given Coulomb's law and the principle of relativity, all the rest of Maxwell's equations follow not as independent experimental results but as logically necessary consequences. Relativity provides for us the total unification of electricity and magnetism that we have hinted at since we began our study of these phenomena.

33.6
GENERAL RELATIVITY

The special theory of relativity is special because it is restricted to uniform motion. Following the development of special relativity, Einstein attempted to formulate a theory that would express the laws of physics in the same form in all frames of reference, including those in accelerated motion. But Einstein recognized that it is impossible to distinguish the effects of uniform acceleration from those of a uniform gravitational field (Fig. 33–30). Consequently, Einstein's general theory became a theory of gravity. Building on the notion of four-dimensional spacetime, Einstein introduced geometrical curvature of spacetime to account for gravity. The predictions of the **general theory of relativity** (1916) differ significantly from those of Newton's theory of gravity only in regions of very strong

general theory of relativity

*See Purcell, *Electricity and Magnetism* (Berkeley Physics Course, vol. 2, McGraw-Hill, 1963, 1985), section 5.9, for a detailed analysis of the force from both frames of reference.

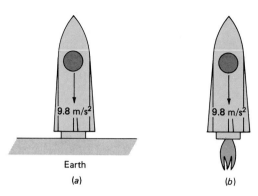

Fig. 33-30
Why general relativity is a theory of gravity. *(a)* In a spaceship at rest on earth, objects accelerate downward at $g = 9.8$ m/s^2. *(b)* Far from any gravitating body, a spaceship accelerates at 9.8 m/s^2. With respect to the accelerating frame of reference of the spaceship, objects accelerate toward the back of the ship at 9.8 m/s^2. Situations *(a)* and *(b)* are impossible to distinguish, so that a theory dealing with nonuniform motion must also consider gravity.

gravitational fields or when the overall structure of the universe is considered. By very strong fields we mean those of objects whose escape speed is comparable to that of light (see Problem 32). Because we have no direct laboratory experience of such fields, the general theory of relativity is not as solidly established as is the special theory. Nevertheless, general relativity is a cornerstone of modern astrophysics, playing a crucial role in the physics of such bizarre objects as neutron stars and black holes. General relativity also addresses cosmological questions of the origin and ultimate fate of the universe. Research in astrophysics and cosmology, in turn, is increasingly confirming the predictions of general relativity.

SUMMARY

1. The **ether** was a hypothetical medium whose properties were supposed to explain the propagation of electromagnetic waves. In particular, such waves were supposed to have speed c relative to the ether.

2. The **Michelson-Morley experiment** and the observation of the aberration of starlight led to a contradiction in physics: earth's motion through the ether could not be detected, yet the earth did not drag ether with it.

3. Einstein's **special theory of relativity** (1905) resolved the contradiction by asserting that uniform motion is undetectable by any experiment, mechanical or electromagnetic. Einstein did away with the ether, declaring simply that **the laws of physics are the same in all frames of reference in uniform motion.** Mechanics and electromagnetism alike are included in Einstein's theory, so that Maxwell's prediction of electromagnetic waves moving at speed c is correct in all frames of reference.

4. The simple statement that the laws of physics are the same in all frames of reference requires profound changes in our common-sense notions of space and time. These changes are described by the **Lorentz transformations,** which relate space and time measurements made in different frames of reference:

$$y' = y$$
$$z' = z$$
$$x' = \gamma(x - vt)$$
$$t' = \gamma(t - vx/c^2),$$

where

$$\gamma = \frac{1}{\sqrt{1 - v^2/c^2}}.$$

Particular manifestations of these transformations include **time dilation** and **Lorentz contraction.**

5. The Newtonian concepts of mass, momentum, and energy are modified by relativity. No longer a constant, mass increases with the speed u of an object:

$$m = \frac{m_0}{\sqrt{1 - u^2/v^2}},$$

where m_0 is the **rest mass,** measured in a frame at rest with respect to an object. Energy and mass become interchangeable, related by Einstein's equation $E = mc^2$. Even the mass of a stationary object represents energy that can be liberated through appropriate reactions.

6. Relativity links space and time into a single four-dimensional framework called **spacetime.** Although individual space and time measurements are not absolutes, intervals in spacetime are. The spacetime interval between two events is given by

$$(\Delta s)^2 = c^2 (\Delta t)^2 - [(\Delta x)^2 + (\Delta y)^2 + (\Delta z)^2].$$

Many other concepts of Newtonian physics may be combined to form **four-vectors** that describe quantities truly independent of particular observers.

7. Although relativity requires substantial modifications of Newtonian concepts, Maxwell's theory of electromagnetism is relativistically correct. But although electric and magnetic fields in any uniformly moving frame of reference obey Maxwell's equations, the fields themselves are different in different frames. What is an electric phenomenon in one frame may be partly magnetic in another, and vice versa. Relativity imposes a logical interrelationship between electricity and magnetism: neither one is possible without the other.

8. The **general theory of relativity** is Einstein's generalization of special relativity to cover all frames of reference, including those in accelerated motion. In its assertion that the effects of acceleration are locally indistinguishable from those of gravity, general relativity manifests itself as a theory of gravity. Although its experimental verification is less complete than that of special relativity, the general theory plays a central role in modern astrophysics and cosmology.

QUESTIONS

1. List as many different kinds of waves as you can, and identify the medium, if any, in which each propagates.
2. Why was the Michelson-Morley experiment a more sensitive test of motion through the ether than independent measurements of the speed of light in two perpendicular directions?
3. Why was it necessary to repeat the Michelson-Morley experiment at different times throughout the year?
4. The entire Michelson-Morley apparatus was floated in a bath of mercury so it could rotate. The experiment was performed for a number of orientations of the apparatus. Why was this necessary?
5. Why do we reject the idea that the ether frame of reference is the earth frame?
6. What is special about the special theory of relativity?
7. Does relativity require that the speed of sound be the same for all observers? Why or why not?
8. How would our world be different if the speed of light were 160 km/h (100 miles per hour)? Would our "common-sense" notions change?
9. A friend argues that the speed of light cannot be the same for all observers, for if one of them is moving toward a light source then that one will clearly measure a higher speed. How can you refute this argument?
10. Time dilation is sometimes described by saying that "moving clocks run slow." In what sense is this true? In what sense does the statement violate the spirit of relativity?
11. If you are in a spaceship moving at 0.95c relative to earth, do you perceive time to be passing more slowly than it would on earth? Think! Is your answer consistent with the theory of relativity?
12. In our light-box example for time dilation, we found that a time interval between two events measured in frame S' was shorter than in frame S. But you could equally well say that frame S is moving relative to frame S', so that clocks in S should "run slow" compared with those in S'. An observer in each frame should judge the clocks in the other frame to "run slow." Is this a contradiction?
13. Is Lorentz contraction just an illusion, or does it "really happen"?
14. To try to circumvent the difficulty of accelerating an object to the speed of light, you build a series of conveyor belts, all running in the same direction, and each moving 10 m/s relative to the one next to it (Fig. 33–31). You step from the ground onto the first conveyor belt, then to the next, and so forth. By the time you reach the 3×10^7th conveyor belt, you should be moving at c relative to the ground. Why doesn't this scheme work?

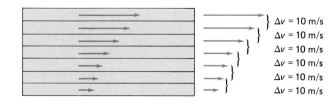

Fig. 33–31 A series of conveyor belts, each moving at 10 m/s relative to its neighbors (Question 14).

15. Does $E = mc^2$ apply to a nuclear reactor? A burning candle? A woodstove? A person metabolizing food?

16. An unstretched rubber band is weighed on an extraordinarily sensitive scale. It is then stretched and weighed again. Is there a difference in the weight? Why or why not?

17. The rest energy of an electron is 511 keV. What is the approximate speed of an electron whose total energy is 1 GeV ($=10^9$ eV)? You need not do any calculations!

18. An atom in an excited state emits a burst of light. What happens to the mass of the atom?

19. In some of the hottest parts of the universe, the thermal energy of particles may be many millions of electron volts. At such temperatures, the number of particles within a closed volume may vary. How is this possible? *Hint:* The rest mass of the electron is equivalent to about 0.5 MeV.

20. The electric field is not invariant, but changes from one frame to another. Is this a violation of relativity? Relativity requires that the laws of physics be the same in all frames of reference. Does this mean that all physical quantities must be the same?

21. The quantity **E·B** is invariant. What does this say about how different observers will measure the angle between **E** and **B** in a light wave?

PROBLEMS

Section 33.2   *Matter, Motion, and the Ether*

1. Consider an airplane flying at 800 km/h airspeed between two points 1800 km apart. What is the round-trip travel time for the plane (a) if there is no wind? (b) if there is a wind blowing at 130 km/h perpendicular to a line joining the two points? (c) if there is a wind blowing at 130 km/h along a line joining the two points? Ignore relativistic effects. (Why are you justified in doing so?)

2. What would be the difference in light travel times on the two legs of the Michelson-Morley experiment if the ether existed and if earth moved relative to it at (a) its orbital speed relative to the sun (Appendix F)? (b) $10^{-2}c$? (c) $0.5c$? (d) $0.99c$? Assume each light path is exactly 11 m in length, and that the paths are oriented parallel and perpendicular to the ether wind.

3. Figure 33–32 shows a plot of James Bradley's data on the aberration of light from the star γ Draconis, taken in 1727–1728. (a) From the data, determine the magnitude of earth's orbital velocity. (b) The data very nearly fit a perfect sine curve. What does this say about the shape of earth's orbit?

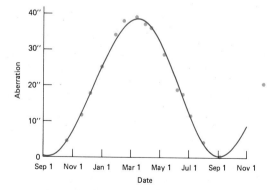

Fig. 33–32 Bradley's data on aberration of starlight from the star γ Draconis (Problem 3). Aberration is measured in seconds of arc (1″ = 1/3600°).

4. Show that the time of Equation 33–2 is larger than that of Equation 33–1 as long as $0<v<c$.

Section 33.3   *Special Relativity*

5. The Andromeda galaxy is two million light-years from earth, measured in a frame at rest with respect to earth and Andromeda. (a) Assuming that you cannot travel faster than light, could you get yourself to Andromeda in one human lifetime (say 70 years)? If so, how fast would you have to go? (b) Once you got there, could you get a radio message back to the friends you left behind on earth? How old would they be when they received the message?

6. Our galaxy is 10^5 light-years across. A high-energy electron crosses the galaxy in 10 years, as measured on a clock in the electron's frame of reference. (a) What is the approximate speed of the electron relative to the galaxy? (b) For an observer at rest with respect to the galaxy, approximately how long does it take the electron to cross the galaxy? (c) Measured in the electron's frame, approximately how wide is the galaxy? (d) Obtain a more precise answer than in (a) for the speed of the electron relative to the galaxy.

7. How fast would you have to move relative to a meter stick for its length to measure 99 cm in your frame of reference?

8. The nearest star beyond our sun is about 4 light-years away. If we built a spaceship that could get to the star in 5 years, as measured by a clock on earth, (a) how long would the pilot of the ship judge the journey to take? (b) How far from earth would the pilot find the star to be?

9. Two twins, A and B, live on earth. On their twentieth birthday, twin B climbs into a spaceship and journeys at $0.95c$ to a star 30 light-years distant, as measured in the earth-star frame. Upon reaching the star, B immediately turns around and comes back to earth at $0.95c$. How old is each twin when the two are reunited?

10. You wish to see what life will be like on earth two million years hence. To do so, you hop a fast space-

ship and travel to another galaxy, then turn around and return to earth. Find a suitable distance to the galaxy and a speed of travel so that two million years elapse on earth, but only 10 years on your spaceship.

11. Two civilizations are evolving on opposite sides of our galaxy (diameter 10^5 light-years). At time $t = 0$ in the galaxy frame of reference, civilization A launches its first interstellar spacecraft. 50,000 years later, measured in the galaxy frame, civilization B launches its first spacecraft. A being from a more advanced civilization C is travelling through the galaxy at $0.99c$, on a line from A to B. Which civilization does C judge to have first achieved interstellar travel, and how much in advance of the other?

12. Repeat the above problem, except assume that civilization B lags A by 1 million years in the galaxy frame.

13. Are there observers who would judge the events in the two preceding problems to be simultaneous? If so, how must each observer be moving relative to the galaxy?

14. Derive the Lorentz transformations for time, Equations 33–10 and 33–11, from the transformations for space, Equations 33–7 and 33–8.

15. Consider the light box of Fig. 33–8. Let event A be the emission of the light flash, and B the return of the light to its source. Assign suitable space and time coordinates to these events in the frame with respect to which box moves at speed v. Apply the Lorentz transformations to show that the time between the two events in the box frame is given by the time dilation Equation 33–5.

16. Use the Lorentz transformations to show in general that if two events are separated in space and time so that a light signal leaving one event could not reach the other, then there is an observer for whom the events are simultaneous. Show that the converse is true as well: that if a light signal could get from one event to the other, then no observer will find them simultaneous.

17. As we look at the distant galaxies, we find them receding from us at speeds proportional to their distances. (This is our primary evidence that the universe is expanding.) Suppose two galaxies located in opposite directions are each moving away from us at $0.50c$. How fast do astronomers in one of the galaxies measure the other galaxy to be moving?

18. Two spaceships are having a race. The "slower" one moves past earth at $0.70c$, and the "faster" one moves at $0.40c$ relative to the slower one. How fast does the faster ship move relative to earth?

19. Muons travelling vertically downward at $0.994c$ relative to earth are observed from a rocket travelling upward at $0.25c$. What speed do passengers in the rocket measure for the muons?

20. Use relativistic velocity addition to show that if an object moves at speed $v < c$ relative to some uniformly moving frame of reference, then its speed relative to any other uniformly moving frame of reference must also be less than c.

21. Electrons in a color TV tube are accelerated through a potential difference of 30 kV. What is the mass of the electrons moving down the tube?

22. Among the most energetic cosmic rays detected are protons with energies around 10^{20} eV. What is the mass of such a proton?

23. What is the kinetic energy of an electron moving at (a) $0.001c$? (b) $0.6c$? (c) $0.99c$? Use suitable approximations where possible.

24. In a nuclear fusion reaction, two deuterium nuclei (^{2}H) combine to form a helium nucleus (^{3}He) plus a neutron. The energy released in this reaction is 3.3 MeV. By how much do the combined rest masses of the helium nucleus and neutron differ from the combined rest masses of the two deuterium nuclei?

25. A large city consumes electrical energy at the rate of 10^9 watts. If you converted all the rest mass of a raisin (~1 g) to electrical energy, for how long could it power the city?

26. The sun has a mass of 2×10^{30} kg and radiates energy at the rate of 4×10^{26} W. This energy arises from the fusion of hydrogen into helium. Estimate the lifetime of the sun, assuming that it will convert about 10% of its hydrogen to helium before it dies, and that 0.7% of the mass is transformed into energy in each reaction.

Section 33.4 *What Is Not Relative*

27. Show from the Lorentz transformations that the spacetime interval of Equation 33–17 has the same value for all observers.

28. Use Equation 33–17 to calculate the square of the spacetime intervals (a) between the events in Problem 11 and (b) between the events in Problem 12. Comment on the signs of your answers, and on the possibility of a causal relation between the events.

29. A light beam is emitted at an event A and arrives at an event B. Show that the spacetime interval between the events is zero.

30. Show that the quantity $E^2 - p^2 c^2$ for a particle is invariant (that is, has the same value in all frames of reference) and is equal to the rest energy of the particle. Here E is the total energy and p the particle's momentum.

Section 33.5 *Electromagnetism and Relativity*

31. Consider a line of positive charge with line charge density λ, as measured in a frame S at rest with respect to the charges. (a) Show that the electric field a distance r from this charged line has magnitude $E = \lambda/2\pi\epsilon_0 r$, and that there is no magnetic field (no relativity needed here).

   Now consider the situation in a frame of reference S' moving at speed v parallel to the line of charge. In this frame, distances along the line are Lorentz contracted, but since charge is invariant the line charge density is therefore increased. (b) Show that the line charge density measured in S' is $\lambda' = \gamma\lambda$, with $\gamma = 1/\sqrt{1 - v^2/c^2}$. (c) Using the result of (b), show that the electric field in S' has magnitude $E' = \gamma\lambda/2\pi\epsilon_0 r$.

(d) In the frame S', the moving line of charge constitutes an electric current. Show that this current is given by $I = \gamma \lambda v$. (e) In S', there must be a magnetic field associated with the moving line charge. Show that this field has magnitude $B' = \mu_0 \gamma \lambda v / 2\pi r$. (f) Show that, although the fields $\mathbf{E}$ and $\mathbf{B}$ differ in the two frames S and S', the quantity $\mathbf{E} \cdot \mathbf{B}$ is invariant (that is, that $\mathbf{E} \cdot \mathbf{B} = \mathbf{E}' \cdot \mathbf{B}'$). (g) Show also that the quantity $E^2 - c^2 B^2$ is invariant.

Although you have demonstrated the invariance of $\mathbf{E} \cdot \mathbf{B}$ and of $E^2 - c^2 B^2$ for a special case, these quantities are true invariants, the same in all frames of reference for any configuration of electric and magnetic fields.

Section 33.6   *General Relativity*

32. Using Equation 8–7, estimate the size to which you would have to squeeze each of the following objects before escape speed at its surface approached the speed of light: (a) the earth; (b) the sun; (c) the Milky Way galaxy, containing about 10^{11} solar masses. Your answers show why general relativity is not needed in most ordinary astronomical calculations.

34

GEOMETRICAL OPTICS

Fig. 34–1
Refraction is a familiar optical effect;
it makes lenses possible.

geometrical optics

physical optics

Much of our contact with the universe around us is through light and images made with light. Even the act of reading this book involves images made within our eyes. The fundamental sixteenth-century observations of the planets made by Tycho Brahe, subsequently explained by Kepler and ultimately interpreted by Newton, were based on studies in the visible part of the electromagnetic spectrum. The development of the telescope and microscope extended our knowledge into realms of very large and very small. These inventions and observations all involve the branch of physics known as **optics**—the study of light and its behavior (Fig. 34–1).

Recent technological developments are providing a renaissance in optics. We can now send telephone calls and other messages as pulses of light along thin strands of glass known as "optical fibers" instead of as electrical signals along copper wires. The advantages of this method are transforming fields of communications into branches of optics.

In the previous chapters, we have shown that electromagnetic waves such as light are a natural consequence of Maxwell's equations. But it proves to be too cumbersome for most purposes to explain optical phenomena from such fundamental principles. It often turns out to be sufficient, when treating the formation of images made with light, to assume that light travels in straight lines within homogeneous substances. We make this assumption here, limiting ourselves to the field of **geometrical optics.** We can use this approximation of geometrical optics whenever any obstacle or aperture in the path of a wave is much larger than the wavelength, so long as we do not study carefully effects of the edges of the obstacles or aperture or the edges of their shadows and images. In the following chapter, we shall consider some ideas of **physical optics,** which requires a more physically complete description of the phenomena involved, including the fact that light consists of electromagnetic waves.

34.1

REFLECTION AND REFRACTION

The fundamental nature of light has been discussed for hundreds of years, and whether light is a particle or a wave has proved to be one of the most stimulating questions of twentieth-century science. We shall discuss that issue in Chapter 36; for the purposes of geometrical optics, it is sufficient to consider light as a wave phenomenon.

Let us begin our discussion of geometrical optics by considering light waves incident on a plane surface that marks the boundary between two transparent substances, each usually called a "medium." Later, we can extend our analysis to curved surfaces; doing so has interesting consequences.

reflected
refracted
rays

In Fig. 34–2, we see that light approaching a surface can bounce off the surface—be **reflected**—or can pass through the surface and be bent—be **refracted.** It is often convenient to talk of the light as composed of straight **rays;** let us first consider the refracted ray.

wavefront

In Fig. 34–3, we consider the refraction of a set of waves moving in the same direction. It is convenient to carry out the analysis first in terms of the **wavefront,** a line drawn through the corresponding crests (or troughs) of the waves. Let v_1 be the speed of light in the top medium, which is often air. Let v_2 be the speed of light in the bottom medium, which might be glass. It is known experimentally that the speed of light depends on the particular medium through which the wave is propagating. Thus in a time interval Δt, the front travels distances $v_1 \Delta t$ in the air and $v_2 \Delta t$ in the glass. The frequency of the wave stays the same (if the frequency were different, for example, at the far side, the waves would pile up at the surface or be depleted there). Since the frequency times the wavelength is equal to the wave speed, the wavelength must change when the speed changes. Since the part of the wavefront that has passed the surface is travelling at a different speed from the rest of the wavefront, the wavefront must bend at the surface.

Fig. 34–2
The beams of light incident from the left split at the surfaces into reflected and refracted beams. The angles at which the reflected and refracted beams go off depend on the angle at which the incoming beams are incident.

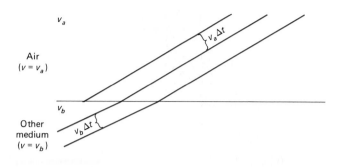 

Fig. 34–3
(Left) Wavefronts bend at a surface because they travel at different speeds in different media. (Right) Refraction in a water tank. The water is shallower to the lower right, which makes the speed of water waves lower. The waves are travelling from top to bottom in this top view.

angle of incidence
angle of refraction

Let θ_i be the **angle of incidence** with which the front approaches the surface, and θ_r be the **angle of refraction** with which the front leaves the surface (Fig. 34–4). Note that the hatched triangle (air) and the shaded triangle (glass) have the same hypotenuse of length d. We have, for incident velocity v_i and velocity after refraction v_r,

$$\sin\theta_i = \frac{v_i\Delta t}{d},$$

and

$$\sin\theta_r = \frac{v_r\Delta t}{d}.$$

From these, we can eliminate the unknown $\Delta t/d$ to find that

$$\frac{\sin\theta_i}{v_i} = \frac{\sin\theta_r}{v_r}, \tag{34–1}$$

where θ_i is the angle of incidence, v_i is the speed of the incident ray, θ_r is the angle of refraction, and v_r is the speed of the refracted ray. For our current example, $v_i > v_r$, so $\sin\theta_i > \sin\theta_r$ and therefore $\theta_i > \theta_r$. Consequently, when passing from air into a medium with a lower wave speed, such as glass, light is deflected so as to make the *wavefronts* more parallel to the

Fig. 34–4
The derivation of Snell's law.

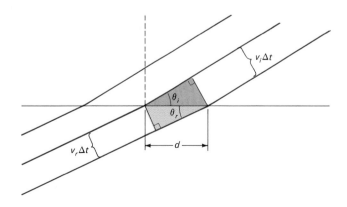

interface. Alternatively, we could say that the *light rays*—which are perpendicular to the wavefronts—are bent toward the normal to the surface. The direction of the rays defines the new direction of propagation.

Suppose that the top surface were a vacuum instead of air; then $v_i = c$. In this case,

$$\sin\theta_i = \left(\frac{c}{v_r}\right)\sin\theta_r.$$

index of refraction   We define the **index of refraction** n of the bottom medium as c/v_r, giving

$$\sin\theta_i = n\sin\theta_r.$$

Since v_r is less than c, n is greater than 1 and light is bent toward the normal when passing from a vacuum into a substance, given the geometrical limit that the angle of incidence $\theta_i < 90°$. Of course, $\theta_i = 0$ implies $\theta_r = 0$, regardless of n, which is physically reasonable because the refracted beam should not be deflected when the incident beam is normal to the surface. Furthermore, if $n = 1$ then $\theta_i = \theta_r$; there is no deflection since there is no interface.

Figure 34–5 shows that the angles of incidence and refraction, respectively, are the same whether measured between the wavefronts and the surface, or between the light rays and the normal to the surface. The latter **Snell's law**   viewpoint is more commonly found. Expressed in this way, we have **Snell's law,**

$$\sin\theta_i = n\sin\theta_r, \tag{34–2}$$

where θ_i is the angle of incidence, measured between the incoming light ray and the normal, and θ_r is the angle of refraction, measured between the outgoing light ray and the normal. This law was discovered in 1621 by Willebrord van Roijen Snell (1591–1626) in the Netherlands as a geometrical construction; he didn't explicitly state what we now call "Snell's law," but the proportionality was implicit. Snell's construction went unpublished, and was noticed only when discussed by Huygens in 1703. In the meantime, René Descartes (1596–1650) had discussed the relation analytically in the 1630's in France, where it is known to this day as Descartes' law.

The indices of refraction are shown for several common substances in Table 34–1. Note that we can rewrite Equation 34–2 in terms of indices of refraction for two media to give the general case

$$n_i\sin\theta_i = n_r\sin\theta_r \tag{34–3}$$

(see Problem 1).

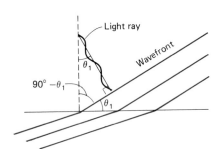

Fig. 34–5
Simple geometry shows that Snell's law holds whether we measure angles from the normal to rays or from the surface to wavefronts.

Example 34–1
Index of Refraction

You are standing at the edge of a 1.00-m-deep reflecting pool, whose top is flush with the ground, and want to pick up your glasses, which have fallen in. The glasses appear to be ahead of you at a 60° angle to the vertical (Fig. 34–6); your eyes are 1.70 m above the ground. At what angle should you reach in? How much closer to you do the glasses appear than they actually are?

TABLE 34–1 ▬▬▬▬▬▬
Indices of Refraction

Substance	State	Temperature (pressure = 1 atm)	Index of refraction
Air	Gas	0°C	1.000293
Carbon dioxide	Gas	0°C	1.00045
Ice	Solid	20°C	1.309
Water	Liquid	0°C	1.333
Ethyl alcohol	Liquid	0°C	1.361
Fluorite (CaF_2)	Solid	20°C	1.434
Fused quartz (SiO_2)	Solid	20°C	1.458
Benzene	Liquid	20°C	1.501
Glass, crown	Solid	20°C	1.52
Salt (NaCl)	Solid	10°C	1.544
Glass, flint	Solid	20°C	1.6–1.9
Diamond	Solid	20°C	2.419

Indices of refraction measured for the yellow line of sodium at 589.0 nm. Fluorite, fused quartz, and glass are commonly used for optics, and diamond is used for optics in special high-pressure situations.

Solution

Since the glasses appear at an angle of 60°, $\theta_1 = 60°$. From Snell's law ($\sin\theta_1 = n\sin\theta_2$), since $\sin\theta_1 = \sin 60° = \sqrt{3}/2$, and $n_{water} = 1.333$, we have

$$\sin\theta_2 = \frac{\sin 60°}{n_{water}} = \frac{\sqrt{3}/2}{1.333} = 0.650,$$

so

$$\theta_2 = 40.5°.$$

To determine the apparent distance to the glasses, we must compute the hypotenuses of the two right triangles shown with colored lines in the diagram. In the air,

$$\cos 60° = \frac{1.70 \text{ m}}{d_{air}};$$

$$d_{air} = 3.40 \text{ m}.$$

Similarly, in the water,

$$\cos 40.5° = \frac{1 \text{ m}}{d_{water}};$$

$$d_{water} = \frac{1.00 \text{ m}}{0.760} = 1.32 \text{ m}.$$

Fig. 34–6
Example 34–1

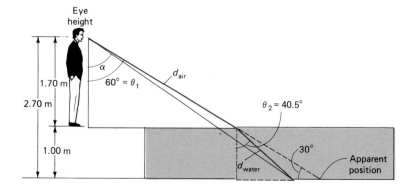

Thus the glasses appear 3.40 m + 1.32 m = 4.72 m from you. If you were focusing a camera, you should set its range to 4.72 m to get the clearest image.

To find out how far away the glasses actually are, we must first compute how far out in the pool they really are horizontally: In the air,

$$\tan 60° = \frac{x_{air}}{1.70 \text{ m}};$$

$$x_{air} = (1.70 \text{ m})\tan 60° = 2.94 \text{ m},$$

the length of the solid, colored horizontal line.

In the water,

$$\tan 40.5° = \frac{x_{water}}{1.00 \text{ m}};$$

$$x_{water} = (1.00 \text{ m})(\tan 40.5°) = 0.85 \text{ m},$$

the length of the dashed, colored horizontal line.

Thus the glasses are actually 2.94 m + 0.85 m = 3.79 m forward horizontally. That distance is one side of a right triangle whose other side is 1.70 m + 1.00 m = 2.70 m. The actual distance to the glasses is the hypotenuse, given by

$$d_{actual}^2 = (3.79 \text{ m})^2 + (2.70 \text{ m})^2 = 21.7 \text{ m}^2,$$

so

$$d_{actual} = 4.65 \text{ m}.$$

The glasses appear 4.72 m − 4.65 m = 0.07 m farther away than they actually are. The angle α up from the vertical at which you should reach is

$$\alpha = \sin^{-1}\left(\frac{3.79}{4.65}\right) = 55°.$$

Example 34–2
Refraction through a Block

You shine a light ray through air at an angle θ from the normal onto a glass block of thickness D. At what angle does the ray emerge into the air at the other side of the block, and what else happens to the ray?

Solution

In Fig. 34–7a, you can see that the ray travels at an angle in the block such that $\sin\alpha = (n_{air}/n_{block})\sin\theta$. This angle does not change, no matter how thick the

Fig. 34–7
Example 34–2

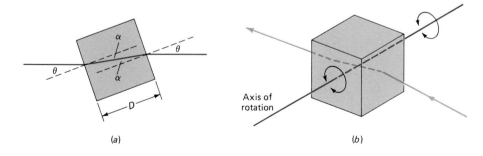

(a)

Axis of rotation

(b)

block is. At the exit surface, the ray strikes at angle α, and is transformed to a final angle $(n_{block}/n_{air})\sin\alpha$. Thus the sine of this angle is

$$\left(\frac{n_{block}}{n_{air}}\right)\left(\frac{n_{air}}{n_{block}}\right)\sin\theta \;=\; \sin\theta,$$

and the ray emerges at the same angle at which it entered. It is displaced, however.

A glass block that is designed to rotate on a fixed axis rapidly (within a fraction of a second) is sometimes used in telescopes to compensate for motion of the incoming light rays caused by atmospheric turbulence. A rotation of the block moves the incoming light ray over without changing its direction. (See Problem 7.)

When a light wave moves into a medium with higher index of refraction, its speed decreases. But physically, nothing has occurred that could change its frequency, which depends on the rate that electrons are oscillating to produce the wave. So from $\lambda f = v$ and $v = c/n$ we have

$$\lambda \;=\; \frac{c}{nf}. \tag{34–4}$$

Thus the wavelength is reduced in the new medium, in agreement with our earlier picture of the propagating wavefronts.

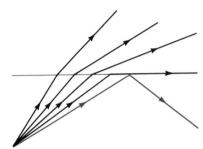

Fig. 34–8
At the critical angle the ray is bent so that it travels along the surface. For greater angles of incidence (color), there is total internal reflection.

34.2
THE CRITICAL ANGLE

critical angle

In Fig. 34–8, we see that a light ray going from a substance of a higher index of refraction to one of a lower index of refraction, as from water to air, is bent farther from the normal as the angle of incidence increases. For some angle, the light ray is bent so that it travels along the surface. For greater angles, the ray is reflected rather than passing through the surface. This **critical angle** corresponds to $\theta_r = 90°$ in Equation 34–3. With $n_r = 1$ for air, Equation 34–3 at the critical angle θ_c becomes

$$n\sin\theta_c \;=\; \sin 90° \;=\; 1,$$

or

$$\theta_c \;=\; \sin^{-1}\!\left(\frac{1}{n}\right), \tag{34–5a}$$

total internal reflection

where n refers to the medium from which the light is incident. For $\theta_i > \theta_c$, the light ray undergoes **total internal reflection** rather than refraction, as shown in color in Fig. 34–8.

More generally, Equation 34–3 shows that

$$\theta_c \;=\; \sin^{-1}\!\left(\frac{n_2}{n_1}\right) \qquad n_1 > n_2 \tag{34–5b}$$

for light propagating from a medium with index of refraction n_1 to a medium with a lower index of refraction n_2.

Example 34–3

Total Internal Reflection in a Prism

Show that the light ray shown in the upper left part of Fig. 34–9 follows the path indicated, if it enters perpendicularly from air into a right-angle prism with two 45° angles. The prism is made of glass whose index of refraction is 1.55.

Solution

The light ray is undeviated as it enters the prism, since $\theta_i = 0$, and therefore $n \sin\theta_r = 0$ and $\theta_r = 0$. The ray strikes the next glass-air interface at an angle of 45°. The critical angle, given that the index of refraction of air is (to the accuracy of this example) indistinguishably close to that of a vacuum, is

$$\theta_c = \sin^{-1}\left(\frac{1}{1.55}\right) = 40.2°.$$

The ray hits at a larger angle than this, so is totally reflected. Since, as we shall soon see, it is reflected at the same angle at which it was incident, the ray then passes undeflected out of the prism.

Taking advantage of total internal reflection, prisms are often used instead of mirrors to provide precision reflection in optical systems. High-quality binoculars invariably use prisms instead of mirrors to reduce the physical size of the binoculars while maintaining the long optical paths needed (Fig. 34–9 lower left).

Example 34–4

Grazing Incidence

If an x-ray hits a mirror straight on, it passes through instead of bouncing off. But x-rays can be bent by a surface if total internal reflection takes place. In the Einstein X-Ray Observatory that studied celestial objects from space, a set of mirrors (Fig. 34–9 center) focused x-rays with reflections at such a low angle that the x-rays were essentially grazing the surface, a phenomenon known as grazing incidence (Fig. 34–9 right). Given that the index of refraction of the nickel surface of the mirror for an x-ray of wavelength 0.4 nm is $1 - \delta$ where $\delta = 1.8 \times 10^{-4}$*, what is the largest possible angle at that wavelength for which reflection occurs?

Solution

We can generalize the derivation of Equation 34–5 to find that, for $n_1 > n_2$, the critical angle

$$\theta_c = \sin^{-1}\left(\frac{n_2}{n_1}\right).$$

For the vacuum of space in which the telescope operated, $n_1 = 1.00000$, and we are given that $n_2 = 1 - \delta = 1.00000 - 0.00018 = 0.99982$, so

$$\theta_c = \sin^{-1} 0.99982 = 89°.$$

Thus the x-ray must be at a greater angle than 89° from the normal, or within 1° from the surface.

---

*The index of refraction can be less than 1 when the frequency of radiation is higher than the natural frequency of oscillation of the electric charges in a material containing free charges. The circumstances for which $n < 1$ are treated in higher-level optics courses. It is nevertheless interesting to see here that we can still apply our formula for the critical angle.

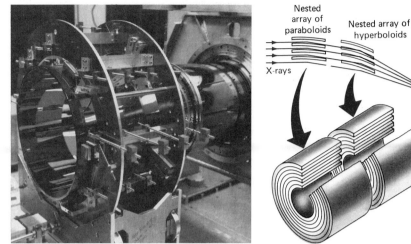

Fig. 34–9
(Above left) Total internal reflection in a 45° right-angle prism of glass whose $n = 1.55$. (Left) Binoculars use total internal reflection in prisms. (Center) The Einstein Observatory, launched by NASA in 1978, focused x-rays with a set of paraboloids followed by a set of hyperboloids, both used at grazing incidence—that is, at very low angles. Since the cross section exposed to x-rays is small at grazing incidence, four nested mirrors were used. (Right) The paraboloids alone would have focused on-axis parallel x-rays but the hyperboloids are necessary to improve the focusing of off-axis x-rays. Without the hyperboloids, off-axis x-rays from a point would have made a small circle around the on-axis image of that point, blurring the image.

The ability to use grazing incidence to focus x-rays allowed scientists to make images in the x-ray region of the spectrum that were comparable in resolution (detectable detail) with the best optical resolution obtainable from telescopes on earth.

Example 34–5
Materials for Total Internal Reflection

What is the minimum index of refraction for total internal reflection in a 45° right-angle prism such as those of Example 34.2, and what materials satisfy this condition?

Solution

$$n_{\min} = \frac{1}{\sin 45°} = \frac{1}{\sqrt{2}/2} = \sqrt{2} = 1.41.$$

All the solid materials shown, including all the glasses, meet this criterion, and 45° right-angle prisms made out of them will provide total internal reflection.

Application
Fiber Optics

Reflection and refraction are now at the basis of one of the greatest technological transformations of our times. Figure 34–10 shows how a series of internal reflections keep a light beam moving along a thin tube. A similar series of reflections optical fiber can keep a light beam moving along a thin pure transparent **optical fiber.** The fiber, often only a few micrometers thick, is made by drawing out molten glass. A typical

Fig. 34–10
Fig. 34–10
(Left) A beam of light undergoes a series of total internal reflections when it is shined along a flow of water, even though the flow is bent. (Right) Some optical fibers guide light through total internal reflection in this way.

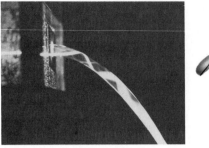

fiber for long-distance communication consists of a small (8 μm) glass core, surrounded by a larger (125 μm) glass cladding whose index of refraction is lower than that of the core. In some fibers, light is reflected abruptly at the core-cladding boundary; in others, gradual refraction in the cladding guides light along the fiber. Laser beams now used for communication are in the infrared region of the spectrum, at wavelengths of 1300 or 1550 nm. At these wavelengths, glass in typical fibers is so transparent that light can travel about 10 km before its intensity is halved. In some systems, the distance signals can travel is also limited by dispersion—the degradation of a signal that occurs because the index of refraction, and therefore propagation speed, varies slightly with wavelength. In undersea fiber-optic cables, propagation distance is limited primarily by attenuation; the first such cables, laid in the mid-1980's, have repeater stations about every 60 km to restore signal intensity.

Communication with light signals travelling along optical fibers has several advantages over satellite communication. One advantage is that communications are more difficult to jam or intercept. Fiber-optic cables are now being laid beneath the oceans, with the idea being that intercontinental communications should be split about equally between satellite and fiber-optic systems.

Optical fibers are also superior to electrically conducting cables. Fiber-optic systems are not sensitive to electrical noise, so they can be used for data transmission in electrically "noisy" environments like the fusion experiment shown on this book's cover. Furthermore, the rate at which a communications system can carry information depends on the frequency of the information-carrying signals. Since the frequency of light signals used in fiber-optic systems is many times the frequency of electrical signals propagating in conducting cables, fiber-optic systems can carry information at a much greater rate. Current technology allows 24,000 simultaneous telephone conversations on a single optical fiber. Since optical fiber cables, each containing many fibers, are much smaller than copper cables, they can be laid in existing rights-of-way under urban streets without the need for extensive excavation.

In addition to optical fibers themselves, fiber-optic engineers must deal with devices for switching optical signals, and for interfacing those signals to electronic circuits. The increasing use of fiber-optic communication has spawned the burgeoning new field of **electro-optics**.

electro-optics

Dielectrics and the Index of Refraction

We have seen (Section 32.5) that the speed of light c in a vacuum follows immediately from Maxwell's equations:

Fig. 34–11
(Left) Light propagates along a coiled optical fiber, guided by internal reflection. Most of the light entering the fiber is guided to the far end, but a small fraction escapes out the sides, so the entire fiber appears to glow. (Right) A modern fiber-optic cable can carry over half a million telephone conversations.

$$c = \frac{1}{\sqrt{\epsilon_0 \mu_0}},$$

where ϵ_0 is the electric permittivity of the vacuum and μ_0 is the magnetic permeability. Suppose, however, that light is propagating in a medium of dielectric constant κ. Though we do not derive it here, Maxwell's equations show that light moves in such a medium at speed $v = 1/\sqrt{\kappa \epsilon_0 \mu_0}$. (See Problem 38 in Chapter 32.) If we recall that the ratio $c/v = n$ defines the index of refraction of the medium, we can make the identification

$$n = \sqrt{\kappa}. \tag{34–6}$$

That an optical property of a dielectric medium, the index of refraction n, can be related to its electrical characteristics is a simple but very important consequence of the fact that light is, indeed, an electromagnetic phenomenon. (The speed of light also varies with the relative magnetic permeability, κ_M, introduced in Chapter 27. But for nonconducting materials in which electromagnetic waves can propagate, κ_M is so close to 1 that magnetic effects are negligible. The dielectric constant κ is generally frequency-dependent, so the index of refraction n does not follow from the low-frequency dielectric constants tabulated in Chapter 24.)

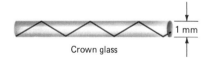

Crown glass

1 mm

Fig. 34–12
Example 34–5; an optical fiber of diameter 1 mm with $\kappa = 1.51$.

Example 34–6

Light Propagation in an Optical Fiber

A light wave of frequency $f = 5.0 \times 10^{14}$ Hz is travelling through a section of optical fiber of diameter 1.0 mm (Fig. 34–12). Suppose the optical fiber has dielectric constant $\kappa = 1.51$. If the ray is incident upon the inner surface of the fiber at 11° and, as is always the case, reflects at the same angle, how many wavelengths are there between successive reflections?

Solution

$\kappa = 1.51$ implies that the index of refraction $n = \sqrt{1.51} = 1.23$. The number of reflections N is thus the distance travelled divided by the wavelength in the dielectric. The distance travelled is the length d along the fiber divided by the cosine of the angle at which the ray travels. Thus

$$N = \frac{d/\cos 11°}{c/nf} = \frac{dnf}{c \cos 11°} = \frac{(10^{-1} \text{ cm})(1.23)(5.0 \times 10^{14} \text{ Hz})}{(0.982)(3.0 \times 10^8 \text{ cm/s})} = 210{,}000 \text{ wavelengths.}$$

Because the number of wavelengths is huge, our geometrical-optics approximation is highly accurate. Note that the wavelength in the dielectric is smaller than that in the vacuum by a factor of $n = 1.23$.

34.3
REFRACTION AND DISPERSION

The experimental fact that light of different wavelengths is bent by different angles at an interface, which corresponds to n being a function of wavelength, is both an advantage and a disadvantage of refracting optical systems. Later on, we shall discuss the disadvantages in connection with such systems as telescopes and cameras. Here we dwell on the important scientific developments that have used the differential bending of light by wavelength.

Figure 34–13 shows the index of refraction of light in glass as a function of wavelength. The variation over the optical spectrum is only about 2 per cent, but this variation is enough to enable us to separate the different wavelengths in a light beam and to study them individually. This separation by wavelength is called **dispersion.**

dispersion

Figure 34–14 shows the dispersion of light in a prism. Note that all the light is bent at each interface, but that the violet light, of wavelength approximately 400 nm, is bent more than the red light, of wavelength approximately 650 nm. After passing through both interfaces and back into air, the net result is that the light is **dispersed,** spread out into its component colors. Isaac Newton showed that if the light were again combined, the light beam would be restored to its original aspect (Fig. 34–15), proving that white light is made up of the entire spectrum of colors.

dispersed

Fig. 34–13
The index of refraction generally depends on wavelength. (The horizontal axis is not to scale.)

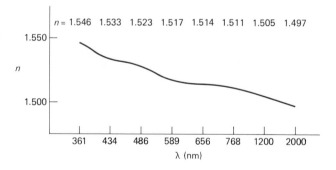

Fig. 34–14
(Left) A prism bends violet light more than red light. (Right) This drawing of the dispersion of sunlight is Newton's own sketch. Note that once a pure color is isolated, it does not change color on further refraction, a point that Newton considered crucial for verifying his theory of colors.

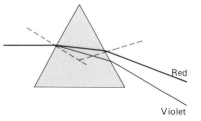

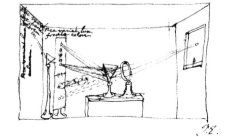

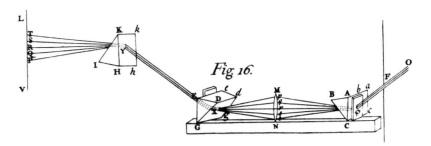

Fig. 34–15
Newton showed that light, though dispersed by a prism, coud be reconstituted. White light comes from point O at right, is separated into colors by prism ABC, recombined by prism DFG, separated again by prism HIJ, and falls as a spectrum on the screen LV at left. The picture is from Newton's *Optics*.

spectral analysis

The dispersion of a prism has led to the discipline known as **spectral analysis,** in which the light from an object is broken down into its component colors. Some substances, such as hot solids, radiate a continuous band of wavelengths (Fig. 34–16a). Others, such as diffuse gases, radiate only a few specific wavelengths (Fig. 34–16b). From studying which wavelengths are present, and their relative intensities, we can determine the constituents of the gas as well as its temperature and pressure. This ability is our link with the universe, for astronomers use spectral analysis to tell the state and composition of the distant stars and galaxies even though we cannot investigate them directly. We shall explore the link between atoms and light in Chapter 36. See the spectra in color plates 7–9.

The actual wavelengths present in light are determined by spectral analysis. The perceptions of the human eye are somewhat different; several different combinations of wavelengths can give us the perception of the same color. This fortunate circumstance allows the manufacture of color

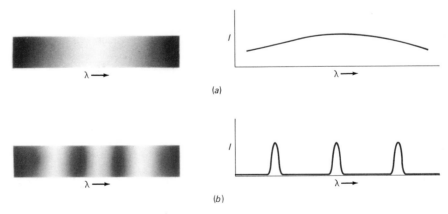

Fig. 34–16
(a) Hot solids radiate a continuous band of color while (b) dilute gases radiate only at a few discrete wavelengths (bottom) known as spectral lines. Dispersion in wavelength goes from left to right. At left, greater intensity is shown as a whiter tone. The axis from top to bottom on these images is either a spatial dimension or else the energy at that wavelength from a point source spread out uniformly from top to bottom to make it easier to see. At right, intensity is shown on the y-axis.

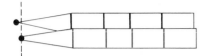

Fig. 34–17
If two objects slightly separated in position are dispersed by the same apparatus, the spectra are aligned with the same separation.

spectral lines

bright lines

emission lines

dark lines

absorption lines

film, in which combinations of three different emulsion layers, each sensitive to a different color, allow our eyes and brains to be fooled into thinking that a whole range of colors is present.

In Fig. 34–17, we see that a point source of light slightly displaced from another point source will have a spectrum slightly displaced from the spectrum of the first point source. If the sources were horizontally aligned, the spectra from each would overlap, and we could not clearly distinguish them. Thus for spectral analysis, we usually choose to place a thin slit in front of the light source, allowing only a line of light to pass into the prism. The effect would be to choose only one of the two sources in Fig. 34–17. The result is a single band of colors. In some cases, there is no continuous band, but only a series of lines perpendicular to the dispersion, each one at a different wavelength. Such a set of individual wavelengths are known as **spectral lines.** Sometimes we have lines at different colors; they are known as **bright lines** or **emission lines.** Sometimes we have **dark lines** crossing a continuous band of colors; the dark lines are also known as **absorption lines** (Fig. 34–18 top).

We can think of the spectral lines as individual images of the slit, dispersed by an amount appropriate to the wavelength. A straight slit (as is most common) gives straight spectral lines; a curved slit (as is used when studying curved objects, such as the gas just above the sun's curved surface) gives curved spectral lines (Fig. 34–18 center). An apparatus consisting, usually, of a slit, optics (lenses or mirrors) to direct the light passing through the slit at the prism as parallel light, the prism, and optics to focus the dispersed light onto film or onto an electronic detector is known as a

Fig. 34–18
(Top) The solar spectrum is a continuum of radiation crossed by dark Fraunhofer lines; the dispersion here runs from left to right. (Center) The emission lines from the curving brightest part of the sun's outer atmosphere show during an eclipse as a set of arcs. A curved slit would also give a curved set of arcs. In this ultraviolet view taken from a rocket, the bright circle near 120 nm is the "Lyman-alpha" line of hydrogen. (Bottom) In a spectrograph, light enters through a slit, and is made parallel before it falls on the dispersing element (here a prism). The two-dimensional image then falls on a detector.

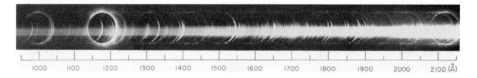

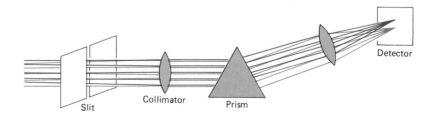

spectrograph

spectrometer

spectrograph or **spectrometer** (Fig. 34–18 bottom). In the following chapter, we shall see that an object known as a "diffraction grating" is often used instead of a prism to disperse light.

Example 34–7

Dispersion in a Prism

Hydrogen has spectral lines that range from the red line at 656.3 nm to violet lines at wavelengths of approximately 360 nm. By what angle do they deviate from each other if a beam of hydrogen light enters at a 30° angle to the normal into a 60° equilateral crown-glass prism (Fig. 34–19)?

Solution

Consider ψ (psi) to be the angle at which the light enters the prism, here 30°. The following angles will depend on the index of refraction for the given wavelength: θ the angle from the normal within the glass, β the angle of internal incidence on the far side of the prism, and ϕ (phi) the angle with the normal at which the light ray exits the prism. Subscripts v and r represent the violet and red rays, respectively.

From Snell's law,

$$\sin\psi = n\sin\theta.$$

From the triangle at the top of the figure, given that the sum of the angles of a triangle is always 180°,

$$180° = (90° - \theta) + 60° + \alpha.$$

Thus

$$\alpha = 30° + \theta$$

and

$$\beta = 90° - \alpha = 60° - \theta.$$

Finally, from Snell's law again, as we exit the glass,

$$\sin\phi = n\sin\beta.$$

From Fig. 34–13, n is 1.546 for the violet light and 1.515 for the red light. Since $\sin\psi = 0.5$,

$$\sin\theta_v = \frac{0.5}{1.546} = 0.3234 \qquad \sin\theta_r = \frac{0.5}{1.515} = 0.3300;$$

$$\theta_v = 18.87°, \qquad\qquad \theta_r = 19.27°,$$

and so

$$\beta_v = 60° - 18.87° = 41.13°. \qquad \beta_r = 60° - 19.27° = 40.73°.$$

Fig. 34–19
Example 34–6

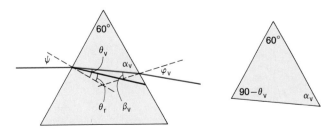

Thus

$$\sin\beta_v = 0.6578, \qquad \sin\beta_r = 0.6525,$$

and

$$\sin\phi_v = \frac{0.6578}{1.546} = 0.4255. \qquad \sin\phi_r = \frac{0.6525}{1.515} = 0.4310.$$

Thus

$$\phi_v = 25.18°. \qquad \phi_r = 25.53°.$$

The rays not only exit the prism at slightly different angles, but are also displaced from each other.

Dispersion in the earth's atmosphere gives rise to an atmospheric phenomenon known as the "green flash," easy to see if (as rarely happens for most people) you are in a location with a perfectly clear sky on the horizon. Near sunset, the sunlight hits the earth's atmosphere at a low angle (Fig. 34–20); we can approximate the atmosphere as a thin layer of air. All the sunlight is bent; at sunset, the sun is actually below the geometrical horizon because of this bending. Let us consider first the top edge of the sun. The red light from this top edge is bent the least and the blue light the most, so we actually have to look slightly higher above the horizon to see its image in blue light than in red light. A similar consideration goes for the bottom edge of the sun. If not for other effects in the earth's atmosphere, we would actually see an overlapping set of images of the sun, with the red image lowest, followed by the other colors higher: ROY G BIV (red orange yellow green blue indigo violet). But, as Lord Rayleigh discovered in the nineteenth century, the scattering of light waves in air decreases as the fourth power of the wavelength; that is, the shorter-wavelength light is scattered by air much more efficiently than longer-wavelength light. Thus the blue, indigo, and violet light scatter so much in the atmosphere at sunset that they do not penetrate far enough to reach us; they make blue skies for people below them. We are left with red, orange, yellow, and green. But water vapor in the earth's atmosphere preferentially absorbs orange and yellow, and at the oblique angle near sunset, there is enough water vapor in the path to eliminate those colors. We are thus left with two images of the sun: a lower one in red and an upper one in green. When the red image sets, the green image remains alone on the horizon for a second or so. The "green flash" can be striking, but you need perfectly clear atmospheric conditions and a horizon right down at 0° or even below from your horizontal (usually available when looking over an ocean) to see it.

Fig. 34–20
The green flash is caused by a combination of dispersion of sunlight and scattering and absorption of certain wavelengths in the earth's atmosphere.

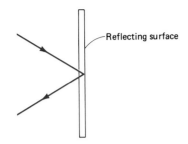

Fig. 34–21
An ordinary mirror reflects from its rear surface, which is usually coated with a silver compound.

specular reflection

diffuse reflection

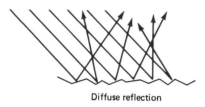

Diffuse reflection

Fig. 34–22
No image is formed in diffuse reflection.

34.4
REFLECTION

Though refraction in your eyes is fundamental to seeing, reflection is the most common optical effect we notice. When we see an image reflected off a water surface or a shiny car or spoon, the rays are reflected from the front of the surface. In ordinary mirrors, though, the light passes through a glass sheet and is reflected at the rear interface with an opaque material (Fig. 34–21), usually silver in the form of a mixture of ammonia, tartaric acid, and nitrate of silver. The glass merely protects the silver from scratching. Only in high-quality optical instruments, such as telescopes, are front-surface mirrors used; to make a front-surface mirror, you coat a smooth material with a reflective substance like silver or aluminum.

In this section, we discuss **specular reflection,** reflection that preserves the relative angles of incoming rays. Smooth surfaces provide specular reflection. An alternative is **diffuse reflection,** in which a rough surface reflects rays at such a wide variety of angles that no image is seen (Fig. 34–22).

Consider (Fig. 34–23) a series of wavefronts approaching a smooth surface. In Fig. 34–23a, we see point F_1 of a front touching the surface. Now let us ask what happens to this point on the front a time interval Δt later when the next front reaches the same position. Since we are assuming that the wave is reflected, it remains in the same medium and its speed doesn't change. Therefore, it is somewhere on a circle of radius $v\Delta t$ centered on G_1. The rest of the wavefront is on the line connecting the new position of F_1 with F_2, which is just striking the surface. Since F_1 was a single point, it must still be a single point; since the line F_1F_2 can only intersect the circle at one point, F_1 must be the tangent point, as shown in Fig. 34–23b. $F_2F_1G_1$ is thus a right triangle. Now draw the perpendicular from F_2 to the second wavefront G_2; G_2 would reach the surface in another Δt. $F_2G_2G_1$ is also a right triangle. Indeed, since $F_2G_2 = F_1G_1 = v\Delta t$, and G_1F_2 is shared by both right triangles, the right triangles must be congruent. Thus the angles $\theta_{incident}$ and $\theta_{reflected}$ are equal.

As we did for refraction, we choose to define angles between the rays and the normal. These angles of incidence, and reflection, respectively, are the same as those defined between the wavefronts and the surface. We have

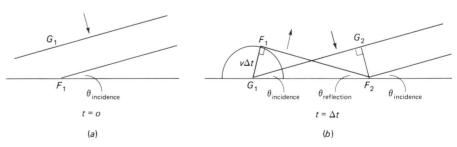

Fig. 34–23
Figure for the derivation of the law of reflection.

thus shown the **law of reflection:**

> The angle of incidence is equal to the angle of reflection,
>
> $$\theta_{\text{incident}} = \theta_{\text{reflected}}.$$ (34–7)

Plane Mirrors

We have thus far dealt only with the directions of the light rays and not with the intensity of the resulting image. Real mirrors ordinarily reflect about 95 per cent of the light, making a considerable dimming with multiple reflections (Fig. 34–24).

image

We know that a mirror can form an **image,** a representation of an original object. In Fig. 34–25, we trace light rays emanating from a bright point. They diverge, and continue to diverge after reflection off a plane mirror. If we look back along these rays, our brains assume that the light has been travelling in a straight line, and reconstruct an image as though it were at the point shown with the dotted lines. Since no light rays actually emanate

virtual image

from this point, we say we have a **virtual image.**

In a flat mirror, the image formed of an extended object (Fig. 34–26) is of the same size and shape as the original. It appears to be behind the mirror the same distance that the original is in front of the mirror.

At first glance, the image in a mirror may appear to be reversed left to right but not top to bottom. But this seemingly asymmetric situation is merely an illusion; actually, the image is reversed front to back. If you write on a transparent sheet of plastic, the writing is reversed in the same way if you look through the back of the plastic or if you hold the paper up to a mirror. Indeed, the reflection in a mirror of the view through the back of the plastic reads normally (Fig. 34–27).

Fig. 34–24
Multiple reflections—with no dimming—unlike real mirrors. Drawing by Charles Addams; © 1957, 1985 The New Yorker Magazine, Inc.

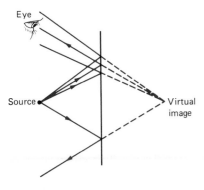

Fig. 34–25
We see a virtual image when we look in a plane mirror.

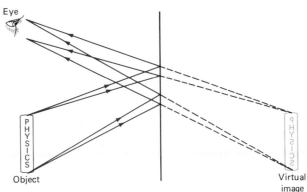

Fig. 34–26
The virtual image in a plane mirror is the same size as the object, is upright, and is reversed back to front.

Fig. 34–27
By photographing in a mirror, we can reverse the reversed image.

A top that spins in one sense will appear to be spinning in the opposite sense when observed in a mirror. If you try to shake hands with your reflection in a plane mirror, you cannot because of the back-to-front reversal. So the view you see of yourself in the mirror is different from the view that others have of you. You can provide yourself with the view that others have by looking in a pair of mirrors set at right angles to each other (Fig. 34–28). You can even shake hands with this image. You may notice that you look a little strange to yourself; this is not the normal view you have. But it matches the view you have of yourself in a TV monitor that displays the image taken by a camera pointing at you.

The double reversal you see in the right-angle mirrors is the basis of the **corner reflector** (Fig. 34–29). No matter what the incoming angle θ, the reflection off the first surface is also at angle θ, so the reflection off the second surface is at angle $90° - \theta$. But the second surface is at a $90°$ angle to the first surface, so the light comes out of the corner with the same angle at which it went in. This phenomenon holds true in the three-dimensional case as well, with three plane reflecting surfaces making up a corner (Prob-

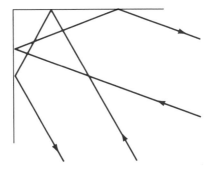

Fig. 34–28
(Left) Light comes out of a perpendicular set of mirrors at the same angle it went in. (Right) Looking in a corner with perpendicular mirrors. Notice that the image in the corner is not reversed, while the image to the side is reversed.

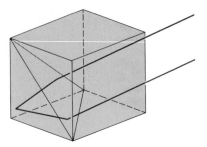

Fig. 34–29
A corner cube reflects light so that it returns in the direction from which it came.

Fig. 34–30
(Left) Corner reflectors are on the Lageos satellite in earth orbit. (Center) A laser beam leaving the Lunar Ranging Observatory (dome near bottom) in Hawaii, heading for corner reflectors on the moon (overexposed at top). (Right) Corner reflectors on the moon.

lem 20). Four sets of such corner reflectors (Fig. 34–30) were left on the moon by American astronauts and by a Soviet spacecraft. From sites in Texas and Hawaii, laser beams are regularly sent to the moon. The beams spread from about 1 cm-diameter at the laser to 5 km-diameter at the moon, so that only a small fraction of the beam is reflected. An even smaller fraction is detected after the light's 2.5-s round-trip journey to the moon by the sensitive detecting apparatus that is located on adjacent telescopes at the laser sites. Monitoring the round-trip time to an accuracy of 1 nanosecond gives 15-cm accuracy in the measured distance to the moon. With this accuracy, we can monitor changes in the earth-moon distance as energy dissipation caused by tidal forces makes the moon spiral slowly away from the earth. Certain earth-orbiting satellites also bear corner reflectors. Detailed analyses of laser light bounced off the moon and off these earth satellites also give accurately the lasers' positions on earth, and can be used for monitoring motions of the earth's crust. We can now detect continental drift directly.

34.5
LENSES AND MIRRORS

We use mirrors and lenses to form images in optical instruments such as eyeglasses, binoculars, telescopes, and microscopes. The point to which *focus* rays from a point source are made to converge is, in general, the **focus.** Focusing light enables us to funnel collected light into a relatively small area to make an image brighter, or, when the image is extended rather than being a point, to magnify or diminish its scale.

For distant objects like stars and planets, the light is diverging so little *parallel light* that the rays are almost parallel; we speak in that case of **parallel light** (Fig. 34–31). In a reflecting system, each parallel ray must be reflected at a dif-

Fig. 34–31
Sufficiently far away from a source of light, the rays are essentially parallel.

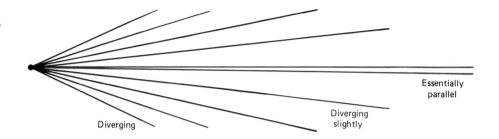

Diverging

Diverging slightly

Essentially parallel

ferent angle to bring all the rays to the focus (Fig. 34–32). A paraboloid, the curve you get by rotating a parabola around its axis, brings on-axis rays of parallel light to a focus (Fig. 34–33).

It is more difficult to make mirrors in the shape of a paraboloid than in the shape of a part of a sphere. But a spherical shape is less good than a paraboloidal shape because the sphere brings only light from its center of curvature to a focus. Parallel light is never quite focused by a spherical mirror, a defect known as **spherical aberration.**

spherical aberration

Fig. 34–32
Reflecting several incoming beams of light with flat mirrors arranged at suitable angles and locations shows that a concave mirror can be duplicated by a large number of individual flat mirrors.

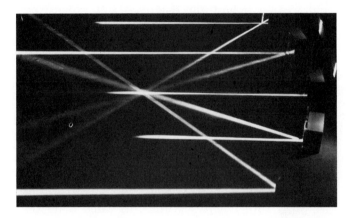

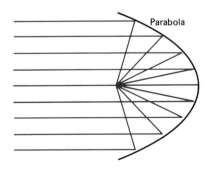

Parabola

Fig. 34–33
(Left) A parabolic mirror focuses to a point parallel light that enters parallel to the mirror's axis. (Right) This large-scale recreation duplicated the reported feat of Archimedes in 212 B.C., who arranged to set ships in the harbor on fire by concentrating solar energy from many plane mirrors set up to duplicate the effect of a paraboloidal reflector.

In a refracting system, light can also be focused. A magnifying glass, for example, can concentrate sunlight so well that it can quickly set fire to paper (Fig. 34–34). In this section, we discuss how simple lenses and mirrors make images. In the following section, we apply these results to optical instruments.

Thin Lenses

focal point

thin lens

Consider a lens shaped so that on-axis parallel light is focused to a point, the **focal point** (Fig. 34–35). In this section, we make the approximation of a **thin lens,** whose thickness is small compared with the radii of curvature of its surfaces. Usually the lens surfaces are sections of spheres.

concave

convex

Each surface of a lens can be either **concave,** like the inside of a spoon, or **convex,** like the outside of a spoon. The whole lens can cause incoming light either to converge or to diverge. Let us consider the following four basic cases (Fig. 34–36):

L1   Double-convex lens, object farther from the lens than the focal point;
L2   Double-convex lens, object closer to the lens than the focal point;
L3   Double-concave lens, object farther from the lens than the focal point; and
L4   Double-concave lens, object closer to the lens than the focal point.

ray tracing

To study image formation in these lenses with the rules of geometrical optics, we carry out **ray tracing.** Using the thin-lens approximation, we also assume that the objects are sufficiently small and close to the axis that no spherical or other aberrations exist. (We treat the more general case of thick lenses in the following section.)

Certain rays are simpler to trace than others, and we always choose to trace these rays for a first look. In particular, any ray from the object that is parallel to the axis is bent so that it goes through a focal point. And the ray from the object that is directed directly at the center of the lens goes through undeviated. Wherever these two rays cross, the image is formed. All other rays from the same point on the object must join them there, given our assumption that aberrations are absent.

Fig. 34–34
A magnifying glass easily concentrates sunlight enough to start a fire.

Fig. 34–35
A suitably shaped lens focuses parallel light to a point.

Consider first the case L1, the double-convex lens with an object farther from the lens than the focal point. Let us consider an arrow as our object. Since it rests on the axis, the base of the arrow must be imaged onto the axis. In the absence of aberrations, it will be on the axis at the same distance from the lens as the rest of the arrow, so we need to find in addition only the image of the top of the arrow. Choosing the rays that are easiest to trace, we first follow a ray from the top of the arrow parallel to the axis. Once it reaches the lens, we draw a straight line from its intersection with the lens through the focal point beyond the lens, since the definition of focal point says that such a ray passes through it. Then draw the "center" ray from the object passing undeviated through the center. You see that it meets the "parallel" ray below the axis on the far side of the lens, beyond the focus. We can then raise a perpendicular to the axis to provide the rest of the arrow.

real image

The image formed in this case is a **real image,** since the rays of light from the object actually reach this point. If we placed a paper or screen at

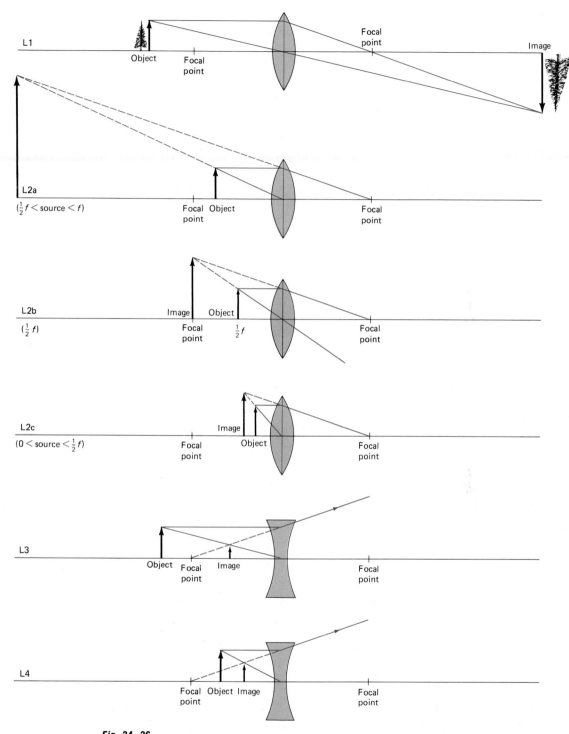

Fig. 34–36

Refraction. L1 shows an object beyond the focal point of a double-convex lens. L2 shows an object inside the focal point of a double-convex lens. In L2a, the source is between the focal point and one-half the focal distance from the lens; in L2b, the source is at one-half the focal distance from the lens, and the image is formed at the focal point; in L2c, the object is closer to the lens than one-half the focal distance. L3 and L4 show objects imaged by a double-concave lens. Only L1 makes a real image.

the real image, we would see the image of the object. Notice that the image is inverted (upside down), a general property of real images formed by single lenses or mirrors.

Now let us consider case L2, the double-convex lens with an object within the focal point. Following the same procedure as above, we draw the parallel ray through the lens and down through the focal point, and we draw the center ray undeviated through the lens. But now we see that these two rays are diverging on the far side of the lens. To make them meet, we must extend them back toward the left. Their intersection defines a virtual image, since no rays of light actually pass through it. If we were to place a paper or screen at this location, no image would be seen; we could view the image only by looking through the lens. Notice that the image is erect (right side up), a general property of virtual images formed by single lenses or mirrors.

If the object is at half the focal distance from the lens, the image appears at the focal point. If the object is closer to the lens, the image is also closer; if the object is farther from the lens than half the focal distance but still within the focal point, the image is beyond the focal point (see Problems 37 to 43).

For case L3, a double-concave lens with the object beyond the focal point, the "parallel" ray diverges, so we find its direction beyond the center of the lens by drawing a line from the near-side focal point through the "parallel" ray's intersection with the lens. This ray intersects with the "center" ray to form a virtual image right side up inside the focal point.

For case L4, a double-concave lens with the object within the focal point, we again find that a virtual image closer to the lens than the object is formed. The image, again, is right side up.

We now derive an equation that links the focal length, f, with the distances from the object to the lens, o, and from the lens to the image, i. In Fig. 34–37, from a point a distance A above the axis, we draw the parallel ray, the center ray, and the ray through the near focal point. This last ray must be parallel to the axis on the far side of the lens, a distance B from the axis. Using the central ray and the parallel ray plus its straight extension, the two right triangles outlined in solid color are similar, so that

$$\frac{A}{A+B} = \frac{o}{o+i}.$$

Further, using the shaded triangles defined by the refracted parallel ray, the lens, and either the axis or the refracted near-focal-point ray, we find that

$$\frac{A}{A+B} = \frac{f}{i}.$$

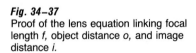

Fig. 34–37
Proof of the lens equation linking focal length f, object distance o, and image distance i.

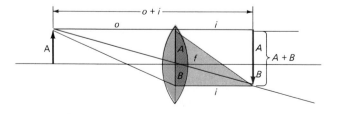

Fig. 34–38
Image distance *i* as a function of object distance *o*, given focal length *f*.

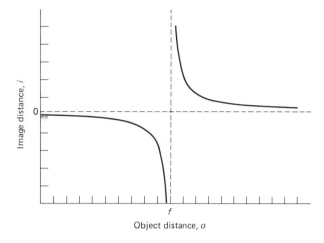

Setting the right sides equal gives

$$\frac{o}{o+i} = \frac{f}{i},$$

or

$$f = \frac{oi}{o+i}.$$

Though this is an adequate formula for *f*, its significance is more readily grasped if rewritten

$$\frac{1}{f} = \frac{1}{o} + \frac{1}{i}. \quad \text{(lens equation)} \tag{34–8}$$

We can solve for $1/i$, to give

$$\frac{1}{i} = \frac{1}{f} - \frac{1}{o} = \frac{o-f}{of}. \tag{34–9}$$

The image distance *i* is graphed in Fig. 34–38. We see that for $o>f$, the image distance is positive, which corresponds to real images. For $o<f$, the image distance is negative, which corresponds to virtual images. When the object distance is equal to the focal length, $1/i=0$, which places the image at infinity; this indeed corresponds to parallel light, as it should.

Similar considerations with *f* negative correspond to concave lenses. So, with proper sign conventions, the lens equation holds for all types of lenses.

Example 34–8

Image Formation by a Thin Double-Convex Lens

Both graphically and with Equation 34–9, (*a*) describe the position and type of image that result from an arrow 20 cm from a thin double-convex lens of focal length 10 cm; and (b) describe the position and type of image if the arrow is 5 cm from the lens.

Solution

(a) The problem is solved graphically in Fig. 34–39a. Or from Equation 34–9,

$$\frac{1}{i} = \frac{o-f}{of} = \frac{(20 \text{ cm}) - (10 \text{ cm})}{(20 \text{ cm})(10 \text{ cm})} = \frac{1}{20 \text{ cm}},$$

and

$$i = 20 \text{ cm}.$$

The image is an equal distance on the opposite side of the lens from the object, and is real and inverted (as all real images are).

(b) The problem is solved graphically in Fig. 34–39b. Or from Equation 34–9,

$$\frac{1}{i} = \frac{o-f}{of} = \frac{(5 \text{ cm}) - (10 \text{ cm})}{(5 \text{ cm})(10 \text{ cm})} = -\frac{1}{10 \text{ cm}},$$

and

$$i = -10 \text{ cm}.$$

The image is virtual and upright (as all virtual images are), and is located at the focal point on the same side of the lens as the object.

Magnification

The image formed by a lens is a different size from that of the object. In Fig. 34–40, we see that the shaded triangles are similar, so the ratio of the sizes of the image and object, respectively, is the same as the ratio of

Fig. 34–39
Example 34–8

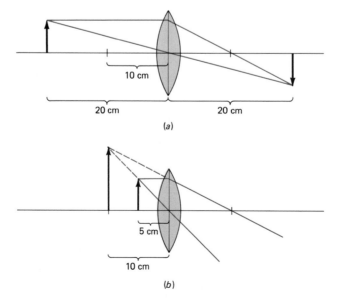

(a)

5 cm

10 cm

(b)

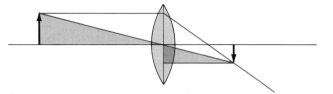

Fig. 34–40
Magnification. The shaded triangles are proportional, so the sizes of the object and image are in the same ratio as the distances from the lens to the object and image, respectively.

the distances of the image and object, respectively, from the lens. From Equation 34–9, we can derive

$$\frac{i}{o} = \frac{f}{o-f}. \quad \text{(linear magnification)} \qquad (34\text{–}10)$$

Magnification is commonly exploited to aid in reading or in examining small objects (Fig. 34–41).

Example 34–9
Magnifying Glass

You hold a magnifying glass of focal length 30 cm a distance of 10 cm from your book. By how much is the image of each word magnified?

Solution

Using Equation 34–10, the linear magnification is

$$\frac{i}{o} = \frac{30 \text{ cm}}{10 \text{ cm} - 30 \text{ cm}} = -1.5.$$

The image is magnified 1.5 times, with the minus sign signifying that it is a virtual image.

Fig. 34–41
A magnifying glass is held close to the object being inspected.

Curved Mirrors

Just as images can be formed by refraction in lenses, images can be formed by reflection off mirrors. If the mirrors are curved instead of flat, distortions may take place (Fig. 34–42).

For parallel light close to the axis of a spherical mirror (to avoid spherical aberration), we note in Fig. 34–43 that the light is reflected off the mirror at any point at an angle that corresponds to the law of reflection for the normal to the surface at that point. For a concave spherical mirror, this normal points radially inward. (For the region near the axis, a sphere approximates a parabola, which indeed focuses on-axis parallel light to a point.) The reflected ray intersects the axis at point f, the focal point. For small angle of reflection θ, the sides Cf and fP total CP, the radius of the circle, which we call R. But the angle PCf is equal to the angle of incidence, since both are interior angles of a line joining parallel lines. The shaded triangle is isosceles, and $Cf = fP$. Thus each is approximately $\frac{1}{2}R$, and

$$f = \tfrac{1}{2}R.$$

Similarly to the four cases we considered for lenses, let us consider four cases for mirrors:

Fig. 34–42
Distortion in funhouse mirrors.

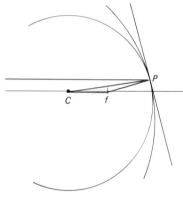

Fig. 34–43
For small angles (smaller than those
drawn here), the focal length of a
spherical mirror is half its radius of
curvature.

M1   Concave mirror, object farther from the mirror than the focal point;
M2   Concave mirror, object closer to the mirror than the focal point;
M3   Convex mirror, object farther from the mirror than the focal point; and
M4   Convex mirror, object closer to the mirror than the focal point.

Following the same ray-tracing methods we used for lenses, we find
the results of Fig. 34–44. The ray from the top of the object that hits the
center of the mirror bounces back so that, when it is back at the distance
of the object, it is as far below the axis as the top of the object is above it.
The ray from the bottom of the object that travels along the axis reflects
back along the axis; there are no other simple rays to follow from the bottom of the object, and we take the distance from the mirror to be the same
for the bottom of the object as for the top of the object. Only the case for
an object outside the focal point of a concave mirror gives a real image.

We could derive the equations relating the focal length and the object
and image distances for mirrors by using geometrical arguments similar to
the ones we used for lenses. But it is sufficient to note that the situation is
identical to that for lenses, if folded over at the lens. So our equation derived for lenses still holds, if we consider the image distance to be positive
if it is on the near side of the mirror and negative if it is on the far side.
Further, f is positive for a concave mirror and negative for a convex mirror.
The sign conventions for o, i, and f are crucial to keeping convex and
concave mirrors and lenses straight. Any answer from the lens equation
must be consistent with the geometrical construction you can get by ray
tracing. In solving problems in geometrical optics, it is advisable to use ray
tracing to check whether the results you obtain from the lens equation are
physically reasonable.

Example 34–10
Images in Curved Mirrors

(a) A lighted arrow is located 40 cm in front of a concave mirror of focal length
12.5 cm. Where is the image? (b) The arrow is moved to 10 cm in front of the mirror.

Fig. 34–44
Reflection in curved mirrors. In M1, the object is beyond the focal length of a concave mirror, and a real image is formed. In M2, the object is within the focal length of a concave mirror, and a virtual image is formed. In M3 and M4, virtual images are formed by convex mirrors.

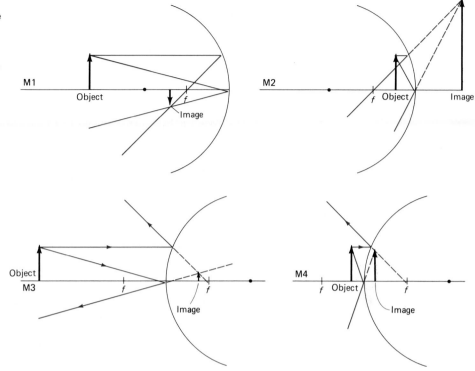

Where is the image? (c) The arrow is now placed 10 cm in front of a convex mirror of focal length 20 cm. Where is the image?

Solutions

Geometrical constructions appear in Fig. 34–45.

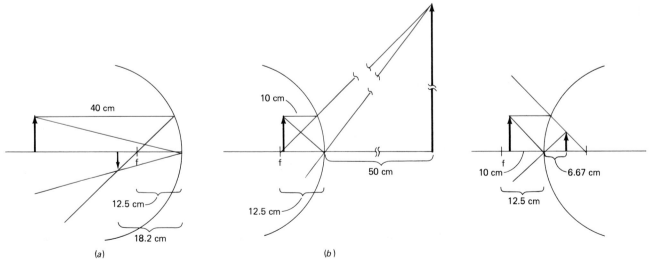

Fig. 34–45
Example 34–10

(a) Although we could solve Equation 34–8 to give Equation 34–9, it is often easiest to remember the former and thus to work directly from it:

$$\frac{1}{f} = \frac{1}{o} + \frac{1}{i},$$

$$\frac{1}{12.5 \text{ cm}} = \frac{1}{40 \text{ cm}} + \frac{1}{i},$$

$$\frac{1}{i} = (0.080 - 0.025) \text{ cm}^{-1} = 0.055 \text{ cm}^{-1},$$

so $i = 18.2$ cm. The image is on the same side as the object, a real image.

(b) Again using Equation 34–7,

$$\frac{1}{f} = \frac{1}{o} + \frac{1}{i},$$

$$\frac{1}{12.5 \text{ cm}} = \frac{1}{10 \text{ cm}} + \frac{1}{i},$$

$$\frac{1}{i} = (.080 - 0.100) \text{ cm}^{-1} = -0.020 \text{ cm}^{-1},$$

and $i = -50$ cm. The image is a virtual image 50 cm behind the mirror.

(c) Again using Equation 34–7,

$$\frac{1}{f} = \frac{1}{o} + \frac{1}{i},$$

$$-\frac{1}{20 \text{ cm}} = \frac{1}{10 \text{ cm}} + \frac{1}{i},$$

$$\frac{1}{i} = (-0.050 - 0.100) \text{ cm}^{-1} = -0.150 \text{ cm}^{-1},$$

and $i = -6.67$ cm, a virtual image 6.67 cm behind the mirror.

34.6

THE LENSMAKER'S FORMULA AND ABERRATIONS

We have considered only thin lenses, in which the light is bent by the lens but then quickly emerges, returning to the same medium in which it originated. Moreover, we have been able to use the single term f, the focal length, instead of physical quantities of the lens such as its radii of curvature and index of refraction. In fact, we can relate these equations via the

lensmaker's formula   **lensmaker's formula,**

$$\frac{1}{f} = (n-1)\left(\frac{1}{r_1} + \frac{1}{r_2}\right), \tag{34–11}$$

where n is the index of refraction of the lens material, and r_1 and r_2 are the radii of the first and second surfaces of the lens, respectively, and are positive for surfaces that curve outward (are convex).

Let us derive this formula for the general case of a lens of thickness t. Later, we can limit ourselves to thin lenses by taking $t \ll r_1$ and $t \ll r_2$. In Fig. 34–46, a pointlike object is at point A, a distance o from the first surface

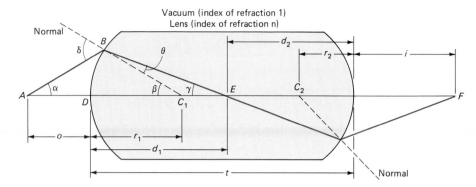

Fig. 34–46
Derivation of the lens maker's equation, showing the geometry of a lens of arbitrary thickness.

of the lens and with a light ray incident on the lens at an angle δ to the normal to the surface.

From Snell's law, $\sin\delta = n\sin\theta$, so $\delta = n\theta$ in the small-angle approximation. From triangles BC_1E and ABC_1, we see that $\theta = \beta - \gamma$ and $\delta = \alpha + \beta$. Combining these results to eliminate θ and δ, we find

$$\alpha + n\gamma = (n-1)\beta.$$

Further, in the small-angle approximation, $\alpha \simeq \tan\alpha = BD/AD = BD/o$, $\beta \simeq BD/r_1$, and $\gamma \simeq BD/d_1$. Since the arc length BD is common to all these angles, the above formula becomes

$$\frac{1}{o} + \frac{n}{d_1} = \frac{n-1}{r_1}, \tag{34–12}$$

which gives us the location of the intermediate image formed at point E. Note that this intermediate image may be real ($d_1 > 0$) or virtual ($d_1 < 0$), depending on the actual physical parameters of the problem.

Next, using the intermediate image as the "object" for surface 2 of the lens yields the analogous formula

$$\frac{n}{d_2} + \frac{1}{i} = \frac{n-1}{r_2}, \tag{34–13}$$

where the final image is formed at point F, a distance i from the surface of the second lens.

Combining these last two equations, we obtain a formula valid for rays from on-axis objects incident at small angles on an arbitrarily thick lens:

$$\frac{1}{o} + \frac{1}{i} + \frac{n}{d_1} + \frac{n}{t-d_1} = (n-1)\left(\frac{1}{r_1} + \frac{1}{r_2}\right), \qquad \text{(thick lenses)} \quad \text{(34–14)}$$

where we used the fact that the lens thickness $t = d_1 + d_2$. Note that Equation 34–12 involves the quantity d_1, which we cannot measure. We can, however, derive d_1 from Equation 34–12, in which case i is given by Equation 34–13.

In the limit of a thin lens, $t \ll d_1$, the undesired terms cancel. Invoking Equation 34–8, we are left with the lensmaker's formula, Equation 34–11.

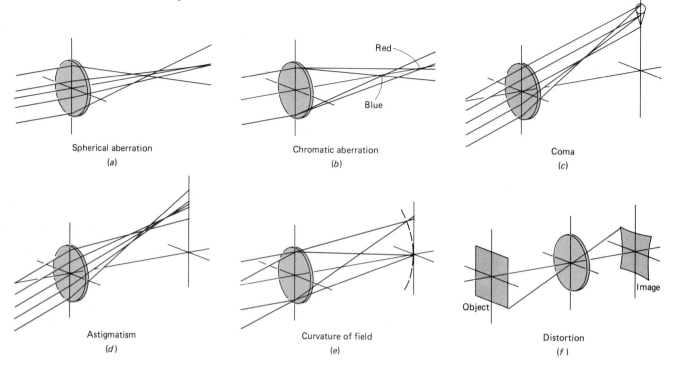

Fig. 34–47
Lens aberrations. *(a)* Spherical aberration. Light passing through the lens at different distances from the axis is focused at different distances. *(b)* Chromatic aberration. Light of different wavelengths is focused at different distances. *(c)* Coma. Off-axis rays passing far from the center of the lens are focused in a different location than light passing near the lens's center. *(d)* Astigmatism. Light that passes along a line across the lens is focused in a different location than light that passes a perpendicular line across the lens. *(e)* Curvature of field. An extended object is in focus along a curved surface rather than on a plane. *(f)* Distortion. Straight lines are distorted into curves because of the shape of the lens.

Nowadays, high-quality lenses are made in complicated nonspherical shapes to limit aberrations. Complicated shapes and combinations of lenses can give reasonably undistorted images over a wider field than the simple lenses we have discussed here. Optical designers still trace rays, but now they do so with computers.

A few standard types of defects—aberrations—are often found in optical systems (Fig. 34–47). We have discussed spherical aberration, the deviation from perfect focus caused by the fact that a spherical mirror does not focus light to a point. **Chromatic aberration,** the focusing of different colors to different points, stems from the fact that the index of refraction of glass is different for different wavelengths of light. It can be limited by assembling compound lenses of differing materials arranged so that several different wavelengths wind up focused at the same point. When two different wavelengths are adjusted, the result is an **achromat.** When three different wavelengths are adjusted, the result is an **apochromat.** Achromats and apochromats still do not guarantee that the colors are well focused in between their special wavelengths, but the deviations are not usually large.

Coma is an aberration that gives images of points slight tails, like comets. **Astigmatism** is an aberration that results from unsymmetric focusing. Simple astigmatism, a common sight disorder, can be compensated for by

chromatic aberration

achromat
apochromat

coma
astigmatism

Fig. 34–48
Astigmatism causes blurring in one direction but not necessarily in the perpendicular direction. (Left) Horizontal blurring. (Right) Vertical blurring.

providing correcting lenses with different radii of curvature in different directions. Many people have eyeglasses that correct for astigmatism (Fig. 34–48).

34.7
OPTICAL INSTRUMENTS AND PHENOMENA

Let us apply our understanding of thin lenses to simplified versions of several common optical instruments.

The Eye

The human eye (Fig. 34–49) is more like a thick lens than a thin lens. It is essentially a ball 2.3 cm in diameter filled with fluid. Light enters the eye through the hard, transparent cornea into a fluid called the aqueous humor. It then passes through an opening whose size is controlled by an adjustable iris. Then the light passes through the eye's lens and the vitreous humor. The combination of cornea, aqueous humor, lens, and vitreous humor focuses the incoming light on the back surface of the eye, the retina. The retina is covered with tiny light-receptor cells. The cones come in types sensitive to different colors, and the rods are sensitive at lower levels

Fig. 34–49
(Left) The human eye. (Right) Rods and cones in a monkey retina, seen in a scanning electron microscope.

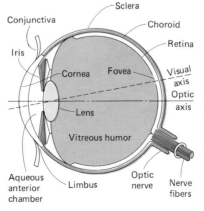
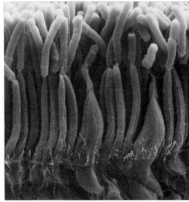

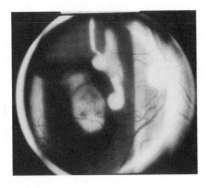

Fig. 34–50
A photograph showing an image on an eye's retina; we see the inverted real image of a girl using a telephone.

Fig. 34–51
Eyeglasses provide convex, concave, or astigmatic lenses to provide correction for aberrations in the eye.

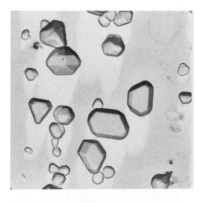

Fig. 34–52
Silver nitrate crystals in a film emulsion, magnified 50,000 times in a scanning electron microscope. Light hitting these crystals makes a chemical change; adding a chemical called a developer blackens the crystals to make a photographic negative.

of light intensity (Fig. 34–49 right). The real image formed on the retina (Fig. 34–50) is transformed into electrical signals that are sent to the brain.

Contrary to general belief, the eye's lens does not provide most of the refraction. Most of the refraction is actually provided by the cornea and the vitreous humor. But the focal length of the eye's lens is adjustable, as the ciliary muscles pull on the lens or relax. Changes in the lens's focal length provide a fine adjustment in our vision.

When the ciliary muscles are relaxed, the lens is relatively flat and has a long focal length. But for nearsighted (myopic) people (that is, people who see well only for nearby objects), the image still forms in front of the retina, and the retinal image is blurred. Diverging (concave) lenses in eyeglasses move the image back so that it is in focus. The extent of the correction is measured in diopters, where the number of diopters is the inverse of the focal length of the lens in meters. Thus a 1-m-focal-length lens provides a 1-diopter correction. A more strongly curved lens with a 0.5-m focal length provides a 2-diopter correction (Fig. 34–51).

In the eyes of farsighted people, on the other hand, the image would form somewhat behind the retina, even when the ciliary muscles are contracted. Converging (convex) lenses provide the correction necessary for the image to form on the retina. The ciliary muscles weaken with age, so stronger converging lenses become necessary.

When neither full relaxation provides clear distant vision nor full contraction provides clear near vision, bifocals often help. They provide different corrections in different parts of the lens. Corrections for astigmatism, if any, can also be included in eyeglasses.

The iris of the human eye can vary the size of the opening from a minimum of about 2 mm to a maximum of about 8 mm. The brain forms an image about 30 times a second, an interval that allows motion pictures and television to give the illusion of motion by forming images every 1/16, 1/24, or 1/30 s.

Eyes of other animals often have different characteristics from human eyes. Some animals, for example, can see light and shadow but not form clear images. Other animals have multiple small systems making up each eye, providing a wider field of view than the human eye.

The Camera

The eye is an excellent imaging device, but it can store images only as a mental picture. In a camera, an image is made to fall onto a surface that is sensitive to light. In a still camera, the surface is ordinarily "film," a plastic transparent base covered with a light-sensitive "emulsion." Tiny particles in the emulsion (Fig. 34–52) undergo chemical reactions when hit by light; these changes can be made permanent by "developing" the film with special chemicals. In a television camera, the light-sensitive surface sends electrical signals to be stored on video tape or recorded in a computer.

In both cases, the essential optical process forms an image on the light-sensitive surface. Though in Fig. 34–53 we draw the image with a simple lens, complex lenses (Fig. 34–54) are usually used to limit aberrations. Further, changing arrangements of optical elements can provide a zoom capa-

Fig. 34–53
Schematic of a camera with a simple lens.

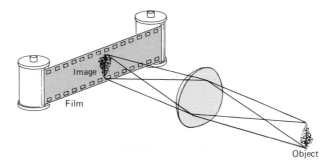

bility, in which the focal length of the lens is adjustable while keeping the image in focus on the film plane.

A major advantage of a camera, of course, is that the image can be permanently recorded in a way that can be shown to many people. Further, the image can be accumulated on film or on certain types of video devices, especially light-sensitive silicon chips known as CCD's (charge-coupled devices). The ability to accumulate light allows us to record fainter images than can be seen with the eye.

Moreover, the larger the lens, the more light it gathers. If that light is focused to make a small image, fainter objects show up than if that same total amount of light is spread out more. The amount that the light is spread out depends on the focal length of the lens; a longer focal length (a telephoto lens) provides a larger image scale and thus spreads the light from a given region on the subject over a greater area.

The sensitivity of a camera to light emitted by an extended object, then, depends on the ratio of the area of the lens through which light is passing and (because of the inverse-square spreading of radiation) on the square of the focal length: sensitivity $\propto A/f^2$. (The measure of sensitivity is known as "speed"; "fast lenses" give brighter images than "slow lenses.") Camera lenses are marked in steps of factors of 2 in the inverse of this ratio, so that lower numbers correspond to higher sensitivity. The numerical markings are in terms of the **focal ratio,** often called the **f-ratio** or **f-stop,** the ratio of the lens's focal length to its available diameter: f/d. For a given focal length f, as you make the diameter d smaller, the f-stop gets larger. The f-stops

focal ratio
f-ratio
f-stop

Fig. 34–54
Lenses in real cameras are much more complicated than simple lenses, in order to correct for aberrations.

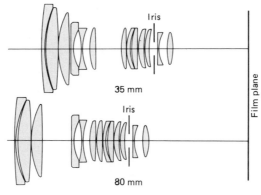

Fig. 34–55
The barrel of each lens shows f-stops and focusing distance.

marked on a camera lens (Fig. 34–55) are inversely proportional to the square roots of the lens's available area; a camera lens shows f-stops 1.4, 2, 2.8, 4, 5.6, 8, 11, 16, 22, 32; these are actually the (rounded) square roots of 2, 4, 8, 16, 32, 64, 128, 256, 512, and 1024. The f-stops are commonly written with a slash: f/1.4 for f-stop 1.4.

In a camera, the focal length of a given lens (or a given zoom setting on a zoom lens) is constant, and the f-stop is changed by varying an iris—a diaphragm—that passes light (Fig. 34–56). When only a small central portion of the lens is used, the light that passes is close to the axis. Light from a wide range of distances strikes the film in a narrow cone, so objects over this wide range of distances are close to forming a sharp image and thus being in focus. This process is the cause of a photographer's common dilemma: to use a wider lens opening to gather more light or to use a narrow lens opening to get a wider range of distances in focus, known as larger depth of field.

pinhole camera When the lens is closed down to almost a point, only undeviated rays pass through the lens, and the lens is unnecessary. A **pinhole camera** is made of just such a pinhole and film. It is relatively insensitive, but can be

Fig. 34–56
(Left) Two lenses of the same focal length but different maximum apertures. (Center) One of the lenses closed part way, changing the f-stop. We see the shape of the diaphragm, which approximates a circle with straight segments. (Right) The same lens closed all the way, giving the highest f-stop.

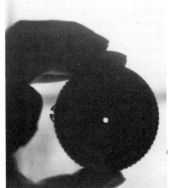

Fig. 34–58
Two pinhole images by Ansel Adams, one through a relatively large hole ($\frac{1}{8}$ inch) and one through a smaller hole ($\frac{1}{50}$ inch), with a longer exposure to compensate (6 s instead of $\frac{1}{5}$ s). Since more light rays from each point on the object could pass through the larger hole, that image is fuzzier.

Fig. 34–57
The small gaps between the leaves in the trees act as pinhole cameras, projecting the crescents of a partially eclipsed sun onto the ground. When the sun is not eclipsed, its pinhole image is round and is generally not noticed.

used to make images by either photographing bright objects (Fig. 34–57) or by taking long exposures (Fig. 34–58).

Telescopes

Astronomers use telescopes to collect light. Large lenses or mirrors collect much more light than the human eye, and focus it onto a CCD, film, or other light-sensitive surface. Certain telescopes are also used to provide magnification, principally when observing the sun or planets.

The simplest form of a refracting telescope is a long-focal-length lens that forms an image on a light-sensitive surface. No other lens is necessary; to take a photograph with a camera through a telescope, one ordinarily removes the camera lens and allows the telescope's image to fall directly on the film. The largest refracting telescope in the world, the telescope at the Yerkes Observatory in Wisconsin (Fig. 34–59), with its lens of 1-m diameter and 12-m focal length, can be thought of as a 12,000-mm telephoto lens.

Fig. 34–59
The 1-m refractor at the Yerkes Observatory in Williams Bay, Wisconsin. In this 1921 photograph, Albert Einstein is right of center.

Fig. 34–60
Magnification in an astronomical
refractor.

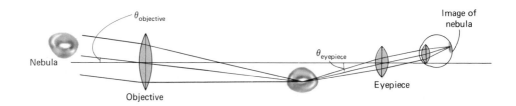

For visual use, a secondary "eyepiece" lens is used as a magnifier to
examine the image that falls in the focal plane of the first lens, the "objective." Since astronomical objects are so far away that their light is essentially parallel, we speak of the objects as being "at infinity." The image of
the objective is thus formed at the distance of the focal point; it is a real
image. The eyepiece is usually used to send parallel light to the eye, allowing the eye to be in the comfortable position that it takes for objects at
infinity. To do so, the eyepiece is placed so that its focal point is at the
focal point of the objective. The eye then sees a virtual image. Eliminating
the distance w that appears in Fig. 34–60, we can see that the angular magnification, the ratio by which the angle subtended by the object appears
larger, is

$$\frac{\theta_{eyepiece}}{\theta_{objective}} = \frac{\tan\theta_{eyepiece}}{\tan\theta_{objective}} = \frac{w/f_{eyepiece}}{w/f_{objective}} = \frac{f_{objective}}{f_{eyepiece}}, \qquad (34-15)$$

where we have used the small-angle approximation $\theta \simeq \tan\theta$. The magnification of a telescope is thus easily changed by changing the eyepiece. It is
generally meaningless to ask the magnification of a telescope, since it is so
readily variable.

Similar considerations apply to reflecting telescopes. Reflecting telescopes have several advantages over refracting telescopes: They are free of
chromatic aberration. The light does not have to pass into the glass, so the
interior structure of the backing for the reflective surface does not have to
be as uniform. The light does not have to pass through the entire surface
and out the back, so the mirror can be supported across its back. Because
of these advantages, a telescope mirror as large as 508 cm (200 inches)
could be made over 50 years ago; it has been in use since 1948 at the
Palomar Observatory on Palomar Mountain in California. Three of the
world's dozen largest telescopes are now located on the high volcanic peak
Mauna Kea in Hawaii (Fig. 34–61 top), where the air is especially clear and
steady and there is less water vapor to absorb the infrared part of the spectrum. A compound mirror made of segments is now being constructed for
that site; it is to be 10 m in diameter (Fig. 34–61 bottom). And new technologies are allowing single mirrors 7.5 m across to be made.

Fig. 34–61
(Above) More optical glass than anywhere else in the world is located on top of the 4200-m-high
volcano Mauna Kea on the island of Hawaii in Hawaii. The largest current visible-light/near-
infrared telescopes there have mirrors with 3.8-m, 3.6-m, and 3.0-m diameters. (Left) A 10-m-
diameter telescope is now under construction, to be erected at Mauna Kea. This Keck Ten-
Meter Telescope will be the world's largest optical telescope. Its mirror is being made of many
individual segments. Each is being stressed by a calculated amount, ground into part of a
sphere, and then allowed to "relax" to an unstressed position that makes it part of the giant
parabolic design.

Fig. 34–62
The 2.4-m-diameter mirror of the Hubble Space Telescope is the most precise large telescope mirror known publically to have been made.

The maximum magnification available with a telescope on earth is actually set not by the limits of optical engineering but rather by the turbulence of the earth's atmosphere; it does no good to magnify blurs. The finest details that can be seen, as set by the atmosphere, are about 0.5 arc sec, corresponding to 1 km on the surface of the moon and equivalent to the diffraction limit for a 25-cm telescope, as we shall discuss in the following chapter. Only the 2.4-m Hubble Space Telescope (Fig. 34–62), by being above the atmosphere, would be able to detect finer details.

Although a paraboloidal mirror (a mirror in the shape of a parabola rotated around its axis) focuses parallel on-axis light to a point, the region of the sky in which images are in clear focus is narrow in angle. Accordingly, most modern telescopes are made of the Ritchey-Chrétien design, in which the aberrations introduced by a spherical main mirror are compensated for by an appropriately shaped secondary mirror.

The Magnifier and the Microscope

To increase the detail we can see in a nearby object, we want to bring the object as close to the eye as possible so that it fills as large an angular field as possible. Our eyes do not focus much closer than 25 cm, but a converging lens near an eye allows us to bring the object closer (Fig. 34–63 left). Let us derive the magnification for such a lens; the situation is different from the use of an ordinary magnifying glass, which is usually placed near the object instead of near the eye.

angular magnification

The **angular magnification** of a magnifier is defined as the ratio of the angle subtended by the image we see to the angle subtended by the object as seen by the naked eye when the object is at a standard distance, set at 25 cm. In Fig. 34–63 (right), we see that the angles are small, and that therefore

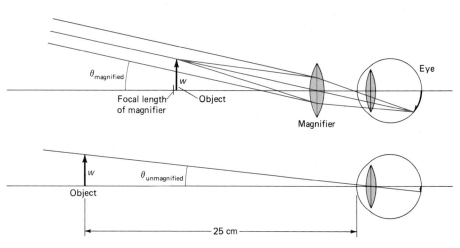

Fig. 34–63
(Left) A jeweler's loupe is held close to the eye. (Right) To compute the magnification of a simple magnifier, we compare the size an object would have if held at the 25-cm near point (bottom) with the angular size it apparently has when viewed through the lens.

$$\theta_{\text{unmagnified}} \simeq \tan\theta_{\text{unmagnified}} = \frac{w}{25 \text{ cm}}$$

and

$$\theta_{\text{magnified}} \simeq \tan\theta_{\text{magnified}} = \frac{w}{f}.$$

Thus the angular magnification is

$$\text{magnification} = \frac{25 \text{ cm}}{f}.$$

In a microscope, the objective lens has short focal length, and forms an enlarged real image of the object. Similarly to a telescope, an eyepiece lens is then used to form a virtual image at infinity of the focal plane.

The magnification of the microscope is equal to the product of the magnification of the objective and the magnification of the eyepiece. We have previously shown the magnification of the objective is i/o. The magnification of the microscope is thus

$$\text{magnification} = \left(\frac{25 \text{ cm}}{f_{\text{objective}}}\right)\left(\frac{i}{o}\right). \tag{34--16}$$

The best "light microscopes" (microscopes using light to illuminate the object) use very short focal length objectives held close to the object to provide the maximum magnification. For higher magnifications, we must use shorter wavelengths of radiation than visible light, or illuminate with particles, as in electron microscopes.

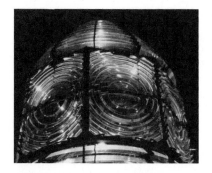

Fig. 34–64
A Fresnel lens installed in 1850 in the Sankaty Lighthouse on Nantucket, Massachusetts.

Application _____

The Fresnel Lens

Large-diameter lenses become too thick to use. In particular, lighthouses cannot concentrate their light to narrow beams because too much light would be absorbed in the lens. To solve this problem, the French scientist Augustin Fresnel, in 1822, worked out a way of making a large lens thinner. Essentially, a Fresnel lens consists of a central "bull's-eye" surrounded by concentric rings. The surfaces of each ring and the bull's-eye are shaped to focus light at an angular distance from the center, but each is as thin as it can be. It is as though a thick disk were removed from each. For a lighthouse (Fig. 34–64), five times more light is transmitted with a Fresnel lens than with an ordinary lens. Today Fresnel lenses made out of plastic often appear on home, office, or car windows to provide a magnified or wide-angle view in spite of being a thin sheet. They also appear in overhead projectors.

34.8
THE RAINBOW

The reflection and refraction by water droplets in the sky causes the most beautiful optical phenomenon: the rainbow. The circular arc of a rainbow is always formed in a well-specified direction with respect to the observer and the location of the sun in the sky (Fig. 34–65).

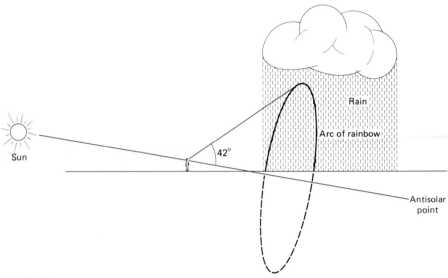

Fig. 34–65
Descartes found that the rainbow is a circle of approximately 42° centered on the antisolar point, the point in the sky opposite the sun.

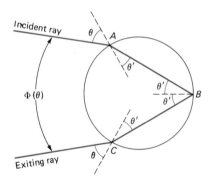

Fig. 34–66
The path of a light ray in a spherical water droplet. The ray enters at angle θ to the normal; the other angles follow from the laws of refraction and reflection.

To derive the angle of elevation of a band of color in a rainbow above the "antisolar point," the point in the sky 180° away from the location of the sun, consider a light ray of wavelength λ incident on a raindrop. For mathematical simplicity, consider the raindrop to be spherical (Fig. 34–66). θ denotes the angle of incidence on the surface of the raindrop, and θ' is the angle of refraction. Note θ' is also the angle of incidence and reflection on the back inner surface of the raindrop, and θ' is the angle of incidence on the exiting point from the rainbow. Because of the symmetries involved, the ray exits at angle θ. (Draw your own figure to verify this.)

Note that the angle Φ (capital Greek phi) between the incident ray and the reflected ray is (Problem 69)

$$\Phi = 4\theta' - 2\theta = 4\sin^{-1}\left(\frac{1}{n\sin\theta}\right) - 2\theta. \qquad (34–17)$$

It was René Descartes (1546–1650) who first realized that for a given value of n, a special angle of incidence θ_c existed at which the scattered rays formed a concentrated beam (Fig. 34–67), giving rise to a band in the sky. The angle corresponds to the minimum possible angle of deviation of the ray as it goes through the drop.

To calculate θ_c, set $d\Phi/d\theta = 0$ (Problem 70), which implies that all those rays within an infinitesimal range $d\theta$ of θ_s will suffer the same deflection. The result, from Equation 34–17, is

$$\cos^2\theta_c = \tfrac{1}{3}(n^2 - 1). \qquad (34–18)$$

Unfortunately, Descartes did not understand that sunlight was actually composed of the whole spectrum of colors. Neither did he appreciate the fact that the index of refraction of water was wavelength-dependent. Consequently, his theory predicted only a "white" rainbow, though he had the angle (40°) approximately correct. Newton, some years later, revealed the role of dispersion in creating the rainbow's beautiful colors (Fig. 34–68).

Fig. 34–67
The paths of a dozen parallel light rays hitting a spherical water droplet. Path 7 corresponds to θ_c, and paths 6 through 11 form the concentrated beam that we see as a color of the rainbow.

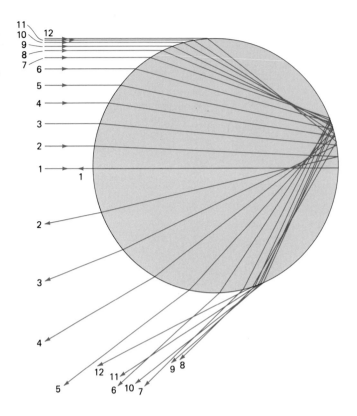

Fig. 34–68
The original drawing from Newton's *Optics* showing the formation of the primary and secondary rainbows (Book I, Part II, Plate IV). The primary rainbow is at an angle of approximately 42° from the antisolar point, and the secondary rainbow is at an angle of approximately 51° from the antisolar point (which is on the extension to the lower right on the line *OP*).

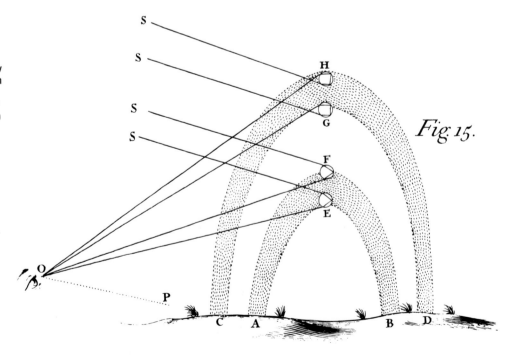

Using the experimentally determined values $n_{red} = 1.330$ and $n_{violet} = 1.342$ gives $\theta_{c,red} = 59.58°$ and $\theta_{c,violet} = 58.89°$. Thus you must gaze up at an angle with respect to the antisolar point $\Phi_{red} = 42.53°$ to see the red band and $\Phi_{violet} = 40.78°$ to see the violet band (Problem 71). Note that the red band is above the violet band.

The secondary rainbow is larger than the primary rainbow. It is formed by a double internal reflection within each raindrop (Problem 72), and is fainter than the primary rainbow. The order of the colors is reversed.

34.9
FERMAT'S PRINCIPLE

We conclude this chapter with a discussion of Fermat's principle, a physical law of great aesthetic appeal that enables us to solve certain problems in geometrical optics. Moreover, Fermat's principle can be generalized and then used to improve our understanding of the propagation of light throughout the universe. Pierre de Fermat (1601–1665) was a French mathematician (Fig. 34–69), often known best as the author of Fermat's last theorem, still one of the major unsolved problems of mathematics.

> *Fermat's principle*: The path of a light ray between any two points is the one that takes the least time with respect to infinitesimally differing paths.

We can use Fermat's principle to verify Snell's law of refraction, which we derived geometrically in Section 34.1. Consider a light ray travelling from a point in a medium with index of refraction n_1, across a straight interface, to a point in a medium with index of refraction n_2 (Fig. 34–69). Since the speed of light in a medium is $v_i = c/n_i$, the time the light ray spends in medium i is $t_i = D_i/v_i$. Thus the total length of time the light takes to travel between the points is

$$t = t_1 + t_2 = \frac{D_1}{v_1} + \frac{D_2}{v_2} = \frac{\sqrt{d_1^2 + y^2}}{c/n_1} + \frac{\sqrt{d_2^2 + (Y-y)^2}}{c/n_2},$$

where y and Y are shown in the figure. We find the minimum value of t by differentiating t with respect to the independent variable y, and setting the derivative equal to 0:

$$\frac{dt}{dy} = \frac{n_1}{c}\frac{d}{dy}(d_1^2 + y^2)^{1/2} + \frac{n_2}{c}\frac{d}{dy}[d_2^2 + (Y-y)^2]^{1/2}$$

$$= \frac{n_1}{c}\frac{1}{2}\frac{2y}{\sqrt{d_1^2 + y^2}} + \frac{n_2}{c}\frac{1}{2}\frac{2(Y-y)(-1)}{\sqrt{d_2^2 + (Y-y)^2}}$$

$$= \frac{n_1}{c}\frac{y}{\sqrt{d_1^2 + y^2}} + \frac{n_2}{c}\frac{(-1)(Y-y)}{\sqrt{d_2^2 + (Y-y)^2}}.$$

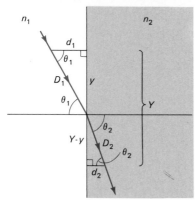

Fig. 34–69
(Above) Fermat. (Below) Diagram for the proof of Snell's law using Fermat's principle: the path of a light ray between any two points is the one that takes the least time.

But

$$\sin\theta_1 = \frac{y}{\sqrt{d_1^2 + y^2}}$$

and

$$\sin\theta_2 = \frac{Y-y}{\sqrt{d_2^2 + (Y-y)^2}},$$

so

$$n_1\sin\theta_1 - n_2\sin\theta_2 = 0,$$

which is equivalent to Snell's law.

Fermat's principle holds even in the curved spacetime that is a feature of Einstein's general theory of relativity. In this theory of gravity, the presence of mass warps the four-dimensional spacetime in which we live. Einstein realized that a light ray passing near the sun would follow a path that takes the minimum time of travel, as in Fermat's principle. This path appears to change the direction to stars near the sun, compared with their position when the sun is not present in that part of the sky (Fig. 34–70). The observational verification of this fact at the total solar eclipse of 1919, when the stars were briefly visible in the daytime near the sun, was the evidence that convinced scientists and the world that Einstein's theory was valid.

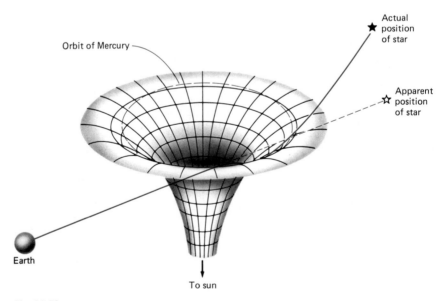

Fig. 34–70
The equivalent of Fermat's principle explains why light seeking the shortest path in warped spacetime makes the positions of stars appear to be slightly displaced when viewed near the sun, as predicted by Einstein's general theory of relativity. The effect is greatly exaggerated in the figure. Einstein's theory also explained an observed rotation of the orientation of the orbit of the planet Mercury.

SUMMARY

Optics is the branch of physics dealing with the behavior of light. When light interacts with objects much larger than its wavelength, it travels in straight lines and we can use geometrical constructions—**geometrical optics**—instead of analysis based on waves and Maxwell's laws of electromagnetism.

1. We treat light as **rays,** and study their **reflection** or **refraction** at a surface. For refraction, we have Snell's law:

$$n_i \sin\theta_i = n_r \sin\theta_r,$$

where n is the index of refraction of a medium.

2. Light moving from a substance of a higher index of refraction n_1 to a lower index of refraction n_2 is refracted more (has a greater angle of refraction) at greater angles of incidence. For the critical angle,

$$\theta_c = \sin^{-1}\left(\frac{n_2}{n_1}\right),$$

the light ray travels along the surface. For larger angles, the light ray undergoes **total internal reflection,** and the law of reflection is obeyed.

3. Communications signals are now often carried in long **optical fibers,** kept in their path by total internal reflection or refraction in fibers whose index of refraction varies radially.

4. It can be shown that

$$n = \sqrt{\kappa},$$

where κ is the dielectric constant of the medium. This result is a consequence of the fact that light is fundamentally an electromagnetic phenomenon.

5. The index of refraction of most materials depends on wavelength. Thus a prism spreads light into its component wavelengths—the light is **dispersed.** This phenomenon is at the basis of **spectral analysis,** the study of the continuous radiation and the **spectral lines** that are typical of individual elements under given conditions of temperature and density. Studies are made with the **spectrograph,** also called **spectrometer.**

6. We deal with **specular reflection,** which preserves the relative angles of incoming rays. This situation is to be contrasted with **diffuse reflection,** in which rays are reflected at all angles because of the roughness of the surface.

   The law of reflection is

$$\theta_{\text{incidence}} = \theta_{\text{reflection}}.$$

7. A mirror can form an **image,** a representation of an original object. We can find the location and orientation of the image by tracing individual rays of light leaving the object. A plane mirror forms a **virtual image,** in which it appears that an image is present at a location even though no light rays actually reach that location. Images in plane mirrors are reversed back to front.

Plane mirrors at right angles to each other make a corner reflector, which can be in either two or three dimensions. A corner reflector sends back light rays in the direction from which they came. A two-mirror reflector provides unreversed images.

8. The point to which rays are made to converge to form an image is the **focus.** Light from a faraway object diverges so slowly that it may be treated as **parallel light.** A paraboloidal mirror focuses parallel light to a precise point; a spherical mirror doesn't quite do so, giving **spherical aberration.**

9. On-axis parallel light is focused to the **focal point.** We consider **thin lenses,** in which the lens's thickness is small with respect to the radius of curvature of its surfaces. Each lens surface can be either **concave** or **convex.**

   We can study image formation with **ray tracing.** We find that an object beyond the focal point of a double-convex lens forms a real image, while an object inside the focal point of a double-convex lens or at any distance from a double-concave lens forms a virtual image.

   The object distance o, the image distance i, and the focal length f are linked by

$$\frac{1}{f} = \frac{1}{o} + \frac{1}{i}.$$

10. The magnification, the ratio of the apparent sizes of the image and object, is $f/(o-f)$.

11. Objects farther from a concave mirror than its focal point have real images, while objects closer than its focal point or objects reflected in convex mirrors have only virtual images.

12. Valid for thin lenses, the lens maker's formula relates the focal length, the index of refraction, and the radii of curvature of the lens surfaces:

$$\frac{1}{f} = (n-1)\left(\frac{1}{r_1} + \frac{1}{r_2}\right).$$

Aberrations of optical systems include **spherical aberration, chromatic aberration, coma,** and **astigmatism.** Lenses that correct for chromatic aberration by providing a common focal length at one and two wavelengths, respectively, are **achromats** and **apochromats.**

13. Several common optical devices involve lenses. These include:

   a. *The eye.* The cornea, aqueous humor, and lens refract light so that it forms an image on the retina; the lens can make adjustments in the focusing. The iris adjusts the aperture. When the focus falls in front of the retina (nearsightedness) or would fall behind the retina (farsightedness), diverging or converging lenses, respectively, can move the image so that it falls on the retina. The number of **diopters** of correction is the inverse of the focal length of the lens in meters.

b. *The camera.* A lens forms an image that falls upon a surface that allows the image to be recorded.

In a **pinhole camera,** no lens is used but images are formed; the hole is equivalent to the central region of a lens.

c. *The telescope.* Refracting telescopes use lenses as objectives and reflecting telescopes use mirrors as objectives. An "eyepiece" lens can be used as a magnifier to examine the focal plane in which the image of the first lens falls. The magnification *m* of a telescope is the ratio of the focal lengths *f* of the objective and the eyepiece,

$$m = f_{objective}/f_{eyepiece,}$$

and can therefore be easily changed by simply changing eyepieces.

d. *The magnifier and the microscope.* A converging lens near the eye, like a jeweler's loupe, is a magnifier, and is different from a magnifying glass, which is used close to the object. Since 25 cm is the standard distance that represents the closest the eye can focus, the angular magnification of a magnifier is (25 cm)/*f*.

The magnification of a microscope is the product of the magnification of the objective and the magnification of the eyepiece:

$$\text{magnification} = \left(\frac{25 \text{ cm}}{f_{objective}}\right)\left(\frac{i}{o}\right).$$

14. The rainbow is formed by symmetric refraction as a light ray enters and leaves a water droplet and by reflection inside the droplet. In the primary rainbow, light rays undergo a single reflection inside the droplet; in the secondary rainbow, light rays undergo a double reflection inside the droplet.

15. Results in geometrical optics can be derived using **Fermat's principle**: the path of a light ray between any two points is the one that takes the least time with respect to infinitesimally differing paths.

QUESTIONS

1. If you notice a piece of coral on the bottom of a fish tank and try to direct your hand to it while looking at the coral horizontally through the water, why won't your hand find the coral where it appears to be?

2. You are using a camera to photograph a hat you are wearing by shooting its reflection in a plane mirror. If you are standing 1 m from the mirror, at what distance must your lens be focused?

3. Using a geometrical construction, show the path of a ray of light in an enlarged model of an optical fiber 0.5 cm in diameter bent in a curve whose radius is 10 cm. Show the reflections for 90° of the curve.

4. While reeling in a fish, you put a net into the water to catch it. Where should you put the net with respect to the image you have of the fish?

5. Describe why a spoon looks bent when you place it in a glass of water.

6. Describe how the critical angle for diamonds affects internal reflections and discuss the way diamonds should be cut and polished for use as jewelry.

7. Describe any color effects that might arise from refraction as you look at fish in a fish tank.

8. You send monochromatic light into an equilateral prism (all sides equal and all angles 60°) held with a flat side down and apex up, and then place a second equilateral prism upside down in the dispersed beam (with apex down and flat side up). What do you see?

9. Would a right-angle 45° prism or an equilateral 60° prism disperse white light more? Explain.

10. Describe the change in your own image in the inside of a spoon as you move the spoon away from you, starting 2 cm from your eye.

11. Describe the change in your own image in the outside of a spoon as you move the spoon away from you, starting 2 cm from your eye.

12. The image in a flat mirror is the same size as the original. Under what circumstances can the image in a concave mirror be the same size as the original?

13. Imagine that you use an optical system to form a real image of a penny. What would occur if you shined a flashlight on that real image (holding the flashlight near your head, just as you might if you were illuminating an actual object), and why?

14. Can a scout troop use a double-concave lens to concentrate sunlight to light a fire?

15. Describe the shapes of the mirrors that make the images shown in Fig. 34–71.

Fig. 34–71 Question 15

16. At the movies, are you watching a real image or a virtual image of the film?

17. How does the image through a thin-lens magnifying glass differ when you hold the magnifying glass near your eye instead of near a printed page?

18. The astronaut in Fig. 34–72 is under water in order to semi-simulate weightlessness in a practice session. Explain why her head looks so small.

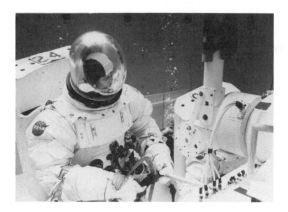

Fig. 34–72 Dr. Shannon Lucid of NASA working underwater on a model of the Hubble Space Telescope (Question 18).

19. In Fig. 34–73, we see a ball inside a transparent sphere inside an aquarium tank. The transparent sphere and the aquarium tank can each be filled with air or water. Explain which combination is occurring in each part of the figure.

Fig. 34–73 Question 19

20. What is the effect if you place a 1-diopter lens immediately behind a 2-diopter lens? Explain.

21. If the same total amount of light affects film the same way no matter what the time interval over which it arrives (the law of reciprocity), what f-stop would correspond to a 1/1000-s exposure if 1/125 s at f/11 also gives the correct exposure?

22. Where in a flashlight would you place the bulb with respect to the focal point of the reflector?

23. To send out a parallel beam of light from a searchlight, you place the light source at the focus of a parabolic mirror. If you wanted a converging beam, would it be helpful to place the light source slightly ahead or behind the focus? Explain.

24. If the total amount of light hitting film has less effect if it comes over a longer time interval ("reciprocity failure"; see Question 21), what would be the effect on long exposure times for night time-exposures?

25. Many photographers use a filter that blocks ultraviolet light to which film but not the human eye is sensitive. How might that improve the photographs?

26. Why is it usually not meaningful to ask the "power" of a telescope?

27. What is the major reason that a large telescope is superior to an ordinary camera lens in studying stars, whose images are points?

28. To avoid a concentration of sunlight, space-based solar telescopes make their secondary reflection with a mirror on the side of the focus nearer the sun and away from the mirror (Gregorian design) instead of with a mirror within the focus (Cassegrain design). Describe for each case whether the secondary mirrors are concave or convex, and whether the images are inverted. Sketch the systems.

29. Contrast a magnifier and a magnifying glass.

30. For the highest magnification with a light microscope, an "oil immersion" system is used in which the objective and the object are linked by oil instead of being separated by air. Explain the advantages of this system.

PROBLEMS

Section 34.1   *Reflection and Refraction*

1. Derive the general form of Snell's law, Equation 34–3:

$$n_1 \sin\theta_1 = n_2 \sin\theta_2.$$

2. From a point at which a wavefront meets a surface, draw circles representing locations of wavefronts travelling at the speed of light inside and outside the medium. Elaborate the geometrical construction to demonstrate Snell's law; this is the form in which Snell presented his law.

3. A fish is located 0.55 m under water. If you are looking directly down at the fish with your camera 1.00 m above the water's surface, at what distance should you focus your camera?

4. You spot a coin at an angle of 45° in 0.75 m of water. Your eyes are 0.78 m above the water. At what angle should you hold a stick so that a magnet at the end of the stick touches the coin?

5. You are looking down at a 30° angle from the horizontal onto a clear lake and see the bottom of a 1-m-long vertical branch, half of which is below and half

above the water. At what angle from the vertical does the light ray that reaches your eyes leave the bottom of the branch?

6. At what angle must you tilt a 2-cm-thick flint-glass block ($n = 1.89$) to make a ray of light deviate by 0.15 cm? *Hint:* Use the small-angle approximation $\sin\theta = \theta$.

7. At one of the National Solar Observatory's telescopes, the edge of the sun appears to oscillate up and down because of turbulence in the earth's atmosphere. A 5.0-cm-thick crown-glass block ($n = 1.52$) is placed in the beam; rotating the block around an axis perpendicular to the incoming beam displaces the light beam without changing its direction. Suppose that the cube can rotate a maximum of 10°. For how much vertical displacement can it compensate? Such a device, which is in common use, is called a "limb guider." (See Example 34–2.)

Section 34.2   *The Critical Angle and Prisms*

8. What is the critical angle for diamond?
9. How does the critical angle change when a quartz prism ($n = 1.46$) is immersed in water instead of air?
10. What is the critical angle in a lens between surfaces of flint glass ($n = 1.89$) and crown glass ($n = 1.52$)?
11. If a scuba diver sets off a strong flash a depth h below the water's surface, the light rays striking the lake surface will undergo total reflection outside a certain radius. Within what distance of the point over the light source would an observer on a ship be able to see the flash?
12. Derive a formula for the critical angle limiting the escape of a light ray from a medium of index of refraction n_2 to a medium of index n_1, if $n_1 < n_2$.

Section 34.3   *Refraction and Dispersion*

13. If the index of refraction in a fused-quartz prism is 1.47 for blue light and 1.45 for red light, calculate the path of a mixture of blue and red light that enters a 60° equilateral prism parallel to one of its surfaces.
14. Spectral lines from hydrogen appear in the visible as the Balmer series: Hα at 656.3 nm, Hβ at 486.1 nm, Hγ at 434.0 nm, Hδ at 410.1 nm, and Hε at 397.0 nm. By how much are these lines separated on a screen placed 1 m to the right of a 30°-60°-90° fused-quartz prism resting on its shortest side if the hydrogen light is incident parallel to the base? Refer to Fig. 34–74 for the indices of refraction.
15. How far from a prism must a screen be placed to see the sodium D lines at 589.0 and 589.6 nm, respectively, separated from each other? The prism has an equilateral cross section; it is hollow but filled with water at a temperature such that the indices of refraction at the wavelengths are 1.3331 and 1.3330, respectively. You may assume that the light enters parallel to one of the faces and that the prism is small compared with the distance to the screen. (If this were not true, the actual dimensions of the prism would have to be specified. Why?) Consider that the

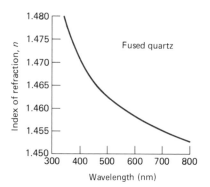

Fig. 34–74   Problem 14

lines can be seen as separated if there is 1 mm between their nominal central wavelengths.

16. Assuming that the earth's atmosphere is equivalent to a flat rectangle 10 km thick and of index of refraction 1.0005, by how much is the angle of the sun above the horizon different from the sun's true position for angles above the horizontal of 30°, 10°, and 5°? Ignore the curvature of the earth. (A more realistic calculation would use many thin layers.)
17. Assuming reasonable parameters, by how much does ordinary window glass affect your estimate of the distance to an object 1 m tall that is actually 10 m away? (*Hint:* Consider displacement of rays by passage of light through the glass slab that is the window. How does the window alter the apparent size of the object? How does it affect the apparent distance?)
18. For safety, workers using radioactive materials at the Nevada Proving Grounds look through lead-glass windows of index of refraction $n = 2$ and thickness 120 cm. Determine how their view of a fuel pellet 1 m distant is different than it would be if the glass were absent.
19. You are looking nearly perpendicularly through a crown-glass ($n = 1.52$) wall of thickness 0.50 cm into a fish tank and see an angel fish swimming at an apparent distance of 0.25 m inside the water. How far must you reach horizontally with a dipper to catch the fish? *Hint:* You may use the small-angle approximation. Can either the glass or the water be ignored?

Section 34.4   *Reflection*

20. Demonstrate that a three-dimensional corner mirror reflects a light ray back in its original direction. *Hint:* Consider a light ray propagating in the direction of the vector $q = q_x\hat{\mathbf{i}} + q_y\hat{\mathbf{j}} + q_z\hat{\mathbf{k}}$. How is q altered by reflection off a mirror in the x-y plane? Under what conditions will the ray bounce off all three mirrors?
21. To what angular accuracy must two mirrors be aligned perpendicular to each other in order to return a light ray within 1° of its original direction?
22. You are standing in a dressing room looking into a three-part mirror whose wing mirrors are at 120° angles with the central mirror. From what angles do you

see yourself when you look into the right-hand mirror?

23. How tall must a plane mirror be to allow you to see your entire reflection, head to toe? How must it be mounted on the wall? Assume that you are 1.62 m tall and that your eyes are 0.12 m below the top of your head. Draw a diagram showing your results.

24. Derive a general formula for the height of a plane mirror as a function of your height that allows you to see your entire reflection, head to toe. How must it be mounted on the wall?

25. The members of your family range from 1.2 to 1.8 m tall. How tall must a plane mirror be, and where must it be mounted on the wall, to allow each of you to see your entire reflection in it?

26. Three plane mirrors are arranged so that a cross section of their inner surfaces is an equilateral triangle. Describe the images of an object at the triangle's center.

Section 34.5   *Mirrors and Lenses*

27. By what angle is light from a bulb in a window of a house 100 m distant diverging as it reaches our eyes? By what angle is light from a point on the sun 1.5×10^8 km distant diverging as it reaches us? Consider in each case two rays entering your eye on opposite sides of your pupil, and assume that your pupil is 2 mm in diameter.

28. Use ray tracing to draw the following images made by reflection in a concave spherical plane mirror of 10-cm radius: an arrow 1 cm high whose base is on the mirror's axis at distances from the mirror (*a*) 3 cm, (*b*) 8 cm, and (*c*) 12 cm.

29. Consider a convex mirror of radius 6 cm. Employ the geometric method of ray tracing to determine the image location of an arrow 2 cm high whose tail is on the mirror's axis at distances (*a*) 2 cm, (*b*) 4 cm, and (*c*) 8 cm from the mirror. Check your result against the analytic formula relating the focal length, image distance, and object distance. Convince yourself that $f<0$ for a convex mirror. Why are the images always virtual?

30. Within what range of object distances does a concave mirror give real images?

31. Use ray tracing to show the spherical aberration in a concave plane spherical mirror of 10 cm radius by imaging of several points along an arrow 6 cm long whose base is on the mirror's axis.

32. Describe the characteristics of a vanity mirror that is to enlarge your face by a factor of 2 when you sit 30 cm from it.

33. Describe (concave or convex, plus focal length) an external truck mirror that is to show a 140° region in its 20-cm diameter when you look from a distance of 1.6 m.

34. The Canada-France-Hawaii Telescope's main mirror is 3.6 m in diameter and has a focal length of 8.5 m. How large is the primary image of the moon, given that the moon subtends an angle of $\frac{1}{2}°$ in the sky?

35. A slide projector has a double-convex lens of focal length 104 mm. (*a*) At what distance must you hold the slide from the lens so that it projects in focus on a screen 4.5 m away? How great will the magnification be? (*b*) You replace the lens with one of focal length 78 mm. How far from the lens should the slide be located, and how large will the magnification be?

36. Two double-convex lenses of focal length 15 cm are placed 25 cm from each other. If an object is 20 cm beyond the first lens, where does the image formed by both lenses fall? Describe the image (real/virtual, right-side-up/inverted, reduced/enlarged).

37. A double-convex lens of focal length 25 cm is followed in an optical system by a double-concave lens of focal length 35 cm. The lenses are separated by 75 cm. If an arrow 50 cm to the left of the convex lens is imaged first by the convex and then by the concave lens, describe the resulting image and its location. Solve the problem both by ray tracing and by using equations.

38. Use ray tracing to show the image formed by a magnifying glass of focal length 5.5 cm held 2.5 cm above a printed page.

39. Your eyes are 0.65 m from a book, and 10 cm up from the book you hold a magnifying glass of focal length 15 cm. What is that magnification?

40. The letters of the Compact Version of the Oxford English Dictionary are 1.5 mm high. How far must you hold a magnifying glass of focal length 15 cm to make an image whose linear magnification is 3?

41. The largest refracting telescope in the world, the 1-m telescope at the Yerkes Observatory, has a focal length of 12 m. If it could be used to observe an airplane passing 1 km overhead, where would the focus occur with respect to the focus for stars?

42. Consider a double-convex thin lens of focal length 5.0 cm. By ray tracing, show the image formation for an arrow on axis (*a*) 3.0 cm, and (*b*) 9.0 cm from the lens.

43. Repeat the preceding problem using the formula for the distances of image and object.

44. What are the magnifications for the preceding problem?

45. By considering whether the image distance is positive or negative, prove that a double-concave lens can never form a real image.

Section 34.6   *The Lensmaker's Formula and Aberrations*

46. What is the focal length of a lens made of crown glass with opposite convex surfaces of radius 5.0 cm?

47. By what factor must you change the radius of a pair of identical convex surfaces making up a crown-glass lens in order to double the focal length?

48. If you make a double-convex lens out of crown glass instead of out of heavy flint glass ($n=1.65$), by how much must you change the 8.5-cm radii of the two surfaces to keep the focal length constant?

49. Two lenses have the same focal length. Both are made

out of crown glass. The two surfaces of one each have radius 12.5 cm. One surface of the other has radius 8.5 cm; what must its other radius be?

50. A plano-concave lens has a focal length of 23 cm and a radius of curvature 10 cm (both f and $r<0$). What material is this lens made of? If the same lens had been cut from a sapphire-like material ($n=1.77$), what would the focal length be?

51. Derive a formula giving the focal length of a system of two thin lenses held next to each other in terms of the focal lengths of the component lenses. *Hint:* Consider each lens to have one flat side and one curved side.

52. What is the equivalent radius of the convex side of a plano-convex fused-quartz ($n=1.458$) lens to equal the 15-cm focal length of a double-convex lens?

53. Using ray tracing, demonstrate the spherical aberration for a spherical lens of 4-cm focal length if an object arrow 2 cm in height is placed a distance of 8 cm from the lens.

54. A double-convex lens has surfaces of radii 20 cm. It is made from glass whose index of refraction is 1.52 for blue light and 1.50 for red light. Describe the chromatic aberration, specifying the distance in focus between red and blue rays.

55. As is apparent from the lens maker's formula, a wavelength-dependent index of refraction n will result in a wavelength-dependent focal length for a lens of given surface radii. This fact is responsible for the chromatic aberration in lenses, as illustrated in the preceding problem. Prove that this variation in focal length of a thin lens is linked to the variation in the index of refraction by $df/f = -dn/(n-1)$.

Section 34.7   *Optical Instruments*

56. Prove that the strength of eyeglass lenses in diopters is additive. For example, prove that placing a pair of 2-diopter eyeglasses in front of a pair of 1-diopter eyeglasses gives the same effect as a single pair of 3-diopter eyeglasses.

57. How many diopters must eyeglasses have to correct an eye whose focal length is 2.0 cm instead of the 2.2 cm required for sharp focus onto the retina? Is the individual nearsighted or farsighted?

58. What is the angular magnification of a jeweler's loupe (simple magnifier) of focal length 4 cm for an object (a) when viewed at the focal point of the lens, and (b) when viewed at the near point 25 cm from you?

59. How much less total light reaches the film in a given time when a camera lens of 135-mm focal length has its aperture closed from f/8 to f/16?

60. Over what distance must a camera lens of 50-mm focal length be adjustable so that it can focus objects at distances between 1.2 m and infinity?

61. How far from a 50-mm-focal-length lens must a 1.8-m person stand to form an image 30 mm high on film? How far must the person stand if a 135-mm telephoto lens is substituted?

62. What is the apparent angular size of the planet Jupiter, which covers 50 arc sec of sky, when viewed through a 1-m-focal-length telescope with an eyepiece of focal length 40 mm?

63. A microscope's objective has a focal length of 0.75 cm. If the object is 0.85 cm from the objective, what are the distance and magnification of the intermediate image? An eyepiece of focal length 1.5 mm is put in place. What is the angular magnification of this eyepiece? Finally, what net magnification results from this compound lens system?

64. In a Cassegrain telescope, a convex mirror reflects the image formed by a first "objective" mirror before its focal point (Fig. 34–75). Using a ray diagram, show the light path for an objective mirror of focal length 1.00 m, with the convex mirror placed 0.15 m before the focal point. What must the focal length of the convex mirror be to place the final focal point of the system 0.12 m behind the front surface of the objective mirror? (A hole is cut in the objective mirror to allow the light to pass through it.)

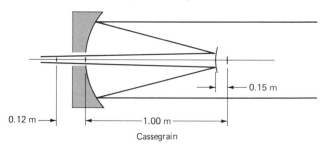

Fig. 34–75   A Cassegrain telescope (Problem 64).

65. In a Newtonian telescope, a flat mirror placed before the focus of a first "objective mirror" reflects the focus out to the side of the tube (Fig. 34–76). If the objective mirror's focal length is 1.20 m and is 20 cm in diameter, where must the flat mirror be placed to put the focus just outside the telescope tube, assuming the tube exactly surrounds the mirror? What shape should the flat mirror have to cut out the minimum amount of light?

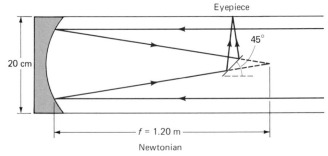

Fig. 34–76   A Newtonian telescope (Problem 65).

66. In the type of telescope used by Galileo in 1610, the eyepiece is a double-concave lens placed slightly in front of the focus of the double-convex objective. With ray diagrams, show that this system gives an

upright image, which makes the design especially useful for terrestrial observing.

Supplementary Problems

67. Use Fermat's principle to prove the law of reflection by finding the point of incidence and the plane of the incident and reflected rays travelling between points $(0, -y_0, z_0)$ and $(0, y_0, z_0)$ in Fig. 34–77. *Hint:* Prove that $x = 0$ and $y = 0$.

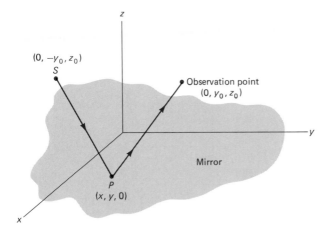

Fig. 34–77 Problem 67

68. Explain the optical effects in the self-portrait (Fig. 34–78) painted by the sixteenth-century Italian Mannerist artist Parmigianino (1503–1540). Fascinated by his own reflection in a barber's curved mirror, Parmigianino painted this picture in Rome in 1524 to demonstrate boldly his skill in "the subtleties of art." Was the artist left- or right-handed? Did he use a convex or a concave mirror to view himself? Why does his hand appear too large for the body of which it is part? Could the image of his hand be bigger than lifesize? If not, why? Finally, how is it that the windows

and the ceiling line in the distance appear curved, not straight as they were in Parmigianino's study?

Fig. 34–78 The self-portrait (1524) of the Italian Mannerist artist Parmigianino. Mannerism, in the wake of the rationalist outlook of the Renaissance, often involved unusual spatial situations. The portrait shown is both painted of and mounted on a curved surface. (Problem 68).

69. Prove that

$$\Phi = 4\theta' - 2\theta = 4\sin^{-1}\left(\frac{1}{n_{\text{water}}}\sin\theta\right) - 2\theta$$

for a rainbow (Section 34.7).

70. Prove that $\cos^2\theta_c = \frac{1}{3}(n^2 - 1)$ for a rainbow (Section 34.8).

71. Evaluate the angles for red and violet bands in the primary rainbow (Section 34.8).

72. Evaluate the angles for red and violet bands in the secondary rainbow, and discuss why the order of colors is reversed from that in the primary rainbow (Section 34.8).

35

PHYSICAL OPTICS

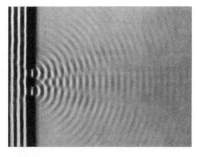

Fig. 35–1
Diffraction in water waves, when the wavelength is comparable with the size of a barrier. Note how the waves spread out beyond the barrier—diffract—and then interfere.

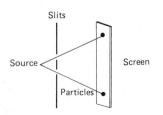

Fig. 35–2
If you didn't know about the wave nature of light, you might consider that light (or particles) passing through a pair of slits would project to a pair of adjacent single bright regions on a distant screen.

In the preceding chapter we explained the behavior of lenses and mirrors by considering that light travels in straight rays. Yet we know that light is an electromagnetic phenomenon, described by Maxwell's equations. In this chapter, we relax the assumption of geometrical optics: that the wavelengths of light are very short compared with the sizes of objects the light encounters. We see that light actually undergoes the same phenomena of interference and diffraction (Fig. 35–1) as other waves; we have already discussed wave interference in Chapter 14. To distinguish it from geometrical optics, the treatment of light and other electromagnetic radiation as wave phenomena is known as **physical optics.**

35.1
INTERFERENCE

The phenomenon of interference played an important role in the history of light. A question asked hundreds of years ago concerned the fundamental nature of light: Is light a particle or is light a wave? The surprising answer that it actually has properties of both was not realized until the twentieth century, as we shall discuss in the next chapter. Although Isaac Newton had argued that light is composed of particles, an experiment carried out in 1801 by the English physicist Thomas Young (1773–1829) seemed to provide conclusive evidence that light is a wave.

    Young's experiment involved the interference of two beams of light. Young allowed light from a single source to pass through two thin slits, and then to fall on a distant screen. If light were but a series of particles, then we would expect that the distribution of particles would result in bright areas on the screen opposite the slits (Fig. 35–2). But Young found that bands of dark and light formed on the screen (Fig. 35–3). Only if light were composed of waves could such a variation result, for then we could have the interference phenomenon described in Chapter 14. Wherever

960

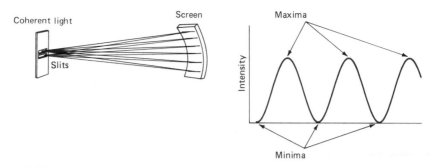

Fig. 35-3
Bands of dark and light formed on the screen of Young's experiment, showing that light is a wave phenomenon.

constructive interference

destructive interference

waves were both at their peaks, we would have bright regions: **constructive interference.** At points where waves had opposite phases, we would have dark regions: **destructive interference** (Fig. 35–4).

Two-Slit Interference

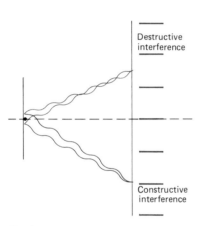

Fig. 35-4
In constructive interference, wave amplitudes are both positive so the resultant amplitude is greater than the amplitude of either individual wave; in destructive interference, the resultant amplitude is diminished.

Consider a pair of thin slits separated by a distance that is not too many times larger than the wavelength of light. Since visible light is approximately 0.4 μm to 0.68 μm in wavelength, a separation considerably smaller than 1 mm is necessary (since 1 mm is approximately 2000 wavelengths). Consider a beam of waves of light all in phase (that is, whose oscillations are all in step); such a beam is termed **coherent.** The first thing to note is that such a coherent beam is necessarily monochromatic; different colors in a beam would correspond to different wavelengths, and the waves in such a beam would soon be out of phase, that is, become **incoherent.** (Beams that *begin* with different waves out of phase are also incoherent.) Until recent decades, coherent beams were generated by filtering ordinary light and using a small hole to isolate a small region of an original light source, since we cannot hope to have coherent light coming from different parts of an ordinary light source. More recently, we use lasers to generate coherent light; among the extremely useful properties of laser light is that all its light waves are both coherent and very nearly monochromatic.

The light passing through each slit is diffracted—spread out as it passes. (We shall analyze diffraction in the following section.) The pattern resulting on the screen (Fig. 35–5) consists of alternate bands of light and

Fig. 35-5
Interference fringes photographed by exposing film to monochromatic coherent light coming from two nearby slits.

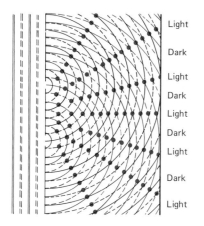

Fig. 35–6
Wave crests are shown in black as they spread out following the passage of coherent light beams through adjacent slits. The wave troughs are in between the crests.

dark; the bands are known as **interference fringes.** Figure 35–6 demonstrates the positions of wave crests following the diffraction at the slits. Where wave crests meet, we have constructive interference. In between, where a wave crest meets a wave trough, we have destructive interference. Note that the screen intersects several instances of constructive interference and several of destructive interference.

In Fig. 35–7, we analyze the situation in which the fringes were created. We consider the screen to be sufficiently far away from the slits that the lines S_1P and S_2P are essentially parallel, where S_i represents slit i and P represents a point on the screen. Then the small triangle at left caused by dropping a perpendicular from line S_1P to line S_2P separates the latter into two parts: one part is equal in length to S_1P and the other part represents the path difference between S_1P and S_2P.

To have maximum intensity, the light rays must travel either the same distance to the screen, which occurs only if $\theta = 0$, or a distance to the screen that is different by a whole number of wavelengths, that is, a distance that is an integer times the wavelength. For this set of distances, the same phase of the wave hits the screen—that is, crest meets crest, trough meets trough, and so on. Thus for maximum intensity, we must have

$$d\sin\theta = n\lambda, n = 0, \pm 1, \pm 2, \ldots \qquad \text{(maximum intensity)} \qquad (35\text{–}1)$$

For minimum intensity, we must have

$$d\sin\theta = (n+\tfrac{1}{2})\lambda, n = 0, \pm 1, \pm 2, \ldots \qquad \text{(minimum intensity)} \qquad (35\text{–}2)$$

In between, we have intermediate intensities.

Example 35–1

Interference from Two Slits

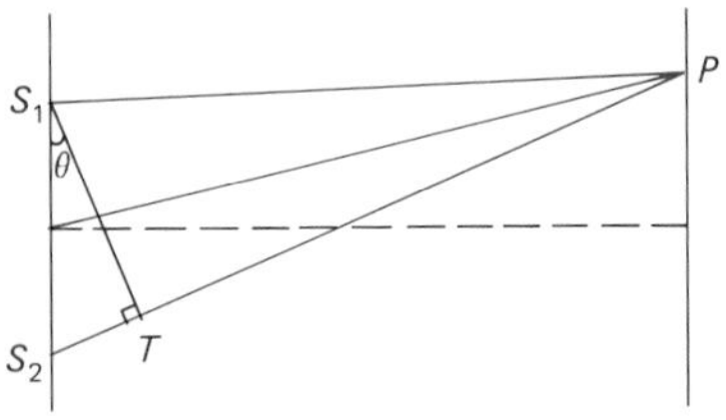

A pair of slits separated by 0.10 mm is completely illuminated by a beam from a helium-neon laser, whose wavelength is 632.8 nm. What is the separation of the central maximum from the first maximum on a screen placed 1500 mm from the slits? How far to the second maximum? By how much is the 60th maximum displaced on this flat screen from its position on a screen curved to make the spacing between maxima equal?

Solution

Solving Equation 35–1 for $\sin\theta$, with $n=1$ to give the first maximum, we have

$$\sin\theta = \frac{\lambda}{d} = \frac{632.8 \text{ nm}}{(0.10 \text{ mm})(10^6 \text{ nm/mm})} = 0.0063.$$

Fig. 35–7
If the screen is sufficiently far away, S_1P and S_2P are essentially parallel, and $S_2T = d\sin\theta$ is the difference that the light from the slits separated by a distance d has to travel at an angle θ before hitting a point on the screen. Physically, the screen could be brought closer if a lens is used, but the essence of the situation is unchanged. Further, the pattern is regularly spaced only if the screen is curved, but this effect is not significant if the screen is sufficiently far away from the slits.

Thus, using the small-angle approximation that $\sin\theta \approx \theta$, $\theta = 0.0063$ radian. [Since π radians $= 180°$, $\theta = (0.0063 \text{ rad})(180°/\pi \text{ rad}) = 0.36°$]. Figure 35–8 shows that the distance P_0P_1 on the screen, either direction from P_0, is $(1500 \text{ mm})\sin\theta$ $= (1500 \text{ mm})(0.0063) = 9.5$ mm.

The second maxima are an additional 9.5 mm from the first maxima. The minima are exactly in between.

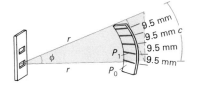

Fig. 35–8
Spacing of maxima (Example 35–1).

To find how far off the separations become if we use a flat screen instead of a curved screen, consider the shaded triangle in Fig. 35–8. From the large triangle marked by two radii and the angle ϕ, we determine that

$$\tan\phi = \tan(60\theta) = 0.396$$

and that

$$\tan\phi = \frac{\beta}{r}.$$

Thus

$$\beta = r\tan\phi = 594 \text{ mm}.$$

In comparison, the distance along the curved screen can be found from the fact that ϕ makes up the same proportion of $180°$ that the length c of the arc makes of half the circumference of the circle. Thus

$$\frac{\phi}{180°} = \frac{c}{\pi r},$$

or

$$c = \pi r\left(\frac{\phi}{180°}\right) = \pi r\frac{(60)(0.36°)}{180°} = 565 \text{ mm}.$$

Evidently the distance between adjacent maxima on a flat screen spreads with distance from the perpendicular between the slits.

Application

Arrays of Radio Telescopes

resolution

It is often important to determine the scale of the finest detail that we can observe in an object. The **resolution** of an optical system is the minimum angular separation between two objects or parts of an object that we can distinguish. Empirically, and for theoretical reasons related to the results we shall study in the following section, the angular resolution of a single-dish radio telescope (Fig. 35–9) is limited to approximately $1.22\lambda/d$. The formula is related to the interference pattern calculated for points as far apart as possible on the dish, so we use it here.

For the 305-m-diameter telescope shown, the largest in the world, at a wavelength of 10 cm (in the radio part of the spectrum), the resolution is 1.22(10 cm)/(30,500 cm) = 0.0003 radian = $0.017° = 1$ minute of arc. To achieve higher resolution, we use arrays of several radiotelescope dishes—"interferometers."

Under construction is the Very-Long-Baseline Array, VLBA. It will be a set of telescopes whose greatest separation is about 8000 km (Fig. 35–10). Actually, two-dimensional coverage is also provided, but here we deal only with one dimension. To allow the coherent radiation arriving at the different individual antennas to be compared, the data are recorded on tape along with a time signal from an atomic clock. Later, in the laboratory, the tapes are played together and synchronized with the clock signals.

We can calculate the maximum resolution of the VLBA by considering it as a two-slit system with the slits separated by 8000 km; these two "slits" correspond to the pair of antennas separated by the greatest distance. (Each other pair has inferior resolution compared with that of this most separated pair.) Radiation from a source directly overhead (Fig. 35–11) will reach each of the antennas simultaneously. Radiation from a second source an angle θ to the side will be resolvable

Fig. 35–9
The 305-m-diameter radio telescope at Arecibo, Puerto Rico, the largest single-dish radio telescope in the world.

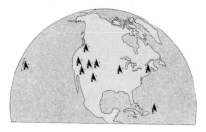

Fig. 35–10
Prospective sites for the ten 25-m-diameter radio telescopes that together will make up the Very-Long-Baseline Array, now under construction. In 1986, a satellite was included in a network for the first time, extending the baseline to 1.4 times the earth's diameter.

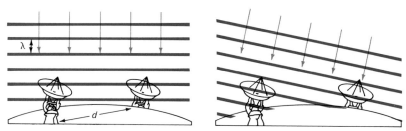

Fig. 35–11
A small change in angle in the sky can be measured by a radio interferometer consisting of several radio telescopes.

from the first source if the maximum of its intensity—its first fringe—falls on the minimum of the intensity of the overhead source. Applying Equation 35–2 for radiation of wavelength 10 cm, we find

$$\sin\theta = \frac{\lambda}{d} = \frac{10 \text{ cm}}{(7500 \text{ km})(10^5 \text{ cm/km})} = 1.3 \times 10^{-8}.$$

This phenomenally high resolution of 10^{-8} radian is finer than can be obtained from any optical telescope, and even much higher than can now be obtained from smaller radio arrays such as the Very Large Array (VLA), Fig. 35–12. With high resolution, astronomers can distinguish as separate different parts of objects such as quasars that are quite far away. To give a terrestrial equivalent, resolution of 10^{-8} radian would allow you to resolve different parts of a person standing in Los Angeles if you were in New York; even at the huge distances of quasars, billions of light years away, it would allow us to detect motions within weeks. In the VLBA and in the VLA, the telescopes are not spaced evenly; as a result, different path differences between them can be measured simultaneously. A measurement with a long path length gives high resolution for small objects, as the above equation shows, but covers only a small region of sky and so turns out not to be sufficient for unscrambling complicated celestial systems at much coarser resolution. So lesser spacings, which provide coarser resolution on a larger scale, must be used simultaneously to give a complete picture of the source. Thus all the different pairs of telescopes in an array are employed; some are at greater distances and some at lesser distances from each other.

Arrays of optical telescopes are under development, but the need to provide simultaneous signals and to maintain the distances constant given the short wavelengths involved have prevented routine measurements from being made. The European Southern Observatory has plans to build an array of four optical telescopes, each 8 m in diameter, a few hundred meters from each other along a line on a mountaintop in Chile. The first telescope is to be completed in 1993, with the entire system operational by about the year 2000.

Using geometric arguments, we have been able to find the positions of interference maxima and minima. But these arguments do not give us a quantitative prediction of the intensity at these points and at locations in between. To find the intensity, we must consider the electric fields. Emanating from each slit i, we may consider that we have electric field (Chapter 32)

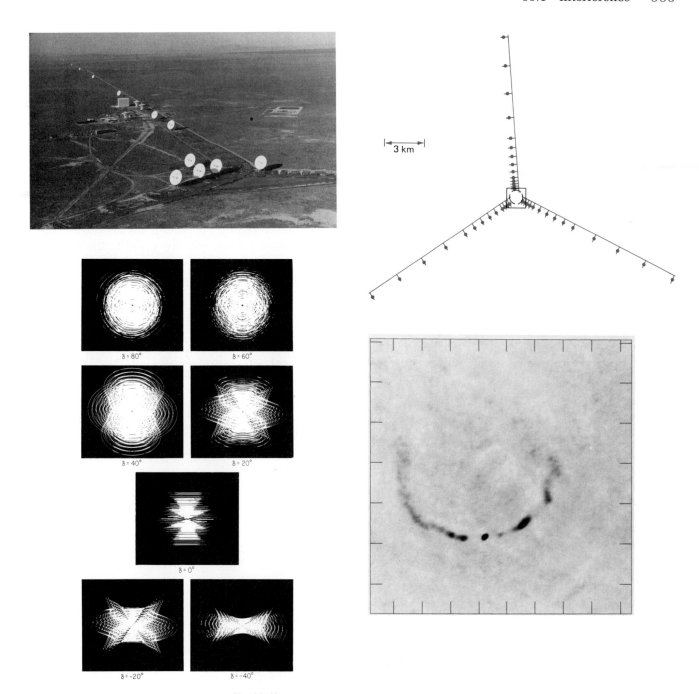

Fig. 35-12
(Upper left) The Very Large Array, VLA, in a plain near Socorro, New Mexico, consists of 27 antennas, each 12 m diameter, spread in a Y shape that covers a circle of approximately 27 km. (Upper right) The individual telescopes of the VLA are logarithmically spaced from each other, to give the maximum amount of information when considering the results from each pair of telescopes. (Lower left) Computer simulations of the sensitivity of the telescope for different angles in the sky. Each frame consists of all or part of 351 ellipses, each ellipse being the projection of a different antenna-pair separation on the sky. (Lower right) Sample data from the VLA: a radio source at a resolution comparable with resolution of ground-based optical telescopes shows that it emits from two lobes on either side of a central source. The minimum size and shape that can be resolved is shown by the oval-shaped center of this radio galaxy. The field of view is 1.75 arc min on each side.

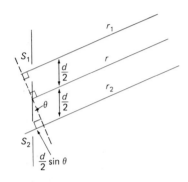

Fig. 35–13
Calculating the intensity of an interference pattern. Lines r_1, r, and r_2 meet at the screen off the figure at upper right, but for $r \gg d$, these lines are nearly parallel and the approximations of Equation 35–4 can be made.

$$E_i = \frac{A}{r_i} \cos(kr_i - \omega t), \tag{35–3}$$

where A is a constant related to the strength of the light source. From Fig. 35–13, we see that, for small θ (which is the case for $r \gg d$),

$$r_1 = r - \left(\frac{d}{2}\right) \sin\theta \tag{35–4a}$$

and

$$r_2 = r + \left(\frac{d}{2}\right) \sin\theta, \tag{35–4b}$$

where r is a mean length defined in the figure. Now if we substitute Equations 35–4a and 35–4b into Equation 35–3, we can simplify the result by noting that we can expand, using the binomial expansion from Appendix A,

$$\frac{1}{r_1} = \frac{1}{r - (d/2)\sin\theta} = \frac{1}{r}\frac{1}{1 - (d/2r)\sin\theta}$$

$$\approx \frac{1}{r}\left(1 + \frac{d}{2r}\sin\theta\right) = \frac{1}{r} + \frac{d}{2r^2}\sin\theta$$

and

$$\frac{1}{r_2} = \frac{1}{r + (d/2)\sin\theta} = \frac{1}{r}\frac{1}{1 + (d/2r)\sin\theta}$$

$$\approx \frac{1}{r}\left(1 - \frac{d}{2r}\sin\theta\right) = \frac{1}{r} - \frac{d}{2r^2}\sin\theta.$$

Thus

$$E_1 = \frac{A}{r}\cos\left[k\left(r - \frac{d}{2}\sin\theta\right) - \omega t\right] + \frac{Ad}{2r^2}\cos\left[k\left(r - \frac{d}{2}\sin\theta\right) - \omega t\right]\sin\theta$$

and

$$E_2 = \frac{A}{r}\cos\left[k\left(r + \frac{d}{2}\sin\theta\right) - \omega t\right] - \frac{Ad}{2r^2}\cos\left[k\left(r + \frac{d}{2}\sin\theta\right) - \omega t\right]\sin\theta.$$

Since $d \ll r$, the second term is negligible. Thus we need consider only the first term in each. Adding the electric fields gives

$$E_{\text{total}} = E_1 + E_2$$

$$= \frac{A}{r}\left[\cos\left(kr - \frac{kd}{2}\sin\theta - \omega t\right) + \cos\left(kr + \frac{kd}{2}\sin\theta - \omega t\right)\right]. \tag{35–5}$$

We can simplify this equation by making use of the trigonometric identity

$$\cos(\alpha + \beta) = \cos\alpha\cos\beta - \sin\alpha\sin\beta.$$

Since $\cos(-\beta) = \cos\beta$ and $\sin(-\alpha) = -\sin\alpha$, we also have

$$\cos(\alpha - \beta) = \cos\alpha\cos\beta + \sin\alpha\sin\beta.$$

Adding these two equations gives

$$\cos(\alpha + \beta) + \cos(\alpha - \beta) = 2\cos\alpha\cos\beta.$$

Using this identity in Equation 35-5 then gives

$$E_{\text{total}} = \frac{2A}{r}\left[\cos(kr - \omega t)\cos\left(\frac{kd}{2}\sin\theta\right)\right]. \qquad (35\text{-}6)$$

Note that although there are two cosine terms, only one of them, $\cos(kr - \omega t)$, represents a propagating wave. The second cosine term, $\cos[(kd/2)\sin\theta]$, is composed of constants and the angle θ. It represents a variation of the field amplitude with angle.

The intensity—power per unit area—of light falling on the screen is given by the magnitude of the Poynting vector that we introduced in Section 32.6. Since we are not interested in variations in intensity on the extremely rapid timescale of the wave period, we use Equation 32-31 for the time-averaged intensity $\overline{S}$:

$$\overline{S} = \frac{E_0^2}{2\mu_0 c},$$

where E_0 is the amplitude of the electric field.

In Equation 35-6, the field amplitude consists of all terms other than the time-dependent cosine. So the intensity at the screen becomes

$$\overline{S} = \frac{2A^2}{\mu_0 c r^2}\cos^2\left(\frac{kd}{2}\sin\theta\right), \qquad (35\text{-}7a)$$

or, since $k = 2\pi/\lambda$ (Equation 32-14),

$$\overline{S} = \frac{2A^2}{\mu_0 c r^2}\cos^2\left(\frac{\pi d}{\lambda}\sin\theta\right). \qquad (35\text{-}7b)$$

Equations 35-7 reflect both the expected $1/r^2$ variation in intensity with distance from the source and the variation with angle arising from the interference pattern (Fig. 35-14). Do these equations make sense? The difference between adjacent maxima represents a cycle of the cosine-squared function. Thus maxima occur at $\pi d\sin\theta/\lambda = 0$ and $\pi d\sin\theta/\lambda = \pi$. The latter reduces to $d\sin\theta = \lambda$, which is indeed the criterion we previously deduced for the first maximum.

Example 35-2
Intensity for Double-Slit Interference

What is the intensity of light falling on a screen at a position halfway between the central maximum and the first minimum of a two-slit interference pattern with $d = 10\lambda$, as was graphed above? Express the intensity as a fraction of the central intensity.

Solution

From Equation 35-7b, the central intensity

$$\overline{S}_{\text{central}} = \frac{2A^2}{\mu_0 c r^2},$$

so the intensity we want to find is

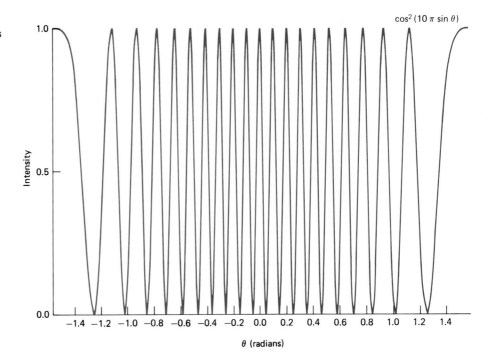

$$\overline{S}/\overline{S}_{\text{central}} = \cos^2\left(\frac{\pi d}{\lambda}\sin\theta\right) = \cos^2(10\pi\sin\theta).$$

Since $\cos(\pi/2) = 0$, the first minimum occurs for $10\pi\sin\theta = \pi/2$, or $\sin\theta = 1/20 = 0.05$. Thus $\theta = \sin^{-1}0.05 = 2.87°$. We wish to find the intensity at $\theta' = \frac{1}{2}\theta = 1.44°$.

$$\overline{S}/\overline{S}_{\text{central}} = \cos^2[10\pi(\sin 1.44°)]$$
$$= \cos^2[(10\pi)(0.025)]$$
$$= \cos^2[0.78 \text{ radian}]$$
$$= (0.707)^2$$
$$= 0.50.$$

Thus the intensity halfway to the first minimum is half the intensity of the central maximum. Similar calculation gives the intensity at any point on the screen (Fig. 35–14).

Multiple-Slit Interference

We can easily generalize our results for interference from two slits to three or more slits. The result is important in helping us design and build instruments for spectroscopic analysis.

In Fig. 35–15 we see the crests emerging from three adjacent slits. The pattern that results appears in Fig. 35–16. We have constructive interference whenever all three crests reach a point at the same time. We assume that the slits are relatively close together, allowing us to ignore small differences in angle. If the waves from two of the slits are both cresting at a given point on the screen, then the third wave has a crest there too. Thus

Fig. 35–15
Calculating interference from three adjacent slits.

Fig. 35–16
Three-slit interference pattern for
$d = 10\lambda$.

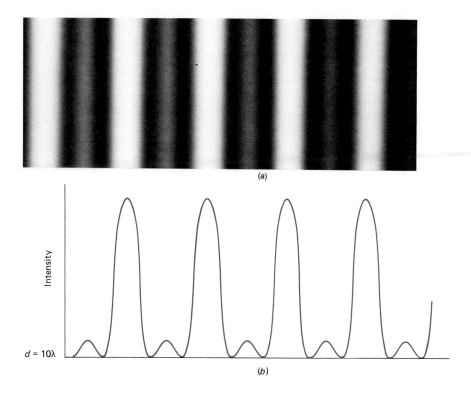

(a)

(b)

$d = 10\lambda$

Fig. 35–16
Three-slit interference pattern for
$d = 10\lambda$.

the criterion for an intensity maximum is the same for three or more slits as it is for the two-slit system:

$$d \sin\theta = n\lambda, \, n = 0, \, \pm1, \, \pm2, \, \ldots \quad \text{(maxima for N slits)} \quad (35\text{–}8a)$$

But note that the criterion for destructive interference is no longer simply that the waves be 180° out of phase, since now we have three waves. The criterion for destructive interference is rather that the waves vary in such a way that their sum is zero. The way to keep the sum of three waves to zero is to have each out of phase from each of the others by one-third of the complete 360° cycle. Thus $d\sin\theta$ must be $(n+\frac{1}{3})\lambda$ or $(n+\frac{2}{3})\lambda$ (Fig. 35–17). Note that $(n+\frac{3}{3})\lambda$ is excluded, because a maximum occurs there.

Fig. 35–17
Waves from three slits must be $\frac{1}{3}\lambda$ out of phase with each other to provide destructive interference.

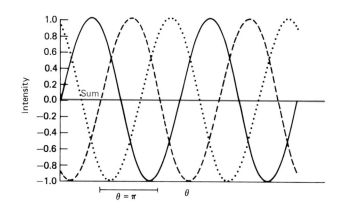

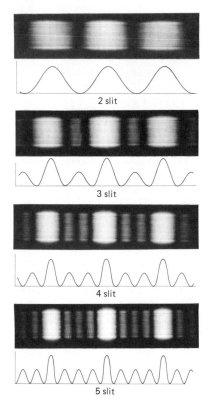

Fig. 35–18
The interference pattern falling on film for two, three, four, and five slits.

For N waves, the minimum intensity (zero field) occurs if the waves are out of phase by $360°/N$. Thus the criterion for minimum intensity for N slits is (Fig. 35–18):

$$d \sin\theta = \frac{m}{N}\lambda,$$

$m = \pm 1, \pm 2, \pm 3 \ldots$, but excluding $m = (\text{integer})(N)$.

(minima for N slits)   (35–8b)

Gratings and Spectrographs

Note the dependence on wavelength in Equation 35–7; the maximum intensity falls at a different location for each wavelength. If we have a great many slits—that is, N is large—there are many minima for each major maximum. The intensity never gets very high between the maxima. Thus we have one major maximum for each value of n. The value of n is the **order.**

But the position of this maximum varies with wavelength. We thus get a series of spectra (Fig. 35–19), one for each value of n. The 0th order—the central maximum, with $n = 0$—is not spread out in wavelength. The other orders are spread out—we say that they are dispersed. The nth-order spectrum is spread out n times as much as the 1st-order spectrum. The different orders appear symmetrically on each side of the 0th order.

A device that disperses a spectrum to be seen by the eye is a spectroscope; when the spectrum is recorded by film or electronically, the device is a spectrograph or spectrometer.

Example 35–3
Dispersion

On a 1st-order spectrum, the sodium D_1-line at 589.6 nm is separated from the sodium D_2-line by 0.6 mm. By how much are they separated in the 2nd-order spectrum? In the 3rd-order spectrum?

Solution

The nth-order spectrum is spread out n times as much as the 1st-order spectrum, so the separation of the two lines in the 2nd-order spectrum is (2)(0.6 mm) = 1.2 mm. In the 3rd-order spectrum, the separation is (3)(0.6 mm) = 1.8 mm.

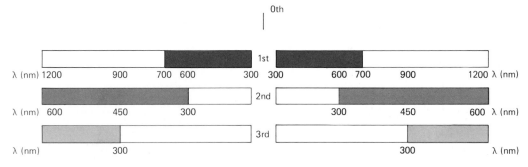

Fig. 35–19
The spectrum from a grating has an infinite number of orders overlapping, though they have different intensities. In the drawing, the orders are displaced from top to bottom for clarity, but on a screen the orders would overlap each other. The visible region of the spectrum is shaded. Note that for a given horizontal position, the order times the wavelength is constant. For example, (900 nm)(1) = (450 nm)(2) = (300 nm)(3).

Example 35–4
Overlapping Orders

Consider a solar observatory with a spectrograph that gives a spectrum whose first-order dispersion is 0.10 nm per millimeter of distance on the electronic detector. The detector is 1 cm wide. How can we use overlapping orders to record simultaneously the spectrum of the dark line of ionized calcium at 393.3 nm (known as the K-line) and the sodium D_1-line at 589.6 nm?

Solution

We can accomplish our task if we allow the K-line in one order to correspond roughly to the D_1-line in another order, so that the two will fit on the same detector. Since the angular displacement of the line from on-axis depends on $n\lambda$, we can approximately equalize $n\lambda$ if we observe the K-line in 6th order [(393.3)(6) is approximately (400)(6)=2400] and the D_1-line in 4th order [(589.6)(4) is approximately (600)(4), which is also equal to 2400]. After we make this estimation, our task is then to see if both lines will actually fit on the detector.

More precisely, $N\lambda$ for the K-line is (393.3)(6)=2360, while $n\lambda$ for the D_1-line is (589.6)(4)=2358.4. So obviously the lines will fit on the same detector. Indeed, the position of the D_1-line seen in fourth order will fall on the detector at the same location as a 2358.4/6=393.1 nm line (very close to the K-line) in sixth order (Fig. 35–20). Since the dispersion is 0.10 nm/mm (that is, nanometers of spectrum per millimeter of detector) in first order, the dispersion is 0.60 nm/mm in sixth order. The difference between 393.1 nm and 393.3 nm is only (0.2 nm)/0.60 nm/mm=0.3 mm on the detector, which is well within the bounds of the detector.

The technique described here is often used in obtaining solar spectra, permitting the astronomer to record two spectral lines for the price of one.

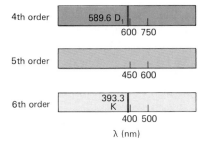

Fig. 35–20
Using overlapping orders to obtain spectra of a pair of desired lines simultaneously on a detector of limited physical size (Example 35–4).

diffraction grating

transmission grating

A set of slits very close together is known as a **diffraction grating** ("grating," for short), and is more frequently used in optical instruments than are prisms for dispersing light. It is common to have 600 slits per mm across a grating several centimeters across. Such gratings can be made in several ways. If the grating passes light, it is a **transmission grating,** and is often made by photoreducing parallel lines. A holographic grating uses interference to generate the parallel lines. Often gratings are used to reflect light, though the effect of multiple slits is the same. Such gratings are often "ruled" on blanks of glass coated with a thin aluminum deposit; a diamond stylus is used to rule many parallel grooves since only diamond is sufficiently hard not to wear down enough to cause the lines to vary. Such a grating is a **reflection grating.*** An original ruled grating can be used as a mold to make a replica grating.

From the position of the minima for a multiple-slit situation that we derived above, we see that the larger N is, the narrower the maximum must be. After all, a minimum is separated from a maximum by only $d \sin\theta = \lambda/N$. So the larger N is, the smaller the difference in wavelength we will be able to detect. In particular, we can tell two wavelengths apart if the maximum of one falls on the minimum of another (Fig. 35–21). But the difference between the maximum and its adjacent minimum for either is

$$\Delta\theta = \frac{1}{N}\frac{\lambda}{d},\qquad (35\text{–}9)$$

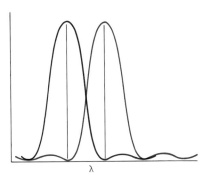

Fig. 35–21
To separate two wavelengths cleanly, the maximum of one can fall on the minimum of the other.

*See Horace W. Babcock, "Diffraction gratings at the Mount Wilson Observatory," *Physics Today*, vol. 39, July 1986, pp. 34–42.

if θ is small so that we can approximate $\sin \theta = \theta$. We can find the difference in the position of the maximum as we change from wavelength λ to wavelength $\lambda + \Delta\lambda$ from Equation 35–1 by noting that for a change in wavelength $\Delta\lambda$, we have $\sin(\theta + \Delta\theta) - \sin\theta = (n/d)[(\lambda + \Delta\lambda) = (n/d)\Delta\lambda]$, and that for a small change in angle $\Delta\theta$ we have $\sin(\theta + \Delta\theta) - \sin\theta \simeq (\theta + \Delta\theta) - \theta = \Delta\theta$. Thus

$$\Delta\theta = n \frac{\Delta\lambda}{d}. \qquad (35\text{–}10)$$

Combining Equations 35–9 and 35–10 gives

$$\frac{\lambda}{\Delta\lambda} = Nn. \qquad \text{(resolving power of a grating)} \qquad (35\text{–}11)$$

resolving power $\lambda/\Delta\lambda$ is the **resolving power** of the grating. It is a dimensionless quantity. Note that the ability to distinguish a small difference in wavelength $\Delta\lambda$ improves as either the number of grooves in the grating increases or as we use a higher order (which spreads out the spectrum more). Only the total number of grooves counts; the groove separation d does not appear in the equation, though more grooves can be fit on a grating of a given size only by ruling them closer together.

In an actual reflection grating, the grooves are often ruled at an angle, so that light is preferentially reflected at a range of angles. Thus certain wavelengths in certain orders have higher intensity than those wavelengths in other orders. The higher the intensity, the shorter the exposure necessary to photograph the spectrum, or the better the ability to make electronic images.

Example 35–5
Resolving Power of a Grating

The largest grating in a spectrograph at the 5-m telescope of the Palomar Observatory has 600 grooves/mm and is 20 cm across. (It is actually a mosaic of four smaller gratings.) What is the finest wavelength resolution that it can provide in first order? In fifth order?

Solution

$N = (600 \text{ grooves/mm})(200 \text{ mm}) = 120,000$ grooves. Thus in first order, the resolving power is 120,000. For a spectral line at 600 nm in the yellow, the grating could potentially resolve $600/120,000 = 1/200$ nm. In fifth order, the resolving power is 600,000, which corresponds to resolving 1/1000 nm. Astronomical objects other than the sun are not sufficiently bright to allow the telescope/spectrograph combination to reach this spectral resolution in a reasonable time. But such resolutions are reached with solar telescopes in combination with their spectrographs.

Spectroscopists are used to the typical range of values represented by resolving power, even though resolving power is dimensionless. A spectroscopist would understand that you are talking about a grating good for high-resolution work if you were to say "the resolving power is 700,000."

Example 35–6
A Grism

grism   A **grism**, also called Carpenter's prism, is a transmission grating ruled onto a prism. The two optical elements are arranged in such a way that the grating order N is transmitted undeviated because the bending angle of the prism compensates for it at a selected wavelength. (We switch to κ instead of *n* for grating order in this problem because we will use *n* to mean index of refraction.) Grisms are useful in astronomy, for example, because if a grism is removed from the system, your image becomes—undisplaced—the star field that is being studied instead of the spectra of the stars in that field, as it was before. It becomes easier to check where you are pointing your telescope.

What is the equation that links the separation of grooves of the grating, the wedge angle of the prism, the order of the spectrum, the index of refraction, and the wavelength that passes through the grism undeviated? What is the angular dispersion?

Solution

If the groove separation of the grating (Fig. 35–22) is *d* mm, the prism angle is α radians, and the diffraction angle from the grating β radians, the grism equation is

$$nd\sin\alpha - d\sin(\alpha-\beta) = N\lambda, \tag{35–12}$$

where *n* is the index of refraction of the material that makes up both the glass and whatever the grating is ruled into and N is the order. The requirement that a given wavelength λ_0 has to pass undeviated ($\beta = 0$) gives

$$(n-1)d\sin\alpha = N\lambda_0 \tag{35–13}$$

and defines α for a given *n*, *d*, and N.

We can express the grism dispersion as the change in wavelength for a change in angle of the exiting beam, $d\lambda/d\beta$. We can find $d\lambda/d\beta$ by differentiating Equation 35–12, noting that *n*, *d*, α, and N are constant. Thus

$$-d\cos(\alpha-\beta)\left(-\frac{d\beta}{d\lambda}\right) = N.$$

But we can eliminate N by using Equation 35–13, which gives

$$N = \frac{(n-1)d\sin\alpha}{\lambda_0}.$$

For small β and β ≪ α, we can approximate cos (α − β) ≃ cosα to give

$$\frac{d\lambda}{d\beta} = \frac{\lambda_0}{(n-1)\tan\alpha}, \tag{35–14}$$

independent of *d* and N, as is the case also for reflection gratings. For a typical grism, one may have α = 30°, *n* = 1.5, N = 2. From Equation 35–13, *d* = $8\lambda_0$ or 4μm for λ_0 = 500 nm, or 250 grooves/mm on the grating.

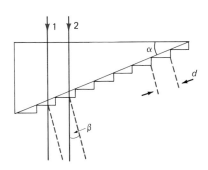

Fig. 35–22
A grism.

Application

Michelson Interferometer

A. A. Michelson (1852–1931), who was to be the first American scientist to win a Nobel prize, invented a type of interferometer over a hundred years ago that is still used for high-precision measurements. Michelson designed his interferometer for fundamental experiments on the speed of light and its constancy, independent of motion of the observer or the source of light; these experiments are relevant to our current understanding of space and time and led indirectly to Albert Einstein's special theory of relativity. We described that application of the Michelson interferometer in Chapter 33.

The basic design of the Michelson interferometer is shown in Fig. 35–23.

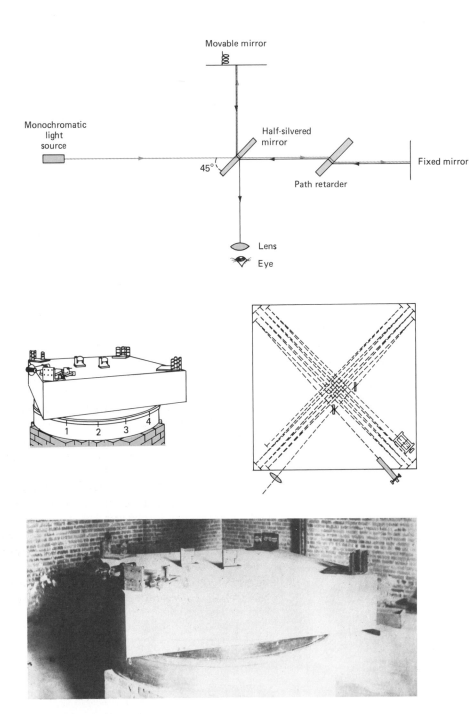

Fig. 35–23
(Top) The simple form of a Michelson interferometer. A half-silvered mirror provides reflected and transmitted rays that travel perpendicular paths. Eventually the rays are combined, allowing them to interfere. Since the ray reflected up to the top and back down traverses the silvered glass plate twice, a transparent glass plate of the same thickness is placed in the other beam to equalize the paths travelled. (Center) The Michelson interferometer as it was used in the Michelson-Morley experiment, which found no variation in the speed of light with changing direction relative to that of the earth's motion. The light bounced back and forth several times to lengthen its path. (Bottom) The original apparatus.

The key idea is that light from a single source is separated into two beams by a half-silvered mirror set at a 45° angle. The beam that is reflected travels a path that is perpendicular to the beam that is transmitted. Both beams are reflected off flat mirrors, and then recombined at the original half-silvered mirror. The resulting interference fringes reveal even slight differences of variations in the length of the paths of the two beams with respect to each other.

If the difference in distance traveled by the two beams is an integral number of wavelengths ($n\lambda$), we have constructive interference. If the distance difference is $(n+\frac{1}{2}\lambda)$, we have destructive interference. In a Michelson interferometer, the position of one of the mirrors is adjustable back and forth. The change in the path length travelled by the beam reflecting from this mirror is twice the distance the mirror moves; if the distance is d_m, then $\Delta d = 2d_m$. Thus a motion of the mirror by $\frac{1}{4}$ wavelength corresponds to the change from constructive to destructive to constructive interference; we see this as a motion of fringes as we look into the interferometer through a small telescope. In this way, the Michelson interferometer allows us to measure distances that are a mere fraction of a wavelength of light.

The Michelson interferometer has found wide use in physics. The ability to measure distances very precisely with it has led to the definition of the meter in terms of wavelengths of light. And Michelson interferometers are used to measure the index of refraction of gases (Problem 33).

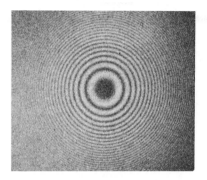

Fig. 35–24
Newton's rings between two pieces of glass that do not quite have the same shape but are touching.

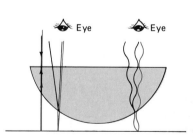

Newton's rings

Fig. 35–25
Newton's rings arise from a variation in the separation of two surfaces.

Newton's Rings and Interference in Thin Films

Isaac Newton noticed that when one glass surface was placed on another, he sometimes saw a series of dark rings (Fig. 35–24). He could not explain these **Newton's rings** with the particle theory of light he had espoused, but we can explain them as interference, a wave phenomenon. Figure 35–25 shows how if one surface is flat and the other is not, Newton's rings arise. The beam reflected off the top of the lower surface is out of phase with the beam reflected off the bottom of the top surface. The fact that the central spot we view is dark indicates that the reflection off one of the surfaces but not the other changes the phase of the wave by 180°; reflections off the top of the lower surface are at a location where the index of refraction becomes higher, while reflections off the bottom of the upper surface are at a location where the index of refraction becomes lower. The former changes the phase by 180°; we have discussed such phase changes in Chapter 14.

Newton's rings are an important tool used by opticians in making optical surfaces. To make an "optically flat" mirror (that is, a mirror that is flat to some precision standard, often $\frac{1}{8}$ or $\frac{1}{20}$ the wavelength of light), one must compare the mirror with a known optical flat and polish the surface until no Newton's rings are visible.

A phenomenon similar to Newton's rings occurs in soap bubbles and in other thin films. Interference between waves reflected off opposite surfaces of the film can interfere (Fig. 35–26). Again, there is no phase change at the first surface, as we go from air to a substance with a higher index of refractory, but there is a phase change of 180° at the second surface, where we return to air. Allowing for this second phase change, we see that the condition for constructive interference is that twice the film thickness δ

Fig. 35-26
In a thin film, light reflected off the top surface interferes with light reflected off the bottom surface. The film must be sufficiently thin that, for a given angle, the two rays overlap; only then can they interfere.

(since the wave makes a round trip through the film) must be an integer multiple of the wavelength in the medium λ_{medium} plus an additional one-half the wavelength to account for the phase change:

$$2\delta = (m + \tfrac{1}{2})\lambda_{medium}, \qquad m = 0, 1, 2, 3, \ldots . \qquad (35\text{--}15)$$

But if the medium has index of refraction n, we know that $\lambda = n\lambda_{medium}$, assuming that the film is suspended in air, so that

$$2n\delta = (m + \tfrac{1}{2})\lambda,$$
$$m = 0, 1, 2, 3, \ldots, \qquad \text{(constructive interference)} \qquad (35\text{--}16)$$

where λ is the wavelength of the illuminating radiation in air. The situation for destructive interference is thus

$$2n\delta = m\lambda,$$
$$m = 0, 1, 2, 3, \ldots . \qquad \text{(destructive interference)} \qquad (35\text{--}17)$$

Example 35-7
Interference in a Soap Bubble

You illuminate a soap bubble with sodium light at a wavelength of 589.0 nm. If the index of refraction in the soap bubble is $n = 1.45$, what thickness of the soap film will result in constructive interference?

Solution

$$2n\delta = \tfrac{1}{2}\lambda, \tfrac{3}{2}\lambda, \ldots$$

so

$$\delta = \frac{\lambda}{4n}, \frac{3\lambda}{4n}, \ldots$$
$$= 102 \text{ nm}, 305 \text{ nm}, \ldots .$$

Camera lenses and other optical surfaces are often coated with a thin film that provides destructive interference for a certain range of light.* If the index of refraction n of the coating is less than that of glass, the incoming wave strikes a medium of greater refractive index at both outer and inner surfaces of the coating, so the reflected waves do not reverse phases (Section 14.6). To provide destructive interference, the phases must differ by one-half wavelength, so the coating must have thickness $\lambda/4n$. The film's thickness usually minimizes reflection in the middle of the optical spectrum. The red and violet extremes of the visible spectrum thus reflect preferentially, giving a purple to amber appearance to the overall reflection, depending on the wavelength of the minimum.

A detailed analysis based on Maxwell's equations shows that exactly zero reflection is achieved if the index of refraction of the coating is given by $\sqrt{n}$, where n is the index of refraction of the glass lens element on which the coating is deposited.

---

*See Philip Baumeister and Gerald Pincus, "Optical Interference Coatings," *Scientific American*, December 1970, pp. 59–75.

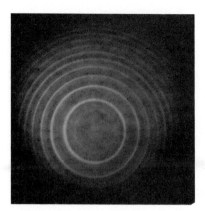

Fig. 35–27
Regular fringes seen in a Fabry-Perot interferometer that is uniformly illuminated.

Fabry-Perot interferometer

Example 35–8
Lens Coatings

How thick should a magnesium fluoride coating on a glass lens be to minimize reflection at 550 nm if the index of refraction of the magnesium fluoride is 1.38 and the index of refraction of the glass in the lens is 1.55?

Solution

For light of wavelength 550 nm in air, its wavelength in the coating is $\lambda_{coating} = \lambda_{air}/n = (550 \text{ nm})/(1.38)$. Since there is no phase change at either surface, Equation 35–15 now provides destructive interference. The thickness of the coating should be $(550 \text{ nm})/(1.38)(4) = 100$ nm, or an odd multiple thereof (that is, 300 nm, 500 nm, etc.). Note that to the accuracy of the method, the index of refraction of the glass does not matter, as long as it is greater than the index of refraction of the coating. Analysis using Maxwell's equation shows that the index of refraction of the glass does affect how low the minimum gets.

A common tool in optics is a **Fabry-Perot interferometer,** in which two plates are separated by a small adjustable distance. By adjusting the distance and monitoring the interference fringes, the wavelengths of the incoming light can be measured accurately (Fig. 35–27).

35.2
DIFFRACTION

We have seen that interference phenomena take place when coherent light passes through two or more slits. A phenomenon similar to interference, also dependent on light having wave properties, takes place even when only a single obstacle is in the path of light. This phenomenon, **diffraction,** results from any obstacle that is roughly the size of the illuminating wavelength. The edge of a shadow, if studied in detail, is not sharp (Fig. 35–28).

diffraction

Diffraction in water waves is relatively easy to see (Fig. 35–29). Diffraction in sound waves allows sound to be heard around a corner. The fact

Fig. 35–28
Diffraction fringes of light passing a straight edge. (Left) Actual fringes. (Right) A plot of light intensity as a function of position.

Fig. 35–29
Diffraction in water waves occurs in a harbor or in a ripple tank, as shown here, when the waves pass through an aperture or when an obstacle blocks the path of the waves.

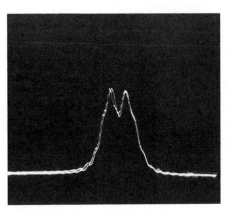

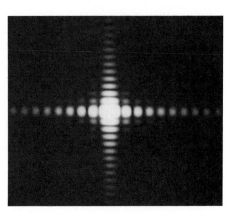

Fig. 35–30
This view of the sun taken from the Solar Maximum Mission spacecraft shows the sun blocked out by a disk placed over the image of the bright sun itself. We see rings of diffracted light around the disk. Beyond them, we see the outer part of the sun, the solar corona, much fainter than the part of the sun that was blocked.

Fig. 35–31
(Left) Fresnel diffraction by a single slit. (Right) Fresnel diffraction pattern of a rectangle.

Fresnel diffraction

Fraunhofer diffraction

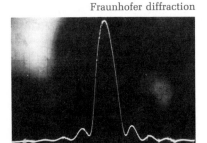

Fig. 35–32
The diffraction pattern of light passed through a single slit at a substantial distance, that is, the Fraunhofer diffraction pattern of a single slit. (Above) Plot of light intensity. (Below) Actual diffraction pattern.

that light waves cast sharp shadows instead of turning around corners means that any diffraction of light is smaller in extent than diffraction of sound; after all, the wavelengths of light are much shorter than the wavelengths of sound. But diffraction of light nevertheless occurs (Fig. 35–30).

Two ranges of diffraction can be considered, depending on the approximations we choose to make. Close to a slit or obstacle, we have **Fresnel diffraction,** after the French physicist Augustin Fresnel (pronounced freh-nel′), who provided mathematical explanations of some aspects of diffraction in 1818. Fresnel diffraction patterns are striking (Fig. 35–31) but their mathematical explanation is beyond the level of this text. At sufficiently great distances from the slit or obstacle, the rays reaching our eyes or a screen are essentially parallel and we have **Fraunhofer diffraction.** This effect is named after Joseph von Fraunhofer, the German optician who is perhaps best known for his work on the solar spectrum, also in the second decade of the nineteenth century. In the following subsection, we limit our explanations to the Fraunhofer approximation.

Diffraction by a Single Slit or Hole

Figure 35–32 shows the image formed by light passing through a single slit and then falling on a distant screen.

Note that not all the light goes straight ahead. Though most goes essentially straight, there is spreading of the image, and the spreading shows the fringes characteristic of an interference phenomenon. The central peak is surrounded by a series of diffraction fringes.

As we shall derive shortly, the positions of the minima are given simply by

$$d\sin\theta = n\lambda, \quad n = \pm 1, \pm 2, \ldots. \tag{35–18}$$

Perhaps surprisingly, the positions of the maxima are not readily stated, though obviously they are near the midpoints between minima. Only the

Fig. 35–33
The diffraction pattern of a slit combines with the two-slit interference pattern for sufficiently narrow slits.

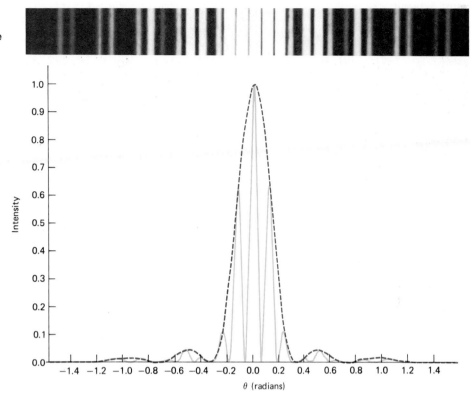

central maximum at $\theta = 0$ has a mathematically simple position. The form of Equation 35–18 shows that interference is probably taking place, and indeed the equation can be derived by calculating the interference that would result between light passing through different parts of the slit.

Note that the first minimum of the diffraction from a single slit takes place at

$$\theta = \sin^{-1}\left(\frac{\lambda}{d}\right). \qquad (35\text{–}19)$$

The smaller d is, the larger the corresponding diffraction angle θ. In our derivations of two-slit interference, we implicitly assumed that the diameter of each individual slit was sufficiently small that its corresponding θ was large. Thus we assumed that the light that was interfering was in the central maximum of each individual slit. Actually, as the nonzero width of slits is accounted for, both interference and diffraction phenomena are simultaneously present (Fig. 35–33).

Figure 35–34 shows the pattern formed by light passing through a small hole. The intensity distribution is very similar to light that passed through a slit, but the effects of the circular geometry change the position of the first minimum to

$$\theta = \sin^{-1}\left(1.22\frac{\lambda}{d}\right), \qquad (35\text{–}20)$$

diffraction ring

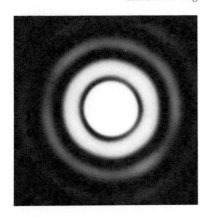

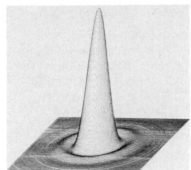

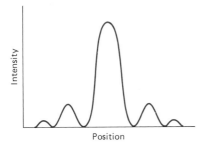

Intensity

Position

Fig. 35–34
(Top) Film records the light falling on it after it passed through a small hole. (Center) The intensity distribution plotted in pseudo-three-dimensions. (Bottom) The radial intensity distribution.

where d is the diameter of the hole. The first bright ring, the first **diffraction ring,** is necessarily farther than this value.

Note that the smaller the hole, the larger the radius of the first minimum and thus of the first diffraction ring. In a camera, if you make your aperture too small (perhaps smaller than f/32), your image may appear less sharp than at a somewhat larger aperture, since diffraction effects begin to become noticeable.

Example 35–9
Diffraction in an Optical Telescope

Astronomers using telescopes of moderate and large aperture are limited in their observations by the earth's atmosphere rather than by diffraction in their telescopes. The atmosphere shimmers and shakes, making varying ray paths that limit the sizes of even the finer star images to about 0.5 arc sec. Below what diameter telescope would diffraction in the telescope rather than the shakiness of the earth's atmosphere become the limiting effect? Consider yellow sodium light at a wavelength of 589.0 nm.

Solution

We can consider the size of the image of a point source like a star to be the distance between the first minima on either side of the center of the image. The size 0.5 arc sec is 0.5 arc sec/(3600 arc sec/°)(π rad/180°) = 2.4×10^{-6} rad, so we can approximate $\sin\theta = \theta$. The distance from the first minimum to the center is

$$a = \frac{1.22\lambda}{\sin\theta} = \frac{(1.22)(589.0 \text{ nm})(10^{-9} \text{ m/nm})}{(2.4 \times 10^{-6})} = 0.30 \text{ m}.$$

Thus only for telescopes smaller than 0.30 m ($\approx$12 inches) is diffraction an important source of spreading stellar images. All moderate and large telescopes are limited by the earth's atmosphere.

The first large diffraction-limited telescope is to be the Hubble Space Telescope (Fig. 35–35). It is to be launched into earth orbit to get above the blurring effects of the earth's atmosphere. One of the major tasks of its designers has been to make a system to control its pointing (the overall direction in which it is pointing) and jitter (motion to and fro around the normal direction in which it is pointing) precisely enough that it can achieve the diffraction limit of approximately 0.1 arc sec provided by its 2.4-m-diameter mirror.

Example 35–10
Diffraction Limitations of Single-Dish Radio Telescopes

The largest fully steerable radio telescope in the world is the 100-m diameter at the Max Planck Institut für Radioastronomie in Bonn, West Germany (Fig. 35–36). At its shortest useful wavelength of 1 cm, what is the narrowest resolution it can obtain?

Solution

$$\frac{1.22\lambda}{d} = \frac{(1.22)(1 \text{ cm})}{(100 \text{ m})(100 \text{ cm/m})} = 1.22 \times 10^{-4} \text{ rad}$$

$$= 0.007° = 25 \text{ arc sec}.$$

Fig. 35–35
The Hubble Space Telescope.
Astronomers hope that the telescope
system will allow imaging of
astronomical objects up to the mirror's
diffraction limit of 0.1 arc sec.

Fig. 35–36
The 100-m radio telescope near Bonn, West Germany, the largest fully steerable radio telescope
in the world. The 305-m Arecibo telescope in Puerto Rico (Fig. 35–9) is built into a natural bowl
in the ground, and can only observe objects that pass nearly overhead.

This result shows that single-dish radio telescopes have resolution substantially
inferior to the resolution of optical telescopes. This deficiency led to the development
of radio interferometers, which provide far superior resolution to that of optical telescopes. At longer radio wavelengths, the limitations of single-dish radio
telescopes caused by diffraction is even more severe.

Astronomers look at much more than single stars in the sky. Indeed,
most stars are members of systems of two or more stars. Further, galaxies
and nebulae are extended objects, and we can approximate our ability to
see detail in them to our ability to resolve (tell apart) point sources such as
a pair of double stars. In other words, if we can tell that two stars 1 arc sec
apart are indeed two instead of one, then we can see 1 arc-sec detail in an
extended source. A hundred years ago, Lord Rayleigh set forth the idea that
we could distinguish the adjacent images of two points if their angular
separation is such that the central maximum on one falls on the first minimum of the other (Fig. 35–37), that is

$$\sin\theta = \frac{1.22\lambda}{d}. \quad \text{(Rayleigh criterion)} \qquad (35\text{--}21)$$

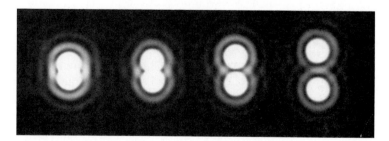

Fig. 35–37
A series of images of adjacent circular apertures corresponding to different angular separations.
In the left part, the sources are not resolved. They are barely resolved—the Rayleigh criterion—
in the second part, and are increasingly resolved as you move to the right.

35.3
HUYGENS' PRINCIPLE

In this section, we discuss a physical principle of great historical interest, advanced by the seventeenth century Dutch scientist Christian Huygens (1629–1695). We employ Huygens' principle to derive the formulae relevant to the phenomenon of single-slit Fraunhofer diffraction. Though only approximate, Huygens' principle affords us substantial calculational convenience, while permitting us to illustrate most of the important physics of diffraction. An exact treatment of the single-slit diffraction problem would entail a complete solution of Maxwell's equations, subject to the precise boundary conditions associated with the slit, including actual material properties of the screen. This would be a difficult task indeed, and has been carried out for only a few specific slit geometries (including rectangular and circular openings), assuming idealized perfectly conducting surfaces.

Fortunately, when source and observation points are quite distant from the slit (the Fraunhofer diffraction approximation), Huygens' principle provides us with a fairly accurate description of the angular-dependent intensity of the diffracted light. Huygens' principle does poorly, though, for the Fresnel diffraction approximation, when either source or observation point is near the slit. For this reason, we restrict ourselves to a quantitative analysis of Fraunhofer diffraction and do not discuss the details of Fresnel diffraction.

Huygens' principle

> *Huygens' principle:* All points of the given waterfront act as point sources of spherically propagating "wavelets" that travel at the speed of light appropriate to the medium. A time Δt later, the new wavefront is given by the unique surface that is tangent to all the secondary waves as they propagate forward.

The principle is easily understood visually (Fig. 35–38). Note that the choice $\Delta t = 1/f$, where f is the frequency of the light waves, results in a spacing of wavefronts that is the wavelength of light in the medium. Since Huygens' construction is assumed valid rather than proved, the following use of Huygens' principle is a plausibility argument and not a rigorous proof.

Looking back at the proofs we made of the law of refraction in Section 34.1 and the law of reflection in Section 34.4, you can see that the geometric constructions we made are equivalent to using Huygens' principle.

We see the relevance of Huygens' principle to single-slit Fraunhofer diffraction in Fig. 35–39, where we see a plane wave of wavelength λ (perhaps emitted from some very distant point source) incident on a slit of width D. Huygens' principle tells us that each small segment of plane wave *in the slit* is the source of propagating spherical secondary wavelets that will constitute the diffracted beam of light. From a mathematical standpoint, we can take the view that the slit is a continuous distribution of point sources, all oscillating with the same phase and amplitude. (Note, however, that there are no actual electric charges there, so Huygens' principle does not correspond to physical sources.) We next integrate over the

Fig. 35–38
Huygens' construction. The new wavefront is drawn along the tangents to the wavelets expanding around points on the old wavefront.

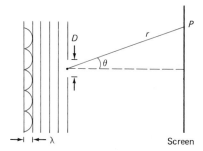

Fig. 35–39
How Huygens' wavelets propagate from a slit, for the derivation of diffraction.

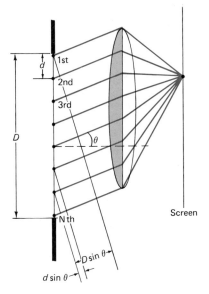

Fig. 35–40
Considering the slit as made up of N smaller Huygens' centers of wavelets.

slit to obtain the net electric field, which gives the light amplitude, at points far from the slit. We consider the distribution of point sources in the slit to be discrete rather than continuous; let N be the finite number of Huygens' centers (the centers of Huygens' wavelets) separated equally by a distance d (Fig. 35–40). Subsequently, we can derive single-slit Fraunhofer diffraction by letting the number of Huygens' centers become infinite ($N \rightarrow \infty$) while the spacing vanishes ($d \rightarrow 0$), but noting that the size of the slit, which is the product $Nd = D$, remains constant.

In our derivation using Huygens' principle, the interference effects that give rise to diffraction are caused by the different path lengths from the Huygens' centers to the observation point. At angle θ, each wavelet has the same amplitude E_θ but differs in phase by $\Delta\phi$ from its neighbor. Since the path length difference between the screen and each of the Huygens' centers is $d\sin\theta$,

$$\Delta\phi = \frac{2\pi}{\lambda} d\sin\theta. \tag{35–22}$$

The intensity at any point in the screen is proportional to the total wave amplitude there. Thus we must first find the total wave amplitude by adding the wave amplitudes from the individual wavelets. We can describe the amplitudes and their phase differences using phasors, which we defined when discussing alternating current in Section 31.2 and used in Section 31.4. Each phasor is an arrow of fixed length but varying angle whose projection on an axis gives an amplitude. Adding the wave amplitudes is equivalent to adding the phasors.

We add a series of phasors in Fig. 35–41a for a point away from the center of the screen; the resultant E_θ is the vector drawn from the beginning of the first phasor to the end of the last. The phase difference ϕ across the entire slit is thus the number of Huygens' wavelets times the phase difference between each:

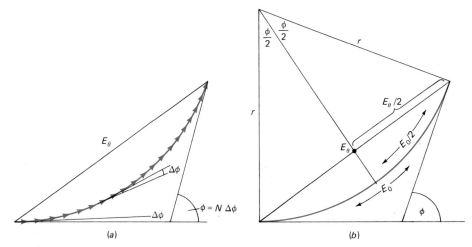

Fig. 35–41
(a) Phasor diagram for amplitude E_θ. (b) The limit as the number of Huygens' wavelets approaches ∞. E_0 is the length of the arc formed by the little arrows in a line.

Fig. 35–42
Phasor diagram for amplitude E_0.

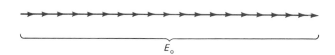

Fig. 35–42
Phasor diagram for amplitude E_0.

$$\phi = N\Delta\phi = \frac{2\pi}{\lambda} Nd\sin\theta, \qquad (35\text{–}23)$$

but since Nd is the slit width D,

$$\phi = \frac{2\pi}{\lambda} D\sin\theta. \qquad (35\text{–}24)$$

To find the resulting wave amplitude, consider the limit as the number of wavelets $N \to \infty$ (Fig. 35–41b). We see that the angle $\phi/2$ defines both half the chord and half the arc. For the chord, the shaded triangle gives

$$\sin(\phi/2) = \frac{E_\theta/2}{r}. \qquad (32\text{–}25)$$

For the arc, the definition of an angle in radians is the arc divided by the radius. And note that if the phase difference $\phi = 0$, the phasors would be in a straight line (Fig. 35–42) whose length is E_0, since the amplitudes would then all be in phase and we could add them directly. Since the arrows that form the arc in the previous figure are just the arrows of this figure curved around, the length of the arc is E_0 as well. Thus

$$\phi/2 = \frac{E_0/2}{r}. \qquad (35\text{–}26)$$

Eliminating r by dividing Equation 35–25 by Equation 35–26 gives

$$\frac{\sin(\phi/2)}{\phi/2} = \frac{E_\theta}{E_0},$$

and

$$E_\theta = E_0 \frac{\sin(\phi/2)}{\phi/2}. \qquad (35\text{–}27)$$

Since the intensity is proportional to the square of the wave amplitude, we have

$$I_\theta = (\text{constant})E_0^2 \left[\frac{\sin(\phi/2)}{\phi/2}\right]^2. \qquad (35\text{–}28)$$

But as $\phi \to 0$, $\sin(\phi/2) \to \phi/2$, so $\lim\limits_{\phi\to 0}\frac{\sin(\phi/2)}{\phi/2} = 1$. Then, with I_0 the value for I_θ for $\phi = 0$, Equation 35–28 becomes

$$I_0 = (\text{constant})E_0^2. \qquad (35\text{–}29)$$

Substituting Equation 35–29 into Equation 35–28 gives

$$I_\theta = I_0 \left[\frac{\sin(\phi/2)}{\phi/2}\right]^2. \qquad (35\text{–}30)$$

Using Equation 35–24 for ϕ, we then have

$$I_\theta = I_0 \left\{ \frac{\sin[(\pi D/\lambda)\sin\theta]}{(\pi D/\lambda)\sin\theta} \right\}^2, \qquad (35\text{–}31)$$

the desired equation for the intensity in terms of angle for a given wavelength and slit width.

The intensity distribution of light on the screen is graphed in Fig. 35–43. Note that minima occur whenever $(\pi D/\lambda)\sin\theta = \pi m$, which implies that

$$D\sin\theta = m\lambda \qquad m = \pm 1, \pm 2, \dots \text{ (but not 0)} \qquad (35\text{–}32)$$

in agreement with Equation 35–12.

Using the phasors of Fig. 35–44, the first maximum corresponds to the phasors completing a semicircle, and the first minimum corresponds to phasors completing a circle. Similarly, the second maximum corresponds to $1\frac{1}{2}$ turns, and the second minimum corresponds to 2 full turns.

Example 35–11

Intensity of the First Diffraction Peak

What is the intensity of the first diffraction peak to the side of the central maximum?

Solution

The first and second minima are at $(D/\lambda)\sin\theta = 1$ and 2, respectively, so the first maximum is approximately midway between them at $(D/\lambda)\sin\theta = \frac{3}{2}$. (We can see this fact also in Fig. 35–47.) Thus, from Equation 35–31

$$I_\theta = I_0 \left\{ \frac{\sin[(\pi D/\lambda)\sin\theta]}{(\pi D/\lambda)\sin\theta} \right\}^2 = \left[\frac{\sin(3\pi/2)}{3\pi/2} \right]^2 = \frac{1}{9\pi^2/4} = 0.045.$$

Thus the intensity at the peak of the first maximum is only 4.5 per cent of the intensity of the central maximum of the diffraction pattern.

Now let us consider the angular width of the central maximum. Its half-width, $\Delta\theta/2$, the distance from the vertical $D\sin\theta = 0$ axis horizontally to the curve, is approximately $\lambda/2$. Since $\Delta\theta$ is small, $\sin(\Delta\theta/2)$ is approximately equal to $\Delta\theta/2$. Thus $D\Delta\theta/2 = \lambda/2$ and $D\Delta\theta = \lambda$. The central maximum is therefore spread out over an angular width

$$\Delta\theta = \frac{\lambda}{D}, \qquad (35\text{–}33)$$

which leads (with a factor of 1.22 for circular geometry) to the Rayleigh criterion. As the slit width gets smaller, $\Delta\theta$ grows larger; narrower slits correspond to wider diffraction patterns. Conversely, if the slit is very wide compared with the wavelength, the effects of diffraction are very small.

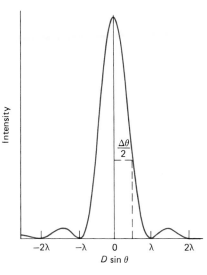

Fig. 35–43
The diffraction pattern of a single slit. The width of the central maximum is shown.

Fig. 35–44
Primary and secondary maxima and minima.

1st maximum   1st minimum   2nd maximum   2nd minimum

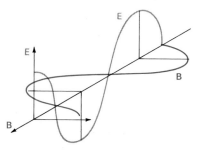

Fig. 35–45
The electric field **E** and the magnetic field **B** of electromagnetic waves are perpendicular.

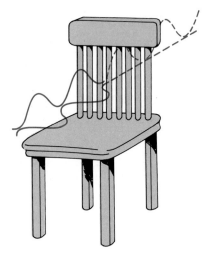

Fig. 35–46
Using the rungs of a chair to polarize waves on a rope.

35.4
POLARIZATION

As we saw in Chapter 32 for all electromagnetic waves, the electric field **E** and the magnetic field **B** of light oscillate in perpendicular directions (Fig. 35–45). For any given light ray, each of these fields oscillates in a given plane. When all electromagnetic waves oscillate in the same plane, the wave is *plane polarized.* We define the direction of the plane of polarization as the direction of oscillation of the electric field. When a mixture of waves is present, each oscillating randomly so that there is an equal distribution of oscillating planes, the wave is *unpolarized.*

All waves can be polarized. The vertical slats on the back of a chair, for example, will allow only vertically oscillating waves on a rope to pass through, and will block horizontally oscillating waves (Fig. 35–46). Analogously, a row of parallel wires will polarize a beam of microwaves (Fig. 35–47). Here, however, electric fields parallel to the wires set up electric currents and the wave energy is absorbed (and subsequently reradiated as a reflected wave). Consequently, only the field component *perpendicular* to the wires passes through, so the polarization is perpendicular to the wires.

Polarizing light waves is slightly more complex. Historically, certain types of natural crystals could be arranged to provide polarization, as we shall discuss later on. The discovery in 1932 by Edwin H. Land of a way to make polarizing material in long sheets, which he called by the tradename Polaroid, made polarization a much more useful phenomenon. (Land, then an undergraduate, took a leave of absence from Harvard College to do his work.) We can think of a sheet of Polaroid as containing parallel bars, though actually it is long chains of molecules that provide the polarization. The incoming radiation does work on these long molecules by making them oscillate along their long axes. Only the electric field oriented parallel to these axes does the work; this component is absorbed. The electric field oriented perpendicularly cannot do work, so passes freely.

In Figure 35–48, we see that light passing through the first Polaroid, the *polarizer,* can sometimes pass through the second Polaroid, the *analyzer.* If the planes of polarization of the polarizer and analyzer are parallel, all that light that passed through the polarizer can pass through the analyzer. (The polarizer itself absorbs slightly over 50 per cent of incident unpolarized light, as in the picture.) If the planes of polarization of the polarizer and analyzer are perpendicular (we say they are "crossed"), then the light that hits the analyzer is polarized perpendicularly to the analyzer's plane of polarization. Essentially none of the light that passed through the polarizer passes through the analyzer in this case.

In the intermediate case, in which the analyzer is oriented at angle θ with respect to the plane of polarization of the polarizer, the amplitude of the electric field that passes is

$$E_2 = E_1 \cos\theta. \tag{35–34}$$

Since the intensity of radiation is proportional to the square of its electric field,

$$I_2 = I_1 \cos^2\theta, \tag{35–35}$$

which is known as the *Law of Malus* (Equation 32–26).

Fig. 35–47
Using straight wires to polarize microwaves.

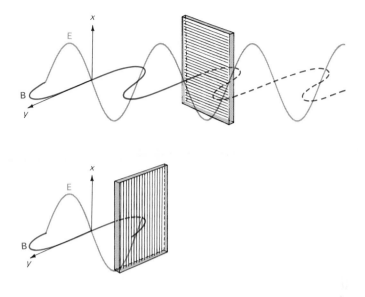

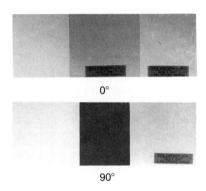

0°

90°

Fig. 35–48
Views through two polarizers with angles of polarization at different angles with respect to each other. (Above) 0°. Almost all the light that passes through one passes through the other. (Below) 90°. The light that passes through one is blocked by the other.

Example 35–12

Polarization at Different Angles

What is the intensity of radiation detected after unpolarized radiation passes through a polarizer and then an analyzer whose planes of polarization differ by 60°? Assume that each provides 50 per cent transmission of unpolarized light.

Solution:

If the incident radiation on the polarizer is I_0, I_1 is incident on the analyzer and I_2 is the intensity after the analyzer, then $I_1 = \frac{1}{2}I_0$, and

$$I_2 = I_1\cos^2\theta = \frac{1}{2}I_0\cos^2 60° = \frac{1}{2}I_0\left(\frac{1}{2}\right)^2 = \frac{1}{8}I_0.$$

Crossed polarizers pass no light. What happens if we place a third polarizer at an intermediate orientation between the crossed polarizers (Fig. 35–49)? Then we can consider the intermediate polarizer to be the analyzer for the first polarizer. It is not perpendicular to the first polarizer, so some light passes. The linearly polarized light from this intermediate po-

Fig. 35–49
The addition of an intermediate polarizer can change the net transmission of the original two polarizers, showing that light of a given polarization can be considered a superposition of two different polarizations.

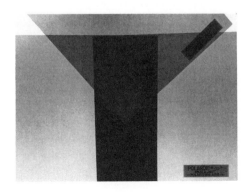

Fig. 35-50
Putting a stressed material between polarizers displays the stress.

Fig. 35-51
Polaroid sunglasses contain sheets of polarizing material to block glare. Note in the lower photograph that the image of a tree reflected off the car is blocked; it was polarized in the reflection.

larizer next hits the last polarizer. It is now not crossed with this last polarizer, so again some light passes. Thus by adding a third polarizer between two polarizers, we allow some light to pass through the entire system. Essentially, the intermediate polarizer has rotated the plane of polarization of the light. (We explore the situation further in Problems 58 and 59.)

The phenomenon of an intermediate polarizer rotating the plane has many practical uses. For example, stressed plastic provides polarization that varies with the degree of stress. Thus using stressed plastic as the intermediate polarizer provides a map of the stress (Fig. 35–50), which is useful in many industrial applications. Crossed polarizers are also used in liquid crystal displays (LCDs), as shown in Fig. 22–20.

Polarization by Reflection

Polaroid sunglasses contain sheet Polaroid that is oriented vertically. By rotating them, you can readily see that reflections from roads or from water are polarized horizontally, and so are blocked by the sunglasses when worn normally (Fig. 35–51). Thus Polaroid sunglasses cut glare preferentially to other incoming radiation.

They do so because reflection can polarize light. Some polarization is provided by reflection at all angles except 0°. In particular, the component of the electric field of incident light that is parallel to a surface reflects more completely than the component that is perpendicular to the surface. Indeed, there exists in dielectrics an angle θ_p, the *polarizing angle*, for which the reflected beam is completely polarized. The angle is known as *Brewster's angle* after Sir David Brewster, who discovered it in 1812. It occurs when the reflected beam and the refracted beam are perpendicular. The refracted beam is partly polarized.

In Fig. 35–52, we can see, from the right side of the normal, that $\theta_p + 90° + \theta_2 = 180°$. Thus $\theta_2 = 90° - \theta_p$. From Snell's law,

$$n = \frac{\sin\theta_p}{\sin\theta_2} = \frac{\sin\theta_p}{\sin(90° - \theta_p)} = \frac{\sin\theta_p}{\cos\theta_p} = \tan\theta_p. \quad \text{(Brewster's law)} \quad (35\text{--}36)$$

Example 35-13

Brewster's Law

What is Brewster's angle for crown glass?

Solution:

For crown glass, n = 1.52, so

$$\theta_p = \tan^{-1}(1.52) = 57°.$$

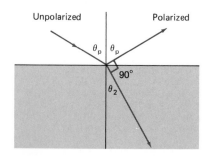

Fig. 35-52
The reflected beam is completely polarized at the Brewster angle, which corresponds to perpendicular incident and reflected beams.

Birefringence and Polarizing Devices

In glass and most other transparent materials, the speed of light is the same in all directions. But in calcite (also called "Iceland spar") and a few other crystals and solutions, the speed of light in different directions is different. These crystals and solutions are *birefringent*.

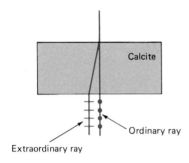

Fig. 35–53
The ordinary and the extraordinary ray passing through a birefringent crystal are polarized at right angles to each other.

When a light ray enters a birefringent crystal, it splits into an *ordinary ray* and an *extraordinary ray*. The ordinary ray passes through following Snell's law. The extraordinary ray, on the other hand, is deviated (Fig. 35–53). Both are plane polarized, though in perpendicular directions. For only one direction in a birefringent crystal, its *optic axis*, do the ordinary and extraordinary rays coincide.

Birefringence can be explained by differing speeds of light in different directions with respect to the material's optic axis. Consider light incident perpendicularly to a birefringent crystal (Fig. 35–54). The ordinary ray is undeviated. The extraordinary ray, though, appears to rotate around the ordinary ray as you rotate the crystal. This phenomenon can be explained as the result of its speed being different, causing it to bend.

Pre-Polaroid devices to polarize light, invented in the early 18th century, used the phenomenon of birefringence. Such devices are still in use to provide more complete polarization when needed in science or industry. In the *Nicol prism* (Fig. 35–55a), one of the rays formed in a calcite crystal is made to be less than the critical angle at the second surface, while the other is greater than the critical angle. Thus only one polarization passes through. In the *Wollaston prism* (Fig. 35–55b), two birefringent crystals with perpendicular optic axes are placed together. The two polarizations are available at different locations on the far side, and can be selected by positioning your detector properly. Certain crystals, called *dichroic*, absorb one of the polarizations much more severely than the other, allowing only one polarization to emerge.

Birefringence also occurs in magnetized plasmas (ionized gases), and its effects provide diagnostic tools for astrophysical plasmas that cannot be probed directly.

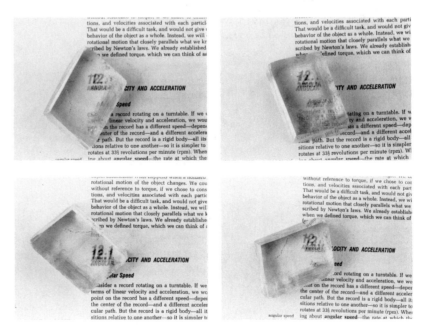

Fig. 35–54
Birefringence results from different light speed in different directions for the extraordinary ray.

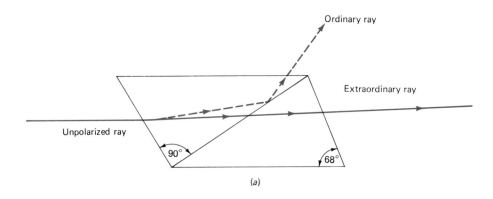

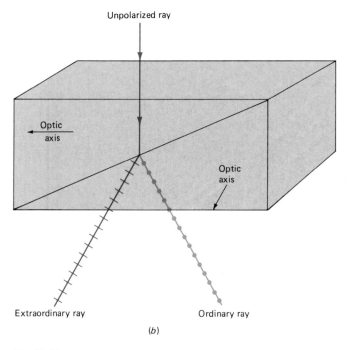

Fig. 35–55
(a) The Nicol prism, two pieces of calcite cemented together with Canada balsam, whose index of refraction makes the critical angle less than the angle of one ray and greater than the angle of the other. *(b)* The Wollaston prism, two pieces of quartz with optic axes perpendicular to each other. Both beams are deviated and are polarized perpendicularly to each other.

Circular Polarization

A plane-polarized beam contains waves whose electric fields oscillate in a given direction. A *circularly-polarized beam* contains waves whose electric fields change in orientation with time, rotating as they pass a given point.

We have seen that the ordinary and extraordinary rays in a birefringent crystal have different speeds. Thus one is delayed in phase with respect to the other. The amount of the phase delay depends not only on the difference between the speeds but also on the thickness of the crystal. Consider a *quarter-wave plate,* a crystal of the suitable thickness to delay one ray $\frac{1}{4}\lambda$

Fig. 35–56
A quarter-wave plate creates a difference of $\frac{1}{4}\lambda$ between the phases of the ordinary and extraordinary rays. $E_\parallel$ is parallel to the optic axis and $E_\perp$ is perpendicular to it. Consider the electric field of the superposition of the ordinary and extraordinary rays—the sum of their individual electric fields. Notice that the direction of this resultant field is rotating as you follow from line A to lines B, C, and D. The light is now circularly polarized.

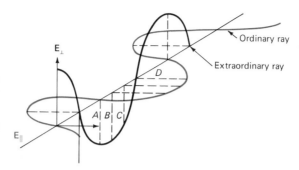

with respect to the other. In Fig. 35–56, we see that at point A the electric field is oriented upward. At point C it is oriented horizontally, and at point B, it has an intermediate orientation. Thus if we allow this wave to pass us, its plane of polarization will rotate. As a result, a quarter-wave plate provides circularly-polarized light.

The direction of the rotation matters. In Fig. 35–57 (top), we see what happens when a sheet containing first a quarter-wave plate and then a linear Polaroid is placed over a coin. The surface of the coin is shiny, and reflects the light that hits it without changing its linear polarization. This linear-polarized light passes back through the quarter-wave plate. But in the lower photograph in Fig. 35–57, the sheet is flipped over so that the light passes first through the Polaroid and then through the quarter-wave plate. The linearly-polarized light that passed through the Polaroid is changed into circularly-polarized light by the quarter-wave plate. But the direction of circular polarization is flipped over by the coin, and the resulting light cannot pass back through. The coin appears dark.

Fig. 35–57
(Above) A quarter-wave plate affixed to a sheet of Polaroid placed over a coin. (Below) The quarter-wave plate now follows the Polaroid sheet. The circular polarization is reversed in the reflection, and the light does then not pass back through.

Polarization by Scattering

When light is reflected off a substance much smaller than its wavelength, it follows *Rayleigh scattering*, which varies as $1/\lambda^4$. Lord Rayleigh worked out the theory in 1871. The circumstance is met by visible light, of wavelength 400 to 650 nm, scattering off oxygen and nitrogen molecules, only about 0.2 nm across, in the earth's atmosphere. Rayleigh's theory also shows that the scattered light is polarized. When we look at the sky in a direction that corresponds to Brewster's angle, so that the light from the sun is scattered toward us at a right angle, we see maximum linear polarization. At other angles, we see lesser degrees of polarization. You can verify this outdoors with Polaroid sunglasses.

The internal reflections in water droplets that contribute to the formation of a rainbow also polarize light. Thus rainbows are highly polarized, and you can miss seeing one if you are wearing Polaroid sunglasses.

Rayleigh scattering is so highly dependent on wavelength that blue light scatters much more efficiently than red light. Thus the daytime sky is blue, as red light passes through to make reddish sunsets for people farther along the globe. When water droplets grow large enough, Rayleigh scattering ceases and scattering becomes uniform at all wavelengths. Thus clouds are white and are unpolarized.

SUMMARY

Interference and diffraction are wave phenomena, and are the subject of **physical optics.**

1. Young's experiment (1801) detected interference from a double slit, and thus showed that light has wave properties. For a double slit,

$$d\sin\theta = n\lambda, n = 0, \pm 1, \pm 2, \pm 3, \ldots$$
(maximum intensity)

$$d\sin\theta = (n + \tfrac{1}{2})\lambda, n = 0, \pm 1, \pm 2, \pm 3, \ldots$$
(minimum intensity)

The time-averaged intensity on a screen is

$$\overline{S} = \frac{2A^2}{\mu_0 c r^2}\cos^2\left(\frac{\pi d}{\lambda}\sin\theta\right).$$

For three or more slits,

$$d\sin\theta = n\lambda, n = 0, \pm 1, \pm 2, \pm 3, \ldots$$
(maximum intensity)

$$d\sin\theta = (m/N)\lambda, m = \pm 1, \pm 2, \pm 3, \ldots$$
excluding $m = $ (integer)(N).   (minimum intensity)

Diffraction gratings, actually interference gratings, come in both reflecting and transmitting types. They use many narrow rulings close together to disperse light. The resolving power of a grating is

$$\frac{\lambda}{\Delta\lambda} = Nn.$$ (resolving power)

2. Diffraction is the spreading of a wave when it hits an obstacle approximately its own size. Fresnel diffraction is the pattern seen close to the obstacle; Fraunhofer diffraction is the pattern seen far away. Fraunhofer diffraction is simpler to treat mathematically.

3. For diffraction of a single slit, the positions of the minima are

$$d\sin\theta = n\lambda, n = \pm 1, \pm 2, \ldots.$$ (straight slit)

The positions of the maxima are roughly in between, but not precisely so.

4. For diffraction of a circular aperture, the position of the first minimum corresponds to

$$d\sin\theta = 1.22n\lambda, n = \pm 1, \pm 2, \ldots$$
(circular aperture)

The **resolution** of a single-aperture telescope depends on its aperture. The Rayleigh criterion is that two point sources are resolved if the first minimum of one falls on the maximum of the other.

5. **Huygens' principle** is an approximation of calculational convenience. It holds that all points of a given wavefront act as point sources of spherically propagating secondary "wavelets" that travel at the speed of light appropriate to the medium. A short time Δt later, the new wavefront is given by the unique surface that is tangent to all the secondary waves.

6. Polarization of light follows Malus's law,

$$I_2 = I_1\cos^2\theta.$$

Reflected light is completely polarized at Brewster's angle.

QUESTIONS
---

1. Why does an oil slick on the ground sometimes show colored bands?
2. Why doesn't sound cast sharp shadows; that is, how does it manage to bend around corners and obstacles?
3. Contrast light and sound from the points of view of whether they undergo reflection, refraction, interference, and diffraction.
4. Do tennis balls show interference phenomena when many of them are thrown against a wall? Why or why not?
5. Does a marching band represent a coherent or an incoherent phenomenon? Explain.
6. Why don't you see interference when you look out between your eyelids?
7. Why don't you see interference effects whenever you look out the window through venetian blinds?
8. On a hot summer evening, as you sit in your screened porch staring at a distant straight lamp, you observe the phenomenon of diffraction because the wire mesh of the screen acts as a rectangular array of slits. Why can you think of the screen, instead, as a rectangular array of point sources of light? (*Hint:* Recall Huygens' principle.)
9. What would you see if one of the pair of sodium D lines fell on the second maximum of the other sodium D line, as imaged through two slits?
10. Why can't we simply take the intensity that results from a two-slit interferometer and calculate how it interacts with an adjacent slit to derive the intensity that results for a three-slit interferometer?
11. In a dark fringe, where has the energy of the interfering waves gone?
12. Our equations show that the intensity at interference maxima is four times the intensity of the individual waves. Explain why this fact does not violate the law of conservation of energy.
13. Suppose you rotate the piece of metal into which two adjacent slits have been cut, while a laser beam is illuminating them and projecting an interference pattern straight ahead. If you rotate the metal around an axis parallel to the slits, what will the effect be on the projected interference pattern?
14. When installing loudspeakers, you are advised to keep the signals in phase. If they are out of phase, would the resulting interference be dependent on the wavelength of the sound radiated?

15. When an approaching car is relatively far away at night, do you see an interference pattern caused by its twin headlights? Explain.

16. Why don't you see interference effects between the front and the back surfaces of your eyeglasses?

17. How can overlapping orders that come from a diffraction grating be put to good use?

18. What is the difference between diffraction and interference?

19. Sketch and compare the maxima and minima in the interference patterns of systems of 4, 5, 6, 7, and 8 slits.

20. Since we can explain the dispersion of light from a grating as an interference effect, do "diffraction gratings" depend on diffraction or interference?

21. A coarsely ruled grating, called an "echelle," will give spectra of low dispersion. What would the result be of passing the resulting overlapping orders through a prism oriented so that it disperses perpendicularly to the echelle?

22. In what way is a pair of radio telescopes separated by 1000 km superior to the 100-m Bonn telescope? In what way is the pair inferior?

23. What is an advantage of spacing an array of radio telescopes irregularly instead of evenly?

24. Describe the change in the diffraction pattern of a single slit as the slit is narrowed.

25. If you lay out a series of radio telescopes along a line, how might the rotation of the earth help you gain increased resolution in various directions?

26. What would the result be if you omit the compensating plate from a Michelson interferometer?

27. Why did Michelson and Morley effectively lengthen their interferometer by reflecting the light back and forth?

28. When the moon passes in front of a star, the intensity of light from the star fluctuates instead of dropping off abruptly. Explain.

29. Why does a total eclipse of the sun, in which the moon comes between the earth and the sun, show smaller diffraction effects than the coronagraph image shown in Figure 35–33, in which a disk in the telescope hides the solar surface?

30. Reflecting telescopes must have secondary mirrors in the light path, so that the image is accessible to view. Refracting telescopes don't have such secondary mirrors. Explain how diffraction might affect image quality in both kinds of telescopes.

31. Why would it be desirable to observe double stars in the blue region of the spectrum instead of in the red?

32. What would be the effect on a Michelson interferometer if it were placed under water instead of in air?

33. What is the total physical length of grooves that must be cut by a stylus on a diffraction grating 10 cm square if it carries 1200 grooves/mm? Comment on why diamond styluses are necessary.

34. The dispersion of light by a prism varies with wavelength. Compare with the dispersion of light by a grating.

35. Sunspots are dark regions of the sun surrounded by the bright solar surface. Why would you prefer to observe sunspots in the infrared than in the visible?

_____ **PROGRAMS** _____

PROBLEMS

Section 35.1   *Interference*

1. A glass plate covered with aluminum has two thin slits ruled in it. The slits are separated by 0.35 mm. The plate is illuminated by yellow sodium light at a wavelength of 589.0 nm. Calculate the angular position of the maxima and of the minima that result.

2. A glass plate covered with aluminum has two thin slits ruled in it. The slits are separated by 0.25 mm. The plate is illuminated by light from a tube filled with hydrogen gas; its radiation at wavelengths of 656.3 nm in the red and 486.1 nm in the blue hits the plate. Sketch the overlapping positions of intensity maxima on a screen 2.5 m distant.

3. Derive Equation 35–6 from 35–5 by expanding the cosine functions.

4. From two slits separated by 0.37 mm, we see an interference pattern emerge with maxima spaced every 0.065°. What is the wavelength of the illuminating radiation?

5. Consider an experiment using film that can resolve two objects' images only when they are separated by more than 0.25 μm. Can you distinguish that the sodium D lines at 589.0 nm and 589.6 nm are double rather than single by their interference through twin slits? The slits are separated by 0.50 mm and the film is 1.50 m away.

6. The green line of gaseous mercury at 546.0 nm falls on a double-slit apparatus. If the fifth dark fringe is at an 0.113° angle to the axis, what is the separation of the slits?

7. Light from a helium-neon laser, 632.8 nm, falls on a Young's experiment (double-slit system) with slits separated by 0.65 mm. Calculate the separation of the second fringe from the first, and of the eighth fringe from the seventh, as measured on a screen 1.7 m away from the slits. By what percentage do these separations differ?

8. For a double-slit experiment with slit spacing 0.35 mm and wavelength $\lambda = 500.0$ nm, what is the phase difference between two waves arriving at a point on a screen 1.5 m distant for $\theta = 0.45°$?

9. For a double-slit experiment with slit spacing 0.25 mm and wavelength 600.0 nm, at what angle is the path difference one-fourth wavelength?

10. For interference through a Young's double-slit experiment, if the illuminating light is a helium-neon laser emitting at 632.8 nm and the spacing of the slits is 0.55 mm, at what angles is the average intensity on the screen 60 per cent of the maximum?

11. What is the minimum phase difference in a double-slit experiment if the intensity on a screen is 15 per cent of its maximum value?

12. The Very Large Array has a maximum spacing between antennas of 27 km. If the antennas used at a radio wavelength of 21 cm, corresponding to the interstellar hydrogen emission, what is the available resolution?

13. A type of interferometry known as "speckle interferometry" is carried out using radiation hitting the extreme opposite sides of a large optical telescope mirror. For the Hale Telescope at the Palomar Observatory, the diameter of whose mirror is 5.1 m, what is the maximum resolution that can be obtained at a wavelength of 500.0 nm?

14. Compare the resolution that will be obtainable from the European Southern Observatory's proposed line of four 8-m-diameter telescopes separated center-to-center by 100 m, used at an optical wavelength of 486.1 nm, and the 8000-km maximum spacing of the Very-Long-Baseline Array of radio telescopes used at a radio wavelength of 21 cm?

15. Your portable radio is in a dead spot caused by direct reception from a radio station being interfered with by indirect reception from the signal bouncing off a wall 1.50 m behind you. The FM band ranges from 88 to 108 MHz. How far forward should you move the set to be at an interference maximum instead of a minimum?

16. Derive that the full angular width of the interference fringes in a Young's experiment is $\Delta\theta = \lambda/2d$ at half the maximum intensity, for small θ.

17. For a two-slit system, sketch the interference pattern on a curved screen as a function of angle from the screen's center, given constant spacing between slits of 0.33 mm for hydrogen-beta radiation at 486.1 nm.

18. For a three-slit system, sketch the interference pattern on a curved screen as a function of angle from the screen's center, given constant spacing of slits of 0.40 mm and sodium-D radiation at 590.0 nm.

19. For a four-slit system, sketch the interference pattern on a curved screen as a function of angle from the screen's center, given constant spacing of slits of 0.33 mm and hydrogen-beta radiation at 486.1 nm.

20. At what angles do minima occur for a 12-slit system illuminated by 632.8-nm laser light if the slits are separated by 0.125 mm?

21. On a screen 1.25 m away from a series of slits, we see a pattern with bright maxima separated by 0.86° and with six intermediate maxima. How many slits are present and what is their separation if the incident light has wavelength 656.3 nm?

22. For a 1200-groove/mm grating, in what orders can we get Hα at 656.3 nm and the K-line of ionized calcium at 393.3 nm both falling on a 4.0-cm-distant detector 2.5 cm across?

23. Consider an 80-groove/mm echelle grating with an eighteenth-order spectrum centered on 600.0 nm falling from side to side across the center of a 2.5 cm × 2.5 cm detector with a dispersion of 10 nm per mm on the detector. A second grating oriented perpendicularly to the echelle grating disperses adjacent spectra crosswise, allowing 10 spectra to the top and 10 spectra to the bottom of the eighteenth-order spectrum to fall on the detector. What are the longest and shortest wavelengths detected by this system?

24. Compare the resolving power of an 80 groove/mm echelle grating used in twelfth order and a 1200 groove/mm grating used in second order, assuming the gratings to be of equal width.

25. What is the closest spacing of spectral lines that could be resolved by a 102 groove/mm echelle grating (of width 2 cm) used in twelfth order in the ultraviolet at 155 nm? (Such a grating is on the International Ultraviolet Explorer spacecraft.)

26. You want to resolve the center of the calcium H-line at 396.85 nm from hydrogen's Hε-line at 397.05 nm in first order ($m = 1$). What is the minimum number of grooves your grating must have? Ignore the intrinsic width of the spectral lines themselves.

27. When viewed through a spectrograph, hydrogen's Hβ-line at 486.1 nm viewed in sixth order is flanked at a position corresponding to 484.3 nm in that sixth order by another line that appears in an unknown order. What are the possible visible wavelengths of this additional spectral line?

28. Derive the formula for the "dispersion" of a grating, $d\lambda/d\theta$ (the derivative of λ with respect to θ), for mth order, in terms of the grating spacing d and the angle θ.

29. As you move one of the mirrors on a Michelson interferometer, you observe 530 bright fringes to move from side to side before your eyes. If you are using sodium light at 589.0 nm, by how much have you moved the mirror?

30. You move one of the mirrors of a Michelson interferometer by 0.12 mm; the interferometer is illuminated with hydrogen's Hβ light at 486.1 nm. How many bright fringes pass before you?

31. You are measuring the wavelength of illuminating radiation with a Michelson interferometer, and see 550 fringes as you move the mirror 0.15 mm. What is the wavelength of the light?

32. One arm of a Michelson interferometer is enclosed in a box that can be evacuated. At the start, both arms are in air, whose index of refraction is 1.000 293. If light of wavelength 641.6 nm travels a round-trip total path of 0.850 m on each arm before evacuation, by how many bright fringes do the fringes shift if one of the arms is evacuated?

33. A thin glass slide, 0.10 mm thick, is placed in one arm of a Michelson interferometer, each of whose beams travels a 1-m-long path from partially reflecting mirror to end mirror and back to partially reflecting mirror. For sodium light at 589.0 nm, if a shift of 55 bright fringes is seen, what is the index of refraction of the glass?

34. In their famous 1887 experiment, Michelson and Morley found no fringe shift to an accuracy of 0.01 fringe. If they were observing at 500.0 nm, and the mirror spacing in their interferometer was $d = 11$ m,

to what percentage accuracy did they determine that the speed of light is constant?

35. What is the minimum thickness of a soap bubble, given that the index of refraction is 1.40, if constructive interference of the sodium line at 589.0 nm is seen?

36. A lamp directly above a 300-nm-thick oil slick ($n = 1.40$) makes an interference pattern. If you are looking straight down, what bright wavelengths do you see reflected?

37. Light reflected from a thin film of acetone on a glass plate shows a maximum at 400 nm and a minimum at 600 nm. If the index of refraction of acetone is 1.25 and the index of refraction of the glass is 1.55, how thick is the film? Assume perpendicular reflection.

38. One day in Hawaii, the sun is directly overhead and shines light onto a 300-nm-thick oil slick characterized by an index of refraction $n = 1.42$. (The sun is never directly overhead in the continental United States.) If you are looking along the illuminating sunbeam, what bright wavelengths do you see reflected?

39. You use a lamp to illuminate a peacock feather, and see light of 411 nm, 459 nm, 520 nm, and 600 nm reflected. What is the thickness of the $n = 1.3$ reflecting part of the peacock feather, treating it as a thin film? What orders provide the light?

40. How thick a coating of index of refraction 1.35 should you use on a glass lens (index of refraction 1.55) to minimize reflection at 500 nm (in the middle of the optical spectrum)?

41. You have two square optical flats that are 20 cm on a side and have $n = 1.52$. They are stacked on a table, with a shim raising one end of the upper flat 0.015 mm on one side, making a wedge of air between them. If you shine 500-nm light on the top surface of the top flat, how many fringes do you see across it?

42. A film of soap on the inside of a fish tank filled with water (index of refraction 1.3) is causing an intensity maximum at 550 nm when you look through the fish tank from the other side, seeing through the water to the film and the glass on its far side. How thick is the soap film?

Section 35.2   *Diffraction*

43. For a slit 0.50 mm wide and Hβ radiation at 486.1 nm, what is the angular location of the first three dark fringes of the diffraction pattern?

44. On a screen 80 cm from a 0.15-mm-wide slit, you notice diffraction fringes separated by 3.0 mm. What is the wavelength of the illuminating radiation?

45. What is the angular separation between dark diffraction rings from a 0.85-mm aperture illuminated by laser light at a wavelength of 632.8 nm?

46. At what angle in seconds of arc is the first minimum in the diffraction pattern of the 3.6-m Canada-France-Hawaii Telescope when observing Hβ radiation at 486.1 nm? Compare with the 0.5 arc sec minimum image size set by the earth's atmosphere.

47. Laser light of wavelength 556.2 nm is incident on a circular hole of diameter 1.0 mm. At what distance from the hole should you place a screen to have the first minimum of the diffraction pattern fall 24 mm from the pattern's center?

48. Calculate the diffraction limit for the 2.4-m Hubble Space Telescope for ultraviolet light at a wavelength of 300.0 nm and for red light at a wavelength of 656.3 nm using the Rayleigh criterion.

49. What size does a telescope have to be to distinguish, by the Rayleigh criterion, the two members of a double star separated by 0.8 arc sec if observed through a filter passing light at a wavelength of 500 nm? What is the distance between the stars if they are a distance of 15 light-years from earth?

50. Describe color variations across a diffraction ring if a filter between the illuminating source and the aperture passes light between 450 and 550 nm. In particular, for the first two diffraction rings of 450-nm light, give the positions of the diffraction rings of the 550-nm light that also passes through the filter.

51. The distance from the center of a circular diffraction pattern to the first minimum on a screen 0.85 m distant is 15,000 wavelengths. What is the size of the aperture?

52. The human eye's iris can expand to 8 mm in near darkness. Given that the eye's index of refraction is 1.35, what is the minimum angle it can resolve for the middle of the spectrum at 500 nm when observing in faint light at night? Saturn's rings can cover as much as 1 arc min of sky. Can we possibly resolve them with the human eye?

53. Galileo's original telescope in 1609 had a front lens that was 2 cm wide. What was its resolution in arc seconds? Jupiter can be as much as 50 arc sec wide, and its Great Red Spot is ⅕ the diameter of Jupiter. Would the Great Red Spot have been resolvable? By what factor?

54. If your eye has an aperture of 2 mm at night with headlights from an oncoming car pointed at it, at what distance can you see as distinct two headlights separated by 1.45 m? Assume a wavelength of 500 nm, near the middle of the visible spectrum.

55. A home dish to receive television programs at a wavelength of about 500 MHz is 1.5 m in diameter. What is the size of its beam at the distance of a satellite orbit 42,000 km from the earth's center? Assume we are located at a latitude of 40° on the surface of the earth, 6400 km from the earth's center. What distance along the equatorial orbit of the satellite must the next communications satellite be to be in the first minimum of the satellite we are observing?

56. Rumor has it that telescopes of the same 2.4-m aperture as the Hubble Space Telescope are flown in low earth orbit (100 km above the surface) pointing downward. What would the resolution of such a telescope be if it were diffraction limited? Compare this resolution to the 0.5 arc sec limitation imposed by the earth's atmosphere when looking outward. The limitation is less when looking downward because of the location relatively near the ground of the turbulent atmospheric layers.

Section 35.4 *Polarization*

57. An unpolarized light beam of intensity I_0 is incident on the first of three polarizers. Each polarizer passes 50% of the intensity. The plane-polarized light from the first polarizer falls on the second polarizer, whose angle of polarization is rotated 45° from that of the first polarizer. Whatever passes through this second polarizer then passes through the third polarizer, whose plane of polarization is rotated another 45°. What is the intensity that exits the system?

58. Repeat Problem 57 with the middle polarizer set with its plane of polarization 60° from that of the first polarizer and 30° from that of the second polarizer.

59. The maximum polarization by a substance is found to occur at an angle of reflection of 55°. What is the substance?

60. By what factor is scattering at 434 nm (blue) enhanced over scattering at 650 nm (red) in the sky?

61. How much less scattering is there in the infrared at 2.2 μm than in the violet at 400 nm?

Supplementary Problems

62. Calculate the separation in angle of a Young's double-slit experiment with slits separated by 0.25 mm for hydrogen radiation at 486.1 nm if (a) the system is in air, and (b) the system is filled with transparent Jell-O, $n = 1.45$.

63. In a Young's double-slit experiment, a thin glass plate of index of refraction 1.56 is placed over one of the slits. The fifth bright fringe appears where the second dark fringe previously appeared. How thick is the plate if the incident light has wavelength 480 nm?

64. The southwest arm of the Very Large Array—a set of radio telescopes used as an interferometer—has telescopes spaced the following distances from a central telescope: 484 m, 1590 m, 3188 m, 5223 m, 7659 m, 10,473 m, 13,644 m, 17,157 m, and 21,000 m. Each two telescopes used together makes a different pair. How many different pairs are there? What is the resolution for the longest and for the shortest spacings, if used at the wavelength of 92 cm?

65. The International Ultraviolet Explorer spacecraft contains a spectrograph that spreads out the ultraviolet spectrum in 53 different strips, each in a different order of diffraction. The strips are overlapping orders, and are dispersed perpendicularly to the main dispersion of the spectrum (Fig. 35–58) by a second grating. In the middle strip, the wavelength is 150.0 nm, and the strip covers 2.5 nm. What are the central wavelengths and wavelengths covered by the 70th, 94th, 96th, and 120th strips if the middle strip is in 95th order? Is there a gap in the wavelength coverage between the 94th- and 95th-order spectra?

66. Derive the formula for constructive interference of a thin film for light incident at an angle θ from the normal.

67. A glass wedge ($n = 1.50$) has a triangular cross section. From the top it is a rectangle 10 cm long by 5 cm wide, and from the side it is a right triangle 10 cm long with the vertical side 1.0 mm high. Describe

Fig. 35–58 Astronomers in the control room of the International Ultraviolet Explorer at NASA's Goddard Space Flight Center in Greenbelt, Maryland. On the screen, they see several of the adjacent bands of spectra recorded by the spacecraft in synchronous orbit and radioed down to earth. Each strip covers 2.5 nm of the ultraviolet spectrum on a diagonal from top left to lower right; 53 adjacent strips, each in a different order, have filled the field of view of the TV-like detector (Problem 65).

the fringes you see if it is illuminated by a 500.0-nm light directly overhead.

68. Consider a plano-convex lens with radius of curvature R, resting curved side down on an optically flat mirror. Derive the formula $d = \sqrt{(m + \frac{1}{2})\lambda R}$ for the distance d of the bright fringes radially from the point of contact, given index of refraction n for the lens.

69. Given the result of the preceding problem, how many bright rings are seen if the radius of curvature is 7.5 cm, the illuminating light is of wavelength 500 nm, and the lens is 2.5 cm across?

70. How would the answer in the preceding problem change if the system were bathed in water, for which the index of refraction is 1.33?

71. An arrangement known as Lloyd's mirror allows a single source to provide coherent illumination equivalent to two slits by reflecting the lone source in a horizontal mirror, and allowing the wave coming directly from the source to interfere with its reflection (Fig. 35–59). If the distance of the source to the screen is D, and the mirror is placed a distance d from the normal drawn from the source to the screen, derive a formula for the separation δ of bright fringes. The wavelength of the light is λ. *Hint:* Locate the second source as the virtual image of the first source.

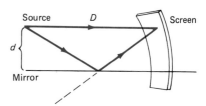

Fig. 35–59 Lloyd's mirror. Note that the reflected wave undergoes a 180° change in phase at the reflection (Problem 71).

36

INSIDE ATOMS AND NUCLEI

The science of mechanics has been at the basis of many triumphs of physics and engineering, and the unification of electricity and magnetism by Maxwell, followed by the discovery of radio waves by Hertz, has certainly transformed our world. Our text has taken you through many of the fundamental concepts in these fields and others, and has given you practice and understanding in solving problems in them. Further, we have at various points brought in contemporary applications of mechanics and of electricity and magnetism. But though the applications may be contemporary, the basic physics is still **classical physics.** We have already seen in the theory of special relativity that Albert Einstein presented in 1905 how our classical ideas of motion can be wrong. And we have alluded to how Einstein's general theory of relativity, advanced in 1916, explained gravity as a geometrical effect in curved space-time. But all of classical physics, and the two relativity theories, appear to be approximations that hold only when we do not look at the detailed structure of matter.

classical physics

A major theme of twentieth-century physics has been the structure of matter and energy on a small scale. In this chapter, we describe some aspects of the theory of **quantum mechanics,** one of the seminal ideas of our time. Quantum mechanics explains the small-scale structure of matter and energy in ways that are often counterintuitive but that nonetheless have been carefully verified by experiment. We go on to indicate some of the current ideas on the structure of matter; in particular, we shall describe how basic particles such as neutrons and protons are made of quarks. The quark theory has been around for over twenty years and is generally accepted. We will finish our brief survey of contemporary physics by showing how the study of quarks and other elementary particles has merged with the study of cosmology to give us a new picture of the evolution of the universe as a whole.

quantum mechanics

This selective survey covers several topics of contemporary interest but slights others. To study any of the fields in this chapter requires much more time and space than we can allot in this basic course. We hope the following whets your appetite.

997

36.1
LEADING TO THE QUANTUM THEORY

The foundations of twentieth-century physics go back to the nineteenth century and earlier. We have shown in the preceding two chapters how optical methods have allowed us to study the spectra of light sources. Newton, indeed, broke down light into a spectrum of colors. In 1804, William Wollaston noticed apparent lines of demarcation between some of the colors. Ten years later, Josef von Fraunhofer, a German optician, succeeded in dispersing the solar spectrum sufficiently that he could discover hundreds of dark lines crossing the spectrum, the **Fraunhofer lines** (Fig. 36–1). We now know of millions of such lines in the solar spectrum, each a specific wavelength where energy from a continuous spectrum is absorbed before it reaches our eyes. Late in the last century, Gustav Kirchhoff showed that sometimes spectral lines could be brighter than their surrounding wavelengths—**emission lines**—and sometimes they could be darker—**absorption lines.**

In 1884, Johann Balmer, a Swiss schoolteacher who was fond of numerology, was trying to figure out an easy way to characterize the wavelengths of the hydrogen lines. We now call these lines Hα, Hβ, Hγ, etc.; Hα is what Fraunhofer labelled the C line in his spectrum below. Balmer realized that the wavelengths of the first four hydrogen lines could be described by the equation

$$\frac{1}{\lambda_{H\alpha}} \propto \left(\frac{1}{2^2} - \frac{1}{n^2}\right), \tag{36–1}$$

where $n = 3, 4, 5,$ and 6. This simple formula, painstakingly discovered by trial and error, eventually led Niels Bohr (Fig. 36–2) to give physical significance to the individual terms (that is, to the $1/2^2$ and to the $1/n^2$). In 1911, Rutherford had presented the idea of the nuclear atom, with a small, positively charged nucleus. In 1913, Bohr explained the hydrogen spectrum with the **Bohr atom,** in which electrons circled hydrogen's nucleus at

Fraunhofer lines

emission lines
absorption lines

Bohr atom

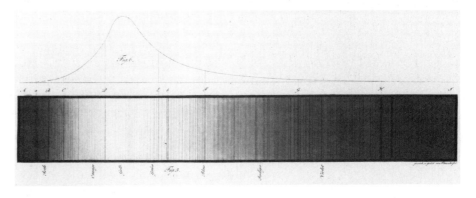

Fig. 36–1
Fraunhofer's drawing of the solar spectrum, made in 1814 and published in 1817. The dispersion in wavelength is from side to side; dark vertical lines indicate the absorption of light at certain wavelengths.

certain radii. At each radius, an electron had a fixed energy. A Bohr atom would not give off energy unless an electron jumped from one energy level to another. If energy level E_n had energy proportional to $-1/n^2$, then the differences in energy would account for the Balmer lines that make up hydrogen's spectrum in the wavelength region visible to the eye. (The minus sign indicates that the electron is bound in the atom; it would take the addition of energy to set the electron free.) The energy of an electron jump would be given off as a bundle of energy, a **quantum.** A quantum of light energy is known as a **photon.** The value n is a "quantum number" describing the energy level.

quantum

photon

The reality of photons as particles of light had been demonstrated in 1905 by Einstein in his explanation of the photoelectric effect. Little was known of the effect in 1905, but Einstein's explanation proved valid for the more detailed observations that came later. In the photoelectric effect, light shining on surfaces made of certain materials would cause the surface to build up a certain potential typical of that type of surface. As ꞏe went to shorter wavelengths of incident light, a higher potential was built up. Stronger light did not cause a higher potential. Experimental limitations at that time kept more detailed observations from being made. We now know specifically that stronger light leads to more electrons given off from a surface, but not electrons of higher energy, while shorter-wavelength light leads to the emission of electrons with higher energy.

Einstein suggested that light energy came only in bundles of certain size, quanta. In his model, a brighter beam of light contains more quanta. Each quantum, though, still has the same energy and so can only knock out electrons of the same energy as before. There are more quanta, though, so more electrons are knocked out. The experimental verification of the relation of energy of incoming photons and emitted electrons was not accepted until the work of Robert Millikan in 1915, about the time Bohr presented his hydrogen model.

Fig. 36–2
Niels Bohr and Albert Einstein.

The top curve in Fraunhofer's drawing (Fig. 36–1) shows the distribution of intensity with wavelength. Though Franhofer's distribution was for light from the sun, a similar distribution exists for any hot gas. Radiation from a perfect hot gas is known as "black-body radiation" (Problems 22 to 25). By the time of Einstein's work, the formula for the distribution of black-body radiation with wavelength had been suggested by Max Planck (Fig. 36–3) in 1900, at first without theoretical basis. But the formula explained the curve so well that Planck went on to find a theoretical basis. His derivation dealt with radiation in a "cavity," the inside of a closed container. (One has to make a small hole in a cavity to allow black-body radiation to escape.) Planck wound up requiring the idea that the energy of the atoms oscillating in the walls of a cavity is emitted only in finite bits ("quanta") rather than as a continuous flow. Further, the size of the steps was proportional to frequency f; the constant of proportionality h is now known as **Planck's constant.** The energy of a quantum of light (which we now call a photon) is given by

Fig. 36–3
Max Planck.

Planck's constant

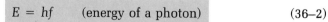

$$E = hf \qquad \text{(energy of a photon)} \qquad (36\text{–}2)$$

where f is the frequency of the light. Einstein's explanation of the photoelectric effect, surprisingly, incorporated Planck's constant. Thus the wave theory and the particle theory of light are linked, given that a simple formula links the wavelength, frequency, and velocity, for any light wave: $\lambda f = c$. Einstein received the 1921 Nobel Prize for his explanation of the photoelectric effect; his special and general theories of relativity were still a little too controversial to be mentioned in the Nobel citation. Planck had received the 1918 Nobel Prize for discovering quanta.

Example 36–1 _____

Energy of a Photon

How much energy in joules and in electron-volts does a photon of sodium D-radiation (wavelength 589.0 nm) have?

Solution

From $\lambda f = c$, we have $f = c/\lambda$, which we can substitute in Equation 36–2 to give

$$E = hf = \frac{hc}{\lambda} = \frac{(6.63 \times 10^{-34} \text{ J·s})(3.0 \times 10^8 \text{ m/s})}{(589.0 \text{ nm})(10^{-9} \text{ m/nm})} = 3.4 \times 10^{-19} \text{ J}.$$

Since from Appendix D, 1 eV = 1.6×10^{-19} J, the photon's energy is

$$E = \frac{3.4 \times 10^{-19} \text{ J}}{1.6 \times 10^{-19} \text{ J/eV}} = 2.1 \text{ eV}.$$

This is a very small amount of energy on a terrestrial scale.

_____

Other series of hydrogen spectra were found in laboratory experiments early in this century. On the basis of the Balmer series, with $n_1 = 2$, and the Paschen series, with $n_1 = 3$, Balmer's formula was generalized:

$$\frac{1}{\lambda} = R\left(-\frac{1}{n_2^2} + \frac{1}{n_1^2}\right), \qquad \text{(Balmer series)} \qquad (36\text{–}3)$$

where R is the *Rydberg constant* and n_2 and n_1 are, respectively, integers describing the upper and lower levels of the transition in the Bohr atom (Fig. 36–4). (R in Equation 36–3 is the value for hydrogen; Problem 3 shows that R has different values for different atoms.) Since $1/\lambda \propto f \propto E$, Balmer's formula simply expresses the law of conservation of energy. In 1916, Theodore Lyman discovered the series for $n_1 = 1$, the Lyman series. Its spectral lines are in the ultraviolet, in a region of the spectrum that does not pass through air.

Example 36–2 _____

Rydberg Atoms

Hydrogen transitions in the optical region of the spectrum are in the Balmer series, but radio astronomers can study higher n_2 and n_1 states of hydrogen. In particular, they now map our galaxy using the 109α line, which is a transition from $n_2 = 110$ to $n_1 = 109$. (It is observed as an emission line; the corresponding absorption line would be at the same wavelength and would represent 109→110.) The transition 272α, the transition from $n_2 = 273$ to $n_1 = 272$ or vice versa, is at one of the longest wavelengths observed. Using the knowledge that the Hα line of the Balmer series is at 656.3 nm, calculate the wavelength and frequency of 272α. Atoms in such high-energy states are known as Rydberg atoms.

Fig. 36–4
Transitions in the Bohr atom, a model that explains hydrogen's spectrum. The energy of an energy level is proportional to $-1/n^2$, where n is its principal quantum number. When an electron jumps from one energy level to another, a photon whose energy is equal to the energy difference between the levels is emitted or absorbed.

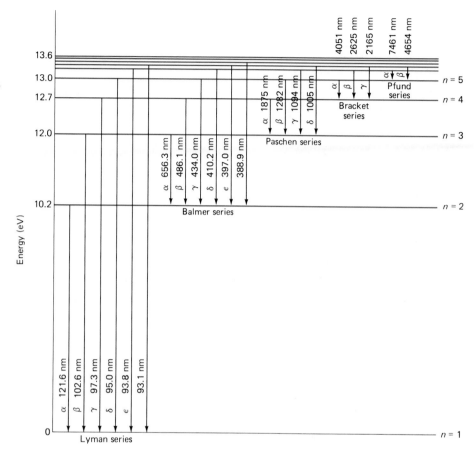

Solution

Inverting Equation 36–3, we solve for R to find

$$R = \frac{1}{\lambda}\,\frac{1}{(-n_2^{-2}+n_1^{-2})} = \frac{1}{656.3\ \text{nm}}\,\frac{1}{(-3^{-2}+2^{-2})} = 1.097 \times 10^{-2}\ \text{nm}^{-1}.$$

Knowing R, we substitute $n_1 = 272$ and $n_2 = 273$ in Equation 36–2 to get

$$\frac{1}{\lambda} = R\left(-\frac{1}{n_2^2} + \frac{1}{n_1^2}\right) = (1.097 \times 10^{-2}\ \text{nm}^{-1})\left(-\frac{1}{273^2} + \frac{1}{272^2}\right)$$

$$= (1.097 \times 10^{-2}\ \text{nm}^{-1})(9.884 \times 10^{-8}) = 1.084 \times 10^{-9}\ \text{nm}^{-1}.$$

Thus $\lambda = (9.22 \times 10^8\ \text{nm})(10^{-7}\ \text{cm/nm}) = 92.2$ cm. The 272α line thus falls in the radio part of the spectrum, with a wavelength just under 1 meter.

Bohr's theory was astonishing, for his assumption that electrons could orbit only in fixed energy levels had no theoretical basis. But it seemed to work, and it gained increasing acceptance in 1914 when an experiment of James Franck and Gustav Hertz showed that electrons could only lose certain discrete amounts of energy when they collided with atoms. Bohr's theory explained hydrogen better than it explained other atoms. Indeed, Bohr

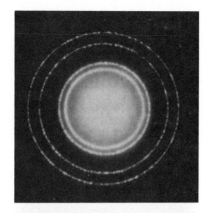

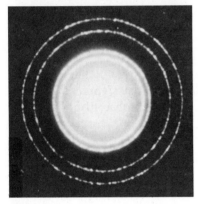

Fig. 36–5
De Broglie's wave theory of matter is proven by the diffraction of an electron beam (above), here shown as the electrons pass through an aluminum-foil target. The pattern is similar to that of diffraction by x-rays (below).

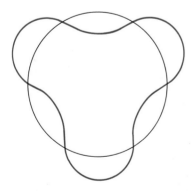

Fig. 36–6
De Broglie showed that the only orbits that exist are those in which an integral number of wavelengths fits.

was tempted not to publish his theory until he had done better for other atoms, but was finally persuaded that he had already made enough of an advance.

Ten years later, in 1923, Louis de Broglie (pronounced "de Broy"), then only a graduate student, generalized the concept that light sometimes had properties of a wave and sometimes had properties of a particle. He suggested that electrons, known to be particles, might also have wave properties. In particular, he proposed that the wavelength of a particle of matter would be linked to its momentum by the same formula that links these quantities for photons: $p = h/\lambda$. His startling new ideas were soon verified experimentally, with an electron of a suitable velocity having the expected very short equivalent wavelength, only about a nanometer. This wavelength is so short that the wave properties of an electron are not ordinarily noticed, but they can be detected (Fig. 36–5). In 1927, diffraction effects were measured for electrons, and diffraction, as we saw in Chapter 35, implies a wave. De Broglie used his idea of matter waves to explain why Bohr's electrons existed only in certain orbits. He suggested that allowed orbits contained standing waves, so that the only orbits that exist are those whose circumference is equal to an integer times the wavelength (Fig. 36–6). De Broglie's ideas explained, finally, why electrons didn't spiral into the nucleus of an atom. After all, on a classical picture, electrons accelerating in circular orbits should radiate, making the atom collapse in a fraction of a second.

Example 36–3
De Broglie Wavelengths

What is the de Broglie wavelength of (a) an electron moving at 1.0×10^6 m/s, and (b) an 0.20-kg baseball pitched at 25 m/s? Relate the results to the 0.1-nm approximate size of a hydrogen atom or to normal terrestrial distances.

Solution

$$\lambda = \frac{h}{p} = \frac{h}{mv,}$$

so

(a) For the electron,

$$\lambda = \frac{(6.6 \times 10^{-34} \text{ J·s})}{(9.1 \times 10^{-31} \text{ kg})(1.0 \times 10^6 \text{ m/s})}$$

$$= 7.3 \times 10^{-10} \text{ m} = 7.3 \times 10^{-1} \text{ nm} = 0.73 \text{ nm,}$$

a few times larger than a hydrogen atom. The de Broglie wavelengths of electrons are sufficiently close to atomic scale that they can be detected by interference effects caused by atoms.

(b) For the baseball,

$$\lambda = \frac{(6.6 \times 10^{-34} \text{ J·s})}{(0.20 \text{ kg})(25 \text{ m/s})} = 1.3 \times 10^{-34} \text{ m.}$$

This distance is so many orders of magnitude smaller than we are able to see that it is irrelevant in a game of baseball. It is even so much smaller than atomic scale that we cannot detect the wave properties of macroscopic objects like baseballs.

36.2
QUANTUM MECHANICS

Schrödinger equation
wave mechanics
wave function

In 1926,* Erwin Schrödinger (Fig. 36–7) succeeded in giving a mathematical theory that did for de Broglie's electron waves what Maxwell had done for Thomas Young's light waves. The **Schrödinger equation,** which describes **wave mechanics,** is written in terms of a **wave function** (usually called ψ, the Greek letter psi) of position and time. This second-order differential equation is not easy to solve; indeed, it can be solved exactly only for the one-particle case except for a few specific multiparticle models. Even for the one-particle case, solutions turn out to exist only for certain discrete values of the energy. The wave function is related to a **probability amplitude,** meaning that (as Max Born later discovered) the square of the wave function ($|\psi^2|$) gives the probability that a particle will be found in any unit volume of space. (The wave function ψ can be a complex number, so we take the absolute value of its square to get a positive number, signified by the vertical bars.) Thus an important, nonintuitive idea arose: An electron is neither here nor there; we can only say that it has a certain probability of being here and a certain probability of being there. Schrödinger's model of the hydrogen atom, then, shows the probability of an electron's being at certain locations (Fig. 36–8). The discrete orbits of the old Bohr atom correspond, on Schrödinger's model, to regions of relatively high probability.

probability amplitude

Fig. 36–7
Erwin Schrödinger.

Einstein and some others did not like these probabilistic ideas; as Einstein wrote Born in 1926, "The theory says a lot, but does not really bring us any closer to the secret of the Old One. I, in any case, am convinced that He does not throw dice." The quotation—or perhaps other similar comments that Einstein made later—is often paraphrased, "I cannot believe that God plays dice with the universe." In any case, years of experimentation leave little doubt on the correctness of quantum mechanics (Table 36–1). Einstein's instincts, in this case, were apparently wrong.

Slightly earlier, in 1925, Werner Heisenberg (Fig. 36–9) had presented another way of calculating the energy of an electron. His method represented variables such as momentum and position as matrices. (Matrices are mathematical devices; in this case, they linked numbers representing the initial and final states.) Heisenberg's theory was thus known as **matrix mechanics.** Though the mathematics of wave mechanics and matrix mechanics looked very different, they were soon proved to be identical. Heisenberg's formulation emphasized the role played by an observer in measuring quantities.

Fig. 36–8
A sample probability distribution of finding an electron at a given position, according to Schrödinger's model of hydrogen. Light areas correspond to regions of higher probability.

Heisenberg's Uncertainty Principle

Many of the ideas of quantum mechanics were not mere mathematical or physical results; some challenged our basic notions of how we think about the universe. Indeed, quantum mechanics contains the idea that the

---

*We refer here to the dates of discovery, not the dates of publication.

TABLE 36–1 ▰▰▰▰

Some Dates in Atomic Physics

Date	Recipient
1900	Max Planck—Discovery of quanta of energy (1918)
1905	Albert Einstein—Theory of the photoelectric effect (1921)
1913	Niels Bohr—Theory of atomic structure (1922)
1914	James Franck and Gustav Hertz—Experimental work on electron-atom collisions (1925)
1915	Robert A. Millikan—Experimental work on the charge on an electron and on the photoelectric effect (1923)
1923	Arthur Holly Compton—Wavelength change of x-ray photons in an electron collision (1927)
1923	Louis-Victor de Broglie—Theory of the wave nature of electrons (1929)
1924	Wolfgang Pauli—The exclusion principle (1945)
1925	Walther Bothe—Experiments on the Compton effect (1954)
1925	Werner Heisenberg—Theory of matrix mechanics (1932)
1926	Erwin Schrödinger—Theory of wave mechanics (1933)
1926	Max Born—Statistical study of wave functions (1954)
1927	Paul A. M. Dirac— Theory of relativistic quantum mechanics (1933)
1930	S. Chandrasekhar—Quantum-mechanical explanation of white dwarf stars (1983)
1931	Ernst Ruske—Invention of the electron microscope (1986)
1948	Sin-Itiro Tomonaga, Julian Schwinger, and Richard P. Feynman—Theory of quantum electrodynamics (1965)

Year that Nobel Prize was awarded is in parentheses.

act of observing a physical system can play a role in determining its condition.

Heisenberg's concentration on observable quantities led to such a role in quantum mechanics for the inherent uncertainties in measurements. In 1927, Heisenberg presented his **uncertainty principle,** which stated that certain related quantities could not both be known with unrestricted accuracy. (Heisenberg's German word can also be called "indeterminacy," which perhaps makes it clearer that a range of values is possible.) In particular, if Δx is the uncertainty in position and Δp is the uncertainty in momentum in the x-direction,

uncertainty principle

Fig. 36–9
Werner Heisenberg.

$$\Delta x \Delta p \geq \frac{h}{2\pi,} \tag{36–4}$$

where h is Planck's constant. Further, if Δt is the least length of time it takes to measure an energy to accuracy ΔE,

$$\Delta E \Delta t \geq \frac{h}{2\pi.} \tag{36–5}$$

These uncertainties are fundamental limits to the precision with which we know the world. Heisenberg's uncertainty principle is a consequence of the wave nature of electrons and other matter.

Heisenberg formulated his quantum-mechanical theory with an emphasis on **observables,** quantities that are determined by experiments. We can illustrate the nature of Heisenberg's uncertainty principle by considering one such experiment: how we locate the position of a particle. Heisenberg suggested a "gedanken" (thought) experiment, with a special kind of microscope that could locate an electron using a single photon. Inside the microscope, the diffraction of the light used to illuminate the electron would lead to an uncertainty in the electron's position. So to locate the electron

observables

more precisely, we would have to use light of shorter wavelength. But shorter-wavelength light corresponds to photons of higher energy and momentum. When this light hits the electron, it changes its momentum, and we don't know the photon's final momentum to allow us to know how much momentum the photon lost. So higher positional accuracy for the electron corresponds to lower accuracy in momentum, as in Equation 36–4.

Example 36–4

Heisenberg's Uncertainty Principle

If we measure the momentum of (a) a 0.20-kg baseball travelling at 40 m/s to an accuracy of 0.1 per cent, or (b) an electron travelling at 1.0×10^6 m/s, what are the fundamental limits on the precision to which we can measure their positions?

Solution

From Heisenberg's uncertainty principle,

$$\Delta x \Delta p \geq \frac{h}{2\pi}$$

so

$$\Delta x \geq \frac{h}{2\pi \Delta p.}$$

(a) For the baseball, the momentum $p = (0.20 \text{ kg})(40 \text{ m/s}) = 8$ kg·m/s, and so $\Delta p = 0.001p = 0.008$ m·kg/s. Thus

$$\Delta x \geq \frac{(6.63 \times 10^{-34} \text{ J·s})}{(2\pi)(8 \times 10^{-3} \text{ kg·m/s})} = 1.3 \times 10^{-32} \text{ m.}$$

The uncertainty in position for macroscopic objects that results from Heisenberg's uncertainty principle is much too small to measure. (b) For the electron, the momentum $p = (9.1 \times 10^{-31} \text{ kg})(1.0 \times 10^6 \text{ m/s}) = 9.1 \times 10^{-25}$ kg·m/s, and $\Delta p = 9.1 \times 10^{-28}$ kg·m/s. Thus $\Delta x \geq (6.63 \times 10^{-34} \text{ J·s})/(2\pi)(9.1 \times 10^{-28} \text{ kg·m/s}) = 1.2 \times 10^{-1}$ m, many times larger than the size of an atom. We would have to settle for much less precision in an electron's momentum to localize its position to an atomic scale.

Pauli's Exclusion Principle and Electron Spin

quantum number

We have seen how different states of the atom in Bohr's model could be distinguished by the values of n (which is always an integer); a quantity like n that distinguishes states is known as a **quantum number.** Three quantum numbers were known at the time of the work of Schrödinger and Heisenberg: n for the principal quantum number that determines the energy of an electron, ℓ for the orbital angular momentum, and m, which was subsequently identified with the orbit's orientation in space. Detailed observation of the hydrogen spectrum had revealed that hydrogen's spectral lines were split in two—they were "doublets." Also in 1925, Wolfgang Pauli (Fig. 36–10) suggested that a fourth quantum number was necessary to explain this splitting. He had a year earlier empirically advanced the

exclusion principle

exclusion principle, which states in general that no two electrons can be in exactly the same state, and then reformulated it based on quantum me-

Fig. 36–10
(Left) Wolfgang Pauli and Bohr looking at a special kind of spinning top whose motion elucidates certain concepts of spin. (Right) George Uhlenbeck (left) and Samuel Goudsmit (right), discoverers of electron spin, with Hendrik A. Kramers between them.

chanics. The existence of the fourth quantum number allowed a second electron to exist with the same n, ℓ, and m quantum numbers.

At the end of that magic year 1925, Samuel Goudsmit and George Uhlenbeck (Fig. 36–10) realized that the splitting could be explained by an electron having an intrinsic spin that could be identified with this new quantum number, which we now call s for spin. The spin angular momentum could have only two values: $\pm\frac{1}{2}\hbar$, where $\hbar = h/2\pi$, giving $s = \pm\frac{1}{2}$. The fact that the spin's values are not integral (compared with the angular orbital momentum, which is always an integral multiple of $\hbar$) shows that the electron is not really a spinning ball, but the classical analogy nonetheless remains useful for visualizing the system.

We can understand how the existence of the spin quantum number allows two electrons to exist on the n, ℓ, and m level: one of the electrons has spin up ($s = \frac{1}{2}$) and the other has spin down ($s = -\frac{1}{2}$), satisfying the exclusion principle. Because of the exclusion principle, a third electron in an atom must go into a state that has at least one different value for n, ℓ, or m.

In the $n = 1$ state, ℓ and m can only be 0. (Though classically, the angular momentum of such a moving entity would not be zero, no actual orbiting electron is present in quantum mechanics and 0 is an acceptable value for the orbital angular momentum ℓ.) The $n = 2$ state allows the electron to have discrete, that is, quantized, amounts of orbital angular momentum ℓ. Values of ℓ for $n = 2$ can be 0 or 1; in general $0 < \ell < (n-1)$. Spectroscopists have a series of letters to give the value of ℓ: s for $\ell = 0$, p for $\ell = 1$, d for $\ell = 2$, f for $\ell = 3$ (based on observed series of spectral lines that had been known, respectively, as sharp, principal, diffuse, and fundamental), and then alphabetically with g and so on. So the lowest energy level of hydrogen, $n = 1$, has only $\ell = 0$, making the sole electron in the "configuration" be in the $1s$ state: the 1 is the value of n, and the s stands for $\ell = 0$. (Helium has the ground configuration $1s^2$, since there can be two $1s$ electrons, each with a different spin; the superscript 2 indicates the two $\ell = 0$ electrons.) If an electron goes to the higher-energy state with $n = 2$ and $\ell = 1$, the configuration will be $2p$. The value of angular momentum affects the shape of the probability distribution of electrons (Fig. 36–11), but does not affect the value of energy, similarly to the way that the energy of a planet in its orbit depends only on the semimajor axis of its orbit and not on its orbital eccentricity.

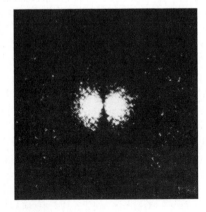

Fig. 36–11
Probability distributions of finding an electron in a hydrogen atom for which the orbital angular momentum ℓ is not zero.

Fig. 36–12
An energy-level diagram for hydrogen, showing that selection rules allow only certain transitions.

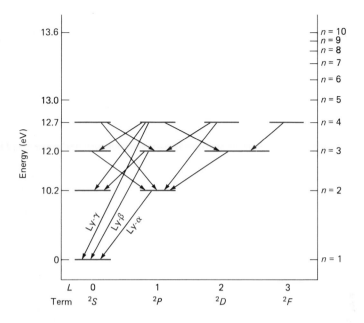

selection rules
allowed

forbidden

electron shell

In quantum mechanics, certain **selection rules** are predicted that specify which transitions between energy levels are **allowed.** For example, $\Delta\ell = \pm 1$ is a selection rule (Fig. 36–12). Transitions that are not allowed are called **forbidden;** forbidden transitions actually can occur, but are relatively rare. Still, they are seen; the spectral lines we detect in the visible from the solar corona and from gaseous nebulae are forbidden (Fig. 36–13).

The total number of states that can exist in a multielectron atom for $n = 2$ is 8: $1s^2 2s^2 2p^6$. A second **electron shell** forms around the shell consisting of the two $n = 1$ electrons. Beyond 8 additional electrons, electrons must go into a third shell. Filled shells are stable, and do not easily undergo chemical reactions; they are found in the noble gases (helium, neon, argon, etc.). Atoms missing one or two electrons from a shell combine eas-

Fig. 36–13
The spectrum of the 2,000,000 K solar corona shows certain "forbidden lines" from highly ionized states of iron and other elements (arrow). The presence of this forbidden radiation reveals the corona's low density, only about 10^9 particles per cubic centimeter since atoms would be deexcited by collisions before they could radiate forbidden lines in a higher density gas. Also visible in this spectrum, and indeed brighter and more obvious, are lines emitted from lower-temperature gas in the solar chromosphere and prominences.

ily with atoms having one or two electrons in their outermost shell. Therefore the exclusion principle is at the basis of chemical structure, and explains the grouping of the elements in the periodic table (Appendix E).

Wave-Particle Duality

principle of complementarity

How do we deal with the idea that light and matter can be sometimes a wave and sometimes a particle? Bohr dealt with this conundrum with his **principle of complementarity,** that the wave picture and the particle picture are complementary. He pointed out that we design experiments to find out, for example, whether light is waves or particles, and that the experiment is different in each case. He showed that being waves or particles is not a property of light itself but rather of light in an experimental situation that we create with waves or particles specifically in mind. So we will never find the notion of light as a particle in conflict with the notion of light as a wave, and cannot ever say that light is one or the other.

Fig. 36–14
P. A. M. Dirac.

Relativistic Quantum Mechanics

positron

antimatter

In 1927, Paul A. M. Dirac (Fig. 36–14) reformulated quantum mechanics to make it consistent with the special theory of relativity. His interpretation of his theory predicted the existence of a particle with the same properties as an electron except that its charge would be positive instead of negative. This **positron** was discovered experimentally in 1932 (Fig. 36–15), verifying Dirac's ideas. The positron was the first known case of **antimatter.** Nonrelativistic quantum mechanics allows us to compute accurately the spectra of atoms, of the interaction between electrons and atoms, and of chemical bonds. Because the associated energies are ordinarily low, Dirac's theory is rarely used in atomic physics.

Fig. 36–15
The cloud-chamber photograph on which the positron was discovered by C. D. Anderson in 1932.

Application _____

Masers and Lasers

stimulated emission

Work on quantum-mechanical ideas has been steady in the decades since the pioneering work of the 1920's and 30's, and we cannot in limited space do it justice. Certainly the development of masers and lasers followed from the particle picture of light. Even the names maser ("<u>m</u>icrowave <u>a</u>mplification by <u>s</u>timulated <u>e</u>mission of <u>r</u>adiation") and laser ("<u>l</u>ight <u>a</u>mplification by <u>s</u>timulated <u>e</u>mission of <u>r</u>adiation") refer to the particle picture in the words "stimulated emission." In particular, in discussing the particle nature of light, there is a certain probability that an atom in one state will spontaneously change to a lower-energy state, giving off a photon. It will do so whether or not any light is shining on it. There is also a certain probability that an atom in one state will gain energy from light shining on it. Einstein showed in 1917 that there is a third possibility (Fig. 36–16): **stimulated emission,**

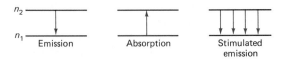

Fig. 36–16
Einstein showed in 1917 that atomic transitions included emission, absorption, and stimulated emission (that is, stimulated by an incident photon).

Fig. 36–17
Theodore Maiman, who made the first working laser, with his device.

Fig. 36–17
Theodore Maiman, who made the first working laser, with his device.

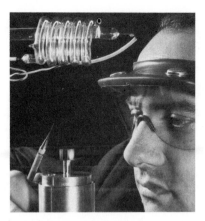

Fig. 36–18
The NOVA laser at the Lawrence Livermore Laboratory is being used to test methods of fusing deuterium and tritium in small pellets. Here we see the target area, where multiple laser beams converge.

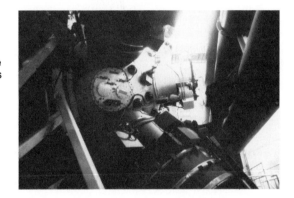

in which light of a suitable wavelength stimulates atoms to change state by the amount that gives off light of that wavelength .

maser

laser

The process of stimulated emission was first exploited in the late 1940's by Charles Townes in the United States and by Soviet scientists to make a **maser.** Later, Townes and Arthur Schawlow worked out ways to have a similar process of stimulated emission affect light, making a **laser** (Fig. 36–17). Laser light can be made very strong, and is monochromatic and coherent. It has thus found many practical uses (Fig. 36–18).

QED

quantum electrodynamics (QED)

Feynman diagram

Another major highpoint in the development of quantum mechanics was the development in 1948 by Sin-Itiro Tomonaga, Julian Schwinger, and Richard P. Feynman of **quantum electrodynamics (QED)** to explain electricity and magnetism in quantum form (Fig. 36–19). QED had been begun by Dirac in nonrelativistic form, but certain infinities occurred in the equations. Tomonaga, Schwinger, and Feynman worked out ways of dealing with the infinities in relativistic calculations. Feynman worked out methods of visualizing the exchange of particles that, QED holds, takes place in all reactions. The **Feynman diagram** in Fig. 36–20 shows how the electric force is carried in electron-proton scattering. The photon is the carrier of the electromagnetic force, so a photon travels between the parti-

Fig. 36–19
(Left) Richard P. Feynman (at right), who discovered how to deal with the infinities in quantum electrodynamics, with H. Yukawa (at left), discoverer of the meson. Feynman, Sin-Itiro Tomonaga (center), and Julian Schwinger (right) independently developed quantum electrodynamics.

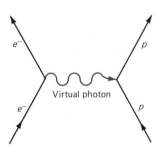
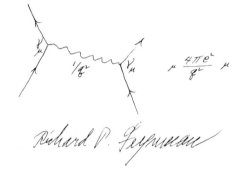

Fig. 36–20
(Left) A sample Feynman diagram, showing electron-proton scattering with the exchange of a virtual photon. (Right) A Feynman diagram in Feynman's own hand. It shows the interaction between any two charged particles.

virtual photon

cles. Since the photon cannot be seen, it is a **virtual photon,** with the word "virtual" being used similarly to its use in virtual images in optics to indicate an image that is not physically present. Experiments have verified QED to extremely high accuracy, making it our best-verified physical theory.

Fig. 36–21
Degenerate matter: the white-dwarf star Sirius B is visible as a faint dot alongside the much brighter star Sirius A; because of the imaging technique, fainter images of Sirius A appear along a horizontal line. White dwarfs are small and faint, and are thus hard to observe. Ultraviolet observations from space confirm that their surface temperatures are of the order of 100,000 K, substantially hotter than normal stars.

Application

Quantum Mechanics and White Dwarf Stars

One large-scale object that can be explained only by the theory of quantum mechanics is a white-dwarf star, a star containing the mass of the sun that has stopped its internal nuclear fusion and shrunk dramatically. Eventually, as was worked out by S. Chandrasekhar in 1930, the star becomes so small that the exclusion principle dominates. The resistance of electrons to being further compressed because of the exclusion principle creates a pressure that counterbalances gravity. When this pressure is present, we say that the matter is "degenerate." White dwarfs are thus in a stable state, in that they do not ever change into a different state; they are degenerate stars, with about as much mass as the sun in a volume equal to that of the earth (Fig. 36–21). We expect our sun to meet this fate one day, in about 5 billion years.

Fig. 36–22
Ernest Rutherford.

36.3
THE STRUCTURE OF MATTER

The quest for the smallest particles of matter has been going on since at least the work of Democritus in ancient Greece two thousand years ago. The Roman philosopher Lucretius reported on this notion of atoms in a long poem in the first century B.C. The notion of an atom—a smallest particle with certain chemical properties—persists, and was given more modern form by John Dalton in 1808, but we can now delve deeply inside individual atoms.

Though the electron was found by J. J. Thomson in 1897, it was not known how electrons and the equivalent positive charge were distributed in atoms. Thomson's own model had the electrons distributed through positive charge that was spread evenly throughout individual atoms, the **plum-pudding model.** But experiments in 1909 to 1911 suggested by Ernest Rutherford (Fig. 36–22) changed this picture. Hans Geiger and his undergraduate assistant Ernest Marsden were shooting alpha particles into thin sheets of gold. They thought—and we have since confirmed—that alpha particles were helium atoms that had lost their electrons. They expected the alpha particles to pass through the gold atoms or to be somewhat deflected. Surprisingly, though most of the alpha particles passed through, a few were bounced back in nearly the direction from which they came. It was, said Rutherford, as though you shot an artillery shell at a piece of tissue paper and had the shell bounce back. Apparently, Rutherford reasoned, the atom must have a small, very massive core: the **nucleus.** In Rutherford's **nuclear model,** the postively charged nucleus was surrounded by an equivalent negative charge.

But the mutual repulsive force of many positive charges in the nucleus would be strong. By 1920, Rutherford suggested that there were other particles in the nucleus in addition to protons. Such a particle, the **neutron,** was discovered by James Chadwick in 1932.

Still, known nuclear reactions would not conserve energy or angular momentum. In 1930, Pauli thus suggested the existence of another particle, very nonreactive, that escapes easily from a nucleus and even travels through the surrounding apparatus without being observed. It is now called the **neutrino** ("little neutral one"). It would have the necessary energy and momentum to make the reactions obey the expected conservation laws. It has generally been thought that the neutrino would have no rest mass, which would allow the neutrino to travel at the speed of light. It was not until 1957 that neutrinos—actually antineutrinos—were discovered, since neutrinos interact so weakly with matter.

There is some possibility that neutrinos do have a very small rest mass. If so, they may even make up most of the matter in the universe. The experiments that indicated that neutrinos have a small rest mass have not been confirmed, though.

The proton and the neutron both follow Pauli's exclusion principle. Thus, similarly to the way the shells of electrons are understood, physicists began to understand the arrangement of particles in nuclei. Just as filled electron shells make elements that do not easily interact chemically, filled

plum-pudding model

nucleus
nuclear model

neutron

neutrino

shell model
liquid-drop model

strong force

binding energy

radioactivity

nuclear shells make nuclei that do not react readily. Maria Goeppert Mayer (Fig. 36–23) and J. Hans Jensen worked out many of the details of this **shell model.** An alternative theory of nuclei treats the nuclear particles as we would treat a liquid in the **liquid-drop model.** Often we must consider both shell model and liquid-drop model simultaneously to interpret nuclear experiments.

We have seen that nuclei are held together by the **strong force.** To pull nuclear particles apart, we must do work. The amount of work we must do to separate a nucleus into its component protons and neutrons is the nucleus's **binding energy.** Some nuclei are much more stable than others, in that they have higher binding energies (Fig. 36–24).

As we all know in this nuclear age, some nuclei are not stable. They spontaneously emit particles, making **radioactivity,** a phenomenon discovered by Henri Becquerel in 1896. He had noticed that nearby uranium salt had fogged a wrapped photographic plate; some scientists would have thrown away the fogged plate but Becquerel interpreted the phenomenon. His student Marie Curie and her husband Pierrre Curie (Fig. 36–25) went on to discover the radioactive elements radium and polonium. Later Mme Curie isolated radium.

Fig. 36–23
Maria Goeppert Mayer with her daughter.

Fig. 36–24
The higher the mass excess, the less stable a nucleus is. The most stable nuclei are those of isotopes in the "valley of stability."

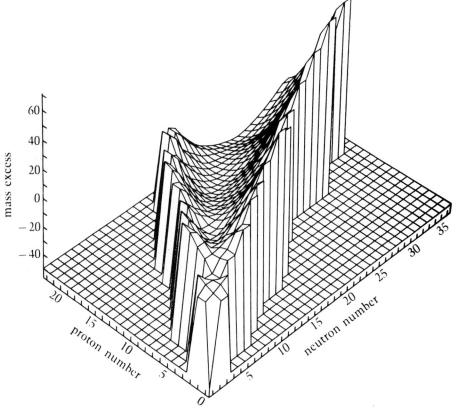

Fig. 36–25
Marie Curie and Pierre Curie.

fission   It was not realized for many years that some elements could spontaneously split—**fission.** The fact that uranium could fission was discovered by Lise Meitner and Otto Frisch in 1939, as they interpreted the experimental results of Otto Hahn and Fritz Strassmann the previous year (Fig. 36–26). Their work, coming at a dismal time in world history with the rise of Hitler in Germany, led to the Manhattan Project for the construction of the atomic bomb. Even Einstein, with his pacifistic tendencies, thought that there was no choice about developing the bomb, for if we didn't do so, the Germans might.

Radioactivity has many peacetime uses. Techniques for using radioactive isotopes as tracers in the human body won a Nobel Prize in Physiology or Medicine for Rosalyn Yalow (Fig. 36–27) in 1977. Radioactivity in car-

Fig. 36–26
Lise Meitner and Otto Hahn. After Hahn and Fritz Strassmann discovered experimentally that the light element barium was present in the decay of uranium, Meitner and her nephew Otto Frisch interpreted the result as a fissioning of the uranium.

Fig. 36–27
Rosalyn Yalow, who received the Nobel Prize in Physiology or Medicine for her work with radioisotopes in the human body.

bon and other atoms is used for dating wood or fossils up to forty thousand years old. Radioactivity in still other atoms can provide dates even farther back.

The history of fission, fusion, and radioactivity is interesting and complicated, and we shall leave it for another course. Here we instead delve deeper into the structure of matter.

36.4
SUBNUCLEAR PARTICLES

By 1932, there were three nuclear particles known: the proton, the neutron, and the neutrino. They were the "elementary particles." Three years later, Hideki Yukawa reasoned that just as the electromagnetic interaction is carried by a particle (the photon), the strong interaction should also be carried by a particle. The more massive this particle, the shorter its range; since the strong force is limited to a nuclear range, Yukawa's particle must lie between the electron and proton in mass. It was called a "meson," from the Greek for "in between." Yukawa's meson was discovered in 1946 in cosmic rays. We now know of many kinds of mesons with different masses.

In the following decades, many additional "elementary particles" were discovered, until the list was over 100. So many particles couldn't be elementary. Various schemes were proposed for classifying them and for predicting the masses of particles yet to be found. Predicting the masses is important, for these masses have equivalent energies $E = mc^2$ and can be created in collisions in particle accelerators if enough energy is made available. Thus high-energy physicists keep pushing for accelerators of higher and higher energies.

Some of the particles did not act as predicted. For example, some lived longer than expected before disintegrating. These were called "strange particles," and were said to have the property "strangeness." A few particles did not seem to decay as quickly as expected. T. D. Lee and C. N. Yang (Fig. 36–28) in 1956 could explain the time scale of these decays if **parity**— that the laws of physics should be the same if reflected in a mirror—was not conserved. Chien Shiung Wu (Fig. 36–29) and her co-workers demonstrated experimentally that parity is not conserved.

In 1964, Murray Gell-Mann and, independently, George Zweig suggested a set of three particles that could be combined in different ways to

Fig. 36–28
T. D. Lee (above) and C. N. Yang (below), who explained theoretically that parity need not be conserved.

parity

Fig. 36–29
Chien Shiung Wu, who demonstrated experimentally that parity is not conserved in certain elementary-particle decays.

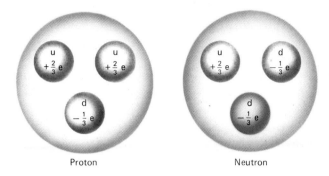

Fig. 36–30
Making baryons out of combinations of quarks.

quarks

form the known nuclear particles. Gell-Mann called these particles **quarks,** from a quotation in James Joyce's *Finnegan's Wake.* One surprising thing about quarks is that they had fractional charge. Two seemed particularly basic: the "up" quark (u) with charge $+\frac{2}{3}e$ and the "down" quark (d) with charge $-\frac{1}{3}e$, where e is the charge on an electron. One of the rules seemed to be that quarks could be combined only in ways in which no fractional charge showed. For example, two up quarks plus one down quark (charge $+\frac{2}{3}e+\frac{2}{3}e-\frac{1}{3}e = +1e$) make a proton (charge $+1e$). Two down quarks plus one up quark (charge $-\frac{1}{3}e-\frac{1}{3}e + \frac{2}{3}e$) make a neutron (charge 0). Other combinations of these quarks and of the third quark, the "'strange" quark (charge $-\frac{1}{3}e$), could account for many of the nuclear particles (Fig. 36–30).

baryons

In particular, all the particles known as **baryons,** which include the proton and the neutron, are made of three quarks. All the mesons contain a quark and its antiquark (Fig. 36–31). A third class of particle—the **leptons**—including the electron and particles called a muon and a tau, may be elementary particles themselves.

leptons

Fig. 36–31
Making mesons out of quark-antiquark combinations. Here we see the charmed quark c and its antimatter counterpart c̄.

As accelerators of higher and higher energy were built, more and more "elementary particles" were discovered. Too many existed, and with particular properties, to be formed out of the up, down, and strange quarks. New quarks had to exist to account for them. In particular, the discovery of the particle that implied the existence of the fourth quark, known as the "charmed" quark, was a triumph of the theory and of its extension by the theorist Sheldon Glashow. The nearly simultaneous discovery of the particle implying charm, called J by Samuel C. C. Ting and his MIT/Brookhaven group and ψ by Burton Richter and his group at Stanford, was particularly noteworthy (Fig. 36–32). The difference in names has never been resolved, and the particle is still often called J/ψ by outsiders.

Fig. 36–32
(Left) The experimental apparatus at Brookhaven National Laboratory with which the particle generally known as J/ψ was discovered. (Center) Burton Richter at Stanford. (Right) Samuel C. C. Ting at Brookhaven.

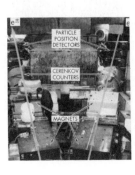

TABLE 36–2
Quarks and Leptons

Name	Symbol	Spin	Charge	Corresponding Leptons
up	u	$\frac{1}{2}$	$+\frac{2}{3}e$	electron (e), electron neutrino (ν_e)
down	d	$\frac{1}{2}$	$-\frac{1}{3}e$	
strange	s	$\frac{1}{2}$	$-\frac{1}{3}e$	muon (μ), muon neutrino (ν_μ)
charmed	c	$\frac{1}{2}$	$+\frac{2}{3}e$	
truth	t	$\frac{1}{2}$	$+\frac{2}{3}e$	tau (τ), tau neutrino (ν_τ)
beauty	b	$\frac{1}{2}$	$-\frac{1}{3}e$	

Two more quarks have since had to be invoked, the "top" (or "truth") quark and the "bottom" (or "beauty") quark. Obviously, the names do not imply the meaning of the words we are used to on a terrestrial scale.

We now realize that quarks come in pairs, and each pair corresponds to a lepton and its associated neutrino (Table 36–2).

color

The Pauli exclusion principle does not allow three particles in the same state, so a new quantum number for quarks must exist. It is known as **color**; the three basic colors are known as red, green, and blue. Each of the three quarks in each baryon has a different color, making the baryon colorless. The force holding quarks together is known as the **color force**, and the theory is thus known as **quantum chromodynamics (QCD)**. QCD is our current theory of matter. In QCD, the particles mediating the color force are known as **gluons**, for they provide the glue that holds elementary particles together.

color force

quantum chromodynamics (QCD)

gluons

To find new particles, elementary-particle physicists need higher and higher energies. Among the accelerators with the highest energies now available are the Fermi National Accelerator Laboratory near Chicago (Fig. 36–33) and CERN, the Centre Européen de Recherche Nucleaire, on the France-Switzerland border. The higher the energy reached, the higher the magnetic field necessary to bend the particles into a given circular orbit. So to reach higher energies, one either has to provide stronger magnets or to build a bigger accelerator ring to provide a more gentle curve. The current

Fig. 36–33
The upper level contains magnets that bend the beam in the main ring of Fermilab; the lower level contains a more recent circle of magnets that constrain a second beam moving in the opposite direction. Allowing these oppositely circulating beams to collide permits physicists to reach higher energies than formerly possible.

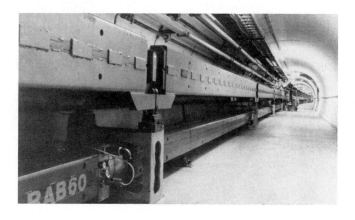

Fig. 36–34
Steven Weinberg, Abdus Salam, and Sheldon Glashow (left to right), who worked out the unification of the electromagnetic force and the weak force.

proposal for a 100-km-diameter Superconducting Super Collider in the United States is under serious discussion, as are projects for huge accelerators in Europe and elsewhere.

36.5
UNIFICATION AND COSMOLOGY

In Chapter 4, we discussed the fundamental forces and some theories about unifying them. In particular, the theory of Steven Weinberg, Abdus Salam, and Sheldon Glashow (Fig. 36–34) predicted the existence of a family of particles to be called W^+, Z°, and W^-; these particles carry ("mediate") the electroweak force. The discovery of the W and Z particles by a huge international consortium headed by Carlo Rubbia, using advances in accelerator physics by Simon van der Meer (Fig. 36–35), seems to have established a unification of the electromagnetic force with the weak force.

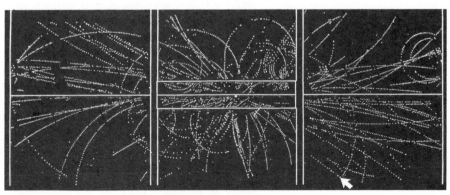

Fig. 36–35
(Left) Carlo Rubbia with the apparatus at CERN with which he and Simon van der Meer discovered the W and Z particles. (Right) The computer output shows the discovery of the first W particle; the arrow at lower right marks a high-momentum decay product. The other tracks also result from the W's decay.

Fig. 36–36
A proton-decay experiment, showing the room prepared to receive 4000 tons of water. No decays have yet been observed, in spite of several ongoing experiments.

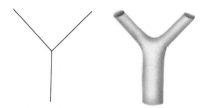

Fig. 36–37
At left, we see an interaction shown as a Feynman diagram for point particles. At right, we see the same interaction shown for superstrings; the path of a superstring through space-time is a surface. Note that at right there is no single point in space-time at which the interaction takes place.

superstring

compactification

TABLE 36–3
Nobel Prizes for Interpreting the Structure of Matter

Date	Nobel Prize Recipient
1903	Antoine H. Becquerel—Discovery of radioactivity (1896)
	Pierre Curie and Marie Curie—Research in radioactivity
1906	J. J. Thomson—Discovery of the electron and other studies
1908	Ernest Rutherford—Investigation of the disintegration of the elements
1935	James Chadwick—Discovery of the neutron (1932)
1936	Victor F. Hess—Discovery of cosmic rays (1911)
	Carl D. Anderson—Discovery of the positron (1932)
1938	Enrico Fermi—Production of new radioactive elements (1934–1937)
1949	Hideki Yukawa—Theoretical prediction of mesons (1934)
1951	John D. Cockroft and Ernest T. S. Walton—Transmutation of nuclei (1932)
1957	C. N. Yang and T. D. Lee—Theoretical prediction of parity violation (1956)
1959	Emilio G. Segrè and Owen Chamberlain—Production of the antiproton (1955)
1963	Eugene P. Wigner—Application of symmetry to elementary-particle theory
	Maria Goeppert Mayer and J. Hans D. Jensen—Nuclear shell structure (1949)
1968	Luis Alvarez—Contributions to particle accelerators (1946)
1969	Murray Gell-Mann—Theory of classification of elementary particles (1964)
1975	Aage Bohr, Ben Mottelson, James Rainwater—Collective and particle motion in nuclei
1976	Burton Richter and Samuel C. C. Ting—Discovery of J/ψ particle, first charmed particle (1973–1974)
1979	Sheldon L. Glashow, Abdus Salam, and Steven Weinberg—unification of the electromagnetic and weak forces (1961–1972)
1980	Val L. Fitch and James W. Cronin—Experimental proof of violation of conservation of CP (charge-parity) (1964)
1984	Carlo Rubbia and Simon van der Meer—Discovery of W particles, predicted by the electroweak force (1983); stochastic cooling for focusing beams (1968)

Year of discovery appears in parentheses.

A further step, **grand-unified theories** (GUT's), try to unify the electroweak force with the strong force. A prediction of grand unification is that the proton is not stable, that it should decay on some very long time scale. The simplest version of the theory says that that time scale may be 10^{31} years. We cannot wait that long, but we can put 10^{33} protons in one place, in the form of the hydrogen in 4000 tons of water molecules, and watch for one year. This experiment is being carried out (Fig. 36–36); no proton decays have been discovered as yet. The simplest version of GUT's seems to have been ruled out, but there are more complex versions that remain viable. Still, it had been hoped that proton decay would have been seen by now. Several proton-decay experiments are now under way. Two other GUT's predictions—that magnetic monopoles exist and that neutrinos have rest mass—are also being investigated. (See Table 36–3 for a summary of investigations of the structure of matter.)

A still newer idea now under intense investigation by many theorists is the notion that what we have been observing as 0-dimensional elementary particles are really tiny 1-dimensional **superstrings** only 10^{-35} m long (Fig. 36–37). In some cases, the superstrings form closed loops. The particles, in this model, correspond to fundamental modes of vibrations of the superstrings. The model is set not in the four-dimensional space-time to which we have become accustomed, but rather in 10-dimensional space-time. To account for why we see only four dimensions, the theory holds that, essentially, the remaining six dimensions are rolled up into a little ball. This **compactification** implies the existence of such little balls containing the extra dimensions at each point of space-time.

The Expansion of the Universe

These experimental and theoretical studies of the smallest units of our universe—elementary particles—are increasingly being linked with studies of the universe as a whole. It has been known since the work of Edwin Hubble in 1929 (Fig. 36–38) that the universe is expanding. Hubble painstakingly observed the Doppler shifts of many galaxies, and discovered that (except for a few nearby ones) they were all receding from us. That is, their spectra are shifted to longer wavelengths, "redshifted." Further, he established what we now know as **Hubble's law,**

Hubble's law

$$v = H_0 d,$$

Hubble constant

where v is the velocity of recession, d is the distance, and H_0 is the symbol we now use for the **Hubble constant** of proportionality. Work over the decades since has established the form of Hubble's law very well; the linear relation between distance and velocity seems to hold as far as we can establish distance independently. Then we extrapolate, using Hubble's law, to find the distances to objects with even greater redshifts.

Hubble's law holds in a universe that is uniformly expanding; since each distance expands by an amount proportional to its length, greater distances expand linearly farther. The expansion of the universe thus need have no center; the whole universe can be in expansion, like slowly rising raisin bread that extends indefinitely far in each direction. No matter what raisin we are on, all the other raisins are moving away from us, and more distant raisins are moving away faster.

As astrophysicists observe objects farther and farther into space, now using techniques not only of optical astronomy but also of radio astronomy, infrared astronomy, and x-ray astronomy, they depend on Hubble's law to assign distances. For example, the objects with the greatest known redshifts—the "quasars"—must be the most distant.

If we invert Hubble's law, we can ask when all the galaxies in the universe would have been together if the universe's rate of expansion is constant. Roughly, they would have been together at $1/H_0$ (Fig. 36–39). We say that there would have been a "big bang" at that time, with continual expansion ever since. The actual theoretical solutions used to describe big-bang cosmology are solutions found by Alexander Friedmann and others in the 1920's to equations of Einstein's general theory of relativity. If astrophysicists could agree on the value of H_0, we would know the age of the universe. But H_0 is uncertain to a factor of about 2, leading to estimated ages for the universe roughly between 10 and 20 billion years.

Fig. 36–38
Edwin Hubble at the prime focus of the 5-meter Palomar telescope.

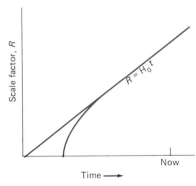

Fig. 36–39
Tracing the universe back in time if it is expanding with slope H_0 shows that its age would be $1/H_0$, if we ignore the effect of gravity slowing the expansion.

The Background Radiation

If we look out into space far enough, we are looking back into time because of the time that radiation took to reach us. When we get back far enough, we are seeing beyond the time when galaxies formed. Indeed, the universe is bathed in a faint glow of microwaves that correspond to a temperature of 3 K (Fig. 36–40). They were formed in a hot early universe, about a million years after the big bang, and have been showing cooler temperatures ever since as wavelengths were redshifted.

Fig. 36–40
Current observations of the microwave background radiation, establishing its temperature at 2.74 K. Essentially we see a plot of intensity versus wavelength.

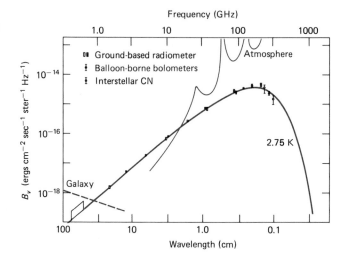

We do observe that the 3 K "background radiation" is very homogeneous and isotropic—that is, it seems the same at all locations in space and in all directions. How did it get so smooth, and how did regions greatly distant on one side of us get to be the same temperature as regions greatly distant on the other side? In ordinary big-bang cosmologies, these regions would have always been separated by too great a distance for light to have travelled between them, so how could they have reached equilibrium with each other?

The Inflationary Universe

inflationary universe

The current solution to these conundrums is the **inflationary universe** theory, first advanced by Alan Guth and then modified by Andrei Linde, Andreas Albrecht, and Paul Steinhardt. The theory incorporates ideas of grand unification. In the earliest stages of the universe, the temperature would have been so high that all the basic forces were unified; they were indistinguishable. In the inflationary universe, starting at 10^{-35} s after the big bang and lasting for 10^{-32} s, the universe expanded by a huge factor: 10^{100} in volume more than it would have grown by the former big-bang theory of unchanging expansion (Fig. 36–41).

Because of this tremendous expansion, distant locations would have been originally closer together than we had thought. Thus they could have reached equilibrium and come to the same temperature. As the universe inflated, it did not immediately undergo a phase transition. (Water undergoes a corresponding phase transition when it cools enough to change from its liquid form to ice. In some cases—including many clouds—water can be "supercooled"; that is, it remains in its liquid state even below its freezing temperature. (The water remains in its liquid super-cooled state until solidification begins; once begun the solidification takes place rapidly.)

Similarly, in the inflationary-universe model, the universe remained in a state in which the basic forces were symmetric longer than they ordinarily would have been. The forces can be seen to be symmetric—aspects of

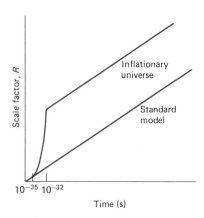

Fig. 36–41
Comparing the inflationary and standard big-bang universes over time.

the same force—only at very high temperatures and energies. They remained symmetric as the universe cooled even beyond the point at which its energy dropped below the state in which the symmetry among the forces is "broken"; such broken symmetries give the apparently separate forces we now observe.

In its supercooled state, the universe had excess energy. It is a [nonintuitive] property of such excess energy that it remained constant as the universe expanded, since the energy was associated with the universe's state rather than with individual particles. This excess energy led to the inflation in a way that can be calculated. When the symmetry was broken, the excess energy was transformed into the matter and energy we observe today (Fig. 36–42).

cosmic strings

There is much discussion at present as to whether **cosmic strings,** "vacuum defects" left over from the interactions of the different phase states of the early universe, exist. They would be submicroscopically thick, though of cosmic length.

It has long been wondered why the amount of mass in the universe is within a factor of 100 of the amount necessary to provide the gravity (or, alternatively, the curvature of spacetime) to "close the universe," that is, to stop its expansion. Observational evidence has seesawed for years over whether the universe is open or closed. If the inflationary-universe theory is valid, then a small volume of preinflation space-time has been spread out into the universe we now observe. It has thus flattened space-time, leading to our being right on the borderline between an open universe that will expand forever and a closed universe that will eventually contract.

The theories of elementary particles are now interacting with theories of cosmology to improve our predictions of the future of the universe and our understanding of its distant past. We still cannot consider what the universe was like in its first 10^{-43} s, the "Planck time," because we have no quantum theory of gravity. Advances in elementary-particle physics and cosmology may go hand in glove. Already observations of the nature of the universe have constrained the number of types of neutrinos that can exist.

Fig. 36–42
A distant cluster of galaxies; we now can account for the creation of all the matter in the universe—both visible and invisible—as a result of the delayed phase change in the inflationary stage.

In years to come, as new accelerators come on line and as new telescopes become available on the ground and in space, we hope to understand our universe better on all scales, from the largest to the smallest.

SUMMARY

1. Fraunhofer discovered spectral lines in the solar spectrum. Balmer explained the hydrogen spectrum numerologically in a way that can be generalized to

$$\frac{1}{\lambda} = R\left(-\frac{1}{n_2^2} + \frac{1}{n_1^2}\right),$$

where R is the Rydberg constant and n_2 and n_1 are integers. In 1913, the Bohr atom assumed that electron orbits are quantized. This model could account for the hydrogen spectrum (with n_2 and n_1 corresponding to the upper and lower levels, respectively, of a transition), if $E = -1/n^2$, given that $E = h\nu = hc/\lambda$. The Franck-Hertz experiment endorsed the model known as the Bohr atom.

2. In 1923, de Broglie realized that electrons have wave properties. In 1926, Schrödinger formulated wave mechanics shortly after Heisenberg formulated matrix mechanics; they were soon shown to be identical. Solutions to the wave equation exist only for quantized levels of energy. Born realized that the square of Schrödinger's wave function corresponded to the probability of finding an electron in a given location.

3. Heisenberg's uncertainty principle (1927) held that $\Delta x \Delta p \geq h/2\pi$.

4. Pauli's exclusion principle states that no two electrons can have the same four quantum numbers. Goudsmit and Uhlenbeck identified the fourth quantum number with the spin angular momentum of the electron, $\pm\frac{1}{2}\hbar$. Bohr's principle of complementarity held that light and electrons had both wave and particle characteristics, and that which set of characteristics showed depended on our experiment.

5. Dirac formulated a relativistic quantum mechanics in 1927, which led to the discovery in 1932 of the positron. The current theory, quantum electrodynamics was formulated by Tomonaga, Schwinger, and Feynman in 1948.

6. The structure of matter has been under study for millennia. J. J. Thomson discovered the electron in 1897. Rutherford presented a nuclear atom in 1911, based on scattering experiments. The neutron was discovered in 1932. Pauli suggested the neutrino in 1930 from conservation arguments.

7. Maria Goeppert Mayer and J. Hans Jensen worked out details of the shell theory of the nucleus. The liquid-drop model must also be considered. Some nuclei are stable, while others are radioactive, as discovered by Becquerel in 1896. Marie and Pierre Curie discovered additional radioactive elements. Lise Meitner and Otto Frisch realized in 1939 that uranium could fission.

8. Over 100 elementary particles were discovered. Lee and Yang could explain some unusual decays if parity is not conserved, as was confirmed experimentally by Wu. In 1964, Gell-Mann and Zweig suggested the quark theory; quarks have fractional charge and make up all baryons. Leptons, including the electron, the muon, and the tau, are also fundamental particles.

9. Weinberg, Salam, and Glashow worked out the unification of the electromagnetic and the weak forces. Rubbia and colleagues, including van der Meer, confirmed the unification by finding the W and Z particles. Grand-unified theories also incorporate the strong force. They predict proton decay, which has not yet been observed.

10. Hubble discovered in 1929 from observations of red-shifts that the universe is uniformly expanding, following Hubble's law:

$$v = H_0 d.$$

We find the distances to the most distant objects from Hubble's law. Big-bang cosmology explains the expansion of the universe on the basis of solutions to Einstein's field equations in general relativity. A homogeneous isotropic 3 K background radiation has been found, confirming the big-bang model. The inflationary universe theory shows why the universe is so homogeneous and isotropic, and why it is on the borderline between being open and closed. We can currently consider the universe no closer to its beginning than the Planck time, 10^{-43} s.

SUGGESTED READINGS

Richard P. Feynman, *QED: The Strange Theory of Light and Matter* (Princeton, NJ: Princeton University Press, 1985).

Alan H. Guth and Paul J. Steinhardt, "The Inflationary Universe," *Scientific American*, May 1984, pp. 116–128.

Abraham Pais, *Inward Bound: Of Matter and Forces in the Physical World* (Oxford: Oxford University Press, 1986).

Melba Newell Phillips, ed., *Physics History from AAPT Journals* (American Association of Physics Teachers, 1985).

Emilio Segrè, *From X-Rays to Quarks* (New York: W. H. Freeman and Co., 1980).

Spencer P. Weart and Melba Phillips, eds., *History of Physics* (American Institute of Physics, 1985). Readings from *Physics Today*.

Steven Weinberg, *The Discovery of Subatomic Particles* (New York: Scientific American Library, 1983).

Steven Weinberg, *The First Three Minutes* (New York: Basic Books, 1977).

QUESTIONS

1. Can you explain the telephone using the laws of classical physics? Explain the laser? Why or why not?
2. Why do the dark lines that Fraunhofer drew cross the spectrum vertically? What is a circumstance in which the lines would appear at an angle?
3. How would the argument that the Balmer series results from jumps between energy levels change if $E_n = +1/n^2$ instead of $-1/n^2$?
4. Imagine an atom that, unlike hydrogen, has only three energy levels. If those levels are equally spaced, how many spectral lines will result? How will their wavelengths compare?
5. If you knew the probability that an electron was in a given volume as a function of radius r from the center of an atom, what can you say about the integral of that probability from $r=0$ to $r=\infty$?
6. Explain how the photoelectric effect shows particle properties rather than wave properties of light.
7. How is the concept of orbital angular momentum different in quantum mechanics from its meaning in classical physics or, indeed, in the Bohr atom?
8. Why is it attractive to have Goudsmit and Uhlenbeck's interpretation of spin instead of merely Pauli's previous suggestion that two electrons could exist with the same n, ℓ, and m quantum numbers?
9. Discuss the principle of complementarity with respect to the photoelectric effect and Young's double-slit experiment.
10. How could a neutrino have no rest mass yet still exist? Describe in terms of Einstein's special theory of relativity. *Hint:* How fast would it go?
11. From the fact that protons repel each other by the electromagnetic force, which drops as $1/r^2$, describe the strength and range of the strong force.
12. Analyze Figure 36–9 in terms of binding energy.
13. Describe a possible difference between uranium and helium in terms of binding energy.
14. Why is it important to have accelerators capable of reaching higher and higher energies?
15. Why would matter-antimatter annihilation be an efficient way of producing energy?
16. Why was the strange quark necessary?
17. In current theory, how are baryons fundamentally different from leptons?
18. Why was the discovery of the J/ψ particle so significant?
19. Compare the significance of the discovery of the J/ψ particle with the discovery of the W particle.
20. How can Hubble's law hold without the universe having a center?
21. How does the existence of the background radiation show that the universe is evolving with time?
22. Discuss how the inflationary universe has helped with two problems of cosmology.
23. Given the tremendous factor of inflation, why doesn't the inflationary model greatly change our estimate of the age of the universe based on tracing the observed expansion back in time?

PROBLEMS

Section 36.1   *Leading to the Quantum Theory*

1. Using the Rydberg constant given in Example 36–2, calculate the wavelength of the first three lines of the Lyman series of hydrogen.
2. Use the fact that Paschen α, the transition between levels 3 and 4 of the hydrogen atom, is a spectral line at 1875 nm to calculate the Rydberg constant. Compare your result with the Rydberg constant generated in Example 36–1.
3. For carbon, the frequencies of spectral lines resulting from a single electron jump are increased over those for hydrogen by the ratio of the Rydberg constants: $R_C/R_H = (1 - m_e/m_C)/(1 - m_e/m_H)$, where the hydrogen mass is approximately 1800 times the electron mass m_e and the carbon mass is 12 times the hydrogen mass. By how many MHz is the 272α carbon line shifted from the analogous hydrogen line?
4. The Rydberg constant for hydrogen is 109,678 cm^{-1}. What is the wavelength at which the Lyman continuum starts, that is, the wavelength at which the Lyman series of spectral lines ends for arbitrarily high n?

Section 36.2   *Quantum Mechanics*

5. Using 6.63×10^{-34} J·s for Planck's constant, what is the minimum uncertainty in momentum that results if you localize a particle to 0.1 nm?

6. With Planck's constant given in the preceding problem, what is the minimum uncertainty in momentum that results if you localize a particle to 1 cm? Consider a proton, in particular, and compare qualitatively your result with the result of the previous problem for $\Delta x = 0.1$ nm.

7. If an electron is travelling 10^8 m/s + or − 10%, what is the uncertainty in localizing its position?

8. List all the allowed (n, ℓ, s) sets of values for $n = 3$.

9. What is the lowest-energy configuration of electrons for nitrogen, which has 7 electrons?

10. What is the lowest-energy configuration of electrons for iron, which has 26 electrons?

Section 36.3   *The Structure of Matter*

11. The half-life of radioactive carbon is 5600 years, meaning that after every period of 5600 years only half the radioactive carbon atoms remain. How long is it before a sample with 10^6 radioactive carbon atoms decays enough so that fewer than 100 such atoms remain?

Section 36.4   *Subnuclear Particles*

12. What are all the groupings of baryons that can be made out of the up, down, and strange quarks? How many are there? Give the total charge of each.

13. What are all the groupings of baryons that can be made out of the up, down, strange, and charmed quarks? How many are there? Give the total charge of each.

14. What are all the groupings of mesons that can be made out of the up, down, and strange quarks? How many are there? Give the total charge of each.

Section 36.5   *Unification and Cosmology*

15. At 1 g/cm³, what is the volume of the 4000 tons of water used in the proton-decay experiment cited?

16. Approximately how many superstring loops side by side would cover 0.1-nm wavelength of an x-ray?

17. Hubble's constant is approximately 17 km/s/megalight-year. What is the distance to a galaxy receding at 2×10^6 km/s?

18. Using the Doppler effect (Section 14.8), what is the distance to a galaxy whose light is redshifted by 0.5%, given $H_0 = 17$ km/s/megalight-year?

19. Using the Doppler effect (Section 14.8), and 17 km/s/megalight-year for Hubble's constant, what is the distance to the nearest quasar, 3C 273? Its spectral lines are redshifted by 16%.

20. If the universe expanded uniformly for all time, its age would be approximately $1/H_0$. Simplify the units of $H_0 = 17$ km/s/megalight-year to find the age of the universe in years.

Supplementary Problems

21. Wien's displacement law applied to a black-body curve says that $\lambda_{maximum} T = 0.289789$ cm·K, where $\lambda_{maximum}$ is the wavelength at which the curve has its maximum value. At what wavelength does the 3 K background radiation have its maximum?

22. Planck's law for the intensity I as a function of wavelength λ and temperature T is

$$I = \frac{2\pi h c^2}{\lambda^5 (e^{hc/\lambda kT} - 1)}.$$

Expand the exponential in a series to derive the (previously known) Rayleigh-Jeans law, the form of Planck's law at long wavelengths.

23. Using the form of Planck's law given in the preceding problem, use a computer to graph Planck's law for $T = 3$ K and $T = 6$ K.

24. Differentiate Planck's law (Problem 22) to derive Wien's law (stated in Problem 21).

25. Using Planck's law (Problem 22), what is the ratio in intensity at a wavelength of 1 mm that corresponds to the difference in temperature between 2.74 K and 2.96 K, two recent values derived for the cosmic background radiation?

ALGEBRA AND TRIGONOMETRY

Quadratic Formula

If $ax^2 + bx + c = 0$ then $x = \dfrac{-b \pm \sqrt{b^2 - 4ac}}{2a}$

Circumference, Area, Volume

Where $\pi \simeq 3.14159$:

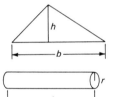

circumference of circle	$2\pi r$
area of circle	πr^2
surface area of sphere	$4\pi r^2$
volume of sphere	$\frac{4}{3}\pi r^3$
area of triangle	$\frac{1}{2}bh$
volume of cylinder	$\pi r^2 \ell$

Trigonometry

definition of angle (in radians): $\theta = \dfrac{s}{r}$

2π radians in complete circle
1 radian $\simeq 57.3°$

Trigonometric Functions

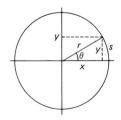

$\sin\theta = \dfrac{y}{r}$

$\cos\theta = \dfrac{x}{r}$

$\tan\theta = \dfrac{\sin\theta}{\cos\theta} = \dfrac{y}{x}$

Values at Selected Angles

$\theta\rightarrow$	0	$\dfrac{\pi}{6}$ (30°)	$\dfrac{\pi}{4}$ (45°)	$\dfrac{\pi}{3}$ (60°)	$\dfrac{\pi}{2}$ (90°)
$\sin\theta$	0	$\dfrac{1}{2}$	$\dfrac{\sqrt{2}}{2}$	$\dfrac{\sqrt{3}}{2}$	1
$\cos\theta$	1	$\dfrac{\sqrt{3}}{2}$	$\dfrac{\sqrt{2}}{2}$	$\dfrac{1}{2}$	0
$\tan\theta$	0	$\dfrac{\sqrt{3}}{3}$	1	$\sqrt{3}$	∞

Graphs of Trigonometric Functions

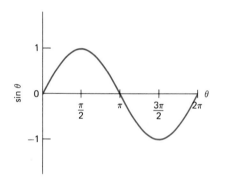 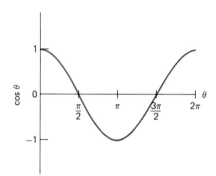 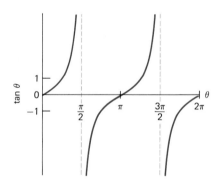

Trigonometric Identities

$$\sin(-\theta) = -\sin\theta$$

$$\cos(-\theta) = \cos\theta$$

$$\sin\left(\theta \pm \frac{\pi}{2}\right) = \pm\cos\theta$$

$$\cos\left(\theta \pm \frac{\pi}{2}\right) = \mp\sin\theta$$

$$\sin^2\theta + \cos^2\theta = 1$$

$$\sin 2\theta = 2\sin\theta\cos\theta$$

$$\cos 2\theta = \cos^2\theta - \sin^2\theta = 1 - 2\sin^2\theta = 2\cos^2\theta - 1$$

$$\sin(\alpha \pm \beta) = \sin\alpha\cos\beta \pm \cos\alpha\sin\beta$$

$$\cos(\alpha \pm \beta) = \cos\alpha\cos\beta \mp \sin\alpha\sin\beta$$

$$\sin\alpha \pm \sin\beta = 2\sin\left[\frac{1}{2}(\alpha \pm \beta)\right]\cos\left[\frac{1}{2}(\alpha \mp \beta)\right]$$

$$\cos\alpha + \cos\beta = 2\cos\left[\frac{1}{2}(\alpha + \beta)\right]\cos\left[\frac{1}{2}(\alpha - \beta)\right]$$

$$\cos\alpha - \cos\beta = -2\sin\left[\frac{1}{2}(\alpha + \beta)\right]\sin\left[\frac{1}{2}(\alpha - \beta)\right]$$

Laws of Cosines and Sines

Where A, B, C are the sides of an arbitrary triangle and α, β, γ the angles opposite those sides:

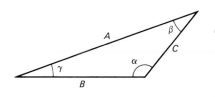

Law of Cosines

$$C^2 = A^2 + B^2 - 2AB\cos\gamma$$

Law of Sines

$$\frac{\sin\alpha}{A} = \frac{\sin\beta}{B} = \frac{\sin\gamma}{C}$$

Exponentials and Logarithms

Graphs

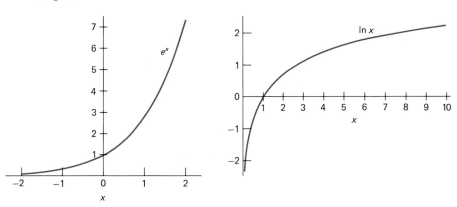

Exponential and Natural Logarithms Are Inverse Functions

$$e^{\ln x} = x, \quad \ln e^x = x \quad e = 2.71828\ldots$$

Exponential and Logarithmic Identities

$$a^x = e^{x\ln a} \qquad \ln(xy) = \ln x + \ln y$$

$$a^x a^y = a^{x+y} \qquad \ln\left(\frac{x}{y}\right) = \ln x - \ln y$$

$$(a^x)^y = a^{xy} \qquad \ln\left(\frac{1}{x}\right) = -\ln x$$

$$\log x \equiv \log_{10} x = \ln(10)\ln x \simeq 2.3\ln x$$

Expansions and Approximations

Series Expansions of Functions

Note: $n! = n(n - 1)(n - 2)(n - 3) \cdots (3)(2)(1)$

$$e^x = 1 + x + \frac{x^2}{2!} + \frac{x^3}{3!} + \cdots \qquad \text{(exponential)}$$

$$\sin x = x - \frac{x^3}{3!} + \frac{x^5}{5!} - \cdots \qquad \text{(sine)}$$

$$\cos x = 1 - \frac{x^2}{2!} + \frac{x^4}{4!} - \cdots \qquad \text{(cosine)}$$

(x in radians)

$$\ln(1 + x) = x - \frac{x^2}{2} + \frac{x^3}{3} - \cdots \qquad \text{(natural logarithm)}$$

$$(1 + x)^p = 1 + px + \frac{p(p - 1)}{2!} x^2 + \frac{p(p - 1)(p - 2)}{3!} x^3 + \cdots \qquad \text{(binomial, valid for } |x| < 1)$$

Approximations

For $|x| \ll 1$, the first few terms in the series provide a good approximation; that is,

$$\begin{aligned}
e^x &\simeq 1 + x \\
\sin x &\simeq x \\
\cos x &\simeq 1 - \tfrac{1}{2}x^2 \\
\ln(1 + x) &\simeq x \\
(1 + x)^p &\simeq 1 + px
\end{aligned} \right\} \quad \text{for } |x| \ll 1$$

Expressions that do not have the forms shown may often be put in the appropriate form. For example:

$$\frac{1}{\sqrt{a^2 + y^2}} = \frac{1}{a\sqrt{1 + \dfrac{y^2}{a^2}}} = \frac{1}{a}\left(1 + \frac{y^2}{a^2}\right)^{-1/2}.$$

For $y^2 \ll a^2$, this may be approximated using the binomial expansion $(1 + x)^p \simeq 1 + px$, with $p = -\tfrac{1}{2}$ and $x = y^2/a^2$:

$$\frac{1}{a}\left(1 + \frac{y^2}{a^2}\right)^{-1/2} \simeq \frac{1}{a}\left(1 - \frac{1}{2}\frac{y^2}{a^2}\right).$$

Vector Algebra

Vector Products

$\mathbf{A}\cdot\mathbf{B} = AB\cos\theta$

$|\mathbf{A}\times\mathbf{B}| = AB\sin\theta,$

with direction of $\mathbf{A}\times\mathbf{B}$ given by right-hand rule:

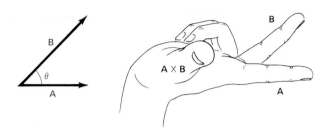

Unit Vector Notation

An arbitrary vector $\mathbf{A}$ may be written in terms of its components A_x, A_y, A_z and the unit vectors $\hat{\imath}, \hat{\jmath}, \hat{\mathbf{k}}$ that have length 1 and lie along the x, y, z axes:

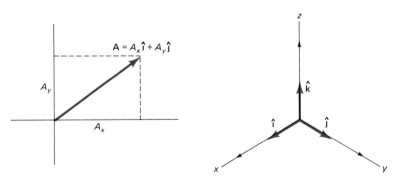

In unit vector notation, vector products become

$\mathbf{A}\cdot\mathbf{B} = A_xB_x + A_yB_y + A_zB_z$

$\mathbf{A}\times\mathbf{B} = (A_yB_z - A_zB_y)\hat{\imath} + (A_zB_x - A_xB_z)\,\hat{\jmath} + (A_xB_y - A_yB_x)\hat{\mathbf{k}}$

Vector Identities

$\mathbf{A}\cdot\mathbf{B} = \mathbf{B}\cdot\mathbf{A}$

$\mathbf{A}\times\mathbf{B} = -\mathbf{B}\times\mathbf{A}$

$\mathbf{A}\cdot(\mathbf{B}\times\mathbf{C}) = \mathbf{B}\cdot(\mathbf{C}\times\mathbf{A}) = \mathbf{C}\cdot(\mathbf{A}\times\mathbf{B})$

$\mathbf{A}\times(\mathbf{B}\times\mathbf{C}) = (\mathbf{A}\cdot\mathbf{C})\mathbf{B} - (\mathbf{A}\cdot\mathbf{B})\mathbf{C}$

CALCULUS

Derivatives

Definition of the Derivative

If y is a function of x [$y = f(x)$], then the **derivative of y with respect to x** is the ratio of the change Δy in y to the corresponding change Δx in x, in the limit of arbitrarily small Δx:

$$\frac{dy}{dx} = \lim_{\Delta t \to 0} \frac{\Delta y}{\Delta x}.$$

Algebraically, the derivative is the rate of change of y with respect to x; geometrically, it is the slope of the y versus x graph—that is, of the tangent line to the graph at a given point:

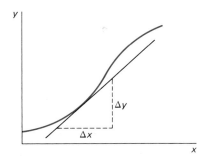

Derivatives of Common Functions

Although the derivative of a function can be evaluated directly using the limiting process that defines the derivative (see Section 2.2 and Example 2–3), standard formulas are available for common functions:

$$\frac{da}{dx} = 0 \quad (a \text{ is a constant})$$

$$\frac{dx^n}{dx} = nx^{n-1} \quad (n \text{ need not be an integer})$$

$$\frac{d}{dx} \sin x = \cos x$$

$$\frac{d}{dx} \cos x = -\sin x$$

$$\frac{d}{dx} \tan x = \frac{1}{\cos^2 x}$$

$$\frac{de^x}{dx} = e^x$$

$$\frac{d}{dx} \ln x = \frac{1}{x}$$

Derivatives of Sums, Products, and Functions of Functions

1. Derivative of a constant times a function

$$\frac{d}{dx}[af(x)] = a\frac{df}{dx} \qquad (a \text{ is a constant})$$

2. Derivative of a sum

$$\frac{d}{dx}[f(x) + g(x)] = \frac{df}{dx} + \frac{dg}{dx}$$

3. Derivative of a product

$$\frac{d}{dx}[f(x)\,g(x)] = g\frac{df}{dx} + f\frac{dg}{dx}$$

Examples

$$\frac{d}{dx}(x^2\cos x) = \cos x\frac{dx^2}{dx} + x^2\frac{d}{dx}\cos x = 2x\cos x - x^2\sin x$$

$$\frac{d}{dx}(x\ln x) = \ln x\frac{dx}{dx} + x\frac{d}{dx}\ln x = (\ln x)(1) + x\left(\frac{1}{x}\right) = \ln x + 1$$

4. Derivative of a quotient

$$\frac{d}{dx}\left[\frac{f(x)}{g(x)}\right] = \frac{1}{g^2}\left(g\frac{df}{dx} - f\frac{dg}{dx}\right)$$

Example

$$\frac{d}{dx}\left(\frac{\sin x}{x^2}\right) = \frac{1}{x^4}\left(x^2\frac{d}{dx}\sin x - \sin x\frac{dx^2}{dx}\right) = \frac{\cos x}{x^2} - \frac{2\sin x}{x^3}$$

5. Chain rule for derivatives

If f is a function of u and u is a function of x, then

$$\frac{df}{dx} = \frac{df}{du}\frac{du}{dx}.$$

Examples

a. Evaluate $\dfrac{d}{dx}\sin(x^2)$. Here $u = x^2$ and $f(u) = \sin u$, so that

$$\frac{d}{dx}\sin(x^2) = \frac{d}{du}\sin u\frac{du}{dx} = (\cos u)\frac{dx^2}{dx} = 2x\cos(x^2).$$

b. $\dfrac{d}{dt}\sin\omega t = \dfrac{d}{d\omega t}\sin\omega t\dfrac{d}{dt}\omega t = \omega\cos\omega t.$ (ω a constant)

c. Evaluate $\dfrac{d}{dx}\sin^2 5x$. Here $u = \sin 5x$ and $f(u) = u^2$, so that

$$\frac{d}{dx}\sin^2 5x = \frac{d}{du}u^2\frac{du}{dx} = 2u\frac{du}{dx} = 2\sin 5x\frac{d}{dx}\sin 5x$$

$$= (2)(\sin 5x)(5)(\cos 5x) = 10\sin 5x\cos 5x = 5\sin 2x.$$

Second Derivative

The second derivative of y with respect to x is defined as the derivative of the derivative:

$$\frac{d^2y}{dx^2} = \frac{d}{dx}\left(\frac{dy}{dx}\right).$$

Example

If $y = ax^3$, then $dy/dx = 3ax^2$, so that

$$\frac{d^2y}{dx^2} = \frac{d}{dx}3ax^2 = 6ax.$$

Partial Derivatives

When a function depends on more than one variable, then the partial derivatives of that function are the derivatives with respect to each variable, taken with all other variables held constant. If f is a function of x and y, then the partial derivatives are written

$$\frac{\partial f}{\partial x} \quad \text{and} \quad \frac{\partial f}{\partial y}.$$

Example

If $f(x,y) = x^3 \sin y$, then

$$\frac{\partial f}{\partial x} = 3x^2 \sin y \quad \text{and} \quad \frac{\partial f}{\partial y} = x^3 \cos y.$$

Integrals

Indefinite Integrals

Integration is the inverse of differentiation. The **indefinite integral,** $\int f(x)\,dx$, is defined as a function whose derivative is $f(x)$:

$$\frac{d}{dx}[\int f(x)\,dx] = f(x).$$

If $A(x)$ is an indefinite integral of $f(x)$, then because the derivative of a constant is zero, the function $A(x) + C$ is also an indefinite integral of $f(x)$, where C is any constant. Inverting the derivatives of common functions listed in the preceding section gives some common integrals (a more extensive table appears at the end of Appendix B):

$$\int a\,dx = ax + C$$

$$\int x^n\,dx = \frac{x^{n+1}}{n+1} + C, \qquad n \neq -1$$

$$\int \sin x\,dx = -\cos x + C$$

$$\int \cos x\,dx = \sin x + C$$

$$\int e^x\,dx = e^x + C$$

$$\int x^{-1}\,dx = \ln x + C$$

Definite Integrals

In physics we are most often interested in the **definite integral,** defined as the sum of a large number of very small quantities, in the limit as the number of quantities grows arbitrarily large and the size of each arbitrarily small:

$$\int_{x_1}^{x_2} f(x)\,dx \equiv \lim_{\substack{\Delta x \to 0 \\ N \to \infty}} \sum_{i=1}^{N} f(x_i)\Delta x,$$

where the terms in the sum are evaluated at values x_i between the limits of integration x_1 and x_2; in the limit $\Delta x \to 0$, the sum is over all values of x in the interval.

The definite integral is used whenever we need to sum over a quantity that is changing—for example, to calculate the work done by a variable force (Chapter 6), the entropy change in a system whose temperature varies (Chapter 19), or the flux of an electric field that varies with position (Chapter 21).

The key to evaluating the definite integral is provided by the **fundamental theorem of calculus,** which relates the definite and indefinite integrals and which we outlined without naming it in Box 2–1. The theorem states that, if $A(x)$ is an *indefinite* integral of $f(x)$, then the *definite integral* is given by

$$\int_{x_1}^{x_2} f(x)\,dx = A(x_2) - A(x_1) \equiv A(x)\Big|_{x_1}^{x_2}.$$

Geometrically, the definite integral is the area under the graph of $f(x)$ between the limits x_1 and x_2:

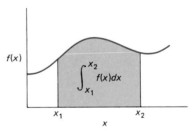

Evaluating Integrals

The first step in evaluating an integral is to express all varying quantities within the integral in terms of a single variable. For example, in evaluating $\int E\,dr$ to calculate an electric potential (Chapter 23), it is necessary first to express E as a function of r. This procedure is illustrated in many examples throughout this text; Example 20–5 provides a typical case.

Once an integral is written in terms of a single variable, it is necessary to manipulate the integrand—the function being integrated—into a form whose integral you know or can look up in tables of integrals. Two common techniques are especially useful:

1. **Change of variables**

   An unfamiliar integral can often be put into familiar form by defining a new variable. For example, it is not obvious how to integrate the expression

   $$\int \frac{x\,dx}{\sqrt{a^2 + x^2}},$$

   where a is a constant. But let $z = a^2 + x^2$. Then

   $$\frac{dz}{dx} = \frac{da^2}{dx} + \frac{dx^2}{dx} = 0 + 2x = 2x,$$

so that $dz = 2x\,dx$. Then the quantity $x\,dx$ in our unfamiliar integral is just $\frac{1}{2}dz$, while the quantity $\sqrt{a^2 + x^2}$ is just $z^{1/2}$. So the integral becomes

$$\int \tfrac{1}{2}z^{-1/2}\,dz = \frac{\tfrac{1}{2}z^{1/2}}{(1/2)} = \sqrt{z},$$

where we have used the standard form for the integral of a power of the independent variable. Substituting back $z = a^2 + x^2$ gives

$$\int \frac{x\,dx}{\sqrt{a^2 + x^2}} = \sqrt{a^2 + x^2}.$$

2. **Integration by parts**

The quantity $\int u\,dv$ is the area under the curve of u as a function of v between specified limits. In the figure below, that area can also be expressed as the area of the rectangle shown minus the area under the curve of v as a function of u.

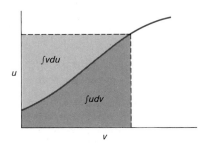

Mathematically, this relation among areas may be expressed as a relation among integrals:

$$\int u\,dv = uv - \int v\,du. \qquad \text{(integration by parts)}$$

This expression may often be used to transform complicated integrals into simpler ones.

Example

Evaluate $\int x\cos x\,dx$. Here let $u = x$, so that $du = dx$. Then $dv = \cos x\,dx$, so that $v = \int dv = \int \cos x\,dx = \sin x$. Integrating by parts then gives

$$\int x\cos x\,dx = (x)(\sin x) - \int \sin x\,dx = x\,\sin x + \cos x,$$

where the $+$ sign arises because $\int \sin x\,dx = -\cos x$.

Table of Integrals

[More extensive tables are available in many mathematical and scientific handbooks; see, for example, *Handbook of Chemistry and Physics* (Chemical Rubber Co.) or Dwight, *Tables of Integrals and other Mathematical Data* (Macmillan).]

In the expressions below, a and b are constants. An arbitrary constant of integration may be added to the right-hand side.

$$\int e^{ax} dx = \frac{e^{ax}}{a}$$

$$\int \sin ax \, dx = -\frac{\cos ax}{a}$$

$$\int \cos ax \, dx = \frac{\sin ax}{a}$$

$$\int \tan ax \, dx = -\frac{1}{a} \ln (\cos ax)$$

$$\int \sin^2 ax \, dx = \frac{x}{2} - \frac{\sin 2ax}{4a}$$

$$\int \cos^2 ax \, dx = \frac{x}{2} + \frac{\sin 2ax}{4a}$$

$$\int x \sin ax \, dx = \frac{1}{a^2} \sin ax - \frac{1}{a} x \cos ax$$

$$\int x \cos ax \, dx = \frac{1}{a^2} \cos ax + \frac{1}{a} x \sin ax$$

$$\int \frac{dx}{\sqrt{a^2 - x^2}} = \sin^{-1}\left(\frac{x}{a}\right)$$

$$\int \frac{dx}{\sqrt{x^2 \pm a^2}} = \ln\left(x + \sqrt{x^2 \pm a^2}\right)$$

$$\int \frac{dx}{x^2 + a^2} = \frac{1}{a} \tan^{-1}\left(\frac{x}{a}\right)$$

$$\int \frac{x \, dx}{\sqrt{a^2 - x^2}} = -\sqrt{a^2 - x^2}$$

$$\int \frac{x \, dx}{\sqrt{x^2 \pm a^2}} = \sqrt{x^2 \pm a^2}$$

$$\int x e^{ax} \, dx = \frac{e^{ax}}{a^2} (ax - 1)$$

$$\int \frac{dx}{a + bx} = \frac{1}{b} \ln (a + bx)$$

$$\int \frac{dx}{(a + bx)^2} = -\frac{1}{b(a + bx)}$$

$$\int \ln ax \, dx = x \ln ax - x$$

THE INTERNATIONAL SYSTEM OF UNITS (SI)

This material is from the United States edition of the English translation of the fifth edition of "Le Système International d'Unités (SI)," the definitive publication in the French language issued in 1985 by the International Bureau of Weights and Measures (BIPM). The year the definition was adopted is given in parentheses.

unit of length (meter): The meter is the length of the path traveled by light in vacuum during a time interval of 1/299 792 458 of a second. (1983)

unit of mass (kilogram): The kilogram is the unit of mass; it is equal to the mass of the international prototype of the kilogram. (1889)

unit of time (second): The second is the duration of 9 192 631 770 periods of the radiation corresponding to the transition between the two hyperfine levels of the ground state of the cesium-133 atom. (1967)

unit of electric current (ampere): The ampere is that constant current which, if maintained in two straight parallel conductors of infinite length, of negligible circular cross section, and placed 1 meter apart in vacuum, would produce between these conductors a force equal to 2×10^{-7} newton per meter of length. (1948)

unit of thermodynamic temperature (kelvin): The kelvin, unit of thermodynamic temperature, is the fraction 1/273.16 of the thermodynamic temperature of the triple point of water. (1957) Also, the unit kelvin and its symbol K should be used to express an interval or a difference of temperature.

unit of amount of substance (mole): (1) The mole is the amount of substance of a system that contains as many elementary entities as there are atoms in 0.012 kilogram of carbon 12. (1971) (2) When the mole is used, the elementary entities must be specified and may be atoms, molecules, ions, electrons, other particles, or specified groups of such particles.

unit of luminous intensity (candela): The candela is the luminous intensity, in a given direction, of a source that emits monochromatic radiation of frequency 540×10^{12} hertz and that has a radiant intensity in that direction of (1/683) watt per steradian. (1979)

SI Base Units

Quantity	SI Unit	
	Name	Symbol
length	meter	m
mass	kilogram	kg
time	second	s
electric current	ampere	A
thermodynamic temperature	kelvin	K
amount of substance	mole	mol
luminous intensity	candela	cd

Examples of SI Derived Units Expressed in Terms of Base Units

Quantity	SI Unit	
	Name	*Symbol*
area	square meter	m^2
volume	cubic meter	m^3
speed, velocity	meter per second	m/s
acceleration	meter per second squared	m/s^2
wave number	reciprocal meter	m^{-1}
density, mass density	kilogram per cubic meter	kg/m^3
specific volume	cubic meter per kilogram	m^3/kg
current density	ampere per square meter	A/m^2
luminance	candela per square meter	cd/m^2

SI Derived Units with Special Names

Quantity	SI Unit			
	Name	*Symbol*	*Expression in Terms of Other Units*	*Expression in Terms of SI Base Units*
frequency	hertz	Hz		s^{-1}
force	newton	N		$m \cdot kg \cdot s^{-2}$
pressure, stress	pascal	Pa	N/m^2	$m^{-1} \cdot kg \cdot s^{-2}$
energy, work, heat	joule	J	$N \cdot m$	$m^2 \cdot kg \cdot s^{-2}$
power	watt	W	J/s	$m^2 \cdot kg \cdot s^{-3}$
electric charge	coulomb	C		$s \cdot A$
electric potential, potential difference, electromotive force	volt	V	J/C	$m^2 \cdot kg \cdot s^{-3} \cdot A^{-1}$
capacitance	farad	F	C/V	$m^{-2} \cdot kg^{-1} \cdot s^4 \cdot A^2$
electric resistance	ohm	Ω	V/A	$m^2 \cdot kg \cdot s^{-3} \cdot A^{-2}$
electric conductance	siemens	S	A/V	$m^{-2} \cdot kg^{-1} \cdot s^3 \cdot A^2$
magnetic flux	weber	Wb	$V \cdot s$	$m^2 \cdot kg \cdot s^{-2} \cdot A^{-1}$
magnetic field	tesla	T	Wb/m^2	$kg \cdot s^{-2} \cdot A^{-1}$
inductance	henry	H	Wb/A	$m^2 \cdot kg \cdot s^{-2} \cdot A^{-2}$
Celsius temperature	degree Celsius	°C		K
luminous flux	lumen	lm		$cd \cdot sr$
illuminance	lux	lx	lm/m^2	$m^{-2} \cdot cd \cdot sr$

SI Derived Units with Special Names Admitted for Reasons of Safeguarding Human Health

Quantity	SI Unit			
	Name	*Symbol*	*Expression in Terms of Other Units*	*Expression in Terms of SI Base Units*
activity (of a radionuclide)	becquerel	Bq	1 decay/s	s^{-1}
absorbed dose	gray	Gy	J/kg, 100 rad	$m^2 \cdot s^{-2}$
dose equivalent	sievert	Sv	J/kg, 100 rem	$m^2 \cdot s^{-2}$

Examples of SI Derived Units Expressed by Means of Special Names

Quantity	SI Unit		Expression in Terms of SI Base Units
	Name	Symbol	
dynamic viscosity	pascal second	Pa·s	$m^{-1} \cdot kg \cdot s^{-1}$
torque	newton meter	N·m	$m^2 \cdot kg \cdot s^{-2}$
surface tension	newton per meter	N/m	$kg \cdot s^{-2}$
heat flux density, irradiance	watt per square meter	W/m²	$kg \cdot s^{-3}$
heat capacity, entropy	joule per kelvin	J/K	$m^2 \cdot kg \cdot s^{-2} \cdot K^{-1}$
specific heat, specific entropy	joule per kilogram kelvin	J/(kg·K)	$m^2 \cdot s^{-2} \cdot K^{-1}$
specific energy	joule per kilogram	J/kg	$m^2 \cdot s^{-2}$
thermal conductivity	watt per meter kelvin	W/(m·K)	$m \cdot kg \cdot s^{-3} \cdot K^{-1}$
energy density	joule per cubic meter	J/m³	$m^{-1} \cdot kg \cdot s^{-2}$
electric field strength	volt per meter	V/m	$m \cdot kg \cdot s^{-3} \cdot A^{-1}$
electric charge density	coulomb per cubic meter	C/m³	$m^{-3} \cdot s \cdot A$
permittivity	farad per meter	F/m	$m^{-3} \cdot kg^{-1} \cdot s^4 \cdot A^2$
permeability	henry per meter	H/m	$m \cdot kg \cdot s^{-2} \cdot A^{-2}$
molar energy	joule per mole	J/mol	$m^2 \cdot kg \cdot s^{-2} \cdot mol^{-1}$
molar entropy, molar specific heat	joule per mole kelvin	J/(K·mol)	$m^2 \cdot kg \cdot s^{-2} \cdot K^{-1} \cdot mol^{-1}$
exposure (x and γ rays)	coulomb per kilogram	C/kg	$kg^{-1} \cdot s \cdot A$
absorbed dose rate	gray per second	Gy/s	$m^2 \cdot s^{-3}$

SI Supplementary Units

Quantity	SI Unit		Expression in Terms of SI Base Units
	Name	Symbol	
plane angle	radian	rad	$m \cdot m^{-1} = 1$
solid angle	steradian	sr	$m^2 \cdot m^{-2} = 1$

Examples of SI Derived Units Formed by Using Supplementary Units

Quantity	SI Unit	
	Name	Symbol
angular velocity	radian per second	rad/s
angular acceleration	radian per second squared	rad/s²
radiant intensity	watt per steradian	W/sr
radiance	watt per square meter steradian	$W \cdot m^{-2} \cdot sr^{-1}$

Rules for Writing and Using SI Unit symbols

1. Roman (upright) type, in general lower case, is used for the unit symbols. If however, the name of the unit is derived from a proper name, the first letter of the symbol is in upper case.
2. Unit symbols are unaltered in the plural.

3. Unit symbols are not followed by a period.

a. The product of two or more units is indicated as follows, for example:

$$N \cdot m.$$

b. A solidus (oblique stroke, /), a horizontal line, or negative exponents may be used to express a derived unit formed from two others by division, for example:

$$m/s, \quad \frac{m}{s}, \quad or \ m \cdot s^{-1}.$$

c. The solidus must not be repeated on the same line unless ambiguity is avoided by parentheses. In complicated cases negative exponents or parentheses should be used, for example:

$$m/s^2 \qquad or \ m \cdot s^{-2}$$
$$m \cdot kg/(s^3 \cdot A) \quad or \ m \cdot kg \cdot s^{-3} \cdot A^{-1}$$

but not:

$$m/s/s$$
$$m \cdot kg/s^3/A.$$

SI Prefixes

Factor	Prefix	Symbol	Factor	Prefix	Symbol
10^{18}	exa	E	10^{-1}	deci	d
10^{15}	peta	P	10^{-2}	centi	c
10^{12}	tera	T	10^{-3}	milli	m
10^9	giga	G	10^{-6}	micro	μ
10^6	mega	M	10^{-9}	nano	n
10^3	kilo	k	10^{-12}	pico	p
10^2	hecto	h	10^{-15}	femto	f
10^1	deka	da	10^{-18}	atto	a

Non-SI units in Use with the International System

Name	Symbol	Value in SI Units
minute	min	1 min = 60 s
hour	h	1 h = 60 min = 3 600 s
day	d	1d = 24 h = 86 400 s
degree	°	1° = $(\pi/180)$ rad
minute	′	1′ = $(1/60)° = (\pi/10\ 800)$ rad
second	″	1″ = $(1/60)' = (\pi/648\ 000)$ rad
liter	L	1 L = 1 dm^3 = 10^{-3} m^3
metric ton (tonne)	t	1 t = 10^3 kg

Units used with the International System Whose Values in SI Units Are Obtained Experimentally

Name	Symbol	Definition
electron volt	eV	(a)
unified atomic mass unit	u	(b)

[a]The electron volt is the kinetic energy acquired by an electron in passing through a potential difference of 1 volt in vacuum; 1 eV = 1.602 19 $\times$ 10^{-19} J approximately.

[b]The unified atomic mass unit is equal to (1/12) of the mass of an atom of the nuclide ^{12}C; 1 u = 1.660 57 $\times$ 10^{-27} kg approximately.

Units in Use Temporarily with the International System

Name	Symbol	Value in SI Units
nautical mile[a]		1 nautical mile = 1 852 m
knot		1 nautical mile per hour = (1852/3600) m/s
angström	Å	1 Å = 0.1 nm = 10^{-10} m
are[b]	a	1 a = 1 dam^2 = 10^2 m^2
hectare[b]	ha	1 ha = 1 hm^2 = 10^4 m^2
barn[c]	b	1 b = 100 fm^2 = 10^{-28} m^2
bar[d]	bar	1 bar = 0.1 MPa = 100 kPa = 1000 hPa = 10^5 Pa
gal[e]	Gal	1 Gal = 1 cm/s^2 = 10^{-2} m/s^2
curie[f]	Ci	1 Ci = 3.7×10^{10} Bq
roentgen[g]	R	1 R = 2.58×10^{-4} C/kg
rad[h]	rad	1 rad = 1 cGy = 10^{-2} Gy
rem[i]	rem	1 rem = 1 cSv = 10^{-2} Sv

[a]The nautical mile is a special unit employed for marine and serial navigation to express distances.
[b]This unit and its symbol were adopted in 1879 and are used to express agrarian areas.
[c]The barn is a special unit employed in nuclear physics to express effective cross sections.
[d]Pressure
[e]The gal is a special unit employed in geodesy and geophysics to express the acceleration due to gravity.
[f]The curie is a special unit employed in nuclear physics to express activity of radionuclides.
[g]The roentgen is a special unit employed to express exposure of x or γ radiations.
[h]The rad is a special unit employed to express absorbed dose of ionizing radiations. When there is risk of confusion with the symbol for radian, rd may be used as the symbol for rad.
[i]The rem is a special unit used in radioprotection to express dose equivalent.

The table below lists some specially named units of the centimeter-gram-second (CGS) system. It is in general preferable not to mix these with the units of the International System. (Although the SI system is generally preferred for scientific work, in some specialized fields, there are good reasons for using CGS or other systems.)

CGS Units with Special Names

Name	Symbol	Value in SI Units
erg	erg	1 erg = 10^{-7} J
dyne	dyn	1 dyn = 10^{-5} N
poise	P	1 P = 1 dyn·s/cm^2 = 0.1 Pa·s
stokes	St	1 St = 1 cm^2/s = 10^{-4} m^2/s
gauss[a]	Gs, G	1 Gs corresponds to 10^{-4} T
oersted[a]	Oe	1 Oe corresponds to (1000/4π)A/m
maxwell[a]	Mx	1 Mx corresponds to 10^{-8} Wb
stilb[a]	sb	1 sb = 1 cd/cm^2 = 10^4 cd/m^2
phot	ph	1 ph = 10^4 lx

[a]This unit is part of the so-called electromagnetic 3-dimensional CGS system and cannot strictly speaking be compared to the corresponding unit of the International System, which has four dimensions when only mechanical and electric quantities are considered.

As regards other units outside the International System, it is in general preferable to avoid them, and to use instead units of the International System. Some of those units are listed in the following table.

Other Units Generally Deprecated

Name	Value in SI units
fermi	1 fermi = 1 fm = 10^{-15} m
metric carat[a]	1 metric carat = 200 mg = 2×10^{-4} kg
torr	1 torr = (101 325/760) Pa
standard atmosphere (atm)[b]	1 atm = 101 325 Pa
kilogram-force (kgf)	1 kgf = 9.806 65 N
calorie (cal)[c]	
micron (μ)[d]	1 μ = 1 μm = 10^{-6} m
x unit[e]	
stere (st)[f]	1 st = 1 m^3
gamma (γ)	1 γ = 1 nT = 10^{-9} T

[a]This name was adopted in 1907 for commercial dealings in diamonds, pearls, and precious stones.

[b]The designation "standard atmosphere" for a reference pressure of 101 325 Pa is still acceptable.

[c]Several "calories" have been in use:
Calorie labeled "at 15°C": 1 cal_{15} = 4.185 5 J (value adopted in 1950)
A calorie labeled "IT" (International Table): 1 cal_{IT} = 4.186 8 J (5th International Conference on the Properties of Steam, London, 1956)
A calorie labeled "thermochemical": 1 cal_{th} = 4.184 J

[d]The name of this unit and its symbol, adopted in 1879, were abolished in 1967.

[e]This special unit was employed to express wavelengths of x rays; 1 x unit = 1.002×10^{-4} nm approximately.

[f]This special unit employed to measure firewood was adopted by the CIPM in 1879 with the symbol "s." The symbol was changed to "st" in 1948.

Source: Extracted from The International System of Units (SI), David T. Goldman and R. J. Bell, eds., National Bureau of Standards Special Publication 330, 1986 edition. Only American alternatives are listed. Omitted for this reason are l for liter; litre, metre, and deca as spellings; and N.m or N m as alternate ways of writing N·m.

APPENDIX D
CONVERSION FACTORS

The day of extensive conversion tables has passed, now that we have calculators small enough to carry in a pocket or wear on a wrist. For example, instead of using a table of conversions from square meters to square feet, students need only remember (or look up) 1 m = 39.37 inches (and the definition 1 foot = 12 inches). Then

$$\# \text{ m}^2 = (\# \text{ ft})^2 (12 \text{ in}/1 \text{ ft})^2 (1 \text{ m}/39.37 \text{ in})^2.$$

This method can be followed at any time and anywhere in the world, even when you don't happen to have your physics text along. In any case, following are a few conversions; $\equiv$ signifies a definition.

Length

Based on 1 m = 39.37 in; 1 ft = 12 in; 1 mi = 5280 ft.

Metric	English System
1 m = 100 cm = 10^{-3} km	= 39.37 in = 3.281 ft = 6.214×10^{-4} mi

Note: to find conversions from cm, simply divide any term in the preceding row by 100; to find conversions from km, multiply any term by 1000.

1609 m = 1.609×10^5 cm = 1.609 km	= 1 mi
1 cm	= 0.3937 in
2.54 cm	$\equiv$ 1 in

Other Length Conversions

Astronomical

1 light year (ly) = 9.46×10^{15} m

1 astronomical unit (AU) = 1.50×10^{11} m (the average distance from the earth to the sun

1 parsec (pc) = 3.26 ly = 3.09×10^{16} m (The parsec is defined as the distance at which, looking back, 1 AU would subtend 1 second of arc; it is a unit that arises naturally in astronomical measurements.)

Atomic

1 angstrom (Å) $\equiv$ 0.1 nm $\equiv$ 10^{-10} m

1 fermi $\equiv$ 10^{-12} m

Area

Metric	English System
1 m^2 = 10^4 cm^2 = 10^{-4} hectares (ha)	= 10.76 ft^2 = 1550 in^2
9.290×10^{-2} m^2	= 1 ft^2
6.452×10^{-4} m^2	= 1 in^2

Other Area Conversions

Normal Scale

1 acre $= 4.356 \times 10^4$ ft^2 $= 4047$ m^2 $= 0.4047$ ha

1 ft^2 $= 144$ in^2

Atomic

1 barn $\equiv 10^{-28}$ m^2

1 shed $\equiv 10^{-2}$ barn

Volume

Metric   *English System*

1 m^3 $= 10^6$ cm^3 $= 10^3$ L $= 35.31$ ft^3 $= 6.102 \times 10^4$ in^3 $= 264.2$ gal

2.832×10^{-2} m^3 $= 1$ ft^3

1.639×10^{-5} m^3 $= 1$ in^3

3.786×10^{-3} m^3 $= 1$ gal (U.S.) $= 0.1337$ ft^3 $= 231$ in^3

Other Volume Conversions

1 barrel (oil) $= 42$ gal (U.S.) $= 0.159$ m^3

1 gal (British) $= 277.42$ in^3 (based on volume of 10 pounds of water) $= 4.55$ L

1 fluid ounce (U.S.) $= 1/128$ gal (U.S.) $= 29.6$ mL

Mass

Metric   *Atomic*   *English System*

1 kg $= 1000$ g $= 6.022 \times 10^{26}$ u $= 6.852 \times 10^{-2}$ slug

1.661×10^{-27} kg $= 1$ u $= 1.38 \times 10^{-28}$

1.459 kg $= 8.786 \times 10^{27}$ $= 1$ slug

Note: The slug is the unit of mass in the English system. A 1-slug mass has a weight $mg = (1 \text{ slug})(32 \text{ ft/s}^2) = 32$ lb. The pound (lb) is often used (incorrectly) as a mass unit. To say that an object has a mass of 1 lb means that its weight at earth's surface is 1 lb. In this usage,

$$1 \text{ kg} = 2.20462 \text{ lb and } 1 \text{ lb} = 0.454 \text{ kg}.$$

Other Mass Conversions

1 ton $\equiv 2000$ lb $=$ the weight of 907.2 kg

1 tonne (t; metric ton) $\equiv 1000$ kg

1 ounce (oz) $\equiv 1/16$ lb $=$ the weight of 28.35 g

Time

1 year $= 365.2422$ d $= 8766$ h $= 5.259 \times 10^5$ min $= 3.156 \times 10^7$ s

1 day $\equiv 24$ h $\equiv 1440$ min $\equiv 8.640 \times 10^4$ s

1 hour $\equiv 60$ min $\equiv 3600$ s

1 min $\equiv 60$ s

Note: the length of the year (the earth's orbital period) changes slowly. Currently, one tropical year (the time between two successive passages of the sun across the point among the stars known as the vernal equinox, defined as the location where the celestial equator crosses the sun's path) is approximately 365.2422 d $= 3.1557 \times 10^7$ s, conveniently close to $\pi \times 10^7$ s.

Force

Metric	English System
1 newton (N) = 10^5 dyne	= 0.2248 lb
4.448 N	= 1 lb

Note: The gram (g) and kilogram (kg) are sometimes (incorrectly) used as force units. A 1-kg force means the gravitational force on a 1-kg mass where g = 9.80665 N/kg. Thus a 1-kg force corresponds to 9.807 N.

Pressure

1 pascal (Pa) $\equiv$ 1 N·m^{-2} $\equiv$ 10 dyne·cm^{-2} = 7.501×10^{-3} mm Hg (torr) = 9.869×10^{-6} atm = 4.015×10^{-3} in H$_2$O = 1.450×10^{-4} lb/in^2 (psi)

1 atmosphere = 1.013×10^5 Pa

1 in H$_2$O = 249.1 Pa

1 mm Hg (torr) = 133.3 Pa

1 lb/in^2 (psi) = 6.895×10^3 Pa

Energy

Metric	Atomic	English System
1 joule (J) $\equiv$ 10^7 erg	= 6.242×10^{18} eV	= 0.2390 cal = 9.85×10^4 Btu
= 2.778×10^{-7} kWh		= 0.7376 ft·lb
1.602×10^{-19} J	= 1 eV	= 1.520×10^{-22} Btu
		= 1.182×10^{19} ft·lb
4.184 J = 1.162×10^{-6} kWh		= 1 cal* = 3.968×10^{-3} Btu
		= 3.086 ft·lb
1054 J		= 1 Btu
1.356 J		= 1 ft·lb
3.6×10^6 J = 1 kWh		

*We have supplied the thermochemical calorie, 4.184 J. The International Table calorie = 4.1868 J.

Other Energy Conversions

1 megaton (explosive yield) = 4×10^{15} J

1 kg (energy equivalence via $E = mc^2$) = 9.0×10^{16} J

Energy Content of Fuels

Energy Source	Energy Content	
coal	2.9×10^7 J/kg	= 7300 kWh/ton = 25×10^6 Btu/ton
oil	43×10^6 J/kg	= 39 kWh/gal = 1.3×10^5 Btu/gal
gasoline	44×10^6 J/kg	= 36 kWh/gal = 1.2×10^5 Btu/gal
natural gas	55×10^6 J/kg	= 30 kWh/100 ft^3 = 1000 Btu/ft^3
uranium (fission)		
normal abundance	5.8×10^{11} J/kg	= 1.6×10^5 kWh/kg
pure U-235	8×10^{13} J/kg	= 2×10^7 kWh/kg
hydrogen (fusion)		
normal abundance	7×10^{11} J/kg	= 3.0×10^4 kWh/kg
pure deuterium	3.3×10^{14} J/kg	= 9.2×10^7 kWh/kg
water	1×10^{10} J/kg	= 1×10^4 kWh/gal = 400 gal gasoline/gal H$_2$O
100% conversion, matter to energy	9.0×10^{16} J/kg	= 931 MeV/u = 2.5×10^{10} kWh/kg

Power

Metric	English System
1 watt (W) $\equiv 10^7$ erg/s	$= 1.341 \times 10^{-3}$ hp $= 0.7376$ ft·lb/s $= 3.413$ Btu/h
745.7 W	$= 1$ hp $= 550$ ft·lb/s $= 2545$ Btu/h
1.356 W	$= 1.818 \times 10^{-3}$ hp $= 1$ ft·lb/s $= 4.628$ Btu/h
0.2930 W	$= 3.929 \times 10^{-4}$ hp $= 0.2161$ ft·lb/s $= 1$ Btu/h

Note: In engineering, Btu/h is often written Btuh.

Other Power Conversion

solar power, direct sunlight at earth's surface: about 1 kW·m^{-2}

Magnetic Field

1 tesla (T) $\equiv 10^4$ gauss $= 10^9$ γ

10^4 T $= 1$ gauss $= 10$ γ

10^{-9} T $= 10^{-5}$ gauss $= 1$ γ

APPENDIX E
THE ELEMENTS

Periodic Table of the Elements

Legend:

1	Atomic number
H	Symbol
1.008	Atomic mass (u)[a]

Periodic table (showing atomic number, symbol, atomic mass):

1 **H** 1.008

2 **He** 4.003

3 **Li** 6.941 · 4 **Be** 9.012 · 5 **B** 10.81 · 6 **C** 12.01 · 7 **N** 14.01 · 8 **O** 16.00 · 9 **F** 19.00 · 10 **Ne** 20.18

11 **Na** 22.99 · 12 **Mg** 24.31 · 13 **Al** 26.98 · 14 **Si** 28.09 · 15 **P** 30.97 · 16 **S** 32.07 · 17 **Cl** 35.45 · 18 **Ar** 39.95

19 **K** 39.10 · 20 **Ca** 40.08 · 21 **Sc** 44.96 · 22 **Ti** 47.88 · 23 **V** 50.94 · 24 **Cr** 52.00 · 25 **Mn** 54.94 · 26 **Fe** 55.85 · 27 **Co** 58.93 · 28 **Ni** 58.69 · 29 **Cu** 63.55 · 30 **Zn** 65.39 · 31 **Ga** 69.72 · 32 **Ge** 72.61 · 33 **As** 74.92 · 34 **Se** 78.96 · 35 **Br** 79.90 · 36 **Kr** 83.80

37 **Rb** 85.47 · 38 **Sr** 87.62 · 39 **Y** 88.91 · 40 **Zr** 91.22 · 41 **Nb** 92.91 · 42 **Mo** 95.94 · 43 **Tc** (98) · 44 **Ru** 101.07 · 45 **Rh** 102.91 · 46 **Pd** 106.42 · 47 **Ag** 107.87 · 48 **Cd** 112.41 · 49 **In** 114.82 · 50 **Sn** 118.71 · 51 **Sb** 121.75 · 52 **Te** 127.60 · 53 **I** 126.90 · 54 **Xe** 131.29

55 **Cs** 132.91 · 56 **Ba** 137.33 · 57–71 Lanthanide series* · 72 **Hf** 178.49 · 73 **Ta** 180.95 · 74 **W** 183.85 · 75 **Re** 186.21 · 76 **Os** 190.2 · 77 **Ir** 192.22 · 78 **Pt** 195.08 · 79 **Au** 196.97 · 80 **Hg** 200.59 · 81 **Tl** 204.38 · 82 **Pb** 207.2 · 83 **Bi** 208.98 · 84 **Po** (209) · 85 **At** (210) · 86 **Rn** (222)

87 **Fr** (223) · 88 **Ra** (226) · 89–103 Actinide series† · 104 **Unq** (261) · 105 **Unp** (252) · 106 **Unh** (263) · 107 **Uns** (262) · 108 **Uno** (265) · 109 **Une** (266)

*Lanthanide Series:
57 **La** 138.91 · 58 **Ce** 140.12 · 59 **Pr** 140.91 · 60 **Nd** 144.24 · 61 **Pm** (145) · 62 **Sm** 150.36 · 63 **Eu** 151.97 · 64 **Gd** 157.25 · 65 **Tb** 158.93 · 66 **Dy** 162.50 · 67 **Ho** 164.93 · 68 **Er** 167.26 · 69 **Tm** 168.93 · 70 **Yb** 173.04 · 71 **Lu** 174.97

†Actinide Series:
89 **Ac** (227) · 90 **Th** 232.04 · 91 **Pa** (231.04) · 92 **U** 238.03 · 93 **Np** (237) · 94 **Pu** (244) · 95 **Am** (243) · 96 **Cm** (247) · 97 **Bk** (247) · 98 **Cf** (251) · 99 **Es** (252) · 100 **Fm** (257) · 101 **Md** (258) · 102 **No** (259) · 103 **Lr** (260)

[a] Atomic mass is average over abundances of stable isotopes. For radioactive elements, mass is that of the most stable important (in availability, etc.) isotope.

Elements and Isotopes (scaled to $^{12}C = 12$)

The atomic weights of stable elements reflect the abundances of different isotopes, and therefore may vary slightly, depending on the origin and treatment of the material. The values given here apply to elements as they exist naturally on earth. For stable elements, parentheses express uncertainties in the last decimal place given. For elements with no stable isotopes (indicated in boldface), sets of most important isotopes selected by the IUPAC Commission on Atomic Weights and Isotopic Abundances are given.

Atomic Number	Names	Symbol	Atomic Weight
1	Hydrogen	H	1.00794 (7)
2	Helium	He	4.002602 (2)
3	Lithium	Li	6.941 (2)
4	Beryllium	Be	9.012182 (3)
5	Boron	B	10.811 (5)
6	Carbon	C	12.011 (1)
7	Nitrogen	N	14.00674 (7)
8	Oxygen	O	15.9994 (3)
9	Fluorine	F	18.9984032 (9)
10	Neon	Ne	20.1797 (6)
11	Sodium (Natrium)	Na	22.989768 (6)
12	Magnesium	Mg	24.3050 (6)
13	Aluminum	Al	26.981539 (5)
14	Silicon	Si	28.0855 (3)
15	Phosphorus	P	30.973762 (4)
16	Sulfur	S	32.066 (6)
17	Chlorine	Cl	35.4527 (9)
18	Argon	Ar	39.948 (1)
19	Potassium (Kalium)	K	39.0983 (1)
20	Calcium	Ca	40.078 (4)
21	Scandium	Sc	44.955910 (9)
22	Titanium	Ti	47.88 (3)
23	Vanadium	V	50.9415 (1)
24	Chromium	Cr	51.9961 (6)
25	Manganese	Mn	54.93805 (1)
26	Iron	Fe	55.847 (3)
27	Cobalt	Co	58.93320 (1)
28	Nickel	Ni	58.69 (1)
29	Copper	Cu	63.546 (3)
30	Zinc	Zn	65.39 (2)
31	Gallium	Ga	69.723 (4)
32	Germanium	Ge	72.61 (2)
33	Arsenic	As	74.92159 (2)
34	Selenium	Se	78.96 (3)
35	Bromine	Br	79.904 (1)
36	Krypton	Kr	83.80 (1)
37	Rubidium	Rb	85.4678 (3)
38	Strontium	Sr	87.62 (1)
39	Yttrium	Y	88.90585 (2)
40	Zirconium	Zr	91.224 (2)
41	Niobium	Nb	92.90638 (2)
42	Molybdenum	Mo	95.94 (1)
43	**Technetium**	**Tc**	**97, 98, 99**
44	Ruthenium	Ru	101.07 (2)
45	Rhodium	Rh	102.90550 (3)
46	Palladium	Pd	106.42 (1)
47	Silver	Ag	107.8682 (2)
48	Cadmium	Cd	112.411 (8)
49	Indium	In	114.82 (1)
50	Tin	Sn	118.710 (7)
51	Antimony (Stibium)	Sb	121.75 (3)
52	Tellurium	Te	127.60 (3)

53	Iodine	I	126.90447 (3)
54	Xenon	Xe	131.29 (2)
55	Cesium	Cs	132.90543 (5)
56	Barium	Ba	137.327 (7)
57	Lanthanum	La	138.9055 (2)
58	Cerium	Ce	140.115 (4)
59	Praseodymium	Pr	140.90765 (3)
60	Neodymium	Nd	144.24 (3)
61	**Promethium**	**Pm**	**145, 147**
62	Samarium	Sm	150.36 (3)
63	Europium	Eu	151.965 (9)
64	Gadolinium	Gd	157.25 (3)
65	Terbium	Tb	158.92534 (3)
66	Dysprosium	Dy	162.50 (3)
67	Holmium	Ho	164.93032 (3)
68	Erbium	Er	167.26 (3)
69	Thulium	Tm	168.93421 (3)
70	Ytterbium	Yb	173.04 (3)
71	Lutetium	Lu	174.967 (1)
72	Hafnium	Hf	178.49 (2)
73	Tantalum	Ta	180.9479 (1)
74	Tungsten (Wolfram)	W	183.85 (3)
75	Rhenium	Re	186.207 (1)
76	Osmium	Os	190.2 (1)
77	Iridium	Ir	192.22 (3)
78	Platinum	Pt	195.08 (3)
79	Gold	Au	196.96654 (3)
80	Mercury	Hg	200.59 (3)
81	Thallium	Tl	204.3833 (2)
82	Lead	Pb	207.2 (1)
83	Bismuth	Bi	208.98037 (3)
84	**Polonium**	**Po**	**209, 210**
85	**Astatine**	**At**	**210, 211**
86	**Radon**	**Rn**	**211, 220, 222**
87	**Francium**	**Fr**	**223**
88	**Radium**	**Ra**	**223, 224, 226, 228**
89	**Actinium**	**Ac**	**227**
90	Thorium	Th	232.0381 (1)
91	Protactinium	Pa	231.03588 (2)
92	Uranium	U	238.0289 (1)
93	**Neptunium**	**Np**	**237, 239**
94	**Plutonium**	**Pu**	**238, 239, 240, 241, 242, 244**
95	**Americium**	**Am**	**241, 243**
96	**Curium**	**Cm**	**243, 244, 245, 246, 247, 248**
97	**Berkelium**	**Bk**	**247, 249**
98	**Californium**	**Cf**	**249, 250, 251, 252**
99	**Einsteinium**	**Es**	**252**
100	**Fermium**	**Fm**	**257**
101	**Mendelevium**	**Md**	**255, 256, 258, 260**
102	**Nobelium**	**No**	**253, 254, 255, 259**
103	**Lawrencium**	**Lr**	**256, 258, 259, 261**
104	**Unnilquadium**	**Unq**	**257, 259, 260, 261**
105	**Unnilpentium**	**Unp**	**260, 261, 262**
106	**Unnilhexium**	**Unh**	**259, 260, 261, 263**
107	**Unnilseptium**	**Uns**	**261, 262**
108	**Unniloctium**	**Uno**	**264, 265**
109	**Unnilennium**	**Une**	**266**

Note: The systematic names for elements 104–109 use the following code for each digit with an "ium" ending to indicate that they are metals: 1—un; 2—bi; 3—tri; 4—quad; 5—pent; 6—hex; 7—sept; 8—oct; 9—enn; 0—nil. (Element 104 was called Rutherfordium (Rf) by the American discoverers and Kurschatovium (Ku) by the Soviet discoverers. Element 105 was called Hahnium (Ha) by the American discoverers and Nielsbohrium (Ns) by the Soviet discoverers.)

Atomic weights and selected isotopes through element 105 from the IUPAC Commission on Atomic Weights and Isotopic Abundances, courtesy of I. Lynus Barnes, National Bureau of Standards. Isotopes for elements 106–109 from Peter Armbruster, Gesellschaft für Schwerionenforschung mbH.

Sun, Planets, Principal Satellites[a]

Body	Mass (10^{24} kg)	Mean Radius[b] (10^6 m except as noted)	Surface Gravity (m/s^2)[c]	Escape Speed (km/s)	Sidereal Rotation Period[d,f] (d)	Mean Distance from Central Body[e] (10^6 km)	Orbital Period	Orbital Speed (km/s)
Sun	7.99×10^6	696	274	618	36 at poles 27 at equator	2.6×10^{11}	200 My	250
Mercury	0.330	2.44	3.70	4.25	58.6	57.9	88.0 d	48
Venus	4.87	6.05	8.87	10.4	−243	108	225 d	35
Earth	5.97	6.37	9.81	11.2	0.997	150	365.3 d	30
Moon	0.0735	1.74	1.62	2.38	27.3	0.385	27.3 d	1.0
Mars	0.642	3.38	3.74	5.03	1.03	228	1.88 y	24.1
Phobos	9.6×10^{-9}	9–13 km	0.001	0.008	0.32	9.4×10^{-3}	0.32 d	2.1
Deimos	2×10^{-9}	5–8 km	0.001	0.005	1.3	23×10^{-3}	1.3 d	1.3
Jupiter	1.90×10^3	69.1	26.5	60.6	0.414	778	11.9 y	13.0
Io	0.0889	1.82	1.8	2.6	1.77	0.422	1.77 d	17
Europa	0.478	1.57	1.3	2.0	3.55	0.671	3.55 d	14
Ganymede	0.148	2.63	1.4	2.7	7.15	1.07	7.15 d	11
Callisto	0.107	2.40	1.2	2.4	16.7	1.88	16.7 d	8.2
and at least 13 smaller satellites								
Saturn	569	56.8	11.8	36.6	0.438	1.43×10^3	29.5 y	9.65
Tethys	0.0007	0.53	0.2	0.4	1.89	0.294	1.89 d	11.3
Dione	0.00015	0.56	0.3	0.6	2.74	0.377	2.74 d	10.0
Rhea	0.0025	0.77	0.3	0.5	4.52	0.527	4.52 d	8.5
Titan	0.135	2.58	1.4	2.6	15.9	1.22	15.9 d	5.6
Iapetus	0.0019	0.73	0.2	0.6	79.3	3.56	79.3 d	3.3
and at least 12 smaller satellites								
Uranus	86.6	25.0	9.23	21.5	−0.65	2.87×10^3	84.1 y	6.79
Ariel	0.0013	0.58	0.3	0.4	2.52	0.19	2.52 d	5.5
Umbriel	0.0013	0.59	0.3	0.4	4.14	0.27	4.14 d	4.7
Titania	0.0018	0.81	0.2	0.5	8.70	0.44	8.70 d	3.7
Oberon	0.0017	0.78	0.2	0.5	13.5	0.58	13.5 d	3.1
and at least 11 smaller satellites								
Neptune	103	24.0	11.9	23.9	0.768	4.50×10^3	165 y	5.43
Triton	0.134	1.90	2.5	3.1	5.88	0.354	5.88 d	4.4
and at least 1 other satellite								
Pluto	0.015	1.3	0.4	1.2	−6.39	5.91×10^3	249 y	4.7
and at least 1 satellite								

[a]Except for Mars, principal satellites are those with radii greater than 500 km.

[b]Mean radius is calculated as average of equatorial and polar radii.

[c]Surface gravity is calculated from mean radius and mass as $g = GM/r^2$, except where measured directly.

[d]Rotation period: Minus sign indicates retrograde rotation, in the opposite sense from the orbital motion. Note that all the moons given are in synchronous rotation, with their period of rotation matching their period of revolution.

[e]Central body is galactic center for the sun, sun for the planets, and planet for the satellites.

[f]Periods given are sidereal—that is, the time for the body to return to the same position or orientation relative to the distant stars rather than the sun.

See page 1050 for additional data.

Astrophysical Data

Astronomical Unit, A.U.	$1.495\,978\,70 \times 10^{11}$ m
Light year	$9.460\,530 \times 10^{15}$ m
Tropical year (1900)	$365.242\,198\,78$ days
Sidereal (by the stars) year	$365.256\,366$ days
Mass of sun	$1.989\,1 \times 10^{30}$ kg
Radius of sun	6.96×10^5 km
Solar constant	$1368\ \mathrm{W \cdot m^{-2}}$
Luminosity of sun (from Solar Maximum Mission, 1980–1986 interval)	3.85×10^{26} W
Mass of earth	$5.974\,2 \times 10^{24}$ kg
Radius of earth (equator)	$6\,378.140$ km
Hubble constant	$50 - 100$ km/s/megaparsec $= 15 - 30$ km/s/Mly $= 1.6 - 3.2 \times 10^{-18}\ \mathrm{s^{-1}}$

CURRENT MOST PRECISE VALUES FOR PHYSICAL CONSTANTS

Parentheses indicate uncertainties in the last decimal places given.

speed of light, c	299 792 458 m/s (exact)
gravitational constant, G	$6.672\ 59(85) \times 10^{-11}$ N$\cdot$m$^2\cdot$kg^{-2}
Planck's constant, h	$6.626\ 075\ 5(40) \times 10^{-34}$ J$\cdot$s
Boltzmann constant, k	$1.380\ 658\ (12) \times 10^{-23}$ J$\cdot$K^{-1}
Stefan-Boltzmann constant, σ	$5.670\ 51(19) \times 10^{-8}$ W$\cdot$m$^{-2}\cdot$K^{-4}
Rydberg constant, R_∞	$1.097\ 373\ 153\ 4(13) \times 10^7$ m^{-1}
Fine-structure constant, α^{-1}	$137.035\ 989\ 5(61)$
Bohr radius, a_0	$5.291\ 772\ 49(24) \times 10^{-11}$ m
Avogadro's number, N_A	$6.022\ 136\ 7(36) \times 10^{23}$ mole^{-1}
mass of hydrogen atom, m_H	$1.673\ 534\ 0(10) \times 10^{-27}$ kg
mass of neutron, m_n	$1.674\ 928\ 6(10) \times 10^{-27}$ kg
	$939.565\ 63(28)$ MeV
mass of proton, m_p	$1.672\ 623\ 1(10) \times 10^{-27}$ kg
	$938.272\ 31(28)$ MeV
mass of electron, m_e	$9.109\ 389\ 7(54) \times 10^{-31}$ kg
	$0.510\ 999\ 06(15)$ MeV
proton-electron mass ratio, m_p/m_e	$1836.152\ 701(37)$
elementary charge, e	$1.602\ 177\ 33(49) \times 10^{-19}$ C
charge to mass, e/m_e	$1.758\ 819\ 62(53) \times 10^{11}$ C$\cdot$kg^{-1}
permeability of vacuum, $\mu_0 = 4\pi \times 10^{-7}$	$12.566\ 370\ 614 \ldots \times 10^{-7}$ H/m
permittivity of vacuum, $\epsilon_0 = 1/\mu_0 c^2$	$8.854\ 187\ 817 \ldots \times 10^{-12}$ F/m

Mathematical Constants

$\pi = 3.141\ 592\ 653\ 6$
$e = 2.718\ 281\ 828\ 5$

Values from *The 1986 Adjustment of the Fundamental Physical Constants*, Report of the CODATA Task Group on Fundamental Constants, prepared by E. Richard Cohen and Barry N. Taylor. Committee on Data for Science and Technology CODATA Bulletin No. 63, Pergamon Press.

NOBEL PRIZES

All Nobel Prizes in physics are listed (and marked with a P), as well as relevant Nobel Prizes in Chemistry (C) and Physiology or Medicine (M). The key dates for some of the scientific work are supplied; they often antedate the prize considerably.

1901   (P) *Wilhelm Roentgen* for discovering x-rays (1895).

1902   (P) *Hendrik A. Lorentz* for predicting the Zeeman effect and *Pieter Zeeman* for discovering the Zeeman effect, the splitting of spectral lines in magnetic fields.

1903   (P) *Antoine-Henri Becquerel* for discovering radioactivity (1896) and *Pierre and Marie Curie* for studying radioactivity.

1904   (P) *Lord Rayleigh* for studying the density of gases and discovering argon.

      (C) *William Ramsay* for discovering the inert gas elements helium, neon, xenon, and krypton, and placing them in the periodic table.

1905   (P) *Philipp Lenard* for studying cathode rays, electrons (1898–1899).

1906   (P) *J. J. Thomson* for studying electrical discharge through gases and discovering the electron (1897).

1907   (P) *Albert A. Michelson* for inventing optical instruments and measuring the speed of light (1880s).

1908   (P) *Gabriel Lippmann* for making the first color photographic plate, using interference methods (1891).

      (C) *Ernest Rutherford* for discovering that atoms can be broken apart by alpha rays and for studying radioactivity.

1909   (P) *Guglielmo Marconi* and *Carl Ferdinand Braun* for developing wireless telegraphy.

1910   (P) *Johannes D. van der Waals* for studying the equation of state for gases and liquids (1881).

1911   (P) *Wilhelm Wien* for discovering Wien's law giving the peak of a blackbody spectrum (1893).

      (C) *Marie Curie* for discovering radium and polonium (1898) and isolating radium.

1912   (P) *Nils Dalén* for inventing automatic gas regulators for lighthouses.

1913   (P) *Heike Kamerlingh Onnes* for studying materials in low temperatures and liquefying helium (1908).

1914   (P) *Max T. F. von Laue* for studying x-rays from their diffraction by crystals, showing that x-rays are electromagnetic waves (1912).

      (C) *Theodore W. Richards* for determining the atomic weights of sixty elements, indicating the existence of isotopes.

1915   (P) *William Henry Bragg* and *William Lawrence Bragg*, his son, for studying the diffraction of x-rays in crystals.

1916   (P) *No prize offered.*

1917   (P) *Charles Barkla* for studying atoms by x-ray scattering (1906).

1918   (P) *Max Planck* for discovering energy quanta (1900).

1919   (P) *Johannes Stark,* for discovering the Stark effect, the splitting of spectral lines in electric fields (1913).

1920   (P) *Charles-Édouard Guillaume* for discovering invar, a nickel-steel alloy of low coefficient of expansion

Five Nobel Prize winners: (left to right) Nernst, Einstein, Planck, Millikan, and von Laue.

(C) *Walther Nernst* for studying heat changes in chemical reactions and formulating the third law of thermodynamics (1918).

1921 (P) *Albert Einstein* for explaining the photoelectric effect (1905).

(C) *Frederick Soddy* for studying the chemistry of radioactive substances and discovering isotopes (1912).

1922 (P) *Niels Bohr* for his model of the atom and its radiation (1913).

(C) *Francis W. Aston* for using the mass spectrograph to study atomic weights, thus discovering 212 of the 287 naturally occurring isotopes.

1923 (P) *Robert A. Millikan* for measuring the charge on an electron (1911) and for studying the photographic effect experimentally (1914).

1924 (P) *Karl M. G. Siegbahn* for his work in x-ray spectroscopy.

1925 (P) *James Franck* and *Gustav Hertz* for discovering the Franck-Hertz effect in electron–atom collisions.

1926 (P) *Jean-Baptiste Perrin* for studying Brownian motion to validate the discontinuous structure of matter and measure the size of atoms.

1927 (P) *Arthur Holly Compton* for discovering the Compton effect on x-rays, their change in wavelength when they collide with matter (1922), and *Charles T. R. Wilson* for inventing the cloud chamber, used to study charged particles (1906).

1928 (P) *Owen W. Richardson* for studying the thermionic effect and electrons emitted by hot metals (1911).

1929 (P) *Louis Victor de Broglie* for discovering the wave nature of electrons (1923).

1930 (P) *Chandrasekhara Venkata Raman* for studying Raman scattering, the scattering of light by atoms and molecules with a change in wavelength (1928).

1931 (P) No prize offered.

1932 (P) *Werner Heisenberg* for starting quantum mechanics (1925).

1933 (P) *Erwin Schrödinger* and *Paul A. M. Dirac* for developing wave mechanics (1925) and relativistic quantum mechanics (1927).

1934 (P) No prize offered.

(C) *Harold Urey* for discovering heavy hydrogen, deuterium (1931).

1935 (P) *James Chadwick* for discovering the neutron (1932).

(C) *Irène* and *Frédéric Joliot-Curie* for synthesizing new radioactive elements.

1936 (P) *Carl D. Anderson* for discovering the positron in particular and antimatter in general (1932) and *Victor F. Hess* for discovering cosmic rays.
(C) *Peter J. W. Debye* for studying dipole moments and diffraction of x-rays and electrons in gases.

1937 (P) *Clinton Davisson* and *George Thomson* for discovering the diffraction of electrons by crystals, confirming de Broglie's hypothesis (1927).

1938 (P) *Enrico Fermi* for producing the transuranic radioactive elements by neutron irradiation (1934–1937).

1939 (P) *Ernest O. Lawrence* for inventing the cyclotron.

1940–1942   No prizes offered.

1943 (P) *Otto Stern* for developing molecular-beam studies (1923), and using them to discover the magnetic moment of the proton (1933).

1944 (P) *Isador I. Rabi* for discovering nuclear magnetic resonance in atomic and molecular beams.
(C) *Otto Hahn* for discovering nuclear fission (1938).

1945 (P) *Wolfgang Pauli* for discovering the exclusion principle (1924).

1946 (P) *Percy W. Bridgman* for studying physics at high pressures.

1947 (P) *Edward V. Appleton* for studying the ionosphere.

1948 (P) *Patrick M. S. Blackett* for studying nuclear physics with cloud-chamber photographs of cosmic-ray interactions.

1949 (P) *Hideki Yukawa* for predicting the existence of mesons (1935).

1950 (P) *Cecil F. Powell* for developing the method of studying cosmic rays with photographic emulsions and discovering new mesons.

1951 (P) *John D. Cockcroft* and *Ernest T. S. Walton* for transmuting nuclei in an accelerator (1932).
(C) *Edwin M. McMillan* for producing neptunium (1940) and *Glenn T. Seaborg* for producing plutonium (1941) and further transuranic elements.

1952 (P) *Felix Bloch* and *Edward Mills Purcell* for discovering nuclear magnetic resonance in liquids and gases (1946).

1953 (P) *Frits Zernike* for inventing the phase-contrast microscope, which uses interference to provide high contrast.

1954 (P) *Max Born* for interpreting the wave function as a probability (1926) and other quantum-mechanical discoveries and *Walther Bothe* for developing the coincidence method to study subatomic particles (1930–1931), producing, in particular, the particle interpreted by Chadwick as the neutron.

1955 (P) *Willis E. Lamb, Jr.* for discovering the Lamb shift in the hydrogen spectrum (1947) and *Polykarp Kusch* for determining the magnetic moment of the electron (1947).

1956 (P) *John Bardeen, Walter H. Brattain,* and *William Shockley* for inventing the transistor (1956).

1957 (P) *T. -D. Lee* and *C. -N. Yang* for predicting that parity is not conserved in beta decay (1956).

1958 (P) *Pavel A. Čerenkov* for discovering Čerenkov radiation (1935) *and Ilya M. Frank* and *Igor Tamm* for interpreting it (1937).

1959 (P) *Emilio G. Segrè* and *Owen Chamberlain* for discovering the antiproton (1955).

1960 (P) *Donald A. Glaser* for inventing the bubble chamber to study elementary particles (1952).
(C) *Willard Libby* for developing radiocarbon dating (1947).

1961 (P) *Robert Hofstadter* for discovering internal structure in protons and neutrons and *Rudolf L. Mössbauer* for discovering the Mössbauer effect of recoilless gamma-ray emission (1957).

1962 (P) *Lev Davidovich Landau* for studying liquid helium and other condensed matter theoretically.

1963 (P) *Eugene P. Wigner* for applying symmetry principles to elementary-particle theory and *Maria Goeppert Mayer* and *J. Hans D. Jensen* for studying the shell model of nuclei (1947).

1964 (P) *Charles H. Townes, Nikolai G. Basov,* and *Alexandr M. Prokhorov* for developing masers (1951–1952) and lasers.

1965 (P) *Sin-itiro Tomonaga, Julian S. Schwinger,* and *Richard P. Feynman* for developing quantum electrodynamics (1948).

1966 (P) *Alfred Kastler* for his optical methods of studying atomic energy levels.

1967 (P) *Hans Albrecht Bethe* for discovering the routes of energy production in stars (1939).

1968 (P) *Luis W. Alvarez* for discovering resonance states of elementary particles.

1969 (P) *Murray Gell-Mann* for classifying elementary particles (1963).

1970 (P) *Hannes Alfvén* for developing magnetohydrodynamic theory and *Louis Eugène Félix Néel* for discovering antiferromagnetism—alternate opposing magnetic domains—and ferrimagnetism—unequally opposing domains (1930s).

1971 (P) *Dennis Gabor* for developing holography (1947).
(C) *Gerhard Herzberg* for studying the structure of molecules spectroscopically.

1972 (P) *John Bardeen, Leon N. Cooper,* and *John Robert Schrieffer* for explaining superconductivity (1957).

1973 (P) *Leo Esaki* for discovering tunneling in semiconductors, *Ivar Giaever* for discovering tunneling in superconductors, and *Brian D. Josephson* for predicting the Josephson effect, which involves tunneling (1958–1962).

1974 (P) *Anthony Hewish* for discovering pulsars (with no mention of Jocelyn Bell, his graduate student, who first noticed the signals in 1967) and *Martin Ryle* for developing radio interferometry.

1975 (P) *Aage N. Bohr, Ben R. Mottelson,* and *James Rainwater* for discovering why some nuclei take asymmetric shapes.

1976 (P) *Burton Richter* and *Samuel C. C. Ting* for discovering the J/psi particle, the first charmed particle (1974).

1977 (P) *John H. Van Vleck, Nevill F. Mott,* and *Philip W. Anderson* for studying solids quantum-mechanically.
(C) *Ilya Prigogine* for extending thermodynamics to show how life could arise in the face of the second law.
(M) *Rosalyn S. Yalow* for developing radioimmunoassay methods.

1978 (P) *Arno A. Penzias* and *Robert W. Wilson* for discovering the cosmic background radiation (1965) and *Pyotr Kapitsa* for his studies of liquid helium.

1979 (P) *Sheldon L. Glashow, Abdus Salam,* and *Steven Weinberg* for developing the theory that unified the weak and electromagnetic forces (1958–1971).
(M) *Allan M. Cormack* and *Godfrey N. Hounsfield* for developing CAT scanners.

1980 (P) *Val Fitch* and *James W. Cronin* for discovering CP (*charge-parity*) violation (1964), which possibly explains the cosmological dominance of matter over antimatter.

1981 (P) *Nicolaas Bloembergen* and *Arthur L. Schawlow* for developing laser spectroscopy and *Kai M. Siegbahn* for developing electron spectroscopy (1958).

1982 (P) *Kenneth G. Wilson* for developing a method of constructing theories of phase transitions.

1983 (P) *William A. Fowler* for theoretical studies of astrophysical nucleosynthesis and *Subramanyan Chandrasekhar* for studying physical processes of importance to stellar structure and evolution, including the prediction of white dwarf stars (1930).

1984 (P) *Carlo Rubbia* for discovering the W and Z particles, verifying the electroweak unification, and *Simon van der Meer*, for developing the method of stochastic cooling of the CERN beam that allowed the discovery (1982–1983).

1985 (P) *Klaus von Klitzing* for the quantized Hall effect, relating to conductivity in the presence of a magnetic field (1980).

1986 (P) *Ernst Ruska* for inventing the electron microscope (1931), and *Gerd Binnig* and *Heinrich Rohrer* for inventing the scanning-tunneling electron microscope (1981).

ANSWERS TO ODD-NUMBERED PROBLEMS

Chapter 1

3. 10^6

5. 8.6 m^2/L

7. yes, by 7 mi/h

9. 10^8

11. 2.5×10^6 m

13. 7.4×10^6 m/s^2

15. (a) 2.5×10^{-4} mm^2; (b) 1.6×10^{-2} mm

Chapter 2

1. 10.14 m/s

3. 12.27 mi/h

5. (a) 24 km north; (b) 9.6 km/h; (c) 16 km/h; (d) 0; (e) 0

7. 13.11 mi/h

9. 48 mi/h

11. 1 m/s = 2.24 mi/h

13. 51 ft/s = 35 mi/h

15. (a) 2d, 17h; (b) 70 km/h

17. 2.6 h later, 1800 km from New York

19. $v_x = 3bt^2$

21. (a) t = 0 s, 0.13 s, 2.5 s; (b) $v_x = 3bt^2 - 2ct + d$; (c) $v_{xo} = 1.0$ m/s; (d) t = 0.065 s, 1.7 s

23. 0

25. 2×10^{11} m/s^2

27. (a) 28 m/s; (b) 23 m/s^2; (c) 9.4 m/s; (d) 11 m/s^2

29. (a) 2.0 m/s^2; (b) 150 m

31. 27 ft/s^2

33. (a) $t = 2v_0/a$; (b) $v = v_0$

35. 22 m/s

37. (a) 0.42 m/s^2; (b) toward Chicago; (c) 1.1 km

39. $a = 125$ g = 1200 m/s^2

41. yes; $a = 370$ m/s^2

43. no collision; they stop 10 m apart

45. 4.6×10^{-3} m/s^2

47. 36 ft/s

49. (a) 27 m; (b) 4.7 s

51. Venus

53. Torey's, by 120 ms

55. 2.7 m/s faster

57. (a) 7.88 m/s and 7.67 m/s, respectively; (b) the upward-jumping diver hits first, by 0.16 s

59. (a) 2.4 m; (b) 47 cm before the chute

61. (b) 3.8 s; (c) 19 m; (d) 100 m

65. (a) 33 m; (b) 8%

67. 1.19 m

Chapter 3

1. 260 m, 7.9° N of W

3. (a) $A_x = 8.7$, $A_y = 5.0$; (b) $A_x = 9.7$, $A_y = -2.6$; (c) $A_x = 10$, $A_y = 0$

5. 18.9 units long, 7.1° E of S

7. $C = -15\hat{\imath} + 9\hat{\jmath} - 18\hat{k}$

9. $v_x = -14$ m/s; $v_y = -12$ m/s

11. 13 mi/h/min 19° W of S

13. 5.1 m/s^2, 49° W of S

15. (a) $10\hat{\imath} - 21\hat{\jmath}$ m/s; (b) 13 m/s; (c) 89°

17. (a) $v(t = 0) = c\hat{\imath} = 6.7\hat{\imath}$ m/s; (b) $t = \sqrt{c/3b} = 1.7$ s; (c) 8.2 m/s

19. (a) t = 18 s; (b) 300 m; (c) 22 m/s at 120° to x-axis

21. (a) 2.6×10^{17} cm/s^2, upward; (b) parabolic

23. (a) 1.4 s; (b) 10 m

25. 1.23 m/s

27. 8.3 m/s at 61°

29. (a) 6640 m/s (14,850 mi/h); (b) 16 min; (c) 8280 m/s

31. 1.1 s

33. 1060 m

37. (a) 8.8 m; (b) 0.5 m

41. 31.2° or 65.7°

43. 2.8×10^{-3} m/s^2

45. 53 min

47. 20 cm

49. 3.44×10^{-10} s

51. $a_t = 3.8$ cm/s^2; $a_r = 95$ cm/s^2

53. (a) 18 km/h, 15 km/h, 10 km/h, 15 km/h; (b) 21 km/h, 17 km/h, 12 km/h, 17 km/h

55. $a = 1000$ m/s^2 in all cases; direction is (a) upward, (b) downward, (c) backward. Velocities in car frame: (a) 19 m/s backward; (b) 19 m/s forward; (c) 19 m/s downward. Velocities in ground frame: (a) 0; (b) 38 m/s forward; (c) 27 m/s at 45° below the forward direction

59. 7.2 m/s at 77° to horizontal

61. 1.2×10^4 m/s^2

65. 26,000 km/h at 26° to equator

Chapter 4

1. 3.8×10^6 N

3. (a) 11 m; (b) 24 m; (c) 43 m; (d) 53 m

5. 2.4×10^5 N

7. 10^{-4} N

9. Uranus

11. $0.75g = 7.4$ m/s^2

13. 98 N in horizontal string; $98\sqrt{2}$ N = 140 N in diagonal string

15. (a) 3.1×10^7 N; (b) 1.6

17. 530 N, 3.6 times the weight

21. (a) 5260 N; (b) 1080 N; (c) 494 N; (d) 589 N

23. 1.3×10^{-21} cm

27. (a) 132 cm; (b) 127 cm; (c) 120 cm; (d) 40 m/s^2

29. 1.9 m/s^2

31. 14 N

35. $\ell + nm(a+g)/k$, where n is the spring number measured from the bottom
37. $Mg + mg(1-y/\ell)$
39. 7.2 m
41. (a) 0.40 mg; (b) 2.40 mg; (c) 1.40 mg
43. (a) 1.6×10^4 N; (b) 850 N
45. (a) 7.1 m/s^2; (b) 48 m/s
47. (a) $a = \dfrac{m_f - m_s}{m_s}\, g$; (b) $y = \dfrac{a_s h}{(1 \div m_s/m_f)\,(a_s + g)}$

Chapter 5

1. 220 N
3. 880 N
5. (a) 6.3 m/s^2; (b) 0.44 s
7. 43 cm
9. 26 s
11. (a) 13 m/s^2; (b) 3.8 m/s
13. 340 N, 1.6 times the weight
15. (a) 8.0°; (b) 0.50 m/s^2
17. 0.38 s
19. 4.1 m/s^2, 4.7% of accident deceleration
21. (a) 1.6 m/s^2; (b) 3.2 N
23. 4.2 m/s^2
25. 95 km/h
27. 0.12
29. (a) yes; (b) yes
31. $T = mg\left(\dfrac{\mu_s \cos\theta - \sin\theta}{\cos\theta + \mu_s \sin\theta}\right)$
33. (a) 9.6 cm; (b) no
37. 12 cm
39. 3.6 m/s
41. 8.2 m/s
43. $h = \frac{1}{2}\ell$
45. 3.45 rev/min
47. (a) 35°; (b) 22°
49. 37°
51. 0.55
53. (a) 63 km/h (17 m/s); (b) 23 km/h (6.3 m/s)
55. $v = \sqrt{gh}$
57. $\mu_k = \dfrac{v_0^2}{2gx_1} - \dfrac{x_2^2}{4x_1 h}$
63. 28 cm

Chapter 6

1. 1300 J
3. 5.9×10^4 J
5. (a) 400 J; (b) 31 kg
7. 5.9×10^6 N
9. (a) 370 J; (b) 0.26
13. (a) 14; (b) −12; (c) −16
17. (a) 45°; (b) 111°; (c) 66°
19. 25°
21. (a) 360 J; (b) 350 J; (c) 357.5 J; (d) 359.375 J
23. $k_B = 8k_A$
25. 190 J
27. (a) 30 J; (b) 56 J; (c) 72 J
29. $F_0\left(x + \dfrac{x^2}{2\ell_0} + \dfrac{\ell_0^2}{\ell_0 + x} - \ell_0\right)$
33. 90 J

35. (a) 1.4×10^{10} J; (b) 3.3×10^6 J; (c) 28 J
37. 2.3×10^7 m/s, 0.077c
39. 2300 J
41. 7400 J
43. (a) 24 J; (b) 18 m/s
45. 4.1 m
47. 85 cm
49. (a) no work is done
51. (a) 60 kW; (b) 1 kW; (c) 0.04 kW
53. (a) 36 MW; (b) 1.1 MW
59. (a) $P = Mgv \cos(vt/R)$, with the argument of cos in radians
63. (b) 2.0 GJ; (c) 45 GJ; (d) 52 GJ
65. (a) $v = \sqrt{2Pt/M}$; (b) $x = (8P/9m)^{1/2}t^{3/2}$

Chapter 7

1. (a) $2\mu mg\ell$; (b) $\sqrt{2}\,\mu mg\ell$
7. $U = F_0\left(x + \dfrac{x^2}{2\ell} + \dfrac{\ell^2}{\ell + x} - \ell\right)$
9. (a) 107 J; (b) 112 J
11. no, the force is not conservative
13. $U = -\frac{1}{3}ax^3 - bx$
15. (a) 1.8 J; (b) +3.6 J; (c) −1.8 J
17. 50 m/s (180 km/h)
23. 2.6×10^7 N/m
25. 15 km/h
29. 2.6 m/s
31. 10^{10} W, the equivalent of 10 1000-MW nuclear plants
33. (a) 2.53×10^5 m/s; (b) 2.91×10^5 m/s; (c) 2.93×10^5 m/s
37. $F = \left[\dfrac{2ax^3}{c^2} - \left(2a + \dfrac{2b}{c^2}\right)x\right]e^{-x^2/c^2}$
41. (a) 18 m/s; (b) 11.5 m/s
43. (a) 41 cm; (b) 1.3 m/s
45. $v = R\sqrt{2\pi e\dot{B}/m}$
47. 2.6 m/s
49. 0.059
51. 62 cm from left end of frictional zone
53. 1000 MW, or twice the output of the coal-burning plant
55. (a) $U = \frac{1}{2}ky^2 - mgy$; (b) $y = mg/k$; (c) $y = 2mg/k$
59. 35 cm
61. (a) 3.3 cm; (b) 6.2×10^7 m/s; (c) 0.76 cm
63. (a) $v = \left[\dfrac{2}{m}\left(\dfrac{ax^2}{2} - \dfrac{bx^4}{4}\right)\right]^{1/2}$; (c) $v_{max} = \dfrac{a}{\sqrt{2mb}}$
65. $v = [2ax^{b+1}/m(b+1)]^{1/2}$

Chapter 8

1. $R_e/\sqrt{2}$
3. 10^{-7}
5. 443 m
7. 1.87 years
9. 16 million years
11. 1.0 hour
15. 9 s
19. $\sqrt{2}$
21. $\frac{1}{99} R_e = 64$ km; underestimate

23. 6.2 km/s

27. $v = \left[2GM\left(\dfrac{3}{R} + \dfrac{1}{r} \right) \right]^{1/2}$

29. (a) $U = -\dfrac{2GMm}{[(\ell/2)^2 + x^2]^{1/2}}$; (c) $F = -\dfrac{2GMx}{[(\ell/2)^2 + x^2]^{3/2}}$

35. (b) 8.8 mm; (c) 3.0 km

41. 1400 km lower

Chapter 9

1. $X = 50$ cm, $Y = 69$ cm, with origin at lower left
3. $\ell/2\sqrt{3}$ along the perpendicular bisector of any side
5. $X = 44$ cm, $Y = 55$ cm, with origin at lower left corner
7. $0.115a$ above the vertex of the missing triangle
9. 0.065 Å from the oxygen, along the symmetry axis
11. $R = (t^2 + \frac{10}{3}t + \frac{7}{3})\hat{\imath} + (\frac{2}{3}t + \frac{8}{3})\hat{\jmath}$;
    $V = (2t + \frac{10}{3})\hat{\imath} + \frac{2}{3}\hat{\jmath}$; $A = 2\hat{\imath}$
13. about 1 Å (10^{-10} m, or the diameter of a hydrogen atom)
15. $m_{\text{mouse}} = \frac{1}{4}m_{\text{bowl}}$
17. 30 m
19. 8.4 km/h
21. (a) 0.14 N/m²; (b) 0.014 mm
23. (a) 0.99 m; (b) 3.9 m/s
25. 3.9 km/s
27. (a) 2×10^4 kg·m/s; (b) 2×10^8 J; (c) yes
29. $v_1 = \left(\dfrac{m_2 kx^2}{m_1^2 + m_1 m_2} \right)^{1/2}$; $v_2 = \left(\dfrac{m_1 kx^2}{m_2^2 + m_1 m_2} \right)^{1/2}$
31. $26\hat{\imath} + 16\hat{\jmath}$ m/s
33. 1100 kg
35. (a) 3.3×10^6 N; (b) 3.4×10^5 kg
37. 0.22
39. $K_{\text{cm}} = 67$ keV before and after; $K_{\text{int}} = 0$ before, $K_{\text{int}} = 80$ keV after
41. $K_{\text{cm}} = 8 \times 10^{10}$ J, largely protons; $K_{\text{int}} = 5 \times 10^{10}$ J, all electrons
43. (a) 37.7°; (b) 0.657 m/s
45. 5.8 s after explosion

Chapter 10

1. (a) $I = 150$ N·s, upward; (b) 3000 N, about 5 times the gravitational force
3. about 6.8×10^{-3} N·s
5. (a) 7.3×10^6 N·s; (b) 5.6×10^6 N
7. $\Delta P/P = 2\%$
9. (a) 6.2 mi/h; (b) 12%
11. 19 kg
15. (a) 10^{15} kg; (b) 10^{12} kg
19. 1.3×10^{-6} J
21. 120°
23. 46 m/s
25. $v_A = -\frac{1}{3}v$, $v_B = \frac{2}{9}v$, $v_C = \frac{8}{9}v$
31. 22°
33. $v_m = 0.32$ m/s, $v_{3m} = 1.3$ m/s
    or $v_m = 1.7$ m/s, $v_{3m} = 0.87$ m/s
35. 13 m/s at 27° to the horizontal
37. $v_A = -\frac{1}{5}v_0\hat{\imath}$; $v_B = \frac{3}{5}\hat{\imath} + \frac{1}{5}\sqrt{3}\hat{\jmath}$; $v_C = \frac{3}{5}\hat{\imath} - \frac{1}{5}\sqrt{3}\hat{\jmath}$
39. 0.88

43. 44
45. (b) $v_1 = 0.28v_0$, $v_2 = 0.48v_0$; (c) 3, $0.26v_0$, $0.31v_0$
47. $v_{1200} = 2.2$ km/h, $v_{1800} = 18$ km/h

Chapter 11

1. 1.4 N·m, into page
3. 30 lbs
5. 420 N
7. 0
9. (a) $-7\hat{k}$; (b) $-8\hat{k}$; (c) $10\hat{k}$; (d) $30\hat{\imath} - 20\hat{\jmath}$;
   (e) $14\hat{\imath} - 28\hat{\jmath}$; (f) 0
11. 72 cm from the center, on the side opposite the worker
13. 16 m from wall
15. 480 N
17. vertical forces both 73.5 N, downward horizontal forces both 33.6 N, away from door jamb at top, toward door jamb at bottom
19. 5.0×10^3 N, tension
21. 0.87
23. 500 N
25. 5.0×10^4 N
27. maximum height of CM is at sphere center; lower for clown (b)
29. stable equilibria at $\theta = 0$ and $\theta = 90°$; unstable equilibrium at $\theta = 90° - \tan^{-1}(\frac{1}{2}) = 63°$
31. There are many possible pairs of forces; one such pair is $F_1 = -0.64\hat{\imath} - 1.9\hat{\jmath}$ applied at $r_1 = 1.0\hat{\imath} + 1.53\hat{k}$, and $F_2 = 1.4\hat{\imath} + 1.4\hat{\jmath} - 1.4\hat{k}$ applied at the origin.
35. (a) $0.44mg$, at 12° to earth's polar axis; (b) $0.036mgR_e$, out of the plane of Fig. 11–61; (c) center of mass is midway between pods; center of gravity is $0.084\,R_e$ to left of center of mass in Fig. 11–61

Chapter 12

1. (a) 7.27×10^{-5} rad/s; (b) 1.75×10^{-3} rad/s;
   (c) 1.45×10^{-4} rad/s; (d) 31.4 rad/s
3. (a) $v = (\pi/30)\,\omega r$; (b) $v = 2\pi\omega r$; (c) $v = (\pi/180)\,\omega r$
5. (a) 66 rpm; (b) 3.7 s
7. (a) 2.0 s; (b) 1.0 rev
9. (a) 15.6 kg·m²; (b) 0.30 sec
13. (a) $\frac{2}{3}m\ell^2$; (b) $\frac{2}{3}m\ell^2$; (c) $\frac{4}{3}m\ell^2$
15. $m_{\text{pulley}} = 0.49$ kg; $m_1 = 0.41$ kg; $m_2 = 0.58$ kg
17. (a) 430 min; (b) 1900 rev
19. 1900 N·m
21. 170 rpm
23. (a) 15 MW; (b) 41%
25. increases by a factor $\omega_2/\omega_1 = 10^7$
27. 17%
29. hollow
31. $\frac{27}{10}(R - r)$
33. 7.1×10^{33} kg·m²/s; 2.7×10^{40} kg·m²/s
35. sun's rotation accounts for 3% of the angular momentum; Jupiter's orbital motion for 62%
37. (a) 45 kg; (b) 22 J
41. $I = mgd/2\omega\Omega$
43. $v = \left[\dfrac{8(m+M)g\ell}{m^2} \left(\frac{1}{4}m + \frac{1}{3}m \right) \right]^{1/2}$

45. Both wheels have stopped rotating about their axes, but the whole contraption is now rotating about the centerline between the two wheels in the same direction as the original disk rotation. The angular speed is

$$\omega_{final} = \frac{MR^2\omega}{\frac{2}{3}m\ell^2 + 3MR^2}$$

Chapter 13

3. (a) $x_0 = 20$ cm; $\omega = \pi/2$ s^{-1}; $\phi = 0$ (b) $x_0 = 30$ cm; $\omega = 2.0$ s^{-1}; $\phi = -\pi/2$; (c) $x_0 = 40$ cm; $\omega = \pi/2$ s^{-1}; $\phi = \pi/4$
5. 1.7×10^3 N/m
7. 0.69 s
9. (a) $\pi\sqrt{m/k}$; (b) $v_0\sqrt{m/k}$
15. 0.11 N·m/rad
17. (a) $2\pi\sqrt{\ell/g}$; (b) $2\pi\sqrt{2\ell/3g}$; (c) $2\pi\sqrt{2\ell/g}$; (d) infinite
19. 0.34 s
21. $R = \sqrt{2\kappa/k}$
23. within 1 μ (10^{-6} m)
29. 5.0 g
31. $\omega^2 = \dfrac{k_1k_2}{m(k_1 + k_2)}$
35. 7 oscillations in x and 4 oscillations in y
37. 20 cm
39. 1.7 rad; 2.4 rad/s
43. $2\pi/\sqrt{2ga}$
45. 0.25 s
47. 18 m/s
51. 6.9 s
53. 45° to x-axis
55. 1.7% at 30°; 4.0% at 45°
57. $R/\sqrt{2}$ above center
61. $T = 2\pi\sqrt{R_e^3/GM_e} = 84$ min

Chapter 14

1. $\lambda = 34$ cm; $T = 1.0$ ms; $\omega = 6300$ Hz; $k = 18$ m^{-1}
3. (a) 0.2 mi/s; (b) 2 mi
5. 0.29 s
7. $y = 2.5 \cos(0.39x + 14t)$ cm
9. $y_0 = 25$ cm; $f = 0.37$ Hz, $\lambda = 12$ m; $V = 4.4$ m/s
11. (a) 37 m/s; (b) 3.3 m/s
13. 590 g
17. $v = \sqrt{gh}$
19. 0.049 W/m^2
21. $4\pi^2 y_0^2 F_0/\lambda$
23. 2.7×10^4 W; drops by $1/\sqrt{2}$ when μ doubles
25. 7.2 m
31. $y_s = 2y_0 \cos(\phi/2)$
33. (a) 0.12 s; (b) 1.1 cm; (c) 0.082 s; (d) 0.86 cm
35. 7
41. (a) 9.75 cm (leaving 20.25 cm of string); (b) 429 Hz − 451 Hz
43. 0.33 Hz
45. (a) 2.9×10^5 m/s; (b) 10^7 K
47. 3.0 cm/s

51. 7.3 km
53. $\lambda_1 = 5.0$ m, $f_1 = 0.56$ Hz; $\lambda_2 = 2.5$ m, $f_2 = 0.79$ Hz

Chapter 15

1. 1 inch H$_2$O = 249 Pa, 1 inch of mercury = 3390 Pa, 1 torr = 133 Pa
3. (a) 176 kg/m^3; (b) 6.8 m^3
5. 3.4×10^5 Pa, or 2.4×10^3 Pa gauge pressure
7. 208 cm^2
9. 5.4×10^{18} kg
11. 10 Pa
13. about 300 ft
15. 7.4×10^3 Pa, or 7%
17. 52 cm higher in hurricane eye
19. 59 g
21. 27 m
23. $\dfrac{\rho\ell}{2} (h_0^2 - h_1^2) \tan\left(\dfrac{\theta}{2}\right)$
27. $y = \frac{4}{5}h$
29. 0.11 gal/min
31. (a) 1650 Pa; (b) 17 cm
33. (a) 79 m/s; (b) 26 mpg
37. $h = h_0 \ln 2 = 5.7$ km

Chapter 16

1. 5
3. They are not equal.
5. (a) 1.37 atm; (b) 0.330 atm; (c) 2.31 atm
7. (a) −40; (b) none; (c) none; (d) 0; (e) none
9. 16×10^6°C; 29×10^6°F; 29×10^6°R
11. 100 W
13. (a) 113 J/K; (b) 630 J/kg·K
15. (a) aluminum; (b) 84 W
17. 68°C
19. 0.70 K
21. 56.2°C
23. 5.6, 0.14, 3.4, 0.18, 5.0, 1.3, all in h·ft^2°F/Btu
25. $R = 12.3$ ft^2·h/btu
27. 75°F
29. 0.80
31. monthly heating cost is now $116; solar gain is worth $13/month
33. 3.3 cm
37. 277 K, or 4°C

Chapter 17

1. (a) volume increases by factor of 473/373 = 1.27; (b) volume doubles
3. 1×10^{15} particles/m^3, about 4×10^{-11} of earth's atmospheric particle density
5. 6.5×10^3 N, outward
7. (a) 235 mol; (b) 5.65 m^3
9. (a) 6.8×10^{26} molecules; (b) 4.1×10^6 J; (c) thermal energy is 11 times greater
11. 3.1×10^3 m/s
13. (a) 1.3×10^{-6} m^3/mol; (b) 175 atm
17. 143 kJ
19. mixture is all water (1050 g) at 10°C
21. 4.9 C

23. 1.2 kg of ice, 0.80 kg of water, all at 0°C.
25. 0.80 kg
29. 0.40°C
31. 285 k
35. 2.101 L, compared with 2.094 L using constant-β approximation
37. on earth, 3.5 km/s for H_2; 0.88 km/s for O_2; compared with escape speed of 11 km/s. On Jupiter, 1.9 km/s for H_2; 0.48 km/s for O_2; compared with escape speed of 60 km/s. More particles in the high-energy tail of the Maxwell-Boltzmann distribution will therefore escape earth, leaving earth with less hydrogen.
39. 36 days; too fast

Chapter 18

1. 5.6 kJ, 8.4 kJ
3. 0.02°C
5. $\frac{4}{3}P_1V_1$
7. 220 J
11. 130 atm
13. (a) $T=300$ K, $W=1.5$ kJ; (b) $T=326$ K, $W=430$ J; (c) $T=336$ K, $W=0$
17. 2.5 kJ
19. 93 J
21. (b) T_0; (c) $T_{van} < T_0$
23. (b) 39 W
25. (a) 2.31 kJ for both gases; (b) 1.8 kJ for H_2; 1.5 kJ for He
27. (a) diatomic; (b) 1.3
29. $W=172$ J, $Q=2700$ J

Chapter 19

1. $12!/6!$, or 6.7×10^5: of these, $6!$, or about one in a thousand correspond to all 6 eggs in the left half
3. assuming oceans cover 75% of earth, with average depth 3 km, energy extracted is 5×10^{23} J, about 2000 times annual use
5. 52% winter, 48% summer
7. 10
11. decrease in minimum temperature
13. 3.4%
17. (a) maximum efficiency is 42%; (b) 2100 MW; (c) 180,000
19. (a) $1-5^{1-\gamma}$; (b) $3T_C(5^{\gamma-1})$; (c) Carnot efficiency is $1 - \frac{1}{3}(5^{1-\gamma})$
21. (a) 7.7; (b) 5.3; (c) 2.8
23. 2×10^7 kg/s, slightly more than the Mississippi River flow
25. (a) 55%; (b) 36%; (c) 32%
27. 717.75 K
33. $\Delta S_{AB} = 1.5nR\ln3$, $\Delta S_{BC} = nR\ln3$, $\Delta S_{CA} = -2.5nR\ln3$
39. $T_h = T_{h0}e^{-[P_0t/mc(T_{h0}-T_c)]}$, $P = 0$ at $t = \frac{mc(T_{h0}-T_c)}{P_0}\ln\left(\frac{T_{h0}}{T_c}\right)$

Chapter 20

1. about 3 C
5. 8×10^8 N, more than 10^6 times your weight

7. electrical force is larger by factor of 2.2×10^{39} at all distances
9. -3.3 μC; $a \gg g$, so okay to ignore gravity; no—acceleration increases
11. 33 μC; 52 μC
15. $\mathbf{F} = 1.6\mathbf{i} - 0.33\mathbf{\hat{j}}$ N; $F=1.7$ N; 11° below x-axis
17. no
19. put $-q$ midway between the others; equilibrium is unstable
21. $\frac{a^2}{\ell + 2a}$ from the right end of the rod

Chapter 21

1. 2.0×10^{-4} N
3. (a) 5.5×10^{11} N/C; (b) 8.9×10^{-8} N
5. 1.6×10^{-12} N
7. 4.1 cm to right of 1 μC charge
13. $\frac{q}{\pi^2\epsilon_0 a^2}$, to right
15. $+3.0$ μC
19. $\pi R^2 E$
21. (a) $-q/\epsilon_0$; (b) $-2q/\epsilon_0$; (c) 0; (d) 0
23. 2.9×10^5 N·m²/C; no
25. $40\epsilon_0 = 3.5 \times 10^{-10}$ C/m³
27. -4.5×10^5 C
29. (a) $E = 0$; (b) $E = \frac{\rho r}{3\epsilon_0} - \frac{\rho a^3}{3\epsilon_0 r^2}$; (c) $E = \frac{7\pi a^3 \rho}{3\epsilon_0 r^2}$
31. $E=0$ between sheets; $E=\sigma/\epsilon_0$ outside sheets, directed away from sheets
33. -0.25 C
35. 3.6×10^{-3} C/m³
37. 0.39 μs later
39. (a) 5.1×10^6 N/C; (b) 34 N/C
41. (a) 7.2×10^6 N/C, from positive to negative plate; (b) 1.4 N/C, directed oppositely from the field of part (a)
43. 41 μC
45. $E = 0$ for $r<a$
$E = \frac{ac}{\epsilon_0 r^2}(e^{-1} - e^{-r})$ for $a<r<b$
$E = \frac{ac}{\epsilon_0 r^2}(e^{-1} - e^{-b})$ for $r>b$

Chapter 22

1. electric force is about 2×10^{12} times stronger
3. $v_{min} = \ell\sqrt{\dfrac{eE}{md}}$
5. (a) 2.8×10^6 m/s; (b) helical
9. (a) zero; (b) 4.0×10^{-3} C/m²; (c) other charges could destroy the spherical symmetry, making the surface charge density nonuniform
11. (b) $-Q$; (c) no
15. (a) 14 μC; (b) -0.18 μC/cm²
17. (a) 3.0×10^{-3} N·m; (b) 1.1×10^{-2} J
19. 1.5×10^{-27} N·m
21. (a) $F=0$, $\tau=0.036$ N·m; (b) $F=1.3$ N, $\tau=0$; (c) 0.036 J
25. 7.8×10^3 N/C

Chapter 23

3. (a) 1.3×10^{-7} J, about the same kinetic energy as a raisin moving at 1 cm/s!; (b) 2.1×10^4 W

5. 9.4×10^7 m/s

7. $6.1 \ \mu C$

9. (a) $V = -E_0 x^2/2x_0$; (b) $4eE_0 x_0$

11. 0 for all three paths

13. 40 V/m

17. (a) $\mathbf{E} = -ay\hat{\mathbf{i}} - ax\hat{\mathbf{j}}$; (b) V/m^2

21. (a) $V(r) = \dfrac{\rho R^3}{3\epsilon_0 r}$; (b) $V(0) - V(R) = \dfrac{\rho R^2}{6\epsilon_0}$;

    (c) $V(r) = \dfrac{\rho}{6\epsilon_0}(3R^2 - r^2)$

23. $Q/2\pi\epsilon_0 R$

25. (a) $d/(\sqrt{2} - 1)$ to right of $-q$; (b) d to right of $-q$ and $d/3$ to left of $-q$

27. 1.2×10^{11} m

29. (b) $V = \dfrac{1}{4\pi\epsilon_0} \dfrac{Q}{R}$

31. 3 kV

35. (a) 2.9×10^5 V; (b) $0.11 \ \mu C$ from smaller to larger sphere

39. 5.8×10^5 V

41. 1.7 cm

43. $y = \frac{1}{3}(d^2 + 2xd - 3x^2)^{1/2}$

Chapter 24

1. $\dfrac{3}{4\pi\epsilon_0} \dfrac{q^2}{a}$

3. $v = q/\sqrt{\pi\epsilon_0 ma}$

5. $U = -7.8 \times 10^{18}$ J or -49 eV; energy released during formation is $-U$

7. $W = \dfrac{Q^2}{8\pi\epsilon_0}\left(\dfrac{1}{a} - \dfrac{1}{b}\right)$

11. 2×10^{11} m^3

13. 2.4×10^{-5} J

15. $Q^2/40\pi\epsilon_0 a$, or 20% more

17. $\dfrac{Q^2}{4\pi\epsilon_0 R}(2^{2/3} - 1) = 3.0 \times 10^{-4}$ J

21. (a) 7600 pF; (b) 1000 V

23. 64 V on each

25. increases by 31 J (doubles)

27. $2\pi\epsilon_0 a$

29. 0.0051 to 0.0082 μF, or 0.0067 μF $\pm$ 24%

31. 0.55 μF, 0.83 μF, 1.3 μF, 1.5 μF, 6.0 μF

33. 75 V

37. (a) 7.5×10^{-2} F; (b) 3.4×10^9 J

41. (a) 15 kJ; (b) 20 kV

43. (a) $C = \dfrac{\kappa + 1}{2}C_0$; (b) $U = \dfrac{C_0 V_0^2}{1 + \kappa}$;

    (c) $F = \dfrac{2C_0 V_0^2}{\ell}\dfrac{\kappa - 1}{(\kappa + 1)^2}$, into capacitor

51. 9 pF/m

Chapter 25

1. 4.8 mA, toward back of tube

3. 2.9×10^5 C

5. (a) 480 m; (b) 50 mA; (c) 14 μA

7. v_d is 8.6 times greater in the copper

9. (a) 4.2×10^6 A/m^2; (b) 3.3 A

11. 7.0×10^{-15} s

13. 2.0×10^{-10}

15. diameter of aluminum should be 1.3 times greater

17. (a) 2.1 cm; (b) 2.7 cm; (c) copper $4.72/m, aluminum $2.07/m

19. 4.0×10^{-4} Ω, 1.0×10^{-6} Ω, 2.5×10^{-7} Ω

21. (a) 69 μA; (b) 82 A

23. 96 W

25. about 10^{12} W, approximately equal to earth's electric power consumption

27. the resistor that dissipates most power has $\sqrt{2}$ times greater diameter

29. (a) 150 A; (b) 1.7 km (remember that there are 2 wires!)

33. 2700 m/s, significantly different from the empirical result

35. 0.63 A

Chapter 26

5. 6.0 V

7. 0.02 Ω

9. 45 Ω

11. 30 A

13. $\mathscr{E} = 1.4$ V; $R_{\text{int}} = 3.7 \ \Omega$

15. 24

19. $R_1\mathscr{E}/(2R_1 + R)$

21. $I_2 = \dfrac{\mathscr{E}_1 R_3 - \mathscr{E}_3 R_1 - \mathscr{E}_2 R_3}{R_2 R_3 + R_1 R_2 + R_1 R_3}$

23. R_1 in all three cases

25. $\mathscr{E}_3 = \dfrac{\mathscr{E}_2 R_1 - \mathscr{E}_1 R_2}{R_1 + R_2}$

27. 1.7% low

29. (a) 13 V; (b) 2.2 mA; (c) 12 V; (d) 2.2 mA

31. 1200 W

33. 0.050 Ω, in parallel

39. (a) $I_1 = \mathscr{E}/R$; $I_2 = 0$; $I_3 = 0$; (b) $I_1 = \mathscr{E}/2R$; $I_2 = \mathscr{E}/2R$; $I_3 = 0$

41. 7.6 s

43. (a) 10 W; (b) 7.5×10^{-3} J

45. (a) $I_1 = 25$ mA; $I_2 = 0$; $V_C = 0$; (b) $I_1 = I_2 = 10$ mA; $V_C = 60$ V; (c) $I_1 = 0$; $I_2 = 10$ mA; $V_C = 60$ V; (d) $I_1 = I_2 = 0$; $V_C = 0$

49. 100 V

51. 83 μs

53. (a) 0; (b) 1.0 A; (c) 0.75 A; (d) 0; (e) 1.0 A; (f) 3.0 A; (g) 3.0 A; (h) 1.0 A

Chapter 27

1. (a) 16 G; (b) 23 G

3. (a) 2.0×10^{-14} N; (b) 1.0×10^{-14} N; (c) 0

5. (a) $(-1.1\hat{\mathbf{i}} + 1.5\hat{\mathbf{j}} + 1.7\hat{\mathbf{k}}) \times 10^{-3}$ N

7. $\mathbf{B} = 0.13\hat{\mathbf{k}}$ T; $\mathbf{v}_2 = -1.4 \times 10^4 \hat{\mathbf{i}}$ m/s

11. 15 G

13. (a) 8.7 km; (b) no change

17. 300 G; yes

19. 0.65 μs

21. (a) double the frequency to 30 MHz; (b) no change; (c) 38 MeV; (d) 38 MeV
23. Ne, CO, NO_2
25. 0.38 N
27. 0.12 T
29. $IA \ln(r_2/r_1)$, out of page for I flowing radially outward
33. (a) 0.35 A·m²: (b) 4.2×10^{-2} N·m
35. 24 A
37. 5.6×10^{-27} J $= 3.5 \times 10^{-8}$ eV
39. $\mu = I\ell^2/4\pi N$
41. 501.8 G
43. $\mathbf{E} = 3.0 \times 10^4 \hat{\jmath}$ V/m; $\mathbf{B} = 0.71(\hat{\jmath} + \hat{k})$ T
45. if bottom of magnet is N then $\boldsymbol{\mu}$ points down; if bottom is S then $\boldsymbol{\mu}$ points up

Chapter 28

1. q/ϵ_0, $-q/\epsilon_0$, 0, 0, 0, 0
3. $\mu_0 I/4a$, into page
5. (a) $\sqrt{2}\mu_0 I/\pi a$; (b) $\mu_0 Ia^2/2\pi x^3$
7. 0.62 G
9. 2.8×10^9 A
11. 5.0×10^{-6} N
13. forces are attractive
19. (a) 0; (b) 3.6 G; (c) 19 G; (d) 5 G
21. $B = \mu_0 I/\ell$ parallel to pipe axis on inside; $B = 0$ outside
23. $(\sqrt{2}/2)\mu_0 i$ both inside and outside (but internal and external fields are mutually perpendicular)
25. (a) $B = \mu_0 J|z|$; (b) $B = \frac{1}{2}\mu_0 Jh$
27. (a) 2300; (b) 3.3 kW; (c) 0
29. (a) 0.016 T; (b) 0.014 T
33. (a) $B = \mu_0 I/2w$; (b) $B = \mu_0 I/2\pi r$
37. (a) $I = \frac{1}{3}\pi J_0 R^2$; (b) $B = \mu_0 J_0 R^2/6r$; (c) $B = J_0 \left(\dfrac{r}{2} - \dfrac{r^2}{3R} \right)$

Chapter 29

3. (a) 0.20 mA; (b) increases to 0.44 mA
5. 1.5×10^{-2} T
7. 3.0 s
11. (a) top to bottom; (b) $B\ell v/R$; (c) $B^2\ell^2 v^2/R$
13. (c) $v = \mathscr{E}/B\ell$; R affects time to reach final speed
15. (a) 6.7 mA, counterclockwise; (b) 6.7 mA, counterclockwise; (c) 0.44 mW, both cases
17. (a) $v = \mathscr{E}/2B\ell$; (b) 0
19. (a) $\phi = a^2(B_0 + \frac{1}{2}ba + bx)$; (b) counterclockwise; (c) a^2bv/R; (d) $a^4 b^2 v^2/R$
21. mgv
23. (a) $\dfrac{-B\ell^2\omega \sin\omega t}{R}$; (b) $\dfrac{B^2\ell^4\omega^2 \sin^2\omega t}{R}$; (c) $\dfrac{-B^2\ell^4\omega \sin^2\omega t}{R}$
25. $E = -bR^2/2r$
27. (a) 140 mV; (b) 0.14 eV; (c) 150 mV/m
29. (a) 530 V/m; (b) 8.0 keV
31. $v = mgR \cos\theta/B^2\ell^2 \sin^2\theta$
33. (a) 2.4 km; (b) 1.6×10^{-5}; (c) \$6.90
35. (a) $J = \dfrac{r}{2\rho}\dfrac{dB}{dt}$; (b) $P = \dfrac{\pi a^4 h}{8\rho}\left(\dfrac{dB}{dt}\right)^2$; (c) 7.0 W

Chapter 30

1. 120 V
3. (a) $2\pi fMI_0 \cos(2\pi ft)$; (b) 130 mH
5. $\dfrac{\mu_0 \ell}{2\pi} \ln \left(\dfrac{a+w}{w} \right)$
7. 400 mH
9. $\mathscr{E} = -(12t + 30)$ V
15. (a) 76 mA; (b) 4.4 V; (c) 7.6 V; (d) 2.2 A/s; (e) 0.58 W
17. about 25 s
19. (a) $I_1 = I_2 = 1.0$ A; $I_3 = 0$; (b) $I_1 = 2.1$ A; $I_2 = 0.43$ A; $I_3 = 1.7$ A; (c) $I_1 = 0$; $I_2 = I_3 = 1.7$ A, flowing clockwise through R_3, L, and R_2; (d) $I_1 = I_2 = I_3 = 0$
21. 3.1 kJ
23. (a) 0.82 A; (b) 3.0 s
25. 2500 W; no—the current decreases steadily, but the energy is proportional to the square of the current, so the power is greatest at first
29. smaller by factor $1/4n^2R^2$
31. (a) 10^{24} J; (b) no; effect is one part in 10^6
33. (a) $u = \mu_0 I^2/8\pi^2 r^2$; (b) $\dfrac{\mu_0 \ell I^2}{4\pi r} dr$
35. (a) $\mu_0 I/w$; (b) $\frac{1}{2}\mu_0 I^2 d/w$; (c) $\mu_0 d/w$
37. (a) 6.7×10^{-10} J/m³; (b) $u_B = \mu_0 \mu^2/32\pi^2 r^6$; (c) 5.8×10^7 m $= 9.1$ R_e

Chapter 31

1. 380 s⁻¹, 170 V
5. $V_{rms} = V_0$
7. $I_1 = I_0 \cos\phi$; $I_2 = I_0 \sin\phi$
9. 45 mA
11. (a) $\omega = 500$ s⁻¹ or $f = 80$ Hz; (b) 2.0 H; (c) $X_L = 4X_C$
13. 4.9 pf to 42 pf
15. 0.20 μH
19. (a) 32 mH; (b) 1.0 V
21. 0.22 μF
23. 20 V
25. (a) 43 μH; (b) R>16 Ω
29. (a) above; (b) $\phi = \tan^{-1} \left(\dfrac{V_{C0} - V_{L0}}{V_{R0}} \right)$
31. (a) $\frac{1}{3}$; (b) 4.0 W
33. (a) 5.5%; (b) 9.1%; (c) large
35. (a) 140 W; (b) 1.2 A; (c) 1.1 kWh; (d) 13¢
37. (a) 280 Ω; (b) 5900 μF
45. (a) $\epsilon_0 w/d$; (b) 3.0×10^8 rad/sec
47. $\dfrac{k}{m} = \dfrac{1}{LC}$

Chapter 32

5. $-x$
11. 2.6 s
13. 5 km
15. (a) 300 m; (b) 6.3×10^6 s⁻¹; (c) 0.021 m⁻¹; (d) 0.39 V/m; (e) 1.3×10^{-9} T
17. 5000 km
19. x-direction
23. 2700 W/m²
25. 3.0×10^{-6} W/m²

27. (a) 1.4×10^{-17} W/m^2; (b) 4.6×10^{-34} W/m^2;
(c) to within 6.7×10^{-7} s
29. fields drop like $1/r$, compared with $1/r^2$ for stationary
point charge
31. (a) 1600 W/m^2; (b) 770 V/m; (c) 2.6×10^{-6} T
33. 1.1×10^8 m^2 if sail is perfect reflector; 2.2×10^8 m^2 if
perfect absorber
35. 6.7×10^{-6} N
39. 0.3 μm

Chapter 33

1. (a) 4.50 h; (b) 4.56 h; (c) 4.62 h
3. (a) $v/c = \tan\theta$, where θ is the amplitude of the
sine curve. Here $\theta = 20''$ (20/3600 degree), giving
$v = 29$ km/s. (b) The orbit is nearly circular.
5. (a) $v = (1 - 6 \times 10^{-10})c \approx 0.999999999c$; (b) no; they
would be 4 My old
7. $0.14c$
9. A is 83 years old; B is 40 years old
11. C judges civilization B first, by 350,000 years
13. Problem 11: An observer moving from A to B at $0.5c$
will judge the events to be simultaneous. Problem 12:
There is no such observer because the events could
be causally related.
17. $0.80c$
19. $0.996c$
21. 9.6×10^{-31} kg
23. (a) 0.26 eV; (b) 130 keV; (c) 3.1 MeV
25. 1 day

Chapter 34

3. 1.41 m
5. $\theta = 40.5°$
7. 3.0 mm
9. air $\theta_c = 43.2°$; water $\theta_c = 65.6°$
11. $d = 1.14\ h$
13. deviations are red 38.2°, blue 41.3°
15. 6.1 m
17. for thickness of 3 mm, $n = 1.6$, appears 1.0001 m
tall $\approx 0.01\%$
19. 0.33 m
21. 0.5°
23. 0.81 m tall, top is 0.06 m below top of your head
25. 1.2 m tall, bottom is 0.6 m off the floor less $\frac{1}{2}$ distance
from eyes to top of head
27. 2×10^{-5} rad $= 0.001°$, 1.3×10^{-14} rad $= 7.6 \times 10^{-13°}$
29. (a) -1.2 cm; (b) -1.7 cm; (c) -2.2 cm
33. convex, $R \approx 0.18$ m
35. (a) 107 mm, 42 times; (b) 79 mm, 57 times
37. virtual image 14.6 cm to left of convex lens
39. 2 times
41. 0.15 m back
43. (a) -7.5 cm; (b) 11 cm
45. magnification $= i/o < 0$, so image is virtual
47. 2
49. 23.6 cm
51. $\dfrac{1}{f} = \dfrac{1}{f_1} + \dfrac{1}{f_2}$
57. -4.5 diopters, nearsighted
59. $\frac{1}{4}$

61. 3.0 m, 8.1 m
63. 6.4 cm; 7.5 times, 170 times, 1300 times
65. 110 cm from the objective mirror, flat and elliptical
71. derive 42.53° for red, 40.78° for violet

Chapter 35

1. maxima: 0°, 0.096°, 0.193°. . . ; minima: 0°, 0.048°,
0.144°. . .
5. yes, since spacing is 1.8 μm
7. 2.484 mm $-$ 0.828 mm $=$ 1.656 mm; 1.657 mm; 0.06%
9. 0.034°
11. 103°
13. 9.8×10^{-8} rad $= 6 \times 10^{-6°}$
15. 0.75 m
17. maxima at $\theta = 0.000°$, 0.084°, 0.169° . . .
19. problem 17 plus secondary maxima at 0.028°, 0.056°,
0.113°, 0.141° . . .
21. 8 slits, 0.044 mm
23. 1058 nm, 142 nm
25. $\Delta\lambda = 0.006$ nm
27. 415 nm, 484 nm, 581 nm
29. 0.156 mm
31. 545 nm
33. 1.324
35. 105 nm
37. 240 nm, 720 nm. . .
39. 1500 nm; $m = 6, 7, 8, 9$
41. 59
43. 0.056°, 0.111°, 0.067°
45. 0.00074 rad $= 0.034''$
47. 35.4 m
49. 0.16 m, 6×10^{-5} ly
51. 0.7 mm
53. 6'', yes, 1.7
55. 19,000 km
57. $\frac{1}{8}$
59. $n = 1.43$, fluorite
61. 920 times
63. 3 μm
65. 70th: 202.7–204.5 nm; 94th: 150.3–152.8 nm; 95th:
148.7–151.2 nm; 96th: 147.1–149.6 nm; 120th: 117.2–
120.4 nm; no gap
67. $0.017(m + \frac{1}{2})$ mm, $m = 0, 1, 2$. . .
69. 4166
71. $\delta = D\lambda/2d$

Chapter 36

1. 122 nm, 103 nm, 97.2 nm
3. 0.17 MHz
5. 10^{-24} kg·m/s
7. 10^{-11} m $= 10$ pm
9. $1s^2 2s^2 2p^3$
11. 74,000 years
13. 20; $+2e$, $+e$, 0, $-e$
15. 3.6×10^9 cm^3
17. 10^{11} ly $= 10^5$ Mly
19. 2.8 Gly
21. 0.097 cm
25. 0.67

CREDITS

We thank Elizabeth Stell and Kimberly S. Levin for their help in obtaining photographs. We thank Hugh Kirkpatrick, Adam Witten, and Crispin Butler for setting up laboratory demonstrations to be photographed. We thank Ernest and John LeClaire of LeClaire Custom Color, as well as Erik Borg, for their expert and expeditious photographic developing and printing. We thank Nikon, Inc., for the loan of photographic equipment.

Cover

Sandia National Laboratories photograph

Color Plates

Plate 1 Swanke Hayden Connell Architects
Plate 2 MTS Systems Corp.
Plate 3 M. Horsmann
Plate 4 NASA/Jet Propulsion Laboratory
Plate 5 Sandia National Laboratories
Plate 6 Jay M. Pasachoff
Plate 7 Dennis di Cicco
Plate 8 Alfred Leitner, Rensselaer Polytechnic Institute
Plate 9 R. Giovanelli and H. R. Gillett, CSIRO National Measurement Laboratory, Australia

Chapter 1

Fig. 1–1 Underlying photo by Alan Dressler and Roger Windhorst, Mt. Wilson and Las Campanas Observatories, Carnegie Institution of Washington
Fig. 1–2 Sprague Electric/Worcester, R. Morrison
Fig. 1–3 Courtesy of Lotte Jacobi
Fig. 1–4 Perkin-Elmer Corporation and Lockheed Missiles & Space Company, Inc.
Fig. 1–5 Heather Couper
Fig. 1–6 National Bureau of Standards
Fig. 1–8 National Bureau of Standards
Fig. 1–9 International Bureau of Weights and Measures
Fig. 1–10 University of California, Lawrence Berkeley Laboratory
Fig. 1–11 National Optical Astronomy Observatories/Kitt Peak
Fig. 1–12 Dale C. Flanders, MIT Lincoln Laboratory, Lexington, MA
Fig. 1–13 Boeing Commercial Aircraft Co.
Fig. 1–14 NASA/JSC, Cosmic Dust Program

Chapter 2

Fig. 2–12 Dr. Harold E. Edgerton, Massachusetts Institute of Technology
Fig. 2–13 NASA/JSC
Fig. 2–15 Jay M. Pasachoff
Fig. 2–16 Jay M. Pasachoff

Chapter 3

Fig. 3–1 National Cente_
Fig. 3–17 *PSSC Physics*, 3r_ & Company, with E_ Center, Inc.
Fig. 3–18 Hanon Isachar/Sygma
Fig. 3–19 Laffont and Pierce/Sygma

Chapter 4

Fig. 4–1 NASA
Fig. 4–2 Courtesy of Fletcher Watson, *Project P_*
Fig. 4–4 Courtesy of Lord Portsmouth and the Tr_ ees of the Portsmouth Estates
Fig. 4–5 Owen Gingerich
Fig. 4–6 From the collection of David Park
Fig. 4–7 Copyright © 1984 by The New York Times Company. Reprinted by permission
Fig. 4–8 Photo by Ralph Crane
Fig. 4–12 NASA/JSC
Fig. 4–14 NASA/JSC
Fig. 4–30 Jay M. Pasachoff

Chapter 5

Fig. 5–1 European Space Agency
Fig. 5–9 Copyright © David Scharf 1977, from *Magnifications* (Schocken Books, Inc.)
Fig. 5–20 By permission of Johnny Hart and News America Syndicate
Fig. 5–28 Michelin Tire Corp.
Fig. 5–31 Jay M. Pasachoff

Chapter 6

Fig. 6–2 American Institute of Physics, Niels Bohr Library
Fig. 6–5 Jay M. Pasachoff
Fig. 6–6 Jay M. Pasachoff
Fig. 6–25 © Lucasfilm Ltd. (LFL) 1981. All rights reserved. From the motion picture *Raiders of the Lost Ark*, courtesy of Lucasfilm Ltd.

Chapter 7

Fig. 7–23 William Kirk, Stanford University
Fig. 7–25 Photo by Erik Borg

Chapter 8

Fig. 8–1 AIP Niels Bohr Library
Fig. 8–2 Chapin Library, Williams College
Fig. 8–3 Jay M. Pasachoff
Fig. 8–4 Chapin Library, Williams College
Fig. 8–7 New Mexico State University Observatory
Fig. 8–9 (above) Courtesy Robert Burger and Marshall Schalk, Smith College

(below) Christopher Duncan, Middlebury
College
Jay M. Pasachoff from the collection of John
E. Gaustad
NASA
European Space Agency
Fig. 8-6 Jay M. Pasachoff
Fig. 8-19 © 1986 Max-Planck-Institut für Aeronomie, Lindau/Harz, FRG, taken by the Halley Multicolour Camera on board ESA's spacecraft Giotto, courtesy H. U. Keller
Fig. 8-26 NASA
Fig. 8-29 Royal Greenwich Observatory ; from A. Eddington, *Nature*, vol. 83, May 26, 1910, pp. 372–373; courtesy of Ruth Freitag, Library of Congress
Fig. 8-42 NASA/JPL
Fig. 8-44 (left) Mark R. Showalter, then at Cornell University
Fig. 8-44 (center and right) NASA/JPL
Fig. 8-47 John F. Hawley and Larry Smarr, University of Illinois at Urbana-Champaign
Fig. 8-49 J. Anthony Tyson, AT&T Bell Laboratories

Chapter 9

Fig. 9-1 Franklin Miller, Jr., from *Physics* by Edward R. McCliment, © 1984 by Harcourt Brace Jovanovich, Inc. Reproduced by permission of the publisher
Fig. 9-9 Weber and Worel, BfG: Hauptverwaltung
Fig. 9-10 Jennifer Davis, Pittsburgh Ballet Theater; photo by Michael Friedlander
Fig. 9-13 Jay M. Pasachoff
Fig. 9-14 National Optical Astronomy Observatories/Cerro Tololo Inter-American Observatory
Fig. 9-17 Copyright © 1920 by The New York Times Company. Reprinted by permission
Fig. 9-19 NASA

Chapter 10

Fig. 10-1 Reproduced by permission of Encyclopaedia Britannica, Inc., illustration by Mark Paternostro, courtesy of Charles Cegielski
Fig. 10-2 Lick Observatory
Fig. 10-3 Dr. Harold E. Edgerton, Massachusetts Institute of Technology
Fig. 10-5 U.S. Dept. of Transportation
Fig. 10-6 Dr. Harold E. Edgerton, Massachusetts Institute of Technology
Fig. 10-7 (left and right) Jay M. Pasachoff

Chapter 11

Fig. 11-29 Swanke Hayden Connell Architects
Fig. 11-30 After Swanke Hayden Connell Architects
Fig. 11-40 Jay M. Pasachoff

Chapter 12

Fig. 12-29 Palomar Observatory Photograph
Fig. 12-30 Jay M. Pasachoff
Fig. 12-31 Palomar Observatory Photograph
Fig. 12-33 © 1978 Anglo-Australian Telescope Board
Fig. 12-42 NASA/JSC
Fig. 12-43 NASA/Goddard Space Flight Center

Chapter 13

Fig. 13-2 AP/Wide World Photographs
Fig. 13-3 After Bart J. Bok and Priscilla F. Bok, *The Milky Way*, courtesy Harvard University Press. Reprinted by permission
Fig. 13-11 (left) Jay M. Pasachoff
Fig. 13-11 (center and right) MTS Systems Corp.
Fig. 13-13 Jay M. Pasachoff
Fig. 13-14 Photo by Erik Borg
Fig. 13-33 Jay M. Pasachoff
Fig. 13-34 Photos by Erik Borg
Fig. 13-36 Jay M. Pasachoff
Fig. 13-37 (left and center) AP/Wide World Photos
Fig. 13-37 (right) Nathan A. Unterman

Chapter 14

Fig. 14-3 Courtesy of Caroll and Nancy Taylor
Fig. 14-4 Jay M. Pasachoff
Fig. 14-17 Jay M. Pasachoff
Fig. 14-27 *PSSC Physics*, 3rd ed. © 1971 by D. C. Heath & Company with Educational Development Center, Inc., Newton, MA
Fig. 14-29 Jay M. Pasachoff
Fig. 14-34 Dr. Daniel I. Becker, North Adams Regional Hospital
Fig. 14-37 After D. C. Miller, *Sound Waves, Their Shape and Speed*, Macmillan, New York, NY
Fig. 14-38 Adapted from Boyce, *Journal of the Fisheries Research Board of Canada*, 31, 1974.
Fig. 14-39 National Film Board of Canada Photothèque ONF, © 1968 Harvard *Project Physics*, courtesy of Fletcher Watson
Fig. 14-40 Courtesy of N. E. Molin, H. Sundin, and E. Jansson, with the assistance of Donald White
Fig. 14-41 Jørgen Christensen-Dalsgaard
Fig. 14-48 Jay M. Pasachoff
Fig. 14-56 Photograph from Transonic Range, U.S. Army Ballistic Research Laboratory; courtesy of Milton Van Dyke, *An Atlas of Fluid Motion*

Chapter 15

Fig. 15-1 (left) Photo by J. Demmer, National Severe Storms Laboratory, National Oceanic and Atmospheric Administration
Fig. 15-1 (center) Big Bear Solar Observatory, California Institute of Technology
Fig. 15-1 (right) Jay M. Pasachoff
Fig. 15-6 Sandia National Laboratories

Fig. 25–19   Jay M. Pasachoff
Fig. 25–20   Courtesy of AT&T Bell Laboratories
Fig. 25–22   Fermi National Accelerator Laboratory

Chapter 26

Fig. 26–1   Adapted with permission from R. Markson, *Nature*, vol. 273, No. 5658, © 1978 Macmillan Journals Limited
Fig. 26–7   Photo by Erik Borg
Fig. 26–17   (left and right) Jay M. Pasachoff
Fig. 26–21   Jay M. Pasachoff
Fig. 26–28   (left) Jay M. Pasachoff
Fig. 26–28   (right) Jay M. Pasachoff
Fig. 26–35   Photo by Erik Borg
Fig. 26–36   Photo by Erik Borg

Chapter 27

Fig. 27–1   Jay M. Pasachoff
Fig. 27–2   (left) D. Balkwill and D. Maratea
Fig. 27–2   (right) R. Blakemore
Fig. 27–4   (right) From *PSSC Physics*, 3rd ed., copyright © 1971 by D. C. Heath and Co., with Educational Development Center, Inc.
Fig. 27–6   Stanford University
Fig. 27–10   Francis Bitter National Magnet Laboratory, Massachusetts Institute of Technology
Fig. 27–14   Palomar Observatory Photograph
Fig. 27–18   High Altitude Observatory and NASA
Fig. 27–20   (center) Photo by Joseph L. Montani © 1978
Fig. 27–20   (right) L. A. Frank and J. D. Craven, University of Iowa, and NASA/GSFC
Fig. 27–21   After Frank Press and Raymond Siever, *Earth*, 4th ed., W. H. Freeman and Co., 1986
Fig. 27–23   Central Design Group, Superconducting Super Collider, Lawrence Berkeley Laboratory, University of California
Fig. 27–24   Taken for *Fortune* by Ivan Masaar, Black Star
Fig. 27–25   Fermi National Accelerator Laboratory
Fig. 27–27   Princeton Plasma Physics Laboratory
Fig. 27–30   Jay M. Pasachoff
Fig. 27–32   Walker Scientific, Inc.
Fig. 27–43   (left, center, and right) Jay M. Pasachoff
Fig. 27–45   General Electric Research and Development Center
Fig. 27–47   Photo by William L. Masterton, from Masterton/Slowinski/Stanitski: *Chemical Principles*, Sixth Edition, courtesy of Saunders College Publishing
Fig. 27–50   William Kirk, Stanford University

Chapter 28

Fig. 28–2   Courtesy of American Association of Physics Teachers
Fig. 28–8   (left) Big Bear Solar Observatory, California Institute of Technology
Fig. 28–8   (right) National Geophysical Data Center, NOAA, D. C. Wilkinson
Fig. 28–9   (left and right) High Altitude Observatory
Fig. 28–14   From *PSSC Physics*, 3rd ed. © 1971 by D. C. Heath & Company, with Educational Development Center, Inc.
Fig. 28–24   (left) High Altitude Observatory and NASA
Fig. 28–24   (right) Adapted from R. Wolfson, *Astrophysical Journal*, v 288, 1985, p. 776, published by the University of Chicago Press; © 1985 The American Astronomical Society
Fig. 28–33   From *PSSC Physics*, 3rd ed. © 1971 by D. C. Heath & Company, with Educational Development Center, Inc.
Fig. 28–36   (left and right) Jay M. Pasachoff

Chapter 29

Fig. 29–19   Con Edison
Fig. 29–20   Jay M. Pasachoff
Fig. 29–21   Earl Richardson, *Topeka Capital-Journal*
Fig. 29–22   Photos by Erik Borg
Fig. 29–23   Jay M. Pasachoff
Fig. 29–24   Jay M. Pasachoff
Fig. 29–25   Photo by Erik Borg
Fig. 29–30   Courtesy of the Princeton Plasma Physics Laboratory
Fig. 29–38   Jay M. Pasachoff

Chapter 30

Fig. 30–3   Jay M. Pasachoff
Fig. 30–6   Photo by Erik Borg
Fig. 30–12   Jay M. Pasachoff
Fig. 30–14   High Altitude Observatory

Chapter 31

Fig. 31–12   Electronics Associates, Inc.
Fig. 31–15   Jay M. Pasachoff
Fig. 31–16   (below) Photo by Erik Borg
Fig. 31–30   Jay M. Pasachoff

Chapter 32

Fig. 32–4   (left) AIP Niels Bohr Library
Fig. 32–4   (right) U.S. Geological Survey
Fig. 32–11   Deutsches Museum, Munich
Fig. 32–12   Jay M. Pasachoff
Fig. 32–13   Vanscan™ thermogram by Daedalus Enterprises, Inc.
Fig. 32–16   (left and right) National Radio Astronomy Observatory, operated by Associated Universities, Inc. under contract with the National Science Foundation
Fig. 32–17   (left and left center) NASA/JPL
Fig. 32–17   (right center) TRW, Inc.
Fig. 32–17   (right) F. R. Harnden, Jr., and colleagues at the Harvard–Smithsonian Center for Astrophysics
Fig. 32–18   Painting by Lois Cohen, Griffith Observatory
Fig. 32–19   American Science and Engineering, Inc./NASA
Fig. 32–20   NASA Landsat/4 Thematic Mapper, processed by the IBM Palo Alto Scientific Laboratory/Ralph Bernstein

Fig. 15–7   Jay M. Pasachoff
Fig. 15–9   AP/Wide World Photos
Fig. 15–12 Courtesy of H. W. Jannasch, Woods Hole Oceanographic Institution
Fig. 15–16 After Karen Arms and Pamela S. Camp, *Biology*, Saunders College Publishing
Fig. 15–17 Tony Korody/Sygma
Fig. 15–18 (left) American Petroleum Institute
Fig. 15–18 (right) Mobil Oil Corp.
Fig. 15–19 J. Guichard/Sygma
Fig. 15–22 Mercedes-Benz of North America
Fig. 15–23 (left) A. E. Perry & T. T. Lim 1978, courtesy of Milton Van Dyke, *An Atlas of Fluid Motion*
Fig. 15–23 (right) NASA/JPL
Fig. 15–26 Richard K. Dean, courtesy of Ausable Chasm Company
Fig. 15–29 Jay M. Pasachoff
Fig. 15–31 ONERA photograph, H. Werlé 1974, courtesy of Milton Van Dyke, *An Atlas of Fluid Motion*
Fig. 15–32 Boeing Commercial Airplane Co.
Fig. 15–33 Peter Senesky, Arizona Sonora Desert Museum
Fig. 15–34 Jay M. Pasachoff
Fig. 15–37 Painting by Lois Cohen, Griffith Observatory
Fig. 15–39 Paul E. Dimotakis, R. C. Lye & D. Z. Papantoniou 1981, courtesy of Milton Van Dyke, *An Atlas of Fluid Motion*
Fig. 15–44 Jay M. Pasachoff

Chapter 16

Fig. 16–1   Kim Vandiver and Dr. Harold E. Edgerton, Massachusetts Institute of Technology
Fig. 16–3   Jay M. Pasachoff
Fig. 16–4   Photo by Erik Borg
Fig. 16–14 Courtesy of E. L. Koschmieder 1974, from Milton Van Dyke, *An Atlas of Fluid Motion*
Fig. 16–15 Photograph obtained by Gören Scharmer at the La Palma Astrophysics Research Station of the Royal Swedish Academy of Sciences
Fig. 16–20 Courtesy of Stephen H. Schneider, after Starley L. Thompson and Stephen H. Schneider, "Nuclear Winter Reappraised," *Foreign Affairs*, Summer, 1986.

Chapter 17

Fig. 17–9   Lou Gilliam, National Optical Astronomy Observatories/National Solar Observatory
Fig. 17–11 The Open University, S272, *The Physics of Matter*
Fig. 17–13 Ken Karp
Fig. 17–16 Jay M. Pasachoff
Fig. 17–21 National Bureau of Standards, courtesy of Charles Tilford
Fig. 17–22 AP/Wide World Photos

Chapter 19

Fig. 19–1   Becor Western, Inc., and American Electric Power System
Fig. 19–5   Jay M. Pasachoff
Fig. 19–6   Environmental Protection Agency, courtesy of Scott, Foresman and Co.
Fig. 19–15 Watts Bar Nuclear Plant
Fig. 19–17 Pennsylvania Power & Light Company
Fig. 19–18 Lockheed Missiles and Space Company, Inc.
Fig. 19–19 Southern California Edison Company
Fig. 19–23 Data from A. Lovins, *Soft Energy Paths*, Ballinger Publishing Co., Cambridge, MA, 1977
Fig. 19–24 Photo by Erik Borg

Chapter 20

Fig. 20–6   University of California, Lawrence Berkeley Laboratory

Chapter 21

Fig. 21–3   Jay M. Pasachoff
Fig. 21–16 *PSSC Physics*, 3rd ed. © 1971 by D. C. Heath & Company, with Educational Development Center, Inc., Newton, MA
Fig. 21–17 The Franklin Institute, Philadelphia, PA, photo by © Richard Dunhoff, courtesy of Donald A. Cooke

Chapter 22

Fig. 22–1   Jay M. Pasachoff
Fig. 22–2   Courtesy of Research-Cottrell, Inc.
Fig. 22–18 Photo by Erik Borg
Fig. 22–24 Jay M. Pasachoff
Fig. 22–27 Object courtesy of SLAC and Sidney Drell, photo by Oscar Kapp, University of Chicago

Chapter 23

Fig. 23–25 Jay M. Pasachoff
Fig. 23–27 Jay M. Pasachoff

Chapter 24

Fig. 24–6   National Center for Atmospheric Research photo
Fig. 24–10 Photo by Erik Borg
Fig. 24–12 Photo by Erik Borg
Fig. 24–13 Photo by Erik Borg
Fig. 24–14 (left) Jay M. Pasachoff
Fig. 24–14 (right) Los Alamos Scientific Laboratories

Chapter 25

Fig. 25–1   From a painting by Edgard Maxence, Institution of Electrical Engineers, courtesy of American Institute of Physics, Niels Bohr Library
Fig. 25–3   High Altitude Observatory and NASA
Fig. 25–11 Jay M. Pasachoff
Fig. 25–16 (left) Smithsonian Institution
Fig. 25–16 (right) Jay M. Pasachoff

Chapter 35 (cont.)

Fig. 35–23   (center and bottom) Michelson Museum, U.S. Naval Weapons Center, China Lake, California/U.S. Naval Academy, Nimitz Library, Special Collections
Fig. 35–24   Christopher C. Jones, Union College
Fig. 35–26   Jay M. Pasachoff
Fig. 35–27   Jay M. Pasachoff
Fig. 35–28   Christopher C. Jones, Union College
Fig. 35–29   From *PSSC Physics*, 3rd ed. © 1971 by D. C. Heath & Company, with Educational Development Center, Inc.
Fig. 35–30   High Altitude Observatory and NASA
Fig. 35–31   (left) G. B. Bianchetti and Salvatore Ganci
Fig. 35–31   (right) Christopher C. Jones, Union College
Fig. 35–32   (top) G. B. Bianchetti and Salvatore Ganci
Fig. 35–32   (bottom) Christopher C. Jones, Union College
Fig. 35–33   (top) Christopher C. Jones, Union College
Fig. 35–34   (top) Christopher C. Jones, Union College
Fig. 35–34   (center) David Stolzman
Fig. 35–35   Perkin-Elmer Corp.
Fig. 35–36   Jay M. Pasachoff
Fig. 35–37   Christopher C. Jones, Union College
Fig. 35–43   Christopher C. Jones, Union College
Fig. 35–48   Jay M. Pasachoff
Fig. 35–49   Jay M. Pasachoff
Fig. 35–50   Jay M. Pasachoff
Fig. 35–51   Jay M. Pasachoff
Fig. 35–54   Jay M. Pasachoff
Fig. 35–57   Jay M. Pasachoff
Fig. 35–58   Jay M. Pasachoff

Chapter 36

Fig. 36–1   Deutsches Museum, Munich
Fig. 36–2   Photo by Paul Ehrenfest; restoration of the negatives and production of the prints by William R. Whipple; AIP Niels Bohr Library
Fig. 36–3   AIP Niels Bohr Library, W. F. Meggers Collection
Fig. 36–5   (top and bottom) From *PSSC Physics*, 3rd ed., © 1971 by D. C. Heath & Co., with Educational Development Center, Inc.
Fig. 36–7   AIP Niels Bohr Library, Landé Collection
Fig. 36–8   A. F. Burrand and Robert Fisher, New Mexico State University, and American Association of Physics Teachers
Fig. 36–9   AIP Niels Bohr Library, Bainbridge Collection
Fig. 36–10   (left) Courtesy of the Institute of Theoretical Physics, Lund, Sweden
Fig. 36–10   (right) AIP Niels Bohr Library, Goudsmit Collection
Fig. 36–11   A. F. Burrand and Robert Fisher, New Mexico State University, and American Association of Physics Teachers
Fig. 36–13   Tokyo Astronomical Observatory
Fig. 36–14   AIP Niels Bohr Library; Fankuchen Collection
Fig. 36–15   C. D. Anderson, California Institute of Technology
Fig. 36–17   Carl Byoir & Associates, Inc., courtesy of Hughes Aircraft Company
Fig. 36–18   Jay M. Pasachoff
Fig. 36–19   (left) AIP Niels Bohr Library, *Physics Today* Collection
Fig. 36–19   (center) Photo CERN
Fig. 36–19   (right) Harvard University Archives
Fig. 36–20   (right) Courtesy of Richard P. Feynman, inscription for Jay M. Pasachoff
Fig. 36–21   Irving Lindenblad, U.S. Naval Observatory Photograph
Fig. 36–22   AIP Niels Bohr Library
Fig. 36–23   AIP Niels Bohr Library, Stein Collection
Fig. 36–24   Jef A. Poskanzer, University of California, Lawrence Berkeley Laboratory
Fig. 36–25   Courtesy National Archives
Fig. 36–26   From Otto Hahn, *A Scientific Biography*, Charles Scribner's Sons, New York, NY, 1966.
Fig. 36–27   Rosalyn S. Yalow
Fig. 36–28   (top) AIP Niels Bohr Library, Meggers Gallery of Nobel Laureates
Fig. 36–28   (bottom) AIP Niels Bohr Library
Fig. 36–29   Chien Shiung Wu, Columbia University
Fig. 36–32   (left) Brookhaven National Laboratory
Fig. 36–32   (center) Chuck Painter, News and Publication Service, Stanford University/Stanford Linear Accelerator Center
Fig. 36–32   (right) Brookhaven National Laboratory
Fig. 36–33   Fermi National Accelerator Laboratory
Fig. 36–34   (left) Photo CERN
Fig. 36–34   (center) Photo CERN
Fig. 36–34   (right) Photo CERN
Fig. 36–35   (left and right) Photo CERN
Fig. 36–36   Brookhaven National Laboratory
Fig. 36–38   Mt. Wilson and Las Campanas Observatories, Carnegie Institution of Washington (Mt. Wilson Observatory photograph)
Fig. 36–40   Courtesy of David Wilkinson and R. Bruce Partridge
Fig. 36–42   National Optical Astronomy Observatories/Cerro Tololo Inter-American Observatory

INDEX

References to illustrations, either drawings or photographs, are identified by i; references to tables by t. Page numbers marked s refer to an entry in the chapter summary. Page numbers marked b (for "basic") identify a definition.

INDEX OF SYMBOLS

PHYSICAL CONSTANTS

Speed of light	c	3.00×10^8 m/s
Elementary charge	e	1.60×10^{-19} C
Electron mass	m_e	9.11×10^{-31} kg
Proton mass	m_p	1.67×10^{-27} kg
Gravitational constant	G	6.67×10^{-11} N·m^2/kg^2
Permittivity constant	ϵ_0	8.85×10^{-12} F/m
Permeability constant	μ_0	1.26×10^{-6} H/m
Boltzmann's constant	k	1.38×10^{-23} J/K
Universal gas constant	R	8.31 J/K·mol
Stefan-Boltzmann constant	σ	5.67×10^{-8} W/m^2·K^4
Planck's constant	h	6.63×10^{-34} J·s
Avogadro's number	N_A	6.02×10^{23} mol^{-1}

SI DERIVED UNITS

newton	$1\ \mathrm{N} \equiv 1\ \mathrm{kg \cdot m/s^2}$
joule	$1\ \mathrm{J} \equiv 1\ \mathrm{N \cdot m}$
watt	$1\ \mathrm{W} \equiv 1\ \mathrm{J/s}$
pascal	$1\ \mathrm{Pa} \equiv 1\ \mathrm{N/m^2}$
volt	$1\ \mathrm{V} \equiv 1\ \mathrm{J/C}$
ampere	$1\ \mathrm{A} \equiv 1\ \mathrm{C/s}$
ohm	$1\ \Omega \equiv 1\ \mathrm{V/A}$
farad	$1\ \mathrm{F} \equiv 1\ \mathrm{C/V}$
tesla	$1\ \mathrm{T} \equiv 1\ \mathrm{N \cdot s/C \cdot m}$
henry	$1\ \mathrm{H} \equiv 1\ \mathrm{V \cdot s/A}$

For more extensive tables of physical constants and conversion factors, see the Appendices at the back of this book.